An Introduction to
Physical Geography
and the Environment

Visit the An Introduction to Physical Geography and
the Environment, Third Edition, Companion Website at
www.pearsoned.co.uk/holden to find valuable **student**
learning material including:

- Video-clips that allow you to see landscapes and processes
  in action
- Multiple choice questions to test your learning
- Models for practical learning
- Annotated weblinks for every chapter
- Comprehensive bibliography
- An online glossary to explain key terms
- Flashcards to test your understanding of key terms

An Introduction to

Physical Geography and the Environment

Third edition

Edited by

Joseph Holden

School of Geography, University of Leeds

PEARSON

Harlow, England • London • New York • Boston • San Francisco • Toronto • Sydney • Auckland • Singapore • Hong Kong
Tokyo • Seoul • Taipei • New Delhi • Cape Town • São Paulo • Mexico City • Madrid • Amsterdam • Munich • Paris • Milan

Pearson Education Limited
Edinburgh Gate
Harlow
Essex CM20 2JE
England

and Associated Companies throughout the world

Visit us on the World Wide Web at:
www.pearson.com/uk

First published 2005
Second edition 2008
Third edition 2012

ISBN 978-0-273-74069-8

British Library Cataloguing-in-Publication Data
A catalogue record for this book is available from the British Library

Library of Congress Cataloging-in-Publication Data
A catalog record for this book is available from the Library of Congress

10 9 8 7 6 5 4 3 2 1
16 15 14 13 12

Typeset in 9.75/13 pt Minion Pro by 73
Printed and bound by Grafos S.A. Spain

BRIEF CONTENTS

Contents

Contents

Contents

Supporting resources

Visit www.pearsoned.co.uk/holden to find valuable online resources

Companion Website for students
- Video-clips that allow you to see landscapes and processes in action
- Multiple choice questions to test your learning
- Models for practical learning
- Annotated weblinks for every chapter
- Comprehensive bibliography
- An online glossary to explain key terms
- Flashcards to test your understanding of key terms

For instructors
- Fully annotated PowerPoint slides for each chapter
- Example field exercises and associated practicals

Also: The Companion Website provides the following features:
- Search tool to help locate specific items of content
- E-mail results and profile tools to send results of quizzes to instructors
- Online help and support to assist with website usage and troubleshooting

For more information please contact your local Pearson Education sales representative or visit www.pearsoned.co.uk/holden

It has been a great delight to co-ordinate the writing of this third edition. The 20 contributors, who are all international experts in their field, have produced excellent chapters which have been updated with new research discoveries and ideas. I have been involved with geography at university level for 17 years and this book still imparts on me a feeling of great enthusiasm and passion for the subject. I hope that readers of the book also feel the same. When the work of the contributors is put together we see that they have researched in detail every environment on the planet. The chapters provide an unrivalled source of rich information from around the world for all budding physical geographers.

There are three new chapters in this edition. Aquatic ecosystems are important components of the biosphere and have been impacted by humans both directly through modification of landscapes and water bodies, and indirectly through pollution or changes to water flow across landscapes. The new chapter written by Lee Brown provides critical insights into the functioning of freshwater ecosystems. There is also a new chapter on weathering which not only outlines key weathering processes but the contributor, Bernie Smith, has presented the information in a thought-provoking way in order to engage the reader with the topic and assist learning. The third completely new chapter, written by Ian Lawson, focuses on the last 11 700 years of Earth history, which form an epoch known as the Holocene. Understanding environmental change during this period as well as human interactions with the environment enables us to properly interpret the landscapes we see around us today and to place current climate change into context.

The focus of this book is the understanding of geographical processes. The book looks at the inter-linkages between different processes, places and environments. Hopefully the book addresses many questions that you have about the way the physical environment works. I hope that the book engages and inspires you and makes you ask new questions about the environment. I encourage you to use your geographical skills to seek answers to those questions and to share your discoveries with others.

Scope of the book

Physical geography is of wide interest and immense importance. It deals with the processes associated with climate, landforms, oceans and ecosystems of the world. The Earth has always been subject to changes in these systems and studying physical geography allows us to understand how Earth systems have come to operate as they do today. It also provides us with insights into how they may operate in the future. The impacts that humans have made on the Earth's environments are ever increasing as the world's population approaches seven billion. Thus the Earth's systems will change in the future both naturally and in a forced way through human action. However, it will be crucial to understand, manage and cope with such change and this can be achieved only by understanding the processes of physical geography. This text is aimed at those embarking on a university course and provides an introduction to the major subjects of physical geography. The book is comprehensively illustrated to aid process understanding and should be of value throughout a university degree. The book provides a baseline of understanding and additionally it directs the reader to resources that encourage them to develop their studies further.

Tools used in the book

In addition to providing a rich source of information the book uses a number of educational tools to aid understanding:

- The book is split into six parts, each with a **part opener** that describes the main themes of that part of the book and the links between the chapters within that part.
- **Learning objectives** clearly outline the purpose and aims of a particular chapter to help locate the reader within the book.
- **Boxed features** explore and illustrate topics and concepts through real-world examples. Scattered throughout every chapter, these insightful applications are differentiated into the following types:
  - fundamental principles;
  - techniques;
  - case studies;

- environmental change;
- new directions;
- hazards.
- **Reflective questions** invite the reader to think about, and further explore, what they have just read. Useful for consolidating learning, these questions are found at the end of each major section of every chapter.
- A **summary** draws together the key ideas of the chapter.
- An **annotated list** of further reading aims to inspire and enable deeper exploration into a topic. The reading lists include important papers as well as textbooks.
- The **comprehensive glossary** serves as an additional resource to help clarify concepts discussed within the book. Key words defined in the glossary are highlighted in the text.

Companion Website

The book also has a dedicated **website** at www.pearsoned.co.uk/holden on which there is a suite of other educational resources for both students and lecturers alike.

Lecturer resources

- **PowerPoint slides:** a set of slides for every chapter comprising bulleted outlines of core topics and the key figures and images from the main text.
- **Field exercise ideas:** suggestions for a variety of field trips and associated activities.

Student resources

- **Annotated weblinks:** several hundred annotated additional websites for students to further explore a topic. There are weblinks listed for each chapter.
- **Digital video-clips:** a unique set of short films allowing students to see the landscapes and processes in action.
- **Interactive models for practical learning:** these models give students the opportunity to explore and understand environmental processes and the principles of modelling.
- **Multiple choice questions:** a set of interactive questions for every chapter that allow students to test and consolidate their understanding.
- **Comprehensive glossary:** unique to this text, a handy resource to assist learning.
- **Bibliography:** an annotated list of further reading material.
- **Flashcards:** designed for students to test their knowledge of key terms for each chapter.

I hope that you are able to use the rich interactive resources that this book provides to further your learning and exploration of the subject of physical geography and the environment.

Joseph Holden
April 2012

NAVIGATION AND SETTING THE SCENE

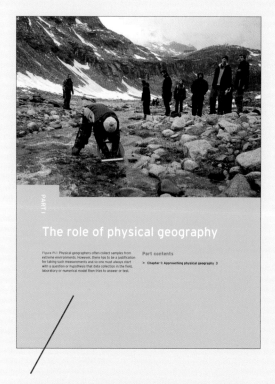

PART 1

The role of physical geography

Figure P1.1 Physical geographers often collect samples from extreme environments. However, there has to be a justification for taking such measurements and so one must always start with a question or hypothesis that data collection in the field, laboratory or numerical model then tries to answer or test.

Part contents

► Chapter 1: Approaching physical geography 3

CHAPTER 21

Freshwater ecosystems

Lee E. Brown
School of Geography, University of Leeds

Learning objectives

After reading this chapter you should be able to:

► understand some of the key scientific concepts underpinning the study of life in rivers and lakes
► recognize some of the major groups of freshwater plants and animals
► appreciate the major differences between flowing and still water ecosystems
► describe some of the ways in which freshwater ecosystems have been altered by human environmental change

21.1 Introduction

Although rivers and lakes constitute only an estimated 0.01% of the world's water resources and cover approximately 0.8% of the Earth's surface, these habitats have a disproportionately high diversity of plants and animals with *at least 6%* (or >100 000) of known species estimated to be found in freshwaters (Dudgeon *et al.*, 2006). Over 10 000 fish species are known from freshwaters with some 90 000 species of invertebrates described, the richest groups being the insects, crustaceans, molluscs and mites. Other well-studied groups of freshwater organisms include amphibians, mammals, birds, macrophytes (plants) and algae (e.g. diatoms, phytoplankton) (Figure 21.1). A diverse assemblage of bacteria, fungi, protozoa and rotifers is also found. However, knowledge of freshwater diversity remains incomplete and new species are identified every year.

Organisms inhabiting freshwaters can be grouped according to their role within aquatic food webs. For example, producers or **autotrophs** are the plants and algae that synthesize biomass from inorganic compounds and light. Producers can be attached to surfaces such as rocks or other plants (e.g. filamentous algae or macrophytes), be rooted in loose sediments (macrophytes), or be free living in the water column (e.g. phytoplankton). Those species that exist within the aquatic ecosystem and directly provide energy to the aquatic food web are termed autochthonous producers. **Heterotrophs** are organisms that obtain organic matter by consuming autotrophs, other heterotrophs or detritus. Members of this group can be considered either as herbivores (consumers of attached algae, plants and phytoplankton), detritivores (consumers of dead organic matter) or predators (consumers of living heterotrophs), although some organisms are omnivorous, feeding on a variety of resources. The diets of many aquatic heterotrophs are also subsidized with organic materials from adjacent terrestrial ecosystems (e.g. leaf litter from

The book is divided into six **parts**, each with a part opener, describing the main themes and links between chapters within that part.

Learning Outcomes introduce topics covered and help you to focus on what you should have learnt by the end of the chapter.

AIDING YOUR UNDERSTANDING

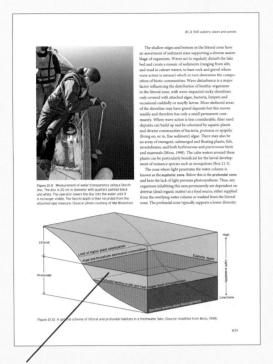

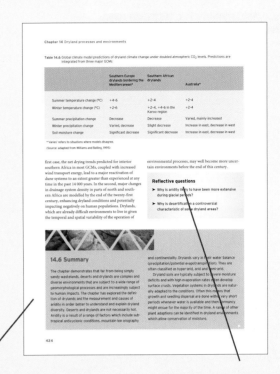

Figures, diagrams and photos feature throughout the text to illustrate key points and clarify topics discussed.

Chapter summaries recap and reinforce the key points to take away from the chapter. They also provide a useful revision tool.

Reflective Questions encourage further thought about topics under discussion and can be used to consolidate learning.

PHYSICAL GEOGRAPHY AND ENVIRONMENTAL ISSUES IN ACTION

Six categories of **boxed features** explore and illustrate topics and concepts through real world examples.

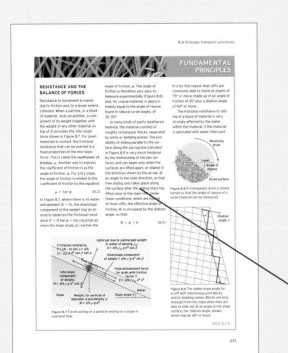

Fundamental principles offer further explanation of core concepts.

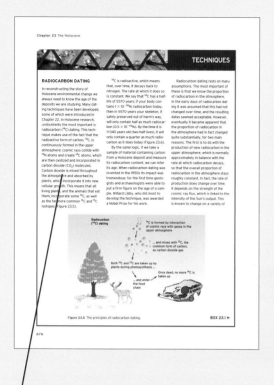

Techniques give detailed coverage of techniques and tools used within physical geography.

Environmental change explores environmental issues that shape the world around us.

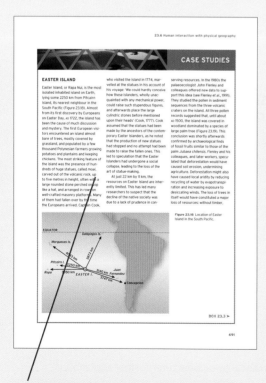

Case studies include a range of international examples and illustrations to add a real world relevance to topics discussed.

Chapter 19 Biogeographical concepts

NEW DIRECTIONS

RESILIENCE AND PANARCHY

The biosphere is dynamic and often has the ability to 'bounce back' or to recover from stresses applied to it. This capacity to return to its previous state is described as resilience. In Figure 19.20, the area of grassland on the headland has remained for many years, despite the stressors of the coastal environment and from trampling by visitors along the coastal path. Unlike straightforward physical resilience, such as within a rubber band, ecological resilience is highly complex and may involve a range of short- and longer-term adaptations. It is problematic to measure or to pre-

dict the ecological limits of any one locality. With increasingly intensive human land use, coupled with climate and other environmental changes, it is important to be able to measure ecological resilience and to find the tipping point. In other words it is important to determine how much stress or disruption can be accommodated before a change to a new state is inevitable. This new state may result in biogeographical boundary changes and the processes involved in the adaptive capacity of ecosystems are therefore important.

The term 'resilience', in this context, was first used by C.S. Holling in 1973 and the measurement of

resilience is becoming an important tool in ecosystem management. The concept should be treated with care, however, as it can be problematic if taken too literally. It is sometimes equated, if on a simplistic level, with 'sustainability' or with successful environmental management. Ecological succession (see Chapter 20) can be regarded as a series of stable states, separated by dynamic, transitional conditions. Some workers, who regard ecological resilience as both a functional and philosophical tool, consider it to be the determinant of movement between stable states and that a system's adaptive capacity can act as a buffer against movement

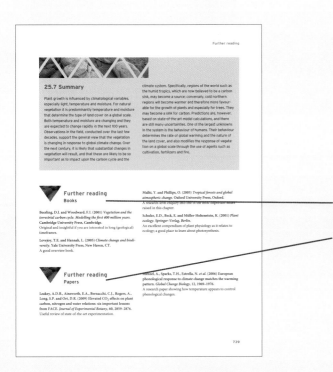

Figure 19.20 Tintagel, Cornwall, England. An example of an area where habitats are displaying their adaptive capacity to both environmental and human-related stresses.

BOX 19.4 ▶

580

New directions introduce the latest thinking and discoveries.

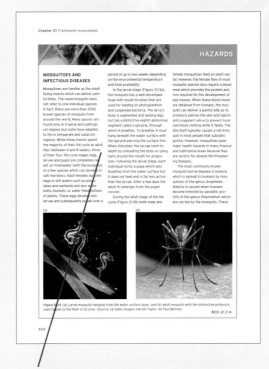

Chapter 21 Freshwater ecosystems

HAZARDS

MOSQUITOES AND INFECTIOUS DISEASES

Mosquitoes are familiar as the small flying insects which can deliver painful bites. The name mosquito does not refer to one individual species; in fact, there are more than 3500 known species of mosquito from around the world. Many species are found only in tropical and subtropical regions but some have adapted to life in temperate and subarctic regions. While these insects spend the majority of their life cycle as adult flies (between 4 and 8 weeks), three of their four life-cycle stages (egg, larvae and pupa) are completely reliant on freshwater (with the exception of a few species which can develop in salt marshes). Adult females lay their eggs in still waters such as ponds, lakes and wetlands and also water butts, buckets, or water-filled hollows of plants. These eggs develop into larvae and subsequently pupae over a

period of up to two weeks depending on the environmental temperature and food availability.

In larval stage (Figure 21.13a), the mosquito has a well-developed head with mouth brushes that are used for feeding on phytoplankton and suspended bacteria. The larva's body is segmented and lacking legs but has a distinctive eighth abdominal segment called a spiracle, through which it breathes. To breathe, it must hang beneath the water surface with the spiracle piercing the surface film. When disturbed, the larvae swim to depth by undulating the body or using hairs around the mouth for propulsion. Following the larval stage, each individual forms a pupa which also breathes from the water surface but it does not feed and is far less active than the larvae. After a few days the adult fly emerges from the pupal cocoon.

During the adult stage of the life cycle (Figure 21.13b) both male and

female mosquitoes feed on plant nectar. However, the female flies of most mosquito species also require a blood meal which provides the protein and iron required for the development of egg masses. When these blood meals are obtained from humans, the mosquito can deliver a painful bite as its proboscis pierces the skin and injects anti-coagulant saliva to prevent localized blood clotting while it feeds. The bite itself typically causes a red itchy spot in most people that subsides quickly. However, mosquitoes pose major health hazards in many tropical and subtropical areas because they are vectors for several life-threatening diseases.

The most commonly known mosquito-borne disease is malaria which is spread to humans by mosquitoes of the genus Anopheles. Malaria is caused when humans become infected by parasitic protists of the genus Plasmodium which are carried by the mosquito. These

(a)   (b)

Figure 21.13 (a) Larval mosquito hanging from the water surface layer, and (b) adult mosquito with the distinctive proboscis used to pierce the flesh of its prey. (Source: (a) Getty Images: Harold Taylor, (b) Paul Bertner)

BOX 21.3 ▶

632

Hazards apply the topics discussed to contemporary environmental problems.

GOING FURTHER

Further reading

25.7 Summary

Plant growth is influenced by climatological variables, especially light, temperature and moisture. For natural vegetation it is predominantly temperature and moisture that determine the type of land cover on a global scale. Both temperature and moisture are changing and they are expected to change rapidly in the next 100 years. Observations in the field, conducted over the last few decades, support the general view that the vegetation is changing in response to global climate change. Over the next century, it is likely that substantial changes in vegetation will result, and that these are likely to be so important as to impact upon the carbon cycle and the

climate system. Specifically, regions of the world such as the humid tropics, which are now believed to be a carbon sink, may become a source; conversely, cold northern regions will become warmer and therefore more favourable for the growth of plants and especially for trees. They may become a sink for carbon. Predictions are, however, based on state-of-the-art model calculations, and there are still many uncertainties. One of the largest unknowns in the system is the behaviour of humans. Their behaviour determines the rate of global warming and the nature of the land cover, and also modifies the response of vegetation on a global scale through the use of agents such as cultivation, fertilizers and fire.

▼ Further reading
Books

Beerling, D.J. and Woodward, F.I. (2001) *Vegetation and the terrestrial carbon cycle. Modelling the first 400 million years.* Cambridge University Press, Cambridge.
Original and insightful if you are interested in long (geological) timeframes.

Lovejoy, T.E. and Hannah, L. (2005) *Climate change and biodiversity.* Yale University Press, New Haven, CT.
A good overview book.

Malhi, Y. and Phillips, O. (2005) *Tropical forests and global atmospheric change.* Oxford University Press, Oxford.
A research-level enquiry into one of the most important issues raised in this chapter.

Schulze, E.D., Beck, E. and Müller-Hohenstein, K. (2001) *Plant ecology. Springer-Verlag, Berlin.*
An excellent compendium of plant physiology as it relates to ecology; a good place to learn about photosynthesis.

▼ Further reading
Papers

Leakey, A.D.B., Ainsworth, E.A., Bernacchi, C.J., Rogers, A., Long, S.P. and Ort, D.R. (2009) Elevated CO_2 effects on plant carbon, nitrogen and water relations: six important lessons from FACE. *Journal of Experimental Botany,* 60, 2859–2876.
Useful review of state-of-the-art experimentation.

Menzel, A., Sparks, T.H., Estrella, N. *et al.* (2006) European phenological response to climate change matches the warming pattern. *Global Change Biology,* 12, 1969–1976.
A research paper showing how temperature appears to control phenological changes.

739

Further reading offers sources of additional information in both **books** and academic **papers** for those who wish to explore a topic further.

Dr Lee E. Brown, University of Leeds

Dr Pippa J. Chapman, University of Leeds

Dr David J. Gilvear, University of Stirling

Professor John Grace, University of Edinburgh

Dr Kate V. Heal, University of Edinburgh

Professor Joseph Holden, University of Leeds

Dr Timothy D. James, Swansea University

Dr Richard Jeffries, Scottish Environment Protection Agency

Professor Mike Kirkby, University of Leeds

Professor Michael D. Krom, University of Leeds

Dr Ian Lawson, University of Leeds

Dr John G. Lockwood, formerly of University of Leeds

Dr Gerhard Masselink, University of Plymouth

Professor John McClatchey, University of the Highlands and Islands

Professor Adrian T. McDonald, University of Leeds

Professor Tavi Murray, Swansea University

Professor Bernie Smith, Queen's University Belfast

Professor Kevin G. Taylor, Manchester Metropolitan University

Professor David S.G. Thomas, University of Oxford

Professor Hilary S.C. Thomas, University of Glamorgan

EDITOR'S ACKNOWLEDGEMENTS

I have been inspired and guided by many people throughout my career at the places where I have studied and worked. Many of these figures probably do not realize it but they influenced me a great deal. I would particularly like to thank the following: Alan Angus, Adrian Bailey, Tim Burt, John Charlton, Nick Cox, Colin Duke, Robin Glasscock, Manuel Gloor, Frank Hodges, Christopher Keylock, Mike Kirkby, Stuart Lane, Adrian McDonald, Vincent McNeany, Sue Owens, Keith Richards, Ian Simmons, Ian Snowdon, Tom Spencer, Derek Talbot, Steve Trudgill and Jeff Warburton. I would also like to mention Patrick Rolfe, a talented, inspirational and thought-provoking individual, who was conducting a PhD under my supervision when he died at the age of just 24. We all miss you Patrick and I am very grateful to have known you.

I am grateful to the contributors for their efforts and outstanding expertise in producing this third edition. I am indebted to many family members for help of various sorts.

These include Eve Holden, Patricia Holden, Henry Holden, Vincent Holden, Clare Holden and a number of people who provided photographs and are acknowledged in the figure captions. Thanks also to Kathryn Smith for help with work on some of the supplementary materials.

I would also like to thank all the reviewers who read and commented in detail upon the chapters and suggested revisions that could be incorporated within this edition. Ian Lawson and I thank Katy Roucoux for her feedback on Chapter 23. My kind appreciation also goes to the whole editorial and production team at Pearson Education including Tim Parker, Maggie Wells, Lauren Hayward and Helen Leech, with particular thanks to Pat Bond who provided significant experience, expertise and support in helping to produce this third edition.

Finally I thank Eve and our beautiful daughters Justina, Mary, Stephanie and Alice for joining me in exploring the physical geography of the Earth.

We are grateful to the following for permission to reproduce copyright material:

Figures

Figure 1.15 after *Journal of Geology*, 60, Fig 1 (Wolman and Miller 1960); Figure 2.5 Martin Jakobson; Figure 2.11 from *Scientific American*, Aug. 1982, p.122 (Bass); Figure 2.12 from Israel, Jordan River Valley, Coast, http://eol.jsc.nasa.gov/sseop/EFS/photoinfo.pl?PHOTO=NM23-755-505; Figure 3.3 adapted from *Sverdrup et al (1942) The Oceans, in Pinet (1996) Invitation to Oceanography*, 1 ed., Pearson Education, Inc. Fig 4.11a, p.141; Figure 3.8 after *Introduction to the World's Ocean*, McGraw Hill Education (Sverdrup et al. 2003) Fig. 6.11, p.180; Figure 4.1 from *Climate Change 2001*, IPCC (2001) Fig 1.1, p.88; Figure 4.12 from *Contemporary Climatology*, 2 ed., Longman (Henderson-Sellars and Robinson 1999) Fig 3.7, p.47; Figure 4.23 from *Climate Change 2007: The Physical Science Basis. Contribution of Working Group 1 to the Fourth Assessment Report of the Intergovernmental Panel on Climate Change*, IPCC (ed. R Alley et al 2007) Fig. WG1-AR4; Figure 5.3 after *Environmental Systems: An Introductory Text*, 2 ed., Taylor & Francis (White et al. 1992) Fig 4.5, p.84; Figure 10.6 from Soils and their use in Northern England, *Soil Survey of England and Wales*, Fig 14, p.47 (Jarvis et al 1984); Figure 10.13 from Soils and their use in Northern England, *Soil Survey of England and Wales*, Fig 16, p.52 (Jarvis et al 1984); Figure 10.20 adapted from *Fundamentals of Physical Environment*, 3 ed., Routledge (Smithson et al 2002) Fig 18.7, p.389; Figure 10.22 adapted from *The Nature and Properties of Soil*, 13 ed., Pearson Education, Ltd. (Brady, N.C. and Weil, R.R. 2002) Fig 9.3, p.347, Brady, Nyle C., Weil, Ray R., The Nature and Properties of Soils, 13th Ed., c2002.Reprinted and Electronically reproduced by permission of Pearson Education, Inc., Upper Saddle River, New Jersey.; Figure 11.30 after *Principles of Hydrology*, 4 ed., McGraw Hill (Ward, R.C. and Robinson, M. 2000) Fig 7.14; Figure 13.9 adapted from *Contemporary Hydrology*, John Wiley & Sons Ltd. (Soulsby in Wilby (ed.) 1997); Figure 14.31

from Sand deserts during the last glacial maximum and climatic optimum, *Nature*, 272, Fig 1b, p.44 (Sarnthein, M. 1978); Figure 15.5 adapted from *Coastal evolution : Late Quaternary shoreline morphodynamics* Cambridge University Press (edited by R.W.G. Carter, C.D. Woodroffe 1994); Figure 15.14 from *Journal of Geophysical Research*, 73 (Bowen et al 1968); Figure 15.18 from THE DISASTER IN THE NETHER-LANDS CAUSED BY THE STORM FLOOD OF FEBRUARY 1, 1953 258 P. J. Wemelsfelder Proceedings of the 4th Coastal Engineering Conference; Figure 15.20 adapted from *Geographical Variation in Coastal Development*, Pearson Education (Longman) (Davies, John Lloyd 1980); Figure 15.24 adapted from *Geographical variation in coastal development*, 2 ed., Longman (Davies, J.L. 1980), Acknowledgement: Title, author, Pearson Education Limited and Copyright line as it appears in our publication (note where figure and diagrams are reproduced, this acknowledgement is to appear immediately below them; Figure 15.28 from *Handbook of Beach and shore Morphodynamics*, John Wiley & Sons Ltd. (Aggard, T. and Masselink, G. in Short A.D. (ed.) 1999); Figure 15.39 from *Fundamentals of Physical Environment*, 2 ed., Routledge (Briggs et al 1997); Figure 16.30 from *Surface processes and landforms*, 1 ed., Pearson Education, Ltd. (Easterbrook, D.J. 1993), EASTERBROOK, DON J., SURFACE PROCESSES AND LAND FORMS, 1st Ed., c1993. Reprinted and Electronically reproduced by permission of Pearson Education, Inc., Upper Saddle River, New Jersey.; Figure 17.5 adapted from *The Frozen Earth, Fundamentals of Geocryology*, Cambridge University Press (Williams and Smith 1989) Fig 1.6, p.14; Figure 17.9 from Proceedings of the Second International Conference on Permafrost, *Yakutsk, USSR* (Molochuskin 1973); Figure 18.10 adapted from *Fundamentals of Biogeography*, Routledge (Huggett 1998), Credit Christen Raunkiaer; Figure 19.18 from CSIRO Australia (Sutherst, B. 1991); Figure 20.12 from A biodiversity concept map, *Woodrow Wilson National Fellowship Foundation* (Leadership Programme for Teachers 1999); Figure 20.13 from Mechanisms of succession in natural communities and their role in community stability and organisation, *American Naturalist*, 111, p. 1121 (Connell, J.H. and

Slatyer, R.O. 1977), University of Chicago Press; Figures 22.9, 25.10 from Global response of terrestrial ecosystem structure and function to CO2 and climate exchange: results from six dynamic global vegetation models, *Global Change Biology*, 7, 357-373 (Cramer, W., Bondeau, A., Woodward, F.I. et al. 2001); Figures 24.2, 24.3, 24.5, 24.7, 25.16b from Climate Change 2007: The Physical Science Basis, *Contribution of Working Group 1 to the Fourth Assessment Report of the Intergovernmental Panel on Climate Change* (Alley, R. et al. 2007), IPCC, Climate Change 2007: The Physical Science Basis. Working Group I Contribution to the Fourth Assessment Report of the Intergovernmental Panel on Climate Change, Figure SPM.6. Cambridge University Press. Figure 24.10 from New directions: rich in CO2, *Atmospheric Environment*, 40, p. 3220 (Korbetis, M., Reay, D.S. and Grace, J. 2006), with permission from Elsevier; Figure 24.11 from *Centre for Atmospheric Science, Dept of Chemistry, University of Cambridge;* Figure 25.1 from Global climate and the distribution of plant biomes, *Philosophical Transactions of the Royal Society*, 359, 1465-1476 (F. I. Woodward, M. R. Lomas and C. K. Kelly 2004); Figure 25.4 from Millennial-scale dynamics of Southern Amazonian rain forests, *Science*, 290, 2291-2294 (Mayle, F.E., Burbridge, R. & Killeen, T.J.), From Millenial-scale dynamics of Southern Amazonian rain forests, Science, 290, 2291-2294 (Mayle, F.E., Burbridge, R, and Killeen, T.J.) Reprinted with permission from AAAS; Figure 25.5 from Height growth response of tree line spruce to recent climate warming across the forest-tundra of eastern Canada, *Journal of Ecology*, 92, 835-845 (Gamache, I. and Payette, S. 2004), British Ecological Society; Figure 25.6 from A global change-induced biome shift in the Montseny mountains (NE Spain), *Global Change Biology*, 9, 131-140 (Peñuelas, J. & Boada, M. 2003); Figure 25.7 from European phenological response to climate change matches the warming pattern, *Global Change Biology*, 12, 1969-1976 (Menzel, A., Sparks, T.H., Estrella, N. et al. 2006); Figure 25.11 from Climate-carbon cycle feedback analysis: results from the C4MIP model intercomparison, *Journal of Climate*, 19, 3337-3353 (Friedlingston, P., Cox, P., Betts, R. et al. 2006), (c)American Meteorological Society. Reprinted with permission, (c)American Meteorological Society; Figure 25.12 from The human footprint in the carbon cycle of temperate and boreal forests, *Nature*, 447, 848-850 (Magnani F, Mencuccini M, Borghetti M, et al. 2007); Figure 25.15 from Positive feedbacks among forest fragmentation, drought and climate change in the Amazon, *Conservation Biology*, 15, 1529-1535 (Laurence, W.F. & Williamson, B. 2001).

Maps

Figure 3.14 (a) adapted from *Chemical Oceanography*, 2 ed., Academic Press (Davies and Gorsline in Riley and Chester (eds.) 1976); Figure 5.4 from *Meteorology Today: Introduction to weather, climate and the environment*, 6 ed., Brooks/ Cole (Ahrens 2000) Fig 18.6, p.484; Figures 5.5, 5.7 adapted

from *General Climatology*, 3 ed., Pearson Education, Ltd. (Critchfield 1983) p.453, CRITCHFIELD, GENERAL CLIMATOL-OGY, 3rd Ed., c1974. Reprinted and Electronically reproduced by permission of Pearson Education, Inc.,Upper Saddle River, New Jersey.; Figure 5.8 from *Meteorology Today: Introduction to weather, climate and the environment*, 6 ed., Brooks/ Cole (Ahrens 2000) Fig 16.8, p.426; Figure 5.37 from *Winter 2010 Mean Temperature 1971-2000 Anomaly*, Met Office (2011), Contains public sector information licensed under the Open Government Licence v1.0; Figure 5.38 from *Winter 2010 Rainfall Amount % of 1971-2000 Average*, Met Office (2011), Contains public sector information licensed under the Open Government Licence v1.0; Figure 6.28 from Haze pollution across South-East Asia on 19th October 2006, www.nea. gov.sg/cms/mss/gif/19.gif, National Environment Agency of Singapore; Figure 10.14 from The distribution of the major soil groups in England and Wales, *National Soil Map* (2004), Cranfield Univeristy; Figure 10.26 from Status of UK Critical Loads, *Critical Loads Methods Data and Maps*, February 2003, Fig 4.1 (CEH, NSRI, Macaulay Institute and DardNI); Figure 15.23 after *The Physical Geography of Western Europe*, Oxford University Press (Hofstede, J. in Koster E.A. (ed.) 2005), and The Common Wadden Sea Secretariat (CWSS); Figure 15.41 adapted from *A Celebration of the World's Barrier Islands*, Colombia University Press (Pilkey, O.H. 2003). Reprinted with permission of the publisher; Figure 16.2 from Bamber J, and Luckman, A., British Antarctic Survey, Jonathan Bamber, University of Bristol.; Figure 17.4 adapted from *The Frozen Earth, Fundamentals of Geocryology*, Cambridge University Press (Williams and Smith 1989) Fig 1.6, p.14; Figure 18.3 from http://glcfapp.umiacs.umd.edu:8080/esdi/ preview?size=browse&granule_id=12531, Hansen et al. 1998; Figure 19.13 from *Proceedings of the International Ornithological Congress*, 10, pp.537-44 (Van Tyne 1951); Figure 22.18 from Large-scale patters of dune type, spacing and orientation in the Australian continental dunefield, *Australian Geographer*, 19, 89-104 (Wasson et al. 1988); Figures 24.4, 24.6 adapted from Climate Change 2007: The Physical Science Basis, *Contribution of Working Group 1 to the Fourth Assessment Report of the Intergovernmental Panel on Climate Change* (Alley, R. et al. 2007), IPCC; Figure 25.16a adapted from Biodiversity hotspots for conservation priorities, *Nature*, 403, 853-858 (Myers, N., Mittermeier, R.A., Mittermeier, C.G., et al. 2000); Figure 27.7 from http://www. cbc.ca/bc/features/floodwatch/pinebeetle.html, British Columbia Flood Watch; Figure 27.9 from Pan-European Soil Erosion Risk Assessment, Irvine, B.

Tables

Table 10.8 adapted from *The Nature and Properties of Soils*, 12 e., Pearson Education, Ltd. (Nyle, C. and Weil, R.R.) Table 4.1, p.125; Table 10.10 adapted from World Map of the Status of Human Induced Soil Degradation - An Explanatory Note,

Global assessment of soil depredation (GLASOD), 2 ed. (Oldeman et al. 1991); Table 13.2 adapted from *Global Environment: Water, Air and Geochemical Cycles*, 1 ed., Pearson Education, Ltd. (Berner, R.A. 1996); Table 14.4 adapted from *Arid Zone Geomorphology: Process, Form and Change in Drylands*, John Wiley & Sons Ltd. (Thomas (ed.) 1997) Table 1.5, p.9; Table 15.1 from Climate Change 2007: The Physical Science Basis, *Contribution of Working Group 1 to the Fourth Assessment Report of the Intergovernmental Panel on Climate Change* (Alley, R. (ed.) 2007), IPCC; Table 15.2 from Climate Change 2007: The Physical Science Basis, *Contribution of Working Group 1 to the Fourth Assessment Report of the Intergovernmental Panel on Climate Change* (Alley et al. (eds.) 2007), IPCC; Table 16.2 from *Physics of Glaciers*, 4 ed., Academic Press (Cuffey, K.M., and Paterson, W.S.B. 2010); Table 17.2 adapted from *Geografiska Annaler*, 42 (Rapp 1960) pp.71–200; Table 18.1 from *Die okozonen der Eerde*, 2 Aufl. UTB1514 (Schultz 1995); Table 19.2 from *An Introduction to Applied to Biogeography*, Cambridge University Press (Spellerberg and Sawyer 1999) Table 5.2, p.112; Table 19.3 adapted from *Biodiversity and Conservation*, Routledge (Jeffries 1997) p.111; Table 25.1 from Global climate and the distribution of plant biomes, *Philosophical Transactions of the Royal Society*, 359, 1465–1476 (Woodward, F. I., Lomas M. R. and Kelly C. K. 2004); Table 26.3 from *Remote Sensing and Image Interpretation*, p.379, John Wiley & Sons Ltd. (Lillesand and Keifer 2000); Table 26.5 adapted from http://modis.gsfc.nasa.gov/about/specifications.php, NASA; Table 26.6 from http://asterweb.jpl.nasa.gov/characteristics.asp, NASA

Photos

Photo on page 1 courtesy of Steve Carver; Figure 1.18 Natalie Suckall; Photo on page 27 Shutterstock.com: Dirkr; Figure 2.3 Bob Finch/School of Earth & Environment; Figure 2.4 Fotolia.com: Jenny Thompson; Figure 2.9 Getty Images: Werner Van Steen; Figure 2.10 USGS: R G McGimsey; Figure 2.17 Fotolia.com; Figure 2.18 NASA: Jet Propulsion Laboratory; Figure 2.19 iStockphoto: Paul Cowan; Figure 2.21 USGS: D W Peterson; Figure 2.22 Courtesy of Steve Carver; Figure 3.1 Planetary Visions; Figure 3.4 Shutterstock.com: Vic Spacewalker; Figure 3.5 USGS; Figure 3.6 James Davies; Figure 3.7 Getty Images; Figure 3.14 (b) and (c) NASA: Jacques Desloitres / LODIS Land Rapid Response Team; Figure 3.16 Plymouth Marine Laboratory: Remote Sensing Group; Figure 3.19 Jon Copley; Figure 3.22 Alamy Images: Louise Murray; Figure 3.25 Corbis: Steve Terrill; Photo on page 75 NOAA; Figure 5.11 Press Association Images: David J Phillip; Figure 5.17 Shutterstock.com: Iafoto; Figure 5.18 Corbis: Ron Kuntz; Figure 6.8 Alamy Images: P B Images; Figure 6.9 DK Images; Figure 6.12 Getty Images: AFP; Figure 6.20 Getty Images: Betty Wiley; Figure 6.23 Getty Images: A J James; Figure 6.27 Alamy Images: Hemis; Figure 7.10 STILL Pictures The Whole Earth Photo Library: Ron Giling; Figure 7.22 Tony Waltham/ Geophotos; Figure 8.12 Satellite image of the Saidmareh Landslide by Geology.com using Landsat GeoCover, data provided by NASA; Figure 8.13 Reuters: Ho New; Figure 8.14 James Davies; Figure 8.24 PhotoDisc: Alan & Sandy Carey; Figure 8.27 Science Photo Library Ltd: Simon Fraser; Figure 9.6 NASA: SeaWiFS Project / Goddard Space Flight Center & ORBIMAGE; Figure 9.7 Andrew Thomas; Figure 9.14 Chris Perry; Figure 9.20 Courtesy of Steve Carver; Figure 9.25 Cathy Delaney; Figure 9.26 Alamy Images: China Images; Figure 9.30 Philip N Owens; Figure 10.7 John Conway; Figure 10.8 Chris Evans; Figure 10.9 E A Fitzpatrick; Figure 10.10 Phil Haygarth; Figure 10.24 (a) Science Photo Library Ltd: Sinclair Stammers, (b)Ardea: Steve Hopkin, (c) FLPA Images of Nature: Thomas Marent; Figure 10.25 Shutterstock.com: Neil Bradfield; Figure 11.21 Kathy Gell; Figure 11.22 Martin Dawson; Figure 11.26 Getty Images: Universal Images Group; Figure 12.4 Tory Milner; Figure 12.5 Science Photo Library Ltd: Bernard Edmaier; Figure 12.28 Charles Perfect; Figure 13.5 STILL Pictures The Whole Earth Photo Library: Mark Edwards; Figure 13.6 M. R. Heal; Figure 14.4 NASA; Figure 14.19 NASA; Figure 14.27 NASA: GSFC / MITI / ERSDAC / Jaros & US / Japan Aster Science Team 2002; Figure 15.12 Rob Brander; Figure 15.16 Rob Brander; Figure 15.17 Martin Austin; Figure 15.31 Anthony Priestas; Figure 15.32 Roland Gehrels; Figure 15.37 Roland Gehrels; Figure 15.40 NASA; Figure 15.44 (c) Aart Kroon; Figure 16.4 The Ohio State University; Figure 16.5 (a) NASA: GRACE; Figure 16.7 (a) and (b) Adam Booth, (c) Mike Crabtree; Figure 16.12 USGS; Figure 16.15 A Gow; Figure 16.17 Tim James; Figure 16.18 (a) and (b) Norwegian polar institute, (c) and (d) Danish Lithospheric center; Figure 16.26 (a) Sarah J Fuller, (b) Andy J Evans;. Figure 16.31 James Davies; Figure 16.37 John Shaw; Figure 17.1 University Centre, Svalbard; Figure 17.2 Tony Waltham/Geophotos; Figure 17.6 Tony Waltham/Geophotos; Figure 17.14 Dr Julian Murton/ University of Sussex; Figure 17.15 NASA; Figure 17.16 (b) Alfred-Wenger- Institute; Figure 17.18 USGS; Figure 17.19 (a) Science Photo Library Ltd: Paolo Koch, (b)University Centre, Svalbard; Figure 17.22 C Fogwill; Figure 17.24 R Shakesby; Figure 18.8 Nick Berry; Figure 18.9 Nick Berry: Despina Psarra; Figure 18.11 Shutterstock.com: Oleg Znamenskiy; Figure 18.13 Shutterstock.com: Patrick Poendi; Figure 18.15 Shutterstock.com: Baldovina; Figure 18.23 Getty Images: Christopher Gruver; Figure 18.24 Alamy Images: Phil Degginger; Figure 18.27 US Fish and Wildlife Service; Figure 18.29 Wouter Buytaert; Figure 19.9 Ronald A. S. Johnston; Figure 19.15 Alamy Images: Mambo; Figure 19.16 Getty Images: Nigel Cattlin; Figure 20.18 Kenfig Nature Reserve; Figure 20.19 Anne Goodenough; Figure 21.1 (d) Getty Images. Mark Ledger; Figure 21.2 (e) Sandy Milner; Figure 21.3 (a) Jonathan Carrivick; Figure 21.11 Mel Bickerton; Figure 21.13 (a) Getty Images: Harold Taylor, (b) Paul Bertner; Figure 21.14 Mel Bickerton; Figure 21.15 Fisheries & Oceans, Canada: E. Debruyn / Reproduced with the permission of Her Majesty the Queen in Right of Canada 2011; Figure 22.3 Alice Milner; Figure 22.4 USGS: Thomas M Gibson;

Figure 22.8 Getty Images: Fred Hirschmann; Figure 22.13 Chris Clarke; Figure 22.14 Ken Hamblin; Figure 22.16 Tony Waltham/Geophotos; Figure 22.17 Tony Waltham/Geophotos; Figure 23.2 Tom Foster; Figure 23.12 John Corr; Figure 23.13 Getty Images: DEA / C Sappa; Figure 23.16 (a) Creatas, (b) Shutterstock.com: Erik Lam, (c) Shutterstock.com: Sue C , (d) Shutterstock.com: Eric Isselee; Figure 23.20 Ty Parker; Figure 23.21 Katy Roucoux; Figure 23.22 Mike Church; Figure 26.2 Environment Agency copyright. All rights reserved.; Figure 26.9 Environment Agency copyright. All rights reserved.; Figure 26.10 Lillesand & Keifer (2000); Figure 26.11 Lillesand & Keifer (2000); Figure 26.16 Courtesy of Landsat.org; Figure 26.18 Tavi Murray & Adrian Luckman; Figure 27.8 Forestry Images: Ron Long.

Cover images: *Front:* Bernhard Edmaier; *Back:* Altitude

Every effort has been made to trace the copyright holders and we apologise in advance for any unintentional omissions. We would be pleased to insert the appropriate acknowledgement in any subsequent edition of this publication.

The role of physical geography

Figure PI.1 Physical geographers often collect samples from extreme environments. However, there has to be a justification for taking such measurements and so one must always start with a question or hypothesis that data collection in the field, laboratory or numerical model then tries to answer or test.

Part contents

Scope

Science, and with it the subject of physical geography, has evolved over time. Part I contains one chapter which deals with the development of physical geography. The chapter provides context that explains why we approach the subject in particular ways today. It describes the basic frameworks for studying science and explains the roles of fieldwork, laboratory work and modelling. The chapter describes the advantages and disadvantages of a range of approaches that we should be aware of when studying physical geography. It therefore sets the scene for the rest of the book by providing the reader with an appropriate grounding in the nature of the subject.

What do we mean by physical geography?

Physical geography is about understanding interactions of processes involving the Earth's climate system, oceans, landforms, animals, plants and people. This understanding requires linking the physical systems together and relating human actions to the physical environment. Of interest to physical geographers are the mechanisms that maintain flows of energy and matter across the Earth. There are components of study which include processes associated with plate tectonics, geomorphology, climatology, glaciology and hydrology that shape the surface of the Earth; the collection of climatic and atmospheric processes acting as one of the ultimate controls on the landscape and biosphere; and the ecological and biogeographical patterns that characterize the living portion of the Earth. Physical geography involves the application of technology to study these components and changes within them. For example, remote sensing provides an aid to monitoring the world's constantly changing natural and human landscapes, the oceans, atmosphere and biosphere.

Geographers often say that they study the 'why of where'. By this they mean that they seek to explain the underlying processes that result in the patterns of natural phenomena and the ways in which humans interact with, and alter, these processes and patterns. In addition to a spatial context, change over time is also a central theme to physical geography.

It is important to be aware of the ways in which physical geographers study physical geography. Some kind of philosophical basis of enquiry is essential in order to allow fair comparisons of results and interpretation of conclusions between different research areas. The scientific methods discussed in Chapter 1 help to form this philosophical foundation. The underlying method does not necessarily mean that all research is done using the same techniques; indeed physical geography utilizes a variety of tools to help understand, measure, observe and predict environmental processes. However, by maintaining a philosophical basis, it reminds us to question the approach we take. In recent years, emphasis has shifted from a position where science represents the ultimate authority informing society, to a realization that science is itself influenced by society, and that many other sources of knowledge must be equally considered. Consideration of the advantages and limitations of a given approach is therefore vital so that we can assess the reliability and usefulness of the conclusions attained.

CHAPTER 1

Approaching physical geography

Joseph Holden

School of Geography, University of Leeds

Learning objectives

After reading this chapter you should be able to:

➤ describe the historical basis of physical geography

➤ understand basic scientific methods

➤ evaluate scientific methods for different types of research in physical geography

➤ appreciate the advantages and limitations of different approaches to physical geography

1.1 Introduction

Physical geography affects most aspects of our daily lives. It is fundamental to human existence. For example, it determines water availability and water quality, weather and climate, soil systems, potential for agriculture, the risk of landslides or other hazards, and if and how we can travel from one place to the next. Figure 1.1 illustrates the major components of physical geography. It deals with the Earth's climate system, which results from a combination of atmospheric, oceanic, land, ice and ecological processes. It also deals with a wide range of processes that affect the landscape of the Earth. For example, plate tectonic processes

are responsible for mountain building, the movement of the continents, ocean floor spreading, ecological isolation and changing climate. In addition the landscape is worn down by weathering and erosion processes, many of which are driven by gravity and water (in solid, liquid and gas form). Water also transports nutrients from soils to plants and from rocks and soils to rivers and into the oceans. It transports nutrients and energy around the globe through the oceans and the atmosphere. It moves sediments across hillslopes, catchments and seas. Understanding the variety of processes that link the components shown in Figure 1.1 (atmosphere, oceans, landforms and biosphere) at global and small scales enables improved prediction of future change of the Earth's environmental systems.

A range of tools are available to physical geographers in order to help us understand, measure, observe and predict environmental processes. These include tried and tested methods along with new technologies such as advanced probes and laboratory methods or geophysical and remote sensing tools that allow us to measure the Earth's features and processes remotely. For example, there are now laser-based imaging velocity probes that allow us to collect data on the speed of water in three directions at the same time (vertically up and down, laterally upstream and downstream and laterally across a river or water body), taking measurements thousands of times per second. This allows

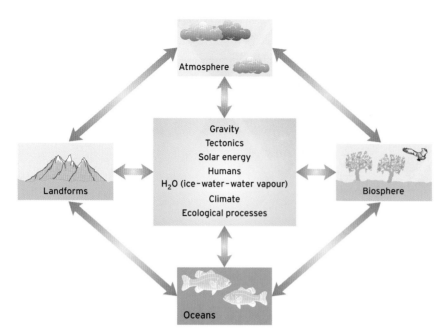

Figure 1.1 Components of the changing environment. The atmosphere, oceans, biosphere and landforms all interact with each other. Important links within the system include H_2O in its different forms, tectonics, ecological processes and humans.

Atmosphere

Landforms

Gravity
Tectonics
Solar energy
Humans
H_2O (ice – water – water vapour)
Climate
Ecological processes

Biosphere

Oceans

us to understand how water actually flows around obstacles and thereby to develop better engineering structures such as bridges or flood protection that are less likely to be eroded or damaged. Another example of a recently developed tool now available to physical geographers is **light detection and ranging** (lidar). This tool uses instruments either on the ground or on board aircraft to measure very accurately the topography of the Earth's surface and also the structure of the vegetation or built environment on top of the Earth's surface. This can be done with a resolution of just a few centimetres and therefore allows lots of important data to be collected over very large areas without the need for a time-consuming field survey, which is particularly advantageous in areas that are difficult to access on foot (see Chapter 26).

Fieldwork, laboratory work and numerical modelling are all important components of the method of physical geography today. However, each particular approach and method has its limitations. No matter what type of measurement device or approach is used, it is often how it is used and why it is being used in that way that are important. In other words, scientific approaches have a philosophical underpinning which can be evaluated. There are a range of approaches to science and physical geography and each approach has advantages and disadvantages. It is therefore necessary to understand these methods and their limitations so that we can: (i) evaluate which are the most appropriate methods to use for a given environmental investigation and (ii) fully evaluate the implications of any given research finding in physical geography.

The approaches that physical geographers have used have varied through time as the subject has developed. In order to understand contemporary practice in physical geography it is therefore necessary to know something about the history and development of the subject. This chapter will briefly describe the way physical geography has developed. It will then move on to discuss how the scientific method has been applied by physical geographers to studies of the environment. The remaining parts of the chapter will look at the principles of and approaches to (i) fieldwork, (ii) laboratory research and (iii) numerical modelling, all of which are important methods of physical geography.

1.2 Historical development of physical geography

1.2.1 Physical geography before 1800

The ancient Greek and Roman geographers produced maps and topographic descriptions of places and their history, performed calculations about the circumference of the Earth, and had a philosophical interest in the relations between humans and the environment. Between the time of the Roman Empire and the sixteenth century, European science progressed very slowly. Often scholars rejected anything that seemed to go against the teachings of the Christian Church. In the Middle East, however, Arab geographers such as Al Muqaddasi (who lived between AD 945 and 988) were pioneering fieldwork whereby observations were given precedence. Indeed Al Muqaddasi stated that he would not present anything unless he had seen it with his own eyes. Such Arab geographers maintained the Greek and Roman techniques and developed new ones. Arab traders travelled throughout Asia, Africa and the Indian Ocean and added a great deal of geographical knowledge to update the classical sources. Any European geographical work was trivial in comparison with the huge amount published by Islamic writers of the Middle Ages. Exploration and learning also flourished in China with advanced triangulation techniques allowing exceptionally good quality maps of the region to be produced from the first century AD onwards. Indeed official Chinese historical texts contained a geographical section from this period onwards, which was often an enormous compilation of changes in place-names and local administrative divisions controlled by the ruling dynasty, descriptions of mountain ranges, river systems, taxable products and so on.

While science was slow to progress in Europe before the sixteenth century, with the **Renaissance** came a renewed interest in the geographical knowledge of the ancients (which the Arab and Chinese scientists had already advanced significantly) and a willingness to test and refine their theories. The European explorations of the fifteenth and sixteenth centuries were part of a major period of invention and discovery. Table 1.1 lists some of the important discoveries made during this period. Improvements in measuring devices such as timekeepers and in mapping and printing techniques were combined with a new geographical knowledge about the world. Indeed many of these new technologies had roots in the pursuit of geographical knowledge. For example, methods for accurately keeping time were developed when stable navigation systems that could determine the longitude (east–west position) of a ship were invented. As the Earth is constantly rotating, knowing the time while making an altitude measurement to a known star or the Sun provided data to accurately calculate longitude. The experiences of the explorers had begun to overturn traditional views of those thought to be authority figures (such as leaders of the Christian Church and the theories of the ancient Greeks). For example, new continents were being discovered and the layout of land masses across the Earth was being determined. A fundamental importance (as recognized much earlier by Al Muqaddasi) was beginning to be placed on the role of real-world experience. This meant that determining whether or not there was a Southern Ocean land mass could only be established through experience and not by just reading the works of Aristotle. This triumph of experience over authority was a central theme of the development of science during this

Table 1.1 Major discoveries of the fifteenth and sixteenth centuries

1410	A translation of Ptolemy's *Geography* was published in Europe (Ptolemy was an Egyptian astronomer and geographer who lived from AD 87 to 150)
1418	Prince Henry the Navigator established the Sagres Research Institute
1455	Gutenberg invented the printing press
1492	Columbus discovered the New World (although some suggest that there were earlier Norse settlements in North America and that there were original migrations from Asia and Europe around 14 000 years ago)
1498	Vasco da Gama sailed around the Cape of Good Hope to India
1500	Cabral discovered Brazil
1504	Columbus correctly predicted the total eclipse of the Moon
1505	Portugal established trading posts in East Africa
1510	Henlein of Nuremberg invented spring-powered clocks permitting smaller (and portable) clocks and watches
1519	Magellan's ships began a circumnavigation of the Earth
1543	Copernicus suggested the Earth and other planets revolved around the Sun
1543	Vasalius produced a detailed description of human anatomy
1556	Tartaglia (Venetian mathematician) showed how to fix position and survey land by compass bearing and distance
1569	Mercator created his map of the world using a projection technique that bears his name
1581	Galileo concluded that the time for a lamp (pendulum) to swing does not depend on the angle through which it swings. This observation eventually led to the development of pendulum clocks
1590	Zacharias and Hans Janssen combined double convex lenses in a tube, producing the first telescope
1592	Galileo developed a type of thermometer based on air

period. However, it was because geography was inextricably linked to exploration, patriotism and colonization that it was considered an important subject by the society of the time. Geographers were making the key advances in discovering new lands, mapping them, changing people's perception of the shape and size of features of the Earth and bringing potential 'wealth' to nations that conquered and colonized others.

1.2.2 Physical geography between 1800 and 1950

1.2.2.1 Uniformitarianism

Prior to the early nineteenth century the prevailing belief of the western world had been that the Earth was created in 4004 BC. The landscapes of the Earth were thought to be a result of catastrophic events. For example, it was thought that river valleys were scoured out during the biblical flood and that peatlands were remnants of the slime left behind after the flood receded (Turner, 1757). However, the increasing scientific knowledge acquired between the sixteenth and the end of the nineteenth centuries began to lead to different views developing. One new idea that emerged, for example, was that the Earth's landscapes gradually changed over time rather than being affected by sudden catastrophic events. Indeed one of the most persistent and influential themes to affect physical geography and especially geomorphology was the *Theory of the Earth* published by James Hutton in 1795 and clarified by Playfair (1802) in his *Illustrations of the Huttonian theory of the Earth*. Hutton and Playfair were scientists who examined the Earth's landscapes and tried to understand their formation. Hutton's theory rejected catastrophic forces as the explanation for environmental features and gave rise to a school of thought known as **uniformitarianism** (Gregory, 1985). The central component of this concept is that present-day processes that we can observe should be used to inform our understanding of past processes that we cannot observe. In other words, many of the processes we can see today are probably the same as those that occurred in the past and so we can infer what went on in the past from understanding contemporary environmental processes. Uniformitarianism propagated the idea that 'the present is the key to the past'. Although this idea was very satisfactory in terms of the processes for understanding the past, of course it cannot be assumed that the rates at which processes operate today (e.g. weathering of rock) are the same as those that occurred in the past. Nevertheless it was still recognized that given enough time a stream could carve a valley, ice could erode rock, and sediment could accumulate and form new landforms. It would take tens of thousands of years to weather the rock to produce the features shown in Figure 1.2(a)–(d) and indeed Hutton speculated that millions of years would

have been required to shape the Earth into its contemporary form. It was not until the early 1900s and the discovery of radioactivity that estimates of the age of the Earth became more reliable. Radioactive elements such as uranium and strontium are unstable and decay at a steady rate. Uranium-238, for example, decays into lead-206. Comparing the ratio of these two elements allows us to determine how much time has passed since the uranium sample was pure when the rock solidified. Radioactive decay also gives off heat and we can determine the rate of Earth cooling to determine a time when it formed. The Earth is in fact around 4.6 billion years old. The oldest rocks that have been found on the Earth date to about 3.9 billion years ago.

1.2.2.2 Darwin, Davis and Gilbert

Charles Darwin was a brilliant scientist who collected and organized specimens. He read some of the writings on uniformitarianism and extended these ideas to biology. Darwin's *The origin of species* published in 1859 was hugely influential in the field of science and in society in general. Indeed it has often been referred to as the 'book that shook the world'. The book outlined how there could be a relatively gradual change in the characteristics of successive generations of a species and that higher plants and animals evolved slowly over time from lower beings. This evolution occurred as a result of competition within local interacting communities (see Chapters 19, 20 and 21). Darwin's book helped throw the idea that there was a complete difference between humans and the animal world into turmoil as he reinforced the suggestion that humans evolved from lower beings. Prior to this it was believed in the western world, based on biblical works, that humans were created superior to other beings. With the idea that humans could have evolved from lower beings came the undermining of traditional religious opinions. However, although some religious leaders did embrace Darwinism at the time, the theories were very different from those that had come before. These ideas radically shook a society where, because of the increasing availability of printed books and papers, intellectual knowledge was being transferred in greater quantity than ever before.

Darwin's ideas influenced most areas of science at the time. The idea of 'change through time' was reflected in evolutionary attitudes to the study of landforms following Darwin's own 1842 study of the evolution of coral islands which was particularly influential in relation to the 'cycle of erosion' idea promoted by W.M. Davis (Gregory, 1985). The approach recommended by Davis, who was a very revered geomorphologist, dominated approaches to

(a)

(b)

(c)

(d)

Figure 1.2 Eroded landforms: (a) a weathered rock face; (b) rugged cliffs eroded by coastal processes; (c) an isolated rock weathered by water, wind, ice and chemical action over thousands of years; (d) an eroding mountain.

physical geography from the late nineteenth century through until the 1950s. Davis suggested in 1889 that the normal cycle of erosion could be used to classify any landscape according to the stage that it had reached in the erosion cycle. He termed this the 'cycle of life', which was a rather biological metaphor for landform development. Figure 1.3 shows the **Davisian cycles of erosion**. A youthful uplifted landscape begins to be dissected by rivers. As the landscape matures these valleys become wider and more gently sloping until eventually all that remains is a flat, old landscape (a **peneplain**). The great success of the Davisian

approach, dominating popular physical geography for 60 years, was due to the fact that it was simple and could easily be applied by people to a wide range of landscapes. As a result of these ideas people then tried to determine the history of an area by establishing which stage of the Davisian cycle it was in. This approach was also known as **denudation chronology**. While Davis had based his ideas on the case study of the Appalachians in the United States, the Davisian ideas were applied by many to help interpret landscapes across the globe (e.g. Cotton, 1922, applied the ideas to parts of New Zealand and Wooldridge and Linton, 1939,

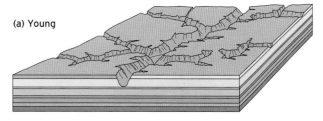

(a) Young

Recently uplifted with new incision

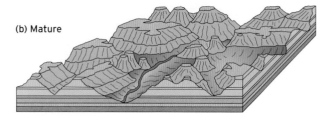

(b) Mature

Deep and widespread valley incision

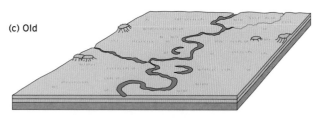

(c) Old

Almost flat, featureless peneplane, with the landscape eroded away

Figure 1.3 The Davisian cycles of erosion: (a) young uplifted stage with very limited incision; (b) a mature stage with deep valley incision and complex topography; (c) an old eroded landscape with few topographic features. (Source: after Davis, 1889)

produced a Davisian interpretation of south-east England). In plant geography and ecology a similar influence was being expressed by Clements in his concept of succession (see Chapter 18). It is notable, however, that two themes of Darwin's work (struggle and selection/randomness and chance) did not have an immediate impact on physical geography (Stoddard, 1966). Indeed the unique contribution of Darwin's theory, which was 'random variation' whereby random change could occur to species from one generation to the next, did not really appear in work by physical geographers until the 1960s (Gregory, 1985). Nevertheless the theme of evolution provided an historical perspective to physical geography which still dominates geomorphology, studies of soils, biogeography and climatology.

An alternative approach that was advocated at the same time as the Davisian approach was that of G.K. Gilbert. Gilbert was an explorer of the American West. Gilbert

wanted to understand *why* particular landforms developed rather than just classify them as being youthful or mature. In order to understand landform development he recognized the importance of describing physical processes and deriving systems of laws that determined how a landform could change. He attempted to apply quantitative methods to geological investigations. His ideas, however, were not taken on board during an era dominated by the descriptive techniques offered by Davis. It was not until the 1950s that physical geography came to revisit his approach and that Gilbert's ideas finally won favour. Until the 1950s, therefore, physical geography was largely descriptive and was concerned with regions. It was concerned with the evolution of environments and their classification. There were virtually no measurements of environmental processes involved and if you look at geography books from that period you will see that they are structured by regions and simply describe regional climates, landscapes, resources and trade (e.g. L. Dudley Stamp's 1949 book *The world: A general geography*).

1.2.3 Physical geography since 1950

1.2.3.1 The quantitative revolution

In the 1950s European and North American geography was forced to change. It was realized that describing places and putting boundaries around them, where in fact real boundaries did not actually exist, was no longer a useful approach. The 1950s were a time of increasing globalization when more people began to travel by air to far-flung destinations and when television began to show programmes made around the world, thereby opening up people's experience and views of the world. Global trade was increasing and mass-produced items such as refrigerators, cars and plastic became much-wanted goods. It became more common for people to own goods that were made in other countries (e.g. Europeans buying Ford cars made in the United States). It therefore became evident that there were increasing human and physical interlinkages between regions. It was also a period of modernity in which there was a societal commitment to order and rationality, and to science as the driving force for future developments and improvements in infrastructure and lifestyles. Physical geography needed to maintain its academic status and it could no longer do so within a society that now had a 'professional' science (see below). The Davisian cycles of erosion could not be verified from a scientific perspective and furthermore they did not *explain* observations. It was too difficult to measure such slow processes over such large spatial scales. Arthur Strahler, a geomorphologist particularly interested in rivers and landform change, proposed that a new dynamic basis for physical geography should be developed based on physical real-world measurements. It was also at this time that Hack (1960), a physical geographer, went into the Appalachians

(coincidentally the very heart of the Davisian theory) and realized that landscapes were more delicately adjusted and that there was some form of equilibrium between rivers and landscapes. Box 1.1 describes these equilibrium approaches and their limitations.

FUNDAMENTAL PRINCIPLES

EQUILIBRIUM CONCEPTS IN PHYSICAL GEOGRAPHY

When Hack (1960) completed a field visit to the Appalachian Mountains he realized that rather than there being one long Davisian erosion process whereby rivers wore away mountains over time, there is in fact a more dynamic set of processes operating. He rejuvenated Gilbert's concept of 'dynamic equilibrium'. He suggested that every slope and every channel in an erosional system are adjusted to each other and that relief and form can be explained in spatial terms rather than historic ones. This work suggested that river profiles were never exactly concave. Instead, when sediment from a hillslope builds up in a river it has to steepen itself in order to move that sediment. Once removed, the river may become less steep in profile. In other words, the rivers and slopes would adjust to each other in an attempt to be in equilibrium.

Of course, the nature of equilibrium investigated depends on the timescale under investigation. Figure 1.4 shows forms of equilibrium

over three timescales (Schumm and Lichty, 1965). Note that over short timescales it may be possible to identify a static equilibrium (no change over time) or a steady-state equilibrium (short-term fluctuations about a longer-term mean value) while over longer time periods the equilibrium might be dynamic (shorter-term fluctuations with a longer-term mean value that is changing).

However, the concept of equilibrium has always been somewhat confusing because different people have chosen to identify different types of equilibrium and because the precise meaning is time dependent. Indeed, equilibrium may be just as generalized and untestable as the Davisian cycle of erosion it was meant to replace. Often it depends on where and when you measure something as to whether it will show equilibrium. Figure 1.5 illustrates this very simply for two systems that in the long term are behaving differently. Because the measurements were done at the times shown in Figure 1.5, it was not possible to identify the nature of the long-term trend and in fact different

trends have been determined from those that are actually occurring in the long term.

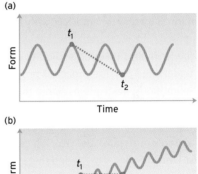

Figure 1.5 The timescale for human measurement makes it very difficult to identify long-term trends and the nature of equilibrium. Here the measurements are taken at two times (t_1 and t_2) for each case. However, because of the timing of the measurements we have incorrectly identified the nature of the long-term change in each case. In (a) we have established a downward trend where there is no long-term trend and in (b) we have identified no change while the long-term trend is upward.

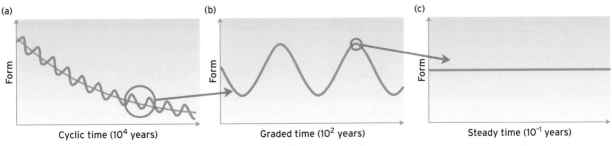

Figure 1.4 Equilibrium over three timescales: (a) dynamic equilibrium; (b) steady-state equilibrium; (c) static equilibrium.

BOX 1.1

It was also during this time that the work of G.K. Gilbert was revisited and his approach eventually embraced. This was largely due to the pioneering studies of the hydrologist Robert Horton and the development of his ideas by Strahler and his graduate students, who included Stanley Schumm, Marie Morisawa, Mark Melton and Richard Chorley. The 1950s are often referred to as the time of a quantitative revolution in geography due to the move away from description and towards measurement. Work began to concentrate on smaller spatial scales where processes could be measured during short-term studies.

1.2.3.2 Functional geomorphology

However, although quantitative techniques were being employed these were not necessarily those that Gilbert had proposed. The measurements that were being performed in the 1950s and 1960s often did not allow us to evaluate or understand physical processes properly. They tended to be quantitative descriptions rather than the measurement of processes. For example, in 1953 Leopold and Maddock, physical geographers who studied rivers, published results of a survey of streams and rivers in the central and south-west United States. They found that stream width, depth and velocity increased in proportion to the discharge to the power of a given number (e.g. width is proportional to discharge to the power of 0.5; see Chapter 12). As the discharge increased downstream the equations suggested that channel width, mean depth and mean velocity should all increase. These equations could be used to make predictions about the discharge or **hydraulic geometry** of rivers across the world (see Chapter 12).

There are two problems with this approach. The first is that the relationships determined are purely statistical relationships (or functional relationships). In other words, they are just a result of the average value of the width, depth, velocity and discharge of all the rivers that were measured, but this does not explain *why* channel dimensions vary in such a way with discharge. These sorts of statistical relationship do not explain the physical processes. The second problem is that such functional geomorphological approaches are often not applicable to areas other than the area for which they were determined. This is because local factors can influence the development of a landform (such as geology or tree roots on a river bank holding the bank together and preventing it from eroding) so that it does not conform to the statistical average. Indeed sometimes it is the unusual cases that we are more interested in rather than the average. It was for these reasons that Yatsu (1992), a Japanese physical geographer with expertise in rock

weathering, accused Strahler himself of 'crying wine and selling vinegar'. This means that he thought Strahler had advocated a new physical geography founded in physically based process measurement (the 'wine') but that Strahler was actually not measuring the physical processes. Hence Strahler was actually advocating a poorer type of physical geography (he was selling only 'vinegar'). Strahler and others around him made measurements but these were measurements of the wrong type. They were not physically based process measurements. It was Yatsu in 1966 who stated: '[Physical geographers] have often asked the what, where and when of things and they have seldom asked how . . . and they have never asked why. It is a great mystery why they have never asked why.' It was not until the mid-1970s that physical geography more fully adopted the idea of measuring processes in order to understand and explain the world. It was during this period that physical geography managed to embrace 'basic scientific methods' (see below). Nowadays the emphasis is very much on understanding processes (e.g. how vegetation interacts with soils and soil development – Chapters 8 and 18; how vegetation and climate interact – Chapter 25; or how glaciers erode their beds and move over the land surface – Chapter 16). However, as the following sections will demonstrate, although a basic scientific method may be adopted, there are actually many types of scientific method proposed and used, and each method can be criticized: this should be borne in mind when reading through the rest of this textbook and during your studies in general.

Reflective questions

➤ How did Darwin's *Origin of species* impact physical geography?

➤ What were the main differences between physical geography before 1950 and that since 1950?

➤ Can you explain what is meant by 'functional geomorphology'?

1.3 Scientific methods

Science uses a number of techniques and has adopted several philosophical approaches. It is important for physical geographers to be aware of such methods so that we can evaluate the advantages and limitations of the way in which we are approaching a research topic. The following section provides information on the key principles of scientific method.

1.3.1 The positivist method

Modern science is grounded in observation. It places special emphasis on empirical (derived from experiment and observation) measurements over theoretical statements (human ideas about how a system might work). Standard scientific method involves the formulation of hypotheses (e.g. 'the Earth's climate has warmed over the past 100 years') and the collection of information to test these hypotheses (e.g. comparing temperature measurements). This helps us explain why physical geographers (and scientists in general) are preoccupied with experimentation and measurement.

One of the key elements of scientific method is that of **causal inference**. This is the idea that every event has some sort of cause and so causal inference is the process by which we link observations under this assumption. However, it is rare for causal inference not to be affected by what we know and think already. Therefore there must be some theoretical basis for research ideas. At the same time, however, the theories are tested so that observations still have greater status than theories.

One of the key approaches for linking theory and observation is known as **positivism**. Positivism is a traditional philosophy of science which has its origins with the philosopher Auguste Comte in the 1820s. Its idea was simply to stick to what we can observe and measure and ensure that science was separated from religious explanations for phenomena. It uses repeatable research methods so that the same tests can be performed again as a quality control. There are different forms of positivist approach. Two important ones are **logical positivism** (or logical empiricism) and **critical rationalism**.

In logical positivism scientists use **inductive** reasoning whereby jumbled knowledge is defined, measured and classified into ordered knowledge. Once it is ordered, then regularities may be discovered and these might suggest the existence of a natural law. Logical positivism uses experiments to acquire knowledge. For example, we might measure water table depth and find that it seems to control the release of carbon dioxide from peat. If we find that in peatland areas with deep water tables, carbon dioxide release is high compared with areas where the water table is shallow, then we may develop a theory that water table depth controls the rate of carbon dioxide release from the peat.

The logical positivism approach was criticized, however, by the philosopher Karl Popper, who suggested that you could not derive a law based on this technique. This is because the failure to do infinite experiments means that there could always be a case that does not fit the law. It may just be that you have not yet found the case that does not obey the law (e.g. the peatland where carbon dioxide release from the peat is greater when water tables are shallow). Instead Popper argued that critical rationalist approaches should be adopted whereby you start with a theory that leads to the formation of a hypothesis. Then you test the hypothesis to the limit in an attempt to falsify the hypothesis. In other words, you try to disprove the hypothesis (rather than try to prove it). This is based on **deductive** reasoning. Figure 1.6 provides a conceptual diagram of this approach. If a theory survives falsification after testing it can still not be proved. You can only say that it has not been disproved. This is the basis for most scientific approaches today. For instance, one theory may be that all sheep have four legs. You test this by observing sheep. If you find every sheep has four legs then you can say that the evidence corroborates the theory and *probably* all sheep have four legs. Many statistical analyses of data result in a calculation of the probability that a theory is correct and has not just occurred by random chance. You cannot be certain, however, that your theory is 100% correct and it is impossible to verify the hypothesis (particularly as not all sheep have been born yet). However, if you find a three-legged sheep then you can say that not all sheep have four legs. In this case you

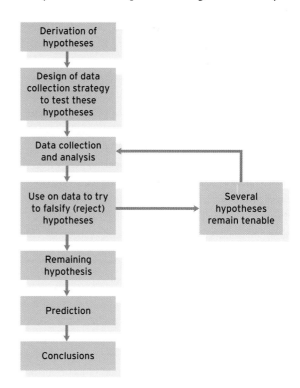

Figure 1.6 Schematic diagram of the typical experimental approach used by critical rationalists.

have falsified the hypothesis and now you must refine the hypothesis and test a new theory. The data collection, analysis and hypothesis testing strategy can involve field measurements, laboratory tests or even numerical modelling.

1.3.2 Critique of the positivist method

While physical geography has now had a 'scientific' basis since the 1950s there are a number of criticisms of the use of traditional science within environmental subjects. These include the following:

1. The idea that it is impossible to do completely objective science: there is always a human who is doing the experiment and that person will always apply some sort of meaning to the research. In other words, there is an interaction between the enquirer and the research they are carrying out. Indeed, sometimes a scientist might be motivated by a range of external forces when doing their experiments. Some scientists are funded by a specific organization with a vested interest in showing that results indicate a particular thing. This may remove independence and objectivity from research (perhaps even only subconsciously) and is why scientists are asked, ethically, to declare their funding source when they publish results.

2. 'Context stripping': this means that although experiments might be well controlled and repeatable, because they involve simplification (e.g. manipulation of water table causing a change in carbon dioxide release) the fact that there may be more complex interactions is often ignored. In this way positivism is **reductionist** because it assumes one can 'close' a system and look at the relationship between two variables in that system while holding all others constant. For example, if we think about the peatland carbon dioxide example discussed above then water tables may themselves be controlled by the vegetation (which may influence how much water is lost to the atmosphere via **evapotranspiration**). The vegetation in turn also affects carbon capture and release from the peatland ecosystem and the types of microbes present within the peat which may also control decomposition and hence carbon dioxide release. Thus, simple positivist experiments may lack the environmental context required for understanding the processes that are operating.

3. Positivist approaches tend to separate out grand theories from local contexts: positivist approaches often produce statistical generalizations that are statistically meaningful but that are inapplicable to individual places (e.g. a general equation based on data from 100 sites that links slope and the frequency of meanders on a river, but which does not seem to work for a particular site being studied).

Figure 1.7 An anemometer which is used to measure wind speed. When there is wind the cups rotate and a counter can record the number of revolutions per second.

4. By actually measuring something we may be changing it: for example, if we wanted to measure the erosion on a hillslope (see Chapter 9) we might dig a pit to collect the sediment. However, by digging that pit we may be changing the rate of erosion on the slope. Similarly if we wanted to measure wind speed we might place an anemometer (Figure 1.7) into the air to measure it. However, by placing that device into the air we may be changing the velocity of the wind around it compared with when the anemometer is not there.

5. Not all things are necessarily measurable, yet they may still be important.

1.3.3 Realism as an alternative positivist approach

In the 1990s realism gained popularity in physical geography as an alternative positivist approach that tried to get around some of the critiques discussed above. This approach attempts to reintroduce philosophy into method. In one version of realism that is often cited in the physical geography literature (e.g. Richards *et al.*, 1997) there are three levels into which any phenomenon can be structured (Figure 1.8): (i) mechanisms (underlying processes that cause things to happen); (ii) events that they produce depending on circumstances; and (iii) the empirical observations of those events made by humans. This approach accepts that in complex open environmental systems, interacting mechanisms may

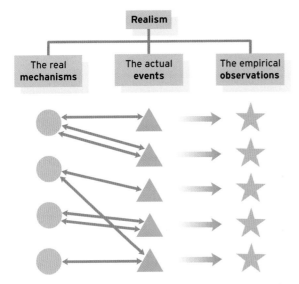

Figure 1.8 Components of realism. Different combinations of mechanisms result in different events, some of which can then be observed or measured.

not always produce an event. Events occur only when the mechanisms interact at the right place at the right time.

Figure 1.8 shows how events are associated with the coming together of mechanisms, which are themselves influenced by other events. An event may be the shattering of a piece of rock. The underlying mechanisms to produce that event will include fluctuating temperatures above and below freezing, the presence of water in rock, and expansion and contraction of water within the rock upon freezing and thawing. However, the rock shattering will occur only if the rock contains water and if the rock has been weakened sufficiently by previous weathering. Therefore, while there may be fluctuating temperatures that cause freezing and thawing of water within a rock, it will only shatter (the event) when the right combination of processes (mechanisms) occurs at the right place and right time.

Different combinations of mechanisms result in different events. For example, a meander bend may form at a given point only if: (i) there is a river flowing past the point; (ii) the right turbulent flows, sediment erosion and transport mechanisms are operating; (iii) the soil material is readily erodible to allow a meander bend to form; and (iv) local vegetation material or geology does not restrict the meander bend formation. In this case the exact size, shape and location of the meander bend will be a result of general meander-bend-producing processes and local factors. In addition this also suggests that the existing landform itself changes the way processes interact and thereby influence how the landform will develop in the future.

This approach brings together the idea of general laws and local events. It helps us understand how local factors influence the mechanisms to produce an individual example.

In this way it is different from traditional positivist approaches which are just interested in generalizations and iron out any irregular or unusual cases. In realism the case study itself becomes of interest in helping us to understand the world around us. This is because what causes change is the interaction of mechanisms with particular places. This is particularly useful for geography, which has a tradition of examining case studies (see below). In addition, the realist approach allows us to evaluate different results at different scales. Often great scientific progress can be made when rules that emerge at one scale of investigation are used to question and critique those that emerge at other scales (Lane, 2001).

1.3.4 No such thing as a single scientific method in physical geography?

More recently, there have been moves by physical geographers to state that while it matters what approach they decide to take, it should not matter that their approach is different from another person's approach. In fact Lane (2001) argued that one of the ways science moves forward is by trying to solve disagreements between one set of findings and another set of findings that have been produced by a different method. It tries to make sense of disagreements. One of the important elements of physical geography is that it is examined by using a wide variety of techniques and approaches which makes for a rich subject and a diversity of findings. If everyone had the same method and approach to physical geography then this might actually hinder scientific progress, discovery and innovation. However, Brown (2004) argued that all physical geographers must assess and present the uncertainty of any data or findings in a common format.

A good example of the benefits of taking multiple approaches is provided by Chapters 22 and 23 which discuss the environmental changes during the past 2.6 million years. In order to investigate such changes it has been necessary to develop a wide range of techniques including climate, ocean circulation and ecosystem modelling, sediment dating techniques, analysis of different types of evidence (e.g. pollen records, fossils, gas bubbles trapped deep inside ice sheets and the contents of ocean floor sediments) and landform interpretation. Indeed much of physical geography is based on 'historical science'. Historical science in this sense means taking measurements that involve historical inference and developing an explanation of phenomena where it is not possible to measure directly the processes involved. An example would be determining the location of an ice sheet 23 000 years ago. To do this would require 'proxy measures' (physical measurements that are based on present evidence of past conditions). The approach is often to build up as much evidence as possible from many different types of sources (e.g. landform shape, radiocarbon dating, fossil plants) so that the hypothesis becomes more acceptable as the multiple lines of evidence are compiled.

It is often the mixture of backgrounds that physical geographers have, and the mixture of ideas about what they want to study, how much they want to achieve, and who they have been collaborating with, that defines how they will approach a new problem given to them. What is important is that you are aware of the advantages and limitations of the various approaches to physical geography so that you can be ready to incorporate them into the way you and others interpret the research findings of investigations in physical geography. The following section describes some of the advantages and disadvantages of field, laboratory and modelling approaches.

Reflective questions

➤ What is a positivist approach to science? What are its benefits and weaknesses?

➤ What are the differences between logical positivism and critical rationalism?

➤ Why is realist theory so attractive to physical geographers?

1.4 The field, the laboratory and the model

Fieldwork, laboratory investigation and numerical modelling are not necessarily independent of each other. Laboratory analysis often requires fieldwork, and numerical models typically require laboratory or field data for validation. Throughout this textbook the findings that are discussed and the processes that are explained will have been investigated using a mixture of field, laboratory and numerical modelling approaches. The following sections identify some of the main types of issues involved with choosing a particular method.

1.4.1 Approaching fieldwork

In physical geography, fieldwork is a fundamental component of research. It is a means of obtaining experience of the environment. However, there is a more specific set of reasons for doing fieldwork rather than other types of investigation. Harré (2002) suggested there were 12 uses of an experiment. These are listed in the first column of Table 1.2. We can see that there are a wide variety of reasons and purposes behind the experimental method. However, the type of experiment and purpose of the experimental method will vary depending on whether the research is extensive or intensive.

Table 1.2 Harré's purposes for an experiment and the role of intensive and extensive fieldwork and of modelling. ✓ and ✗ indicate whether they fit this role

Harré's role of an experiment	Fieldwork Intensive/case study	Extensive case study	Numerical models
As formal aspects of method			
1 To explore the characteristics of a naturally occurring process	✓ Detailed environmental processes examined	✓ General environmental relationships	✓ Traditional role of modelling to examine a range of scenarios
2 To decide between rival hypotheses	✓ Choose between potential processes	✓ Choose between potential influences	✓ Parts of model are tested to see which are the best hypotheses
3 To find the form of a law inductively	✗ Hard to find laws from case studies	✓ An important aspect	✗ Not possible because you need laws before you can get a model to work

Table 1.2 (*continued*)

Harré's role of an experiment	Fieldwork		Numerical models
	Intensive/case study	Extensive case study	
4 As models to simulate an otherwise unresearchable process	✗	✗	✓ One of the most useful aspects of modeling
5 To exploit an accidental occurrence	✓ Good for investigating unusual cases	✗ Not included as by definition it is not accidental	✓ Used more to understand potential accidental occurrences such as a landslide
6 To provide negative or null results	✓ We can falsify from one case study and the advantage of this approach is understanding why it is false	✓	✗ Not possible as we do not know whether the negative result is just because we use models or whether it is a real falsification
In the development of the content of a theory			
7 Through finding the hidden mechanism of a known effect	✓ Main justification for case studies	✗ Hard to investigate mechanisms behind data	✗ We cannot find hidden mechanisms if they are not included in the original model
8 By proving something exists	✓ Demonstrating something exists	✗ Demonstrates relationships but not phenomena	✓ They can corroborate findings of other methods
9 Through the decomposition of an apparently simple phenomenon	✓ Main justification for case studies	✗ Generally not enough detailed knowledge available	✗ Not possible
10 Through demonstration of underlying unity within apparent variety	✗ More likely to identify difference	✓ Identifies generalizations	✓ They can identify general patterns
In the development of technique			
11 By developing accuracy and care in manipulation	✓ Design for particular applications	✓ Statistical methods	✗ Not possible
12 Demonstrating the power and versatility of apparatus	✓ Develop/test new technique at case study site	✗	✗ Not possible

(Source: after Harré, 2002; Lane, 2003)

1.4.1.1 Extensive research

Extensive research consists of a large number of samples over many places with empirical observations as the basis for theoretical development. Extensive research methods involve measuring a large number of samples so that we can make statements about the entire population. For example, we might measure 1 000 000 of the sand grains on 1000 beaches. From this we then attempt to make a statement about all the sand grains on beaches across the world. In other words, we infer things from the sample for the entire population. It is just like asking 1000 people which party they would vote for in an election and then assuming that those 1000 people were representative of the entire country. However, we qualify our statements by saying what the probability is of a statement being correct. For example, we may state that we are 95% certain that the results have not occurred purely by chance. It is not possible to be 100% certain unless the entire population has been measured, which in physical geography is generally not easy! There is a great deal of literature on extensive sampling methods (e.g. Burt *et al.*, 2006).

1.4.1.2 Intensive research and case studies

Intensive research may be restricted to one or two places in which theoretical developments are used as the context for empirical observations and whereby process and mechanisms can be investigated. Case studies are intensive field research methods and are used widely in physical geography. They can involve sets of measurements taken of natural conditions, or indeed of conditions that humans have disturbed (e.g. an agricultural field compared with a natural hillside). It is very common for researchers to do field experiments at case study sites whereby you manipulate a factor (e.g. you change the vegetation cover) and measure how this affects other factors (e.g. water infiltration). In many ways the approach adopted in a field experiment could be the same as in a laboratory (see below) but the environmental context would be different as you have less control over the other factors that operate, but at the same time the situation is likely to be more representative of the environment you are studying (Figure 1.9).

There are a number of problems with using case studies. Firstly, if you spend a lot of time in just one place it must be asked whether the findings are relevant to other places. In most cases they will not be. Therefore, we have to ask what is actually the point of doing intensive research at just one place (e.g. on one glacier in New Zealand)? Case studies provide us with a better way of looking at how things are related. In extensive studies it is often hard to establish whether particular variables are really related to each other

Figure 1.9 A field experiment to test whether distance from a pool in a peat bog influences gas release from the peat. The work could have been conducted in the laboratory, but the field environment may provide more realistic conditions, albeit with more complications.

or whether other factors have affected them. In small-scale case studies, however, we may be able to see how one variable affects another. Furthermore, a case study fits into the realist perspective outlined above which states that we are interested in how general relationships occur in particular places. Therefore case studies are used to focus on individual events in order to identify how the processes are coming together to create that event (Richards, 1996). This helps us identify the causes of an event (e.g. a landslide). In other words, by using a case study in the field we can establish how processes operate and how they come together at a particular place and time to produce an event.

Table 1.2 shows how case studies and extensive field studies fit with Harré's roles for experiments. By using this table it is possible to identify the reasons why we might do fieldwork. However, it should be noted that Harré's table is heavily biased towards the experimental sciences, and of course much of physical geography is based on historical science.

1.4.2 Approaching laboratory work

Physical geographers use the laboratory for two reasons: (i) analysis of something sampled in the field in order to derive its properties (e.g. concentration of carbon dioxide in an air sample); and (ii) experimentation in order to see how something behaves under controlled conditions. Of course with laboratory work care is needed in order to ensure that samples are not contaminated and that they are stored correctly so that they are not affected by laboratory conditions (e.g. storing ice cores in a freezer at the correct temperature). Laboratory work also requires very careful research design so that we can be sure that a 'fair test' has been carried out.

A good example comes from determining whether sulphate deposited in rainwater controls the release of dissolved organic carbon from soils. Sulphate has historically been released in many parts of the world where industrialization has occurred due to the burning of fossil fuels (Chapters 10 and 23). This led to rainwater with high concentrations of sulphate resulting in more acidic rain ('acid rain'). More recently, however, in western Europe and North America the concentrations of sulphate in rainwater have been falling. At the same time the concentration of dissolved organic carbon in river waters has been increasing (Worrall and Burt, 2004). Dissolved organic carbon is a hazard because it leads to the water becoming discoloured (brown) and also makes it difficult to treat to make it safe for drinking. In order to test whether the decrease in sulphate deposition is a causative factor in dissolved organic carbon release, a laboratory experiment can be set up. This can be done for three soil types as some soils might behave differently from others.

In the laboratory this requires a number of samples of each of the soil types in order that we have enough replicates to be sure that one result was not just an unusual result compared with others. It also requires that we establish whether sulphate in rainwater has any effect at all. Therefore we would need to put some of the soils into distilled water and some into rainwater with a sulphate solution. The distilled water tests act as a control with zero sulphate. We could then measure how much dissolved organic carbon is released from the soil samples. A weak sulphate solution and a strong sulphate solution could also be tested. In this case, there would be nine **treatments** (three sulphate solutions × three soil types). We would also need replicates of each of the treatments in order to establish how much variation there was. So if we said we wanted eight replicates per treatment this would result in 9 × 8 samples being required, a total of 72 (Figure 1.10). Thus, the number of samples can increase dramatically if you want to include more variables (e.g. another soil type, test the effect of water pH on dissolved organic carbon release and so on). Therefore in order to avoid a lot of wasted resources it is very important to make sure that the research design is carefully considered.

A range of critiques of laboratory work have been put forward, including the difficulty in scaling up results from the laboratory to the field and, like models (see below), the simplicity of the experiments ignores complex feedback processes and interlinkages that occur in the real world.

	Peat	Podzol	Brown earth
Distilled (control)			
Low-concentration sulphate solution			
High-concentration sulphate solution			

Figure 1.10 Treatments and replicates for a laboratory soil test. Note that there are eight replicates for each treatment. Treatments include distilled water, low-sulphate-concentration water and high-concentration water each on three soil types. There are therefore 72 separate tests for this simple experiment.

1.4.3 Approaching numerical modelling

Table 1.2 shows how numerical models fit into Harré's reasons for experimentation. Models can be used to help us understand whole systems, tell us which parts of the system are most important (by changing each part of a model separately you can see which changes result in the biggest differences overall), help us make predictions and simulate what might happen if we varied something in the landscape or climate (e.g. the effect of changing land management on catchment runoff and water quality). A good example of this type of modelling approach is given in Box 1.2. Models are ways of simplifying reality. For example, a map is a model of part of the Earth. There are distinctions between different types of model:

- conceptual;
- statistical;
- probabilistic;
- deterministic.

Conceptual models (e.g. a map, a flow chart) express the ideas about how processes in a system work. This is the starting point for any modelling, as without an idea about how things work you cannot do anything else. It is important to ensure that the concepts are correct before embarking on a numerical model.

Statistical models are used where data have been collected that allow relationships to be statistically established and then predictions to be made. For example, by measuring the diameter of 100 trees in each of 20 plots of a tropical rainforest it may be possible to predict how altitude or aspect influences tree growth. It may therefore be possible to make predictions about tree size in other areas of the rainforest just by determining the aspect or altitude. Note that this is a classic example of functional research.

CASE STUDIES

MODELLING PROCESSES WE CANNOT MEASURE

Gildor and Tziperman (2001) wanted to experiment with causes of past climatic change using a model. They developed a simple box model which is shown schematically in Figure 1.11 which represents the oceans, land and atmosphere. The land and oceans could be covered with variable amounts of ice. When there is more ice, the global climate (atmosphere) is cooler and also more of the Sun's energy is reflected back out to space by the ice. Rocks and water absorb more energy than ice thereby making the Earth's climate warmer (see Chapter 4).

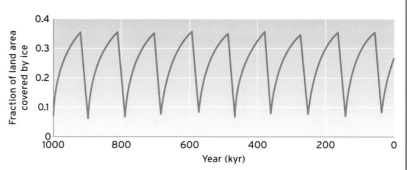

Figure 1.12 Land ice extent predicted by Gildor and Tziperman's (2001) model in the northern polar box as a proportion of total land area. (Source: after Gildor and Tziperman, 2001)

The ocean model was made up of four surface and four deep-water boxes as shown in Figure 1.11(b). The two polar boxes of the ocean may be covered by sea ice to variable extent. Each of the four atmosphere boxes had land, ocean, land ice or sea ice at its base, each with a specified reflection of the Sun's energy. The land part of the polar boxes may be covered by glaciers that can be of variable extent. Each model then used simple energy and mass balance equations.

When Gildor and Tziperman ran the model, which involved working through all of the equations at different time steps, they obtained an interesting variability on a timescale of 100 000 years. A result is shown in Figure 1.12 for 1 million years of model predictions. This result shows the land ice in the northern polar box as a fraction of land area. There is a sawtooth pattern whereby land ice cover gradually grows and then quickly melts every 100 000 years. This is a similar pattern to the real ice cover over the past 2.6 million years as determined by records from deep-ocean sediments (see Chapter 22).

Figure 1.13 shows the results of different parts of the model for a 200 000 year time series. Initially, the ocean has no ice cover (Figure 1.13b) and the atmosphere and ocean temperature in the northern box are mild (Figure 1.13c and g). However, glaciers start growing because there is enough moisture available to make snow accumulation greater than melting. The temperature of the northern atmosphere box is, while mild, still below freezing (Figure 1.13c). Thus, the reflection of the Sun's energy begins to increase and so the atmospheric and ocean temperatures begin to decline slowly (Figure 1.13c and g). Eventually, at 90 000 years, the sea surface temperature reaches the critical freezing temperature and sea ice begins to form very rapidly (Figure 1.13b). This further increases reflection of energy and decreases atmospheric temperatures. Sea ice can now grow more quickly. In less than 20 years the sea ice cover in the polar box is almost total. However, the sea ice growth is self-limiting as it insulates the ocean below it from the cold atmosphere and thus there are no more net losses from the ocean and thus no more sea ice growth.

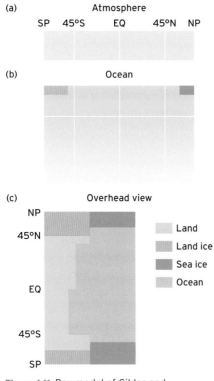

Figure 1.11 Box model of Gildor and Tziperman (2001): (a) side view of atmospheric boxes; (b) side view of ocean boxes with shaded regions representing sea ice cover; (c) top view of land and ocean with sea and land ice cover. (Source: after Gildor and Tziperman, 2001)

BOX 1.2 ➤

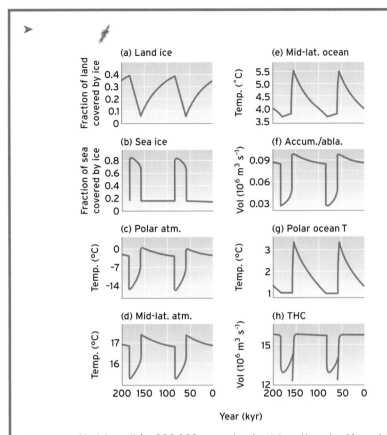

Figure 1.13 Model result for 200 000 years showing (a) northern land ice extent as a fraction of the polar box area (e.g. 0.1 = a tenth of the polar box is covered in land ice); (b) northern sea ice extent as a fraction of the polar ocean box area (e.g. 0.1 = a tenth of the polar box has sea ice cover); (c) northern atmospheric mean annual temperature, °C; (d) mean mid-latitude atmospheric temperature, °C; (e) deep-ocean mid-latitude temperature, °C; (f) source term and sink term (dashed) for northern land ice, 10^6 m^3 s^{-1}; (g) upper-ocean polar box temperature, °C; (h) thermohaline circulation discharge through the northern polar boxes, 10^6 m^3 s^{-1}. (Source: after Gildor and Tziperman, 2001)

At this point air temperature is at a minimum, and land and sea ice cover at a maximum. However, because of the low air temperatures there is a significant reduction in moisture flux to the poles (i.e. evaporation of water from oceans and land is reduced, thereby reducing precipitation). Furthermore, the sea ice prevents evaporation from the polar ocean and further reduces moisture sources for the land. This reduces precipitation for

glaciers and now melting is greater than accumulation. The glaciers recede and reflection decreases. Hence atmospheric temperature starts to rise. The oceans begin to warm slowly, too. However, because there is still sea ice cover, not enough moisture is being delivered to land glaciers and they melt fairly rapidly. Once the glacier coverage decreases significantly, atmospheric temperatures rise rapidly and the sea ice begins

to melt owing to heat from the mid-latitude oceans. The sea ice melting is slow at first (Figure 1.13b; 90 000 to 70 000 years) until the deep ocean has warmed sufficiently (Figure 1.13e; 70 000 years). At this point sea ice melt is rapid and within 40 years the ocean is free of ice. The deep ocean warms during this phase because it is affected by vertical mixing from the upper ocean, which is affected by global warming due to the melting of land ice. Hence the deep ocean produces a delay in the response of sea ice. Now the lack of land and sea ice increases temperatures dramatically and with it the amount of precipitation available for land ice growth. Here the cycle reaches its starting point again and the land ice slowly starts to grow.

This model suggests that sea ice plays a critical role in climate switching and that ocean circulation changes and instabilities are not the trigger of rapid climate changes. Instead the ocean circulation responds to rapid changes in sea ice. Obviously the model suffers from low resolution because there are only eight ocean boxes and four atmospheric ones. Nevertheless, Gildor and Tziperman (2001) showed that the mechanism the model predicts is independent of other theories of global climate change such as changes in the Sun's energy reaching the Earth's surface (orbital forcing; see Chapter 22) and yet produces similar patterns of ice growth and retreat to that seen in the ocean sediments. Hence, this sort of simple modelling approach can point us in new directions for research. The next step after these sorts of model findings is to find evidence for the hypotheses they raise.

BOX 1.2

Probabilistic models assume that there is some sort of random behaviour that is part of the system and assign a likelihood to events or data (e.g. 95% probability that this is the correct output). This is often expressed by a ranked numerical value or an estimate of best case, worst case or most likely.

Deterministic models assume a cause and effect link via a process mechanism. The models assume there is only one possible result (which is known) for each alternative course or action. The model contains no random components and hence each component and input are determined exactly. Both deterministic and probabilistic models are based on physical laws and obey rules of storage and energy transfer. There are a number of stages in developing these numerical models and these can be summarized as follows:

1. Start with a conceptual model based on theory and what is relevant.
2. Convert the theory into mathematical equations and rules.
3. Incorporate the equations into a simulation model (by connecting the equations together and giving them some sort of numbers as input data).
4. Apply the model to a real-world situation.
5. Compare the model predictions with real-world observations.
6. The model may now need some fine tuning by tweaking it so that it gives the best results for the majority of cases – note that this does not mean that it will work perfectly for all cases. It may even mean that while at first the model might work really well for one or two cases, the final model might not work as well for those cases. On average, however, it will work better overall when all cases are taken into account.
7. Now use the model to simulate and predict (as long as we know the limits of its predictive capability).

Of course numerical modelling is subject to a range of problems, including those associated with simplifying spatial and temporal scales in order to keep the model simple enough for a computer to run it. For example, if you wanted to run an atmospheric model that predicted how the air moved across the Earth's surface in response to pressure gradients, the surface topography and other forces (see Chapter 4), some topographic data would be needed. A decision would also have to be taken on how often predictions should be made. However, once topographic and air mass detail becomes very fine such as around 2 m × 2 m × 2 m resolution then it becomes computationally difficult to do all of the calculations that move air from one 8 m^3 cell to the next and also to do it for every cell at the same time. Furthermore, repeating this every minute will lead to enormous amounts of computer power being required. Therefore we have to make decisions about which processes to include and what spatial and temporal scales to adopt. We must always remember when analysing numerical model predictions that the model itself is just a simplification of reality and that often many real-world processes have been removed.

It is also often difficult to apply models developed at one scale to situations at a very different scale (e.g. applying a valley glacier model to the whole of the Antarctic; see Chapter 16). There are often problems in dealing with the net effect of very small-scale processes (e.g. how does a catchment runoff model deal with rapid water movement through soil cracks? See Chapters 11 and 13). Other problems include the fact that models can become so tweaked that they no longer reflect the processes they were originally intended to represent (Beven, 1989). One of the most frequent problems with numerical modelling is in not having enough real-world data (or good enough quality data) to test the model and thus the need for field and laboratory work in order to help provide this (e.g. Siebert and Beven, 2009). At the same time models can also help determine new areas for research. They often show which areas are the most important or which we know least about. This may be shown, for example, by changing the numbers in different parts of the model and investigating what effect that has on the overall model output. The part of the model that produces the biggest change in output overall can be called the most sensitive and therefore this part may be the most important to get right. Other parts of the model may have very little impact when they are changed and therefore are less important. Models may also be used to help us to predict or explain something that is actually impossible to measure (e.g. Box 1.2). In some cases it is not possible to compare the model predictions to real-world observations. Nevertheless, these sorts of models still provide us with useful information and understanding about the working of physical systems that may have not been possible to envisage without the numerical model.

Reflective questions

➤ What are the advantages and disadvantages of setting up an experiment in field conditions rather than in the laboratory?

➤ In what ways have the various field, laboratory and modelling approaches discussed above been typical of logical positivist, critical rationalist or realist philosophies?

➤ Why is numerical modelling a useful tool to physical geographers?

1.5 Using physical geography for managing the environment

Many of the world's environmental problems lie in unsound human management of the environment (see Chapters 24, 25 and 27). Humans have the potential to recognize and respond to opportunities and to threats that are natural or caused by humans, and to perhaps avoid or mitigate them (see Chapters 24 and 27). Physical geography provides a long-term understanding of environmental change within which to place contemporary environmental change. Physical geographers study past and present climates (Chapters 4–6, 22–25), tectonics (Chapter 2), oceans (Chapter 3), glaciology (Chapters 16 and 17), landform development (Chapters 7–17) and ecological and biogeographical processes (Chapters 18–21, 25). They also study the interacting nature of these global and local processes. Physical geographers are involved with measuring environmental change. This can be done using a wide range of tools, many of which are discussed throughout this textbook (e.g. remote sensing – Chapter 26; water and sediment monitoring – Chapters 7–17; ecological survey – Chapters 19–21). In addition to monitoring environmental change we are able to test and refine theories about how processes actually operate. This allows us to understand more readily whether changes we see today are part of normal dynamic Earth processes or whether they are the result of human interaction with the environment.

Through fieldwork, laboratory investigation and numerical modelling geographers are able to understand a wide variety of environmental processes and how these processes interact. This improved understanding helps us to predict the likely effects of future environmental changes. This is because our predictive models rely on us first having a good conceptual model of how the system operates. Geographers are able to contribute to the debates surrounding global climate change and can help provide policy-makers with some answers as to how best to deal with certain problems. Rather than just investigating whether the increased flooding in a particular catchment is due to climatic change or land management, physical geographers are able to take a more holistic approach (looking at how the whole system responds rather than just one bit of it). A traditional approach would simply explore a problem (e.g. flooding) in relation to possible causes (e.g. land management, climatic change). This approach fails to address the linkages between problems that emerge within particular catchments and where any one solution to a problem (e.g. blocking land drains) may have positive (enhanced biodiversity) and negative (release of dissolved organic carbon into rivers) impacts upon other parts of the environment. In other words, rather than thinking about just one thing at a time in environmental management, we ought to be seeking to look at the full range of possible effects.

Furthermore, because of the history of physical geography, geographers have had experience of bringing together large-scale approaches with small-scale approaches, linking case studies with general context. In this way physical geographers can readily adapt to the relevant scale of enquiry required by policy-makers. They can also link together scales of approach through process understanding. This is important because sometimes cause and effect are not clear and the alternative framework for thinking about environmental change is one in which form–process feedback interactions are considered. For example, Clifford (1993) suggested that water could undulate over an individual pebble in a river. This undulation in flow causes small-scale erosion upstream of the pebble and deposition downstream thereby changes the shape of the river bed around the pebble leading to an exaggeration of the undulating flow, which in turn makes a bigger feature on the river bed, eventually leading to a large pool and **riffle** (Figure 1.14).

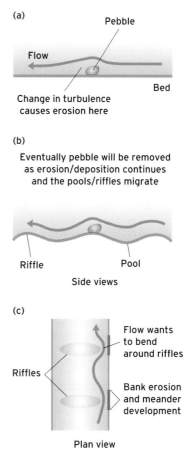

Figure 1.14 Flow around an obstacle on the river bed causes erosion and deposition around the obstacle (a). This in turn affects the flow at a larger scale which in turn affects the turbulent flow structure of the water, the erosion and the deposition processes, causing pool and riffle development and meander bend development (b) and (c). The whole process then becomes self-maintaining.

In turn this larger-scale feature will interact with the flow of water, which in turn influences meander development and the way sediment is eroded, transported and deposited. Thus there are feedbacks between processes and landforms at lots of different spatial scales. On a temporal scale, too, there have been approaches to physical geography that illustrated how the individual location, the history of that location, and therefore the exact timing of an event can be important in determining how a landscape will respond. Box 1.3 provides an example of this.

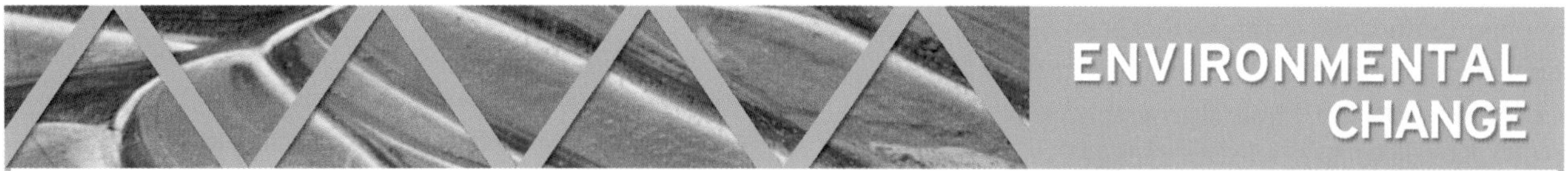

ENVIRONMENTAL CHANGE

MAGNITUDE-FREQUENCY CONCEPTS IN PHYSICAL GEOGRAPHY

In the environment there will be events that occur either very rarely or very frequently and those that occur somewhere in between. Typically small-magnitude and very large-magnitude events are very rare and so intermediate-sized events are most frequent. Wolman and Miller (1960) suggested that it is these intermediate-sized events that, in the long term, will do the most work (Figure 1.15). So for a river carrying sediment it might be the river discharge event that occurs twice a year that performs most of the sediment transport rather than the 1 in 100 year flood. At the time of the big 100 year flood it might seem as if there

has been a great deal of sediment transport. However, in the long term it is the twice yearly events that have carried the most sediment downstream.

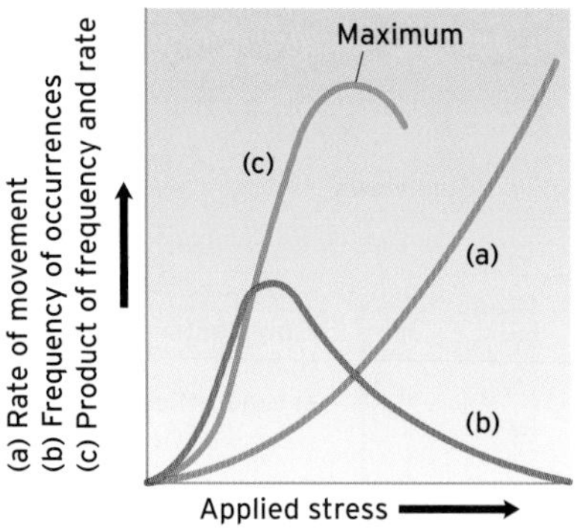

Equally the response of a particular site to a flood event may vary depending on how long it has been since the last big flood event. This means that there may be a flood of a particular magnitude in a river but it may not carry as much sediment as an earlier flood of the same magnitude. This may be because the first flood flushed out all of the available sediment that had been building from the hillslopes and so the second flood did not have much sediment available to carry. In Figures 1.16 and 1.17 there is a plentiful supply of in-channel sediment so that the next flood in these cases will be able to transport large amounts of sediment down the valley. Therefore the timing of an event in relation to

Figure 1.17 A river has undercut the steep bank resulting in a small landslide and a plentiful supply of local sediment which can now be transported downstream.

other events matters. Furthermore, different environments take different amounts of time to recover from a major event (such as a landslide or a flood). This is known as the **relaxation time**. In ecology, relaxation time is an important concept where disturbance can cause the population of species to decline. If there are a number of disturbances then the population will only recover to original levels if there is sufficient time between the disturbances for the recovery to take place (e.g. see Figure 20.2d in Chapter 20). Some environments are also more sensitive to change than others. Hence the environmental response to a particular process or event depends on where you are and at what time the event occurs in relation to other events.

Figure 1.15 The magnitude-frequency concept. In the long term the intermediate-sized events that occur perhaps only once or twice a year will do the most work. (Source: after Wolman, M.G. and Miller, J.P., 1960, Magnitude and frequency of forces in geomorphic processes, *Journal of Geology*, **68**, pp. 54-74, Fig. 1. Reprinted by permission of the University of Chicago Press)

Figure 1.15 axis labels:
(a) Rate of movement
(b) Frequency of occurrences
(c) Product of frequency and rate
Maximum
(c)
(a)
(b)
Applied stress

Figure 1.16 An upland river channel with lots of sediment available to be mobilized. How a landscape will respond to an event may depend on how long it has been since the last big event and how quickly a landscape recovers from events.

BOX 1.3

The fact that geographers have such a variety of approaches can be of enormous benefit to policy-makers. For example, while evaluating a river reach subject to erosion near houses, some geographers might examine the river sediment distribution to establish sediment transport processes and hence design some coarse stone block protection for the river banks. Other physical geographers might take a longer-term view of the river system and start to look for evidence of how the river has behaved over time (e.g. Lane, 2001). This might suggest a different management strategy. For example, if the geographers find evidence that the river has been subject to historically frequent avulsions (where the channel suddenly changes its position), then perhaps progressive abandonment of the area by those who live there would be a better strategy since no matter how much river bank engineering there is, that particular river may be prone to major channel change and homes would never be safe from the threat of flooding and erosion. Therefore, by using different skills and different scales of approach (long term and short term) geographers can aid environmental management decision-making and planning.

As a society we are just starting to learn about the problems with scientific research (as discussed above) and how it is approached. We are therefore moving away from a society that simply accepts science as a dominant source of knowledge to a position where society is just one of many different sources usefully contributing knowledge (Lane, 2001). Successful environmental management requires an enhanced engagement with the relationship between science and society and in human–environment interactions. This is where physical geographers should really be making headway in the world. Box 1.4 describes an example of where this has recently occurred. The last chapter of this book describes the considerations and processes involved in environmental management.

Reflective question

➤ Why is it beneficial to use a range of approaches and look at problems at different scales when considering environmental management?

CASE STUDIES

STAKEHOLDERS AT THE CENTRE OF SCIENCE – DEVELOPING A STRATEGY FOR MOORLAND MANAGEMENT

Many upland environments such as the moorlands of the Paramo in the Andes or the hills of the United Kingdom are very sensitive to environmental disturbance but have been managed intensively over the past few hundred years (Holden *et al.*, 2007). Often there are conflicting interests between those who live and work in these moorland areas and those who have different views on how the landscape should be used and managed (Reed *et al.*, 2010). There is often also a lack of scientific evidence about the impacts of some types of management and a lack of communication between scientists and the stakeholders involved. Furthermore, some stakeholders such as water companies are worried that what other stakeholders (e.g. farmers, conservation bodies) might do to the land could impact their water quality.

In order to tackle moorland management in a way that fully engaged geographical scientists with stakeholders while at the same time enabling the different stakeholders to discuss the issues with each other, a new project was initiated in UK moorlands and is described at www.env.leeds.ac.uk/sustainableuplands. The project began by identifying the stakeholders for case study areas and then asking them (both independently and in groups) what they thought the major problems were and how they would like the moorlands to be managed in the future (Figure 1.18). This way the stakeholders, and not the scientists, decided the important topics to be examined. Then the geographers collected new data, used existing data and developed integrated models to understand how the management scenarios that stakeholders had suggested would alter the vegetation cover, the water quality, the flood risk, the carbon balance, biodiversity and economic livelihoods of the area. The results were then shown to the stakeholders, who then reconsidered their management options in light of the results. Simplified versions of results were developed into short films which are available on the project website

BOX 1.4 ➤

provided above. The films involved stakeholders and scientists. The stakeholders have been redesigning their management plans and influencing the future research that geographers need to do. In this way the stakeholders have been at the heart of the science being done and were at the heart of the dissemination of that science.

Figure 1.18 Stakeholders and scientists discussing potential future management scenarios in the UK moorlands. (Source: photo courtesy of Natalie Suckall)

BOX 1.4

1.6 Summary

This chapter has shown that some of the early roots of physical geography belonged to the time of exploration and colonization. Since then a number of key figures have influenced how physical geography was done, and these include Darwin and Davis. Until the 1950s geography was predominately a subject of inventory and classification. It was rather descriptive in nature. However, during the 1950s this position could no longer be upheld and physical geography began to embrace positivist science. Many positivist studies of the 1950s and 1960s lacked a focus on processes and instead were interested in statistical relationships and regularities. The aim was to generate generalizable equations and laws that could be used across the world. However, by the 1970s it was realized that this did not help us understand the world. Only by understanding processes could we explain the cases that did not fit the normal statistical relationships. Thus, experimental methods were adopted that tried to determine cause–effect relationships. Inductive methods were replaced by deductive methods following the Popperian ideal of falsification. In the 1990s realism was one of the philosophical approaches that physical geographers adopted in order to get around the problems with earlier forms of positivism. This allowed geographers to recognize that all places are unique, yet there are underlying physical mechanisms that come together to produce an event. It is the place itself and the way the processes come together (and have come together at the place in the past) that matter. It is through case studies that we more adequately understand the operation of environmental processes. In a complex world where many processes are non-linear it is important to have the right framework for investigation. A framework that allows both intensive and extensive study and that appreciates that there are a range of feedbacks between a landform or an event and processes seems to be appropriate.

Fieldwork, laboratory work and numerical modelling are all important components of physical geography and many studies will adopt all three approaches. Each approach needs to be carefully designed and there are clear reasons for requiring each type of study. At the same time each approach has a number of problems and it is these that make science fallible. Behind every result presented and every graph plotted there was a human idea and a human reason for obtaining that result. Thus physical geography (and all science) has an element of subjectivity about it. A numerical model may produce huge amounts of data but that model can only be roughly as good as the original conceptual idea that was developed and can only produce results that are of equivalent or worse standard than the real-world input data it was given. Of course models can be used to simulate events that would otherwise be very difficult to measure and observe and they can be used to make predictions and guide future research directions.

This chapter has provided a somewhat different introductory chapter from those you will see in other physical geography textbooks. Rather than saying this is how we do things in physical geography it has tried to explain why we do certain things in the way that we do them. The aim of this chapter was to encourage readers to think critically about the subject of physical geography and the approaches to it. In some ways, therefore, I am encouraging you to read the remaining chapters in this textbook with a critical eye. While the contributors have put together a comprehensive set of material on physical geography to provide you with a baseline knowledge and understanding of the physical environment, it is also evident that there is still much we do not understand about the natural environment and human interactions with it. In particular the environment is always changing and is dynamic in nature. Sometimes we can predict fairly well how a system will respond to change, but in most cases we cannot – particularly those systems that are global in nature and consist of a complex set of process interactions all operating on different spatial and temporal scales. I hope that this book will be the first step towards inspiring you to be one of those that helps us to understand the way in which the world works. Because there are so many new questions, so many processes we do not understand and so many hypotheses that are constantly being revised, there is a huge opportunity out there for you to make an impact on our understanding of the world. Will you take this opportunity and engage with physical geography?

Further reading
Books

Clifford, N.J., Holloway, S., Rice, S.P. and Valentine, G. (eds) (2008)*Key concepts in geography*. Sage Publications, London.
There is a lot of relevant material in this book which further develops the material covered in the chapter and also covers integrating concepts with human geography.

Clifford, N.J., French, S. and Valentine, G. (eds) (2010) *Key methods in geography*. Sage Publications, London.
This book contains useful and clear chapters on a range of topics including data collection, data analysis and numerical modelling.

Gregory, K.J. (2000) *The changing nature of physical geography*. Arnold, London.
This is a very good short introduction to the history of physical geography over the past 150 years; particularly focus on chapters 2 and 3.

Inkpen, R. (2004) *Science, philosophy and physical geography*. Routledge, London.
This book describes science–philosophy interactions and how there is not one simple single scientific method. Some complex material is described in a very accessible way by the author.

Jones, A., Duck, R., Reed, R. and Weyers, J. (2000) *Practical skills in environmental science*. Prentice Hall, Harlow.
This book has lots of great information about field and laboratory techniques, approaches to sampling and designing projects.

Montello, D.R. and Sutton, P.C. (2006) *An introduction to scientific research methods in geography*. Sage Publications, London.
This book provides a basic outline of scientific approaches and data analysis techniques.

Rhoads, B.L. and Thorn, C.E. (eds) (1996) *The scientific nature of geomorphology*, Proceedings of the 27th Binghampton Symposium in Geomorphology, 27–29 September 1996. John Wiley & Sons, Chichester.
This is a really excellent edited volume which explores the main issues that surround approaches to physical geography. Chapters 1, 2, 5 and 7 are particularly useful. Chapter 7, for example, outlines the role of case studies in physical geography.

Schumm, S.A. (1991) *To diagnose the Earth; ten ways to be wrong*. Cambridge University Press, Cambridge.
Schumm's essay is a thoughtful discussion of methods in geomorphological work.

Woolgar, S. (1988) *Science: The very idea*. Methuen, London.
This is an important book on the problems of scientific method.

Further reading
Papers

Beven, K. (2007) Towards integrated environmental models of everywhere: uncertainty, data and modelling as a learning process. *Hydrology and Earth System Sciences*, 22, 460–467.
This journal paper is freely available online and presents relevant arguments around the use and limits of models.

Gildor, H. and Tziperman, E. (2001) A sea-ice climate switch mechanism for the 100-kyr glacial cycles. *Journal of Geophysical Research*, 106, 9117–9133.
This is the full paper providing details of the model described in Box 1.3.

Wolman, M.G. and Miller, J.P. (1960) Magnitude and frequency of forces in geomorphic processes. *Journal of Geology*, 68, 54–74.
The paper which has been highly influential in understanding the relationship between the size of events and their reoccurrence as described in Box 1.3.

PART II

Continents and oceans

Figure PII.1 White Island or Whakaari (in Maori) off the South Island of New Zealand is a volcanic island that emerged from the water surface around 16 000 years ago, which is very recent in geological terms. It is thought that the volcano itself is around 200 000 years old as this is when it started growing on the floor of the sea and it has since risen 1600 m in height from its base. In fact the largest mountains on Earth are volcanoes that have formed on the ocean floor with Mauna Kea being 10 203 m tall. New volcanic islands can form at any time when an underwater volcano grows to reach the surface. Sometimes an island can appear for a few years before being eroded by wave action and disappearing below the water surface again until the next eruption. This is the case for Home Reef volcanic island in the Tongan South Pacific.

Part contents

Scope

The previous part introduced many of the concepts that underlie approaches to the study of physical geography. A theme emerged which is that physical geography has shifted over the years to an emphasis on understanding the processes and mechanisms that drive the Earth's environmental systems. Part II offers a foundation for gaining such an insight by describing the processes involved in some of the most important underlying elements of physical geography: plate tectonics and the oceans.

The distribution of the land surface and the shape and nature of the oceans, which themselves cover 71% of the Earth's surface, are ultimately controlled by plate tectonics. The realization in the 1960s that large-scale crustal plate movements were occurring over the Earth revolutionized geological and biogeographical study; it became possible to understand why hazards such as volcanoes and major earthquakes are confined to particular areas and to explain the spatial distribution of mountain belts and flat plains. It is now possible to explain the three main types of rock we see on the Earth's surface, the topography of the ocean floor and the reasons behind geographical patterns of evolution. The development of plate tectonic theory and the concepts involved are approached in Chapter 2.

Throughout history, the oceans have often been perceived as a powerful and sometimes spontaneous natural force; only experienced fishermen had a sense of the behaviour of the sea. Although many of the secrets of the oceans remain guarded, our inquisitiveness as scientists has led to some spectacular discoveries. These include the pinnacle role played by the oceans in controlling the Earth's global and local climate system and the discovery of teeming life around volcanic vents in the ocean floor where there is no sunlight. The oceans are the major driving mechanism by which energy is circulated around the planet and their nature and importance are described in Chapter 3.

CHAPTER 2

Earth geology and tectonics

Michael D. Krom
Earth and Biosphere Institute, School of Earth and Environment, University of Leeds

Learning objectives

After reading this chapter you should be able to:

➤ describe how the theory of plate tectonics developed from ocean basin research

➤ understand the structure of the Earth and how plates are able to move across the Earth

➤ describe the nature of the three main types of rock found on Earth

➤ explain what happens at the boundaries of plates when they move apart or together in different ways

➤ describe how the theory of plate tectonics has enabled earth scientists to understand the processes of continental drift and the many geological features preserved on continents and underneath the oceans

2.1 Introduction

Until the mid-1960s, although geologists had assembled much detailed information on geological processes both on continents and in the ocean, there was no comprehensive theory to explain the interrelationship between these processes. It is no exaggeration to say that with the development of plate tectonics, the subject of geology was revolutionized. It is now possible to understand and explain such apparently widely divergent subjects as the location of volcanoes, where and why earthquakes occur and why mountain belts are located only in certain confined areas. We now understand the structure of the ocean including why the deepest part of the ocean is near the continents and not, as was once thought, in the centre of the ocean and why the crust under the ocean is much younger than most of the rocks on the continents. Many of the patterns seen in the evolution of fossils can now be explained and it has even been possible to explain why the mammals of Australasia are dominated by marsupials; the Americas have a few marsupials and Africa and Asia have none at all.

The purpose of this chapter is to explain our present knowledge of the structure of the Earth, including both its interior and its outer layers, which we now know are divided into plates. The history of the development of global plate tectonic theory will be explained and the chapter concludes with a section explaining the location and development of many of the major features of the continents as we see them today. An understanding of plate tectonics is fundamental to understanding a wide range of other subjects in physical geography including biogeography, oceans and global landform development.

2.2 The Earth's structure

The Earth is roughly spherical, as even some of the ancient Greeks realized. It is now known that it is an oblate spheroid, somewhat flattened at the poles with a radius of 6357 km at the poles and 6378 km at the equator. The flattening of the Earth is very small but the shape of the Earth is very close to that expected of a fluid rotating with the same velocity that the Earth travels through space. This suggests that, over a long timespan, the Earth behaves in some ways like a fluid.

2.2.1 The interior of the Earth

If the Earth was cut in half and the interior structure exposed, the layers would appear as in Figure 2.1. At the centre of the planet is a core that has a radius of 3470 km. The inner core is solid and very dense ($\sim$13 g cm^{-3}) with a thickness of 1390 km. It is magnetized, and has a temperature probably in the region of 3000°C. The solid inner core is surrounded by a transition zone about 700 km thick, which in turn is surrounded by a 1380 km thick layer of liquid material that together form the outer core. It is somewhat less hot than the inner core and also less dense ($\sim$11.5 g cm^{-3}). The core is thought to be made up of iron and nickel with a small amount of lighter elements such as silicon. The pressure at the core of the Earth is up to 3 million times the Earth's atmospheric pressure. The next layer is called the **mantle**, and contains the largest mass of any of the layers of the Earth, approximately 70% of the total mass. The outer part (180 km) of the mantle along with the overlying crust is called the **lithosphere** and is rigid; this floats on the more mobile **asthenosphere**. Beneath the asthenosphere is a transition zone (350–700 km deep) to the lower mantle, which is located from 700 km to 2900 km below the surface. The mantle is composed principally of magnesium–iron silicates and has a density of 3.35 g cm^{-3}. When the rocks in the mantle are subjected to differential pressure and heat, they flow slowly, rather like ice flows in a glacier (see Chapter 16).

2.2.2 The outer layers of the Earth

The Earth's outermost layer is the cold, rigid, yet thin layer that we are directly familiar with, known as the crust. The large continental land masses are formed primarily of granite-type rock, which has a high content of aluminium and silica. Quartz and sodium-rich feldspar are the two principal minerals present. The crust is comparatively thick ($\sim$35–70 km) and has a density of 2.8–2.9 g cm^{-3}. By

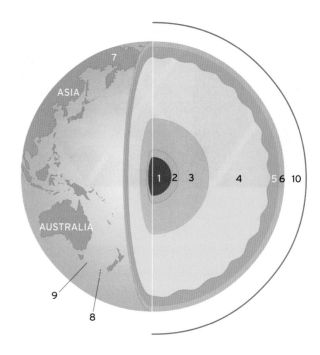

1 Inner core, solid, 1390 km radius
2 Transition zone, 700 km thick
3 Outer core, liquid, 1380 km thick
4 Lower mantle, semi-rigid, ~1200 km thick
5 Transition zone, 350–700 km deep
6 Asthenosphere, rigid, 180 km thick
7 Continental crust, rigid, 35–70 km thick
8 Oceanic crust, rigid, 6–10 km thick
9 Hydrosphere, ~4 km deep
10 Atmosphere, ~30 km+

Figure 2.1 The internal structure of the Earth showing the major layers and their known thicknesses. The core consists of an inner solid centre, a transition zone and an outer liquid zone. The mantle contains approximately 70% of the mass. The upper mantle consists of the asthenosphere and a transition zone. Above the asthenosphere lies the lithosphere, which is made up of the continental and ocean crust above which there is the hydrosphere and atmosphere.

contrast the rock underlying the oceans is mainly basalt, which is lower in silica and higher in iron and magnesium. The principal minerals in basalt are olivine, pyroxenes and feldspars rich in calcium. The oceanic crust is much thinner than the continental crust, being only 6–10 km thick, but has a higher density ($\sim$3 g cm^{-3}).

The boundary between the crust and the mantle is called the **Mohorovičić discontinuity**, named after its discoverer, and usually called simply the Moho. At one time it was thought that the Moho was the layer where the Earth's rigid crust moved relative to the mantle. However, it is now thought that the crust and the upper mantle together form the rigid upper layer known as the lithosphere (60–100 km thick) which floats on the asthenosphere and moves as the Earth's plates.

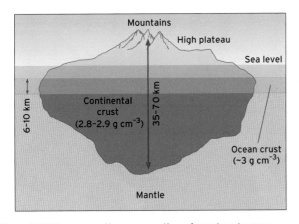

Figure 2.2 Diagrammatic cross-section of crust and upper mantle showing the principle of isostasy. This shows the less dense continental crust 'floating' on the upper mantle in a similar way to the manner in which an iceberg floats in seawater.

The difference between the average elevation of the continents and the oceans is determined principally by differences in the thickness and density of continental and oceanic crust (Figure 2.2). This is the principle of **isostasy**. A common analogy used to describe isostasy is that of an iceberg. The top of an iceberg is above sea level and is supported by the buoyancy of the displaced water below the surface. The deeper an iceberg extends below the surface, the higher the same iceberg reaches above the surface. Isostasy is, however, a dynamic situation. For example, when the major ice sheets, which until 10 000 years ago covered Europe and North America, melted (see Chapter 22) a significant weight was removed from the continents. These continents are still rebounding upwards as a result of slow flow processes in the upper mantle. An extreme example of this is in areas of west–central Sweden that are still moving upwards at a rate of 2 cm yr^{-1}. Areas can remain elevated where tectonic plates are colliding, as the crust thickens where two continental plates come together. This has occurred in the Tibetan Plateau for example. It can also occur in areas of volcanic activity such as Yellowstone National Park, USA.

Reflective questions

➤ What is the internal structure of the Earth?

➤ Imagine that the Earth was the size of a large (spherical) orange or apple. Based on the relative thicknesses of the different Earth layers discussed above, calculate what the thickness of the continental and oceanic crust would be for that apple. How does that compare with the average thickness of an orange skin (0.2 cm) or an apple skin (0.2 mm)?

2.3 Rock type and formation

The rocks that are found at or near the Earth's surface can broadly be divided into three types: **igneous rocks**, **sedimentary rocks** and **metamorphic rocks**. There are many varieties of each of these three main types of rock and the exact type of rock that is formed can depend on factors such as temperature, pressure and the minerals present at the time of formation.

2.3.1 Igneous rock

Igneous rocks are formed from **molten rock** (rock that has melted) which cools and hardens. If the molten rock is erupted at a volcano, then the subsequent cooled and hardened rock is very fine grained with crystals that can only be seen under a microscope. There are a number of fine-grained rocks erupted in this way including basalt, andesite and rhyolite (depending on the amount of silica in the magma) and volcanic glasses which are so fine grained they look and behave like glass. This type of rock generally cools quickly because it is exposed to the cooling air or water at the Earth's surface upon eruption. Basalt covers the ocean floor (which covers over two-thirds of the Earth's surface). If the rock cools more slowly then there is time for crystals to grow within the molten rock, and a coarse-grained rock is formed. Granites or gabbros are igneous rocks of this type containing large crystals (Figure 2.3a). The Sierra Nevada mountains, New Mexico, and Dartmoor, UK, are made up primarily of granite. In addition to grain size, igneous rocks are often divided into acid rocks (often light coloured) which are rocks formed from the melting of continental rocks and basic rocks (often dark coloured) formed most commonly by the melting of oceanic rocks.

2.3.2 Sedimentary rock

Sedimentary rocks are formed from the products of the chemical and/or physical weathering of rocks exposed at the Earth's surface. The sediment produced from such weathering can accumulate over time and eventually build up a deposit which, over time, can harden to form rock. Many sedimentary rocks are formed after the weathering products such as sand, silts or clays are transported by rivers and deposited downstream in coastal regions (see Chapter 9). Sediments are cemented together, and compacted and hardened over time by the weight and pressure of the sediments above them and by the precipitation of chemical cements such as calcium carbonate or silica. These processes result

(a) Granite

(b) Sandstone

(c) Mica schist

Figure 2.3 Photographs and thin sections of three different rocks. (a) Granite is an igneous rock. Note the large crystals of pink and white feldspar (which can be seen in both the thin section and the hand specimen) which form as the molten rock cools slowly within the Earth's crust. (b) Sandstone is a sedimentary rock. The hand specimen shows layering formed when the sand was laid down under the sea. In the thin section it is possible to see the individual sand grains which show up as black or white grains under cross-polarized light. (c) Mica schist is a metamorphic rock. Note the fine layering seen in both the thin section and the hand specimen brought about by the recrystallization of the original minerals under conditions of high temperature and pressure. (Source: photos courtesy of Bob Finch, School of Earth and Environment, University of Leeds)

in rocks such as sandstones, siltstones or mudstones (Figure 2.3b). Sedimentary rocks often contain structures that represent a record of the physical conditions present when the rocks were deposited. In addition there are sedimentary rocks formed by the accumulation of the remains of either the skeletons or the organic remains of microscopic animals. These can be as fossils within other rocks or can represent most of the rock itself such as chalks, which are almost entirely the remains of coccoliths (dead microscopic plants called algae). It is amazing to think that in some areas such as the south-east of England there are thick layers of the Earth's surface that are made up almost entirely from the skeletons (tests) of coccoliths (Figure 2.4). Other sedimentary rocks are formed when the concentration of a dissolved mineral in water is so great that mineral precipitates are formed. This can often happen when the water evaporates to leave behind the solid minerals. Halite or gypsum are formed, for example, when seawater is evaporated towards dryness.

2.3.3 Metamorphic rock

Metamorphic rocks form as a result of partial melting and recrystallization of existing sedimentary or igneous rocks. These changes usually take place where there is also high pressure such as under hundreds of metres of bedrock or where rock is crushed at the junction of tectonic plates (see below). As a result, many metamorphic rocks have a layered structure caused by this external pressure (Figure 2.3c). Metamorphic rocks tend to be harder than sedimentary rocks and are more resistant to weathering and erosion. For example, limestone and mudstone change to marble and slate when metamorphosed.

Figure 2.4 A white chalk cliff on the south coast of England. (Source: Fotolia.com: Jenny Thompson)

2.3.4 The rock cycle

Over time all rock types can convert into other forms and this has often been termed the rock cycle. Igneous and sedimentary rocks can become metamorphic rocks under pressure and heat. All rock types can erode to form the layers of sediment that can eventually become sedimentary rocks, and all rocks can be completely melted. When molten rock eventually cools and hardens at or near the Earth's surface it will form igneous rock.

> **Reflective question**
>
> ➤ How are igneous, sedimentary and metamorphic rocks formed?

2.4 History of plate tectonics

2.4.1 Early ideas of global tectonics

As maps of the world became more complete a number of individuals began to examine in detail the shapes of the continents and their apparent relationships to one another. In a paper published in 1801, the German explorer Alexander von Humboldt noted that the bulge of South America seemed to fit into the bight of Africa (Figure 2.5). Francis Bacon, a British philosopher of the late sixteenth and early seventeenth centuries, is often erroneously credited with first noticing this fit. In fact he commented on the similar shape of the *west* coasts of Africa and South America in his *Novum Organum* written in 1620. Antonio Snider-Pelligrini in 1858 created a map that showed for the first time the positions of the American and European–African continents before they broke up. It was not, however, until the early twentieth century that Alfred Wegener and Frank Taylor independently proposed that this fit was not coincidental and that the continents were in fact slowly drifting about the Earth's surface (Wegener, 1915, 1966). Taylor soon lost interest, but Wegener continued to develop and expand on his theory until his death in 1930. He suggested that starting in the Carboniferous period (250 million years ago), a large single continent, which he called **Pangaea**, slowly broke up and drifted apart creating the continents as we know them today (see Figure 2.22 below).

A wide range of evidence, apart from the fit of continents across the Atlantic, was advanced to support the theory of continental drift. Rare, identical fossils, such as the Cambrian sponge-like organism *Archaeocyatha*, were found in rocks

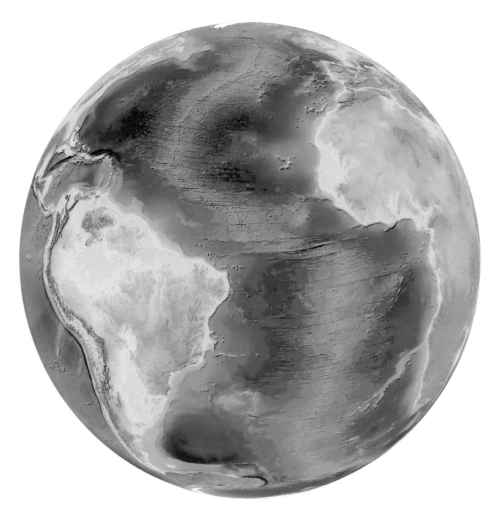

Figure 2.5 Bathymetric map of the Atlantic Ocean showing how Africa and Europe have drifted apart from the Americas. Transform faults can be seen crossing the mid-ocean ridge. (Source: courtesy of Martin Jakobsen who created the image using data from GEBCO; www.ngdc.noaa.ngg/gebco/gebco.html)

on different continents, now separated by thousands of kilometres of ocean. Glaciations that were known to have occurred in Carboniferous times appeared to have affected contiguous areas but only if the continents were fitted back together. Likewise, mountain belts of similar age, rock types and tectonic history such as the mountains of Scandinavia, the Highlands of Scotland and the Appalachians of North America could all be fitted together within a reconstructed supercontinent.

Wegener's theory provoked considerable debate in the 1920s and for a time was quite widely accepted. However, most geologists simply could not believe it was possible to move the continents through the rigid crust of the ocean basins. Since there was no mechanism to drive this continental drift, the theory was not treated very seriously. It became a footnote in geology textbooks and was seldom taught seriously to students of the earth sciences.

2.4.2 Evidence that led directly to plate tectonic theory

The situation changed after the Second World War as a result of the major increase in our understanding of the **bathymetry** of the deep ocean and the rocks that underlie these basins (see Box 2.1 for the reason behind this increased knowledge of the deep ocean). For the first time scientists were able to construct detailed bathymetric maps of the sea floor beneath the oceans. It was found that there was a large and continuous mountain range running through the centre of many of the world's oceans. Detailed bathymetric surveys found that there was a valley in the centre of these mid-ocean mountain ranges that had the same shape as valleys caused by rifting on land. This was interpreted as evidence for the pulling apart of the ocean basins at their centre and thus provided some of the first direct evidence in support of continental drift (Figure 2.5). It was also

CASE STUDIES

HOW THE EVIDENCE FOR THE THEORY OF PLATE TECTONICS DEVELOPED FROM THE COLD WAR BETWEEN THE SOVIET UNION AND THE UNITED STATES

It is a sad reflection on human civilization that many of the major advances in technology and knowledge have come about as a result of waging, or planning for, war. One dramatic example of this is evidence for the theory of plate tectonics. At the end of the Second World War, the wartime alliance between the (capitalist) United States and Britain and the (communist) Soviet Union (USSR) came to an abrupt end. The US Government became worried about the threat posed by the growing fleet of Soviet submarines, particularly because only a few years previously attacks on allied shipping by U-boats in the Atlantic had so nearly won the war for Germany. So the United States developed new sophisticated instruments and technology to study various aspects of the deep-ocean basins. It also set up new oceanographic institutions and paid for the education and research expenses of a whole generation of oceanographers and marine scientists.

Detailed surveys were carried out of the bathymetry of the deep oceans to enable US submarines to navigate their way safely in the oceans. It was also necessary to know the most likely routes of passage for Soviet submarines and where they might hide. These surveys were carried out using the new, improved, echo-sounding equipment that was also used to find submarines underwater. The detailed maps of the oceans that show the presence of the mid-ocean ridges and deep trenches were made during this period (e.g. Figure 2.6).

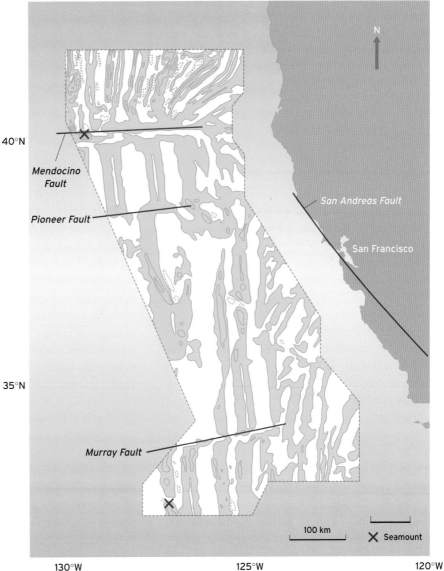

Figure 2.6 Magnetic anomalies of total magnetic field for the area off the western coast of California. Positive anomalies are shown in orange and negative anomalies in white. This was the first such map produced. (Source: after Mason and Raff, 1961)

BOX 2.1 ➤

> During this explosion of oceanographic information, it was found that there was a depth zone in the ocean that had sound properties that allowed submarines to 'hide' from ships using echo sounders. Thus, at the same time, other technologies not based on sound waves were developed to search for submarines. One of these was underwater towed magnetometers. These very sensitive instruments were used to try to detect the distortion of the Earth's magnetic field caused by steel (magnetic) submarines. These magnetometers, however, were then used by marine geologists to examine the magnetic properties of the rocks underlying the ocean. These surveys found magnetic stripes across the ocean floor that Vine and Matthews (1963) interpreted as crucial evidence for the formation of new oceanic crust at the mid-ocean ridges (Figure 2.6).
>
> Furthermore, during the 1950s the United States and the Soviet Union tested their nuclear weapons in explosions at ground level, resulting in large amounts of radioactivity being released into the atmosphere. This radioactive fallout spread over the entire Earth and was found to be contaminating everything, including such sensitive substances as food. In particular it was found that the milk fed to children was contaminated with strontium-90. A campaign was launched by the Mothers' Union of North America and other groups to stop this atmospheric testing of nuclear weapons. It is a tribute to the success of this campaign that a treaty banning the atmospheric testing of nuclear weapons was passed in 1963. From then on, during the remainder of the Cold War, all nuclear weapons testing was carried out underground. Naturally the US Government was keen to gain as much information as possible about the underground nuclear testing being carried out by the Soviet Union and later by China and other countries. So it set up a global network of seismic stations to monitor these tests. These seismic stations were able to locate and measure not only nuclear tests but also natural earthquakes. It is the data from these seismic stations that were crucial in defining the detailed geometry of the plate boundaries and the movement of the plates.
>
> **BOX 2.1**

found that the deepest parts of the world's oceans were not, as had been expected, in the middle of the ocean but instead were located very close to the edge of the ocean, particularly within the Pacific. Prevailing theories had not predicted these major features and were unable to explain them.

Then in the 1960s, Harry Hess at Princeton University proposed that deep within the Earth's mantle, there are currents of low-density molten material that are heated by the Earth's natural radioactivity (Hess, 1962). These form convection cells within the mantle. When the upward-moving arms of these convection currents reach the rigid lithosphere, they move along it, cooling as they go until eventually they sink back into the Earth's interior. In the regions where the upward-moving limb of mantle material breaks through the crust of the sea floor, underwater volcanoes would appear, developing a mountain range supported by the hotter mantle material below. Lava from these volcanoes erupted and then hardened to form new crust along the underwater mountain chain. As the limbs of the convection cell moved apart, they dragged the overlying ocean crust with them. Since the size of the Earth has remained constant, it is necessary to remove old crust at the same rate as it is being produced at the mid-ocean ridges. The great deep trenches around the Pacific were proposed as areas where the old crust dips down and disappears back into the Earth's mantle.

The second piece of evidence to explain these sub-oceanic features came from the study of **palaeomagnetism**. Most igneous rocks contain some particles of magnetite (an iron oxide, Fe_3O_4) which is strongly magnetic. Volcanic lavas such as the basalt which erupts at the mid-ocean ridges are high in magnetite and erupt at temperatures in excess of 1000°C. As the lava cools below 600°C, the **Curie point**, the particles of magnetite become oriented in the direction of the Earth's magnetic field, recording that field permanently relative to the rocks' location at the time they were erupted. If the Earth's magnetic field changes subsequent to the formation of the igneous rock, the alignment of these particles will not be affected. When geologists began to study this palaeomagnetism, they discovered to their surprise that the direction of the Earth's magnetic field reversed itself periodically. A compass needle that points towards the North Pole today would point south during a period of reversal. It is still not known why these reversals take place but they have occurred once or twice every million years for at least the past 75 million years. In fact, there is evidence of magnetic reversals into the Pre-Cambrian (i.e. more than 600 million years ago). A magnetic reversal is not sudden but is very quick in geological time, taking about 20 000 years for completion.

In 1960 **magnetometers** were towed over the sea floor off the western coast of North America, by scientists from Scripps Oceanographic Institute. A clear pattern of stripes appeared, caused by changes in the polarity of the Earth's magnetic field locked into the crust of the sea floor (Figure 2.6). However, at that time, the bathymetric maps of that region of the Pacific Ocean were considered classified information by the US Government (see Box 2.1), and the scientists involved were not able to relate the magnetic patterns to the mid-ocean ridge, which was subsequently found to run through the area. It was not until Vine and Matthews (1963) proposed that these stripes represented a recording of polar magnetic reversals frozen into the sea floor that we had the evidence needed to demonstrate ocean floor spreading. As the molten basalt flowed out at the mid-ocean ridge, it solidified and retained the magnetic field existing at the time of formation. Sea floor spreading then moved this material away from the ridge on both sides to be replaced by more molten material (see Figure 2.8 below). Each time the Earth's magnetic field reversed it was preserved in the new basaltic lava crust. Vine and Matthews (1963) suggested that in such a case, there would be a symmetrical pattern of magnetic polarity stripes centred on the ridges and which would become progressively older as you travelled away from the ridge.

The third piece of evidence that appeared to create the new theory of plate tectonics came from the study of earthquakes. The global seismic network set up to monitor underground nuclear testing (see Box 2.1) gave geophysicists much detailed information on the epicentres of earthquakes and the direction of their movement. When maps were constructed of the worldwide distribution of earthquakes, it was found that there were narrow bands which were subject to frequent (and often destructive) earthquakes and relatively large areas of stable lithospheric crust in between. Isacks *et al.* (1968) looked not only at the precise epicentre of earthquakes (position and depth), but also at the direction of movement of the two layers of rock that caused these earthquakes. As a result of this work, it became obvious that there were a series of relatively rigid plates which were moving about on the Earth's surface colliding with one another. The continents, far from being crucial to this, appeared to be merely passengers being carried about on these moving plates. McKenzie and Parker (1967) put all this information together on a stringent geometrical basis and thus the theory of plate tectonics was born.

The final piece of evidence that confirmed plate tectonics came from the ocean drilling programme (ODP). Cores were drilled in a variety of locations in the ocean basins. It was found that the thickness of sediment increased from the mid-ocean ridges towards the edge of the basin. The thickest (and oldest) sedimentary cover was found nearest to the continents. However, at no place in the ocean were

sediments found that were over 200 million years old. Thus the ocean floor is young compared with the continents.

Reflective question

➤ What are the key pieces of evidence that were used to develop the theory of plate tectonics?

2.5 The theory of plate tectonics

2.5.1 Lithospheric plates

The theory of plate tectonics states that the Earth's lithosphere is divided into a series of rigid plates which are outlined by the major earthquake belts of the world (Figure 2.7). The boundaries of the plates coincide with **mid-ocean ridges**, **transform faults**, trenches and actively growing mountain belts. At present, seven major lithospheric plates have been identified: the Pacific, Eurasian, African, Australian, North American, South American and Antarctic – as well as numerous smaller plates off South and Central America, in the Mediterranean area and along the north-west United States (Figure 2.7).

The geological features found at a given plate boundary depend on the relative motion of the two plates that are in contact. Plate boundaries that are moving apart are called **divergent plate boundaries**. This is the place where new oceanic lithosphere is being created, the mid-ocean ridges. Plates slide past each other along major strike–slip faults called transform faults, such as the San Andreas Fault in California. When two plates move towards one another, at a **convergent plate boundary**, different features are formed. If one of the plates slides underneath the other, a **subduction** zone (the location where one plate is forced below another) is formed. Alternatively a mountain belt such as the Alps or Himalayas can be formed. Transform faults and convergent plate boundaries are the sites of major destructive earthquakes (e.g. Box 2.2 below).

It is now known that the driving forces that cause the motion of the plates are convection cells within the mantle. These convection cells transfer heat (and thus energy) from the very hot Earth's core towards the Earth's surface. The exact size and location of these cells are still matters of scientific debate. The maximum size of each cell should correspond to the size of the plates, although there may also be smaller cells. The upward-moving limbs of convection cells are called plumes or **hot spots** (see Section 2.6.4). Although these plumes ultimately cause the

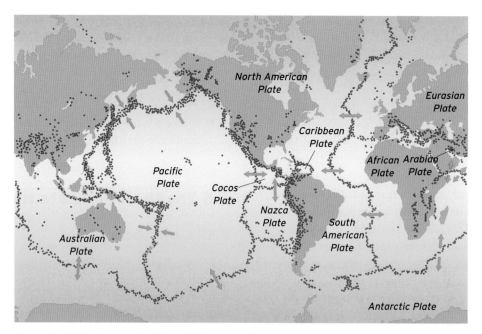

Figure 2.7 Major lithospheric plates of the world. The plates are delineated by the distribution of earthquakes over a period of nine years. The names of the plates and their relative motion are shown in the figure. (Source: based on data from NOAA)

motion of the plates, the geometric relationship between the hot spots and the plates is not as simple as Hess first suggested. While there are examples where plumes are directly under spreading centres (e.g. Iceland), there are also hot spots that are far from present-day active divergent boundaries, such as Hawaii and Yellowstone. The motion of these convection cells moves the plates by viscous forces dragging the base of the lithosphere. In addition the upwelling magma at the mid-ocean ridges forces the plates apart as new crust is formed, while the plates are further dragged apart by the weight of the older, dense and cold part of plates which are furthest from the mid-ocean ridge and which also have a thick load of sediment as they sink back into the mantle.

2.5.2 Rates of plate movement

Vine and Matthews (1963) were the first to calculate both the rate and direction of plate motion over the past several million years. Knowing the age of the magnetic reversals and the position of the stripes relative to the mid-ocean ridges, it was possible to determine the rate of sea floor spreading. Using this method, it has been calculated that the sea floor moves away from the ridge system at a rate of $1-10$ cm yr^{-1}. More recently we have been able to locate positions on the Earth with extreme accuracy using laser beams fired from orbiting

satellites (see Chapter 25). It has therefore been possible to confirm these estimated rates of plate motion by direct measurement of their present rates of movement determined by satellites. The direction of plate motion can be expressed only as a motion of one plate relative to the adjacent plate since all the plates on the Earth are moving and there is no fixed reference point that can be used.

While rates of $1-10$ cm yr^{-1} may appear to be slow, over geological timescales they are most definitely not. Consider, for example, the opening of the North Atlantic Ocean which, on the basis of geological evidence, started 200 million years ago. The present rate of movement of the North American Plate away from the Eurasian Plate is approximately 2.5 cm yr^{-1}. That corresponds to 1 km of movement in 40 000 years, resulting in a 5000 km separation over 200 million years. This distance corresponds quite closely to the actual present separation of these two plates. Plates, however, do not necessary move at a constant rate or indeed in a constant direction. If, for example, one plate with continental crust at its leading edge bumps up against a similar plate boundary on the adjacent plate, mountains are formed and the plates eventually stop moving relative to one another. This cessation of movement not only affects the two plates involved in the collision, but also has consequences on the motion of plates over a considerable fraction of the world.

2.6 Structural features related directly to motion of the plates

2.6.1 Divergent plate boundaries

2.6.1.1 Mid-ocean ridges

Mid-ocean ridges are the spreading centres where the lithosphere is created (Figure 2.8). The faster the spreading rate, the broader is the mountain range associated with that spreading. The best example of this is the East Pacific Rise, which is the mid-ocean ridge located between the Pacific Plate and the Nasca Plate where spreading rates are as high as 16.5 cm yr^{-1}. By contrast the Mid-Atlantic Ridge, with a spreading rate of 2–3 cm yr^{-1}, has steeper slopes characteristic of a slow spreading centre.

The axis of the mid-ocean ridge consists of a central valley or **graben**. Lava flows out onto the sea floor via long fissures that are orientated parallel to the ridge axis rather than from the more circular volcanic craters we see associated with other types of volcanism. As the oceanic crust is carried away from the active (and hot) central ridge, it gradually cools and sinks. Eventually, over the millions of years that the oceanic crust takes to be carried across the ocean basin, it becomes covered with deep-ocean sediments and the original primary volcanic and tectonic features are blanketed over.

Mid-ocean ridges are where most of the Earth's volcanism occurs, producing about 20 km^3 of magma per year that solidifies to produce new ocean floor. The eruption of this magma occurs mostly underwater. The average depth of the mid-oceanic ridges is 2.5 km. This makes it difficult to study such eruptions directly. As a result, much of the detailed knowledge we have of the processes that occur at the mid-ocean ridges has been obtained from studies of Iceland, the one location where there is a comparatively large island straddling the mid-ocean ridge (Figure 2.9). The rock that is erupted at these divergent plate boundaries is called mid-ocean ridge basalt. When mid-ocean ridge basalt erupts it is generally hot (1000°C) and very runny. Large volumes of basalt lava are formed at such volcanoes and the ease with which the lava flows when erupted on land leads to the formation of **shield volcanoes**. These are volcanoes with such gentle slopes that sometimes they are barely discernible as mountains at all. Eruptions from these volcanoes are rarely explosive because any gas that is in the magma can easily escape through the runny liquid. Sometimes larger bubbles of gas do burst out of the magma and can produce spectacular fountains of fire. Eruptions can occur out of circular vents or quite frequently as walls of molten lava issuing from a linear crack in the Earth. When lava erupts underwater it cools rapidly, forming rounded structures called **pillow lavas**. Under such conditions the lava does not flow far from the vent where it was erupted (Figure 2.10).

The lava for these volcanoes comes from the partial melting of the upper mantle. When mantle rocks, which are themselves solid and move by **elastic creep**, move up towards the surface as the plates separate, pressure is released. This release of pressure causes the outer edges of the minerals to melt, forming a microscopic network of interconnected tubes and veins. The resulting fluid is

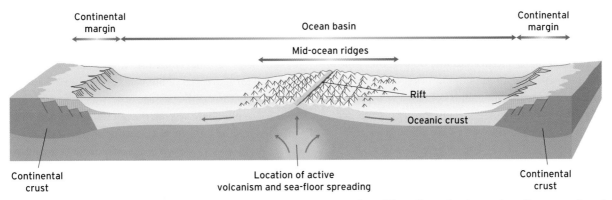

Figure 2.8 Diagrammatic cross-section of a mid-ocean ridge showing the location of the active volcanism and sea floor spreading at the ridge crest.

Figure 2.9 Runny lava flowing from Eyjafjallajökull volcano, Iceland. (Source: Getty Images: Werner Van Steen)

Figure 2.10 Pillow lava, which has this characteristic shape, is formed when basaltic magma is erupted underwater. The photo shows pillow lava from Fifty Mountain in Montana. (Source: R.G. McGimsey and the US Geological Survey)

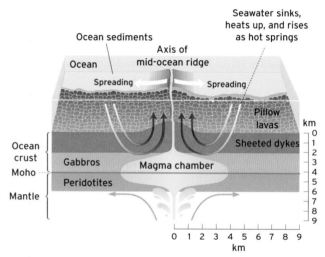

Figure 2.11 The characteristic layers of ocean crust shown here are found on land preserved within ophiolite suites such as in the Troodos Mountains in Cyprus. Most of the seawater enters at cracks on the ridge and circulates parallel to the ridge axis. (Source: From *Scientific American*, Aug. 1982, p. 122 (Bass))

so runny and the pressure so great that the lava is rapidly squeezed out and passes upwards into the lava chambers found below the mid-oceanic volcanoes and eventually out onto the surface in eruptions. The composition of this liquid is derived from, but not identical to, the mantle from which it was ultimately formed.

Our knowledge of these layers is derived to a large extent from study of those areas on land where oceanic crust has become exposed. These layers are called **ophiolites** and they are found in a number of locations. One of the best preserved sections of ophiolites is on the island of Cyprus. The Troodos Mountains contain a well-preserved geological section (Figure 2.11). On top are oceanic sediments (mainly chalk). Below this are pillow lavas and other volcanic rocks underlain by extensive areas of sheeted dykes which represent evidence for the cracks through which the lava flowed. The deepest rocks are **gabbros** and other rocks, which represent the lowest ocean crust and upper-mantle rocks.

Apart from this volcanic activity, the mid-ocean ridges are also the site of widespread hydrothermal activity. Seawater which seeps into the cracks and fissures in the basaltic lavas becomes superheated. As a result it rapidly rises and forms a series of hot vents along the rift valley of the ridge. During the passage of seawater through this subsurface plumbing system, the hot water interacts with the surrounding basaltic rock, causing chemical changes to occur. Sulphate ions, which are present in large quantities in seawater, are chemically reduced to sulphide, while at the same time trace metals including manganese and iron are dissolved. The result of all these chemical changes is that the water re-emerges as a jet of hot water and carries with it a black precipitate of metal sulphides, which gives these features their popular name of **black smokers**. For a discussion of the unique biological **communities** associated with these hot vents see Chapter 3.

2.6.1.2 Rift zones on land

Rift zones are not limited to the ocean. They also occur on land and have in the past been responsible for the break-up of land masses. For such a feature to form on land, however, it is necessary for there to be considerable thinning of the continental crust first. The best example of this today is the Syrian–African Rift Valley (Figure 2.12). Initially localized faulting produced a sunken rift zone or graben. As the fault continued to pull apart there was a further thinning of the crust and deepening of the rift valley until eventually magma was erupted at the surface covering much of the Golan Heights. The magma in turn forced the crack to widen. As the rift continued to deepen, it eventually sank below sea level, as in the Dead Sea, which is at –425 m. In this example the rift to the south opened to the ocean, and seawater entered. This formed the present-day Red Sea. The floor of the Red Sea has both oceanic basalt and subsided blocks of continental-type crust on its floor. If this process continues, the crust will continue to thin and eventually magma will well up into the centre of the rift, forming a new mid-ocean ridge spreading centre.

The lava that is extruded from this type of volcano results in 'rivers and lakes' of lava which solidify into ropey lava also known by its Hawaiian name of pahoehoe (Figure 2.13) and a granular lumpy lava known technically as a-aa. The lava from such eruptions can flow for considerable distances and tends to infill existing landscapes. Once it begins to weather it can form very fertile soil.

2.6.2 Transform faults

Transform faults are found at the plate boundary where two plates slide along next to one another (Figure 2.14). Here the lithosphere is being neither created nor destroyed. The faults can exist as long boundaries between plates such as the Northern Anatolian Fold in Turkey and the Altyn Tagh Fault in China. These boundaries are characterized by frequent major and destructive earthquakes. The movement can be from a few centimetres in a comparatively small earthquake to 1–2 m in a large one.

Transform faults can also occur in small sections in the ocean between spreading centres. The boundary of the divergent plate is not necessarily linear. For mechanical reasons the central graben and spreading centre are almost

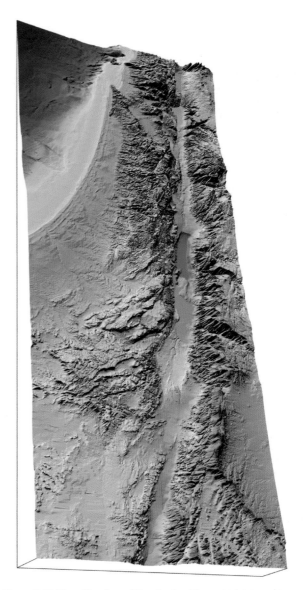

Figure 2.12 The rift valley of the Jordan River can be seen in this image. The country of Israel is to the left of the river valley and the country of Jordan is to the right of the river valley. The Jordan River is 322 km long and originates in the Anti-Lebanon Mountains. The river flows southwards through the drained Hula Valley Basin into the freshwater Lake Tiberias (Sea of Galilee) which is the northernmost depression in this figure. The Dead Sea which is in the centre of this image is 70 km long and presently is 399 m below sea level. The Jordan River Rift Valley is a small part of the 6500 km long Syrian-African Rift. (Source: John Hall; Image Science and Analysis Laboratory, NASA-Johnson Space Center. 'The Gateway to Astronaut Photography of Earth,' http://eol.jsc.nasa.gov/sseop/EFS/printinfo.pl?PHOTO=NM23-755-505, accessed 14 March 2008)

straight. Thus, in order to accommodate curves in the plate boundary, the central graben is displaced by transform faults. The opposite sides of a transform fault are two different plates moving in opposite directions, and the movement

Figure 2.13 Photograph of ropey lava also known as pahoehoe.

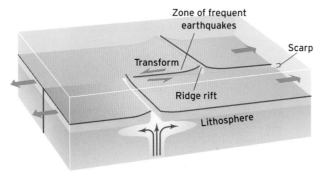

Figure 2.14 Diagrammatic cross-section of an oceanic transform fault showing the region of the fault with the most active seismicity. Where such transform faults occur on land they are the sites of the destructive earthquakes.

results in frequent shallow and often severe earthquakes. Where this fault zone extends beyond the active spreading centre, it is now within the boundary of a single plate and thus the two sides of the fault are moving in the same direction. The earthquakes that occur on this extended section of the fault are much smaller in magnitude and

less destructive. The best known example of an active and dangerous transform fault is the San Andreas Fault. It extends from where the actively spreading East Pacific Rise intersects the west coast of Mexico, through southern California and the San Francisco Bay region, to the Juan de Fuca Rift off the coast of Oregon and Washington states.

2.6.3 Convergent plate boundaries

2.6.3.1 Subduction and volcanism

When two plates consisting of oceanic lithosphere at their leading edge converge, one of the plates is driven under the other one. This subduction causes a deepening of the ocean at the boundary and a deep-ocean trench generally forms. The only occasion when a trench is not formed at a subduction zone is when it is very close to the mid-ocean spreading centre. This is because the lithosphere will still be relatively warm and therefore less dense and more buoyant than the surrounding lithosphere. Hence there will be no propensity for a deep trench to form. At most subduction zones, however, the lithosphere is cool and dense and will readily subduct.

The **Wadati–Benioff** zone is a band of rock, 20 km thick, which dips from the trench region under the overlying plate. It contains the location of all the earthquake foci associated with the descending lithospheric plate. Initially the angle of dip is 45°, becoming steeper as the plate descends to greater depths. Deep-focus earthquakes associated with the descending plate have been observed to depths of 400 km in slowly moving plates and up to 700 km in faster plates. The major destructive earthquake and resulting **tsunami** that struck Indonesia and the eastern shore of the Indian Ocean occurred at a subduction zone (Box 2.2).

CASE STUDIES

EARTHQUAKE IN THE INDIAN OCEAN ON 26 DECEMBER 2004

On 26 December 2004 at 8 a.m. local time, the section of the plate boundary between the Indian and Burmese plates situated off the west coast of Sumatra, Indonesia (Aceh province),

gave way catastrophically. This resulted in one of the largest earthquakes ever felt, with a magnitude of 9.3. It was the longest-duration earthquake ever recorded (500–600 seconds) and was large enough to cause the entire surface layers of the planet to vibrate by over a centimetre of vertical displacement. In total

1200 km of the fault moved. The slip did not happen instantaneously but took place over a period of several minutes. Seismograph and acoustic data (Figure 2.15) indicate that the first phase involved a rupture about 400 km long, 100 km wide at a depth of 30 km below the seabed. This is the longest rupture ever known to have

BOX 2.2 ➤

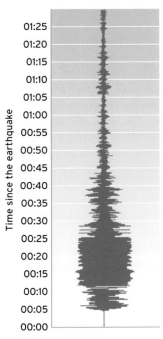

Figure 2.15 Seismograph of the magnitude 9.3 earthquake recorded on 26 December 2004. (Source: Modified from Rapid Earthquake Viewer, http://rev.seis.sc.edu/earthquakes/2004/12/26/00/58/50)

been caused by a single earthquake. This first rupture, which was centred off Aceh, propagated at a speed of about 2.8 km s^{-1} or 10000 km h^{-1} and lasted about 100 seconds. After a pause of about 100 seconds, a second major rupture occurred, continuing northwards towards the Andaman Islands. This rupture occurred more slowly than that in the south (2.1 km s^{-1}) and continued north for another 5 min to the point where the subduction boundary intersects with a transform fault. In addition there were several other sub-events, the largest of which would normally have been ranked as major earthquakes in their own right, with magnitudes of over 7.5. Subsequently when scientists examined the area of the fault line using high-resolution side scan sonar, they found major disruption including several large landslides. In total there was 10 m of lateral movement and 4–5 m of vertical movement along the fault line. It has been estimated that the plate movements during the earthquake displaced ~30 km^3 of seawater radiating outwards along the entire 1200 km length of the rupture. The volume of sea floor upthrust raised global sea level by 0.1 mm. The total energy generated in this earthquake was 3.35 × 10^{18} joules (equivalent to 0.8 gigatonnes of TNT).

The largest earthquakes, such as the Sumatra–Andaman event, are almost always associated with major thrust events at subduction zones. This earthquake was the third largest in the past 100 years but because of the location compared with other events, it had by far the largest loss of life. To put this into perspective, in Table 2.1 you can see the magnitude of several major famous and destructive earthquakes.

The reason for such a large loss of life in this earthquake was not the earthquake itself but the tsunami generated by the earthquake. Tsunami is a Japanese word for harbour wave. Japan is adjacent to a similar subduction zone as Indonesia and is thus liable to similar earthquake-generated phenomena. As with all tsunamis, this one behaved differently in deep water compared with shallow water. In deep water the maximum wave height 2 h after the earthquake was a very modest 60 cm. The wave moved, however, at very high speed (500–1000 km h^{-1}). In shallow water, near coastlines, the wave slows down but in doing so forms large destructive waves. In this case, a wave 24 m high struck the Indonesian coast, destroying towns

Table 2.1 A list of some of the largest earthquakes during the past 100 years, the estimated number of fatalities and whether a major tsunami was produced. Note that although the largest earthquakes are all associated with subduction zones and produce major tsunamis, the number of fatalities is most closely associated with the population density in the area affected

Earthquake location	Magnitude and date	Tsunami	Number of fatalities
Chile[+]	9.5, 1960	Yes	>2 000
Sumatra–Andaman[+]	9.1–9.3, 2004	Yes	283 000
Prince William Sound, Alaska[+]	9.2, 1964	Yes	125
San Francisco[+]	7.8, 1906	No	~3 000
Kashmir[+]	7.6, 2005	No	80 000
Tangshan, China	8.2, 1976	No	242 000[*]
Gansu, China	7.8, 1920	No	c.200 000[*]

[*]Government estimates. Aid agencies estimated the casualties as much higher.
[+]Data from www.earthquake.usgs.gov

BOX 2.2 ➤

➤

and villages in its path. In total, 130 700 people were killed in Indonesia. The wave also propagated across the Indian Ocean and killed people in Sri Lanka (35 300), India (12 400), Thailand (5400) and many other countries. The tsunami even killed 8 people in Africa some 5000 miles to the west.

Could anything have been done to prevent such a large loss of life? Scientists at an international geophysics meeting a few months prior to the earthquake had forecast a major earthquake on this section of the fault but, like all such forecasts, it was impossible to say when it would occur or indeed how severe such an earthquake would be, although it was expected to be a 'big' one.

Very little could have been done for the people of Aceh province. They had very little time between the earthquake being felt and the tsunami striking. In areas with a subduction fault situated off the coast, it is necessary for the local population simply to know that if there is a major earthquake felt, then they must make rapidly for high ground because the tsunami may be much more destructive than the earthquake that caused it. However, such major earthquakes happen in the same region very infrequently, possibly once every several hundred or even thousand years, so information about such events cannot generally be handed down by word of mouth. There is therefore a requirement for education of people by teachers and others in the area. In fact one of the few coastal areas to evacuate before the tsunami arrived was on the Indonesian island of Simuelue. Here island folklore recounted an earthquake and tsunami in 1907. As a result of the 1907 story being well known by the islanders, they fled inland immediately after the initial shaking, knowing that a tsunami might be on its way.

Further away from the epicentre, it would have been possible to generate a tsunami warning. Such a system is in place around the Pacific Ocean (see Figure 2.7 for earthquake locations around the Pacific). If there is a major earthquake around the Pacific, the Tsunami Warning Center in Hawaii contacts regional authorities across the Pacific, who then have to get a warning to local residents. On 26 December 2004, the Tsunami Warning Center did realize very quickly that a major tsunami was being generated in the Indian Ocean. However, the problem was that it simply did not have any mechanism to warn the people on the coasts of the Indian Ocean around Thailand, India and Sri Lanka of the danger that they faced.

So one is left with stories such as the one where Tilly Smith, a 10-year-old British girl, was on Maikhao Beach in northern Phuket, Thailand. She was on the beach with her parents and saw the sea retreating dramatically from the beach. She had learned about tsunamis in geography lessons in school and recognized the warning signs of the retreating sea and frothing bubbles. As a result she and her parents warned others on the beach and they were able to evacuate to safety.

BOX 2.2

As the oceanic lithosphere is carried down into the mantle, it is heated. Water and other volatiles which were carried down with the plate are freed, producing a low-density mixture that rises to the surface on the overriding plate side of the trench. If the overriding plate is oceanic, then basaltic volcanoes are produced, which form into island arcs. The islands are generally situated about 100 km behind the oceanic trench.

When an oceanic plate collides with a plate that has continental crust at its leading edge, the oceanic plate is always the one that subducts beneath the continental one. The oceanic plate carries with it oceanic sediments, which are oldest and thickest at the plate boundary. Some of these sediments are scraped off the oceanic plate and are added directly to the continental plate (Figure 2.16). The remaining sediment gets carried down with the descending plate. The oceanic crust contains water that was trapped within it during its formation and from the pore waters of oceanic sediments. As the plate descends into the mantle, the water causes the basaltic rock to melt. The rising magma is also hot enough to start to melt the continental crust of the overlying plate. The magma that results from this process is very sticky and

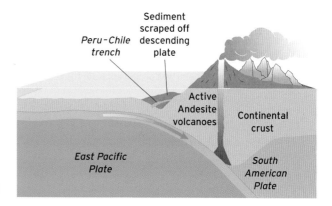

Figure 2.16 Diagrammatic cross-section of a convergent plate boundary showing the subduction of the oceanic plate at the continental margin producing a deep-sea trench and a volcanic belt close to the continental margin.

would not flow at all were it not for the water it contains, which reduces its viscosity to the point where it continues to rise. As the ascending magma approaches the Earth's surface, the pressure is reduced. The water and other gases trapped in the magma begin to expand but cannot escape.

As a result, a considerable overpressure builds up, which cannot escape because the volcanic vent is blocked with a plug formed from the cooling of the lava from a previous eruption. Eventually, the rising lava reaches the point where it breaks through this lava plug and erupts at the surface. The gases have expanded to such a degree that the lava is not so much a liquid, as you find in a mid-ocean ridge eruption, but in effect an incandescent mixture of foam and ash. The result can be a great and destructive explosion that can destroy large areas and kill many people.

Examples of these volcanoes that have resulted in great destructive eruptions include Santorini (Greece; see Box 2.3), Krakatoa (Indonesia) and more recently Mount St Helens (USA). The lava can flow down the hillside as a boiling mud flow or as a stream of incandescent foam (called a **nuée ardente**). It can form pillars of gas and ash that can ascend as much as 40 km into the stratosphere. The dust and gases put into the atmosphere can have widespread effects on climate. After the Mount Tambora eruption in Indonesia in 1815, the world's temperature was lowered for over a year and the following year was known as the year without a summer because dust in the atmosphere reduced the amount of sunlight that reached the Earth. Once the initial destructive phase has passed, the sticky lava flows out of

Figure 2.17 Mount Fuji is an example of an andesitic volcano which is formed behind a subducting oceanic plate. These volcanoes are typically highly destructive when they erupt. (Source: Fotolia.com)

the mountain and builds its shape up into the characteristic steep-sided volcanoes such as Mount Fuji (Figure 2.17) and Mount Vesuvius. Once the lava ceases to flow, it cools in the volcanic vent. This volcanic plug allows a considerable overpressure to build up within the volcano, repeating the cycle until the next catastrophic eruption occurs.

CASE STUDIES

THE DESTRUCTIVE POWER OF SANTORINI

Located 120 km north of Crete is the island of Thira, which is also known as Santorini (Figure 2.18). In 1628 BC it was the site of what was probably the largest volcanic eruption to take place during recorded history. The island is situated above the descending African Plate which sinks below the island of Crete. As such it is similar in type to the many volcanoes around the Pacific, such as Mount St Helens, Washington, USA, or Krakatoa, Indonesia. When the eruption of Santorini occurred it began as a typical but very large andesitic eruption. It has been calculated that during the eruption between 30 and 40 km^3 of molten rock and ash were ejected. The ash cloud was blown into the sky to a height of 35 km or more. Ash from the eruption of Santorini has been found throughout much of the eastern Mediterranean Basin including the islands of Kos and Rhodes, western Turkey and as far as the Nile Delta in Egypt, 800 km away (Stanley and Sheng, 1986).

What made this particular eruption so destructive was that, once the magma chamber was partially emptied, the crater (also known as a **caldera**) collapsed and the sea rushed in. The resulting interaction between seawater and the hot magma caused an explosion that was probably heard over much of the eastern Mediterranean. By analogy with similar modern eruptions, particularly those in Indonesia, it is expected that the eruption would have resulted in a vast tsunami which has been calculated to have been possibly 200 m high. This tsunami would have devastated any coastal communities in the region. In particular, the explosion of Santorini may be largely to blame for the destruction of the Minoan civilization in Crete (Figure 2.19). All that remains of the original island after the cataclysmic explosion are the thin

BOX 2.3 ➤

➤

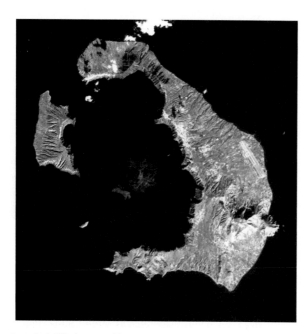

Figure 2.18 Satellite image of the island of Thira (Santorini) showing the extent of the caldera which was formed during the eruption of 1628 BC. The central cone is appearing again above the water in the centre of the caldera. (Source: NASA Jet Propulsion Laboratory (NASA JPL))

Figure 2.19 The ruins of Zakros Palace, Crete, which was an important part of the Minoan civilization. This civilization was essentially destroyed by the eruption of Santorini in 1628 BC. (Source: iStockphoto: Paul Cowan)

crescent-shaped islands around the original caldera. The small island in the middle of the caldera is the site of the volcanic vent itself, which is gradually growing out of the sea (Figure 2.18).

Such high-altitude ash clouds as produced by the Santorini eruption typically have severe effects on global climate. Locally this would have blocked out the sunlight almost completely and reduced global temperatures. Regionally it has been suggested that the ash cloud may have caused a failure in the monsoonal rains in East Africa, while globally it is likely to have caused widespread crop failure and hardship as global temperatures were lowered. The Santorini explosion has been dated by a prominent sulphuric acid peak in a south Greenland ice core, which was dated as 1645 BC (±20 years) (Hammmer *et al.*, 1987), and high-altitude bristlecone pines in California and pines in Ireland suffered severe frost damage in the year 1672 BC (±2 years) (LaMarche and Hirschboeck, 1984; Baillie and Munroe, 1988). It has even been suggested that the explosion of Santorini may have been responsible for the biblical plagues in Egypt (Wilson, 1985).

BOX 2.3

The composition of the lava that erupts behind an oceanic trench depends on the precise nature of the continental rocks that are partially melted during their formation together with partially melted mantle material. One widespread form of this lava is called andesite, which is named after the Andes Mountains. It is intermediate in composition between basalt and granite. This process, over many cycles, is the ultimate origin of the rocks that make up the continents.

2.6.3.2 Mountain building

Although plate tectonics is very good at describing the principal features of the sea floor, and the changing distribution of the continents through geological time, it is not very good at describing the details of what happens when two continental plates collide (Figure 2.20). Indeed it might be argued that this is the reason why it took until the 1960s before plate

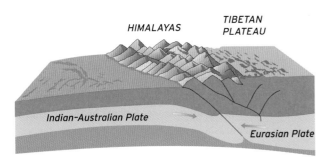

HIMALAYAS
TIBETAN PLATEAU

Indian-Australian Plate

Eurasian Plate

Figure 2.20 Diagrammatic cross-section showing the type of plate boundary that is formed when two continental plates collide. Typically this results in multiple thrusting and folding of the rocks, double thickening of the continental crust, and high mountains and plateaux such as those found in the Himalayas and Tibet today.

tectonic theory was properly developed. There is no subduction when two plates carrying continental lithosphere at their leading edge collide. Examples of such boundaries today are in southern Europe where Italy has moved north to collide with Europe to form the Alps and in Asia where India has collided to form the Himalayas. In these regions there is no obvious plate boundary which can be defined either by earthquake epicentres or indeed geologically.

When two continents collide, both being buoyant, neither can sink under the other. The crust is thickened because it has been shortened and compressed as if squeezed in a vice. As a result the rocks are folded and deformed, crumpled and faulted and most dramatically uplifted to form great mountain belts. The thick crust that is a result of such plate collisions sticks up like an iceberg with high mountains and an even deeper root below floating on the dense mantle. As the high mountains are eroded, the root at the bottom moves up and exposes rocks and minerals that have been metamorphosed by the high temperatures and pressures which are found in the core of the mountain belts.

In addition to the high mountains that form at the boundary of the two plates, a number of other features are seen when two continents collide. These can include major horizontal faults. Faults of this type do not make mountains. They allow blocks to rotate or to move sideways out of the principal collision zone.

2.6.4 Hot spots

In addition to the comparatively simple pattern of volcanism associated with plate boundaries, there are areas where there are volcanoes which are not associated directly with plate boundaries. These features are called hot spots. They represent the surface expression of mantle plumes and

are related to convective processes which originate at the core–mantle boundary deep in the Earth. At present 41 of these hot spots have been identified (Sverdrup *et al.*, 2004). Some are associated with ocean spreading centres and the large amount of lava that they provide has created islands such as Iceland. Hot spots can also form in mid-plate locations. In cases where there is a volcano situated immediately above the hot spot, then, as the plate moves over the hot spot, a series of volcanic islands or **seamounts** are formed. The best example of such a system is the Hawaiian island chain. The island of Hawaii is furthest to the east and is the site of the most active volcano on Earth at present (Figure 2.21). There are a series of islands, Maui, Molokai, Oahu and Kauai, in a line stretching more than 500 km to the west. Each island is made up of an extinct volcano, with the volcanoes getting progressively older as you travel west. Further west the islands are so old and eroded that they disappear below the ocean water surface to become seamounts. This also happens because, as the plate moves west, it cools and the ocean floor becomes deeper towards the ocean trench. During July and August 1996, there were a series of

(a)

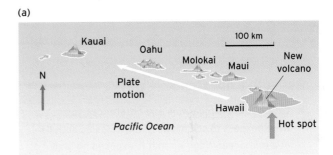

Kauai

Oahu

Molokai
Maui

New volcano

N

Plate motion

Hawaii

Pacific Ocean

Hot spot

100 km

(b)

Figure 2.21 (a) The islands of Hawaii are located above an oceanic hot spot. The island chain is caused by the interaction of that hot spot with the Pacific Plate, which is moving to the north-east. (b) Night-time 50 metre high burst of lava on Mauna Ulu, Kilauea volcano in Hawaii. (Source: D.W. Peterson and US Geological Survey)

Figure 2.22 Bubbling mudpots and billowing steam from hot vents in Yellowstone National Park. (Source: photo courtesy of Steve Carver)

volcanic eruptions and earthquakes on the sea floor 30 km to the south-east of the big island of Hawaii called Loihi, which seems to be associated with a new island forming.

Hot spots are also found in the middle of continental plates. One such hot spot is beneath Yellowstone National Park, and is responsible for the spectacular series of geysers, mudpots and other hydrothermal and volcanic features seen there (Figure 2.22).

Reflective questions

➤ What are the similarities and differences you would expect to see between the volcanic activity of a mid-ocean ridge when it is exposed on land (such as in Iceland) and the same feature at the bottom of the ocean?

➤ A majority of the most destructive earthquakes occur on transform faults. List the locations of all the earthquakes you have ever heard of and then look at Figure 2.7 to see what type of plate boundary they are associated with.

➤ What landforms would you associate with divergent and convergent plate boundaries and transform faults?

2.7 The history of the continents

Plate tectonics provides a theory to explain the nature and location of tectonic features on the present Earth. It also enables geologists to reconstruct how the present arrangement of continents came about. Using the

magnetic record found within the oceanic crust and other direct evidence, it has been possible to reconstruct accurately the history of continental drift over the past 200 million years. When this reconstruction was carried out it was found that all the present continents were once bound together in a single land mass called Pangaea (Figure 2.23). This land mass was formed by the northern continent of Laurasia (which consisted of present-day North America and Eurasia) combined with the southern continents of South America, Africa, India, Australia and Antarctica (Gondwanaland) across the ancient Tethys Sea. This is the supercontinent proposed by Wegener (1966). The land masses appear to join most effectively if they are joined not at the present seashore but at the boundary of the continental slope at about 2000 m water depth.

When the continents moved together in this way, many hitherto inexplicable geological features become understandable. For example, it is possible to use the magnetism that is locked in rocks when they are formed to calculate how the position of the North Pole changed with time. These so-called polar wandering curves could be simply explained as the record of the tracks of the continents over the globe of the Earth. This movement of the continents resulted in their being found not only in different geometrical orientations but also at different latitudes. As a result, geological features such as **evaporites**, sandy desert features and coral reefs, which are common in the geological record of western Europe and North America in the Palaeozoic and Mesozoic (see Chapter 22), make sense since they were deposited when those continents were at lower latitudes than they are today.

One additional consequence of the break-up of Gondwana is that a series of passive continental margins were created. Passive margins are coastlines that are not directly related to present plate boundaries. Such coastlines, while not involved in earthquake or other tectonic activities, are important as being the base levels for continental denudation. They represent the locations where sediments eroded from the various continents that were part of Gondwana (e.g. Africa, India, South America, Australia and Antarctica) are deposited and build up.

The Earth is 4.6 billion years old (Eicher, 1976). It is thus a reasonable question to ask whether plate tectonics and continental drift were processes active earlier than 200 million years ago and, if so, how far back in time have these processes operated? There is no reason to believe that these processes were not active well before 200 million years ago. However, the evidence we have available to us becomes rapidly more fragmentary the earlier we go

CHAPTER 3

Oceans

Michael D. Krom

Earth and Biosphere Institute, School of Earth and Environment, University of Leeds

Learning objectives

After reading this chapter you should be able to:

➤ describe the vast scale of the oceans and how the processes that control the shape of ocean basins are different from those that shape landscapes

➤ understand the processes that control and drive both the surface and deep-water flows in the oceans

➤ understand the processes that control the nature and amount of sediment deposited within the oceans

➤ understand the factors that control primary productivity in the surface layers of the ocean and the distribution of animals in the sea including those that make up commercial fisheries

➤ outline the factors involved in the development of a commercial marine agriculture

3.1 Introduction

When astronauts first looked back on planet Earth from space (Figure 3.1) they were struck by its colour. The Earth looks blue from space because most of its surface is covered by seawater. The oceans not only cover most of the area of the planet but also contain almost all of the water on Earth. Furthermore, much of the oxygen we breathe is produced by phytoplankton that live in the surface layers of the ocean. These interactions between the ocean and the atmosphere play a large part in controlling both the climate and the weather at the Earth's surface (see Chapters 4, 5, 6, 22 and 24). The oceans transfer energy across the planet in the form of ocean currents and waves. Much of the excess carbon dioxide resulting from human emissions is taken up by oceanic processes either temporarily or permanently. The study of oceanic processes is therefore crucial if we are to understand global climate change.

We also use the oceans as a source of food. However, present fishing methods have become so efficient that we are in danger of depleting fish stocks over large parts of the ocean. We are only just beginning to develop the technology to grow marine organisms as an agriculture in the same way as we farm the land. Indeed this is probably the last major untapped food resource on the planet. At the same time we use the oceans as a repository for our waste, often assuming the oceans are so large that they have an infinite capacity to absorb our pollutants. Yet it is clear from the increased incidence of toxic plankton blooms and other undesirable effects that this is not true. For all these reasons it is important that we study the oceans and understand how they operate.

Figure 3.1 A view of the cloudless Earth from space. This view is centred over the Pacific and highlights how much of the Earth's surface is water covered. (Source: Planetary Visions)

3.2 The ocean basins

3.2.1 The scale of the oceans

There are a number of ways of expressing the total amount of water in the oceans. Seawater covers 361 million square kilometres (361×10^6 km²), which represents 71% of the surface of the globe. The total volume of water is enormous: 1.37 thousand million cubic kilometres (1.37×10^9 km³). Most of this water is contained in the three great oceans of the world: the Pacific, Atlantic and Indian Oceans (Table 3.1). Such numbers are so large that they are difficult to conceptualize. One way of looking at this number is to consider a sphere the size of the Earth completely covered with water. The ocean water would then cover such a sphere

to a uniform depth of almost 2700 m. That amount of water represents almost 93% of the total amount of water on the planet. Glacial and land ice make up a further 1.6% and groundwater 5% (see Chapter 4). Freshwater, in the form of rivers and lakes, the water that we as land animals require for our very existence, represents only 0.04% of the total amount of water on the planet. Even the largest lakes in the world (i.e. Lake Superior and Lake Baikal) are very small in area and particularly in volume compared with the world's oceans (Table 3.1). This freshwater is derived entirely from evaporation from the surface of the oceans.

3.2.2 Geological structure of the ocean basins

Geologists have suggested that oceans have been present on Earth for much of its 4.6 billion year history. Yet it came as a considerable surprise to discover in the 1960s that although there are rocks on land that were formed very early in the history of the Earth, the rocks that underlie the ocean are all comparatively recent. The oldest rocks in the ocean are only 200 million years old. Furthermore, the oldest rocks in the ocean were not found in the centre, as might be expected, but at the ocean margins. The explanation for this observation is plate tectonics (see Chapter 2). The ocean basins form when two plates spread apart and new crust is formed at the mid-ocean ridge. As the plates continue to diverge, the ocean basin gets wider (and older). As the oceanic crust cools it sinks, so the depth of water increases as one travels away from the mid-ocean ridges. The edge of the ocean basin, and generally its deepest part, is the bottom of the continental rise which is also where the oceanic lithospheric plate, composed mainly of basalt, is joined to the continental (granitic) plate. Eventually, as a result of the movement of plates over the Earth's surface, one or both of these passive margins becomes a subduction zone and the oceanic crust plunges back into the Earth's mantle to be remelted and returned whence it came. Chapter 2 describes these tectonic features in more detail.

Table 3.1 The amount of water in each ocean and its depth with some large freshwater lakes

	Atlantic Ocean	Pacific Ocean	Indian Ocean	Lake Superior	Lake Baikal
Area (km²)	106×10^6	180×10^6	75×10^6	82×10^3	32×10^3
Volume (km³)	346×10^6	724×10^6	292×10^6	12×10^3	23×10^3
Mean depth (m)	3332	4028	3795	150	740

3.2.3 The depth and shape of the ocean basins

The deep ocean is nearly as rugged in its bathymetry as the terrain we see on land. There are high mountains, deep valleys and canyons as well as areas of flatter plains, with hills of varying heights rising from the plains below. Indeed the undersea mountains are in general longer, the valley floors wider and the canyons often deeper than the equivalent features on land. The principal difference is that while the dominant feature on land is erosion, with rain, wind and ice acting to erode the landscape and often to accentuate features (see for example Chapters 7, 9 and 16), the net process underwater is sedimentation, which acts to fill in the valleys and to cover these features. However, the rate of sedimentation tends to be very slow so that topographic features formed near the mid-ocean ridge remain largely intact as the floor spreads away from the ridge.

Extending from the shoreline is the **continental shelf** (Figure 3.2). The width of the continental shelf can vary from a few tens of metres in areas such as parts of the western coast of North and South America to hundreds of kilometres in the North Sea and off the northern coast of Siberia. It is usually less than 150 m deep and is the area of the seabed where most of the coarse-grained sediment (sand and silt) derived from the physical erosion of the land is deposited. Because it is shallow and close to land, dissolved nutrients, which fertilize the sea and are needed by **phytoplankton** to grow, are relatively abundant (see below). It is often an area of rich fisheries. Also, because it is underlain with continental rocks, it has been exploited for oil and other minerals.

The continental slopes beyond the shelf break are similar in relief to mountain ranges on continents. Their total height is generally 1–5 km above the **abyssal plain** below

(Sverdrup *et al.*, 2004). The continental slope, and to a lesser extent the continental shelf, are often dissected by large submarine canyons. These canyons, which may originally have been formed as erosional features on land during the recent Ice Age when sea level was much lower (up to 130 m lower) than it is now, are often the locations for turbidity currents. These are flows of sediment-laden water, which periodically move down the slope at speeds in excess of 20 km h^{-1} and can cause considerable erosion on their way. Once these currents reach the continental rise, they form deep-sea fans, which have many of the characteristics of deltaic fans found in shallow water (see Chapter 15).

The continental slope and, beyond it, the abyssal plain, where depths often exceed 4000 m, are entirely underlain by basaltic oceanic crust. It is, however, untrue to think of these regions as flat and featureless. Seamounts (such as guyots or abyssal hills), all of which are volcanic in origin, dot the deep-sea floor. For example, in the Pacific Ocean alone, there are more than 20 000 seamounts, a few of which are tall enough to reach the surface as oceanic islands or **atolls**.

Mid-ocean ridges are a continuous set of mountain chains that extend for some 65 000 km. They are tectonically active with frequent volcanic eruptions and earthquakes. They are the origin of the basaltic rocks which underlie all the ocean basins (see Chapter 2). In many areas of the mid-ocean ridges there are also submarine hot springs (see Box 3.3 below).

The deepest valleys on Earth are the narrow, steep-sided oceanic trenches which characterize particularly the edge of the Pacific Ocean. These are the locations where oceanic plates plunge down into the mantle of the Earth. The Challenger Deep, which is the deepest part of the Marianas Trench in the western Pacific, is 11 020 m deep. By comparison Mount Everest is only 8848 m high.

Reflective questions

➤ What is a continental shelf?

➤ Why are the oldest rocks on the ocean floor not very old in geological terms?

3.3 Physical properties of the ocean

3.3.1 Salinity

In 1872, HMS *Challenger* sailed from Portsmouth on a voyage lasting three and a half years. This was the first systematic scientific investigation of the world's oceans and was as

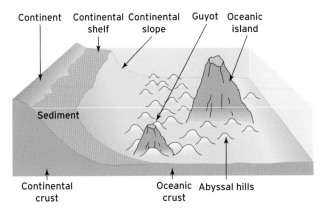

Figure 3.2 An idealized cross-section of the floor of an ocean basin from the coast of the continent. Abyssal hills and guyots are known to be volcanic in origin.

Table 3.2 The concentration of major elements present in seawater

Element	Concentration in seawater (mg L^{-1})
Chloride	19 500
Sodium	10 770
Magnesium	1 290
Sulphur (as sulphates)	905
Calcium	412
Potassium	380

much of a voyage into the unknown as space probes to Mars and Venus are today. One of the most surprising things that the scientists on board discovered was that wherever you go in the world, the chemical composition of the ocean is *approximately* the same. The major salts that make up most of the salts in seawater are present in the same ratio to one another. Seawater is made up principally of sodium chloride and magnesium sulphate with significant amounts of potassium, calcium and bicarbonate (Table 3.2).

The chemicals that make up the salt in seawater were originally derived from the chemical weathering of rocks on land. They flow as a dilute solution in river water into the sea. Once these chemicals reach the sea, they are involved in a variety of chemical, geological and biological reactions which ultimately remove them from seawater and deposit them in the sediments at the bottom of the ocean. So, for example, calcium can be removed by marine organisms to form their shells and skeletons. These are then deposited, sometimes in enormously thick and extensive deposits of calcium carbonate, such as the chalk deposits of south-east England. It is the balance between the processes of supply of salts by the rivers and their removal into the sediments of the oceans that maintains the particular chemical composition of the oceans that we find today.

Although the ratio of most chemical elements in seawater is constant over all the world's oceans, the actual concentration varies depending on where you are (Pinet, 2000). Over large parts of the ocean, seawater salinity is 35.5 g of solid matter per kilogram of water. The concentration of salt in the ocean is generally expressed as parts per thousand (ppt) or in salinity units. However, surface water in the ocean can be diluted by the addition of freshwater from rain, river inflow or ice melting. Surface waters can also be concentrated by evaporation. The salinity of surface water in the climatic bands of the dry desert latitudes (20°N and S) is higher than in the wetter regions of the equator or the temperate latitudes (Figure 3.3) because there is the greatest net evaporation in those latitudes. It is perhaps

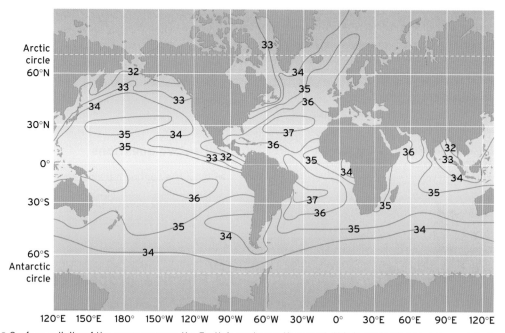

Figure 3.3 Surface salinity of the oceans across the Earth in parts per thousand. This is controlled by net precipitation (rainfall - evaporation) with a maximum at the same latitude as the major deserts and minima at the equator and towards the higher latitudes. (Source: Sverdrup, H.U., Fleming, Richard H., Johnson, Martin W. *Oceans*, 1st edition, © 1942. Adapted by permission of Pearson Education, Inc., Upper Saddle River, NJ)

CASE STUDIES

SAVING THE DEAD SEA: SHOULD WE BUILD A RED SEA-DEAD SEA CANAL?

The Dead Sea is the lowest land spot on the planet (−425 m and still sinking). It is a terminal lake sitting in the Syrian-African rift valley between Jordan and Israel. It is the saltiest lake in the world with a salinity of ~275 ppt compared with ~35 ppt for normal seawater. It is 25% denser than ordinary seawater with a density of 1250 kg m^{-3} compared with 1025 kg m^{-3} for seawater. This density means that it is difficult to swim in the Dead Sea but you can simply sit and float in it (Figure 3.4).

As a result of this extreme salinity, the Dead Sea is (almost) literally dead. There are no plants or animals which can survive in its waters. The only organisms which are found are photosynthetic bacteria and Achaea and then only in significant numbers after the occasional floods or rain events when the surface waters are diluted by 10% or more. The Dead Sea has a unique chemistry being rich in magnesium, calcium and crucially potassium which is extracted for use as

Table 3.3 Major element chemistry of (1) present-day Dead Sea water, (2) Red Sea water of the type which will be carried in the proposed Red Sea–Dead Sea conduit and (3) discharge brine from a reverse osmosis plant of the type which will be discharged into the Dead Sea if the proposed project is built.

Type of water	Na	K	Ca	Mg	Cl	SO$_4$
Present Dead Sea water	1.11	0.16	0.38	1.57	5.25	1.42
Red Sea water	0.53	0.012	0.012	0.06	0.64	0.26
Discharge brine from a reverse osmosis plant	1.03	0.023	0.023	0.011	1.24	0.54

(Source: Data from Tahal and GSI Dead Sea conveyance study mid-term report 2010, http://web.worldbank.org).

fertilizer by the Israeli and Jordanian Dead Sea Chemical Works (Table 3.3). These unusual properties are because Jordan River water, which is perfectly ordinary freshwater, used to flow into the Dead Sea at its northern end and then largely evaporate. Pure water evaporated as a result of the extreme hot arid conditions in the region and the small amount of salts in the freshwater were concentrated and accumulated in the Dead Sea, building up the unusual chemistry found today.

The Dead Sea is not only interesting geochemically but also culturally

and historically. There are many interesting biblical locations at or close to the Dead Sea including Jericho, the longest continually inhabited town in the world, the oasis of Ein Gedi where David hid from King Saul and Qumran where the Dead Sea Scrolls were found, which provide a unique detailed record of life in the desert in the period immediately before the birth of Christ. The fortress of Masada which was the location of the last stand of the rebellious Jews against the Roman Empire in AD 79 is also along the shores of the Dead Sea.

This area of the Middle East is desperately short of water. In 1963, Israel completed a dam across the southern outlet of the Sea of Galilee and simultaneously constructed the National Water Carrier, which is a canal and water conduit from the northern end of that lake through to the southern part of Israel. This was part of the dream of David Ben-Gurion, a founder of the State of Israel, to make the desert bloom. The water from the National Water Carrier is a vital part

Figure 3.4 The density of the Dead Sea is 25% greater than normal seawater, allowing people to float in it without the aid of inflatables. (Source: Shutterstock.com: Vic Spacewalker)

BOX 3.1 ➤

➤

of the economy of modern-day Israel. Jordan also diverts the water from the Dead Sea catchment, the River Yarmuk which flows into the River Jordan just south of the Sea of Galilee. If anything the water needs of the Kingdom of Jordan are even more acute than those of Israel since Jordan has significantly less rainfall and only a very small outlet to the sea to enable desalination to take place. Jordan is at present 'mining' groundwater thousands of years old from under the desert sands.

The result of this combined water diversion is that the Dead Sea is shrinking (Figure 3.5) at a rate of ~1 m per year with the level dropping from −407 m in 1993 to −422 m in 2009. It is predicted that, at present rates, the level of the Dead Sea will stabilize at a depth ~150 m below present lake level (i.e. ~−570 m). As a result of this shrinking, the tourist beaches are now a significant walk from the hotels (Figure 3.6). In addition a series of sinkholes have appeared which can even represent a hazard to walkers in the area

(Figure 3.7). The entire southern basin of the Dead Sea has disappeared (Figure 3.5) and is now entirely occupied by evaporation pans operated by the Israeli and Jordanian Dead Sea Chemical Works. They pump water from the northern basin into a series of pools in which first gypsum ($CaSO_4$), then halite (NaCl) and finally salts of magnesium and potassium crystallize out and are purified and used for fertilizers.

A Possible Red Sea–Dead Sea Canal

In response to this problem the World Bank has commissioned a feasibility study to examine whether it might be possible to build a water conduit from the Red Sea, 150 km to the south, to the Dead Sea to solve or at least alleviate these environmental problems. The proposal is to build six parallel pipes to take water from the eastern side of the Gulf of Aqaba, pump it up 200 m to pass over the high point in the Syrian–African rift valley and then allow it to flow down into the Dead Sea.

Figure 3.6 Coastal retreat along the shores of the Dead Sea. (Source: James Davies)

Figure 3.7 Sinkhole around the Dead Sea caused by lowered groundwater levels. (Source: Getty Images)

A key part of this plan would be to desalinate the water close to the high point using reverse osmosis technology. This would represent a source of freshwater which would be used mainly within the Kingdom of Jordan. The waste brine would then flow down to the Dead Sea. The elevation drop of ~600 m would then drive a hydroelectric power station close to the shores of the Dead Sea before the water would be discharged into the Dead Sea. This discharge would reverse the current drawdown of the Dead Sea level. In total this plan would cost ~US$10 billion (2010 values) to construct. It is estimated that a little more than half of the cost will be recouped by the sale of the freshwater produced and the hydroelectricity generated.

The feasibility study is designed to look at not only the engineering and technical aspects of this proposed project, but also the potential environmental consequences of building and

(a)

(b)

Figure 3.5 Changes in the Dead Sea over time showing the retreat of the sea: (a) 1972 when the waters of the northern and southern basin were directly connected; (b) 2010 when only the northern basin remains; the southern basin has become entirely industrial evaporation pans fed by water pumped up from the northern basin. The development of industrial evaporation pans can also be seen in the southern part of the Dead Sea. (Source: US Geological Survey)

BOX 3.1 ➤

operating such a conduit. In particular the study is looking into the possible effects of pumping 2 million m³ per year of seawater from the top of the Gulf of Aqaba. These effects might include changes in circulation in the northern gulf as well as potential ecological effects. The northern Gulf of Aqaba is the most northerly coral reef in the world. It is already under considerable environmental pressure as a result of tourism and other anthropogenic activities. It is important to be sure that water extraction

into a proposed conduit would have insignificant further effects on the coral reefs or other aspects of the marine ecosystem.

The other major potential environmental effect of this proposed conduit is the effect on the Dead Sea itself. What is proposed is adding reject seawater brine to the Dead Sea in place of the Jordan freshwater that used to flow into the lake. These two waters have very different chemistries. Detailed experimental and theoretical modelling studies are

being carried out to determine what will be the changes in the nature of the Dead Sea, its circulation, chemistry and biology caused by this new water addition. It is only once these environmental studies are complete, and interested parties have had their say, that a decision can be made on whether to build this ambitious water project or not. Updates on the project can be found on the World Bank website (see http://go.worldbank.org/MYWJ6T5R50).

BOX 3.1

not surprising that the salinity of enclosed basins can vary by more than that of the open ocean. The highest salinities in the ocean today are found in the Red Sea where the salinity reaches 41.5 ppt due to high evaporation (Box 3.1). By contrast the waters of the upper Baltic Sea near Finland, which has large amounts of freshwater both raining upon it and flowing in from the adjacent land, can get as low as 5 ppt.

3.3.2 Temperature structure of the oceans

Oceans are very important in controlling the climate of the Earth. The surface of the ocean gains heat by radiation from the Sun, particularly in lower latitudes, and by conduction and convection from warm air flowing over the waves. Heat is lost by evaporation, by reradiation to space and by conduction to cold air above. Measurements made over the Earth's surface show that more heat is gained than lost at the low latitudes, while more heat is lost than gained at the high latitudes. Because water has a very high specific heat (see Chapters 4 and 5), it acts as a major store of the energy of the Sun. The movement of ocean currents from the low latitudes to the high latitudes is very important in transferring energy to the colder regions and thus maintaining the Earth's temperatures in the present range, which is basically conducive to life as we know it. When ocean circulation changes occur, such as during El Niño events, they can have a dramatic effect on the climate of the Earth. El Niño is the situation where unusually the entire central Pacific from Australia to Peru is covered with a relatively thin layer of warm water (see Chapter 4). As a result of this change, weather patterns alter not only adjacent to the current, such as rainfall and floods in the

Atacama Desert and snowfall in Mexico City, but also over much of the world.

For most areas of the ocean, there is a warm surface layer. The thickness of this surface mixed layer varies with both season and location but is generally of the order of 50–300 m thick. This layer is underlain by waters of decreasing temperature (and often increasing salinity). The depth zone with rapidly changing temperature is called the **thermocline** (Figure 3.8). Below this deep thermocline,

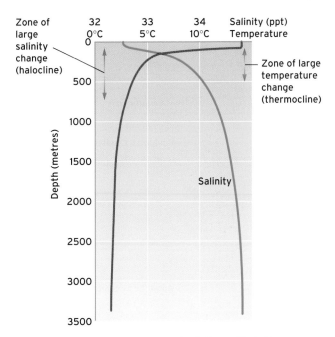

Figure 3.8 Vertical temperature and salinity profile in the ocean based on the north-east Pacific. There are large changes near the surface. (Source: after Sverdrup, K.A. *et al.*, *Introduction to the world's oceans*, 7th edition, 2003, McGraw-Hill, Fig. 6.11. Reproduced with permission of the McGraw-Hill companies)

the temperature is relatively uniform, often showing only a small further decrease to the ocean bottom. Figure 3.8 also shows a **halocline**, which is the layer where salinity concentration also rapidly changes.

Temperature and salinity together control the density of seawater. If the density of seawater increases with depth, then the water column is stable. If, however, there is more dense water on top of less dense water, then the situation is unstable. Vertical mixing will take place until the water column has a similar density over that depth interval. This is the process that drives the three-dimensional circulation of the oceans (see below).

Reflective questions

➤ Why does ocean water contain salt?

➤ What factors lead to differences in salinity across the oceans?

➤ Why would differences in water temperature or salinity encourage water movement?

3.4 Ocean circulation

3.4.1 Surface currents

Early information on surface water currents of the oceans was crucial for any voyages by ships driven by wind or oar power. The surface currents are driven by the prevailing winds. Except for the dramatic seasonal changes in those parts of the northern hemisphere affected directly by the monsoons (see Chapter 5), there is a nearly constant pattern for the winds blowing over the ocean that drives the large-scale surface currents. The trade winds that blow out of the south-east in the southern hemisphere and out of the north-east in the northern hemisphere (see Chapters 4 and 5) drive the Northern and Southern Equatorial Currents, which move in a westerly direction parallel to the equator (Figure 3.9). These currents are deflected by the continents when they reach the coastal areas.

In addition to the continents deflecting the water, the **Coriolis effect** (see Chapter 4) also causes water to deflect to the right in the northern hemisphere and to the left in

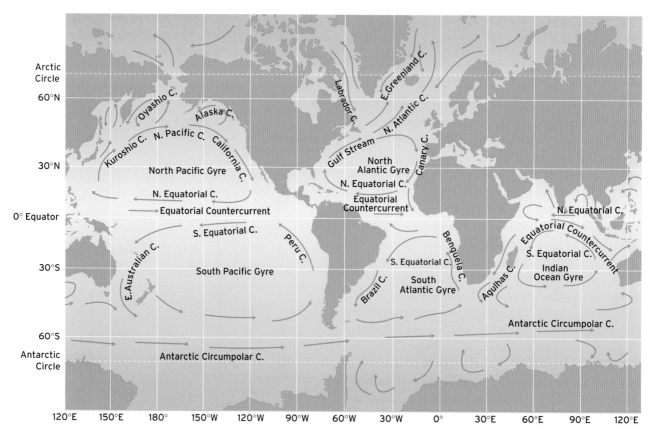

Figure 3.9 Map of the surface currents of the world's oceans. Wind-driven gyres rotate clockwise in the northern hemisphere and anticlockwise in the southern hemisphere. (Source: after Pinet, P.R. *Invitation to oceanography*, 1st edition, 1996: Jones and Bartlett Publishers, Sudbury, MA, www.jbpub.com. Reprinted with permission)

the southern hemisphere. This forms the warm western boundary currents along the eastern coasts of North and South America, eastern Australia, eastern Asia and eastern Africa. In the northern hemisphere in the Atlantic, this warm western boundary current is better known as the Gulf Stream. The Gulf Stream forms the western and northern section of the North Atlantic subtropical **gyres**, which is one of the most dominant surface features of the world's oceans (Figure 3.9). These subtropical gyres are so called because their centres are located approximately 30°N or S of the equator, in the subtropics. In all of the world's oceans there is a westerly current in the temperate latitudes and the gyre is completed by a cold current flowing back towards the equator on the western side of the continents. The centre of some of the gyres contains large amounts of debris such as plastic which has been carried there by the surface ocean currents after being dumped into the oceans by humans. There may be as much as 100 000 tonnes of plastic waste in the centre of the North Pacific Gyre with an average of 330 000 pieces of plastic per square kilometre (Moore *et al.*, 2001). The plastic can have harmful affects on sea creatures either by entangling them or by poisoning them since small toxic chards of plastic, resembling plankton, can be mistakenly eaten by small fish.

The other major feature of the Earth's surface circulation is the circumpolar currents which flow around Antarctica with a westerly current below 60°S and an easterly drift above 60°S. Because most of the land

on Earth at present is in the northern hemisphere, there is no equivalent circumpolar current in the northern hemisphere.

3.4.2 The deep currents of the oceans

The large-scale deep circulation of the world's oceans is not driven by wind but by density variations. The density of seawater is controlled by its salinity and temperature. The resulting circulation is called the **thermohaline circulation**. When a body of water becomes denser than the water surrounding it, it sinks. In the world at present, the two most important regions where deep-water currents form are the North Atlantic–Arctic Ocean (see Box 3.2) and the Antarctic oceans (Figure 3.10). In the North Atlantic Ocean, relatively saline water from the Gulf Stream moves rapidly north into the Norwegian Sea. There, in winter, it cools to a temperature of 2–4°C and has a salinity of 34.9 ppt. This water, being denser than the surrounding water, then sinks away and forms the **North Atlantic Deep Water (NADW)**. This water flows south and forms the major fraction of the deep water in the whole of the Atlantic (Dickson *et al.*, 1990). The other major source of deep cold water is formed along the edge of the Antarctic continent where a mass of even denser water with a temperature of −0.5°C and a salinity of 34.8 ppt is formed in winter. This **Antarctic Bottom Water (ABW)**, which is the densest water formed in the oceans at present, then flows north under the NADW. Eventually the two water masses mix before

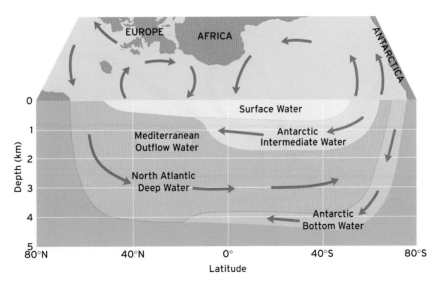

Figure 3.10 Vertical structure of water masses in the Atlantic Ocean. The Antarctic Bottom Water is densest and flows north from Antarctica. The North Atlantic Bottom Water flows south from Greenland above the Antarctic Bottom Water. Above these are intermediate water masses that are formed somewhat nearer to the equator. The surface layers extend to 300–500 m depth and are controlled by winds and other factors (see Figure 3.9).

FUNDAMENTAL PRINCIPLES

THE FORMATION OF NORTH ATLANTIC DEEP WATER AND THE OCEAN CONVEYOR BELT

Figure 3.11 shows the ocean conveyor belt. The Gulf Stream starts in the Caribbean and flows initially as a rather narrow, rapidly flowing stream of water off the coast of Florida and the southern states of the United States up to Cape Hatteras, North Carolina. From Cape Hatteras it leaves the coast and spreads out to form the North Atlantic Drift, which forms the northern part of the North Atlantic Subtropical Gyre (see also Figure 3.9). As it flows north, it gradually cools but the flow is still rapid enough to mean that the Gulf Stream and its equivalent in the western Pacific, the Kuroshio Current, are extremely important in the global transfer of heat from equatorial regions towards higher latitudes. The Gulf Stream flows around Iceland and into the Norwegian and Greenland Seas.

Here in winter, this relatively saline (34.9 ppt) and warm water cools to the point where it becomes denser than surface waters and falls away into the depths. Initial studies of the rate of North Atlantic Deep Water (NADW) current formation were carried out in 1972–1973 during the Geochemical Ocean Section Study (GEOSECS) programme. Radioactive pollutants produced both in the atmosphere by nuclear bomb testing (^{3}H and ^{14}C) and by the discharge of nuclear waste at Windscale (Sellafield) in north-west England (^{137}Cs) were measured in the water column. Since it was known when these radioisotopes were first released into the environment, it was possible to

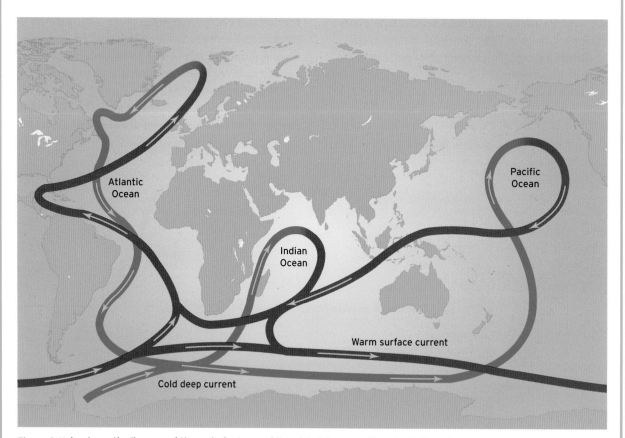

Figure 3.11 A schematic diagram of the main features of the global thermohaline circulation, also known as the oceanic conveyor belt.

BOX 3.2 ➤

determine from their distribution how far and fast they had moved, and thus the rate and amount of the NADW produced (Schlosser *et al.*, 1999). Scientists were surprised to discover that water downwelled as fast as an average of 0.1 to 1.5 m day^{-1}. More recently direct current measurements suggest that most of the NADW passes rather close to the Greenland coast as a western boundary current and then into the deep North Atlantic (Dickson *et al.*, 1990).

It has been suggested that global warming may cause sufficient freshwater to be released from the melting of ice in the northern latitudes to slow or even to stop the formation of the NADW by reducing salinity to the level where it is insufficiently dense to sink. If this occurred it would stop the Gulf Stream flowing north and thus cause the oceanic conveyor belt, which is transferring heat from the equator, to cease. This would have a dramatic cooling effect on the climate of north-west Europe, making it resemble that experienced at present in areas of north-east Canada such as Labrador. Such a scenario has been used to explain the re-cooling that occurred during the Younger Dryas (11.5–12.8 kyears BP) (Broecker,

1998) as the northern hemisphere warmed up after the end of the last Glacial Maximum (see Chapter 22) as well as the global climate anomaly 8200 years ago (Barber *et al.*, 1999). A series of current measurement and tracer studies are being carried out by a joint international programme of the US National Science Foundation and UK Natural Environment Research Council RAPID-WATCH programme. The aim of this programme is to determine whether measurable and significant changes are taking place in the northward flow of water in the North Atlantic Drift.

BOX 3.2

flowing eastwards into the Indian and Pacific Oceans where the mixture forms a major fraction of their deep water. The major deep-water masses in the oceans at present are formed in the Atlantic and/or the Southern Oceans. That is because there is no free access to the Arctic Ocean from either the Indian or Pacific Ocean.

Subsurface water can also be formed nearer to the equator. Near the equator, the upper boundary of the NADW is formed by water produced at about 40°S. This Antarctic Intermediate Water (AIW) is warmer (5°C) and less salty (34.4 ppt) than the NADW and hence remains above it. The other source of intermediate water (water that sinks to a depth of over 1000 m in the Atlantic) is Mediterranean Intermediate Water. This water is formed in the eastern basin of the Mediterranean. It has a much higher salinity (37.3 ppt) and temperature (13°C) than other water masses in the Atlantic and therefore once it flows out into the Atlantic at Gibraltar it remains recognizable as a water mass as far as Iceland and the West Indies. It is also of geological importance, since it is believed that water formed in this manner (by evaporation and then cooling) was the source of the deep waters of the world's oceans when there were no ice caps present.

Water that sinks away into the deep has to be balanced by water upwelling to the surface. This occurs many thousands of kilometres from the regions where deep water is formed. There are regions in the world, principally along the eastern margins of the major continental masses, where the surface currents are driven directly offshore. In order

that volume is preserved, this water is replaced by upwelling water from below. These regions of upwelling are important because the upwelled water contains an abundance of plant nutrients, nitrate and phosphate, which, when they reach the surface layers where there is sufficient light, result in a vast bloom of phytoplankton which in turn sustains major fisheries. One example of such an upwelling region is off the coast of Peru. This is the area affected by El Niño, which results in the upwelling being interrupted by the covering of the warm surface waters from the west and the collapse of the fisheries.

3.4.3 The weather of the ocean

From 1971 to 1973 an international expedition called the Mid-Ocean Dynamics Experiment (MODE) was mounted to study intensively an area of ocean several hundred kilometres across in the western Atlantic. To the surprise of the scientists involved in this experiment, it was found that the pattern of currents was much more complex than the comparatively simple ocean currents described above. What we have described so far might be called the climate of the ocean. What the MODE scientists found was that the ocean also has 'weather'. They found that there are a series of eddies and fronts in the ocean in much the same way as there are cyclones, anticyclones and warm and cold fronts in the atmosphere (see Chapters 4 and 5). These eddies are characteristically of the order of 100 km in diameter. They can spin either clockwise or anticlockwise and can have either a

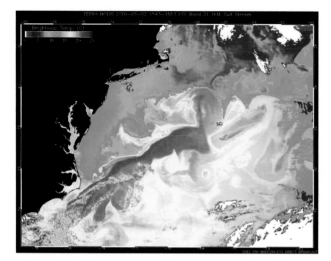

Figure 3.12 Colour satellite image of the western Atlantic showing eddy formation. The colours indicate mixing of waters of different temperatures with warmer waters in yellow-red and cooler waters in blue-green. The mixing of these waters occurs in 'swirls' or giant eddies. (Source: image courtesy of Liam Gurmley, MODIS Atmosphere Team, University of Wisconsin, Madison)

warm or a cold core. Some eddies are formed at the boundary between two major currents. Figure 3.12 shows eddies that form at the boundary between the warm Gulf Stream and the adjacent cold Labrador Current. These rings, formed at this boundary, spin off and can survive for a few months to a couple of years before either dissipating or being incorporated back into the original current. Other eddies are caused by the interaction of a current with a promontory of land, such as the Alboran Gyre at the entrance to the Mediterranean Sea (Figure 3.13), or by a current flowing over a seamount (e.g. Cyprus Eddy). Other eddies have been observed in the middle of the ocean. The cause of these is simply unknown at present. In practical terms these eddies can be important in transferring energy from one part of the ocean to another as well as being sites for enhanced or reduced biological productivity (Smith *et al.*, 1996).

Reflective questions

➤ How are the surface and deep-ocean currents different? What drives these currents?

➤ Where are the key locations where deep oceanic water rises to the surface?

➤ How could climate change alter deep oceanic circulation?

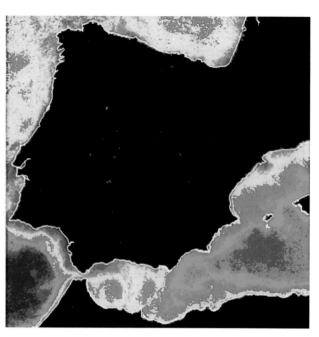

Figure 3.13 Colour satellite image of the Strait of Gibraltar with Portugal and Spain to the north and Morocco to the south. The figure shows an eddy being formed as incoming water from the Atlantic 'relaxes' after passing through the narrow strait.

3.5 Sediments in the ocean

The continental shelves and other near-shore areas are usually underlain by sands and silts. Because we are living in a warm period with ice caps melting and sea level rising (see Chapter 15), much of the new sediment eroded from the continents is being deposited in estuaries, while the shelf areas such as the North Sea are mainly underlain with relic sediments from the melting of glaciers and ice caps after the last Glacial Maximum. However, there are regions, particularly opposite major rivers such as the Bramaputra–Ganges and the Amazon, where sediments are being actively deposited at present (Goodbred and Kuehl, 1999). These sediments can deposit such that they become unstable and become sources for turbidity currents, which carry the sand and silt onto the continental rise and beyond.

There are four types of sediments that are found in the deep ocean. These are: **biogenous**, the remains of the skeletons of marine organisms; **lithogenous**, particles derived from the physical and chemical breakdown of rocks and minerals; **hydrogenous**, sediments derived from geochemical processes involving organic matter, and particularly iron and manganese oxides and seawater; and finally **cosmogenous**, particles derived from outer space.

The most abundant sediments on the ocean floor are sediments derived from biological remains (Figure 3.14).

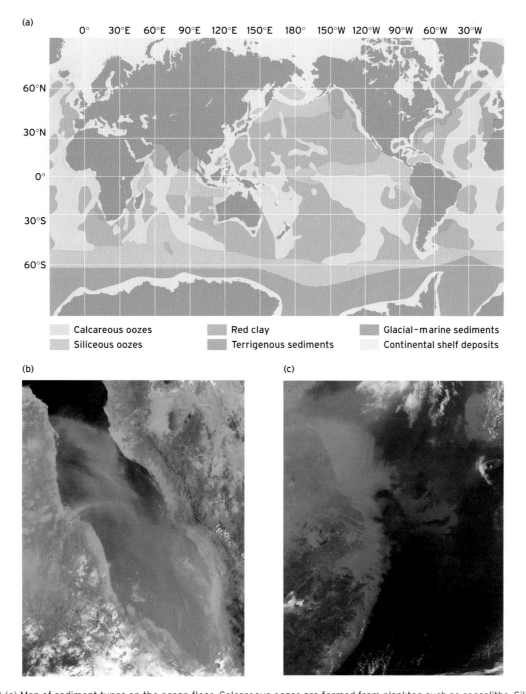

Figure 3.14 (a) Map of sediment types on the ocean floor. Calcareous oozes are formed from plankton such as coccoliths. Siliceous oozes are formed mainly from diatoms (Southern Ocean) and radiolaria (Tropical Ocean). Red clays are the slowly depositing sediment far from land. The deposited material is derived mainly from atmospheric inputs carried far out to sea by dust storms such as that shown in (b) over the Red Sea. Terrigeneous and continental shelf sediments are formed from land-derived material such as that seen entering the oceans from rivers such as the Yangtze in (c) which are then redistributed by oceanic currents; and also glacial marine sediments mainly derived from melting icebergs. (Source: (a) adapted from Davies and Garsline, 1976; (b) and (c) Jacques Desloitres, MODIS Land Rapid Response Team, NASA/GSFC)

Calcareous oozes are found over large parts of the oceans and particularly the Atlantic Ocean. They are made up of the remains of minute organisms called plankton that live in the surface waters of the oceans. The most abundant organisms found in calcareous oozes are coccoliths,

foraminifera (see Chapter 22) and pteropods. These oozes, which are made up principally of calcium carbonate, are found to a depth of 3500 m in the Pacific Ocean and to a somewhat greater depth (~5500 m) in the Atlantic Ocean. Below that depth, which is called the calcite compensation depth (CCD), the increasing concentration of carbon dioxide in the water and hydrostatic pressure cause the water to become undersaturated with respect to calcium carbonate, and the shells dissolve.

In some parts of the ocean, particularly around the Antarctic and central Pacific, the principal planktonic organisms growing in the surface waters do not have calcareous skeletons. They are diatoms in the Antarctic Ocean and radiolaria in the central Pacific. These organisms have skeletons made of opaline silica. As a result, the sediments below these regions are called siliceous oozes.

Over the remaining areas of the deep ocean, the sediments are mainly lithogenous in origin. Red clays are derived mainly from dust transported by the prevailing winds and then deposited. As might be imagined, the rate of deposit of such sediments is very slow, ~1 mm per 1000 years. In some of these areas of slow deposition, manganese nodules form. We still know very little about exactly how these nodules form. We do know that they grow exceedingly slowly, at a rate of between 1 and 200 mm per million years. It is probable that at least part of the manganese and iron which forms the nodules is supplied by black smokers at the mid-ocean ridges (Box 3.3 below). The nodules are usually concentric and in addition to manganese contain significant amounts of iron, cobalt, copper and zinc. See also Chapter 9 for further information on ocean sediments.

Reflective question

➤ What are the main types and sources of sediments found in the deep ocean?

3.6 Biological productivity

3.6.1 Photosynthesis in the ocean

The basic building blocks of all life in the sea, as on land, are photosynthesizing plants. It may appear to the casual observer, whose main contact with the sea is a stroll along the beach, that the only plants in the sea are seaweeds, but that is far from true. By far the largest biomass of plants in the sea are phytoplankton. Phytoplankton, the grasses of the sea, are mainly single-celled plants known as algae. Although some plankton have the ability to move to some degree, they make no purposeful motion against the ocean currents and are carried from place to place suspended in the water. Phytoplankton, which are able to photosynthesize, vary in size from ultraplankton, which are less than

5 µm (0.005 mm) in diameter, to net plankton, which can be found in chains between 70 and 1000 µm (1 mm) in length. Groups of organisms belonging to the phytoplankton include diatoms, **dinoflagellates** and coccolithopores. Because they reproduce mainly by asexual division, which can occur every 12–24 h, under favourable conditions phytoplankton can rapidly bloom to form vast numbers of individuals. These can discolour the water and even, in the case of coccolith blooms, be seen from orbiting satellites.

The basic equation for photosynthesis is:

$$106CO_2 + 16NH_3 + H_3PO_4 + 106H_2O$$
$$\rightarrow (CH_2O)_{106}(NH_3)_{16}(H_3PO_4) + 106O_2 \qquad (3.1)$$

An additional requirement for this photosynthetic reaction to take place is light energy. It is the chlorophyll within the plant that is used to convert light energy into chemical energy, into the organic matter plants require for growth. The rather curious elemental ratio of 106C : 16N : 1P (carbon, nitrogen, phosphorus) is used because it has been found that almost all marine organic matter has this particular elemental ratio (Redfield *et al.*, 1963). The reverse reaction, respiration, involves the breakdown of organic matter to release energy. This is the energy all living organisms need for their life processes. Plants both photosynthesize and respire. This provides the food for animals. However, a healthy plant population will produce more organic matter by photosynthesis than it will respire. All animals only respire using the organic matter that they consume as fuel.

Plants are restricted to the surface layers of the ocean where there is sufficient light for them to photosynthesize. This layer is called the **photic zone**. As a rough guide, plankton can still grow in light that is 1% of the light intensity that reaches the sea surface. Some species of ultraplankton can successfully photosynthesize at a tenth of this light level. The penetration of sunlight in clear and turbid seawater is compared in Figure 3.15. In clear oceanic water the photic zone stretches down to approximately 100 m and in extreme cases such as the eastern Mediterranean to 200 m. As the turbidity of the water increases, owing to the presence of sediment or other particles, the depth of the photic zone decreases.

The amount of light available for photosynthesis varies with time of day and with the weather. Clouds reduce not only the amount of light but also the depth to which that light will penetrate. Light also varies with season and with latitude. These variations in light affect the amount and timing of productivity in the ocean. In the North Atlantic all the other requirements for primary productivity are present throughout the winter except for the fact that there

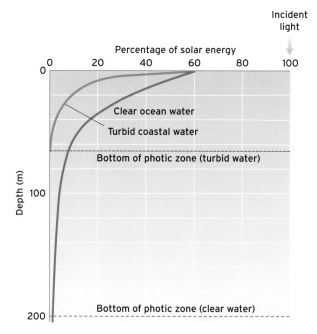

Figure 3.15 Depth of light penetration in clear and in turbid water. The photic zone represents that part of the water body into which solar radiation can penetrate allowing photosynthesis.

Figure 3.16 A near-true colour Landsat image of a coccolith bloom off the south-west coast of England. (Source: courtesy of the Remote Sensing group at the Plymouth Marine Laboratory)

is simply not enough light for the plankton to grow. In March, the light flux increases sufficiently for the plankton to start to grow. Plankton grow very fast, doubling their numbers over a 24 h period; this results in the annual plankton bloom (Figure 3.16).

3.6.2 Importance of nutrient supply to primary productivity

The presence of light does not guarantee plant life. Phytoplankton also need dissolved nutrients. The most important nutrients for plants in the sea are nitrate (NO_3^-) and phosphate (PO_4^{3-}). They are the fertilizers of the sea and are stripped from the surface layers of the ocean by the plankton, which incorporate them into their tissues. When the plankton die, the organic matter is broken down and

these nutrients are released back into the water. This can occur at depth or they can be returned to the water in the form of waste products of herbivores and carnivores. Figure 3.17 illustrates a typical depth profile for nutrients in ocean water. Vertical mixing brings these nutrients back to the surface.

The highest productivity in the oceans occurs in those areas where the supply of dissolved nutrients is greatest. The most productive areas of the ocean are the areas of coastal upwelling which have abundant life. Dissolved nutrients are supplied naturally from land by chemical weathering and by the breakdown of land plants and animals. Hence estuaries are also areas of high productivity. This natural process is often amplified by humans, resulting in **eutrophication** (areas of unnaturally high primary productivity induced by high nutrient loading). The high productivity on the shallow

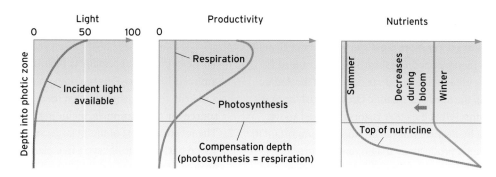

Figure 3.17 Typical nutrient and productivity profile for ocean water during spring plankton bloom.

coastal shelves is sustained not only by coastal runoff but also by mixing and recycling from the shallow sea floor. By contrast large areas of the world's oceans (e.g. the subtropical gyres) that do not have a direct supply of dissolved nutrients are, for all practical purposes, a biological desert.

Plant populations are limited by the lack of any essential nutrient: if the concentration of such a nutrient falls below the minimum required, the population growth stops until the nutrient is supplied. Most of the world's oceans have been found to be essentially limited equally by nitrate and by phosphate. Both need to be added in a ratio of 16 : 1 for natural growth to occur. There are some areas of the ocean, most notably the eastern Mediterranean, which are phosphorus limited. Apart from the major nutrients, N and P, plants also need micronutrients such as iron, copper, zinc and molybdenum. Recently it has been shown that large areas of the eastern central and northern Pacific and in the Southern Ocean, which are so far away from land that they are affected neither by runoff from land nor by dust input from the atmosphere, are iron limited (Boyd *et al.*, 2000).

Life in water requires carbon dioxide and oxygen, as does life on land. Carbon dioxide is required by plants for photosynthesis. It is contributed by respiration processes and also is absorbed by water from the atmosphere. Because seawater has the capacity to absorb large quantities of carbon dioxide, under natural conditions the amount of carbon dioxide in water is never limiting. Oxygen is also needed by all organisms that liberate energy for organic matter by normal aerobic respiration. There are groups of bacteria that are able to liberate energy from the breakdown of organic matter using other compounds such as nitrate or sulphate as the oxidizing agent. These bacteria respire **anaerobically**. Oxygen is supplied to surface water by exchange with the atmosphere and also as a by-product of photosynthesis. There are, however, no processes that supply oxygen to deep water, while processes of respiration consume oxygen. Thus in most areas of the ocean there are lower concentrations of dissolved oxygen than at the surface, as shown in Figure 3.18.

Plants and marine animals other than birds and mammals do not control their body temperatures. Their physiology is regulated by the temperature of the water and within limits metabolic processes occur more rapidly in warm water than in cold water. It is therefore perhaps surprising that some of the most productive areas of the world's oceans are in the Antarctic Ocean and in the northern Atlantic Ocean in early spring. The reason for this is that the other factors discussed above, availability of light, nutrients and so on, are more important in controlling primary productivity than is temperature.

There is, however, one group of creatures that does not rely on light energy for survival of their food chain in the

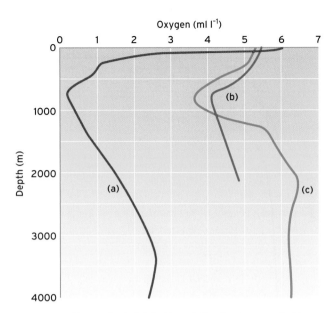

Figure 3.18 The vertical distribution of dissolved oxygen in three ocean waters: (a) the Pacific south of California; (b) the eastern Mediterranean; and (c) the Gulf Stream. (Source: adapted from Open University, 1989)

oceans. Instead they survive by chemical energy produced by inorganic reactions at the mid-ocean ridge. This is also the only ecosystem on the planet that remains unaffected by anthropogenic inputs and change. These systems are described in Box 3.3.

3.6.3 Animals of the sea

3.6.3.1 Food chains

In all the waters of every ocean, one organism preys on another. At the base of this food chain (see Chapter 20 for further explanation of food chains) are the phytoplankton, the primary producers, while at the top of the pyramid are the carnivores: sharks, tuna and of course now also humans. On the first rung of this ladder are the herbivores which eat plant life directly (Figure 3.20). In the oceans by far the most numerous are the zooplankton. Within the zooplankton there are two types of organism:

1. Those animals that remain as floating animals throughout their life cycle. This includes animals such as salps and krill, the tiny shrimps that form the major food source for many animals, including the blue whale.
2. The juvenile forms of many larger marine organisms spend part of their life cycle in a planktonic form. This enables **sessile** or quasi-sessile organisms to spread their offspring into new areas.

HYDROTHERMAL VENTS

During the 1970s, geologists and geo-chemists who studied the processes occurring at the mid-ocean ridges hypothesized that apart from the volcanic eruptions and earthquakes, which they knew took place at these spreading centres, it was likely that there should be hot springs there as well. If such springs existed, they were expected to have unusual chemistries and possibly to have a significant effect on the chemical composition of the ocean as a whole. In March 1977, an expedition of scientists from Woods Hole Oceanographic Institute with the research submersible *Alvin* arrived over the Galapagos Ridge in the eastern Pacific to locate and investigate the ridge area at a depth of 2000 m look-ing for vents. What they found when they finally located a hot spring, and dived down to investigate it, surprised and delighted them. Subsequently similar hydrothermal vents have been found at all of the mid-ocean ridges, at other spreading centres and at arc vol-canoes (e.g. Pedersen *et al.*, 2010).

They did indeed find water issuing from the vents at temperatures of 17°C, which was significantly warmer that the 2°C found in the surrounding water. The hot water was seen to rise from the vent, producing shimmering upward-flowing streams rich in silica, barium, lithium, manganese, hydro-gen sulphide, methane, hydrogen and sulphur. Since that time studies have been carried out of many other vents that have been found in the mid-ocean ridges in all of the world's oceans. Temperatures as high as 380°C have been recorded for the water issuing from these vents. This temperature was high enough to melt the plastic coating on the probes that were first used to measure temperature in these streams. It was also hot enough to soften the Perspex windows of the submersible *Alvin* itself! At many of these areas, mounds and chimney-shaped vents were found which were tens of metres high and which ejected a continuous stream of black particles. These black smokers contain mainly particles of manganese, iron together

with small amounts of copper, lead, cobalt, zinc, silver and other metals, as well as sulphur and sulphide (Figure 3.19a). When such deposits are uplifted on to land, they result in com-mercially important massive sulphide deposits such as at Troodos, Cyprus.

However, probably most remark-able and surprising of all the discov-eries at the vents was the presence of dense communities of large animals (Figure 3.19b). Many of these animals looked very different from animals found elsewhere on Earth including giant tube worms, 3 m in length, with red tips, clams with red blood and blind shrimps.

It was immediately clear that these vent communities were unlike any other biological community on Earth. They were not dependent on organic matter derived from the sur-face layers of the ocean and hence on food that was ultimately derived from the Sun's energy. The vent communi-ties derive their metabolic energy from the chemical energy contained within the hot water in the springs.

(a)

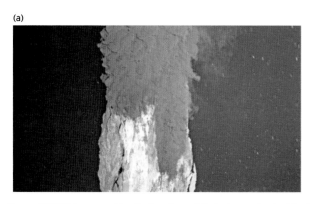

(b)

Figure 3.19 Mid-ocean ridge hot springs: (a) photograph of a black smoker vent from a deep-sea diving vehicle; this is the deepest vent yet photographed at 4960 m deep on the Mid-Cayman Spreading Centre. (b) Communities of animals including tube worms around a deep-sea hydrothermal vent. A large cluster of small white crabs can be in the bottom centre of the photo. (Source: photos courtesy of Jon Copley)

BOX 3.3 ➤

> It has subsequently been shown that at the base of this remarkable food chain are groups of bacteria that derive their metabolic energy from the oxidation of the hydrogen sulphide, methane or hydrogen emitted with the vent water.

Detailed studies at the mid-ocean ridge in the centre of the equatorial Atlantic have found that the most abundant animal found at these vents is the blind vent shrimp, *Rimicaris*. These shrimps live very close to the hottest (and most toxic) vents. They live both by grazing chemosynthetic bacteria and by growing these bacteria within their shells (rather like humans growing potatoes and carrots on their backs). Because of this symbiotic relationship with these bacteria, the shrimps have to live right next to the outpouring waters where the highest concentrations of hydrogen sulphide are found. Although the shrimps are blind, they do have a heat-sensing organ on their backs which they use to avoid becoming boiled.

One intriguing question still to be answered is how the individuals in these communities manage to spread from one vent to the next. It is known that each individual vent has only a quite limited lifespan (~100 years) before the waters switch off and the vent ceases to flow. Since the next vent may be several kilometres away, it is necessary for the organisms to spread their offspring in such a way as to reach the new vent wherever it is. Scientists have now shown that for many vent organisms such as the tube worms, the larvae are simply broadcast into the surrounding waters together with seed bacteria. The few to survive and grow are those that land by chance on a hot spring. By contrast the young of organisms such as shrimps start life close to the 'home' hot spring. They then swim off and live off organic debris in the deep sea until they find a new vent to colonize, or die in the attempt to find one. Once they find a new vent, they lose their eyes, which are replaced by a heat-sensitive organ that enables them to live adjacent to the hot water coming out of the vent.

Recently similar communities of animals which are dependent on chemical rather than solar energy for their energy source have also been found at cold seeps. These cold seep communities are often found in much shallower water. They depend generally on hydrocarbons such as methane or hydrogen sulphide. As with deep-sea vents, these organisms live in symbiotic relationship with bacteria which oxidize these reduced compounds for energy. For example, cold seeps have been found in the Gulf of Mexico and off the coast of California.

BOX 3.3

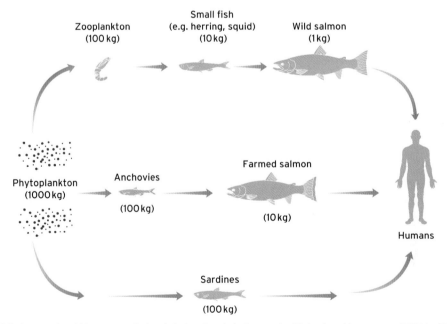

Figure 3.20 A simplified example of three oceanic food chains. Food chains are inefficient and lose around 90% of their energy between different food levels (see Chapter 10). Therefore 1000 kg of phytoplankton (photosynthetic cost) can produce 1 kg of wild salmon, 10 kg of farmed salmon and 100 kg of sardines.

There may be many or few links in the food chain from the primary producers to the top carnivore (Figure 3.20). Food chains in the ocean are rarely simple. They are most likely to show complex branching and interactions. As a result, such interactions are better described as a food web.

In general as one moves upwards from the first **trophic level** (see Chapter 20 for information on trophic levels), the size of the organism increases and the number of organisms decreases. The larger numbers of small organisms at the lower trophic levels collectively have a much larger biomass than the smaller numbers of large organisms at the upper levels. It is estimated that the overall efficiency of energy transfer up each layer of an open-ocean trophic pyramid is about 10%. Thus in order to gain 1 kg of weight, a salmon must eat 10 kg of smaller fish, and the fish need to consume 100 kg of carnivorous zooplankton which in turn require 1000 kg of phytoplankton. Thus it takes 1000 kg of primary producers to produce 1 kg of the top carnivore. The 90% energy loss at each level of the pyramid goes into the metabolic needs of the organisms at that level, such as energy required for moving, breathing, feeding and reproduction (see Chapter 20).

3.6.3.2 Fisheries

At present we harvest marine organisms principally by hunting them in boats using nets (Figure 3.21). The catch depends on custom, culture, economics and availability. Possibly more than with any other food, what one culture considers a great seafood delicacy, another will not eat either because of taste or in some cases because of religious taboos. Many commercial fisheries catch from both the higher trophic levels (e.g. cod, salmon, halibut and tuna)

Figure 3.21 Fishery catch depends upon custom, culture, economics, technology and availability.

and the lower levels (e.g. herring, shellfish, anchovies and sardines). Harvesting fish high in the trophic pyramid is energy inefficient. Because of the complex interactions within the marine food web as well as our very poor knowledge of the marine biological system, depleting the stocks of fish at one trophic level has unpredictable effects on a number of organisms elsewhere in the food web. What is clear, however, is that with modern fishing methods, large boats and nets, sonar and other efficient locating equipment, we are succeeding in severely depleting many of the world's fish stocks (e.g. at Grand Banks, off Newfoundland, and the North Sea).

There is, however, hope. Most marine organisms have a very high **fecundity**. For example, each female gilthead seabream, a fish at the higher levels of the trophic pyramid in the Mediterranean, produces 1 million eggs per year. In nature only two of those offspring need to survive every several years to ensure the continuation of the species. In captivity, survival rates of 20% or more have been achieved from egg to adult fish, so that 200 000 offspring have survived. Thus it is possible, once overfishing has stopped, for marine stocks to rebound to previous natural levels much faster than it is possible to restore natural systems on land.

3.6.3.3 Mariculture

Mariculture, the culturing of marine organisms as in agriculture, is probably the last major untapped food resource on Earth. Freshwater aquaculture began in China with the farming of carp some 4000 years ago. In medieval times in Europe almost every monastery had its own fishponds to provide the monks with fresh fish for Fridays. Despite this, the farming of marine fish is a comparatively recent technology. There are three main types of mariculture. Sea ranching is where the marine organism is restrained in cages or in the case of shellfish grown on ropes hung from rafts. No artificial food is provided in this type of culture and thus the overall yield of the system is constrained by the natural productivity of the area. An example of this type of mariculture is shown in Figure 3.22 which shows ropes seeded with mussels. The shellfish grow over a period of several months and are then harvested and sold.

By far the most widespread form of mariculture carried out at present is growth of fish in pens or nets. In such systems, shown in Figure 3.23, fish are put into a net slung beneath a set of floats when they are fingerlings. They are then fed an artificial diet and allowed to grow until they are large enough to be harvested. Sea-cages are now common in the sea lochs of Scotland, where the principal species being cultured is salmon. Salmon is also grown in Scandinavia,

Figure 3.22 Ropes seeded with shellfish which grow in the water. (Source: Alamy Images: Louise Murray)

Figure 3.23 Fish-cages in a sea loch off the Isle of Skye, Scotland. Fish are grown inside the cages.

the United States and Canada. In the Mediterranean, gilthead seabream is the major species grown, while in Japan and South-East Asia, where most of the world's mariculture is carried out, a number of different species are cultured. In such systems the waste products of the culture, the 70–80% of the food that is not used for growth, is simply discharged into the environment and washed away. This can cause local problems of pollution such as the degradation of the coral reef in Eilat and possibly also encourage blooms of 'red tide' toxic dinoflagellates (red microscopic algae).

The third type of fish culture system is fishponds. These ponds, which are adjacent to the sea, are used to grow fish and shrimps, particularly in South-East Asia. Areas of natural lagoons have also been netted off and used to grow fish. The principal problem with such systems is that the accumulated metabolic wastes, particularly ammonia, are toxic to the fish. The resulting water quality problems limit the number of fish that can be grown and hence the commercial success of the fish farms. They are also vulnerable to natural disasters such as tsunamis (see Box 2.2 in Chapter 2

and Section 15.4.3 in Chapter 15) and tropical storms. An example of integrated mariculture systems that attempt to deal with the problems of pollution is provided in Box 3.4.

3.6.4 Pollution

Until recently it was considered relatively normal practice to discharge domestic and industrial wastes of various types into the ocean. This was done because the oceans are so vast that it was considered that the input would result in no significant change in the concentrations already there. This was the so-called 'dilution is the solution' method of waste disposal. However, we now realize that even the most remote parts of the world have been changed by our careless discharge of wastes. The snows in Greenland now contain measurable lead from car exhausts while the penguins of Antarctica contain DDT residues in their body fats. Pollution of ocean waters can occur via direct discharge into the oceans of sewage and industrial effluent, from shipping (e.g. oil spills) and from changing rainfall chemistry. Pollution can also occur indirectly via runoff from rivers and land (e.g. sewage, fertilizers and pesticides, industrial effluent, sediment). This pollution can alter the nutrient balance in the water, add toxic chemicals to the system and alter the turbidity.

A further set of problems related to pollution results from the interactions of the oceans and atmosphere (see Chapter 4). Many atmospheric gases are taken up by the oceans via precipitation which contains the chemicals or via living matter near the surface of the oceans. Therefore the ocean chemistry may change because we have altered the composition of the atmosphere. Box 3.5 describes the process of acidification of ocean waters through this process.

Reflective questions

➤ A turbid river containing abundant plant nutrients discharges into a coastal sea. The particles drop out of the water column in decreasing grain size until eventually only the finest particles remain in the water column. How might such a system control the primary productivity distribution on the coastal shelf?

➤ Under what conditions might each of the terms in the photosynthetic equation become limiting?

➤ What is mariculture and what types of mariculture are there?

CASE STUDIES

INTEGRATED MARICULTURE IN FISHPONDS

The farming of fish in fishponds, while being technically more difficult than culture in cages, principally because the fish swim about in their own metabolic waste products, has several advantages. It is possible to reduce the pollution into the environment by recycling or treating the waste. It is also possible to mimic the natural food chains by producing an integrated fish culture. Such systems are more energy efficient and can result in a greater profit since more than one species can be cultured.

The National Mariculture Centre, Eilat, Israel, has been in the forefront of developing such integrated fishpond culture systems. In these systems seabream, a fish that lives naturally in the Mediterranean and is much sought after by the French and Italians, is grown. The first sys-tem developed was a semi-intensive system in which the fish were grown in earthen ponds. The nutrients excreted by the fish caused phytoplankton to grow. The phytoplankton removed most of the toxic ammonia from the fishpond water and added life-giving oxygen. Because of the particular conditions of the pond culture in Eilat, a large number of these phytoplankton were large benthic diatoms. These phytoplankton were then fed to oysters, which became a second successful cultured product. The commercial success of this system was, however, limited because it was not possible to prevent the phytoplankton blooming and crashing, which caused large fluctuations in the water quality in the ponds and occasional mass mortalities of the fish.

In order to solve these problems a fishpond–seaweed system was developed. In this system fish are grown at high density in a concrete tank. The water from the fish tank is cleaned of the waste metabolic nutrients, most significantly ammonia and phosphates, by being passed through a tank in which seaweed is grown. It requires approximately three tanks of seaweed to clean the waste from one tank of fish. Seaweed, though edible by humans, does not have great commercial value. However, in this system the seaweed is fed to abalone, which at present retail at $50 per kg. The waste faeces from the fish tank are put into a sedimentation tank and the water flowing over that is used to support a successful oyster culture. This system uses the natural sunlight that is abundant in Eilat to convert waste nutrients into seaweed, abalone and oysters, all of which represent additional commercial products from this aquaculture system (Figure 3.24).

An alternative system for an environmentally friendly intensive mariculture system that does not require

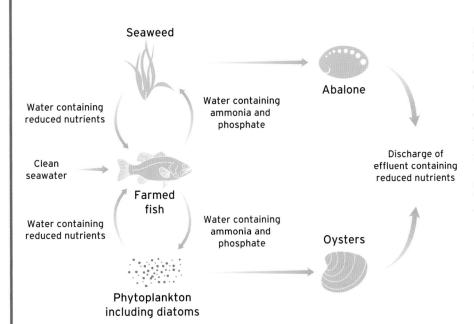

Figure 3.24 Integrated mariculture systems. In the lower system waste nutrients (ammonia and phosphate) excreted by the fish are taken up by phytoplankton including diatoms which are then fed to oysters. In the upper system, the waste water is passed over seaweed (often ulva) which removes the nutrients. The seaweed is then fed to abalone. The treated waste water is then recycled to the fishponds or discharged. Such systems produce abalone, seaweed and oysters in addition to fish as commercial products while reducing the amount of nutrient waste being discharged into the environment.

BOX 3.4 ➤

➤
abundant sunlight involves the use of bacterial biofilters to treat the wastes from the fishponds. In this system (Figure 3.25), the water from the fishpond containing ammonia and oxygen is passed over a nitrifying filter, which contains bacteria that convert ammonia to the nitrate in the presence of air. Although the nitrate is far less toxic than ammonia to fish, it is still undesirable to let it accumulate in the system. To remove the nitrate, some of the water is passed over a sedimentation pond containing anaerobic denitrifying bacteria that convert the nitrate to nitrogen gas. These bacteria also remove phosphates from the reaction stream at the same time. Although this system produces only one commercial product, the fish, it does have several practical advantages. It is a completely closed system, which means that it can be operated far from the sea and potentially near the market for the fish. Also, because the system is closed, it is commercially viable to heat the system. This means that fish can be grown in climates such as in Britain and northern Europe that are too cold to grow fish fast enough to be commercially viable under normal temperatures.

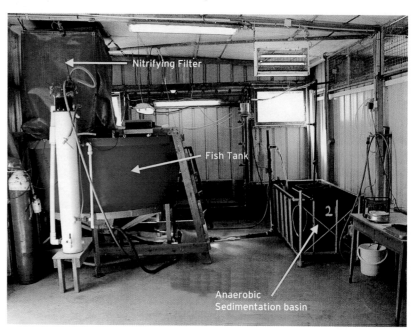

Figure 3.25 This pilot mariculture system uses bacterial biofilters to clean the waste nutrients from the water and enable the system to operate with no waste water discharge at all. (Source: Corbis: Steve Terrill)

BOX 3.4

ENVIRONMENTAL CHANGE

ACIDIFICATION OF THE SURFACE OCEAN

It is known that until approximately 2006, 50% of the carbon dioxide added to the atmosphere by anthropogenic activities was removed from the atmosphere by natural buffer systems. The ocean system formed a very important part of this buffer system with excess carbon dioxide being transferred into the deep ocean with deep-water formation in places such as the North Atlantic and Southern Ocean. In addition some excess carbon dioxide was taken up by biological productivity in surface waters and then transferred to the deep water via waste products from surface-grazing organisms.

There is now evidence that this natural buffer system is being used up and the carbon dioxide we emit into the atmosphere is largely accumulating

BOX 3.5 ➤

➤

in the atmosphere. One of the effects of this increased carbon dioxide in the atmosphere is that it dissolves in the surface waters of the ocean and makes them more acidic. Scientists are very concerned about this effect and are actively studying the possible consequences of acidification of the surface ocean.

One effect that has been predicted concerns the ability of marine organisms to build their shells from calcium carbonate. This is important not only for the seashells that are commonly found at beaches but also for coral reefs and for phytoplankton such as coccoliths which can form blooms in vast numbers under favourable conditions (see Figure 3.16). At present the surface waters of the ocean are close to the saturation point of calcium carbonate and natural processes such as photosynthesis can locally make the system supersaturated and facilitate the precipitation of calcium carbonate. However, if the surface water becomes more acid, then it will be more difficult for corals to grow. Indeed it has been predicted that all coral reefs will cease to grow and start to be dissolved within 30 to 50 years based on current estimates for increased carbon dioxide in the atmosphere (Figure 3.26).

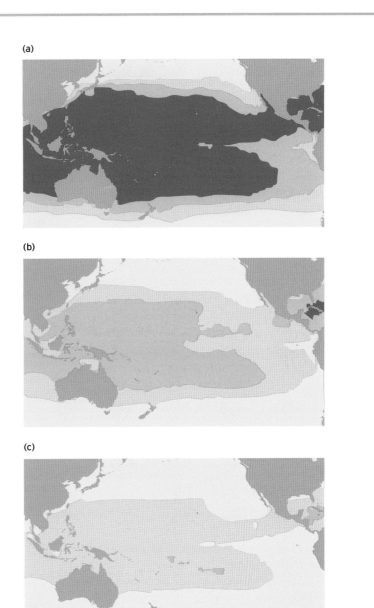

(a)

(b)

(c)

■ >4 (optimal)   ■ 3.5−4 (adequate)   ▦ 3−3.5 (marginal)

□ <3 (extremely low)   ■ Land

Figure 3.26 Changes in growth potential of Pacific coral reefs as related to rising atmospheric CO_2 levels and declining pH and saturation state of surface ocean waters: (a) calculated pre-industrial values for 1870 with atmospheric CO_2 concentrations of 280 ppm; (b) projected values for 2020-2029 with CO_2 concentrations of 415 ppm; (c) projected values for 2060-2069 with CO_2 concentrations of 517 ppm. (a) was high pH conditions while (c) will be low pH conditions. (Source: after Guinotte *et al.*, 2003)

BOX 3.5

3.7 Summary

Oceans cover 71% of the surface of the Earth and a very considerable higher percentage of the habitable niches on planet Earth. The mean depth of the oceans is approximately 3500 m. The bottom of the ocean is far from flat. Most of the basic structure is controlled by plate tectonic processes. At the centre of the oceans are mid-ocean ridges. Away from the centre, the abyssal plains gradually deepen towards the edge of the ocean basin where in some locations they plunge into the ocean trenches. The continental slopes and shelves are of varying widths from hundreds of kilometres to less than a kilometre. In most areas of the ocean, the hills and valleys are covered by a drape of sediments derived from the continent or from biological or chemical processes in the ocean basins themselves.

While the surface currents of the world are driven by prevailing winds, the deeper circulation is controlled by the density of the water, which in turn is controlled by its temperature and salinity. The two most important deep-water masses in the world are formed in the North Atlantic in winter (North Atlantic Deep Water) and at the edge of the Antarctic (Antarctic Bottom Water).

Most of the primary productivity in the world occurs in the surface waters of the oceans. The total amount is controlled by the depth to which usable light penetrates, the photic zone, and by the supply of dissolved nutrients (e.g. nitrogen and phosphorus). All organisms in the ocean, except those living at oceanic hot springs, depend on these phytoplankton via a food chain. We are just beginning to develop marine agricultures as a sustainable alternative to our current over-exploitation of fish stocks. In addition we are only beginning to realize and deal with the problems associated with pollution of the ocean waters.

Further reading
Books

Pinet, P.R. (2009) *Invitation to oceanography*, 5th edition. Jones & Bartlett, Sudbury, MA.
This book contains good sections on geology, chemistry and biology as well as current processes. There are also sections on tides and waves which are relevant to coastal processes discussed in Chapter 15.

Segar, D.A. (2007) *Introduction to ocean sciences*, 2nd edition. W.W. Norton, New York.
Another well-written and colourful textbook with useful questions and good discussion of the history of oceanographic study.

Sverdrup, K.A. and Armbrust, E.V. (2008) *An introduction to the world's oceans*. McGraw-Hill, London.
This book develops the themes discussed in this chapter in more depth. It is a very well-illustrated text containing lots of weblinks and suggested further reading.

Thurman, H.V. (2004) *Introductory oceanography*. Prentice Hall, Upper Saddle River, NJ.
This is a clearly written oceans textbook containing some useful reflective questions and exercises.

Further reading
Papers

Hoegh-Guldberg, O., Mumby, P.J., Hooten, A.J. *et al.* (2007) **Coral reefs under rapid climate change and ocean acidification.** *Science*, 318, 1737–1742.
A review of some ocean acidification impacts.

Pedersen, R.B., Rapp, H.T., Thorseth, H.I. *et al.* (2010) **Discovery of a black smoker vent field and vent fauna at the Arctic Mid-Ocean Ridge.** *Nature Communications*, 1, 126.
Discoveries of hydrothermal vent systems are still being made.

PART III

Climate and weather

Figure PIII.1 Humans have the ability to deliberately and inadvertently alter atmospheric processes and the climate. Normally invisible, wind wakes take shape in the clouds behind the Horns Rev offshore wind farm west of Denmark. (Source: Vattenfall)

Part contents

Scope

The Earth's climate and weather system forms the focus of Part III. The climate is a major environmental influence on the natural vegetation, the landscape and human activity at any given location. Chapter 4 offers an understanding of the science behind the processes occurring in the atmosphere, the most unstable and rapidly responding element of the climate system. Chapter 5 then outlines how these processes result in the nature of the climatic zones that cover the Earth. These broad global zones contain within them substantial variation at the local and regional scale and these smaller-scale processes form the focus of Chapter 6.

Climate change is an important topic and while some of the processes are introduced in Chapter 4, the topic of climate change is dealt with in more detail in four chapters (Chapters 22 to 25) within Part VI of this book. Nevertheless Chapters 4, 5 and 6 still discuss the interactions of humans with climate and weather systems. Humans adapt to local and regional features of the Earth's climate system as well as modify them. For example, urban areas often have different local climates to the surrounding rural zones and humans have developed techniques for locally inducing rainfall by seeding clouds, and for reducing wind speed or snowfall by building shelter belts.

The complexity of the Earth's climate system sometimes feels overwhelming. Weather forecasters struggle to accurately predict the weather only two or three days ahead in many parts of the world. The physical mechanisms interact at a variety of scales within an open system, exchanging energy and matter in ways that often seem hidden from our grasp. However, it is possible to approach these mechanisms at a variety of levels. At the most simple level, we can observe a basic driving force behind these systems. This is the need to dissipate the energy that circulates through the system and that results in processes that transport energy and matter. It is possible to think of a system in terms of an energy or matter budget, with inputs, outputs, stores and transfers. As an example the global climate system principally reacts to the input of energy from the Sun. This input is concentrated at the equator and the large-scale atmospheric and oceanic circulation of the Earth is the mechanism by which this energy is transported away from the equator in an attempt to create a global balance. The Sun's energy is eventually released back out of the atmosphere into space so that the Earth's average temperature remains relatively stable. The above illustrates how we can simplify systems in order to help understand what drives the Earth system processes we pursue and how the same concepts can be used for a variety of subjects and over different scales of space and time.

Atmospheric processes

John G. Lockwood

Formerly of School of Geography, University of Leeds

Learning objectives

After reading this chapter you should be able to:

➤ describe the major components of the climate system and their most significant interactions

➤ understand the basic properties of the atmosphere and the various flows of energy and matter that connect components of the climate system

➤ describe general wind circulation processes and understand concepts of Rossby waves, jet streams, El Niño and North Atlantic Oscillation

➤ understand the greenhouse effect and predict the likely impact of enhanced greenhouse effect on climate and be aware of some positive and negative feedback effects

4.1 Introduction

The term 'climate system' is often used to refer to an interactive system consisting of five major components: the atmosphere, the hydrosphere (e.g. oceans, lakes, rivers), the cryosphere (e.g. ice, snow, glaciers), the land surface and the biosphere (e.g. vegetation) (Figure 4.1). The climate system is driven by various external mechanisms, the most important of which is the Sun. In this chapter the main concern is with the atmosphere, which is the unstable and rapidly changing part of the climate system. In particular its composition, which is continually changing, together with the resulting climate changes have become of major international concern. The Stern Review (Stern, 2006) and the IPCC report (2007a, b) commented that an overwhelming body of scientific evidence indicates that the Earth's climate is rapidly changing, predominantly as a result of increases in so-called greenhouse gases caused by human activities. Greenhouse gases trap heat near the Earth's surface and cause the surface temperature to increase. Carbon dioxide (CO_2) and methane (CH_4) are among the most important anthropogenic greenhouse gases.

The Earth's atmosphere has a composition that is strongly influenced by the biosphere (vegetation and animals); conversely, atmospheric composition is of importance for the biosphere. The present-day atmosphere consists of 78.1% nitrogen, 20.9% oxygen and 0.035% CO_2 with a surface pressure of 1 atmosphere. If the effect of life were removed, then it is estimated that the atmospheric composition would be 1.9% nitrogen, nil oxygen and 98% CO_2, with a surface pressure of 70 present-day atmospheres – an atmosphere similar to that found on present-day Venus, a planet without life or surface water. This is important because it illustrates the close relationship

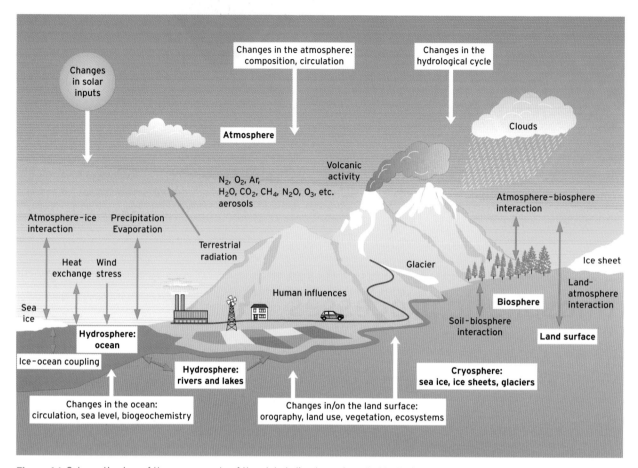

Figure 4.1 Schematic view of the components of the global climate system (bold), their processes and interactions (thin orange arrows) and some aspects that may change (thick white arrows). (Source: IPCC, 2001)

between life and atmospheric composition. The present atmosphere of the Earth is created by life, particularly plant life. Oxygen concentrations in the absence of life are at least 1000 times smaller than they are in the contemporary atmosphere. Using solar radiation to provide the energy for photosynthesis, vegetation converts CO_2 into carbon and frees oxygen gas. The carbon is combined with other elements to build plant tissue, such as stems, leaves and wood. This material is often known as biomass. Carbon dioxide from the atmosphere is absorbed by biomass throughout its lifetime and is then released into the atmosphere again when burnt. This is why biomass is often referred to as carbon neutral. Since carbon dioxide is a greenhouse gas, this suggests a relationship between life and atmospheric temperature. The total amount of carbon dioxide is not much different on Earth from that on Venus, it is just the distribution between reservoirs that differs. In the absence of water on Venus, carbonate rocks have probably never been formed and most of the carbon dioxide remains in the atmosphere. On Earth most of the carbon dioxide is

contained in solid rocks such as limestones, chalk, coal, etc., formed initially by the action of life. It is essential for present life that the vast majority of 'buried carbon' remains out of the atmosphere. If just 0.03% of the 'buried carbon' was returned to remain in the atmosphere, the atmospheric carbon concentration would double, with damaging results for climate change. The problem with fossil fuels is not that they will run out, but that we cannot use the majority of the reserves because we will overload the atmosphere with carbon dioxide.

The emergence of historically rapid climate change in the twenty-first century due to increasing atmospheric greenhouse gas concentrations adds new and urgent dimensions to the age-old challenges of, among others, poverty, inequality, infection and environmental stress (see Chapter 24). Concern about the greenhouse effect arises over two issues: how human activities might enhance the natural effect, and the likely impacts of such an enhancement. Since the Industrial Revolution, human activities have increased atmospheric trace gases such as carbon dioxide.

Before the start of the industrial era (around 1750), atmospheric CO_2 concentration had been approximately 280 parts per million (ppm) for several thousand years. It has risen continuously since then, reaching 383 ppm in 2007. The annual CO_2 growth rate was larger during the last 10 years (1.9 ppm/year from 1995 to 2005) than it has been since the beginning of continuous direct atmospheric measurements (1.4 ppm/year from 1960 to 2005). This unexpected increase in CO_2 concentration is important, because the current suites of emission scenarios informing the international and national climate change agenda seldom include emissions data post 2000 (Anderson and Bows, 2008). In terms of global demand, preliminary data from 2009 (International Energy Agency (IEA), 2010) suggest that China has overtaken the United States to become the largest user of energy, and that China is driving much of the change predicted by the IEA. However, even after years of growth to 2035, China will still use less energy per person than countries in the Organization for Economic Co-operation and Development (OECD). Chinese growth will push demand for fuels, contributing 36% to the predicted growth in global energy use. China is also on its way to becoming a global leader in renewable energies, although most of these technologies will require ongoing government subsidies. Amid all this discussion of energy demand, the IEA (2010) emphasizes another of its estimates: 1.4 billion people across the world lack any access to electricity and 2.7 billion still use wood or other biomass to cook. While governments argue over energy policy, a huge proportion of the world population is still stuck in energy poverty.

Before human activities became a significant disturbance, and over periods that are short compared with geological timescales, the exchanges between the carbon cycle reservoirs (atmosphere, ocean, vegetation, soil) were remarkably constant. For several thousand years before the beginning of industrialization around 1750, a steady balance was maintained, such that the mixing ratio of CO_2 in the atmosphere, as measured from air bubbles trapped in ice cores, kept within about 10 ppm of a mean of about 280 ppm. Air trapped in the Antarctic and Greenland ice sheets also indicates that the atmospheric concentration of CO_2 did not exceed 300 ppm for the 650 000 years before the beginning of the industrial era. The Industrial Revolution disturbed this balance, with fossil fuel burning providing the main cause of the increase of over 30% in CO_2 concentration from about 280 to about 383 ppm in 2007. The fossil fuels largely consist of carbon compounds where the carbon is largely from CO_2 withdrawn from the atmosphere at the time by vegetation or ocean surface plankton. Burning the fossil fuels restores this CO_2 to the atmosphere.

At the present rate of increase, atmospheric CO_2 concentration will reach 400 ppm by 2020.

MacKay (2009) commented that climate change sceptics claim that the recent increase in atmosphere CO_2 concentration is a purely natural phenomenon. Yet something happened around AD 1800 which left an imprint in both atmospheric composition and temperature not seen for at least 600 000 years. This something was the Industrial Revolution, which started in a serious manner in 1698 with Watt's steam engine. From 1769 to 1800, Britain's annual coal production doubled. After another 30 years (1830), it had doubled again. The next doubling happened within 20 years (1850), with another doubling within 20 years of that (1870). This increase in coal production led to a great increase in prosperity in Britain. British coal production peaked in 1910, but meanwhile world coal production continued to double every 20 years. From 1769 to 2006, world annual coal production increased 800-fold, and is still increasing today. Most of this coal was burnt and produced CO_2, a significant amount of which has remained airborne. Much of western prosperity is fossil fuel based, and is thus responsible for increasing atmospheric CO_2 and global warming.

Largely as a result of increasing atmospheric greenhouse gas concentrations, global mean temperatures have increased by 0.7°C since around 1900. Over the past 30 years, global temperatures have risen rapidly and continuously at around 0.2°C per decade, bringing the global mean temperature to what is at or near the warmest level reached in the current interglacial period, which began around 11 700 years ago. By considering long averaging periods, the effects of year-to-year variability are reduced. It is noteworthy that the decade 2000 to 2009 was the warmest in 160 years, significantly warmer than the 1990s, which were warmer in turn than all earlier decades (Kennedy and Parker, 2010). IPCC (2007a) reported that a global assessment of data since 1970 has shown that it is likely that anthropogenic warming has had a discernible influence on many physical and biological systems. Most climate model calculations show that a doubling of pre-industrial levels of greenhouse gases is very likely to commit the Earth to a rise of between 2 and 5°C in global mean temperatures (IPCC, 2007a). This level of greenhouse gases will probably be reached between 2030 and 2060.

Over the past 2.6 million years there have been numerous advances and retreats of major ice sheets across the northern parts of Europe and North America (see Chapter 22). It has often been assumed that since the last retreat of these ice sheets, around 10 000 years ago, global climate has undergone few variations. However, examination of both proxy climate indicators (such as plant and animal remains) and direct climate observations indicates that the recent climate shows a surprisingly high degree of variability. It can therefore no longer be assumed that the global climate during the past 10 000 years has been either stable or benign (see Chapter 23). Furthermore, humans have interacted with atmospheric processes by burning fossil fuels and adding important trace gases to the atmosphere. These contribute to global and local climate changes.

Predicted changes in global temperature due to human-induced global warming may seem small, but they need to be compared with the change in global average temperature between the present and the depth of the last ice age at around 18 000 years ago. The difference is only 5 to 8°C, making a warming of 2 to 4°C by the end of the twenty-first century of great significance. A temperature increase of 2°C is now considered to represent the threshold between dangerous and extremely dangerous climate change. Risks rise rapidly and non-linearly with temperature increases. Once temperature increase rises above 2°C up to 4 billion people could be experiencing growing water shortages. Agriculture could cease to be viable in parts of the world, particularly in the tropics, and millions more people will be at risk of hunger. This rise in temperature could see 40–60 million more people exposed to malaria in Africa. The warmer the temperature, the faster the Greenland ice sheet could melt, accelerating sea-level rise. Above 2°C, the risk of a disintegration of the West Antarctic ice sheet rises significantly. A 2°C temperature rise above pre-industrial levels is a clear limit that should not be exceeded. Current energy use and CO_2 emission trends run directly counter to the repeated warnings sent by the United Nations. The Intergovernmental Panel on Climate Change (IPCC, 2007a, b) concludes that reductions of at least 50% in global CO_2 emissions compared with 2000 levels will need to be achieved by 2050 to limit the long-term global average temperature rise to between 2.0 and 2.4°C. Recent studies suggest that climate change is occurring even faster than previously expected and that even the '50% by 2050' goal may be inadequate to prevent dangerous climate change. While the 2009 UNFCCC Conference of the Parties in Copenhagen during December 2009 failed to deliver any formal 'climate deal', the non-binding Copenhagen Accord recognized the scientific view 'that the increase in global temperature should be below 2 degrees Celsius' (UNFCC, 2010). The adoption of this target occurred despite increasing evidence that for at least some nations and ecosystems, the risk of severe impacts is already significant at 2°C; hence, the Accord includes an intent to consider a lower 1.5°C target in 2015. Therefore the 2009 Copenhagen Accord recognized

the scientific view 'that the increase in global temperature should be below 2 degrees Celsius' (UNFCC, 2010) despite growing views that this might be too high. At the same time, the continued rise in greenhouse gas emissions in the past decade and the delays in a comprehensive global emissions reduction agreement have made achieving this target extremely difficult, arguably impossible, raising the likelihood of global temperature rises of 3 or 4°C within this century.

This chapter therefore seeks to explore the Earth's fundamental global atmospheric processes so that we may better understand, predict and mitigate the effects of such climate changes. It will start by introducing the basics of atmospheric processes before moving on to examine energy systems and moisture circulations. The chapter will then return to a more detailed treatment of atmospheric motions including processes linked with inter-annual variations in climates such as those associated with El Niño. Finally the important processes associated with the greenhouse effect will be discussed.

4.2 The basics of climate

Global atmospheric circulation consists of wind systems that can show marked annual and seasonal variations. They are one of the principal factors determining the distribution of climatic zones. The two major causes of the global wind circulation are inequalities in the distribution of solar **radiation** over the Earth's surface, particularly in a north–south direction (more received at the equator), and the Earth's rotation. Global wind and ocean current circulations redistribute heat from equatorial regions where it is in surplus to polar regions where it is in deficit. The solar radiation received mainly at the equatorial Earth's surface and reradiated back out from the whole surface provides the energy to drive the global atmospheric circulation, while the Earth's rotation determines its shape (Lockwood, 2009).

Fundamental causes of seasonal differences in climate across the globe are the approximately spherical shape of the Earth and the way the Earth's axis of rotation is tilted at 23.44° in relation to a perpendicular to the plane of the Earth's orbit round the Sun. This tilt is probably the result of the Earth being hit by another large object in space several billions of years ago. The gravitational attraction of the Sun, the Moon and the other planets causes the tilt of the Earth's axis of rotation to vary between at least 21.8° and 24.4° over a regular period of about 40 000 years. At present it is decreasing by about 0.000 13° a year. The greater the tilt of the axis, the more pronounced the difference between

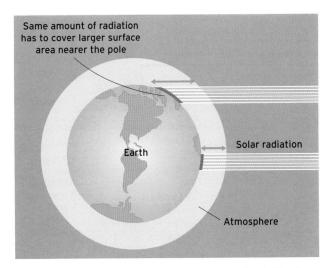

Same amount of radiation has to cover larger surface area nearer the pole

Earth

Solar radiation

Atmosphere

Figure 4.2 The effect of apparent altitude of the Sun in the sky upon solar radiation received at the Earth's surface. The Earth's 23.44° tilt means that the Sun appears overhead at midday at the Tropic of Cancer (23°27′N) on 21-22 June and the Tropic of Capricorn (23°27′S) on 22-23 December.

winter and summer. If the axis tilt were zero, the present seasonal differences between winter and summer would not exist. The spherical shape of the Earth results in sharp north–south temperature differences. This is because, as Figure 4.2 shows, the same amount of solar radiation is spread over a larger area nearer the poles and also has a thicker layer of atmosphere to penetrate because of the angle at which the solar radiation is received. The tilt in the Earth's axis of rotation is responsible for month-by-month changes in the amount of solar radiation reaching each part of the planet, and hence the variations in the length of daylight throughout the year at different latitudes and the resulting seasonal weather cycle. With zero axial tilt the length of daylight would be approximately 12 hours everywhere throughout the year. The 23.44° tilt accounts for the position of the tropics: the Tropic of Cancer at 23°27′N and the Tropic of Capricorn at 23°27′S. At the tropics the Sun is overhead at midday on the solstices, 21–22 June (northern summer) and 22–23 December (northern winter) respectively. The length of daylight throughout the year does not vary significantly from 12 hours over the whole of the area between the tropics. At the spring (21 or 22 March) and autumnal (22 or 23 September) equinoxes, when the noon Sun is vertically overhead at the equator, day and night are of equal length everywhere across the Earth. This is not so at the winter and summer solstices, since each year the areas lying polewards of the Arctic (at 66°33′N) and Antarctic (at 66°33′S) Circles have at least one complete 24 hour period of darkness at the winter solstice and one complete 24 hour

period of daylight at the summer solstice. The Arctic Circle marks the southernmost latitude (in the northern hemisphere) at which the Sun can remain continuously above or below the horizon for 24 hours. Similarly, the Antarctic Circle marks the northernmost latitude (in the southern hemisphere) at which the Sun can remain continuously above or below the horizon for 24 hours. At the poles themselves there is almost six months of darkness during winter followed by six months of daylight in summer.

In the tropical world the Sun is nearly always almost overhead at midday, with little seasonal variation in day length from about 12 hours. Seasonal temperature variations in the humid tropics are therefore generally small, particularly near the equator. The annual temperature range (mean January minus mean July) is very small at 3°C or less over the oceans in the equatorial zone, and is only 2 or 3°C greater at 30°N and 30°S. While values of annual temperature range are equally small over the equatorial continents, values increase rapidly towards the subtropics to reach, for example, 20°C in the Sahara Desert and 15°C in central Australia (see Chapter 5).

The interiors of the northern hemisphere mid-latitude continents are far removed from oceanic influences and because of their location experience extreme continentality of climate. In continental climates, the influence of seasonal variations in daylight length and incoming solar radiation on temperature is greatly exaggerated, producing large seasonal temperature variations. Thus Bergin (60°24′N, 5°19′E, 44 m) on the Atlantic coast of Norway has a mean annual temperature of 7.8°C and an annual range of 13.7°C, while at approximately the same latitude the central Asian city of Omsk (54°56′N, 73°24′E, 105 m) has a mean annual temperature of 0.4°C and a range of 38.4°C.

By definition, the positions of the Tropic of Cancer, Tropic of Capricorn, Arctic Circle and Antarctic Circle all depend on the tilt of the Earth's axis relative to the plane of its orbit around the Sun. However, this angle is not constant, but has a complex motion determined by the superimposition of many different cycles with short to very long periods. As the axial tilt varies, so do the positions of the tropical and polar circles. This causes the tropical circles to drift towards the equator by about 15 metres per year, and the polar circles to drift towards the poles by the same amount. As a result of the movement of the tropical circles, the area of the tropics decreases worldwide by about 1100 km^2 per year on average.

The major controls of very long-term climatic change include palaeogeography, greenhouse gas concentrations, changing orbital parameters and varying ocean heat transport. On timescales of a few years, variations in ocean heat transport and atmospheric greenhouse gas concentrations are particularly effective in changing climate. The greenhouse

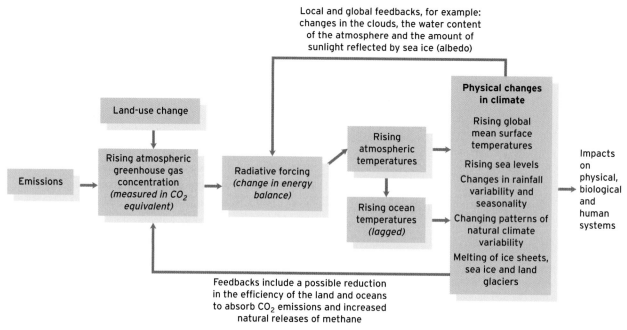

Figure 4.3 The link between greenhouse gases and climate change. (Source: after Stern, 2006)

effect is a natural process that keeps the Earth's surface around 30°C warmer than it would be otherwise. Without this effect, the Earth would be too cold to support life. Fourier realized in the 1820s that the atmosphere was more permeable to incoming solar radiation than outgoing infrared radiation and therefore trapped heat (Fourier, 1824). The link between greenhouse gases and climate change is illustrated in Figure 4.3, and is explored further later in the chapter.

Reflective question

> ➤ Why are there seasonal differences in climate across the globe?

4.3 The global atmospheric circulation

The excess of incoming solar radiation over the tropics, compared with the polar regions, drives the atmospheric circulation. Incoming solar radiation is mainly absorbed by the Earth's surface in low-latitude regions, while infrared long-wave radiation is continuously radiated to cold outer space from across the whole of the Earth's surface. This creates the north–south temperature gradients that drive the atmosphere/ocean circulation. Modern observations of the atmosphere support a three-cell circulation model in each hemisphere as shown in Figure 4.4. Air rises near the equator where there is a heat surplus, flows poleward at high levels,

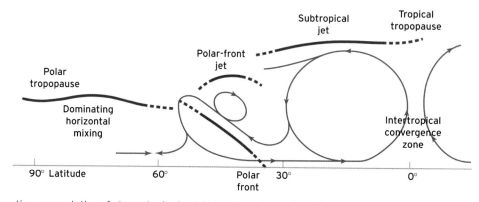

Figure 4.4 Schematic representation of atmospheric circulation cells and associated jet stream cores in winter. The tropical Hadley, the middle-latitude Ferrel and the polar cells are clearly visible. Surface winds are not directly north or south because of the Earth's rotation that forces the air to move in an easterly or westerly direction.

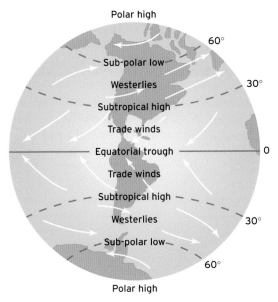

Figure 4.5 Idealized mean surface wind circulation.

convergence zone (ITCZ) where conditions are favourable for ascending motion, condensation of water vapour, cloudiness and high precipitation rates. It is only a few hundred kilometres in width and its position varies from day to day. Thus the ascending limb of the Hadley cell is limited to a few areas of cloud and rain in the equatorial trough, while the descending motion covers much of the subtropics giving widespread desert conditions. The Hadley cells are relatively stable; therefore Figure 4.4 is a reasonable representation. Compared with the Hadley cells, the middle-latitude atmosphere is highly disturbed and the Ferrel cell circulation shown in Figure 4.4 is highly schematic. At the surface the predominant features are a series of eastward-moving closed cyclonic and anticyclonic circulation systems. At high levels in the atmosphere these features are associated with large-scale open waves with **jet streams** imbedded in them. These waves and jet streams are discussed in more detail in Section 4.6.4.

sinks at about 30°N and S and returns at low levels to the equatorial regions. Climatologists know this simple circulation cell as the Hadley cell; it is restricted to tropical regions by the Earth's rotation (see Section 4.6.2). Over both poles where there is a heat deficit, a similar circulation with subsiding air and surface-level winds flowing back towards the equator exists, but they are diverted in an easterly direction by the Earth's rotation. Between the tropical Hadley cell and the polar cell there is, to complete the three-cell structure, a reverse cell with sinking air around 30°N and S, poleward-flowing westerly winds, and rising air along the boundary (the **polar front**) between the mid-latitude and polar cells. This middle-latitude cell is known as the Ferrel cell.

Between 20° and 30° latitude from the equator are areas of high pressure, such as the Azores anticyclone, where air from the tropical Hadley cells descends (Figure 4.5). Since descending air warms (see Section 4.6.1) there is little chance of moisture condensation and generally these are regions of clear skies with high pressure and light winds. Almost all the major deserts of the world lie in these latitudes (see Chapters 14 and 18). Recent research by Qiang Fu *et al.* (2006) suggests a widening of the tropical circulation zone and a poleward shift of the subtropical jet streams and their associated subtropical dry zones. Since 1980, the area defined climatically as 'tropical' has expanded by 277 km in either direction away from the equator. The equatorward-flowing easterly winds on the equatorial sides of the subtropical anticyclones are known as **trade winds**. The trade winds meet in a trough of low pressure near the equator. This trough contains a convergence zone near the surface known as the **intertropical**

Reflective question

➤ Which major circulation zone do you live in? What are its main characteristics?

4.4 Radiative and energy systems

4.4.1 The nature of energy

Many meteorological processes involve flows of energy; it is therefore essential to have some understanding of the nature of energy. Energy supply is the very life-blood of existence. Energy may formally be defined as the capacity for doing work. It may exist in a variety of forms including heat, radiation, potential energy, kinetic energy, chemical energy and electromagnetic energies. It is a property of matter capable of being transferred from one place to another, of changing the environment and is itself susceptible to change from one form to another. Except at the sub-atomic scales, energy is neither created nor destroyed and from this it follows that all forms of energy are exactly convertible to all other forms of energy, though not all transformations are equally likely. This is because of something called **entropy**, which is explained in more detail in Box 4.1. When energy is used, it is still there, but normally cannot be used over and over again, because only *low-entropy* energy is 'useful'. Once energy is used even once, it converts to a

ENERGY EXCHANGES AND ENTROPY

In very general terms it can be stated that used drinks bottles can be recycled while used energy cannot. Used energy still exists in a high-entropy form but cannot be reused, because only low-entropy energy is useful. Taken one step further, it can be noted that organized systems seem to have an annoying tendency to become disorganized. Both these very simple statements are underpinned by the concepts of entropy and thermodynamics.

Energy transformations are best understood in terms of the first and second **laws of thermodynamics** and the concept of entropy (Lockwood, 1979). The first law of thermodynamics is a statement of the law of conservation of energy for a thermodynamic system (see Chapter 26). Entropy, however, is a measure of the unavailability of a system's heat energy for conversion into mechanical work. The lower the entropy, the more energy the system has for conversion into mechanical energy. It is also a measure of the degradation or disorganization of the system: the higher the entropy, the more disorganized the system. Friction continually degrades and destroys atmospheric motions by converting them back into heat, but this heat is in the form of low-grade, high-entropy energy, which cannot create fresh motion systems.

The second law of thermodynamics asserts that the entropy of an isolated system increases with time so that it becomes more disordered. The whole universe is progressing from a state of simple order to states of increasing disorder. So if the system starts off in a state with some kind of organization, this organization will, in due course, become degraded, and the special features (e.g. wind motions) will become converted into 'useless' disorganized particle motions. Thus, over time, all organized atmospheric motions, unless continually renewed, will be destroyed by friction and vanish.

However, observation suggests that many natural systems do not increase their entropy through time. On medium and large scales, atmospheric motions show a high degree of stable organization. Since entropy is always tending to increase in these systems, they must be receiving new supplies of low entropy to maintain the stable, low-entropy conditions. It is therefore necessary to enquire about the source of this supply of low entropy. The answer to this may be illustrated by a consideration of the energy flow in plants. Green plants take atmospheric CO_2, separate the oxygen from the carbon using photosynthesis, and then use the carbon to build up their own substances and grow leaves, roots, fruits and so on. In doing this they supply free oxygen gas to the atmosphere; indeed a test of the presence of carbon-based life on a planet is free oxygen in the atmosphere. The plant biomass produced can be burnt later using atmospheric oxygen and producing CO_2, releasing the stored energy. Photosynthesis results in a large reduction of entropy in the vegetation system. Green plants achieve this entropy reduction by using **short-wave radiation** from the Sun to drive the chemistry of the photosynthetic process. The light from the Sun brings energy to the Earth in a comparatively low-entropy form, namely in the photons of visible light. The Earth does not retain this energy, but after some time reradiates it all back into space. If this were not so the temperature of the Earth would continually increase. However, the reradiated energy is in a high-entropy form of relatively useless radiant heat or infrared photons. The sky is in a state of temperature imbalance: one region of it (occupied by the Sun) is at a very much higher temperature than the rest. The Earth receives energy from the hot Sun in a low-entropy form and reradiates it to the cold regions of the sky in a high-entropy form. The existence of the cold sky is essential to make the entropy flow system work. The total amount of entropy exported by the Earth-atmosphere system to space is 22 times the amount of entropy imported by the incoming solar radiation at the top of the atmosphere. The gains in entropy represent the export, in the form of **long-wave radiation**, of the disorder continually being created in atmospheric flow patterns and surface ecological systems.

BOX 4.1

useless low-grade, high-entropy form. Thus hydrogen is a low-entropy fuel; on combining with oxygen it produces heat energy and high-entropy water. Water is completely useless as a fuel. The transformation of light into heat is common, and if sunlight falls on green plants it can be transformed into chemical energy via the formation of new compounds. The direct transformation of sunlight into kinetic energy of atmospheric motion is extremely uncommon, since this normally takes place via heat energy. It is possible for any particular system to produce an exact energy budget, in which energy gained exactly equals the energy lost plus any change in storage of energy in the system.

Stated in simple climatological terms, heat is a form of energy and the indirect transformation of heat energy into various forms of mechanical energy is the process that drives the global circulation of the atmosphere. Thus the transformation of heat into mechanical energy is responsible for the formation of weather systems whose cumulative effects define the climate of a particular region. Friction continually degrades and destroys atmospheric motions by converting them back into heat, but this heat is in the form of low-grade, high-entropy energy, which cannot create fresh motion systems. Throughout Section 4.4 and Box 4.1 it will be demonstrated that the source of low-entropy energy, which drives the atmospheric circulation, is sunlight, while the low-grade, high-entropy energy produced as a result of the motions is lost to space as long-wave infrared radiation. At the most fundamental level, the important fact is that the Sun is a hot spot in an otherwise cold dark sky. Had the entire sky been at the same temperature as the Sun, then its energy would have been of no use in driving atmospheric and oceanic circulations because long-wave loss would not have been possible. One final question is why is the night sky dark since the universe is of immense size and full of radiating stars? The answer is that the universe is expanding, and this expansion causes the wavelength of light from distant stars to increase (reddening) and move away from the visible due to the so-called Doppler effect. So without the expansion of the universe atmospheric circulations (and vegetation as well) would not be possible!

4.4.2 Distinguishing between temperature and heat

It is important to distinguish between temperature and heat. Temperature is a measure of the mean kinetic energy (speed) per molecule of an object, while heat, or sensible heat as it is often called, is a measure of the total kinetic energy of all the molecules of that object. Thus the temperature of the air is simply a measure of the 'internal energy' of the air. This internal energy is associated with the random motion of the molecules. If two equal masses of gas are brought into intimate contact, this internal energy is rapidly shared between them and their temperatures become equal.

As the temperature decreases it is possible to imagine a state being reached when the molecules are at complete rest and there is no internal energy, a point on the temperature scale known as **absolute zero**. This has been determined as 273.15 degrees Celsius below the melting point of pure ice (0°C), and the Kelvin temperature scale is measured upwards from absolute zero in Celsius units, making 0°C equivalent to 273.15 K. The Kelvin temperature scale is often used in basic physical equations, particularly those concerned with radiation.

Heat is transferred from high- to low-temperature objects and this alters either the temperature or the state of the substance, or both. Thus, a heated body may acquire a higher temperature and this is known as **sensible heat** gain. Alternatively a heated body may change to a higher state (solid to liquid, or liquid to gas) and therefore acquire latent (or hidden) heat. For example, ice at 0°C can be heated and melt to form water. Once the melting process is complete the temperature of the water may still be 0°C. The extra heat was simply used to change the state of the solid to a liquid and this is known as **latent heat**.

One or more of the processes of **conduction**, **convection** or radiation affects the transfer of heat to or from a substance. Conduction is the process of heat transfer through matter by molecular impact from regions of high temperature to regions of low temperature without the transfer of matter itself. It is the process by which heat passes through solids (e.g. soils, rocks). The effects of conduction in fluids (liquids and gases) are usually negligible in comparison with those of convection. Convection is a mode of heat transfer in a fluid (e.g. atmosphere, oceans), involving the movement of substantial volumes of the substance itself. Examples are wind systems in the atmosphere and ocean currents.

Radiation is the final form of heat transfer and is transmitted by any object that is not at a temperature of absolute zero (0 K). This type of heat transfer is discussed in the following section.

4.4.3 Radiation

4.4.3.1 The nature of radiation

Radiation does not require any medium through which to travel and so it can pass through a vacuum with the speed of light. This is how radiation from the Sun reaches the Earth. Radiation can be regarded as having both a wave structure and a particle (**photon**) structure (see Box 26.1 in Chapter 26).

Sometimes its properties are best described in terms of waves, while at other times in terms of a particle structure. Radiation is characterized by wavelength, of which there is a wide range or spectrum extending from very short X-rays through ultraviolet and visible light to infrared, microwaves and long radiowaves (see Figures 26.4 and 26.5 in Chapter 26).

The radiation laws contained in Box 4.2 describe the emission of radiation by an object. Further information about electromagnetic energy (radiation) is also provided in Chapter 26. These laws show that the amount of radiant energy emitted by any object is much greater as temperature increases. At the same time hotter objects also produce more radiation at shorter wavelengths than cooler objects because the maximum wavelength emitted by an object is inversely proportional to absolute temperature (K). The

high temperature of the radiating surface of the Sun (about 6000 K) results in over 99% of the solar energy being at wavelengths of less than 4 μm (μm stands for micrometre and is equivalent to one-millionth of a metre), whereas the much lower temperature of the Earth's atmosphere and surface (generally around 300 K or lower) yields most radiative energy in the 4–100 μm region.

It is therefore convenient to divide the entire atmospheric radiative regime into two parts: the solar (or short-wave) regime and the terrestrial (or long-wave) regime. A division of the spectrum at about 4 μm effectively separates the two. The atmosphere is nearly transparent to short-wave radiation from the hot Sun, of which large amounts reach the Earth's surface. However, the atmosphere readily absorbs longer-wave infrared radiation.

FUNDAMENTAL PRINCIPLES

RADIATION LAWS

A perfect **black body** is one that absorbs all the radiation falling on it and that emits, at any temperature, the maximum amount of radiant energy. This definition does not imply that the object must be black, because snow, which reflects most of the visible light falling on it, is an excellent black body in the infrared part of the spectrum. For a perfect all-wave black body the intensity of radiation emitted and the wavelength distribution depend only on the absolute temperature, which is the temperature measured in degrees Celsius from absolute zero, and in this case the Stefan–Boltzmann law (equation 4.1) applies. This law states that the flux of radiation from a black body is directly proportional to the fourth power of its absolute temperature, that is:

$$F = \sigma T^4 \qquad (4.1)$$

where F is the flux of radiation (watts per square metre), T is the absolute temperature (kelvin) and σ is a constant ($10^{-8}\,W^{-2}K^{-4}$).

A black body emits radiation with a range of frequencies, but with a maximum at frequency λ_{max}. It can be shown by the Wien displacement law that the wavelength of maximum energy λ_{max} is inversely proportional to the absolute temperature:

$$\lambda_{max} = \alpha/T \qquad (4.2)$$

where α is a constant ($1.035 \times 10^{11}\,K^{-1}\,s^{-1}$).

Objects on the Earth's surface are commonly assumed to emit and absorb in the infrared region as a grey body: that is, as a body for which the Stefan–Boltzmann law takes the form:

$$F = \varepsilon \sigma T^4 \qquad (4.3)$$

where the constant of proportionality ε is defined as the infrared emissivity or, equivalently, the infrared absorptivity (one minus the infrared albedo). Typical infrared emissivities are in the range 0.90 to 0.98.

At normal temperatures many gases are not black bodies since they emit and absorb radiation only in selected wavelengths. Many molecules in the atmosphere possess spectra that allow them to emit and

absorb thermal infrared radiation (4–100 μm); such gases include water vapour, CO_2 and ozone, but not the main constituents of the atmosphere, oxygen or nitrogen. These absorption properties are directly responsible for the natural greenhouse effect. Vigorous vertical mixing takes place in the **troposphere** as explained in Section 4.6.1. Therefore it is not the change in thermal infrared radiation at the Earth's surface that determines the strength of the greenhouse warming. Instead it is the change in the irradiance or radiative flux (F) at the top of the troposphere (**tropopause**) that expresses the forcing of the climate system by radiation. Unless a gas molecule possesses strong absorption bands in the wavelength region of significant emission from the ground surface, it can have little effect on the radiative flux at the tropopause (Houghton, 1997). Thus global climate change is strongly affected by the atmospheric composition. The roles of the troposphere and tropopause are discussed in Section 4.6 (see Figure 4.12).

BOX 4.2

4.4.3.2 Global radiation

Global radiation is the sum of all short-wave radiation received, both directly from the Sun (direct solar radiation) and indirectly (diffuse radiation) from the sky and clouds, on a horizontal surface (radiation pathways are shown in Figure 4.6). Direct solar radiation casts shadows, and its intensity on a horizontal surface varies with the angle of the Sun above the horizon (solar angle). Diffuse radiation, in contrast, does not cast shadows, is usually much less intense than the direct component and only varies slightly throughout daylight. Since diffuse radiation is the result mainly of scattering and absorption in clouds, its intensity depends on cloud structure and thickness, not just on solar angle (see Figure 26.6 in Chapter 26).

Global radiation with a large direct component shows a strong diurnal variation in intensity (e.g. Figure 4.7), with maximum values when the Sun is at its greatest angle above the horizon around midday. This diurnal variation nearly vanishes on cloudy days when the diffuse component dominates the global radiation. The actual worldwide distribution of global radiation reflects astronomical factors and the distribution of cloud, which in turn is determined by the global wind circulation. Thus Figure 4.8 shows that the areas receiving most global radiation are found in the subtropics where there are clear skies because of the prevailing anticyclonic conditions associated with the descending air from the Hadley cell circulations illustrated in Figures 4.4 and 4.5.

4.4.3.3 Albedo

Radiation falling on a solid surface may be partly reflected, partly absorbed and partly transmitted. In contrast, water is translucent and light penetrates the surface layers of the oceans, while the atmosphere is nearly transparent to short-wave radiation. Radiation received from the surface of an object may result from either reflection by that surface, or radiation emitted (according to the radiation laws in Box 4.2) from the surface, or indeed both. **Albedo** is a measure of the reflecting power of a surface, being the fraction of the short-wave radiation received that is reflected by

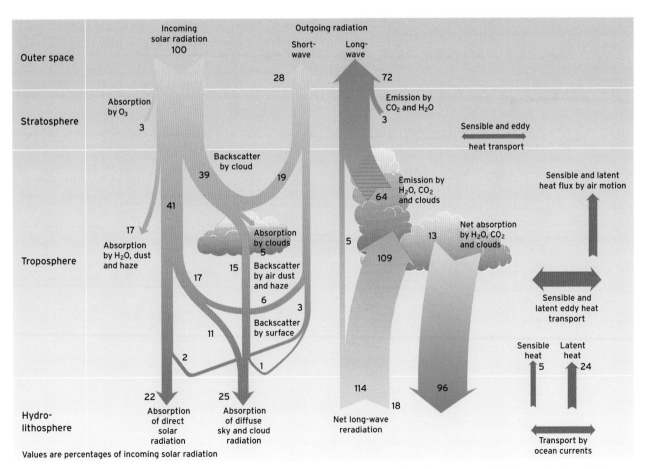

Figure 4.6 Schematic diagram showing the interactions that radiation undergoes in the atmosphere and at the land surface. Values are percentages of incoming solar radiation. The stratosphere is the upper atmosphere and the troposphere the lower atmosphere. The hydro-lithosphere refers to the land or oceans. (Source: after Rotty and Mitchell, 1974)

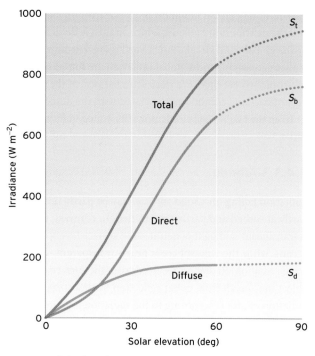

Figure 4.7 Solar irradiance (*F*) on a cloudless sky (16 July 1969) at Sutton Bonington (53°N, 1°W): S_t total flux; S_b direct flux on horizontal surface; S_d diffuse flux. Solar elevation is the angle of the Sun above the horizon. (Source: after Monteith and Unsworth, 1990)

a surface. Albedo is sometimes called **reflectivity**, but this term properly refers to the reflected : received ratio for a specific wavelength. Reflectivity therefore varies with wavelength; for example, grass is green because it reflects much of the green light in solar radiation and absorbs most of the energy in other colours.

Albedo is important because it determines the fraction of the incoming solar radiation that is absorbed by the surface. This absorbed radiation is then available to heat the surface. Typical albedo values are contained in Table 4.1. The greater the albedo, the less solar radiation is absorbed. Albedo varies from 0.08 for dark, moist soils, to 0.86–0.95 for fresh snow. In the latter case nearly all the incident solar radiation is reflected and the solar warming of the surface is small. Clouds also have very high albedos, and therefore greatly restrict the amount of sunlight falling on the ground below them. The planetary albedo of the Earth as seen from space is made up of three main components. These are the light reflected from the actual land and sea surfaces of the Earth, the light reflected by clouds, and light scattered upwards by the atmosphere (Figure 4.6).

The global distribution of surface albedo is shown in Figure 4.9, where the extremely high albedo values of the ice fields of Antarctica should be noted and compared with

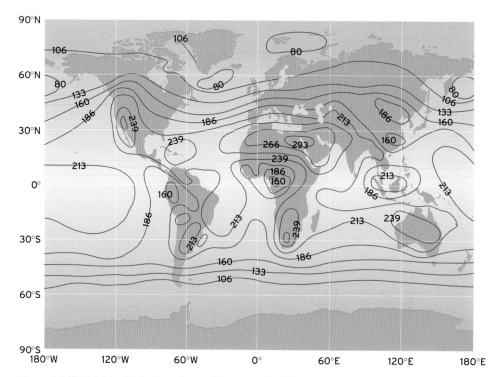

Figure 4.8 Worldwide distribution of annually averaged global (direct plus diffuse) surface-received solar radiation (W m^{-2}). (Source: after Lockwood, 1979 and Budyko, 1974)

Table 4.1 Values of albedo for various surfaces

Surface	Albedo (percentage of incoming short-wave radiation which is reflected)
Fresh, dry snow	80–95
Sea ice	30–40
Dry light sandy soils	35–45
Meadows	15–25
Dry steppe	20–30
Coniferous forest	10–15
Deciduous forest	15–20

the low albedo values over the equatorial rainforests. Particularly high gradients of surface albedo are seen in both the Arctic and Antarctic where ice fields replace open water or ice-free land. High albedo values also occur over the dry sands of the subtropical deserts. Albedo over the oceans is, except for the high latitudes, nearly uniform at around 0.10.

4.4.3.4 Net radiation and energy balances

The full energy balance of, for example, a sample land surface (Figure 4.6) involves not only radiative fluxes, but also fluxes of energy in the form of sensible heat (heat that can be detected by a thermometer) and latent heat (e.g. energy contained in evaporated water vapour). The sensible heat fluxes may be both downwards into the soil and upwards into the atmosphere. The energy available for the sensible and latent heat fluxes depends on the net radiation. The **net radiation** at a surface is the difference between the total incoming and the total outgoing radiation. The net radiation indicates whether net heating (positive) or cooling (negative) is taking place; the net radiation will normally be negative at night, indicating cooling, but during the day it may be positive or negative depending on the balance of the incoming and the outgoing radiation. Low cloud blankets the Earth's surface and at night restricts the long-wave radiation loss, thereby causing only small net radiative losses and temperature decreases. If the sky is clear at night, the long-wave radiative loss is large, so is the net radiative loss, resulting in a large nocturnal temperature decrease. The term 'available energy' is often used to describe net radiation minus the sensible heat transfer into the soil.

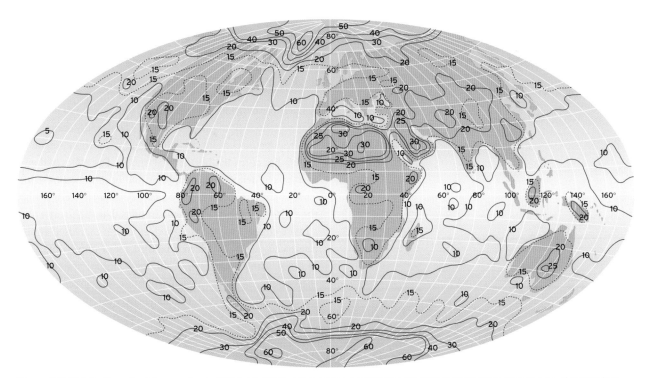

Figure 4.9 Distribution of minimum albedo (%) from Nimbus 3 satellite measurements during 1969-1970. This is a good approximation to the albedo of the surface. The assumption is that the albedo of the Earth-atmosphere system is higher over each area in the presence of clouds than for a cloud-free atmosphere. Therefore estimates of the albedo of the surface and also locations of persistent cloud fields and of ice and snow can be made. The effects of changing cloud fields are removed by displaying only the lowest observed satellite albedo value in each area. (Source: after Raschke *et al.*, 1973)

Net radiation is defined as:

$$R_n = (1 - \alpha)Q - L_{out} + L_{in} \tag{4.4}$$

where R_n is the net radiation, α is the albedo of the surface, Q is the incoming global radiation, L_{out} is the outgoing long-wave radiation from the surface and L_{in} is the incoming long-wave radiation from the atmosphere. For a normal surface (land or ocean) the sensible and latent heat fluxes are balanced by the net radiation. This gives an equation of the form:

$$R_n = \lambda E + H + S \tag{4.5}$$

where λ is the latent heat of vaporization, E is the evaporation rate, H is the sensible heat flux into the atmosphere and S is the sensible heat flux into the soil or ocean.

4.4.3.5 The Bowen ratio – the partition of net radiation

The key issue over land is the partition of the net radiation into sensible and latent heat fluxes. One approach to this problem is to consider the Bowen ratio, which is defined as $H/\lambda E$ (sensible heat flux divided by latent heat flux). For normal moist surfaces, values of the ratio are in the range 0 to 1 (Figure 4.10). The value of the Bowen ratio varies widely over land. The Bowen ratio decreases with increasing temperature and the ratio reaches zero at about 32–34°C over moist surfaces. The ratio also depends on the availability of water for evaporation, which is controlled by soil moisture,

vegetational controls, and input of dry air from above. Drier conditions result in an increase in the Bowen ratio.

If water is readily available for evaporation, because it is on the soil surface or the roots of vegetation have access to water in the soil, then as much as 75–80% of the total daytime transport of energy to the atmosphere occurs as latent heat. This high evaporation rate from a moist surface cools the surface and keeps daytime maximum temperatures below about 32–34°C. At this temperature all the net radiation is used for evaporation. Higher evaporation rates can occur only by extracting heat from the atmosphere and thereby cooling it. At the other extreme of the desert, where water for evaporation is very limited, most of the net radiation during daylight is transferred to the atmosphere as sensible heat, which warms the lower atmosphere. Under these circumstances the Bowen ratio has values above 1 and daylight surface temperatures can reach values considerably above 34°C since the surface is not cooled by evaporation. Because the desert surface is hot in daytime, the long-wave emission can be very large, significantly reducing the net radiation. The degree of wetness of a surface therefore exerts, through the Bowen ratio, a strong control over surface temperature and near-surface atmospheric temperature. For example, maximum temperatures observed over tropical oceans or moist tropical land surfaces rarely exceed 32–34°C, while over dry tropical deserts they may reach 50°C.

4.4.4 Thermal inertia

A scientific consideration of critical importance to policy-makers who seek to mitigate the effects of undesirable anthropogenic climate change is that the Earth's climate system has considerable thermal inertia. The effect of the inertia is to delay the atmospheric and oceanic response to various climate forcings. Of particular importance is that the existence of thermal inertia implies that still greater climate changes will be in store for present increased atmosphere CO_2 levels, which may be difficult or impossible to avoid. The primary symptom of the climate system's thermal inertia, in the presence of increasing greenhouse gas concentrations, is an imbalance between the energy absorbed and emitted by the planet. Hansen (2005) finds using a climate model, driven mainly by increasing human-made greenhouse gases and aerosols, among other forcings, that the Earth is now absorbing 0.85 ± 0.15 W m^{-2} more energy from the Sun than it is emitting to space. This imbalance is confirmed by precise measurements of increasing ocean heat content over the past 10 years. This implies an expectation of additional global warming of about 0.6°C

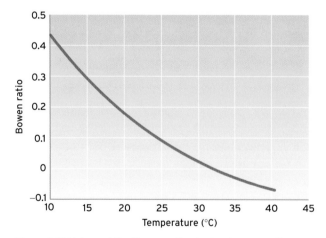

Figure 4.10 Values of the Bowen ratio at a freely evaporating surface for various surface temperatures. This diagram applies only to a freely evaporating surface. If the surface becomes dry, the value of the Bowen ratio will increase to above that appropriate for the surface temperature. (Source: after Lockwood, 1979)

without further change of atmospheric composition. IPCC Working Group 1, Fourth Assessment Report (2007a) (WG1 AR4) estimated a further 0.6°C unavoidable warming by the end of the century relative to 1980–1999 even if atmospheric greenhouse gas concentrations remain at 2000 levels. Working Group 2, Fourth Assessment Report (2007b) (WG2 AR4) commented that this leads to the conclusion that there are some impacts for which adaptation is the only available and appropriate response.

4.4.5 The atmospheric energy balance

The atmosphere continually loses more heat to space in the form of long-wave radiation than it gains from the absorption of short-wave radiation, and this is equivalent to a cooling of 1 or 2°C per day. The heat lost to space in the form of long-wave radiation is largely compensated for by the latent heat released by atmospheric water vapour that has evaporated from the surface, condensing to form precipitation. Typical annual global values for the energy balance of the atmosphere are:

- net radiative loss = -99.4 W m^{-2} (this must be balanced by the sensible and latent heat gained by the atmosphere);
- sensible heat gain = 20.4 W m^{-2};
- latent heat gain = 79.0 W m^{-2}.

Annual global precipitation is therefore linked closely to the atmospheric energy balance and can change only if there are associated changes in the global energy balance. The latent heat gain of 79 W m^{-2} is equivalent to an average annual global precipitation of about 1004 mm yr^{-1}, which has shown yearly variations of about ±50 mm (±5%) over the past 100 years.

On a global mean basis, to a good approximation, a balance between radiative cooling and condensational heating maintains the mean equilibrium atmospheric temperature. This has interesting and important consequences for the relationship between changes of global atmospheric temperature and associated changes in global precipitation. For example, a temperature increase initiated by a reduction in radiative cooling in the lower atmosphere, because of an increase in atmospheric CO_2, and therefore the greenhouse effect, can be achieved with a corresponding decrease in condensational heating and therefore a reduction in precipitation. In contrast, for a temperature increase initiated by an enhancement in condensational heating (increase in global precipitation), for instance, due to increased sea surface temperatures, and therefore evaporation, the atmosphere can adjust itself to a new steady state through an increase in

radiative cooling. For both cases, although the new steady-state temperature is higher, the changes in global precipitation, and therefore in the intensity of the hydrological cycle, are in opposite directions. Thus an increase in temperature due to global warming may decrease or increase the rainfall. This is explored further in the next section and it is to the role of moisture in the atmosphere that we now turn our attention.

Reflective questions

➤ Where are the main areas of latent heat release in the global atmosphere?

➤ What is the visual evidence of convection in the atmosphere?

➤ How would you expect daily totals of global radiation to vary through the year at (a) the South Pole, (b) Singapore and (c) London?

➤ Why are snow and ice fields able to exist on high tropical mountains?

➤ Will the transfer of sensible heat to the atmosphere over a moist grass surface be greater at higher or lower temperatures?

➤ Many hot desert areas have very high surface albedos. Do they warm or cool the atmosphere in comparison with a similar but vegetation-covered area?

4.5 Moisture circulation systems

4.5.1 Moisture in the atmosphere and the hydrological cycle

The full cycle of events through which water passes in the surface–atmosphere system is best illustrated in terms of the hydrological cycle, which describes the circulation of water from the oceans, through the atmosphere back to the oceans, or to the land, and then to the oceans again by overland and subsurface routes. Water in the oceans evaporates under the influence of solar radiation and the resulting water vapour is transported by the atmospheric circulation (see Section 4.3) to the land areas, where precipitation may occur in a variety of forms including rain, hail and snow. Some of this precipitation will infiltrate into the soils and rocks below the surface from where it flows more slowly to river channels or sometimes directly to the sea (see Chapter 11). The water remaining on the surface and in the upper layers of the soil will partly evaporate back to the atmosphere. Since evaporation is linked to the surface energy balance, the hydrological cycle is directly linked to, and indeed is part of, global energy exchanges.

It is simplest to regard the hydrosphere as a global system consisting of four reservoirs linked by the hydrological cycle. These are the world ocean, polar ice, terrestrial waters and atmospheric waters; estimates of the amount of water stored in these reservoirs are summarized in Table 4.2. The average residence times (Table 4.3) of water in the various reservoirs are of particular interest to climatologists. The water vapour in the air at any one time represents about one-fortieth part of the annual rainfall or the supply of

Table 4.2 Estimated volumes of principal components of the hydrosphere

	Surface area covered (10^6 km^2)	Present amount of water resources (10^3 km^3)	(%)
World ocean	361	1 370 000	93
Polar ice	16	24 000	2
Terrestrial waters	134	64 000*	5
Atmospheric waters	510	13	0.001

*Includes groundwater (almost 64 000), lakes (230), soil moisture (82).

Table 4.3 Average residence times of water

Water source	Residence time
Atmospheric waters	10 days
Terrestrial waters	
Rivers	2 weeks
Lakes	10 years
Soil moisture	2–50 weeks
Biological waters	A few weeks
Groundwaters	Up to 10 000 years
Polar ice	15 000 years
World oceans	3600 years

water for about 10 days of rainfall. The Earth's atmosphere contains on average an amount of water vapour which, if it were all condensed and deposited on the surface of the Earth, would stand to a depth of about 25 mm. Therefore to maintain the precipitation amounts observed in many parts of the world, there must be a continual flow of very moist air into the area from the surroundings. The general wind circulation (Section 4.3) is therefore very important in explaining the observed global precipitation patterns. Particularly high precipitation regions are found where moist winds from oceanic areas either flow onto windward continental coasts, Europe for example, or converge into low-pressure troughs such as that found in the equatorial zone.

4.5.2 Global distribution of precipitation and evaporation

While on a global scale over an annual period, evaporation approximately equals precipitation, this is not so for either particular regions or seasonal timescales. Figure 4.11 shows estimated annual distributions of precipitation and evaporation. It is clear that the main areas of evaporation, mostly over the oceans, do not coincide with the major precipitation regions. This implies an atmospheric flux of water vapour from the oceans to the land masses. Indeed, the water forming precipitation over the continental land masses can be divided into two components. Firstly, there is the water component of the precipitation that originated from distant oceans and has been circulated over the land by low-level winds. Secondly, there is the component that is formed of locally evaporated water. In the middle-latitude continents, the first component dominates in winter and is associated with the middle-latitude westerlies and active frontal depressions, while the second is more important in summer, and tropical/equatorial areas, and is associated with convective storms. Evaporation rates (Figure 4.11b) over the land masses are relatively low, only about 50% of the global average (Brutsaert, 1982). There are a number of reasons for this, mostly linked to the moisture-holding capacity of the land surface soils, and the annual distribution of precipitation and net radiation. Many land surfaces are able to sustain typical evaporation rates of 2–3 mm day^{-1} for less than a month. In climates with marked wet and dry seasons, found particularly in the tropics, soil moisture soon becomes exhausted in the dry season and evaporation rates fall to very low values. Here vigorous evaporation is restricted to the wet season. In many areas in the subtropics, the wet season is short or non-existent, and annual evaporation rates are very low. Similar arguments apply in many middle-latitude regions, where precipitation is

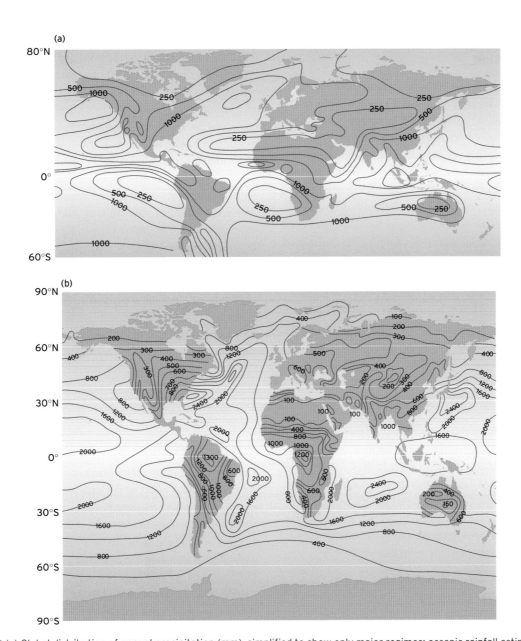

Figure 4.11 (a) Global distribution of annual precipitation (mm), simplified to show only major regimes; oceanic rainfall estimated. (Source: after Riehl, H., *Introduction to the atmosphere*, 1965, McGraw-Hill, Fig. 2.16. Reproduced with permission of the McGraw-Hill Companies) (b) Global distribution of annual evaporation (mm) based upon a number of recent estimates. Oceans are almost infinite moisture sources, while land areas frequently become dry, so there is a marked discontinuity between adjacent land and ocean regions. Oceanic evaporation shows a dependence upon ocean surface temperature. (Source: after Brutsaert, 1982)

often plentiful in winter, but because of a Sun low in the sky associated with short daylight hours, net radiation is restricted and hence evaporation rates are low. Much of the winter precipitation either recharges soil moisture or forms river runoff. In contrast, in summer with long daylight hours and a Sun high in the sky, the net radiation values are high but soil moisture becomes seriously depleted, because of low precipitation rates, leading to restricted evaporation rates.

In many summer and dry-season cases the actual land surface evaporation is considerably less than predicted by the energy balance equations with assumed moist land surfaces. The evaporation may be considered as being restricted by some form of resistance in the evaporative pathway from the soil to the atmosphere. When the soil is moist the resistance is relatively low, but as the soil moisture decreases, the resistance increases in response and limits the evaporative loss.

4.5.3 The influence of vegetation on evaporation

Evaporative losses from land surfaces have three components, known as transpiration, **interception** loss and soil surface evaporation. While they are governed by the same physical laws, they do not necessarily behave in the same manner under similar conditions. Water is extracted from the soil by the roots of actively growing vegetation and transpired through small holes in the vegetation leaves called stomata. Transpiration is therefore strongly controlled by the nature and condition of the vegetation. During rainfall, water is intercepted in vegetation canopies and evaporated back to the atmosphere during and just after the rainfall event. This re-evaporated rainwater is termed interception loss. Provided there is rainwater on the plant leaves, the rate of interception loss is independent of the vegetation state but strongly dependent on atmospheric conditions. Lastly, water is evaporated directly from moisture held on or just below the soil surface. The sum of transpiration, interception loss and soil surface evaporation is termed evapotranspiration. Further discussion of evapotranspiration is provided in Chapter 11.

Evaporation from the ground surface depends on the net radiation, which supplies the energy for the latent heat of vaporization, and also the vertical gradient of water vapour pressure. Equations describing evaporation are normally expressed in terms of conceptual resistances, using Ohm's law for electricity as a direct analogue. Ohm's law gives the relationship between current (amps) in a circuit to the electrical potential (volts) and the resistance of the wire:

$$\text{Current} = \frac{\text{potential difference}}{\text{wire resistance}} \qquad (4.6)$$

For entities transported from the surface through the atmosphere such as water vapour, heat and CO_2 this may be rewritten to read:

$$\text{Flux rate} = \frac{\begin{array}{c}\text{concentration} \\ \text{difference of property}\end{array}}{\begin{array}{c}\text{resistance to flow} \\ \text{exerted by the system } (r)\end{array}} \qquad (4.7)$$

where r represents the appropriate system resistance with units of seconds per metre. In the case of evaporation the concentration difference may be expressed as the vertical gradient of vapour pressure.

In the case of **transpiration**, which is the removal of water from plants, two resistances are frequently used: one applies to the passage of water though the vegetation from the soil to the atmosphere, and the other applies to the passage of water vapour from the canopy surface to a standard level (often at 2 m) in the overlying atmosphere. The value of the first (the bulk canopy resistance) depends on such factors as the vegetation type, soil moisture, sunlight and atmospheric humidity. The value of the second (the aerodynamic resistance) depends on such factors as vegetation height, wind speed and so on.

Stomatal pores in leaves are necessary for the uptake of CO_2 from the ambient air, but at the same time water vapour escapes via transpiration since there is a common physical pathway of water vapour and CO_2 through the stomatal pores. The degree of opening of the pores can be considered as a compromise in the balance between limitation of water loss and admission of CO_2. The generally observed closure of stomata when CO_2 increases is an expression of this compromise. Thus for many vegetation types the bulk canopy resistance increases as atmospheric CO_2 increases, resulting in a fall in transpiration rates. As a result vegetation uses water more efficiently in an enhanced CO_2 atmosphere. Therefore some vegetation types could grow in drier climates than they do at present (Lockwood, 1993, 1999).

4.5.4 Drought

Droughts rank among the world's costliest natural disasters because they affect a very large number of people each year. Drought is a recurring phenomenon that has plagued civilization throughout history. A drought is considered to be a period of abnormally dry weather that causes serious hydrological imbalance in a specific region. However, the definitions of 'serious' and 'abnormally dry' depend on the nature of the local climate and the impact of the drought on the local society. A dry spell in a humid climate may be classified as a drought, while similar conditions in a semi-arid climate would be considered a wet period. This makes it difficult to produce a definition of drought that applies in a variety of climates.

The American Meteorological Society (1997) grouped drought definitions into four categories: meteorological or climatological, agricultural, hydrological and socio-economic. If atmospheric conditions result in the absence or reduction of precipitation over several months or years the result is a meteorological drought. A few weeks' dryness in the surface layers (vegetation root zone), which occurs at a critical time during the growing season, can result in an agricultural drought that severely reduces crop yields, even though deeper soil levels may be saturated. The onset of an agricultural drought may lag that of a meteorological drought, depending on the prior moisture status of the surface soil layers. **Precipitation deficits** over a prolonged period that affect surface or subsurface water supply, thus reducing stream flow, groundwater, reservoir and lake levels, will result in a hydrological drought, which will persist long after a meteorological drought has ended. Socio-economic drought associates the supply and demand of some economic good with elements of meteorological, agricultural and hydrological drought.

There are a number of quantitative definitions of drought based upon knowledge of precipitation, soil moisture, **potential evapotranspiration** or some combination of the above. A commonly used and widely accepted meteorological drought index is the Palmer Drought Severity Index (PDSI). It originated in the United States and is obtained from a simplistic model of the cumulative anomaly of moisture supply and demand at the land surface, which requires knowledge of both precipitation and potential evapotranspiration. Model results are adjusted to allow for the local climate before the final index is calculated. Regardless of the simple nature of PDSI, it was found by Dai *et al.* (2004) to be significantly correlated with the observed soil moisture content within the upper 1 m of soil during warm-season months. They also found that PDSI was a good proxy for stream flow. Using a global data set of PDSI, Dai *et al.* (2004) found that very dry regions over global land areas have increased from ~12% to 30% since the 1970s, with a large jump in the early 1980s due to an El-Niño-induced precipitation decrease and enhanced surface warming. IPCC (2007a) commented that it is likely that droughts have increased in many areas since the 1970s, and that it is more likely than not that there is a human contribution to this trend. During the past two to three decades, there has been a tendency for more extreme (either very dry or very wet) conditions over many regions, including the United States, Europe, east Asia, southern Africa and the Sahel.

The UK Hadley Centre's global climate model (HadCM3) has been used, together with PDSI, to study drought on a global basis (Burke *et al.*, 2006). The model shows that between 1952 and 1998 on average 20% of the land surface was in drought at any one time. At decadal timescales, on a global basis, the model reproduced the observed drying since 1952 reported by Dai *et al.* (2004) and analysis showed that there was a significant influence of anthropogenic emissions of greenhouse gases and sulphate aerosols in the production of this drying trend. Future projections of drought in the twenty-first century using HadCM3 show regions of strong wetting and drying with a net overall global drying trend. The proportion of the land surface in extreme drought is predicted to increase from 1% for the present day to 30% by the end of the twenty-first century.

Reflective questions

➤ What are the fundamental differences in precipitation and evaporation likely to be during the year between northern and southern Europe?

➤ What reasons are there that actual evaporation from a desert surface might be lower than predicted by typical energy balance equations?

4.6 Motion in the atmosphere

4.6.1 Convective overturning

Convection is the dominant process for transferring heat, mostly in the form of latent heat, upwards from the surface. During daylight the surface of the Earth is warmed by short-wave radiation from the Sun. Air close to the surface is heated and rises, because on warming its density becomes lower than that of the surrounding cooler air. In this discussion it is useful to consider an isolated parcel of air which is warmed at the surface. Because pressure falls with height in the atmosphere, as the parcel rises it also expands so that the internal pressure of the parcel and the external pressure at the same level in the atmosphere are always equal. An expanding parcel of air chills because its molecules lose a little of their internal energy, which as explained earlier we sense as temperature, as they bounce off the retreating parcel walls.

Vertical motions in the atmosphere are usually rapid enough to make temperature changes occur much more quickly than any interchange of heat, by conductive or radiative processes, between the parcel and its surroundings (see Figure 6.2). The temperature changes may then be said to be **adiabatic**: that is, there is no interchange of heat between the parcel and its surroundings. Adiabatic changes also imply constant entropy throughout the changes. The rate of temperature fall with height (**dry adiabatic lapse rate**) in the rising air parcel under these conditions is 9.8°C km^{-1}. A similar rate of warming is observed in a parcel of air sinking under adiabatic conditions. Cooling in a rising air parcel may cause the air to become saturated with water vapour, condensation to take place and clouds and precipitation to form. When water vapour condenses to form water droplets in the rising air parcel, it releases latent heat which warms the air slightly and therefore causes the rate of temperature decrease with height to fall below the dry adiabatic lapse rate. The new **lapse rate**, in the presence

Chapter 4 Atmospheric processes

of condensation, is known as the **saturated adiabatic lapse rate**. Its exact value is variable since it depends on the rate of condensation. See Box 6.1 in Chapter 6 for further information on lapse rates.

Considerations of mass balance dictate that as some air masses rise, other air masses descend, so the lower atmosphere is continually turning over. This layer of the atmosphere where there is large-scale mixing, and which also contains most of the atmospheric water vapour as well as most clouds and weather phenomena, is known as the troposphere (Figure 4.12). The upper boundary of

the troposphere is known as the tropopause, which has an altitude of about 5–6 km over polar regions and 15–16 km over equatorial regions. Temperature in the troposphere falls from around 15°C at the surface to around −57°C at the tropopause at a rate determined by these convective processes. Since condensation (to form clouds and rainfall) and the release of latent heat occur during the convective processes, the observed fall of temperature with height in the troposphere is less than the dry adiabatic lapse rate. Observations indicate that it is about 6°C km^{-1}, but it varies with location in the troposphere. The **stratosphere** extends from the top of the tropopause to about 50 km height (the stratopause). In the stratosphere the temperature either varies very little with height or increases with height with warmer air overlying cooler air. This is because the maximum temperatures are associated with the absorption of the Sun's short-wave radiation by ozone which is found at this altitude. In the thermosphere temperatures increase with altitude because of the absorption of extreme ultraviolet radiation (0.125–0.205 μm) by oxygen but really these 'temperatures' are theoretical because there is so little air (so few molecules – and remember that temperature is simply a measure of the mean kinetic energy of the molecules of an object).

4.6.2 The Earth's rotation and the winds

Over timescales of around a year or longer, the complete Earth–atmosphere system is almost in thermal equilibrium. Consequently the total global absorption of solar radiation by the Earth's surface and atmosphere must be balanced by the total global emission to space of infrared radiation. While most of the solar radiation is absorbed during daylight in the tropical regions of the world, infrared radiation is lost continually day and night to space from the whole of the Earth's surface and atmosphere, including the middle and polar latitudes. Heat energy is therefore transferred from the tropics, where it is in surplus, to middle and high latitudes, where it is in deficit, by the north–south circulations in the atmosphere and oceans. These circulation patterns were described earlier and take the form of a number of cells in the atmosphere, with rising air motion at low and middle latitudes, and sinking motions in subtropical and high latitudes. This very simple circulation both implies a rotating Earth and is further modified by the Earth's rotation.

The simplest atmospheric circulations would arise if the Earth always kept the same side towards the Sun. Under these circumstances the most likely atmospheric circulation would probably consist of rising air over an extremely hot, daylight face and sinking air over an extremely cold, night face. The diurnal cycle of heating and cooling obviously

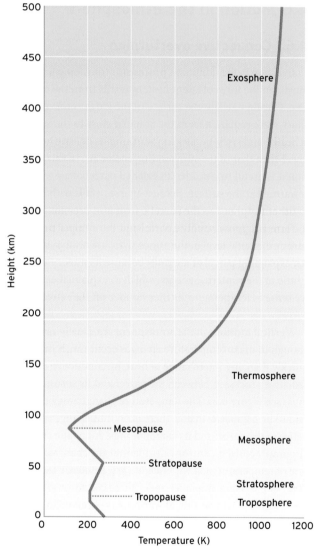

Figure 4.12 Temperature variation with height (lapse rate) in the atmosphere. Note the three regions of increased temperature where absorption occurs: (i) the surface (visible and near-infrared absorption); (ii) lower stratosphere (ultraviolet absorption by ozone); (iii) thermosphere (high-energy absorption). 0 K = −273.14°C, 273.14 K = 0°C. (Source: from Robinson and Henderson-Sellers, 1999)

would not exist since it depends on the Earth's rotation. Surface winds everywhere would blow from the cold night face towards the hot daylight face, while upper flow patterns would be reversed. The climatic zonation would be totally different from anything observed today. Theoretical studies suggest that if this stationary Earth started to rotate, then as the rate of rotation increased, the atmospheric circulation patterns would be progressively modified until they approached those observed today (Lockwood, 1979). Even with the present rate of rotation two different atmospheric circulation patterns are possible. This is because if the Earth rotated towards the west instead of towards the east as at present, then the mid-latitude westerlies would become easterlies and depressions and anticyclones would rotate in the opposite sense. Under these circumstances, Europe would have a climate similar to that of present-day eastern Canada, while eastern Canada would enjoy the present-day, mild, western European climate.

Observations of air flow in the atmosphere show that, except very near to the surface and in equatorial regions, it is almost parallel to the **isobars** (contours of equal pressure). This seems odd because air should move at 90° to isobars from high to low pressure. Instead the Earth's rotation acts on the atmosphere through the so-called **Coriolis effect** (see Box 4.3). The force due to the horizontal pressure gradient balances the Coriolis effect and so the wind blows parallel to isobars (Figure 4.13). This wind, which exists above about 1 km from the surface, is known as the **geostrophic wind**. The speed of this wind is proportional to the pressure gradient (i.e. wind speeds are greater when the isobars are closer together) but is also determined by the strength of the Coriolis effect. As the Coriolis effect varies with latitude (Figure 4.14b) the geostrophic wind for the same pressure gradient will decrease towards the poles. The direction is

such that the air flow is anticlockwise around low pressure in the northern hemisphere. The converse applies in the southern hemisphere where air flow is clockwise around low pressure. For high-pressure anticyclones the flow in both hemispheres is the opposite way round.

There are added complexities in atmospheric air flow which make long-term weather forecasting a challenging exercise. These complexities can be related to feedback effects that are not straightforward or linear as described in Box 4.4.

4.6.3 Rossby waves

Compared with the Hadley cells discussed in Section 4.3, the middle-latitude atmosphere is highly disturbed and the suggested circulation of the Ferrel cells in Figure 4.4 is largely schematic. At the surface, the predominant features are irregularly shaped, closed, eastward-drifting, cyclonic and anticyclonic systems, while higher up, smooth wave-shaped patterns are generally found. These waves are often termed **Rossby waves**, after C.G. Rossby who first investigated their principal properties in the late 1930s and early 1940s. Rossby waves are important because they strongly influence the formation and subsequent evolution of surface weather features. Middle-latitude frontal depressions tend to form and grow rapidly just downwind of upper troughs of Rossby waves, while surface anticyclones tend to develop just downwind of upper ridges. Surface depressions in particular tend to move in the direction of the upper flow with a speed which is directly proportional to the upper flow.

In a simple atmosphere, Rossby waves could arise anywhere in the middle-latitude atmosphere, as is observed in the predominantly ocean-covered southern temperate latitudes. In contrast, the northern temperate-latitude Rossby waves tend to be locked in certain preferred locations. These preferred locations arise because the atmospheric circulation is influenced not only by the differing thermal properties of land and sea, but also by high north–south-aligned mountain ranges such as the Rockies and Andes. In January, at about 5.5 km in the northern hemisphere, the dominant troughs are found near the eastern extremities of North America and Asia, while ridges lie over the eastern parts of the Pacific and Atlantic Oceans (Figure 4.15). Climatologically, the positions of the two troughs are associated with cold air over the winter land masses and the position of the ridges with relatively warm sea surfaces. Atlantic Ocean frontal depressions form just downwind of the American trough, move eastwards with the upper winds and decay south of Iceland in the upper trough. In July, the mean 5.5 km flow pattern found in January over the Pacific

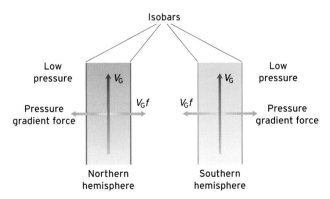

Figure 4.13 The geostrophic wind: V_G is the geostrophic wind; f is the horizontal component of the Coriolis force. The pressure gradient force tends to balance the Coriolis force. The pressure gradient force acts from high to low pressure.

FUNDAMENTAL PRINCIPLES

THE CORIOLIS EFFECT

The concept of the Coriolis effect may be briefly explained as follows. Large-scale, horizontal winds flowing across the Earth's surface tend, if they are not deflected by horizontal pressure gradients, to move in a straight path in relation to a reference frame fixed in relation to the distant stars. The movement of winds over the Earth's surface is usually made with reference to the latitude and longitude grid which rotates with the Earth and thus continually changes orientation in relation to the distant stars. Thus to an observer on the rotating Earth, who is unaware of the Earth's rotation, it appears that a force is acting on the winds that causes them to be deflected to the right (left) in the northern (southern) hemisphere (Figure 4.14). The observer will make the same 'mistake' whatever the initial wind direction. The observer may conclude that the deflection to the right has been brought about by a force acting to the right of the wind. In the nineteenth century the French physicist Coriolis formalized this concept of an apparent force caused by the Earth's rotation, later called the Coriolis effect. Thus winds blowing towards the equator are deflected by the Coriolis effect towards the west (generating easterly winds), while poleward-flowing winds are deflected towards the east (generating westerlies). The magnitude of the Coriolis effect varies with latitude as shown in Figure 4.14(b).

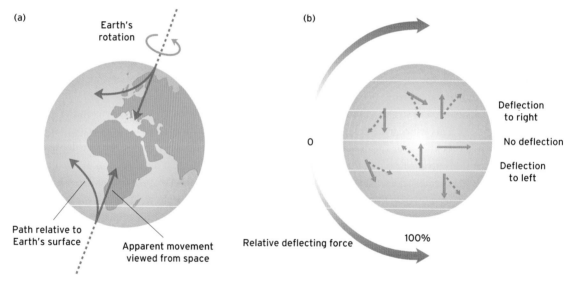

Figure 4.14 (a) The influence of the Coriolis effect on air movement. (b) The changing magnitude of the Coriolis effect with latitude.

BOX 4.3

has moved about 25° west and now lies over the warm North American continent, while there is a definite trough over the eastern Pacific (Figure 4.16).

4.6.4 Jet streams

Embedded in the Rossby waves are fast-moving bands of air known as jet streams. They are caused by sharp temperature gradients associated with frontal depressions.

Separate powerful westerly jet streams, not associated with the middle-latitude frontal jet streams, are also found along the poleward borders of the tropical Hadley cells. Normally a jet stream is thousands of kilometres in length, hundreds of kilometres in width and several kilometres in depth. An arbitrary lower limit of 30 m s^{-1} (about 60 mph) is assigned to the speed of the wind along the axis of a jet stream. The global distribution of jet streams is briefly described below.

NON-LINEAR EFFECTS

The climate record shows rapid step-like shifts in climate variability that occur over decades or less, as well as climate extremes (e.g. droughts) that persist for decades. The variability of climate can be expressed in terms of two basic modes: the forced variations which are the response of the climate system to external forcing, and the free variations due to internal instabilities and feedbacks to non-linear interactions among the various components of the climate system. The external causes operate mostly by causing variations in the amount of solar radiation received or absorbed by the Earth, and comprise variations in both astronomical (e.g. orbital parameters) and terrestrial forcings (e.g. atmospheric composition, aerosol loading). The internal free variations in the climate system are associated with both positive and negative feedback interactions between the atmosphere, oceans, cryosphere and biosphere. These feedbacks lead to instabilities or oscillations of the system on all timescales, and can operate independently or reinforce external forcings.

What all this means is that the climate system belongs to a particular set of mathematical systems known as non-linear systems. They have unusual, and perhaps unfamiliar, properties that are of great importance to understanding climate change. Such systems are very common in the environment. Examples of non-linearity in the climate system include the following. The Asian monsoon shows an abrupt change or seasonal jump in the northern spring characterized by a sudden shift northwards of the subtropical westerly jet stream in early June. This is associated with nearly simultaneous changes in atmospheric circulation parameters over large regions of southern Asia. On a longer timescale, throughout the Alps, the last 3000 years were characterized by repeated glacial events at intervals of 200–400 years, which, in some, were comparable to the Little Ice Age of around 1700. The advances were quite sudden and rapid, with advances in the later ones obscuring the earlier events. The sparse long records of Sahel rainfall suggest that the conditions of the last couple of decades may be unprecedented in the context of the last several centuries, and also that shifts from one variability state (e.g. 'wet') to another (e.g. 'dry') may occur over only a couple of years.

The inherent non-linear nature of the atmosphere is manifest in interactions and fluctuations on a wide range of space- and timescales. Even on the annual scale, abrupt circulation changes are observed as seasonal changes progress. Many climatic switches take place on timescales that are relatively short, and make them of significant societal relevance. There are a number of types of behaviour found in dissipative, highly non-linear systems, under non-equilibrium conditions, such as the climate system. In particular, as a system evolves, due for example to climate change, it may approach a so-called **bifurcation point** where the natural fluctuations become abnormally high, as the system may 'choose' among various regimes. These bifurcation points are often known as 'tipping points' in the popular literature. Under non-equilibrium conditions, local events have repercussions throughout the whole system, with long-range correlations appearing at the precise point of transition from equilibrium to non-equilibrium conditions. Long-range correlations (teleconnections) are indeed observed in the atmosphere, and the strength of these teleconnections is observed to vary with time.

A very important property of developing weather systems or climate systems is that they are very sensitive to initial conditions. Slight changes in the initial conditions can lead to very different development pathways. This is very noticeable in the development of weather systems, and makes weather forecasting very difficult for more than a few days ahead. Indeed it places a limit of about 20 days on the length of detailed weather forecasts. Weather forecasts are normally produced with the aid of numerical atmospheric models. A standard technique is to vary very slightly the meteorological data used to start the model. If the resulting forecasts are very similar, the atmospheric conditions are stable and the weather forecast can be made with some confidence. In contrast, if the forecasts diverge rapidly, the resulting weather forecasts show little confidence. It is this great sensitivity to initial conditions that makes seasonal climate forecasting difficult, if not impossible. Similarly it is not possible to give detailed climate data for some point in the future; rather the climate data have to be expressed in terms of probabilities.

BOX 4.4

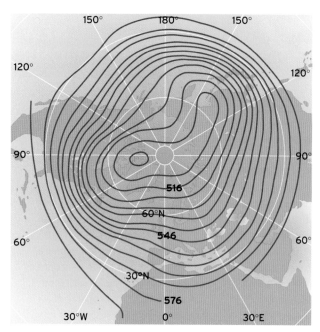

Figure 4.15 Monthly mean 500 millibar contours in the northern hemisphere for July, based on data from 1951 to 1966. (Source: after Moffit and Ratcliffe, 1972)

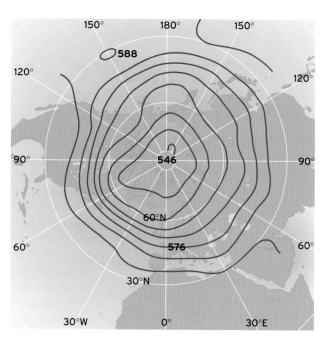

Figure 4.16 Monthly mean 500 millibar contours in the northern hemisphere for January, based on data from 1951 to 1966. (Source: after Moffit and Ratcliffe, 1972)

4.6.4.1 The subtropical westerly jet streams

The subtropical westerly jet streams are found near 30°N and S, corresponding to the poleward boundaries of the tropical Hadley cells, and often form continuous belts around the globe (Figures 4.4 and 4.17). The jet stream cores with the highest wind speeds are normally found near the 12 km level, just below the tropopause. The subtropical jet streams draw their high wind speeds partly from the circulations associated with the Hadley cells and are associated with the descending limbs of the cells. In winter the northern subtropical westerly jet stream divides north and south around the Tibetan Plateau. The two branches merge to the east of the plateau and form an immense upper convergence zone over China. In May and June, in association with the development of the south-west monsoon, the subtropical jet stream over northern India slowly weakens and disintegrates, causing the upper westerly flow to move northwards into central Asia. While this is occurring, an easterly jet stream, mainly at about 14 km altitude, builds up over the equatorial Indian Ocean and expands westward into Africa (Figure 4.17).

4.6.4.2 The equatorial easterly jet stream

During summer, the northern subtropical westerly jet stream is less well developed, especially over Asia, where it moves polewards and is replaced by an easterly jet stream at about 14 km altitude over the equatorial Indian Ocean.

The formation of the equatorial easterly jet stream is connected with the formation of an upper-level high-pressure system over Tibet. In October the reverse process occurs: the equatorial easterly jet stream and the Tibetan high disintegrate, while the subtropical westerly jet stream reforms over northern India.

The equatorial easterly jet stream forms over Indonesia and disintegrates over northern Africa. The disintegration of the jet stream creates a region of subsidence around North Africa, which reinforces the subsiding limb of the Hadley cell and the associated Sahara Desert conditions. When it is well developed the equatorial easterly jet stream therefore has the effect of intensifying the summer aridity over the Sahara Desert and the Middle East.

4.6.4.3 Polar front jet streams

The polar front jet streams are associated with the warm and cold fronts of temperate-latitude depressions, and are found at approximately 10 km altitude. They show considerable day-to-day variation as surface-level fronts form and develop.

Reflective questions

➤ How does the Coriolis effect vary with latitude?

➤ To what extent does convection play a role in atmospheric processes?

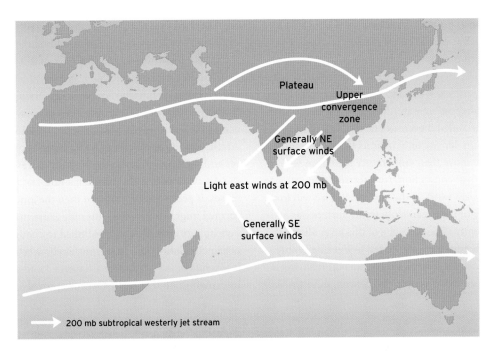

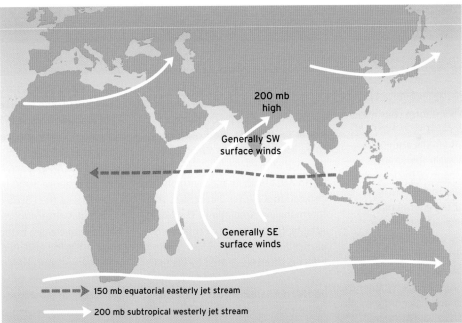

Figure 4.17 Schematic illustration of the major upper-troposphere features of the Asian monsoon in the northern winter (top panel) and summer (bottom panel). The 200 mb pressure level in the atmosphere is around 12 km altitude; the 150 mb level is at around 14 km. (Source: after Lockwood, 1965)

4.7 The influence of oceans and ice on atmospheric processes

We have seen how the land surface can interact with atmospheric processes via albedo effects, the supply or retardation of moisture and topographic interactions. However, the

oceans also influence atmospheric processes, not only via albedo and moisture circulation but via other energy transfer processes too. Most substances become denser as they get colder. However, freshwater is unlike any other substance because it is densest at about 4°C. Therefore when the surface is cooled below that temperature, the coldest water stays on top as it is less dense (and hence more buoyant) than the water at 4°C. When water freezes, the ice floats on the top. This is because water expands when it freezes and becomes less dense. All other substances contract when they freeze.

By contrast, when saline water, which is normal for the world's oceans, is cooled it becomes denser, but does not reach its maximum density until near its freezing point, at about −2°C. Just above that temperature the ocean water will undergo marked convection, transferring cold water to the depths and warm water to the surface. Sea ice will form only when a layer of the ocean close to the surface has a relatively low salinity. The existence of this layer allows the temperature of the surface water to fall to freezing point, and ice to form, despite lower levels of the ocean having a higher temperature. For example, the perennial Arctic sea

ice is formed in a relatively freshwater layer formed by run-off from rivers draining the surrounding continents. This is important because the river drainage of much of both Arctic Canada (Mackenzie River system) and Siberia (e.g. Lena River system) is into the Arctic Ocean, thus giving it a positive water balance and causing a net outflow of ocean water across 70°N. However, precipitation variations across either Arctic Canada or Siberia may cause variations in river run-off and thereby influence the amount of freshwater reaching the Arctic Ocean. Thus, through a range of feedback processes, if precipitation changed then the amount of sea ice formed might change and this may cause major northern hemisphere climatic changes linked to changes in albedo, ocean circulation and heat transfers between deep and shallow layers of the ocean (see Chapters 3 and 22). Further information on the Arctic Ocean is provided in Box 4.5.

CASE STUDIES

THE ARCTIC OCEAN

The radiation budget of the polar surfaces is nearly always negative because of the absence in winter of any solar radiation, the low angle of solar incidence in the summer, and the high albedo of the ice fields. Indeed, the polar regions serve as global sinks for the energy that is transported polewards from the tropics by warm ocean currents and atmospheric circulation systems, particularly travelling cyclones and blocking anticyclones (see Chapter 5). However, there are major differences between the Arctic and Antarctic regions in terms of the distribution of land, sea and ice. The Arctic is a largely ice-covered, landlocked ocean, whereas Antarctica consists of a continental ice sheet over 2000 m thick surrounded by ocean.

In winter the southern limits of the Arctic sea ice are constrained by the northern coastlines of Asia and North

America, but pack ice extends into middle latitudes off eastern Canada and eastern Asia where northerly winds and cold currents transport ice far southwards. West of Norway there is open water to about 78°N because of the warm waters of the North Atlantic Drift and thermohaline circulation (Chapter 3 describes the thermohaline circulation). The North Atlantic thermohaline circulation is particularly vigorous and its climatic effect is often illustrated by comparing the surface temperature of the northern Atlantic with comparable latitudes of the Pacific, since the former is 4 to 5°C warmer. During summer the sea ice melts back towards the pole, but much of the Arctic ice survives at least one summer melt season, and as a consequence its mean thickness is 3-4 m. The thickness is highly variable locally, with narrow openings (leads) throughout the pack ice even in winter, and larger openings

(polynyas) adjacent to coastal areas where winds blow offshore. While there appears to have been no significant long-term trend in Antarctic sea ice extent, Arctic sea ice extent has decreased by about 2.5% per decade since 1970. An extensive ice sheet that rises above 3000 m in elevation covers more than 80% of Greenland.

Simulations from even the earliest global climate models have shown that the effects of loading the atmosphere with greenhouse gases would be seen first and be especially prominent in the Arctic, largely due to feedbacks involving surface air temperature and the loss of sea ice and snow cover. Snow and ice have a high albedo, while that of seawater is low. Therefore particularly important are the ice−albedo feedbacks, where absorption of solar radiation increases when snow melts, and the loss of sea ice cover allowing for strong heat transfers from the ocean to the atmosphere.

BOX 4.5 ➤

➤ The Arctic Ocean and its rim is becoming the object of major political importance to the surrounding countries. The Arctic rim will be transformed by climate change into a new economic powerhouse. As the ice recedes, ecosystems extend and minerals and fossil fuels are discovered and exploited, the Arctic will become a place of 'great human activity, strategic value and economic importance'. The eight nations of the Arctic rim – the USA, Canada, Russia, Greenland, Iceland, Finland, Sweden and Norway – will become increasingly prosperous and powerful (Smith, 2010). This is because there is a high possibility that there are extensive mineral and oil deposits under the Arctic Ocean. Also if the Arctic ice retreats further during the northern summer, it will open sea routes between the Atlantic and Pacific Oceans. These sea routes, to the north of Siberia and Canada, the North-east and North-west Passages, could be of major commercial importance.

In October 2007, surface temperature showed very large positive anomalies (10–12°C) over areas of the Arctic Ocean experiencing record sea ice loss. This basic pattern of autumnal warming has been emerging over the past seven years or so. Numerical climate model simulations suggest that conditions are ripe for a 'tipping point' where summer sea ice vanishes once the spring ice thickness averaged across the Arctic Ocean thins to about 2.5 m, close to the value estimated for spring 2007 from IceSat, NASA's satellite altimeter system. In one of the model simulations, the tipping point occurred in 2024 when 1.8 million km^2 of ice was lost. However, modelling of Arctic sea ice by the UK Met Office Hadley Centre shows that ice invariably recovers from extreme events, but that the long-term trend of reduction is robust, with the first ice-free summer expected to occur between 2060 and 2080.

Associated with this Arctic warming, the extent and intensity of summer surface melt over the Greenland ice sheet has shown a general upward trend. The Greenland ice sheet will start to melt more quickly than snow accumulates once regional temperatures are high enough. It is not yet known what this exact threshold is, but it could be a global temperature rise of somewhere between 1.9 and 4.6°C above pre-industrial values, with resulting sea-level increases of up to 1 m. There is evidence that some of this meltwater is reaching the base of the large Greenland glaciers, lubricating them and increasing the iceberg discharge. This raises concern that projections of sea-level rise through the twenty-first century are too conservative. What is surprising climatologists is the rapid rate of change in Arctic conditions.

BOX 4.5

Western Europe receives an almost continuous supply of heat and moisture from the North Atlantic. This is most noticeable in winter when north-western Europe is significantly milder than the temperature norm for its latitude. Indeed, the mildness of the winter climate of western Europe is one of the more spectacular latitudinal anomalies in the world climatic pattern; in January on the Norwegian coast the temperature anomaly may be as high as 22–26°C, gradually decreasing to 1°C southwards across the continent. This winter temperature anomaly arises not only from the prevailing westerly winds but also because at comparable latitudes the Atlantic Ocean is 4–5°C warmer than the Pacific. The extension into very high latitudes and the northward narrowing of the northern North Atlantic have consequences on the Atlantic Ocean circulation which in turn has a series of unique effects on the climatic system. This is in complete contrast to the much more benign Pacific Ocean. Warm, saline surface water flows into the northern North Atlantic, after travelling from the Caribbean Sea, via the Gulf Stream and the North Atlantic Drift (see Chapter 3), warming western Europe. This is part of the so-called thermohaline circulation, which is strongly developed in the Atlantic but completely absent from the Pacific. Thus the atmosphere forms just one part of a complex interacting climate system that also incorporates the land, the oceans, ice cover and vegetation. It is because of such a system that there can also be inter-annual variations in atmospheric processes and it is to examples of these that we now turn.

Reflective questions

➤ Why is western Europe mild compared with eastern Canada?

➤ Why does ice form on the sea surface rather than on the sea floor?

4.8 The Walker circulation

Satellite imagery and rainfall data clearly show (Figure 4.11) three equatorial regions of maximum cloudiness and rainfall: the so-called 'Maritime Continent' of the Indonesian Archipelago, the Amazon river basin in South America and the Zaire river basin in Africa. The rest of the equatorial region is comparatively dry, and some, such as the coasts of Peru, even desert. These longitudinal variations in rainfall are associated with east–west regional circulations along the equator, the most important being the so-called Walker circulation, which involves rising air motion over the Indonesian Archipelago and sinking over the eastern Pacific (Figure 4.18). The rising air motion takes place mostly in deep convective clouds and is associated with intense convective rainfall and therefore the wet humid climates of the Indonesian Archipelago. The subsiding air suppresses cloud formation and rainfall, giving rise to the coastal deserts of Peru. The Hadley cell circulation refers to the north–south component of these circulations: equatorward motion at low levels, rising in the convective regions near the equator, and poleward flow aloft. The Walker circulation refers to the east–west component, which is particularly prominent in the equatorial plane. Both circulations are driven by the release of latent heat in deep convective shower clouds.

The Walker circulation is closely coupled with the sea surface temperature distribution over the Pacific, with relatively cool water in the east and warm in the west. When the Pacific Ocean off the coast of South America is particularly cold, the air above is too stable to take part in the ascending motion of the Hadley cell circulation. Instead, the equatorial air flows westwards between the Hadley cell circulations of the two hemispheres to the warm West Pacific where, having been heated and supplied with moisture from the warmer waters, the equatorial air can take part in large-scale ascent (Figure 4.18). The easterly winds that blow along the equator and the north-easterly winds that blow along the coast of Peru and Ecuador both tend to drag the surface water along with them. The Earth's rotation then deflects the resulting surface currents towards the right (northwards) in the northern hemisphere and to the left (southwards) in the southern hemisphere. The surface waters are therefore deflected away from the equator in both hemispheres and also away from the coastline. Where the surface water moves away under the influence of the trade winds, colder, nutrient-rich water upwells from below to replace it. Since the newly upwelled water is colder than its surroundings, its signature in infrared satellite images takes the form of a distinctive 'cold tongue' extending westwards along the equator from the South American coast.

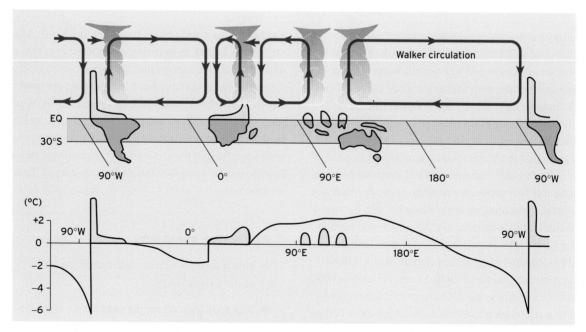

Figure 4.18 Schematic representation of the Walker circulation along the Equator during non-ENSO (El Niño Southern Oscillation) conditions. The sea surface temperature departures from the zonal mean along the equator are shown in the lower part of the figure. (Source: adapted from Peixoto and Oort, 1992; Wyrtki, 1982)

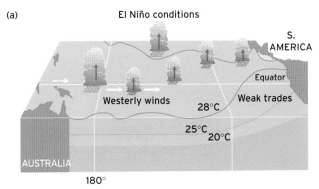

(a) El Niño conditions

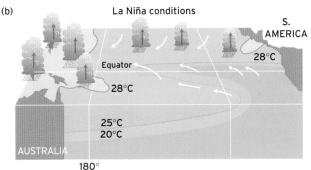

(b) La Niña conditions

Figure 4.19 A schematic view of (a) El Niño and (b) La Niña. During La Niña, intense trade winds result in a cold equatorial Pacific in the east and warm waters in the west where moist air rises. The air subsides in the east, a region with little rainfall. During El Niño the trade winds along the equator relax, as does the temperature gradient of the ocean waters, and the warm surface waters of the west flow eastwards. (Source: after Philander, 1998)

4.8.1 El Niño Southern Oscillation

Since the Sun is nearly overhead for the whole year near the equator, the seasonal cycle there is weak. This allows other cycles to dominate the equatorial ocean–atmosphere system. The Southern Oscillation is dominated by an exchange of air between the south-east Pacific high and the Indonesian equatorial low, with a period that varies between roughly 1 and 5 years (Philander, 1998). For several years during one phase of this oscillation, the trade winds are intense and converge into the warm-water regions of the western tropical Pacific, including the Indonesian islands and north-eastern Australia, where rainfall is plentiful and sea-level pressure is low (Figure 4.19). At such times the ocean surface in the eastern tropical Pacific is cold and the associated atmosphere over the ocean and coastal equatorial South America is also cold and dry. More extreme forms of this 'normal' condition are known as La Niña. Every few years, for several months or longer, the trade winds relax, the zone of warmer surface waters and heavy

precipitation shifts eastwards, and sea-level air pressure rises in the west while it falls in the east (Figure 4.19). This is El Niño. Under these conditions Indonesia experiences drought and associated forest fires, while coastal equatorial South America experiences heavy rainfall and floods. These ocean and atmospheric changes are part of an oscillating system known as the **El Niño Southern Oscillation** (ENSO), consisting of the warm ocean pool near the Indonesian Islands, the cold Pacific Ocean pool off equatorial South America, and the atmospheric circulation over the equatorial Pacific. The Southern Oscillation may be defined in terms of the difference in sea-level pressure between Darwin in Australia and Tahiti. Records are available, with the exception of a few years and occasional months, for this pressure difference from the late 1800s. While the El Niño phenomena are associated with extreme negative Southern Oscillation values, most of the time there are continuous transitions from high to low values, with most values being positive. ENSO is important climatologically for two main reasons. Firstly, it is one of the most striking examples of inter-annual climate variability on a global scale. Secondly, in the Pacific it is associated with considerable fluctuations in rainfall and sea surface temperatures and also extreme weather events around the world (Figure 4.20).

Australian rainfall is more variable than could be expected from similar climates elsewhere in the world, mainly due to the impact of ENSO. Australian rainfall fluctuations, as well as being more severe because of ENSO's influence, also operate on a very large spatial scale. High rainfall totals in Australia occur when the Southern Oscillation Index (SOI) is large and positive (La Niña events). The high rainfalls and floods give an example at the end of December 2010. In contrast when the SOI is strongly negative (El Niño years) drought occurs over much of the continent. Thus the continental scale of the 1982–1983 drought is typical of many years, although it was more severe than most. Extended periods of drought or extensive rains in Australia do not occur randomly in time, in relation to the annual cycle. The ENSO phenomenon, and Australian rainfall fluctuations associated with it, are phase locked with the annual cycle. Thus the heavy rainfall of an anti-ENSO event tends to start early in the calendar year and finish early in the following year. The dry periods associated with ENSO events tend to occupy a similar time period. For example, the 1982–1983 drought started about April 1982 and broke over much of the country in March and April 1983.

Australian vegetation should be suited to an environment of highly variable rainfall with frequent severe droughts or wet periods. Among the characteristics of

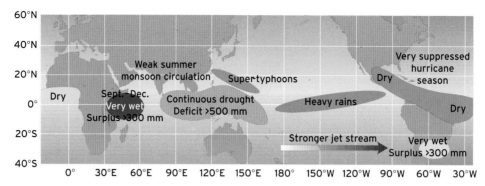

Figure 4.20 Schematic diagram showing the major impacts of El Niño during the period June to December 1997. (Source: after Slingo, 1998)

Australian vegetation that may be, at least in part, attributable to ENSO's influence on climate are the following:

- *Absence of succulents* (see Chapter 18): succulents are almost totally absent from Australian arid and semi-arid regions, because although adapted to arid climates and requiring little moisture, they need regular rainfall. Such plants are therefore unsuited to the high variability ENSO produces over much of Australia.
- *Fire resistance/dependence*: much of the Australian flora is fire resistant or even dependent on fire for successful reproduction. Fires are common during drought periods, which are often associated with ENSO events.
- *Fluctuating climax*: the high inter-annual variability of annual rainfall in arid and semi-arid parts of Australia affected by ENSO means that vegetation appears adapted to the climate in such a way that demographic components are in a state of unstable equilibrium.

4.8.2 North Atlantic Oscillation

A major source of inter-annual variability in the atmospheric circulation over the North Atlantic and western Europe is the North Atlantic Oscillation (NAO), which is associated with changes in the strength of the oceanic surface westerlies. It is often measured by the difference of December to February atmospheric pressure between Ponta Delgado in the Azores (37.8°N, 25.5°W) and Stykkisholmur, Iceland (65.18°N, 22.7°W). When the values are greater than the average the index is said to be positive. Statistical analysis reveals that the NAO is the dominant mode of variability of the surface atmospheric circulation in the

Atlantic and accounts for more than 36% of the variance of the mean December to March sea-level pressure field over the region from 20° to 80°N and 90°W to 40°E between 1899 and 1994.

The oscillation is most marked during the winter. There can be great differences between winters with high and low values of the NAO index. Typically, when the index is high the Icelandic low pressure is strong, which increases the influence of cold Arctic air masses on the north-eastern seaboard of North America and enhances the westerlies carrying warmer, moister air masses into western Europe. Low values of the index are associated with large stationary or slow-moving anticyclones over the North Atlantic or north-western Europe. These block the normal progression of low-pressure systems from the North Atlantic over north-western Europe, and are therefore often termed blocking anticyclones. Blocking anticyclones can occur at any time of the year; in summer they can give hot dry conditions, but in winter with the warm westerlies absent, they often lead to very low temperatures. The record cold December 2010 in the United Kingdom was partly due to a continuous series of blocking anticyclones over the eastern Atlantic. Thus, NAO anomalies are related to downstream wintertime temperature and precipitation across Europe, Russia and Siberia. High-index winters are anomalously mild while low-index winters are anomalously cold.

Reflective question

➤ How can ENSO affect human activity across the world?

4.9 Interactions between radiation, atmospheric trace gases and clouds

4.9.1 The greenhouse effect

The gases nitrogen and oxygen which make up the bulk of the atmosphere neither absorb nor emit long-wave radiation. It is the water vapour, CO_2 and some other minor gases present in the atmosphere in much smaller quantities which absorb some of the long-wave radiation emitted by the Earth's surface and act as a partial blanket. To examine the effect of these absorbing trace gases on surface temperature it is useful to consider an atmosphere from which all cloud, water vapour, dust and other minor gases have been removed, leaving only nitrogen and oxygen. As the Sun's short-wave radiation passes through the atmosphere, about 6% is scattered back to space by atmospheric molecules and about 10% on average is reflected back to space from the land and ocean surfaces. The remaining 84% heats the surface. To balance this incoming radiant energy the Earth itself must radiate on average the same amount of energy back to space in the form of long-wave radiation. It can be calculated from radiation laws (Box 4.2) that, to balance the absorbed solar energy by outgoing long-wave radiation, the average temperature of the Earth should be $-6°C$. This is much colder than is actually observed (15°C).

This calculation error occurs because the atmosphere readily absorbs infrared radiation emitted by the Earth's surface with the principal absorbers being water vapour (absorption wavelengths of 5.3–7.7 μm and beyond 20 μm), ozone (9.4–9.8 μm), CO_2 (13.1–16.9 μm) and all clouds (all wavelengths). Only about 9% of the infrared radiation from the ground surface escapes directly to space. The rest is absorbed by the atmosphere, which in turn reradiates the absorbed infrared radiation, partly to space and partly back to the surface. This blanketing effect is known as the **natural greenhouse effect** and the gases are known as greenhouse gases. It is called 'natural' to distinguish it from the **enhanced greenhouse effect** due to gases added to the atmosphere by human activities such as the burning of fossil fuels and deforestation. It needs to be stressed that the natural greenhouse effect is a normal part of the climate of the Earth and that it has existed for nearly the whole of the atmosphere's history. Concern about the greenhouse effect arises over two issues: how the natural greenhouse effect may vary with time, and how human activities might modify and enhance the natural effect.

Human activities since the Industrial Revolution have increased atmospheric trace gases such as CO_2. Before the start of the industrial era, around 1750, atmospheric CO_2 concentration had been 280 ± 10 parts per million (ppm) for several thousand years. It has risen continuously since then, reaching 379 ppm in 2005, with a rate of increase over the past century that is unprecedented over at least the past 20 000 years (see also Chapter 21). The annual CO_2 growth rate was larger during the last 10 years (1995–2005 average 1.9 ppm per year) than it has been since the beginning of continuous direct atmospheric measurements (1960–2005 average 1.4 ppm per year). The present atmospheric CO_2 concentration has not been exceeded during the past 650 000 years, and possibly the past 20 million years. Several lines of evidence confirm that the recent and continuing increase of atmospheric CO_2 content is caused by human CO_2 emissions, and in particular fossil fuel burning. These human-induced CO_2 emissions enhance the already existing greenhouse effect, causing global warming and fundamental changes in climate (see Chapter 24). The 100-year linear global temperature trend is 0.74°C and the linear warming trend over the last 50 years of 0.13°C per decade is nearly twice that for the past 100 years (IPCC, 2007a).

Although CO_2 is the most important greenhouse gas, other gases also make a contribution to climate change. The combined effect of the increases to 1990 of the minor greenhouse gases, methane, nitrous oxide and tropospheric ozone, is to add a warming forcing equivalent to that of an additional 60 ppm or so of CO_2. Even if there were no further increase in these minor gases, the 1990 forcing would still require to be added to future projections of change driven by CO_2. As an example, the effect of this, if turned into equivalent amounts of carbon dioxide (CO_2e), would be that the 450 ppm CO_2 only level would become about 520 CO_2e ppm and the 550 level would become about 640 CO_2e ppm. Note that, although the amount of forcing from the minor gases is the same, when turned into equivalent CO_2, the amounts added increase with the CO_2 concentration to which the amount is added. This is because the relationship between radiative forcing and concentration is non-linear. It should also be noted that it is not always clear in the literature as to whether the authors are using simple CO_2 concentration or the equivalent CO_2 value.

4.9.2 A simple climate model of the enhanced greenhouse effect

It is possible to construct a simple climate model to illustrate the enhanced greenhouse effect due to the addition to the atmosphere of radiatively active trace gases such as CO_2. Figure 4.21 shows a mean temperature profile for the troposphere, calculated assuming convective equilibrium, which

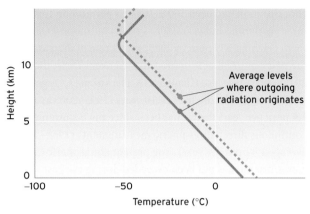

Figure 4.21 The distribution of temperature in a convective atmosphere (solid line). The dashed line shows how the temperature increases when the amount of carbon dioxide present in the atmosphere is increased (for doubled carbon dioxide in the absence of other effects the increase in temperature is about 1.2°C). Also shown for the two cases are the average levels from which thermal infrared radiation leaving the atmosphere originates (about 6 km for the unperturbed atmosphere). (Source: after Houghton, 1994)

gives a lapse rate of 6°C km^{-1}. The upper atmospheric temperature is estimated by assuming that the stratosphere is in radiative equilibrium (its temperature is controlled by its radiation balance and not by vertical convection). The height at which these two straight-line temperature profiles connect is the tropopause. Viewed from space, the Earth is observed to have a temperature of about −18°C. This is because most of the infrared radiation to space takes place from the middle atmosphere where the temperature is around −18°C; most of the greenhouse gases are below this level. As the amount of infrared-absorbing greenhouse gases mixed into the atmosphere is increased, it becomes more likely to absorb infrared radiation. Thus the effective level at which outgoing infrared radiation originates must rise to allow the radiation to escape to space at the same rate as before. However, because the tropospheric lapse rate does not change significantly from 6°C km^{-1} then the temperatures at a given height in the troposphere must be dragged up by the changes. Furthermore, since the height

NEW
DIRECTIONS

GREENHOUSE GAS STABILIZATION AND GLOBAL TEMPERATURE INCREASES

Most pollutants released into the atmosphere have very short residence times, often just a few days. Clouds or rainfall washes out the common ones, such as sulphur dioxide, or the larger particles settle out rapidly. The distribution of such pollutants is therefore restricted to near their source regions, usually large urban or industrial regions. Their extent is rarely global; even the sulphur compounds emitted by erupting volcanoes (Box 4.7) rarely stay in the atmosphere for more than a few years. Long-term concentrations of gases such as sulphur dioxide are usually maintained by continuous emissions. If the emissions cease, atmospheric concentrations drop rapidly and the gases often vanish from the atmosphere. This is the common experience with sulphur dioxide, where closing down pollution sources causes a rapid improvement in atmospheric conditions.

Carbon dioxide concentrations in the atmosphere depend on the interactions between the various carbon reservoirs forming the so-called carbon cycle. Consider an emission of anthropogenic CO_2 into the atmosphere. Different processes, with very distinct timescales, are responsible for determining the lifetime of anthropogenic CO_2. For timescales of particular human interest, ranging from decades to centuries, the responses to excess atmospheric CO_2 include ocean uptake (see Box 3.5 in Chapter 3), changes in land surface carbon uptake, CO_2 fertilization and alterations to vegetation cover. For scales of centuries to about 5000-10 000 years, ocean uptake becomes dominant. At present, the terrestrial biosphere appears to be a net sink of carbon, in spite of anthropogenic deforestation predominately in the tropics. Long-term numerical modelling studies predict a reversal of present-day net CO_2 uptake by the biosphere as the Earth warms, resulting in a net release of carbon to the atmosphere by the end of the century. Studies also indicate that roughly 80% of an anthropogenic CO_2 input into the atmosphere has an average perturbation lifetime in the atmosphere of approximately 300-450 years. The remaining 20% could remain in the atmosphere for more than 5000 years after

BOX 4.6 ➤

emissions cease. Archer (2005) commented that care is required in using 300 years as a lifetime of anthropogenic CO_2, because it misses the immense longevity of the tail in the CO_2 lifetime. He considers that a better approximation of the lifetime of fossil fuel CO_2 for public discussion might be '300 years, plus 25% that lasts forever'.

The long life of the anthropogenic CO_2 released into the atmosphere makes stabilization difficult, and at present nearly impossible in the short term, for levels below the present atmospheric concentration. The global carbon cycle has a strong one-way arrow in the short term directed to increases in atmospheric CO_2 concentration. This is because a substantial fraction of the CO_2 emitted into the atmosphere by human activity remains there, in effect, for centuries to millennia. This is a very important point to understand: without the use of large-scale atmospheric carbon capture technologies, it is not possible to reduce atmospheric CO_2 concentrations in the short term.

If the above properties of the carbon cycle are real and enduring, then it is likely that bringing future anthropogenic carbon emissions to zero will not reduce global average surface temperatures except in the very long term. Rather, once temperatures have peaked, they will remain almost steady (see Section 4.4.4 on thermal inertia). Several recent studies have sought to exploit this observation in order to provide a simple link between levels of cumulative anthropogenic carbon emissions and future warming. Allen *et al.* (2009), for example, considered the cumulative carbon emissions summed between pre-industrial times and 2050, linking them to peak global

warming. It can be argued that warming by a given date is proportional to cumulative CO_2 emissions to that date. The recent Copenhagen Accord contains the aim of limiting warming to no more than 2°C, drawing on earlier targets from the EU and G8. Though not specified in the Copenhagen Accord, this 2°C warming limit is usually presumed to be relative to pre-industrial levels. Using the results in Allen *et al.* (2009), a 2°C limit on the most likely peak CO_2-induced warming could be achieved by limiting cumulative CO_2 emissions to one trillion tonnes of carbon (1 TtC). Present-day human activities, such as burning fossil fuels and changing land use, are estimated to put about 26 000 million tonnes of CO_2 a year into the atmosphere.

Bowerman *et al.* (2010) commented that cumulative emission targets represent the sum of emissions over time, and therefore these cumulative emissions could be distributed over time in a number of ways. For example, an early peak in emissions could be followed by a relatively slow rate of post-peak decline, or a later peak could be followed by a much more rapid decline. It may not be technically or politically feasible, or economically desirable, to decrease emissions at rates much in excess of 3 or 4% per year, so that peaking later may not be viable, assuming a 2°C warming target.

The so-called A1FI emissions scenario is considered by the IPCC to be one of a number of equally plausible projections of future greenhouse gas emissions from a global society that does not implement policies to limit anthropogenic influence on climate. It assumes that the world is market oriented with fast per capita growth and fossil-fuel-intensive energy production. It also assumes that the

global population peaks in 2050 and then declines. Previously, this scenario has received less attention than other scenarios with generally lower rates of emissions. However, there is no evidence from actual emissions data to suggest that the A1FI scenario is implausible if action is not taken to reduce greenhouse gas emissions, and hence it deserves closer attention than has previously been given. The evidence available from recent simulations, along with existing results presented in the IPCC AR4, suggests that the A1FI emissions scenario would lead to a rise in global mean temperature of between approximately 3 and 7°C by the 2090s relative to pre-industrial levels, with best estimates being around 5°C. A temperature rise of 4°C is predicted by the 2070s, but if carbon cycle feedbacks are strong, 4°C could be reached in the early 2060s – this latter projection appears to be consistent with the upper end of the IPCC's likely range of warming for the A1FI scenario.

The Committee on Climate Change (CCC, 2008) has analyzed eight greenhouse gas emission trajectories. Three trajectories had global emissions peaking in 2028, with subsequent reductions in total CO_2e emissions of 1.5%, 2% and 3% per annum. Five of the trajectories have global emissions peaking in 2016 with subsequent reductions in total emissions of 1.5%, 2%, 3% and 4%. None of the three trajectories with emissions peaking in 2028 would keep atmospheric greenhouse gas concentrations below 550 ppm CO_2e by the end of the century, and all would give a global average warming of 2.5-2.8°C this century. The CCC (2008) does not believe that a global policy which leaves emissions peaking as late as 2028 is adequate

BOX 4.6 ➤

for control of global warming. All the trajectories with emissions peaking in 2016, except the one with 1.5% annual reduction after 2016, would keep concentrations below 550 ppm CO_2e by the end of the century. Only emission reductions after 2016 at 3 or 4% per year would limit the chance of reaching 4°C to very low levels, with central model estimates indicating a 2.2°C rise this century with the 3% trajectory, and a 2.1°C rise with the 4% one. Even in these cases the CCC (2008) commented that the chances of exceeding 2°C by 2100 would be 63% and 56% respectively.

It is often argued that it may not be technically, economically or politically feasible to eliminate emissions of all greenhouse gases while, for example, preserving global food security. Crop and animal farming can produce large amounts of CO_2 and methane. This limit has been referred to as an 'emissions floor'. It is difficult to estimate a compelling emissions floor, either in terms of its size (in gigatonnes of carbon per year (GtC yr^{-1})), or in terms of the extent to which it can reduce over time as new technologies become available. Nevertheless, Bowerman *et al.* (2010) commented that it makes sense to consider the possibility that it may prove prohibitively expensive to reduce emissions beyond some positive level. They use the following conventions: if the emissions floor is constant, it is referred to it as a 'hard floor'. If, on the other hand, society is able to continue to reduce residual CO_2 emissions, eventually to the point where net emissions are zero, then it is called a 'decaying floor'.

The above discussion strongly suggests that it is unlikely that global temperature increases can be kept at relatively safe levels of 2°C or below without major reductions in greenhouse gas emissions in the next few years and preferably by 2020. Further discussion of contemporary climate change is provided by Chapter 24.

BOX 4.6

of the tropopause is increased, the stratosphere must cool slightly. In particular, and of great importance, the surface temperature is also increased.

From this simple model, the doubling of the CO_2 content of the atmosphere in the absence of other effects would increase the tropospheric temperature by about 1.2°C. In reality various other effects could increase the predicted temperature changes (Box 4.6).

4.9.3 Radiative interactions with clouds and sulphate aerosols

The terms 'forcing' and 'feedback' are frequently used in climatology. Forcings are processes that act as external agents to the climate system, such as changes in solar input to the Earth, the loading of the atmosphere with volcanic ash and aerosols (see Box 4.7), or rising levels of CO_2 gas. As a result of temperature changes caused by increasing concentrations of greenhouse gases, other changes may take place that could in turn influence temperature. For example, an increase in temperature could result in greater evaporation from the ocean, enhancing atmospheric humidity. Since water vapour is an active greenhouse gas, the increase in atmospheric humidity causes a further increase in temperature. This is an example of a positive feedback process.

It also demonstrates that the intensity of the hydrological cycle is closely linked to the greenhouse effect.

4.9.3.1 Clouds

Conventionally, the radiative effect of clouds is discussed in terms of 'cloud radiative forcing', even though cloud effects are actually feedback processes. The radiative forcing of the Earth's climatic system is in part determined by the distribution of cloudiness, since clouds strongly influence the distribution of both short-wave and long-wave radiative fluxes within the atmosphere. At short wavelengths, clouds generally increase planetary albedo and so cool the planet by reflecting more solar radiation to space. However, for long-wave radiation, clouds generally add to the greenhouse effect as they absorb upward-moving infrared radiation and reradiate it back downwards. Thus cloudy days are normally cooler than clear-sky days, particularly in summer, because the short-wave radiative input to the surface is restricted. In contrast, cloudy nights are warmer than clear nights, particularly in winter, because the long-wave radiative loss from the surface to space is restricted (Figure 4.22). However, observations from satellites show that the albedo cooling effect dominates, and the net effect of clouds at present is to cool the Earth.

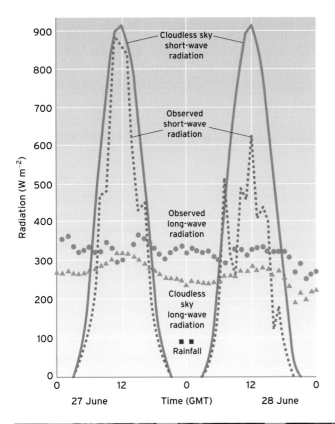

Figure 4.22 Values of incoming short- and long-wave radiation for actual and cloudless sky conditions for a site near Pateley Bridge, North Yorkshire, for two days in June, 1984. During the night of 27/28 June, 2 mm of rain fell in two events of 1 mm each. It is seen that, while cloud makes effectively no contribution to the downward-moving long-wave radiation across the midday period of 27 June (a relatively cloudless day with near-maximum values of incoming short-wave radiation), it contributes around 90 W m^{-2} throughout the rainfall of the night of 27/28 June and only slightly less during the relatively cloudy 28 June (with much reduced incoming short-wave radiation). (Source: after Lockwood, 1993)

HAZARDS

VOLCANOES AND CLIMATE

Volcanic eruptions are of two main types: effusive eruptions and explosive eruptions (Chapter 2). Effusive eruptions are those in which local lava flows predominate, and are of little meteorological interest. Explosive eruptions can throw vast quantities of rocks, dust and gases to great heights. The volume of material blown into the atmosphere during an explosive eruption can amount to 1 km^3 or more. During the eruption of Hekla, Iceland, in 1947, dust fell on ships in the Atlantic Ocean up to 800–1000 km away to the south and east.

Major volcanic eruptions can be highly disruptive, mainly due to five significant effects: tsunamis may inundate surrounding low-lying land; sulphate **aerosols** circulating in the stratosphere may cause globally lowered temperatures for several months or years; toxic elements such as fluorine deposited on the ground or in water may cause death if ingested; tropospheric sulphates, which precipitate within a few days as acid rain, may stunt or kill vegetation over huge areas downwind from the eruption; and volcanic dust may disrupt aircraft movements by damaging jet engines.

The gases emitted by eruptions are largely water vapour, carbon dioxide, sulphur dioxide, and small quantities of other gases. Volcanoes emit two gases that can have an impact on global temperatures: sulphur dioxide and CO_2. They each have very different effects and work on different timescales.

Sulphur dioxide

When this gas is emitted to high altitudes (about 12–14 km or above) it enters the stratosphere. Here it can form acid droplets that partially scatter and reflect sunlight away from the Earth, causing surface cooling. The droplets have a fairly immediate impact, and in sufficient concentrations may cool the climate for a few months, or even a year or two, but then the droplets fall out of the stratosphere and temperatures return to normal. Large-scale winds in the stratosphere tend to flow towards the east with a small polar component. Thus stratospheric dust clouds tend

BOX 4.7 ➤

to move eastwards and polewards; they also tend to be restricted to the particular hemisphere in which they formed. Typically, volcanic dust takes 2 to 6 weeks to circuit the Earth in middle or lower latitudes, and from 1 to 4 months to become a fairly uniform veil over the whole of the latitude zone swept by the wind system into which it is injected. Thus volcanic eruptions near to the equator can spread stratospheric dust over much of one or, if they are almost on the equator, both hemispheres. In contrast polar eruptions often have only a restricted influence. The final stage of a worldwide dust veil from an equatorial eruption is probably a concentration of the last remaining airborne dust over the two polar caps.

Because of their relatively large size compared with the wavelength of the incident short-wave radiation, sulphuric acid droplets and volcanic dust particles predominantly scatter solar radiation in the direction of the incident beam. The result is that scattering reduces the direct radiation much more substantially than is the global solar radiation, since the bulk of the radiation scattered from the direct beam reappears at the surface as diffuse radiation. For example, after the eruption of Mount Agung (Bali) in 1963, the stratospheric dust veil was the most effective since 1902–1903 and possibly since 1883–1886. A marked drop in the direct beam radiation at Aspendale, Melbourne, Australia, was observed which was largely compensated by a similar large increase in the diffuse components. The changes in the direct and diffuse components almost completely cancelled out, and only a very slight fall was detectable in the global radiation curve. The consequence is that the cooling effect of volcanic eruptions is generally small. Large volcanic eruptions of the calibre of Krakatoa (1883) or Agung (1963) are required to cause atmospheric dust loadings that would affect surface solar radiation.

Carbon dioxide

This is a greenhouse gas, so when it is emitted in large enough quantities it can have a warming impact on the global climate. The last remnants of CO_2 emissions have an atmospheric lifetime of several hundred years, so any impact will be felt over a long timescale. Even so, it is still much smaller than that produced from human activity, such as burning fossil fuels and changing land use.

Examples of volcanic eruptions

1. Tambora volcano, Sumbawa, Indonesia: The Tambora volcano on the island of Sumbawa in Indonesia erupted on 10 April 1815, sending a massive cloud of aerosols into the stratosphere. This was one of the largest eruptions of the past 500 years. The year 1816 is often known as the 'year without a summer'. Snow every month of the year started a mass migration from the US East Coast across the Appalachian Mountains to the Midwest. The effects of the 1816 summer on agricultural productivity in New England were mainly due to a series of killing frosts that reduced the growing season. This, along with a severe drought, reduced agricultural output to record low levels. In Europe, by contrast, there were record low temperatures accompanied by above average rainfall and cloudiness, the combination of which slowed the growth of crops and produced fungus and moulds. Thus, food production was negatively affected in both regions, but by different mechanisms. This shows the complex relationship between climate and its impacts.

2. Mount Pinatubo, Philippines: Recent history has produced a good case study of the impact of a large volcanic eruption. Mount Pinatubo in the Philippines erupted in 1991, making it one of the biggest volcanic events of the twentieth century. It put about 20 million tonnes of sulphur dioxide into the stratosphere. As expected, this impacted global temperatures. The average for the following year was reduced by between 0.1°C and 0.2°C, although temperatures quickly recovered the year after. In addition, an estimated 250 million tonnes of CO_2 was put into the atmosphere. This is a significant amount, but is still much smaller than that produced from human activity, such as burning fossil fuels and changing land use, which is estimated to put out about 26 000 million tonnes of CO_2 a year.

3. Eyjafjallajökull eruption, 14 April –23 May 2010: Volcanic eruptions are not uncommon in Iceland (Petersen, 2010). This eruption is of interest because the upper winds were northerly and spread clouds of volcanic ash over north-western Europe, severely disrupting air travel. This was because the ash clouds can damage jet engines and cause them to fail.

4. Santorini, eastern Mediterranean: The eruption of Santorini (Thera) was one of the largest volcanic events of the last four millennia (see Box 2.3 in Chapter 2). Some

BOX 4.7 ➤

30 cubic kilometres of rock was erupted, and ash from the eruption has been identified across more than 2 million square kilometres extending from the Black Sea in the north through Turkey and the south-eastern

Mediterranean to the Nile Delta in the south. The best date for this eruption, consistent with all the data, is 1628 BC.

5. Toba, Sumatra, Indonesia: The largest known eruption of the past 100 000 years was the great Toba

eruption about 71 000 years ago, which occurred close to the beginning of a major glacial advance, although a causal relationship has yet to be established.

BOX 4.7

4.9.3.2 Aerosols

Over the last few years, it has become evident that, when averaged over the global atmosphere, part of the enhanced greenhouse effect has been offset by a negative forcing due to the human-induced emission of aerosols (Figure 4.23). Aerosols have two effects, a direct one and also by modifying cloud properties. Aerosol particles reflect short-wave radiation and therefore increase atmospheric albedo. Aerosols also increase cloud albedo and influence cloud lifetime and precipitation. Sulphate aerosols, formed from sulphur dioxide emitted during the combustion of fossil fuels and partly responsible for producing acid rain, are regarded as particularly important. Sulphate aerosols located in the troposphere are rapidly washed out of the atmosphere by rainfall; hence they do not travel far from their industrial sources. The IPCC Report (2007a) suggested that anthropogenic contributions to aerosols (primarily sulphate, organic carbon, black carbon, nitrate and dust) together produce a cooling effect, with a total direct radiative forcing of -0.5 W m^{-2} and an indirect cloud albedo forcing of -0.7 W m^{-2}. The combined radiative forcing due to increases in the greenhouse gases of CO_2, methane and nitrous oxide is estimated to be $+2.30$ W m^{-2}. Therefore IPCC (2007a) reports with a very high confidence that the globally averaged net effect of human activities since 1750 has been one of warming, with a radiative forcing of between $+0.6$ and $+2.4$ W m^{-2} (Figure 4.23).

Reflective question

➤ Sulphate aerosols originating from industrial areas cool the atmosphere but cause acid rain. Attempts to remove such aerosols from the atmosphere will accelerate global warming. What therefore should be the policy on industrial pollution?

4.10 Geoengineering

Scientists have identified a range of engineering techniques, collectively called geoengineering, to address the control of atmospheric greenhouse gases and reduce the risks of climate change. Basically they attempt to modify the various atmospheric processes described in this chapter. None are fully operational at present, but there is growing interest in developing some of them. One class removes greenhouse gases from the atmosphere after they have been released; the other involves solar radiation management, such as deflecting sunlight away from the Earth. The first approach makes use of biological agents, such as land plants or aquatic algae, to produce, for example, biofuels. There is growing interest in this approach, and the planting of forests is already considered in international climate negotiations. The second approach could involve increasing the albedo of land surfaces so that they reflect more sunlight and thus cool the Earth. It is not yet clear how this could be done on a scale large enough to be effective.

Reflective question

➤ Are there any planned geoengineering projects intended for your area to control greenhouse gas concentrations? Do some web searches to find out.

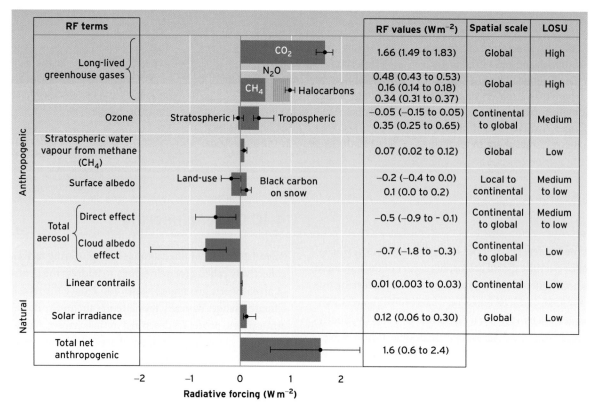

RF terms		RF values (W m^{-2})	Spatial scale	LOSU
Long-lived greenhouse gases	CO_2	1.66 (1.49 to 1.83)	Global	High
	N$_2$O, CH$_4$, Halocarbons	0.48 (0.43 to 0.53) 0.16 (0.14 to 0.18) 0.34 (0.31 to 0.37)	Global	High
Ozone	Stratospheric, Tropospheric	−0.05 (−0.15 to 0.05) 0.35 (0.25 to 0.65)	Continental to global	Medium
Stratospheric water vapour from methane (CH$_4$)		0.07 (0.02 to 0.12)	Global	Low
Surface albedo	Land-use, Black carbon on snow	−0.2 (−0.4 to 0.0) 0.1 (0.0 to 0.2)	Local to continental	Medium to low
Total aerosol — Direct effect		−0.5 (−0.9 to − 0.1)	Continental to global	Medium to low
Total aerosol — Cloud albedo effect		−0.7 (−1.8 to −0.3)	Continental to global	Low
Linear contrails		0.01 (0.003 to 0.03)	Continental	Low
Solar irradiance		0.12 (0.06 to 0.30)	Global	Low
Total net anthropogenic		1.6 (0.6 to 2.4)		

Figure 4.23 Global-average radiative forcing (RF) estimates and ranges in 2005 for anthropogenic carbon dioxide, methane, nitrous oxide and other important agents and mechanisms, together with the typical geographical extent of the forcing and the assessed level of scientific understanding (LOSU). (Source: IPCC, 2007a)

4.11 Summary

The climate system consists of a closely coupled atmosphere-ocean-ice-land-vegetation system. Short-wave radiation from the Sun is the main energy input that drives the climate system, while long-wave radiation from the atmosphere to the cold of space is the main energy lost. The processes of absorption, reflection and reradiation of energy, however, are of fundamental importance to atmospheric processes. There is a net radiation surplus at the surface near the equator and a net radiation deficit at the poles, leading to energy transfers. These transfers to rectify the equator/pole energy imbalance take place through atmospheric and oceanic motions.

The basic circulation of the atmosphere is dominated by a three-cell structure in each hemisphere that is controlled by a pressure gradient force. The Earth's rotation, however, has a profound influence on air movements across its surface, making circulation systems much more complex than they would otherwise be. In the tropical world there are mean north-south circulations, known as Hadley cells, with rising air over the equator and sinking air in the subtropics. There are also east-west circulation cells along the equator known as the Walker circulation. At the surface in middle latitudes, the predominant features are closed, eastward-drifting, cyclonic and anticyclonic systems, while higher up in the troposphere smooth wave-shaped patterns are the general rule. There are normally five to eight upper (Rossby) waves circling the poles. They are important because frontal depressions tend to form and grow rapidly just downwind of upper troughs while surface anticyclones tend to develop just downwind of upper ridges. Embedded within the Rossby waves are strong, narrow currents of air known as jet streams.

Hydrological processes play an important part in atmospheric and climatic processes. For example, condensation is important in controlling the convective overturning of the atmosphere and clouds act as both cooling and warming agents across the planet. Atmospheric circulation patterns are naturally variable on an inter-annual scale. Among the more important are the El Niño Southern Oscillation associated with the Walker circulation, and the North Atlantic Oscillation with its influence on winter climate in western Europe. On longer timescales there is also natural climate variability.

A natural greenhouse effect operates because certain trace gases in the atmosphere such as water vapour and carbon dioxide act to absorb long-wave radiation emitted from the Earth's surface and reradiate it back downwards. They act as a blanket. This effect is being enhanced by increased atmospheric concentrations of such greenhouse gases as carbon dioxide due to the use of coal and oil as energy sources. However, there are also negative feedback effects from atmospheric aerosol pollution produce by various industrial processes. Increased clouds and sulphate aerosols can increase atmospheric albedo and reflect more of the Sun's energy directly back into space and thus partly offset the greenhouse effect. Nevertheless, increasing atmospheric greenhouse gas concentrations are now dominating global climate change, and unless restricted in the present decade have the potential to cause massive damage by the end of the century.

Further reading
Books

Barry, R.G. and Chorley, R.J. (2009) *Atmosphere, weather and climate.* Routledge, London.
This book provides excellent further coverage of the processes discussed in this chapter and has been very popular with students for many years.

Houghton, J. (2009) *Global warming: The complete briefing,* 4th edition. Cambridge University Press, Cambridge.
This book describes the uncertainties surrounding predictions of the effects of global warming including positive and negative feedbacks.

IPCC (2007a) *Climate change 2007: The physical science basis.* Contribution of Working Group 1 to the Fourth Assessment Report of the Intergovernmental Panel on Climate Change. Solomon, S., Qin, D., Manning, M. *et al.* (eds). Cambridge University Press, Cambridge.
The latest assessment of the Intergovernmental Panel on Climate Change demonstrating the lastest scientific understanding.

IPCC (2007b) *Climate change 2007: Climate change impacts, adaptation and vulnerability.* Contribution of Working Group 2 to the Fourth Assessment Report of the Intergovernmental Panel on Climate Change. Adger, N. *et al.* (eds). Cambridge University Press, Cambridge.
The latest assessment of the Intergovernmental Panel on Climate Change predicting the potential magnitude of change and impacts of climate change.

Lockwood, J.G. (2009) The climate of the Earth. In: Hewitt, C.N. and Jackson, A.V. (eds) *Atmospheric science for environmental scientists.* Wiley-Blackwell, Chichester, 1–25.
A detailed introduction to the circulation and climate of the Earth's atmosphere.

MacKay, D.J.C. (2009) *Sustainable energy – without the hot air.* UIT, Cambridge.
An excellent introduction to the physics of energy use. Free copies are available on the Internet at www.withouthotair.com.

Robinson, P.J. and Henderson-Sellers, A. (1999) *Contemporary climatology.* Pearson Education, Harlow.
The authors provide a good overview of climate basics with some excellent diagrams.

Stern, N. (2006) *The economics of climate change: The Stern Review.* Cambridge University Press, Cambridge.
The Stern Review is an independent, rigorous and comprehensive analysis of the economic aspects of climate change.

Further reading
Papers

Allen, M. R., Frame, D.J., Huntingford, C. *et al.* (2009) Warming caused by cumulative carbon emissions towards the trillionth tonne. *Nature*, 458, 1163–1166.

Bigg, G.R. (2001) Back to basics: the oceans and their interaction with the atmosphere. *Weather*, 56, 296–304.
This short paper provides a straightforward introduction to the principal ocean–atmosphere interactions.

Hansen, J., Sato, M., Kharecha, P. *et al.* (2008) Target atmospheric CO_2: where should humanity aim? *The Open Atmospheric Science Journal*, 2, 217–231.
Good paper about uncertainties and targets for atmospheric emissions and carbon uptake measures.

Lockwood, J.G. (1999) Is potential evapotranspiration and its relationship with actual evapotranspiration sensitive to elevated atmospheric CO_2 levels? *Climatic Change*, 41, 193–212.
A paper providing further information on Section 4.5.3.

Matthews, H. D. and Caldeira, K. (2008) Stabilizing climate requires near-zero emissions. *Geophysical Research Letters*, 35, L04705, doi: 10.1029/2007GL032388
Some interesting calculations and thoughts on the climate inertia problem.

CHAPTER 5

Global climate and weather

John McClatchey
Environmental Research Institute, University of the Highlands and Islands, and Honorary Research Fellow, University of Nottingham

Learning objectives

After reading this chapter you should be able to:

➤ understand how major features of the general atmospheric circulation play a role in climate

➤ describe the types of climates that exist across the Earth

➤ identify the type of weather associated with different types of climate

➤ recognize the importance of seasonality in different climates

➤ understand why there is a distinct geographical distribution of climates

➤ explain why there are differences between the climates of the northern and southern hemispheres

5.1 Introduction

The climate at any place is the most important environmental influence on the natural vegetation, the landscape and human activity. For the natural world, three components of climate are crucial, these being the precipitation, the thermal conditions and the wind conditions. For example, in desert climates the landscape is usually dominated by **aeolian** (wind-related) features, while in much of the middle latitudes, fluvial (water-related) features dominate the landscape. The vegetation cover is also strongly controlled by the thermal conditions and by the availability of moisture.

The climate at any one location arises from the types of weather that are experienced there over a period of years. However, the climate is not just about average values of weather elements such as temperature or precipitation but includes the range and extremes of those elements and the frequencies of types of weather. For example, in desert regions the average monthly precipitation total may be almost meaningless, as precipitation might occur only on one or two days a month. Of much more importance is the number of days on which precipitation falls and the amounts that fall in each event. Of course in all climates there can be substantial local and regional variations from the average conditions (see Chapter 6) but despite such local variations, the types of weather and the weather systems in each major climate regime will be similar. Therefore in order to understand why a location has a particular climate, the starting point should always be to understand what types of weather are a feature of that climate. This chapter therefore starts by outlining the general controls of weather conditions and climate across the Earth, before discussing the features of the major global climate zones.

5.2 General controls of global climates

The Sun drives both weather systems and the global climate and, as a result, the latitude of any location plays an important role in determining the climate. The decline of the solar radiation input that takes place in moving from the equator to higher latitudes eventually leads to a negative net radiation balance at the top of the atmosphere (the balance between the incoming solar and outgoing terrestrial radiation). This occurs at close to 45°N and 45°S and would lead to a cooling at higher latitudes if energy were not transported there from lower latitudes by ocean currents and the general circulation of the atmosphere. Therefore, although important, solar radiation alone does not control the climate. The distribution of the oceans and continents, ocean currents and the general circulation of the atmosphere also play an important role (see Chapter 4). Hence despite having a similar solar radiation input, places at the same latitude may have mild winters in one case and very cold winters in another. For example, Scotland has much milder winters than Labrador at the same latitude in north-east Canada. Climate is often assessed by using a network of observation stations that record meteorological and climatological data. There are a number of types of such records that can be used to assess climate and these are discussed in Box 5.1.

A key factor determining global climate is the presence of zones of ascent or descent of air. In areas of the world subject to high-pressure anticyclonic conditions, air is subsiding (descending) and is being warmed by compression. That

TECHNIQUES

METEOROLOGICAL AND CLIMATOLOGICAL OBSERVING NETWORKS

It is important to be aware of the availability and type of climatological records that can be used to assess climate. There are a number of different types of observing station but the three principal types are synoptic stations, climatological stations and precipitation stations. They are not spread uniformly across the planet. There are very few in oceanic locations or in less developed countries, deserts, polar regions and other harsh or inaccessible places. Such stations tend to be concentrated in developed countries.

Synoptic stations

Synoptic observing stations provide surface weather element observations as an input into numerical weather forecasting models. The principal elements recorded are temperature, humidity, wind speed, wind direction and atmospheric pressure (and change over past 3 hours). These

elements are recorded at all synoptic stations, including automatic stations. For stations at which meteorological observers are based, observations of cloud cover, cloud heights, visibility, precipitation, current weather and past (over past 3 hours) weather are also recorded but recent developments in equipment now allow visibility and some cloud information to be recorded at some automatic weather stations (Figure 5.1).

Synoptic stations record weather elements at least every 6 hours at the same time at every station in the world (at 00:00, 06:00, 12:00 and 18:00Z, the Z denoting Universal Time, equivalent to Greenwich Mean Time) and many stations make hourly observations. These synoptic stations are located to provide measurements representative of a wide area. Many are at airports or military aviation bases and therefore the local area surrounding the station is generally flat and open.

Solar radiation is recorded at only a few stations, mostly at what are termed 'agrometeorological stations'

(used to provide information specifically for agriculture). At some of those stations both the direct and diffuse solar radiation are recorded and at a few stations there will also be an observation of the net radiation, which is the balance between the short- and long-wave radiation (see Chapter 4).

Climatological stations

Climatological observations are generally taken at least once per day, although at some sites it is twice per day. The time at which observations are taken at an 'ordinary' climatological station is decided by each country (e.g. 09:00 GMT in the United Kingdom). All climatological stations make an observation of the maximum and minimum temperatures and precipitation total over the previous 24 h. The 'average' temperature is just the sum of the maximum and minimum temperatures divided by 2. Note that for synoptic stations, however, a more representative mean can be obtained from the hourly values. Some stations record soil temperatures (at various depths) and some record a night 'grass' (surface)

BOX 5.1 ➤

➤

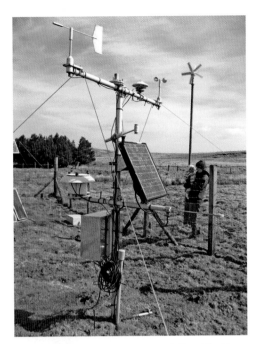

Figure 5.1 An automatic weather station. The blue solar panel is used to charge the battery supply.

Figure 5.2 A Stevenson screen used for housing thermometers.

minimum temperature. Some non-instrumental observations of weather phenomena are also recorded.

There are far more climatological stations than synoptic stations (which also record climatological observations), as they are used to provide an indication of the variation of climate within a region (Chapter 6). Across the United Kingdom there is about one climatological station per 500 km^2, although the density is much less in the Scottish Highlands and much greater in lowland England. This is a high density of stations compared with most countries. The number of stations required to show the local variations of climate within a region depends on the topography and the proximity to the sea or very large lakes (such as the North American Great Lakes).

Precipitation stations

In addition to climatological stations there are also a large number of precipitation stations at which only precipitation is recorded. In areas of frequent winter snowfall, special snow gauges are used to obtain a rainfall equivalent measure. The United Kingdom has around 5000 precipitation stations (about one per 30 km^2), a high density compared with most countries.

World Meteorological Organization standards

Both synoptic and climatological stations have to conform to certain standards set out by the World Meteorological Organization (WMO). These ensure that observations taken at any place in the world are directly comparable. For example, thermometers are sited in a screen (Figure 5.2) at between 1 and 2 m above a grass surface (this may be a snow surface during winter). This screen has to be above the surface as at night there can be differences of over 10°C between the (colder) grass (surface) minimum and screen minimum temperatures as a result of surface radiational cooling. The height at which wind is recorded is also important as the friction of the surface slows the wind and causes the wind direction to change. The standard height for an anemometer (wind speed recorder) and wind vane is 10 m (although a range between 8 and 12 m is acceptable).

A 'climate normal' (average) is established over a 30 year period for a standard station updated every 10 years (e.g. 1951–1980; 1961–1990; 1971–2000), which is the 30 year average of the individual monthly values. Often, however, meteorological and climatological observations are made as part of micrometeorological, ecological, hydrological and even geomorphological investigations. Such measurements may not conform to WMO standards but may be appropriate in terms of the particular investigation. However, non-standard observations cannot be directly compared with standard measurements and therefore they need to be used with care if used in any local climate assessment.

BOX 5.1

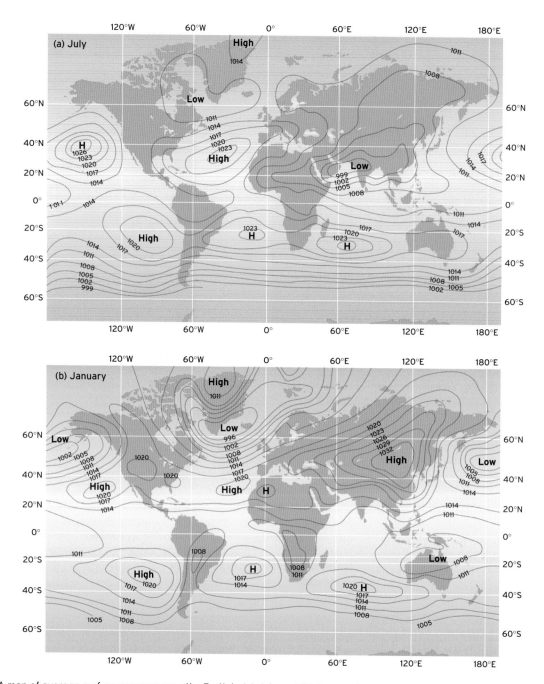

Figure 5.3 A map of average surface pressure over the Earth in (a) July and (b) January in hPa (or millibars). High-pressure areas represent zones of descending air (air at the surface is diverging away from that point) while low-pressure areas are zones of ascending air (air at the surface is converging towards that point). (Source: after White *et al.*, 1992, *Environmental Systems: An Introductory Text*, 2nd edition, by D.N. Mottershead, S.J. Harrison and I. White, Fig. 4.5, p. 84. Published by Taylor and Francis Books Ltd (Nelson Thornes) 1992. Reproduced by Permission of Taylor & Francis Books UK)

warming leads to the formation of a **temperature inversion** at a few hundred to perhaps 2000 m above the surface whereby warmer air overlies colder air. This prevents any further rise of the lower colder air because it is more dense and less buoyant than the overlying warm air. This inversion therefore inhibits the growth of clouds, and as clouds need to be deep to give substantial falls of precipitation, areas in which anticyclones are located tend to be dry. Conversely, in areas of the world with

low pressure, air is ascending (as a result of convergence in the lower atmosphere) and clouds can often grow to considerable depths. This convergence can enhance thermal convection or encourage **slantwise convection** (movement of air upwards and north or south) if there are noticeable thermal contrasts between regions of air (e.g. at fronts).

A map of average surface pressure (Figure 5.3) therefore provides an initial indication of likely areas of ascent (relatively wetter areas) and descent (relatively drier areas). In addition, because the middle and high latitudes have an overall energy deficit (more radiation is lost to space at the top of the atmosphere than is gained from solar radiation) the weather systems help transport energy from lower to higher latitudes. Therefore the average position of the weather systems can give an indication of regions for which air is generally being brought into from lower latitudes. These regions will be milder than might otherwise be expected and the circulation of air around pressure systems means that these regions will be on the southern and eastern sides of low-pressure systems in the northern hemisphere (northern and eastern sides in the southern hemisphere).

The type of weather experienced at a particular location depends on latitude, the position and movement of pressure systems and the presence or absence of areas of ascent or descent (subsidence) of air. The thermal climate will depend on latitude (height of the Sun in the sky), the principal prevailing wind directions (from warmer or colder areas) and the cloudiness (thick clouds reduce daytime maximum temperatures but keep night-time minimum temperatures higher). Cloud and precipitation depend on the presence or absence of ascending air (giving cloud) or subsidence (giving clear skies). Therefore, although the following discussion will be separated into latitudinal zones, there will be substantially different climates experienced within each latitudinal zone. A number of classifications of climate have been suggested, with probably the most commonly used being Köppen's classification that was developed in the early twentieth century. The classification scheme was modified by Köppen's students (Geiger and Pohl, 1953) and is shown in Table 5.1. It is, however, important to realize that the boundaries between different types of climate are not distinct and that climates merge into each other. Nevertheless a very general map of climatic zones is provided by Figure 5.4 based on Köppen's classification.

Reflective question

➤ What are the main factors controlling the location of climate zones?

5.3 The tropics and subtropics

5.3.1 Equatorial regions

Horizontal air movement arises from the pressure gradient force that is created if there is a pressure difference between different places. Over much of the Earth, once air moves as a result of a pressure gradient force, the Coriolis effect becomes apparent (see Chapter 4) and the winds deviate

to the right in the northern hemisphere and to the left in the southern hemisphere. Close to the equator the Coriolis effect is negligible and therefore air follows the pressure gradient from high to low pressure. This is important as it means that near the equator, the weather is not dominated by the movement of large circulatory weather systems such as anticyclones and depressions. However, outflow from the large subtropical anticyclones (the north-east and south-east trade winds) does play an important role leading to convergence of air into a region of somewhat lower pressure sometimes termed the equatorial trough.

This trough can be seen as a distinct **intertropical convergence zone** (ITCZ), particularly over the oceans, and is part of the Hadley cell circulation (see Chapter 4 for further details on the formation of the ITCZ). The ITCZ is not located along the equator but moves quite well north of the equator in some places in the northern hemisphere summer and a little south of the equator in the southern hemisphere summer, with an average position somewhat north of the equator (Figure 5.5). This movement is in response to seasonal differences in pressure gradients caused by changes in solar energy received during the year. The movement of the ITCZ north and south is greater over land because the seasonal cooling and warming are more pronounced over land (the oceans warm and cool at a much slower rate than land). The ITCZ is typically from 4° to 8°N (in March and September respectively) in the Pacific. There are, however, extensions of the ITCZ further south of the equator (Linacre and Geerts, 1997), particularly in the southern hemisphere summer. These are into southern Africa, east of the Andes in South America and a larger and more consistent extension called the South Pacific convergence zone, roughly south-south-west from the north-east of Australia (Figure 5.5).

Low-level convergence such as at the ITCZ leads to rising air and the formation of clouds. Therefore the weather close to the equator is dominated by frequent convectional clouds and rain. This convectional activity may be relatively random with many individual cumulus clouds being formed but there are also more organized cloud clusters. For example, in some tropical coastal areas, more organized features sometimes occur diurnally as part of a land and sea breeze circulation (see Chapter 6). Equatorial climates are therefore dominated by the movement of the ITCZ with many places close to the equator having no distinct dry season. However, some equatorial areas show a clear peak in the monthly precipitation totals at each of the equinoxes with lower totals when the ITCZ is at its northern and southern limits (summer in the northern and southern hemisphere respectively). Total annual precipitation is high and in some parts of the Amazon and West Africa totals can be from 2500 to over 4000 mm.

Table 5.1 Köppen's climate classification with additional modifications by Geiger and Pohl (1953)

1st	2nd	3rd	Basic description	Classification criteria[†]
A			Humid tropical:	
	f		tropical wet (rainforest)	Wet all year
	w		tropical wet and dry (savanna)	Winter dry season
	m		tropical monsoon	Short dry season
B			DRY:	PET > ppn
				BS/BW boundary is one-half dry/humid boundary*
	S		semi-arid (steppe)	Mean annual $T \geq 18°C$
	W		arid (desert)	Mean annual $T < 18°C$
		h	hot and dry	Mean annual $T \geq 18°C$
		k	Cold and dry	Mean annual $T < 18°C$
C			Moist with mild winters:	Coolest month $T < 18°C$ and $-3°C$
	w		dry winters	Wettest summer month 10 times rain of driest winter month
	s		dry summers	Driest month < 40 mm and wettest winter month ≥ 3 times driest month
	f		wet all year	Criteria for w or s not met
		a	summers long and hot	Warmest month $>22°C$ with more than 4 months $> 10°C$
		b	summers long and cool	All months below 22°C with more than 4 months $> 10°C$
		c	summers short and cool	All months $< 22°C$ with 1 to 3 months $> 10°C$
D			Moist with cold winters:	Coldest month $\leq -3°C$ warmest month $> 10°C$
	w		dry winters	Same as Cw
	s		dry summers	Same as Cs
	f		wet all year	Same as Cf
		a	summers long and hot	Same as Cfa
		b	summers long and cool	Same as Cfb
		c	summers short and cool	Same as Cfc
		d	as c with severe winters	Coldest month $< 38°C$
E			Polar climates:	Warmest month $< 10°C$
	T		tundra	Warmest month $> 0°C$ but $< 10°C$
	F		ice cap	Warmest month $\leq 0°C$

*ppn = annual precipitation; T = mean annual temperature; PET = potential evapotranspiration. Potential evapotranspiration greater than annual precipitation (PET > ppn); dry/humid boundary given by ppn = $2T + 28$ when 70% of rain in summer half-year; ppn = $2T$ when 70% of rain in winter half-year; ppn = $2T + 14$ when neither half-year has 70% of rain.

[†]Temperature limits given for warmest and coldest months in classification section are for average monthly temperatures.

(Source: Goiger and Pohl, 1953)

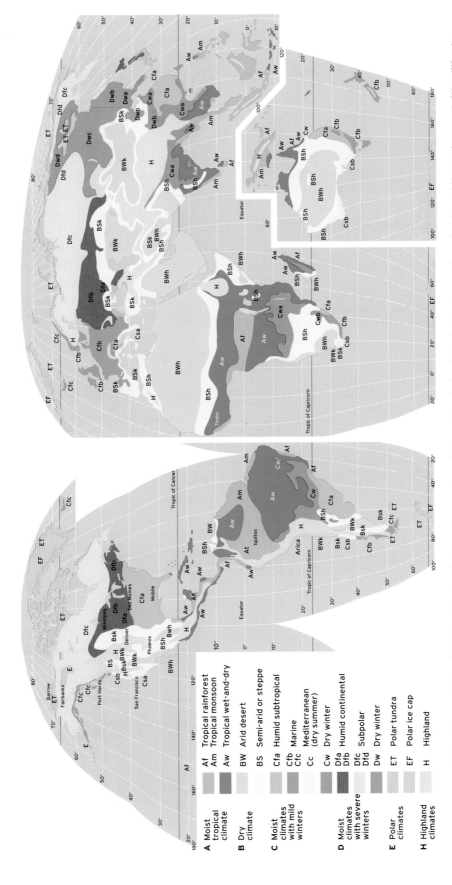

Figure 5.4 A map of the Köppen–Geiger–Pohl climate classification. (Source: from *Meteorology today: Introduction to weather, climate and the environment*, 6th edition by AHRENS, 2000. Reprinted with permission of Brooks/Cole, a division of Thomson Learning: www.thomsonrights.com, Fax 800-730-2215)

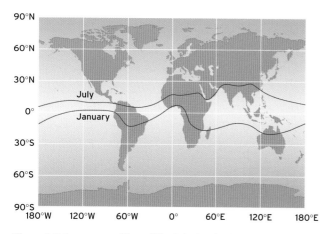

Figure 5.5 Average position of the intertropical convergence zone in January and July. North-east trade winds and south-east trade winds converge at the ITCZ. Here solar heating and convergence of air results in ascension and instability. Thus precipitation rates are high. (Source: Critchfield, H.J., *General climatology*, 3rd edition, 3rd © 1974. Adapted (or electronically reproduced in case of e-use) by permission of Pearson Education, Inc., Upper Saddle River, NJ)

As the Sun is always high in the sky, average temperatures in the equatorial regions are fairly constant throughout the year, although maximum temperatures may be a little lower during the wetter periods. Humid conditions keep night temperatures high, giving a relatively small diurnal range with average annual temperatures around 27°C. Typical monthly temperature and precipitation values for equatorial stations are given in Figure 5.6. Equatorial climates are not exactly alike in every place, particularly over the continents of Africa and South America, as topography and proximity to the sea do play a role in altering thermal and precipitation regimes (see Chapter 6).

Moving north and south away from the equator into the trade wind belt, between about 5° and 20° latitude, the ITCZ still plays a key role in the climate but as the ITCZ is only close to these areas in the summer (in that hemisphere), there is now a distinct rainy (summer) and dry (winter) season. This greater seasonality gives a greater range of temperatures through the year, but still much less than in many middle and high latitudes, with the highest temperatures tending to be observed just before the rainy season. There can be quite large average

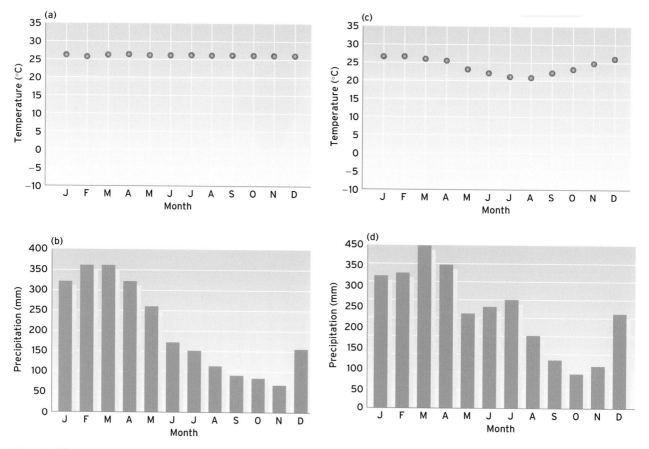

Figure 5.6 Example temperature and precipitation graphs for equatorial stations. Mean monthly temperatures (°C) and mean monthly rainfall (mm) for (a) and (b) Belem, Brazil (1.5°S), and (c) and (d) Tamavate, Madagascar (18°S).

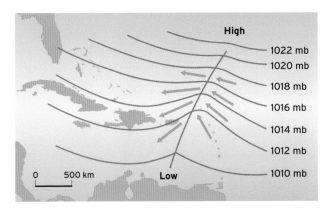

Figure 5.7 An easterly wave in the subtropical easterly flow. Easterly waves occur within the trade wind belt and represent waves of air (moving north and south) for air travelling in an easterly direction. (Source: after Critchfield, H.J. *General climatology*, 3rd edition, 3rd © 1974. Adapted (or electronically reproduced in case of e-use) by permission of Pearson Education, Inc., Upper Saddle River, NJ)

diurnal ranges of temperature in places (15°C or more), particularly in the dry season. The trade winds also mean that there is a fairly steady but not particularly severe wind regime.

5.3.1.1 The easterly wave and tropical cyclones

One important type of more organized cloud is related to easterly waves in the trade winds. Such waves represent moving troughs of slightly lower pressure denoting zones of low-level convergence. This convergence is concentrated along the trough line resulting in a zone of heavier rain. These wave features (Figure 5.7) are important as they can develop into tropical depressions and tropical storms (Malkus, 1958) and some of those become tropical cyclones. These cyclones are called hurricanes in the Atlantic and typhoons in the west Pacific (Boxes 5.2 and 5.3). The WMO has produced a classification system that allows us to define when tropical depressions become severe enough to be called tropical storms or tropical cyclones (Table 5.2).

Tropical cyclones require high sea surface temperatures (at least 27°C) and do not form over land as the energy of the system comes from the release of latent heat when the clouds are formed. The moisture evaporating from the sea is therefore a key factor in sustaining tropical cyclones and when that source of moisture is cut off (when hurricanes move inland) the intensity declines. Although tropical cyclones occasionally move inland, the majority are confined to the oceans. Even across the oceans the tracks of tropical cyclone systems show a distinct geographical distribution (Figure 5.8). Tropical cyclones do not occur close to the equator as there needs to be a sufficient Coriolis effect to help develop the circulation of the initial depression. As well as requiring high sea surface temperatures, the mid-troposphere has to be moist in order to allow tropical cyclone development. There must also be no strong temperature inversion inhibiting mixing of the warm moist surface air with the mid-troposphere air. This is why tropical cyclones are extremely rare in the South Atlantic. In the South Atlantic there is a noticeable extension of the trade wind inversion from south-west Africa towards Brazil and hence there is insufficient opportunity for the build-up of a deeper moist layer. A strong temperature inversion also extends out into the Pacific from northern Chile and Peru and the sea surface temperatures are also relatively cool. This again inhibits the development of tropical cyclones. However, the Pacific is much wider than the Atlantic and therefore a deeper moist layer has time to build up and as a result tropical cyclones are found in the central and west Pacific south of the equator (Brasher and Zheng, 1995). Figure 5.8 shows the tracks of tropical cyclones in both the northern and southern hemispheres. It is also of note that there are fewer tropical cyclones in the southern hemisphere as the summers are somewhat cooler than those in the northern hemisphere and therefore the area of the ocean with high sea surface temperatures is smaller and confined to north of 20°S. In the northern hemisphere, high sea surface temperatures of 27°C and above extend to 30°N.

Table 5.2 World Meteorological Organization classification of tropical storms

	Wind speed		Beaufort scale
	m s^{-1}	knots	
Tropical depression	<17.2	<34	7
Moderate tropical storm	17.2-24.4	41-47	8-9
Severe tropical storm	24.5-32.6	48-63	10-11
Hurricane (tropical cyclone)	>32.6	>63	12

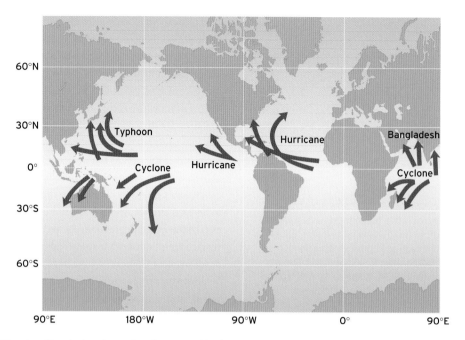

Figure 5.8 Typical tracks of tropical cyclones, hurricanes and typhoons. Note that hurricanes and typhoons are also tropical cyclones but that the names depend on their location (called hurricanes in the Atlantic, typhoons in the west Pacific). (Source: from *Meteorology today: Introduction to weather, climate and the environment*, 6th edition by AHRENS, 2000. Reprinted with permission of Brooks/Cole, a division of Thomson Learning: www.thomsonrights.com, Fax 800-730-2215)

ENVIRONMENTAL CHANGE

ARE HURRICANES ON THE INCREASE?

The official hurricane season runs from 1 June to 30 November. In the US National Oceanic and Atmospheric Administration (NOAA) report on the 2005 Atlantic hurricane season, a record 28 named storms were reported. Prior to this, the number of reported storms in any year since reliable records began in 1944 did not exceed 19 (although it is believed that there would have been 21 named storms in 1933 and there may have been other years before the 1940s when 20 or more storms might have occurred). Of the 28 named storms in 2005, 15 were hurricanes and 7 of these were major hurricanes (categorized as 3 or more on the Saffir–

Simpson scale – see Table 5.3). In addition, 4 of those 7 major hurricanes reached category 5 status, the highest number of category 5 hurricanes in any year since records began. The 2005 season started early and as well as having the most costly and one of the most deadly hurricanes (Katrina) the season also had hurricane Wilma (another category 5 storm) in October which had the lowest central pressure ever recorded (882 hPa) for any Atlantic hurricane (the previous record was 888 hPa in hurricane Gilbert in 1988).

An alternative to counting the number of storms to determine how stormy a year has been is to use the NOAA Accumulated Cyclone Energy Index. This is based on the cumulative strength and duration of each storm.

Using that index, 2005 was not the most active season, but the third most active season on record, behind 1950 and 1995.

Multi-decadal signal and the El Niño Southern Oscillation signal

Tropical cyclone activity in the Atlantic has been above normal since 1995. This has been largely in response to the active phase of the 'multi-decadal signal'. The multi-decadal signal relates to cycles of about 20-40 years in monsoon rains over West Africa and the Amazon Basin and in North Atlantic sea surface temperatures. The current phase of the cycle has involved lower **wind shear** (change of wind speed and/or direction with height) and warmer sea surface temperatures

BOX 5.2 ➤

Table 5.3 The Saffir-Simpson hurricane category scale

Hurricane category	Wind speed				Storm surge	
	mph	km h^{-1}	knots	m s^{-1}	ft	m
1	74-95	119-153	64-82	33-42	4-5	1.2-1.5
2	96-110	154-177	83-95	43-49	6-8	1.8-2.4
3	111-130	178-209	96-113	49.5-58	9-12	2.7-3.7
4	131-155	210-249	114-135	58.5-69	13-18	4.0-5.5
5	>155	>249	>135	>69	>18	>5.5

across the tropical Atlantic. In addition there have been weak low-level easterly (trade) winds and a westward expansion of upper-level easterly winds from Africa, the African Easterly Jet. These conditions are favourable to the development of hurricanes. This phase is expected to continue for the next decade or perhaps longer.

As well as the multi-decadal signal, El Niño Southern Oscillation (see Chapter 4) episodes occur roughly every three to five years, and generally last 9 to 15 months. El Niño events involve reduced upwelling and therefore lead to a warming of the surface ocean waters over the central equatorial Pacific. Strong upwelling gives La Niña events, cooling the surface waters. Changes in sea surface temperatures in this region alter the

patterns of tropical convection across the central and east-central equatorial Pacific with warmer temperatures (El Niño) increasing convection and colder temperatures (La Niña) reducing convection (see Chapter 4). The La Niña episodes encourage upper-atmosphere easterly winds and reduced wind shear at lower levels in the tropical Atlantic which as noted above are conditions that are favourable to Atlantic hurricanes. In contrast, El Niño events produce upper westerly winds and increased wind shear in the same region.

Although ENSO events can encourage (La Niña) or act against the formation of Atlantic hurricanes (El Niño), the tropical multi-decadal signal is the dominant feature and can mask any influence of ENSO

events. A multi-decadal signal favourable to hurricanes can be enhanced by a La Niña event and diminished by an El Niño. The influence of an El Niño was evident in 1997, 2002 and 2006, the only years with below-average hurricanes in the 11 years to 2006. However, the relationship is not simple, as the exceptional hurricane season of 2005 was not a La Niña year.

The average number of named storms per year since 1995 has been 13.0, compared with 8.6 during the preceding 25 years when the multi-decadal signal was in an inactive phase (Table 5.4). An average of 7.7 hurricanes and 3.6 major hurricanes per year since 1995 compares with 5 hurricanes and 1.5 major hurricanes per year between 1970 and 1994 (Table 5.4).

Table 5.4 Number of Atlantic hurricanes per year

	Average 1970-1994	Average 1995-2005	2005	2006
Storms named	8.6	13.0	28	9
Hurricanes	5.0	7.7	15	5
Major hurricanes	1.5	3.6	7	2

BOX 5.2 ➤

This box has indicated that there are many factors that combine to determine the numbers of hurricanes that form. However, the International Panel of Climate Change report has warned that global climate change will lead to more hurricanes over the next century (IPCC, 2007b). In fact IPCC suggests that there is a greater than 66% chance that the intensity of hurricanes will also increase over the next few decades. However, as noted in Table 5.4, the number of hurricanes in 2006 was typical of the 1970–1994 average and was less than the average over the most recent period (1995–2005). Potential future increases in hurricanes and hurricane intensity and of typhoons in the Pacific must therefore be put into the context of multi-decadal and ENSO signals that can overlap and mask potential long-term change. There will be considerable variability in the frequency and intensity of tropical cyclones (hurricanes and typhoons) from year to year and it will therefore be difficult to discern a definite increase in either intensity or frequency of tropical cyclones for many years to come.

BOX 5.2

CASE STUDIES

HURRICANE KATRINA

In 2005, hurricane Katrina was the most costly in US history and also caused the highest number of deaths from a single hurricane since 1928. Tropical storm Katrina was designated on 24 August 2005 at which time it was located in the central Bahamas (Figure 5.9). Katrina began strengthening rapidly and became a category 1 hurricane 24 km east-north-east of Fort Lauderdale at 17:00 EDT (21:00 UTC) on 25 August. At 18:30 EDT (22:30 UTC), the hurricane made landfall between Hallandale and North Miami Beaches with sustained winds estimated at 36 m s^{-1} gusting to over 40 m s^{-1}. Katrina moved south-west across the tip of the Florida Peninsula during the night but the landfall did little to reduce the intensity as the storm reintensified as it moved back to sea over the warm waters of the Gulf. The sustained winds over Florida were never higher than 36 m s^{-1} but the heavy rain and gusty winds caused substantial damage and flooding, and 14 people lost their lives. By way of comparison, the 1987 October storm in southern England also had maximum sustained winds of around 36 m s^{-1}.

Katrina moved west after entering the Gulf of Mexico and then over the next few days gradually turned to the north-west and then north. The high sea surface temperatures and an upper-level anticyclone over the Gulf encouraged the rapid intensification, which led to Katrina attaining 'major hurricane' (category 3) status on the afternoon of 26 August. Katrina continued to strengthen and by 07:00 CDT (12:00 UTC) on 28 August, hurricane Katrina reached category 5 status with wind speeds of 72 m s^{-1} or more and a pressure of 908 hPa with the maximum sustained wind speeds of close to 78 m s^{-1} being reached at 10:00 CDT, remaining at that speed until the afternoon. At 16:00 CDT (21:00 UTC), Katrina's minimum central pressure dropped to 902 hPa, one of the lowest pressures ever recorded. At this time Katrina was at its peak strength with hurricane force winds extending outwards up to 168 km from its centre and tropical storm force winds (up to 33 m s^{-1}) extending outwards up to nearly 370 km. Sustained tropical storm force winds were already battering the south-east Louisiana coast and the 16:00 CDT (21:00 UTC) Bulletin from the National Hurricane Center warned of coastal storm surge flooding of 5.5 to 6.7 m above normal tide levels, locally as high as 8.5 m, and stated 'some levees in the Greater New Orleans area could be overtopped'.

At 04:00 CDT (09:00 UTC) on 29 August the hurricane's centre was 144 km south-south-east of New Orleans with winds of 67 m s^{-1} near the centre and gusts to hurricane force (33 m s^{-1}) along the coast. Just over 2 h later Katrina made landfall in Plaquemines Parish just south of Buras (between Grand Isle and the mouth of the Mississippi River) as a strong category 3 hurricane (wind speeds about 57 m s^{-1} and a central pressure of 920 hPa). By 08:00 CDT (13:00 UTC), Katrina was only 64 km south-east of New Orleans with hurricane force winds extending up to 200 km from the centre of the storm. In the right front quadrant of the storm, which is where the strongest winds are generally found as there is an additive effect from the winds circulating

BOX 5.3 ➤

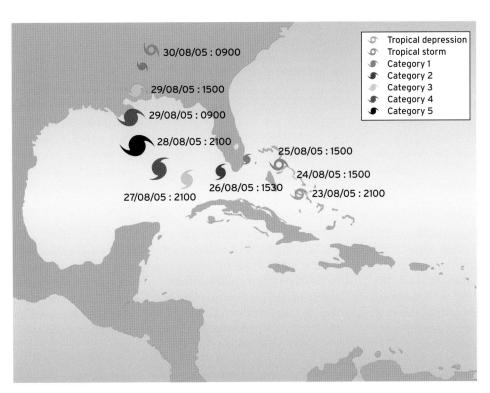

Figure 5.9 The storm track of hurricane Katrina.

round the hurricane with the direction of movement (Figure 5.10), Pascagoula Mississippi Civil Defense reported a wind gust to 53 m s^{-1} and Gulfport Emergency Operations Center reported sustained winds of 42 m s^{-1} with a gust to 45 m s^{-1}. By 10:00 CDT (15:00 UTC), the eye of Katrina was making its second northern Gulf coast landfall near the Louisiana–Mississippi border. The northern eyewall (Figure 5.10) was still reported to be very intense by WSR-88D radar data and the intensity was estimated to be near 54 m s^{-1}.

Katrina caused enormous damage to homes and businesses in both Louisiana and Mississippi estimated at around US$125 billion. The loss of human life was even more catastrophic with a death toll of 1833 with several hundred people still listed as missing (Graumann et al., 2005). The majority of the deaths were in Louisiana (1577) and Mississippi (238) with 14 deaths in Florida and 2 each

in Alabama and Georgia. This made Katrina the third deadliest hurricane since 1900, after the Galveston hur-

ricane of 1900 (at least 8000 deaths) and the Lake Okeechobee Hurricane of 1928 (over 2500 deaths).

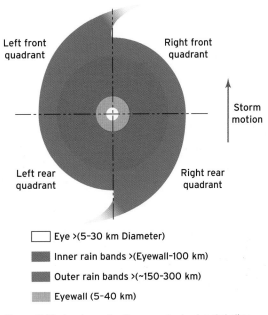

Figure 5.10 A schematic diagram of a horizontal slice through the structure of a hurricane.

BOX 5.3 ➤

129

➤

A detailed account of the hurricane with satellite images and a discussion of the historical perspective has been produced as a NOAA Technical Report (Graumann *et al.*, 2005) and much of the above summary of the progression of the hurricane is based upon that report.

While the wind damage caused by Katrina was significant, the bulk of the devastation was caused by flooding, largely due to the very substantial storm surge which peaked at 8.5 m at Pass Chritian, Mississippi. A surge of 7.3–8.5 m was estimated along the western Mississippi coast across a path of about 32 km. The surge was 5.2–6.7 m along the eastern Mississippi coast, 3.0–5.8 m along the Louisiana coast and 3.0–4.6 m along the Alabama coast.

A number of factors contributed to the extreme storm surge:

- the massive size of the storm;
- the strength of the system (category 5) just prior to landfall;
- the 920 hPa central pressure at landfall; and
- the shallow offshore waters.

In the delta country south-east of New Orleans, a number of towns were completely flooded, with Plaquemines and St. Bernard Parishes particularly badly affected. The levee system protecting New Orleans was put under severe pressure due to the rise in the level of Lake Pontchartrain caused by the surge. As reported by Graumann *et al.* (2005) the damage and high-

water marks indicate that the surge reached up to 19 km inland in some areas, especially along bays and rivers, and in New Orleans there were significant failures in the levee system on 30 August on the 17th Street Canal, Industrial Canal and London Avenue Canal levees. As much of New Orleans lies below sea level, the failure of the levees led to drainage of water into the city, leading to 80% of the city being underwater to depths of over 6 m (Figure 5.11). Graumann *et al.* (2005) also noted that while much of the flood waters had been cleared by 20 September, the storm surge from hurricane Rita on 23 September caused a new breach in the repaired Industrial Canal levee and many of the areas of the city were flooded again.

Figure 5.11 Flooding in New Orleans, following hurricane Katrina, August 2005. (Source: PA Photos: David J Phillip)

BOX 5.3 ➤

Tropical cyclones give very high rainfall totals and intensities. The rainfall falling over two to three days from a single hurricane can even be close to the average annual total in some places. The relatively low frequency of tropical storms, and the even lower frequency of tropical cyclones at any individual location, mean that the standard climate statistics do not provide any real indication of the importance of these events in the climate of these areas. They are high-magnitude, low-frequency events. In any one year, two locations that have similar mean rainfall totals may well have widely differing totals if one of the locations had experienced a severe tropical storm or tropical cyclone.

5.3.1.2 Monsoons

One other important feature of the climate in the north-east and south-east trade wind areas is that there are regions that experience an exceptionally wet rainy season with a very distinct dry season (Figure 5.12). These are the regions that experience **monsoon** conditions (monsoon is from

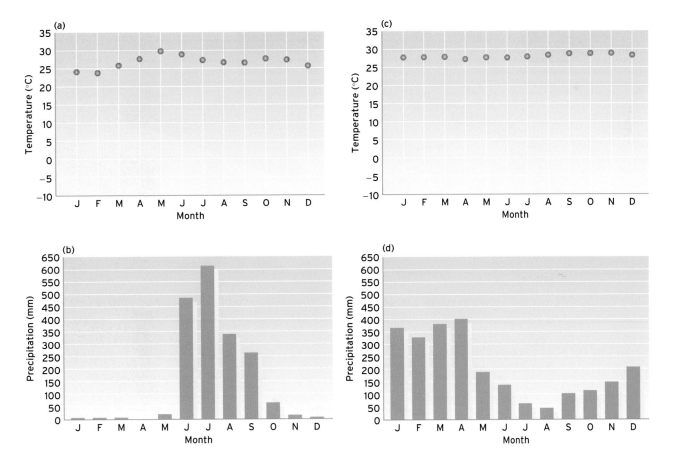

Figure 5.12 Example temperature and precipitation graphs for monsoon climate stations. Mean monthly temperatures (°C) and mean monthly rainfall (mm) for (a) and (b) Bombay, India (19°N), and (c) and (d) Manaus, Brazil (3°S).

the Arabic, meaning season). The largest and most intense monsoon is in Asia but there are monsoon-type climates in West and East Africa, and in Australia (Webster, 1981). There is a much less well-defined monsoonal climate in South America, partly as a result of the relatively cool sea surface temperatures to the west of South America that increase stability in the lower layers of the atmosphere. The Andes also play a role as they obstruct the trade winds and air descent in the lee of the mountains increases atmospheric stability.

Monsoon regions are all subject to the switching of the wind direction as the ITCZ moves north and south. During the Asian winter, the winds are generally north-easterly (although they may be north-westerly over western India). At high levels in the troposphere (above 11 000 m) there is a distinct westerly jet stream (see Chapter 4) located to the south of Tibet. This jet stream is the southern and stronger branch of the subtropical westerly jet stream, the northern branch being located to the north of Tibet. In spring, this southerly branch of the jet stream weakens but remains south of Tibet while the northern branch strengthens and

becomes extended. At the same time, the north of India is warming, with temperatures reaching a maximum in May. This warming creates a 'heat low' beginning a process that encourages the inflow of warm moist air from the south (Robinson and Henderson-Sellers, 1999).

In summer the upper-level westerly jet stream to the south of Tibet breaks down and then moves north across Tibet. As it moves across the mountains and the Tibetan Plateau, the high ground blocks the flow and the lower portion of the jet is deflected and re-established to the north. The strong convection over India (enhanced by the heat low) creates an outflow of air aloft and the southerly outflow develops into an easterly jet (under the influence of the Coriolis force). This upper air flow reversal from a westerly to an easterly jet is associated with the onset of the monsoon season in India and South-East Asia (Figure 5.13). As noted earlier, the south-east trades are located to the south of the ITCZ. In some parts of the world these winds remain basically south-easterly as the ITCZ moves north during the northern hemisphere summer. Over India and Asia where the ITCZ moves much further from the equator than in

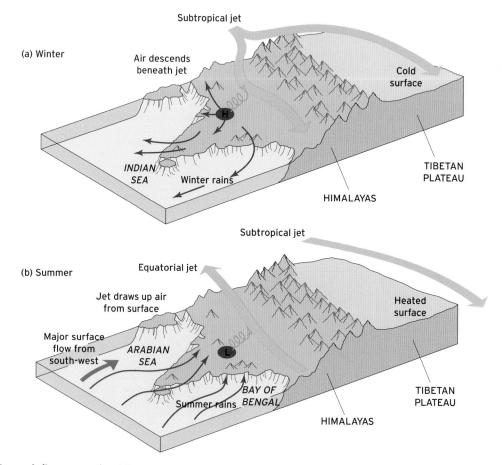

Figure 5.13 Upper air flow reversal and the onset of the Asian monsoon. (Source: after Robinson and Henderson-Sellers, 1999)

most regions, the south-east trades move far enough north to become affected by the Coriolis force. As a result, they are deflected to the right and become south-westerly. The warm moist winds of the monsoon are therefore south-westerly. The return of the upper westerly jet takes place in October but the cessation of the monsoon rains is less distinct than their start. October and November also have the most frequent occurrences of tropical cyclones in the Bay of Bengal, and the rains from those cyclones give rise to a rainfall maximum at this time of year in south-east India.

Monsoon rains are not continuous throughout the monsoon summer period, as there are breaks between more active phases. In some places there is **orographic** enhancement of rainfall (forcing air to rise leading to further condensation, see Chapter 6) and the alignment of hills can also lead to increased low-level convergence (creating zones of ascending air). In such places the rainfall can be exceptionally high, as for example in Assam (north-east India) where a number of places have exceptionally high rainfall totals (annual totals in excess of 10 000 mm). Cherrapunji, at 1340 m above sea level, is the most famous of the wet places. These exceptionally high totals compare with more typical values of between 1500 and 2000 mm close to the Bay of Bengal coast (~300 km to the south of Cherrapunji).

Despite such examples of orographic and convergence-induced enhancement of rainfall, the dominating influence on rainfall in monsoon regions is the large-scale circulation

system creating the monsoon itself. The Asian monsoon therefore comes about as a complex interaction of the formation of the heat low, the changes in the upper air flow patterns, the movement of the ITCZ and the topographical barrier of the Himalayas and the Tibetan Plateau. It is also important to note that there is great variability from year to year. For example, El Niño years can be associated with the failure of the monsoon rains, resulting in food shortages (Kumar *et al.*, 2006).

There is a less intense monsoon circulation over West Africa but the lack of a major mountain barrier allows a more steady movement of the ITCZ. However, as with Asia, the northward movement is sufficient to allow the Coriolis force to deflect the winds round to the south-west, bringing warm moist air in from the Atlantic Ocean. Northern Australia also has some monsoon rains but there is no topographical barrier to the south, and the land mass is not as large as in Asia and so the 'heat low' is not as intense.

5.3.2 The Sahel and desert margins

The influence of the poleward movement of the ITCZ declines further away from the equator. The effects of this are seen most clearly in the Sahel region of Africa that lies on the southern side of the Sahara Desert (Figure 5.14). Most of the year the region is under the influence of the north-east trade winds blowing out of the subtropical anticyclone to the north. Temperatures are high during the day (possibly higher than 40°C

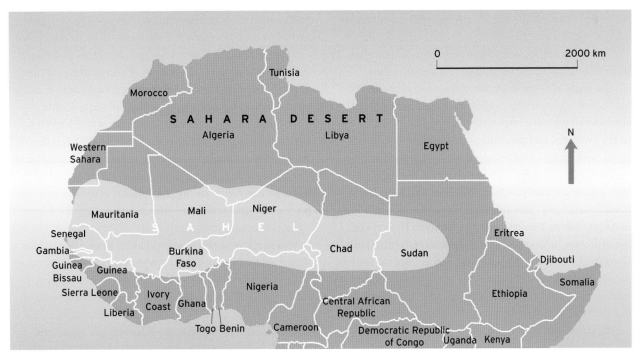

Figure 5.14 The Sahel and Sahara regions of Africa.

in early summer) and there is a substantial diurnal range of temperature (10–15°C). The Sahel region still has a rainy season (when the ITCZ is at its furthest north), but the amounts are generally modest (300–600 mm) and in some years the rains fail to come (Nicholson *et al.*, 1996). During normal years, the rainfall is usually sufficient to sustain the vegetation and any grazing by animals. In dry years, however, there is an encroachment of the desert from the north (although that encroachment may be partly due to over-exploitation by people – see Chapter 14). There are several regions of the world with a similar climate to that of the Sahel, but they are not as extensive and the problems of drought are less severe. For example, in northern Argentina and south-east New South Wales in Australia the water supply is better because rivers flow into them from more humid areas.

5.3.3 Subtropical deserts

To understand the geographical pattern of these deserts, it is necessary to examine the positioning and intensity of the subtropical anticyclones that form the descending part of the Hadley cells north and south of the equator. These anticyclones are large features extending across the North and South Atlantic and Pacific Oceans centred at about 30°N and S. The intensity of the temperature inversions is greater on the eastern side of these anticyclones and the inversions are at lower altitudes: 300–500 m, as opposed to 1500–2000 m on the western extension of the anticyclones. In part this is due to stronger subsidence in the east but it is also due to the circulation of the ocean currents, with cool currents being present under the eastern end of these anticyclones. The lower sea surface temperatures in these currents help to increase atmospheric stability and reduce convection.

The driest hot deserts are therefore found in the western coastal regions of the continents where the subtropical anticyclones are most intense (see Figures 5.4 and 14.7). In the southern hemisphere they are found in Namibia in south-west Africa (Namib Desert), in western Australia (Great Sandy and Gibson Deserts) and in northern Chile (the Atacama Desert). In the northern hemisphere they are found in southern California (Sonoran and Mojave Deserts), in Africa (Sahara Desert), in Arabia (Arabian Desert) and in north-west India and southern Pakistan (Great Indian Desert). The extent of the hot desert region from the western Sahara, through to the Arabian Desert east of the Red Sea and then again in southern Iran, Pakistan and north-west India, is not mirrored elsewhere in the world, except in Australia.

This very large extent of the hot desert region in Africa, the Middle East and southern Pakistan arises as the large land mass allows a much greater eastward elongation of the

subtropical anticyclone from the east Atlantic. This helps damp down convection over these regions. In addition, unlike the western side of the Pacific and Atlantic where warm ocean currents bring warm moist air to the east of the continents, warm currents do not move polewards in the Indian Ocean but move along the equator and then cross the equator off the east coast of Africa. The Australian deserts have greater annual precipitation totals than some of the other deserts at similar latitudes; the driest regions have annual totals close to 90 mm, while stations in the Sahara may have less than 15 mm. Part of the reason for greater precipitation is that the Australian anticyclone is not a constant feature but an average of individual anticyclones moving eastwards across the continent, allowing occasional inflow of moister air from the oceans to the north and south of the continent.

In North and South America there is only a relatively small area of desert due to major mountain barriers to the east: the southern Rockies and Mexican mountain ranges in the north and the Andes in the south. However, as the deserts in North and South America are situated to the lee of the mountains, the descent of the trade winds as they cross the mountains further dries the air, thereby intensifying the aridity. It is also important to note that variations in weather patterns over time can lead to extension or contraction of desert regions (Tucker *et al.*, 1991).

The main features of the weather in most desert regions are the wind and the high daytime temperatures (typically over 45°C in the summer in Libya). The wind increases aridity and causes considerable aeolian erosion (by sand and other particles carried in the wind) in some places (see Chapters 9 and 14). The dry air and clear skies of the anticyclones give large diurnal ranges of temperature (as much as 20°C in some places). During winter, night temperatures can even drop below freezing in parts of these deserts.

In South America the northern parts of the desert are narrow and rainfall increases rapidly inland. This is due to the sea breeze circulation that can trigger thunderstorms if it reaches the edge of the Andes where forced ascent of the moist air can penetrate the temperature inversion and trigger the potential instability aloft. South of 10°S the desert widens and even the western Andes are dry. The aridity is most noticeable in the Atacama Desert in northern Chile. While it does rain on a few days per year in most deserts (even in the Sahara), rainfall is very rare in the Atacama Desert which has the lowest annual precipitation totals of any place in the world (with the possible exception of the central Antarctic). Close to the coast, typical maximum temperatures in summer may be 25°C with diurnal ranges of 5–10°C and although temperatures do increase somewhat inland they are never extreme owing to the moderating influence of the ocean. The desert areas in southern California are similar, but inland from the Californian coast, in the lee of both the coastal ranges and the Sierra Nevada, temperatures in Death Valley do reach the very high values found in Libya. Examples of temperature and precipitation data from desert climates are given in Figure 5.15 which provides monthly mean temperature and precipitation totals for Khartoum (Sudan) and Baghdad (Iraq).

5.3.4 Humid subtropics

On the western side of the subtropical anticyclones there is a deep moist layer and convective activity is stronger than on the eastern side giving a higher likelihood of the development of rain clouds. A particular feature of this type of climate is the hot (often over 32°C), very humid summers associated with the tropical maritime air. These uncomfortable conditions can occasionally be interrupted if cooler air moves in from higher latitudes as part of the return flow of the Hadley or Ferrel circulation cells (see Chapter 4). However, if a ridge in

the upper westerlies becomes established, very hot and humid conditions can last for weeks. The winters are generally mild, although there can be outbreaks of cold polar air into these regions. These unusually cold periods can cause major damage to sensitive crops such as citrus fruits and coffee beans (e.g. in Florida or south-east Brazil). Further into the continental interiors winters become more severe and at the same latitude winters in China are colder than winters in the United States. Annual precipitation totals are typically between 1100 and 1700 mm in this climate regime. Examples of areas with this type of climate include south-east Australia (Figure 5.16a and b), Taiwan and south-east Brazil, Paraguay, Uruguay and north-east Argentina in South America and the south-east states in the United States (e.g. Florida, Georgia, Alabama, Louisiana; Figure 5.16c and d), North and South Carolina. There is also a small zone of this type of climate in the east of southern Africa. The climate of eastern China is similar except that in central and southern China there is a winter minimum of precipitation (increasingly evident towards the north) rather than the more evenly spread rainfall in, for example, Georgia and Alabama. This winter minimum is a result of

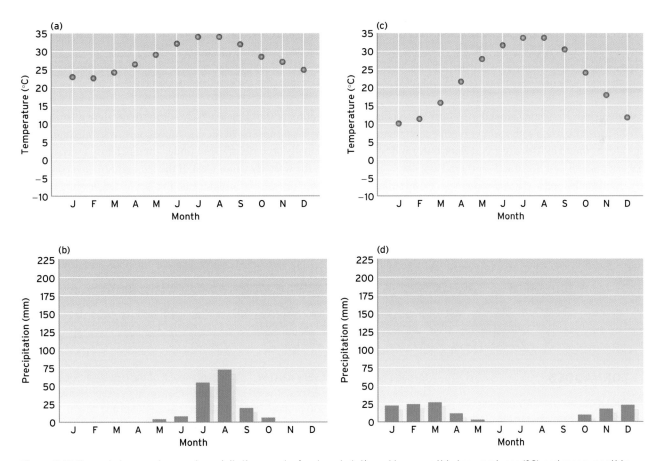

Figure 5.15 Example temperature and precipitation graphs for desert stations. Mean monthly temperatures (°C) and mean monthly rainfall (mm) for (a) and (b) Khartoum, Sudan (15.5°N), and (c) and (d) Baghdad, Iraq (33.5°N).

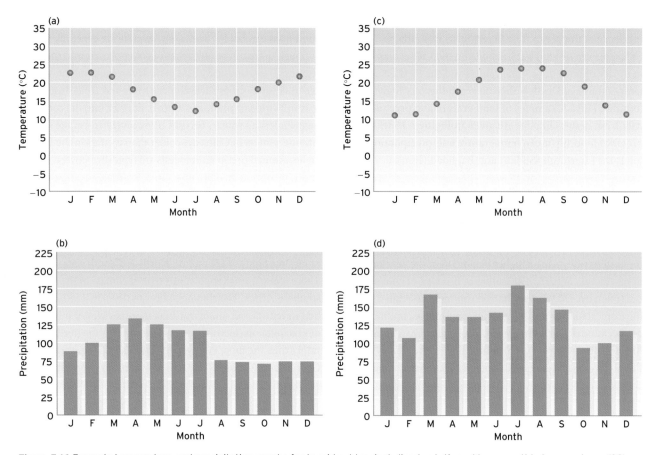

Figure 5.16 Example temperature and precipitation graphs for humid subtropical climate stations. Mean monthly temperatures (°C) and mean monthly rainfall (mm) for (a) and (b) Sydney, Australia (34°S), and (c) and (d) New Orleans, USA (30°N).

cool dry winds circulating around the winter Siberia high-pressure region. Areas with a winter precipitation minimum can also be found in Africa (e.g. east Zimbabwe) and South America (south-east Brazil and northern Paraguay).

Humid subtropical climates do not generally suffer from severe winds but the more coastal areas (in both the United States and China) can be hit by tropical cyclones (hurricanes) as they turn northwards and westwards. In Texas, Oklahoma and Kansas, tornadoes are another localized

feature of the climate which is important. Here warm moist air from the Gulf of Mexico moves northwards inland and initially becomes trapped under a temperature inversion in the westerly winds aloft. If the temperature inversion is penetrated, substantial instability is released, leading to the growth of very substantial storm clouds, some of which will have associated tornadoes (Box 5.4). Although rare at any one place, these tornadoes are an important feature of the climate of this part of the United States.

HAZARDS

TORNADOES

Tornadoes are violently rotating columns of air extending to the ground (Figure 5.17). They are capable of causing great damage with wind speeds of over 300 km h^{-1} (Figure 5.18). A

tornado can be either very narrow and only a few metres across or very large with some over 500 m wide. They can often travel long distances, with some causing damage over 75 km. Tornadoes are often thought of as a

phenomenon of the mid-west United States. However, they occur all over the world (e.g. Holden and Wright, 2004) and even in the United Kingdom the Tornado and Storm Research Organisation (TORRO) reports an

BOX 5.4 ➤

➤

Figure 5.17 A tornado in Texas rampaging across fields. Tornadoes form when two air masses of different temperatures and humidity meet. If the lower layers of the atmosphere are unstable, a strong upward movement of warmer air is formed. This starts to spiral as it rises, and intensifies. Only a small percentage of these systems develop into the narrow, violent funnels of tornadoes. Wind speeds can reach up to 400 km h^{-1} and they can damage an area 1 mile (1.6 km) wide and 50 miles (80 km) long. Tornadoes come in many shapes and sizes. (Source: Shutterstock: Iafoto)

Figure 5.18 Damage to homes on the end of a cul-de-sac demolished by a tornado in Grandview, Missouri in 2003. (Source: Ron Kuntz/Corbis)

average of 33 tornadoes per year (mainly in southern Britain, rarely in Northern Ireland or Scotland). Nevertheless, the largest ones tend to be concentrated in the Plains of the United States where atmospheric conditions often occur that suit their formation. On 25–28 April 2011 a series of tornadoes, described by the US National Weather Service (NOAA) as the most severe in recorded history, tore across the southeastern United States leading to a state of emergency in seven states. More than 300 tornadoes were reported killing more than 320 people and leaving a million people without power.

Tornadoes form when temperature and wind flow patterns in the atmosphere can cause enough moisture, instability, lift and wind shear for tornadoes to form in association with thunderstorms (Figure 5.19). The most destructive and deadly tornadoes occur from supercells which are rotating thunderstorms with a well-defined radar circulation called a mesocyclone. As well as tornadoes, supercells can also produce damaging hail and strong winds. All thunderstorms tend to produce lightning and heavy precipitation and in supercells the lightning is often more frequent and the precipitation can lead to flash floods. The rotating in the storms is due to wind shear which is when the wind direction changes and the wind speed increases with height. This kind of wind shear and instability usually exists only ahead of a cold front and depression system. The rotation of a tornado partly stems from updrafts and downdrafts caused by the unstable air interacting with the wind shear. Cyclonically flowing air which is already slowly spinning to the left (in the northern hemisphere) converges towards the centre of the thunderstorm, causing it to spin faster due to the conservation of angular momentum. This is a similar process to one you can try for yourself. If you sit on an office chair and spin round with your arms outstretched you spin slowly. If you then pull your arms towards your chest you will suddenly start spinning faster. It is this conservation of angular momentum that creates the very high wind speeds within tornadoes.

BOX 5.4 ➤

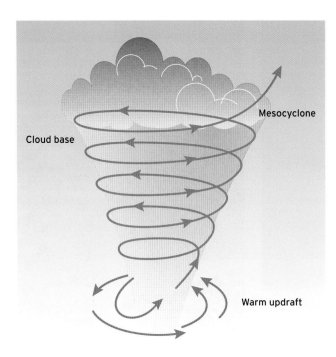

Figure 5.19 Formation of a tornado. Warm rising air meets cooler air aloft creating turbulence. These winds start to rotate because of wind shear which is when the wind direction changes and the wind speed increases with height. The larger mesocyclone which then develops aloft can generate sufficient strength to extend a funnel cloud down to the ground.

Most tornadoes rotate *cyclonically* (counterclockwise in the northern hemisphere and clockwise in the southern hemisphere) but the Coriolis force does not play any real part in the rotation as the size of tornadoes is too small for the Coriolis force to have any real effect. In fact there have been observations of anticyclonic rotating tornadoes, usually in the form of waterspouts (which are essentially tornadoes over water and also mainly have a cyclonic rotation), non-supercell land tornadoes, or anticyclonic whirls around the rim of a supercell's mesocyclone.

Tropical cyclones (hurricanes and typhoons) can also spawn tornadoes and while not all tropical cyclones that move across land will have tornadoes, some have major tornado outbreaks. The size of the tornado outbreak does not appear to be associated with the intensity of the tropical cyclone.

Tornado intensity is measured by the Fujita scale (F-scale named after Theodore Fujita), which was replaced in February 2007 by the enhanced Fujita scale (EF-scale) (Table 5.5).

Table 5.5 Enhanced F-scale for tornado damage (1 mph = 1.6 km h^{-1})

	Fujita scale		Derived EF-scale		Operational EF-scale*	
F number	Fastest 1/4-mile (mph)	3 second gust (mph)	EF number	3 second gust (mph)	EF number	3 second gust (mph)
0	40-72	45-78	0	65-85	0	65-85
1	73-112	79-117	1	86-109	1	86-110
2	113-157	118-161	2	110-137	2	111-135
3	158-207	162-209	3	138-167	3	136-165
4	208-260	210-261	4	168-199	4	166-200
5	261-318	262-317	5	200-234	5	Over 200

*Note that the EF-scale still is a set of wind estimates (not measurements) based on damage. It uses 3 s gusts estimated at the point of damage based on a judgement of eight levels of damage related to 28 indicators.

BOX 5.4

Reflective questions

➤ What is the main feature influencing the climate in equatorial regions and how does it affect temperature and rainfall?

➤ What are the necessary conditions for the formation of tropical cyclones?

➤ Why are there not as many tropical cyclones in the southern hemisphere as the northern hemisphere?

➤ What factors make the Asian monsoon a much more marked feature than in other parts of the world at similar latitudes?

➤ Why are the deserts located where they are and what factors help increase aridity in many of these deserts?

➤ What is the dominant feature of summer weather in humid subtropical regions?

5.4 Mid- and high-latitude climates

In the tropics and subtropics the weather is often relatively predictable but the middle latitudes are dominated by weather systems that move across the planet. This makes the weather both much more variable and also much less predictable, particularly in areas close to the oceans. The middle latitudes of the southern hemisphere are dominated by oceans. The only land in the southern hemisphere middle latitudes is the southern tip of South Africa, the most southern parts of Australia and New Zealand, and central and southern Argentina and

Chile. In the northern hemisphere, however, as well as the North Atlantic and Pacific Oceans, there are the major land masses of the North American, European and central Asian continents. These span all of the middle latitudes and extend into the polar regions, and (in the case of North America and Asia) into the subtropics.

5.4.1 Depressions, fronts and anticyclones

The equatorial and subtropical climates are dominated by the weather systems associated with the thermally direct Hadley cell circulation (convection along the ITCZ and subsidence in the subtropical anticyclones) that transfers energy from lower latitudes. In the middle latitudes, however, there is also an energy transfer from lower latitudes to higher latitudes but this is not achieved by direct convection via heating at the surface. Instead the transfer is accomplished through the movements of large weather systems. However, just as in the tropics and subtropics, the key to understanding the development, movement and dissipation of mid-latitude weather systems is not what happens at the surface but what happens in the middle and upper troposphere.

The key factor is the positioning and movement of the westerly polar front jet stream. The jet stream is a 'thermal wind' related to sharp thermal gradients in the atmosphere. In the case of the polar front jet stream, the thermal gradient is created by the temperature difference between polar and tropical air where the two **air masses** meet (the polar front). A range of types of air masses exist and it is the interaction and modification of these air masses that may determine the weather conditions experienced in the middle and high latitudes. Further explanation and examples of air masses are discussed in Box 5.5. The polar front is just a steeper part of the normal low- to high-latitude temperature

FUNDAMENTAL PRINCIPLES

AIR MASSES

The term 'air mass' is given to a body of air that has a very large horizontal extent and in which its potential temperature and moisture content are similar through most of the troposphere (close to the surface there may be differences). An air mass develops over a source region where

it has remained for a period of days (Barry and Chorley, 2003). There are four basic types of air mass according to their source regions. These are tropical maritime, tropical continental, polar maritime and polar continental.

Essentially the tropical air masses are from low latitudes or the subtropics and polar air masses are from high latitudes. There are also extreme ver-

sions of polar air masses called Arctic maritime and Antarctic continental. Continental air masses are relatively dry and maritime air masses are relatively humid. The term 'relatively' has to be used as warm air holds much more moisture than cold air. A polar maritime air mass with 90% relative humidity holds 3.9 g of water vapour per kg of air at 0°C while a tropical

BOX 5.5 ➤

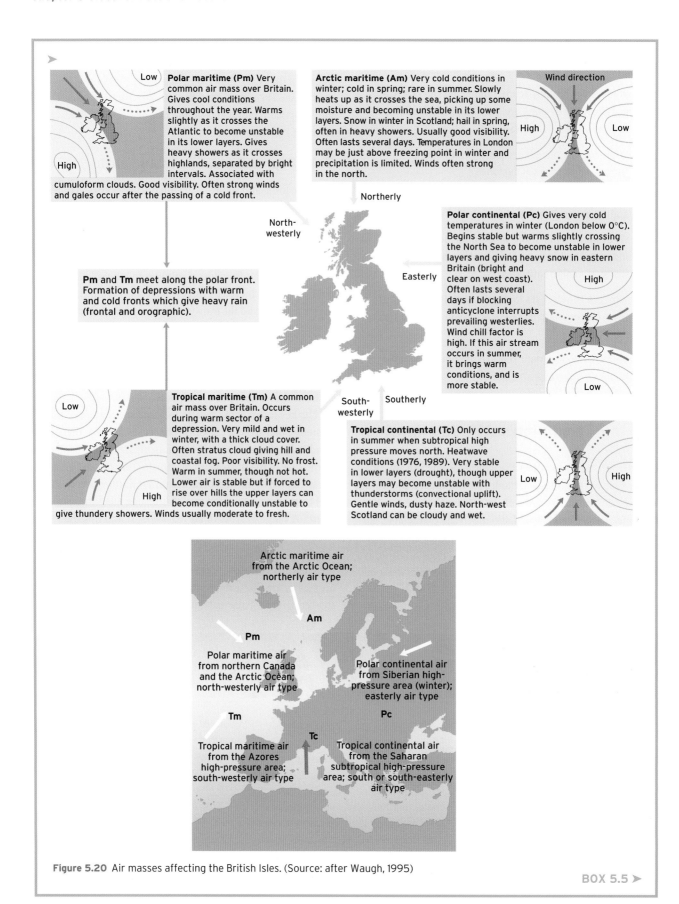

Polar maritime (Pm) Very common air mass over Britain. Gives cool conditions throughout the year. Warms slightly as it crosses the Atlantic to become unstable in its lower layers. Gives heavy showers as it crosses highlands, separated by bright intervals. Associated with cumuloform clouds. Good visibility. Often strong winds and gales occur after the passing of a cold front.

Arctic maritime (Am) Very cold conditions in winter; cold in spring; rare in summer. Slowly heats up as it crosses the sea, picking up some moisture and becoming unstable in its lower layers. Snow in winter in Scotland; hail in spring, often in heavy showers. Usually good visibility. Often lasts several days. Temperatures in London may be just above freezing point in winter and precipitation is limited. Winds often strong in the north.

Wind direction

Northerly

North-westerly

Polar continental (Pc) Gives very cold temperatures in winter (London below 0°C). Begins stable but warms slightly crossing the North Sea to become unstable in lower layers and giving heavy snow in eastern Britain (bright and clear on west coast). Often lasts several days if blocking anticyclone interrupts prevailing westerlies. Wind chill factor is high. If this air stream occurs in summer, it brings warm conditions, and is more stable.

Easterly

Pm and **Tm** meet along the polar front. Formation of depressions with warm and cold fronts which give heavy rain (frontal and orographic).

Tropical maritime (Tm) A common air mass over Britain. Occurs during warm sector of a depression. Very mild and wet in winter, with a thick cloud cover. Often stratus cloud giving hill and coastal fog. Poor visibility. No frost. Warm in summer, though not hot. Lower air is stable but if forced to rise over hills the upper layers can become conditionally unstable to give thundery showers. Winds usually moderate to fresh.

South-westerly

Southerly

Tropical continental (Tc) Only occurs in summer when subtropical high pressure moves north. Heatwave conditions (1976, 1989). Very stable in lower layers (drought), though upper layers may become unstable with thunderstorms (convectional uplift). Gentle winds, dusty haze. North-west Scotland can be cloudy and wet.

Arctic maritime air from the Arctic Ocean; northerly air type

Am

Pm

Polar maritime air from northern Canada and the Arctic Ocean; north-westerly air type

Polar continental air from Siberian high-pressure area (winter); easterly air type

Tm

Tc

Pc

Tropical maritime air from the Azores high-pressure area; south-westerly air type

Tropical continental air from the Saharan subtropical high-pressure area; south or south-easterly air type

Figure 5.20 Air masses affecting the British Isles. (Source: after Waugh, 1995)

BOX 5.5 ➤

continental air mass with a 30% relative humidity has 8 g of water vapour per kg of air at 30°C.

Tropical maritime air is common in both hemispheres, but tropical continental air is less common as the only really large land mass in the subtropics is northern Africa and to a lesser extent in Australia. India is a source region for tropical continental air in winter but the intense winter high pressure over Siberia and the mountains to the north act as a barrier. In summer, central Asia can be a source for tropical continental air although strictly speaking it is in the middle latitudes. Polar maritime air is common in both hemispheres with source regions in the high-latitude oceans.

As with all air masses, there is not a single set of characteristics defining this air mass as the values change according to the actual source region and the time of year. Polar continental air is found only in the northern hemisphere as there is no continent at higher latitudes in the southern hemisphere other than the Antarctic where the air is classed as being Antarctic continental.

It is important to note that all air masses are modified by the underlying surface. If the surface is colder than the air mass then low-level stability will be increased. If it is warmer than the air mass, low-level stability will be decreased. The best example of this is Arctic maritime air moving

south across the North Atlantic. The sea surface is much warmer than the air mass and this warms the lowest layers, decreasing stability and encouraging convection. This leads to the formation of frequent instability showers, a characteristic of this type of air mass. By studying source areas for air and tracking their movement over a particular region and predicting how the air masses might be modified by local conditions and how they might interact with other air masses, it is possible to provide short-range weather forecasts. Figure 5.20 provides an example of how air masses can affect the British weather.

BOX 5.5

gradient (it is where there is a sharp transition between warm and cold air). The polar front jet stream is not fixed both in terms of its location as it moves north and south and there are marked waves along its length which vary in amplitude with shallower waves giving what is called a zonal flow (west to east) and larger amplitude waves giving a meridional flow (a more marked north to south component as shown in Figure 5.21).

Air accelerates into anticyclonically curved waves creating a zone of upper-level divergence while the flow slows down as it enters a cyclonically curved wave creating upper-level convergence. If the upper-level divergence is greater than any low-level convergence this leads to a fall in surface pressure and creates a zone of ascending air. Such a situation is one of **cyclogenesis** (development of a depression). Conversely, if there is strong upper-level convergence and weaker lower-level divergence, there is a zone of subsiding air and this situation is one of **anticyclogenesis**. The thermal gradient in the upper westerlies is therefore of more significance in the development of weather systems than temperatures close to the surface.

Waves along the upper westerly polar front jet stream (Figure 5.22) can develop into frontal depressions. Air generally rises at fronts (by slantwise convection, as warm air is forced to rise above the cooler air it meets), leading to the formation of clouds (and precipitation) as the air cools on ascent. Fronts therefore mark areas of general precipitation

(which may fall as snow in winter), although there are bands of more intense precipitation embedded within those areas. Some less active fronts may produce little or no precipitation.

Jet streams are not simple continuous features. They have marked entrances where the flow becomes more concentrated into a stronger jet. They also have exits where the jet spreads out and the flow rate reduces. Associated with jet entrances and exits are marked zones where the formation of depressions and anticyclones (or anticyclonic ridges) is favoured through divergence or convergence of the upper winds (leading to cyclogenesis and anticyclogenesis respectively) (Figure 5.23). It is important to remember that if there is an upper-level outflow (divergence) the pressure will fall at the surface and air will therefore converge at lower levels and rise up to replace the ouflow. As a result there is a general (slow) upward motion of air in depressions in the conveyor belts moving through the system. In anticyclones or ridges the air flows inwards in the upper atmosphere, leading to an increase in pressure with descent and outflow at lower levels.

The descent of air in anticyclones increases the temperature (adiabatically) and dries out the air, which means anticyclones bring clear conditions. However, the descending air does not fall all the way to the land surface and due to the adiabatic warming a temperature inversion forms. While the air above may be clear, on some occasions the inversion can trap moister air below and a layer of cloud can form below the inversion, giving a condition termed

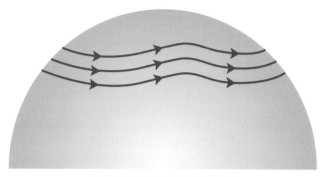

Zonal flow (generally west to east)

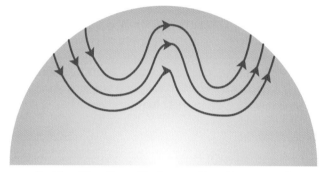

Meridional flow (flow with a large north-south component)

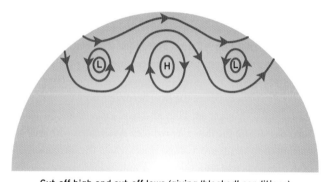

Cut-off high and cut-off lows (giving 'blocked' conditions)

Figure 5.21 Zonal and meridional flow.

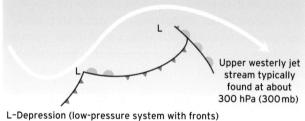

L-Depression (low-pressure system with fronts)

Figure 5.22 Jet stream development. Waves along the jet stream can develop into frontal depressions.

'anticyclonic gloom'. This is more common in winter than summer. This is because during summer solar radiation warms the air below the inversion sufficiently to cause the cloud to dissipate. On occasions the upper westerly flow

Confluent thermal trough		Diffluent thermal trough
CON		DIV
Entrance	Jet core	Exit
DIV		CON
Confluent thermal ridge		Diffluent thermal ridge

CON – Convergence (leads to anticyclogenesis and descent of air)
DIV – Divergence (leads to cyclogenesis and ascent of air from below)

Figure 5.23 Jet stream entrance and exit, cyclogenesis and anticyclogenesis.

can be interrupted by what is termed a blocking anticyclone generally formed by an intensification of a ridge in the upper westerlies into a closed circulation. This can lead to quite long periods of anticyclonic weather in areas where normally the weather would be characterized by the passage of mid-latitude depressions.

The ascent of air in depressions is concentrated to some extent along the warm and cold fronts. Once aloft, the air rotates to become more parallel with the upper-level flow (Figure 5.24). The ascent of the air is at perhaps 20 cm s^{-1} compared with the 5–20 m s^{-1} that is typically found in large convective clouds. An ascent of air will eventually produce cloud as water vapour condenses out of the atmosphere in cooler conditions. The ascent of warm moist tropical air will quickly lead to the formation of layers of cloud (stratiform cloud) and continued ascent of air is likely to lead to precipitation. The ascent of air in depressions takes place over a much wider area and over a much greater time than is the case for a convective cloud, and therefore although precipitation rates (e.g. how heavy it is raining) may be less than those found with convective clouds, there can be substantial amounts of precipitation arising from a frontal depression. It is important to note that even within a frontal precipitation zone there will be areas of more intense precipitation so that fronts do not produce simple areas of steady precipitation.

Fronts tend to slope gently at a rate of 1 m vertical rise for every 80–150 m of lateral distance (slope of 1 : 80 to 1 : 150) with cold fronts being steeper than warm fronts. Over time the cold front tends to overtake the warm front, leading to what is termed an **occluded front**, which is classified as warm or cold depending on whether the air ahead of the warm front is colder or warmer than the air following the cold front (Figure 5.25). The complex three-dimensional nature of depressions and fronts is important and is one reason why weather forecasting is a very complicated science.

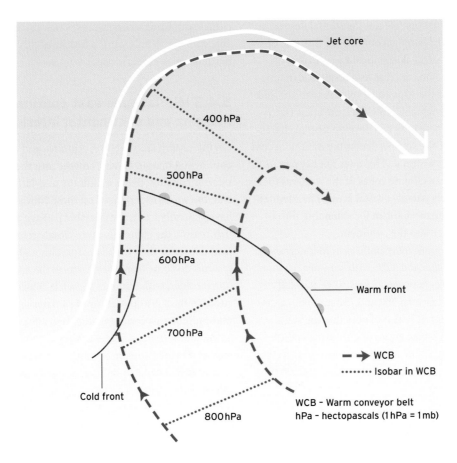

Figure 5.24 Conveyor belts in depressions. Slantwise convection (i.e. with a strong horizontal motion as well as conductive ascent) in the warm conveyor belt carries sensible and latent heat polewards.

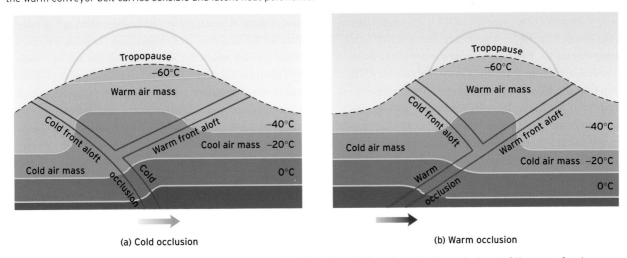

(a) Cold occlusion

(b) Warm occlusion

Figure 5.25 Occluded fronts: (a) cold occlusions are where air behind the cold front is colder than air ahead of the warm front; (b) warm occlusions are where air behind the cold front is less cold than air ahead of the warm front.

5.4.2 Mid-latitude western continental margins

The Mediterranean-type climate is the first subtype of climate found polewards of the subtropical desert regions and is characterized by a mild wet winter half-year and a hot dry summer half-year. It is found in the far south-west of South Africa, in central Chile, on south-west-facing coastlines in the south of Australia, in California, as well as in the Mediterranean itself. In winter, mid-latitude frontal depressions bring rain to these areas, although in the Mediterranean itself, most of the depressions are not the frontal depressions of the Atlantic, the latter accounting for only 9% of Mediterranean depressions. A significant proportion of Mediterranean depressions develop as a result of dynamic effects on air flow over the Alps and Pyrenees that can lead to the formation of 'orographic' low-pressure areas (Barry and Chorley, 2003). These lows can develop frontal characteristics, particularly if the air flow across the mountains has a cold front embedded within it. This does not happen in the other Mediterranean climate zones of the world as they comprise only relatively narrow coastal areas. The Mediterranean Sea provides the mechanism for extending the climate type much further into the continent.

Typically average winter temperatures in Mediterranean climates will be between 5 and 12°C with summer daytime maximum temperatures between 25 and 30°C. Rainfall totals will typically be between 400 and 750 mm with a distinct summer minimum. It is also of note that the summers in the Mediterranean climate zones of California and central Chile are drier than in the Mediterranean owing to the upwelling of cold water off those coasts. This stabilizes the air and inhibits convection.

Further polewards, mid-latitude western continental margins are most extensive in the northern hemisphere, although a similar climate regime is experienced by southern Chile, Tasmania and New Zealand. These climates have unusually mild winters for their latitude (e.g. the British Isles, western Europe and the west of Norway). These mild winters are particularly marked in the north-east Atlantic as the North Atlantic Drift pushes relatively warm water a long way north. There is a similar climate in much of New Zealand where again there is a warm current off the western coast. Generally temperatures in the southern hemisphere are lower than those at similar latitudes in the northern hemisphere owing to the large extent of the southern oceans and the paths of their ocean currents. As well as unusually mild winters for their latitude, these climates have a remarkably small range of annual temperature, have precipitation distributed throughout the year and there is considerable orographic enhancement of precipitation in the coastal mountain ranges (see Chapter 6). The mountain ranges of North and South America keep this climate confined to a relatively narrow coastal strip while in Europe the relatively low-lying ground from the Netherlands to Russia allows this climate type to extend to Poland in the east (Robinson and Henderson-Sellers, 1999). Average winter temperatures are typically between 2 and 8°C with average summer maximum temperatures between 15 and 25°C. Precipitation totals are generally in the range 500–1200 mm. The mid-latitude depressions that are a feature of this

climate can bring strong winds which can cause considerable damage. Windiness is a feature of this type of climate, particularly in coastal areas.

5.4.3 Mid-latitude east continental margins and continental interiors

On the eastern side of North America and Asia, the eastern continental margin climates merge into the continental interior climates. Being within the mid-latitude westerly belt, the winds experienced on these continental margins have generally had a considerable passage across land. For that reason, the climate is more closely related to the continental interior than to the oceans, although some weather systems do come from the oceans to the east. Winters are much colder, with frequent snowfall, than those experienced in the subtropical humid climates to the south. These mid-latitude east continental margin climates do not exist in the southern hemisphere as they require large land masses. Furthermore, in South America east of the Andes, where such a climate might exist, the Andes act as a block to the westerlies and descent in the lee of the mountains dries the air and creates a climate that is more like a mid-latitude continental interior (semi-arid) climate.

The main extent of the humid continental type of climate is in North America, China and eastern Russia. In China and North America this climate merges from the humid subtropical into a humid continental maritime margin, with increasingly severe winters, although at lower latitudes the summers are hot and long. As noted earlier, in China the cool winds circulating around the winter Siberian high pressure mean that there is less precipitation in winter than at other times of the year. The three winter months (December, January, February) have a total of close to 13 mm precipitation in Beijing while in New York the total for the same period is over 230 mm.

Summers are similar in both Asia (eastern China, Korea and central Japan) and the eastern United States (south of New England), being hot and humid. Average temperatures in July for Beijing, China (40°N) are 26°C and those in New York (41°N) 24.5°C. Mean winter temperatures are and 0°C respectively. In summer the June, July, August precipitation total is over 460 mm in Beijing while it is less than 320 mm in New York. Overall Beijing is drier (annual total close to 620 mm) than New York (over 1110 mm) owing to the aridity of Asia north and east of the Tibetan Plateau. Some examples of conditions experienced at mid-latitude climate stations are given in Figure 5.26.

Winters become increasingly colder as you move north or west into the mid-latitude continental interiors and summers

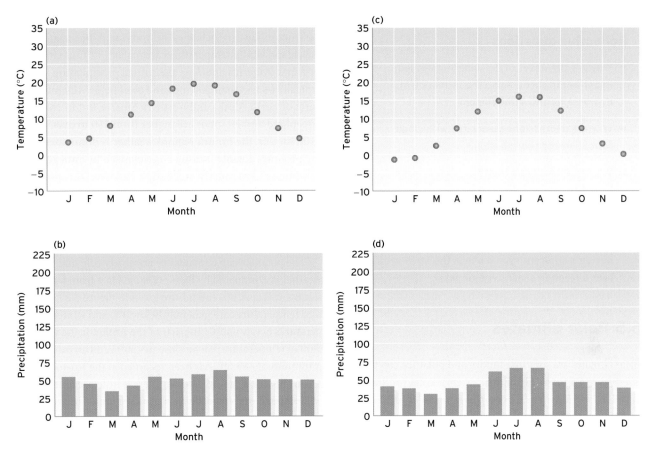

Figure 5.26 Example temperature and precipitation graphs for mid-latitude climate stations. Mean monthly temperatures (°C) and mean monthly rainfall (mm) for (a) and (b) Paris, France (49°N), and (c) and (d) Berlin, Germany (52.5°N).

also become milder and less humid (north of 45°N). Further north still (north of 50°N) the climate becomes subpolar with severe winters and relatively short summers (only three, or fewer, months with average temperatures above 10°C). For much of the mid-latitude continental interiors, precipitation is distributed throughout the year but generally with a distinct summer maximum. This summer precipitation is mainly in the form of convective showers although there are some weak frontal systems. Winter precipitation tends to fall as snow and as temperatures are low it often does not melt until the spring thaw. Total precipitation amounts are relatively low (below 500 mm), but the cold winter period and summer maximum of precipitation ensure that in most years there is sufficient moisture for plant growth. The main wheat-growing areas of North America have this type of climate. In Asia, however, the southern part of these mid-latitude continental interiors is semi-arid (as are those states in the United States just east of the Rockies). East of the Caspian Sea the climate becomes truly arid. These are cold desert regions in which winters are cold, although summers may still be warm. The Gobi Desert is an example of this type of desert.

The further north you go the shorter the summer and the growing season. Winters are cold with average temperatures below −12°C in the coldest month and below −25°C in northern regions. In the coldest regions of Siberia the average temperatures in the coldest month can even be as low as −50°C. There is a very large range in temperatures with the warmest months having average temperatures of over 21°C, and even in the coldest parts of Siberia, July average temperatures reach over 13°C (an annual range of over 60°C). In North America there can also be some extreme diurnal ranges in temperature, particularly in areas prone to Chinook winds (see Chapter 6) or if warm moist air from the south pushes much further north than usual. Diurnal changes in temperature have even exceeded 50°C. These regions are influenced by mid-latitude weather systems. However, in winter, high pressure dominates, especially in Siberia where pressures can reach over 1080 mb. Precipitation totals tend to fall as you move north within the mid- and high-latitude continents of the northern hemisphere, with totals below 400 mm in the northern United States and southern Canada and falling below 300 mm further north. In eastern Siberia, annual totals can even be less than 150 mm.

5.5 Polar climates

Polar climates are split into tundra and polar ice cap types (see Figure 5.4 for a map of their extent). Polar tundra is found in North America, in northern Labrador, the far north of Quebec, North West Territories (north and east of the Great Slave and Great Bear Lakes), part of central and northern Yukon and north Alaska. Polar tundra can also be found in Europe and Asia in northern Sweden, Finland and Russia and also in northern Iceland. Coastal Greenland is also classed as having a polar tundra climate as is the northern part of Graham Land in the Antarctic. The polar ice caps are found in central Greenland and all of the Antarctic (except the northern part of Graham Land).

In the polar tundra climate the temperature of the warmest month will be above 0°C but below 10°C. Winter temperatures are generally extremely low (average temperature in January below −25°C), although in coastal Greenland winter temperatures are higher (−7°C in January). In North America and Asia the annual precipitation will typically be less than 300 mm and even below 120 mm in parts of north Alaska and northern Siberia, but in coastal Greenland annual totals are higher from 750 to over 1100 mm. In the tundra regions weather is dominated by the prolonged winter season characterized by anticyclonic conditions (particularly in Siberia). However, this provides a sharp contrast to the short growing season, which although not very warm does provide an opportunity for the local flora and fauna to survive, if not actually flourish (Chapter 8).

The polar ice cap climates are extremely cold and weather is dominated by high pressure. Data from a polar station at Ivigut, Greenland, are given in Figure 5.27. Summer temperatures are generally below 0°C and winter temperatures below −40°C. In parts of the Antarctic the mean annual temperature can be close to −50°C and an extreme minimum of −89.6°C has been recorded at the Vostok research station (21 July 1983). There is little precipitation with annual totals typically less than 100 mm. These areas can actually be classified as cold deserts. In the central Antarctic there is almost no precipitation. Air with a temperature below −40°C contains almost no water vapour (even when saturated the amount of water vapour held at such low temperatures is very small), and therefore even if clouds form there is unlikely to be any precipitation. Polar ice caps cool the air in contact with them and as a result there can be strong winds blowing off the centre of the ice caps towards the coasts. This climate type is therefore dominated by the cold and by the frequent strong winds that produce extreme wind chill.

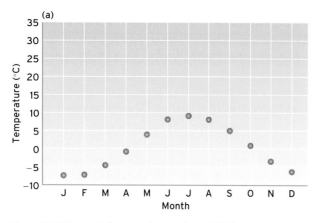
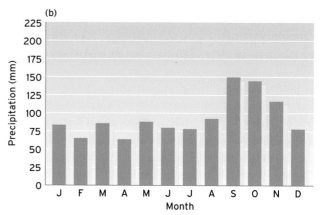

Figure 5.27 Example temperature and precipitation graphs for a polar climate station at Ivigut, Greenland (61°N).

5.6 A global overview

The previous sections have outlined the major climate regions of the world but it is worth taking a global overview to consider how the various parts of the global climate interact and how there can be changes in opposite directions at the same time in different regions of the world. Chapter 4 provided an introduction to the El Niño Southern Oscillation (ENSO) and the North Atlantic Oscillation (NAO) but there are other teleconnections and indices which try to characterize variability or differences across large regions through time. Key indices described by the IPCC (2007a) are:

- *Southern Oscillation Index (SOI)*: Mean sea-level air pressure anomaly difference for the pressure at Tahiti minus the pressure at Darwin, normalized by the long-term mean and standard deviation of the mean sea-level pressure difference. Data are available from the 1860s. Data from Darwin can be used alone, as these data are more consistent than those for Tahiti prior to 1935. An El Niño event involves warming of tropical Pacific surface waters from near the International Date Line to the west coast of South America. This limits the upwelling of cold water near South America and typically occurs every 3 to 7 years. Periods of below-average temperatures in the eastern tropical Pacific are called La Niña events. El Niño events reduce the sea surface temperature gradient across the equatorial Pacific and are linked with the atmospheric Southern Oscillation, which brings changes in trade winds, tropical circulation and precipitation. El Niño Southern Oscillation events are coupled ocean–atmosphere phenomena with global implications that have extratropical teleconnections characterized by changes in the jet streams and storm tracks in mid-latitudes (particularly in winter months) as well as mean sea-level pressure anomalies.
- *North Atlantic Oscillation (NAO) Index*: The difference in normalized mean sea-level air pressure anomalies between Lisbon in Portugal and Stykkisholmur in Iceland has become the most widely used NAO Index and extends back in time to 1864. The data can go further back to 1821 if Reykjavik is used instead of Stykkisholmur and Gibraltar instead of Lisbon. The NAO has a strong link to the alternation of westerly and blocked flow across the Atlantic and is present from the surface up into the stratosphere.
- *Northern Annular Mode (NAM) Index*: The amplitude of the pattern defined by a mathematical term known as the 'leading empirical orthogonal function' of winter monthly mean northern hemisphere mean sea-level air pressure anomalies polewards of 20°N. The NAM has also been known as the Arctic Oscillation (AO), and is closely related to the NAO.
- *Southern Annular Mode (SAM) Index*: The difference in average mean sea-level air pressure between the southern hemisphere middle and high latitudes (usually 45°S and 65°S), from gridded or station data (Gong and Wang, 1999; Marshall, 2003), or the amplitude of the leading empirical orthogonal function of monthly mean southern hemisphere 850 hPa height polewards of 20°S. This was formerly known as the Antarctic Oscillation (AAO) or High Latitude Mode (HLM). The principal mode of variability of the atmospheric circulation in the southern hemisphere extratropics is the SAM Index. It is essentially a zonally symmetric structure, but with a zonal wave pattern, and reflects changes in the main belt of subpolar westerly winds. Enhanced Southern Ocean westerlies occur in the positive phase of the SAM.
- *Pacific-North American pattern (PNA) Index*: The mean of normalized height at which the air pressure is equal to 500 hPa at 20°N, 160°W and 55°N, 115°W minus those at 45°N, 165°W and 30°N, 85°W. As with the NAO, the PNA appears related to periods of blocked flow, particularly in the Gulf of Alaska, and periods of stronger westerlies. It is associated with changes in the Aleutian Low, the Asian jet, and the Pacific storm track, and affects precipitation in western North America and the frequency of Alaskan blocking events and associated cold-air outbreaks over the western United States in winter.
- *Pacific Decadal Oscillation (PDO) Index and North Pacific Index (NPI)*: The NPI is the average mean sea-level pressure anomaly in the Aleutian Low over the Gulf of Alaska (30°N–65°N, 160°E–140°W) and is an index of the PDO, which is also defined as the pattern and time series of the first empirical orthogonal function of sea surface temperature over the North Pacific north of 20°N. The PDO broadened to cover the whole Pacific Basin is known as the Inter-decadal Pacific Oscillation (IPO). Decadal to inter-decadal variability of the atmospheric circulation is most prominent in the North Pacific, where fluctuations in the strength of the winter Aleutian low-pressure system co-vary with North Pacific sea surface temperature in the PDO. These are linked to decadal variations in atmospheric circulation, sea surface temperature and ocean circulation throughout the whole Pacific Basin in the IPO.

One other index that has been noted (but not by the IPCC) is the Atlantic Multidecadal Oscillation (AMO), which is related to North Atlantic sea surface temperatures that show a 65–75 year variation (0.4°C range), with a warm phase during 1930–1960 and cool phases during 1905–1925 and 1970–1990.

Some of the above indices and their teleconnections provide a means of making long-range weather predictions. Note that these are not forecasts which use numerical weather prediction models, but they give an indication of very generalized patterns of weather that might be expected, for example, for a winter seasonal prediction. The SOI and NAO do provide useful information when making such predictions but they also have an influence in global climate, as can be seen in Figure 5.28 where the warmest year (1998) coincided with a strong El Niño event (Figure 5.29).

Although there does seem to be some relationship between variations in the indices and global climate it is important to remember that temperature is not the only climate variable and it may be, for example, that precipitation

has been affected (see Box 5.6). It is also important to recognize that there is considerable inter-annual variability in global climate and that inter-annual variability will be even greater at a regional level and variations in precipitation are much greater than those for temperature. The extent of inter-annual variability can clearly be seen in the Central England Temperature (CET) series (Figure 5.30). Although the CET series goes back as far as 1659, it has only been plotted from 1731, as the period from 1731 is where the temperatures are to the nearest 0.1°C (before that many years had temperatures only recorded to the nearest 0.5 or 1.0°C). The CET data can be found at: http://hadobs.metoffice.com/hadcet. It is also important to note that this inter-annual variation in global as well as the CET series is less than the variation at a seasonal or monthly scale. There are two particularly cold years in the CET series, 1740 and 1879, with average temperatures of 6.84 and 7.42°C respectively. There is no obvious explanation as to why those two years were so much colder than the years immediately before and after. In Figure 5.31 the CET data have been

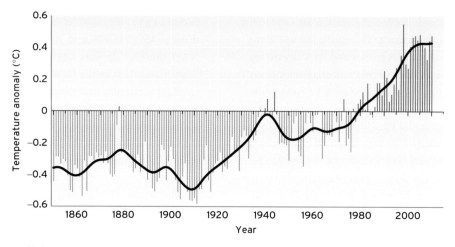

Figure 5.28 Global mean air temperature 1850-2010. (Source: Climate Research Unit, University of East Anglia)

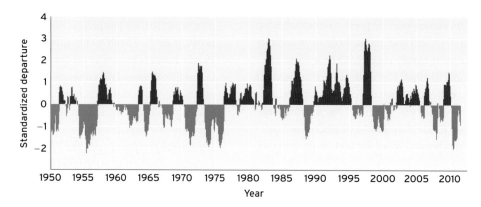

Figure 5.29 Multivariate El Niño Southern Oscillation Index 1950-2010. (Source: Klaus Walter, www.esrl.noaa.gov/psd/enso/mei/)

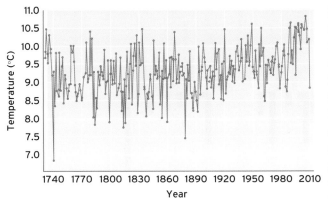

Figure 5.30 Central England temperatures 1731-2010.

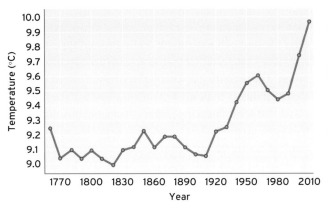

Figure 5.31 Central England 30 year average temperatures (updated every 10 years).

plotted on a 30 year average starting in 1760 (as the first point at which 30 years have passed with temperatures recoded to 0.1°C). However, it appears that, overall, the period at the end of the seventeenth century (which is not shown as the precision of the measurements was only to the nearest 0.5 or 1°C) was the coldest in the full CET series, with the coldest year being 1695 (7.25°C), a little colder than 1879 but still more than 0.4°C warmer than 1740.

This inter-annual variability is why it is important not to assume that a few years when the average temperature increases or decreases is evidence of a warming or cooling trend. There is the danger of 'cherry-picking' data and the standard practice is to produce a 'climate normal', which is an average calculated over a 30 year period that is updated every 10 years. Hence, in 2000 the average temperature was that for the period 1961–1990 but from 2001 until 2010 the average used was over the period 1971–2000. From 2011 the average is over the 1981–2010 period. A plot of these climate normals (such as in Figure 5.31) is much smoother than the annual data but it should be noted that as there is a 20 year overlap each value is strongly correlated with the

previous value and also correlated with the next earlier value (when there is a 10 year overlap). Given the variable nature of climate it should not be unexpected to find a period of 30 or even 40 years in which there is a slight increase or decrease in the decadally updated 30 year averages, but given the smoothing effect it would not be expected that the overall increase or decrease would be large. The plot of global 30 year average temperatures is shown in Figure 5.32 from 1860. In the global series in the first 40 years (1880–1920) there is only a variation of just over 0.1°C in the decadal 30 year averages, while from 1920 to 2010 (90 years) there is a continuous increase in the averages of over 0.67°C, which strongly suggests that there is a real increase in global temperature. Some critics have suggested that there was a cooling after 1940, which does appear to be the case even with the smoothed line shown on Figure 5.28. However, the standard 30 year averages in Figure 5.32 do not show a cooling at that time, just a period of little or no change from 1960 to 1980. That period was characterized by more frequent negative ENSO indices (see Figure 5.29) and a more frequent occurrence of a negative winter NAO Index (Figure 5.33). That period of negative winter NAO was marked by colder winters in northern Europe.

Of course the global temperature series only extends back to 1850, but the CET series starts in 1659 and temperatures were recorded to the nearest 0.1° C from the 1720s. The CET 30 year averages (updated every 10 years) in Figure 5.31 show a period of 150 years (1760–1910) with only a small variation of about 0.25°C from the warmest (1760) to the coldest (1820) 30 year average but a very clear increase totalling nearly 0.92°C from 1910 to 2010. There is a short period of decline in the 30 year averages of about 0.16°C from 1960 to 1980, but as noted earlier that is related to the frequent negative winter NAO indices during that period.

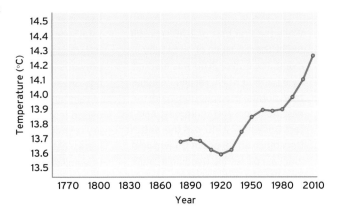

Figure 5.32 Global 30 year average temperatures. Data commence in 1850 and so the first point on the plot starts in 1880. The average 30 year value is updated every 10 years.

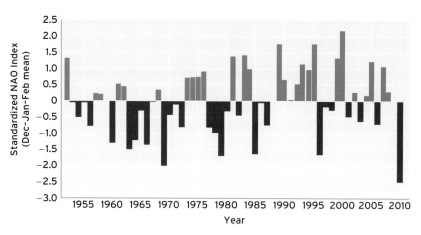

Figure 5.33 The winter (December, January, February) NAO Index.

The global temperature series has been criticized as it may be that some stations have been affected by urbanization which will increase local temperatures (see Chapter 6), although those developing the series have tried to eliminate any such effect by changing the stations used to compile the series. In the case of the CET series there has been very careful quality control to ensure the values are not affected by urbanization. Therefore, although there is considerable inter-annual variability the increase in both series observed since the start of the twentieth century would appear to be a genuine change, rather than one influenced by urbanization. The preceding discussion has demonstrated that climate indices can be used to measure some of the processes that are influencing global climate in individual years. However, there is a clear signal that overall the world has warmed during the last 100 years and it is important not to be misled by individual years (e.g. Box 5.7) or by relatively short-term changes.

CASE STUDIES

THE QUEENSLAND FLOOD DECEMBER 2010-JANUARY 2011

While the first part of 2010 was still under the influence of an El Niño event, the end of 2010 was marked by a La Niña which results in higher than normal precipitation in north-east Australia and south-east Asia. In fact December 2010 was the wettest December on record in Queensland and caused severe flooding in much of southern and central Queensland. There was substantial rainfall from storms in south-east Queensland

and a tropical cyclone made landfall south of Cairns. Heavy rainfall events continued into January 2011. Some examples of the December rainfall totals are given in Table 5.6 and the differences from the average and previous wettest Decembers indicate how severe the rainfall was. A map of rainfall across Australia for the period from 28 November to 17 January is shown in Figure 5.34 and illustrates just how much rain fell across Queensland.

The heavy rainfall led to flooding with thousands of people being evacuated and damage estimated at

around A$1 billion, as well as 35 flood-related deaths. This event is an illustration of how large-scale ocean and atmospheric features such as ENSO can result in substantial regional impacts that can be of a sufficient magnitude and extent to influence global averages for that season or year. What the Queensland floods also illustrate is that the large-scale ocean and atmospheric changes that the ENSO Index reflects can have a large influence not only on temperature but also on precipitation.

BOX 5.6 ➤

Table **5.6** Rainfall at selected stations in Queensland for December 2010

Station	Total rainfall for December 2010 (mm)	Previous wettest for December (mm)	Average for December (mm)
Mount Perry, The Pines	584.8	365.4	118.8
Pittsworth	433.6	297.5	98.7
Cambooya Post Office	325.6	298.9	100.9
Gin Gin Post Office	803.7	411.0	127.4
Biggenden Post Office	558.9	441.4	123.9
Howard Post Office	631.4	543.0	123.4
Hillgrove Station	399.4	254.6	72.2
Laura Post Office	410.4	406.6	149.2
Kenilworth Township	547.1	481.1	151.4
Byfield Childs Road	770.0	651.9	175.5
Thangool Airport	374.4	344.4	94.2
Rewan Station	587.2	408.1	123.1
Monto Township	499.0	248.5	95.9
Wiseby	603.2	339.0	107.5
Carnarvon Station	535.8	238.9	103.4
Elphinstone Pocket No 1	698.2	559.6	164.1
Bundaberg Aero	573.2	480.3	128.9

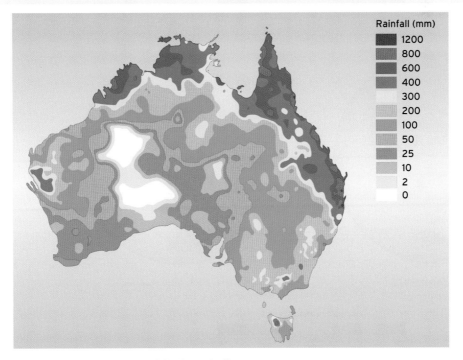

Figure 5.34 December 2010 rainfall in Australia.

BOX 5.6

CASE STUDIES

THE WINTER OF 2009-2010 IN NORTH-WESTERN EUROPE

The winter of 2009-2010 was unusual in the northern hemisphere with cold and snow affecting the eastern United States and northern Europe but with much milder conditions than normal in Canada which caused problems with a lack of snow for the 2010 Winter Olympics in Vancouver. This pattern arose due to a strongly negative phase of the NAO (Figures 5.35 and 5.36). At the same time there was a positive phase of the ENSO and this led to a different pattern than would have been the case with a negative NAO alone, as shown in Figure 5.35.

In the United Kingdom the winter was particularly cold in central and northern Scotland (Figure 5.37), but it was not only temperature that was affected as can be seen from the map of precipitation anomalies for

Figure 5.35 The combination of ENSO and negative NAO.

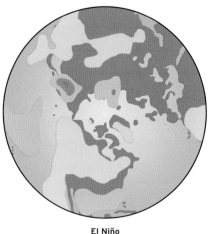

El Niño

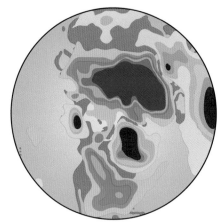

Negative NAO

Temperature anomalies (°C)

−3　　　0　　　3

= Land mass

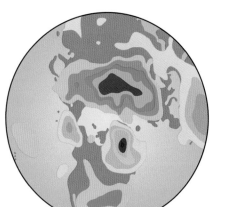

El Niño + negative NAO

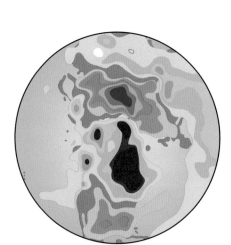

Winter 2009-2010 observed

Temperature anomalies (°C)

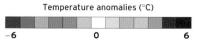

−6　　　0　　　6

BOX 5.7 ➤

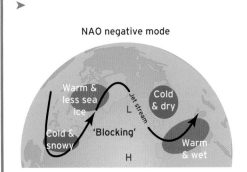
NAO negative mode

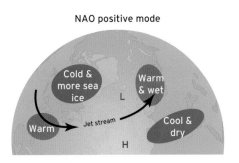
NAO positive mode

Figure 5.36 Phases of the NAO.

the winter (Figure 5.38). The precipitation map shows a reversal of the normal pattern of precipitation in the United Kingdom, which would be for greater precipitation in the west especially over the mountains of north-west Wales, England and Scotland. This reversal of the normal pattern occurred as much of the precipitation fell as snow and came from a generally easterly or north-easterly direction rather than the usual westerly direction. The map also illustrates how the mountains, particularly in the eastern Scottish Highlands, acted to reduce precipitation in the west (again a reversal of the usual situation).

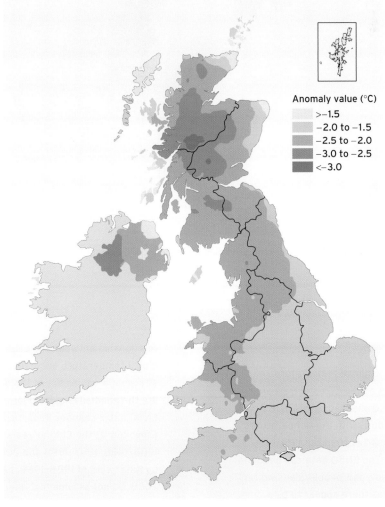

Anomaly value (°C)
>−1.5
−2.0 to −1.5
−2.5 to −2.0
−3.0 to −2.5
<−3.0

Figure 5.37 Winter 2010 mean temperature anomaly from the 1971–2000 average. (Source: Met Office (2011))

BOX 5.7 ➤

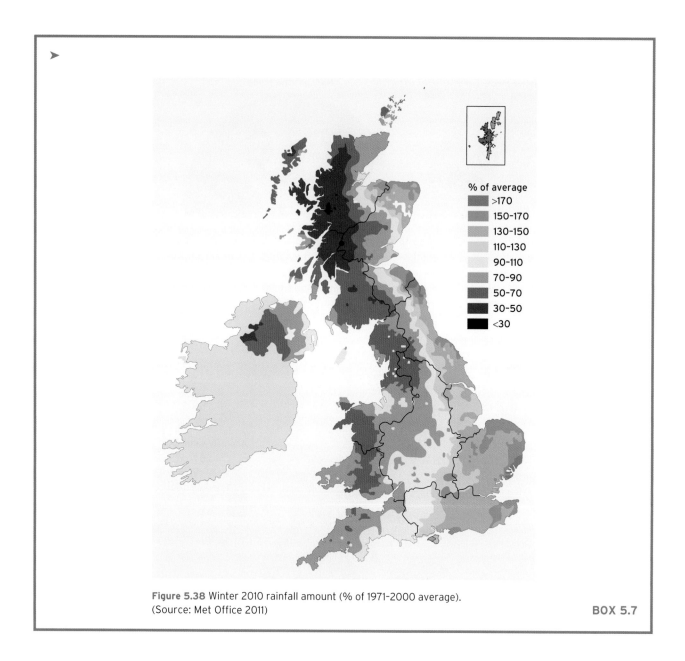

Figure 5.38 Winter 2010 rainfall amount (% of 1971-2000 average).
(Source: Met Office 2011)

BOX 5.7

Reflective question

➤ The raw data for the global temperature series can be found on the Climatic Research Unit website (www.cru.uea.ac.uk/cru/data/temperature/#datdow) with the temperatures being given as anomalies from the 1961-1990 average of 14.0°C. The ENSO data can be found at www.esrl.noaa.gov/psd/people/klaus.wolter/MEI/table.html. Does there appear to be a relationship between the ENSO values and the global temperature anomalies? (Hint: do not look at absolute values but at the direction of change: are the anomalies getting larger or smaller as the ENSO Index changes in particular years? Look especially at the El Niño years of 1982-1983, 1986-1988, 1997-1998 and 2009-2010 and in the La Niña years of 1988-1989, 1998-2001 and 2006-2008. Do not assume that any signal will be clear.)

5.7 Summary

This chapter has shown that the climate of any region is a result of the type and frequency of the weather systems found in that region. Zones of ascending and descending air such as the ITCZ (ascending), the subtropical anticyclones (descending) and the slantwise convection at the polar front (ascending) play an important role in climate. Equatorial climates are dominated by movements of the ITCZ whereas at higher latitudes in the tropics the easterly wave, monsoonal conditions and tropical cyclones may be more important, although it is still the movement of the ITCZ that partly controls these. Thus, some regions may have a fairly constant climate but are subject to occasional extreme events that have a substantial impact (e.g. regions prone to tropical cyclones). At still higher latitudes in the tropics, desert conditions are prevalent associated with the anticyclonic conditions related to the descending limb of the Hadley cell.

The distribution of land masses and oceans play an important role in global climates. Thus the southern hemisphere experiences different mid- and high-latitude climate conditions to the northern hemisphere owing to the lack of land masses within these regions. Air masses are important in the middle and high latitudes. They are distinguished by the source area from which they originate (continental or maritime, polar or tropical). The boundaries between air masses are known as fronts and represent sharp contrasts in temperature and moisture contents of the air. At fronts the warm air rises above the cooler air, often resulting in condensation and precipitation. Continental interiors tend to have different climates from those close to oceans even at the same latitude.

Individual climate elements and their changes through the year vary in their importance between different climates. For example, the distribution of precipitation during the year as well as annual precipitation totals is important because a region may not be arid if there is a winter rainfall maximum, even if it has a fairly low annual total and a hot summer. In some regions the heat and humidity of the summer are the dominating features of the climate while in others it may be precipitation totals or winter cold. There are also some climate types where there may be large differences between individual locations in one or more climate element (this is discussed further in Chapter 6). It is also important to recognize that climate types do not have distinct boundaries, as unless there is a major mountain barrier one type of climate usually merges gradually into another.

Finally, it is clear that far from being constant, the climate, even as measured over the normal 30 year climate period, is not constant. During the instrumental record before the twentieth century these climate changes were relatively small, but the last 100 years have been marked by a period of increasing temperatures.

Further reading

Books

Ahrens, C.D. (2003) *Meteorology today – An introduction to weather, climate, and the environment.* Brooks/Cole, Pacific Grove, CA.
This is an American textbook which provides a good clear overview and is very nicely illustrated with colourful figures. There are lots of reflective and essay-style questions and a useful interactive CD.

Barry, R.G. and Chorley, R.J. (2009) *Atmosphere, weather and climate*, 9th edition. Routledge, London.
This book contains useful chapters on air masses, fronts and depressions and on climates of temperate and tropical zones. It has been a very popular book over the years and is now in its ninth edition.

Linacre, E. and Geerts, B. (1997) *Climates and weather explained.* Routledge, London.
The presentation is good on the general principles and there is a large amount of material on winds at different scales.

McIlveen, J.F.R. (2010) *Fundamentals of weather and climate.* Oxford University Press, Oxford.

This textbook goes into great depth and there is a lot of science (equations!). This should suit a range of interests, but those keen to get into real detail should look at this text.

O'Hare, G., Sweeney, J. and Wilby, R. (2005) *Weather, climate and climate change.* Prentice Hall, Harlow.

Excellent and accessible introduction to the area.

Robinson, P.J. and Henderson-Sellers, A. (1999) *Contemporary climatology.* Pearson Education, Harlow.

This book contains good sections on tropical and mid-latitude climates.

Further reading
Papers

There are a number of papers about the northern European winter of 2009–2010 described in Box 5.7 in a special issue of the journal *Weather* (January 2011, Vol. 66, Issue 1).

CHAPTER 6

Regional and local climates

John McClatchey

Environmental Research Institute, University of the Highlands and Islands, and Honorary
Research Fellow, University of Nottingham

Learning objectives

After reading this chapter you should be able to:

➤ understand how local factors can modify regional climate

➤ understand how altitude and topography control local and regional climates

➤ describe how large water bodies influence local and regional climates

➤ recognize how human activity can have a deliberate or inadvertent impact on local climate

6.1 Introduction

There is substantial variation at the local and regional scales within the global climate zones discussed in Chapter 5. In some places the variation is part of a relatively gradual change from one climate type to another. For example, in the mid-west states of the United States, moving north from Tennessee through Kentucky to Illinois (Figure 6.1) there is a steady change in climate. The whole of this large area north of the southern half of Missouri, Illinois and Indiana and north of the whole of Ohio (i.e. the coverage of Figure 6.1) is within the humid continental climate type. However, there are more gradual regional variations. Tennessee has a humid subtropical climate but from Illinois to the north into Canada the climate is described as humid continental (see Chapter 5). In moving northwards the winters become increasingly severe. Northern Iowa has summers that are noticeably cooler than the hot humid summers experienced in the south of Missouri. This cooling of the summers continues into Canada. North of Winnipeg the summers become much shorter with less than four months having temperatures above 10°C. Iowa, Wisconsin and Michigan are within the same climate subtype with at least four summer months with temperatures above 10°C. However, even within this area, there are marked regional variations in climate on top of the gradual change northwards. The most obvious of these differences are associated with the areas bordering the Great Lakes of Michigan and Superior.

In any climate region there can be very marked local variations in certain climate elements. For example, over a distance of little more than 100 km across Scotland there are places in the west where the total annual precipitation is nearly 10 times greater than that in parts of the east. However, on a global or even European scale the whole of Scotland falls well within the limits of a single climate type. The classification or description of a climate at a particular place therefore depends on the scale at which that climate

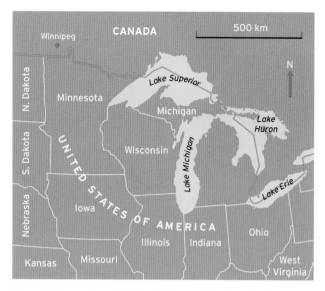

Figure 6.1 The mid-west United States and the Great Lakes.

is being considered. The smaller the geographical extent of the area of interest, the more important it becomes to have detailed climatological observations and statistics (see Box 5.1 in Chapter 5) in order to specify the climate of an individual location. It is also important to note that regional and local climates are a feature of the land and not the oceans because altitude, topography and proximity to the sea (or large bodies of water) are the key features that lead to the development of local and regional climates. These factors and associated processes will be discussed in this chapter as it is important to understand how both local and global processes interact at different scales. This interaction helps to explain the nature of local and regional climate across the Earth's surface.

6.2 Altitude and topography

The climate of mountains has always been of interest to scientists. Early studies helped establish that pressure and temperature fell with height and in the latter part of the nineteenth century a number of mountain observatories were established in Europe and North America to support astronomical studies and weather forecasting. Examples include Mount Washington (New Hampshire, established 1870), Sonnblick (Austria, 1886) and Ben Nevis (Scotland, 1883). Some of these mountain observatories closed after 10–20 years of observations (e.g. Ben Nevis, 1883–1904) but many are still in existence. The rate of fall of temperature with altitude (the **lapse rate**, see Box 6.1) varies in different parts of the world. The amount of solar radiation that can potentially be received by the ground surface actually increases with height. This is because less radiation has been absorbed or reflected by components of the atmosphere back into space. The lower the altitude, the thicker the layer of atmosphere that can reflect solar radiation back into space. However, whether solar radiation received at the ground surface actually increases with height in any particular mountain range depends upon local cloudiness. Harding (1979) reported that in the mountains of the United Kingdom, which tend to be cloudy, solar radiation decreased by 2.5 to 3 million $J\,m^{-2}\,day^{-1}\,km^{-1}$, which was regarded as typical by Grace and Unsworth (1988). Nevertheless, even in those areas where received solar radiation increases with altitude, temperature is still likely to decline upwards because of adiabatic processes (see Table 6.1). In many places wind speed and precipitation increase with altitude but this is not true everywhere. Substantial mountain ranges

Table 6.1 Predicted change of pressure and temperature with height based on a lapse rate of 6.5°C per 1000 m (pressure P in hPa or millibars and temperature T in °C)

Height (m) (above mean sea level)	Tropical		Mid-latitude		High-latitude	
	P	T	P	T	P	T
0	1013	25.0	1013	15.0	1013	5.0
1000	902	18.5	899	8.5	894	−1.5
2000	801	12.0	795	2.0	788	−8.0
3000	710	5.5	701	−4.5	691	−14.5
4000	627	−1.0	616	−11.0	605	−21.0
5000	552	−8.5	541	−17.5	527	−27.5

LAPSE RATES

The rate at which temperature falls with increasing altitude is known as the **environmental lapse rate**. An air parcel will rise if it is warmer than the surrounding environment. Once the air parcel reaches the same temperature as the surrounding environment it will stop rising (Figure 6.2). When air rises (ascends) it expands. This is because the air pressure decreases. Conversely if air sinks (descends) it is compressed as the pressure increases. If no energy is added to or lost from that air as it rises (or falls), the changes in pressure and temperature are the result of what is termed an adiabatic process. When air rises and the pressure falls, the energy for the expansion of the air comes from the air itself. As temperature is a measure of the energy

of air, if energy is removed by expansion of the air then the temperature of the air decreases. The reverse is true if air sinks and is compressed, in which case the temperature of the air increases.

If air expands adiabatically as it ascends the temperature of the air falls at a constant rate of 9.8°C km⁻¹ (Figure 6.2). If air is compressed adiabatically as it sinks the temperature of the air increases at the same rate. This rate of temperature change is called the dry adiabatic lapse rate. The term 'dry' gives a clue as to why this is not the only rate of change of temperature with height, as it applies only if the atmosphere remains unsaturated. Air can hold only a certain amount of water vapour at any temperature and if the temperature falls or if more water is evaporated

into the atmosphere then once saturation is reached water vapour will condense out of the atmosphere. When water condenses out of the air, energy is released. This energy is the latent heat of vaporization and is the energy released when gaseous water vapour condenses into liquid water. When water evaporates or ice melts, this uses up energy in order to change the state of the water but without changing the temperature of the water (see Chapter 4). When the reverse occurs energy is released. This energy release reduces the cooling rate of the air as it rises and expands. In the absence of any loss of total water content (gaseous, liquid or solid water) from the air, the rate of change of temperature with height is given by what is termed the saturated adiabatic lapse rate.

However, unlike the dry adiabatic lapse rate, the saturated adiabatic lapse rate depends on the amount of water vapour the air can hold at any temperature. As can be seen in Table 6.2 the amount of water vapour air can hold more than doubles for each 10°C increase in temperature. This means that the amount of latent heat released is much less at low temperatures than at high temperatures as less water vapour will condense out of the air at lower temperatures. The saturated adiabatic lapse rate therefore varies from around 0.3°C per 100 m close to the surface in the tropics, where air temperatures are over 30°C, to close to the dry adiabatic rate at temperatures below −40°C. Temperatures of below −40°C are normally found at heights of between 5 and 10 km in mid-latitude regions.

When saturated air ascends, the temperature decreases at the

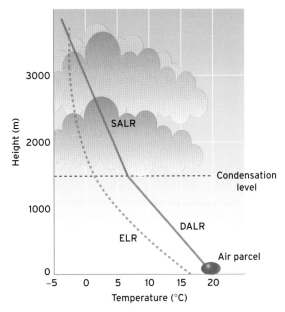

Figure 6.2 Air parcel buoyancy. A parcel of air warmed at ground level will rise if it is warmer than its surroundings and will cool at the environmental lapse rate (ELR). Note that top of the cloud occurs where the air parcel temperature is the same as its surroundings. DALR - dry adiabatic lapse rate (9.8°C/1000 m); SALR - saturated environmental lapse rate.

BOX 6.1 ➤

➤

Table 6.2 Water vapour saturation vapour pressures

Temperature (°C)	−40	−30	−20	−10	0	10	20	30
Water vapour pressure (hPa)	0.19	0.51	1.25	2.86	6.11	12.27	23.37	42.43

saturated adiabatic lapse rate and the water that condenses out of the air forms a cloud. However, when saturated air descends, it will warm at the saturated adiabatic rate only if all of the liquid or solid water (cloud droplets or ice crystals) is re-evaporated back into the air. If any of the water has fallen out of the cloud as precipitation there will be less water to evaporate and the air will therefore warm more on descent than it cooled on ascent.

The difference between the dry and saturated adiabatic lapse rates is crucial in understanding why some atmospheric conditions are unstable (or **conditionally unstable**) (Figure 6.3). This instability occurs because if rising air becomes saturated, any further ascent will cause the air to cool at the saturated adiabatic rather than dry adiabatic lapse rate. Such saturated air will therefore be warmer than the surrounding air and as warmer air is less dense than colder air, the saturated air will

ascend further as a result of this density difference. This leads to strong upward convection and the growth of shower clouds. A situation is described as conditionally unstable if the environmental lapse rate is steeper than the saturated adiabatic lapse rate through the lower atmosphere (around 10 km). If clouds form and grow into this layer with a steep lapse rate they are then going to continue to grow to great depths. The instability is 'conditional' as clouds have to form and reach the height at which the cloud temperatures become warmer than the surrounding air. Although adiabatic processes are common in the atmosphere, particularly when there is widespread ascent or descent of air such as in frontal ascent or anticyclonic subsidence, energy can be gained or lost from air by a number of processes including the loss of energy associated with precipitation that falls out of the atmosphere.

Radiative exchanges can also be important in the atmosphere. While

air is largely transparent to solar radiation, certain atmospheric gases absorb terrestrial (long-wave) infrared radiation (see Chapter 4). Liquid or solid water completely absorbs and reradiates long-wave radiation. This means that clouds play an important role in radiative exchanges. Radiational losses at the top of clouds can cause localized cooling. Radiational heating from the ground surface is also important. Strong heating can create steep lapse rates in the lowest layers of the atmosphere. These steep lapse rates can be greater than even the dry adiabatic lapse rate and are termed **super-adiabatic**. Such steep lapse rates cause rapid local convection which tends to mix the atmosphere. This effect is greatest in summer and at lower latitudes.

At night when long-wave infrared radiational emission from the surface cools the ground, the lowest few hundred metres of the atmosphere can be cooled as a result. This creates

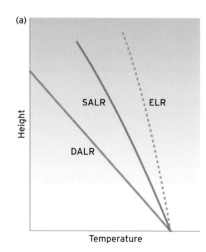

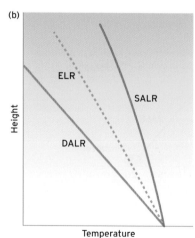

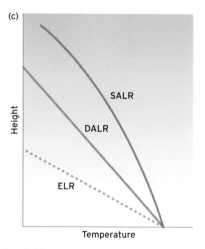

Figure 6.3 Atmospheric stability relationships between the ELR, DALR and the SALR: (a) stable; (b) conditional instability; (c) absolute instability.

BOX 6.1 ➤

a temperature inversion perhaps a few hundred metres deep. This is where warmer air overlies cooler air so that it is no longer the case that temperature declines with altitude. Once warmer air overlies cooler air, the lower layer is trapped because it is denser. Thus, pollutants from

fossil fuel combustion (e.g. fires, car engines) may not be able to escape from the lower air layer and thus a long-lasting 'smog' can develop. During these times public health can be at severe risk.

However, while adiabatic lapse rates are important in the atmos-

phere, other processes, such as mixing of air, affect temperatures. If rising air mixes with its surroundings its energy will be shared with the surroundings and the temperature changes will no longer be adiabatic.

BOX 6.1

can also act as a barrier to the movement of weather systems and even smaller ranges of mountains and hills can give rise to noticeable differences in the weather (and hence in climate) on the lee side of those mountains. Hills and mountains are therefore important as they can create substantial regional and local modifications to the general climate of that part of the world. The following discussion will explain how individual climate elements are modified by hills and mountains, and how a specific regional climate may be developed on the leeward side of upland areas.

6.2.1 Pressure

The fall of air pressure with height is the most consistent feature of mountain climate. Up to around 3000 m the fall in pressure is close to 10 mb per 100 m. The rate of fall is more rapid in colder (denser) air and therefore pressures are higher at the same altitude in the tropics compared with the middle and high latitudes (Table 6.1).

6.2.2 Temperature

The change of temperature with altitude is known as the lapse rate. Box 6.1 describes the important characteristics of lapse rates and should be read in order to understand fully the following section. The values in Table 6.1 give an indication of the differences in the fall of pressure with altitude in tropical, mid-latitude and high-latitude regions, but they are based on an assumed lapse rate of 6.5°C per 1000 m. However, there can be substantial diurnal variations in lapse rates and these will influence altitudinal pressure gradients. During the day strong solar heating warms the air close to the ground, steepening the lapse rate in the first few hundred metres of the atmosphere. In the first few tens of metres the lapse rate can exceed the dry adiabatic lapse rate of 9.8°C per 1000 m (see Box 6.1), although the lapse rate will depend on how well mixed the atmosphere is. The stronger the wind, the greater the turbulent mixing and the

closer the lapse rate will be to the dry adiabatic rate (as long as there is no condensation of water vapour). In cloudy conditions with strong winds, the atmosphere is so well mixed in the lowest few hundred metres that below the cloud base the lapse rate will approximate the dry adiabatic lapse rate. Within the cloud the lapse rate will be at the saturated adiabatic rate. At night, light winds and clear skies allow the ground to cool through long-wave radiation loss and a strong temperature inversion can form close to the ground (Figure 6.4). **Katabatic drainage** can further strengthen the night-time inversions (see Box 6.4 below). These diurnal variations in the fall of temperature with height are influenced by cloud cover and the presence or absence of vigorous mixing caused by strong winds. Together these play an important role in controlling temperatures of mountain regions.

As well as diurnal variations, the lapse rates also vary according to air mass. As air masses move they can be warmed or cooled by the underlying surface, leading to a steeper or a shallower lapse rate respectively. Hence Arctic or Antarctic maritime air masses, which are always warmed from below as they move into lower latitudes, have the steepest lapse rates, which are often close to the dry adiabatic lapse rate. Tropical air masses, however, are cooled from below as they move away from lower latitudes. This reduces the lapse rate. The affect of warming and cooling from below is illustrated in Figure 6.5, which shows an air mass being cooled as it moves across a cool lake and one being warmed as it moves from a cold lake to a warm land area. Temperature inversions are temporarily formed as the air is warmed or cooled from below. Polar maritime air may also have steep lapse rates if there is a large contrast between the air and sea surface temperatures. As the contrast between the air and sea surface temperatures varies with the seasons, there are some seasonal differences in lapse rates. The average lapse rates in mountains of the British Isles are steeper than in most mountainous regions as a result of the frequency of polar and Arctic air masses and the relatively

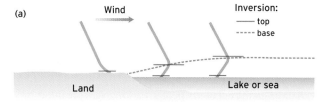

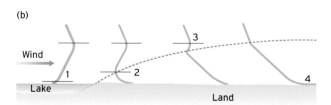

Figure 6.5 Temperature inversion: (a) the modification of an unstable temperature profile to give a surface-based inversion over a cold surface (e.g. lake/ice); (b) elevated inversion due to the advection of stable lake air across the shoreline to a warmer land area on a spring afternoon. (Source: after Oke, 1987)

warm surrounding sea surface temperatures (Harding, 1978). Air with steep lapse rates is conditionally unstable (see Box 6.1) and instability showers are therefore common. Air with a shallow lapse rate is stable, damping down convective activity. However, such air can be subject to forced ascent over mountains or dynamically induced ascent such as in a mid-latitude depression. If the ascent is sufficient to cool the air to its dew point temperature, clouds will form and this can lead to enhancement of precipitation. Tropical continental air moving towards higher latitudes is cooled from below but in summer that cooling may only extend through a relatively shallow layer. Tropical continental air

is dry and is formed over an area in which there was strong heating from below. Therefore above the layer of air that is cooled from below, the lapse rate will be much steeper. This means that there can be strong conditional instability. Convection that rises above the cooled layer can trigger the rapid formation of deep convectional clouds.

Anticyclonic conditions are also important in determining temperature variations with altitude. The large-scale descent of air in anticyclones creates a dry adiabatic lapse rate aloft, above a strong temperature inversion. For example, in the trade wind belt flowing out of the subtropical anticyclones, there is a marked temperature inversion, generally at a few hundred metres above the surface on the eastern side of the anticyclone rising to 2000 m in the west. Mountain ranges such as the Atlas in Africa, the Andes in South America and the Sierra Nevada in California often penetrate well above these subtropical inversions. This leads to distinct changes in temperatures and therefore local climate on ascending such mountain ranges.

Subsidence inversions as shown in Figure 6.6 are a feature of all anticyclones. Air is forced to descend under the anticyclonic conditions but part of this descending air may be warmer than the air below it. Thus the temperature might vary with altitude as shown in Figure 6.6. Depending on the height of the inversion, mountain ranges in any part of the world may be above the height of the inversion base. This will lead to unusual conditions on those mountains with different lapse rates above and below the inversion. There may also be a very low humidity. The descending air in anticyclones leads to exceptionally low values of humidity close to the base of the inversion. However, as anticyclones found in the middle and high latitudes are

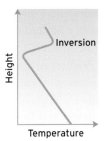

Approximately 10 km

Dry adiabatic lapse rate

Subsidence inversion

Altitude

Temperature

Figure 6.6 Lapse rate under anticyclonic conditions with a subsidence inversion.

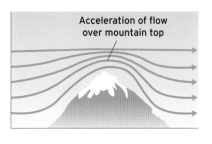

Figure 6.7 Acceleration of winds over ridges and mountain tops where the air is compressed by a temperature inversion aloft. Because the space for the air to pass through is narrower then air is forced through at greater speeds.

short-lived compared to subtropical anticyclones, on some days there will be a sharp change in the weather conditions experienced on ascending the mountains. Such changes will be incorporated into the averages of the climate elements.

6.2.3 Wind

As well as changes in temperature with height, the wind regime in mountains can be quite different from that at lower levels. It is not, however, altitude that is necessarily the key factor. It is the topography itself that is important. For example, the wind can be funnelled through valleys or even gaps between individual peaks, particularly if the orientation is in the direction of the prevailing wind. This will result in much greater local wind speeds. In addition, individual peaks and exposed ridges will experience higher winds as there will be less surface friction acting to reduce wind speeds as the wind approaches. Friction reduces surface winds by about 30% compared with the 'free atmosphere'. In some circumstances, the compression of air between mountain summits and a temperature inversion aloft can lead to wind speeds above free atmosphere values (Figure 6.7). This frequently occurs around the Cairngorm Mountains in Scotland and at Mount Washington in New Hampshire (Figure 6.8), where maximum gusts have been recorded at 76 and 103 m s^{-1} respectively. The wind regimes in mid-latitude mountains are also influenced by the westerly winds. The westerly winds are generally faster aloft. However, in the tropical and subtropical trade wind belts, the north-east and south-east trade winds generally weaken with height. Therefore mean wind speeds can be low on tropical and subtropical mountains. For example, typical wind speeds are 2 m s^{-1} during the period December to February at 4250 m in New Guinea and an

annual mean of 5 m s^{-1} at 4760 m in Peru. In the Himalayas the monsoon circulation gives strong westerly winds through the winter half-year (October–May) with more moderate easterly winds in the summer (June–September). The westerly winds decrease from over 25 m s^{-1} at 9 km in the winter half-year to only 10 m s^{-1} by the end of May (being replaced by easterlies in the second half of June). These wind speed changes in the Himalayas emphasize the importance of the weather systems experienced in mountain regions in determining the wind regime.

One other local and regional climate feature of winds is a warm and dry wind that blows down lee slopes of hill and mountain ranges. These winds tend to warm and dry as a result of the compression and adiabatic warming of the air in the lee of the hills and mountains as it descends from higher levels. It is called a **Föhn wind**, although it has other names in different parts of the world such as in Canada where it is called the **Chinook**. The onset of the wind is

Figure 6.8 Strong winds at the top of a mountain ridge known as Glyder Fawr in Snowdonia National Park, North Wales, UK, resulting from compression of air by a temperature inversion above and lack of friction from surrounding terrain. (Source: Alamy Images: P B Images)

typically accompanied by a sharp rise in temperature often with a substantial decrease in relative humidity. In Canada, in the lee of the Rockies, temperature rises of over 20°C have been recorded in just a few minutes. Evidence of smaller rapid rises in temperature has been found in Scotland and even with winds across the Pennines in England, where the hills are typically only 500–700 m high (Lockwood, 1962). Föhn winds in the Alps and other mountainous regions can cause rapid snow melt, greatly increasing avalanche risk and flooding (Barry, 1992).

It has already been noted that surface cooling can lead to katabatic drainage into valley bottoms. Over glaciers and particularly over the Antarctic ice sheets more substantial katabatic winds can form due to local cooling (e.g. Renfrew and Anderson, 2002). These winds, which can be extreme, will flow into hollows and valley bottoms and are a special case of mountain winds (see Box 6.4 below).

6.2.4 Precipitation

The amount of moisture the air can hold is strongly dependent on temperature and as temperatures fall with height, generally the moisture content of air does so too. It might therefore be expected that precipitation would also decrease with height as the moisture content declines. However, in the lowest 3000 m of the atmosphere this is certainly not the case. Air forced to rise over mountains cools at the dry adiabatic lapse rate until the dew point temperature is reached. At this point clouds form and temperature then decreases at the saturated adiabatic lapse rate if there is any further ascent of the air. Therefore, while temperature is reduced by adiabatic expansion as the pressure falls, the moisture content of the air does not change until saturation is reached. Hence, even in a very dry region, if air is forced to rise, it will eventually reach saturation. For example, if dry air (say 30% relative humidity) at 30°C is forced to rise, it will become saturated at about 2000 m and cloud will form. Unless there is no wind, the forced ascent of air over hills will therefore provide a supply of moisture from lower levels, even though there may be a decrease of vapour pressure with height in the free atmosphere. This forced ascent of air means that in moist airstreams, clouds will form over relatively low hills. Even in dry airstreams clouds will form if mountains are sufficiently high (Figure 6.9). The mere

Figure 6.9 Cloud formation over mountains at Applecross, in western Scotland. As the air is forced to rise over the mountain it expands adiabatically and saturation of the air occurs aloft. The water vapour can then condense to form clouds. (Source: Dorling Kindersley)

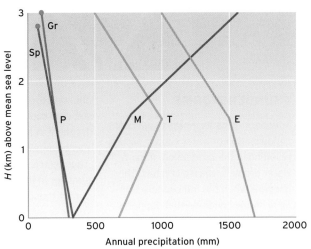

Figure 6.11 Precipitation changes with height in different parts of the world: E, equatorial; T, tropical; M, middle latitude; P, polar (Sp, Spitzbergen; Gr, Greenland). Only in the middle latitudes does precipitation increase with altitude over the 3000 m range shown. For equatorial and polar areas precipitation decreases with altitude and for tropical areas precipitation increases to about 1500 m and then declines above this height. (Source: after Lauscher, 1976)

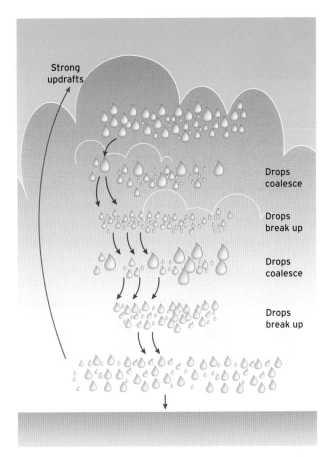

Figure 6.10 Raindrop formation where there are strong updrafts.

formation of clouds will not, however, lead to precipitation. Precipitation has to be initiated. This occurs either through the formation of ice crystals in the upper parts of a cloud which then fall through the cloud leading to aggregation of crystals and production of supercooled water (**Bergeron process**), or by coalescence of smaller droplets onto larger droplets falling more quickly through the cloud. In a convective cloud, the maximum rate of precipitation will be close to the cloud base, as once rain falls out of the cloud the raindrops begin to evaporate. If there are strong updrafts even the raindrops may be transported upwards and, if that is happening, the zone of maximum precipitation may be above the cloud base (Figure 6.10).

Thunderstorms can often occur when the atmosphere is very unstable, producing very rapid falls of precipitation. Box 6.2 provides more details of these hazardous features of the atmosphere.

In the tropics and subtropics, precipitation is often as a result of convective activity and therefore the highest rainfall totals are found at typically between 1000 and 1500 m, at or just above the average cloud base. This is very common in the trade wind belts where the air above the trade wind inversion is very dry. For example, rainfall on Mauna Loa in Hawaii is over 5500 mm at 700 m but only 440 mm on the summit at 3298 m, well above the inversion (Barry, 1992). In the moister equatorial regions, rainfall generally tends to decrease with height. For example, in equatorial Africa rainfall on mountains above 3000 m is only 10–30% of the highest totals which are observed lower down the mountains. In the middle latitudes, however, precipitation totals increase with altitude above 3000 m. Thus, there are distinct latitudinal differences in the change of precipitation with height in mountains (Figure 6.11).

The presence of mountains in the middle latitudes enhances precipitation in a number of ways. The most important effect is that low-level cloud is formed as air is forced to rise over the mountains (Box 6.3). Although convective precipitation can form a significant proportion of the rainfall totals in some mid-latitude locations, much of the precipitation arises from frontal activity associated with depressions. Orographic enhancement through the **feeder–seeder mechanism** (Box 6.3) can be substantial at warm fronts and in warm sectors, and to a lesser extent with cold fronts. The forced ascent of air over hills and mountains may also intensify vertical motions in depressions and troughs or even trigger conditional instability in polar or arctic airstreams. The general increase of wind speeds with height also ensures there is a supply of moist air brought in to replace any loss of water content through precipitation.

HAZARDS

THUNDERSTORMS

Thunderstorms form when significant condensation of water vapour occurs, resulting in the production of many water droplets and ice crystals. This happens when the atmosphere is in an unstable condition that supports fast upward motion. Although thunderstorms often happen during warm weather when heating of the ground surface causes sufficient moisture to accumulate in the lower atmosphere, and the warm surface causes there to be a steep adiabatic lapse rate, what is required for thunderstorm formation is an unstable atmosphere (see Box 6.1) through a considerable depth of the atmosphere (possibly right up to the tropopause). The unstable atmosphere means that considerable energy is released as water condenses out of the atmosphere to form the deep cumulonimbus clouds that characterize thunderstorms. That energy can range from the equivalent of a small nuclear bomb (say 10 kt of TNT equivalent) in a small thunderstorm to over 100 times more in a severe thunderstorm. Thunderstorms occur anywhere in the world but are most frequent in the tropics where they can be an almost daily occurrence. Thunderstorms are more common in summer in the mid-latitudes, although winter thunderstorms can occur, due to low-level convergence along a cold front, while at high latitudes thunderstorms are fairly rare, and form only in the summer. At high latitudes in winter, the air is so cold at the surface and throughout the atmosphere that there is insufficient moisture in the air to provide enough energy for a thunderstorm. Thunderstorms are usually accompanied by heavy rainfall, often with strong winds and possibly hail (Figure 6.12). Thunder is caused by the explosive expansion of a narrow column of air which is heated by a lightning discharge. Therefore lightning precedes all thunder. Large cumulonimbus clouds form, often extending to great heights during thunderstorms, although the storm cloud is not normally larger than a few kilometres in diameter. Thunderstorms can be single or multi-cellular and a series may form a squall line with an associated gust front of strong winds. Supercell storms are severe storms characterized by wind shear with height creating a rotating updraft or mesocyclone. Severe tornadoes are associated with supercell storms (see Box 5.4).

Lightning occurs when a large charge is built up within a cloud and is then discharged. In tall cumulonimbus clouds electrical charge is built up as water droplets, hail and ice crystals collide with one another in the strong air movement (a bit like when you rub a balloon on your jumper you can create a charge that, when you put the balloon near your hair, makes your hair stand on end). The positive and negative electrical charges in the cloud separate from one another, the negative charges dropping to the lower part of the cloud and the positive charges staying in the middle and upper parts (Figure 6.13). Positive

Figure 6.12 Large hailstones which fell in Les Esserts near Lausanne, on 23 July 2009. Such large hailstones can be very damaging and have been known to smash car windscreens and cause damage to buildings. (Source: Getty Images: AFP)

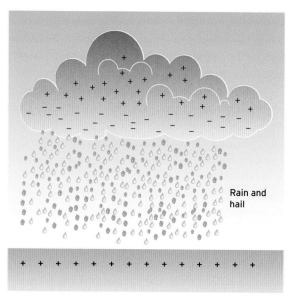

Rain and hail

Figure 6.13 Electrical charge in a thunderstorm.

BOX 6.2 ➤

➤

electrical charges also build upon the ground below. When the difference in the charges becomes large, a flow of electricity occurs in a discharge event. In-cloud lightning is most common but lightning does strike the ground or strikes from the ground to the cloud. A lightning strike occurs in less than a millionth of a second. The temperature of a lightning bolt can be hotter than the surface of the Sun.

Although the lightning is extremely hot, the short duration means it is not necessarily fatal: hundreds of people are struck by lightning every year, but not all die. Lightning at the ground is the most common natural cause of forest fires, which can cause considerable damage and present a major hazard to both people and the environment, particularly after long periods of drought.

It is worth noting that large hail (Figure 6.12) (which in supercell storms can be up to 10 cm in diameter, although large hail is more typically 2-4 cm across) can cause major damage to crops, property and vehicles. NOAA notes that hail causes damage to crops and property amounting to $1 billion per year in the United States.

BOX 6.2

As a result of orographic enhancement of precipitation, mid-latitude hills and mountains have unusually high annual precipitation totals (Box 6.3). In New Zealand, average precipitation totals on the windward side of the Southern Alps can reach 10 000 mm yr^{-1}.

As well as enhancing precipitation on the windward side of mountains, there can be substantial reductions of precipitation in lower-lying areas to the lee of the mountains. These 'rain shadow' areas are found in many places such as northern Chile (south-east trade wind belt), Patagonia in Argentina and east of the Rocky Mountains in the United States (mid-latitude westerlies) and on a smaller scale in many hilly locations. This occurs because the air which can now descend down the lee side of the mountains warms adiabatically. This in combination with the earlier loss of moisture through precipitation on the windward and summit parts of the mountains makes the air less saturated and thus less likely to produce precipitation.

The other feature of mountains is that precipitation may fall as snow which can accumulate over time. Over the winter substantial amounts of snow can accumulate in mountains. Often snow melt in the spring can produce large river flow peaks downstream even when there is no precipitation at the time. Precipitation in mountains is therefore not just enhanced but its hydrological impact may occur a number of months (or years) later. Given sufficient depth of snow (many metres) surviving over many summers it is possible for a glacier to form. At present nearly all mid-latitude mountain glaciers are retreating as annual melt is greater than snow accumulation. Despite that retreat, as the glacier surface ice is at 0°C even in mid-summer, glaciers have an impact on their local summer climate as they act as heat sinks. Glaciers cool air in contact with the surface and depending on the moisture content of the air they can act

as either a local moisture source or sink. If air is dry then water vapour can **sublimate** (change directly from the solid to gaseous state) into the air above the glacier. If the air is moist then water can condense out of the air onto the surface of the glacier which is then acting as a moisture sink.

Even in mountains where no glaciers exist, snow patches can survive over the summer. The Observatory Gully snow patch on Ben Nevis in the Scottish Highlands is an example of this, having only melted completely a few times over the past 120 years. Although snow may last for many years before melting, the Observatory Gully patch is too shallow to form a glacier. Like glaciers, such snow patches can also modify the microclimate close by. However, this is to a much smaller extent than glaciers. The snow can also have local ecological impacts creating a niche for certain alpine species. Again the local topography is important with snow patches forming and surviving in local shaded depressions or gullies in the mountains.

6.2.5 Frost hollows

Cold air tends to move downhill by katabatic drainage. This drainage of cold air into lower-lying areas can give unusually high occurrences of frost and low temperatures in certain locations. Such frost hollows are much more a feature of middle and high latitudes than other locations. This is because low temperatures occur when there is strong radiational cooling at the surface resulting in a surface temperature inversion. As water vapour is a strong absorber of long-wave radiation, a significant amount of which it re-emits down to the surface, the relatively moist air in most subtropical and tropical regions (deserts are the exception) reduces the amount of this surface radiational cooling. Geiger (1965) described an extreme example of local surface

OROGRAPHIC ENHANCEMENT OF PRECIPITATION

Precipitation totals increase with altitude in the tropics (up to around 1500 m) but the increase with altitude is greatest in the middle latitudes. This mid-latitude orographic enhancement of precipitation is related to the feeder–seeder mechanism. As air is forced to rise over the mountains the adiabatic cooling reduces air temperature to the point at which the air becomes saturated with water vapour. Further ascent leads to the formation of cloud, a common feature of mid-latitude mountains. This cloud acts to increase precipitation falling from higher 'seeder' clouds because the lower 'feeder' cloud droplets are swept into the precipitation falling through the cloud (Browning and Hill,

1981). This feeder–seeder enhancement of precipitation as illustrated in Figure 6.14 can be substantial at warm fronts and in warm sectors and to a lesser extent with cold fronts. This is most apparent in coastal mountain ranges such as the New Zealand Alps, the Coast Mountains of British Columbia and the western Highlands of Scotland.

Orographic enhancement of precipitation is difficult to measure as in the middle latitudes winter precipitation often falls as snow and the turbulent flow around mountains means that the deposition of snow can be into gullies and hollows. Standard measurements of precipitation using rain gauges are therefore impossible (see Chapter 11). In addition, orographic enhancement is not constant in all precipitation events. Enhancement tends to be greatest

at warm fronts and in warm sectors and therefore as weather events are not distributed evenly through time, there are seasonal differences in enhancement. In addition, even across a relatively narrow mountain range such as the Scottish Highlands, typically 80 km wide and only up to around 170 km at their widest (from the west coast to the east of the Cairngorms), there is a rapid decline in enhancement from west to east. While enhancement rates can be over 4.5 mm m^{-1} in the western Highlands, in the east of the Cairngorms the rate is only of the order of 1.33 mm m^{-1} (McClatchey, 1996). This enhancement gives annual totals of just over 2000 mm on the highest tops of the eastern Cairngorms, as compared with 6000 mm in parts of the western Highlands.

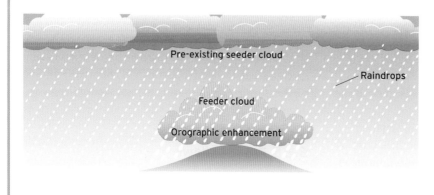

Figure 6.14 Schematic diagram of the feeder–seeder mechanism. The frontal cloud aloft produces precipitation which falls and hits the water droplets in the orographic cloud. These then combine to increase the overall precipitation totals on the mountain/hilltop.

BOX 6.3

cooling at the Gstettneralm sinkhole in Austria where temperature inversions from the bottom to the top (about 150 m) can be over 27°C and extreme minima of below −50°C have been recorded in the valley bottom. An example of an area prone to low temperatures in Europe is given in Box 6.4. Low-lying areas with well-drained soils such as sands or gravels, or thin soils on chalk, are more likely to experience increased frequencies of frosts or

unusually low temperatures. Outside the tropics and subtropics, the lowest surface air temperatures are, however, almost always recorded when the ground is snow covered as the snow insulates the air from the soil heat flux (Robinson and Henderson-Sellers, 1999). In addition the cold air above snow surfaces contains less water vapour and therefore the surface radiational cooling will be greater.

CASE STUDIES

KATABATIC DRAINAGE AND EXTREME TEMPERATURE MINIMA

The density of air is inversely proportional to temperature ($\rho \propto 1/T$; density increases as temperature falls). In light winds, mechanical mixing of the air is very limited, and at night there is a lack of convective turbulence that is normally created by solar radiation warming the ground during the day. As a result the air close to the surface is cooled during the night as the surface temperature drops. This cooling is greatest when the sky is cloud free and the air is dry. As this cooled air is now denser than the air aloft, if it is on a slope, the air can start to move downslope in what is called katabatic drainage. This flow of air is not fast and is not like that of water but more like the flow of something like porridge. Very rarely genuine katabatic winds can occur but this really only happens in the Antarctic when cold air flows off the main ice sheets (which can be at over 3000 m above sea level) down to the coastal ice shelves. In most parts of the world katabatic drainage is slow and the cold air can pond up behind restrictions in the flow such as walls (on a small scale) and on a larger scale where a wide valley becomes constricted at a lower point in the valley.

Anticyclonic conditions are most likely to give rise to stronger katabatic drainage, as winds tend to be light and the sky cloud free (calm and clear conditions). Such night-time conditions are sometimes called radiation nights as the surface has its maximum radiational loss under such conditions. Anticyclones have a marked temperature inversion aloft formed by the subsidence of air from higher levels and on such radiation nights the surface temperature inversion, formed by the cooling at the surface, can extend up to the anticyclonic inversion aloft.

An example of an exceptionally deep surface temperature inversion was observed in the Cairngorm Mountains, Scotland, in 1982. On 8 January 1982, a surface minimum temperature of −31.3°C was recorded at Grantown-on-Spey (the minimum air temperature was −26.8°C) while at the summit of Cairngorm (1247 m above Grantown-on-Spey) the minimum temperature was −12.6°C, a temperature inversion of over 18°C from the valley surface. Even lower temperatures were recorded in the Spey Valley on 10 January 1982 but no observation was available from Cairngorm summit on that date. A map of the temperatures on 10 January 1982 (Figure 6.15) using remote sensing (see Chapter 23) shows clearly how there are particular areas (frost hollows) that have the lowest minimum temperatures (McClatchey et al., 1987).

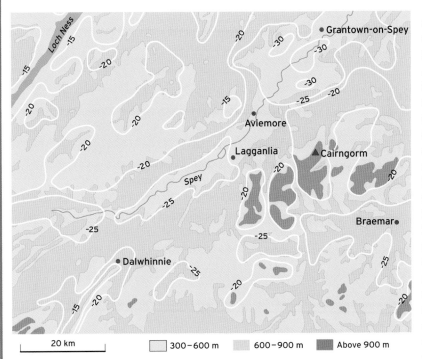

Figure 6.15 Satellite-derived temperatures in the Scottish Highlands for 10 January 1982. Severe low temperatures can be seen in the low-lying hollows whereas warmer temperatures are found on the mountain tops. This is an example of katabatic drainage. (Source: after McClatchey et al., 1987)

BOX 6.4

Reflective questions

➤ While an average temperature lapse rate for the troposphere may be around 6.5°C km⁻¹, under what circumstances would the lapse rate in mountains (i) show an increase in temperature with height from the valleys to well up the hillsides; (ii) be close to the dry adiabatic lapse rate; (iii) show a fall in temperature followed by a rise and then a further steep fall in temperature on ascent?

➤ Why is it generally windier on the top of mid-latitude mountains than at lower elevations?

➤ Why is the change of wind speed with height in tropical and subtropical mountains different from that in mid-latitude mountains?

➤ Why do rainfall totals in the middle latitudes increase much more greatly with height than those in the tropics?

➤ What conditions are needed for thunderstorms to form and what hazards do they pose?

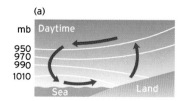

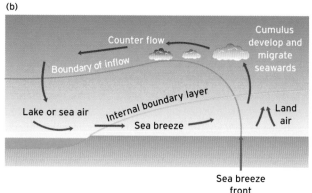

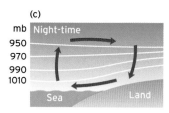

Figure 6.16 Circulation of sea and land breezes for (a) and (b) day time and (c) night-time. (Source: after Robinson and Henderson-Sellers, 1999)

6.3 Influence of water bodies

Unlike land surfaces, water bodies have little diurnal change in surface temperature except in very shallow water close to the water's edge. Surface temperatures are fairly constant for a number of reasons. Solar radiation is transmitted through water to a considerable depth and is not absorbed at the surface as is the case for land surfaces. The high **specific heat** of water (the energy required to increase water temperature) means that it requires more energy to be absorbed for any given temperature change than other substances. Furthermore the surface layers of water bodies tend to be well mixed, which helps spread any temperature change through a substantial depth of water. In addition, energy at the surface is used largely for the latent heat needed for evaporation rather than sensible heat that would cause a change in water temperature.

Over the land there are much more substantial diurnal changes in air temperature particularly in the summer half-year when solar radiation is stronger. In the middle latitudes, sea surface temperatures (and the air in the layers close to the surface) are therefore cooler than land surfaces

during the day in the summer half-year. They are warmer than land surfaces at night. The same is true in the high latitudes but in the winter half-year the temperature of the snow-covered land may remain colder than sea surface temperatures during both day and night. Such differences in local temperature result in sea and land breezes as shown in Figure 6.16. Features like this can also develop where there are large inland bodies of water, such as the Great Lakes in North America, when they are called lake breezes.

Sea breezes form only when there are light wind conditions (typically anticyclonic conditions) as the stronger winds of more active systems help reduce land–sea temperature differences through vigorous mixing of air. Summer sea breezes are a feature of many coastal areas and typically have speeds between 2 and 5 m s⁻¹. Sea breezes exist from the surface to 2 km above ground and may penetrate 30 km or more inland (occasionally as far as 100 km). The sea (or lake) breeze brings cooler (occasionally up to 10°C cooler) more humid air inland and a shallow sea breeze front may be evident. Uplift takes place along this front and can trigger the development of cumulus cloud (Figure 6.17).

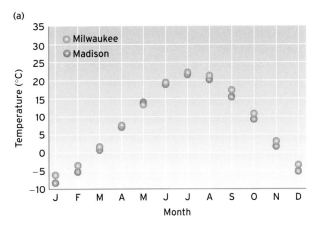

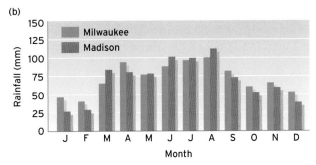

Figure 6.18 Mean monthly temperature (a) and precipitation (b) at Milwaukee (43°N) and Madison (43°N). Milwaukee is close to Lake Michigan and so has milder winters owing to the influence of the warmer water. The lake water in winter is warmer than the surrounding land because it cools more slowly than land. Thus air over the lake becomes warmed by the lake water. The lake also acts as a moisture source and so winters in Milwaukee are wetter than those in Madison that is too far away from Lake Michigan to be affected.

Figure 6.17 Cumulus clouds develop at the coast where air rises over the land and migrates seawards.

These clouds can be carried seawards by the counterflow present aloft. Occasionally sea breezes from different directions can converge enhancing uplift, which can lead to convective showers. The sea breeze dies off as night falls and is often replaced by a weak land breeze (Figure 6.16).

In the introduction, the regional change in climate across the mid-west United States was highlighted as an example of how, although within a single climate type, there would be differences in local climate. A further illustration of this can be seen in the temperatures and rainfall of Milwaukee and Madison in Wisconsin, USA (Madison is about 120 km west of Milwaukee). Both are in the same climate type but Milwaukee comes under the influence of Lake Michigan, giving it slightly milder winters and slightly drier summers than Madison. This is illustrated by climate data shown in Figure 6.18.

Another feature of middle- and higher-latitude large inland water bodies such as the Great Lakes in North America is lake-effect snow. When cold polar or Arctic air moves across a warmer underlying water body the temperature of the lowest layers of the air is increased, which causes instability and the formation of convective activity. As well as warming the lowest layer of the air, some additional moisture is also added, but the main effect is the relative warming (the air may still be very cold, just not as cold as upwind of the water). The convective activity causes precipitation which, as the air is cold, falls as snow. The convective activity can be enhanced at the edge of lakes as the greater surface friction over the land can cause slowing down of air and hence some convergence which then forces air to rise. Convection can be also enhanced with across-lake flow if there is an upslope (or orographic) effect when the air reaches land. Lake-effect snow results in annual snowfall totals of over 250 cm to the lee of the lakes, and exceeds 500 cm in the Tug Hill Plateau in New York (to the lee of Lake Ontario) and on the Keweenaw Peninsula of northern Michigan (to the lee of Lake Superior).

In areas where lakes become frozen (often with snow cover) the lake-effect season ends at this point. For example, at Lake Winnipeg there can be lake-effect snow in November but then the lake freezes and the effect ends. For Lake Erie the lake-effect snow season often ends in late January or early February when the lake freezes over. The Lake Ontario lake-effect snow season, however, continues into March as it does not freeze completely.

As well as occurring over lakes a similar snow effect can be found with polar or Arctic air flow over bays, some more enclosed seas and oceans. Examples include the eastern Black Sea region, the Aegean and Athens, eastern Italy and the Adriatic, and in the United Kingdom cold, fairly dry, easterly air flow from Europe can be warmed and moistened over the North Sea giving snow over eastern England and eastern Scotland. An ocean effect is also seen in the Sea of Japan and can occur in Nova Scotia and Cape Cod and in northern Scotland on the other side of the Atlantic. In strong northerly Arctic maritime air, the effect of warming by the north-east Atlantic can be enhanced by the development of small cyclonic features called polar lows, which, for example, can bring snow to northern Britain even as late as June (e.g. snowfall at the start of June 1975 led to the cancellation of cricket matches).

Reflective questions

➤ Why do sea and land breezes form?

➤ Why is there more snow in the vicinity of some large lakes?

6.4 Human influences

6.4.1 Shelter belts

It was noted earlier that coastal areas are subject to stronger winds than inland areas when winds are blowing off the sea. This is a result of reduced friction over the relatively smooth sea surface. This suggests that any alterations in surface roughness have an impact on the local wind. However, unless there is a permanent change to a new surface (as from sea to land) individual roughness elements such as trees or buildings will have an effect for only a relatively short distance downwind. If a line of trees or hedges is planted upwind of a field sown with sensitive crops it is possible to reduce the local wind speeds in the field to provide some protection. The same thing can be done around a

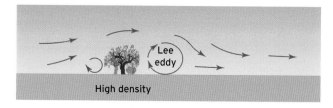

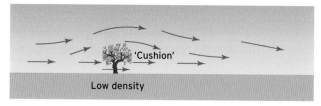

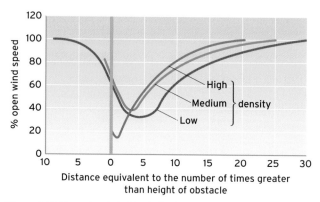

Figure 6.19 The role of shelter belts in reducing wind speed downwind. Low- and medium-density belts offer better shelter than high-density belts (e.g. walls) for a greater distance away from the belt. This is because the small amount of air flowing through the low-density belts cushions the air flowing over the top of the belt. Without this the air flowing over the top immediately subsides causing fast-moving eddies to form and thus rendering high-density belts less useful. (Source: after Nägeli, 1946)

garden to provide shelter. The ideal shelter belt is slightly permeable as the lower layer of air acts somewhat like a cushion and extends the reduction in wind speed over a longer distance as shown in Figure 6.19. An impermeable shelter belt will produce a greater reduction in wind speed close to the belt. However, recovery of the wind speed to upwind values is more rapid for impermeable belts, taking place over a distance of about 10–15 times the height of the barrier. Low-density shelter belts have an impact 15–20 times the height of the barrier downwind with medium-density shelter belts having the greatest impact of up to 20–25 times the height of the barrier.

The use of shelter belts is best when there is a particular wind direction from which damaging winds come. Examples can be found in southern France where shelter belts are

Figure 6.20 A snow fence near Mayflower Beach, Dennis, Massachusetts. A build-up of snow can be seen around the fence which reduces the amount of snow reaching the adjacent path on the left. (Source: Getty Images: Betty Wiley)

planted to protect crops from winds that can come down the Rhône Valley from the north. Snow fences set back from roads and railway lines are also a type of shelter belt as they are used to reduce air flow to allow snow to fall to the ground before the air reaches the road. This keeps roads and rails more clear of snow than would otherwise be the case (Figure 6.20). In many ski resorts snow fences are also used to reduce local air flow and help allow snow to accumulate on the pistes.

6.4.2 Urban climates

The fabric of towns and cities substantially alters surface characteristics compared with surrounding rural areas. The urban surface is much rougher than most vegetation. An indication of roughness can be given by what is called the **roughness length**. This is of the order of 5–20 cm for agricultural crops but up to 10 m for tall buildings. There are also important changes to the radiation and energy fluxes in urban areas. The urban fabric (stonework, road materials, roofs and so on) strongly absorbs solar radiation. There is also a substantial release of energy into the urban atmosphere as a result of humans heating their environment (domestic and industrial). This is especially the case for

mid- and high-latitude cities in winter. The urban atmosphere is also affected by air pollution with increased levels of carbon monoxide, oxides of nitrogen and various hydrocarbons. Although the local climatic impact of urban areas is always present, it is reduced in strong wind conditions as the vigorous mixing spreads any impact through a greater depth of the atmosphere and rapidly transports effects away from the urban area. The greatest impact of urban areas on the local climate is therefore found during light wind conditions. While the most important impact of urban areas on climate is linked to air pollution, there are local climate changes in both the temperature and wind regimes experienced by towns and cities.

The most commonly discussed climate modification in urban areas is the **urban heat island** effect. This is so called as the urban area is an 'island' of warmer air within the surrounding cooler rural air. The urban heat island occurs mainly at night when the urban atmosphere cools more slowly than that in the surrounding rural areas. It is more strongly developed in generally light wind conditions with clear skies when long-wave radiational loss is greatest (at night). There appear to be critical wind speeds above which the urban heat island disappears. Examples are 12 m s^{-1} for London, 11 m s^{-1} for Montreal and 4–7 m s^{-1} for Reading (England) and these critical speeds are related to size of the urban area.

Oke (1976) identified two parts to the modification of the urban atmosphere: the **urban canopy layer** and the **urban boundary layer** (Figure 6.21). Oke (1987) also suggested how an urban heat island would develop these two layers. In the case of the urban canopy layer he suggested that the following would all play a role: (i) greater absorption or direct solar radiation due to 'canyon geometry' (width of the road plus the height of the buildings on either side);

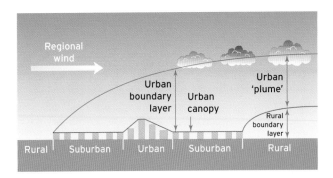

Figure 6.21 The urban canopy and boundary layer. The canopy layer consists of the spaces around the local buildings beneath the mean height of the buildings, whereas the boundary layer is the layer affected by the urban environment. (Source: after Oke, 1976)

(ii) greater daytime heat storage due to properties of urban materials; (iii) anthropogenic heat release from buildings (largely due to heating losses); and (iv) decreased evaporation. In the urban boundary layer, however, he suggested entrainment of air from the canopy layer, anthropogenic heat from roofs and chimneys, and downward flux of sensible heat (see Chapter 4) from the overlying stable layer would be the principal causes of the urban boundary layer heat island.

Even in relatively light wind conditions, the turbulence caused by air flow over buildings is enough to maintain a well-mixed atmosphere (up to an altitude of a few hundred metres in the largest cities). This mixing establishes an adiabatic lapse rate and as a result a stable layer is formed aloft. The heat island formed in this well-mixed urban boundary layer is strongly related to the city size and types of building in the city and therefore the largest urban boundary layer heat islands are found in cities such as New York (e.g. Gedzelman and Austin, 2003). The failure to differentiate between the local urban canopy layer and the

more general urban boundary layer heat islands can lead to erroneous conclusions being drawn in observational studies.

Oke and East (1971) recorded a maximum heat island of up to 12°C in Montreal in winter, which is unusual for mid-latitude cities where the maximum heat island is normally observed in summer. However, in high-latitude cities the anthropogenic heat losses in winter can be very large (owing to the extra internal heating switched on in offices and homes), leading heat island maxima to occur at that time of year. Even in London, a mid-latitude city with relatively mild winters, anthropogenic heat losses reach over 200 W m^{-2} in winter which is greater than typical solar radiation amounts at that time of year. Infrared technology is now being attached to aircraft to examine which buildings emit most heat and to determine those areas most in need of additional insulation (Figure 6.22).

While urban heat islands are at their maximum under light wind conditions, in strong winds the buildings can create powerful gusts as the wind is forced to flow round

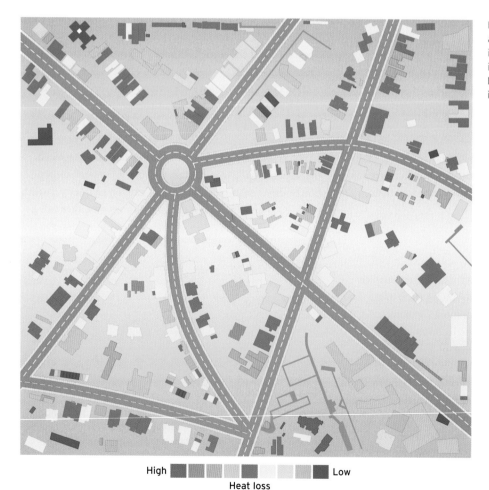

Figure 6.22 Heat loss from an urban area detected using infrared sensors on an aircraft. It is possible to see which individual buildings are in need of increased insulation.

High   Low

Heat loss

Figure 6.23 Wind speeds can be increased in some places in urban areas by funnelling or the Venturi effect. (Source: Getty Images: A J James)

tall buildings (Figure 6.23). In a zone close to tall buildings gusts may reach 2.5 or even 3 times the mean wind speed upwind of the building. The **Venturi effect** occurs when winds are forced to funnel between two buildings increasing localized wind speeds (Figure 6.24a). These winds can cause difficulty in walking and opening doors and may put severe stress on the buildings. Transverse currents can be generated when buildings are at right angles to the wind. Here pressure differences between the upwind and downwind sides of the buildings can lead to unexpected strong gusts (Figure 6.24b).

Overall, however, large urban areas tend to reduce overall mean wind speeds as their greater roughness slows wind speeds to below their rural upwind values. This modification of wind flow around buildings also plays a role in the dispersion of air pollution and can lead to areas when the local air pollution is unusually high as a result of the trapping of pollution between buildings. Although there is some evidence that convectional rainfall can be enhanced by urban areas (but perhaps downwind of the city), there is no real indication that urban areas have any influence on precipitation events. Therefore, the climate modifications created by urban areas are largely the canopy and urban boundary layer heat islands and the modification of the wind regime. Box 6.5 describes techniques that you can use for measuring the urban heat island. In addition there is increased air pollution in cities but levels at any one place are related to emissions, dispersion and local geography and not just to any change in the local climate. Urban areas do also have some influence on local atmospheric moisture conditions related to lower evapotranspiration (Deosthali, 2000).

6.4.3 Atmospheric pollution and haze

Most air pollution events are localized in urban or industrial areas where traffic emissions or pollutants from factories occur. There can be periods when the atmospheric conditions

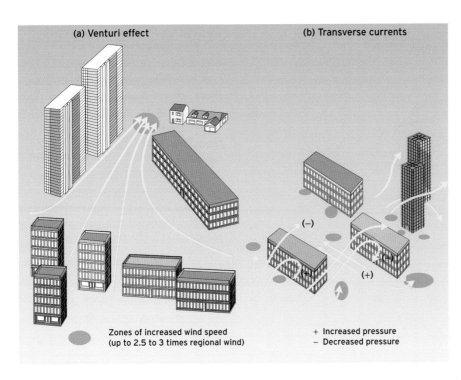

(a) Venturi effect (b) Transverse currents

Zones of increased wind speed
(up to 2.5 to 3 times regional wind)

+ Increased pressure
− Decreased pressure

Figure 6.24 Wind regime around buildings: (a) flow can be funnelled into narrow passages between buildings resulting in the Venturi effect; (b) transverse currents can develop when buildings are at right angles to the wind. (Source: after Thurow, 1983)

TECHNIQUES

OBSERVING URBAN HEAT ISLANDS

Urban heat islands provide students with an opportunity to make their own observations of human influence on the local climate. The urban heat island (measured by the difference between temperatures in the surrounding rural areas and those within the town or city) is most clearly marked at night and as such is not a warming of the urban atmosphere but rather a reduced cooling rate in comparison with surrounding rural areas (Figure 6.25).

The best conditions under which to observe the urban heat island are on nights with light winds and clear skies. The light winds are important as stronger winds will mix the atmosphere and the surface cooling which takes place at night (through long-wave radiation loss) will be spread through a greater depth of the atmosphere. The lack of cloud cover is important as the water droplets in clouds mean that clouds act as black body radiators at terrestrial temperatures and so the long-wave radiation loss from the surface at night will be largely compensated for by the long-wave radiation emitted from the cold base. In fact, well-mixed air in the **surface boundary layer** of the atmosphere will tend to have an adiabatic lapse rate in the absence of strong surface heating (through solar radiation warming) or cooling (long-wave radiation loss to the atmosphere). It is therefore the case that on nights with strong winds and a complete low-level cloud cover, the lapse rate up to the cloud base will be close to the dry adiabatic lapse rate (this is also the case during the day if the low-level cloud cover is fairly dense limiting strong solar heating).

In making observations on what are termed 'radiation nights' (nights with light winds and clear skies), it is important to be aware that certain lower-lying spots are often prone to record somewhat lower temperatures than those in the surrounding areas (see Section 6.2.5). Thus in taking an observation at any site to assess the urban heat island effect it is important to note that the relative elevation of the site and the opportunity for colder air to drain away from, or towards, the site will have an impact on the observation.

Early research on the urban heat island made use of thermometers attached to vehicles. These vehicles were usually driven on a route into the town or city and then returned by the same route. Assuming the weather conditions remained the same over the period of the two transects (inwards and outwards), the first set of observations on the inward transect would record higher temperatures at each location than the return transect as cooling would be taking place throughout the period. Observations of the urban heat island typically show a relatively sharp rise in the value of the urban heat island in moving into the built-up area of the town or city. There then tends to be a slower rise into the city centre.

It is important to note that in making observations of the urban heat island any individual, or a group of individuals, will generally be making their observations within the urban canopy layer (that layer of air within what are termed the 'street canyons') and not the urban boundary layer. Observation of the urban boundary layer requires the use of meteorological masts or

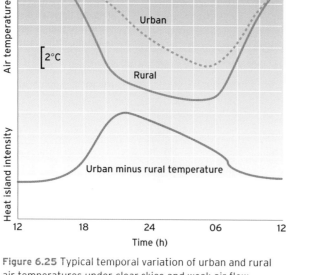

Figure 6.25 Typical temporal variation of urban and rural air temperatures under clear skies and weak air flow. (Source: after Oke, 1987)

BOX 6.5 ➤

➤
Table 6.3 Suggested causes of canopy layer urban heat island

Urban feature	Effect on canopy layer urban heat island
Air pollution	Increased absorption and re-emission
Canyon geometry	Reduced view of sky, reduced wind speed
Buildings	Direct heating of the local environment
Vehicles	Direct heating of the local environment
Construction materials	Increased thermal admittance and water proofing

remote sensing techniques. There are a number of suggested causes of the canopy layer heat island (Table 6.3) and the importance of these will vary depending on the particular building, the building structure, orientation with respect to the wind, traffic volume and the canyon geometry (topography of the buildings and urban structures).

Rather than studying a long transect from the rural surroundings to the centre of the town or city, an interesting study can be made by taking observations at locations which have clear differences in urban features. For example, quite rapid changes in temperatures can be observed in moving from urban parks into built-up areas (Figure 6.26). Urban heat island investigations can make interesting projects but the research design, detailed observation of the environment and proper and consistent exposure of the thermometer(s) are vital. Examples of investigations could include:

- observing any temperature differences between relatively narrow and broad street canyons;

- looking at differences between deep and shallow street canyons of the same width;

- recording differences between temperatures in parkland (on lit paths) and nearby built-up areas;

- vehicle traverses from rural areas into and out of the town or city.

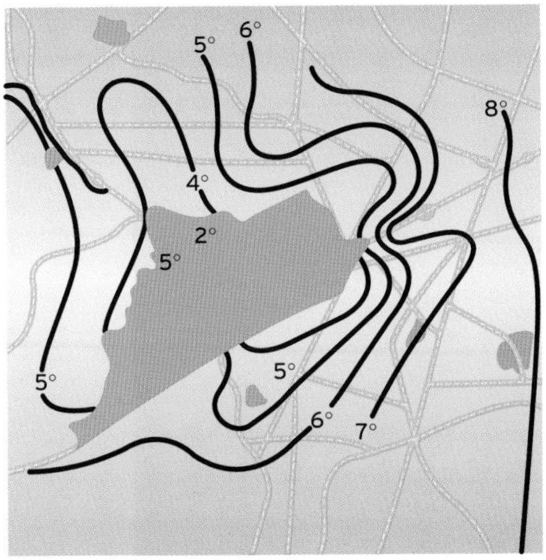

In undertaking any study of the urban heat island canopy layer, it is important to make detailed observations of the local environment (e.g. building type, materials, canyon geometry) as well as taking temperature observations. Any change in weather conditions will also influence observed temperatures and it is crucial that there is a proper and consistent exposure of the thermometer. Thermometer exposure refers to the way in which the thermometer is presented to the environment. At fixed sites thermometers are in screens designed to ensure that the thermometer records the air temperature as it is shielded from solar and terrestrial radiation (see Figure 5.2 in Chapter 5). Ideally all thermometer exposures should provide radiation shielding and the observations can be improved if air is drawn past the thermometer (**aspiration**). Errors in exposure can invalidate observations. An interesting discussion on exposure is provided by Perry *et al.* (2007). The temperature observations must also be made at the same height above the surface, as in some places temperature lapse rates can be steep at night. The difference between the ground surface and the minimum temperature recorded by a thermometer inside a Stevenson screen at 1.2 m height can be over 10°C on a calm clear night with a snow-covered surface.

Before undertaking any urban heat island investigations there are certain safety issues that need to be addressed. Personal safety is important and as investigations will take place at night it is important that

Figure 6.26 Isotherms (heat contours) around an urban park, Mexico City, with clear and calm air. (Source: after Jauregi, 1990–91)

BOX 6.5 ➤

➤

such a study should not be undertaken by one person on their own. Observations made using thermometers mounted on a vehicle provide increased safety but it is important that the driver is not involved in making any observation of either temperatures or the nature of the environment through which the vehicle is being driven. The driver's attention must be fully given to the road, pedestrians and to all types of other vehicles. If a vehicle is not used, high-visibility clothing should be worn and observations should be made by two or more persons remaining together. Again for safety, when on foot the observers should keep to reasonably well-lit areas and should avoid areas which are known to be unsafe.

In assessing the results it will be important to examine whether the geographical location (how close to the centre of the town or city) and the nature of the local environment at each location (Table 6.3) help provide any explanation for observed differences. It is also important to be aware that the strength of the urban heat island varies over the course of a night and temperature differences between different locations may be the result of the different times at which the observations were made (Chow and Roth, 2006). Any differences in temperatures between different locations may also provide some indication of the rate at which any pollution is dispersed as poor mixing of air both vertically and horizontally will encourage greater pollution concentrations and a higher local air temperature can be related to areas of calm air where mixing is reduced. Air pollution in towns and cities is largely due to that from vehicle exhausts and high vehicle volumes combined with poor local mixing can lead to higher pollution concentrations in certain urban streets.

BOX 6.5

of urban areas are a danger to human health (Figure 6.27). In regions with high solar radiation such as Los Angeles, Athens and Mexico City, the ultraviolet radiation reacts with the uncombusted hydrocarbons from vehicle emissions and produces a photochemical smog which irritates the eyes, nose and throat. However, often these conditions are localized and do not persist for long periods. Recently, however, in some places there have been haze pollution events that have lasted months and spread over many hundreds of kilometres.

Many of the large haze pollution events have resulted from forest fires. Forest fires can occur both through natural action and by human intervention. Forest fires in Indonesia are very largely the result of human activity as fire is used to clear land for agricultural purposes. Fire is cheap and, as well as reducing vegetation cover, enriches what are often very poor soils. Indigenous tribes, such as the Dayak people in Kalimantan, have traditionally used **shifting cultivation** (slash and burn) techniques and their use has been in tune with the natural environment with strict traditional rules of

Figure 6.27 Smog over Mexico City. Smog blocks out the sunlight and promotes respiratory and other health problems. (Source: Alamy Images: Hemis)

using fire. Unfortunately, the large number of settlers who came from other islands and new plantation companies do not follow rules that help the long-term maintenance of the environment. Plantation companies (or people hoping to profit from providing services to them) have a particular responsibility as they are largely aware of potential environmental damage and yet place a higher value on their own profits. The Indonesian Government has banned the use of fire for clearing land for a number of years but fires continue to be lit as Indonesia expands its wood pulp, palm oil and rubber industries. In addition to fires, the extensive logging of the rainforests, particularly selective logging, plus other land-use changes have played an important role in making the Indonesian rainforests more susceptible to fire. Rainforests are humid and fires do not naturally take hold. Selective logging and other agricultural land uses open up the forest and allow it to dry out more easily. The forest fires have caused massive damage within Indonesia but due to the smoke from the fires, damage has also been caused to neighbouring countries, such as Malaysia and Singapore. The haze from the fires in 1986 covered the South-East Asian region for weeks, particularly affecting Malaysia and Singapore, causing health problems, disruption of shipping and aviation, and also caused the temporary closure of international airports. As a result there are significant economic losses as well as ecological damage. A useful history of Indonesian fires is presented by Gellert (1998) and the Indonesian fires of 2006 are covered in more detail in Box 6.6.

CASE STUDIES

SOUTH-EAST ASIAN HAZE OF 2006

A severe haze occurred from September to late October 2006 across South-East Asia (Figure 6.28). This event was caused by uncontrolled burning associated with land clearance in Indonesia. The haze affected Mayalsia, Singapore and southern Thailand and may have reached South Korea. The haze added to any local sources of pollution and so was an even greater problem in industrial and high-density urban areas. As with

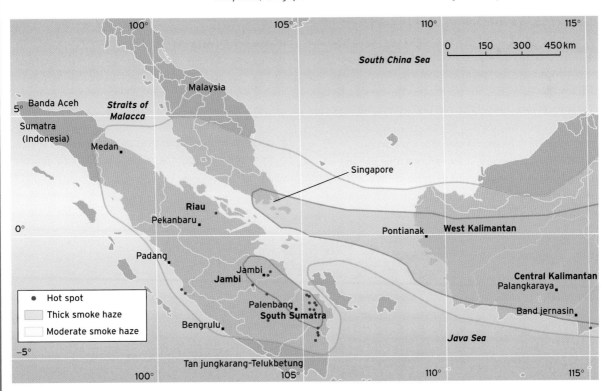

Figure 6.28 Haze pollution across South-East Asia on 19 October 2006. (Source: National Environment Agency of Singapore)

BOX 6.6 ➤

➤

many air pollution events, the local topography played a role, especially in the highly urbanized and industrialized Klang Valley of Malaysia. Air quality improved in late October after heavy rainfall helped put out the forest fires.

Singapore has a good air quality monitoring network and there is a telemetric network of air monitoring stations strategically located in different parts of Singapore. These stations measure sulphur dioxide, nitrogen dioxide, ozone, carbon monoxide and particulate matter called PM10 (particulate matter of 10 microns or smaller in size). The Singapore Government uses the 'Pollutant Standards Index' (PSI), an index developed by the US Environmental Protection Agency to provide accurate, timely and easily understandable information about daily levels of air pollution.

Table 6.4 Pollution Standards Index

PSI value	PSI descriptor
0–50	Good
51–100	Moderate
101–200	Unhealthy
201–300	Very unhealthy
Greater than 300	Hazardous

The index is based on the pollutants listed above. The PSI value is calculated and graded according to Table 6.4. A PSI value is assigned to each pollutant through a linear function relating the pollutant level to its PSI sub-index. Once all sub-indices are calculated, the maximum sub-index is given as the overall PSI value.

During the 2006 haze event the PSI value in Singapore reached 150 on 7 October (well into the unhealthy level) as a result of the PM10 from the Indonesian forest fires. Even higher values were recorded in Malaysia. The pollution led the Singapore Ministry of Education to recommend the suspension of outdoor activities.

BOX 6.6

As well as human activity, natural climate cycles can increase the susceptibility of forests to fires. For example, during El Niño events there tends to be reduced rainfall over Indonesia which leads to drier than usual vegetation. Severe fires in Indonesia took place during the two strongest El Niño events of recent years, 1982–1983 and 1997–1998, and hence these fires are commonly cited as impacts of El Niño events. However, while there is some evidence of a connection between El Niño and forest fires for earlier strong El Niño events, there were no fires as severe as those of 1982–1983 and 1997–1998. Additionally, fires now occur every year and are not simply coincident with El Niño events.

Reflective questions

➤ Why is a permeable barrier more effective as a shelter belt than an impermeable barrier?

➤ What is the urban heat island effect?

➤ If the mean wind speeds in cities are slower than in rural areas why are urban wind gusts sometimes faster?

6.5 Summary

Local climate variations vary from extremely localized microclimates to more generalized regional climates within a general climate type. Regional and local geographies are the key factors in determining the magnitude and importance of these climates. A particular climate classification type can therefore be affected more locally by a number of different influences, some of which will depend on particular weather conditions or on the time

➤

of year. The most obvious and rapid climate gradients result from changes in altitude. With increasing altitude atmospheric pressure decreases, and associated with this are decreases in temperature. Precipitation totals often increase substantially with altitude and the wind regime in hills and mountains can be much more severe than that for nearby low-lying ground. Small topographic features can also be important. At night, radiational cooling takes place, reducing the temperature of the air close to the ground. This cooler air is slightly denser and tends to move down slopes to lower ground and into depressions (this is termed katabatic drainage). There are also micro-climate variations introduced by vegetation and these include both forests and deliberately planted shelter belts.

The climate near sea coasts is also modified as a result of the influence of the sea, particularly in middle and high latitudes. Coastal areas are often subject to stronger winds than more inland areas. Winds over the sea are stronger as the frictional reduction of wind speed close to the surface is much less than over land. Therefore, if the wind direction is from the sea to land, coastal areas are subject to stronger winds. In addition, there can often be substantial differences between air temperatures over land and sea partly due to the high specific heat of water

but also due to the ability of solar radiation to penetrate to considerable depths before being fully absorbed. This can lead to cooling sea breezes in coastal regions particularly in summer (see below). Lakes can also have an influence on the climate such as lake-effect snow, but except in the case of very large lakes (such as the North American Great Lakes that behave like inland seas) any effect is limited to a very narrow strip around the edge of the lake.

Human activity can also alter climate on a regional and even on a global scale. Examples include the haze pollution caused by forest fires, and emissions of CO_2 into the atmosphere through the burning of fossil fuels (see Chapters 3 and 24). However, there are more localized changes to climate that are a result of human activity. The most obvious example is the creation of a distinct urban climate in built-up areas. Urban areas affect the urban atmosphere and in lighter winds strong heat islands can develop. The urban atmosphere can be split into a canopy (between building) layer and a boundary (above roof) layer and each will have its own heat island. Buildings also modify wind flow. Towns and cities tend to have lower wind speeds but greater gustiness than surrounding rural areas.

Further reading
Books

Barry, R.G. (1992) *Mountain weather and climate*. Routledge, London.
Excellent book on topographic controls of local, regional and global climate with a whole section dedicated to case studies.

Geiger, R., Aron, R.H. and Todhunter, P. (2003) *The climate near the ground*. Rowman and Littlefield, Oxford.
This is republication of a book from 1965 which is still of great relevance today. It details a whole range of ground–air interactions ranging from soil and vegetation energy balances to the deposition of dew on ponds. It is a very detailed text.

Oke, T.R. (1987) *Boundary layer climates*. Routledge, London.
This is a fairly technical and physically based textbook with good detail and excellent explanations of differences between processes over vegetated and non-vegetated surfaces.

Robinson, P.J. and Henderson-Sellers, A. (1999) *Contemporary climatology*. Pearson Education, Harlow.
Chapters 10 and 11 are particularly relevant to the material discussed above.

Rosenberg, N.J., Blad, B.L. and Verma, S.B. (1983) *Microclimate: The biological environment*. John Wiley & Sons, New York.
Only Chapter 9 covering the subject of shelter belts is relevant here.

Further reading
Papers

Three useful research papers which look at change over time in gradients of rainfall and surface temperature with altitude and examine upland climate and weather are:

Beniston, M. (2006) Mountain weather and climate: a general overview and a focus on climatic change in the Alps. *Hydrobiologia*, 562, 3–16.

Burt, T.P. and Holden, J. (2010) Changing temperature and rainfall gradients in the uplands. *Climate Research*, 45, 57–70.

Holden, J. and Rose, R. (2011) Temperature and surface lapse rate change: a study of the UK's longest upland instrumental record. *International Journal of Climatology*, 31, 907–919.

A paper that deals with the topic covered by Box 6.4 is:

McClatchey, J., Runcres, A.M.E. and Collier, P. (1987) Satellite images of the distribution of extremely low temperatures in the Scottish Highlands. *Meteorological Magazine*, 116, 376–386.

PART IV

Geomorphology and hydrology

Figure PIV.1 Small-scale braided flow patterns, similar to those seen on large rivers, flowing into a pool with a cluster of pebbles on a sandy beach. Many landforms and processes can be seen to develop and operate at both small and large scales with feedbacks operating between the scales. Understanding the interactions of these processes and landforms is the key to understanding geomorphology.

Part contents

Scope

This part of the textbook deals with those processes outside of tectonics that mould and shape landscapes. The processes of weathering are considered in Chapter 7 and the discussion includes consideration of weathering under different environmental conditions including in the urban environment. Weathered materials are transported by erosion. Chapter 8 considers erosion processes and deals with the interactions between the shape of the landscape and erosion processes. Chapter 8 deals with issues of form-process relationships in relation to hillslope profiles and landscape evolution. However, the landform itself is not simply a product of geomorphological or hydrological processes: the processes themselves will be influenced by the landform. In addition, landscapes may incorporate a legacy from the operation of geomorphological processes under past climatic regimes such as dryland and post-glacial environments. Therefore contemporary processes and landscapes are likely to be linked to historic processes and landscapes and there may be time lags in landscape response. Change through time in hillslope form is covered in Chapter 8 ranging from slow creep processes to fast, hazardous landslides. Sediments transported by erosion processes can also form landscapes elsewhere and the nature of sediments and sedimentary landforms is covered by Chapter 9.

The products of weathering and sediment movement often help to form soils. The soil acts as the interface between the atmosphere, lithosphere, hydrosphere and biosphere in which we exist. It performs a wide range of essential functions that sustain life. The properties of soils are determined by the processes that maintain the inputs and outputs of material through the soil system. Soil can be degraded, lost and improved by natural or human activities. An appreciation of these issues is imperative if we want to preserve our soil in a healthy state. Chapter 10 attempts to provide such a basis.

This part of the book also deals with the nature of water (especially Chapters 11, 12, 13 and 15), solutes (especially Chapter 15, but also Chapter 7) and ice movements (especially Chapters 16 and 17) in a range of environments including drylands (Chapter 14), spectacular glacial landscapes (Chapter 16), periglacial environments (Chapter 17) and coastal environments (Chapter 15). It describes such motions of energy and mass on the hillslope, in the floodplain, within channels dominated by water and ice, and in coastal zones. The first seven chapters of this part detail processes that can occur on any landscape on Earth whereas the remaining four chapters describe the processes and nature of particular Earth environments.

Studying the form of the Earth and its various characteristics seems an obvious facet of geography and is commonly perceived as the entirety of the work physical geographers do. However, this refers only to geomorphology. Geomorphologists are interested in trying to ascertain the processes and mechanisms that interact with, and create, the form of the landscapes we study. Often, while most processes may be operating on all land environments, the dominating processes depend on the regional climate characteristics or tectonic context. The chapters in this part of the book highlight that it is often the relative abundance of the various processes, and the balance between the process rates, that determines the nature of the landforms in different environments.

On a rudimentary level, it is the Earth's attempt to create some kind of balance of inputs and outputs of energy that leads to the internal reorganization of sediment and materials, resulting in the changing shape of landscapes. It is through uplift, erosion and sedimentation that landscape forms are maintained. Erosion and sedimentation are dependent upon the physical agent exchanging the energy in the system, which is most commonly a transporting fluid (water, wind or ice), although gravity also plays a significant role. Although erosion and sedimentation are contrasting processes, they are inextricably linked and the same sediments can be reworked many times. Chapters 7, 8 and 9 delve into issues of weathering, erosion, sediments and sedimentation to provide a basis for the following chapters. Sediments (Chapter 9) and solutes (Chapter 13) are transported from hillslopes (Chapter 8) to fluvial environments (Chapter 12), most commonly by hydrological processes (Chapter 11), which also interact with glacial sediment and solute systems (Chapter 16), finally ending at the coastal zone (Chapter 15). This input-output budgeting concept is most commonly used in catchment hydrology and fluvial geomorphology as well as within glaciology where the mass balance (the budget) of glacier ice is studied with respect to its influence on glacier movement and hydrology. Coastal studies are another classic example in which the sediment budget is used as a framework for studying dynamic system adjustments to wave, tide and current processes.

Within every chapter in this part, special consideration is given to the influence of humans on landscape processes and how understanding of the processes can result in improved management strategies.

CHAPTER 7

Weathering

Bernard J. Smith

School of Geography, Archaeology and Palaeoecology, Queen's University Belfast

Learning objectives

After reading this chapter you should be able to:

➤ describe the main weathering processes

➤ explain the role of water in weathering

➤ understand the importance of multiple weathering processes operating in combination or sequence

➤ describe the climatological and geological controls on weathering at different scales

➤ outline the main factors associated with building stone weathering

7.1 Introduction

As the name implies, this chapter will deal with how **geomaterials** (rocks and **regoliths**) react when exposed to weather. This is a study of change or '**metamorphism**', in which materials that were typically formed under conditions of high temperature and/or pressure within the Earth's crust come into equilibrium with the highly variable and often rapidly fluctuating low-temperature, low-pressure and chemically complex conditions at and near its surface. In most cases this involves a progressive simplification of their mineralogy and chemistry and a decrease in their structural integrity.

This chapter will examine one of the most crucial Earth surface processes. Without weathering there would be no soils (Chapter 10), and thus no plants and no people. Without weathering there would also be a dramatic reduction in global erosion and deposition (Chapter 8). Weathering directly provides the solute load of rivers (Chapter 13) and is an essential precursor that unlocks the transport of sediment by all but the most energetic of geomorphological processes (Chapter 9).

Despite this, weathering is something of a neglected field of study in geography and environmental science, perhaps because there are sometimes challenging chemistry and physics equations, but more often because the changes involved in weathering are often individually very small, gradual and to a large extent hidden from view either inside rock or within and beneath a blanket of soil. The physical measurement of weathering effects is therefore complex, time consuming and requires long-term observation using specialist equipment before any noticeable change can be recorded. This is not to say, however, that over geological time the gradual changes brought about by weathering cannot be just as effective in shaping the Earth at the landscape scale as, for example, the most dramatic, but infrequent erosional events. This is because what weathering lacks in terms of magnitude it more than makes up for in terms of persistence and pervasiveness.

Because of its hidden nature and the somewhat limited attention often paid to its understanding, weathering is an area of study in which there remains much debate and uncertainty amongst researchers regarding precisely how different processes operate. One consequence of this is that those not involved in the debate are prone to sweeping generalizations regarding such things as the power of the Sun to crack rocks, which can confuse obvious symptoms with the underlying causes of breakdown and decay. For example, the widespread damage often associated with a severe frost may very well have more to do with its tendency to exploit weaknesses generated over a long period by other weathering processes than the ability of ice crystallization to fracture fresh rock. While there are some processes that act in isolation, it is much more common to find weathering processes operating in combination or in sequence to the extent that some processes may act as essential precursors to the operation of others.

7.2 Environmental and material controls on weathering

The controls on weathering processes represent a competition between two key parameters, those of environment and material, in which many see environment as the main driving force (Figure 7.1). Hence there is often a desire to classify weathering on the basis of climate into, for example, categories such as tropical and periglacial. It is clear, however, that the consequences of the exposure of geomaterials to different environmental conditions are also strongly influenced by a wide range of material properties. Indeed, where these properties are particularly distinctive they may overtake environmental influences in terms of their significance. Hence distinctive features and landscapes can be recognized that are associated with the weathering characteristics and controls exerted by, for example, granitic rocks

and limestones (e.g. see Box 7.4). Where these extreme views break down is when environments are less extreme and/or geological characteristics are less pronounced. It is in these situations (i.e. the majority) that one is required to pay more detailed attention to the complex interactions between the two sets of controls, as well as the precise nature of the processes operating.

There are no weathering processes that can be uniquely ascribed to a particular climatic zone. Hence a weathering classification based on climate may be problematic. There are a limited number of physical, chemical and biological processes that either are able to operate, or are prevented from operating, under certain environmental conditions, with differing degrees of effectiveness and in different combinations with other processes. For example, in the case of 'tropical weathering', we are not referring to a set of unique processes found only in these regions, but rather to particular combinations and intensities of generic weathering processes related to a specific climatic regime. This in turn results in distinctive weathering products and, in limited cases, supposedly definitive landforms. Such 'climatic landforms' are, however, not always what they appear to be. For example, granite tors that are found in many textbooks on tropical environments are in fact a feature of most climatic regimes (Figure 7.2). Their origins may have followed different pathways, but their final appearance often differs only in terms of the detailed morphology of the individual granite blocks. This convergence of shape or form is partly in response to a strong, but largely passive, lithological control on weathering conditioned by the characteristic mineralogy of granites. However, it also owes much to active structural controls that are a response to the release of physical stresses contained within the rock that in turn result in the formation of distinctive patterns of joints and fractures.

The role of surface morphology is itself important since there are multiple feedbacks in the weathering system (Figure 7.1). As weathering progresses it often develops complex surface morphologies such as variously sized hollows. This topography can then influence how processes operate. This might, for example, be through the modification of surface temperature by the creation of shade, by allowing water to accumulate, by allowing salt to accumulate in hollows protected from rainwash on vertical surfaces and by creating micro-environments in which organisms can flourish. These are positive feedbacks that encourage further growth of the hollow.

Weathering processes, which characteristically operate at the scale of individual grains and crystals, are controlled on a day-to-day basis by the complex micro climates experienced at the rock–atmosphere interface and within the pores

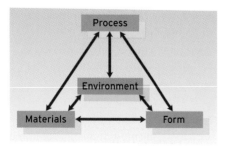

Figure 7.1 The weathering system.

(a)   (b)

Figure 7.2 (a) A granite tor in the Guadarrama Mountains near Madrid, Spain composed of joint bounded boulders. (b) Section in a road cutting near to the tor in (a) showing an 'incipient' tor shaped by differential subsurface weathering controlled by the joint system within the granite. If the weathered regolith were to be removed by erosion the tor would then be revealed.

of fresh rock and weathered regolith (see Chapter 8) and the soils above them (see Chapter 10). At this micro-scale, it is still temperature and moisture that mainly control patterns and rates of weathering but often in complex ways. In the case of temperature, seasonal and diurnal cycles together with erratic variations in surface temperature driven by, for example, variations in cloud cover and wind speed, interact to create complex patterns of temperature change over time. These parameters are particularly significant for a wide range of near-surface physical weathering processes, both directly through influencing responses such as thermal expansion of minerals and indirectly through their influence on variables such as relative humidity and moisture uptake, movement and evaporation. Overall, however, the moisture balance within these materials is driven primarily by the availability of precipitation across different spatial and temporal scales. 'Deep wetting' of materials is, for example, most likely to reflect seasonal rainfall patterns, whereas near-surface wetting and drying within a few centimetres of the surface is more likely to be 'event related' and controlled by factors such as the amount, duration, intensity and frequency of individual storms. Included in these factors is wind direction and velocity, which studies have shown to be crucial in determining the efficiency of driven rain in the wetting of rock (Ashurst and Dimes, 1998). What we do not understand in any detail, however, is how individual wetting events contribute over time to longer-term, deeper wetting and drying nor what role some processes such as direct condensation of water onto rock play in weathering.

The atmosphere is an important environmental control on weathering. It provides oxygen required for many

weathering processes. Air within pores and fractures of geomaterials can differ greatly from that of the open atmosphere depending on the nature of the materials and the presence of any decomposing organic matter capable of producing gases such as methane (CH_4). In a geomorphological context, however, the most important of these other gases is likely to be carbon dioxide (CO_2), because of its role in the formation of carbonic acid which dissolves rocks such as limestones. The most complex reactions between geomaterials and the atmosphere are most likely to occur in the course of urban stone decay. This is primarily the result of the addition to the atmosphere of gases such as sulphur dioxide (SO_2) and various oxides of nitrogen (NO_x) from the burning of fossil fuels. Through a combination of **photochemical oxidation** and reaction with moisture in the atmosphere this results in the formation of a complex chemical cocktail that includes naturally occurring carbonic acid as well as sulphuric and nitric acids and which falls as 'acid rain'. Total 'acid deposition' is, however, more complex than this as acids can be produced directly through the reaction of gases with moist stone surfaces (dry deposition) or precipitated onto surfaces from fog and condensation (occult deposition). While such deposition is, like naturally occurring carbonic acid in the atmosphere, very effective in dissolving carbonate rocks such as limestone, it plays a much more significant role when it reacts with building materials to produce soluble salts. These are responsible for widespread damage to stone through a variety of mechanisms associated with the process of salt weathering (Goudie and Viles, 1997). The most potent of these salts is generally considered to be gypsum (calcium sulphate, $CaSO_4 \cdot 2H_2O$), formed most commonly by the reaction of sulphuric acid with stone containing calcium carbonate, although it can be deposited directly onto building stone where gases have reacted in the atmosphere with carbonate-rich combustion particles (particulate deposition). In this way, and through reactions with carbonate mortars, it is possible for all stone types to be 'contaminated' with complex salts, even if they themselves contain no carbonate fraction. Because of this the urban environment not only ranks alongside deserts and coasts as an environment particularly prone to salt weathering, but is for much of the world's population the environment in which they are most likely to encounter readily identifiable evidence of weathering. In fact, it could be argued that the built environment represents a well-constructed 'weathering experiment', in which friendly architects simultaneously expose different combinations of stone types to the same environmental conditions, conveniently align structures to permit evaluation and comparison of meteorological effects and often provide the researcher with a considerably greater knowledge of the physical characteristics and stress histories of the stone than is normally available for natural rock outcrops. Through this it is possible to address the common complaint of students of natural weathering that explanations are hindered because they have too few controls on the processes operating and know little of the history of the rocks they study.

The following sections explore the salient points raised above by trying, as far as is possible, to link weathering processes to the environmental and material factors that control them and the features and products they produce.

Reflective question

➤ Why is it inappropriate to classify weathering only on the basis of regional climate?

7.3 Some fundamental processes

The processes by which geomaterials weather are normally grouped into three categories: physical, chemical and biological. In reality all three interact with each other and it is equally valid for the physical and chemical consequences of biological activity to be split and considered under the other two headings. This section opts for this split of biological processes under the two main headings and attempts an outline of the principal weathering *mechanisms* that constitute the different weathering *processes*. However, it is firstly important to acknowledge and explain the central role of water in all forms of weathering.

7.3.1 Weathering and the role of water

Alteration of geomaterials occurs in two realms. There are complex changes that take place relatively deep within the Earth's crust, generally in the absence of oxygen, in response to reactions with heated and chemically complex gases and liquids. Often referred to as hydrothermal alteration, these processes are enormously important in, for example, the mobilization and concentration of specific elements or weathering products into zones that can be then economically mined or quarried. This includes extensive deposits of the clay mineral kaolinite formed through the hydrothermal alteration of granites in areas such as the south-west of Britain, and which form the basis of that region's 'china clay' industry. In this chapter, however, we are primarily concerned with weathering that takes place at or near the Earth's surface under mainly **sub-aerial** conditions where they are in close contact with the atmosphere. In these environments the most important control on chemical alteration is the presence of water. This is both as a potential **reagent** and as the principal means by which the products of weathering are removed and weathering allowed to continue. The precise role of water, especially in chemical alteration, depends on a number of key properties.

Water molecules are surprisingly complex. They consist of one oxygen and two hydrogen atoms, but exist as a covalent bond in which electrons are shared between atoms. In water, this bond is uneven and the oxygen atom attracts **electrons** more strongly than the hydrogen ones. This creates polar molecules with an asymmetric charge distribution and partial negative and positive charges at the ends of each molecule. It is this polarity that allows water to separate other polar solute molecules and explains why water is such an effective **solvent** (see Chapter 13). On its own, water can undergo what is termed as self-ionization, in which one of the bonds of the liquid water molecules breaks to create positively charged aqueous hydrogen ions or protons (H^+) and negatively charged hydroxyl ions (OH^-). In pure water the free hydrogen ions react with other water molecules to produce H_3O^+. At normal temperature (25°C) the concentration of free hydrogen ions is extremely low at only one-tenth of a millionth of a gram per litre of water. To bring the figures into a more manageable range, the concentration is expressed as pH with a neutral water solution of pH 7 (see Chapter 10 for an explanation of pH). Any solution with a greater concentration of hydrogen ions is acidic and has a lower pH value, while those with a lower concentration have a higher value and are considered to be **alkaline**. In turn, any substance which, when added to the water, increases the proton concentration is therefore said to be acidic or a proton donor. In contrast, substances that are proton acceptors are said to be alkaline and are referred to as **bases**. As an example, adding a base such as sodium hydroxide to water results in the dissociation of an aqueous sodium ion (Na^+) and an aqueous hydroxyl ion (OH^-), which will then combine with any free hydrogen ions.

In nature perhaps the most common proton donor is carbonic acid, formed by the reaction of atmospheric carbon dioxide with water. When this dissolves in water it dissociates to form hydrogen and bicarbonate ions:

$$H_2O + CO_2 \rightarrow H_2CO_3 \rightarrow H^+ + HCO^{3-} \tag{7.1}$$

The importance for weathering of proton donation is that these protons can react with minerals such as many commonly occurring aluminium silicates (e.g. feldspars) and replace positively charged metal **cations** (e.g. Na^+, K^+ and Ca^{2+}) in their crystal lattices through a process known as hydrolysis. This leaves behind weathering products made up mostly of 'hydrated aluminium silicate minerals' or clays.

In addition, pH is a key control on the solubility of substances, especially weathering products such as metal oxides and hydroxides. The importance of pH in influencing relative solubility is indicated in Figure 7.3.

A further key property of water that controls chemical reactions and determines solubility is its redox potential, or Eh. This is the ability of a chemical species to bring about an oxidation or a reduction reaction. Eh is explained in Chap-

ter 13. In terms of rock weathering the most common oxidation or reduction reactions are those associated with iron oxides. Iron commonly exists in two **valent** forms, either as oxidized, trivalent ferric iron (Fe^{3+}) or divalent, reduced ferrous iron (Fe^{2+}). Ferrous iron is therefore typically found under saturated conditions and most commonly in its hydrated form as minerals such as limonite (FeO(OH) · nH_2O) and goethite (FeO(OH)) that give soils such as gleys (see Chapter 10) their distinctive yellowish green colour. When exposed to the atmosphere these minerals oxidize (rust), most typically into the mineral haematite (Fe_2O_3) that is responsible for the reddish colour of many iron-rich soils. Another important distinction between these two valent forms is their relative solubility. From Figure 7.3 it can be seen that the reduced form of iron is relatively soluble under typical Earth surface conditions, so that it can, for example, be mobilized and concentrated under saturated conditions. If, however, it is allowed to oxidize it will become fixed under all but the most acid of conditions. When associated with repeated wetting and drying this switch from ferrous to ferric states can lead to the progressive concentration of iron as, for example, nodules within a soil profile or more uniformly in distinct soil horizons.

The difference in solubility of iron in relation to its oxidized or reduced states is also a graphic indication that solubility in natural environments is dependent on both pH and Eh, both of which are influenced in turn by temperature. The relationships between the two are typically shown in the form of 'Pourbaix diagrams' for individual elements in which Eh and pH are plotted against each other to establish the range of conditions under which mobilization and fixation are likely to occur. The phase diagram for iron in water is shown in Figure 13.1 (see Chapter 13), indicating its various states in relation to combinations of Eh and pH.

As well as the chemical properties of water that are significant for weathering, water also possesses a range of physical properties that strongly influence both weathering environments and the processes that operate within them. For example, the strong hydrogen bonds in water molecules give water a high cohesiveness and consequently a high **surface tension**, adhesion and **capillarity** (see Chapters 10 and 11). Within large pores the majority of the water is held by a combination of surface tension and capillarity, but very close to grain and crystal surfaces and also in very fine pores, the water is bonded in its ionic state by much stronger forces of electrochemical **adsorption**. This creates a very thin film of water bound very strongly to the underlying adsorbent (see also Chapter 10). The different states at which water is held exert a major influence on processes such as the evaporation of water held within porous

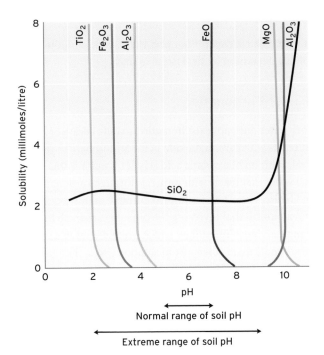

Figure 7.3 Solubility of metal oxides plotted against pH for some common products of chemical weathering. (Source: modified from Douglas, 1979)

materials and explains why initial drying may be quite rapid, whereas thorough drying requires considerably more time and energy and is unlikely ever to completely remove all adsorbed moisture. The same forces act to attract moisture into dry materials by exerting a 'suction potential'. Finer pores exert a greater suction and attract and hold moisture and any dissolved substances for longer. This property is very well understood in terms of soils, but is also a key control on moisture ingress, movement and retention across all geomaterials. It is therefore one of the reasons that porosity has a considerable influence on many weathering processes. Adsorption and capillary attraction are not the only ways in which water can be held and it can also be absorbed into the structure of a material. So-called 'mineral hydration' occurs when hydroxyl and hydrogen ions are absorbed into a substance, leaving the non-water component chemically unchanged. The most common reactions of this type typically involve salts and results in the creation of new minerals or **hydrates**, but the reaction is reversible and on drying the salt reverts to its original anhydrous form. The other area where water absorption is important is that of clay minerals (see Box 10.3 in Chapter 10), and there is an important group of these minerals known as smectites that can absorb water into their crystal lattice. These swelling clays, of which montmorillonite is the best known, form a major soil group, but when found within rocks the potential swelling and contraction associated with wetting and drying could be a contributory factor in their eventual physical breakdown. Indeed wetting and drying, or **slaking**, is recognized by many as a weathering process in its own right, to which certain rock types such as basalts are particularly prone because of the presence of secondary minerals that hydrate and dehydrate. The area where hydration has greatest significance is, however, in the area of salt weathering, where it is recognized as one of the key decay mechanisms. It shares with most mechanisms of physical weathering the key property of a repeated change in volume that generates internal physical stress. One cannot leave the role of water without acknowledging its unique property of having a 'triple point' that allows it to freely change between liquid, gaseous and solid states under normal Earth surface conditions. This facilitates not only wetting and drying, but also the best known of all volume changes, that of water into ice.

7.3.2 Chemical Weathering

Most weathering scientists consider there to be five main chemical processes that act individually and in combination to alter geomaterials, all of which involve water. These are described below.

7.3.2.1 Hydrolysis

Hydrolysis is the chemical reaction between water that is slightly acidic and a mineral. Aluminium silicates, of which the best known are the feldspars, typically combine a metal with silica, aluminium and oxygen atoms. Within the feldspars these range from calcium feldspar or anorthite ($CaAl_2Si_2O_8$) and sodium feldspar or albite ($NaAlSi_3O_8$), both of which belong to the plagioclase group, to potassium feldspar or orthoclase ($KAlSi_3O_8$). Other commonly occurring minerals which are weathered include groups of metal silicates, most notably ferromagnesian minerals containing iron and magnesium such as olivine (($Mg,Fe)_2SiO_4$), and phyllosilicate or sheet minerals such as biotite ($K(Mg,Fe)_3(AlSi_3O_{10})(OH)_2$) that are found particularly in igneous rocks. During the course of hydrolysis the metal cations are selectively replaced within the crystal lattice by a process known as **congruent dissolution**. Typically the first to go are the most mobile, monovalent cations such as sodium and potassium, followed by divalent ones such as calcium and magnesium and finally by the least mobile polyvalent ions such as iron and aluminium. Through this process, these minerals are progressively altered into clay minerals that are themselves phyllosilicates, together with soluble oxides that are usually lost from the system. A typical, simplified, reaction is that for orthoclase:

$$2KAlSi_3O_8 + 2(H^+OH^-) \rightarrow Al_2Sl_2O_5(OH_4) + K_2O + 4SiO_2 \quad (7.2)$$

or orthoclase feldspar + ionized water gives kaolinite + soluble potassium oxide + soluble silica.

7.3.2.2 Carbonation

Pure water does not exist in nature, mainly because rainwater reacts with carbon dioxide in the atmosphere to form a weak solution of carbonic acid (H_2CO_3) and it is this acidified water that reacts with minerals through carbonation and hydrolysis. The best-known example of this sequence of reactions is that involving the calcium carbonate that comprises most limestones, which produces soluble calcium bicarbonate that is then lost in solution (Box 7.1).

It is clear from Box 7.1 that the carbonation process is highly dependent on the continued presence of carbon dioxide in solution, which in turn is initially dependent upon its concentration in the atmosphere. Beneath a soil cover, however, the relative concentration (partial pressure) of carbon dioxide in soil air, and also cave air, is typically greater than that in the open atmosphere. This can lead to a higher content of dissolved carbon dioxide in soil moisture and helps

FUNDAMENTAL PRINCIPLES

CARBONATION EQUATIONS

This dissolution process underlies the formation of **karst** landforms and landscapes (see Box 7.4) and is also referred to as incongruent dissolution, on the basis that any clay minerals found are derived from impurities within the original limestone rather than the alteration of the rock itself. The precise sequence is:

$$CO_2 + H_2O \rightarrow H_2CO_3 \qquad (7.3)$$

$$H_2CO_3 \rightarrow H^+ + HCO_3^- \qquad (7.4)$$

$$Ca^{2+} + CO_3^{2-} + H^+ + HCO_3^- \rightarrow Ca^{2+} + 2(HCO_3^-) \, (7.5)$$

or calcium carbonate + carbonic acid gives soluble calcium bicarbonate.

Carbonation is not restricted to carbonate minerals and carbonation and hydrolysis are also effective in the weathering of silicate minerals such as orthoclase:

$$2KAlSi_3O_8 + 2H_2CO_3 + 9H_2O \rightarrow Al_2Si_2O_5(OH)_4 + 4H_4SiO_4 + 2K^+ + 2HCO_3^- \, (7.6)$$

or orthoclase + carbonic acid gives kaolinite + soluble silicic acid + soluble potassium bicarbonate. Carbonation is also effective for ferromagnesian minerals such as olivine:

$$Mg_2SiO_4 + 4CO_2 + 4H_2O \rightarrow 2Mg^{2+} + 4HCO_3^- + H_4SiO_4 \qquad (7.7)$$

or olivene + carbon dioxide + water gives soluble magnesium carbonate + soluble silicic acid.

BOX 7.1

to explain how limestone is often dissolved more rapidly beneath a soil cover than when exposed to the atmosphere. It also helps explain how limestone can be precipitated from solution (i.e. a solid form produced from a dissolved form) as water flows out from under the ground and carbon dioxide is lost to the atmosphere. An interesting anomaly arises from the partial pressure of carbon dioxide being inversely related to temperature and it falls by approximately 50% between 0 and 20°C. This has lead some geomorphologists to suggest that karst solution (see Section 7.5 below) should, unlike almost all other chemical weathering, be most rapid in areas with cold climates. This ignores, however, a number of key factors, especially in relation to the elevated level of carbon dioxide within soils that derives primarily from biological processes that would be inhibited under cold conditions. It also fails to consider the relative aridity of these regions or the effective aridity associated with the prolonged freezing of what water is available.

7.3.2.3 Solution

Water is a highly effective solvent. There are many naturally occurring minerals, such as salts contained within evaporites, that are highly soluble in water, but this also applies to those that derive from the carbonation of primary minerals, as well as many oxides and hydroxides. The degree of solubility is highly dependent on factors such as the pH and Eh of the water solvent. One important factor is that the rate of dissolution can be rapidly curtailed if there is no water flow and water that is static above a surface becomes saturated with the solute.

7.3.2.4 Oxidation and reduction

Oxidation and reduction have already been dealt with in relation to the role of water in weathering, where they are seen not only to be the controlling factors in the transformation of substances through either the addition or removal of oxygen, but in doing so to exert considerable control on their mobility in solution.

7.3.2.5 Biologically related weathering

There are numerous ways in which organisms act to initiate and enhance rock breakdown and alteration. The most obvious of these is the large quantities of carbon dioxide produced by the respiration of microorganisms and the breakdown (humification) of organic matter within soils, both of which considerably enhance the carbonation process. Certain bacteria may also produce oxygen, which can lead to the formation of inorganic acids. In particular, oxygen from bacteria is central to the nitrification process in soils in which ammonium is converted into nitrate, with the release of two hydrogen ions (protons) leading to the acidification of the soil. As well as inorganic acids, the biodegradation of organic matter produces organic acids, which react directly with and alter the mineral constituents of soils. These are typically complex mixtures of acids containing carboxyl and phenolate groups, the most common groupings being dark brown humic acids that are soluble only above pH 2 and yellower fulvic acids that are soluble in water under all commonly occurring pH conditions. One of the most important roles within soils of organic acids is that they act as effective chelating agents. **Chelation** is the process by which metal ions, in particular, are kept in solution and prevented from precipitating within, for example, plant roots. Chelation is, however, also the process by which these cations can be complexed and removed from the surfaces of fine mineral soil particles and colloids by ion exchange, in which they are typically replaced by hydrogen ions. Such 'cation exchange' (see Chapter 10) is a major route by which plants obtain nutrients, but also a mechanism by which mineral soils are weathered and, importantly, progressively acidified. Organic acids are not the only available chelating agents and microorganisms, especially bacteria, secrete a range of 'siderophores' or 'iron-carrying' compounds which are equally effective in chelating a range of other metals including aluminium, copper, zinc and manganese. In a weathering context, siderophores are especially important in scavenging iron from minerals in wet environments where it can then be removed in solution.

As well as their role in weathering and soil formation, there is growing interest in the role that organisms, particularly bacteria, fungi and algae, play in the surface weathering of rock (e.g. Allsopp *et al.*, 2004). When examined microscopically, all exposed rock surfaces are seen to be colonized by microorganisms. Bacteria, in particular, are able to survive extreme environments. Some of these organisms also penetrate pores and cracks and live successfully, especially where light is still able to penetrate through transluscent grains and crystals.

These micro organisms do not make a major contribution to landscape evolution, but they can be locally important and result in a range of micro-scale phenomena that are often diagnostic of certain environmental conditions. There are three main types of bacteria based on where they obtain their nutrients. Heterotrophic bacteria obtain nutrients from organic compounds, autotrophic from inorganic compounds and mixotrophic from both. In addition to these there is an important sub-group of chemolithotrophic bacteria, which are capable of oxidizing metals from minerals with the aid of an enzyme. These have been used extensively on a commercial basis to leach metals such as copper from ores, but in the field of geomorphology they have also been linked to the metabolism of manganese, which can then be precipitated on rock surfaces as a black crust or

'rock varnish' especially in arid and semi-arid environments where a soil cover is absent and there is limited surface water. A further grouping of bacteria is cyanobacteria or blue-green algae, which obtain their energy through photosynthesis. Colonial forms of these bacteria can often be found living in depressions on rock surfaces and they are commonly associated with locally concentrated weathering and erosion. On sandstones, for example, it has been suggested that photosynthesis by endolithic bacteria can result in more alkaline conditions which could be responsible for the weakening of inter granular cements through the dissolution of silica leading to grain loss. Conversely, it is also proposed that some cyanobacteria are able to mobilize manganese from superficial dust on rock surfaces, which, when precipitated, provides an alternative mechanism of rock varnish formation. Given, however, that other researchers would still argue that the iron and manganese in these varnishes could derive from the inorganic weathering of dust and/or inorganic, outward migration from within the rock, it is clearly a subject around which there is much ongoing debate (Dorn, 1998).

The final category of organisms to be considered here are inter dependent communities of fungi and oxygen-producing algae. Lichens (Figure 7.4) can be linked to a range of chemical and physical weathering processes and yet their presence on a surface is often seen as an indicator of stability and age. This has lead some to suggest that there

Figure 7.4 Lichens growing on the surface of granite in the Namib Desert. When they dry and lift from the surface they can 'pluck' away loosened grains as dust.

are different categories of lichens, some of which are considered to be bioprotectors shielding the surface from the actions of other processes and others that promote weathering (Lisci *et al.*, 2003). Lichens can excrete humic and other organic acids (so-called lichen acids) that are thought to promote both hydrolysis and chelation. At the same time, lichens also produce oxalic acid which, on limestones, can produce a coating of relatively insoluble calcium oxalate that could protect the underlying surface from carbonate dissolution. In the past, builders were known to have coated limestone surfaces with a variety of organic substances, including urine and eggs, to create a surface oxalate crust or patina of 'case-hardened' limestone.

7.3.3 Products of chemical weathering

Although the underlying chemistry of weathering is relatively well researched and understood, there is often uncertainty in how weathering has operated at a site or what might happen in the future. This is because of complex combinations of materials and varied environmental conditions. Aspects of the uncertainty are explored in subsequent sections dealing with geological and climatic controls on weathering, but as a starting point the following section attempts to introduce a degree of order by establishing some broad categories of weathering products derived from primary rock-forming minerals. Broadly speaking these come under four headings as shown in Figure 7.5.

7.3.3.1 Resistates

Resistates are unaltered primary minerals that are effectively immune to a specific process. At any one stage during the weathering of a rock it might include minerals that survive for longer because they are more resistant to alteration than others, even though in the fullness of time and with continued weathering they would also disappear. The most commonly referred to of these minerals is quartz, which survives as recognizable crystals within the weathered residues of rocks such as granite, long after the other minerals, typically feldspars and micas, have turned to clay.

7.3.3.2 Secondary minerals

As the name implies, secondary minerals are the weathering products of primary minerals. The most common alteration is of silicate and aluminium silicate minerals into clay minerals through hydrolysis. The precise composition of the clay mineral depends on the initial mineral and the intensity and/or duration of weathering. The most complex clay minerals (and therefore those found in less intensely weathered environments) are so-called 2 : 1 lattice clays made up of two sheets of corner-sharing SiO_4 tetrahedra, sandwiching a sheet of AlO_4 octahedra. Different groups of 2 : 1 clays can be distinguished by the presence of, for example, interlayer potassium (illite) (see Box 10.3 in Chapter 10) or the substitution of magnesium, iron, nickel and manganese within

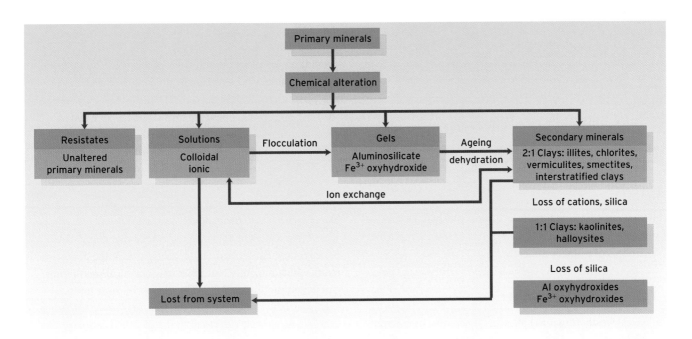

Figure 7.5 Processes and products of chemical weathering. (Source: from Fookes *et al.*, 1988)

the sheets (chlorite). These additions give the clays a rigid, non-swelling structure. The sheets in other 2 : 1 groups such as the smectites are less rigidly attached and allow water to penetrate and the minerals to swell. Continued or more intense weathering can, however, lead to the loss of one of the silica sheets (desilicification) of 2 : 1 clays, and the formation of a 1 : 1 lattice clay, the best known of which is kaolinite (see Box 10.3). Under extreme weathering conditions the second silica sheet may also be lost, to leave only aluminium or iron oxyhydroxides, such as the aluminium hydroxide, gibbsite ($Al(OH)_3$), which is one of the main constituents in the aluminium ore bauxite. The fact that bauxite is commonly associated with highly weathered tropical soils has encouraged researchers to establish links between certain clay mineral assemblages and different climatic regimes and different intensities of weathering (Bland and Rolls, 1998). However, there are likely to be significant variations in weathering intensity related to factors such as local drainage conditions.

7.3.3.3 Solutions

During the course of chemical weathering it is normal for metal cations, in particular, to be oxidized or transformed into salts and lost in solution. This is normally in the form of ionic solutions (e.g. $NaCl_{(s)} \rightarrow Na_{(aq)}^+ + Cl_{(aq)}^-$), but can also be in the form of **colloidal** solutions which contain extremely fine particles between 1 and 1000 nanometres in diameter (a nanometer is a thousandth of a millionth of a metre). Typically, these solutes are lost to the weathered material, although they may be precipitated nearby if environmental conditions (e.g. pH, Eh, temperature or the partial pressure of a gas) change, and they are the source of most terrestrial geochemical sediments.

7.3.3.4 Gels

The final category of weathering products is gels. These are solid, jelly-like substances ranging from soft and malleable to hard and tough. In a weathering context, perhaps the best studied are silica gels that can be formed by the evaporation of aqueous silicate solutions in surface and shallow subsurface waters, themselves the product of, for example, the hydrolysis of primary minerals.

7.3.4 Physical weathering

Physical weathering is the physical breakdown of material though either the application of an external stress or the release of an accumulated internal stress, without the involvement of any chemical alteration. In this way material can be reduced in strength and size to a point where it becomes susceptible to movement by a process of erosion. Where there is an existing relative topography material can be lost directly through gravitational collapse (see Chapter 8). In addition, the creation of fracture networks creates pathways for moisture penetration and biological colonization that facilitate chemical weathering, as well as greatly increasing the available surface area over which chemical reactions can occur. Fracturing can be brought about through a number of different mechanisms, but all of them are ultimately linked to some form of expansion or contraction. This volume change is either of the material itself or of an alien material created within it or absorbed during the process of weathering.

7.3.4.1 Dilatation – pressure release

The most common form of dilatation is that associated with the release of pressure when intrusive igneous rocks (see Chapter 2) such as granite are exposed through the erosion and removal of the overlying rock into which they were intruded. At this point the stresses accumulated during cooling at high pressure are released and result in the physical expansion of the rock. At the margins of an intrusion this expansion is greatest towards the surface leading to the generation of internal shear stresses parallel to the surface. This in turn results in characteristic curvilinear 'sheet joints' parallel to the curved surface of the cooling granite intrusion or **batholith** (Figure 7.6). Such joints are common in newly exposed intrusions and help to give the characteristic dome shape to many granite hills (Figure 7.7). They are also particularly useful in the quarrying of granite as they provide an initial set of joints that can then be exploited to break the stone into manageable blocks. The fact that they are 'pressure release' features is demonstrated by so-called 'A-Tent' structures. These are formed where an isolated sheet of granite arches upwards from the surface of the granite mass (Figure 7.8). Measurement of the sheets always shows them to be longer than the hole from which they came, indicating not only their original compressive loading, but also their expansion once this pressure is removed. Deeper within the intrusion the stresses are resolved into sets of joints at right angles to each other that define large-scale rectangular blocks. These joints provide the pathways along which moisture and ultimately chemical weathering can penetrate (Figure 7.2b). The density of these joints is highly variable, depending on the structure of the intrusion, and on a large scale can result in the so-called compartmentalization of the landscape as areas with a high joint density are weathered and eroded more rapidly than those with a

Figure 7.6 Curved sheet jointing generated by dilatation in granite, Yosemite National Park, California.

Figure 7.7 Half Dome, Yosemite National Park, California, demonstrating the structural control of curvilinear sheet joints on the shape of the dome prior to its truncation by glacial erosion that exploited vertical joint sets within the batholith.

Figure 7.8 An A-Tent structure on a granite batholith, South Australia. The granite slabs are larger than the hole from which they sprang demonstrating the expansion associated with pressure release.

low joint density. Such processes are often cited as central to the formation of isolated granitic hills characterized by widely spaced joint systems.

As well as the development of joints and fractures spaced over metres or tens of metres, pressure release is also now considered to operate at the scale of millimetres and less. In this way it can lead to the development of dense networks of microfractures around and through individual crystals as minerals with different physical characteristics expand at different rates following pressure release. Depending on the minerals present, these fractures represent a secondary porosity along which moisture and chemical alteration can

Figure 7.9 Detail of a weathered granite from south-east Brazil, in which all of the feldspars have been weathered by hydrolysis into white kaolinite.

penetrate leading to the formation of a clay-rich or 'argillaceous' weathered regolith (Figure 7.9). Alternatively, if the minerals present are less susceptible to chemical weathering a sandier or 'arenaceous' regolith can develop in which the original minerals and structures of the rock are retained, but now it is so fragmented that it can be dug out by hand. In granites these sandy regoliths are sometimes referred to as 'grus' after the term coined by quarry workers in the south-west of England, where spreads of sandy material derived from the disaggregation of arenaceous regoliths are a common feature.

7.3.4.2 Thermoclasty

Temperature change always works in conjunction with other weathering factors. Even in the driest desert moisture plays a role in weathering. Hence understanding the role of temperature in weathering has been tricky. To get over these problems researchers have attempted to isolate thermal effects under laboratory conditions. One of the earliest of these experiments was carried out by Griggs (1936) who alternately heated a block of granite in front of an electric fire and allowed it to cool in a dry cellar. The upshot of the experiment was that despite heating the block many times and far beyond any temperature to be found in nature, he observed no physical change until he gave up and started to cool the block by pouring water on it. Since this experiment, laboratory tests have become much more sophisticated and now employ, for example, programmable climatic cabinets that can mimic temperature and humidity conditions from a wide range of climates. However, these experiments have had comparatively little success in inducing breakdown that is uniquely attributable to thermoclasty.

In designing these experiments, researchers normally choose to study one of two potential thermal processes. The first of these is what Yatsu (1988) described as 'thermal shock'. This is where temperature change is so rapid that the stresses generated by expansion or contraction of the rock cannot be accommodated by the required deformation. The most obvious example of thermal shock is that experienced during a fire, and there are many observations of cracked boulders after the passage of bush and forest fires. It seems reasonable to suppose that over time thermal shock could play an important role in the breakdown of debris in, for example, Mediterranean and savanna environments prone to such fires. Other than this special case, it is difficult to imagine how natural cycles of heating and cooling could generate a sufficient shock to fracture fresh rock. Recently, however, there has been interest in the potential role of short-term, near-surface reductions in rock temperature triggered by, for example, the spread of cloud cover over a surface previously exposed to bright sunlight (Smith, 2009). This is particularly the case in polar and high mountain environments where high rock surface temperatures can occur at the same time as low air temperatures, resulting in a very rapid fall in surface temperature when clouds or shade appear. It is most likely, however, that the mechanical stresses associated with any such temperature change will be restricted to a very shallow, near-surface zone.

Adjacent crystals of different minerals heat up and expand at different rates in response to differences of thermal characteristics such as **albedo**, thermal conductivity, heat storage capacity and **coefficient of thermal expansion** as well as in some cases the orientations of their crystal lattices along which expansion can be concentrated. This, in turn, sets up stresses at crystal boundaries that can be translated into stresses across adjacent crystals.

Thermal shock effects are now considered to operate at small scales, as compared with earlier ideas about the splitting of large boulders. It is also recognized that the forces generated are most effective where the expansion is constrained in some way by surrounding crystals or rock (because they have something to push against). This constraint effect has increasing practical significance as architects use more and more stone as thin claddings on buildings and are faced with replacing bowed and cracked panels put in place with inadequate expansion joints.

The alternative to thermal shock as a mechanism for rock breakdown is that of 'fatigue failure'. This is where breakdown occurs through the repeated application of a low-magnitude stress, which on its own is insufficient to initiate a fracture, but cumulatively can weaken the rock to the point where it becomes susceptible to even a low-magnitude stress. This is in much the same way as if you were asked to snap the handle of a metal spoon: you would not attempt it in one go, but instead would bend it backwards and forwards many times until it eventually broke. Fatigue failure is seen by many as a more viable contributory mechanism to rock weathering not least because there is no need to invoke extreme and abnormal temperatures. However, fatigue failure is difficult to test in the laboratory or in nature as it is extremely difficult to isolate and test the effect. In the laboratory, for example, many weathering tests seek to accelerate the process under study, but any attempt either to speed up the heating and cooling process or increase the temperature range would invalidate any link with the fatigue effect. Any such tests would therefore have to run in real time and are thus completely impractical. In nature, meanwhile, the fact that fatigue effects only manifest themselves over very long time periods makes it difficult to dissociate them from the contributions of other weathering mechanisms.

7.3.4.3 Freeze-thaw (frost weathering)

Of all weathering processes, freeze–thaw is the one that the majority of people can most easily comprehend, in part because of its clear link to an easily recognizable environmental condition (water freezing), but also because it is perceived to involve a relatively straightforward mechanism – water expanding on freezing. However, a number of researchers question some of the long-held assumptions about its effectiveness. For example, there has been an assumed relationship between freeze–thaw and the production of angular rock fragments. In reality, however, the conditions required for freeze–thaw to generate significant stresses within a stone are much more complicated and rare than a simple drop in temperature below 0°C and the freezing of water already in the pores of a stone. For example, the maximum expansion associated with the transition of water into ice occurs at −22°C. If this point is actually reached, the pressure generated, theoretically 207 MPa, is more than enough to fracture most rocks. As temperatures drop, however, a number of factors combine to inhibit freezing and also to relieve any pressure generated by ice that has formed within pores. For example, when ice begins to form within pores the pressure on the remaining water increases. This lowers the freezing point and slows down further freezing. The presence of impurities in the water such as salts will also lower the freezing point, and in an unsaturated, open-textured stone the pressures generated by freezing and expansion might be partly absorbed by the compression of air within the pores and feasibly by

displacement and possible expulsion of water via any exposed surface. For this reason the rock should ideally be at or near saturation and frozen from all sides if freeze–thaw is to be most pronounced. The process is accentuated by rapid freezing and so is considered to be most effective at or near the rock surface, leading to the preferential release of individual grains and fine rock fragments. If the expulsion of water is particularly rapid or the surrounding porosity relatively low, the pressure of the displaced water may itself lead to breakdown of the rock by a process known as 'hydrofracture'.

Ice segregation is the process whereby water is gradually attracted towards areas of ice formation by a suction gradient established in adjacent pores as the ice forms (see Chapter 17). In this way, lenses of ice are created that can theoretically continue to grow against a constraining pressure. Such phenomena have been recognized for many years in soils and result in the fracturing and heaving of these and other fine-grained, weak materials, but recent experimental studies have suggested that ice segregation and growth might also be capable of fracturing stronger rocks, especially where pre-existing microfractures can be exploited and extended. Unlike volumetric expansion, the conditions required for ice segregation are not especially exceptional and ice can continue to grow under conditions of relatively slow freezing associated with sustained low temperatures. Experimental and field observations also suggest that disruption associated with ice lens formation could be a much more deep-seated process than volumetric expansion, especially because it can be enhanced by prolonged freezing leading to 'slow ice separation' at depth. This could in turn lead to fracturing at greater depths within rock formations and potentially the production and release of larger and bigger quantities of rock debris.

Debris produced by freeze–thaw seems to be both fine and coarse (Figure 7.10). **Microgélivation** is considered to occur where ice crystallizes within pores and microfractures at the grain and crystal scale, and is responsible for the creation of fine debris (Figure 7.10a). It is also considered by some to be responsible for the rounding of larger debris and rock surfaces. This contrasts with **macrogélivation**, which is principally the exploitation of fractures and potential lines of weakness within larger rock masses to produce coarse, angular debris or clasts (Figure 7.10b). Some researchers have recently questioned this division of weathering products on the basis of size (Hall and Thorn, 2011). The same might also be said about the presumed relationship between freeze–thaw and the angularity of the debris. Weathering associated with any expansive process, be it heating and cooling, wetting and drying, freezing and thawing or, as

(a)

(b)

Figure 7.10 Frost weathering of granitic rocks in a Bronze Age stone circle at Copney, Northern Ireland. The stone was buried for over 2000 years beneath a peat cover before the site was excavated. The stress inheritance of leaching under highly acid conditions considerably weakened the stones, which made them particularly susceptible to freeze-thaw, especially as the site was waterlogged for much of the year: (a) microgélivation and granular disaggregation that is typical of the smaller blocks; (b) macrogélivation and splitting along internal fracture planes typical of the larger blocks. (Source: STILL Pictures The Whole Earth Photo Library: Ron Giling)

described next, salt crystallization and dissolution, is likely to exploit internal structural weaknesses that are often linear in nature. The production of a similar end product from a variety of origins and through a number of different processes is known as **equifinality** and highlights the danger of trying to deduce process based solely on form or shape.

7.3.4.4 Salt weathering

Traditionally salt weathering has been seen as something that is important in coastal and desert environments where salt is available in abundance, but largely irrelevant outside of these areas. However, it is possibly the most important cause of rock breakdown in built environments. Building owners across the world will testify, irrespective of climate, to the effectiveness of salt from sources such as rising groundwater in weathering stone. Salt weathering also occurs in polar and mountain environments that are themselves often arid. Unlike freeze–thaw, salt weathering is not constrained to the precise coming together of a limited and very specific set of environmental parameters for its operation.

Salt weathering is a physical weathering process associated with the growth or expansion of salt crystals in pores and fractures. Of course, the processes driving this expansion are predominantly chemical and relate either directly to crystal growth out of solution or to the absorption of moisture (hydration) by salts that have previously crystallized within the rock. Theoretically, crystal growth alone could result in rock breakdown as, provided that the solution remains supersaturated and a thin liquid film is maintained at the boundary between the crystal tip and the adjoining pore or crack wall, crystals will continue to grow against any confining pressure. The stresses generated within an individual pore are unlikely to be sufficient to cause failure as they act on too small an area (Smith and McGreevy, 1988). Instead, it is the ability of salts to spread and accumulate throughout a zone within the rock that allows them to generate sufficient stress to cause failure by interacting with larger structural flaws that ultimately control rock strength. The exception to this might be at the rock surface, where individual grains and crystals are only partially confined. The effectiveness of salt crystallization in causing breakdown is also closely linked to the nature of the pores and it has been suggested that the controlling factor is the presence of micropores (less than 5 millionths of a metre in diameter) that have the suction potential to absorb moisture and are readily bridged by salt crystals. Such a process is more difficult to accomplish in larger pores. The growth of clearly defined, pore-bridging crystals may not be essential for salt weathering as hydration could be favoured where pores become completely filled with microcrystalline salt. The same might be said of differential thermal expansion as a possible weathering mechanism, where salts are present with coefficients of thermal expansion greater than that of the enclosing rock (Figure 7.11). For further details on salt weathering processes and hazards, see Goudie and Viles (1997).

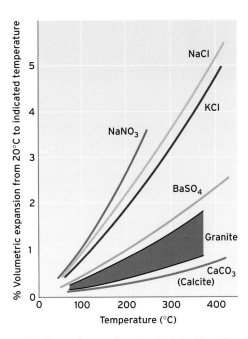

Figure 7.11 The thermal expansion of salts linked to salt weathering, compared with that of granite. (Source: from Cooke and Smalley, 1968)

As with other processes linked to physical breakdown, the importance of salt weathering lies not so much in the shock effect of one-off crystal growth, but in the fatigue effect associated with regular and frequent volume changes linked to repeated solution and recrystallization, hydration and dehydration, and possibly differential thermal expansion and contraction. These changes are largely controlled by alternating heating and cooling and wetting and drying that may in turn be linked to natural diurnal cycles, but may also be event driven in relation to individual storms. It is in this context that hydration and dehydration take on special significance, in that, depending on the salt, these processes can occur when specific combinations of temperature and relative humidity are crossed without the presence of liquid water. For some salts these thresholds could potentially be crossed several times during the course of a day as temperature and relative humidity rise and fall. The hydration pressures generated can be considerable, and theoretically could exceed the **tensile strengths** of many common rocks.

The relative significance of crystallization, hydration and differential thermal expansion is still unclear but it is most likely that in reality all three mechanisms are likely to overlap and work together. It should be stressed, however, that not all salts hydrate or dehydrate under commonly experienced environmental conditions (e.g. sodium chloride), and many commonly occurring salts (e.g. calcium sulphate) can have a low solubility that makes it difficult for them to penetrate

far into the rock. This is where having a mix of salts can be important. The presence of sodium chloride is known to increase the solubility of calcium sulphate, and together they are considerably more effective in causing decay. A mix of salts could also increase the number of hydration and dehydration cycles that occur within a rock. Some salts such as sodium chloride can also absorb moisture from the atmosphere (deliquesce) and promote the operation of other chemical weathering processes. Salinity is also known to enhance the solubility of certain elements, most notably silica. While this is unlikely to play an important role in dissolving the rock mass as a whole, it could be significant in some sandstones where crystalline quartz grains (with a generally very low solubility) are held together by amorphous silica cements with a somewhat higher solubility. In this situation there may be preferential removal of the cement enhancing the susceptibility of the stone to more classic salt weathering, as well as other mechanical weathering processes.

For urban stone decay, one of the diagnostic traits of salt weathering is so-called 'sanding', whereby running one's hand over, say, a sandstone block removes a layer of loose material, which would otherwise collect as the small piles of sand that are often seen along the bottoms of walls. This process occurring on stone is referred to as **granular disaggregation** (Figure 7.12). It is generally agreed that this disaggregation is associated with a near-surface or surface accumulation of salt (efflorescence) and both laboratory experimentation and field observations from innumerable buildings have linked this to the slow evaporation of salts brought to the surface in solution, a process that is enhanced if the salt itself has a high solubility. The complementary process to surface salt accumulation is thought to occur when rock surfaces are rapidly heated, especially

Figure 7.12 Granular disaggregation of granite, arid southern Namibia.

NEW DIRECTIONS

SALT WEATHERING BY ION DIFFUSION

Recently attention has been drawn to ion diffusion as an alternative mode of salt weathering. In this process salts can migrate through stone to concentrate in certain areas. This is where the individual ions (e.g. Na^+

and Cl^-) migrate through a saturated rock from a zone of high concentration to one with a low concentration without any flow of the solution itself. The process could be important in areas where rock is saturated for long periods by rising groundwater, or possibly during long periods of wet-season saturation that could

penetrate deep into a rock. One specialist area of weathering where this process is especially important is that of concrete, where chloride diffusion and subsequent corrosion of iron reinforcing rods is a major source of structural failure as the iron expands.

BOX 7.2

in conjunction with a surface air flow. Under these conditions, rapid evaporation dries out the immediate subsurface zone more rapidly than moisture can be drawn out from the interior. In this way the water connectivity with the surface is broken and the only way that moisture can be lost is as vapour, causing salt to crystallize at depth. This concentration is clearly enhanced if the salt itself has a relatively low solubility. Repeated many times, this process allows salt to accumulate in a subsurface zone, close to a frequent wetting depth, and eventually results in **contour scaling**, in which a complete surface layer up to several centimetres in thickness can first blister before falling away (Figure 7.13). This can sometimes reveal a subsurface accumulation of crystallized salt. Box 7.2 discusses an alternative mechanism for salt weathering.

Granular disaggregation and contour scaling cannot be claimed as uniquely salt weathering in origin, being the possible end points of a range of linked processes, including thermal expansion and contraction that can establish

stresses either between grains or between the rock surface and its interior. However, a more definitive indicator of the operation of salt weathering is the occurrence of cavernous weathering at a variety of scales indicative of the tendency for salt weathering to create hollows protected from rain-wash, in which salts are then preferentially retained. The hollows also establish their own microclimates that are typically cooler and moister than adjacent exposed surfaces, which in turn enhance the direct precipitation of moisture, its retention and the deeper absorption of salt in solution. They thus create a positive feedback that promotes further weathering and the growth of the cavern. Two scales of hollow are generally recognized: honeycombs of the order of a few millimetres to a few centimetres wide and deep (Figure 7.14); and **tafoni** that can be measured in tens of centimetres to metres (Figure 7.15).

The location of these hollows may reflect an initial, localized variation in weathering susceptibility across a rock surface, perhaps related to porosity, grain size, mineralogy or degree of cementation (Turkington, 1998). Sometimes these structural controls are easily recognized as lines of hollows that pick out a particular horizon within a rock, but in other cases the controls may be very subtle and the distribution may appear to be random. As honeycombs grow and begin to intersect each other to cover a surface they can appear to take on a strange degree of uniformity. In other situations it is possible to identify a degree of environmental control so that in hot desert environments, for example, basal tafoni can be found in cliff-foot locations where groundwater seepage is concentrated.

Cavernous hollows of all types are particularly common in coastal and desert locations but their widespread occurrence in urban environments suggests that it is the presence of salt, rather than any specific environmental control, that

Figure 7.13 Surface scaling of granite, arid southern Namibia.

Figure 7.16 Large block of granite dislodged by the growth of a tree root, Erongo, Namibia.

Figure 7.14 Honeycomb weathering of a carbonate cemented quartz sandstone in a salt-rich coastal environment, Yehliu, northeast Taiwan. Weathering of the less consolidated sand beneath the carbonate-rich nodule has produced a 'pedestal rock', which is also a common feature of salt weathering in many rocks.

Figure 7.15 Cavernous hollow or tafoni in sandstone, Petra, Jordan.

is the key to their formation. It should also be remembered that salt weathering is not the only process that favours warm, humid and sheltered environments and the formation of hollows is also associated with various forms of biological weathering in which, for example, bacterial and algal communities create their own micro-environments in otherwise harsh locations.

7.3.4.5 Physical biological processes

The most visual link between biological processes and the physical breakdown of rocks is when the growth of tree roots is seen to split and prise apart rocks (Figure 7.16). As in Figure 7.16 the penetration of the root is greatly aided by the pre-existing presence of a network of joints and a degree of prior granular disaggregation. At a smaller scale, it has been suggested that processes such as grain release can be facilitated through the generation of tensile stresses as fungi grow along micro-cracks. It is also proposed that the expansion of both fungi and lichens as they rehydrate after a dry period can exert significant stresses through exertion of 'turgor' pressure (the pressure of water against cell walls). Some authors have suggested that the volume expansion of some species could be close to 4000% on rehydration.

Reflective questions

➤ Why is water important for weathering?

➤ What are the key chemical weathering processes?

➤ What are the key physical weathering processes?

7.4 Climatic controls on weathering

In terms of links between weathering and generalized climatic parameters, the one that arguably has the strongest justification is that between increased temperature and increased chemical weathering, based as it is on the approximate doubling of most chemical reaction rates with every 10°C rise in temperature. Of course such reactions are also likely to be influenced by the availability of water and the nature of any biological activity, linked in turn to other factors such as available carbon dioxide. When we come to consider other processes such as freeze–thaw, however, the links with mean annual temperature become much weaker and there are many other more meaningful environmental parameters that control its operation and effectiveness. These include absolute temperature below zero and the rate of freezing in relation to the number and duration of frost events and the depth of frost penetration. Thus it is possible to envisage constraints on freeze–thaw activity controlled not just by conditions that are too warm to freeze, but also too cold to thaw and too dry for ice to form. Thermal controls on other weathering processes are similarly complex as described above.

For moisture it is not necessarily the absolute amount of rainfall that is significant but the balance between rainfall and **potential evapotranspiration** which provides a better measure of moisture availability. It is also the case that rainfall distribution can exert a considerable influence on patterns of weathering. In savanna environments, for example, alternate wetting and drying linked to the rise and fall of the water table might favour cycles of leaching and precipitation of certain minerals within soils (especially iron) and ultimately the formation of **duricrusts** such as ferricretes or laterites. Continually wet conditions under rainforests, together with the rapid recycling of organic matter, could favour continuous leaching of soils and weathered regoliths (see Chapter 10). Such assumptions apply specifically to patterns of deep wetting and associated deep weathering. In terms of weathering closer to a rock surface, weathering is much more likely to be related to the characteristics (duration, frequency, intensity and total amount) of individual rainfall events linked to patterns of evaporation immediately following them. There is no minimum annual rainfall below which weathering does not occur. This is because many weathering processes, such as salt weathering and microbial activity, require very little water for their operation and they can be adept at obtaining this directly from the atmosphere or from groundwater.

In his famous diagram, Strakhov (1967) attempted a broad linkage between climate, weathering depth and the nature of weathering products (see Figure 10.11 in Chapter 10). As a generalization this is undoubtedly justified, but it masks a raft of contradictions and complicating factors. The underlying assumption behind the relationship between increased weathering depth and low-latitude environments is that it is the product of increased intensity of weathering associated in turn with high temperatures and abundant all-year-round rainfall. Possibly the first thing to note, however, is a common failing of all such generalizations, which is that in reality global climates do not conform to a perfect zonal model, and it is common, even within the tropics, to find a range of climatic types at any one latitude, including climates associated with, for example, mountain ranges (Chapter 6). Within each of these zones it is also the case that weathering processes and depths are highly dependent on rock type and also upon those rock types that have been most widely studied and reported. Within the humid tropics this most probably means ancient, basement rocks in stable **shield areas** that are likely to be granitic in character and have a propensity for deep weathering because of their combined physical and mineralogical characteristics. Under these circumstances it might also be the case that profiles are deep, not only because of the intensity of processes such as hydrolysis, but because the profiles are old and the rate of surface erosion under a protective forest cover is relatively slow. That is, we can substitute stability and longevity for weathering intensity. There are many areas of the humid tropics in steep, tectonically active areas on weathering-resistant rocks that have relatively thin soils. Conversely, it is also possible to find deep weathering profiles outside of the tropics. In many cases this reflects a long-term climatic inheritance. For example, pre-Quaternary deep weathering profiles have been found in areas such as Scotland and Scandinavia and this has questioned the power of cold-based ice sheets in particular to reshape the landscape (Migon and Bergstrom, 2001; Olvmo *et al.*, 2005).

Like Strakhov (1967) others have attempted the same climate zoning exercise using the occurrence of specific clay minerals. This is based on the assumption that weathering intensity increases with increased temperature and rainfall and that increased weathering intensity is in turn reflected in the progressive simplification of clay minerals from 2 : 1 lattice clays, to 1 : 1 clays, to iron and aluminium oxides and hydroxides. While such sequences do occur it is also true that specific weathering processes (e.g. the carbonation of orthoclase) are capable of bypassing this sequence to directly produce clays such as kaolinite. Likewise, recent

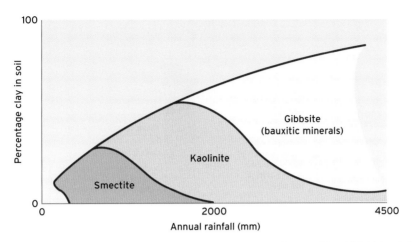

Figure 7.17 Influence of climate on clay mineralogy under a perennially wet climate in Hawaii in which high rainfall and rapid leaching cause the preferential removal of silica and metal cations. (Source: after Sherman, 1952)

studies from countries such as Thailand (Hermann *et al.*, 2007) have suggested that gibbsite, which is usually thought of as the end product of weathering under wet tropical conditions, can appear as an early and direct transformation from micas and feldspars in rocks such as granite and gneiss without any intermediates. The key variable in this case appears to be freely draining soil conditions and high leaching rates, a control previously emphasized in a study

of basalt weathering from Hawaii (Figure 7.17). Sherman's (1952) study on Hawaii is frequently used to demonstrate climatic controls on weathering and clay formation, but given that it is a study of one small island it could more accurately be thought of as a demonstration of variability within one broad climatic zone based upon local conditions. Such variability has also been demonstrated in many other studies from the tropics (see Box 7.3).

CASE STUDIES

DEEP WEATHERING IN RIO

Despite the widespread assumption that the humid tropics are underlain by a uniformly thick blanket of deeply rotted rock reduced mainly to simple clay minerals, recent research has emphasized that there is considerable local variability. A study of deep weathering around Rio de Janeiro by Power and Smith (1994) showed that soils on well-drained crests, subject to enhanced cation and silicate removal, were dominated by gibbsite and kaolinite, whereas complex clays such as montmorillonite were more common in poorly drained foot slopes and valley floor

locations. A study in the same area by Smith and Sanchez (1992) also demonstrated how, under the same climatic conditions and with the same climatic history, subtle variations in mineralogy could drastically affect weathering response. Granitic rocks containing predominantly potassium (orthoclase feldspars) weathered to produce a deep, but predominantly sandy, arenaceous regolith. Adjacent rocks rich in calcium and sodium feldspars (plagioclase), which are more prone to chemical alteration, produced equally deep, but clay-rich, argillaceous regoliths (Figure 7.18). One important consequence of this in terms of potential slope hazard is

that slopes underlain by the clay-rich regoliths are much more likely to fail catastrophically through mudflows and slides (Figure 7.19), whilst the arenaceous profiles are more likely to experience erosion by surface wash. Active erosion also means that, particularly on steep slopes, deep weathering profiles are replaced either by thin soils that are prone to 'slip' off the underlying bedrock after prolonged rainfall, or by large boulders held in place by surrounding finer debris (Figure 7.18). These boulders are themselves prone to rolling downslope, with great destructive potential, if the supporting material is washed away.

BOX 7.3 ➤

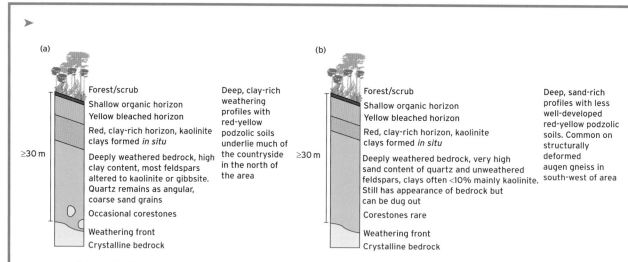

Figure 7.18 Schematic diagrams of (a) clay-rich and (b) sandy deep weathering profiles on granitic rocks in the Rio de Janeiro region of Brazil. Sandy profiles tend to form on rocks comparatively rich in potassium feldspars.

Figure 7.19 Slope failure at Niteroi, Brazil in May 2010 resulting from the saturation of deeply weathered, clay-rich weathered granite regolith.

BOX 7.3

It is important that global models of climatic controls on weathering are recognized for what they are, namely generalizations, and only the broadest of starting points to any understanding of weathering. Thus, while they indicate that the occurrence of certain processes may be favoured in particular climatic zones this is no guarantee of their exclu-sivity or their universal effectiveness. The danger comes if one seeks to employ the relationships they imply to the interpretation of weathering and weathering products at the landform scale. At this scale it is essential that attention is focused appropriately on the scale at which the processes in question operate.

Reflective question

➤ What are the key climatic controls on weathering?

7.5 Geological controls on weathering

Weathering is often portrayed as an assault on the integrity of a rock, and, as such, the majority of weathering investigations have focused on understanding the motives and capabilities of the assailant (i.e. the weathering processes involved). As with any good detective story, a true understanding of the crime must also involve a sound knowledge of the crime scene and the victim, especially their background and history. It is because of this that a theme throughout this chapter has been the linking of process and

environment, primarily climate, to those characteristics of geomaterials that make them particularly susceptible or resistant to specific modes of decay. Weathering resistance is a complex characteristic and materials that are formidably resistant to one form of weathering may be highly susceptible to another. For example, most karstic limestones (see Box 7.4) are not porous and are therefore resistant to salt and frost damage. They are, however, highly soluble. The rate of solubility, like most chemical weathering processes, is itself dependent on complex environmental factors. This complexity has not prevented researchers from coming up with tables of relative susceptibility to chemical weathering. The most widely quoted of these is based on the Bowen reaction series (Figure 7.20a), created in the early twentieth century, which ranked common minerals found in igneous rocks in terms of their temperature of formation. This ranking was later picked up by Goldich (1938) who, on the basis that the Earth's surface is a low-temperature

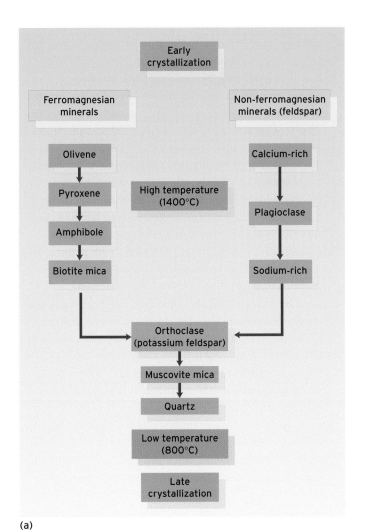

(a)

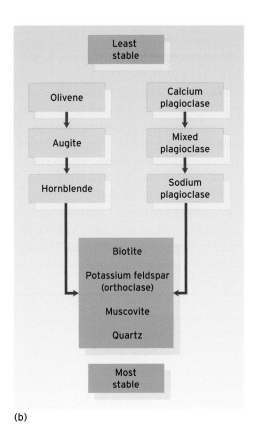

(b)

Figure 7.20 (a) The Bowen reaction series of igneous minerals and (b) the weathering series of Goldich (1938).

environment, used the listing as an indication of relative weathering susceptibility under near-surface conditions, with the most stable and therefore resistant at the bottom of the list (Figure 7.20b). In reality patterns and rates of decay will vary considerably in response to localized variations in the weathering environment and there is no guarantee that the stability sequence could not be reversed under certain circumstances.

It would be convenient if a similar table could be compiled for susceptibility and resistance to physical weathering. The most obvious candidate in terms of the physical properties that could determine weathering resistance is that of the hardness of the material, expressed in terms of its resistance to various kinds of shape change when a force is applied. Unfortunately, however, the same difficulties apply as they do for chemical decay, in that there are many potential combinations of environmental conditions and material properties under which physical breakdown might be induced. There are, for example, a variety of forces to which materials might be subject, to which the same material will have a varying degree of resistance. Included in these are **scratch hardness**, **indentation hardness** and **rebound hardness**. The best known is scratch hardness, by virtue of 'Moh's hardness scale', which is used by geologists as an aid to mineral identification. It does not provide any great insight into weathering in itself, although it might give an indication of abrasion resistance during erosion. Rebound hardness has received some attention in recent years thanks to the use of the 'Schmidt hammer'. This was initially developed to test the hardness of concrete, and fires a spring-loaded metal bolt at the surface to measure its rebound. While of debatable use on rough and weathered surfaces, where it tends to compress the material rather than rebound, it has had some success in, for example, discriminating between rocks that are more or less susceptible to different types of karst formation (Day, 1981). Recently, the development of smaller and more sensitive rebound instruments, such as the 'equotip portable harness tester' has offered greater potential for the characterization of weathered material in a non-destructive manner (see Viles *et al.*, 2010).

The techniques described above can assess susceptibility to weathering rather than providing any insight to the actual nature of rock breakdown. The physical breakdown of rocks tends to result from the application of three main stress types applied either generally across a rock mass or at specific points right down to the micro-scale. The first of these is **compressive stress**, in which an axial 'pushing force' is applied to a material that crushes it once the compressive strength of the material is exceeded. This con-

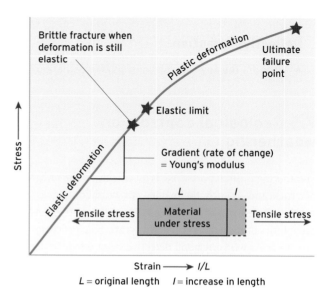

Figure 7.21 Idealized stress–strain curve.

trasts with **tensile stress**, in which an axial 'pulling force' is applied, which results in the distortion of the material, followed by its rupture or fracture once the tensile strength is exceeded. Between these two are **shear stresses**, in which offset forces induce a sliding failure parallel to the direction of the force once shear strength is exceeded. Resistance to these stresses is by no means uniform and the same material will invariably respond differently and more or less readily to the different stress types. The nature of this response can be shown in the form of a stress–strain curve in which the applied stress is plotted against the distortion of the material. Figure 7.21 shows a representative stress–strain curve for material under tensile stress, in which strain is expressed as the ratio of the increase in length of the sample compared to its original length. From the curve it can be seen that there is an initially linear increase in strain, during which the material behaves elastically, and will return to its original dimensions once the stress is removed. The gradient of this straight line is known as the **Young's modulus** for the material. Following this, the material may enter a phase of plastic deformation, having surpassed its **elastic limit**, which is characterized by non-reversible shape change. If the stress continues, eventually a point is reached at which the material fractures or ruptures. This final failure could be an eventual response to the gradual development of networks of micro fractures developed during repeated plastic deformation, and it is this process that lies at the heart of **fatigue failure**. Obviously, the precise nature of this curve will differ greatly between materials. Some materials are, for example, more 'ductile' than others and have a greater ability to deform under an applied stress, although this property is in turn affected by other factors such as temperature and a material typically becomes more ductile when heated. In contrast, other materials (e.g. glass) may be brittle in character and have little if any capacity for deformation, and instead will tend to fracture under the application of a well-directed stress. Therein, however, lie a number of contradictions, in that brittle materials can often be considered as hard and are particularly resistant to, for example, abrasion forces. Likewise there is often confusion over the difference between weak and strong, in that it is often perceived that materials that bend are weak, and that those that do not are strong. Many materials that experience significant elastic and plastic deformation are in fact capable of surviving stresses that would fracture more brittle materials. In the context of weathering, one further complication is that the nature of materials can be altered by their surroundings. Thus, when a rock in the form of a pebble is unconstrained it can expand and contract elastically in response to, say, heating and cooling and absorb the stresses generated. In contrast, the same material, when constrained within a larger rock mass, cannot absorb stress through expansion and will behave inelastically, making it more susceptible to

the development of, for example, micro fractures and ultimately brittle fracture.

The chemical and physical properties described above can be broadly thought of as lithological in character, relating as they do to intrinsic properties within the material. As we will see in the following section dealing with stone in the urban environment, however, by the time that most geomaterials are subject to weathering they will have experienced complex stress and deformation histories. A common result is the compartmentalization of the rock mass into structural units ranging in dimension from sub millimetres to spacings measured in kilometres. These are defined by joints and fractures, some of which are 'real' and others that are 'incipient' and only revealed when exploited by subsequent weathering. As indicated earlier in this chapter, the three-dimensional distribution of these structural properties exerts major controls on both the operation of weathering processes and the morphology of subsequent landforms. Nowhere is this combination of lithological and structural control better illustrated than in the case of the hard, non-porous, well-jointed limestones that form the distinctive karst topographies of the world (Box 7.4).

CASE STUDIES

KARST LANDFORMS

Where rocks are highly soluble, so that most of their mass can be removed in solution, a set of distinctive, or 'karst', landforms may be formed. Karst occurs most commonly in limestone, but may also occur in other soluble rocks, such as gypsum. Some comparable forms can also be generated through melting of ice to form 'thermokarst'. Where the rocks are strong, water flow and solution is concentrated along joints and other porous areas, enlarging them in a complex network of passages, many of them underground, although generally with some connections with the surface. These passages may evolve over very long periods, and some

become enlarged into caves some of which can be hundreds of metres in extent and form networks over many kilometres.

Joint passages evolve most rapidly where there is most water flow, so that they tend to be formed mainly above the water table, although this can alter over time through climate or base-level change. This also means that caves frequently develop along the lines of stream valleys, because the streams have the largest and most continuous water flow. Streams, in this way, can progressively bring about their own demise, as more and more of the water sinks into the bed and follows underground passages, which may only emerge at the boundaries of the soluble rock mass.

Other factors that increase the rate of solution are the presence of soil organic matter and cold temperatures, because carbon dioxide is more soluble in the cold. The combination of these factors with the flow means that karst development is most rapid in humid tropical areas, and slowest in hot deserts.

Rapid karst development lowers the hills until they intersect the water table, so that they eventually form towers rising out of an almost level plain (Figure 7.22). In temperate areas, karst usually develops more slowly, so that cave systems are formed beneath a surface topography of hills and valleys (Figure 7.23). Where the landscape has been glaciated, as in northern Britain, the rock is, in many

BOX 7.4 ➤

Figure 7.22 Towering karst pinnacles above flat-lying paddy fields south of Guilin, Yangshuo, Guangxi, China. (Source: Tony Waltham/Geophotos)

Figure 7.23 Limestone sinkholes overlying a cave system on Sinkhole Plain, near Mammoth Cave, Kentucky, USA.

BOX 7.4 ➤

places, initially smoothed by glacial abrasion. The action of solution and joint enlargement only gradually destroys the initial smooth surface, leaving a very characteristic surface of clints and grikes (Figure 7.24). In unglaciated landscapes, no visible trace of the original surface is left, and a landscape of pinnacles, bounded by joint planes, is equally characteristic (Figure 7.22).

Figure 7.24 Clints (higher rock) and grikes (gaps between) formed in limestone in the Yorkshire Dales, northern England.

BOX 7.4

Reflective questions

➤ What are the main geological controls on weathering?

➤ Under what main weathering controls do karst landforms dominate?

7.6 Urban stone decay and lessons for rock weathering

Placing stone in a building does not immunize it from natural weathering processes (Smith *et al.*, 2008). Outside of the dissolution of limestone by acid rain, the principal cause of urban stone decay is salt weathering. Salts can derive either directly in the form of pollution particles deposited on stone, or indirectly through reactions between different forms of acid deposition and stones and mortars. This applies especially to limestones in sulphur-rich atmospheres and leads to the formation of calcium sulphate in the form of gypsum. In areas of elevated atmospheric pollution, and where this reaction takes place faster than the gypsum can be dissolved and washed away by rain, this can result in a uniform crust across complete facades, turned black by the inclusion of soot and other particles. Where levels of pollution are lower or rainfall amounts higher, such black

gypsum crusts are largely restricted to areas sheltered from rainwash, where they form primarily as a result of dry gaseous and **occult deposition**. While it is the presence of such crusts that drives many owners to expensively clean their buildings, the potential longer-term significance lies in the possibility that the gypsum can be washed into the stone, sometimes during the course of cleaning. There it can join other salts (derived from atmospheric pollution, rising groundwater, marine aerosols, road salt and other sources) to cause breakdown through, for example, repeated crystallization and dissolution, hydration and dehydation and differential thermal expansion. This can result in many of the patterns of decay, such as scaling, flaking, granular disaggregation and honeycombing, that we have come to associate with salt weathering of natural rock outcrops (Figure 7.25).

Because of the financial costs involved in construction, repair and replacement of stone in buildings there has been considerable research into the causes of urban stone decay in recent years, but those involved face the same difficulties as rock weathering researchers in terms of convincing building owners of the differences between symptoms and causes, explaining the underlying complexities of the processes operating and demonstrating the importance of accurate diagnosis for effective conservation. Such a communication process is made even more difficult by the widespread public belief that stone should be immutable, that stone buildings should last forever, and that if they do not it is because of some human failing rather than a natural

Figure 7.25 Honeycombing produced by salt weathering of a clastic limestone used as a building stone in Gozo, Malta. This is a low-pollution environment in which stone continues to weather through 'natural' processes.

process of change. What follows therefore are some suggestions as to what building owners and architects might usefully understand about stone decay.

7.6.1 Stone decay is multifactorial

Even more so than with rock weathering, urban stone decay is the product of multiple processes acting synergistically. This derives in part from the superimposition of anthropogenic factors on top of those associated with natural change. One of these is the wider variety of salt types to be found within urban environments that can exploit an increased range of environmental thresholds in terms of processes such as hydration and dehydration. This also includes the consequences of the construction process itself, in which, for example, constraining stone within hard impermeable mortars (especially Portland cement) increases the likelihood of fracture development as the stone attempts to expand against this restraint in response to heating and cooling. At the same time forcing water to drain through the stone rather than through the surrounding mortar can promote salt weathering that further exploits the microfracture network. The end result of this is a phenomenon known as boxworking in which stones disappear to leave only the hard mortar as an empty framework (Figure 7.26).

Figure 7.26 Boxworking resulting from the rapid weathering and erosion of soft oolitic limestone constrained within a hard cement mortar, Oxford, England.

7.6.2 Rates of stone decay are unpredictable

Research into rates of urban stone decay, which go back to the nineteenth century and the work of the famous Scottish geologist Archibald Gieke on the erosion of dated gravestones, has typically envisaged a gradual rate of change. This is partly because of an emphasis on gradual loss of material in solution, but also because of a desire to justify the extrapolation of long-term rates of decay from short-term observations. There are, however, numerous problems with this approach. Environmental conditions are never stable, especially in urban environments where patterns, types and levels of pollution are known to have varied greatly in historical and recent times. It is also the case that many building stones, such as quartz sandstones, are prone to decay in a very non-linear fashion. Typically they can show no surface evidence of decay for many years, especially if this is also associated with case hardening by, for example, near-surface iron cementation. Then they can lose a complete outer layer almost instantaneously through the falling away of a contour scale. If the newly exposed subsurface layer has been sufficiently weakened, perhaps by the outward migration of iron cements, and if enough salt has penetrated deeply into the stone, a set of positive feedbacks could be initiated that lead to rapid decay and complete loss of the stone block. Such a process is possibly encouraged by the fact that as the stone retreats it creates in front of it a humid, shaded environment in which more salt can accumulate. Conversely, if there is only a shallow subsurface zone of weakening and limited salt penetration below the scale, its loss may be followed by only limited granular disaggregation

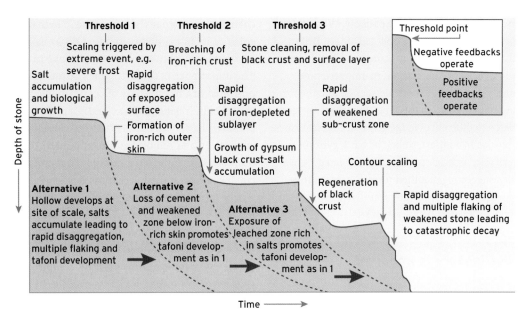

Figure 7.27 Schematic diagram of potential decay sequence of a quartz sandstone used in construction in a polluted environment. (Source: from Smith, 2003)

before a negative feedback is initiated and the newly exposed surface stabilizes. Stabilization can also be brought about if, for example, a black gypsum crust is allowed to develop following scaling. In fact, recent studies of clastic limestones in the still highly polluted city of Budapest and the once heavily polluted city of Oxford have identified multiple crust formation, where stone has gone through many cycles of scaling and stabilization. Interestingly, there is some evidence that a decrease in atmospheric sulphur and particulates (which help to catalyze the rapid formation of gypsum) resulting from clean air legislation could retard this 'scabbing over' and might accelerate the decay of some intrinsically weak stones in the longer term (Smith *et al.*, 2010a). This process of episodic decay is illustrated in Figure 7.27, which also highlights how effects such as scaling and rapid surface loss may also be triggered by human interference such as aggressive surface cleaning. What this also illustrates is the danger and uncertainties associated with any attempt to precisely predict the long-term performance of stone, especially based on one-off observations or only short-term monitoring.

7.6.3 Decay is spatially variable

Casual observation of almost any ancient stone building or structure will reveal stones and groups of stones that have experienced markedly different patterns and rates of weathering and erosion (Figure 7.28). Given that individual walls are likely to be constructed of stone from the same source that was quarried at the same time, the seemingly random distribution of decay that is often seen is a graphic illustration

Figure 7.28 Weathering of quartz sandstone at Bamburgh Castle, north-east England illustrating the variable response of the same stone type based upon subtle differences in porosity, bedding, cementation and other physical and chemical characteristics.

of how quite subtle differences in mineralogy and structure can determine susceptibility to different processes and/or relative resistance to the same one. Alternatively, on some walls there does appear to be a degree of spatial organization of decay. This could possibly be a response to environmental factors, of which the most obvious is decay near to ground level in response to salts derived from groundwater. One way of examining this spatial dimension is to study the degree of connectivity exhibited by particular types of decay across a wall. This can be done by assigning a principal decay type to each block, and counting the number of sides that are adjacent to blocks demonstrating the same type of decay. When this was done for quartz sandstones in a polluted area of Belfast, this showed that stones experiencing scaling and rapid retreat typically had a low level of connectivity, mainly zeros and ones (Smith and Viles, 2006). This could suggest that the factors controlling susceptibility to this type of mechanical decay are mainly related to differences in the physical characteristics of individual blocks. In contrast, soiling by black gypsum and biological crusts were associated with a higher average score for connectivity, suggesting that their formation is controlled to a greater extent by localized environmental factors that override material differences between blocks.

7.6.4 Stress history is important

Following the successful implementation of clean air legislation across much of western Europe, many building owners believed that it was safe to clean buildings without the fear that they would rapidly resoil and safe in the knowledge that stone decay would effectively cease. It came as something of a surprise that despite much cleaner air many buildings continued to decay. In general this has been put down to a 'memory effect', but it is also a tacit recognition that natural rock weathering processes also have a role to play in urban stone decay.

During the course of its lifetime, building stone, or any natural rock or rock outcrop for that matter, will experience a range of environmental conditions and be subject to a variety of forces. For building stone this could include a number of pre-emplacement effects such as dilatation and associated micro fracturing during quarrying, as well as cutting and shaping before it is placed in a building. Post-emplacement, it could be subject to many years of environmental cycling and physical stressing through heating and cooling, wetting and drying and freezing and thawing as well as potential chemical alteration and loading with a range of pollutants, including complex salts, under

changing environmental conditions. These factors could enhance its susceptibility to future decay irrespective of whether atmospheric pollutants continue to be available. The importance of these 'complex histories' has been demonstrated in a recent study of coastal medieval sandstone churches in Ireland and Scotland. Over their lifetime these buildings have been subject to many changes, including the application and removal of plaster renders that have loaded the stone with lime, often multiple fires and a period of enhanced freeze–thaw activity associated with the Little Ice Age of the sixteenth to nineteenth centuries, which itself followed the Medieval Warm Period (see Chapter 23). These changes are in addition to previous campaigns of conservation including repointing. In an experiment to understand the effects of these different factors, samples of fresh sandstone were given complex stress histories by subjecting them in the laboratory to different combinations of freeze–thaw cycles, addition and removal of lime plaster and heating in a wood fire, after which they were artificially salt weathered to replicate the ongoing stresses associated with their coastal location (McCabe *et al.*, 2007). The upshot of this was that different combinations produced very different patterns of weight loss during the course of the salt weathering simulation, perhaps the most interesting being the pattern of loss associated with stones that had been heated by fire. In comparison with other stones, weight loss was initially slow, perhaps because the heating made the stones water repellant for a period, but at a later stage the stones started to rapidly fracture and decay. This possibly reflected the exploitation of a deeper-seated network of microfractures initiated during their rapid heating and cooling.

Reflective question

➤ What key lessons for rock weathering have we learned from studying building stone decay?

7.7 Summary

This chapter has introduced the key chemical, physical and biological processes that drive the weathering of geomaterials. It has also explained the underlying principles that control their operation and interpretation. Weathering is an essential precursor to almost all other geomorphological processes. It is all around us and is happening all the time. It is pervasive, persistent and can operate under normal environmental circumstances. Weathering is often portrayed as a competition between environmental and material controls in which one has to be identified as the victor. The reality is more subtle than this; while one set of controls may appear to be dominant the end product, be it a landform or regolith, is always formed by an interaction involving both factors. The shapes of features produced by weathering have a feedback effect on weathering processes by influencing micro-environmental conditions.

Despite an historical tendency to classify weathering processes, forms and products on the basis of simplified large-scale environmental controls, such as meteorological measures of mean annual rainfall and mean annual temperature, the most important environmental controls on weathering processes tend to act at the meso- and micro-scales at the interface between materials and the environment. Therefore, to understand weathering it is important to focus investigations on the temporal and spatial scales over which the responsible processes operate. Thus, if breakdown is through granular disaggregation one should focus initially on understanding the behaviour of individual grains in response to environmental changes that operate over a few millimetres and in minutes rather than years or even days. Likewise, if salt and ice weather rock by exploiting pores and micro fractures the first thing we need to understand is what goes on in these spaces and what controls it. An implication of such a small-scale approach is that eventually we have to upscale our understanding to explain the formation of landforms and landscapes over much longer timescales.

This has proven to be extremely difficult because rarely do weathering processes operate singly. Weathering is the result of the interaction of two or more processes operating together or in sequence.

All geomaterials carry within them a stress history, even so-called fresh rock will have embedded 'memories' of the conditions under which it was formed, as well as any subsequent tectonics and the processes responsible for its exposure at the Earth's surface. This stress history, especially where it is complicated by, for example, numerous environmental changes over a long period of exposure, can have a major influence on subsequent rates and patterns of weathering. This inheritance effect is particularly significant in the field of urban stone decay. Stress inheritance is especially relevant in situations where decay is episodic in nature and where it is the long-term accumulation of stress and material weakening that typically brings rock close to the threshold beyond which it is subject to rapid and often catastrophic decay. As rock nears this strength–stress threshold it becomes more susceptible to damage from exceptional environmental conditions and processes that did not necessarily contribute to its previous weakening. Hence the common coincidence of weathering with extreme events has led to an over-emphasis of the effects of, for example, extreme high and low temperatures to the detriment and neglect of more mundane conditions and regular environmental cycles that lay the foundations for much of the change and decay that constitutes weathering.

Further reading
Books

Allsopp, D., Seal, K. and Gaylarde, C. (2004) *Introduction to biodeterioration,* 2nd edition. Cambridge University Press, Cambridge.
A thorough overview of the biological processes underpinning rock weathering.

Ashurst, J. and Dimes, F.G. (1998) *Conservation of building and decorative stone.* Butterworth–Heinemann, Oxford.
This text gives a very thorough background to stone types and stone decay and treatment from the perspective of leading stone conservationists.

Bland, W. and Rolls, D. (1998) *Weathering: An introduction to the scientific principles.* Arnold, London.
This is an excellent book for the reader to explore the chemistry of weathering in greater detail.

Dorn, R.I. (1998) *Rock coatings.* Elsevier, Amsterdam.
An extremely thorough overview of the characteristics of rock coatings and the processes responsible for them, by a leading researcher in the field.

Douglas, I. (1979) *Humid landforms.* MIT Press, Cambridge, MA.
Primarily a text on geomorphpology, but some good background on weathering controls.

Ford, D.C. and Williams, P. (2007) *Karst hydrogeology and geomorphology*. John Wiley & Sons, Chichester.
An excellent overview of karst processes and associated landforms.

Goudie, A.S. and Viles, H.A. (1997) *Salt weathering hazards*. John Wiley & Sons, Chichester.
An in-depth study of how salt weathering has turned out to be one of the most pervasive weathering processes that has many implications not just for natural landscapes, but particularly for the durability of materials in the built environment.

Migo, P. (2006). *Granite landscapes of the world*. Oxford University Press. Oxford.
An up-to-date summary of the weathering of one of the world's most distinctive rock types and an excellent demonstration of how weathering contributes to the development of distinctive landforms.

Selby, M. J. (1993) *Hillslope materials and processes*, 2nd edition. Oxford University Press. Oxford.
A very approachable exploration of a core area of geomorphology that emphasizes the underlying processes that contribute to landform development, including excellent explanations of weathering processes.

Thomas, M.F. (1994) *Geomorphology in the tropics: A study of weathering and denudation in low latitudes*. John Wiley & Sons, Chichester.
The most comprehensive overview of geomorphology in those parts of the world that have the strongest links between weathering and landscape development.

Warke, P., Smith, B. and Savage, J. (2010) *Stone by stone*. Appletree Press, Belfast.
This text provides an easy-to-read field guide to building stone weathering which is thoroughly illustrated with photographs. The pocket book includes practical guidance on stone identification, diagnosing stone decay and surveying building condition.

Yatsu, E. (1988) *The nature of weathering: An introduction*. Sozosha, Tokyo.
One of the most detailed accounts of rock weathering.

Further reading
Papers

Hall, K. and Thorn, C. (2011) The historical legacy of spatial scales in freeze–thaw weathering: Misrepresentation and resulting misdirection. *Geomorphology*, 130, 83–90.
This paper challenges traditional assumptions about freeze–thaw and deals with topics outlined in Section 7.3.4.3.

Smith, B.J., Gomez-Heras, M. and McCabe, S. (2008) Understanding the decay of stone-built cultural heritage. *Progress in Physical Geography*, 32, 439–461.
A review of the latest literature and understanding on building stone decay.

CHAPTER 8

Hillslopes and landform evolution

Mike Kirkby

School of Geography, University of Leeds

Learning objectives

After reading this chapter you should be able to:

➤ understand the importance of water in shaping the landscape

➤ describe the key processes of sediment and solute transport

➤ explain how the balance and rate of processes influences the evolving form of hillslope profiles

➤ use slope profiles to interpret the processes shaping them

8.1 Introduction

Over 90% of landscapes that are currently not glaciated consist of hillslopes, and the remainder consist of river channels and their floodplains. Although hillslopes are not generally the most active part of the landscape, they provide almost all of the material which eventually leaves a river catchment through the more active channelways. The processes by which parent material is broken down by weathering (see Chapter 7) and carried to the streams are therefore vital to an understanding of how the catchment works as a geomorphological machine. The weathered debris on hillslopes

(the **regolith**) is also the raw material from which soils are developed (see Chapter 10). Land management such as agricultural practice strongly affects the rate and types of hillslope processes. The way in which farmland is managed can dramatically influence whether soil erosion remains at an acceptable level, or is increased to a rate which leads to long-term and often irreversible degradation of the soil.

Terrestrial landscapes are dominated by erosion, and the material removed is ultimately transported to the oceans where it takes part in the continual slow recycling of the Earth's crust as tectonic plates spread apart and collide to form new mountains (Chapter 2). Geomorphological processes form an essential part of this crustal recycling which periodically renews the surface of our planet in episodes of **orogeny**, erosion and isostatic response. Water helps to break up rocks as part of the process of weathering, and drives sediment transport processes that carry soil materials down to the ocean, progressively eroding the land. This chapter reviews the various hillslope sediment processes, the factors which influence their rates and the ways in which the processes in an area influence the form of the hillslopes. The balance between process rates has a very strong influence on the form of both the landscape and its soils, and plays a large part in the distinctive appearance of landscapes in different climatic regions of the Earth. Weathering and erosion provide the raw material which rivers transport through

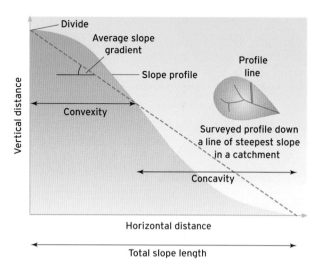

Figure 8.1 A typical hillslope profile.

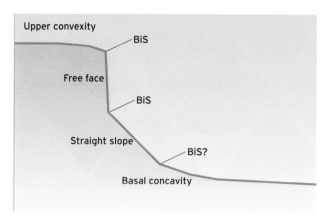

Figure 8.2 Elements of a slope profile: upper convexity, free face, straight slope and basal concavity. BiS is break in slope.

their valleys to the oceans. However, changes in hillslope erosion rate may not be matched by similar changes in river transporting capacity, resulting in either floodplain **aggradation** and widening, or valley incision.

8.2 Slope profiles

The appearance of hillslopes around the world varies considerably (Allen, 1997). Some of these differences are related to the vegetation and soils on the hillslopes, but there are also important differences in the form of hillslope profiles. Most profiles are convex near the divide (top of the slope) and concave near their base (Figure 8.1). The most significant differences between them are in their total length, gradient and convexity.

8.2.1 Slope length

Slope profiles are generally surveyed along a straight line (on a map) from the divide, following a line of steepest descent, to the nearest point at their base (Figure 8.1), which is usually in a river or floodplain, or, for coastal slope profiles, directly into the sea. The length of inland slope profiles is linked to the **drainage density** of channels (see Chapter 12), which varies, depending on gradients, climate and rock type. In temperate areas such as Britain, slope profiles are almost a kilometre in length, whereas some semi-arid areas have slopes that are less than 10 m long. Drainage density generally tends to be low (1–5 km km^{-2}) in humid areas such as north-west Europe, and higher (10–500 km km^{-2}) in semi-arid areas, particularly where the soils are developed from impermeable rocks such as shales, clays or marls, on which intensely dissected **badlands** may occur.

8.2.2 Slope steepness

Slope profiles also differ in their average gradients, from steep cliffs to gentle slopes with almost imperceptible gradients. Figure 8.2 shows the four possible elements of a profile, consisting of an upper convexity, a free face, a straight slope of almost constant gradient and a basal concavity. Not all of these features are found in every slope profile, and some may be repeated more than once within a profile. For example, many profiles do not have a free face at all, but grade smoothly from convexity to concavity, as in Figure 8.1, whereas many steep profiles have a number of free faces, each with a straight slope and/or basal concavity below it.

- Upper convexity: from the divide, there is generally a convex slope of increasing gradient and with slopes from level (0°) to a maximum of up to 35° (although usually less).
- Free face: below the upper convexity there may be a cliff of free face, usually consisting of bedrock at a slope of up to 70° (although exceptionally steeper and/or locally overhanging). The free face may be somewhat stepped if there are rock layers of different resistances.
- Straight slope: there may also be a straight section of almost uniform gradient. Below a free face this usually consists of a scree or **talus** slope, at 30° to 40°, with a surface of loose stones. In some cases the straight slope is cut into bedrock, which is usually visible in patches beneath a thin layer of loose stones, and is called a boulder-controlled slope. Where the rock is layered, the sequence of free face and talus slope may be repeated several times. Straight slopes at 10° to 25°, in fine materials, are also commonly found in sands and clays, although not associated with a free face, but merging directly into the upper convexity.

- Basal concavity: at the foot of the slope there is usually a basal concavity which leads down towards the valley-bottom river and floodplain. This is usually in sand or finer materials.

Between these slope elements, there may be breaks in slope (BiS in Figure 8.2), within which the slope gradient changes relatively abruptly. There are commonly breaks in slope at the top of a free face, and between the free face and the straight slope below it. In arid and semi-arid areas, there is usually another break in slope between the straight slope and the basal concavity, although this is not normally found in humid areas, and is associated with the different balance of processes. Comparing hillslopes of different overall steepness, the proportion of different slope elements changes, and the free face and/or straight slope are commonly absent on lower-gradient slopes. These differences are sketched in Figure 8.3, and this represents one of the very many possible evolutionary sequences over time. On even steeper slopes, the basal concavity and/or the straight slope may also disappear, so that the cliff plunges directly into a river as a gorge.

The slope steepness reflects the relationships between the hillslope and conditions at the slope base. Where there is a river that is cutting rapidly downwards at the slope base, then slopes are inevitably steeper than where the river is stable. The most rapid downcutting is usually associated with tectonic uplift, since the river is generally able to cut down almost as fast as the land is uplifted, while the upper part of the hillslopes is initially little affected. Steep slopes are also associated with coastal areas, owing to undercutting by wave action (Chapter 15). Coastlines are some of the commonest locations for good free face development. Cliffs are also common in formerly glaciated areas, where slopes have been steepened by glacial erosion, and post-glacial processes have not yet (after 10 000–15 000 years) had time to eliminate them.

Over time, and in tectonically stable areas, slope processes progressively erode the landscape to produce more and more gentle slopes. The gentlest slopes are therefore found in tectonically stable shield areas, such as West Africa, central Australia and northern Canada and Eurasia. Where rocks are more easily eroded, slopes flatten more quickly, so that, in any area, the gentlest slopes are usually found on clays and shales. However, there are exceptions which include badland areas where the dissection is severe, promoting steep gully sides.

8.2.3 Slope convexity

The amount of the hillside that is convex in profile is usually expressed as a proportion of the total slope length. Hillslopes vary from almost complete convexity (100% convex) to a narrow convexity and a much broader concavity, although most slopes are mainly convex. Where slopes appear to have long concavities, as in the 'skyline' profile shown in Figure 8.4, the profiles seen from a distance are not the lines of steepest slope, but a line or a series of lines along the divides between channels. These divides follow the concave profiles of the streams between them, whereas the true 'steepest descent' slope profiles run from the divides into the nearest stream and are convex, or convex and straight.

The proportion of convexity is related to both the balance between slope processes and the relationship between the slope and the river at its base. Where the river is cutting down rapidly, or undercutting the base of the hillside, then slopes are not only steeper but also tend to be more convex. Similarly, where the streams are aggrading, more of the slope profile tends to become concave.

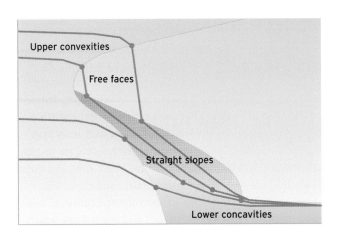

Figure 8.3 Slope elements on steep and gentle gradient hillslopes.

Reflective question

➤ Compare slope profiles you have seen or studied in different areas: how do they differ in the proportions of each slope component (Figures 8.2 and 8.3) – or are they all the same?

8.3 Hillslope transport processes

Slope processes are of two very broad types: (i) weathering and (ii) transport of the regolith. Within each of these types, there are a number of separate processes, which may

'skyline' profile

'steepest descent' profiles

Figure 8.4 Cliffs and gullies south of Tatouin, Tunisia. The gullies are in sandstone cliffs and across the basal concavity in shales. Visually the skyline slope profile follows the divides between the concave gully thalwegs. However, true slope profiles are from the divides into the nearest channel thalwegs.

be classified by their particular mechanisms into groups (Table 8.1), although many of these processes occur in combination. Most slope processes are greatly assisted by the presence of water, which helps chemical reactions, makes masses slide more easily, carries debris as it flows and supports the growth of plants and animals. For both weathering and transport, the processes can conveniently be distinguished as chemical, physical and biological.

Landscapes evolve over time in response to the internal redistribution of sediment, usually with some net removal of material to rivers or the ocean. The way in which landscapes and slopes evolve depends on their initial form, the slope processes operating and the boundary conditions which determine where and how much sediment is removed (Allen, 1997). These relationships between process and form are discussed later in this chapter. Weathering is dealt with in Chapter 7 and so the subsequent section deals with transport processes.

Where there is a plentiful supply of material, and the process which moves it can only move a limited amount for a short distance, the rate of transport is limited by the

transporting capacity of the process, which is defined as the maximum amount of material which the process can carry (Kirkby, 1971). A transport process of this kind, such as **rainsplash** (see below), is described as **transport limited**. Some other processes are limited, not by their capacity to transport, but by the supply of suitable material to transport, and are described as **supply limited**. For example, rockfall (see below) from a cliff has a very large potential capacity to carry material, but is limited, fortunately, by a shortage in supply of freshly weathered material.

There is not always a clear distinction between transport- and supply-limited processes, but it is an important distinction which has a substantial impact on the way in which hillslopes evolve over geological time periods. Landscapes that are dominated by transport-limited removal are generally covered by a good layer of soil and vegetation, and slope gradients tend to decline through time. Landscapes where removal of material is mainly supply limited, however, tend to have sparse vegetation, thin soils and steep slopes which tend to remain steep throughout most of their evolution (Carson and Kirkby, 2009).

Table 8.1 Classification of the most important hillslope processes

	Weathering processes	Transport processes	Type (S/T)
Chemical	Mineral weathering	Leaching	S
		Ionic diffusion	T
Physical	Freeze-thaw Salt weathering Thermal shattering	Mass movements:	
		Landslides	S
		Debris avalanches	S
		Debris flows	S
		Soil creep	T
		Gelifluction	T
		Particle movements:	
		Rockfall	S
		Through-wash	T
		Rainsplash	T
		Rainflow	T
		Rillwash	T
Biological	Animal digestion Root growth	Biological mixing (often included within soil creep)	T

Types: T = transport-limited; S = supply-limited removal (see text).

8.3.1 Chemical transport processes (solution)

The process of solution is closely linked to chemical weathering, as rock and water interact. The chemical reactions that are altering rock material in place are, at the same time, releasing the lost material dissolved in the water. Rocks that weather rapidly therefore lose material in solution rapidly too. In Figure 8.5, rain falls on the soil, where it picks up solutes from the regolith, in proportion to the concentration of each constituent in the regolith, and its solubility. As water in the rock gradually weathers the bedrock, solutes also diffuse out of the upper layers of rock into the soil water. Some water is lost by percolation, containing solutes carried, for example, into limestone cave systems. Some

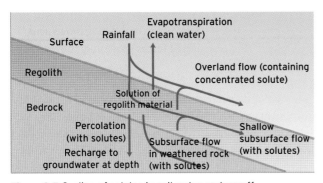

Figure 8.5 Cycling of solutes in soil water and runoff.

water is lost to evapotranspiration, and this carries little or no solutes, so that the remaining overland flow and subsurface flow runoff have an increased concentration of solutes. This concentration effect is only marked in relatively arid climates, where the evapotranspiration is high.

In extreme cases, some of the soluble material reaches its maximum saturated concentration, and any further concentration leads to redeposition of the dissolved material near the surface. This occurs most commonly for calcium, which is often found to form crusts of calcrete near the surface in arid and semi-arid areas. The concentration of solutes is therefore generally highest in dry climates, but the total amounts removed in solution are much less than in more humid areas. Where a flow of water contains dissolved material, the rate at which the solutes are carried away or advected is determined by their concentration in the runoff water. This is called **advective solution** or leaching, and is very effective at removing solutes from regolith near to the surface, both in runoff and in percolating waters. Once material is leached out, it generally travels far downstream, and its rate is supply limited. Leaching is not very effective, however, in removing material from the bedrock–regolith boundary because little water usually flows across this boundary. However, there is a rapid change in solute concentration at this boundary as water remains in contact with material of different composition (see Chapter 13). Close to the bedrock, there is a high concentration of solute ions in the water. Further above the bedrock, there is a lower concentration in the slightly more weathered regolith.

This difference in concentration results in a net upward movement of ions, which means that there will be a movement of solutes away from the bedrock towards the regolith, even though no water is moving. This is because of **ionic diffusion**. Ions move about randomly over short distances and Figure 8.6 compares the number of downward movements from the upper area of low concentration and the number of upward movements from the lower area of high concentration. Even though the movements are random, Figure 8.6 illustrates how random movements in all directions cause a net diffusion of material from areas of higher ion concentration to areas of lower concentration. Around the regolith–bedrock interface the concentration of ions is higher near the unweathered bedrock than in the partially weathered regolith, driving a net upward movement of ions, carrying solutes away from the bedrock, even though little water is moving. In this case the solute load depends not on the flow of water but on the differences in concentration between the layers. This movement of solutes by ionic diffusion is not as fast as by leaching where there is appreciable water moving, but can be very important in the early stages of rock weathering, when little water is able to flow through the almost intact rock (Yatsu, 1988). Because

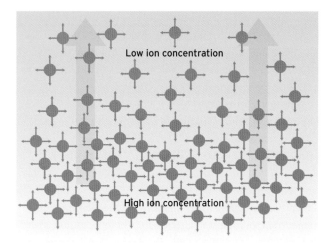

Figure 8.6 Net transport resulting from random diffusion. Because the concentration is greater nearer the bottom of this figure than at the top then there will be more upward movements than downward movements and so the net transport will be upwards.

material moves only a short distance by ionic diffusion, and is not limited by the supply of suitable material, solution by ionic diffusion is a transport-limited process.

Leaching is generally the most important process in carrying solutes down the slope and into the rivers (see Chapters 10 and 13). Both leaching and ionic diffusion, however, play an important role in moving solutes vertically. One particularly significant role of vertical leaching is in carrying plant nutrients down to the roots from decaying leaf litter deposited on the surface, and so completing the nutrient cycle of the vegetation (see Chapter 20).

8.3.2 Physical transport processes

When material is physically transported down a hillslope, it may travel as a mass or as independent particles. In a **mass movement**, a block of rock or soil moves as a single unit, although there may be some relative movement within the block. The movement of the block is mainly determined by the forces on the block as a whole, and the individual rock or soil fragments within the block are in close contact, so that they are moved together, almost irrespective of the properties of the individual constituent grains. The alternative to a mass movement is a **particle movement**, in which grains move one, or a few, at a time, and do not significantly interact with one another as they move. For a particle movement, forces act on each particle separately, and they move selectively, mainly depending on their sizes, but also on other factors such as shape and density (Selby, 1993). Some processes can behave in either way, according to the size of an individual event. For example, in small

rockfalls there is little interaction between the few blocks coming down the cliff face, but larger blocks may break up into fragments which interact as they fall, giving them some of the characteristics of a mass movement.

Both mass movements and particle movements can occur at a range of rates. In general, however, movements driven by large flows of water tend to be more rapid than drier movements. The more rapid movements also tend to carry material farther, and so tend to be supply limited, whereas slower movements tend to be transport limited.

8.3.2.1 Force and resistance

Movement of material is decided by a balance of forces, some of which promote movement and some of which resist movement. For mass movements, the forces act on the block of material which is about to move, and for particle movement on each individual particle. The main forces promoting movement are those of gravity and water detachment. On a slope, there is always a component of the weight of the material that tends to pull it downslope, and this applies equally to particle and mass movements. Flowing water can detach fragments of rock or soil if it passes over them rapidly. It can do this in three ways: (i) flow of water over the surface picks up material from the surface; (ii) raindrops strike at the surface at up to 10 m s^{-1} and jets of water from their impact can return at almost equal speed; and (iii) in a deposit which has enough fine-grained material in it, water can also permeate the entire deposit, and convert it into a mixture of water and sediment which moves as a thick slurry. This process is called **fluidization**, and is able to carry large masses of material in **debris flows** (see below). Friction and **cohesion** provide resistance to movement. Box 8.1 provides details of such resistance processes and the balance of forces operating on hillsides. In simple terms, material begins to move when the forces promoting its movement become larger than the resistances holding it back. The ratio of these forces is known as the **safety factor**:

$$\text{Safety factor (SF)} = \frac{\text{sum of forces resisting movement}}{\text{sum of forces promoting movement}}$$

(8.1)

If SF > 1 then movement will not occur; if SF ≤ 1 then movement will begin. As soon as movement begins, the resisting forces on the material usually decrease, as the moving material detaches itself from the bonds which originally held it in place. The moving material therefore accelerates at first. Material slows down again only when the promoting

FUNDAMENTAL PRINCIPLES

RESISTANCE AND THE BALANCE OF FORCES

Resistance to movement is mainly due to friction and, to a lesser extent, cohesion. When a particle, or a block of material, rests on another, a component of its weight (together with the weight of any other material on top of it) provides the 'into-slope' force shown in Figure 8.7. For given materials in contact, the frictional resistance that can be exerted is a fixed proportion of the into-slope force. This is called the **coefficient of friction**, μ. Another way to express the coefficient of friction is as the angle of friction, ϕ. For a dry slope, the angle of friction is related to the coefficient of friction by the equation:

$$\mu = \tan \phi \qquad (8.2)$$

In Figure 8.7, where there is no water entrainment ($E = 0$), the downslope component of the weight ($mg \sin b$) exactly balances the frictional resistance ($F = N \tan \phi = mg \cos \beta \tan \phi$) when the slope angle, β, reaches the

angle of friction, ϕ. The angle of friction is therefore very easy to measure experimentally (Figure 8.8), and, for coarse material, is approximately equal to the angle of repose found in natural scree slopes, of 30–35°.

In many kinds of partly weathered bedrock, the material consists of roughly rectangular blocks, separated by joints or bedding planes. The possibility of sliding parallel to the surface along the zig-zag line indicated in Figure 8.9 is very much hindered by the interlocking of the two surfaces, and can begin only when the surfaces are lifted apart, or dilated in the direction shown by the arrow, at an angle to the slide direction, so that free sliding only takes place along the surface after the sliding mass has lifted clear of the main cliff. Under these conditions, which are typical of most cliffs, the effective angle of friction, Φ, is increased by the dilation angle, so that:

$$\Phi = \phi + \theta \qquad (8.3)$$

It is for this reason that cliffs are commonly able to stand at angles of 75° or more, made up of an angle of friction of 35° plus a dilation angle of 40° or more.

The frictional resistance to sliding of a block of material is very strongly affected by the water within the material. If the material is saturated with water, then part

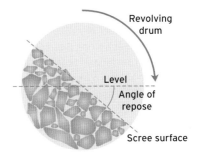

Figure 8.8 A transparent drum is slowly turned so that the angle of repose of a scree material can be measured.

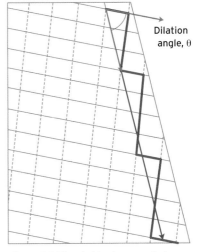

Figure 8.9 The stable slope angle for a cliff with interlocking joint blocks and/or bedding planes. Blocks are only released from the slope when they are able to slide out at an angle to the slope surface, the 'dilation angle' shown, which may be 40° or more.

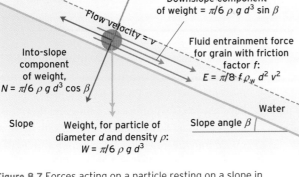

Figure 8.7 Forces acting on a particle resting on a slope in overland flow.

BOX 8.1 ►

of its weight is carried by the water, following **Archimedes' principle** that the upthrust is equal to the weight of water displaced. On a hillslope, the surface of the flowing water is parallel to the surface, and so this upthrust, or weight relief, affects only the into-slope component of the weight and not the downslope component. The effect of saturating the material is therefore to reduce the frictional resistance, which is proportional to the into-slope force. This is a very important factor, and is able to reduce the frictional force to roughly half its value under dry conditions. This means that a slope that becomes saturated from time to time can be stable only at slopes of about half of the angle of friction.

In some materials, especially unweathered rocks and some clays, there is still some resistance to movement even when there is little or no overburden weight. This residual resistance is called the cohesion of the material, and the total cohesive force is equal to the cohesion value for the material multiplied by the area of effective contact. Cohesion is thought to develop in materials that have been consolidated at depth. Hence clays formed close to the surface, such as some tills, have little cohesion, while older clays, particularly those consolidated over geological time periods, have substantial cohesion. When these compressed clays are brought up to the surface, however, weathering along fissures in the clay gradually reduces the cohesion, so that, over a period of 50-100 years, the cohesion becomes very small again.

Material strength is made up of friction and cohesion. Both of these are usually expressed as a **stress**, or force per unit area, and measured in megapascals (equal to millions of newtons per square metre) (see also Box 16.3 in Chapter 16). The total strength, s, exerted to prevent sliding is expressed as a stress, or force per unit area:

$$s = c + \sigma \tan \phi \qquad (8.4)$$

where c is the cohesion, σ is the normal force (N) per unit area and ϕ is the angle of friction. The angle of friction and the cohesion are essentially properties of the material. Some typical values are summarized in Table 8.2. The general pattern in this table is that materials with a lot of clay minerals (clay, till and shale) have lower angles of friction than others, and heavily consolidated materials (limestone, granite, sandstone and shale) have higher cohesion than unconsolidated materials which have been formed closer to the surface (sands, gravels, clay, till and chalk).

Table 8.2 Typical values for soil strength parameters in soil and rock materials

	Sand/gravel	Clay	Glacial till	Chalk	Limestone	Granite	Sandstone	Shale
Cohesion (MPa or million N m^{-2})	0	0–10	0–10	10–20	100–400	100–400	10–30	10–40
Angle of friction (degrees)	30–40	5–20	15–35	30–35	30–35	30–35	30–35	10–25

BOX 8.1

forces also decrease. This usually happens where the material comes down to lower gradients where the downslope component of the gravity force becomes less, or where the water flow carrying the material spreads out and moves more slowly, or drains out into the ground or the surrounding area.

8.3.2.2 Rapid mass movements

There are many names for different types of rapid and slow mass movements. In rapid mass movements, the crucial distinction is between **slides**, in which the moving mass essentially moves as a block, and **flows**, in which different parts of the mass move over each other with differential movement or **shear**. Figure 8.10 shows the difference between velocity profiles for a slide and flow. It is usually found that flows occur in masses with more water mixed into the moving mass, in proportion to the amount of regolith or rock material. In a slide, water is often very important in reducing the frictional resistance and allowing movement to begin, but there is little water within the moving mass. In a flow, there is usually almost at least as much water as solids, and sometimes many times more. Water

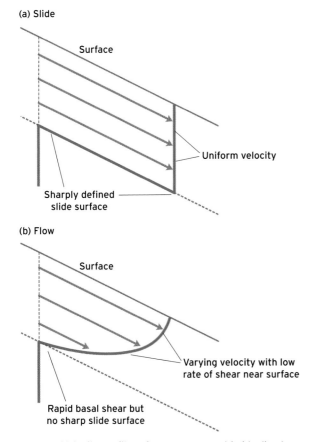

(a) Slide

Surface

Uniform velocity

Sharply defined
slide surface

(b) Flow

Surface

Varying velocity with low
rate of shear near surface

Rapid basal shear but
no sharp slide surface

Figure 8.10 Velocity profiles of mass movement in idealized slides and flows. Shear is the differential movement between layers. In a slide, shear is concentrated at the slide surface. In a flow, shear is spread throughout the moving mass.

and regolith materials can be mixed together in almost any proportions if they are moving fast enough, although coarse materials (sand, gravel and boulders) can only remain suspended through the mixture in the fastest and largest flows. Table 8.3 shows a classification of mass movements based on their water content, and the resulting type of flow.

In a slide, the form of the original block can usually still be seen, particularly at the upslope end, where the **back-scar** is usually clear. The slide mass may show multiple scars and

cracks where it has moved. The downstream end, or toe, shows much more severe deformation producing a hummocky topography where the mass has advanced over the previous surface. Slides may be more or less planar when there are lines of weakness which follow geological structures or are parallel to the ground surface. Cliffs with lines of weakness parallel to the face often fail in this way, creating slab failures in which a flake of rock collapses completely or partially (Figure 8.11a). Sometimes the flake only partly separates and begins to lean progressively outwards until it fails by toppling. Many low-angle (10–20°) clay slopes also fail in planar slides, along surfaces near the base of the weathered regolith (Figure 8.11b). Planar slides are long (in the downslope direction) relative to their depth (measured into the slope), with length : depth ratios of 10 : 1 to 20 : 1 (Skempton and DeLory, 1957).

Slides also occur deep within the mass of a slope. In tills or consolidated clays, these rotational slides (Figure 8.11c) move on surfaces 5–10 m deep, but in strong rock, the slides may be at depths of 50–250 m, in proportion to the much greater cohesive strength of the rock compared to the clay (Table 8.3). Their length : depth ratios are typically 3 : 1 to 6 : 1. The largest slides, in rock, may therefore move entire mountain sides, and may be very destructive, such as the Hope and Turtle Mountain slides in the Canadian Rockies, the prehistoric Saidmarreh slide in south-west Iran (Figure 8.12) and a major recent slide in Taiwan (Figure 8.13).

Where the slope material moves in a wetter mixture, the mass movement becomes more like a flow. In a pure flow, such as a river, debris is lifted into the flow by turbulent eddies and settles to the bed again under its own submerged weight. Material is deposited as a series of bars within channels. As the concentration of solids is increased, many of the moving grains strike other grains before reaching the bottom, and are supported on a bouncing layer of grain-to-grain collisions. The lift provided by these collisions, called the **dispersive grain stress**, becomes more important than turbulence in a **hyper-concentrated flow**, and is maintained by the power of the flow. Along the edges of the flow, the

Table 8.3 Rapid mass movements classified by water content

Water content		Density (kg m^{-3})	Types of flow	Sediment forms
More solids than water	↑	2600	Slides	Back-scar and toe
		1900	Debris flows	Thixotropic forms
		1700	Debris avalanche	Marginal levees and lobes
More water then solids	↓	1000	Fluvial sediment transport	Mid-channel bars

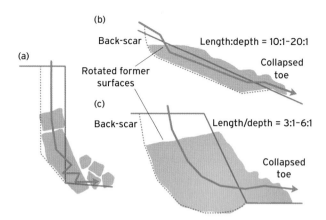

Figure 8.11 Types of rapid mass movement: (a) slab failure on a steep gradient; (b) low-gradient planar slide, length : depth = 10 : 1 to 20 : 1; (c) steep rotational failure, length : depth = 3 : 1 to 6 : 1. Red lines show general path of moving mass.

Figure 8.12 The prehistoric Saidmarreh landslide, south-west Iran. Material from the 15 km wide source area was spread over 64 km², over riding the ridge on the other side of the valley. (Source: Geology.com, using Landsat Geo cover data provided by NASA)

Figure 8.13 Massive landslide, near Keelung, north Taiwan. (Source: Reuters: Ho New)

water drains out sideways, leaving a levee of material, while the centre of the flow continues downslope. When the flow eventually comes to a stop on gentler slopes, the levees are joined by a loop of debris around the front edge of the flow. This kind of flow is relatively common on steep mountain slopes, and is generally known as a **debris avalanche**. In many formerly glaciated valleys, debris avalanches come down the steep side slopes, and their lobes spread over the gentler foot slopes, each flow following the lowest path avail-

able over the lobes of previous flows. Creation of the mixture which behaves as a hyper-concentrated flow is thought to occur through small slides on gully sides which fall into a rapidly flowing stream of water down the centre of the gully.

Where there are sufficient fine-grained materials (silt and clay) in the mixture, then a true **debris flow** can occur, with a still higher ratio of solids to water. Drainage at the edges of the flow is less evident, and grains are supported because they sink only slowly in the mixture, which is both dense and viscous (sticky), so that the flow is much less turbulent and even large rocks sink very slowly. The flow may still move at dangerously high speeds, but also tends to stop suddenly, as drainage lowers the water content to a critical level, and the whole mass suddenly sets into a rigid mass. As water drains out, the viscosity rises sharply as grains are no longer separated by flowing water, but collide with one another and coagulate. This behaviour, in which the viscosity decreases as the rate of shear (relative movement) increases, is called **thixotropic**. This creates more and more rapidly moving masses once the movement has been triggered, and very rapid solidification as the movement slows down, through drainage and/or on lower gradients. Debris flows can be triggered by landslides during intense storms, particularly in semi-arid areas where there is plenty of loose material and in steep mountain areas worldwide. After volcanic

Figure 8.14 Multiple mass movements following the 2008 Sichuan earthquake, China. (Source: James Davies)

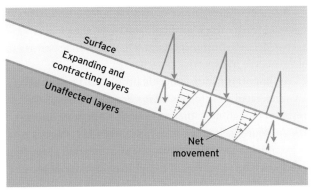

Figure 8.15 Soil creep due to expansion and contraction in a sloping regolith. The net movement is downslope.

eruptions, freshly deposited ash commonly provides a plentiful supply of loose, low-density material, which is particularly prone to debris flows, known as **lahars**. Flows may also be triggered by subsurface seepage of water into the bottom of an unstable human-made deposit, as occurred in the 1966 Aberfan slide in South Wales, where springs saturated a low-density mine spoil heap and coal waste flowed into a school and houses, killing 144 people, of whom 116 were children.

In steeplands, whole areas may be affected by swarms of mass movements. These are usually triggered by rainfall as it wets the soil and reduces the safety factor, but the underlying cause may be exceptional storms, for example linked to hurricanes or typhoons, that lead to unprecedented levels of soil saturation. Swarms of mass movement can also be caused by land-use change, for example where logging removes the trees that help to transpire soil water and add to the cohesion of the soil with their roots. Earthquakes may also be responsible, shaking the ground, and so adding to the forces that promote movement and reducing the safety factor. Figure 8.14 shows shallow landslides and debris flows triggered by the 2008 Sichuan earthquake.

8.3.2.3 Slow mass movements

The essential characteristic of slow mass movements is that they do not involve movement bounded by a discrete slip surface. Failures occur between individual soil aggregates, and not over the whole of an area. Movements are usually driven either by 'heaves' of expansion or contraction, or by apparently haphazard movements between aggregates. Heaves are usually caused by freezing and thawing of soil water, or by wetting and drying of the soil. Haphazard or apparently random movements are usually caused by biological activity,

which mixes the soil in all directions. In all cases, these movements do not cause any net movement when they occur on level ground, but on a slope the steady action of gravity causes more downhill than uphill movement, and there is a gradual transport of regolith material, at a rate that increases with gradient (Figure 8.15). One particular form of slow mass movement, which is intermediate between slow and fast behaviour and is driven by freeze–thaw activity, is **gelifluction**, which occurs in periglacial areas (see Chapter 17).

Wetting or freezing of the regolith has been shown experimentally to produce an expansion that is almost perpendicular to the soil surface. As the soil then dries (after wetting) or thaws (after freezing), it sinks back closer to the vertical, under the influence of gravity on the more open texture of the expanded soil (Figure 8.15). The rate of movement is therefore thought to be roughly proportional to the slope gradient. At different times of year, the expansion penetrates to different depths, so that, totalling over the year, the lateral movement of material is greatest close to the surface, and dies away gradually with depth.

Biological organisms (e.g. earthworms, rabbits and gophers) move material more or less randomly in all directions, but gravity again provides a bias which leads to net downhill movement on a slope at a rate roughly proportional to gradient.

The three main drivers for soil creep, wetting–drying, freeze–thaw and biological mixing, can all be of similar magnitudes, although one or other dominates in any particular site (Selby, 1993). For example, in temperate deciduous forests, biological mixing may dominate, whereas in upland areas, freeze–thaw is probably the most important. Most measurements suggest that rates of near-surface movement by soil creep are typically 1–5 mm per year, dying away to nothing at depths of 300 mm. The total sediment transport capacity is usually estimated, from these measurements, as about:

$$C = 10 \times (\text{tangent slope gradient}) \text{ cm}^3 \text{ cm}^{-1} \text{ yr}^{-1} \quad (8.5)$$

Because soil creep moves the regolith material, and only operates when there is an ample soil cover, the process is considered to be transport limited, so that transport is always at the transporting capacity.

An important anthropogenic process that behaves like an accelerated soil creep is **tillage erosion**, which is the result of ploughing, either up- and downslope or along the contour. Each time the soil is turned over, there is a substantial movement of soil. Up- and downhill ploughing produces a direct downhill component of movement as the turned soil settles back. Contour ploughing can move material either up or down, according to the direction in which the plough turns the soil. Contour ploughing in which the soil is turned downhill moves approximately 1000 times as much material as soil creep. Contour ploughing in both directions (soil turned uphill and then downhill or vice versa) or ploughing up- or downhill produces a smaller net movement, but the overall rate is still about 100 times greater than natural soil creep. Sediment transport is more rapid using modern heavy machinery than with primitive ploughs, but it is clear that tillage erosion may have been responsible for more soil movement in some areas during the past few centuries than natural soil creep during the whole of the last 10 000 years. The accumulated effect is often apparent from the build-up of soil behind old field boundaries and from the infilling of hollows within arable fields.

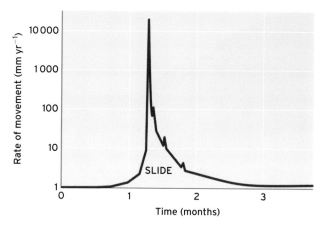

Figure 8.16 Continuous soil creep before and after a slide.

8.3.2.4 Continuous creep

For several days or weeks, both before and after landslides (Figure 8.16), the soil commonly moves slowly along the slide surface in a process which is intermediate between a landslide and soil creep. The rates may be characteristic of soil creep, but the movement is concentrated along a well-defined surface. In some clays, much of the movement on a hillside seems to consist of this continuous creep, which is usually shown up by occasional tears in the vegetation cover, and by evidence of more distinct landslides here and there (Figure 8.17). In this type of movement, the main

Figure 8.17 An earthflow in southern Italy, where continuous creep is occurring over most of a field.

movement is a downslope glide rather than an expansion and contraction, but some sites show a combination of normal soil creep and continuous creep, so that many combinations of heave and glide are possible.

8.3.3 Biological mixing

Plant roots and burrowing by creatures of all sizes produce an overall biological mixing. Because there is more material in dense soil than in uncompacted soil, there is a net diffusion of material from denser to looser soil (Figure 8.18). Small organisms such as bacteria are very abundant, but generally move little material, as they are much smaller than the soil aggregates. Larger organisms such as rabbits may move large quantities of material, but are much less abundant than earthworms. The greatest total effect is generally due to creatures of moderate size, such as earthworms, termites and ants, which are just large enough to move the aggregates. The rates of mixing are much lower when the soil environment is not suitable for them due to waterlogging, cold, lack of nutrients and/or lack of air (usually at depth). However, in some areas, larger rodents, such as gophers and mountain beavers, have the greatest impact.

The regolith is normally loosely packed near the surface, and denser at depth, and this **bulk density** profile is a result of two processes, both strongly driven by biological mixing: (i) the balance between a net upward diffusion of mineral material by biological mixing, and its settlement under the action of gravity, at a rate which is greatest where the bulk density is least; (ii) the balance between the net downward mixing of low-density organic litter and its decomposition to

(a)

(b)

Figure 8.19 Two scree slopes. (a) Boulder-controlled slope in Nevada where cliff and scree are dynamically retreating. (b) Static accumulation on a scree in Yorkshire.

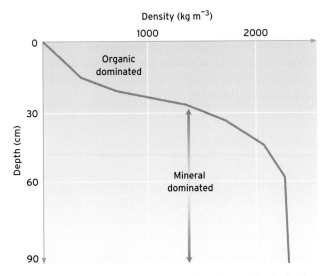

Figure 8.18 A bulk density profile, showing the combined effects of increasing pore space and increasing organic content towards the surface.

carbon dioxide. The combination of these two processes gives a bulk density which is very low at the surface, and gradually increases with depth to a constant value at 0.2–2.0 m according to conditions, where the soil is deep enough (Figure 8.18). Within this layer of biological mixing, the mechanical mixing processes are generally much more rapid than any chemical processes, so that the mineral soil shows only very small chemical differences. If weathering goes to depths beyond the reach of biological mixing, there is a clear distinction between the homogenized mixed upper layers and an undisturbed saprolite in which the detailed rock structures are preserved intact within the weathered regolith.

8.3.4 Particle movements

8.3.4.1 Rockfall and screes

Although cliffs may lose material in large slab failures, they more commonly lose smaller blocks, which fall as they are released from the cliff face by weathering along the joints and/or bedding planes around them (Terzaghi, 1962). These blocks often break into smaller fragments on impact, and the pieces bounce and roll down the **scree slope** at the foot of the cliff, with little or no interaction between them (Figure 8.19). The scree slope is itself constructed by accumulation of the falling blocks as the cliff retreats. As it retreats, the scree covers the base of the cliff, and protects the base from further loss. In this way a bedrock core is established within the scree. This can occasionally be seen in road-cuts, or where the loose scree is quarried away.

The blocks falling onto a scree have a range of sizes, and the scree slope acts as a dynamic sieve which sorts the stones as they bounce, roll and slide down its length. Each time a block makes contact with the scree surface, it may come to a stop, or it may continue downhill. A small block landing on a surface of coarser blocks can readily be trapped between the blocks and stop, whereas a large block tends to slide over the gaps between smaller blocks (Figure 8.20). Small blocks therefore tend to stop near the top of the scree slope, and larger blocks go farther down, creating a slight down-scree coarsening of the grain size, which is maintained as the scree continues to accumulate.

The broken blocks will, in time, also weather away. In arid areas, the boulders are often covered in a tough, dark **desert varnish**, which is produced as the interior of the boulder weathers. When the varnish is broken, perhaps by an impact, the weathered interior of the boulder breaks

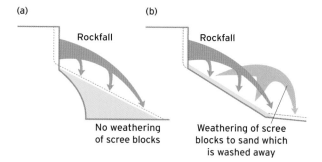

Figure 8.21 Static and dynamic cliff and scree evolution: (a) static accumulation of scree with burial of cliff; (b) continued dynamic retreat of cliff and boulder-controlled slope.

down into sand, which is easily washed off the steep scree slope (Melton, 1965). The way in which the scree slope develops depends on the ratio between the rates of these two stages of weathering: (i) from intact bedrock to boulders and (ii) from boulders to sand. Where the second stage is very slow, the cliff is gradually buried in its own detritus as the scree extends farther and farther up the cliff (Figure 8.21a). This is the normal pattern observed on cliffs in formerly glaciated areas, such as Britain (Fisher, 1866). Beneath such a scree there is usually a convex parabolic rock core. Where the second stage is faster than the first, however, material is removed from the scree as quickly as new material is added to it by rockfall. Here scree is only a thin veneer on a bedrock slope at the angle of repose of the scree material (30–40°), which is called a **boulder-controlled slope** (Bryan, 1922). The landform, consisting of a cliff and boulder-controlled slope (Figures 8.19a and 8.21b), retreats across the landscape, maintaining an almost constant ratio of total to scree heights (equal to the weathering ratio described above). Such forms are also familiar from the American south-west (e.g. Monument Valley), where cliff and scree have retreated until outliers are separated by a broad desert plain. Just below the base of the cliff, the bedrock core can often be seen, preserved at the angle of repose of the scree (Figure 8.22).

8.3.4.2 Wash processes

Water is directly responsible for the other main processes of material transport as particles which move more or less independently. The least significant process is **through-wash**, in which regolith particles are moved through the regolith. The pores between grains of equal size are much smaller than the grains themselves, so that grains can be washed through textural pores only if they are at least 10 times smaller than the grains they are passing through. Through-wash is significant, therefore, only in washing

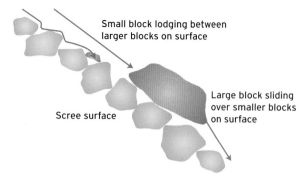

Figure 8.20 Relative roughness for movement of stones which are smaller or larger than the scree surface.

Figure 8.22 Bedrock, at the angle of repose of scree material, exposed at the top of a boulder-controlled slope in Provence, southern France.

silt and clay out of clean sands (e.g. in the sandy layers of podzolic soils – see Chapter 10), and in washing clays into and through structural pores, such as cracks and root holes, which are often lined with **clay skins** which have been deposited there in this way.

The more important wash processes take place at the surface. Material may be detached by two processes, **raindrop impact** and **flow traction**, and transported either by jumping through the air or in a flow of water. Combinations of these detachment and transport processes give rise to the three different processes: rainsplash, **rainwash** and **rillwash**, as indicated in Table 8.4. Raindrops detach material through the impact of drops on the surface. Drops can be as large

Table 8.4 Types of wash processes

	Transportation style	
Detachment by	Through the air	In overland flow
Raindrop impact	Rainsplash	Not applicable
Overland flow traction	Rainwash	Rillwash

as 6 mm in diameter, and fall through the air at a **terminal velocity** which is related to their size. For the largest drops, the terminal velocity is about 10 m s^{-1}, but they attain this only after falling through the air for about 10 m. If their fall is interrupted by hitting the vegetation, drops hit the ground at a much lower speed, and have much less effect on impact. As drops hit the surface, their impact creates a shock wave which dislodges grains of soil or small aggregates up to 10 mm in diameter and projects them into the air in all directions. The total rate of detachment increases rapidly with the energy or momentum of the raindrops, and thus with the rainfall intensity. As a working rule, the rate of detachment is roughly proportional to the square of the rainfall intensity. Where the raindrops fall into a layer of surface water which is more than about 6 mm thick, the impact of the drop on the soil surface is largely lost. Impact through thinner films can still detach aggregates into the water, and other detached grains jump into flowing water films, which then transport grains which they do not have the power to detach.

If water is flowing with sufficient force, it exerts a force on the soil which is sufficient to overcome the frictional and cohesion resistance of soil particles (Figure 8.17). This can be expressed by the safety factor (Box 8.2) as discussed above. An important feature of all particle movements is that different grain sizes are carried selectively. For a surface of mixed grain sizes, the safety factor is generally determined by the average of the coarser grain sizes present, as small grains hide behind and are protected by larger grains, and cannot easily be dislodged on their own. The coarsest material may also be only partially submerged in a shallow flow, increasing its safety factor because the fluid entrainment force and the upthrust (Figure 8.17) are both reduced. The threshold is also influenced by the vegetation cover, which absorbs some of the flow power. A dense grass cover may, in practice, provide an extremely resistant surface that is vulnerable only where there is a bare patch, due, for example, to grazing pressure. At low flows, some fines can be detached from between coarser grains, but at higher flows, the whole surface begins to break up together ('equal mobility'), as coarse material releases trapped fines. Once detached, fine grains generally travel farther, but, as coarse grains settle, they again begin to trap fines in the pockets they create. However, in general, travel distance in an event, and therefore the contribution to total sediment transport, decreases with increasing particle size.

Transportation through the air, in a series of hops, is able to move material both up- and downslope, but there is a very strong downslope bias on slopes of more than about 5°. As a rough guide, the net rate of transportation

FUNDAMENTAL PRINCIPLES

BALANCE OF FORCES ON A STONE IN WATER

For a grain of diameter d, fully submerged in a flow of depth r on slope s, the safety factor SF is:

$$SF = \frac{\Delta d}{rs} \qquad (8.6)$$

where $\Delta = (\rho_s - \rho_w)/\rho_w$ is the ratio of submerged grain density to water density and ρ_s, ρ_w are the grain and water densities respectively.

The ratio Δ has a value of about 1.65 for mineral grains, but may be much lower for aggregates. If the safety factor falls below a critical value, then the driving forces are greater than the resistance, and the particle may be detached by the flow traction. Corrections to this expression for the safety factor are needed to allow for the cohesion force between fine grains, which increases roughly as $1/d$, and for the tendency for particles to roll on their own down very steep slopes which are close to their angle of repose. With these corrections:

$$SF = \frac{\Delta d(1 - s/\tan\phi) + c/d}{rs} \qquad (8.7)$$

where c is a constant, related to the cohesion, with a value of 1-10 mm^2.

The critical value of the safety factor is also influenced by the effect of turbulence in the flow, so that the critical value for entrainment is not 1.0, but lies between 10 and 20 in experimental practice. The resistance and therefore the safety factor has a minimum at around 1 mm for sand grains in water, and has the general form shown in Figure 8.23. The particular values of the safety factor will depend on a number of other variables, particularly slope gradient (s) and flow depth (r). Once the safety factor falls below the critical threshold, the total rate of detachment is proportional to the deficit below this threshold.

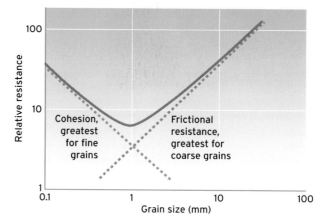

Figure 8.23 Resistance to detachment by overland flow, showing effects of cohesion and grain friction. Safety factor = resisting forces/driving forces.

BOX 8.2

(downhill minus uphill) increases linearly with slope gradient, and inversely with the grain size transported. The gross rates of material transport, for rainsplash, are generally similar to those for soil creep. Rainsplash, however, is strongly particle size selective, and operates only on the surface, whereas soil creep operates over a significant depth of soil, and carries material together as a coherent mass. Protection from raindrop impact, either by vegetation or by stones, strongly suppresses rainsplash by reducing the impact velocity of raindrops. Individual stones may be left capping miniature pillars of soil as shown in Figure 8.24, and microtopography, including tillage features, are gradually smoothed out as rainsplash redistributes material, eroding high points and filling depressions.

Where the surface is not protected from raindrop impact, either by overhanging vegetation or by coarse gravel, the impact of raindrops on soil aggregates leads to **crusting** of the surface (see also Chapter 14). As raindrops strike the surface, some water is forced into aggregates, compressing air inside them, causing them to explode in a process known as **slaking**, and breaking them down to their constituent grains and smaller aggregates (Figure 8.25a and b). According to the grain sizes involved, these are then washed into the pore spaces around intact aggregates, creating an impermeable seal, which changes as each raindrop strikes the surface. Where the soil is mainly silt sized, a structural crust is formed at the surface (Figure 8.25c). Where there is appreciable sand, or stable sand-sized aggregates, the crust forms

result, there is a very strong relationship between vegetation and runoff generation. Plot experiments on silt soils in Mississippi, for example, have showed a 40-fold difference in runoff between a bare crusted field (80% annual runoff) and a densely vegetated plantation (2% annual runoff).

Once there is overland flow, material can be carried in the flow, and some material can move much further than through the air during rainsplash. The presence of overland flow provides a thin layer of water on the soil surface, generally distributed rather unevenly, following the microtopography. This layer of water attenuates the impact of raindrops, and significantly reduces detachment when it is deeper than the raindrop diameter (6 mm). In shallow flows, the combination of detachment by raindrop impact and transport by the flowing water is the most effective transport mechanism, and is known as **rainflow**. This process provides a significant fraction of the material carried into and along rills and larger channels.

When and where the flow is deeper than 6 mm, raindrop detachment becomes ineffective, and detachment is related to the tractive stress of the flowing water. Sediment is detached when the downslope component of gravity and the fluid entrainment forces overcome frictional and any cohesive resistance in the soil (Figure 8.23). Detachment increases with discharge and gradient, and decreases with grain size except where cohesion is significant. Flows powerful enough to detach material generally suppress raindrop detachment, and detached material is also carried by the flow. This combination of processes is called rillwash, and is responsible for most of the erosion by running water in major storms. Much of the material exported from an eroding field is the direct product of enlarging these small rill channels during the storm, and almost all of the material detached by raindrop impact also eventually leaves the area through these channels.

Combining the effects of these three wash processes which are active during storms under a sparse vegetation cover, much of the area is subject only to rainsplash (Figure 8.26), which feeds into areas, some spatially disconnected, with thin films of water where rainflow is dominant. These areas in turn provide sediment to the eroding channels where rillwash is actively detaching material and enlarging the channels. In larger storms, the areas of rillwash and rainflow increase, and become better connected to the channels. The runoff generated per unit area and the area contributing runoff to the outlet both increase, giving a greater than linear response of runoff to increased storm rainfall. Because sediment transport also increases more than linearly with discharge, the non-linearity of the relationship between rainfall and sediment load is even stronger, making the

Figure 8.24 Stones protecting soil pillars from rainsplash. Columns of soil are left intact with a stone sitting on top of the pillar. (Source: PhotoDisc: Alan & Sandy Carey)

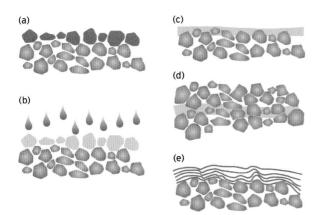

Figure 8.25 Processes of soil crusting: (a) original soil aggregates; (b) break-up of surface layer of aggregates into constituent grains under raindrop impact; (c) structural crust at surface; (d) sieving crust below the surface; and (e) sedimentary crust of in-washed and deposited fines.

below the surface (a sieving crust, Figure 8.25d). Often the fine broken-down material washes into depressions and is redeposited in a layered sedimentary crust (Figure 8.25e).

All of these types of crust create a relatively impermeable surface which strengthens the surface and severely limits infiltration, so that runoff from subsequent storms is increased. Vegetation cover protects the soil from direct raindrop impact and so dramatically reduces this crust formation, and, as a

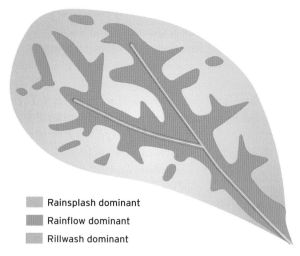

■ Rainsplash dominant
■ Rainflow dominant
■ Rillwash dominant

Figure 8.26 Domains of wash processes in a semi-arid microcatchment.

erosion pattern very sensitive to topography at scales from the catchment to individual soil clods.

Many sparsely vegetated areas develop temporary rills and gullies which are channels formed during storms and destroyed by infilling between storms (Figure 8.27). In agricultural fields, infilling is generally through tillage, sometimes deliberately after each storm and otherwise following the annual cultivation calendar. In uncultivated areas, natural processes of wetting and drying, or freezing and thawing, create a loose surface layer which accumulates downslope along the depressed rill lines, and gradually obliterates them. Rills are small channels, generally 5–10 cm deep, that are formed on a smooth hillside and are not associated with a depression. Over a series of storms,

the rills reform in different locations, and gradually lower the whole hillside more or less evenly. Ephemeral gullies form along shallow depressions, and tend to reform along the same line in each storm, enlarging and deepening the depression, while the infilling processes bring material from the sides and gradually widen the depression.

In a particularly large storm, channels may form that are too large to be refilled before the next event. These channels then collect runoff in subsequent events, leading to further enlargement, and may become permanent additions to the channel network. As material is exported, undercutting of the surface layer can lead to further rapid growth of a linear or branching gully system, which disrupts agriculture and roads, and may be very difficult to restore.

Selective transportation removes fine material from the soil, leaving behind coarser material that 'armours' the surface. As the surface is lowered by erosion, the armour layer consists of the coarsest fraction in the layer of soil that has been eroded, and so develops more and more over time. The coarse armour progressively begins to protect the soil by reducing detachment rates, increasing infiltration and providing an increased resistance to flow. All of these effects reduce the rate of erosion until some equilibrium is approached. In this equilibrium, local differences in sediment transport rate balance differences in armour grain size. One effect of this process, acting over a period, is to establish a relationship between surface grain size and gradient, with coarser material on steeper slopes.

The effects of selective transportation are only evident where the regolith contains some coarse material. This usually consists of weathered bedrock, but may consist of

Figure 8.27 Rill development on an exposed slope.

fragments of calcrete or other **indurated** soil horizons. How-ever, erosion of soils that contain little or no coarse material, for example deep **loess** deposits or some deeply weathered tropical soils, cannot produce an armour layer, so may continue unchecked to great depths, often allowing the formation of extensive gully systems. On the other hand, shallow, stony soils produced by weathering of the bedrock show an enhanced effect of armouring because erosion brings lower layers of the regolith to the surface, containing less and less fines, and the end point of erosion may be a rocky desert, with no prospect of recovery. Some rocks, for example coarse sandstones and granites, produce a discontinuous distribution of grain sizes in their weathering products, dominated by joint-block boulders of weathered rock and the sand grains which are produced as the boulders break down. On these bedrocks, desert slopes often show a sharp break in slope at the base of steep hillsides, between straight slopes close to the angle of rest and the basal concavity (Figure 8.2). If grain size is plotted against gradient for these slopes, the sharp break in slope represents missing gradients which correspond to the gap in the grain size distribution.

8.3.5 The balance between erosion processes

Process rates are affected by topography, particularly slope gradient, and the collecting area for overland flow; by run-off generation and flow paths controlled by climate, soil type and land use; and by the properties of parent materials mediated by the regolith. Each of the processes discussed above in the sections on weathering and transport may be dominant under some circumstances, and in this section some qualitative comparisons are made between the rates of co-existing processes.

Gradient is the strongest and most universal driver of hillslope processes, but processes respond to it very differently (Figure 8.28). Solution rates are only slightly affected by gradient, at least until slopes are so low that little water circulates through the regolith. Several processes, including soil creep, gelifluction (see Chapter 17) and wash processes, increase almost linearly with gradient, although the rate begins to increase more rapidly as they approach the angle of stability (Carson and Kirkby, 2009). Rapid mass movements, including landslides and many debris flows, only begin to move above a fairly sharply defined threshold gradient. Thus on the gentlest slopes, solution may be the dominant process, while the steepest slopes are generally dominated by rapid mass movements. However, the transition gradient from one process to another depends on other factors, such as the regolith materials and climate.

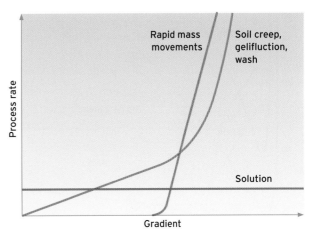

Figure 8.28 Schematic relationships between process rates and slope gradient.

The effect of climate is closely linked to the role of vegetation. Here only areas with near-natural vegetation are considered, and it should be remembered that cultivation, fire and/or grazing can greatly modify these relationships. A rainfall–temperature diagram can be used to sketch the range of conditions. However, conditions generally change through the year at any site, and processes may therefore show a seasonal pattern, in which the vegetation cover responds to monthly changes with some delay. Removal in solution is primarily associated with the amount of subsurface runoff, and is greatest where rainfall is high and temperatures are cool, but not frozen. The pattern of relative rates is sketched in Figure 8.29 on this basis, showing a maximum rate in wet temperate climates. Annual climate loops for south-east Spain (Almeria) and south-east Brazil (Rio), showing monthly mean values of precipitation and temperature for each month of the year,

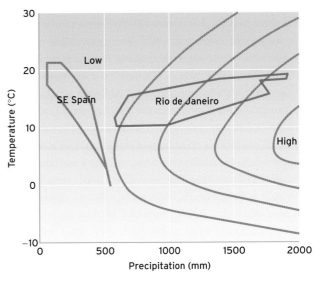

Figure 8.29 Rates of solution of igneous rocks in different climate regimes near Rio de Janeiro, Brazil and south-east Spain.

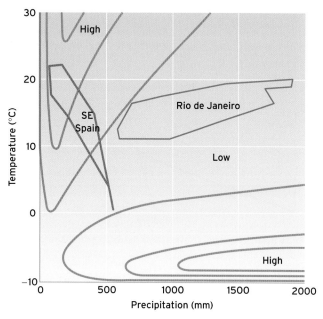

Figure 8.30 Rates of wash erosion for uncultivated land in different climate regimes near Rio de Janeiro Brazil and south-east Spain.

Other processes show less interesting responses to climate. Despite the importance of moisture in driving rapid mass movements, storm events occur in most climatic regimes and are, with earthquakes, the most important triggers of landslides and debris flows. Mass movements therefore show relatively large increases with increasing rainfall and for increasingly frost-prone climates. The contrast between processes in different climates therefore shows several features, although with many exceptions according to local conditions and histories:

- on steep slopes, rapid mass movements are generally dominant;
- on very low slopes, solution is generally dominant;
- on moderate slopes, wash processes are favoured under semi-arid conditions and solution processes under humid conditions.

Hillslopes dominated by wash processes typically show sparse vegetation, thin stony soils, often with a surface armour and poor soil development with little organic matter. Hillslopes dominated by solution show deep soil profiles with strong development of clay minerals, often encouraging mass movement. The soil is fine grained with high organic content and a dense vegetation cover. These soil and vegetation characteristics are also mirrored by differences in hillslope form.

have been included for reference. Although deeply weathered soils are most widespread in humid tropical areas, this distribution mainly reflects the much longer time for development in tropical shield areas (10–100 million years) than in recently glaciated temperate areas (10–20 000 years), more than differences in rates of removal.

Removal by wash shows a more complex pattern, with two regions of high erosion potential (Figure 8.30), both associated with sparse natural vegetation. One high is in areas that are too cold to support vegetation, but warm enough to have at least seasonal runoff. The second high is in semi-arid climates, where sparse vegetation is combined with intense rainfall. At a given temperature, there is an initial rise with rainfall as runoff increases while vegetation remains sparse. Beyond a maximum, erosion declines as the increase in vegetation cover more than compensates for the increase in rainfall. Although not shown in Figure 8.30, there is some evidence for an eventual gradual rise in erosion at very high rainfalls, under a closed forest canopy which can provide no additional protection. This pattern strongly reflects the relationship between climate and natural vegetation. The corresponding pattern for a fixed vegetation cover such as the extreme of a bare surface shows a steady increase with rainfall, almost irrespective of temperature except under permanently frozen conditions. The difference between these two patterns gives some measure of the effect of clearing natural vegetation for agriculture, and shows why forest clearance is particularly damaging in its erosional impact, both for wash erosion and, as noted above, for mass movements.

Reflective questions

➤ Can you summarize and explain the response of each hillslope sediment transport process to slope gradient?

➤ What processes would you expect to be dominant on gentle slopes in North Africa and New England?

➤ How would you expect the surface and subsurface appearance of soils undergoing soil creep and rillwash to differ? Consider their texture, sorting, organic matter content and horizon development.

➤ Can you compare the processes you would expect to find in (a) tectonically active mountains and (b) humid tropical shield areas?

➤ How do wash processes depend on hillslope hydrological processes?

➤ How do climate and vegetation influence the rate and dominance of slope processes acting?

8.4 Evolution of hillslope profiles

Hillslope processes move material around the landscape, primarily, for each climatic zone, in response to gradient and hydrological conditions. Each process discussed above responds to these factors in a distinctive manner, and therefore shows a characteristic distribution over an area, or down the length of a particular slope profile. These differences gradually change the form of the hillslope profile and, if the climate and tectonic regime remain reasonably uniform, lead to a consistent relationship between profile form and the dominant processes acting. This section explores these relationships in order to understand the principles that link process and form. Some simple models are examined that make use of these principles and show how they may be used to interpret real landscapes.

8.4.1 Concepts

The history of hillslopes is primarily one of erosion, and land masses would eventually become rather flat plains close to sea level (after about 10–100 million years) if there were no tectonic uplift. However, few, even cratonic, areas ever reach this stage, partly because few areas are absolutely stable relative to sea level, and because erosion is substantially (~75%) compensated by isostatic upift. More commonly, areas reach an approximate balance, or equilibrium, between erosion and tectonic uplift and such landforms are relatively easy to analyze and understand. The most useful single concept in understanding how hillslopes evolve is the principle of mass balance. When sediment is transported, the loss from the source area exactly balances the addition to the receiving area. In a few landscapes, a sequence of landforms can be seen which represent either the linear progress of a process, or different process rates along a climatic gradient, and it is possible to substitute space for time (**ergodic method**), and interpret the spatial sequence as an evolutionary sequence over time. Where such simplifying assumptions can be made, even approximately, landscapes can be most readily interpreted. Often, however, quantitative models are required to understand how process and landform are related to one another.

8.4.1.1 Mass balance

The most general statement of mass balance is the storage equation:

Input − output = net increase in storage   (8.8)

This expression can be applied to the mass of any identifiable component of a hydrological or geomorphological system. The component may be water, total Earth materials, a chemical element or compound, a sediment fraction defined by grain size or source rock, or a population of tracers (e.g. radioactive or painted pebbles). Budgeting may be done in absolute terms, or with reference to a chosen fixed datum. For example, Earth materials may be budgeted with reference to sea level as a datum, and water may be budgeted as deficit or surplus relative to saturation. Finally the physical space for which a budget is calculated can be whatever is most convenient: it may, for example, be for a one-dimensional balance of vertical fluxes at a point; for a channel reach; for a particular catchment; or for the whole Earth surface. What is important is that inputs and outputs take full account of gains and losses for the component system of interest, and include all transfers of mass across the boundaries of the defined system. Examples of mass balance approaches are given in Box 8.3.

8.4.1.2 Equilibrium and other simple landforms

Although there is a complex interplay between landforms and processes, some understanding of how processes shape landforms is gained by considering simple landscapes in which there is an approximate balance between the rates of processes, and the shape of the hillslope is either constant (Hack, 1960) or evolving in a simple way. A strict equilibrium can generally be achieved only when tectonic uplift is exactly equal to the rate of downcutting at every point in the landscape. In practice, both tectonic uplift (in earthquakes) and erosion (in major storm events) are episodic, so that equilibrium can only be considered by taking long-term averages, and most real landscapes depart even more substantially from a true equilibrium (see Chapter 1). Nevertheless, the concept of equilibrium provides a powerful tool, as in many other branches of science, for simplifying the analysis of a complex system, and offers important insights into the relationship between the set of processes acting and a corresponding characteristic form for the hillslope profile.

Three types of situation are of particular value in approximating to recognized types of landscape behaviour, and in simplifying the relationship between form and process. The first is the constant downcutting form, in which uplift exactly balances vertical downcutting. Such forms are found in areas of strong tectonic uplift, in which slope gradients steepen until slopes and rivers carry away the sediment as fast as uplift raises new material. The second is parallel retreat, in which a hillslope profile migrates

235

FUNDAMENTAL PRINCIPLES

MASS BALANCE

One example of a mass balance is to consider total sediment for a flood-plain reach (Figure 8.31). Using the terms in the figure to expand the basic storage equation:

$$
\begin{aligned}
&\text{(Inputs from upstream +} \\
&\quad \text{hillslope inputs)} - \text{output} \\
&\quad \text{downstream} = \text{net} \qquad (8.9) \\
&\quad \text{increase in (in-channel} \\
&\quad \text{and floodplain) storage}
\end{aligned}
$$

By measuring or estimating the terms in this expression, an estimate can be made of whether the floodplain is aggrading or degrading, and a succession of mass balances can provide an estimate of what is happening in an entire catchment, and how long sediment spends in different parts of the catchment. This residence time is calculated as:

$$
\begin{aligned}
&\text{Residence time} \\
&\quad = \frac{\text{volume in storage}}{\text{average annual flux}} \quad (8.10)
\end{aligned}
$$

In this case the volume is the total volume of floodplain alluvium, and the average flux is the mean of the input and output rates. Estimates have shown that residence times are longest (10 000 years) within the hillslope soil layers, least in small channels (10 years), and gradually increase downstream (100–1000 years) (Dietrich and Dunne, 1978). The sediment budget can be further subdivided, and in this example it may be relevant to separate the budget into grain size fractions. In this case, an additional input for each separate size class is the breakdown from coarser sizes, and an additional output is the breakdown into finer material.

For the hillslope, it is often convenient to break the length of the slope profile into equal sections (Figure 8.32),

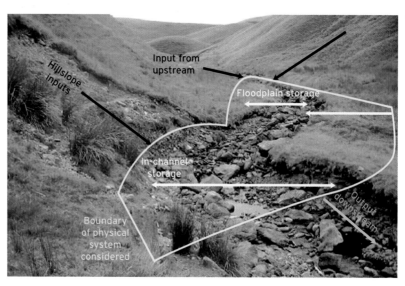

Figure 8.31 Components of a valley floor sediment budget. The physical system is bounded by the edges of the floodplain and the top and bottom of the reach.

and examine the sediment budget for each section. For the section of interest, representing one particular store, the storage equation is:

$$
\begin{aligned}
&\text{Sediment in from} \\
&\quad \text{upslope} - \text{sediment out} \\
&\quad \text{to downslope} = \text{increase} \qquad (8.11) \\
&\quad \text{in section storage}
\end{aligned}
$$

For a section of length Δx over a short time period Δt during which the surface elevation is increased by Δz and there is no addition of, for example, wind-blown material, this storage equation can be put into symbols in the form:

$$
(S_{IN} - S_{OUT})\Delta t = \Delta z \Delta x \qquad (8.12)
$$

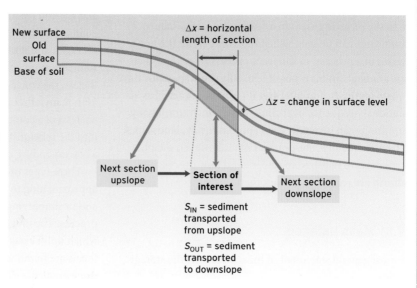

Figure 8.32 A sediment budget for a section of a slope profile.

BOX 8.3 ➤

or

$$\frac{\Delta z}{\Delta t} = \frac{(S_{IN} - S_{OUT})}{\Delta x} \qquad (8.13)$$

where S_{IN} and S_{OUT} are the rates of sediment transport (per unit contour width). The left-hand side of equation (8.13) represents the rate of change of elevation over time, or the rate of aggradation (negative if erosion). The right-hand side represents the current rate of change of sediment transport with position. Thus the

storage equation converts the spatial pattern of erosion (the right-hand side) to a forecast for the rate of change over time (erosion or deposition on the left-hand side). This is the basis for modelling hillslope evolution over time. Each short-term forecast of erosion or deposition changes the form of the hillslope. Each change of the hillslope form changes the rate of sediment transport as it responds to the topography.

If the distances Δz, Δx and the time Δt tend to zero in an appropriate way, the equation becomes the partial differential equation:

$$\frac{\partial z}{\partial t} + \frac{\partial S}{\partial x} = 0 \qquad (8.14)$$

where S is the sediment transport. The change in sign from equation (8.13) is due to the convention that the change in S is taken in the sense of increasing downslope (i.e. as $S_{OUT} - S_{IN}$).

BOX 8.3

laterally across the landscape as it erodes. Landscapes of this general type are found in some semi-arid areas, where steep boulder-controlled slopes and cliffs are maintained in near-horizontal sedimentary rocks during the retreat of escarpments across distances of several kilometres (Figure 8.19a). The third is slope decline, in which the landscape profiles remain the same in form, but become increasingly muted in their vertical relief, and eventually decline to a horizontal plain. These forms are described in areas of long tectonic stability, and more commonly described for humid than for arid areas.

Because of the relatively low density of continental rocks and the flexibility of tectonic plates, the unloading of the mantle by erosion is partially or completely compensated by isostatic uplift, which replaces about 70% of the loss by erosion, and is spread over an area which depends on the rigidity of the slowly flexing plate.

Where erosion is occurring in a sequential fashion, in the migration of a meander cutting into the valley wall, or where a spit grows along the coast progressively to protect cliffs behind from erosion (Chapter 15), then the spatial sequence of visible profiles represents a sequence of passive slope recession over different periods since undercutting was active. Figure 8.33 shows an example of where this space–time substitution may be applicable. As the broad meander bend has migrated downstream from Y to X, the sequence of slope profiles from X to Y can be interpreted as an evolutionary sequence showing the retreat of a cliff by rockfall, with development of an angle of repose slope below it. Although it is often possible to draw slope profiles within an area and arrange them into such a sequence, it is not generally appropriate to do so except in very particular circumstances such as those shown in Figure 8.33. More

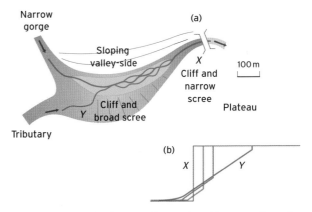

Figure 8.33 The Rio de Aguas Gorge above Turre, south-east Spain, showing space-time substitution: (a) sketch map; (b) schematic sequence of cliff-scree sections from X to Y.

generally, the profiles form a spatial set representing the differences in process rates among the range of topographic (area and slope gradient) settings found within a catchment, and should not be reinterpreted as a time sequence.

8.4.2 Models

The discussion of slope evolution can only be carried forward through the application of mathematical and numerical models (Culling, 1963; Kirkby, 1971; Kirkby *et al.*, 1993). The cornerstone of these models is the mass balance equation for a section of a slope, which converts the spatial pattern of sediment transport into a forecast for local rates of erosion or deposition. To create a model formally requires three other types of information: (i) the functional relationship between topography and sediment transport rates; (ii) the initial form of the profile at some appropriate starting time; and (iii) the boundary conditions, which define how the model slope interacts with the rest of the landscape. In each case, simple assumptions can be used to show some aspects of slope development, but more complex conditions may be needed to match the evolutionary history of a particular hillside.

The relationship between process rate and topography can be developed by summing the rates across the frequency distribution of storm or other events, linking the long-term rates to detailed process mechanics. However, in most slope models, this summation is taken for granted, and results are quoted directly for the long-term average rates. The driving variable of discharge is represented by its topographic surrogate which is often the distance from the divide (or the collecting area in a three-dimensional landscape). Furthermore, assumptions are commonly made about whether removal is transport limited or supply limited, and the discussion here will focus on the simpler transport-limited case. With these assumptions, there have been many attempts to express sediment transport as an algebraic function of distance from the divide, x, and local (tangent) gradient, g. Not all processes can be readily put in this form, and for landslides and other rapid mass movements they have only very limited validity, but the expressions in Table 8.5 provide a useful and relevant basis for comparing form and process. In the literature, there is some range of exponents which appear to give acceptable results, and these values should be regarded only as indicative.

Creep, rainsplash and gelifluction are all driven primarily by slope gradient, operate even on gentle slopes and are not driven by flow processes. They are generally thought to be linearly dependent on gradient over the full range. Rainflow is similarly driven by a uniform detachment, but with material carried by flow, which therefore increases with distance from the divide. Rillwash depends on detachment by the power of the flowing water, and hence depends strongly on both gradient and distance (Schumm, 1956, 1964; Dunne and Aubry, 1986). Landslides only occur above a threshold gradient, g_0, and the distance moved by material increases strongly as the angle of repose, g_T, is approached. These two threshold slopes determine the rather complicated form of its dependence on gradient alone (Scheidegger, 1973). Solution is usually described by a constant rate of denudation, with material accumulating linearly with the collecting area.

Boundary conditions describe the spatial relationship of the profile with the remainder of the landscape. If the profile follows the line of steepest descent, there are no exchanges of material with neighbouring profiles, and the important boundaries are at the top and bottom of the profile. It is normally convenient to take the top of the profile as the divide, and this is defined by no sediment crossing this line, or by considering the profile on the other side of the divide to be a mirror image.

The lower boundary condition usually describes the connection of the profile with the stream or floodplain at its foot. In reality there are interesting and complex interactions at this point, but, for simplicity, it is often adequate to assume that the stream is a passive agent, removing all the sediment delivered to it at an unchanging position. Another simple alternative is to assume that the stream is downcutting at a steady rate.

With these tools it is possible to create a numerical model for the progress of slope evolution for a given process or combination of processes. However, a good qualitative idea of how slope form responds to process can be obtained by analyzing the constant downcutting equilibrium form, in which erosion exactly balances tectonic uplift at every point. For this and other equilibrium assumptions, the unvarying slope form obtained is independent of the initial form of the slope profile, greatly simplifying the range of possible outcomes. Box 8.4 gives an example of how this is done.

A similar approach to that in Box 8.4 can be applied to the parallel retreat of hillslopes at a constant horizontal rate. This can be applied most fruitfully to the combination of creep and mass movements, to give the form of a steadily retreating hillslope. Figure 8.36 illustrates the types of profile generated in this way, with the same slope process rates, but with different rates of slope retreat. Each profile shows a convex section at the top, of gradually increasing gradient, associated with the dominance of creep processes. This convexity becomes sharper as the rate of retreat is increased. If retreat is sufficiently rapid (> 0.04 mm yr^{-1} in this example),

Table 8.5 Indicative long-term sediment transport functions, assuming transport-limited removal

Process	Sediment transport function
Creep, rainsplash, gelifluction	$\sim g$
Rainflow	$\sim x$
Rillwash	$\sim x^2 g^2$
Landslides	$\dfrac{\sim g(g - g_0)}{(1 - g/g_T)}$, valid for $g_0 < g < g_T$
Solution	$\sim x$

x = distance from divide; g = local tangent gradient; g_0, g_T are constants.

TECHNIQUES

MODELLING EQUILIBRIUM FORM OF HILLSLOPE PROFILES

If the slope is eroding (and uplifting) at a constant rate T, then the sediment transported past a point at distance x from the divide must be exactly Tx as this is the area between the new and old surface levels. The form of the equilibrium slope profile can then be derived directly from the sediment transport relationships in Table 8.5. For example, for soil creep, the sediment transport can be expressed both as Tx and through the process relationship as proportional to g, say equal to Ag for a suitable constant A. Putting these two expressions equal to one another, $Tx = Ag$, or the gradient $g = Tx/A$. In other words, the equilibrium slope is a convex parabolic shape in which gradient increases steadily and linearly downslope. The slope profile can be either expressed as a relationship between gradient and distance, or recalculated as the parabolic slope profile (Figure 8.34).

The same procedure can be followed to work out the profiles associated with each of the separate processes in Table 8.5. Thus, for rillwash, $Tx = Bx^2g^2$ for some constant B, which can be re-expressed as $g = (T/Bx)$. This expression shows that the gradient decreases steadily downslope, so that the profile is concave throughout. This procedure can also be applied for landslides, but gives an indeterminate result for solution or rainflow.

In practice, processes generally occur together, and creep or rainsplash are generally the most important processes near to the divide.

(a) $g = T/A\,x$

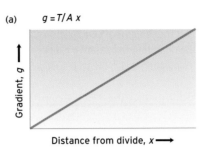

(b) $z = z_0 - \frac{1}{2}T/A\,x^2$

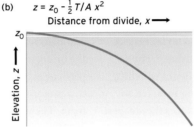

Figure 8.34 Constant downcutting equilibrium profile for soil creep: (a) gradient against distance; (b) elevation against distance.

The same procedure can be applied to a combination of processes, as is shown below for a combination of creep and solution, and for rainsplash and rillwash.

Constant downcutting form for creep plus solution:

- creep: $S = Ag$ for constant A;
- solution: $S = Cx$ for constant C;
- in equilibrium with uplift at rate T: $Tx = Ag + Cx$;
- rearranging, $g = (T - C)/Ax$ which is a uniform convex profile.

Constant downcutting form for rainsplash and rillwash:

- rainsplash: $S = Ag$ for constant A;
- rillwash: $S = Bx^2g^2$ for constant B;
- in equilibrium with uplift at rate T: $Tx = Ag + Bx^2g^2$.

When this is rearranged with g on the left-hand side the equation becomes a quadratic equation:

$$g = \frac{-A + \sqrt{(A^2 + 4TBx^3)}}{2Bx^2} \quad (8.15)$$

Example results are shown in Figure 8.35 for a combination of

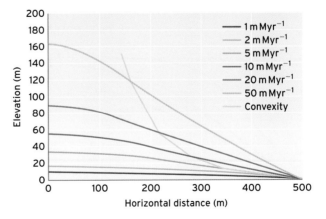

Figure 8.35 Constant down-cutting profiles for rainsplash plus rillwash under different rates of uplift (in m per million years). The width of the convexity is greatest where uplift is slowest. The grey line indicates the transition from convex to concave slope profiles.

BOX 8.4 ➤

> ➤
>
> rainsplash and rillwash keeping the process rates (*A* and *B*) constant and varying the rate of uplift (equal to denudation). The slope profiles developed are convexo-concave in profile and are generally steeper under higher rates of uplift (and matching denudation), and the
>
> convexities tend to be narrower for the steeper slopes. The form of these slopes depends on the values of the process rate constants, which in turn depend on climate and soil controls, but these general conclusions stand. By changing the relative rates of the rainsplash and rillwash transport, it
>
> can also be seen that as rillwash is increased (perhaps due to changed climate or land cover), the concavity becomes broader, and that the convex and concave sections of the slope correspond roughly to the zones where rainsplash and rillwash are respectively dominant.
>
> **BOX 8.4**

as in all the profiles drawn, the convexity continues until the threshold for landslides (g_0 in Table 8.5: 22° or 40% in this example) is crossed. From that point, the slope gradient increases much more slowly downslope, and an increasing proportion of the material is carried by landslides. The profiles in Figure 8.36 can be visualized as a series of sea cliffs, with decreasing severity of wave attack going from left to right. The most aggressive wave attack produces the steepest cliffs, dominated by frequent mass movements. In progressively less exposed situations, the angle of the cliff decreases. Inland, where there is little or no retreat, the same material forms low-angle slopes, with minimal impact of landslides. The model smoothes out the effect of individual landslides, which, in reality, create much less regular profiles, in which the form of individual slides shows up as a series of steps (see Figure 8.11c or 8.13), and this effect is greatest on the steepest slopes.

Models can also be used to generate evolutionary sequences, for profile development over time from a given initial slope form (see the Companion Website for the book for simple slope models you can adjust). Figures 8.37 and 8.38 show two such sequences, both starting from an initial form of a plateau with a stream vertically incised into it, and then remaining stable in position. Figure 8.37 shows the uniform convexities associated with soil creep, rainsplash or gelifluction processes, and the eventual evolution to a uniform parabolic form showing no trace of the initial form. There is a strong similarity between the final forms in Figures 8.36 and 8.37, in both cases showing a hillslope dominated by creep processes. However, the effect of landslides in Figure 8.36 produces much more uniform gradient on the steeper slopes than under creep without slides in Figure 8.37. Figure 8.38 shows, for the same initial conditions, development under rainsplash and rillwash together. Although there is an initial rounding of the sharp plateau edge, the effect of rillwash becomes increasingly evident over time, with the development of a marked concavity in the lower part of the profile. As with creep, the final forms show no trace of the initial form, and appear to be declining smoothly towards a level plain. However, the combination of convexity and concavity is characteristic of the processes acting in each case, and the convex and concave areas roughly correspond to the areas where the rainsplash (or creep) and rillwash are respectively dominant.

Thus models, both simple and complex, are able to make use of current understanding of process rates and mechanisms, and show that these processes are able to produce many of the features of observed landscapes. Three-dimensional models (Ahnert, 1976; Willgoose *et al.*, 1991; Howard, 1994; Tucker *et al.*, 2001) are also able to take account of the interactions between streams and hillslopes, which control the spacing or density of channels in the landscape. Some recent models are also able to incorporate other aspects of hillslope profiles, including the development of soils, armour layers and vegetation patterns. With improved knowledge of the climatic drivers of process rates, and of past climates, there is also scope to understand how

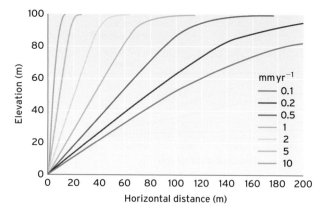

Figure 8.36 Modelled profiles for a 100 m high hillslope in equilibrium with a constant rate of lateral retreat under creep and landslides. Legend shows rates of retreat in mm yr^{-1}. In this example, the threshold gradient for landslides is 40% (22°).

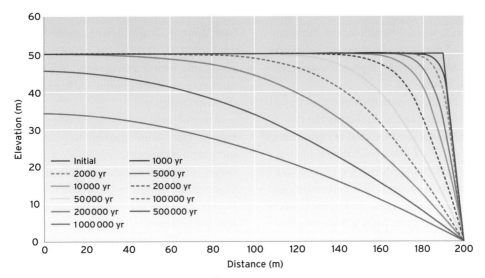

Figure 8.37 Modelled slope evolution by soil creep.

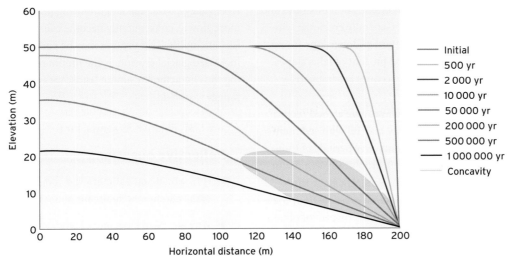

Figure 8.38 Modelled slope evolution by rainsplash and rillwash.

landscapes have evolved and how we are currently modifying them through global changes in land use and climate.

8.4.3 Interpreting landscape form

The hillslopes themselves contain many clues to the processes that formed them, although, as erosional forms, they always tend to destroy rather than preserve the evidence of their formation. The gross form of the profile is generally related to the processes that formed it and partially reflects its history. Many authors once argued about whether this history reflected periods of erosion under tectonically stable conditions (Davis, 1954), the tectonic history (Penck, 1924) or a time-independent form (Hack, 1960). Convex profiles

are generally associated with creep (Gilbert, 1909), rainsplash or gelifluction, but, as Figure 8.36 illustrates, rapid incision can lead to initial convexity almost irrespective of the process. Similarly, concave profiles are generally associated with rillwash or fluvial processes, yet rapid deposition can also lead to initial concavity irrespective of process. Mass movements generally lead to more or less rectilinear slopes of uniform gradient except in situations of exceptional activity. Thus landslides produce a landscape with uniform slopes close to and slightly above the threshold of sliding (g_0 above), and scree slopes form at angles close to their angle of repose. However, very active areas, such as actively retreating coastal cliffs, are dominated by large rotational slides with a much more complex topography of large back-tilted blocks and crumpled toe areas. Under less active erosion, as is generally found inland, the slides become shallower and the slopes straighter, although often retaining a more or less hummocky topography associated with individual slide blocks.

Smaller features are also important in interpreting process activity. The irregular hummocks that are the remnants of landslide back-scars and toe areas are one example of features that can survive for thousands of years in the landscape. Another important feature is the lines of accumulated sediment above contouring field boundaries, and equivalent erosion below them. These may be the result of deliberate terracing, but, in many cases, are accumulated over many centuries of agriculture by tillage erosion and/or wash processes. On a still finer scale, active wash processes can produce small terraces behind each clump of vegetation, and erosional steps a few centimetres high below plants or larger stones. Wash processes also sort and selectively transport surface stones, leading to a concentration of stones and some sorting of surface material. Generally erosional winnowing of fines gives rise to a pattern of downslope fining, while local patches deposition and the foot of talus slopes may show downslope coarsening as fines are trapped by coarser material.

In some areas, stream head hollows give a good record of episodic erosional activity. They may fill with sediment from fast and slow mass movements over periods of thousands of years, and then empty catastrophically in a major event (Dietrich and Dunne, 1993). Similarly, large mass movements may bury former soil surfaces below their toe deposits. Such sites therefore offer some prospect of obtaining a stratigraphic record of slope history.

Interpreting the form of real landscapes and understanding process mechanisms and rates are the two complementary halves of geomorphology, which need to be integrated within a broader view of environmental processes and Earth history. Two of the most exciting areas of current research are into the quantitative relationships between landforms and climate, and between landforms and tectonics. It is clear that different climatic regions have different assemblages of process and form. Humid areas are generally dominated by creep, mass movements and solution under a dense vegetation cover and well-developed soils. Hillslopes are usually mainly convex, typically with a low

($1–5$ km km^{-2}) drainage density. In contrast semi-arid areas have stony, shallow soils with sparse vegetation and surface armouring. These hillslopes evolve under supply-limited conditions, with dominant wash processes, low solution rates, concave slope profiles and high ($10–100$ km km^{-2}) drainage densities.

Areas of rapid tectonic uplift also show distinctive slope morphologies, dominated by steep slopes and the mass movement processes which are most active on steep slopes. Largely irrespective of climate, soils are thin and stony, and processes are limited by the weathering of fresh bedrock to a state where it can undergo mass movements, and by removal processes in largely bedrock-floored steep mountain rivers. The rapid erosion fuelled by tectonics is further enhanced by isostatic uplift – as erosion removes mass from the land, the surrounding area gradually bulges upwards in response, as the flexible continental plate floats higher on the viscous mantle beneath. Erosion is most rapid along the river valleys, while the neighbouring peaks, which erode more slowly, consequently become even higher due to the isostatic uplift. This process may be partially limited by increased glacial erosion as the mountain peaks rise, creating the characteristic alpine landscapes of high mountain ranges worldwide.

Reflective questions

➤ How would you estimate the components of a small catchment sediment budget?

➤ Why are deep soils most commonly found in the tropics and subtropics?

➤ How well do familiar landscapes show a good relationship between current processes and current landforms?

➤ What are the three types of valuable situation when approximating recognized types of landscape behaviour and simplifying the relationship between form and process?

➤ What can modelling approaches tell us about hillslope evolution?

8.5 Summary

Hillslope processes transform and transport parent materials to river channels which eventually deliver almost all sediment and solutes to the sea. Weathering consists of largely *in situ* transformation to regolith; and erosion consists of the net removal of this weathered material. Solution dissolves rock and soil materials, progressively leaving behind the less soluble minerals in the regolith. Physical and biological expansion processes break the regolith into finer grains which can be transported more easily. Mechanical processes, usually aided by water, transport material downhill, in mass movements and surface wash. Although the different processes are all driven by water and gradient, they produce distinctive small-scale features in the landscape. Because processes depend differently on flow and gradient, they also create different and distinctive hillslope forms, which can be analyzed through models. Both small- and large-scale forms can be used to infer the processes acting in the landscape.

Further reading
Books

Allen, P.A. (1997) *Earth surface processes*. Blackwell Science, Oxford.
This is probably the best up-to-date book on hillslope processes. It is moderately demanding, and written from an earth science perspective.

Anderson, M.G. (ed.) (1988) *Modelling geomorphological systems*. John Wiley & Sons, Chichester.
This is a good survey of different approaches to modelling in landscape systems.

Carson, M.A. and Kirkby, M.J. (2009) *Hillslope form and process*. Cambridge University Press, Cambridge.
A classic from 1972 reprinted again in 2009.

Gregory, K.J. (2010) *The Earth's land surface*. Sage, London.
A very well-written guide to key geomorphological themes.

Kirkby, M.J., Naden, P.S., Burt, T.P. and Butcher, D.P (1993) *Computer simulation in physical geography*, 2nd edition. John Wiley & Sons, Chichester.
This is a guide to programming some simple models in geomorphology and hydrology.

Middleton, G.V. and Wilcock, P.R. (1994) *Mechanics in the Earth and environmental sciences*. Cambridge University Press, Cambridge.
This book is a state-of-the-art guide to relevant continuum mechanics for the mathematically able and committed student.

Selby, M.J. (1993) *Hillslope materials and processes*, 2nd edition. Oxford University Press, Oxford.
An excellent introductory text.

Further reading
Papers

Evans, D.J and Kirkby, M.J. (eds) (2004) *Critical concepts in geomorphology, Vol. 2 – Hillslope geomorphology*. Taylor and Francis, London.
A selection of classic papers in geomorphology.

Kirkby, M.J. (2011). Hydromorphology, erosion and sediment. In: McDonnell, J. (ed.), *Benchmark volumes in hydrology*. IAHS Publications, Wallingford.
Another selection of classic papers in geomorphology in relation to hydrology.

CHAPTER 9

Sediments and sedimentation

Kevin G. Taylor

School of Earth, Atmospheric and Environmental Sciences, University of Manchester

Learning objectives

After reading this chapter you should be able to:

➤ **understand the origin, composition and classification of sediments**

➤ **appreciate the products and patterns of sedimentation in surface environments**

➤ **understand the response of sedimentation to environmental change**

➤ **be aware of the role and impact of sedimentation in natural and anthropogenic catchments and basins**

9.1 Introduction

Sediments are an important component of the Earth's surface and their movement and deposition are key processes in environmental systems. In conjunction with weathering and erosion (Chapters 7 and 8), sedimentation shapes the face of the Earth. Sedimentation creates floodplains, builds coastlines and fills the ocean basins. Sediments have a major impact on natural and engineered systems. In this chapter, we will introduce the concepts of sediments and sedimentation, describe the features that

sedimentation produces, and assess the impact that sedimentation has on the environment.

Sediments are collections of grains of pre-existing rocks, fragments of dead organisms or minerals precipitated directly from water at Earth surface temperatures. In Earth surface environments these grains accumulate (through the process of sedimentation) to form sediment packages and sedimentary successions. Sediment is a term restricted to loose, unconsolidated material (e.g. sand). If the sediment is buried and hardened, through the action of heat and pressure, to form a rock (a process that can take from tens to millions of years) it is then termed a sedimentary rock, and the suffix '-stone' is added (e.g. sandstone). Sediment classification is based on a combination of the sediment's constituents and the mode of origin of the sediment. The three major types of sediment are terrigenous **clastic sediments** (simply termed clastic sediments throughout this chapter), **biological sediments** and **chemical sediments**.

Clastic sediments are composed of particles (clasts) derived from pre-existing rocks (which may be igneous, metamorphic or sedimentary in nature). These clasts are derived from the weathering and erosion of bedrock and soil material and comprise both primary and secondary minerals. The material that is formed as a result of this weathering is transported away from the site by water, wind or ice and will ultimately settle out and accumulate in a range of continental or marine environments. Biological sediments are derived from organic materials and can be either the remains of dead organisms (e.g. shells, plants) or build-ups of framework-building organisms (e.g. coral reefs). Shell material and other skeletal fragments are commonly composed of calcium carbonate ($CaCO_3$), in the form of either calcite or aragonite. Biological sediment composed of such material is often referred to as carbonate sediment. Sediments in which organic material (e.g. plant remains) is a significant component are commonly termed organic-rich biological sediments. For example, peat is an organic-rich biological sediment composed of dead plant material accumulated in wet environments. Chemical sediments are those that are produced by chemical processes and are formed predominantly as a result of precipitation of minerals directly from a water body. A good example of this is the precipitation of salt from evaporating seawater or an inland sea, as is happening in the Dead Sea today.

Sedimentation is the process by which sediment is deposited, leading to its accumulation. The most common cause of deposition is the settling out of sediment from a transporting fluid (water, wind or ice). As such, sedimentation is the opposite of erosion and transportation. While erosion removes material from a location, sedimentation leaves material behind. Although they are contrasting processes, sedimentation and transportation are closely interconnected processes. Many sediments can be sedimented and then reworked and transported many times prior to the material permanently accumulating in sediment packages. Sedimentation is therefore a dynamic process, which cannot be considered in isolation from erosion and transportation.

9.2 Clastic sediments

As stated above, clastic sediments are composed of grains of rock, weathered and eroded from pre-existing bedrock material. The dominant grains present within clastic sediments are quartz, feldspar, mica, clay minerals and iron oxides (Nichols, 1999). In sediments those minerals which are most resistant to weathering are concentrated relative to those that are less resistant. This is because more easily weathered material will be more likely to be dissolved into solution (see Chapters 7 and 13), remain in suspension or be transported well away from the source area. In general, sediment that accumulates near its bedrock source bears a greater resemblance, compositionally, to the bedrock

material than does sediment deposited a long way from its source. Quartz dominates the composition of most clastic sediments as this mineral is most resistant to weathering and transport.

9.2.1 Classification of clastic sediments

The most widely used scheme for classifying clastic sediments is based on the size of the clasts, or grains, within the sediment. Three major grain size classes can be recognized. Gravel refers to grains greater than 2 mm in size. Sand refers to grains less than 2 mm but greater than 63 μm (1/16th of a millimetre) in size. Mud refers to grains less than 63 μm in size. These three major classes can be subdivided further using the Udden–Wentworth grain size scale (Figure 9.1). The Udden–Wentworth grain size scale is the scheme most widely used to classify the grain size of sediments. Each grain size class within the Udden–Wentworth scale is a factor of two larger than the previous one, and is therefore a logarithmic scale (logarithmic to base 2). As can be seen from Figure 9.1, each of the three major classes is subdivided into smaller classes (e.g. very fine sand, fine sand, medium sand, coarse sand, very coarse sand). The Φ (pronounced 'phi') scale is a numerical representation of the Udden–Wentworth scale, based on the logarithmic nature of the scale. The Φ value = $-\log_2$ grain size (in mm). This scale is mathematically more convenient to use

mm	Φ	Class terms	
		Boulders	
256	−8	Cobbles	
128	−7	Cobbles	
64	−6		
32	−5		
16	−4	Pebbles	
8	−3		
4	−2		
2	−1	Granules	
1	0	Sand	Very coarse
0.5	1	Sand	Coarse
0.25	2	Sand	Medium
0.125	3	Sand	Fine
0.0625	4	Sand	Very fine
0.0312	5	Silt	Coarse
0.0156	6	Silt	Medium
0.0078	7	Silt	Fine
0.0039	8	Silt	Very fine
		Clay	

Figure 9.1 The Udden-Wentworth grain size classification scheme for sediment grains. The Φ value = $-\log_2$ grain size (in mm).

than fractions of millimetres. Note that a larger value of Φ represents a smaller grain size.

Grain size on unconsolidated sediment is readily estimated by observing the sediment through a hand lens or a binocular microscope. A more accurate determination of grain size can be made by sieving. Sediment is shaken through a stack of sieves of reducing mesh size and the percentage mass of sediment trapped at each sieve is weighed and calculated. Determination of grain size by sieving has the benefit of allowing statistical calculations to be made on the sediment (e.g. the mean and modal grain size). In consolidated sedimentary rocks grain size is commonly determined qualitatively by microscope examination.

9.2.2 Clastic sediment grain shape and texture

A great deal of information about the origin, history, source and environment of deposition of a sediment can be inferred by studying textural properties of the grains within a sediment. The **sorting** of a sediment is a measure of the degree to which the grains in the sediment are clustered around one grain size (in other words, a measure of the spread, or standard deviation, of grain sizes in a sediment). Although this can be determined statistically by sieving analysis, sorting is more usually estimated visually through a microscope by comparison with sorting charts. In such a case, the terms poorly sorted, moderately sorted, well sorted and very well sorted are used. Although a number of factors control sorting within a sediment, in general the further a sediment has been transported from its source, the better the sorting of the grains.

Grain roundness is another textural property that contains environmental information. Grains can be very angular, angular, subangular, sub-rounded, rounded and well rounded (Figure 9.2). Grains within sediments become

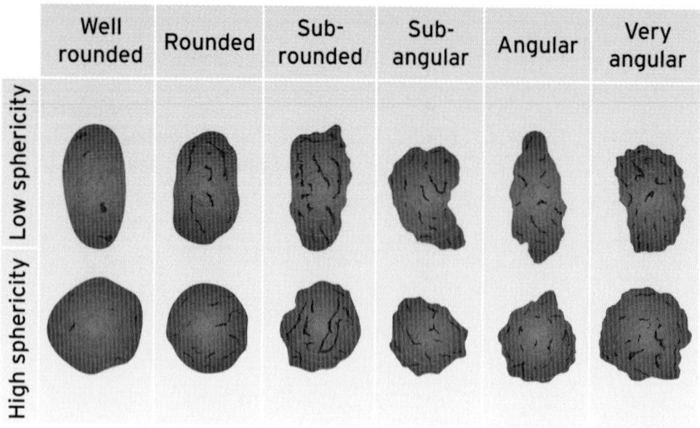

Figure 9.2 Roundness and sphericity scale for sediment grains. (Source: after Pettijohn *et al.*, 1987)

rounded by continual abrasion during transport as a result of the impact of the sediment grains with each other. Well-rounded grains indicate that the sediment has undergone extended transport prior to deposition. Wind-blown dune sands in deserts are commonly very well rounded as a result of continual grain impacts during wind transport. In contrast, sediment grains in **scree slopes** (see Chapter 8 and below) are commonly very angular as they have undergone only limited transport.

9.2.3 Sediment transport and sedimentation

With the exception of *in situ* organic build-ups, such as reefs, and chemical precipitation of minerals directly from water, virtually all sediments are deposited after some element of transport. This is particularly true for clastic sediments. Sediment transport can take place as a result of gravity, but more commonly transport is by water, wind or ice. The density of the transport medium has a major control on the ability of the medium to carry sediment. The higher the density of the medium, the larger the grains that can be transported.

Transport of sediment by gravity alone is only important on steep slopes and can be thought of as the first stage of erosion and transport of weathered material. Material may move down a slope, under the action of gravity, by a number of mechanisms, depending on the grain size and cohesiveness of the material, and the slope angle (see Chapter 8). Major mechanisms are **rockfalls, landslides, soil creep** and **slumping** (see Chapter 8). In rockfalls, consolidated material falls and breaks up into a jumble of material at the base of a cliff or steep slope. In contrast, a landslide is where a large coherent mass of material moves down a slope undeformed. Slumping is similar to landsliding, but contains saturated slope material (pore spaces are full of water) which deforms upon movement. Rockfalls, landslides and slumps are rapid events. Soil creep is the very slow, imperceptible, movement of material down a slope (see Chapter 8). Screes are accumulations of sediments that build up adjacent to mountain fronts, developed through the collection of loose sediment material removed from the mountain by gravity-driven sediment transport. Sediments fall onto the surface of the scree and move down the scree surface. They come to rest at the base of the scree where the slope shallows out. Scree slopes are composed mainly of poorly sorted, angular gravels, and exhibit only crude layering.

Another form of subaerial gravity flow is a debris flow. A debris flow is a slurry-like flow containing both solid material and water but with a high ratio of solids to water (see Chapter 8). Debris flows are highly destructive and are common in all climate regimes. They are often started after heavy rainfall on debris-laden mountain slopes, but can also be initiated by earthquakes, volcanic eruptions and even forest fires. The sediments deposited by debris flows are generally poorly sorted with little, or no, internal stratification.

Water is by far the most common medium for sediment transport. Water moves as a result of flow in channels or as a result of currents generated by wind and tides. If water movement is fast enough it may carry sediment, and in many cases this transport can be for hundreds of kilometres before the sediment grains are deposited. Box 9.1 describes important principles associated with sediment transport and water. Transport of sediment by air (wind) can also be an important mechanism, but its effectiveness is limited by

FUNDAMENTAL PRINCIPLES

SEDIMENT TRANSPORT BY WATER

To understand transport of sediment by water we need to have a basic understanding of the nature of flowing water. Water can flow in one of two ways: **laminar flow** and **turbulent flow**. In laminar flow the water molecules all flow in the same direction, parallel to each other (Figure 9.3). As a result, almost no mixing of water takes place during laminar flow. Laminar flow is uncommon in surface waters, being restricted to low flow velocities and very shallow water. In turbulent flow, water molecules move in many

directions, with an overall net flow in one direction. As a result water undergoing turbulent flow is well mixed.

The Reynolds number (*Re*) is a dimensionless (it has no units) quantity which indicates the extent to which a flowing fluid is laminar or turbulent. The Reynolds number relates the velocity of a flow (*u*), the ratio between the density and viscosity of the fluid (*v*, the fluid **kinematic viscosity**) and the length of the pipe or channel through which the fluid is flowing (*l*). The Reynolds number $(Re) = ul/v$. It has been experimentally determined that when the Reynolds number is low (less than 500) laminar flow dominates, and when the Reynolds number is high (greater than 2000) turbulent flow dominates. The Reynolds number is applicable to both water and air, but the lower viscosity of air results in turbulent flow dominating at lower flow velocities than water.

Sediment grains in water are transported by one of three processes (Figure 9.4). Firstly, grains can be moved along the bed surface by rolling. Secondly, grains may bounce along the bed surface; this process is termed **saltation**. Thirdly, material may be lifted off the bed surface and transported in suspension in the fluid,

(a) Rolling

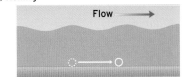

(b) Saltation

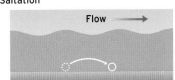

(c) Suspension

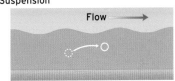

Figure 9.4 The three mechanisms of sediment transport in flowing water: (a) rolling, (b) saltation and (c) suspension.

kept in suspension by turbulent flow in the fluid. Sediment transported by the first and second mechanism is termed **bed load** and that transported by the third process is called **suspended load**. Whether a sediment grain will be transported as bed load or suspended load will depend on the size of the sediment grain and the velocity of the fluid flow. At low current velocities only small sediment grains (clays) will be transported

Laminar flow

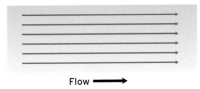

Turbulent flow

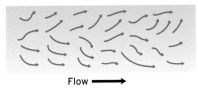

Figure 9.3 Laminar and turbulent flow in water.

BOX 9.1 ➤

in suspension. At higher velocities larger grains may be transported in suspension, but it is rare for grains larger than sand sized to be transported as suspended load.

The **Hjulström curve** (Figure 9.5) shows the nature of flow velocity required to move sediment of different grain sizes in water. This graph shows both the velocity of water required to keep a sediment clast in transport and the velocity required to move a stationary clast. Therefore, as a consequence it also shows the velocity below which sediment of a specific grain size will be sedimented. As can be clearly seen in Figure 9.5, as sediment grain size increases, a higher velocity is required both to keep the grains in transport and to move stationary grains. For fine sediment grain sizes only low flow velocities are required to keep sediment in transport. A consequence of this is that fine-grained sediments will accumulate only under very quiet water conditions. In contrast to this, fine-grained sediments require relatively more energy to be moved from a stationary position. This is due to the fact that clay grains, which dominate fine-grained sediments, are cohesive in nature and clump together. As a result of this, it takes greater flow velocity to transport stationary clay material than sand-sized particles. This means that, although clay-sized particles are deposited only when current velocity effectively falls to zero, once deposited, clay size material is not easily eroded and retransported.

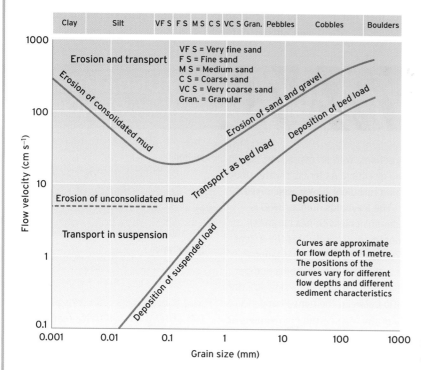

Figure 9.5 The Hjulström diagram, illustrating the relationship between grain size and current velocity for sediment grain transport. The two curves show the energy required to keep sediment in transport (lower curve) and the energy required to transport grains from a stationary position. (Source: after Press and Siever, 1986)

BOX 9.1

the low density of air and as a result only small grains can be transported by the wind. Wind-blown sediments are important indicators of climate change in the past and may also have important impacts on climate itself (Lowe and Walker, 1997). Such issues are addressed in Box 9.2. Although ice is a solid material, it moves and deforms slowly in the form of glaciers and ice sheets and can therefore be thought of as a fluid. As a result it can transport large amounts of sediments slowly over relatively short distances (see Chapter 16; Leeder, 1999).

9.2.4 Products of sedimentation – bedforms

A **bedform** is a morphological feature formed when sediment and flow interact on the sediment bed surface.

Ripples on a sandy beach and sand dunes in a desert are both examples of sediment bedforms. As sediment is moved along the bed surface by the current, small irregularities on the bed surface influence the manner in which sediment is transported and deposited. Bedforms deposited under **unidirectional flow** (currents that are flowing in dominantly one direction such as river currents and wind) are different from those formed under **oscillatory flow** (currents that oscillate backwards and forwards such as wave currents).

Current ripples form under unidirectional flow and are small bedforms (up to 5 cm in height and 30 cm in wavelength) and form predominantly in sand-sized sediment. Current ripples are not symmetric but have a shallow stoss side and a steep lee side as shown in Figure 9.8(a). Sediment is transported up the stoss side and avalanches down

ENVIRONMENTAL CHANGE

WIND-BLOWN TRANSPORT OF SEDIMENT

Wind-blown sediment transport is called aeolian transport. Although such transport can take place in many environments, it dominates in arid and semi-arid environments with little water. As air has a lower viscosity than water, higher flow velocities are needed to move sediment grains. At typical wind speeds, medium sand grains (up to 0.5 mm) are the largest grains that can be transported (this contrasts with the pebbles and boulders that can be transported by water). As a result the deposits built up from wind-blown sediment grains are generally fine grained in nature. Transport distances for such material

can be vast, with dust deposits transported thousands of kilometres from the Sahara of North Africa to the Atlantic Ocean (Figures 9.6 and 9.7). This material settles out on the sea floor, contributing to sedimentation in the oceanic environment. The amount of material removed is huge and for the Sahara approximately 2.6×10^8 tonnes per year. The extent of this dust transport has changed through time, a fact documented through climate and dust records preserved in ice cores and deep-sea sediments. Dust transport from land surfaces was greater during the last glacial period than today. Thompson *et al.* (1995) documented increased dust transport during the Late Glacial Stage in ice cores from glaciers in the

high Peruvian Andes. They concluded that atmospheric dust contents were up to 200 times as high as today as a result of increased aridity.

As well as responding to changes in global climate, the transport of mineral dust through the atmosphere also has direct impacts upon climate and biological systems itself. Very fine-grained mineral dust is highly effective at scattering light and, therefore, may have a cooling effect on climate (see Chapter 4). Recently, it has been clearly documented that the element iron, present within wind-blown dust derived from deserts (Figure 9.6), acts as a nutrient in the surface layers of oceans and helps promote algal growth in the water column (so-called primary production) (Moore and Braucher, 2008). This **primary productivity** results in the uptake of carbon dioxide from the atmosphere and is, therefore, an important component of the global climate system. Such observations have led some researchers to suggest that the artificial addition of iron to the oceans could be one way to lower atmospheric carbon dioxide levels (a process commonly termed iron fertilization). It has also been proposed that dust transported from the Sahara has had a detrimental impact on the development of Caribbean corals. It has been recognized that, since the late 1970s, fluxes of wind-blown dust to the Caribbean from the Sahara have increased and it has been suggested that this has led to environmental stress on Caribbean corals (Shinn *et al.*, 2000).

Loess is a fine-grained (less than 50 μm) sedimentary deposit composed of grains of quartz, feldspar, carbonate and clay minerals

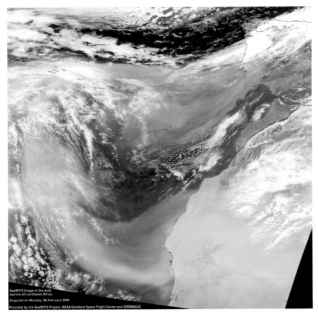

Figure 9.6 Iron-rich wind-blown dust from the Sahara being delivered to the North Atlantic Ocean where it acts as a nutrient stimulating surface water primary productivity. (Source: image courtesy of SeaWiFS Project, NASA/Goddard Space Flight Center and ORBIMAGE)

BOX 9.2 ➤

transported by wind from arid land surfaces and deposited elsewhere, often thousands of kilometres from its source (Pye, 1987). Thick deposits of loess sediments are present in the Czech and Slovak Republics of central Europe, and in central China, in an area known as the Loess Plateau. It is believed that these loess deposits were formed during full glacial conditions. The loess was derived from winds blowing across arid glacial outwash plains. These loess deposits also contain soil layers which are interpreted to reflect times of wetter climatic conditions, with negligible loess sedimentation (Lowe and Walker, 1997).

Figure 9.7 Dust transport by wind in the northern Sahara, Morocco. (Source: photo courtesy of Andrew Thomas)

BOX 9.2

the lee side. As a result the ripple migrates in a downstream direction. In plan view (looking from above) the shape of current ripples can vary from straight to sinuous to linguoid (Figure 9.8b). This variation in ripple shape can be in response to water flow and water depth.

Dune bedforms are larger than current ripples (up to 10 m high) but have similar cross-sections and form in a similar manner. Ripples are more predominant in silt to medium sand-sized sediments, whereas dunes are more predominant in medium to coarse sands (Figure 9.9). A clear relationship has been documented between flow velocity and bedforms (King, 1991) (Figure 9.9). At low flow velocities, bedforms do not form. At greater flow velocities current ripples form within fine to medium sand-sized sediment, dune bedforms forming at higher flow velocity. At very high velocities bedforms do not form owing to the speed of sediment transport. This is known as the upper-plane bed stage.

Wind-generated waves in water bodies (predominantly shallow marine settings, but sometimes also present in lakes) produce circular, oscillatory water motion (see Chapter 15). Beneath the surface of the water body, at the sediment surface, this motion is translated into horizontal oscillatory current movement (Figure 9.10). This motion sweeps grains away from a central zone and deposits particles as symmetrical ripples on the sediment bed. In cross-section and plan aspect, wave-formed ripples are symmetrical in shape and as such can easily be distinguished from current ripples formed under unidirectional flow conditions (Nichols, 1999).

Both sand ripples and sand dunes may be formed by wind-transported sediment. The most predominant environment for wind-produced bedform formation is that of the arid desert environments (Chapter 14), but localized sand dunes may also form in coastal environments as coastal sand dunes (Chapter 15). Arid zone aeolian (wind-blown) bedforms can be highly variable in size, ranging up to 600 m in wavelength and 100 m in height. When sediments are deposited from wind, the most common result is sand dunes (Figure 9.11). Under simple conditions of unidirectional wind patterns, simple dunes may form, with stoss and lee slopes. However, in many desert regions wind directions can change seasonally, which leads to more complicated dune bedforms. Unidirectional dunes can take two forms. **Transverse dunes** are linear features, with a shallow windward side and a steep lee slope. The internal structure is similar to dunes and ripples formed below water. **Barchan dunes** also have a shallow windward and steep lee side, but these are isolated dunes, with a characteristic crescent shape. These dunes are most commonly formed as sediment is moved across a hard substrate, such as a dried-up lake bed. **Stellate** (star-shaped) **dunes** form under conditions in which wind directions are variable, and with no particular direction prevailing. These dunes do not migrate and may

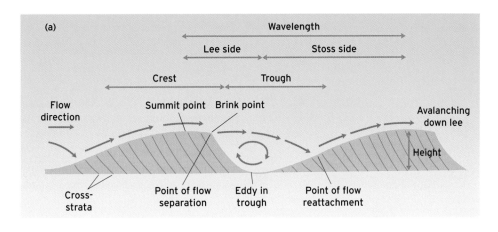

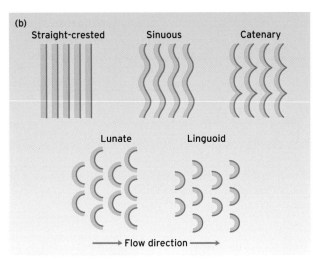

Figure 9.8 (a) Schematic diagram to illustrate the formation of a current ripple under unidirectional current flow. (b) Shape of current ripples in plan view. The change from straight-crested to linguoid is governed by current strength and water depth. (Source: after Tucker, 1981)

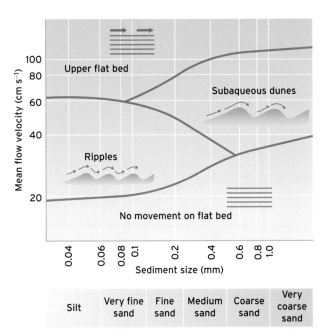

Figure 9.9 A bedform–flow diagram to illustrate the grain size and current velocity regimes under which sediment bedforms are present. (Source: after Nichols, 1999, and King, 1991)

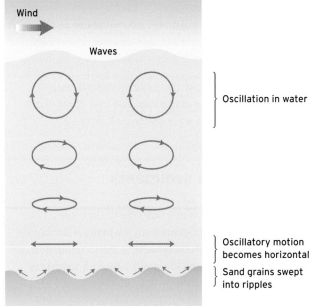

Figure 9.10 The formation of symmetrical wave ripples by oscillation of a water body. (Source: after Nichols, 1999)

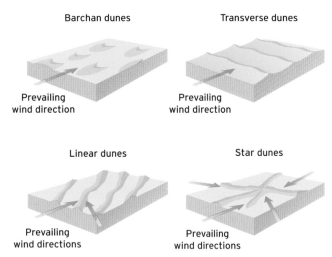

Barchan dunes

Transverse dunes

Prevailing wind direction

Prevailing wind direction

Linear dunes

Star dunes

Prevailing wind directions

Prevailing wind directions

Figure 9.11 Aeolian bedforms in arid environments. (Source: after Nichols, 1999)

be initiated at irregularities in the ground surface. **Seif** (or linear) **dunes** form where two distinct wind directions are present at approximately right angles to each other. Wind-formed ripples are commonly present on the surface of sand dunes. Chapter 14 provides further details on sand-dune forms and processes.

Reflective questions

➤ What grain roundness would you expect for a beach sand?

➤ Looking at the Hjulström curve, what can you say about the transport of boulders within rivers and streams?

➤ How would you use ripples to distinguish between sediment deposited in a river and that deposited in a shallow marine wave-dominated setting?

9.3 Biological sediments

In areas where there is a minimal supply of clastic sediments, other components form the major contribution to sediments. The main component is the accumulation of dead organisms or the build-up of framework structures (**bioherms** such as coral reefs). Two main types of biological sediments can be recognized, carbonate sediments and organic-rich sediments. The major components in carbonate sediments are skeletal fragments of marine or freshwater organisms (e.g. brachiopods, molluscs,

echinoids, corals, foraminifera, calcareous algae), bioherms (e.g. corals) and non-biological components (e.g. ooids, peloids) (Figures 9.12, 9.13 and 9.14). Reefs are the result of framework building by organisms composed of calcium carbonate material (Masselink and Hughes, 2003). Organisms most characteristic of reef build-up are corals, but bryzoa, coralline algae and brachiopods have also produced bioherms in the geological past. Non-biological components form as the result of direct precipitation of calcium carbonate in the form of grains, although some of these grains may form via microbiological processes. Ooids and pisoids are concentrically coated grains whereas peloids are grains with little internal structure (Figure 9.13).

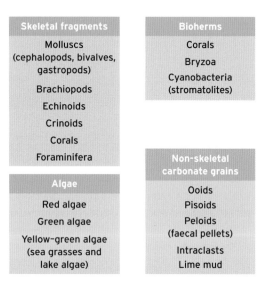

Skeletal fragments	**Bioherms**
Molluscs (cephalopods, bivalves, gastropods)	Corals
	Bryzoa
Brachiopods	Cyanobacteria (stromatolites)
Echinoids	
Crinoids	
Corals	**Non-skeletal carbonate grains**
Foraminifera	
Algae	Ooids
	Pisoids
Red algae	Peloids (faecal pellets)
Green algae	
Yellow–green algae (sea grasses and lake algae)	Intraclasts
	Lime mud

Figure 9.12 Components of carbonate sediments.

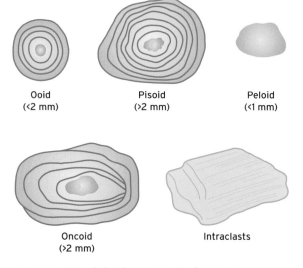

Ooid (<2 mm)

Pisoid (>2 mm)

Peloid (<1 mm)

Oncoid (>2 mm)

Intraclasts

Figure 9.13 Non-skeletal components of carbonate sediments. (Source: after Nichols, 1999)

(a)

(b)

Figure 9.14 Tropical carbonate sedimentary environments. (a) Assemblage from a clear, tropical reef setting at Eleuthera, Bahamas. Note the presence of large coral colonies. All material in such environments becomes transported and deposited as sediment upon the death of the organisms. (Source: photo courtesy of Chris Perry) (b) A reef assemblage forming in a muddy, sediment-dominated reef setting, Queensland, Australia. (Source: photo courtesy of Chris Perry)

Carbonate sediments are most abundant within warm tropical waters, as carbonate production is favoured in strong solar radiation, warm-water environments. Although traditionally viewed as forming in clear waters, free from terrestrial sediment inputs, reefs have also been reported from tropical shallow waters containing high levels of terrestrial sediment inputs (Figure 9.14) (Perry *et al.*, 2008). Carbonate sediments form predominantly in tropical and subtropical shallow marine environments, but calcium carbonate accumulation can also occur within temperate shallow marine environments and lakes. These sedimentary environments are sensitive to inputs of land-derived sedi-

ment and an example is discussed in Box 9.3. Sediments composed of shells of single-celled organisms (e.g. diatoms and foraminifera) also accumulate in deep-sea environments. Such sediment is commonly termed calcareous ooze, or siliceous ooze, depending on the composition of the shell material making up the sediment. When buried and lithified, carbonate sediments are known as limestones. Chalk, a pure white limestone formed throughout northwest Europe 80–65 million years ago (see Figure 2.4 in Chapter 2), is an example of a sedimentary rock composed of the remains of shells of a calcareous blue–green alga (Tucker and Wright, 1990).

ENVIRONMENTAL CHANGE

TERRESTRIAL SEDIMENT IMPACTS UPON TROPICAL CARBONATE SEDIMENTS

Carbonate sediments, composed of the mineral calcium carbonate, commonly accumulate in tropical shallow marine environments. Much of this calcium carbonate is derived from the skeletal remains of organisms. This accumulation of

calcium carbonate is an important sink for carbon dioxide and so is, therefore, an important part of the global carbon cycle. Carbonate sediments accumulate mostly in clear waters, with minimal inputs of land-derived sediment. In many parts of the world changes in land use, deforestation or climate change have led to increased inputs of land-derived sediments. One example where such

sediments have had a major impact is that of Discovery Bay, north Jamaica (Figure 9.15). In this example, bauxite dust, an iron-rich, aluminium-rich material that is mined for aluminium production, has been discharged from a loading terminal into Discovery Bay, a semi-restricted embayment fronted by fringing reefs and dominated by the deposition of carbonate sediments. Since the 1960s these inputs

BOX 9.3 ➤

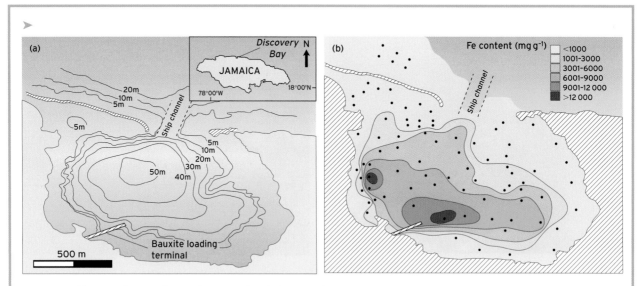

Figure 9.15 Discovery Bay, north Jamaica. (a) Water-depth map of the bay, also showing the location of the bauxite loading point. (b) Distribution of iron within the bay, showing the high levels of bauxite accumulation within the bay.

have led to the bay containing areas where surface sediment is composed of up to 35% bauxite sediment (Perry and Taylor, 2004). As well as significantly altering the composition of the surface sediments, this bauxite material has also led to markedly altered element cycling within the sediment. Iron and phosphorus are released from the sediments into the overlying water column and greater calcium carbonate is buried, compared with sediments in areas of the bay that have not received bauxite material (Taylor *et al.*, 2007).

BOX 9.3

Figure 9.16 Peat formation occurs in waterlogged areas such as in (a) and can build up thick deposits as shown in (b).

The most widely distributed type of organic-rich biological sediment is peat, which forms through the accumulation of dead plant material in waterlogged swamp and bog environments (Figure 9.16). The accumulation of peat is favoured in areas where rates of plant breakdown are low, which is most common in waterlogged stagnant conditions. Under such conditions, low oxygen, coupled with low pH, inhibits the breakdown of organic matter by bacteria and fungus. This leads to the accumulation of organic material. Peat can be cut and dried for use as a burning fuel. If peat layers are buried beneath further sediment layers, water is squeezed from the material and volatiles (water vapour and carbon dioxide) are lost. This process leads to the formation of lignites and coals. Many of the large coal deposits of north-west Europe were deposited in freshwater swamps during the Carboniferous era, 350 to 300 million years ago.

Reflective question

➤ What types of landform are composed largely of biological sediments?

9.4 Chemical sediments

The most important chemical sediment is that termed evaporative sediment. These sediments are formed as a result of minerals precipitated out of lake or seawater as waters are concentrated by evaporation. As seawater is evaporated the least soluble mineral precipitates out first. This is usually calcite (calcium carbonate, $CaCO_3$) followed by gypsum (calcium sulphate, $CaSO_4$) and halite (sodium chloride, $NaCl$) as the water becomes more saturated. If water becomes very concentrated a number of salts of potassium become precipitated (bittern salts). The most commonly encountered evaporite mineral is **gypsum**. Water concentrated to 19% of its original volume will precipitate gypsum. Halite is only deposited once the water has been reduced to less than 10% of its original volume and, therefore, is less common than gypsum. Many water bodies undergoing evaporation are periodically recharged by addition of water, either by rainfall or river water input, making it rare for such concentrated evaporation. The high solubility of $NaCl$ also means that it is readily redissolved on exposure to water. Bittern salts are very rare and form only after complete evaporation of a standing body of water.

Evaporative sediments are common in arid climates (see also Chapter 16) and may form in coastal settings or in standing bodies of water. In the case of coastal settings, the best documented examples are **sabkhas**. These are low-angle tidal flat environments (the coasts of the Persian Gulf being a classic example). Evaporation of groundwater draws in seawater, which upon evaporation precipitates gypsum. However, in these settings the water rarely becomes concentrated enough to precipitate halite. Thick deposits of halite have been formed in the past, such as those extracted in the Cheshire region of northern England. However, such thick deposits are not forming in the present day. To produce such thick accumulations of halite complete evaporation of large water bodies has been invoked, such as the Mediterranean Sea. However, this may become a phenomenon of the future if climate and land management change impact on evaporative processes.

Reflective question

➤ In which environments are chemical sediments more likely to be found?

9.5 Sedimentation in Earth surface environments

Sedimentation takes place in a wide range of Earth surface environments. In general, sedimentary environments can be thought of as a traverse from upland mountainous environments, through lower-lying continental environments, shoreline and shallow marine settings, eventually to oceanic environments, and this is the structure adopted in the following section (Figure 9.17). This passage is particularly true in the case of clastic sediments where the transport of sediment by water, wind or ice moves sediments from continental environments to offshore environments. Given the scope of this chapter, only a brief summary can be given here of the characteristics of sedimentation in Earth surface environments. Other chapters of this book provide further details within specific environments.

9.5.1 Continental environments

9.5.1.1 River environments

The term fluvial is generally used to describe river environments and processes. There are three major types of

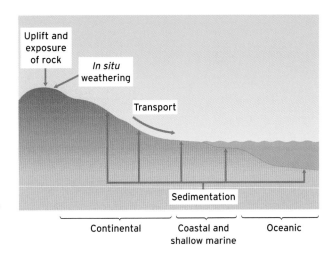

Figure 9.17 A schematic diagram to illustrate the pathways and major sites of sedimentation of clastic sediments in surface environments. (Source: after Nichols, 1999)

fluvial environment that can be recognized: braided (where the river consists of several shifting channels separated by islands, or bars); meandering (where singular sinuous channels are surrounded by low-lying floodplains); and anastomosing (where the river consists of stable, splitting and rejoining channels, thereby possessing elements of both braided and meandering form). For a detailed discussion of fluvial geomorphology see Chapter 12.

Braided channels form most commonly where water flows over loose sand or gravel, often in mountainous or upland areas. A characteristic feature of braided rivers is the mobile nature of the channels and the intervening bars (Figure 9.18; see also Figure 12.4 in Chapter 12). The sediments deposited in braided rivers are most commonly composed of gravels and coarse sands. Fine-grained sediments may be deposited on bars, but they are not common. The predominant sedimentary bedforms present within braided river sediments are dune bedforms, formed as bars accrete and move forward within and between the channels.

In contrast to braided rivers, meandering rivers form in low-lying areas, with a predominance of finer sediment. Meandering is a term used to describe the sinuous nature of the channels (Figure 9.18). Sediments deposited in meandering river environments are quite different from braided rivers, being predominantly finer grained than braided river sediments. Sediments deposited in the channels are coarser than those deposited in floodplains, as the water flowing within the channels transports the finer materials downstream. During periods of high stage (floods) water may overflow the channels and deposit suspended sediment upon the floodplain. As the sediment carried in suspension

is silt and clay sized, the resulting sediment on floodplains is fine grained. Such fine-grained sediments are high in nutrients and as a result floodplains are fertile areas. Indeed, it is the flooding and deposition of fresh sediment that keeps the land fertile, which is why flood management and dam building can have a major negative impact on soil fertility in river areas. In addition to the floodplains, sediments accumulate on the inside of meanders, which leads to the formation of point bars (Figure 9.18). Current ripples, especially on the surface of point bars, are the most common sediment bedform in meandering river environments (Tucker, 2001).

9.5.1.2 Arid environments

Desert environments are those regions of the Earth where potential evaporation exceeds rainfall (often the definition that potential evapotranspiration is more than twice the precipitation is used; see Chapter 14) and in these environments standing bodies of water are rare. The major process of sediment transport and deposition is by wind action, although flood events can also lead to fluvial conditions prevailing under wet seasons. Three major forms of sedimentation take place in arid environments: **sand seas**, **alluvial fans** and **playa** lakes.

Sand seas (also known as **ergs**) are areas of sand accumulations, and large sand seas are present in the Sahara, Namibia, south-western North America and western central Australia. Sand seas are not the same as deserts. Deserts are simply dryland environments and may or may not contain sand seas. These sand seas are composed of dune bedforms deposited from wind-blown sediments (see above).

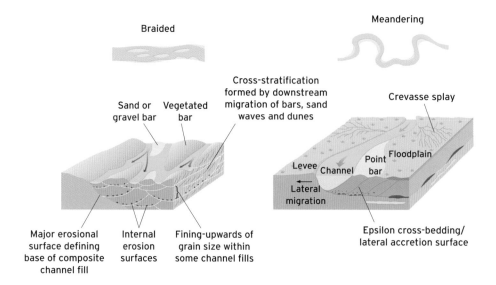

Figure 9.18 Sedimentation associated with braided and meandering river systems. (Source: after Tucker, 1981)

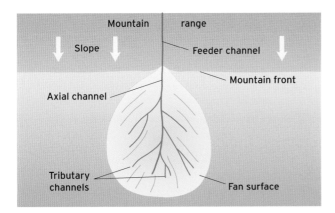

Figure 9.19 Typical features of an alluvial fan. (Source: after Harvey, 1997)

Sediment grains in these environments are typically well rounded as a result of grain–grain collisions during transport, and composed almost entirely of quartz. The red colour of desert sands is the result of a thin coating of iron oxide as a result of the oxidizing conditions in arid environments (see Chapter 14).

Alluvial fans are cones of sediment that accumulate at mountain fronts. The major agent of sediment transport on alluvial fans is flowing water (which is generally present only during wet seasons). This flowing water spreads out and deposits sediment as it slows down in distant parts of the alluvial fan surface. The result is a semicircular fan (Figure 9.19). Sediment deposits on alluvial fans are a mixture of gravels and coarse sands, with a general decrease in grain size away from the mountain front. Alluvial fans are particularly common in arid mountainous settings with well-developed fans present in Death Valley in California (e.g. Figure 9.20), and the Atlas Mountains of Morocco (Harvey, 1997).

Playa lakes are **ephemeral** (seasonal) bodies of water which accumulate during rainfall events. Water flowing into these lakes deposits a layer of silt and clay. As the lake dries up, evaporite minerals are precipitated (gypsum and halite), and desiccation cracks form within the sediment surface. If the water body dries up completely, a salt crust known as a salt pan is produced. At the next wet period further sediment is deposited followed by additional formation of evaporite minerals. This alternation of muds and evaporite minerals is characteristic of playa lakes.

9.5.1.3 Glacial environments

Ice plays a major role in sediment transport and sedimentation and gives rise to a number of characteristic landforms and sediment deposits. However, erosion and

Figure 9.20 An alluvial fan in an arid mountainous region of Death Valley National Park. (Source: photo courtesy of Steve Carver)

sedimentation through glacial activity are significantly different from those resulting from water and air. Detailed information on sediments in glacial systems can be found in Chapter 16.

9.5.2 Coastal and marine environments

9.5.2.1 Delta environments

There have been many definitions for deltas but a broad definition can be given as a discrete shoreline protuberance formed at a point where a river enters an ocean or other body of water. Deltas are sites where sediment supplied by the river is accumulating faster than it is being redistributed in the ocean by waves and tides. These environments vary depending on whether a delta is dominated by river processes (e.g. the Mississippi Delta), wave processes (e.g. the Rhone Delta) or tidal processes (e.g. the Ganges Delta). Further details of delta environments are provided in Chapter 15.

9.5.2.2 Estuaries and salt marshes

An estuary is a semi-enclosed coastal water body where there is a mixture of river and seawater and where there is a mixture of fluvial and marine processes. At the present time estuaries are common as a result of the post-glacial rise in sea level drowning the mouths of rivers. Two major

Figure 9.21 An estuarine environment in a tropical setting, Queensland, Australia. Note the presence of sinuous tidal channels, with areas of vegetated fine-grained sediments stabilized by mangroves.

morphological elements are present in estuaries: tidal channels (Figure 9.21) and tidal mudflats. Tidal channels are major sites of sediment transport and hence consist of sand-sized sediment. Within tidal channels subaqueous dune bedforms and ripples commonly form. In estuaries both flood (incoming tide) and ebb (outflowing tide) currents can be present, but in general either the ebb or the flood current is strongest and this is reflected in the nature of the bedforms. Tidal mudflats are regions away from strong flood or ebb currents. Suspended sediment (silt and mud) is carried over the mudflats during high tide and deposited upon the mudflat as the tide turns and water velocity falls. The frequency of flooding depends on the height of the tide and the elevation of the mudflat. Mudflats gradually build upwards and therefore become flooded less frequently over time. If flood frequency is low, salt-tolerant plant species will colonize the mudflat, and these systems are termed salt marshes. In tropical environments, mangroves form in such environments as a result of similar processes (Figure 9.22). Chapter 15 provides more information on estuary and salt marsh systems.

9.5.2.3 Beaches, barriers and lagoons

Between deltas and estuaries coastlines may be sites of sediment erosion or sediment deposition (see Chapter 15).

Figure 9.22 Details of a tropical coastal mangrove environment in Queensland, Australia. Note the presence of tree roots trapping and stabilizing sediment.

Along coastlines that are sites of deposition, sedimentation may take place on beaches, lagoons or barriers. A beach is an area that is continuously impacted by waves. Sediment accumulating on beaches (which may be supplied by cliff erosion or by **longshore drift**) is continuously reworked and is characteristically well sorted and well rounded. On very shallow-sloped beaches, waves and wind may form ripples, but on steeper-sloped beaches low-angle sediment accumulation may be present, especially on wave-dominated beaches. A barrier island is a beach detached from the main coast to form a ridge of sediment parallel to the coast (see Chapter 15). Such islands are most common along shorelines with a low tidal range and high wave energy. Sediment accumulates along the front of the island in a beach environment, whereas landward of the island, quiet conditions allow the accumulation of fine-grained material, either in salt marshes or lagoons. Lagoons are areas of low energy and are normally formed behind a barrier such as a barrier island. Along clastic sediment shorelines, muds and salt marshes develop, whereas in carbonate-dominated shorelines, fine-grained carbonate mud accumulates.

Within tropical coastal settings in which clastic sediment input by rivers is minimal, the deposition of carbonate biological sediments can dominate. In such environments, sediment may be formed by the build-up of reef-building organisms, the accumulation of skeletal material and the inorganic precipitation of calcium carbonate. Reefs composed of corals accumulate in shallow, high-energy conditions and commonly act as barriers to shallow lagoon environments behind. Within these shallow lagoons, fine-grained sediment composed of precipitated calcium carbonate (lime mud) accumulates.

9.5.2.4 Shallow marine environments

The nature of sediment deposited in shallow marine environments is governed by the strength of currents produced by tides and storms. Sands are the predominant sediment deposited, commonly as sand dunes up to 10 m in height. In wave- and storm-dominated environments (micro-tidal) sands are deposited as wave-rippled and symmetrical hummocks. In tidal-dominated shallow marine environments (macro-tidal) sand waves and sand ribbons form. Sand waves form under lower tidal flow and are aligned perpendicular to the direction of tidal flow. Sand ribbons form under higher tidal flows and are aligned parallel to tidal flow. Fine-grained sediment tends to be deposited in waters deep enough not to be affected by major storms and tides. At the present time, many shallow marine environments (e.g. the North Sea) are covered with a layer of coarse sand and pebbles. These deposits are relict from the time of much lower sea level during the last Ice Age. Finer-grained sediment is currently being trapped in estuaries.

9.5.2.5 Oceanic environments

Oceanic environments include those environments from the edge of the continental shelf into the abyssal plains (see Chapter 3), and span a water depth from 100 m or less to over 8000 m in some of the deep-sea trenches. Sedimentation in oceanic environments takes place via two major processes: **turbidite currents** and **pelagic sedimentation**. Turbidite currents are mixtures of sediment and water which, because of their increased density relative to seawater, flow down and along the bottom surface of the oceans. In this process, they transport sand and clay-sized sediment from shelf slopes to deeper oceanic environments, depositing sediment as a thin bed widely across the sea floor (Figure 9.23). Turbidite flows are commonly triggered by earthquake events, with one of the best documented examples being in the Grand Banks area of Newfoundland in 1929. The resulting turbidite flows broke transatlantic telephone cables on the seabed.

Pelagic sedimentation is the slow background sedimentation of fine-grained material falling through the water column to the seabed. The best developed pelagic sediments accumulate in the deep sea where clastic sediment input from continents is minimal. Three types of pelagic sediments have been documented from deep-sea

Figure 9.23 Sandstone beds deposited by turbidite currents, Tabernas, southern Spain.

environments: brown clay, carbonaceous ooze and siliceous ooze. Brown clay is a sediment accumulation of fine-grained clay grains and glass fragments. The sediment is derived predominantly from wind-blown continental material, volcanic material and micrometeoric grains. Carbonaceous ooze is an accumulation of calcite tests of microscopic organisms living in the water column (e.g. foraminifera and coccoliths), and is widely distributed on the ocean floor. Siliceous ooze is composed of the tests of microscopic organisms made of silica (e.g. radiolaria and diatoms) and has a localized distribution on the ocean floor. See Figure 3.14 in Chapter 3 for further information.

Reflective questions

➤ What are the main differences in the dominant sediment processes between continental, coastal and oceanic environments?

➤ What type of bedforms do you find in rivers?

➤ What type of bedforms do you find in arid environments?

9.6 Response of sedimentation to environmental change

Both natural and anthropogenic activities can have a major impact on the style and rate of sedimentation in surface environments. Therefore it is important that we can measure the rates of sedimentation on different environments and Box 9.4 provides some examples of how this can be done. The natural changes that impact most upon sedimentation are climate change and sea-level change. Climate change leads to changes in rainfall and vegetation which have major impacts upon sediment supply in the catchment, and thereby sedimentation in associated receiving water bodies. Sea-level change results in changes in the base level of sedimentary systems. The result is commonly either the marine inundation of coastal and continental environments, or the exposure of shallow coastal shelves. In general, these natural changes are slow and gradual, although there are numerous examples in the geological past where such changes have produced marked changes in sedimentation style. At the present time sea-level rise, associated with global climate warming, is having marked impacts on low-lying coastal systems (see Chapter 15).

TECHNIQUES

MEASUREMENT OF SEDIMENTATION RATES

The rate of sedimentation is a measure of the thickness of sediment (normally measured in centimetres) that accumulates at a specific location over a specific amount of time. Generally, accumulation rates are quoted in centimetres or millimetres per year, but may also sometimes be quoted as grams per cm² of sediment. However, in many environments sedimentation rate is generally very low, and so rates in centimetres per hundred years, or even per thousand years, are commonly quoted.

Sedimentation rates can be measured for environmental systems using a range of techniques. Short-term measurements can be made by collecting sediment that accumulates in a sediment trap and measuring the amount of sediment deposited over a month or a year. Alternatively, in salt marsh or floodplain environments short-term sediment accumulation rates have been measured by laying down grass mats on the marsh and measuring sedimentation upon the mats over durations of 1–10 years. Longer-term measurements of sedimentation rates can be made using radionuclide determination. Caesium-137 (^{137}Cs) is a radioactive element that was released into the atmosphere by atomic weapons testing in the 1950s and by the Chernobyl nuclear power station incident in 1986 (Figure 9.24). By measuring the vertical location of these two concentration peaks of ^{137}Cs in the sediment, an estimate of annual sedimentation rate can be made. For longer-term sedimentation rate estimations, archaeological artefacts or ^{14}C dating can be used to estimate sediment accumulation rates over hundreds to thousands of years (see Chapter 23). Under special circumstances, yearly layers of sediment (commonly called **varves** in lake sediments) can be recognized, allowing for accurate estimates of sedimentation rate in such cases (Figure 9.25).

Typical sedimentation rates in natural systems display a wide range

BOX 9.4 ➤

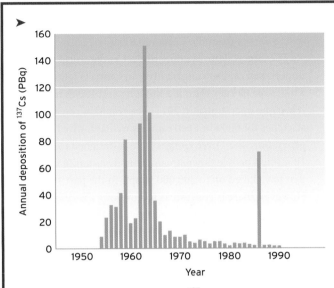

Figure 9.24 The inputs of radioactive ^{137}Cs from the atmosphere to the northern hemisphere since 1950. The peaks in the 1950s and 1960s as a result of atomic weapons testing, and in 1986 as a result of the Chernobyl incident, allow for estimates of sedimentation rate to be made for sediments over the past 50 years, assuming constant sedimentation rates. (Source: after Owens *et al.*, 1996)

of values. In general, sedimentation rates are low, being of the order of less than 1 cm yr^{-1}. Sedimentation rates are very low in oceanic environments (commonly less than 1 cm per 1000 years), as a result of their great distance from sediment sources. Sedimentation rates in lakes, floodplains and coastal settings can be much higher (in the range of 0.1-10 cm yr^{-1}). It should be remembered, however, that sedimentation is a dynamic process and rates of sediment accumulation may vary over time scales from daily to yearly. Longer-term changes in sedimentation rate will also result from natural and anthropogenic changes to the system.

Figure 9.25 Annual sediment layers (varves) deposited in Quaternary Glacial Lake Riada, central Ireland. Core is approximately 25 cm in length. (Source: photo courtesy of Cathy Delaney)

BOX 9.4

Of greater short-term impact and concern are the effects of anthropogenic activities on sedimentation. Such activities can be either direct, through the engineering of water bodies (e.g. dams, reservoirs), or indirect, through changes in catchment characteristics (e.g. mining, urbanization).

9.6.1 Dams and reservoirs

Dams and reservoirs have been constructed since early human history for regulation of water, but in the last 50 years the construction of major dams for water supply and hydroelectric generation has increased markedly. As well as having marked impacts upon water flow within catchments downstream of the dam, they also have a marked impact on sedimentation and sediment transport throughout the catchment. The two most significant impacts are the trapping of sediment behind the dam (see below) and the reduction in the sediment load of the river downstream. Sedimentation in lakes behind dams leads to less floodplain sedimentation downstream, which reduces nutrient supply and increases the potential for erosion (clear-water erosion). Before 1930 the Colorado River, USA, carried up to 150 million tonnes of suspended sediment annually to its head in the Gulf of California. Since that time, a number of dams have been built on the Colorado. The Glen Canyon Dam, for example, was constructed as a sediment trap to prolong the life of Lake Mead behind the Hoover Dam which was completed in the 1930s. Sediment is currently being deposited and trapped behind these structures (e.g. in Lake Mead and Lake Powell) and is no longer

discharged into the sea. Indeed, as a result of dams and water abstraction in southern California, water no longer enters the sea from the Colorado River.

The largest dam-building project in the world is being undertaken on the Yangtze River in China (Figure 9.26). The Three Gorges Dam is 2 km long and 100 m high; it became operational in 2011. The resultant reservoir stretches for 600 km upstream. As well as for hydroelectric generation, the Three Gorges Dam is designed to protect 10 million people downstream from devastating floods that have killed up to 300 000 people in the past 100 years. However, there is concern as to how long the reservoir will last, given that silt carried down by the Yangtze will sediment behind the dam, perhaps eventually filling the reservoir. On the Yellow River, also in China, a reservoir behind the Sanmenxia Dam filled with silt within four years of construction and had to be emptied, dredged and rebuilt. In the year 2000 the reservoir had less than half its original

Figure 9.26 The Three Gorges Dam, Yangtze River, China. (Source: Alamy Images: China Images)

capacity (Chengrui and Dregne, 2001). The Yangtze carries 530 million tonnes of silt through the Three Gorges area each year. A similar, smaller dam on the Yangtze (the Gezhouba Dam), a test run for the Three Gorges, lost more than a third of its capacity within seven years of opening. To minimize this loss of capacity in the Three Gorges area, two approaches have been recommended. The first is to keep the reservoir levels low during high-flow seasons to allow more transported sediment into the lower reaches of the river. The second is to increase tree cover in the catchment in order to decrease erosion of silt into the river.

The construction of water reservoirs in upland environments, especially in north-west Europe, has been a common practice. Soil erosion in such catchments is high, leading to significant sediment supply within the catchment. This erosion has been exacerbated by human land-use practices and pollution. Sediment within upland catchments has major impacts on the water reservoirs within them. The most important is that sedimentation within the reservoir (Figure 9.27) reduces the effective volume of water that can be held by that reservoir. Another impact is that the sediment has a deteriorating effect on the colour of the water as a result of the high organic content of the sediment (mainly peats). Although sedimentation impacts vary between catchments, it has been estimated, for example, that a typical water reservoir in upland United Kingdom loses 10% of its volume as a result of sedimentation over a 100 year lifetime.

9.6.2 Mining

The activity of mining economic deposits from the Earth can have major impacts on sedimentation in both river and coastal systems (Box 9.5). Mining activity can take the form of subsurface mining for metals, or the open-cast mining of

Figure 9.27 An upland water reservoir after drought conditions showing the layers of sediment that have built up. Such sedimentation reduces the capacity of these reservoirs.

metals, coals and aggregate material. In all cases, the mining activity exposes large amounts of material to subaerial and fluvial erosion through the production of piles of spoil and waste material. This increase in erodibility leads to increased sediment loads in rivers and increased deposition of material in downstream environments. In addition to large **sediment yields**, the sediment that is deposited in downstream settings is commonly highly contaminated with metals, which has an impact on organisms living within those environments. In mining activities that are close to the coast, estuaries will experience increased sedimentation. In Cornwall, UK, historical mining of metals, especially tin, released large volumes of particulate mine waste into river systems (Pirrie *et al.*, 2002). This sediment was deposited in the coastal zone, leading to the rapid siltation of small estuaries and the loss of ports.

HAZARDS

TOXIC SEDIMENT DISCHARGES AS A RESULT OF MINING ACTIVITIES

Potentially toxic elements (e.g. arsenic, mercury, zinc, lead, cadmium) are a major concern in the environment, and in many cases

these are associated with sediment in aquatic systems as a result of the affinity of metals for the particulate fraction, and their low solubility in water. Therefore, the dispersion of contaminated sediment in aquatic systems can often have major negative impacts upon ecosystems. The

dispersion of this contaminated sediment can be a consequence of the disposal of overburden or low-grade ore material as unconfined spoil heaps (Figure 9.28) from mines such as the copper mine shown in Figure 9.29, or the release

BOX 9.5 ➤

➤

Figure 9.28 Unconfined spoil heaps of metal-rich material, Copper Flats copper mine, New Mexico, USA.

Figure 9.29 Santa Rita open-cast copper mine, New Mexico, USA.

BOX 9.5 ➤

➤

of fine-grained processed mineral waste (tailings) from an impounded tailings reservoir. In the former case, metal contamination of river sediments is as a result of long-term hydrological remobilization of sediment from the spoil heaps. In the case of the latter, contamination is generally the result of an acute, short-lived event resulting from engineering failure.

An example of a severe pollution event linked to the dispersion of metals associated with a tailings dam spill is the Aznalcóllar copper–silver–lead–zinc mine in Spain,

45 km west of Seville. On 25 April 1998 the tailings dam, which held the fine-grained metal-rich tailings waste from the mining activities, failed, releasing metal-rich sediment into the Agrio and Guadiamar Rivers. Approximately 4600 ha of floodplain land was flooded with an estimated 2 million m^3 of metal-rich tailings, as well as over 5 million m^3 of metal-rich acidic water. This was the worst recorded pollution event in Spanish history. The contaminated sediment was deposited along a length of river 40 km downstream from the mine, and 800 m wide.

The clean-up of the affected rivers took the form of mechanical excavation, and manual clearing of the deposited layer of tailings sediment. Recent research has shown that there are long-term impacts upon the river systems as a result of the deposition of these tailings' sediment on the floodplain and the channel bed (Hudson-Edwards *et al.*, 2005; Kraus and Wiegand, 2006). Similar toxic sediment inputs to river systems as a result of mining have happened recently in Bolivia (Rio Pilcomayo; Hudson-Edwards *et al.*, 2001) and Romania (Tisa Basin; Macklin *et al.*, 2003).

BOX 9.5

9.6.3 Urbanization

Urbanization has a marked effect on sediment sourcing and sedimentation, mainly through the anthropogenic nature of sedimentary material and the engineering of the land surfaces (Taylor and Owens, 2009). There are two main types of urban sediment: aquatic sediments in urban water bodies (e.g. canals, docks), and as street sediments on road surfaces ('street dust'). Urbanization of sediment catchments has a number of effects on sedimentation in water bodies. Watercourses become engineered, commonly by channelization and culverting, and land surfaces are paved over. All this has the effect of increasing the rate and extent of sediment supply to receiving water bodies, while vegetation loss reduces the sediment storage capacity of the system. In addition to sediment yields, sediment quality also markedly decreases as a result of urbanization. Sediment can become contaminated by sewage, industrial pollution and vehicular pollution. As a result of this, sediment pollution is a problem within urban water bodies, and chemical reactions in the sediment can lead to the remobilization of this pollution and the generation of noxious methane gas (Taylor *et al.*, 2003). Sediment accumulates on street surfaces as a result of industrial, vehicular and building activities and these sediments commonly contain high concentrations of lead and other metals (Robertson *et al.*, 2003). These sediments have been implicated in respiratory diseases, as the fine-grained fraction of this sediment can be resuspended in the atmosphere and inhaled by humans living and working in urban environments.

9.6.4 Sediment management

The field of sediment management has rapidly expanded in response to the environment pressures exerted upon sediment systems. As Box 9.6 indicates, many systems suffer from issues of sediment quality (e.g. pollution) and sediment quantity (e.g. sediment requiring dredging). Management of sediment in such cases may take the form of source control by cleaning up contaminated sediment sources such as mine sites, industry and wastewater treatment works, for example. Site-specific remediation, or clean-up, can take the form of dredging, sediment nourishment, sediment capping (to isolate contaminated sediments from overlying water) or *in situ* chemical or biological treatment.

Reflective questions

➤ Thinking of dams, mining and urbanization, which of one or two of these activities are most likely to (a) increase rates of sedimentation; (b) change the type (or quality) of the sediment; and (c) decrease rates of sedimentation?

➤ Can you describe some methods of measuring sedimentation rates?

CONTAMINATED SEDIMENT MANAGEMENT AND TREATMENT, PORT OF HAMBURG, GERMANY

The contamination of sediments with potentially toxic elements (e.g. metals, persistent organic pollutants) is increasingly being recognized as a major problem. In many cases, the presence of this contamination needs management and remediation. Such an example is the Port of Hamburg, Germany, which is a major economic shipping port on the River Elbe, approximately 100 km from the North Sea. Sediment is delivered to the port from upstream sources on the River Elbe and from tidal transport of sediment from the North Sea. Sedimentation rates are high, and sediment accumulation in the port significantly reduces water depth. As a result a dredging programme is needed to maintain water depths in the port to allow continued shipping access (Figure 9.30). Approximately 3 to 4 million m³ of sediment is dredged each year. Once the sediment has been dredged from the port, it cannot be dumped at sea as the sediment is contaminated and must be treated as a controlled waste. The River Elbe upstream of Hamburg flows through industrial and mining areas in Germany and the Czech Republic and for this reason the sediment in Hamburg Port is contaminated (Netzband *et al.*, 2002), containing high levels of arsenic, mercury, chromium, lead and organic pollutants (polychlorinated biphenyls, PCBs, and polyaromatic hydrocarbons, PAHs).

Figure 9.30 Dredging operations in Port of Hamburg. (Source: photo courtesy of Philip N. Owens)

The option taken to deal with this dredged contaminated sediment is to treat it in a specially built sediment treatment plant. The dredged material is passed through rotary screens and sieves to remove the coarser fraction of the sediment, leaving behind the fine sediment fraction (<63 mm). This fine sediment fraction contains the majority of the contaminants as a result of the high capacity of silt and clay minerals to adsorb contaminants. The resulting coarser sediment fraction is lower in contamination and can be used for building material. The fine sediment fraction is dewatered and disposed of to a specially built landfill facility. By removing the coarse fraction from the dredged material the volume of material which requires landfill disposal is significantly reduced.

This dredging of sediment from the port also acts as a pollutant filter to the North Sea, annually removing approximately 30% of metal contaminants from the River Elbe that would otherwise be discharged to the North Sea.

This management of sediment represents a site-specific approach. A more sustainable approach for contaminated sediment management is one that considers sediment management on the river basin scale, identifying and minimizing contaminant inputs at source. It is increasingly being recognized that river basin scale approaches to sediment management are the most effective, from both economic and ecological viewpoints (Owens, 2005).

BOX 9.6

9.7 Summary

Sedimentation is a major process acting to shape the Earth's surface. Sediments are derived from fragments of pre-existing rocks (clastic sediments), the remains of organisms (biological sediments) and the direct precipitation of minerals from seawater (chemical sediments). Sediments are deposited in all surface environments, from continental settings to oceanic settings. Within these environments distinct grain composition, grain shape and sediment bedforms are preserved. This allows the reconstruction of past sedimentary environments through the recent and geological past. Both natural and anthropogenically induced changes in environmental conditions impact upon sedimentation processes. Of these, anthropogenic impacts (e.g. engineered structures and urbanization) have the most marked and rapid effects. It should be clear that a thorough understanding of sediments and sedimentation processes is required for physical geographers and environmental scientists to interpret Earth surface geomorphology.

Further reading
Books

Leeder, M.R. (1999) *Sedimentology and sedimentary basins.* Blackwell Science, Oxford.
This textbook is aimed at higher-level undergraduate students in the earth science and geographical disciplines. Although much of the detail is in depth, the student will find further information on aspects of sedimentology not covered in this chapter. Subjects covered well include the physical transport of sediment by water, wind and ice, the role of tectonics on sedimentation and large-scale sedimentary processes on the Earth's surface. Readers interested in the more physical and mathematical aspects of sediment transport and deposition will be well served by this book.

Lowe, J.J. and Walker, M.J.C. (1997) *Reconstructing Quaternary environments.* Longman, Harlow.
This deals extensively with methods and examples of how Quaternary environments can be reconstructed from the rock record. As such, it has a large amount of information upon sedimentary successions and would give the student a good background to ancient sedimentary deposits. It also provides useful information on loess and wind-blown deposits.

McManus, J. and Duck, R.W. (eds) (1993) *Geomorphology and sedimentology of lakes and reservoirs.* John Wiley & Sons, Chichester.
This edited volume contains a series of papers dealing with the topic of sedimentation within lakes and reservoirs. Of particular interest is a set of papers on the impact of sedimentation in engineered reservoirs. This book takes a more geographical approach than most textbooks on sedimentology.

Nichols, G. (2009) *Sedimentology and stratigraphy,* 2nd edition. Blackwell Science, Oxford.
This book is designed for undergraduates studying sedimentology in the earth sciences. Although latter parts of this text are based on interpreting sedimentary environments from the geological record, the first half of the book provides a good, clear overview of the major processes operating on sediments, the formation of bedforms and the sediments deposited in the full range of Earth surface environments.

Perry, C.T. and Taylor, K.G. (eds) (2007) *Environmental sedimentology.* Blackwell Publishing, Oxford.
This book provides an extensive introduction to the sedimentology of contemporary Earth surface environments and the impacts of climatic and environmental change upon these environments. A large range of terrestrial, coastal and marine environments are covered, and aspects of the biology, physics and chemistry of sediments are included, as well as good case examples of sediment pollution, management and remediation.

Stow, D.A.V. (2005) *Sedimentary rocks in the field: A colour guide.* Manson Publishing, London.
A highly visual guide to the appearance of sedimentary rocks in the field, but with also good illustrations of sedimentary structures.

Tucker, M.E. (2001). *Sedimentary petrology,* 3rd edition. Blackwell Publishing, Oxford.
This book gives a full coverage of the textures and composition of sediments and sedimentary rocks. It has especially good coverage of the minerals that make up sediments and sedimentary rocks and clearly sets out how sedimentary rocks may be classified.

Tucker, M.E. and Wright, V.P. (1990) *Carbonate sedimentology.* Blackwell Scientific Publications, Oxford.
This book gives full coverage of the composition, origin and depositional environments of carbonate sediments. There are very good chapters on the constituents of carbonate sediments and a wide range of carbonate depositional environments are covered. There are also chapters on the geochemistry of carbonate sediments and on carbonate sediments in the geological record.

Further reading
Papers

Moore, J.K. and Braucher, O. (2008) Sedimentary and mineral dust sources of dissolved iron to the world ocean. *Biogeosciences,* 5, 631–656.
A paper on some of the material discussed in Box 9.2.

Taylor, K.G. and Owens, P.N. (2009) Sediments in urban river basins: A review of sediment-contaminant dynamics in an environmental system conditioned by human activities. *Journal of Soils and Sediments,* 9, 281–303.
A paper that brings together research knowledge on urban river sediment contaminants.

CHAPTER 10

Soil in the environment

Pippa J. Chapman

School of Geography, University of Leeds

Learning objectives

After reading this chapter you should be able to:

➤ describe the components that make up soil and understand how they affect the physical, chemical and biological properties of soil

➤ describe the processes of soil formation and the factors that control soil development, and understand how these lead to the features seen in soil profiles

➤ compare and contrast the different methods of soil classification

➤ define the important physical properties of soil and explain important cation-clay mineral interactions and the processes that control soil pH and relate these concepts to soil fertility

➤ understand the important functions soil biota perform and the factors that control their diversity and activity

➤ appreciate the environmental importance of soil and understand the threats posed to soil and soil processes by human activities

10.1 Introduction

Soil is a major component of the Earth's **ecosystems** and forms at the interface of the atmosphere (air), **lithosphere** (rocks), **biosphere** (plants and animals) and hydrosphere (water). Soil can be defined as a complex medium, consisting of inorganic materials (such as sand, silt and clay minerals), organic matter (living and dead), water and air, variously organized and subject to dynamic processes and interactions. As well as being a major component of the natural system, soil has a key role in the use and management of the environment by humans, where it performs a wide range of essential functions:

- It is a medium for plant growth, providing plants with support, essential nutrients, water and air. Plant life, in turn, supports animal life.
- It acts as a reservoir for water, influencing the quantity of water in our rivers, lakes and aquifers (see Chapter 11).
- It has a filtering and transforming role for materials added to the soil (see Chapter 13). Thus, soil is often able to protect the quality of our air and water.
- It recycles dead plants and animals into nutrients needed by all living things.
- It provides a habitat for organisms. A handful of soil may be home to billions of organisms, belonging to thousands of species.
- It provides raw materials such as clays, gravels, sands and minerals as well as fuels such as peat. It also provides a physical base for the foundations of buildings and roads.

- It helps to protect our cultural heritage. Soils preserve a diverse range of archaeological remains which are a vital resource for understanding anthropogenic history.

Soil is, therefore, essential to the maintenance of the environment. Without soil, the biosphere in which we live could not function. As Doran and Parkin (1994) stated: 'The thin layer of soil covering the Earth's surface represents the difference between survival and extinction for most terrestrial life.'

Soil, however, is not an unlimited resource and can be lost, degraded or improved by natural processes and human activities. As most soils take thousands or even millions of years to form, they cannot be replaced if they are washed away or polluted. Understanding the nature and distribution of soils and the processes operating within soil is therefore essential if we want to preserve our soil in a healthy state and improve our understanding of ecosystem dynamics. This chapter begins by describing the components of soil before examining soil formation processes. These processes impact the physical and chemical properties of soil, which are also described. The types of organisms found in soil and the functions they perform are also covered in this chapter. The chapter concludes by examining a range of impacts of human activities on soils and soil processes.

10.2 The components of soil

Soil comprises four major components: the inorganic or mineral fraction, organic matter, water and air. The relative proportion of these four components greatly influences the physical, chemical and biological properties of a soil. In a soil, the four components are mixed in a complex way. Figure 10.1 shows that approximately half the soil volume

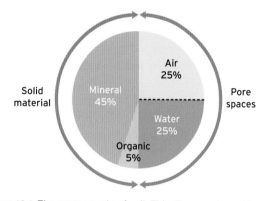

Figure 10.1 The components of soil. This diagram shows the composition by volume of a typical topsoil. The dashed line between water and air indicates that the proportion of these two components varies with soil moisture. Water and air make up the pore spaces whereas the solid material is made up from mineral and organic matter.

of a typical **topsoil** consists of solid material (inorganic and organic); the other half consists of voids or **pore spaces** between the solid particles. Most of the solid material is inorganic mineral matter. However, the influence of the organic component on soil properties is often far greater than its small proportion suggests. Most agricultural soils contain between 1 and 10% organic matter and are referred to as inorganic or mineral soils because of the low organic content.

Air and water fill the pore spaces between solid soil particles. The relative proportion of air and water fluctuates greatly and they are inversely related to each other. Following rainfall, water fills the pore spaces, expelling much of the air. As water gradually drains away or is used by plants, air refills the pores.

10.2.1 Mineral particles

In most soils the mineral fraction predominates. Mineral particles are derived from the **weathering** of **parent material**. Weathering (see Chapter 7) refers to the breakdown of rocks and minerals by the action of physical and chemical processes. The larger mineral particles, which include boulders, stones, gravel and coarse sands, are generally rock fragments, whereas smaller particles are usually made of a single mineral. There are two major types of mineral particles: **primary minerals**, which are minerals that have changed little since they were formed in magma, such as quartz, feldspars and micas; and **secondary minerals**, which are formed from the breakdown and chemical weathering of less resistant primary minerals such as clays and oxides of iron and aluminium. The composition and size of mineral particles have a great influence on the physical and chemical properties of a soil (see Sections 10.5 and 10.6).

10.2.2 Soil organic matter

Soil organic matter can be divided into three categories: (i) decomposing residues of plant and animal debris referred to as **litter**; (ii) resistant organic matter known as **humus**; and (iii) living organisms and plant roots collectively referred to as the **soil biomass**. Fresh plant and animal litter is progressively decomposed by soil microorganisms (see Section 10.7) to a more or less stable end product called humus which is resistant to further decomposition. During the breakdown of organic matter by microorganisms, plant nutrients are released, particularly nitrogen, phosphorus and sulphur. This process is called **mineralization**. The balance between inputs of plant and animal materials and losses by decomposition determines the amount of organic matter in the soil. The organic matter content of soils varies greatly but it usually represents between 2 and 6% by volume, but even a small amount is important as it has such a large impact upon many of the major physical, chemical and biological properties of soil. In arid climates, where less vegetation grows, soil organic matter is low (0.5 to 2%), whereas in environments where decomposition processes are drastically slowed, such as in waterlogged conditions, a surface accumulation of only partially decomposed material builds up to form depths of several metres. These organic soils contain more than 65%, and often more than 90%, organic matter and are known as **peat** (see Chapter 9).

Soil organic matter is a very important component of soils because it: (i) is the main food for soil organisms; (ii) binds mineral particles together and therefore stabilizes the soil's structure and protects it from erosion; (iii) improves water holding capacity; (iv) improves **porosity** and aeration and therefore aids the growth of plants; and (v) is a major source of nutrients and therefore influences soil fertility.

10.2.3 Soil water

Water is essential to the ecological functioning of soils. Plant and soil organisms depend on water to survive. Water is also a major driving force in soil formation as it is required for parent material weathering. All chemical weathering processes depend on the presence of water. Water together with soil air fills up the pore spaces between the mineral and organic components of the soil. Soil water, however, is not pure; it contains dissolved organic and inorganic substances and is known as the **soil solution**. When some compounds dissolve in the soil solution, the atoms become separated as **ions**. An ion is an atom, or group of atoms, bearing an electrical charge. For example, when table salt (sodium chloride), which has the chemical formula $NaCl$, dissolves in water the atoms separate and form ions (see Figure 10.2). The sodium ions have a single positive charge and are indicated by the symbol Na^+, whereas chloride has a single negative charge as indicated by the symbol Cl^-. Positively charged ions are referred to as **cations** and negatively charged ions as **anions**. In soil solution important cations and anions include those in Table 10.1.

An important function of the soil solution is to ensure the continual supply of some of these cations and anions to the plant roots. Soil solution is also the main agent of **translocation**, carrying dissolved ions, including pollutant ions, and small particles through the soil to surface and groundwaters.

Water is held in soil by the attraction of water molecules to each other and to soil particles. Water exists as one of three states in the soil (Figure 10.3). The amount held in

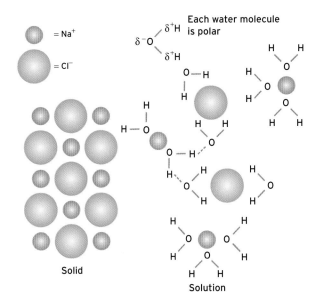

Figure 10.2 The dissolution of table salt (NaCl) in water. Since water (H_2O) is a polar molecule (one end has a different charge from the other end), the negative end is attracted to positive ions and the positive end is attracted to negative ions. Table salt completely dissolves (dissociates) in water as the water molecules keep the sodium (Na^+) and chloride (Cl^-) ions apart and stop them from reforming the solid salt. The figure shows how the water molecules are linked to the sodium and chloride ions.

each state changes over time, which affects the amount of water available to plants and the potential movement of nutrients and pollutants within the soil. When all the soil pores are filled with water from rainfall, the soil is described as **saturated** (Figure 10.3a). The soil, however, does not stay in this state for very long as, under the action of gravity, water will start to drain out of the larger pores and is replaced by air (Ward and Robinson, 2000). This water is called **gravitational water** and when all of it has drained away the soil is said to be at **field capacity** (Figure 10.3b). The small pores retain water against the force of gravity: this water is known as **capillary water** and represents the

Table 10.1 Some important cations and anions in soil solution

Cations		Anions	
H^+	Hydrogen	Cl^-	Chloride
Na^+	Sodium	NO_3^-	Nitrate
K^+	Potassium	SO_4^{2-}	Sulphate
Ca^{2+}	Calcium	HCO_3^-	Bicarbonate
Mg^{2+}	Magnesium	OH^-	Hydroxide
Al^{3+}	Aluminium		

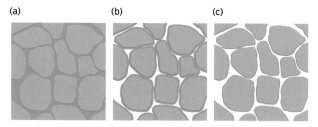

(a)   (b)   (c)

Figure 10.3 Soil water states at: (a) saturation, when all pore spaces are filled with water; (b) field capacity, when the smaller pores are filled with water and the larger pores are filled with air; (c) permanent wilting point, when plants can no longer exert sufficient suction to withdraw the water that is tightly held around the soil particles.

majority of water that is *available* for plant uptake. This water remains in the soil because the combined attraction of (i) the water molecules to each other and (ii) water to the soil particles is greater than the gravitational force. Capillary water moves within the soil from zones of higher potential (wet areas) to lower potential (dry areas) (see Chapter 11). The most common movement is towards plant roots and the soil surface, where it is lost by evaporation and transpiration. The final type of water is **hygroscopic water**, which is held as a tight film around individual soil particles (Figure 10.3c). This water is *unavailable* to plants as the attraction between the water and the soil particles is greater than the 'sucking power' of plant roots. A soil in which all the water is hygroscopic will appear dry although some water still remains. The drier the soil, the harder the plant has to work to obtain the remaining water held in progressively smaller pores. Eventually there comes a point when plants cannot withdraw the tightly held water from the soil and this is known as the permanent **wilting point** (Figure 10.4). The water retained in the soil between the states of field capacity and the wilting point is known as the plant available water or **available water**.

The amount of water held in each state is related to a number of factors including **soil texture**, **soil structure** (see Section 10.5.3) and organic matter content. Since soil water occurs as films around soil particles, if there are many small particles (i.e. clay size particles) in a soil it will hold more water due to the greater soil surface area per unit volume. However, much of the water held in soils with a high proportion of clay size particles is unavailable to plants (i.e. hygroscopic water) because it is held in very small pores. The general relationship between soil texture and soil water availability is shown in Figure 10.4. Soil structure influences the nature and abundance of soil pores and soil permeability and therefore the rate at which water drains through the soil. For example, if a soil has a lot of well-connected pores and is very permeable, water will percolate rapidly through it. Organic matter increases the soil's moisture holding capacity and indirectly affects water content through its influence on soil structure and total pore space (White, 1997).

10.2.4 Soil air

Soil air occupies pores that are not filled with water. Soil animals, plant roots and most microorganisms use oxygen and release carbon dioxide when they respire. In order to maintain biological activity, oxygen needs to move into the soil and carbon dioxide must move out of the soil. This ventilation of the soil is known as **aeration**. Aeration is affected primarily by the pore size distribution, pore continuity, the soil water content and the rate of oxygen consumption by respiring organisms. As soil air is 'compartmentalized' by the presence of water and intervening soil particles, the composition of the soil air differs from that of atmospheric air. Generally it also has higher moisture and carbon dioxide and lower oxygen concentrations than the atmosphere (see Table 10.2). Carbon dioxide concentrations are often several hundred times higher than in the atmosphere.

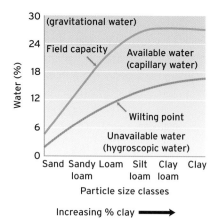

Figure 10.4 The general relationship between soil texture and soil water availability.

Table 10.2 The composition (% by volume) of soil air relative to the open atmosphere. There tends to be less oxygen and more carbon dioxide in soil air than in the atmosphere because of the respiration of microorganisms in the soil which uses oxygen and produces carbon dioxide

	Atmosphere	Soil air
Nitrogen (N_2)	79.01	79.0
Oxygen (O_2)	20.96	18.0–20.8
Carbon dioxide (CO_2)	0.035	0.15–1.0

However, the composition of soil air is constantly changing with marked diurnal and seasonal fluctuations. These changes are often associated with the differences in biological activity between night and day or between summer and winter. There will be less respiration on a cold winter's night in a temperate zone than on a warm summer's day and so carbon dioxide concentrations in soil air may be much lower in winter. The composition of soil air also varies considerably from place to place in the soil.

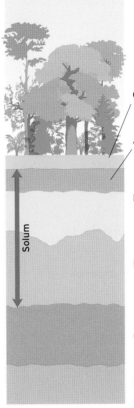

O	Organic horizon – surface layer dominated by the accumulation of organic matter
A	Mineral horizon – mineral material mixed with decomposed organic matter. Dark colour due to presence of organic matter
E	Mineral horizon – depleted in clay, iron and aluminium oxides. Eluvial horizon. Lighter in colour than A and B horizons
B	Mineral horizon – enriched in iron, aluminium and clay minerals. Illuvial horizon
C	Unconsolidated material
R	Bedrock

Figure 10.5 A hypothetical mineral soil profile showing the relative position of the major horizons that may be present in a well-drained soil in the temperate humid region. Not all the horizons described here are present in every soil profile, and the relative depths vary.

Reflective questions

➤ What is the difference between primary and secondary minerals?

➤ Why does soil organic matter have such an important impact on the properties of soil?

➤ What factors affect the soil's moisture holding capacity and why?

➤ Why is the concentration of carbon dioxide in soil air higher than that in the air above the soil?

10.3 Soil profile

Soils are described by the characteristics of their **soil profile**. This consists of a vertical section through the soil from the ground surface down to the parent material. It is made up of a series of distinctive horizontal layers known as **soil horizons**. This horizontal alignment is mainly due to the translocation of materials by the movement of water through the soil. The removal of solid or dissolved material from one horizon is called **eluviation**, while the deposition in another horizon is referred to as **illuviation**.

The soil horizons are given letters according to their genesis and their relative position in the profile. The major horizons are shown in Figure 10.5. Note that not all the horizons described here are present in every soil. The O horizon is a surface layer dominated by the accumulation of fresh or partially decomposed organic matter. The A horizon can occur at or near the surface (beneath the O horizon) and contains a mixture of mineral and organic (mainly humus) material and is therefore usually darker than the horizons below. Beneath this occurs the E horizon or elluvial horizon. As the E horizon is a zone of depletion (e.g. of clay, organic matter, iron) it is usually a pale, ashy colour. E horizons are common in high-rainfall areas, especially in soils developed under forests. The underlying B horizon is often a zone of accumulation (e.g. of clay, iron, organic matter, carbonates) often referred to as the illuvial horizon. In some soils, the accumulation of iron oxides in the B horizon gives it a reddish colour. The A, E and B horizons are sometimes referred to as the **solum** (from the Latin for soil or land). It is in the solum that the soil-forming processes are active and that plant roots and animal life are largely confined. The B horizon usually grades into the C horizon, which largely comprises unconsolidated weathered parent material known as the regolith. Although the regolith is affected by physical and chemical processes it is little affected by biological activity and therefore not part of the soil solum. If unweathered rock exists below the C horizon it is called bedrock and is designated the R horizon.

In some soil profiles, the soil horizons are very distinct in colour, with sharp boundaries, whereas in other soils the colour change between horizons may be very gradual, and the boundaries difficult to locate. However, colour is just one of the many physical, chemical and biological characteristics by which one horizon may differ from the horizon above or below it.

The informal terms 'topsoil' and 'subsoil' are often used to describe soil. Topsoil refers to the upper portion of the soil (usually the A horizon or plough horizon) and is the part most important for plant growth. The subsoil refers to the part of the soil below the topsoil (plough depth) and usually relates to the B horizon.

Reflective questions

➤ Draw a diagram of the typical soil horizons and explain their importance in determining the properties of soil.

➤ What is the difference between eluviation and illuviation?

10.4 Soil formation processes

10.4.1 Pedogenesis

The process of soil formation, called **pedogenesis**, takes place over hundreds and thousands of years. The soil is an open system, which allows input of materials to the soil, the loss of materials from the soil and internal transfers and reorganization of these materials within the system. Soil horizons develop as a result of a number of processes occurring within the soil, which can be classified into the following categories: additions, removals, mixing, translocations and transformations.

The main input of soil material comes from the parent material of the soil. Mineral particles are released from the parent material by weathering at the base of the soil, and contribute to the lower layers of the soil. Significant inputs of material come from surface accumulation, particularly of organic matter. Inputs also include solutes and particles carried by precipitation and the wind, energy from the Sun and gases from the atmosphere.

The main losses from the soil occur through wind and water erosion and **leaching**. Leaching is the removal of soil material in solution and is most active under conditions of high rainfall and rapid drainage. The percolating water carries soluble substances downwards through the soil profile, depositing some in lower layers but removing the most soluble entirely (see Chapter 13). Removals also include the loss of gases and uptake of solutes by plants.

Mixing of organic and inorganic components is an important process that is carried out by soil animals, microbes and plant roots, freezing and thawing of water, and shrinking and swelling of the soil. Humans also cause physical mixing of the soil by ploughing. Chemical and biological processes can also transform soil components. Organic compounds decay and some minerals dissolve while others precipitate. These transformations result in the development of soil structure and a change in colour from that of the parent material. Translocation of material within the soil profile often occurs in response to gradients of water potential (e.g. suction) and chemical concentrations within soil pores. Suspended and dissolved substances may move up or down through the soil profile.

The net result of these processes occurring over a long period of time is the formation of different soil horizons. However, the processes that dominate at a particular site are dependent on the environmental conditions at that site. In areas where rainfall exceeds evapotranspiration, net water movement is down through the soil (Figure 10.6a). The extent of leaching is often indicated by the acidity of the soil (Jarvis *et al.*, 1984). In many freely draining soils, clay is carried from the upper horizons by percolating water to lower horizons and this is known as clay eluviation, or lessivage (Figure 10.6b). The clay is redeposited as skins or coats on the surfaces of aggregates or in pores and around stones. Soil horizons characterized by clay accumulation are described as **argillic**. Clay eluviation tends to produce a group of soils known as acid brown earths or **luvisols** (Figure 10.7).

Podzolization may occur in soils where there is intense leaching and translocation of material (Figure 10.6c). Organic acids complex with iron and aluminium compounds that are transported downwards from the E horizon by percolating water and deposited in the B horizon. Podzolization occurs on freely drained sites under forests and heath plants and the end product of this process is a soil called a podzol (Figure 10.8), the characteristics of which are the presence of an organic layer, a leached E horizon and an accumulation of iron, aluminium and humic material in the B horizon. These soils are not very productive for agriculture because they are acidic and the free drainage results in leaching of fertilizers away from plant roots.

In many locations waterlogging leads to the reduction, mobilization and removal or redeposition of iron compounds in the soil (Figure 10.6d). The reduction of ferric (Fe^{3+}) to the more mobile, grey ferrous (Fe^{2+}) iron compound (see Chapter 15) by microorganisms is known as **gleying**. The soil loses the brown/red colour of ferric oxide and becomes grey or bluish. Alternate phases of reduction and oxidation due to fluctuations in the water content result

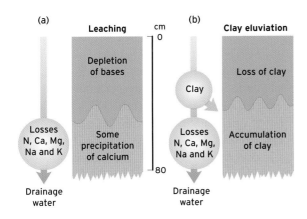

(a) **Leaching**
cm 0
Depletion of bases
Clay
Losses N, Ca, Mg, Na and K
Some precipitation of calcium
80
Drainage water

(b) **Clay eluviation**
Clay
Loss of clay
Losses N, Ca, Mg, Na and K
Accumulation of clay
Drainage water

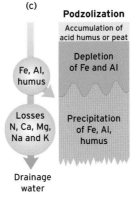

(c) **Podzolization**
Accumulation of acid humus or peat
Fe, Al, humus
Depletion of Fe and Al
Losses N, Ca, Mg, Na and K
Precipitation of Fe, Al, humus
Drainage water

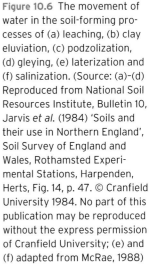

Figure 10.6 The movement of water in the soil-forming processes of (a) leaching, (b) clay eluviation, (c) podzolization, (d) gleying, (e) laterization and (f) salinization. (Source: (a)-(d) Reproduced from National Soil Resources Institute, Bulletin 10, Jarvis *et al.* (1984) 'Soils and their use in Northern England', Soil Survey of England and Wales, Rothamsted Experimental Stations, Harpenden, Herts, Fig. 14, p. 47. © Cranfield University 1984. No part of this publication may be reproduced without the express permission of Cranfield University; (e) and (f) adapted from McRae, 1988)

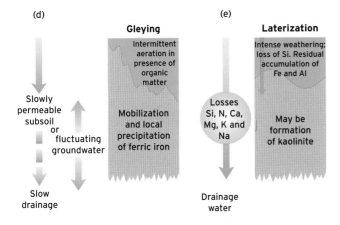

(d) **Gleying**
Intermittent aeration in presence of organic matter
Slowly permeable subsoil or fluctuating groundwater
Mobilization and local precipitation of ferric iron
Slow drainage

(e) **Laterization**
Intense weathering; loss of Si. Residual accumulation of Fe and Al
Losses Si, N, Ca, Mg, K and Na
May be formation of kaolinite
Drainage water

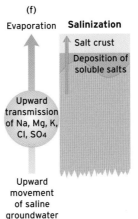

(f) **Salinization**
Evaporation
Salt crust
Deposition of soluble salts
Upward transmission of Na, Mg, K, Cl, SO4
Upward movement of saline groundwater

Figure 10.7 An argillic brown earth from Kent, UK. (Source: photo courtesy of John Conway)

Figure 10.8 A podzol from the Upper Wye catchment, Wales. (Source: photo courtesy of Chris Evans)

Figure 10.9 A stagnohumic gley. (Source: photo courtesy of E.A. Fitzpatrick)

in soil having a mottled appearance with brown/red iron oxide spots or streaks occurring along root channels and larger pores as shown in Figure 10.9.

Laterization (ferralitization) occurs in tropical and subtropical soils where high temperatures and heavy rain result in intense weathering and leaching (Figures 10.6e and 10.10). Almost all the by-products of weathering are leached out of the soil leading to the development of horizons depleted in base cations (e.g. calcium, magnesium, potassium and sodium) and enriched in silica and oxides of aluminium and iron (McRae, 1988). The red colour of these soils is due to the presence of haematite and goethite (Figure 10.10). Conversely in areas where evapotranspiration exceeds rainfall, such as arid and semi-arid areas, water is drawn to the soil surface and as water evaporates salts are precipitated at or near the surface (Figure 10.6f; Chapters 9, 11 (Box 11.5) and 14). This salinization process is almost the complete opposite of leaching.

10.4.2 Factors affecting soil formation

The major processes involved in soil formation described above are controlled by local and regional environmental factors. In the late 1800s, Dokuchaiev, a Russian scientist, was one of the first to recognize that soils do not occur by chance but usually form a pattern in the landscape and develop as a result of the interplay of climate, parent material, organisms and time. Building on this work in the 1930s and 1940s, Hans Jenny suggested that relief was an additional important factor (Jenny, 1941).

Figure 10.10 A red laterite soil (also known as an oxisol) formed under a wet, humid and warm tropical climate, Brazil. (Source: photo courtesy of Phil Haygarth)

10.4.2.1 Climate

Climate is perhaps the most influential factor affecting soil-forming processes as it determines the moisture and temperature regimes under which a soil develops. In addition, climate is influential in determining vegetation distribution (see Chapters 18 and 25). Rainwater is involved in most of the physical, chemical and biological processes that occur within the soil, and particularly weathering and leaching. To be effective, however, water must pass downwards through the whole of the soil profile and into the regolith. The amount of precipitation that percolates downwards through the soil is mainly related to total annual precipitation and rate of evaporation (from vegetation and soil), although topography and permeability of the parent material are also important factors. Overall, percolating water stimulates weathering processes, which helps to differentiate the soil into horizons and influences soil depth.

The main effect of temperature on soils is to influence the rate of soil formation via mineral weathering and organic matter decomposition. For every 10°C rise in temperature, the speed of chemical reactions increases by a factor of two or three; biological activity doubles, up to around 30–35°C, and evaporation of water increases. As rates of chemical weathering are greatest under conditions of high temperature and humidity, soils in tropical areas are often several metres deep while those in polar regions are shallow and poorly developed (Figure 10.11). In addition, soils are influenced by microclimates that are related to altitude and aspect.

10.4.2.2 Parent material

Soils may develop on the weathered surfaces of exposed, consolidated *in situ* rock surfaces, or unconsolidated superficial material that has been transported and deposited by gravity, water, ice or wind. Parent material influences soil formation through the process of weathering and then through the influence of the weathered material on soil processes. Rock types influence the rate of weathering through their mineralogical composition and the surface area of the rock exposed. The larger the exposed surface area, the faster the rate of weathering. Some minerals are more susceptible to weathering than others. Goldich (1938) proposed a 'stability series' for the silicate minerals as shown in Figure 10.12. This arrangement of minerals is identical to Bowen's reaction series, where the silicate minerals are placed in their order of crystallization. The minerals that crystallize first form under much higher temperatures than those that crystallize last. Consequently, the minerals that crystallize first, such as olivine and pyroxene, are not as stable at the Earth's surface, where the temperature and pressure are very different from the environment in which they form. In contrast, quartz, which crystallizes last, is the most resistant to weathering (Figure 10.12).

Knowledge of rock mineralogy allows rocks to be placed in their order of susceptibility to weathering. Hard igneous rocks and Carboniferous and Jurassic sandstones weather slowly to give shallow, stony, coarse-textured soils. In contrast, softer Permo-Jurassic sandstones weather more

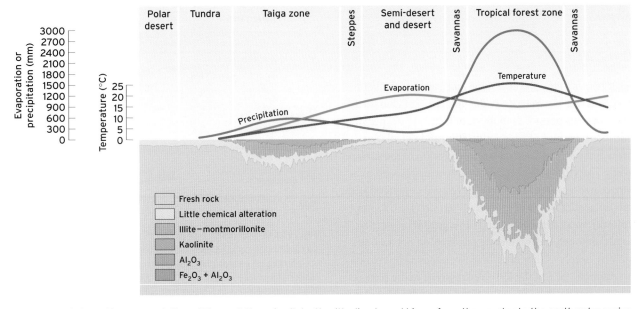

Figure 10.11 Schematic representation of the variation of soil depth with climate and biome from the equator to the north polar region. See Chapter 18 for information on each of the named biomes. Soils are deeper in the wet humid tropics and in the temperate zone and most shallow in dry or very cold locations. The weathering products of aluminium and iron oxides are also shown. (Source: after Strakhov, 1967, as adapted in Birkland, 1999, Fig. 10.5, p. 274)

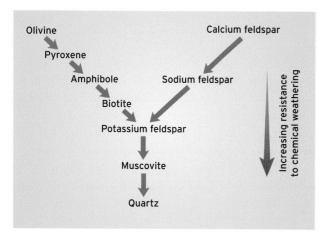

Figure 10.12 Weathering sequence for common rock-forming minerals. The sequence ranks the silicate minerals according to the general ease of weathering.

rapidly to give deeper, less stony, loamy or sandy soils. The soils that develop on all these parent rocks are generally acidic owing to the low base cation content of these rocks or the bases are leached from the soil faster than they are replenished by weathering. Carboniferous, Permo-Triassic and Jurassic clays, siltstones, mudstones and shales are all fine-grained rocks which weather to give silty or clayey soils which are generally slowly permeable. The weathering products of chalk and limestone are very soluble, and therefore soil depths are often shallow, particularly on steeper slopes. At the foot of the slope, where deeper soils form, they are well drained and base rich. Further information about parent material weathering can be found in Chapter 7.

10.4.2.3 Relief

Relief relates to the altitude, slope and aspect of the landscape and can hasten or delay the influences of climatic factors. Slope steepness is an important factor, as steeper slopes reduce the amount of water infiltrating and percolating through the soil and allow increased erosion of the surface layers. Therefore soils formed on steeper slopes tend to be thin, coarse textured and poorly developed compared with soils on gentler slopes or more level terrain (see Chapter 8). However, weathering rates tend to be greater on steeper slopes, although the weathering products do not accumulate very deeply as they are efficiently removed by erosion. For example, 90% of the dissolved material in the rivers of the Amazon Basin comes from the steep Andes Mountains which only cover 12% of the Basin (Gaillardet *et al.*, 1997).

On slopes with less permeable parent material, surface waterlogging causes gleying on flat ground, whereas the

soils on steeper slopes are drier as most precipitation runs off the surface or through the upper horizons to lower ground. This produces the pattern of soil distribution illustrated in Figure 10.13(a). On slopes with very permeable parent material water tends to penetrate to the subsoil, leaving the higher ground and steep slope well drained, whereas soils on the lower slopes and valley bottoms are more likely to be affected by groundwater as shown in Figure 10.13(b). Milne (1935) was the first to use the term soil **catena** for topographically determined soil profiles in East Africa. Where there is no change in the geology along the slope, soil differences in the catena are brought about by drainage conditions, differential transport, eroded material and the leaching, translocation and redeposition of mobile chemical constituents.

Aspect affects the solar energy received at the ground surface. In the northern hemisphere, south-facing slopes receive more and are therefore warmer and generally lower in moisture than north-facing slopes. Consequently, soils on the south slopes tend to be drier, less densely vegetated,

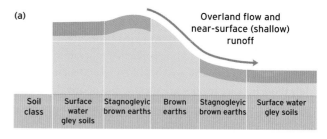

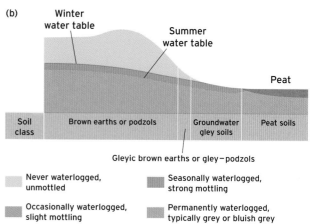

Figure 10.13 Relationship between slope, hydrology and soil formation on (a) slowly permeable parent material and (b) permeable parent material. (Source: Reproduced from National Soil Resources Institute, Bulletin 10, Jarvis *et al.* (1984) 'Soils and their use in Northern England', Soil Survey of England and Wales, Rothamsted Experimental Stations, Harpenden, Herts. Fig. 16, p. 52 © Cranfield University 1984. No part of this publication may be reproduced without the express written permission of Cranfield University)

and thus lower in organic matter. These differences are reversed in the southern hemisphere.

Altitude influences climate (see Chapter 6). Temperature declines with altitude and precipitation tends to increase with altitude in the middle latitudes. This leads to an excess of rainfall over evaporation and as a result leaching rates are high and waterlogging occurs where the drainage is poor. The lower temperatures also lead to a reduction in biological activity and therefore slower decomposition of organic matter. This in turn leads to the accumulation of thick organic horizons at the surface and ultimately to the formation of peat.

10.4.2.4 Organisms

Organisms include plants, animals, microorganisms and humans. Vegetation extracts water and nutrients from the soil and under natural conditions returns most of the nutrients it uses to the soil in litter. The type of vegetation influences the type and amount of litter that is returned to the soil. Different soil types support different vegetation communities. Vegetation also protects the soil from water and wind erosion by intercepting rainfall, decreasing the velocity of runoff, binding soil particles together, improving soil structure and porosity, and providing a litter cover which protects the soil surface against raindrop splash.

Earthworms and other small animals such as moles mix and aerate the soil as they burrow through the soil. Earthworms have been found to increase the infiltration rate of fine-textured soils and contribute towards increasing the stability of the soil structure by intermixing organic matter with mineral particles (Curtis *et al.*, 1976). Soil organisms, including fungi, bacteria and single-celled protozoa, play a major role in the decomposition of organic matter (see Section 10.7). The end product is humus. Humans influence soil formation through manipulation of vegetation, agricultural practices such as drainage and irrigation, the additions of fertilizers, lime and pesticides, and urban and industrial development.

10.4.2.5 Time

Over time soil is continually forming from the parent material, under the influence of the climate, topography, vegetation and soil organisms. Soil genesis is a long process; the formation of a layer 30 cm thick takes from 1000 to 10 000 years. During this time, the properties of the soil continually change. This is manifest by changes in the soil profile including the number of horizons, their depth and their degree of differentiation. When the rate of change of a

soil property with time is negligible, the soil is said to be in steady state. However, in reality soil rarely reaches this state because of changes in one of the environmental factors. For example, changes in the world's climate over geological time accompanied by changes in sea level, erosion and deposition have produced large changes in the distribution of vegetation and parent material. Therefore most soils have not developed under a single set of environmental factors but have undergone successive waves of pedogenesis. The most recent large change in climate resulted in alternating glacial and interglacial periods of the Pleistocene (see Chapter 22). In high and middle latitudes, glaciation removed the majority of soils and covered large areas with drift material. Therefore, soil development in these areas began again on new surfaces after the final retreat of the ice during the Holocene which began about 11 700 years ago (Chapter 23).

10.4.2.6 Combined influences

It can be seen that the five factors influencing soil formation do not operate as single independent factors. Climate influences vegetation and human activities and is itself affected by topography. Vegetation is influenced by climate and parent material. The combined influence of the five factors produces a set of soil-forming processes, which results in the world's distinctive soil profiles. Not all soils develop the same amount or combination of horizons and therefore specific combinations of horizons are used to classify soils. Box 10.1 provides details on soil classification schemes.

Reflective questions

➤ What are the similarities and differences between the soil-forming processes of leaching and clay eluviation?

➤ Can you explain why soils are considered as open systems and how this influences soil formation? (Drawing a diagram may help with your answer.)

➤ Can you describe the process of podzolization? What materials are removed? Why does this happen? Where are they being redeposited?

➤ Why are soils in tropical areas often several metres deep while those in polar regions are shallow and poorly developed?

➤ How does soil development vary down a slope of relatively uniform parent material?

SOIL CLASSIFICATION

Not all soils develop the same amount or combination of horizons and therefore specific combinations of horizons are used to classify soils. There are a number of different soil classification systems used throughout the world, many of which are summarized by FitzPatrick (1983) and Gerrard (2000). The two most commonly used are the soil taxonomy (classification) of the United States Department of Agriculture (USDA) and the system used by the Food and Agriculture Organization – United Nations Educational, Scientific, and Cultural Organization (FAO-UNESCO).

The USDA soil taxonomy scheme is a hierarchical classification with soils divided into: (1) order, (2) suborder, (3) great group, (4) subgroup, (5) family and (6) series. There are 11 orders that are differentiated by the presence or absence of diagnostic horizons, features that show the dominant set of soil-forming processes that have taken place or chemical properties. This is essentially a subjective process as there are no fixed principles involved. A brief summary of the characteristics of the soil orders is presented in Table 10.3. Suborders are differentiated using criteria that vary from order to order. The number of subgroups ranges from two to seven per order. In the differentiation of the great groups, the whole assemblage of horizons is considered, together with a number of diagnostic features. Great groups are divided into subgroups by the addition of adjectives to the great groups' names. Further subdivision into families occurs on the basis of physical and chemical properties. The final level of subdivision, the series, is achieved on the basis of the locality in which that type of soil was first recognized. It has no real value in terms of soil classification but is used in soil mapping at more detailed scales.

The FAO-UNESCO (1974) scheme was designed for the production of the Soil Map of the World. It is now a very widely used scheme. It has 28 major soil groups that are subdivided into 153 units. The major

Table 10.3 Characteristics of the soil orders of the soil taxonomy scheme of soil classification

Order	Characteristics
Alfisols	Generally possess an argillic horizon and a moderate to high base saturation. They are common in humid temperate environments
Andisols	Form on parent material of volcanic origin, especially ash, and are often very fertile
Aridisols	Occur in both cold and hot dry areas where effective rainfall is low and often have a high salt content
Entisols	Recently formed soils with limited development of horizons
Histosols	Organic-rich soils
Inceptisols	Soils of humid regions with altered horizons that have lost material by leaching but still contain some weatherable minerals
Mollisols	Dark-coloured, base-rich soil of temperate grasslands
Oxisols	Red, yellow or grey soils of tropical and subtropical regions. They have strongly weathered horizons enriched in silica, clay and oxides of aluminium and iron and are acidic with a low nutrient status
Spodosols	Podzolic soils that have an organic surface horizon, a bleached grey to white eluvial horizon and a B horizon enriched in organic matter and oxides of iron and aluminium. These soils are coarse textured, highly leached and acidic, with a low nutrient status
Ultisols	Highly weathered, leached, acidic and have a low nutrient status. They usually have an argillic horizon and the B horizons of well-drained soils are red/yellow due to the accumulation of iron oxides. They have developed over a long period of time in humid, warm, temperate and tropical regions
Vertisols	Swelling clay soils with deep, wide cracks

BOX 10.1 ➤

➤

group names come from a number of linguistic routes and many have been used before in other classification schemes, while others have been newly devised. Approximate equivalents with the soil taxonomy orders are shown in Table 10.4. Many of the soil characteristics used to define the major soil groups are morphological, such as texture, structure and colour, while soil processes or chemistry define other groups. The division of groups into units is based on the presence or absence of diagnostic horizons and properties. One advantage of this scheme is that it is less hierarchical than many other schemes.

The Soil Survey of England and Wales classified soils according to broad differences in the composition or origin of the soil material and the presence or absence of specific diagnostic features. At the highest levels soils are divided into a small number of categories known as major soil groups, the distribution of which is shown in Figure 10.14.

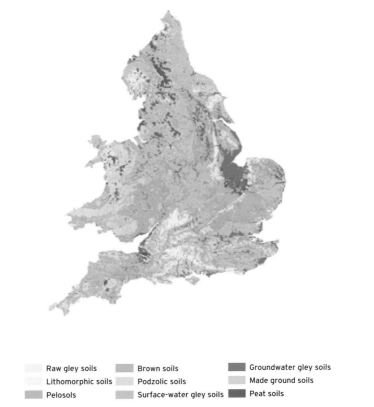

Raw gley soils	Brown soils	Groundwater gley soils
Lithomorphic soils	Podzolic soils	Made ground soils
Pelosols	Surface-water gley soils	Peat soils

Figure 10.14 The distribution of the major soil groups in England and Wales. (Source: Reproduced from National Soil Resources Institute, National Soil Map, 2004. © Cranfield University and for the Controller of HMSO 2004. No part of this publication may be reproduced without the express written permission of Cranfield University)

Table 10.4 Approximate relationship between the USDA soil taxonomy soil orders and FAO-UNESCO major soil groupings

USDA soil taxonomy soil orders	FAO-UNESCO major soil groups
Alfisols	Luvisols
Andisols	Andosols
Aridisols	Calcisols, Gypsisols, Solonchaks, Solonetz
Entisols	Arenosols, Fluvisols, Leptosols, Regosols
Histosols	Histosols
Inceptisols	Cambisols
Mollisols	Chernozems, Greyzems, Kastanozems, Phaeozems
Oxisols	Alsiosl, Ferralsols, Nitosols, Plinthosols
Spodosols	Podzols
Ultisols	Acrisols, Lixisols
Vertisols	Vertisols

BOX 10.1 ➤

➤

Table 10.5 provides a brief description of the main characteristics of these major soil groups. Within each major group, soils are progressively subdivided into soil group, soil subgroup and then into soil series (Avery, 1990). A slightly different system was developed for soils in Scotland (MISR, 1984). However, it is possible to amalgamate the two systems to show the relative occurrence of the major soil groups in the United Kingdom (Table 10.6).

Table 10.5 Description of the major soil groups in the UK

Major soil type	Characteristics
Raw gley soils	Soils that occur in mineral material that has remained waterlogged since deposition. They are chiefly confined to intertidal flats or saltings
Lithomorphic soils	Shallow soils usually well drained and formed directly over bedrock in which the only significant soil-forming process has been the formation of an organic or organic-enriched mineral surface
Pelosols	Slowly permeable non-alluvial clayey soils that crack deeply in dry seasons
Brown soils	Generally free-draining brownish or reddish soils overlying permeable materials
Podzolic soils	Soils with dark brown, black or ochreous subsurface layers resulting from the accumulation of iron, aluminium or organic matter leached from the upper layers. They normally develop as a result of acid weathering conditions
Surface water gley soils	Seasonally waterlogged, slowly permeable soils
Groundwater gley soils	Soils with prominently mottled or grey subsoils resulting from periodic waterlogging by a fluctuating groundwater table
Peat soils	Predominantly organic soils derived from partially decomposed plant material that accumulates under waterlogged conditions
Human-made soils	Predominantly organic soils derived from partially decomposed plant material that accumulates under waterlogged conditions

Table 10.6 The occurrence (%) of the major soil groups in the UK

Soil type	England and Wales	Scotland
Lithomorphic soil	7	10
Brown soils (including pelosols)	45	19
Podzols	5	24
Gley soils (including surface water and groundwater)	40	23
Peat soils	3	24

(Source: Avery, 1990; MISR, 1984)

BOX 10.1

10.5 Physical properties of soil

Soil physical properties are those properties of the soil that you can see, feel, taste and smell. By observing **soil colour**, we can estimate organic matter content, iron content, soil drainage and soil aeration. By feeling the soil we can estimate the kinds and amounts of different size particles present. Soil physical properties have a huge influence on how soils function in an ecosystem and how they can be managed.

10.5.1 Soil colour

Soil colour is easy to observe and although it has little effect on the soil, it is possible to use it to determine the nature of soil properties such as organic matter content, aeration and drainage characteristics. Colour also helps us to distinguish the different soil horizons of a soil profile. Soils with a higher amount of organic matter are black or dark brown in colour. Surface horizons are usually darker than subsequent horizons owing to their higher organic matter content (see Figures 10.7, 10.8 and 10.9). Soil colour can be used to determine the drainage characteristics of a soil because of the colour change that takes place when various iron-containing minerals undergo **oxidation** and **reduction**. In well-drained soils iron is oxidized and imparts a reddish or yellowish colour to the soil. In water-logged soils, the iron minerals are reduced, owing to the anaerobic conditions, and impart a grey or blue colour to the soil.

A standard system for soil colour description has been developed using the **Munsell colour chart**. In this system, three measurable variables determine colour. **Hue** is the dominant colour of the pure spectrum (usually redness or yellowness), **value** is the degree of darkness or light-ness of the colour (a value of 0 being black) and **chroma** is the purity or strength of the colour (a chroma of 0 being natural grey). By using a standard colour book, the observer can express soil colour as a letter numerical code as well as descriptively. For example, in the case of Black (10Y/R 2/1) the hue is 10Y/R, value 2 and chroma 1.

10.5.2 Soil texture

Mineral particles in the soil vary considerably in size from boulders (greater than 600 mm in diameter) and stones (greater than 2 mm in diameter) down to **sand**, **silt** and **clay**. The sand-, silt- and clay-sized particles are

Table 10.7 Soil particle size classification schemes

Size class	Size range (diameter, mm)		
	International or Atterberg System*	USDA-FAO System†	SSEW, BS and MIT System‡
Clay	<0.002	<0.002	<0.002
Silt	0.002–0.02	0.002–0.05	0.002–0.06
Sand	0.02–2	0.05–2	0.06–2
Gravel	>2	>2	>2

*This system subdivides sand into fine and coarse fractions.

†This system subdivides sand into very fine, fine, medium, coarse and very coarse fractions.

‡This system is adopted by the Soil Survey for England and Wales, British Standards and the Massachusetts Institute of Technology; it subdivides sand into fine, medium and coarse fractions and refers to gravel as stones.

often referred to as the fine fraction or **fine earth**, and are usually separated from the larger soil particles by passing through a sieve with 2 mm diameter holes. Within the fine earth fraction, size definitions vary between different systems (Table 10.7). All set the upper limit of clay as 2 μm (2 micrometres; two-millionths of a metre) but differ in the upper limit chosen for silt and the way in which the sand fraction is subdivided (see also Chapter 9, e.g. Figure 9.1).

Soil particles are classified into different size fractions because as particles become smaller they have differ-ent properties (see Table 10.8). In particular, as particles become smaller the total surface area of the soil particles in the soil becomes larger and this has a large influence on water holding capacity, **cation exchange capacity (CEC)** (see Section 10.6.1) and rate of mineral weathering.

It is extremely rare for soils to be composed of a single particle size class. Thus, soil texture refers to the relative proportions of the sand-, silt- and clay-sized fractions in a soil. The classification of texture in terms of particle size distribution is normally shown as a triangular diagram (Figure 10.15). Combinations of different proportions of sand, silt and clay result in 11 main textural classes although the number of classes may vary between different countries. Triangular diagrams can be used to determine a textural class if the particle size distribution is known, or to determine a range of particle size distributions if a textural class is known. The texture of a soil can be determined by measuring the particle size in the laboratory or in the

Table 10.8 Influence of soil separates on some properties and behaviour of soil

Property/behaviour	Sand	Silt	Clay
Water holding capacity	Low	Medium	High
Aeration	Good	Medium	Poor
Drainage rate	High	Slow to medium	Slow
Organic matter decomposition	Rapid	Medium	Slow
Compaction	Resists	Easily compacted	Easily compacted
Susceptibility to water erosion	Low	High	Low
Ability to hold nutrients	Poor	Medium to high	High
Leaching of pollutants	Allows	Moderately retards	Retards

(Source: Brady, Nyle C.; Weil, Ray R., *The nature and properties of soil*, 12th edition. © 1999. Adapted by permission of Pearson Education, Inc., Upper Saddle River, NJ)

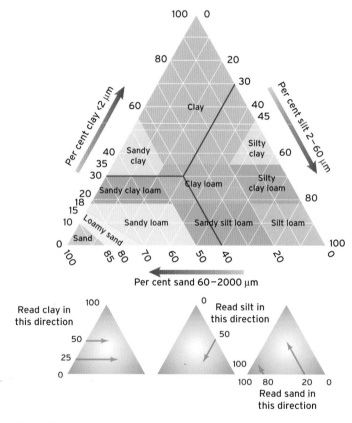

Figure 10.15 The triangular diagram of soil textural classes adopted in England and Wales. For example, a soil with 40% sand, 30% silt and 30% clay is a clay loam, as highlighted in the diagram.

field by working moist soil between finger and thumb as explained in Box 10.2.

Soil texture is an important property as it greatly influences the soil's ability to absorb and retain water. For example, coarse-textured soils have larger pore spaces because the sand-sized particles do not fit as closely together as the smaller silt- and clay-sized particles. As a result, sandy soils have high percolation rates but lower water retention capacities because water passes rapidly through the pores and little sticks to the soil particles. The opposite is observed for fine-textured soils.

TECHNIQUES

DETERMINING SOIL TEXTURE BY FEEL

The simplest and quickest way of assessing soil texture is by working a moist sample of soil in your hand. Begin by removing all stones of more than 2 mm and any large roots, moisten the soil and mould it in your hand for a few minutes, then follow the steps shown in Figure 10.16.

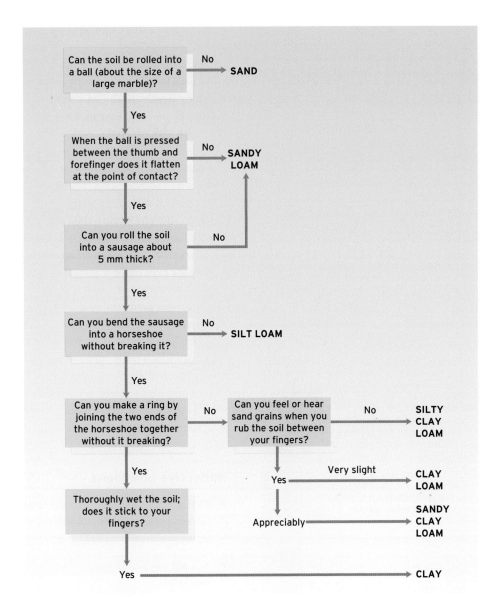

Figure 10.16 Flow chart for determining soil texture by feel.

BOX 10.2

10.5.3 Soil structure

Soil particles normally do not remain detached from one another. Instead, soil particles tend to adhere to each other, forming larger groupings called **aggregates** or **peds**. Soil structure is characterized in terms of the shape (or type), size and distinctness (or grade) of these peds. Each ped is separated from another by voids or natural surfaces of weakness. Soil structure is divided into four principal types: blocky, spheroidal, platy and prismatic as shown in Figure 10.17. The following terms are used to describe the distinctness of the structure: (i) structureless (no observed peds); (ii) weak (indistinct peds; when disturbed breaks into a lot of unaggregated material); (iii) moderate (well-formed peds; little unaggregated material when disturbed); and (iv) strong (distinct peds; remains aggregated when disturbed).

The formation of stable peds requires the action of physical, chemical and biological factors as follows: (i) Freeze–thaw, wet–dry and shrink–swell processes have a dramatic effect on soil structure throughout the whole profile, although the effects are the greatest in the top layer, where the exposure is direct. These processes help to mould the soil into peds, change the volume of the soil and move soil particles from one place to another throughout the soil profile. (ii) The activities of burrowing animals, such as earthworms, lead to the mixing of mineral and organic particles of soil and the formation of stable organo-mineral complexes. They can also produce macropores and channels which in turn influence infiltration rates and hence soil drainage. (iii) The secretions of soil animals can act as nuclei for ped formation while fine roots and microbes produce a range of polysaccharide substances which bind soil particles together and fungal hyphae literally hold mineral and organic particles together. Together, these factors combine to produce peds/aggregates in soil.

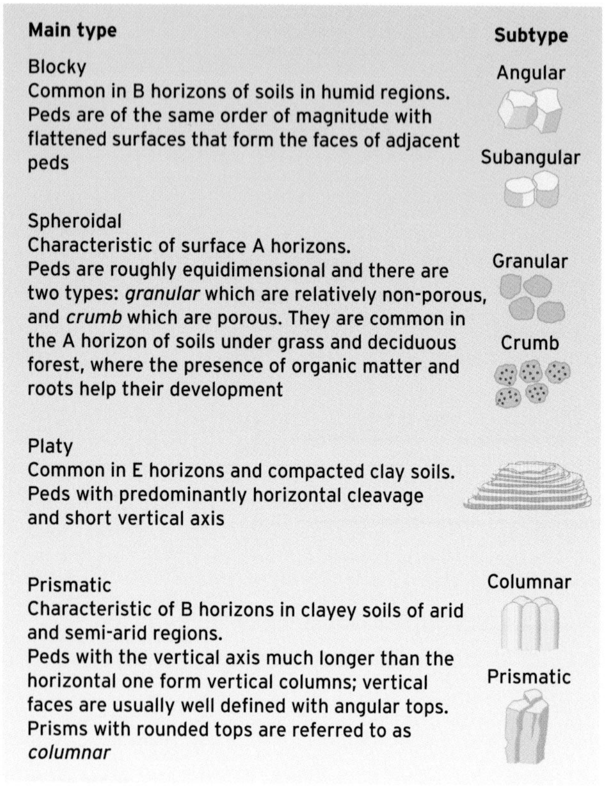

The clay-sized particles and organic compounds largely hold the peds together. As a result, coarse-textured soils tend to have weakly developed structures, whereas fine-textured soils generally have moderate to strong structures. The strength with which the individual peds are held together influences both the soil's resistance to erosion and ease of cultivation. A strong structure holds the soil together and causes it to resist erosion. The same characteristics, however, make it difficult to plough (White, 1997). Ploughing tends to alter and weaken the soil structure, and the passage of farm machinery leads to soil compaction.

Clearly the size, shape and arrangement of the peds determine the pore space or porosity of the soil. A soil with a well-developed structure is typically less compact and has a greater permeability and porosity than does a coarse-grained soil with a poor structure. The size and connectivity of the soil pores are important in determining the ease with which water, air and biota move through the soil. Good structural development is therefore necessary to obtain well-drained and well-aerated soils, to promote free movement of soil biota, allow roots to proliferate and enable aerobic microbial processes to dominate.

Main type	Subtype
Blocky Common in B horizons of soils in humid regions. Peds are of the same order of magnitude with flattened surfaces that form the faces of adjacent peds	Angular / Subangular
Spheroidal Characteristic of surface A horizons. Peds are roughly equidimensional and there are two types: *granular* which are relatively non-porous, and *crumb* which are porous. They are common in the A horizon of soils under grass and deciduous forest, where the presence of organic matter and roots help their development	Granular / Crumb
Platy Common in E horizons and compacted clay soils. Peds with predominantly horizontal cleavage and short vertical axis	
Prismatic Characteristic of B horizons in clayey soils of arid and semi-arid regions. Peds with the vertical axis much longer than the horizontal one form vertical columns; vertical faces are usually well defined with angular tops. Prisms with rounded tops are referred to as *columnar*	Columnar / Prismatic

Figure 10.17 Diagrammatic representation of the main types of soil structure.

Reflective questions

➤ What particles contribute to soil colour and how can soil colour provide valuable insight into the drainage status of a soil?

➤ Why are soil particles classified into different size fractions?

➤ How does soil texture influence soil structure?

➤ In which texture class does a soil with 10% sand, 60% silt and 30% clay fall? (It may help to look at Figure 10.15.)

10.6 Chemical properties of soil

Chemical properties give soils their ability to hold nutrients and create a desirable environment for plant growth. They are strongly influenced by parent material and organic matter content as they control the amount and type of **colloids** in a soil. **Soil colloids** are very small (less than 0.002 mm in diameter) particles that stay suspended in water. The most important soil colloids are clays and humus (organic matter).

10.6.1 Clay minerals and cation exchange

Clay minerals are formed from the weathering products of aluminium and silicate minerals. Box 10.3 explains the structure of clay minerals. Clay minerals are extremely small (<0.002 mm in diameter), and have a large surface area and a negative electrical charge so they are able to attract and hold water and cations. Therefore, they have a fundamental influence on both the physical and chemical properties of the soil. The electrical charge on clay colloids

FUNDAMENTAL PRINCIPLES

THE STRUCTURE OF CLAY MINERALS

The structure of all clay minerals is based on two types of sheets consisting of repeating units of (i) a silicon (Si) atom surrounded by four oxygen (O) atoms in the form of a tetrahedron as shown in Figure 10.18(a) and (ii) an aluminium (Al) or magnesium (Mg) atom surrounded by six oxygen (O) or hydroxy (OH) atoms in the shape of an octahedron (Figure 10.18b). The individual units are linked together by sharing oxygen atoms to form silicon tetrahedral sheets and aluminium octahedral sheets (Figure 10.19). All clay minerals are built from various combinations of these two sheets.

Alternating sheets of one tetrahedral sheet and one octahedral sheet

produce what are known as 1:1 clays. Kaolinite is the commonest 1:1 clay mineral (Figure 10.20). Each pair of sheets is held together by hydrogen ions, making it a relatively rigid and stable structure. In 2:1 clays, the aluminium octahedral sheet is sandwiched between two silicon tetrahedral sheets. There are many different types of 2:1 clays that are distinguished on the basis of how the unit layers are held together and the

spacing between the unit layers. In illite, the unit layers are held together by potassium (K$^+$) ions, which make it a relatively stable clay (Figure 10.20). In contrast, weak oxygen bonds hold the unit layers in **montmorillonite** together. As a result, water molecules can penetrate between the layers enabling it to expand and contract on wetting and drying. The structures of some of the different clay minerals are shown in Figure 10.20.

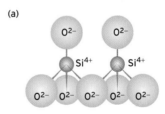

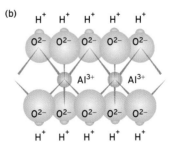

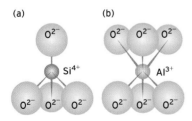

Figure 10.18 The structure of (a) a silicon tetrahedron and (b) an aluminium octahedron.

Figure 10.19 The structure of (a) a silicon tetrahedral sheet and (b) an aluminium octahedral sheet.

BOX 10.3 ➤

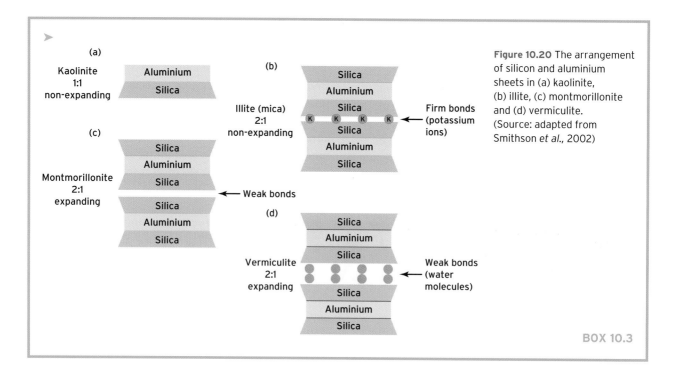

Figure 10.20 The arrangement of silicon and aluminium sheets in (a) kaolinite, (b) illite, (c) montmorillonite and (d) vermiculite. (Source: adapted from Smithson *et al.*, 2002)

BOX 10.3

results from **isomorphous substitution** which occurs during the formation of clay minerals. It is when one atom in the crystal lattice is replaced by another atom of similar size without disrupting or changing the crystal structure of the mineral. For example, Al^{3+} may replace Si^{4+} in the tetrahedral sheet (Figure 10.19a) and Mg^{2+} or Fe^{3+} or Fe^{2+} or Ca^{2+} can replace Al^{3+} in the octahedral sheet (Figure 10.19b). The clay mineral structure is electrically neutral, but as the replacing ion generally has a lower positive charge than the ion it replaces, the clay mineral becomes electrically charged. For example, the silicon ion has a positive charge of 4 (i.e. Si^{4+}). If silicon is replaced by aluminium (Al^{3+}), which has a charge of 3, then the clay has lost one positive charge, which leaves one unsatisfied negative charge on the clay.

The net negative charge on clay minerals is balanced by cations (e.g. Ca^{2+}, Mg^{2+}, K^+, Na^+, Al^{3+} and H^+), which are attracted to the surface of the clay minerals and held (**adsorbed**) there by electrostatic attractions. They are referred to as **exchangeable cations** because cations in the soil solution can displace adsorbed cations on the clay surface. Interchange between a cation in solution and another on the surface of a colloid is known as **cation exchange**. All cation exchange reactions must be chemically balanced. For example, if a clay containing Na^+ is washed with a solution of $CaCl_2$, each Ca^{2+} ion will replace two Na^+ ions as Ca^{2+} has double the charge of Na^+, and Na^+ will be washed out in the solution, as shown in Figure 10.21.

Cation exchange reactions are also rapid and reversible. The distribution of the cations between soil and solution depends on their relative concentrations and the force of attraction to the negatively charged surface of the clay mineral. Cations adsorbed on the exchange sites are in equilibrium with cations in the soil solution. For example, if Ca and Mg are the dominant cations in the soil solution they will also dominate the exchange sites. In general, the strength of adsorption increases as the charge of the cation increases and the size of the hydrated cation decreases. The sequence of preferred adsorption is: $Al^{3+} > Ca^{2+} > Mg^{2+} > K^+ > Na^+$ (Cresser *et al.*, 1993).

Cation exchange capacity, commonly abbreviated to CEC, is the ability or capacity of a given quantity of soil to hold cations. This capacity is directly dependent on the overall net negative charge of the colloids present in the soil. It is usually expressed as milliequivalents (meq) per kg of oven-dried soil. The main factors controlling the CEC of a soil are the number of colloids present (soil texture), the type of colloids present and organic matter content. The CEC of a soil is a very important property as it controls both soil fertility and soil acidity. Soils with a high CEC usually have a high capacity to store nutrients and are therefore potentially more fertile than soils with a low CEC.

The soil cations that are readily adsorbed onto soil colloids can be divided into two groups. Firstly there are the base cations, which include the important plant nutrients Ca^{2+}, Mg^{2+}, K^+ and Na^+. Secondly there are acid cations, which include Al^{3+} and H^+. Related to this distinction in cations is the term **base saturation**, which is defined as the

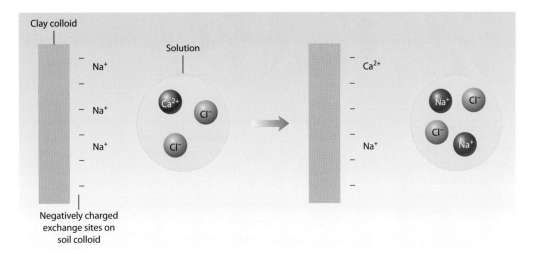

Figure 10.21 Schematic diagram representing cation exchange between calcium ions in solution and sodium ions held on the surface of a negatively charged colloid.

proportion of exchange sites occupied by base cations and is calculated as follows:

Base saturation (%)

$$= \frac{(Ca^{2+} + Mg^{2+} + K^+ + Na^+)}{Ca^{2+} + Mg^{2+} + K^+ + Al^{3+} + H^+} \times 100 \quad (10.1)$$

A soil with a high base saturation (greater than 35%) is more fertile than a soil with a low base saturation.

10.6.2 Soil acidity

The degree of acidity or alkalinity of a soil is an important variable as it affects most soil physical, chemical and biological processes. Whether a soil is acidic, neutral or alkaline is determined by measuring the hydrogen ion concentration

in the soil solution. In pure water at 24°C, water ionizes to give equal concentrations of hydrogen (H^+) and hydroxide (OH^-) ions:

$$H_2O \Leftrightarrow H^+ + OH^- \quad (10.2)$$

The concentration of both H^+ and OH^- ions is 1×10^{-7} (0 .000 000 1) **moles** per litre. As it is inconvenient to use these very small numbers to express the concentrations of H^+ ions, a simpler method of using the negative logarithm of the hydrogen concentration was developed, known as pH (see Chapter 13).

As the pH scale is logarithmic, a change of one unit represents a 10-fold change in hydrogen concentration (Table 10.9). For example, a 10-fold increase in H^+ ion concentration from 1×10^{-5} to 1×10^{-4} moles per litre

Table 10.9 The pH scale

pH	H⁺ concentration (moles per litre)	OH⁻ concentration (moles per litre)	Description
3	0.001	0.000 000 000 01	Excessively acidic
4	0.000 1	0.000 000 000 1	Strongly acidic
5	0.000 01	0.000 000 001	Moderately acidic
6	0.000 001	0.000 000 01	Slightly acidic
7	0.000 000 1	0.000 000 1	Neutral
8	0.000 000 01	0.000 001	Alkaline
9	0.000 000 001	0.000 01	Strongly alkaline
0	0.000 000 000 1	0.000 1	Excessively alkaline

FUNDAMENTAL PRINCIPLES

HOW ALUMINIUM INFLUENCES SOIL ACIDITY

As clay minerals weather and break down, the aluminium in the octahedral layer is released into the soil solution, where it either reacts with water or is adsorbed onto the exchange sites of negatively charged clay minerals. Al^{3+} ions are adsorbed in preference to all the other major cations. The influence that aluminium has on soil acidity is itself dependent on the acidity of the soil. At pH less than 5, aluminium is soluble and exists as Al^{3+}. When Al^{3+} enters the soil solution it reacts with water (it is hydrolyzed) to produce H^+ ions:

$$Al^{3+} + H_2O \Leftrightarrow AlOH^{2+} + H^+ \quad (10.3)$$

Thus the acidity of the soil increases (pH falls). In soils with a pH of between 5 and 6.5, aluminium also contributes H^+ ions to the soil solution but by different mechanisms, as aluminium can no longer exist as Al^{3+} ions but is converted to aluminium hydroxy ions:

$$Al^{3+} + OH^- \Leftrightarrow AlOH^{2+} \quad (10.4)$$

$$AlOH^{2+} + OH^- \Leftrightarrow Al(OH)_2^+ \quad (10.5)$$
aluminium hydroxy ions

These hydroxy aluminium ions act as exchangeable cations, just like Al^{3+}, and are adsorbed by the clay minerals. They are in equilibrium with hydroxy aluminium ions in the soil solution, where they produce H^+ ions by the following reactions:

$$AlOH^{2+} + H_2O \Leftrightarrow Al(OH)_2^+ + H^+ \quad (10.6)$$
$$Al(OH)_2^+ + H_2O \Leftrightarrow Al(OH)_3 + H^+ \quad (10.7)$$

In soils where the pH is above 7, Ca^{2+} and Mg^{2+} dominate the exchange sites and most of the hydroxy aluminium ions have been converted to gibbsite ($(AlOH)_3$), which is insoluble and cannot be adsorbed by the negative clay minerals as it has no charge. The general relationship between soil pH and the composition of cations held on the exchange sites of clay minerals is presented in Figure 10.22. In a neutral soil the exchangeable cations that dominate the cation exchange sites are the base cations, whereas in an acidic soil aluminium and hydrogen ions dominate the exchange sites.

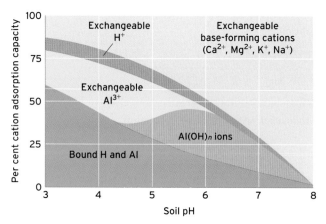

Figure 10.22 Relationship between soil pH and the cations held on the exchange sites of colloids, including clay minerals. (Source: Brady, Nyle C. and Weil, Ray R., *The nature and properties of soil*, 13th edition. © 2002. Adapted (electronically reproduced in case of e-use) by permission of Pearson Education, Inc., Upper Saddle River, NJ)

BOX 10.4

---

is represented by a one-unit decrease in pH from 5 to 4. Table 10.9 also shows the inverse relationship between the concentrations of H^+ and OH^- ions. As one increases, the other must decrease proportionally as the product of the H^+ and OH^- concentrations must always equal 1×10^{-14}.

Although the pH scale ranges from 1 to 14, most soils have a pH of between 3.5 and 9. Very low values are often associated with soils rich in organic matter, whereas high values usually result from the presence of sodium carbonate. Although the reason a soil becomes acidic is because of excess H^+ ions in the soil solution, it is the presence of

aluminium that is largely responsible for producing these H^+ ions as discussed in Box 10.4.

There are a number of natural processes and human-induced changes that result in hydrogen and aluminium becoming the predominant cations in the soil and thus increase its acidity:

- Leaching–percolating water removes base cations (Ca^{2+}, Mg^{2+}, K^+ and Na^+) faster than their rate of release from weathering and therefore cation exchange sites become dominated by aluminium and hydrogen.

- The respiration of roots and microbes, and the decomposition of organic matter, release hydrogen ions.
- The addition of acids such as H_2SO_4 and HNO_3 from the atmosphere in acid rain also increases hydrogen ions in the soil.
- Use of acid-forming fertilizers. For example, the application of ammonium-based fertilizers results in nitrate and H^+ release.
- Harvesting of the crop removes the base cations from the soil as they are not returned to the soil as litter.

Soil pH affects which plants grow well as they vary considerably in their tolerance to soil pH. Soil pH also determines the fate of many pollutants, affecting their breakdown, solubility and possible movement from the soil to surface waters and groundwaters. For example, many heavy metals become more water soluble under acid conditions and can move down with water through the soil to aquifers and surface waters (see Chapter 13).

There are 16 **essential elements** without which green plants cannot grow normally. The availability of these essential nutrients for plant uptake is greatly influenced by soil pH (Figure 10.23) as are the number, species and activities of soil organisms. All 16 essential elements must be present in the correct proportions as too little or too much of any element will result in symptoms of nutrient deficiency or toxicity. On the basis of their concentration in plants they are divided into **macronutrients** (carbon, oxygen, hydrogen, nitrogen, phosphorus, sulphur, calcium, magnesium, potassium and chloride) and **micronutrients** (iron, manganese, zinc, copper, boron and molybdenum).

Strongly acidic soils (pH 4–5) usually have a low or reduced supply of the macronutrients, particularly calcium, magnesium, potassium, nitrogen, phosphorus and sulphur, and high, often toxic concentrations of the micronutrients, especially iron, manganese and zinc. While most nutrients are more soluble in acid soils than neutral or slightly basic soils, phosphorus and molybdenum become insoluble at low pH and unavailable to plants. Soil pH also influences plant growth by the influence of pH on activity of beneficial microorganisms and root activity. In acidic soils the number and activity of many soil organisms are reduced, including worms, bacteria that convert ammonium to nitrate, organisms that break down organic matter and nitrogen-fixing bacteria.

In contrast, high soil pH leads to phosphorus and boron becoming insoluble and unavailable to plants and concentrations of the micronutrients, particularly iron, manganese, zinc and copper, becoming so low that plant growth is

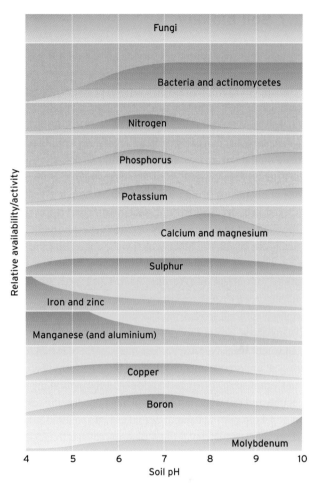

Figure 10.23 The influence of soil pH on the availability of plant nutrients and activity of soil organisms.

restricted. Therefore a pH range of 6 to 7 is generally most favourable for plant growth as most plant nutrients are readily available in this range. However, some plants have adapted to a pH above or below this range.

Reflective questions

➤ Why are cations attracted to clay minerals?

➤ What is the difference between a 1 : 1 clay mineral and a 2 : 1 clay mineral?

➤ What is the meaning of each of the following terms: isomorphous substitution, cation exchange capacity, base saturation?

➤ Aluminium is considered an acid component of the soil. Why should this be true?

➤ Why is the level of soil pH so important for plant growth?

10.7 Soil biology

Although soils may appear to be made up of inert soil particles, in fact, they are home to many different types of soil organism. For example, a single teaspoon of soil may contain thousands of species, millions of individuals and hundreds of metres of fungi (European Commission, 2010). Without the activities of soil organisms, dead vegetation would accumulate on the soil surface, and there would be no recycling of nutrients. Thus a healthy, fertile soil depends on a vibrant range of organisms living within it; from bacteria and fungi to tiny insects, earthworms and small mammals. However, despite the fact that the majority of Earth's species might actually live in soil, soil biota, which is the collective term for all organisms living within the soil, have until recently received little attention (Wardle, 2002). This is because the majority of soil organisms are very, very small, and they are extremely difficult to study and lack the sentimental appeal of many above-ground species (Bardgett, 2005).

In recent years, we have become more aware of the wide variety of important functions that soil biota perform (European Commission, 2010). In addition to decomposing organic materials and releasing nutrients, soil biota regulate the carbon flux, keep pests at bay and decontaminate polluted land, and provide raw material for new pharmaceuticals to tackle infectious diseases. For example, in 1928 Alexander Fleming noticed that a soil fungus growing in his laboratory was inhibiting the growth of nearby cultures of staphylococcus bacteria. From this he deduced that the fungus must be killing the infectious disease by some means, and shortly afterwards isolated penicillin from the fungus, which is used in antibiotics throughout the world today to treat many serious diseases.

10.7.1 The soil biota

The organisms that live in soil can be grouped according to size into: microfauna (body width < 0.1 mm), which includes the microbes, bacteria and fungi as well as algae, nematodes and protozoa; mesofauna (body width 0.1–0.2 mm), which includes microanthropods and enchytraeids; and macrofauna (body width > 2 mm), which includes earthworms, millipedes and termites (Figure 10.24). The primary role of the soil biota is to decompose organic matter, like leaf and plant litter, and in the process release nutrients such as nitrogen and phosphorus. These fauna are often allocated to functional

(a)

(b)

(c)

Figure 10.24 Examples of soil (a) microfauna – protozoa, (b) mesofauna – enchytraeids and (c) macrofauna – termites. (Source: (a) Science Photo Library: Sinclair Stammers, (b) Ardea: Steve Hopkin, (c) FLPA Images of Nature: Thomas Marent)

groups based on their feeding habit. Some feed primarily on microbes (microbial feeders) or litter (**detritivores**), whereas others feed mainly on plant roots (**herbivores**) or other animals (**carnivores**).

Collectively the mass of organisms in a given mass of soil is referred to as the **soil biomass**. The mass of macro- and mesofauna can be measured directly as they can be physically separated from the soil and is usually expressed as kg per hectare (to a certain depth). The soil biomass can range from 2 to 5 t ha^{-1} (White, 1997), with earthworms making up the largest single contribution (~75%; Wild, 1993). This is more than the weight of the cattle grazing upon the soil in a temperate grassland. In contrast, microfauna are intimately mixed with soil particles and organic matter and, being very small, are difficult to isolate for weighing or counting. For microfauna, collectively referred to as the soil microbial biomass, the methods used to measure numbers and/or mass include either direct observation of organisms on agar plates, and physiological or biochemical methods such as extraction of adenosine triphosphate (ATP), substrate induced respiration (SIP) and chloroform fumigation–incubation. These methods give no information about the species of biomass, but give an estimate of biomass size which is valuable in modelling carbon turnover in soil (White, 1997). The most common method used to measure soil microbial biomass is chloroform fumigation followed by incubation or extraction (Jenkinson and Powlson, 1976).

In addition to being grouped by sizes, the soil organisms can be classified into three main groups based on the principal function they perform in the soil: primary, secondary and higher consumers. Together all these soil organisms interact with each other and their surroundings in a complex system to perform a variety of functions that are important in maintaining fully functioning soils that can support human life.

10.7.1.1 The primary consumers

The primary consumers include all the organisms that decompose organic matter. They are able to break down and mineralize complex organic substances, releasing carbon dioxide and nutrients (e.g. NH_4, PO_4 and SO_4) as well as producing new compounds that are less susceptible to decomposition and accumulate as **soil organic matter** in the soil.

Microbes (bacteria and fungi) are the most abundant and diverse members of this group, which also includes algae and viruses. Bacteria are single-celled organisms, are very abundant, with populations ranging from 100 million to 3 billion in a gram of soil, can reproduce rapidly, doubling their population in minutes under favourable conditions, and live in the water-filled pores in soil. However, they are very sensitive to environmental conditions for their survival and growth, including moisture, temperature and porosity. When conditions are not conducive for growth they can enter a dormant phase and come back to life after a period of weeks, months or years in response to changing environmental conditions.

Certain groups of bacteria perform specific functions, particularly in relation to the nitrogen cycle. For example, nitrifying bacteria are responsible for oxidizing ammonium (NH_4), which is produced from the decomposition of proteins, into nitrate (NO_3) which is taken up by plants and once again converted to proteins. This process is known as **nitrification**. Other bacteria are able to convert nitrogen from the atmosphere into nitrogen-containing organic substances in the process of **nitrogen fixation** while different bacteria return equal amounts of nitrogen to the atmosphere through a series of processes called **denitrification**. Chapter 25 discusses the nitrogen cycle (see Box 25.1).

Fungi comprise an eclectic group that varies from single-celled yeasts to complex structures visible to the human eye. A gram of soil can contain around a million fungi, such as yeast and moulds. Fungi are chemo-heterotrophic, which means they require a chemical source of energy rather than being able to use light as an energy source, as well as organic substrates to obtain carbon for growth. Many fungi are parasitic, often causing disease to their living host, but others live on dead or decaying organic matter, thus breaking it down. Fungi that are able to live symbiotically with living plants, creating a relationship that is beneficial to both, are known as mycorrhizae. **Mycorrhizal** fungi obtain the carbohydrates they require from the plant roots, in return providing the plant with nutrients and moisture. In addition to decomposing organic material, fungi bind soil particles together, thereby enhancing soil structure, provide a food source to microbial feeding fauna and act as plant pathogens.

10.7.1.2 Secondary consumers

This group of soil organisms, which is dominated by protozoa, nematodes and microarthropods, exists alongside the microbes, feeding on them and on each other, and on soil organic matter. Protozoa (Figure 10.24a) are single-celled organisms which require a water film around soil particles for both feeding and moving. Hence their activity is restricted to the water-filled pores in soils, but they can withstand drying of the soil by rapidly forming resistant structures called cysts. Nematodes are tiny worm-like creatures which also require water to move, feed and reproduce in.

Nematodes can be grouped according to the type of food they consume: plants, bacteria, fungi and other animals. Macroarthropods are small invertebrates (animals without a backbone) which feed on a combination of bacteria, fungi, other macroarthropods and decaying organic matter.

10.7.1.3 Higher-level consumers

This group of soil organisms includes a wide range of species, such as earthworms, ants, termites (Figure 10.24c), woodlice, millipedes, centipedes, beetles and spiders. Some of these organisms, such as millipedes and woodlice, consume organic matter, whereas others, such as centipedes, spiders and beetles, are the major predators in soil. Earthworms, termites and ants help mix, move and aerate the soil, influencing its structure and porosity. Moles and part-time soil residents, such as voles, snakes, lizards, mice, rabbits, lemmings and badgers, also play a role in maintaining soil biodiversity as they mix plant litter and roots into the soil when burrowing as well as creating airways and macropores through which water can pass.

10.7.2 Factors influencing soil biodiversity

The activity and diversity of soil organisms are controlled by a combination of biotic and abiotic factors. The main abiotic factors are:

- climate, which controls temperature and moisture regimes within soil;
- soil physical properties such as texture and structure;
- soil chemical properties such as pH and salinity.

The growth and activity of soil organisms increase at higher temperatures and moistures, although soil organisms vary in their optimal ranges of temperature and moisture. As climate varies both spatially across the globe and temporally at the same place, due to diurnal and seasonal variations, the climatic conditions to which soil organisms are exposed vary greatly. Soil texture influences the activity and diversity of soil organisms as it determines the ability of the soil to retain water (see Figure 10.4) and nutrients. Soil structure is important because it influences the porosity of soil, which controls both the distribution of water in soil and the extent to which biota are able to enter and occupy pore spaces, which is controlled by pore neck diameter and the size of the organisms. Soil pH influences soil biota as it controls nutrient availability (see Figure 10.23); acid soils are high in soluble aluminum (see Box 10.4) which can be toxic to many soil organisms as well as plants. Most microbes grow within the pH range 4–9 and many soil organisms

are sensitive to acidic conditions. Soil salinity can also cause severe stress to soil organisms, leading to their rapid desiccation. However, the sensitivity towards soil pH and salinity differs between species. For example, earthworms occur in low numbers in acidic soils and their abundance increases as soil pH increases towards pH 7 (Edwards and Boheln, 1996), whereas acid soils tend to be dominated by enchytraeid worms (Figure 9.24b) (Cole *et al.*, 2002) which replace earthworms as the dominant soil animal.

Soil organisms influence plants and organisms that live entirely above ground, and these influences take place in two directions. Plants can exert a strong influence on the activity and composition of soil organisms, especially close to their roots, called the **rhizosphere**. In turn, plant growth may be limited or promoted by these soil organisms. For more information on the linkages between plant and soil biological communities see Bardgett (2005) and Wardle (2002).

Overall, soil organisms are very sensitive to environmental conditions, such as moisture and supply of organic matter, which may be altered as a result of human activities. Thus an increase in soil degradation, as discussed in Section 10.8, is likely to have an impact on soil biodiversity. Therefore, it is important that we seek to protect soil biodiversity.

Reflective questions

➤ What roles do fungi carry out in the soil?

➤ Can you name some of the processes that bacteria carry out that are important in the nitrogen cycle?

➤ What are the main functions that soil biota perform?

➤ What are the main factors that control the activity and diversity of soil organisms?

10.8 Impact of human activities on soils and soil processes

The soil functions described in this chapter are at risk from human activities. Pressures include agriculture, drainage, extraction, application of wastes and urban development. These pressures can lead to soil degradation such as soil erosion, contamination (by heavy metals, organic contaminants such as pesticides, radionuclides (from nuclear waste) and excessive use of nitrogen and phosphorus fertilizers),

Table 10.10 Estimated areas affected by major soil threats in Europe

Threat	Area affected (million hectares)	Percentage of total European land area
Water erosion	115	12
Wind erosion	42	4
Acidification	85	9
Pesticides	180	19
Over-fertilization	170	18
Soil compaction	33	4
Organic matter loss	3.2	0.3
Salinization	3.8	0.4

(Source: Oldeman *et al.*, 1991)

Figure 10.25 Rill and gully development across a sparsely vegetated slope in Spain. (Source: Shutterstock.com: Neil Bradfield)

acidification, soil compaction, loss of organic matter, salinization and loss of biodiversity. All of these threats lower the current and/or future capacity of the soil to support human life. At a global level, the total area of soil that has been degraded by human activities (almost 20 million km^2) exceeds the total area of farmland (15 million km^2) and the main causes are deforestation, overgrazing and poor agricultural management (Oldeman *et al.*, 1991). In Europe, an estimated 633 million hectares (6.33 million km^2) are affected by some kind of degradation process (Table 10.10). In this section emphasis is placed on the most severe soil degradation problems. For each of these the causes, magnitude, impact on soil function and remedies will be briefly discussed.

10.8.1 Soil erosion

Soil erosion is a two-phase process consisting of the detachment of individual particles from the soil mass and their transport (see Chapter 8). It is a natural process but is accelerated by human activities that expose the soil during times of erosive rainfall or windstorms, or that increase the amount and speed of **overland flow**. Farming practices such as overgrazing, removal of vegetation and/or hedgerows, ploughing up and down slopes, abandonment of terraces, compaction by heavy machinery and poor crop management may have these effects.

The GLASOD study estimated that 15% of the Earth's ice-free land surface is afflicted by some form of land degradation (GLASOD, 1990). Of this soil erosion by water is responsible for 56% and wind erosion is responsible for about 28%. During the past 40 years, nearly one-third of the world's arable land has been lost by erosion and continues to be lost at a rate of more than 10 million hectares per year (Pimentel *et al.*, 1995). Soil erosion's most serious impact is its threat to the long-term sustainability of agricultural productivity, which results from the 'on-site' damage that it causes. Erosion by water can quickly remove large volumes of soil as shown in Figure 10.25, which bury or destroy crops in localized areas, and leave channels, **rills** and gullies that in the worst case can inhibit agricultural machinery cultivating the land.

In the long term, soil erosion results in a reduction of soil depth, with fertile topsoil being lost at the rate of several millimetres per year (Morgan, 1986). In the United States, an estimated 3.6×10^9 tonnes of soil and 118×10^9 tonnes of water are lost from the 160 million hectares of cropland each year (Pimentel *et al.*, 1995). In more than one-third of the total land of the Mediterranean Basin, average yearly losses exceed 15 tonnes per hectare (UNEP, 2000) which represents a reduction in productivity of around 8% (Pimentel *et al.*, 1995). In the Russian Federation, it is estimated that the humus content of agricultural soils decreases by about 1% each year (Karavayeva *et al.*, 1991) and that long-term productivity is endangered since the annual erosion rates exceed the rate of humus production. In India, 113 and 38 million hectares are subject to water and wind erosion, respectively, and between 5.37 and 8.4 million tonnes of plant nutrients are lost every year due to soil erosion (Gerrard, 2000).

In addition to 'on-site' effects, the soil that is detached by accelerated water or wind erosion may be transported considerable distances. This gives rise to 'off-site problems', including sediment deposition on roads and in watercourses and reservoirs (see Chapter 9). Another major off-site impact results from agricultural chemicals (fertilizers, pesticides, heavy metals) that often move with eroded sediment. These chemicals can pollute downstream watercourses.

A few studies have attempted to estimate the economic impact of soil erosion. For example, Pimentel *et al.* (1995) estimated that soil erosion in the United States translates into an on-site economic loss of more than $27 billion each year, of which $20 billion is for replacement of nutrients and $7 billion for lost water and soil depth. They also state that the total on- and off-site costs of damage by wind and water erosion and the cost of erosion prevention each year are $44 billion. In 1991, the direct cost impact of soil erosion in Spain was estimated at $212 million per year, including the loss of agricultural production, impairment of water reservoirs and damage due to flooding (ICONA, 1991). In addition, the cost of attempts to fight erosion and restore the soil were estimated at about $3 billion over a period of 15–20 years (ICONA, 1991).

Measures to control erosion include the retention or planting or strips of permanent vegetation (trees or hedges) to form shelter belts to reduce the effects of wind (see Chapter 6), and tillage techniques such as contour ploughing, strip and alley cropping, and use of cover crops to reduce the rate at which water is able to move across the soil surface. Rotation farming, adjusting stocking levels and agro-forestry practices can also be used to help reduce soil erosion. Although severe soil erosion is nearly always irreversible, in less severe cases damage can still be prevented.

10.8.2 Soil acidification

Soil acidification is a natural process that occurs over the long term, but it has recently been enhanced by human action though the emission of sulphur and nitrogen compounds from the combustion of fossil fuels, resulting in acid rain. Humans also influence the acidity of soils by agricultural practices such as the harvesting of crops, the draining of waterlogged soils and the overuse of nitrogen fertilizers. However, in central and western Europe and North America acidic deposition is by far the most important cause of soil acidification.

Soil acidification is not a visually obvious feature although it has a large impact upon the ecosystem. The most sensitive soils to acidification are those derived from base-poor igneous (e.g. granite) and metamorphic rocks. Soils developed on these rocks tend to have low base cation and clay contents. Soils containing carbonates or with higher base cation and clay contents have a greater capacity to buffer acidification. This differential ability of soils to cope with acidification has been examined through use of a 'critical loads' approach as discussed in Box 10.5.

Soil acidification has the effect of increasing the leaching of base cations, such as calcium and magnesium, and depleting the soil's buffering capacity. It also increases the solubility of heavy metals in the soil, such as aluminium, manganese, lead, cadmium and zinc, which can be toxic to plants. This may lead to decreased plant growth or changes in plant communities. For example, forest decline in central Europe is linked with increasing acidity in soils. Populations of soil organisms may also change, with a shift towards more acid-tolerant species. As a result, a number of soil processes can slow down. For example, the decomposition of litter becomes slower, leading to surface accumulation. Soil acidification gradually leads to acidification of waters draining from them (see Chapter 13). Acidity and high concentrations of aluminium can lead to deterioration of aquatic life with losses in the diversity and size of invertebrate and fish populations.

Soil acidification can be slowed down by a reduction in acid deposition. Since the 1980s considerable national and international effort has been made to decrease emissions of acidifying pollutants. This is a difficult problem because those countries suffering most from acid rain are not always the main polluters owing to the movement of the pollution in the atmosphere across the planet. In 1979, 34 European and North American countries adopted the Convention on Long-Range Transboundary Air Pollution, which bound them to reducing emissions. The convention agreed firm targets in 1983, when 21 European countries agreed to reduce their sulphur dioxide (SO_2) emissions by 30% from the 1980 levels by 1993. Twelve countries reached this target by 1988. The United Kingdom, however, did not ratify this agreement, but did eventually declare an intention to reduce SO_2 emissions by 30% in the late 1990s. In the United States, an 'Acid Rain Program' commenced in 1995 that aimed to achieve environmental and public health benefits through reductions in emissions of SO_2 and oxides of nitrogen (NO_x). Since the signing of these agreements there has been a substantial reduction in SO_2 emissions in western Europe, although emissions of NO_x continued to increase until the late 1980s before starting to decline.

Between 1990 and 2006, there has been a decrease in UK SO_2 and NO_x emissions by 82 and 46% respectively, which will result in a decline in the deposition of sulphur and

HAZARDS

CRITICAL LOADS

Some soils are less able to cope with acidification than others. To help quantify the effects of soil acidification and relate them to the acid deposited, an 'effects-based' approach, known as critical loads, has been developed. A critical load is defined as 'a quantitative estimate of exposure to one or more pollutants below which significant harmful effects on elements of the environment do not occur according to present knowledge' (Nilsson and Grennfelt, 1988). Deposition above that limit may lead to harmful effects on the environment. Maps of critical loads and their exceedances (excess over the critical load) have been used to show the potential for harmful effects to systems at steady state as an aid to developing strategies for reducing pollution. For soil acidification the critical loads are based on the rate of release of base cations from soil minerals by weathering, indicating the capacity of the receiving soil to buffer acid inputs. Figure 10.26 shows critical loads of acidity for soils in the United Kingdom. Critical loads have also been developed for soil-plant systems

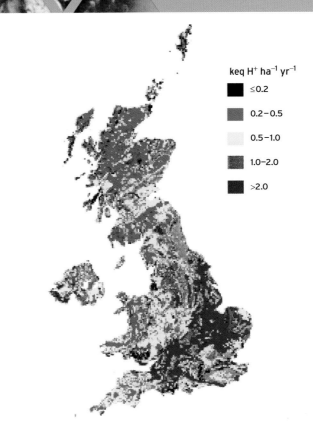

keq H^+ ha^{-1} yr^{-1}

■ ≤0.2

■ 0.2–0.5

□ 0.5–1.0

▨ 1.0–2.0

■ >2.0

Figure 10.26 Empirical critical loads of acidity for soils. (Source: CEH, NSRI, Macaulay Institute and DardNI)

on the basis of biological indicators that reflect the health of the whole system. For example, the critical molar ratio of calcium (or calcium plus magnesium) to aluminium in soil solution is a commonly used criterion to protect the fine roots of trees and may be used in the calculation of critical loads for forest soils.

BOX 10.5

nitrogen to soils. Despite this decline, modelling and experimental data predict that the recovery of some soils will take decades (NEGTAP, 2001). However, there is some recent evidence from soil monitoring that soil pH is increasing in soils across England (NSRI, 2004; Emmett *et al.*, 2010).

10.8.3 Soil pollution

A wide range of substances including heavy metals, pesticides and fertilizers can pollute soils. A distinction is often made between soil pollution originating from clearly

defined sources (local or point source pollution) and that from undefined sources (diffuse or non-point sources). The introduction of pollutants to the soil can result in damage to or loss of soil functions and may result in water contamination.

10.8.3.1 Heavy metals

Although heavy metals such as cadmium, copper, chromium, lead, zinc, mercury and arsenic are present naturally in soil, they can originate from a number of other sources including industry (e.g. atmospheric emissions, waste disposal, effluent disposal), agriculture (e.g. application of sewage sludge, farm wastes and fertilizers), waste incineration, combustion of fossil fuels and road traffic. Long-range transport of atmospheric pollutants can also add to the metal load in an area. In general, the highest concentrations of heavy metals are associated with local sources of contamination such as mining and industrial facilities both in operation and after closure. Thus the largest and most affected areas are concentrated around heavily industrialized regions of the world, such as north-west Europe. Heavy metals accumulate in the soil as they bind to organic matter and clay minerals and are largely unavailable to plants. However, if soil acidity increases, heavy metals are released into soil solution, where they can be taken up by plant roots and *soil organisms*, or leached into surface and groundwaters, thus polluting the food chain and affecting drinking-water quality.

10.8.3.2 Pesticides and organic solvents

Organic compounds such as pesticides, oils, tars, chlorinated hydrocarbons, PCBs (polychlorinated biphenyls; used in the manufacture of electrical appliances) and dioxins are widely used in industry and agriculture and enter the soil through atmospheric deposition, direct spreading onto land, or contamination by wastewaters and waste disposal. Modern agricultural production systems rely on pesticides (mainly fungicides, insecticides and herbicides) for crop protection and disease control purposes. These pesticides are applied either directly or indirectly to the soil. The intensive use of pesticides has occurred since the Second World War. In northern and western Europe pesticide use peaked in the 1980s (Stanners and Bourdeau, 1991), whereas in southern Europe its use is still increasing.

The behaviour of pesticides in soil is influenced by a number of factors including its chemical properties, climate and soil type (particularly soil texture). Many pesticides, particularly the older ones, have a broad activity spectrum. This means they affect organisms that they were not intended to target. Pesticides can affect soil directly by adsorption onto clays and organic matter, by affecting soil microorganisms and plant growth, and by moving through the soil to surface and groundwaters. In Europe, the maximum admissible concentration of pesticides and metabolites in drinking water is set at 0.5 mg L^{-1} (EEC 80/778). The impact of pesticide use on the soil is very much dependent on the specific pesticide used. However, as there are over 1000 different compounds on the market, all behaving differently in the soil, it is very difficult to identify and evaluate all the threats posed by pesticides.

10.8.3.3 Fertilizer use – nitrogen and phosphorus

Soils used for intensive agriculture require additional nutrients, particularly nitrogen, phosphorus and potassium, to maintain optimum plant productivity. Over the past 50 years the use of inorganic fertilizers has increased by between 5 and 10 times to increase crop yield. Nitrates are not adsorbed on soil particles but remain in solution, from where they may be taken up by the plant, leached out in drainage water or denitrified. In contrast phosphorus is adsorbed strongly on the surface of clay particles and to iron and aluminium oxides. Nitrates are therefore leached out in drainage water whenever there is sufficient excess of rainfall. The concentration of nitrates in drainage water depends on the volume of drainage water and the amount of nitrates available for leaching in the soil. Leaching of nitrates is undesirable because in drinking water they are considered to be a health hazard and in marine waters can cause **eutrophication** (Burt *et al.*, 1993).

Although there is a link between the use of inorganic nitrogen fertilizers and nitrate leaching, it is indirect, as long as the recommended amount is used, it is not applied in the autumn and there is no crop failure. The more fertilizer used, the greater the crop yield and the more nitrogen in the plant residue (straw, stubble, roots). This can lead to an increase in soil organic nitrogen and the potential for nitrate leaching after mineralization of organic nitrogen to nitrate. Mineralization of soil organic nitrogen can also occur when permanent grasslands and woodland are brought into cultivation. Leaching of nitrates is most pronounced on free-draining soils during autumn and winter. The ultimate loss depends on the soil texture, land use, rainfall pattern, drainage properties, the presence or absence of vegetation, the amount and availability of nitrogen applied, the timing of applications in relation to crop growth and, for grassland, the intensity of grazing by livestock (Burt *et al.*, 1993).

Until recently, the risk of water pollution from phosphorus was believed to be minimal owing to the fact that the majority of phosphorus applied in fertilizer is quickly bound to the soil. However, long-term inputs of phosphorus from fertilizers and manure to intensive crop and livestock agricultural systems have been made at levels that often exceed outputs in crop and animal produce. Calculations of an annual phosphorus balance for European agriculture indicate that western European countries actually operate an annual phosphorus surplus (Ulén *et al.*, 2007). In the United Kingdom this surplus is 15 kg ha^{-1} yr^{-1} (Withers *et al.*, 2001). For over 20 years, fertilizer phosphorus application rates for individual crops have remained relatively constant (Withers *et al.*, 2001). As a result, the total amount and availability of phosphorus in agricultural soils have increased to a point where some soils can be classified as 'over-fertilized'. There is concern that this may lead to increased phosphorus loading to the aquatic environment, through a combination of leaching and eroded soil material. Although phosphorus losses are small in comparison with nitrate, only a small increase in phosphorus concentration is needed to produce a large change in the ecological dynamics of lakes and rivers.

To reduce phosphorus loss from agricultural land the concentration of soil phosphorus and/or the transport of phosphorus from land to water needs to be reduced. As most phosphorus loss is associated with the movement of fine soil particles, the same measures that are used to reduce soil erosion, such as cover crops and vegetated buffer strips between cultivated land and watercourses, would be beneficial.

Nutrient leaching from agricultural land is mainly a problem in areas of intensive agriculture, where fertilizer use is greatest, such as western Europe and North America. In Europe, the 1991 Nitrates Directive aimed to reduce water pollution caused or induced by nitrates from agricultural sources through the designation of Nitrate Vulnerable Zones (NVZs). Within these zones, farmers are required to undertake measures to reduce nitrate leaching. In addition, codes of good agricultural practice aim to narrow the imbalance between fertilizer input and plant uptake through a combination of measures such as: (i) better adjustment of fertilizer application and crop demands by soil testing and taking account of organic manures as sources of nutrients; (ii) application of fertilizers at the most appropriate times (i.e. when the crop most needs it); (iii) improvement of methods of manure application; (iv) minimizing leaching losses from arable land by sowing autumn crops; and (v) less intense use of grasslands. However, nitrate leaching cannot be completely prevented, especially in areas with high precipitation and where high crop yields are achieved on permeable soils (Burt *et al.*, 1993).

10.8.4 Soil organic matter and carbon

Soil organic matter is a vital component of productive and stable soils; it is the primary energy source for a wide range of soil organisms; it is an important store of global carbon which, if disturbed, can increase greenhouse gas concentrations in the atmosphere; it has a critical role to play in maintaining soil structure; and it influences water retention and regulates nutrient supply. It also acts as a buffer against

many of the threats discussed above. Intensive cultivation in Europe and North America has led to a decline in organic matter due to increased soil organic matter decomposition rates upon cultivation and loss of the organic rich topsoil through erosion (Dawson and Smith, 2007). This is of particular concern in Mediterranean areas, where 75% of the total area has a low (3.4%) or very low (1.7%) soil organic matter content (COM, 2002). Agronomists consider a soil with less than 1.7% organic matter to be in pre-desertification stage. In England and Wales, the percentage of soils with less than 3.6% organic matter rose from 35% to 42% in the period 1980 to 1995. In the same period, in the Beuce region south of Paris, soil organic matter decreased by half (COM, 2002). The ploughing up of grasslands, the abandonment of crop rotation and the burning of crop residues all reduce the amount of vegetation matter returning to the soil and increase carbon dioxide levels in the atmosphere. The reformed EU Common Agricultural Policy requires all farmers in receipt of the single payment to take measures to protect their soil from erosion, organic matter decline and structural damage. Various measures can be taken to increase the organic content of soils. These include the introduction of grass into agricultural rotations, ploughing crop residues back into the soil, increasing land left with a grass cover, and the spreading of animal manure. Recent research in England has shown that farmers can be encouraged to change their soil management when presented with information on the likely benefits to their farm of increasing soil matter (Gaunt *et al.*, 2008).

Soils are a major reservoir of carbon. In the UK, soils store in the order of 10 billion tonnes of carbon (Milne and Brown, 1997; Tomlinson and Milne, 2006). Loss of soil carbon leads to deterioration in soil function as well as contributing to greenhouse gas emissions and thus climate change. However, our knowledge of how soil carbon stocks are changing is limited. Recent work by Bellamy *et al.* (2005) suggested that over the last 25 years there has been a decline in soil organic matter, and therefore soil carbon, in agriculturally managed soils across England and Wales. They reported that carbon has been lost from UK soils at an annual rate of 13 million tonnes, equivalent to about 8% of the UK's current carbon emissions from the burning of fossil fuels. If correct this is a significant contribution to greenhouse gas emissions. However, there is some debate concerning the possible cause of this decline in soil carbon (Smith *et al.*, 2010b) which requires further investigation. Further losses of soil carbon could occur as a result of climate change, owing primarily to changes in temperature and soil moisture speeding up the decomposition of organic matter. However, the increase in decomposition may be counteracted by the higher rate of carbon dioxide uptake by plants, as they grow faster (Kirschbaum, 2000). A recent review for the European Commission (2008) was unable to find strong and clear evidence for either an overall combined positive or negation impact of climate change on soil carbon stocks. Hence there are still great uncertainties regarding the impact of climate change on soil carbon cycling. Current national and international policies are to maintain levels of soil organic matter/carbon and, where appropriate, to increase levels. Box 10.6 describes how soils can be managed to store carbon. In addition, certain habitats such as peatlands that contain large stores of soil carbon need to be protected.

MANAGING SOILS TO STORE CARBON

Soil is a major component in the global carbon cycle, containing about 1500 Pg (1 Pg = 1 Gt = 10^{15} g) of organic carbon (Batjes, 1996), which is about three times the amount in vegetation and twice the amount in the atmosphere. Through photosynthesis, plants convert carbon dioxide (CO_2) into organic forms of carbon and return some to the atmosphere through respiration. The carbon that remains in plant tissue is added to the soil through their roots and as litter when plants die and decompose. This carbon is then stored in the soil as soil organic matter. Carbon can remain stored in the soil for millennia, or be quickly released back into the atmosphere as CO_2. Climate, vegetation type, soil texture and drainage all influence the amount and length of time carbon is stored in the soil. Therefore, soils play a major role in maintaining a balanced global carbon cycle. However, the carbon content of soil is smaller today than a few hundred years ago owing to the intensification and mechanization of agriculture. Agricultural practices have depleted soil organic carbon pools by two main routes:

1. Reducing the amount of carbon returned to the soil in litter by harvesting and removing the crop.

2. Excessive use of tillage practices which breaks up the soil, increasing the decomposition rate of soil organic matter, which leads to an increase in the release of CO_2 from the soil.

Figure 10.27 illustrates the soil carbon changes over time on agricultural land. Estimates of historic soil organic carbon loss range from 40 to 90 Gt (Smith, 2004), of which about one-third is attributed to soil degradation and accelerated erosion and two-thirds to mineralization (Lal, 2004). Conversion of natural ecosystems to fields for crop production and grasslands causes depletion of a soil's carbon content by as much as 75% (Lal, 2004). Severe depletion of the soil organic carbon pool degrades soil quality and leads to a decline in crop production. Currently there is concern that many of the world's agricultural soils are alarmingly depleted of carbon (see Section 10.8.4).

There is, however, a way to reverse the soil carbon release process as research has shown that soils can regain lost carbon by absorbing or 'sequestering' it from the

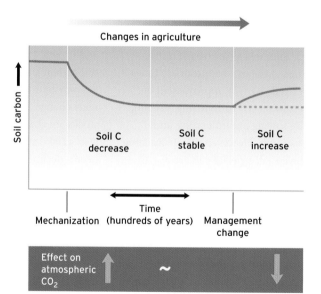

Figure 10.27 Changes in soil carbon stocks resulting from changes in agricultural land use and management.

BOX 10.6 ➤

300

atmosphere. **Carbon sequestration** implies transferring atmospheric CO_2 into long-lived pools and storing it securely so it is not immediately re-emitted to the atmosphere. Thus soil carbon sequestration means increasing soil carbon stocks. This can be best achieved through changes in land-use and management practices in soils that have been depleted in carbon, such as intensively managed agricultural soils and degraded soils. Some of the soil carbon sequestration options available for agricultural land include reduction in tillage, reducing fallow periods, improving efficiency of animal manure use and crop residue use, conversion of arable land to grassland, woodland or bioenergy crops and restoring degraded land. Estimates of the maximum yearly carbon mitigation potential for some of these land management options are shown in Figure 10.28 and compared with the 1990 CO_2 emissions from the United Kingdom. Many of these land management practices also improve soil quality, plant production and water conservation, reduce erosion, and enhance wildlife habitat and species protection, which result in increased biodiversity. Recent research has shown that the largest potential for increasing soil carbon

stocks in England comes from land-use change (e.g. arable to woodland) rather than changes in land management (King *et al.*, 2004). However, large-scale land-use change may result in food production being transferred to other, currently non-arable, areas of the country or the world, in which case the net benefit may be low.

Improvements in measuring, monitoring and verifying changes

in carbon stocks in soils are needed for quantitative economic and policy analysis. Currently, world scientists can combine data on soil carbon, land use and climate to create models that estimate the carbon change related to farm management practices. However, they are continuing to refine measurement methods for greater accuracy.

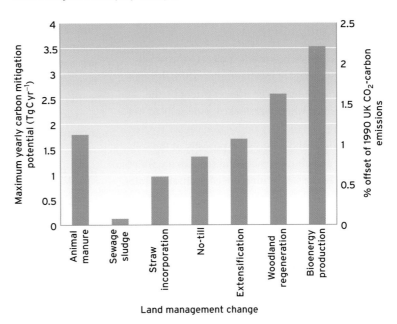

Figure 10.28 Estimates of the maximum yearly carbon mitigation potential, and maximum percentage offset of UK 1990 CO_2 emissions of some different land management changes. (Source: after Smith *et al.*, 2000)

BOX 10.6

10.8.5 Other threats

10.8.5.1 Salinization

Salinization is the accumulation in soils of soluble salts of sodium, magnesium and calcium to the extent that the soil fertility is severely reduced. The process is mainly a problem in arid and semi-arid areas, where there is insufficient rain to leach away soluble salts, and upward soil water movement due to evapotranspiration leads to salt precipitation at or near the soil surface (see Chapters 9, 11 and 14). Irrigation of the land with water of a high salt content dramatically worsens the process. In coastal areas salinization can also

be associated with abstraction of groundwater for drinking and industry. This can cause the intrusion of salty seawater as it attempts to replace lost freshwater in the ground. Human-induced salinization is estimated to affect 50% of all irrigated land (Abrol *et al.*, 1988). Salinization is reversible but reclamation of saline soils is expensive, as it requires complex amelioration techniques. It is estimated that:

- in Syria, 45% of the irrigated land is affected (Ilaiwi *et al.*, 1992);
- in Spain, 3% of the 3.5 million hectares of irrigated land is severely affected, significantly reducing its agricultural potential, and another 15% is under serious risk (COM, 2002);

- in the Hungarian plains, irrigation has caused salinization of more than 20% of the region (Stanners and Bourdeau, 1995);
- in the Russian Federation, 7% of agricultural soils are saline, part of them naturally so, such as the Solonchaks (Stanners and Bourdeau, 1995);
- in Romania, 6% of irrigated land became saline between 1970 and 1995 (Stanners and Bourdeau, 1995);
- in Australia, salinization could devastate up to 12 million hectares of land during the next 100 years and already costs at least A\$500 million a year in Victoria alone (Gowing, 2003).

10.8.5.2 Soil compaction

Soil compaction occurs when soil is subject to the repetitive and cumulative effect of mechanical pressure through the use of heavy machinery, or to a lesser extent by the trampling of livestock when overstocked, especially in wet soil conditions (Figure 10.29). Compaction causes reorientation of soil particles and reduces the porosity of the soil. This slows water and air movement and reduces the water holding capacity of the soil. With reduced infiltration rates,

Figure 10.29 Soil compaction caused by sheep.

compacted soil is likely to produce much greater volumes of overland flow, which increases the risk of soil erosion. Compaction can occur within both the top- and subsoil. Soil compaction mainly affects heavily mechanized agricultural land such as central Europe and North America. In Europe about 33 million hectares (4% of land area) is at risk from soil compaction, mainly in the Russian Federation, Poland, Germany, Belgium, the Netherlands and north-west France (Oldeman *et al.*, 1991). The major impact of soil compaction is a decline of agricultural productivity. Surface compaction may reduce yield by up to 13% (average 5%) and subsurface compaction may reduce yields by between 5 and 35% (Stanners and Bourdeau, 1995). Surface soil compaction can be resolved by reworking the soil. However, subsoil compaction is much harder to resolve. Compaction can be prevented by reducing the amount of tillage, reducing traffic over fields, improving land drainage, increasing organic matter and reducing surface pressure of machinery by increasing the axles and wheels on agricultural machinery, increasing tyre width and reducing tyre pressure.

10.8.5.3 Soil biodiversity

Soil is the habitat for a large variety of living organisms assembled in complex and varied communities (see Section 10.7). Soil biodiversity reflects the variability among living organisms in the soil. Species range from the myriad of invisible microbes, bacteria and fungi to the more familiar macro-fauna such as earthworms and termites. Soil is by far the most biological diverse part of Earth. Soil organisms are known to play a key role in many of the fundamental functions that occur in soil. However, the roles played by most groups of soil organisms are very poorly known. Soil communities are currently exposed to a wide range of impacts, including soil erosion, agricultural intensification and the deposition of acidic pollutants, with poorly documented effects on diversity of the soil biota, and virtually unknown effects on ecosystem processes.

Soil management strongly influence soil biota in agricultural systems (Stockdale *et al.*, 2006). For example, different practices cause a shift in habitat and in substrate availability, which results in changes in abundance of individual species (Van Camp *et al.*, 2004). However, evidence of the threats to biodiversity and opportunities for its conservation and improved management is mainly qualitative and hence there is the need for research that quantifies these threats. Reliable indicators need to be developed so that long-term monitoring programmes can be set up. To date, no comprehensive indicator of soil biodiversity exists (European Commission, 2010). In addition, no legislation or regulation exists that is specifically targeted at soil biodiversity, whether at international, EU, national or regional level. This partly reflects the lack of awareness of soil biodiversity and its value, as well as the complexity of the subject (European Commission, 2010).

10.8.6 Policy and legislation

Soil degradation processes are a major worldwide problem with significant environmental, social and economic consequences. As the world's population continues to grow there is increased need to protect soil as a vital resource. With the addition of a quarter of a million people each day, the world food demand is increasing at a time when per capita food productivity is beginning to decline (Pimental *et al.*, 1995). Growing awareness of the need for a global response has led to national and international initiatives. In 1972, the Council of Europe's Soil Charter called on states to promote a soil conservation policy. The World's Soil Charter and the World's Soils Policy sought to encourage international co-operation in the rational use of soil resources (COM, 2002). In 1992, at the Rio Summit, the participating states adopted a series of declarations of relevance to soil protection. In particular, the concept of sustainable development was agreed and legally binding conventions on climate change, biological diversity and desertification adopted. The aim of the 1994 Convention to Combat Desertification was to prevent and reduce land degradation, rehabilitate partly degraded land and reclaim desertified land. For further information on **desertification** see *Ecosystems and human well-being: Desertification synthesis*, a report of the Millennium Ecosystem Assessment (2005a).

In 2001, the EU indicated soil loss and declining soil fertility as the main threats to sustainable development as they erode the viability of agricultural land (COM, 2002). Although several different EU policies (e.g. on water, waste, pesticides, industrial pollution prevention) contributed to soil protection, they were not sufficient to ensure an adequate level of protection for all soil in Europe. Hence in response to concerns about the degradation of soils in the EU, the European Commission adopted a Communication 'Towards a Thematic Strategy for Soil Protection' in April 2002 and in September 2006 the Commission adopted the Thematic Strategy for Soil Protection. The strategy's objective is to define a common and comprehensive approach, focusing on the preservation of soil functions, and based on the following principles:

1. Preventing further degradation to the soil and preserving its functions.

2. Restoring degraded soils to a level of functionality consistent at least with current and intended use, thus also considering the cost implications of the restoration of soil.

The strategy outlines why further action is needed to ensure a high level of soil protection across the EU and what kind of measures must be taken. There are currently four main elements to the strategy:

1. Measures to address soil erosion, organic matter decline, compaction, salinization and landslides, obliging Member States to identify risk areas and develop associated programmes of measures and targets.

2. Measures to address soil contamination, including actions such as an inventory of contaminated sites, the production of soil status reports, identification of sites on which potentially polluting activities are taking place or have taken place, and the production of a remediation strategy for such sites.

3. Measures on soil sealing, obliging Member States to take appropriate steps to minimize soil sealing, or mitigate its effects, using construction techniques and products which would allow as many soil functions as possible to be maintained.

4. Awareness raising, reporting and exchange of information.

In May 2004, a First Soil Action Plan for England was published, which set out a three-year programme designed to ensure that soil will be used and looked after in ways that get the best out of a vital natural resource. It represented a key milestone in soil policy in that it highlighted the first comprehensive statement on the state of soil in England and how Government and other partners were working together to improve soils. The Action Plan addressed issues that are listed under eight headings: (1) protecting soils in the planning system, (2) minimizing contamination of soils, (3) predicting and adapting to the impacts of climate change on soils, (4) soils for agriculture and forestry, (5) interactions between soils, air and water, (6) soils and biodiversity, (7) soils, the landscape and cultural heritage and (8) soils in mineral extraction, construction and the built environment.

The EU Thematic Strategy for Soil Protection included proposals for a 'Soil Framework Directive' which seeks to harmonize and raise the level of soil protection across Europe. The proposed directive primarily seeks to address seven key threats to European soils: (1) erosion, (2) decline in organic matter content, (3) soil compaction, (4) soil salinization, (5) landslides, (6) contamination and (7) soil sealing.

Through the soil strategy, the Commission is establishing a framework, which may eventually become an official Framework Directive with legislation, based on common principles and objectives to address the different facets of soil degradation. Member States will be obliged to identify where the problems occur, but they are free to decide what to do, and to what extent, in order to address these problems. In 2009, a new Soil Strategy for England was published, a successor to the First Soil Action Plan for England, which addresses many of the threats identified in both the thematic strategy and the draft Soil Framework Directive within an English context. Through the adoption of polices that help protect soil, other environmental media such as air and water will also be improved.

Reflective questions

➤ How does the concept of critical loads work?

➤ What are the main controls on the nature and intensity of soil erosion?

➤ Explain what is meant by the term 'soil compaction' and how the effects of compaction can be reduced in soils.

➤ Why is it important to maintain or even enhance the organic matter content of a soil?

10.9 Summary

Soil is composed of minerals, organic matter, air, water and living organisms in interactive combinations produced by physical, chemical and biological processes. Soil is an essential component of the terrestrial biosphere and performs a wide range of essential functions that sustain life. It supports plant growth on which humans rely for food, fibre and wood for fuel and building materials. It provides a habitat for large numbers of animals and microorganisms that decompose dead plants and animals into the nutrients needed by all living things. It acts as a reservoir for water, and has a filtering, transforming and buffering role. It also provides raw materials and a physical base for the foundations of buildings and roads.

Soil is made up of mineral and organic materials, and contains pore spaces occupied by water and air. The minerals include residues of the parent material and secondary minerals, which are the product of weathering. The organic matter is composed of readily decomposing plant, microbial and animal products, living organisms and roots and resistant organic matter known as humus. The soil water contains solutes and dissolved gases, and is referred to as the soil solution. The composition of soil air differs from that of atmospheric air in that it generally contains more carbon dioxide and less oxygen owing to the respiration of soil organisms and roots. The soil

is an open system, which allows input of materials to the soil, the loss of materials from the soil and internal transfers and reorganization of these materials within the system. It is the processes of additions, removals, mixing, translocations and transformations that are influential in differentiating soil material into a series of horizons that constitute the soil profile, and in determining the nature and properties of soil.

The soil-forming processes that dominate at a site are controlled by five interacting environmental conditions: parent material, climate, topography, organisms and time. These environmental variables are known as the soil-forming factors and they control the direction and speed of soil formation. Climate and organisms determine the rate at which chemical and biological reactions occur in the soil, while parent material and topography define the initial state for soil development and time measures the extent to which reactions will have proceeded.

There are large differences between soil profiles from place to place throughout the world. To describe soil profiles in a coherent manner, various classification schemes have been introduced. The physical properties of soil depend largely on the size of the soil particles (soil texture) and on their arrangement (soil structure) into peds or aggregates. Texture and structure influence the distribution and movement of water and air in the soil and thus greatly affect plant growth. Soil colour is used as

➤

an indicator of organic matter content, drainage and aeration.

Clay minerals are the product of weathering. The structure of all clay minerals is based on two types of sheets: the tetrahedral sheet, which consists of repeating units of a silicon atom surrounded by four oxygen atoms in the form of a tetrahedron; and the octahedral sheet, which consists of repeating units of an aluminium atom surrounded by six oxygen atoms or hydroxy (OH) groups in the shape of an octahedron. Alternating sheets of one tetrahedral sheet and one octahedral sheet produce what are known as 1:1 clays, whereas an aluminium octahedral sheet sandwiched between two silicon tetrahedral sheets produces 2:1 clays. Isomorphous substitution (Al^{3+} replaces Si^{4+} or/and Fe^{2+} or Mg^{2+} replaces Al^{3+}) in the crystal lattice of clay minerals results in an overall net negative charge. This charge is balanced by cations, positively charged ions, which are attracted to the surface of the clay minerals and held there by electrostatic attractions. These cations are known as exchangeable cations as they can be displaced by cations in the soil solution in the process of cation exchange. In most agricultural soils, calcium and magnesium are the dominant exchangeable cations. However, as the acidity of the soil increases, aluminium and hydrogen ions dominate the exchange sites.

There are a number of natural and human-induced changes that increase soil acidity. They include leaching of base cations, respiration of roots and organisms, decomposition of organic matter, deposition of acids from the atmosphere, application of nitrogen fertilizers, removal of base cations in crop harvests and draining of waterlogged land. Soil pH determines the fate of many pollutants, affecting their breakdown, solubility and possible movement from the soil to surface and groundwaters.

The availability of the essential nutrients for plant uptake is also influenced by soil pH as are the number, species and activities of soil organisms. Low soil pH leads to a decline in the number and activities of many soil organisms, an increase in concentrations of aluminium, iron, manganese and zinc (to the extent of toxicity to plants and other organisms), a decrease in the concentration of the macronutrients (to the extent that plants may show signs of deficiencies) and a reduction in root activity. In contrast, high soil pH results in phosphorus and boron becoming insoluble and unavailable to plants and low concentrations of the micronutrients, particularly iron, manganese, zinc and copper, resulting in restricted plant growth.

Soils are home to many different soil organisms that perform a wide variety of important functions such as decomposing organic material, regulating the carbon flux, decontaminating polluted soils and even providing raw materials to tackle infectious diseases. These soil organisms can be grouped according to size or into three main groups depending on the principal function they perform. The diversity and activity of soil organisms are controlled by a combination of biotic and abiotic factors.

Soil, as a resource, is being increasingly exploited throughout the world. The pressures of agriculture, drainage, extraction, application of wastes and urban development have led to soil degradation such as soil erosion, contamination, acidification, compaction, loss of organic matter, salinization and loss of biodiversity. Ensuring that our soil is sustainable, and not damaging to future generations, will involve, as well as regulatory bodies, all sectors of society whose activities and decisions affect soils. This is one of the world's biggest challenges for the future.

Further reading
Books

Bardgett, R. (2005) *The biology of soil: A community and ecosystem approach.* Oxford University Press, New York.
This textbook provides an excellent introduction to soil ecology. It describes the vast diversity of biota that live in soil, discusses the factors that control this diversity across different temporal and spatial scales and considers how biotic interactions in soil influence decomposition and nutrient cycling.

Brady, N.C. and Weil, R.R. (2007) *The nature and properties of soils,* 14th edition. Pearson Education, Harlow.
This is the latest edition of a comprehensive and popular textbook. It covers all the major aspects of soils, and therefore provides an in-depth reference for all areas of soil science and management. The book is well written and often illustrates ideas and concepts with diagrams, graphs, tables and case studies (mainly American), making the subject available and easy to understand.

Cresser, M.S., Kilhma, K. and Edwards, A.C. (1993) *Soil chemistry and its applications.* Cambridge University Press, Cambridge.
This textbook demonstrates the role soil chemistry plays with other areas of soil science and environmental science such as water quality and pollution science.

European Commission (2010) *Soil biodiversity: Functions, threats and tools for policy makers.* European Commission, Paris.
This report, available at http://ec.europa.eu/environment/soil/biodiversity.htm, reviews the state of knowledge of soil biodiversity, its functions, its contribution to ecosystem services and its relevance for the sustainability of human society.

FitzPatrick, E.A. (1983) *Soils: Their formation, classification and distribution.* Longman Scientific and Technical, Harlow.
This textbook includes a comprehensive section on soil classification and a comparison of the many different systems used throughout the world.

Rowell, D.L. (1994) *Soil science: Methods and applications.* Longman, London.
This textbook includes useful examples of practical soil work in the field and laboratory.

Royal Commission on Environmental Pollution (1996) *Nineteenth report – Sustainable use of soil.* HMSO, London.
This report outlines the major uses of soil in the UK and the environmental issues associated with each use.

Stanners, D. and Bourdeau, P. (1995) *Europe's environment: The Dobris assessment.* European Environment Agency, Copenhagen.
Focus on Chapter 7 (soil), pages 146–169. This chapter gives a detailed and comprehensive review of the problems and threats facing soils in Europe.

Toy, T.J., Foster, G.R. and Renard, K.G. (2002) *Soil erosion: Processes, prediction, measurement and control.* John Wiley & Sons, Chichester.
Good for case studies of soil erosion and information on practical methods of measuring and preventing erosion.

White, R.E. (1997) *Principles and practice of soil science: The soil as a natural resource,* 3rd edition. Blackwell Science, Oxford.
This textbook covers all aspects of soil science and is divided into three sections: soil habitat, soil environment and soil management.

Wild, A. (1993) *Soils and the environment: An introduction.* Cambridge University Press, Cambridge.
This textbook has two sections: part A covers soil properties and processes, whilst part B considers soils in relation to the environment.

Further reading
Papers

Batjes, N.H. (1996) Total carbon and nitrogen in the soils of the world. *European Journal of Soil Science,* 47, 151–163.

Bellamy, P.H., Loveland, P.J., Bradley, R.I., Murray Lark, R. and Kirk, G.J.D. (2005) Carbon losses from all soils across England and Wales 1978–2003. *Nature,* 437, 245–248.

Smith, P. (2004) Soil as carbon sinks: the global context. *Soil Use and Management,* 20, 212–218.

Smith, P., Chapman, S.J., Scott, W.A. *et al.* (2010b) Climate change cannot be entirely responsible for soil carbon loss observed in England and Wales, 1978–2003. *Global Change Biology,* 13, 2605–2609.
The above are a set of papers dealing with how to quantify carbon stocks in soils and determine the role that climate change and land management are having on soil carbon stocks.

CHAPTER 11

Catchment hydrology

Joseph Holden

School of Geography, University of Leeds

Learning objectives

After reading this chapter you should be able to:

➤ describe and critically evaluate the measurements involved in measuring the inputs, outputs and stores of water in catchments

➤ understand the main ways water moves across and through the landscape and how water reaches rivers

➤ explain spatial and temporal changes in runoff generation

➤ evaluate the form of river hydrographs and describe the likely processes leading to their form

➤ describe different forms of flooding

➤ understand how land use and climate change may affect catchment hydrological processes, and how these might alter runoff and storm hydrographs

11.1 Introduction

A **catchment** is also known as a **drainage basin** and both define an area of land in which water flowing across the surface drains into a particular stream or river. In North America the same definition also applies to the term **watershed**. The catchment is a convenient unit because it is normally well defined topographically, it can be studied as a series of nested units of increasing size (so that larger catchments are made of many smaller subcatchments), and it is an open system for which inputs and outputs of mass and energy can be defined and measured. Catchments are drawn on the basis of land surface topography. The boundary of a catchment is called a **drainage divide**; water on one side of the divide will flow to one river and on the other side of the divide it will flow to a different river. The proportion of land area compared with the density of river channels or total stream channel length within a catchment may determine how efficiently water can be removed from a catchment since water in channels tends to move much more quickly than water across and through hillslopes.

The pathways by which water travels to a river or lake will often determine how quickly that water will reach the river or lake. The flow paths also determine water quality; water that has been in contact with soil or rock for long periods, for example, often has a very different chemistry from **precipitation** water (see Chapter 13). The dominance of different types of water flow process across or through the landscape is controlled by local climate and catchment features such as geology, topography, soils and vegetation. Knowledge of the relevant mechanisms and their controls is important for determining catchment hydrological response to a precipitation event. The amount of precipitation that reaches the river channel can be very great (almost 100% in some urban areas) or very low (less than 5%), depending on soil or rock water storage and evaporation. Changes in catchment management (or land use) such as urbanization, ploughing, afforestation, deforestation and artificial soil drainage can result in changes to the flow paths for water across hillslopes and therefore to changes in the timing, volumes and quality of water reaching the river channel. Thus, flood risk and water supply can be altered if there are changes in the relative dominance of runoff flow paths occurring within the catchment.

The movement of water on hillslopes occurs in a number of forms including overland flow (of different types), subsurface flow involving micropores, macropores and natural soil pipes, to displacement flow and groundwater discharge. The dominance of these processes varies with climate, topography, soil character, vegetation cover and land use, but may vary at one location (e.g. seasonally) with soil moisture conditions and with storm intensity and duration. The spectrum of runoff processes is related to both the type and intensity of erosion process (in both particulate form and dissolved or solutional form) and to the resulting mode of hillslope and landform evolution (see Chapters 7, 8 and 13). Therefore, while the landscape plays a role in determining runoff processes, the runoff processes themselves help to shape the landscape.

This chapter aims to investigate the main components of a catchment water balance, to examine runoff production processes and to evaluate the role of these processes in generating river flow. This will help explain why there is spatial and temporal variability in the dominance of particular processes and therefore such variability in river flow. The chapter will then be able to demonstrate how environmental change may lead to changes to hillslope runoff processes and river flow.

11.2 Measuring the main components of catchment hydrology

A water balance is often defined by a simple equation which is used to express the idea that the water inputs are equal to the outputs plus or minus changes in water stores:

$$P = Q + E + \Delta(I + M + G + S) \qquad (11.1)$$

where P is precipitation, Q river discharge, E evapotranspiration, I interception and biological water store, M soil water storage, G groundwater storage and S channel and

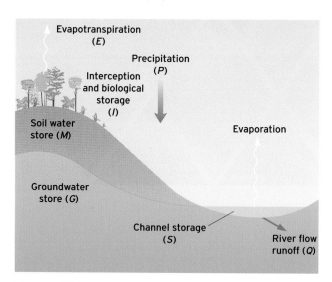

Figure 11.1 Main measurement components of a catchment water budget. The letters used form part of equation (11.1).

surface storage. Seasonal variations in the balance (especially between precipitation and evaporation) help to control the seasonal patterns of river flow. In arid zones such as central Australia or south-east Spain high rates of evapotranspiration may mean that river flows are small (many are completely dry) even after a rainfall event.

Figure 11.1 shows the main components of a traditional catchment water budget. It does not include the transfer processes that will be discussed later. Measuring the main components shown in Figure 11.1 might at first seem straightforward but there are many problems associated with the techniques employed. The following sections discuss each component of the water budget and associated measurement strategies.

11.2.1 Precipitation

The main hydrological inputs to any catchment system are in the various forms of precipitation (e.g. hail, mist, dew, rain, sleet or snow). Precipitation is usually expressed in units of length such as mm or cm (or inches in the United States). This is because it is assumed that the precipitation has fallen uniformly over a given area and so the volume of water is divided by the surface area of the catchment (or gauging instrument) to give a depth of water. Rain gauges are traditionally used to collect inputs of precipitation to the ground surface (Figure 11.2). These gauges funnel water into a collecting device. Many gauges can automatically record the volume of water received (usually by a tipping

(a)

(b)

Figure 11.2 Rain gauges. (a) There are two rain gauges in this image. In the foreground there is a rain gauge at ground level surrounded by a metal grid. The grid reduces potential splash and turbulence effects which might influence the amount of precipitation that enters the gauge. In the background a green rain gauge can be seen in the centre of a sunken circle. The circle around the gauge has been dug to reduce turbulence and splash effects. (b) The internal mechanism of a tipping bucket rain gauge. The water drips through a funnel at the top into the sea-saw buckets. Each time the bucket tips this is recorded by a small datalogger.

bucket or siphon mechanism) and if gauges are remote or flood warnings need to be provided, data can be collected by **telemetry** (e.g. mobile phone).

Rain gauges provide point measurement of precipitation. Often large areas of catchments are only gauged by one or two rain gauges. Yet it is well known that there can be great local differences in rainfall received. Topography, aspect and the localized nature of many storm events may mean that readings from one rain gauge cannot be applied to the whole catchment. If there is a network of gauges then spatial averaging can be done using arithmetic means,

Thiessen polygons (Figure 11.3a) or **isohyets** (contours of equal rainfall; Figure 11.3b). Often topography complicates spatial interpolation between gauges since slight changes in altitude can result in large changes in precipitation. It is sometimes difficult to install and maintain a sufficiently dense network of gauges and therefore remote sensing technology is becoming widely used in precipitation measurement (see Chapter 26). Radar techniques allow rainfall totals to be estimated across catchments through time and space (Campos *et al.*, 2007). Radar can remotely pick up changes in raindrop density across a catchment

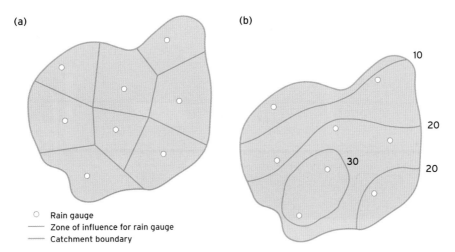

Figure 11.3 Producing spatial data from point rain gauges: (a) using Thiessen polygons, where polygons are constructed around a set of points in such a way that the polygon boundaries are equidistant from the neighbouring points; (b) using contours (isohyets) of rainfall.

allowing improved flood forecasts and catchment water budgets to be developed. Nevertheless it is often necessary to check radar data using ground rain gauges. Satellite imagery can also be used to estimate snow cover and how much water might be stored in snowpacks.

Precipitation in forms other than rainfall, and wind turbulence around gauges, both cause problems for precipitation measurement. Careful siting of a rain gauge, usually at ground level, is needed so that turbulence does not affect precipitation entering the gauge orifice (Figure 11.2a). In temperate and polar regions snowfall is not immediately recorded on most rain gauges since water is not freely available to trickle into the collecting device. Even those gauges that are heated suffer from the problem that snow that has already fallen elsewhere can blow into the gauge, causing precipitation totals to be overestimated. Many mists and dews can precipitate water onto vegetation or the ground surface. Rain gauge surfaces have different textures from surrounding vegetation and may not accurately record these inputs. For example, in Newfoundland, Price (1992) used wire netting to catch fog precipitation and found that 50% of summer precipitation to the vegetation and ground surface came in the form of fog that was not correctly

measured by rain gauges. Therefore it is often necessary to have separate recording devices for different types of precipitation (e.g. using automatic snow depth, pressure or light reflectance recorders).

11.2.2 River flow

River flow is the only phase of the hydrological cycle in which water is confined into well-defined channels allowing 'accurate' measurements to be made of the quantities involved. Good water management is founded on reliable river flow information. Many river flow gauges exist around the world and generally these take the form of a water-level recorder (normally a pressure sensor or a float) housed in a protective well which is fixed into the river at a suitable location (Figure 11.4). In order to get good control over water levels and to allow for small changes in discharge to result in measurable changes in water level around the recording instrument, many gauging sites take the form of weirs or flumes (Figure 11.5). These large in-stream engineering features, however, are not always practical on large rivers.

If measurement of water volumes is required in order to produce a water budget then recorded water levels

(a)

(b)

Figure 11.4 Stilling wells containing a pressure sensor to record water level through time. (a) The data recorder is in the protruding green box while the well is the green piping going down into the ground. The well is connected to the river by tubing so that the water level in the well is the same as in the river. There are two stilling wells directly next to each other in this photo acting as a check against each other. (b) This stilling well is on a steep bank and is therefore some distance from the river so that it does not become an eyesore. Only the green box at the top of the well (which contains the data recorder) can be seen as the rest of the well is buried but connects to the river via tubing.

(a)

(b)

(c)

Figure 11.5 Typical gauging stations: (a) crump weir, (b) flat crested weir, (c) flume-type weir.

must be converted into a discharge reading for the river. This requires a **rating equation** to be determined for each river flow gauging site. Sometimes this is well known for a particular shape of weir but generally field calibration is required. Box 11.1 shows how stream discharge can be measured and how the data can then be used to derive a rating equation. There are inevitable errors involved with using rating equations for river flow, not least those associated with the techniques of measuring discharge itself (e.g. inaccurate measurement of stream cross-section, effect of instruments on local water velocity, inadequate mixing of tracers, vertical changes in water velocity not accounted for by a measuring instrument), the difficulties with obtaining reliable discharge data at high flows and the fact that many river channel cross-sections change shape over time (see Chapter 12).

Often the volume of water that has been discharged by a river is divided by the catchment area in order to allow runoff volumes to be compared with precipitation depths, **infiltration rates** or evaporation losses which are also commonly expressed in units of length. By working out how much water entered the catchment as precipitation and measuring how much came out (by measuring stream discharge) we can determine the 'efficiency' of the catchment. For example:

- Precipitation = 40 mm in 7 days.
- River discharge = 120 000 m^3 of water in 7 days.
- The catchment area is 15 km^2 (this catchment area is equivalent to $15 \times 1000 \times 1000$ m^2 (since there are 1000×1000 m^2 in 1 km^2) = 15 000 000 m^2).
- If we were to spread the discharge evenly over the surface of the entire catchment this would give us a depth of water of 120 000 m^3/15 000 000 m^2 = 0.008 m = 8 mm of water.
- So when compared with the depth of rainfall that fell over the entire catchment, which was 40 mm, a discharge of 8 mm is equivalent to 20% of the precipitation.
- Hence the catchment efficiency was 20%.
- If there was no change in catchment storage then we can say that 80% of the water was lost through evapotranspiration.

In many catchments river flow is not gauged but it is sometimes estimated using the water balance equation (11.1) when the other variables (e.g. precipitation, evapotranspiration) are known. On an annual basis, or over longer periods, it is often assumed that storage of soil and groundwater is

TECHNIQUES

RIVER DISCHARGE MEASUREMENT

Several methods can be adopted to measure river discharge. Three main methods are illustrated below.

Velocity-area method

This method requires estimation of the water velocity to be multiplied by an estimation of the cross-sectional area of water. The river cross-section is surveyed and water and bed level are plotted. Water velocity can be measured using a variety of instruments (see Herschy, 1999). Figure 11.6 shows a river survey with the water velocity being measured using an impeller meter. These are commonly used and work by counting the number of revolutions of an impeller as the water flows past it. It is usual to place the impeller at three-fifths of the water depth to give an estimate of mean velocity in the water column. This is because water will move at different velocities at different heights and it is assumed that three-fifths of the depth provides the best average. Measurements of velocity can be done at several points across the river. Each separate velocity measurement can then be multiplied by a separate area value associated with that reading as indicated in Figure 11.7. The separate discharge values for each segment can then be summed to give the discharge for the whole cross-section.

Acoustic gauging

Ultrasonic discharge gauges (Figure 11.8) work by recording the time for a beam of acoustic pulses to cross a river at different depths. Devices are placed on opposite banks of the river and the travel time for

Figure 11.6 Measuring water velocity using an impeller meter.

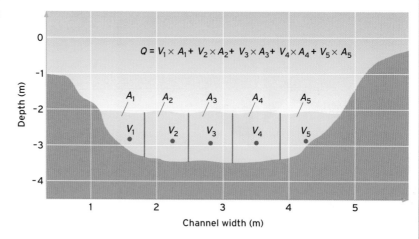

$$Q = V_1 \times A_1 + V_2 \times A_2 + V_3 \times A_3 + V_4 \times A_4 + V_5 \times A_5$$

Figure 11.7 Velocity-area cross-section plot to calculate discharge.

BOX 11.1 ➤

➤

 (a)

 (b)

Figure 11.8 An ultrasonic gauging station. (a) The round objects at the edge of the river measure the travel time for ultrasonic pulses to cross the river. In the photo you can see a pair in the foreground and in the background you can see one of a nearby pair on the far bank. (b) A close-up of one of the rounded ultrasonic gauges with access cover (on the right). There are also water-level recorders on site, one of which can be seen on the left with the stage board.

the pulse is used to calculate the stream flow velocity. These devices can be left *in situ* to give an automatic record. A water-level sensor is also needed and the cross-sectional area has to be surveyed in advance for different water levels. In non-uniform channels multiple path ultrasonic gauges are required which use a series of pairs of devices arrayed vertically to characterize the vertical velocity profile and hence get a better value of discharge across the channel. Ultrasonic gauges cannot be used well where there is dense aquatic vegetation or where high suspended sediment loads are likely.

An alternative acoustic method of river gauging is via the use of Doppler probes which record the change in frequency when acoustic waves are reflected from natural suspended sediment or gas bubbles in river water. The change in frequency is proportional to the velocity of the particle or gas bubble. The Doppler method therefore works well in murky rivers unlike the ultrasonic discharge gauges described above. Sometimes Doppler profilers are deployed from boats or cables across rivers

to provide velocity measurements across the channel.

Dilution gauging

An alternative method is to calculate discharge based on dilution. This involves adding a known concentration of chemical (e.g. sodium chloride) to the river and measuring the amount of dilution in the river water (Figure 11.9). A graph of concentration against time can then be plotted (Figure 11.10). Using the following equation the discharge of the river can then be established:

$$Q = \frac{(C_i - C_b)V}{\int (C_d - C_b)dt} \qquad (11.2)$$

where Q = discharge, C_i = concentration in added water, C_b = background river water concentration, V = volume of water added to river, C_d = measured downstream concentration after water is added and t = time of measurement. The equation simplifies to:

$$Q = \frac{(C_i - C_b)V}{\text{shaded area under graph shown in Figure 11.10}} \qquad (11.3)$$

Further details on how this technique can be used are given in Burt (1988).

Using a rating curve

If water levels are being measured at a point on a river using an automatic gauge these levels can be converted into discharge values. This is done using a rating equation. It is necessary to measure (e.g. by using one of the techniques above) the discharge of the river when the water level is at different heights. The values can then be plotted as shown in Figure 11.11 and the equation for the line of best fit is calculated. This can then be used to infer the discharge from any water height at that point on the river. Care must be taken when inferring discharge values that are beyond the range of water-level values that have been calibrated. This is because the curve can change shape when the river floods or for very low flows. It is also very difficult to measure river discharge using velocity–area techniques or dilution gauging when a river is in flood or for very large rivers.

BOX 11.1 ➤

313

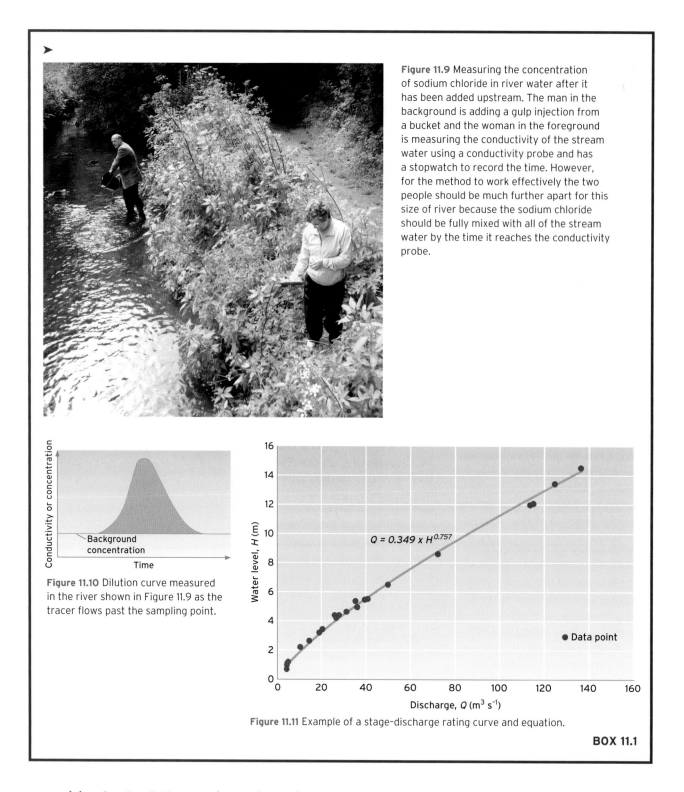

Figure 11.9 Measuring the concentration of sodium chloride in river water after it has been added upstream. The man in the background is adding a gulp injection from a bucket and the woman in the foreground is measuring the conductivity of the stream water using a conductivity probe and has a stopwatch to record the time. However, for the method to work effectively the two people should be much further apart for this size of river because the sodium chloride should be fully mixed with all of the stream water by the time it reaches the conductivity probe.

Figure 11.10 Dilution curve measured in the river shown in Figure 11.9 as the tracer flows past the sampling point.

$Q = 0.349 \times H^{0.757}$

Figure 11.11 Example of a stage-discharge rating curve and equation.

BOX 11.1

zero and thus $Q = P - E$. However, this may be satisfactory for an annual water budget but it is often not sufficient to determine daily discharge for ungauged rivers. A number of techniques have been developed to enable daily discharge estimates for ungauged rivers and an example is described in Box 11.2.

11.2.3 Evapotranspiration

Evapotranspiration is evaporation plus transpiration (the biological process by which water is lost from a plant through its leaves). Evapotranspiration is difficult to measure and is affected by: solar radiation (providing **latent heat**;

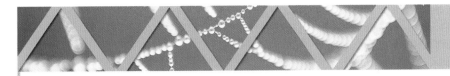

ESTIMATING DAILY DISCHARGE FOR UNGAUGED RIVERS

Daily river flow data are important for water resource management, ecological studies and future downstream discharge. However, many rivers are ungauged. One recent technique was developed by Archfield and Vogel (2010) who selected a gauged stream in New England. They then compared daily flow from this stream with a 50 year record from 27 other streams in the area. Using statistical analysis they then mapped out the area surrounding the gauged stream where daily flow rates were corre-

lated to other streams 98% of the time. They repeated this for streams which correlated 95% of the time and so on. They then repeated the whole exercise for the other gauged streams. The scientists were then able to select ungauged streams and from their maps pick out the gauged stream which was most likely to have a matching pattern of daily discharge. This could then be used to estimate discharge in ungauged streams

Clearly this technique is less reliable where there are few gauged streams within large areas of the world and there remain significant challenges. The use of satellites is therefore being advocated. The SWOT

satellite mission can measure water surface height and has been used for ocean monitoring. However, it has been suggested that this system could be redirected to measure water heights across the land surface at 100 m resolution to within a centimetre accuracy every 10 days. Therefore, water levels in rivers and lakes could be monitored. Discharge can then be inferred (but note that it is not directly measured) over time since water slopes can be estimated and over time more information about the river cross-sections will be derived at low flows and high flows by the SWOT system.

BOX 11.2

see Chapters 4 and 5); temperature of the air and the evaporation surface (these influence the capacity of the air to hold moisture and the rate at which evapotranspiration can occur); wind speed (removing saturated air); humidity; turbulence (and hence surface roughness as determined by topography or vegetation); plant biology (transpiration processes); and availability of water. As for precipitation, evapotranspiration can vary widely over short distances and there are often mosaics of values coinciding with changes in vegetation cover.

Evaporation can be measured using an evaporation pan, which is simply a container of open water in which the water depth is measured. Evaporation is calculated as precipitation minus the fall in water level. However, these measurements tend to be unreliable because of the effect of direct sunlight and heating of the pan material. **Atmometers** can give direct readings of evaporation; a water supply is connected to a porous surface and the amount of evaporation over a given time is measured by the change in water stored. However, measurement of evapotranspiration rather than evaporation is necessary since most catchments of interest are not made up entirely of open water. **Lysimeters** are often used to measure evapotranspiration but they can be very difficult to install and maintain. A lysimeter (Figure 11.12) isolates a block of soil (with its vegetation cover) from its surroundings so its water balance can be

measured. The weight of the block is measured to determine how much water has been lost. If the soil in the lysimeter is kept moist by the addition of water, and well covered by vegetation (a grass sward is ideal), evapotranspiration is

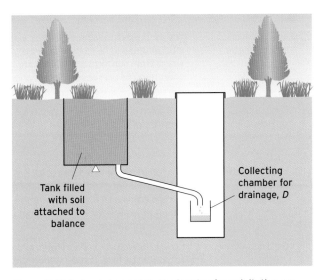

Figure 11.12 A simple lysimeter. The inputs of precipitation are measured as are the outputs from soil drainage out of the block. The block is weighed and evapotranspiration is estimated by determining the change in weight over time once drainage and precipitation are taken into account.

controlled by the weather and is largely independent of the amount (biomass) of the vegetation. This allows measurement of **potential evapotranspiration** (PE), which is the evapotranspiration from a vegetated surface with unlimited water supply. Because of difficulties in directly measuring PE it is often estimated based on equations that take account of solar radiation, air temperature, dew point temperature, wind speed and the vegetation roughness properties (e.g. vegetation height and density). However, as with precipitation and lysimetry measurements, there are problems with spatial variability of evapotranspiration and the applicability of local calculations to entire catchments.

11.2.4 Soil water

The ability of a soil to absorb and retain moisture is crucial to the hydrology of an area. The soil water store will vary seasonally and will depend on soil properties, soil depth and precipitation. Deep permeable soils can store large quantities of water, providing a moisture reserve through times of drought and helping to sustain river flow during dry periods. There are a wide range of techniques for the measurement of soil water content. These are listed in Table 11.1. The most common method is the gravimetric technique whereby soil samples are extracted, weighed

Table 11.1 Some common methods for determination of soil water content

Technique	Method	Main advantages	Main disadvantages
Feel	Estimate by touch	Very quick, cheap	Not quantitative and requires vast experience
Gravimetric	Soil samples weighed and heated in oven at 105°C for 24 h and then reweighed to determine moisture loss by mass	Accurate, little equipment required	Slow, destructive, requires many samples
Electrical resistance blocks	Gypsum blocks installed into soil and electrical resistance measured	Cheap, good data over time which can be continuously recorded	Not good in very wet conditions, gypsum slowly dissolves, slow response to real soil moisture changes
Neutron probe	Radioactive source lowered into hole and fast neutrons released – these are impeded by hydrogen nuclei in soil water. A detector senses scattering of impeded neutrons	Repeatable at any given site, rapid measurement	Must be calibrated for each new test site, radiation hazard, high cost
Time domain reflectometry	Waves of electromagnetic energy sent into ground and reflections of these waves from subsurface features are collected. The presence of water affects the speed of the wave	Rapid collection of spatial data	Must be precisely calibrated. Lots of noise on the signals
Capacitance sensors	Uses dielectric constant of soil as a measure of water content	Small-scale differences between soil layers can be measured as well as values close to the soil surface	Difficult to install and calibrate as air gaps and local inhomogeneities affect readings
Heat dissipation sensors	The temperature in a porous block is measured before and after a small heat pulse is applied to it. The amount of heat flow is proportional to the amount of water contained within the soil	Variation in the conductivity of water is not a problem, easy to automate, non-destructive	Large power requirement, high cost
Carbide method	CO_2 gas is emitted when water is added to calcium carbide. A small wet weight soil sample is mixed with calcium carbide in a sealed container and the pressure of the produced gas is measured	Relatively inexpensive, rapid in field measurements	Destructive, specialized equipment and reagents needed
Remote sensing	Satellite/airborne imaging	Large spatial coverage, fast	Difficult to calibrate the images, coarse scale

and then heated in an oven for 24 h at 105°C to remove the moisture and then reweighed. However, obtaining acceptable point measurements and reliable areal measurements is still difficult. Remote sensing of soil moisture characteristics is possible in some environments either using airborne or satellite instruments or using ground-based geophysical equipment such as **ground-penetrating radar** (see Chapter 26). Airborne or satellite techniques are preferable as they can provide wide spatial coverage in a short period of time and can often illustrate the nature of medium-scale variability in soil moisture (e.g. comparing hillslope to hillslope or hillslope top to hillslope bottom). However, most of these Earth observational techniques are limited to examining soil moisture in the upper 5 cm of the soil profile (Schmugge and Jackson, 1996) which may not be representative of the entire soil profile. A range of techniques exist for examining soil water movement and these will be discussed later in this chapter.

11.2.5 Groundwater

Groundwater is water in the **saturated zone** below the land surface. Therefore groundwater can be held within soils or in bedrock. Nearly all rocks are porous and can hold water within them. The depth at which a soil or rock is fully saturated (all the pore spaces are full of water) is called the **water table** (Figure 11.13). If we dig a hole into the ground (such as a well) then water will flow from saturated rock into the well. The water level within the well will rise to a constant level which demarcates the water table. Therefore

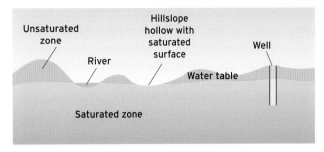

Figure 11.13 Saturated and unsaturated zones and the water table.

measuring the water table is relatively simple and requires an instrument to record the water level in a well (similar to those used in river-level gauging). The water table is often not flat depending on topography and geology. Sometimes the water table will be at the surface, particularly after heavy rain in hillslope hollows.

Water table data from one point in a peat bog are shown in Figure 11.14 over time. The water table was measured by using a pressure sensor that recorded water level in a thin, shallow plastic tube that was perforated and installed into the peat. The water table is close to the surface most of the time, but during dry spells the water table slowly falls owing mainly to evapotranspiration and some drainage of water from the hillslope. When it rains the water table quickly rises towards the surface again as shown by the almost vertical lines on the plot.

The groundwater store is important because in some areas it is the only source of water for humans and it represents virtually all (97%) of the Earth's non-saline and

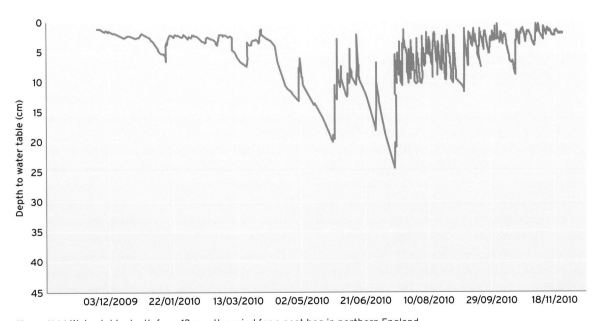

Figure 11.14 Water table depth for a 12 month period for a peat bog in northern England.

non-frozen water resources (Price, 2002). However, measuring exactly how much water is in any groundwater store is very difficult and requires an estimation of the porosity of the rock or soil that is holding the water. If the bedrock contains only a few small pores then it cannot hold much water, yet if all the pores are full then it is still saturated. Water is not just stored as groundwater but moves through saturated soil and rock. This water movement will be discussed later in this chapter. One assumption that is often made in water balance studies is that catchments are watertight. However, bedrock geology may allow water to flow out of a catchment below the surface without reaching the local river system. Often surface topography does not tell us exactly which direction water may be flowing below the surface and so it is sometimes difficult to determine groundwater losses and gains from a catchment.

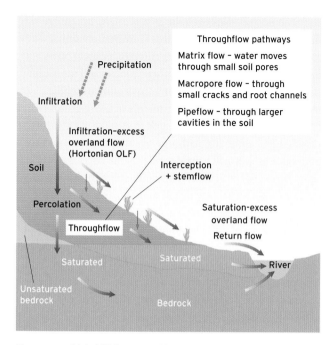

Figure 11.15 Main hillslope runoff pathways.

Reflective questions

➤ What are the main problems with measuring catchment water budgets?

➤ How would you measure discharge if you visited a mountain stream that was strewn with boulders and coarse sediment making it difficult to use an impeller meter?

➤ Why is very little known about peak discharge values of many of the world's largest rivers?

11.3 Hillslope runoff pathways – how water reaches rivers and lakes

Figure 11.15 illustrates the main hillslope runoff pathways for water. Precipitation can either hit the surface of the hillslope directly or be intercepted by vegetation. This intercepted water can be stored on leaves and tree trunks which shelter the ground beneath. This is called **interception storage**. **Stemflow** is the flow of water down the trunk of a tree or stems of other vegetation species allowing water to reach the hillslope. There are then two possibilities for direct precipitation or stemflow once it reaches the hillslope: either to infiltrate into the soil (see Box 11.3) or to fill up any depressions on the surface and flow over the surface as overland flow.

11.3.1 Infiltration

Infiltration is the process of water entry into the surface of a soil and it plays a key role in surface runoff, groundwater recharge, ecology, evapotranspiration, soil erosion and transport of nutrients and other solutes in surface and subsurface waters. Surface water entry is influenced by vegetation cover, soil texture, soil porosity and soil structure (e.g. cracks, surface crusting) and compaction. The infiltration rate is the volume of water passing into the soil per unit area per unit time (e.g. m s^{-1}, mm h^{-1}). The maximum rate at which water soaks into or is absorbed by the soil is the **infiltration capacity**. This is very important in determining the proportion of incoming rainfall that runs off as infiltration-excess overland flow and the proportion that moves into the soil. If infiltration is occurring at less than the infiltration capacity then all rain reaching the soil surface will infiltrate into the soil.

The infiltration capacity of a soil generally decreases during rainfall, rapidly at first and then more slowly, until an approximately stable value has been attained (Figure 11.16). Soil surface conditions may impose an upper limit to the rate at which water can be absorbed, despite a large available capacity of the lower soil layers to receive and to store additional infiltrating water. Often the infiltration capacity is reduced by frost; snowmelt above a frozen surface can lead to rapid generation of overland flow and large flood peaks. Field ploughing can increase soil infiltration

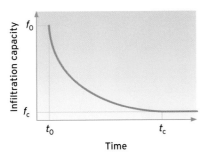

Figure 11.16 The typical decline of infiltration capacity during a rainfall event.

capacity (Imeson and Kwaad, 1990). Soils with well-developed humus and litter layers (such as tropical rainforest soils) tend to have high infiltration capacities (Ward and Robinson, 2000).

Soil water movement continues after an infiltration event, as the infiltrated water is redistributed. Bodman and Colman (1943) suggested that for a uniform soil there would be a series of zones in the wetting part of the soil profile during an infiltration event. The zone nearest the surface is a saturated zone (typically in the upper centimetre of the soil profile). As water penetrates more deeply a zone of uniform water content, the transmission zone, develops behind a well-defined wetting front. There is a sharp change in water content at the wetting front. Figure 11.17 shows the water content with depth for a sandy soil 4 min after ponded infiltration at the surface. Note that the soil below the wetting front still has some pre-event moisture.

11.3.2 Infiltration-excess overland flow

If surface water supply is greater than the rate of infiltration into the soil then surface storage will occur (even in urban catchments in small surface depressions). When the surface depressions are filled they will start to overflow; this is called Hortonian overland flow or **infiltration-excess overland flow**. Horton's (1933, 1945) theory of hillslope hydrology assumed that the only source of storm runoff was excess water that was unable to infiltrate the soil. In this theory infiltration divides rainfall into two parts. One part goes via overland flow to the stream channel as surface runoff; the other goes initially into the soil and then through groundwater flow to the stream or into groundwater storage or is lost by evapotranspiration.

In many temperate areas infiltration-excess overland flow is a rare occurrence except in urban locations. The infiltration capacity of many soils is too high to produce infiltration-excess overland flow (Burt, 1996). Infiltration-excess overland flow is more likely in semi-arid areas where soil surface crusts have developed and rainfall events can be particularly intense. It is also more likely in areas where the ground surface is often frozen, such as northern Canada or parts of Siberia. Often infiltration-excess overland flow will occur only on spatially localized parts of a hillslope such as in tractor wheelings on arable land. This spatially localized occurrence of infiltration-excess overland flow is known as the **partial contributing area concept** (Betson, 1964). This suggests that only parts of the catchment or hillslope will

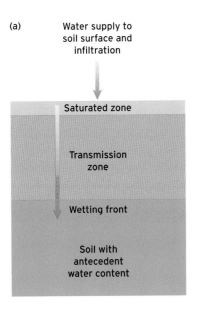

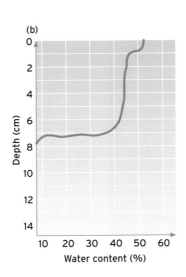

Figure 11.17 Soil water zones during infiltration: (a) theoretical zonation; (b) measured water content for a sandy soil in Iowa 4 minutes after ponded infiltration commenced.

319

contribute to infiltration-excess overland flow rather than the whole catchment as Horton had originally suggested in the 1930s and 1940s.

11.3.3 Saturation-excess overland flow

When water infiltrates a soil it will fill the available pore spaces. When all the pore spaces are full the soil is saturated and the water table is at the surface. Therefore any extra water has difficulty entering the soil because it is saturated. Hence overland flow will occur. This type of overland flow is known as **saturation-excess overland flow**. It can occur at much lower rainfall intensities than those required to generate infiltration-excess overland flow. Saturation-excess overland flow can occur even when it is not raining. This might happen, for example, at the foot of a hillslope (Figure 11.18). Water draining through the soil is known as **throughflow** (see below). Throughflow from upslope can fill up the soil pores at the bottom of the slope and so the soil becomes saturated. Any extra water is then forced out onto the surface to become overland flow. This water is known as '**return flow**' and is a component of saturation-excess overland flow. Therefore saturation-excess overland flow is more likely to occur at the bottom of a hillslope, or on shallow soils where there is restricted pore space for water storage.

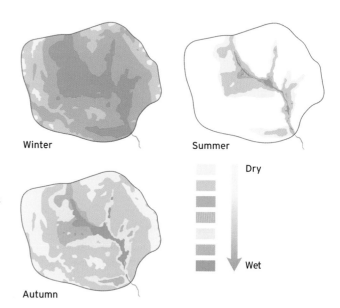

Figure 11.19 Seasonal changes in catchment saturation for a small headwater in Denmark. The likely source areas for saturation-excess overland flow vary throughout the year.

The area of a catchment or a hillslope that produces saturation-excess overland flow will vary through time. During wet seasons, for example, more of a catchment or hillslope will be saturated and therefore able to generate saturation-excess overland flow than during dry seasons. If the catchment starts off relatively dry then during a rainfall event not much of the area will generate saturation-excess overland flow, but as rainfall continues then more of the catchment becomes saturated, especially in the valley bottoms, and therefore a larger area of the catchment will produce saturation-excess overland flow (Figure 11.19). The fact that the area of a catchment in which saturation-excess overland flow occurs tends to vary is known as the '**variable source area concept**' (Hewlett, 1961). The variable source area model has become the dominant concept in catchment hydrology.

The main differences between the two overland flow types are related to the water flow paths. For infiltration-excess overland flow all of the flow is fresh rainwater that has not been able to infiltrate the soil. However, saturation-excess overland flow is often a mixture of water that has been inside the soil (return flow) and fresh rainwater reaching the hillslope surface. Therefore its chemistry will be very different (see Chapter 13).

11.3.4 Throughflow

If water infiltrates the soil several things can happen:

- It can be taken up by plants and transpired (or be lost from the soil by evaporation).

Figure 11.18 Saturation-excess overland flow occurring through grass after rainfall has stopped.

- It can continue to percolate down into the bedrock.
- It can travel laterally downslope through the soil or rock – this is called throughflow.

Worldwide, most water reaches rivers by throughflow, through the soil layers or through bedrock. Throughflow can both maintain low flows (**baseflow**) in rivers by slow subsurface drainage of water and also contribute to peak flows (**stormflow**) through its role in generating saturation-excess overland flow and as an important process in its own right (Burt, 1996). There are different ways that water can move through soil and this affects the timing of water delivery to the river channel. Soils are not uniform deposits as they have cracks and fissures within them. Water can move through the very fine pores of soil as **matrix flow**,

or it can move through larger pores called macropores (**macropore flow**), or even larger cavities called soil pipes (**pipeflow**). Water moving through the soil matrix occurs in a laminar fashion whereas flow within macropores and pipes is turbulent (see Box 9.1 in Chapter 9 for explanation of laminar and turbulent flow).

11.3.4.1 Matrix flow

Flow through the matrix of a porous substance should behave as determined by '**Darcy's law**'. As Box 11.3 indicates, Darcy's law allows us to calculate the likely rate of water movement through a porous medium when it is saturated (the **saturated hydraulic conductivity**). Hence

FUNDAMENTAL PRINCIPLES

DARCY'S LAW

Darcy's law is a mathematical relationship originally determined by Henry Darcy in 1856 that allows us to calculate the amount of water (or other fluid) flowing through a substance. It equates volumetric discharge per unit time (q) to the product of the area of substance being tested (A), the hydraulic gradient, I (which is the difference in water pressure (Δh) between one end of the substance and the other divided by the length (L) of the substance being tested), and a coefficient

(K = saturated hydraulic conductivity). This may be expressed as:

$$q = KIA \qquad (11.4)$$

Figure 11.20 provides a schematic diagram of how the relationship can be established in a laboratory test and helps explain what the letters stand for. The diagram shows a cylinder of soil inside a tube. Water enters the left side of the tube and leaves the right side of the tube after having passed through the soil. The water moves through the tube because the soil is porous. The water pressures

h_1 and h_2 can be measured at each side of the tube and the difference between the water pressures at both ends is called the head difference (Δh). Note that $\Delta h/L$ (length of tube) = I. The rate that water leaves the tube can be measured (q). For a given soil in a tube of cross-sectional area A, the hydraulic conductivity, K, can be determined as it is the only unknown in the equation.

By changing the hydraulic head (the water pressure difference between one end of the tube and the other) and measuring the discharge for these different values it is also possible to confirm that the relationship between head and discharge is linear for most materials (a graph of I against q should be a straight-line plot). It should also be possible to determine the head or discharge conditions for which the relationship deviates from the linear form (when the graph starts to deviate from a straight line) and therefore the threshold beyond which Darcy's law is no longer applicable. The relationship holds only for laminar (non-turbulent)

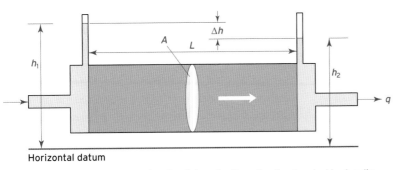

Horizontal datum

Figure 11.20 Laboratory apparatus for determination of soil saturated hydraulic conductivity (K) based on Darcy's law. The letters are as indicated in the text and equation (11.4).

BOX 11.3 ➤

flow of fluids in homogeneous porous media.

The hydraulic conductivity (units of velocity) of a soil or rock is an important parameter and is frequently used for estimating water movement through hillslopes. Techniques for determining K in the field include pumping water out of a well and timing how long it takes for the water level in the well to reach the original level, and timed movement of

tracers (e.g. dye). The factors affecting hydraulic conductivity include those associated with the fluid and those associated with the soil or rock including temperature, salinity, pore space geometry and soil or rock surface roughness. As the scale of approach increases, often the estimation of hydraulic conductivity can be found to increase because of incorporation of ever-larger and more extensive fracture systems. Flow

analysis predictions based upon Darcy's law in the presence of massively fissured soils or rocks such as karstic limestones (see Chapter 7) or highly fractured crystalline rocks can lead to large errors. Flow in such cases cannot be described adequately by a linear relationship such as Darcy's law and more detailed analysis of turbulent flows is required.

BOX 11.3

it is possible to estimate the amount of flow taking place as matrix throughflow. Often the saturated hydraulic conductivity will vary with depth and soil type. Sandy soils typically have a high hydraulic conductivity compared with clay soils that have low hydraulic conductivities. Therefore water will drain through sand more quickly than clay. Lateral throughflow through the matrix will occur in any soil in which the hydraulic conductivity declines with depth. If both soil and bedrock remain permeable at depth, however, then percolation remains vertical and little lateral flow can occur; infiltrating water will serve only to recharge groundwater storage (Burt, 1996).

Water also moves through soils on hillslopes that are unsaturated. Even after a long drought most soils contain some water. This suggests that gravitational drainage and evapotranspiration are not the only forces at work in moving water within soils and that other forces involved must be very strong. Chapters 7 and 10 discuss the processes by which water is held within soil and it was shown that water remains in the soil after gravitational drainage because of the combined attraction of the water molecules to each other and the water to the soil particles. This water that is held in soil against the force of gravity is known as capillary water and it will move within the soil from wet areas to dry areas. It is more difficult to get water out of small soil pores (spaces between the solid particles) than it is to obtain water from larger soil pores. This is why sandy soils are much dryer than clay soils. The attractive forces holding the water to the soil particles and between the thin layers of water are much greater in small pores. This means that large soil pores lose their water before small pores. However, the forces exerted by small empty soil pores mean that when water is added to a soil it will fill small pores first; thus the

'suction' or 'soil water tension' exerted by small pores is greater than that exerted by larger pores. When all of the soil pores are full of water the soil is saturated and there are no suction forces. Instead there will be forces associated with gravity and the pressure of water above a given point. There will be a positive **pore water pressure** caused by the pressure of water from above. When the soil is unsaturated there will be forces associated with gravity and negative forces associated with the suction effect. It is therefore possible to measure these suction or positive pore water forces and to determine in which direction water is likely to move through a hillslope.

11.3.4.2 Macropore flow

Macropores are pores larger than 0.1 mm in diameter and can promote rapid, preferential transport of water and chemicals through the soil, not only because of their size but also because they are connected and continuous over sufficient distances to bypass agriculturally and environmentally important soil layers (Beven and Germann, 1982). Therefore if a field has many macropores, surface fertilizer applications may get washed through the macropore channels and may not enter the main part of the soil. Fertilizer applications could therefore be transported out of the hillslope before they can be taken up by plants, potentially resulting in downstream water quality problems. Macropores can be formed by soil fauna, plant roots and cracking (Figure 11.21). They have been identified by using microscope and visual observation techniques (e.g. dyes) and even using neutron imaging (Box 11.4). However, the occurrence of a macropore does not necessarily mean that there will be preferential flow of water through the channel. The

(a)

Figure 11.21 Macropores (a) caused by surface cracking; (b) caused by crane fly larvae (which can be seen in the photograph) and identified using dye staining. The blue dye is added to the soil surface. It is possible to see areas in the soil where water has quickly percolated down the macropores as these are stained blue. These macropores result in much higher infiltration and percolation rates. (Source: photo (b) courtesy of Katy Gell)

(b)

process of water flow in macropores has three components: water is delivered to macropores, the available water then flows some distance into the macropores, and finally the water may be absorbed through the walls or the base of the macropores. A macropore must be sufficiently connected to a supply of water in order for there to be flow.

Macropores may not take up much space in the soil and if they are open at the soil surface they will often only take up a tiny proportion of the soil surface. Despite their small spatial role, macropores can still have a high impact on runoff and play a large role in throughflow as water can preferentially flow through them. Holden and Gell (2009) showed that macropores caused by crane fly larvae could increase soil percolation rates over areas without larvae. A study in Niger on a crusted sandy soil showed that 50% of infiltrated water moved through macropores (Leonard *et al.*, 2001). Some studies in upland peat catchments have indicated that 30% of throughflow moves through macropores (Holden *et al.*, 2001; Holden, 2009).

11.3.4.3 Pipeflow

Natural soil pipes are subsurface cavities of diameter greater than 1 mm that are continuous in length such that they can transmit water, sediment and solutes through the soil and bypass the soil matrix. Soil pipes are larger versions of soil macropores. Soil pipes are created by a wide range of processes including faunal activity (animal burrows) and by turbulent flow through desiccation (shrinkage) cracks, biotic (e.g. roots) and mass movement cracks enhancing macropores into pipe networks. Climate, biota, human activity, soil chemistry, soil texture, erodibility, soil structure, hydraulic conductivity, clay minerals, cracking potential and dispersivity are all important controls on soil piping (Jones, 1981). Pipes can be up to several metres in diameter and several hundred metres in length (e.g. Holden and Burt, 2002a) and occur in a broad range of environments. Pipe outlets (Figure 11.23) may transmit a large proportion of water to the stream in some catchments. Holden and Burt (2002a) and Smart *et al.* (2012) found 10 to 14% of river discharge in peat catchments had moved through the pipe network on peatland hillslopes while 43% of discharge

NEUTRON IMAGING OF WATER FLOW PATHS

Neutron radiography provides images similar to X-ray images but with the advantage that organic materials or water are clearly visible in neutron radiographs because of their high hydrogen content, while many structural materials such as aluminium or steel are nearly transparent. Neutron imaging experiments have been carried out on soil using the neutron imaging station, Neutrograph, at the Institut Laue-Langevin (France). This is the world's most intense beam of its kind. The samples are placed approximately 11 m from the neutron source (the nuclear fission reactor), and are irradiated with neutrons. As the neutron beam passes through the sample, it is attenuated (the intensity is reduced) depending on the length of the neutrons' path through the sample (the sample's thickness) and the material present along that path. The attenuated beam then passes into a lead-shielded, light-tight box, where a screen converts the neutrons

to photons and a special camera detects their presence.

Projections with an exposure time of between 50 and 150 ms were taken every 0.225° as the sample was rotated through 180°, resulting in 800 unique images. These images were then computationally reconstructed to produce cross-sectional slices, which are perpendicular to the beam and have an in-slice resolution of approximately 160 μm. Figure 11.22(a) shows a neutron image of a whole soil core when

all the slices have been put together while Figure 11.22(b) shows one of the slices. A small macropore can be seen near the bottom and top of the core. It is also possible to see that the macropore near the top has rough walls and may have only recently formed whereas the lower macropore has smooth walls and may have water flow through it to smooth its sides.

Part of the method description in this box was kindly provided by Martin Dawson, University of Leeds.

(a) 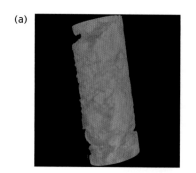 (b)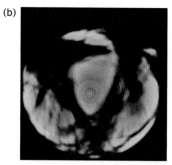

Figure 11.22 Neutron images of a soil core. (a) Macropores near the top and bottom of the core. The core is only 8 cm long and 4 cm in diameter. (b) A thin horizontal slice through part of the core with black areas within the core indicating macropores. (Source: Photos courtesy of Martin Dawson)

BOX 11.4

moved through pipe networks in semi-arid loess soils of China (Zhu, 1997).

Piping is common in arid and semi-arid areas such as south-east Spain and Arizona where shrinking and desiccation cracking are common. Often soil crusting in these environments can result in infiltration-excess overland flow. If a few cracks at the surface exist then infiltration will be concentrated at those points. Turbulent flow may then enlarge cracks to form soil pipes. In these areas pipes can grow so large that they eventually collapse, forming large gullies, and so they are important in shaping the landscape of several regions. Often the networks of pipes can be quite complex and may look something like that shown in Figure 11.24. Pipes can therefore provide rapid

connectivity of water, sediment and solutes throughout the soil profile and form complex meandering and branching networks.

11.3.4.4 Groundwater flow

Groundwater is water held below the water table in both soils and rock. Therefore groundwater flow has, to some extent, already been discussed in the sections on matrix flow, macropore flow and pipeflow. However, further treatment of groundwater flow as a separate component is required because of its worldwide importance. Groundwater holds around 30% of the world's freshwater store. Some countries such as Austria, Hungary and Denmark rely

(a)

(b)

Figure 11.23 Natural soil pipe outlets producing water in a peatland.

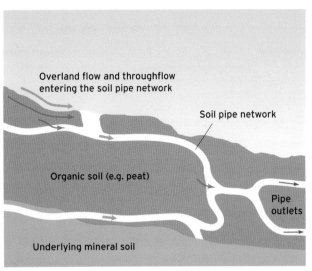

Figure 11.24 Schematic example of a short vertical section through a soil pipe network. The pipe allows water to move quickly through different layers of the soil.

almost entirely on groundwater (Holden, 2011) to provide for people's needs, while half of the water resources in the United States come from groundwater. In many places groundwater has been over-abstracted or the abstraction brings other problems such as around the Tuscon area

of Arizona where ground subsidence has resulted from groundwater abstraction. In parts of Australia groundwater abstraction has led to the soils becoming salty and unsuitable for plant growth (see Box 11.5). In many catchments water is supplied to the stream from groundwater in the bedrock. This is water that has percolated down through the overlying soil and entered the bedrock. Rock has small pores, fractures and fissures. Therefore it is possible to use Darcy's law (see Box 11.3) to investigate flow rates through bedrock where fissures are at a minimum. Where there are large fractures such as in cavernous limestone areas (e.g. Cutta Cutta caves area of northern Australia) then it may not be so useful.

Groundwater may be a large store of water, but in order for it to be available to supply river flow the holding material (rock or soil) needs to be not just porous but permeable. That is to say, a rock (or soil) may be porous but relatively impermeable either because the pores are not connected or because they are so small that water can be forced through them only with difficulty. Conversely a rock that has no voids except one or two large cracks will have a low porosity and therefore a poor store of water. Nevertheless because water will be able to pass easily through the cracks the permeability will be high (Burt, 1996). Layers of rock sufficiently porous to store water and permeable enough to allow water to flow through them in economic quantities are called **aquifers**. Sometimes aquifers can be confined between impermeable rock layers (**aquitards**) and are open only for recharge and discharge at certain locations

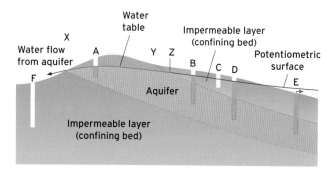

Figure 11.25 A confined aquifer. Between X and Z the aquifer is unconfined and has a water table; to the right of Z the aquifer is confined and has a potentiometric surface. Wells A, B, D and E enter permeable material and strike water. Wells C and F are in impermeable material which will yield water only very slowly. (Source: after Price, 2002)

(Figure 11.25). Sometimes aquifers are recharged from stream and lake bed seepage rather than supplying river flow itself. However, groundwater is generally an important supplier of river baseflow (although glacier ice melt is another source). Depending on the nature of the aquifer, baseflow discharge may be uniform throughout the year or peak discharge may lag significantly behind precipitation inputs.

Reflective questions

➤ Why are the source areas for infiltration-excess overland flow and saturation-excess overland flow likely to be different in any given catchment?

➤ Why might the areas producing overland flow in a catchment vary through time?

➤ What is Darcy's law?

➤ Why might fertilizers and pesticides applied to some fields move through the soil very quickly and bypass the plant roots leading to river pollution?

11.4 River discharge, hydrographs and runoff production processes

Seasonal variation in river flow which tends to be repeated each year is known as the **regime** of the river. Often this is expressed as monthly discharge. Chapter 12 provides some examples of river regimes in different climates. Rivers in equatorial areas tend to have a fairly regular regime, tropical

rivers show a marked contrast between discharge in dry and wet seasons, while in other climatic areas snow may complicate the regime as it does not contribute to runoff until melting occurs. Thus regimes can often be complex and despite important climatological controls may also depend on the catchment geology and soils that control the relative role of hillslope runoff production processes.

There are several important components of river flow that need to be assessed in order to determine flood risk and water availability for reservoirs and abstraction for drinking water and industry. The most important of these is river discharge and its variability over time. A graph of the way discharge varies over time is called a **hydrograph**. Hydrographs can be analyzed for long periods or for short storm events. Long records of river discharge are useful because they can tell us about the likely frequency of particular high- and low-flow events and the seasonality and overall variability in flows. Unfortunately long-term records of river flow are not widely available.

11.4.1 Stormflow

It is often useful to analyze the characteristics of individual storm hydrographs. This is because it gives us information on how quickly rainwater moves from the hillslope to the stream channel. Figure 11.27 identifies the main components of a storm hydrograph. The separation on the hydrograph of baseflow and stormflow is often arbitrary and a variety of techniques have been adopted; these are discussed in Ward and Robinson (2000). The baseflow is the amount of water in the river channel that is derived from groundwater sources and can be considered to be a 'slowflow' component of the hydrograph. In the case of Figure 11.27, baseflow rises only slowly and some time after the rain has fallen. This is because water in the groundwater zone typically travels much more slowly and by longer, more tortuous flow paths than by other processes. The amount of baseflow in a river depends on seasonal variations in precipitation, evapotranspiration and vegetation. However, there is a wide range of baseflow runoff responses. In cavernous limestones water will very quickly move through cracks and fissures providing rapid, peaked groundwater hydrographs. However, it is unusual for aquifers to provide a major contribution to storm hydrographs (Price, 2002).

The hydrograph in Figure 11.27 shows that there is a lag time between the precipitation and the peak discharge of the river. This lag time is affected by the hillslope runoff processes discussed above. Where infiltration-excess overland flow dominates the hillslope runoff response then the

CASE STUDIES

GROUNDWATER ABSTRACTION IN AUSTRALIA

As a vital resource, groundwater supplies over 65% of irrigation water for farming in South Australia, Victoria and New South Wales. Western Australia uses groundwater to supply 72% of its urban and industrial demand. Up to 4 million people in Australia rely totally or partially on groundwater for domestic supply. Such demand puts pressure on groundwater resources. This is because if the inflow into the ground is much smaller than the outflow from the aquifers then the groundwater resource will decrease. While there are large groundwater reserves in Australia, many are in very remote areas or where the water is difficult to access. Therefore the more accessible groundwater reserves may become depleted. In fact, 30% of Australia's groundwater management units are either almost overused or certainly overused.

Additionally there is the problem of salinization. Salinization is the build-up of salt within the soil (Figure 11.26). This causes severe damage to farmland as the land is no longer able to support crops. Water from deep below the surface contains salts. Normally these salts are held deep below the surface where they do not affect plants. However, when the deep groundwater is brought to the surface and evaporated in hot conditions this leaves behind salts as a deposit, killing plants and soil organisms. There are two main ways excess salts are brought to the surface. The first source is from irrigation water which

is abstracted from deep sources. The second source is when land use is changed so that trees are removed. The trees may keep the water table fairly deep by using a lot of water. However, when they are removed, the water table may quickly rise to the surface and mobilize the salts that are stored in the soil. Approximately 5.7 million hectares of Australia are within regions at risk or suffering from dryland salinity. In 50 years this

may increase to 17 million hectares (ANRA, 2001). There are a series of knock-on effects of soil salinization. For example, the high concentrations of salt at the soil surface can also pollute streams when runoff occurs and the native vegetation also becomes damaged. Damage to river ecology due to salt pollution is also associated with loss of streambank vegetation, which can then also lead to exacerbated bank erosion.

Figure 11.26 Salinization of soils in Western Australia.
(Source: Getty Images: Universal Images Group)

BOX 11.5

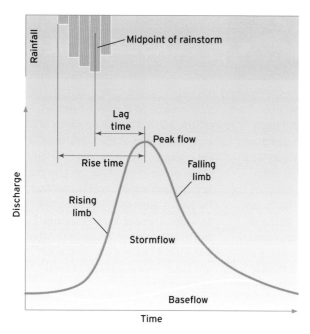

Figure 11.27 Components of a storm hydrograph.

hydrograph is likely to have a short lag time and high peak flow (Figure 11.28a). If throughflow in the small soil pores (matrix flow) dominates runoff response on the hillslopes then the hydrograph may look something like that in Figure 11.28(b). The lag time is long and peak flow low. However, since throughflow contributes to saturation-excess overland flow then throughflow can still lead to rapid and large flood peaks. In some soils or substrates only a small amount of infiltration may be needed to cause the water table to rise to the surface. There may even be two river discharge peaks caused by one rainfall event. This might occur where the first peak is saturation-excess overland flow dominated (with some precipitation directly in the channel). The second peak may be much longer and larger and caused by subsurface throughflow accumulating at the bottom of

hillslopes and valley bottoms before entering the stream channel. Throughflow may also contribute directly to storm hydrographs by a mechanism called piston or displacement flow. This is where soil water at the bottom of a slope (old water) is rapidly pushed out of the soil by new fresh infiltrating water entering at the top of a slope.

The proportion of precipitation that is produced as stormflow in a river may vary from storm to storm. During a wet season saturation-excess overland flow may be more common since the area of the catchment over which it is generated is greater. As a result river discharge peaks will be greater and lag times shorter than during drier antecedent conditions.

The occurrence of hillslope flow processes in the catchment and their relative dominance affect the speed at which water is delivered to the stream (Table 11.2). Catchments dominated by infiltration-excess overland flow have the shortest lag times and the highest peak flows. This is why urbanization can lead to increased flood risk downstream. Dekker and Ritsema (1996) showed for a clay soil in the Netherlands that flow through shrinkage cracks when the soil was relatively dry produced higher peak flows than when the soil was wet and lacked macropore channels.

11.4.2 Flow frequency

In very large catchments it is often difficult to spot individual storm hydrographs on the river at the lower end of the catchment. This is because the length of time for flood peaks to travel down the various tributaries and through the system may be many days or weeks. Therefore longer-term analysis of daily or weekly flows is required. Furthermore, over long periods it is possible to identify important differences in catchment response through seasons. Figure 11.29 shows hydrographs for two adjacent catchments in Dorset, England. There is a great difference in hydrograph response

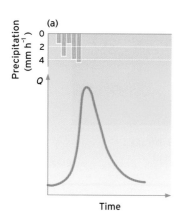

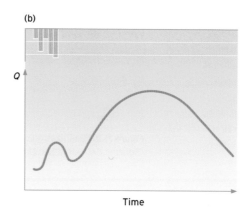

Figure 11.28 Example hydrographs for a single storm: (a) for an infiltration-excess overland flow dominated catchment; (b) for a throughflow dominated catchment.

Table 11.2 Relative flow speeds for runoff processes

Pathway	Relative speed of delivery
Infiltration-excess overland flow	Very fast
Saturation-excess overland flow	Fast
Throughflow:	
Matrix flow	Slow
Macropore flow	Medium
Pipeflow	Fast

which is not related to different climates. The Sydling Water catchment is dominated by baseflow and lies on permeable chalk limestone with high soil infiltration capacities. Therefore little storm runoff is generated within the catchment and groundwater flow dominates river flow. For the River Asker, however, there are many events when stormflow is produced and the catchment is more responsive to rainfall. There are many storm hydrographs during the year and it is

likely that overland flow or rapid throughflow mechanisms (through macropores and soil pipes) generate most of the runoff in the Asker catchment. The infiltration capacities of the soils are quite high and so infiltration-excess overland flow is unlikely to be a frequent occurrence. However, the substrate is impermeable clay in the Asker catchment and so there is little chance for percolating water to penetrate deeper subsurface layers. Therefore the soils above the bedrock are more easily saturated and can rapidly generate saturation-excess overland flow. The soils are also quite thin and there are steep slopes.

When analyzing long-term discharge records it is sometimes useful to compare **flow duration curves**. Hourly or daily flows are grouped into discharge classes and the percentage of time that any particular flow is equalled or exceeded is plotted. Often these curves are plotted on a probability scale (this is the scale on which a normal distribution plots as a straight line) as shown in Figure 11.30. This allows the lower and higher ends of the curves to be examined and compared with other curves in more detail. Generally, if rivers are being compared it is usual to plot the other axis as the discharge divided by mean discharge to allow fair comparison between large and small catchments.

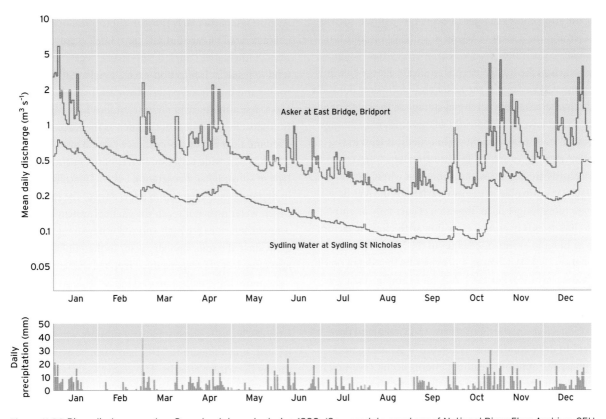

Figure 11.29 River discharge on two Dorset catchments during 1998. (Source: data courtesy of National River Flow Archive, CEH Wallingford, and the Environment Agency, South Wessex area)

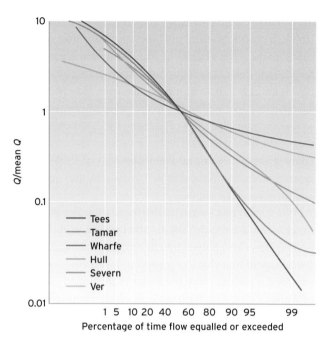

Figure 11.30 Flow duration curves for some British rivers. (Source: after Ward, R.C. and Robinson, M., *Principle of hydrology*, 4th edition, 2000, McGraw-Hill, Fig. 7.14. Reproduced with the kind permission of the McGraw-Hill Publishing Company)

Examples of flow duration curves are plotted in Figure 11.30. Flow duration curves that plot steeply throughout, such as the Tees and the Tamar (UK), denote highly variable flows with a large stormflow component, whereas gently sloping curves such as the River Ver (UK) indicate a large baseflow component. The slope at the lower end of the flow duration curve characterizes the perennial storage in the catchment so that a flat lower end indicates a large amount of storage.

Flow duration curves and regime analysis tend to suggest that river flows are stable. Unfortunately long-term flow records are not always available, but where they are available, it is possible to identify variations in river flow that are more longer term. Research is currently attempting to link trends in annual river flow with climatic factors such as El Niño events or changes in the North Atlantic Oscillation (NAO). Chapter 4 provides further detail on El Niño and the NAO. For the NAO, for example, during a positive phase northern Europe and eastern North America have warmer and wetter conditions while southern Europe, the Mediterranean, northern Africa and Greenland have cooler but drier conditions. When the NAO is in a negative phase the opposite weather patterns occur, resulting in northern Europe and eastern North America having cooler but drier conditions while southern Europe, the Mediterranean, northern Africa and Greenland having warmer but wetter conditions. Therefore the behaviour of a particular

NAO phase can result in a potential increase in frequency of flooding in certain areas and drought in others.

Figure 11.31 highlights the opposing relationships that exist between river flow and the NAO. The River Agueda in Portugal and River Guadalquivir in Spain both lie in close proximity to the Atlantic Ocean and hence have a good negative relationship with the NAO (when the NAO is positive, little precipitation falls in southern Europe and drought is frequent). The River Kent located in north-east England lies in close proximity to the North Atlantic and therefore has a strong positive relationship with the NAO (when the NAO is in a positive phase, rainfall increases, therefore flooding is more frequent). The Loire River in France, while benefiting from having a long time series, is also a heavily developed river system both historically and presently. Therefore the relationship between the NAO and river discharge is somewhat masked. However, a negative relationship is evident. Current research interest is focused on trying to locate where the boundaries of the different relationships between river discharge and the NAO lie and whether they are changing with climate change.

11.4.3 River flow in drylands and glacial regions

Arid zones, particularly those in subtropical drylands, are characterized by rare but intense rainfall events. Such high rainfall intensities coupled with sparse vegetation cover tend to result in infiltration-excess overland flow, rapid runoff and high flood peaks. However, many dryland soils are coarse and sandy in nature and thus have high infiltration capacities. In these areas, such as the Kalahari in Africa, overland flow is a rare occurrence (see Chapter 14). There can thus be a wide variation in response depending on the type of substrate and vegetation cover. Typically river flows in drylands will cease within a few days of the rainstorm. Much water does not reach the channel system because evaporation rates are high and soil water storage capacity is great since the soils were dry prior to the rainfall. Water is also often lost via seepage into river beds. Chapter 14 provides more information on dryland hydrology.

River discharge in glacial regions tends be dominated by two features. Firstly a seasonal control, which means that river discharge is greatest in early summer when snow and ice melt is at a maximum. Discharge can be extremely low during the winter months. The glacier ice acts as a long-term water store so that annual precipitation inputs to the catchment do not necessarily match the outputs. Changes in the size of the glacier will be a main determinant of catchment water balance. Secondly there is a diurnal control so

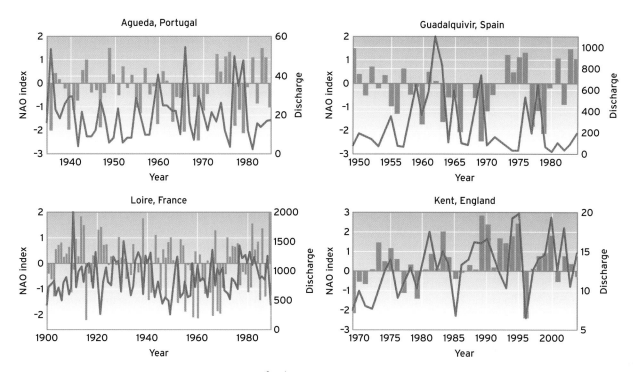

Figure 11.31 Mean annual discharge (blue, measured in m³ s⁻¹) of four rivers in Europe and corresponding NAO index (red). The strength of the NAO phase, and therefore the associated impacts, is shown by their distance from the zero axes. Note that the length of the available data record for each river is different. (Source: Figure based upon river data kindly supplied from Global Runoff Data Centre (GRDC). Courtesy of Sara Alexander, University of Leeds)

that night-time discharge tends to be much lower than that of the mid-afternoon. This reflects melting cycles of the glacier ice during daylight hours. Typically peak discharge lags a few hours behind peak temperature as it takes time for the energy to melt the ice and then for the meltwater to be routed through the river network. Further detail on glacial and periglacial hydrological processes is provided in Chapters 16 and 17.

Reflective questions

➤ What factors might affect the lag time between peak rainfall and peak river discharge?

➤ What are the main ways stormflow can be generated?

➤ Why might the dominance of particular runoff processes vary over time and space?

➤ What shape do you think the flow duration curves of Sydling Water and the River Asker should be? (Both catchments are discussed above and their annual hydrographs are shown in Figure 11.29.)

11.5 Flooding

There are different types of flood events. Most people think of occasions when rivers overtop their banks and spill out onto surrounding land. This is known as fluvial flooding. However, there are other types of flooding. Pluvial flooding can occur when there is heavy rainfall which leads to concentrated overland flow inundating an area. A town can be flooded in this way without a river overtopping its banks because urban zones often have impermeable surfaces and so overland flow can be quickly generated. Groundwater flooding can also occur in some areas where there are concentrated zones of saturation-excess overland flow. Coastal flooding caused by tidal surges, storm surges or tsunamis are also major problems in many locations.

It should be noted that flooding is a natural phenomenon and is to be expected on all rivers. The severity of any flooding will depend on the response of hillslope runoff production processes to heavy or prolonged rainfall or snowmelt and to the nature of the area being flooded. Often fluvial flood frequency is analyzed by examining flow duration curves. However, it is more common to analyze the flood record for the number of times water level in a river peaked above a given (often critical) level. Sometimes this water-level value is not the same as a discharge value because the lower parts of many catchments are affected by tides. If high storm discharge from upstream coincides with a spring high tide or a sea-level surge then flooding can be exacerbated. If long-term river-level records are available then simple return period calculations are possible. For example, if a water level greater than 20 m occurred 10 times in 10 years we would say that the return frequency of the 20 m flood at a given point is on average once per year. This helps us determine how often certain flood events might take place. Of course we could get a 20 m flood occurring three times in one year and the return frequency is just an average value. However, inferring what might be expected in the future from flood events that have happened in the past may not be reliable given that land management change and climate change might impact hydrological processes operating within a given catchment.

The magnitude of floods has been both increased and decreased by human action in different locations. Additionally humans have decided to live in low-lying areas subject to flooding. These areas tend to be where there are fertile soils (often made more fertile by regular flooding) suitable for crops and where navigation of rivers by boats allows transport of goods and people. So floods should be expected if people live on floodplains and it is possible to produce maps of areas prone to flooding. Nevertheless people still choose to develop land and live in zones where flooding is likely. Humans have modified the landscape of many regions to change both deliberately and inadvertently the size and frequency of floods. Covering more of the landscape with concrete and tarmac, which are impermeable, and then channelling flow into drains that feed streams, will inevitably lead to increased flood risk. Large-scale change across entire catchments could change flood risk too, such as deforestation or overgrazing which compacts the soil, leading to reduced infiltration.

Additionally it is possible that human modification of the climate may also lead to changes in flood frequency and flood magnitude. For example, it is forecast for some areas that a warmer climate could mean more intense rainfall events. This could lead to more floods. However, predicting flood risk is very difficult, especially in a changing climate. If a flood of a particular size has occurred twice in the past hundred years it might be called the 1 in 50 year flood. However, that does not mean it will occur only twice in the next 100 years. With changing climate (or land management) the same size flood could occur 20 times (1 in 5 year flood). The difficulties of predicting the risk from flooding are of major importance to humans because often designs of flood defence are made based on flood return periods (e.g. to protect against the 1 in 100 year flood).

In some locations flooding has increased in recent years due to both changing precipitation regimes (e.g. positive NAO in northern Europe) and land management activity which results in enhanced production of stormflow. One of the problems with building larger and better flood defences around our towns and cities in order to counter greater river discharges is that the flood water has to go somewhere. Many solutions to flooding have involved either building taller levees or embankments next to rivers or straightening the river channel and clearing out the sediment and vegetation to allow faster, more uniform river flows. However, there are many examples from around the world where these techniques have led to worse flooding downstream (e.g. the Mississippi had its worst (and most disastrous) flooding in 1993 after many years of river engineering works). This is because sending the water more quickly through one part of the river system simply reduces the lag time downstream and increases the overall flood peak. Floodplains normally have a function: they act as a temporary store of water, which means that the flood peak downstream is not as great as it would otherwise be. However, the demand for building or farming on the flat fertile floodplains of the world means that fewer and fewer of them are freely available for a river to store its water. Thus flooding in lowland areas is becoming an ever-increasing problem as the extra water is brought downstream more quickly and in greater quantities than ever before.

Changes in flood risk caused by land management change are not always as simple as would be first expected. For example, it is not possible to say that if you cover 20% of a catchment with trees to take up more water that you will reduce the flood risk by 20%. Indeed, such an activity that reduces flooding in a small area might actually lead to more flooding downstream. Figure 11.32 shows a catchment with hydrographs at different points down the system. Within this larger catchment a small tributary catchment is highlighted. If land management were to change in this catchment that resulted in a lower flood peak and delayed lag times (e.g. dense trees were planted) it would be expected that this would be beneficial to those downstream. However, the figure shows that because the peak from the small tributary has been delayed, it now peaks at the same time as the peak in the main river channel. Therefore it is contributing most discharge at the same time as the main river. The effect of this synchronous peaking is an increase in the flood peak in the main river channel. This will therefore have a negative impact on those living downstream and increase flood risk. This example highlights the importance of considering whole catchments and not just single parts of catchments. Flood risk mitigation strategies must examine whole catchments and models are needed that allow examination of the

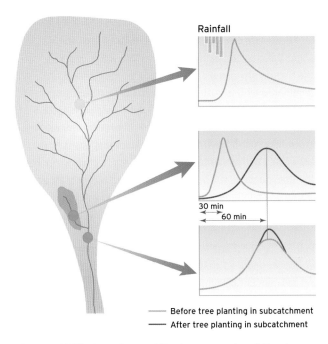

Rainfall

30 min
60 min

—— Before tree planting in subcatchment
—— After tree planting in subcatchment

Figure 11.32 The importance of flood wave synchronicity when considering impacts of land management change on flood risk. Despite the tree planting activity causing the small tributary catchment to have a lower flood peak, this has still resulted in a higher overall flood peak in the main channel. This is because the timing of the flood peak from the tributary has been delayed so that it now matches that in the main channel and so it makes the overall flood larger.

timing and volume of water moving through the drainage network across whole catchments. Some parts of a catchment might be more sensitive to change than others and the same change in one part of a catchment (e.g. tree planting) might not have the same impact if that change had occurred in another part of the catchment.

Because the hydrograph shape and relative size of the peak flows are often determined by hillslope runoff processes, human activities can affect the form of a storm hydrograph. Soil erosion, deforestation, afforestation,

grazing intensity (compacting the soil surface) and other agricultural activities (such as ploughing, irrigation, crop growth) or the development of towns and drain systems are ways in which the hydrograph shape may be affected. Building dams and changing or diverting stream courses are other ways the hydrograph shape could be altered. If the soil surface or the subsurface structure of the soil is altered in any way then it is likely that the flow paths for water from the hillslope will alter and the shape of the hydrograph will change. Many of these changes may not be simply reversible. For example, many organic soils such as peats change their structure when dried. If they are rewet, their structure does not revert to its original form and is permanently damaged. Therefore in some cases, changing the land management and causing damage to soils might lead to permanent change in flow paths, which could lead to enhanced flooding. If land management was altered back again to its original use, this may not reduce flood risk because the damage to the soil has already occurred.

Reflective questions

➤ Based on climate change predictions for your area how might your local river flow change over the next 50 years?

➤ What factors might increase or decrease flood risk for a river system?

➤ For a river near your home find out what land management changes have taken place within its catchment: what are the likely impacts of these land-use changes on the hydrology of the catchment in terms of water balance, runoff production processes, shape of the storm hydrograph and long-term river flow?

11.6 Summary

This chapter has illustrated the main components of catchment water budgets and shows how each of these can be measured. The main problems with measurement techniques are related to extrapolating point data over an

entire catchment and errors associated with the measurement techniques themselves. Advances are being made on collecting spatial data such as precipitation radar and remote sensing of soil moisture but these still need to be calibrated using point data.

Water reaches the river channel in a range of ways, some by direct precipitation on the channel but mostly by movement through soil or bedrock. Water that runs over the land surface as overland flow is often water that is being returned to the surface from the soil except where infiltration-excess overland flow occurs or where fresh rainwater meets an already saturated land surface and mixes with return flow. Throughflow occurs in a variety of forms, each of which affects residence time of water in the soil and contact with soil or bedrock constituents. Throughflow can contribute to river stormflow and to river baseflow. The dominance of particular runoff processes in contributing to river flow both changes over time (e.g. during a storm or over seasons) and varies spatially across a catchment.

The nature of river discharge is controlled by the water balance and by the ways in which water reaches the river in any given catchment. Some catchments are dominated by flashy stormflow with minimal baseflow (sometimes none) whereas others with deep permeable soils and rocks are dominated by baseflow with limited stormflow runoff. There is a spectrum of responses in between. Environmental change can lead to changes in the dominance of particular runoff production processes within any catchment or to timing of delivery of water to the river channel and through river channel systems. This leads to changes in catchment response to precipitation in both the short and long term. Some land management or climate-induced changes to runoff production processes may not be simply reversible.

Further reading
Books

Anderson, M.G. and Burt, T.P. (eds) (1985) *Hydrological forecasting*. John Wiley & Sons, Chichester.
This text contains a wide range of important chapters covering hillslope hydrology; overland flow, subsurface flow, infiltration and channel routing are all covered very well.

Arnell, N. (2002) *Hydrology and global environmental change*. Pearson Education, Harlow.
This is an excellent companion to this chapter. The book is full of great examples and explains terms and processes clearly. There are many useful comments and diagrams about how environmental change and hydrological processes interact.

Herschy, R.W. (ed.) (1999) *Hydrometry, principles and practice*. John Wiley & Sons, Chichester.
This book contains an overview of river gauging techniques and instruments. It is a little limited on some aspects of the water budget, but written very clearly with lots of useful photographs.

Jones, J.A.A. (2010) *Water sustainability: A global perspective*. Hodder Education, London.
This book provides a large-scale perspective on hydrology and water resource management.

Lehr, J.H. (ed.) (2005) *Water encyclopedia*. John Wiley & Sons, New York (5-volume set).
This is a comprehensive reference work with excellent articles on all aspects of hydrology. I contributed eight articles to this encyclopedia.

Petts, G.E. and Calow, P. (eds) (1996) *River flows and channel forms*. Blackwell Scientific. Oxford.
There are lots of good examples of runoff production processes and catchment management in this book.

Price, M. (2002) *Introducing groundwater*. Nelson Thornes, Cheltenham.
This is a clear, well-illustrated and nicely written book on groundwater. The best introduction to groundwater you can read. It also contains material on analysing river flows.

Shaw, E.M., Beven, K.J., Chappell, N.A. and Lamb, R. (2010) *Hydrology in practice*. Taylor & Francis.
This is a detailed overview of hydrology with practical material and case studies.

Ward, R.C. and Robinson, M. (2000) *Principles of hydrology*. McGraw-Hill, London.
These authors provide one of the best general textbooks on hydrology.

Further reading
Papers

Clifford, N.J. (2002) Hydrology: the changing paradigm. *Progress in Physical Geography*, 26, 290–301.
A review article on hydrology.

O'Connell, E., Ewen, J. O'Donnell, G. and Quinn, P. (2007) Is there a link between agricultural land-use management and flooding? *Hydrology and Earth Systems Sciences*, 11, 96–107.
The search for farming impacts on flood risk.

Uchida, T., Asano, Y., Onda, Y. and Miyata, S. (2005) Are head-waters just the sum of hillslopes? *Hydrological Processes*, 19, 3251–3261.
A paper that discusses the scale of approach to hydrological measurement.

Walker, J.P., Houser, P.R. and Willgoose, G.R. (2004) Active microwave remote sensing for soil moisture measurement: a field evaluation using ERS-2. *Hydrological Processes*, 11, 1975–1997.
An article that tests techniques for soil moisture determination.

Wood, S.J., Jones, D.A. and Moore, R.J. (2000) Accuracy of rainfall measurement for scales of hydrological interest. *Hydrology and Earth System Sciences*, 4, 531–543.
A paper that raises issues around water budget calculations.

CHAPTER 12

Fluvial geomorphology and river management

David J. Gilvear[1] and Richard Jefferies[2]

[1]School of Biological and Environmental Science, University of Stirling
[2]Scottish Environment Protection Agency, Edinburgh

Learning objectives

After reading this chapter you should be able to:

➤ understand basic controls on the processes of flood generation and sediment delivery to rivers

➤ define and measure the size and shape of river channels
and drainage networks

➤ explain how water and sediment move in river channels

➤ describe the environmental controls on the size and shape
of river channels and rates of channel change

➤ show an awareness of how an understanding of river
channel behaviour can be used to manage rivers in a more
sustainable and environmentally sensitive way

12.1 Introduction

Rivers vary greatly in appearance with changes both from
source to mouth and between individual rivers. It is this
morphological diversity that is the fascination for many
people. In the case of Europe, contrast the quietly flowing
small chalk-fed streams shaded by overhanging willows
of southern England with the turbulent milky-coloured
torrents draining glacial regions of the Alps. Similarly,
contrast the raging torrents in the Alps that form the
headwaters of the River Rhine with the more tranquil
character of the same river as its passes through the 'polder' landscape of the Netherlands and discharges into
the North Sea. At a local scale, salmon fishermen identify
reaches and 'pools' of unique character to improve their
chances of a successful catch. The size, shape and location
of a river can also be transformed overnight by a single
large flood, by depositing sediment in some areas and
reactivating other reaches by erosion. Every year there
are damaging floods in most countries. Extreme flooding
occurs every year somewhere in the world. The extensive
flooding in Pakistan in 2010 cannot have escaped anyone.
The United Nations estimates that 21 million people were
displaced and structural damage is estimated as exceeding $4 billion. At a smaller scale, flash floods in Jeddah in
Saudi Arabia in 2009 caused massive destruction and loss
of life. In England in 2009 flooding also caused extensive
damage to land and property and unfortunately loss of
life at Workington when a major road bridge collapsed

owing to erosion of the foundations. Floods and the associated fluvial processes of sediment erosion and deposition are a major threat to humankind and one likely to get worse under global warming scenarios. Of lesser human consequence, the salmon fishermen may return following a flood to find that the location and nature of their favourite pools have changed. Indeed, on the River Tay in Scotland, salmon fishermen pay large amounts of money to obtain lifetime fishing rights to a particular 'pool' only to return following a very large flood to find it infilled with sediment.

Rivers are dynamic landscape features that adjust their morphology, in both time and space. They are a significant hazard to humans and yet have also been vital to human civilization. They provide landscapes that have a very high nature conservation value via the habitats they provide for plants and animals. Describing and explaining variability in space and time of in-channel sediment mobilization, transport and deposition processes together with the landforms they create are encompassed within the scientific discipline called fluvial geomorphology. The movement of sediment from catchment-wide sources via hillslope runoff, together with subsequent deposition on floodplains, also forms part of the subject of fluvial geomorphology but is the focus of other chapters within this book (Chapters 8 and 9). This chapter will examine the geomorphology of river channels.

Fluvial geomorphology, until relatively recently, was a subject practised by a few academics in their quest to understand the evolution of river landscapes. Modern fluvial geomorphology involves detailed field study of processes using precise measurements, often involving sophisticated equipment, statistical analysis of the field data obtained, mapping and monitoring change on rivers using airborne and spaceborne remote sensing and modelling using large and powerful computers and numerical models. Fluvial geomorphologists now work alongside civil engineers, water resource managers, planners and ecologists in a common quest to utilize rivers for water supply, power and navigation, to alleviate flooding and destruction of houses and transport networks, and yet to maintain the nature conservation value of rivers. Fluvial geomorphologists are also leading the way with regard to river restoration, which focuses on reversing the historical legacy of environmental degradation of rivers caused by inappropriate **channelization**, bank protection and river flow regulation methods.

12.2 Catchment processes: energy and materials for rivers

12.2.1 Runoff, river regimes and floods

Catchment runoff is controlled by regional climate and catchment characteristics such as topography, geology, soils, vegetation and land use. The percentage of rainfall that reaches the channel may vary from more than 90% to less than 10% depending on the water balance (see Chapter 11). Seasonal variability in the water balance, together with a number of other variables, controls the pattern of streamflow throughout the year. Such patterns are known as runoff regimes and may be revealed simply by plotting daily flow against time, but more often are shown using monthly flow data, sometimes expressed for each month as a ratio to mean annual flow. At the global scale four major river regimes can be identified (Figure 12.1). The first type of runoff regime is one dominated by snow and ice melt which produce a major peak of streamflow during the late spring in the case of snow melt or early summer with glacial **ablation** (see Chapter 16). In temperate, oceanic areas precipitation occurs all year with a winter or autumn maximum, but the runoff regime is more the reflection of the marked peak of evapotranspiration during the summer months. Tropical, non-equatorial, river systems receive high precipitation

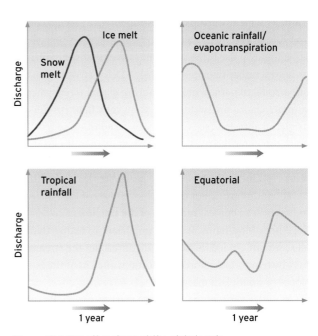

Figure 12.1 Runoff regimes at the global scale.

during the summer but experience a marked dry season during the winter. Evapotranspiration is high at all times so that the streamflow mirrors the seasonal pattern of rainfall. Finally equatorial rivers have more complex regimes because precipitation has a bimodal distribution with two clear maxima.

At the regional scale, although river regimes may show broad similarity, closer inspection of flow fluctuations reveals distinct differences. In steep, impermeable mountainous catchments with intense precipitation episodes, regimes with rapid rates of water-level change and large flood peaks are apparent. These are known as **flashy regimes**. In contrast, lowland catchments with permeable geologies have slower rates of water-level change and smaller flood peaks. These are often termed 'subdued' river regimes. Such differences, particularly in the size of flood peaks, have important implications for river channel morphology. Many river regimes have been highly modified by human river flow regulation; in most cases this results in a reduction in the size and frequency of floods and an increase in low flows.

12.2.2 Sediment sources and delivery

Sediment is transferred to river systems from a variety of catchment sources, including surface erosion on hillslopes, by rill and gully erosion and landsliding (see Chapters 7, 8 and 9), usually during or following intense precipitation and runoff. Sediment delivery to river systems is therefore often highly pulsed (the sediment arrives in distinct phases or pulses, rather than continuously). Delivery will often be seasonal in nature, most notably in glacial rivers where summer glacial meltwater flushes out **rock flour** produced by glacial abrasion (see Chapter 16). In addition sediment can enter the river directly when river banks erode floodplain materials or adjacent valley sides. The bulk of a river's sediment comes from the headwater located within mountainous areas. For example, over 80% of the Amazon's sediment load comes from 12% of the catchment covering the Andes Mountains. It is also important to realize that not all sediment eroded in the catchment reaches the stream. It may be temporarily stored in locations such as at the base of slopes or floodplain margins.

As with river regimes, there are marked variations in the volume of sediment eroded at the global and local scales. At the global scale the highest natural **sediment yields** are observed for catchments receiving between 250 and 350 mm precipitation annually with a sparse vegetation cover. With increasing precipitation, vegetation cover increases and

Table 12.1 Average sediment yield (t km^{-2} yr^{-1}) for a variety of rivers

Kosi, India	3130
Brahmaputra, Bangladesh	1370
Colorado, USA	424
River Ystwyth, UK	164
Orinoco, Venezuela	90
Zambesi, Mozambique	75
Amazon, Brazil	67
Danube, Russia	27
River Exe, UK	24
Volga, Russia	19

sediment release from the catchment markedly declines (Langbein and Schumm, 1958). Sediment yield thus tends to reach a minimum with a forest cover at 750 mm precipitation annually. Further increases in annual precipitation appear to offset any further vegetation biomass increase and slightly elevate sediment yields beyond the 750 mm maximum (Douglas, 1967). Given that precipitation varies widely at global and regional scales and the Earth's surface varies from horizontal plains to steep, sometimes near vertical, mountainous terrain the average sediment yield of rivers varies widely (Table 12.1). The sediment yield per unit area is highest for small rivers although obviously the total load of large rivers usually exceeds that of smaller ones. In the case of the United Kingdom, the average value lies somewhere in the range 30–40 t km^{-2} yr^{-1} but values range from less than 1.0 to 500 t km^{-2} yr^{-1}. It is important to realize that human activities can also release vast amounts of sediment to river systems and modify the natural sediment yield. Construction activity and deforestation are two activities that are known to inject large pulses of sediment to river systems. Poor agricultural practices, such as ploughing downslope, lead to sheet wash, rilling and gullying and can also result in elevated sediment input to rivers.

Apart from the total load the **calibre** of sediment also varies widely and this has important implications for channel morphology. Tropical weathering tends to result in the majority of sediment delivered to river systems being silts and clays. In temperate areas silts and clays are also prevalent but in mountainous terrain, sands and gravel and even boulders enter the stream system. In Arctic and alpine areas coarse sediments are well represented but glacial erosion also produces vast amounts of silt-sized material. Knowledge of river sediment loads and the calibre of the material is important in river engineering. It can determine the size of sediment traps at water intakes and the lifetime of a flood storage reservoir, for example.

> **Reflective questions**
>
> ➤ Apart from water what else does a river transport in the natural environment?
>
> ➤ What are the main controls on how much sediment reaches the mouth of a river?

12.3 River channel morphology: measuring rivers

Critical to understanding the geomorphic behaviour of river systems is the ability to measure their morphological attributes precisely (Kondolf and Piégay, 2003). Stream channels consist first and foremost of a drainage network whereby each time two river networks join, the downstream flow channel dimensions nearly always increase. Discharge can diminish downstream in arid areas owing to evapotranspiration, such as in the Okavango River in southern Africa. For a given river channel reach there are five basic variables or **degrees of freedom**: slope (which is also influenced by the nature of channel network), channel width, depth, **channel planform** and bed-roughness. Reliable measurement of these variables needs to take into account the fact that their values change not only at the regional scale but also locally. Furthermore, precise definition of the terms in a field situation can be complex.

12.3.1 Channel networks and slope

Despite the variable size and nature of stream patterns, each network has been found to have an ordered internal composition. Horton (1945) used this observation to promote the **stream order** as a way of describing drainage networks in a numerical manner. In his ordering system each source stream was designated as a first-order stream; two first-order channels meet to generate a second-order stream, and so on. He then drew a line from the highest-order stream to the headwaters along the channel, which involved least deviation from the line of the main stem of the channel network. He then related this line to changes in altitude. Therefore it is along this line that channel slope changes from source to mouth are measured. Measurement is undertaken

either by using contour levels and distance measurements from a map or by field survey. When the values are plotted against distance from the stream source this is known as the channel **long profile**.

Ordering methods have been refined subsequent to the work of Horton. Strahler's (1957) modification of Horton's ordering system has been most widely adopted (Figure 12.2a). Strahler also designated headwater tributaries as first-order streams with their meeting producing a second-order stream and so on, but he omitted the second reordering procedure so that all headwater tributaries are designated first order and the main channel is not redesignated along its length. Other variables used to quantify the stream network include: **main stream length** (MSL, km) which is the distance of the main river channel from source to mouth and equates with the length over which the long profile is measured (Figure 12.2b); **total stream length** (TSL, km) which is the combined length of all components of the channel network; and **drainage density** (DD, km) which is the drainage area (DA, km^2) divided by the TSL (Figure 12.2b). These variables are important

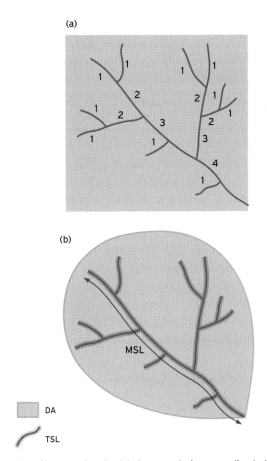

Figure 12.2 Stream networks: (a) stream ordering according to the systems of Horton and Strahler; (b) measuring mainstream length (MSL) and total stream length (TSL) and drainage area (DA).

because they reflect the combined effects of topographical, geological, pedological and vegetation controls on catchment hydrology. Indeed, these variables have been utilized as a surrogate for river flow in the analysis of channel morphology and provide clues as to the nature of the flood hydrology. High drainage densities, for example, often mean flashy flow regimes.

12.3.2 Channel cross-section: width, depth

River channels are three-dimensional linear features. To describe their reach morphology, in terms of size and shape, their form is broken down into channel cross-section and planform. Channel size and shape vary locally within a reach. In order for reliable measurements to be made at one location that can be compared with measurements elsewhere on the river, the cross-section should be measured on a straight reach and located consistently with reference to river bed landforms (see section below on channel bed morphology). Channel size and shape are usually quantified by measuring **bank-full** channel dimensions in cross-section. Bank-full conditions occur when the discharge just fills the channel, and this is significant because it is the water level at which flow resistance (via friction of the bed and banks) tends to be at a minimum so the conveyance of water at this level is most efficient. Efficient transfer of water is sometimes termed hydraulic efficiency. Identification of the position of bank-full stage in the field, however, is often difficult. A simple uniform channel cross-section with two straight banks that intersect a floodplain at a sharp angle is rare in nature. The valley floor on either side of the channel may be at different elevations, the banks have irregular sides that gradually merge with the floodplain and vegetation may overhang and colonize banks masking their presence.

Measurement of the channel cross-section can be used to determine channel bank-full width (W), channel depth (d), length of the **wetted perimeter** (length of the two river banks plus channel bed) at bank-full (P) and channel cross-sectional area (A) (Figure 12.3). These variables can also be used to calculate two other key variables: the width–depth ratio (W/d) and the **hydraulic radius** (R), which is A divided by P.

12.3.3 Channel planform

The planform, or pattern, of a channel can be divided into three broad types, namely braided (Figure 12.4), meandering (Figure 12.5) and straight. In reality many channels are intermediate types and in addition to these three broad types, **anastomosing** and wandering planforms have been identified. Variables quantifying the form of undivided or

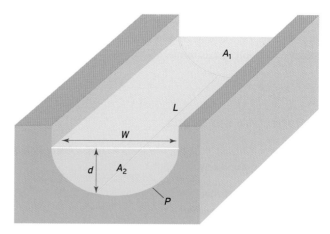

Figure 12.3 Morphological measurements of stream channel cross-section: channel bank-full width (W), channel depth (d), length of the wetted perimeter (length of the two river banks plus channel bed) at bank-full (P) and channel cross-sectional area (A). L is the distance from one cross-section to the next where the area, depth, width, etc., may be different.

single channels are numerous. The most common is **channel sinuosity** (Figure 12.6). This is the ratio of the length of river between two points to the length of the valley between these two points. Rivers display a continuum of sinuosities from a value of 1 to more than 5. Straight channels are defined as having a sinuosity of less than 1.5. A meandering channel refers to a single channel with a number of bends which result in a channel sinuosity in excess of 1.5. However, other variables can be used to describe meandering channels further. These include the meander width, meander wavelength and radius of curvature. In the case of divided channels, channel multiplicity can be calculated by measuring the total length of the perimeters of sand and gravel islands (known as **bars**) and vegetated islands in the river and dividing by the length of the channel reach over which this process was undertaken. One problem with this approach, however, is that the value obtained is not a constant and varies with the water level at the time of survey. Bars can become 'drowned' during high water levels and emerge during low flows.

12.3.4 Channel boundary materials

Channel boundaries can consist of cohesive sediments, sands and gravels, bedrock or vegetation. Bedrock or vegetation boundary types need only be recorded as such, but as a basis for analysis of sediment transport, detailed field sampling and subsequent analysis of bed sediments are required. For sand and silts, or mixed sands and gravels, a bulk sample or grab sample is abstracted and analyzed for

Figure 12.4 A braided river in Turkey. Note that some of the gravel bars appear to be bulldozed and that gravel extraction is taking place at the top of the picture. (Source: courtesy of Tory Milner)

Figure 12.5 A meandering tributary before it enters the larger Yukon River, Alaska. (Source: Science Photo Library Ltd: Bernard Edmaier)

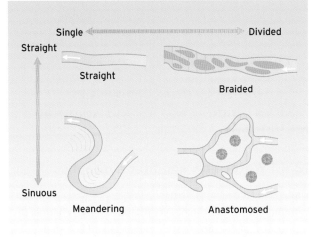

Figure 12.6 The four main river planform types defined according to channel sinuosity and channel multiplicity: braided, meandering, anastomosing and wandering.

particle size. Sediment mineralogy and roundness can be measured using a hand lens and used to help determine sediment sources. For coarse channel bed sediments, which typically have a great range of sizes, often 100 particles are selected randomly and their size measured individually with calipers or a pebble sizing plate (Bunte and Abt, 2001). If all the particles are roughly of the same size the sediment is referred to as well sorted, whereas if they are highly variable in nature they are poorly sorted. In general, riverine sediments tend to be well sorted but can show a bimodal distribution. Roundness is judged visually. **Imbrication** of particles, whereby their long axes tend to be aligned along a common vector due to flow direction, can also be noted and used to understand flow paths during floods.

Movement of stream bed material can be monitored using tracers. Traditionally bed material was painted and the loss and occasionally subsequent discovery of the pebbles could be used in the first case to identify movement and in the second track movement distances (Kondolf and Piégay, 2003). More recently use has been made of magnetized pebbles, which enables relocation using a metal detector, or pebbles with a radio-tracking device inserted (Sear *et al.*, 2000). Sediment load can also be calculated by recording the amounts of material in sediment traps and stored in reservoirs (see Chapter 9).

Reflective questions

➤ Imagine yourself by your local river. Try to decide where you would define the bank-full limit to be on the channel.

➤ How do you overcome the problem of measuring sediment size when the river bed comprises pebbles with a wide range of sizes?

12.4 River channel processes: understanding water and sediment movement

12.4.1 Water flow and flow hydraulics

Understanding what governs the velocity of water flow and amount of flow that can be accommodated by a channel of a given size, shape, **boundary roughness** and bed slope is central to fluvial geomorphology. Water within a channel is subject to two principal forces: the force of gravity, which induces water flow, and frictional forces. Hence, a channel

in which gravitational forces are high and channel boundary roughness is low will have a high average water velocity. A similar channel, however, with a channel boundary consisting of boulders, and thus high frictional forces, will have a slower average water velocity, even though the turbulence of the water may give the impression of fast-moving water. Mean water velocity in an open channel can be estimated using the Manning equation:

$$V \, (\text{m s}^{-1}) = \frac{R^{0.66} \, S^{0.5}}{n} \tag{12.1}$$

where V is mean water velocity, S channel bed slope (expressed as a gradient in m per m such as 0.005), R hydraulic radius (see above) and Manning's 'n' is a measure of channel roughness (Manning, 1891; Chow, 1959; Simon and Castro, 2003). Indices of bed grain size are usually used to quantify channel roughness although in-channel vegetation and the overall shape of the channel (form roughness) also need to be taken into account. Defining a representative value of bed roughness for heterogeneous channel beds is problematic, particularly since larger particles have a greater effect on roughness than small particles. Values will thus only be an approximation of roughness.

For a given cross-section the way in which water velocity, depth and width increase with a rise in water level has been termed 'at-a-station hydraulic geometry' (Langbein, 1964). Water velocity, width and depth express themselves as power functions of discharge:

$$w = aQ^b \tag{12.2}$$

$$d = cQ^f \tag{12.3}$$

$$v = kQ^m \tag{12.4}$$

where Q is stream discharge in $\text{m}^3 \, \text{s}^{-1}$, w is water width in metres, d is water depth in metres and v is water velocity in m s^{-1}. Since wdv is equal to Q it can be established that the sum of the exponents b, f and m is 1 and the sum of the intercept values a, c and k is equal to 1. In most river channels, velocity increases more rapidly than water depth and width. Typical values for b, f and m are 0.1, 0.4 and 0.5. Stream power is another useful measure of river flow hydraulics and is described in Box 12.1.

Within a river channel cross-section, considerable variation in water velocity and flow characteristics occurs. The velocity distribution in cross-section can be shown by **isovels**, lines of constant velocity. Generally water velocities are greatest in mid-channel and much lower at channel margins and close to the stream bed (Figure 12.7). At river bends centrifugal forces give rise to an excess fluid pressure on the

TECHNIQUES

STREAM POWER

One of the most important geomorphic variables is **stream power**, which describes the energy available in a particular area of channel. Total stream power (Ω), measured in watts, is calculated at any given cross-section in a river by:

$$\Omega = \rho g Q S \qquad (12.5)$$

where Q is stream discharge (m³ s⁻¹), S is channel slope, g is acceleration due to gravity (9.8 m s⁻¹) and ρ is water density (1000 kg m⁻³). Discharge is a measure of the volume of water passing through a particular cross-section during one second, so total stream power is the amount of energy available within a single

second and within a given cross-section on a channel to do geomorphic work, such as eroding, moving or depositing sediment.

Total stream power is a useful variable to calculate in order to understand the driving force available for morphological change, and can be related to rates of bank or bed erosion or to understand channel dynamics and channel type.

Unit stream power (P) (sometimes known as specific stream power) provides a more detailed insight into morphological processes because it describes the energy available per unit width of channel:

$$P = \frac{\rho g Q S}{w} \qquad (12.6)$$

where w is water width in metres and the other variables are as in equation (12.5). Because this calculation is absolute (W m⁻¹), it can be used to compare the energy regime between cross-sections meaningfully, and to predict when particles are entrained and the rate at which they move downstream. Unit stream power can, therefore, be used to assess the likely morphological impact when one or more independent fluvial variables are altered (e.g. change to channel slope, width, depth, bed or bank erodibility or discharge), such as an increase in channel slope during river diversions or an increase in discharge under future climate change scenarios.

BOX 12.1

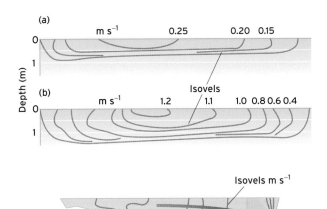

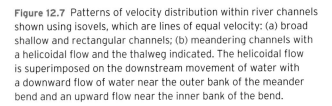

Figure 12.7 Patterns of velocity distribution within river channels shown using isovels, which are lines of equal velocity: (a) broad shallow and rectangular channels; (b) meandering channels with a helicoidal flow and the thalweg indicated. The helicoidal flow is superimposed on the downstream movement of water with a downward flow of water near the outer bank of the meander bend and an upward flow near the inner bank of the bend.

outer bank and a deficit on the inner bank. This creates a surface water gradient between the inner bank (lower elevation) and outer bank (higher elevation). The deeper water on the outside of the bend, having a greater hydraulic head, falls to the bottom of the outer bank (into the pool) and then moves up to the inside of the bend. The water therefore circles outwards as it moves downstream, following a spiral or **helicoidal flow** path. The area of maximum water velocity, known as the **thalweg**, moves towards the outer bank in response (Figure 12.7). This idealized model of flow occurs in sinuous channels, but the pattern of flow is often altered or dominated by local controls on channel morphology such as funnelling of flow between boulders or trees.

12.4.2 Sediment movement

River beds contain sediments of mixed size and these may be classified in terms of how they are transported: in suspension (wash load), or along the bed (bed load). Transport of fine sediments, which are typically sourced from soil erosion and river bank erosion, is termed wash load and sediments are predominantly suspended in the water and transported downstream. Coarse sediments such as gravels and cobbles, which are typically sourced from river bank or bed erosion and hillslope input, are dominantly moved by rolling, hopping or bouncing (known as saltation).

The type of transport – and where and when it occurs – depends on flow hydraulics. A particle can be mobilized or entrained if the forces applied to it are greater than the resisting forces. Applied forces include the fluid drag from moving water (which depends on velocity, large-scale turbulence and the kinematic viscosity of the water) and the down slope potential energy of the particle's submerged mass. Resistance to **entrainment** is provided by packing and by gravity (the submerged weight of the particle). Taken as an average over time, **shear stress** may be quite low, but erosion can still occur because of turbulence from eddies. Most channels usually show considerable variation in the levels of turbulence and eddies in the flow create short-lived (seconds) and localized (metres) peaks in shear stress that entrain larger particles (Bridge, 2003). Once in motion, particles transported as bed load roll, slide or saltate (hop) along the bed in a shallow zone only a few grain diameters thick (see Chapter 9). If the particle is small (typically < 2 mm), or entrainment forces are sufficiently high (due to increased discharge, or localized turbulence), it may be carried upwards into the main body of water and transported in suspension. Deposition and cessation of bed load movement for an individual particle occurs when velocity falls below the critical conditions. Hence, assuming no change in hydraulic conditions, a particle will come to rest only if it becomes lodged against an obstruction or falls into an area sheltered from the main force of the water by a larger particle.

Hydraulic sorting is an inevitable consequence of selective transport whereby, in non-cohesive sediments, finer particles are preferentially moved downstream. Another useful concept is that of **stream competence**. This is the largest size of particle that a stream can carry as bed load at a single time or position. Stream capacity refers to the maximum volume of sediment that a stream can carry.

In contrast to bed load, suspended load transport is determined not only by the discharge of the river and nature of flow, but also by its rate of supply from the drainage basin. In many cases the suspended load is 'supply limited'. This means that the sediment transport capacity of the river exceeds the rate of supply of sediment from the catchment. Supply limitation for the finest particles of suspended load, especially clay and silt, occurs most of the time in rivers because bank and hillslope sediment sources are activated only during precipitation events and high flows. However, many channels have flow conditions that are capable of transporting the wash load over a wide range of flows. Thus, although there is a broad correlation between discharge and suspended sediment transport, a plot of the two can exhibit a wide scatter of points which relate to temporal variations in hillslope controls on sediment mobilization. Sediment movement, in both suspended and bed load form, is highly pulsed, with 'waves' of sediment moving through the river system. As a result sediment transport in rivers has been likened to a jerky conveyor belt (Ferguson, 1981).

Within a given river channel reach, if net erosion and deposition are equal, the river bed will remain at approximately the same elevation. Channels that are neither aggrading nor degrading may remain vertically stable but often undergo planform changes as the river migrates naturally across the floodplain. These changes are usually relatively slow and are normally evident if observed over a period of years. If erosion exceeds deposition, a lowering of the river bed (degradation) and subsequent bank collapse usually occurs. Depositing channels occur when the input exceeds output, resulting in vertical aggradation of the bed and triggering lateral deposition along channel margins. Often this results in an increase in the rate of bank erosion and channel change. Channel and floodplain landforms and their stability, in this context, can also be viewed in terms of the balance between sediment input and output and a natural tendency to migrate across the floodplain. Identifying which situation occurs in a reach of interest is of critical concern to river management. A knowledge of sediment budgets and the possibility of sediment starvation because of trapping and removal of sediment upstream is also of paramount importance in terms of the stability of infrastructure and, in particular, bridges. In recent years a number of notable bridge collapses have been caused owing to undermining of bridge foundations. Unparalleled flood discharges could also cause excessive scour around the base of bridge piers. Geomorphologists have a responsibility to be able to predict the possibility of altered sediment budgets and the effects of changes in sediment budgets on channel morphology. Stable channels, especially bedrock channels, are less likely to be a problem although short-term fluctuations in channel form caused by flooding or sediment pulses moving down the river can bring about erosional or depositional change. The role of floods and discharge of varying magnitude in controlling sediment transport is discussed in Box 12.2.

FUNDAMENTAL PRINCIPLES

THE DOMINANT DISCHARGE CONCEPT

Rivers erode their beds and banks and receive inputs of sediment from the surrounding landscape during floods. In general, floods with higher flows have greater potential to erode and transport sediment per unit time. Medium-sized floods, however, although transporting less sediment per unit time, occur more frequently and can thus transport greater sediment loads in the long term. In the long term the total amount of geomorphic work is the product of the sediment transported during medium-size floods and their frequency (Figure 12.8). The flood discharge that achieves the greatest total geomorphic work is referred to as the dominant discharge and is often considered to equate to bank-full discharge, although this is not necessarily the case (Emmett and Wolman, 2001). Small floods are generally ineffective in mobilizing coarse river beds and causing bank erosion but can transport wash load introduced to the river by hillslope runoff. In many large temperate rivers the majority of sediment transport is accomplished by flood events which occur between twice each year and once in every five years. However, while this concept is useful, it is also overly simplistic because morphological change usually lags behind hydrological change. Also, dominant discharge does not equate to bank-

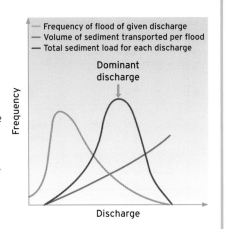

Figure 12.8 Illustration of the derivation of dominant discharge.

full in some rivers, particularly in highly active river channels (braided or wandering).

BOX 12.2

Reflective questions

➤ Do you think that water moves through a reach fastest in upland boulder-bed rivers or gentle lowland rivers?

➤ Why do river channels sometimes usually have clear water during low flows but are turbid at high flows?

➤ Do you think that river channels adjust their size and shape to optimize their carrying of water or sediment or both?

12.5 River channels: linking channel processes and morphology

Linking channel size and shape to water and sediment transport processes would be relatively easy if channels responded proportionally and instantaneously to the size of floods. This does not occur, because different elements of a channel's morphology have different susceptibilities to change and they may change over different timescales (Figure 12.9). Rivers have been described as having historical hangovers because of the lag between process change and landform response.

When considering fluvial processes, it is useful to consider over what timescales change occurs. These include:

- instantaneous timescales (10^{-1}–10^{0} years) (hours/days);
- very short timescales (10–10^{1} years) (1–10 years);
- short timescales (10^{1} and 10^{2} years) (10–100 years);
- medium timescales (10^{3} and 10^{4} years) (1000–10 000 years);
- long timescales ($>10^{5}$ years) ($>$100 000 years).

Discharge and sediment loads will vary in instantaneous time as a result of individual flood events. Cross-sectional form parameters adjust over instantaneous and short timescales, planform and local-scale profile change over short and medium timescales, and the overall longitudinal river profile only changes over the medium and long term. Figure 12.9 illustrates this diagrammatically by showing how different features of varying size (or length scale) change over different timescales.

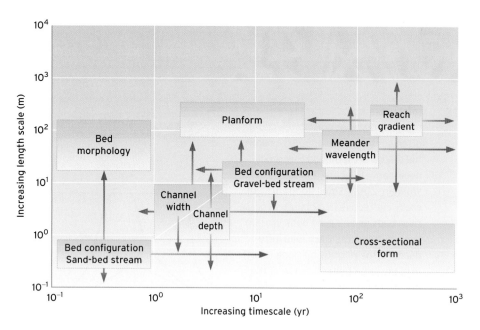

The concept of equilibrium is central to linking channel processes and morphology. True equilibrium could occur only if all morphological variables responded in instantaneous time, but as stated above this is rarely the case. Negative feedback is an important component of equilibrium between form and process. For example, if a channel is too narrow for a given flow and sediment concentration during an exceptional flood event, bank erosion may enlarge the channel and reduce flow velocity. Channel form thus oscillates around an average and this is referred to as dynamic equilibrium (see Chapter 1, especially Figure 1.4).

Of course no adjustment will occur if the channel boundary prevents erosion because it is resistant bedrock or strengthened by a vegetation root mat, concrete or artificially placed bank protection such as **gabions** (wire cages filled with rocks; Figure 12.10). In these cases the discharge will be accommodated in the existing channel or by inundation of the valley floor. Valley floor inundation occurs during a flood even if channel adjustment is occurring, because the rate of change of discharge is more rapid than the rate at which the channel can adjust its dimensions and bed-roughness. If there is a large flood that enlarges the channel, followed by a period of years with smaller floods, flow velocities will be reduced in the enlarged channel, resulting in sediment deposition and a return towards the morphology that existed before the exceptionally large flood event. However, if there is no sediment delivery from upstream for deposition then little or no adjustment is likely to occur in channel dimensions. The time taken for a channel to return to its original form after a flood-induced change is

Figure 12.10 Flood engineering through a village on Inchewan Burn, a tributary of the River Tay, Scotland. The bed and banks were made up of gabions but the coarse bed material during flood events has ripped some of the gabions on the bed and so for much of the year the flow moved through the gabions rather than over them, preventing fish movement. Since this photograph was taken in 2009 this reach has now been restored by removing the gabions and placement of large boulders in some cases fixed with steel pins. Water now flows permanently over the bed.

termed the relaxation time or recovery period. In temperate rivers this is usually fast in relation to the return periods of floods. Rapid re-vegetation is an important component of the recovery. Human activity that alters fluvial processes, such as channelization or river flow regulation, may alter the rate of change so that the river system is no longer in equilibrium. Processes of feedback (negative or positive) may then create a new (and possibly undesirable) equilibrium state, such as a wandering channel, which is characterized by accelerated rates of bed incision, bed deposition or bank erosion. An understanding of equilibrium is therefore important for river management.

12.5.1 Long profile

The **long profile** (slope of a river from its source to mouth) is typically concave with progressively lower gradients downstream. The degree of concavity, however, varies among rivers according to a host of factors. These include inherited landscape form (which is dominated by glaciation in most of the UK), geology, tectonics and variability in runoff. Indeed many rivers that drain passive continental margins (see Chapter 2) have a significant convexity. Natural variations to the concave downstream model also occur owing principally to interruption by lakes or resistant rock bands. Where particularly resistant rock bands exist, waterfalls or rapids result (Figure 12.11). At the reach scale the long profile may locally steepen or be reduced in gradient owing to localized aggradation and degradation processes. At the larger scale such aggradation or degradation can be caused by a rise or fall in base level (e.g. from sea-level change or dredging within a reach).

12.5.2 River channel cross-sections

Channel cross-sections adjust, as described above, to accommodate the discharge and sediment load from the drainage basin, as permitted by the constraints of bed and bank erodibility, sediment dynamics within the section, and channel slope. Consider how the channel bank-full cross-section varies along a river from source to mouth. Change in channel form from upstream can be summarized by stating that with distance downstream there is an increase in bank-full cross-sectional area. However, there is not a linear relationship between increase in discharge and channel cross-sectional area. A given increase in channel cross-sectional area will result in a proportionally greater increase in bank-full discharge capacity because larger channels are typically more efficient at transporting water. Secondly, boundary roughness generally decreases downstream,

(a)

(b)

Figure 12.11 Waterfalls (a) and rocky rapids (b) often occur at a particularly resistant rock band. In this case the rocky rapids shown in (b) are frozen during winter in a boulder-bed river in North Korea.

which promotes faster flow for an otherwise unchanged cross-section. This reduced roughness is able to offset the decrease in transport capacity caused by a decrease in the channel slopes that usually occur downstream. The relationship between downstream increase in discharge and bank-full indices of channel morphology is known as 'downstream hydraulic geometry' (Leopold and Maddock, 1953) and average values are:

$$w^b = aQ^{0.5} \qquad (12.7)$$

$$d^b = cQ^{0.4} \qquad (12.8)$$

$$v^b = kQ^{0.1} \qquad (12.9)$$

where w^b, d^b and v^b are channel bank-full cross-sectional width, depth and velocity values, respectively. Such relations imply an adjustment of channel shape whereby channels become broader and shallower in proportional terms while velocity only increases slowly downstream. These equations describe the change in channel form downstream in uniform sediments. However, bed and bank erodibility and sediment load are important in controlling channel cross-sectional geometry. Channels with a high percentage of silt/clay in their banks, and rivers transporting much of the sediment load in suspension, tend to be narrower and deeper than sand and gravel-bed rivers. This is due to flow hydraulics. Channels can carry material in suspension most effectively under certain optimal hydraulic conditions: typically, semicircular channels are hydraulically most efficient at transporting water and, therefore, suspended sediment. Many river engineering projects, such as straightening and alteration of the cross-section shape over long distances, have sought to create a hydraulically efficient channel. Semicircular channels are, however, liable to bank collapse, so a trapezoidal cross-section is used instead (flat bed and sloping banks). Although trapezoidal channels are hydraulically efficient, they are often poor at transporting coarse bed sediment, which requires high bed shear stresses and a wide, shallow channel over which the material can be mobilized and transported. When engineering river channels it is important to maintain a mobile bed and to ensure that the downstream continuity of bed sediment transport through the reach from upstream is maintained.

Vegetation can also be important in controlling cross-sectional form by influencing bank resistance and flow hydraulics. Root systems increase bank strength, so vegetation-lined channels tend to be narrower (Zimmerman *et al.*, 1967) and more stable (Abernethy and Rutherfurd, 2000). A reduction in the vegetation cover often results

in bank erosion which, once initiated, is difficult to stop. Large wood within the channel provides extra roughness and hydraulic complexity and has a significant effect on cross-section morphology (Gregory, 1992; Keller and Macdonald, 1995) and can dissipate energy effectively. The most effective river management is that which works with these natural processes rather than against them.

12.5.3 Channel planform

When viewed from above, channels vary greatly in appearance. They range from those with tortuous bends that snake through the landscape to straight and multiple-channel rivers. The processes controlling the development of these channel patterns through time have long been the focus of fluvial geomorphology (e.g. Leopold and Wolman, 1957; Kellerhals and Church, 1989; Van der Berg, 1995). A conceptual model of morphological types in relation to some controlling process variables is shown in Figure 12.12. Studies have consistently demonstrated that channel slope and discharge are important controlling variables. Early work by Leopold and Wolman (1957) based on numerous field studies of gravel-bed rivers in the United States suggested that for any given discharge there is a threshold slope above which channels will meander. They also found another higher threshold above which they will braid. They also found that these critical slope thresholds decreased with increasing discharge (Figure 12.13a). Thus, braided planforms tend to be found on large rivers, or on small rivers with steep slopes (where stream power is high). Braided streams have several channels which except at high flood are divided by active coarse-grained bars. Each channel may be sinuous but overall a fairly straight planform exists and the width–depth ratio is large. At the lower end of the energy spectrum, anastomosing rivers have a network of semi-permanent and very slowly changing interconnected sinuous channels with cohesive banks, although a variety of forms have been identified in temperate, tropical and Arctic environments. The relationship between planform and the primary control of stream power is complex owing to other influencing variables; Simpson and Smith (2001) working on the Milk River in North America found that it failed the Leopold and Wolman (1957) slope–discharge test in that it was braided when its discharge and slope values suggested it should be meandering. They suggested that this was because the river had a sandy bed whereas Leopold and Wolman had worked mainly on gravel-bed rivers in the United States. Clearly bed sediment size is important in controlling river planform.

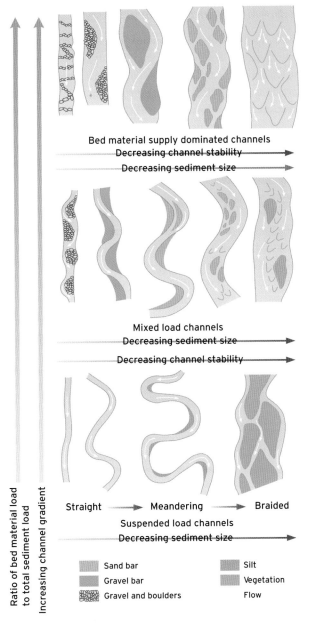

Figure 12.12 Conceptual model of morphological types of channels indicating the conditions (sediment size, channel gradient, ratio of bed load to suspended load) under which river channels will be straight, meandering or braided.

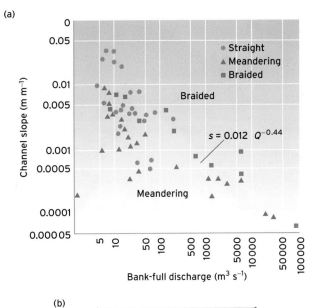

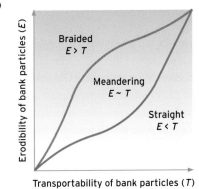

Figure 12.13 (a) The relationship between braided and meandering planforms and the controlling variable of slope and discharge. (b) Straight, meandering and braided patterns defined in terms of erodibility and transportability of bank particles.

Where the bed material is coarse, the bank material highly erodible, or the dominant mode of sediment transport is by bed load movement, there will be a tendency for the river to be laterally active, as in active meandering, wandering or braided rivers. River channel planforms can be differentiated using the ratio between the transportability of bank particles to the erodibility of bank particles (Figure 12.13b). Braided and **wandering gravel-bed rivers** tend to have uncohesive coarse floodplain sediments that provide high rates of sediment input (from bank erosion and bank collapse) that travel only

for short distances (typically less than 1 km) during floods. In areas where bank erosion cannot occur, such as in bedrock-lined rivers, channels tend to be straight or follow the pattern of geological structure (e.g. faults). If there is sufficient energy in the system, rivers are likely to meander, although research in undisturbed, forested rivers in the Pacific North-west in the United States has shown that rivers containing very large volumes of large dead wood tend also to have a lower sinuosity because the wood dissipates the majority of the energy in such systems (Keller and Swanson, 1979), which is usually used to create the secondary flow forms required for meandering.

It is important to understand the potential response of channel planform when river management alters a process control such as width, depth, sinuosity, slope, or water or sediment discharges. If the activity takes a stream across a threshold then the river can change its form and level of stability. Furthermore, if channels are wholly constructed and

12.5.4 Channel bed morphology

The morphology of the stream bed changes downstream. Often there are bedrock channels in the upper section of the long profile with an exponential decline of bed material particle size downstream. This downstream fining is due to particle sorting and abrasion. Particle sorting occurs because smaller particles are preferentially entrained and transported further downstream (Ferguson *et al.*, 1996). Abrasion, involving grinding and chipping, causes the bed material to become more rounded downstream. Perturbations to this general pattern are caused by injections of sediment from tributaries, hillslope inputs or bank collapse, by lithological changes along the long profile or by reach variability in stream power. Lakes and artificial structures such as weirs will also interrupt the pattern of sediment transport.

Superimposed on the long profile and cross-sectional geometry are a variety of erosional and depositional bedforms. The type of forms depends upon whether the bed is composed of predominantly sands, gravels or bedrock and upon channel gradient. Within gravel-bed rivers the most common bedforms are pool–riffle sequences. Pools are characterized during low flow by relatively slow flowing water with fine bed material. **Riffles** are formed by accumulation of relatively coarse material and are characterized by shallow, more rapidly flowing water (Figure 12.15). The spacing of pools and riffles is often five to seven times the

Figure 12.14 The Arve River between Argentiere and Chamonix in the French Alps following bulldozing and straightening.

a planform is adopted which is not in equilibrium with the processes it is unlikely to be a success for long. Bulldozing the Arve River between Argentiere and Chamonix in the French Alps (Figure 12.14) into a straight, narrow and deep channel is unlikely to be a long-term solution to channel instability and flooding. The Arve River's natural tendency is towards a braided pattern and so it will try to return to this form following the channelization, causing major local instability.

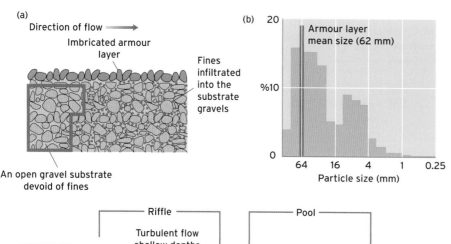

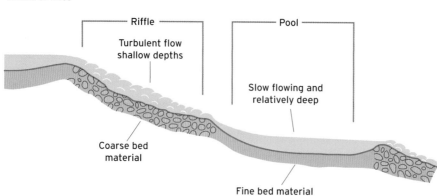

Figure 12.15 A pool-riffle sequence showing the main morphological attributes of pools and riffles (viewed from the side) including (a) a schematic diagram of stream bed armouring with and without infiltrated fine sediment and (b) a particle size distribution of the bed sediments on the gravel-bed River Tryweryn, Wales.

Figure 12.16 Large sand dunes formed during the high-flow season, Cinaruco River, Venezuela.

channel width but a variety of spacings have been found in natural channels. In mountain streams where the size of bed materials is large relative to the size of the channel and/or the channel slope is steep (typically greater than 10%), a step–pool morphology, with steps of coarse material such as boulders (Montgomery and Buffington, 1997) or large dead wood (Keller and Swanson, 1979), sometimes replaces pools and riffles.

The channel bed of sand-dominated reaches is commonly composed of ripples (less than 40 mm in height and 600 mm in wavelength) and larger **dunes** forming traverse sand bars (see Chapter 9). During periods of high discharge sand dunes can be formed, as seen in Figure 12.16. The size and shape of these bedforms change with time since their form is directly related to flow velocity. Downstream migration of dunes and ripples occurs as material is carried up the stoss side and avalanches over the crest and down the lee side (see Chapter 9). At very high flow velocities a flat river bed can be formed or **anti-dunes** created. Anti-dunes migrate upstream as erosion from the downstream side of the anti-dune throws material into saltation and suspension more rapidly than it can be replenished from upstream.

Such sedimentary structures can be related directly to stream power and the diameter of the bed material. In bedrock channels a variety of sculptured forms can be found, in part controlled by the rock type and structure. The most well-known feature is the **pothole** caused by **corrasion** (mechanical wearing and grinding), **cavitation** (pressure changes due to bubble collapse in turbulent flow) and **corrosion** (chemical weathering). In some bedrock rivers rhythmical forms similar to pools and riffles have also been identified. Thus, the rapids within the Grand Canyon of the Colorado River, which are well known to white-water rafters, are spaced at intervals of 2.6 km with deeper sections in between.

Sizeable accumulations of bed load material are referred to as bars (Figure 12.17). For example, riffles are **lobate** gravel bars but are usually studied independently of other bar forms owing to their regular spacing. Bars may consist of coarse or fine material, be relatively stable or highly mobile, and be attached or detached from the river banks. In general, mid-channel bars are found on braided rivers, lateral and diagonal bars on wandering gravel-bed rivers, whereas point bars form on the inside of bends within **meandering rivers** (Figure 12.17). Most bar forms consist of an upstream portion of

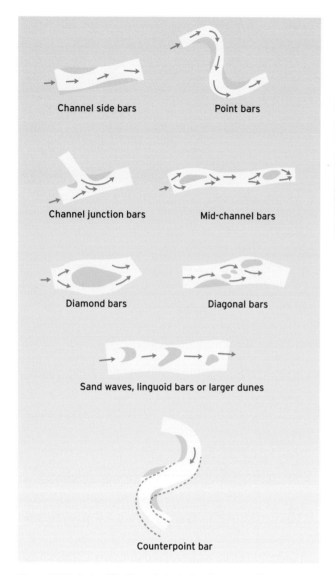

Figure 12.17 A classification of simple bar forms and flow patterns.

Figure 12.18 A vertical 'freeze core' of a gravel substrate. The core has been collected by inserting a tube into the sediment and then filling the tube with liquid nitrogen. This freezes the water around the tube and the nearby sediment thereby allowing it to be pulled out with the sediment intact for investigation. The core shows mixed grain sizes below the coarser 'armour' layer that is characteristically found at the surface.

Reflective questions

➤ Why are roof gutters normally semicircular in cross-section?

➤ If the river were also to carry larger gravel-sized material would you design it with a semicircular cross-sectional form?

12.6 River channel changes: rates and types of channel adjustment

sediment known as the bar head, with a tail of finer material. Associated with bar forms and channel bed topography is variability in grain size. Grain size tends to be greatest in the centre of channels and finer at the channel margins (relating to the normal pattern of flow velocity across the channel). Within meandering channels, because water velocity is greatest close to the outer bank, grain size tends to be at a maximum here and slowly decreases towards the inner bank. Exceptions to these generalizations will reflect complex flow hydraulics. In natural channels with non-uniform channel morphology, patterns of flow velocity during floods are highly variable and complex. The vertical structure of bed forms in gravel-bed rivers typically consists of a surface layer of stones with finer material beneath. The coarser layer is referred to as an **armoured layer** as it protects the finer material from being transported (Figure 12.18).

River channels adjust their channel slope, cross-section, planform, bed morphology or stream network over a range of timescales in relation to natural and anthropogenic changes (Figure 12.19). This chapter is focusing on short timescale changes because of their direct relevance to river management. Some changes take place slowly and gradually while others can occur almost instantaneously in response to individual flood events. Channel change can also be distinguished according to whether it is **autogenic** or **allogenic**. Autogenic change refers to fluctuations of channel form about an equilibrium condition. Allogenic change refers to adjustments of channel form in response to a change through time in the sediment and water regime of the river. Box 12.3 provides a simple way of determining how channel shape might change if discharge or sediment input changes.

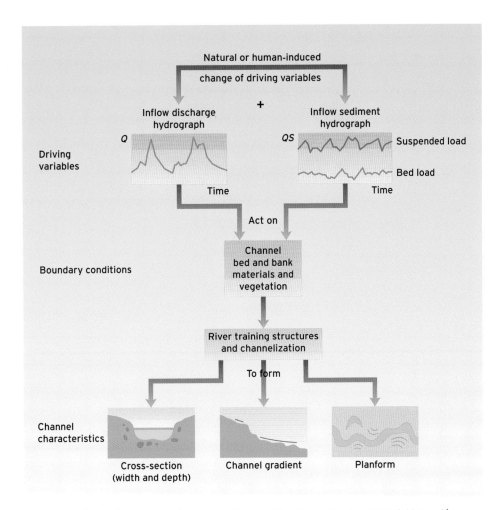

Figure 12.19 Principles of channel form and change. Alteration of the driving variables and/or boundary conditions causes changes in channel characteristics.

TECHNIQUES

PREDICTING CHANNEL RESPONSE TO CHANGES IN DISCHARGE AND SEDIMENT YIELD

Predicting how channels might adjust to changes in sediment yield, sediment type and discharge is complex. Rivers vary in their sensitivity to discharge or sediment load. Similarly their sensitivity in terms of morphological adjustment to direct human intervention such as chan-

nelization, river margin vegetation removal or gravel extraction, and indirectly via land-use change, varies greatly (Kondolf *et al.*, 2002). The highly eminent American fluvial geomorphologist Stan Schumm, however, produced a set of relationships as to how channels would respond to change in discharge or sediment yield (Schumm, 1977). Using a plus or minus sign to denote an increase or decrease respectively, the effects of a change in discharge (Q) or bed

load (Q_b) on channel forms can be hypothesized:

$Q+ w+, d+, (w/d)+, \lambda+, S-$

$Q- w-, d-, (w/d)-, \lambda-, S+$

$Q_b+ w+, d-, (w/d)+, \lambda+, S+, P-$

$Q_b- w-, d+, (w/d)-, \lambda-, S-, P+$

where w is width, d mean depth, λ meander wavelength, S channel gradient and P channel sinuosity.

The algorithms were based on Schumm's observations of

BOX 12.3 ➤

➤

predominantly sand-bed channels in semi-arid and sub-humid regions. Therefore their wider applicability has yet to be fully tested. Furthermore, changes in discharge or sediment load rarely occur alone because of their interdependence on climatic and catchment variables. Four combinations can be postulated:

$Q+, Q_b+$
   $w+, d+-, (w/d)+, \lambda+, S+, P-$

(such a change may be caused by urbanization)

$Q-, Q_b-$
   $w-, d+-, (w/d)-, \lambda+, S+-, P+$

(such a change may be caused by adoption of good agricultural management practices such as use of buffer zones and water retention ponds)

$Q+, Q_b-$
   $w+-, d+, (w/d)+-, \lambda+-, S-, P+$

(such a situation may occur when a river is in receipt of water from a 'donor' stream)

$Q-, Q_b+$
   $w+-, d-, (w/d)+-, \lambda+-, S+, P-$

(such a situation may occur where construction or mining activity takes place in a catchment).

These simple algorithms can be useful in assessing how a channel might respond to climate change or human activity and they may predict the direction of change. Such simplistic models, however, should be used with caution. Channels with highly variable ratios of bed to bank resistance and complex modes of sediment transport can respond in different ways. Even if the directions of change are predicted, the rate and magnitude of change are still likely to be unknown. The river will alter whichever component is the easiest. For example, if the banks are soft and erodible and the bed is bedrock, the river will widen.

BOX 12.3

12.6.1 Cross-sectional change

The extent of bank or bed erosion will depend upon the resistance of the channel bed and banks. Sand-bed rivers with highly mobile beds, such as the Fraser River in Canada, can experience changes of 5 m in bed elevation during a flood season. On such rivers even greater bed lowering can occur around bridge piers where scour is exacerbated. Knowledge of such bed-level change during floods is critical in creating sound foundations for river structures such as that shown in Figure 12.20. Channel widening usually takes place during extreme floods often resulting in movement of a river by more than a whole channel width. In the United States there is evidence that floods of 100–200 year recurrence interval have resulted in channel widening of between 60 and 600%. A 100 year recurrence interval flood on natural reaches of the River Tummel in Scotland, however, resulted in channel widening of 10% despite the presence of uncohesive sand and gravels.

12.6.2 Planform change

Changes in channel pattern vary according to planform geometry. Meandering rivers shift their position primarily by extension, translation, rotation or enlargement (Hooke, 1980) (Figure 12.21a). However, differences in bank strength, resulting from sediment and vegetation variability, ultimately cause more complex changes in form to occur (Micheli and Kirchner, 2002). Braided river channels change chaotically in response to bar development and shift their position laterally across the floodplain. Wandering gravel-bed rivers exhibit a number of types of movement including meander development and **avulsion**. Avulsion is the process whereby a channel shifts from an old course to a new course, leaving an intervening area of floodplain intact.

Generally, rates of channel shifting are greater for braided rivers than wandering gravel-bed rivers and slowest for meandering rivers. Locally, however, bank erosion on the outside of meander bends can be quite rapid and ultimately result in meander cutoffs (Figure 12.21b). In the United Kingdom, medium-sized meandering rivers have been observed to have migrated across more than more than 50% of the floodplain in less than 200 years. For small and medium-sized meandering rivers (less than around 20 m wide), average bank erosion rates are often less than 0.1 m yr^{-1}, usually less than 0.5 m yr^{-1}, but may reach values of 5 m yr^{-1} or more. In contrast, some meandering channels appear static with little or no change in planform over a few hundred years or more. Mobility appears to relate to the degree of incision and stream power. The meandering Luangwa River in Zambia is an example of a highly mobile meandering river. On the Luangwa River average annual bank erosion rates have varied between 1 and 20 m over the 40 years from 1956 to 1997 (Gilvear *et al.*, 2000). Figure 12.22 shows a reach of the Luangwa in 1956

Figure 12.20 Exposure of bridge piers on a tributary of the Luangwa River, Zambia. The original bed level can be seen by the 'ragged' concrete on the piers.

and 1988 where channel migration and meander cutoff have occurred. Such vast changes in the channel have resulted in safari lodges being swept into the river. The safari lodges are located on the outside of the meanders, despite the erosion threat, because animals come down to the river edge to drink. They do so via the point bars on the opposite side of the river from the safari huts and thus many animals can be easily spotted. Prediction of the rate and direction of meander migration is problematic on meandering rivers. The rate of bank erosion varies around

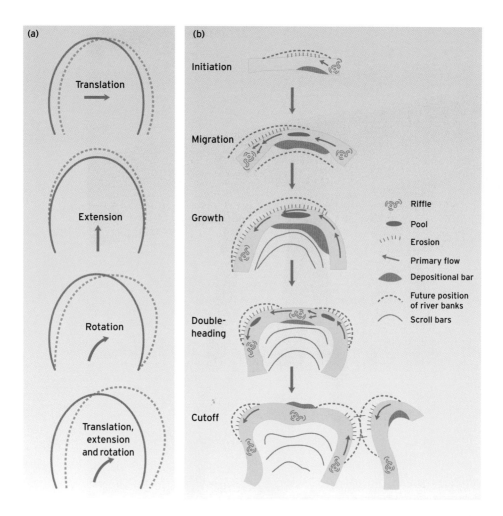

Figure 12.21 Modes of channel planform change on a meandering river: (a) types of change and (b) the development of a straight reach into a meandering reach and then into a straight reach as the meander bend is cut off leaving 'oxbow' lakes on the floodplain.

the meander bend and as the meander develops and alters its sinuosity. The rate of erosion will also be dependent upon the size and timing of floods over proceeding years and on the resistance of river bank sediments encountered as the river migrates across the floodplain.

12.6.3 Human-induced change

Over the last few thousand years human activity has significantly modified the majority of river channels in developed parts of the world. This modification accelerated over the past 200–300 years through channelization, straightening and embanking. In addition there have been changes resulting from human activity altering river flows and sediment yields. Such activities include reservoir construction, urbanization, building construction, mining, land drainage and vegetation

change such as afforestation and deforestation. For example, Wolman (1967) produced a model of the response of channels to land-use change in the Piedmont region of the United States and this model is shown in Figure 12.23. His model of land-use change over the past 200 years deals with the conversion from forest to urban land with interim stages of arable agriculture, reversion to woods and grazing and construction activity. The model suggests episodes of aggradation, scour, stability and bank erosion within the affected river channels. For example, channels below urban areas in the United States have been shown to have channel capacities of up to six times the size of their rural counterparts. In the United Kingdom values up to 150% are more common (Gregory et al., 1992). The nature of channel change in detail is complex and a good example is the case of rivers regulated by dams as discussed in Box 12.4.

Figure 12.22 Channel planform changes on the meandering Luangwa River, Zambia, over a 32 year period.

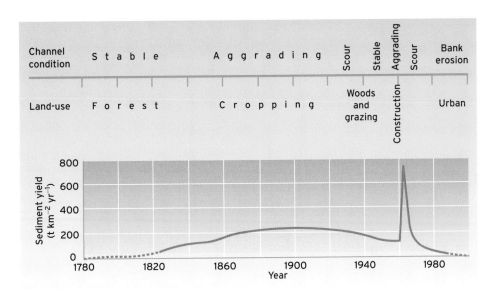

Figure 12.23 A model of variations in sediment yield and channel response over time, Piedmont region, USA.

ENVIRONMENTAL CHANGE

RIVER CHANNELS BELOW DAMS

Dams or weirs typically disconnect river habitats by preventing the upstream migration of fish and the downstream movement of sediment, and large dams often alter hydrology, particularly dominant discharge. A common response, except in bedrock channels, to the release of sediment-free water below dams is degradation of the channel bed mainly because sediment is lost from the reach but not replaced. In time degradation progressively moves downstream. This process often continues unless bedrock is met, the channel becomes armoured to the effects of clear water erosion, or an unregulated tributary injects sediment to the regulated river. Such degradation and erosion can

be prevalent along the whole course of the river downstream of the dam. Increased coastal erosion 965 km downstream of the High Aswan Dam on the Nile has even been blamed on the impoundment.

The primary response of the regulated river to the change in flood regime, however, usually seems to be a decrease in channel capacity. This usually takes the form of a reduction in channel width which can often be over 50%. Width reduction is largely achieved through the formation of depositional bars and **berms** and subsequent vegetation colonization. Shrinkage of over 50% in the width of the River Spey in Scotland, down-stream of the Spey Dam constructed in 1938, occurred over 50 years (Figure 12.24; Gilvear, 2003).

Many structures were built in the early part of the twentieth century

or before and are now reaching the end of their design life. Fluvial geomorphology has increased the general understanding of how such structures affect river stability and habitats. Ownership of such large, old and potentially dangerous structures also carries a significant liability. As a result, many old dams are now being removed or modified to allow fish and sediment to move through the fluvial system. In 2011 the world's largest ever dam removal project took place with the removal of Elwah and Glines Canyon Dams within the Olympic National Park, USA. This opened up 100 km of river to Pacific salmon, and geomorphologists are monitoring downstream dispersion of sediment formerly trapped behind the dam and related channel adjustment.

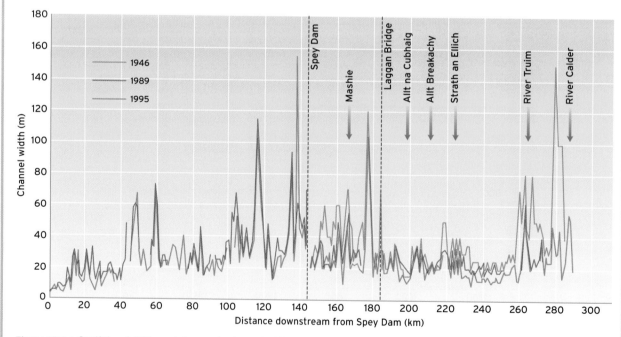

Figure 12.24 Spatial variability and changes in channel width (1945-1989) downstream of the Spey Dam, Scotland.

BOX 12.4

12.7 Fluvial geomorphology and environmentally sound river management: living and working with nature

The above sections have demonstrated that river channels are not static features in the landscape but have high levels of dynamism and are very sensitive to changes in flow or sediment yield or changes in bed and bank resistance. Such temporal variability has to be incorporated into river management if environmentally sound and sustainable solutions are to be reached. This is the focus of this section in the chapter. Rivers can be classified into different types and this often provides a simple way for managers to determine which rivers might be more dynamic than others in particular environmental settings.

12.7.1 River management and the engineering tradition

Historically, many rivers in developed countries have been straightened, or have had major changes made to their cross-section shape (widening, deepening, embankments) in order to drain land for agriculture or other development. This has typically led to significant loss of habitat area and habitat variety and ecological damage such as loss of spawning gravels for salmon. These changes are also unsustainable because river processes often result in a return to the original morphology: river systems can be thought of as having a 'memory'. The memory effect, and associated recovery, frequently leads to morphological instability that is difficult and expensive to manage. A notable example of this is the Mississippi River which regained much of its sinuosity following straightening in the early twentieth century.

However, in many developed countries there are now programmes to assess and improve environmental quality, and these recognize that alterations to morphology have significant impacts on ecosystem health. In Europe, for example,

the European Union Water Framework Directive (European Union, 2000) requires Member States to assess the ecological quality of rivers, lakes and wetlands, and to restore lost functioning where possible. The goal of the directive is for the water environment to achieve good ecological quality. To do this, individual countries have to assess the biological quality, the water quality and, for the first time, the morphological quality of the water environment. The UK has developed a Morphological Impact Assessment System (MImAS; Box 12.5) to assess how significant morphological alterations to the river are. This tool shows that around one-third of all main rivers in Scotland have significant morphological damage, and require restoration to achieve good quality (Scottish Environment Protection Agency, 2007).

Restoration will only occur where the environmental benefits outweigh the potential cost, and the morphology of some rivers is never likely to be fully restored – for instance through a major city. Depending on the river system, a different level of effort may be needed to achieve morphological restoration. Morphologically active parts of a river may recover naturally and the best approach is to leave these alone. In other rivers, where either the changes are very significant, or the energy levels in the river are naturally low, physical restoration, such as re-meandering, may be required. Restoration also needs to take climate change effects on morphology into account. However, rivers will only recover naturally if the fundamental fluvial processes of sediment and water movement through the system are maintained. This can be achieved by regulation of those activities that can interrupt such processes. Similarly, river restoration activities such as re-meandering may also fail if these are not designed with due account for system processes.

The morphological impacts of river engineering are increasingly being assessed and regulated in order to protect the existing quality of the environment. In some countries, environmental assessment and regulation of river engineering, such as sediment extraction, is increasingly used to ensure that processes of sediment and water movement operate naturally within a river system. However, many countries and land managers have yet to adopt this approach, and significant morphological impacts continue to occur. An understanding of river geomorphology is required in order to identify which rivers require specific management.

Over the last decade, therefore, the science of fluvial geomorphology has moved from being a mainly academic discipline to an applied science that is now used regularly to help manage and improve the environment. Many large engineering consultancies now employ fluvial geomorphologists to advise on projects involving large-scale river engineering works.

TECHNIQUES

ASSESSING AND IMPROVING MORPHOLOGICAL QUALITY

Many rivers have been modified by human activity. Developed countries in Europe now have a statutory requirement from the EU Water Framework Directive (WFD; European Union, 2000) to assess how these modifications have affected river morphology and associated ecological quality. In the UK, a Morphological Impact Assessment System (MImAS) has been developed. The inputs to the tool are (a) length and type of engineering modification; (b) length of the river or reach to be assessed; and (c) channel type. Each type of modification has a particular impact rating for each channel type, which is multiplied by the length of channel affected, and scaled to the reach assessment length:

$$\text{Impact} = (M \times (R/L) \times 100) \quad (12.10)$$

where M is the modification impact rating, L is the length of the modification and R is the length of the reach length being assessed. For instance, the impact rating for channel realignment on a meandering river, which is particularly sensitive to morphological change, is 0.62. If the assessment length is 1 km and 500 m of this has been realigned, then:

$$(0.62 \times (500/1000) \times 100) = 31\%$$

The WFD states that rivers should be at 'good status', which means that the ecology is healthy, and the threshold for good morphological status is 25%. In the example above, therefore, an improvement to the morphological quality is needed of 6%. Using the equation above, the tool can then be used to calculate what length of impacted channel requires restoration, which

in this instance is 96.7 m. Improvement requires restoration of the natural channel processes and morphology, and is required unless the area that needs restoration cannot be altered because it has a sustainable human use or would be disproportionately expensive to undertake. In such circumstances, the river may be termed 'Heavily Modified' and a lower-quality morphological condition is considered adequate. In Scotland, around one-third of all rivers do not reach good status for morphology and require restoration. Elsewhere in Europe, modifications to river morphology are more widespread and the requirement for restoration is likely to affect more rivers.

BOX 12.5

12.7.2 Living with rivers

As a result of the failure of a number of river management and engineering schemes there has been a radical shift in the nature of river management (Leeks *et al.*, 1988; see Chapter 27). This shift is towards working with, rather than against, natural processes and accepting the dynamic nature of river channels. This is because the long-term viability of many engineering structures cannot be assured given the highly mobile nature of rivers and the likely impact of a changing climate on river morphology in the future. Flood embankments can be lost to bank erosion, and current deflectors meant to deepen the river and curtail its lateral movement may well be largely ineffective in a river with high stream power. In this context, geomorphologists have a role in: (i) deciding on what type of rivers are activities such as channel straightening or floodplain development permissible; (ii) deciding where and how far from rivers structures can be built; and (iii) designing long-term and environmentally sensitive bank protection and engineering

solutions to river management problems. Such an approach places the emphasis on living with rivers rather than fighting against the forces of nature. The role of the geomorphologist in the three situations above is illustrated here with reference to a number of research projects.

In a study of the success of channelization schemes Brookes (1985) examined channel response in England, Wales and Denmark. The schemes varied in their type, age and extent and were on a variety of river types. Channelization schemes were found to be successful on low-energy streams but on streams with stream powers above 35 W m^{-2} stream channels reverted back towards their natural morphology (Figure 12.25). Such a finding is obviously of direct relevance to assessing where channelization schemes may be an appropriate solution to an environmental management problem (Brookes, 1985).

The role of the geomorphologist in assessing where structures should or should not be built is illustrated by the River Tay in Scotland. Here reconstruction of channel planform evolution over the past 250 years and scrutiny of

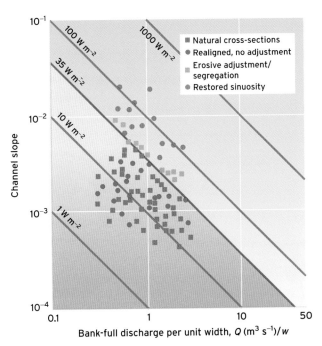

Figure 12.25 The response of river channels to channelization in relation to stream power. (Source: after Brookes, 1985)

Table 12.2 Percentage of flood embankment failures on the River Tay in comparison with their geomorphic setting (1993 flood event)[*]

Overlying relic channels less than 250 years old	14
Overlying relic channels older than 250 years	5
Outside bend of meanders	28
Overlying relic channels and outside of a bend	8
Perpendicular to valley sides	13
Others	32

[*]Subsequent to 1993, 70% of the locations have experienced failure at least once more during floods in 1998, 2003, 2005 and 2007.

the present-day morphology revealed that flood embankment failures along the river occurred more often on the outside of meander bends and where embankments had been constructed over old courses of the river (Gilvear and Black, 1999) (Table 12.2). This suggests that, in the case of the River Tay, and probably more generally (Box 12.6), it is unwise to build flood embankments over areas of relatively recent fluvial activity on high-energy and thus inherently mobile river systems.

HAZARDS

FLOODING AND BANK EROSION ON THE BRAHMAPUTRA RIVER

The Brahmaputra River drains a catchment of 580 000 km², with more than half the area lying within China and Tibet, the remainder being composed of parts of Bhutan, India and Bangladesh. In total the river is 2906 km long with its source being the Kangklung Kang Glacier in Tibet at an altitude of 4877 m. On leaving the Himalayas it flows for 640 km within the state of Assam, before entering Bangladesh and flowing into the Bay of Bengal. At its mouth the mean flow of the Brahmaputra River is 19 830 m³ s⁻¹ making it the fourth largest river in the world in terms of discharge. The Brahmaputra also has one of the highest sediment loads of any river in the world with an estimated annual load of 402 million tonnes.

Once the river meets the alluvial plains of India and Bangladesh it forms a large, wide, braided river system up to 20 km across and has a valley floor width of 70–80 km. Each year in the wet season widespread inundation of the valley floor brings misery to millions of people. To combat this flooding thousands of kilometres of embankments have been built to try to limit the extent of flooding. However, the Brahmaputra, like all braided rivers, shifts its course from year to year, threatening cities and the stability of embankments and destroying agricultural land. It is estimated that 868 km² of land in India

was lost by erosion in the twentieth century. The river is also reported to have widened and become shallower with bank erosion rates almost quadrupling in the twentieth century and bed aggradation rates of 16.8 cm yr⁻¹ having been reported. This increase in erosion and sedimentation has been linked to increased sediment delivery to the river, resulting from deforestation and an upward trend in flood magnitudes possibly linked to climate change.

Understanding how the Brahmaputra River will alter its morphology and course in the future is of crucial importance. However, on such a large river, traditional field-based approaches to collecting geomorphological data have severe limitations. High spatial and spectral resolution

BOX 12.6 ➤

➤ satellites, not to mention Google Earth, however, can now provide important information on channel planform and channel instability. These remote sensing technologies allow the mapping of water surfaces and sand bars. Images can be compared over time to determine channel planform change and the shifting position of the river (Figure 12.26). Wet-season images also allow the extent of inundation to be mapped. Such hazard mapping is important in assessing risk and developing flood mitigation strategies.

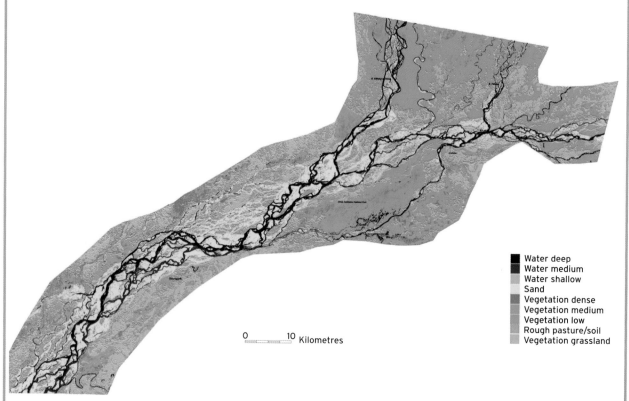

Water deep
Water medium
Water shallow
Sand
Vegetation dense
Vegetation medium
Vegetation low
Rough pasture/soil
Vegetation grassland

0   10 Kilometres

Figure 12.26 A satellite image of the Brahmaputra River, India, classified into major land cover types. The image shows a reach with a large mid-channel island and braided channel patterns. Note the scale.

BOX 12.6

Traditional engineering approaches to bank protection on meandering channels have tended to starve the river of sediment (Kondolf, 1997) and to alter flow patterns such that erosion continues to occur even where rock-filled gabions and concrete have been used to protect the river bank. An example of an alternative geomorphic approach to stemming bank erosion is the use of submerged vanes or hydrofoils. These work on the basis of modifying the flow adjacent to the bank such that the processes of erosion are reduced and bank failure prevented. The structures are located in order to generate a secondary flow cell. This secondary cell has opposite polarity to the main flow cell that results from helicoidal flow in the meander bend. Convergence of faster water occurs in the centre of the channel and causes bed scouring. Eroded material is then deposited towards the outer bank. As shear stresses against the outer bank, toe region and bank height are reduced, bank retreat is prevented. Installation of such structures has been successfully utilized on the River Roding in England (Hey, 1996).

12.7.3 River maintenance

River maintenance may be needed where the channel dynamics are in conflict with human use. For example, river flow regulation by dams can cause sedimentation and siltation of gravels below tributary junctions due to the reduced frequency and magnitude of floods. This, in turn, can have adverse impacts on aquatic organisms such as salmon, which require almost silt-free gravels to spawn.

In recent years there has therefore been a move towards releases of reservoir water down salmon rivers to flush fine sediments from coarser ones, and to encourage fish to migrate. These pulses are termed either flushing flows or channel maintenance flows. The most well-known, and probably largest, example of a channel maintenance flow was the huge one from the Glen Canyon Dam on the Colorado River in 1997 (Kaplan, 2002). However, to a water resource company this water may be seen as a loss that affects its profit margin. The company may wish to know the effective discharge and duration for the release to meet its objectives. This is an area where fluvial geomorphologists can provide advice. For channel maintenance, a general rule of thumb is that sediment within gravel-bed channels begins to be mobilized when flow depth is greater than 80% of channel bank-full depth (assuming that the dominant discharge equates to bank-full flow; see Box 12.2). Flushing flow recommendations in the United States have been either hydrologically determined and based on a percentage of the mean annual flow or geomorphologically determined. The latter involves either direct field measurement of the threshold discharge for sediment transport or modelling of sediment transport to predict the discharge that causes incipient motion of particles. Some geomorphologists argue that the flood that occurs once every 1.5 years is theoretically the most suitable discharge given that, on average, river channels have a channel capacity equivalent to this flow level. However, in reality the dominant discharge in terms of sediment transport is highly variable between reaches and rivers.

Another area where a geomorphological understanding of river systems has led to a change in river maintenance is in the field of dam sedimentation and clear water erosion problems downstream. In many places sands and gravels are now lifted over dams and put into the river downstream. The major hydroelectric company in Scotland, for example, now undertakes this practice after 50 years of removing sediments from behind its diversion dams and stockpiling the material. Alternatively the Grande Dixence hydroelectric power company in Switzerland has traps that automatically purge material downstream when they become full of sediment. This can occur up to 50 times per year owing to the high bed load yield of alpine streams. However, such river maintenance alone may not successfully impact on the downstream river unless there are peak discharges able to redistribute the material along the course of the regulated river. Indeed such schemes may exacerbate sedimenta-

tion problems that are a feature of regulated rivers. This sedimentation can actually increase flooding by reducing the channel capacity (via infilling), despite discharges being lower. It is thus apparent that a good understanding of the relationship between flow processes, erosion, sediment transport and depositional processes is necessary for improved river management.

12.7.4 Building new river channels

The principles of fluvial geomorphology are increasingly being used to help engineers build new river channels. New channels may be required for restoration, or because another human use of the land is needed (Gilvear and Bradley, 1997). As discussed earlier, the traditional approach to diverting channels was to create a straight trapezoidal channel, but these are often unstable and morphologically and ecologically impoverished. From an environmental viewpoint such traditional techniques are no longer acceptable in most circumstances. A channel morphology that replicates the natural regime is preferable. It ought to be in equilibrium with the processes of water movement and transfer and create habitat conditions that conserve in-stream biota. At its simplest the channel can 'mirror' the former natural course but in other situations design criteria need to be based on geomorphic principles and processes while incorporating some hard engineering. For example, a 2.7 km sinuous gravel-bed river diversion was built on the River Nith, Scotland, in 2005 (Figure 12.27). This partially mirrors the old channel and was constructed with reference to geomorphic principles.

Figure 12.27 The River Nith diversion, south-west Scotland, showing its engineered morphology. The point bar features visible on inside of the meander bends were not engineered but developed over the first few flood events to flow through the diversion. The spoil behind is from a coal mine which lies beyond.

(a)

(b)

Figure 12.28 The re-meandered Sinderland Brook near Manchester, England, on the day after completion (a) and three months later (b). (Source: photo (b) courtesy of Charles Perfect)

There was some hard engineering but only at critical locations where prevention of erosion was of paramount importance. The old natural river has previously been lost to valley floor mining for coal. However, before this river was obliterated the fish were rescued and introduced to the new channel. A post-project monitoring programme has shown that 5 years after construction the river has developed a wide range of fluvial features such as riffles and gravel bars, a near-natural invertebrate fauna and fish community, but perhaps still lacks the levels of physical habitat heterogeneity apparent in unmodified streams.

12.7.5 River restoration

River restoration is often undertaken where past human activity has resulted in morphological and ecological damage to river systems. Unsympathetic engineering treatments reduce channel area (for instance, due to straightening) and habitat diversity, which is important to maintain healthy ecology. Restoration can reverse these adverse morphological and ecological impacts. River restoration can also be used as a flood-control measure. For example, to reduce future flooding, the Swiss Government in 2004 announced plans to reinstate meanders on over 100 rivers at a cost of over £80 million. Re-meandering increases channel roughness and therefore reduces velocities, and in turn this reduces water velocities and attenuates the flood wave. The process often includes reinstatement of natural channel dimensions, re-meandering and creation of pools and riffles. Sometimes this is

as simple as mirroring an upstream reach or producing an exact copy of what once existed at that location. However, the original (historic) form may no longer be appropriate, for instance if discharge or sediment supply has changed owing to catchment alterations, climate change or other influences. Also, in many cases the nature of the channel that existed before the straightening occurred is not known. Therefore geomorphologists have to design channels in equilibrium with river regime, sediment load and calibre and must consider what the natural channel pattern ought to be. Empirical equations can be used to identify appropriate channel dimensions in the absence of any historic information (Soar and Thorne, 2002).

A good example of a river restoration project involving re-meandering is the case of Sinderland Brook near Manchester, England. Land was sold off for housing development. A condition of the purchase was that the straightened and highly channelized stream that runs across the land was re-meandered. Figure 12.28 shows the re-meandered morphology immediately following construction and one year later after vegetation colonization of the riparian zone had occurred. The scheme has led to increased in-stream species diversity and reduced flood risk to an existing housing estate. A different type of restoration project was undertaken on the Highland Water in the New Forest, England. Here, in the mid-twentieth century, channel realignment for forestry had caused channel instability (deepening and erosion). Figure 12.29 shows the river following reintroduction of gravel during a process known as substrate replenishment. The success of the project relied upon determining the correct bed material particle size distribution so that

Figure 12.29 The Highland Water in the New Forest, England, following introduction of bed gravels. Prior to substrate replenishment, dredging and subsequent incision had created a channel devoid of gravel and hydraulic features such as pools and riffles.

the material did not undergo large-scale removal during the first large flood but still remained sufficiently mobile so that pools and riffles would be created. Another concern

was that the permeability of the gravels did not result in the complete disappearance of water during low flows; to combat leakage the gravel was laid down in layers, consolidated and some finer material added to create less permeable layers. From these two examples of river restoration the direct and paramount importance of an understanding of fluvial geomorphology is evident.

Reflective questions

➤ Based on your knowledge of fluvial geomorphology, do you think you could design a river to be in equilibrium with a given flood regime and sediment loading?

➤ What are the main problems associated with traditional engineering approaches to river management?

➤ Why is morphological diversity in river channels important to the flora and fauna?

12.8 Summary

This chapter has focused on fluvial processes, the linkage between processes and landforms, and natural and human-induced channel change. It has also examined the relevance of fluvial geomorphology to the needs of river management in the twenty-first century.

Catchments vary in the amount of water and sediment they carry. This variation is a function not simply of catchment size but also of local topography, land management, geology, soils, vegetation, and so on. Catchments with high drainage densities are likely to have high peak flows and a flashy discharge regime. River channel networks can be described and classified in a number of ways including the stream order systems of Horton and Strahler. The size and shape of river channels can be described in terms of channel cross-section and planform which may vary over short distances. River channel planform ranges from braided and anastomosing to meandering and straight river channels. River cross-sections are not uniform and water and sediment flows through a cross-section vary. At river bends a circulatory pattern of flow (helicoidal flow) is superimposed on the downstream movement of the water.

Water within a channel is subject to gravity and frictional forces. Steep, smooth channels will have a high average water velocity. A similar channel with a channel boundary consisting of boulders, and thus high frictional forces, will have a slower average water velocity, even though the turbulence of the water may give the impression of fast-moving water. Mean water velocity in an open channel can be estimated using the Manning equation which accounts for channel slope, hydraulic radius and channel roughness. For a given cross-section the way in which water velocity, depth and width increase with a rise in water level is known as hydraulic geometry. Stream

➤

power is a key parameter in determining rates of erosion, sediment transport and instability. Slight changes in velocity can significantly affect potential stream power.

For a particle to be entrained from the stream bed or bank, a threshold has to be passed whereby a critical velocity or shear stress exceeds the frictional forces that resist erosion. This is dependent upon channel slope, particle size and shape, and immersed weight in relation to the bed shear stress and fluid kinematic viscosity. However, processes such as imbrication may also play a role. Deposition and cessation of bed load movement for an individual particle occur when velocity falls below critical conditions. Hydraulic sorting occurs under these conditions. Sediment can be carried as bed load or as suspended load and bed load transport is almost entirely a function of flow volume, velocity and turbulence. Particles roll, slide or saltate along the bed.

Bed morphology can vary depending on bed material while the nature of a channel's boundary materials may also play a role in channel stability. In coarse sediments pool-riffle sequences may dominate, whereas in fine sandy channels dune structures may be found. Bedrock channels, however, may be subject to cavitation and corrosion processes. River channel change can occur very quickly during a flood event and channels can avulse from one site to another. Channels may take a long time to recover from a large event in terms of adjusting their size and shape back to suit lower flows but this relaxation time varies from river to river.

Many channelized rivers have altered their course or shape following engineering works and this has caused on-site and upstream and downstream problems. Rivers are naturally dynamic and yet humans often require stable river channels. Thus fluvial geomorphology has a very important role to play in modern-day river management. It can identify the causes of management problems, at both reach and catchment scales, and predict the impacts of human intervention on rivers with mobile bed sediments and erodible banks. River restoration, design of new river channels and river maintenance all require fluvial geomorphological insight. However, much remains to be understood about the behaviour of river channels.

Further reading
Books

Chow, V.T. (1959) *Open channel hydraulics.* McGraw-Hill, New York.
A classic textbook on the physics of water movement in channels.

Gregory, K.J. (1992) Vegetation and river channel process interactions. In: Boon, P.J., Calow, P. and Petts, G.E. (eds), *River conservation and management.* John Wiley & Sons, Chichester, 255–269.
This chapter discusses the role of vegetation in controlling river morphology.

Gupta, A. (2007) *Large rivers: Geomorphology and management.* John Wiley & Sons, Chichester.
This book has many very useful case studies and illustrates techniques for managing very large rivers.

Keller, E.A. and Macdonald, A. (1995) River channel change: The role of large woody debris. In: Gurnell, A. and Petts, G.E. (eds), *Changing river channels.* John Wiley & Sons, Chichester, 216–236.
This chapter discusses the role of woody debris in controlling river dynamics.

Kondolf, M. and Piegey, H. (2003) *Tools in fluvial geomorphology.* John Wiley & Sons, Chichester.
This is a volume aimed at providing comprehensive details of the techniques required to map, monitor and investigate the geomorphology of rivers.

Petts, G.E. and Calow, P. (eds) (1996) *River restoration.* Blackwell Science, Oxford.
This is a series of very useful chapters on river engineering and restoration. The chapter by Hey is particularly relevant.

Sear, D.A., Lee, M.W.E., Oakley, R.J., Carling, P.A. and Collins, M.B. (2000) Coarse sediment tracing technology for littoral and fluvial environments: A review. In Foster, I.D.L. (ed.), *Tracers in geomorphology.* John Wiley & Sons, Chichester, 21–55.
A chapter on how to measure and monitor bed particle movement.

Thorne, C.R., Hey, R.D. and Newson, M.D. (1997) *Applied fluvial geomorphology for river engineering and management.* John Wiley & Sons, Chichester.
This is a volume aimed at providing an overview of fluvial geomorphology as a basis for effective management and engineering within rivers.

Further reading
Papers

Abernethy, B. and Rutherfurd, I.D. (2000) The effect of riparian tree roots on the mass-stability of riverbanks. *Earth Surface Processes and Landforms*, 25, 921–937.
This paper details the importance of tree roots in terms of bank stability.

Brookes, A. (1985) River channelisation, traditional engineering methods, physical consequences and alternative practices. *Progress in Physical Geography*, 9, 44–73.
This is a review article that shows how traditional engineering approaches to channelization can lead to geomorphic problems and environmental degradation. It also explains how geomorphic knowledge can be used to design better channelization schemes. Professor Andrew Brookes is well known for his work on channelized rivers.

Bunte, K. and Abt, S.R. (2001) Sampling surface and subsurface particle-size distributions in gravel- and cobble-bed streams for analyses in sediment transport, hydraulics and streambed monitoring. *USGS Rocky Mountain Research Station Gen Tech Report RMRS-GTR-74.*
A comprehensive review detailing methods and their effectiveness in accurately measuring bed material particle size.

Ferguson, R.I., Hoey, T., Wathen, S. and Werritty, A (1996) Field evidence for rapid downstream fining of river gravels through selective transport. *Geology*, 24, 635–643.
This paper presents a case study on a Scottish river of the downstream change in bed particle size due to differential transport of particles of different size.

Gilvear, D.J. and Black, A.R. (1999) Flood induced embankment failures on the River Tay: implications of climatically induced hydrological change in Scotland. *Hydrological Sciences Journal*, 44, 345–362.
This is a paper that notes the importance of past channel changes to flood embankment stability and how future changes in flood hydrology that may occur owing to climate change could heighten levels of flood embankment instability.

Gilvear, D.J. and Bradley, S. (1997) Geomorphic adjustment of a newly constructed ecologically sound river diversion on an upland gravel bed river, Evan Water, Scotland. *Regulated Rivers*, 13, 1–13.
This is a short case study paper examining how a newly constructed river diversion responded to a large flood just a few months after its completion.

Gurnell, A.M., Morrissey, I.P., Boitsidis, A.J. *et al.* (2006) Initial adjustments within a new river channel: interactions between fluvial processes, colonising vegetation and bank profile development. *Environmental Management*, 38, 580–596.
This paper illustrates the role of geomorphology in river engineering and how fluvial processes shape physical habitat and biota.

Keller, E.A. and Swanson, F.J. (1979) Effects of large organic material on channel form and fluvial processes. *Earth Surface Processes and Landforms*, 4, 361–380.
This paper discusses the role of woody debris in controlling river morphology.

Kondolf, G.M. (1997) Hungry water: effects of dams and gravel-mining on river channels. *Environmental Management*, 21, 533–551.
A review of the effects of sediment trapping and removal on downstream channel morphology.

Montgomery, D.R. and Buffington, J.M. (1997) Channel-reach morphology in mountain drainage basins. *Geological Society of America Bulletin*, 109, 596–611.
This paper classifies river types as found in Oregon and Washington and relates the types to fluvial processes and topographic and geological controls – relevant to Box 12.4.

Petts, G.E. (1984) Sedimentation within a regulated river. *Earth Surface Processes and Landforms*, 9, 125–134.
This is an examination of how river regulation caused pronounced sedimentation downstream – relevant to Box 12.5.

Zimmerman, R.C., Goodlet, J.C. and Comer, G.H. (1967) The influence of vegetation on channel form of small streams. In: *Symposium on River Morphology*, International Association of Hydrological Sciences Publication 75, General Assembly at Bern, 255–275.
One of the first papers demonstrating the role of vegetation on bank strength and channel shape.

CHAPTER 13

Solutes

Kate V. Heal

School of GeoSciences, The University of Edinburgh

Learning objectives

After reading this chapter you should be able to:

➤ **understand the difference between solute concentration and solute flux**

➤ **explain how solutes are affected by hydrological processes operating within catchments**

➤ **appreciate the interaction of geology, climate, soil, biotic and human factors that produce the solute characteristics of a catchment**

➤ **understand how spatial and temporal patterns in solutes occur**

➤ **explain why changes in climate and land use affect solutes**

13.1 Introduction

Water is essential for life, and in the natural world it consists of more than pure hydrogen and oxygen. It contains many other substances in dissolved or solid forms. This chapter deals with solutes – substances dissolved in water. Solutes are naturally occurring. However, human activities can discharge solutes into the environment and also alter the processes affecting both the **concentrations** and fluxes of naturally occurring solutes. A contaminant is a chemical compound present in the environment, at a level significantly greater than its natural abundance, owing to anthropogenic activities. An understanding of solute sources and transport processes underpins many topics in physical geography and environmental science including mineral weathering, soil development, biogeochemical cycling, plant and animal nutrition, aquatic ecosystem health, and the quality of water resources for human use.

This chapter will examine solute processes within a catchment framework since these processes are intimately linked with the hydrological processes discussed in Chapter 11. A range of chemical, biological and physical factors affect solutes, and understanding solute transport processes is important because of the implications for devising effective strategies for managing water quality in catchments. The factors that determine spatial and temporal patterns of solutes can be examined at a number of scales and these will be outlined within this chapter. Knowledge of these factors is required in order to predict the impacts of environmental change on solutes in catchments. Such environmental change may include pollution incidents, land use change or changes in atmospheric chemistry.

13.2 Solutes: some key controls

It is vital to appreciate some of the factors affecting solutes before solute processes are explored within catchments. Solute concentrations at a given point and the spatial distribution of solutes are constantly changing as the result of complex interactions of physical, chemical and biological factors. This section briefly discusses the aspects of solute chemistry relevant for understanding solute processes in catchments. Box 13.1 explains the different units that are used to report solute concentrations. For detailed information on solute chemistry see Langmuir (1997).

13.2.1 Solute form

Solutes exist in a number of different forms in the environment. It is important to know the form of a solute as well as the concentration since different forms of solutes have differing toxicities and bioavailabilities (see Chapters 20 and 21) to living organisms. For example, the free aluminium ion is much more toxic to fish than aluminium that is complexed to **organic** compounds. Many nutrients occur in inorganic and organic forms. For example, dissolved carbon occurs as dissolved organic carbon (DOC) and dissolved inorganic carbon (hydrogen carbonate, HCO_3^-, and carbonate, CO_3^{2-}, ions), whilst dissolved nitrogen occurs as

FUNDAMENTAL PRINCIPLES

UNITS OF CONCENTRATION

A variety of units are used for expressing solute concentrations in water which can cause confusion and make it difficult to compare the results of different studies. The simplest way of expressing solute concentration is by mass, as the mass of solute per volume of water, normally as $mg\ L^{-1}$. Thus a concentration of calcium of $5.7\ mg\ L^{-1}$ measured in river water means that there is 5.7 mg of dissolved calcium in every litre of river water. Units of $\mu g\ L^{-1}$ are used instead for solutes that occur at lower concentrations, such as potentially toxic metals. Since freshwater has a density of $1\ g\ cm^{-3}$, solute concentrations are sometimes alternatively expressed by mass as parts per million (ppm) or parts per billion (ppb): 1 ppm means that there is one unit mass of solute dissolved in 1 million unit masses of water. The units of ppm and ppb are the same as $mg\ L^{-1}$ or $\mu g\ L^{-1}$, respectively, when freshwater is the solvent. However, ppm and ppb units can be confusing and are rarely used for solute concentrations in physical geography

and environmental science. A more chemically useful way of expressing solute concentrations by mass is as moles of solute per volume of water $(mol\ L^{-1})$:

$$\text{Concentration (mol L}^{-1}) = \frac{\text{concentration (g L}^{-1})}{\text{relative atomic mass}} \quad (13.1)$$

For example, a concentration of calcium of $5.7\ mg\ L^{-1}$ re-expressed in $mol\ L^{-1} = 0.0057/40.08 = 1.42 \times 10^{-4}\ mol\ L^{-1} = 142\ \mu mol\ L^{-1}$ ($1.42 \times 10^{-4} \times 10^6$) since $5.7\ mg\ L^{-1} = 0.0057\ g\ L^{-1}$ ($5.7\ mg\ L^{-1}/1000$) and the relative atomic mass of calcium = 40.08. The unit $mol\ L^{-1}$ is sometimes abbreviated to M so the concentration of 142 $\mu mol\ L^{-1}$ might be written as 142 μM.

If there is interest in comparing quantities and properties of different compounds or investigating reaction mechanisms or relationships between solute concentration and biological activity, solute concentrations may be expressed in terms of charge equivalent concentrations, calculated as:

$$\text{Charge equivalent concentration (eq L}^{-1}) = \text{concentration (mol L}^{-1}) \times \text{ionic charge} \quad (13.2)$$

For example, the charge equivalent concentration of 142 $\mu mol\ L^{-1}$ of calcium = 284 $\mu eq\ L^{-1}$ (142 $\mu mol\ L^{-1} \times 2$, since the ionic charge on calcium is +2). The units of charge equivalent concentration are also sometimes written as $mol_c\ L^{-1}$.

Finally, care must be taken with the concentration units where a particular species is of interest. For example, concentrations of the two different forms of inorganic nitrogen (N), nitrate, NO_3^-, and ammonium, NH_4^+, were determined to be 15.0 mg $NO_3^-\ L^{-1}$ and 0.532 mg $NH_4^+\ L^{-1}$. To calculate the total inorganic N concentration in the water sample, the NO_3^- and NH_4^+ concentrations must be converted to concentrations of N using the relative atomic masses of N (14.01), NO_3^- (14.01 + (3 × 16.00) = 62.01) and NH_4^+ (14.01 + (4 × 1.01) = 18.05). Thus, 15.0 mg $NO_3^-\ L^{-1}$ = 3.39 mg $NO_3^- - N\ L^{-1}$ (15.0 × (14.01/62.01)) and 0.532 mg $NH_4^+\ L^{-1}$ = 0.413 mg $NH_4^+ - N\ L^{-1}$ (0.532 × (14.01/18.05)). Therefore the total inorganic N concentration in the water sample is 3.80 mg L^{-1} (3.39 + 0.413).

BOX 13.1

organic nitrogen and inorganic nitrogen (nitrate, NO_3^-, and ammonium, NH_4^+, ions). Measurement of both organic and inorganic forms of solutes is required for a complete understanding of nutrient flows in the environment. Studies in both tropical and boreal forests show that organic nitrogen often accounts for at least 50% of the nitrogen transported in streamwater and other hydrological flow paths (Goller *et al.*, 2006; Kortelainen *et al.*, 2006). A further complication is defining what constitutes a solute since **colloids**, very small particles, such as of clay minerals, organic material and iron oxyhydroxides, with diameter 0.1–10 μm may be suspended in water. For routine water sample analysis, 'soluble' is defined as material that passes through a filter with pores of 0.45 μm diameter. In addition, many elements can occur in more than one **oxidation state** in the environment. For example, iron may be present as iron +2 (ferrous, Fe^{2+}) or iron +3 (ferric, Fe^{3+}). The oxidation state of an element exerts an important control on its solubility and availability for transport by water in the environment. The oxidation state of an element at a particular point in the environment is mainly governed by the surrounding pH and **redox potential** conditions.

13.2.2 pH and redox potential

pH is a measure of acidity from the activity of hydrogen ions in a system. The pH scale ranges from 0 (extremely acidic) to 7 (neutral) to 14 (extremely alkaline). Chemically, pH is the negative logarithm to the base 10 of the concentration of hydrogen ions in solution (see Chapter 10):

$$pH = -\log_{10}[H^+] \tag{13.3}$$

Redox potential is a measure of how oxidizing or reducing the environment is in terms of the occurrence of electrons. Redox potential is normally measured as Eh, in units of volts (V). From electrochemical theory, **Eh-pH stability fields** can be plotted for different elements that show the forms expected under particular Eh and pH conditions, if chemical equilibrium is assumed. The Eh–pH stability field for iron is shown in Figure 13.1.

The influence of pH and redox potential on the form and environmental impact of solutes is illustrated by the occurrence of iron in freshwater. The Eh–pH stability field for iron shows that it occurs in highly acidic, but well oxidized, conditions (Eh > 0 V) (e.g. in acid mine drainage), and also in neutral pH, reducing conditions (Eh < 0 V) (e.g. in fens). This is partly why acid mine drainage is so damaging to aquatic ecosystems. Not only is the high Fe^{2+} content of the water toxic, but also soluble Fe^{2+} is converted to insoluble

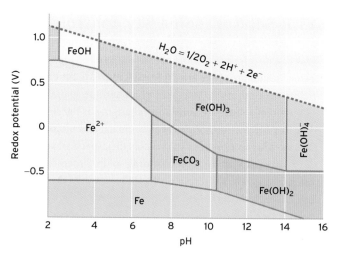

Figure 13.1 Eh–pH stability field for aqueous iron at atmospheric pressure and 25°C. If the water redox potential (Eh) and pH are known the figure shows the form in which iron will occur. For example, at pH 4 and Eh 0 V, iron would be present as Fe^{2+}. (Source: after Morgan and Stumm, 1965)

Fe^{3+} (in the form of iron +3 hydroxide, $Fe(OH)_3$) when the mine drainage mixes with water of a higher pH. Fe^{3+} then precipitates on the banks and bed of the river, smothering plants and animals living in these habitats. Box 13.2 explains the contribution of redox potential in creating the human health hazard of arsenic contamination of groundwater supplies in Bangladesh and West Bengal.

13.2.3 Temperature and pressure

Increasing temperature increases the solubility of some solids in water and speeds up the rate of chemical reactions, but reduces the solubility of most gases. The last is particularly significant for the dissolved oxygen content of rivers and lakes that is necessary for organisms, such as fish, to survive. In contrast, at high pressure, the solubility of gases in water increases. This can influence the pH of water and carbon cycling in different parts of the catchment hydrological system as follows. In groundwater, the solubility of carbon dioxide in water increases because the pressure is higher than atmospheric. Carbon dioxide dissolves in water to form carbonic acid, a weak acid, causing a decrease in pH in groundwater. When the groundwater comes into contact with the surface environment again, in for example a spring, the reverse effect occurs. Carbon dioxide in the spring water can come out of solution, because of the reduced pressure, and the pH of the water increases as it now contains less carbonic acid. Not only does such degassing change water pH, but it may also result in emission of the greenhouse gas carbon dioxide. For example,

THE LARGEST MASS POISONING IN HISTORY: ARSENIC IN GROUNDWATER SUPPLIES IN BANGLADESH AND WEST BENGAL

The world's largest outbreak of arsenic poisoning has been occurring in Bangladesh and West Bengal, India, since the 1990s. The cause of poisoning is the consumption of groundwater containing high concentrations of arsenic. As well as the total arsenic concentration, the chemical speciation of arsenic is important in assessing the impacts on public health. Arsenic has two oxidation states in the environment. Arsenic +3 is the most toxic but is difficult to remove from water. In comparison arsenic +5 is less toxic and easier to remove from water. The health effects of arsenic poisoning appear slowly, starting with skin lesions and followed by skin and internal cancers,

particularly of the bladder, lung and kidney. An estimated 35–77 million people in Bangladesh and 2 million people in West Bengal are at risk of consuming well water contaminated with arsenic.

Until the 1970s, most of the population in the affected area obtained water supplies from surface waters which were contaminated with sewage, resulting in widespread gastrointestinal disease. From the 1970s thousands of low-cost tube-wells were dug to access groundwater supplies with a better microbiological quality. Many of the new wells were sunk into sedimentary aquifers with a naturally high arsenic concentration, from arsenic bound to iron minerals in delta sediments which were deposited 25 000–80 000 years ago. Decomposition of organic matter in the delta sediments consumed all the available oxygen, creating reducing conditions that converted insoluble

Fe^{3+} to the soluble Fe^{2+} state. The solubilization of iron also mobilized arsenic bound to the iron minerals so that arsenic concentrations of 0.05–1 mg L^{-1} have been measured in wells, compared with the World Health Organization recommended maximum concentration for drinking water of 0.01 mg L^{-1}. There is ongoing research into the extent to which human activity, such as the overpumping of groundwater or introduction of anthropogenic organic matter, has affected arsenic concentrations in groundwater in Bangladesh and West Bengal, but, so far, evidence of a significant deleterious effect of human activity on arsenic concentrations is not compelling. Geologists, chemists and engineers are working together to identify alternative water supplies for the affected population and, in the short term, to develop technologies to reduce arsenic concentrations in the well waters.

BOX 13.2

Billett *et al.* (2004) reported that degassing of carbon dioxide from a lowland peatland stream accounted for 13% of the total annual flux of carbon in the stream.

13.2.4 The role of particulates

The movement of particulate material in catchments can affect the distribution of solutes within the catchment hydrological system. For example, eroded phosphorus-rich soil from agricultural areas where applications of manure and fertilizer exceed the crop nutrient requirements is probably a more important source of phosphorus for aquatic systems than the leaching of soluble phosphorus from soils (see Chapter 10). Particles transported in water can also play an important role in the transport of solutes through **adsorption** of dissolved species on the particle surface. Particles can range from large boulders, leaves and litter to small particles of algae, viruses, **colloids** and minerals.

A volume of small mineral particles (10^{-8}–10^{-10} m diameter) has an extremely large surface area (because there are more particles in a given volume than for coarser particles) and a surface charge, resulting in the 'sticking' of dissolved species that have the opposite charge to the particle surface.

13.2.5 Solute fluxes

All the factors discussed above are important controls on solute concentrations and distributions within a catchment, but an understanding of hydrology is vital for determining the transport of solutes between different stores within the catchment. In general solute concentrations are highly dependent on the mass of water passing through a system. Measurement of solute concentrations enables comparison of catchment water quality with environmental quality standards (e.g. for drinking water, freshwater fish). However, the **solute load** or flux is of more interest for some

Figure 13.2 The Bow River and downtown Calgary, Alberta, Canada.

is used in studies of chemical weathering rates from catchments, in biogeochemical studies to determine whether catchments are sinks or sources of a contaminant or nutrient, to investigate the response of catchments to changes in land use and/or precipitation chemistry and, in some countries, for managing pollution in rivers and lakes. Pollutants in wastewater and stormwater from the city of Calgary, Canada, shown in Figure 13.2, are managed using targets based on estimated total fluxes to minimize deleterious impacts on the Bow River which flows through the city.

purposes. The solute flux is simply the solute concentration of the water at a point in time multiplied by the discharge occurring at the same time. However, the calculation of fluxes often involves assumptions and uncertainties, and these are discussed in Box 13.3. Calculation of solute fluxes

Reflective questions

➤ If a large number of fish deaths occurred in your local river and the suspected cause was metal contamination of the river water, what environmental factors would you need to consider in an investigation of this suspected cause?

➤ What is the difference between solute concentration and solute flux?

TECHNIQUES

ESTIMATING SOLUTE FLUXES

The solute flux (sometimes called the solute load) is simply the solute concentration of the water at a point in time multiplied by the discharge occurring at the same time:

$$\text{Solute flux (mg s}^{-1}\text{)}$$
$$= \text{solute concentration (mg L}^{-1}\text{)}$$
$$\times \text{discharge (L s}^{-1}\text{)} \quad (13.4)$$

In this calculation care should be taken with the units of discharge if they are in $m^3 \ s^{-1}$. By applying equation (13.4) for every second over a period of time, such as a year, the solute flux can be estimated by multiplying by the total seconds in the time period of interest. The solute

fluxes calculated in this way are often used in biogeochemical studies of the sources, cycling and stores of

contaminants in the environment. For example, Table 13.1 shows the annual solute fluxes in an upland catchment

Table 13.1 Annual solute fluxes of trichloroacetic acid (TCA) in hydrological pathways in an upland catchment in south-west Scotland

Hydrological pathway	Annual TCA flux ($\mu g \ m^{-2}$)
Rainwater	2100
Cloudwater	300
Throughfall	1400–1600
Stemflow	60–110
Streamwater	2300

(Source: based on data from Stidson *et al.*, 2004a, b)

BOX 13.3 ➤

➤ covered by forest plantation and moorland in south-west Scotland of trichloroacetic acid (TCA), a phytotoxic chemical with human-made and natural sources. In the table the solute fluxes are expressed as mass per catchment area. The calculated solute fluxes in Table 13.1 show that annual TCA inputs (rainwater and cloudwater) to and outputs (streamwater) from the catchment are approximately equal. Comparison of input fluxes with throughfall and stemflow fluxes shows that 30-40% of TCA inputs are removed by the forest canopy. However, the similarity of the TCA output and input fluxes demonstrates that production of TCA within soil and/or vegetation in the catchment is also occurring.

Although it is relatively straightforward to monitor discharge continuously (see Chapter 11) it is difficult and costly to measure solute concentrations constantly, so the calculation of fluxes often involves assumptions and uncertainties concerning the solute concentration values used. The two main methods of estimating loads are interpolation and extrapolation methods. Interpolation methods use the measured concentration values only and combine these with discharge measurements in different ways to estimate flux. In the simplest interpolation method the mean solute concentration is multiplied by the mean discharge measured during the time period of interest to obtain the solute flux. In more complicated

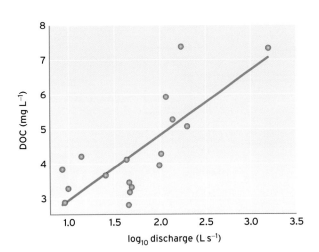

Figure 13.3 Significant relationship between streamwater DOC concentration and discharge in an upland Scottish catchment ($n = 16$, $R^2 = 58.5\%$, $P < 0.001$).

versions of the interpolation method continuous discharge measurements are combined with the solute concentration determined for a specific time period. The main disadvantage of interpolation methods is that fluxes of solutes mobilized during storm events are often underestimated. This is because when water sampling for measurement of solute concentration is intermittent there is a higher probability of sampling low flows, in which solute concentrations are lower. Extrapolation methods address this problem by using the relationship between measured solute concentration and discharge to estimate solute concentration for every time point for which there is a discharge measurement. Using these finer time resolution concentration estimates

with the discharge measurements normally yields more accurate solute flux estimates. Figure 13.3 shows the significant solute concentration-discharge relationship used to estimate dissolved organic carbon (DOC) flux in an upland catchment. From the relationship in the figure, DOC concentrations were calculated for the discharge measurements made every 10 minutes and used to estimate a DOC flux for the 8 month study period of 3.1 g DOC m^{-2}, which is higher than the flux of 2.3 g DOC m^{-2} estimated using the simplest interpolation method (mean DOC concentration × mean discharge for the study period). For further information on different methods of estimating fluxes and their relative merits see Johnes (2007).

BOX 13.3

13.3 Solutes within the catchment hydrological system

Catchments are the scale for assessing biogeochemical cycling and managing water quality. Solute hydrology can be considered at all stages of the catchment hydrological system in a similar manner to water quantity. Catchment hydrology

influences the transport of solutes, and Figure 13.4 shows how processes affecting solute concentrations take place within each store and transport pathway of water in a catchment. The following section explores the processes influencing solute concentrations within each compartment of the catchment hydrological system, from precipitation inputs to outputs in rivers and standing waters.

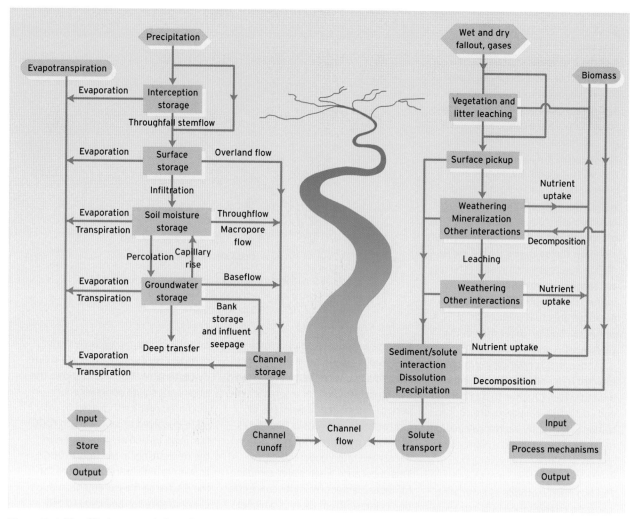

Figure 13.4 Simplified representation of hydrological processes and associated solute processes operating within the catchment hydrological system. (Source: after Walling and Webb, 1986)

13.3.1 Precipitation

Precipitation inputs to catchments take the form of wet deposition, as rain, snow, hail and **horizontal interception**, and dry deposition, as gases and particulates. The natural pH of rainwater of 5.7 is due to the presence of carbonic acid, formed by dissolution of carbon dioxide in atmospheric moisture. Precipitation contains a mixture of cations (positively charged) and anions (negatively charged) (see Chapter 10) derived from natural sources in the ocean and on land and also originating from human activities (Table 13.2). Sulphur, particles, fluorine and potentially toxic elements, such as lead, cadmium and mercury, emitted from volcanoes (see Figure 13.5) can also affect precipitation composition. Concentrations of solutes in wet deposition vary over time and space, depending on the proximity of solute sources, the source of the air mass producing the precipitation, and the nature of precipitation. For example,

in the United States, sodium concentrations in precipitation are highest on the Atlantic and Pacific coasts as sea salts are the main source. In contrast, the main source of calcium is wind-blown dust so maximum calcium concentrations in precipitation occur in the continental interior.

Altitude and the type of precipitation affect the solute input to a catchment as well as the total quantity of precipitation. For example, cloud droplets contain higher concentrations of sulphate and nitrate ions, compared with rain. Ground-level cloud droplets are intercepted by the vegetation canopy and, in some environments, can contribute significantly to solute inputs to catchments. Needle-bearing conifer trees are particularly efficient scavengers of cloud droplets from the atmosphere because of the large surface area of needles. This process is partly why the planting of conifer forests on **base**-poor soils has exacerbated acidification of soils and waters in temperate latitudes receiving acid deposition (see Chapter 10).

Table 13.2 Sources of individual ions in rainwater

Ion	Origin		
	Marine	Terrestrial	Pollution
Sodium	Sea salt	Soil dust	Biomass burning
Magnesium	Sea salt	Soil dust	Biomass burning
Potassium	Sea salt	Biogenic aerosols Soil dust	Biomass burning Fertilizer
Calcium	Sea salt	Soil dust	Cement manufacture Fuel burning Biomass burning
Hydrogen	Gas reaction	Gas reaction	Fuel burning
Chloride	Sea salt	–	Industrial hydrochloric acid
Sulphate	Sea salt Dimethylsulphide from biological decay	Dimethylsulphide, hydrogen sulphide, etc., from biological decay Volcanoes Soil dust	Fossil fuel burning Biomass burning
Nitrate	Atmospheric nitrogen plus lightning	NO_2 from biological decay Atmospheric nitrogen plus lightning	Vehicle emissions Fossil fuels Biomass burning Fertilizer
Ammonium	Ammonia from biological activity	Ammonia from bacterial decay	Ammonia fertilizers Human, animal waste decomposition
Phosphate	Biogenic aerosols adsorbed on sea salt	Soil dust	Biomass burning Fertilizer
Hydrogen carbonate	CO_2 in air	CO_2 in air Soil dust	–
Silica, aluminium, iron	–	Soil dust	Land clearing

(Source: Berner, R.A., *Global Environment: Water, Air and Geochemical Cycles*, 1st edition. © 1996. Adapted by permission of Pearson Education, Inc., Upper Saddle River, NJ)

Figure 13.5 Gaseous emissions from the summit crater of Volcán Villarrica, one of the most active volcanoes in Chile. (Source: STILL Pictures The Whole Earth Photo Library: Mark Edwards)

Solutes dissolved in atmospheric moisture may be transported for thousands of kilometres across continents, whereas dry deposition of gases and particulates is most significant immediately downwind of the source. Dry deposition inputs to a catchment are more difficult to quantify than wet deposition inputs but measurement and modelling studies suggest that they are important. In the United Kingdom in 2006, dry deposition accounted for 20% of the total deposition of sulphur, 52% of oxidized nitrogen deposition and 32% of reduced nitrogen deposition (RoTAP, 2011).

13.3.2 Evapotranspiration and evaporation

Evapotranspiration of water from vegetation and water surfaces in a catchment acts to concentrate the solutes remaining in the system. In arid environments, river flow

Figure 13.6 The Salar de Uyuni on the altiplano in Bolivia was covered by a large lake around 40 000 years ago. The lake water has evaporated away over time to leave an estimated 10 billion tonnes of salt deposits. (Source: photo courtesy of M.R. Heal)

can decrease downstream (see Chapter 14) and solute concentrations may increase with distance from the headwaters. This is due to the effect of evaporation. Figure 13.6, for example, shows the white salt deposits created in the Salar de Uyuni, Bolivia, due to evaporation of the lake that formerly covered the area.

13.3.3 Interception

Precipitation inputs to the catchment are intercepted by vegetation surfaces and then either evaporate back to the atmosphere, or are transported by stemflow or **throughfall** to the ground surface. Contact of precipitation with vegetation surfaces can result in both gains and losses of solutes (Figure 13.7). Solute concentrations can increase in throughfall and stemflow compared with precipitation owing to the washing off of atmospheric aerosols, deposited on vegetation by dry deposition, and also owing to the leaching of solutes exuded by vegetation. The magnitude of solute enrichment depends on the vegetation species and location. For example, in temperate forest plantations, maximum solute concentrations in throughfall and stemflow occur on the edge of the plantation where there is a

larger tree surface area for the interception of precipitation. Interception of precipitation can also result in losses of solutes where some nutrients are absorbed by the biomass. The gain and loss of solutes in throughfall and stemflow can result in localized spatial variability in solute inputs to the ground surface within catchments.

13.3.4 Soil

The ground surface is an important divide in catchment hydrological systems for solute processes as well as for hydrology. Water that does not infiltrate into the ground surface will flow rapidly towards the river channel and has a limited time to react with the soil. In contrast, water infiltrating into the ground surface enters the soil and will interact with its components, such as the alternating mineral and organic layers shown in the soil exposure in Figure 13.8. Chemical, physical and biological processes within the soil alter solute concentrations and significantly influence the composition of surface waters. Some of these important processes are discussed below and examples of their effects are also given in Section 13.3.4.7.

Figure 13.7 Measurements of precipitation, throughfall and stem-flow concentration and flux near this rainforest in southern Chile (Oyarzún *et al.*, 2004) showed uptake of ammonium (NH_4^+) and organic nitrogen by the forest canopy. Higher concentrations of nitrate (NO_3^-), Na^+, K^+, Ca^{2+} and Mg^{2+} in throughfall compared with precipitation were attributed to washoff of dry deposition from the canopy and leaching from leaves.

Figure 13.8 An exposure of soil near Cotopaxi volcano in the Andes Mountains, Ecuador. The soil comprises light-coloured layers of ash, deposited during volcanic eruptions, and dark-coloured organic horizons that have developed between eruptions as vegetation matter has accumulated in the cool, wet upland climate.

13.3.4.1 Weathering

Weathering reactions are key mechanisms by which soils maintain their pH within a given range to counteract the acidity generated by organic acids and decomposition. Solutes are released into soil solution by the weathering of primary and secondary minerals (see Chapter 7), frequently due to dissolution of minerals by carbonic acid or organic acids. An example of this process is the dissolution of orthoclase, a widely occurring mineral in soils formed on acid igneous and metamorphic rocks (see Chapter 7), by carbonic acid to form the secondary mineral kaolinite, with the release of potassium ions:

$$4KAlSi_3O_8 + 4H_2CO_3 + 18H_2O \rightarrow$$
$$Al_4Si_4O_{10}(OH)_8 + 8H_4SiO_4 + 4K^+ + 4HCO_3^- \quad (13.5)$$

Orthoclase, carbonic acid and water react to form kaolinite, potassium ions and other by-products.

13.3.4.2 Cation exchange

Positively charged ions in water percolating through the soil can exchange with other cations adsorbed onto negatively charged clay minerals and humic substances with large surface areas, in the process of cation exchange. For example:

$$\begin{matrix} Ca^{2+} & + & 2H^+ & \leftrightarrow & 2H^+ & + & Ca^{2+} \\ \text{(attached} & & \text{(in soil} & & \text{(atttached} & & \text{(in soil} \\ \text{to colloid} & & \text{solution)} & & \text{to colloid} & & \text{solution)} \\ \text{particle)} & & & & \text{particle)} \end{matrix} \quad (13.6)$$

Thus ions are exchanged from the water to the soil and vice versa. Chapters 7 and 10 provide further detail on cation exchange.

13.3.4.3 Anion adsorption

Negatively charged anions such as sulphate (SO_4^{2-}) and phosphate (PO_4^{3-}) are also adsorbed onto positively charged surfaces in soils. This reaction is particularly important for phosphate because the adsorption of phosphate within soil is probably the reason why it is the limiting nutrient for biological productivity in most freshwater systems.

13.3.4.4 Microbiological activity

Microbiological activity is the action of bacteria and other microscopic organisms within the soil. It affects solute concentrations by increasing the rate of soil and rock weathering, by the production of organic acids and by decomposition of dead plant and animal material to provide soluble nutrients for plant growth (see Chapters 7 and 10). The decomposition of organic nitrogen to form the ammonium ion which can then be oxidized to the nitrate ion is a good example of this process:

$$\text{Organic N} \Rightarrow NH_4^+ + OH^- \tag{13.7}$$

Organic nitrogen is converted to the ammonium ion.

$$NH_4^+ + 2O_2 \Rightarrow NO_3^- + H_2O + 2H^+ \tag{13.8}$$

Ammonium ions are oxidized to form nitrate ions and other by-products.

13.3.4.5 Oxidation and reduction

The redox potential of the soil affects the form of substances with more than one oxidation state and therefore the concentrations of solutes in soil water. In well-aerated soils, substances with multiple oxidation states are present in the oxidized form. However, when the soil pore space is filled with water, the absence of oxygen causes bacteria which function in anaerobic conditions to reduce substances to obtain energy. As soils become progressively waterlogged, nitrate is reduced to nitrogen gas (N_2), Fe^{3+} is reduced to soluble Fe^{2+} and sulphate is reduced to hydrogen sulphide gas. Therefore an increase in iron concentrations and a decrease in nitrate and sulphate concentrations in soil water will occur with progressively more reducing conditions.

13.3.4.6 Chelation

Elevated concentrations of iron and aluminium can occur in soil water because of the formation, with soluble humic substances, of stable complexes known as **chelates**.

13.3.4.7 Examples of the effect of soil processes on solute concentrations

The processes discussed above affect solute concentrations in soil water and frequently result in substantial modifications of solute composition and concentration from the inputs to the ground surface. The variation in soil water processes over space and also with depth in the soil profile makes examination of soil water chemistry very complex. Variations in solute concentrations with depth in two schematic soil profiles are shown in Figure 13.9. In the podzol soil type (see Chapter 10 for discussion on podzols), pH is low and the content of organic substances is high at the top of the profile because of the organic acids produced by decomposition of dead vegetation and animals at the soil surface. The pH and base cation concentrations increase with depth as weathering of minerals releases cations into solution and concentrations of organic acids decrease. Inorganic aluminium concentrations peak in the middle of the soil profile owing to the formation of chelates. Soluble organic substances precipitate lower down the profile as pH increases.

The calcareous brown earth soil (Figure 13.9b) displays different patterns of solute concentrations with depth, largely because of the absence of an organic surface horizon. Here the pH and base cation concentrations in soil water are high at the surface because of the lower concentrations of organic substances and increase with depth as the influence of organic acids declines. Inorganic aluminium concentrations in soil water are negligible throughout the soil profile because the formation of chelates is limited by the low concentration of organic substances.

13.3.5 Groundwater

Solute concentrations in groundwater are commonly the highest of any compartment in the catchment hydrological system owing to the longer residence times of water. The solute concentration in groundwater at any particular point is controlled by three main factors:

1. Residence time: As the residence time of groundwater increases, solute concentrations also increase because there are more opportunities for chemical weathering

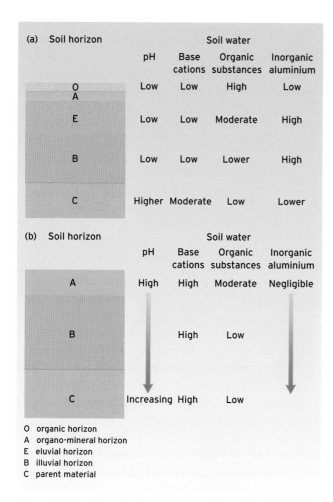

O organic horizon
A organo-mineral horizon
E eluvial horizon
B illuvial horizon
C parent material

Figure 13.9 Variation of soil water composition with depth in two schematic soil profiles: (a) podzol; (b) calcareous brown earth. Chapter 10 provides more details on podzol and brown earth soil types and processes. (Source: adapted from Soulsby, 1997)

Table 13.3 Chemical composition of rainwater and groundwater from different rock formations in western Cuba. All units mg L^{-1} apart from Eh (V).

Parameter	Rainwater	Serpentinite	Limestone/ schist/sandstone
Sodium	4.8	15.4	119
Calcium	0.8	5.2	457
Magnesium	4.5	44.7	50.9
Sulphate	< limit of detection	8.9	1300
Chloride	5.7	16.6	37.3
Total dissolved solids	24	329	2210
Eh	–	0.156	−0.300

(Source: adapted from Fagundo-Castillo *et al.*, 2008)

products to enter solution. Residence time is affected by the rate of **recharge** of groundwater and the composition of recharge water.

2. Rate and composition of recharge water: Groundwater that is recharged rapidly with water of a low solute concentration will have low solute concentrations and will also have a high dilution potential for contamination. In contrast, pollution events will have a longer-term impact on groundwater that is recharged slowly. The rate of recharge will also influence the redox status of groundwater. Reducing conditions occur most commonly in groundwater with a low rate of recharge from oxidized atmospheric waters.

3. Geochemistry of the surrounding geology: Different minerals in rocks weather at different rates and produce different solutes.

The effect of these factors on groundwater composition is illustrated by comparison of the characteristics of rainwater and water samples from different groundwater groups in western Cuba in Table 13.3. Concentrations of all ions in groundwater are higher than in rainwater as the result of mineral weathering, but there is also variation in ionic concentrations between different groundwaters. Concentrations of all solutes are considerably higher in groundwater from the sandstone/limestone/schist formation compared with the serpentinite formation because minerals in the former rock types are less resistant to weathering and also because the water residence time is longer. The longer water residence time in the sandstone/limestone/schist formation is the cause of the lower Eh value compared with groundwater in the serpentinite formation.

13.3.6 Rivers

The role of in-stream processes in altering solute concentrations in catchments has been recognized only in the past few decades. Water in river channels is frequently not in equilibrium with channel sediments because of its short residence time. Consequently the rates of reactions are particularly important in this compartment of the catchment hydrological system. A number of processes have now been identified that may alter solute concentrations once water has entered the river channel.

As for soil, cation exchange can occur between solutes in flowing water and solutes adsorbed onto the channel bed. Solute concentrations in river channels may be further reduced by physical storage within channel sediments. Biological processes in river channels also affect nutrient concentrations, through uptake and temporary retention of nitrate, ammonium and phosphate, followed by mineralization and re-release – a process known as 'nutrient spiralling' (Ensign and Doyle, 2006).

Concentrations of iron, and associated sorbed potentially toxic metals, are particularly subject to alteration when water enters river channels in catchments. The increase in pH that may occur when carbon dioxide comes out of solution as groundwater enters river channels (see Section 13.2.3) can cause precipitation of Fe^{3+} on the channel bed and therefore a decrease in iron concentrations in river water. However, iron concentrations in acidic river water have been reported to increase during the middle of the day due to the **photoreduction** of solid Fe^{3+} in channel sediments to soluble Fe^{2+} (Gammons *et al.*, 2005). Diurnal variations in light levels also affect the activity of photosynthesizing organisms (**autotrophs**) in river channels and have been associated with observations of diurnal variations in nitrate concentrations in river water (shown in Figure 13.10). Maximum nitrate concentrations typically occur in river water in the early morning when the activity of autotrophs and their nitrate uptake is lowest, whilst minimum concentrations occur in late afternoon when autotrophic activity peaks.

13.3.7 Lakes and reservoirs

One of the main processes affecting solute concentrations in lakes and reservoirs is **stratification** (see Chapter 21). Stratification normally occurs because of temperature differences within the water column that cause density differences. It is most common in temperate lakes and reservoirs deeper than 10 m because of seasonal changes in climate. It occurs less frequently in tropical lakes that have nearly constant temperatures all year round. It results in the development of two layers of water with very different chemical properties as shown in Figure 13.11. The surface layer of water, known as the **epilimnion**, is warmer and has an oxidizing environment because it is exposed to the atmosphere. Hence chemical substances with a number of oxidation states occur in the oxidized form in the epilimnion. The epilimnion is separated from the lower layer of water, the **hypolimnion**, by a thermocline (see Chapters 3 and 21). In the hypolimnion the water temperatures are cooler and low dissolved oxygen concentrations can cause a reducing environment due to the isolation of the water from the atmosphere and sunlight. Consequently, substances with multiple oxidation states occur in the reduced form in the hypolimnion and the potent greenhouse gas, methane, may be formed by reduction of organic material in lake sediments. Concentrations of the metals iron and manganese often increase in the hypolimnion owing to reduction of the solid form contained in basal sediments. This process can cause concentrations of iron

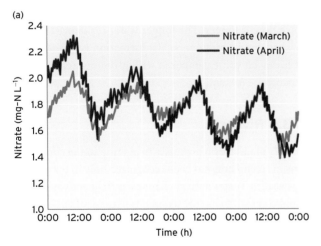

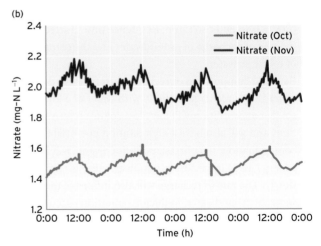

Figure 13.10 Diurnal variation in nitrate concentrations (mg L^{-1} nitrate-N) measured every 15 minutes using an *in situ* ion-selective electrode in a stream draining a forested catchment in south-west Slovenia. The greatest diurnal variation in nitrate concentrations occurs in spring (a – March and April) and is attributed to high light levels because the stream is not yet shaded by deciduous trees and also to high diurnal variation in streamwater temperature. Diurnal variation in nitrate concentrations is lower in autumn (b – October and November). (Source: adapted from Rusjan and Mikoš, 2010)

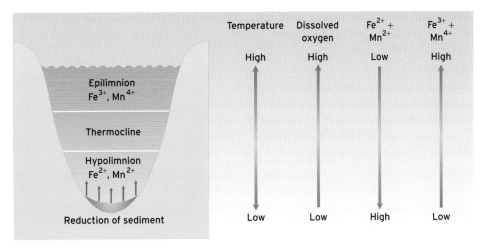

Figure 13.11 Stratification of water in lakes and reservoirs and its effect on solute concentrations.

and manganese in reservoir waters to exceed drinking water quality standards, resulting in the supply of discoloured, metallic-tasting water to consumers.

Reflective questions

➤ Why are the solute concentrations in river water not exactly the same as in the precipitation that fell on the catchment?

➤ Which factors may increase the risk of groundwater pollution?

➤ What are the main processes other than soil processes that alter solute concentrations in catchments?

13.4 The role of hydrological pathways in solute processes

Hydrological pathways, the routes that water takes through the catchment from precipitation inputs to river channel outputs, control which of the processes discussed above influence solute concentrations. Knowledge of hydrological pathways is essential for devising appropriate solutions to **diffuse pollution** in catchments. The influence of hydrological pathways on solute concentrations is summarized in Figure 13.12. Precipitation inputs to catchments normally have low solute concentrations. Solute concentrations are relatively low and unaltered in hydrological

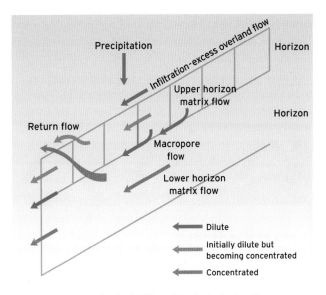

Figure 13.12 The effect of different hydrological pathways on solute concentrations. Water that is unable to infiltrate into the soil will have solute concentrations that are similar to the precipitation. Water that gets into the soil but moves by fast pathways, such as macropores or soil pipes, will have little time in contact with the soil and so will also be quite dilute in terms of solute concentrations. However, water that passes through the matrix of the soil (small pores) will move more slowly and therefore be in contact with the soil for longer. Hence the solute concentrations will be much greater. Some of this soil water can reach the stream either directly from the soil or by first being transferred to the surface as return flow. Return flow occurs when the soil is saturated and so water returns to the surface (e.g. near the foot of a hillslope). See Chapter 11 for further details of hillslope flow pathways. (Source: after Burt, 1986)

pathways with a short residence time within the catchment, such as infiltration-excess overland flow, and macropore or pipeflow (see Chapter 11). In contrast, solute concentrations may be altered in throughflow in the upper soil horizon and in the return flow component of saturation-excess overland flow because there is more time for interaction with the soil. Solute concentrations are initially low in these pathways but increase owing to the addition of solutes from weathering reactions and microbiological activity. By the time soil water has percolated to lower soil horizons in the catchment, solute concentrations are relatively high as the result of interaction with the soil.

Different hydrological pathways in a catchment may therefore have distinct solute concentrations. Solute concentrations measured in runoff reaching the river channel are the outcome of all the different hydrological pathways in the catchment, each with a different solute signature. Consequently, solutes have been widely used in hydrology to estimate the contribution of different hydrological pathways in a catchment to runoff generation. The flow of a particular hydrological pathway, such as throughflow or groundwater flow, can be estimated by measuring the concentration of a suitable solute in the hydrological pathways and the river channel. Along with measurements of river discharge this enables the flow in a hydrological pathway to be calculated using **chemical mixing models**. For two

hydrological pathways Pinder and Jones (1969) suggested that equation (13.9) could be used:

$$Q_2 = \left(\frac{C_T - C_1}{C_2 - C_1} \right) Q_T \qquad (13.9)$$

where C is the concentration of solute, Q the flow, subscripts 1 and 2 refer to two different hydrological pathways and subscript T to the total runoff in the river channel.

Chemical mixing models are based on three main assumptions. Firstly, complete mixing is required in the hydrological pathways of interest so that the solute concentration is uniform in each pathway. Secondly, the solute must mix conservatively when the hydrological pathways combine with no chemical reaction occurring between them (e.g. oxidation, precipitation). Thirdly, the difference in solute concentration between the hydrological pathways must be greater than the internal variation within each pathway. Any solute that meets these criteria can be used in chemical mixing models. The most widely used solutes for this purpose are chloride, which generally behaves conservatively within catchments, and the naturally occurring **stable isotopes** (see Box 13.4) oxygen-18 and **deuterium**. Chemical mixing model calculations can be performed for a number of points in time to separate storm hydrographs for a river into different flow components on the basis of solute concentrations.

TECHNIQUES

STABLE ISOTOPES FOR IDENTIFYING HYDROLOGICAL FLOW PATHWAYS AND SOLUTE SOURCES

Naturally occurring **stable isotopes** of water and solutes are increasingly used as non-invasive tools to identify hydrological flow pathways and the biogeochemical cycling processes in catchments. The different relative atomic masses of the stable isotopes of an element, arising from the different numbers of neutrons in the

nucleus, mean that different isotopes are favoured, or fractionated, by specific hydrological and biogeochemical processes. For example, in water, H_2O, hydrogen occurs in the environment as the stable isotopes ^1H and ^2H (deuterium) which have relative atomic masses of 1 and 2, respectively, whilst oxygen occurs as the stable isotopes ^{16}O and ^{18}O, which have relative atomic masses of 16 and 18, respectively. Precipitation is enriched compared with ocean water in ^1H and ^{16}O because these 'lighter' isotopes are evaporated in prefer-

ence to the heavier isotopes ^2H and ^{18}O. As precipitation water moves through the different flow pathways within the catchment hydrological system, it becomes progressively depleted in the 'lighter' isotopes of ^1H and ^{16}O due to water losses by evapotranspiration. Consequently, soil water and groundwater with long residence times often have different stable isotope signatures compared with precipitation and water that has moved rapidly through the catchment system to the river channel, such as by overland flow or pipeflow.

BOX 13.4 ➤

The abundances of stable isotopes of an element are measured using a mass spectrometer in samples of waters from different hydrological flow pathways and are reported as delta (δ) values with units of parts per thousand (per mil or ‰), calculated as shown in equation (13.10):

$$\delta(\text{‰}) = \left(\frac{R_{\text{sample}} - R_{\text{standard}}}{R_{\text{standard}}} \right) \times 1000 \qquad (13.10)$$

where R is the ratio of heavy : light isotope.

The standard normally used for isotopic analysis of water is SMOW (Standard Mean Ocean Water). Because ocean water contains relatively more heavy ^{2}H and ^{18}O due to preferential evaporation of lighter ^{1}H and ^{16}O, the delta values for all catchment water samples are negative. Once the isotopic compositions of different water types have been quantified they can be used in mixing models, such as shown in equation (13.9) and subject to the same assumptions, to quantify the contribution of different hydrological flow pathways to river discharge. For example, Laudon et al. (2007) used the different δ^{18}O compositions of snowmelt (-15.2 to -18.1‰) and soil water (-13.2‰) samples to identify the processes of streamflow generation during the spring snowmelt flood in boreal catchments.

Stable isotope ratios of solutes in water, such as δ^{18}O and δ^{15}N (based on the ratio of ^{15}N to ^{14}N) in nitrate (NO_3^-), can be used to identify the relative importance of different processes in biogeochemical cycling. Based on the very different values of δ^{18}O in the nitrate present in precipitation inputs and produced by microbial nitrification within catchment soils, several studies, such as by Barnes et al. (2008), have shown that the majority of nitrate exported in river water from forested catchments is derived from within the catchment rather than from direct atmospheric deposition, Such results are important for ascertaining the effect of enhanced atmospheric deposition of nitrogen on ecosystems.

BOX 13.4

Using solutes as tracers has increased the understanding of hydrological processes in catchments. The advantage of using solutes as chemical tracers is that interference with the catchment hydrological system is minimal, whereas direct measurements of hydrological pathways by physical methods can alter the flows being measured. Knowledge of hydrology and hydrological pathways in a catchment is important in explaining and modelling solute concentrations in rivers and also in understanding biogeochemical cycling and temporal and spatial patterns of solutes in catchments. Conversely, solute concentrations and fluxes can be used as tools in understanding catchment hydrology.

Reflective questions

➤ Which combination of hydrological pathways would make a catchment most vulnerable to spillage of a polluting chemical?

➤ Which flow path would you expect to produce the greatest concentration of solutes: infiltration-excess overland flow, return flow or macropore flow? Why?

➤ Explain how solute tracers can be used to identify hydrological flow pathways in catchments.

13.5 Temporal patterns of solutes

Solute concentrations in rivers vary over time because the hydrological processes that generate runoff within catchments are dynamic. Four main factors control temporal patterns of solute concentrations in catchments. Firstly, the quantity and mobility of the solute supply determine its availability for transport to the river channel. For example, increased sulphate concentrations may occur after drought in streams draining organic soils, as shown in Figure 13.13. This is due to lowering of the water table, causing oxidation of reduced

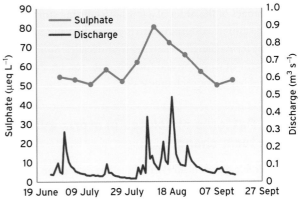

Figure 13.13 Increased concentrations of sulphate in a headwater stream in central Wales after summer drought conditions. Such flushes also result in streamwater acidification episodes. (Source: adapted from Evans et al., 2008)

forms of sulphur compounds that have accumulated over long time periods in wet, anaerobic soils, to sulphate which is flushed into the river channel in subsequent rainfall events.

Secondly, the composition of precipitation inputs to the catchment affects solute concentrations in river water. For example, precipitation in coastal regions is often enriched by sea salt during stormy weather conditions when droplets of seawater are incorporated into precipitation. These **sea-salt events** can have a dramatic effect on river water composition since chloride ions cause an increase in **specific conductance**. If the soil has a low base content, sodium ions in the precipitation exchange with hydrogen and aluminium ions adsorbed onto soil cation exchange sites, resulting in a flush of acidified waters with high concentrations of toxic aluminium ions which can kill fish.

Thirdly, catchment size and heterogeneity affect solute concentrations. In large catchments, solute concentrations in the main river channel represent the sum of solute processes operating in the individual tributary catchments, each of which may differ considerably in their individual concentrations.

Finally, changes in hydrological pathways in catchments over time affect solute concentrations in river water as different pathways have different solute signatures. Groundwater and throughflow deep within the soil are the major source of base cations to river water from mineral weathering. Therefore base cation concentrations in river water are normally higher in drier conditions when these hydrological pathways are the main source of river water. Base cation concentrations decline in wetter conditions since more runoff from macropore flow and overland flow is generated, with lower base cation concentrations, thereby diluting the groundwater inputs.

Temporal patterns in solute concentrations in river water are studied at three different timescales: from short-term changes of the order of a few hours during storm events, to annual patterns occurring over days and months, and long-term changes occurring over decades. Some of the patterns of variation and the processes responsible are examined below for each timescale.

13.5.1 Patterns of solutes in storm events: short-term changes

Considerable research effort in solutes has focused on storm events. Dramatic changes in solute concentrations occur and a significant proportion of solute fluxes are transported during storm events. The relationship between solute concentrations and river discharge is of particular interest for incorporation into hydrological models. Studies of solute patterns in river water during storm events have found significant variations between solutes, and also between individual storm events and between catchments for the same solute. The response of individual solutes during storm events varies because of the different magnitude and location of solute stores within a catchment and the extent to which they are accessed by the hydrological pathways that generate storm runoff.

Nevertheless, three generic relationships have been identified between solute concentrations and discharge: positive, negative and **hysteresis**. Positive relationships between solute concentrations and discharge occur when solutes are washed rapidly into the river channel by overland flow and shallow throughflow. In Figure 13.14 dissolved organic carbon (DOC) concentrations increase with discharge owing to the flushing of organic matter from more organic-rich surface soil horizons in the catchment. Negative relationships between solute concentrations and discharge occur for solutes that are mainly derived from mineral weathering and transported to the river in deep soil throughflow and groundwater flow. During storm events, inputs from these hydrological pathways are diluted by runoff generated from near-surface inputs. This effect is evident in Figure 13.14 where calcium concentrations decrease as discharge increases owing to dilution by near-surface throughflow with lower calcium concentrations.

More complex hysteresis relationships occur between solute concentrations and river discharge during storm events when very different solute concentrations are measured at the same discharge on the **rising** and **falling limbs** of the storm hydrograph. This results in hysteresis loops when solute concentration is plotted against discharge. Figure 13.15 illustrates how anticlockwise and clockwise hysteresis loops can develop for solute concentrations and river discharge. Anticlockwise loops occur when solute concentration is higher on the falling limb of the storm hydrograph compared with the rising limb. This could be caused by displacement into the river channel of soil water containing high solute concentrations by rainfall percolating through the soil. Clockwise hysteresis loops are formed when a higher solute concentration occurs on the rising limb of the storm hydrograph compared with the falling limb. This can arise when solutes are flushed from ground and vegetation surfaces and surface soil horizons at the start of storm events. Clockwise hysteresis loops for manganese concentrations and discharge in the River Nidd are shown in Figure 13.16 and are caused by flushing of manganese from surface soil horizons on the rising limb of the storm hydrograph. Depletion of manganese soil stores during consecutive storm events results in changes to the shape of the hysteresis loops. More information about different shapes

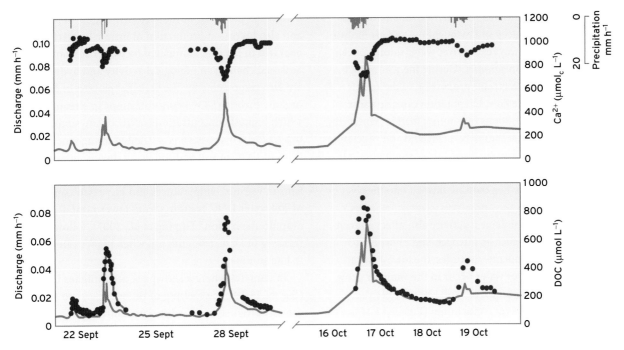

Figure 13.14 Solute concentrations and discharge in storm events in a stream draining a mixed hardwood forested catchment in the Adirondack Mountains, New York State, USA. (Source: adapted from Christopher *et al.*, 2008)

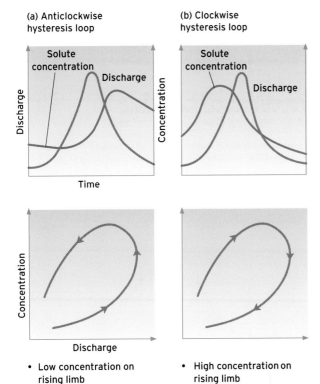

Figure 13.15 Schematic diagram of the formation of (a) anticlockwise and (b) clockwise hysteresis loops for solute concentrations in storm events.

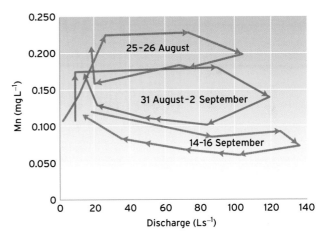

Figure 13.16 Hysteresis relationship between manganese concentrations and discharge during storm events in the Upper Nidderdale catchment, North Yorkshire, England, in 1994.

of hysteresis loops and their role in identifying sources of runoff can be found in Evans and Davies (1998).

13.5.2 Annual patterns of solute concentrations

Annual patterns of solutes in catchments occur because of systematic seasonal changes in hydrological pathways, weather, and biological and human activities. In temperate latitudes, with drier summers and wetter winters, solutes that occur mainly in deep soil throughflow and groundwater hydrological pathways cause river water solute concentrations to be higher in summer. During the winter such solutes are diluted as a greater proportion of the runoff occurs as near-surface flow. Evaporation can also cause higher solute concentrations in summer than in winter. In contrast, solutes that accumulate on ground and vegetation surfaces from dry deposition during drier summer conditions exhibit maximum annual concentrations in river water in the autumn when they are flushed into the river by increasing rainfall.

Annual cycles of biological activity also affect concentrations of nutrients such as nitrogen and phosphorus in river water. In temperate latitudes, the lowest nitrate concentrations in river water occur in the summer growing season because uptake from soil water by plants and microorganisms is at a maximum (Figure 13.17). Maximum stream nitrate concentrations occur in the

autumn as the result of reduced demand from plants and also increased production of nitrate by **nitrifying bacteria** in the soil, stimulated by wetter soil conditions. Seasonal changes in biological activity and hydrology are also responsible for the seasonal pattern in dissolved organic carbon (DOC) concentrations in streamwater, in which highest concentrations occur in late summer (Figure 13.18). This pattern is attributed to higher concentrations of DOC in soil and soil water during summer, due to the stimulation of DOC production by soil microorganisms in warmer temperatures and increased concentration of soil water DOC resulting from lower rainfall and higher evapotranspiration (Tipping *et al.*, 2007), followed by flushing of DOC from the soil when higher rainfall occurs in late summer.

In climates where a snowpack accumulates in winter (Figure 13.19), the melting of the snowpack in spring causes extremely high solute concentrations in river water. This is caused by the release of dry deposited

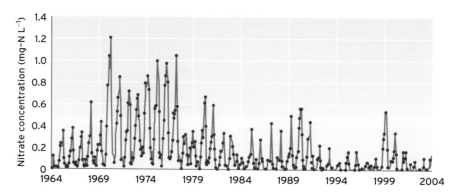

Figure 13.17 Monthly nitrate concentrations in a headwater stream in the Hubbard Brook Experimental Forest, New Hampshire, USA. Stream nitrate concentrations are consistently lower in the growing season (May to September). (Source: adapted from Judd *et al.*, 2007)

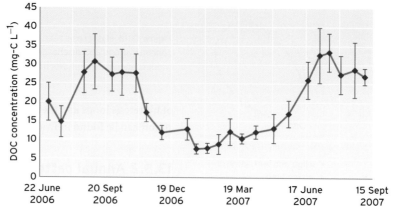

Figure 13.18 The seasonal pattern in dissolved organic carbon (DOC) concentrations in five rivers draining partially forested upland catchments in central Scotland. Each point is the mean ± 1 standard deviation of samples collected on the same day. (Source: adapted from Waldron *et al.*, 2009)

Figure 13.19 In climates where a snowpack accumulates in winter, such as at Lake Agnes in the Rocky Mountains, Alberta, Canada, solutes are released into river water when the spring snowmelt occurs.

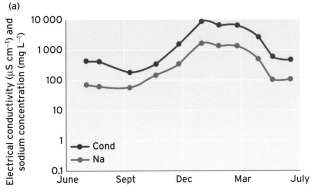

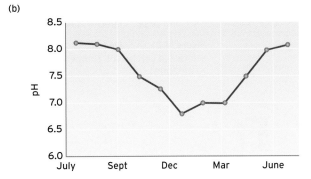

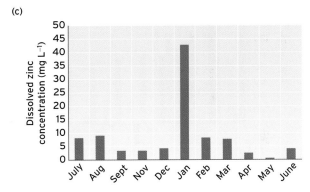

Figure 13.20 Electrical conductivity, pH, and concentrations of sodium and zinc in soil water sampled throughout the year 4 m from a road in central Sweden. (Source: after Bäckström *et al.*, 2004)

solutes, accumulated on the surface of the snow during the winter, and also by flushing of solutes from the soil that have remained in a frozen state during the winter. High levels of aluminium and acidity, released into rivers during snowpack melting, can be particularly damaging to fish populations, although the toxicity of aluminium is lowered in streams rich in dissolved organic carbon (Laudon *et al.*, 2005).

Seasonal variations in human activity such as winter ploughing of fields, spring application of fertilizer and salting of road surfaces in winter may all cause distinct annual patterns of solute concentrations. Some of the effects of winter applications of road salt in Sweden on soil water composition are shown in Figure 13.20. In winter months soil water electrical conductivity and sodium concentrations increased, whilst pH decreased, resulting in higher zinc concentrations in soil water.

13.5.3 Long-term patterns of solute concentrations

Changes in solute concentrations in river water over decades are difficult to evaluate because of the frequent unavailability of records of sufficient quality and length. Figure 13.21 illustrates the difficulty of identifying a long-term trend in river water dissolved organic carbon (DOC) concentrations hidden amongst considerable storm event and annual variability. Graphical analysis techniques show a significant upward trend in DOC

concentrations the cause of which is hotly debated (see Box 13.5).

The main causes of long-term patterns of solute concentrations in river water are climatic variability and climate change, land management change and changes in human polluting activities (e.g. atmospheric pollution). Acidification of freshwaters in many parts of north-west Europe and North America occurred from the eighteenth century as the result of atmospheric pollution from fossil fuel combustion for domestic, industrial and transport purposes. With the decrease in atmospheric emissions and deposition of acidifying sulphur and nitrogen compounds

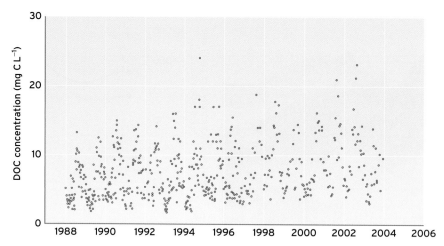

Figure 13.21 Dissolved organic carbon concentrations measured in a stream draining an afforested catchment, southern Scottish Highlands, UK, 1988-2006. (Source: after Dawson *et al.*, 2008)

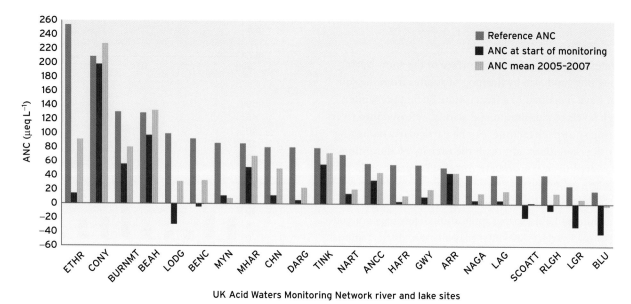

Figure 13.22 Acid neutralizing capacity (ANC) values at UK Acid Waters Monitoring Network upland river and lake sites modelled in 1860 (reference values) and measured at the start of monitoring (1988-1990) and in 2005-2007. ANC indicates the ability of water to resist acidification by a strong acid. The ANC values in this figure were calculated from the alkalinity, dissolved organic carbon and ionic aluminium concentrations of the water samples. (Source: Kernan *et al.*, 2010)

from human activities since the 1980s as the result of international agreements to reduce emissions, some acidified freshwaters have started to recover, with acid neutralizing capacity (ANC) values returning towards earlier modelled values as shown for upland freshwaters monitored in the UK (Figure 13.22).

The precise causes of long-term patterns of solute concentrations in river water are difficult to identify because of the interaction of causal factors. For example, climate change often results in changes to land use and management. It is evident from Figure 13.22 that, although ANC values are returning towards historical values at some sites, the rate of recovery of freshwaters from acidification is variable. Suggested reasons for the slower recovery at some sites include:

1. Limited recovery of the bases in the soil to neutralize acidic inputs.

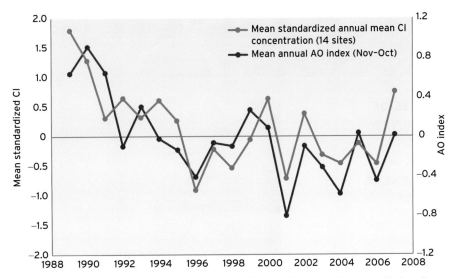

Figure 13.23 Standardised annual mean freshwater chloride concentrations at 14 river and lake monitoring sites near the western coast of the UK and the mean annual Arctic Oscillation (AO) Index for 1989-2007. The two parameters are closely positively associated indicating that the occurrence of sea-salt events in freshwaters is linked with regional circulation patterns. The recovery from acidification of many freshwaters during this time period may be partly explained by the reduced intensity of the AO index, resulting in fewer sea-salt events. (Source: Kernan *et al.*, 2010)

2. The continued release of sulphur from soil, even though deposition of sulphur compounds from human activities has decreased.

3. The release of nitrate from catchment soils offsetting the effects of lower acidifying sulphate concentrations in freshwaters.

4. A slower increase in water pH than expected because of increasing dissolved organic carbon concentrations.

5. Other confounding stresses, such as the uptake of neutralizing base cations from the soil by forestry plantations and climate change. For example, recovery from acidification of some freshwaters in the UK since the 1980s has been associated with the decreased frequency in occurrence of sea-salt events, a natural cause of the freshwater acidification. The frequency of occurrence of sea-salt events affecting freshwaters in the western UK is associated with the Arctic Oscillation, similar to the North Atlantic Oscillation (see Chapter 4), which affects weather patterns in the northern Atlantic Ocean. Figure 13.23 shows the positive association between the strength of the Arctic Oscillation and chloride concentrations in freshwaters in the western UK. It is predicted that, due to climate change, the Arctic Oscillation may become stronger in coming decades, resulting in more frequent sea-salt events and freshwater acidification.

Reflective questions

➤ Thinking about dilution and other processes, do solute concentrations always decline as discharge increases? What are the reasons for your answer?

➤ What effects could climate change have on solute patterns at the storm event, annual and long-term timescales and how would you explain the processes which cause the effects you have identified?

13.6 Spatial patterns of solutes

Solute concentrations and fluxes in rivers and lakes vary over space as well as over time owing to differences in climate, geology, topography, soils, vegetation and land management. The significance of the factors controlling spatial patterns of solutes depends on the scale of observation. At the global scale, climate and geology account for most of the variation in solute concentrations, although human activities are increasingly influential. At the regional scale, the factors of soil, vegetation and land management are more important. The influence of these different sets of controls on spatial patterns of solutes will be examined at the global and regional scales in the following sections.

ENVIRONMENTAL CHANGE

WHAT IS THE CAUSE OF INCREASED DISSOLVED ORGANIC CARBON CONCENTRATIONS IN NORTHERN HEMISPHERE SURFACE WATERS?

Many studies have reported significant increases in the concentration of dissolved organic carbon (DOC) in lakes and rivers in the northern hemisphere since the 1980s. These observations have raised concerns about the degradation of soil carbon stores and also about drinking water treatment and quality because of the formation of potentially carcinogenic trihalomethanes when the standard method of disinfecting drinking water with chlorine is applied to DOC-rich water.

A number of explanations have been proposed for the observed increases in DOC concentrations:

1. Rising air temperatures are creating a larger soil store of DOC due to enhanced microbial decomposition of soil organic matter.

2. Rising atmospheric CO_2 concentrations are stimulating plant growth, resulting in increased production of compounds exuded from plant roots that are rich in DOC.

3. Increasingly aerobic conditions in peatlands, arising from higher air temperatures and/or lower rainfall, remove constraints on the activity of enzymes that decompose organic matter to produce DOC (Freeman *et al.*, 2001).

4. More intensive land management activities, such as land drainage, forest planting and harvesting, burning and livestock grazing (such as in the UK uplands), have resulted in greater DOC mobilization. For example,

Figure 13.24 Area of former forestry plantation which was harvested as part of a wind farm development in central Scotland.

disturbance associated with the harvesting of forestry plantations to clear land for wind farm developments (Figure 13.24) has been associated with increased DOC and phosphorus concentrations in runoff (Waldron *et al.*, 2009).

5. Reductions since the 1980s in the deposition of acidic non-marine sulphur compounds and sea salts have reduced the acidity and ionic strength of soil water, resulting in increased mobility of DOC in organic soils.

6. Sustained atmospheric nitrogen deposition may decrease the activity of DOC oxidizing enzymes in soils, resulting in the greater availability of DOC compounds for export from the soil. It may also increase microbial activity and nitrogen mineralization in organic soils causing an increase in DOC production.

7. Changes in catchment hydrology in which either decreases in water flow could result in increased measured DOC concentrations or changes in hydrological flow paths could mean

that more flow occurs through DOC-rich surface organic soils.

There is much debate and ongoing research concerning the explanation and interpretation of the observed increase in DOC concentrations in rivers and lakes, but it is unlikely that there is a single explanation applicable to all situations. Clark *et al.* (2010) suggest that the explanations proposed are not incompatible, but interact with each other and are influenced by catchment properties so that the causes of increased DOC concentrations vary between catchments. For example, in regions where acid deposition was high in the past, the main cause of increasing DOC concentrations is likely to be decreasing non-marine sulphur deposition. In contrast, in peat-dominated catchments where the natural rate of sulphate reduction in waterlogged anaerobic conditions exceeds sulphur inputs from atmospheric deposition, decreasing sulphur deposition is unlikely to cause increased DOC concentrations.

BOX 13.5

13.6.1 Global patterns of solutes

Global non-anthropogenic controls on solute concentrations and fluxes in river water have been analyzed by Walling and Webb (1986). Solute concentrations in rivers often increase with mean annual temperature because of greater evaporative loss from river systems, thereby increasing river solute concentrations. The total ionic flux, however, increases with mean annual runoff and slope steepness in rivers. This is because it is the rate of water movement at the weathering front that largely determines the rate of solute release. There is also a marked difference in ionic fluxes between rock types. Ionic fluxes in rivers draining catchments developed on different geologies are in the order: sedimentary > igneous extrusive > igneous intrusive and metamorphic. Pristine freshwaters are seldom found today so Meybeck (2003) argues that the main processes affecting river systems globally are anthropogenic, in particular river damming, eutrophication, chemical contamination and decreased river flow due to irrigation.

13.6.2 Regional patterns of solutes

At the regional scale, land use is the dominant control on solute concentrations in river waters. Land use affects solute concentrations by altering the magnitude and mobility of solute sources in the catchment and also by altering the hydrological pathways that transport solutes to the river channel. The effects of conifer plantations, agricultural and urban land uses on solute patterns are discussed below as examples.

13.6.2.1 Conifer plantations

The alteration of solute concentrations in river water by conifer plantations depends on the stage of the forestry cycle. The most significant effect of conifer plantations on nutrient concentrations is an increase in nitrate concentrations if the trees are harvested by clear-cutting. This effect is caused by creation of an improved microclimate for microbial decomposition in forest floor soils, resulting in increased rates of production of nitrate. Elevated nitrate concentrations in river water persist until new vegetation establishes and takes up nitrogen from the soil.

Conifer plantations in catchments developed on base-poor geologies contribute to high levels of acidity and aluminium in river waters. These effects are most marked in mature plantations and are caused by several processes as summarized in Figure 13.25. The major cause of acidification is the interception of acidic atmospheric pollutants by the forest canopy. Conifer trees are more efficient

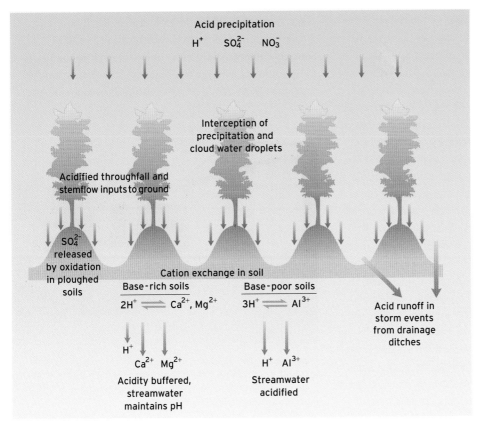

Figure 13.25 Processes of acidification of river water in catchments dominated by conifer plantations.

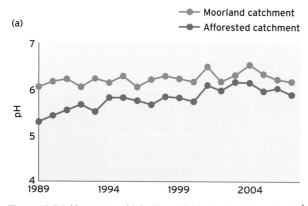
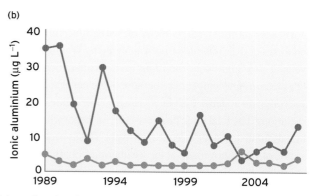

Figure 13.26 Mean annual (a) pH and (b) ionic aluminium (μg L^{-1}) in lake waters in adjacent moorland (blue line) and afforested (green line) sites in the southern Scottish Highlands, UK, 1989-2007. At the start of monitoring, lake water in the afforested catchment had a lower pH and higher ionic aluminium content than lake water in the catchment with a moorland shrub vegetation cover. However, the acidification of the lake water draining the afforested catchment has decreased over time and is attributed to declining acidic atmospheric deposition. (Source: Kernan *et al.*, 2010)

scavengers of atmospheric pollutants than broad-leaved trees and grassland vegetation because the nature of the canopy surface and the large surface area of pine needles encourage the condensation of acidic cloud droplets and deposition of acidic gases and particles. Consequently, high inputs of acidity (hydrogen ions) occur to soils underlying conifer plantations. Acid inputs to forest soils are initially neutralized by the exchange of hydrogen ions with calcium and magnesium ions at cation exchange sites. However, if acid inputs continue faster than base cations are released into the soil from mineral weathering, the soil cation exchange sites become dominated by hydrogen ions and aluminium resulting in increased levels of acidity and aluminium in river water. The neutralizing capacity of the soil can be depleted further by the uptake of base cations by trees which are then harvested and removed. Another factor that contributes to river water acidification in conifer plantations is the preparation of ground for planting. Drainage ditches are often created by ploughing to improve the soil moisture conditions for the trees. The ditches alter hydrological pathways in the catchment by forming a more efficient drainage system, transferring runoff rapidly to river channels during storm events (Holden *et al.*, 2007). As a result, runoff only passes through the acidic surface soil horizon and has limited opportunity to percolate to less acidic deeper soil horizons where neutralization of acidity can occur. Drainage ditch construction also increases the acidity of river water through allowing soil oxidation to occur. This results in production of the acidifying anion, sulphate. However, as acidic deposition has decreased since the 1980s in some parts of the world, there is evidence that the enhanced acidification of freshwaters due to greater interception of acidic atmospheric pollutants by forest

plantations has started to be reversed in some catchments (see Figure 13.26).

13.6.2.2 Agricultural land management

The main effects of agricultural land management on solute concentrations in river water are an increase in concentrations of nutrients and pesticides. Increased concentrations of these may arise from point sources such as leakage of pesticides, silage, slurry and milk wastes from storage facilities, or from diffuse pollution, such as from farmyards (Figure 13.27). Sources of diffuse pollution are numerous and are summarized for nitrogen and phosphorus in Figure 13.28. Where fertilizer or manure applications exceed crop nutrient requirements,

Figure 13.27 One source of diffuse pollution from agriculture is runoff from farmyards and roofs when faecal material from animals and birds, sediment and spillages of oil and hydrocarbons are washed off by rainfall.

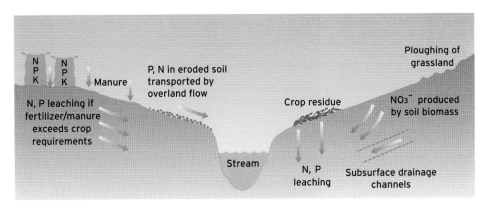

Figure 13.28 Diffuse sources and transport of nitrogen and phosphorus from agricultural land to river channels.

or are applied on frozen ground or just before heavy rainfall, leaching of soluble nitrogen and phosphorus occurs from the soil. Artificial land drainage can increase the percolation of water, water residence time and also improve soil aeration. These factors increase nitrate leaching from the soil because there is more opportunity for leaching to occur and conditions are more favourable for nitrate production by microbial activity. Ploughing grassland for conversion to crops also increases nitrate leaching from the soil for similar reasons. Decomposition of crop residues results in the release of nitrogen and phosphorus in the soil, which may be transported into river courses. Finally, soil erosion transports nutrients attached to

soil particles from agricultural land to river channels. This transport pathway is particularly important for phosphorus.

13.6.2.3 Urbanization

Urbanization typically results not only in increased flood frequency and magnitude but also in increases in concentrations of metals, nutrients, pesticides and organic matter in river water. These elevated concentrations arise from an increased magnitude of solute sources such as metals deposited from tyre breakdown and corrosion of vehicle brake linings, and intensive pesticide use on gardens, public parks and road verges (Table 13.4). They also

Table 13.4 Sources of selected solutes in urban areas

	Vehicle use	Pesticide use	Industrial/other use
Copper	Metal corrosion	Algicide	Paint
			Wood preservative
			Electroplating
Lead	Fuel	-	Paint
	Batteries		Lead pipe
Zinc	Metal corrosion	Wood preservative	Paint
	Tyres		Metal corrosion
	Road salt		
Chromium	Metal corrosion	-	Paint
			Metal corrosion
			Electroplating
Nitrogen	Combustion of fuel	Fertilizers	Cleaning operations
			Animal excrement
Chloroform	Formed from salt	Insecticide	Solvent
PCBs (polychlorinated biphenyls)	-	-	Electrical insulation

(Source: after Novotny and Harvey, 1994, and Makepeace *et al.*, 1995)

result from the dramatic changes in catchment hydrological pathways brought about by urbanization. Traditionally, urban drainage engineering aimed to remove runoff as rapidly as possible from urban surfaces to prevent inundation. This aim was achieved through the use of nearly impervious surfaces for car parks, highways and roofs, from which water drained rapidly into a wastewater treatment works or a nearby river. However, this flushed solutes into urban rivers during storm events, often in toxic concentrations, as dry deposited solutes are washed from impervious surfaces. Solute pollution in urban rivers is worse during medium-sized storm events, occurring two or three times a year, since sufficient runoff is generated to mobilize dry deposited solutes from urban surfaces but there is not enough water to dilute solute concentrations. Urban **Best Management Practices (BMPs)** or Sustainable Urban Drainage Systems (SUDSs) have been introduced to reduce the pollution of urban rivers and also reduce the risk of small and medium-sized floods. These include structures such as ponds, wetlands (Figure 13.29) and porous paving that can increase the storage of runoff in the catchment and allow water quality improvement by physical, chemical and

Figure 13.29 Example of a SUDS: a linear wetland treating diffuse pollution from a commercial development in central Scotland. (Source: photo courtesy of E. Nitschke)

biological processes. A new and increasing water quality concern associated with urban runoff and wastewater, but also with agricultural land uses, is the presence and effect of endocrine disrupting chemicals which are discussed in Box 13.6.

HAZARDS

NEW SOLUTES OF CONCERN: ENDOCRINE DISRUPTORS

In the past few decades there has been increasing production of numerous synthetic organic chemicals, such as pesticides, pharmaceuticals, microbial disinfectants and personal care products, for use in industry, agriculture, medical treatment and household products. While many of these chemicals have undoubted benefits, they are also endocrine disruptors (EDs), which affect sex determination in fetuses, brain development in babies and the growth and activity of organs in adults. The entry of EDs into the water environment has been associated with adverse impacts on health and reproduction in wildlife and

humans, including even in the next generation. Consistent proof of a causal link between exposure to synthetic EDs and effects is difficult. There are hundreds of thousands of different EDs present in the environment, both synthetic and naturally occurring, which may have damaging combined effects even at exposure to very low concentrations over long periods of time. Consequently no standards have yet been developed for maximum allowable ED concentrations in waters. EDs enter the water environment from both diffuse sources, such as surface runoff from intensive agricultural activities in which veterinary products and pesticides are used and leakages from septic tanks and landfill sites, and point sources, in particular

discharge from wastewater treatment works containing EDs originating from personal care products such as soaps, detergents and perfumes, and excreted pharmaceuticals and contraceptives. Although wastewater treatment processes may be effective in removing some EDs, removal rates differ between specific processes and EDs and also vary seasonally. Since EDs typically occur at very low concentrations (nanograms per litre, ng L^{-1}) in water, sophisticated and expensive procedures and instruments are required to detect individual EDs and determination of EDs in water samples is not routine, The first US-wide survey of 100 pharmaceuticals and other organic wastewater contaminants

BOX 13.6 ➤

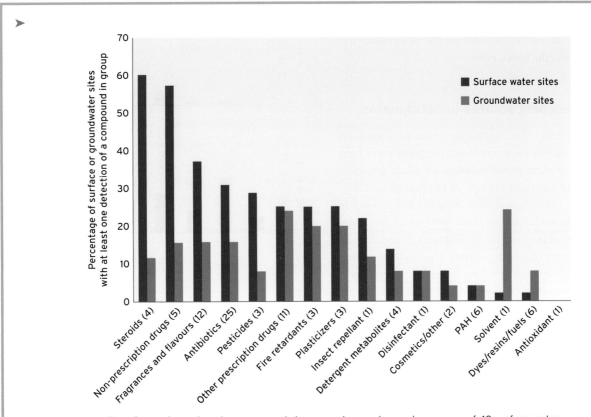

Figure 13.30 Detection of organic wastewater compounds by general use category in a survey of 49 surface water and 25 groundwater drinking water sources throughout the USA in 2001. All steroids analyzed are naturally occurring. (Source: adapted from Focazio *et al.*, 2008)

(many of which are EDs) in drinking water sources detected 63 of the compounds in at least one water sample and a median of 4 compounds at each site. The most frequently detected organic wastewater contaminants by type of product are shown in Figure 13.30 and include steroids, prescription and non-prescription drugs, fragrances, antibiotics, pesticides, fire retardants, plasticizers, insect repellent, detergent by-products, disinfectants and cosmetics. The contaminants were detected more frequently in surface water than groundwater, probably because the main sources are surface waters (into which direct discharge of wastewater effluent occurs) and there is less potential for degradation and removal of contaminants in surface waters (Focazio *et al.*, 2008).

BOX 13.6

Reflective questions

➤ How do the types, sources and dominant transport processes of solutes from urban and agricultural land compare and contrast?

➤ How would you design an investigation to identify the effects of different land uses on water quality?

13.7 Modelling solutes

Models are widely used to predict solute concentrations in catchments, rivers and lakes for catchment management and planning. This may include simulating the effects of pollution incidents on river water quality, for example. As with all models, solute models require field or laboratory measurements for model calibration and validation to evaluate the success of the model (see Chapter 1). Since solute concentrations are controlled by many hydrological, soil, chemical and biotic factors, solute models must also include simulations of these processes. Examples of solute modelling at two different scales are discussed below: the catchment and the watercourse.

13.7.1 Modelling solutes in catchments

Many catchment solute models have been developed. For example, it may be necessary to predict the effect of land-use change or atmospheric pollution on solute concentrations in river water. An example of this type of model is the MAGIC (Modelling Acidification of Groundwater In Catchments) model (Cosby et al., 2001) which was developed to predict the long-term effects on soil and surface water chemistry of acid deposition. The model assumes that the concentration of solutes in surface waters is governed by atmospheric deposition, soil weathering reactions, uptake of solutes by biomass and runoff. A series of chemical equations simulates the reactions of solutes in the soil, including cation exchange and adsorption. The model is a **lumped model** which assumes that the catchment is homogeneous, with uniform soil properties and precipitation inputs.

Application of the MAGIC model to a catchment requires data for soil physical and chemical properties, rainfall/runoff characteristics, precipitation chemistry and base cation weathering rates. The MAGIC model was applied to predict the chemistry of the surface waters from 1860 (prior to industrialization) to 2100 of 59 lakes in south-west Scotland that have been affected by acidification (Helliwell and Simpson, 2010). The simulations took account of changing acid deposition and land use over time. Figure 13.31 shows the distribution of simulated pH values between the lakes for different years from 1860 to 2100. The peak of the frequency distributions is a measure of the average pH of lake waters. The model results show that lake water pHs decreased from 1860 to the lowest values in 1970 when pH values in approximately half of the lakes were less than 5.5 (a critical threshold for salmonid fish). From 1970 surface water pH has increased over time so that by 2100 it is projected that pH values less than 5.5 will occur in only 17% of the lakes.

13.7.2 Modelling solutes in watercourses

It is often necessary to develop models of the processes controlling water quality within rivers and lakes, particularly for operational purposes, for example to respond to a pollution spillage into a river, or to set concentration

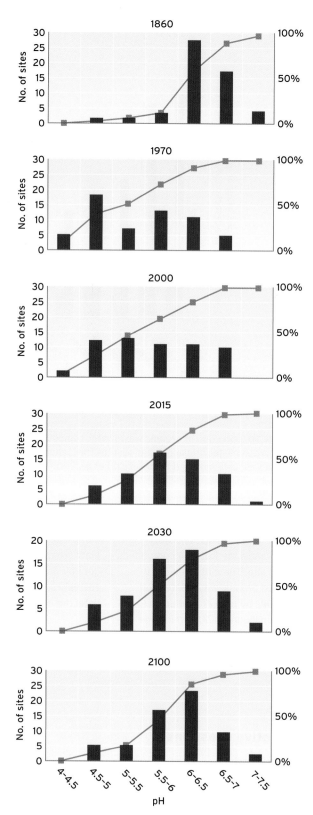

Figure 13.31 Histograms (left-hand y-axis) and cumulative frequencies (right-hand y-axis) of surface water pH for selected years from 1860 to 2100 in 59 lakes in Galloway, south-west Scotland, simulated using the MAGIC model. (Source: after Helliwell and Simpson, 2010)

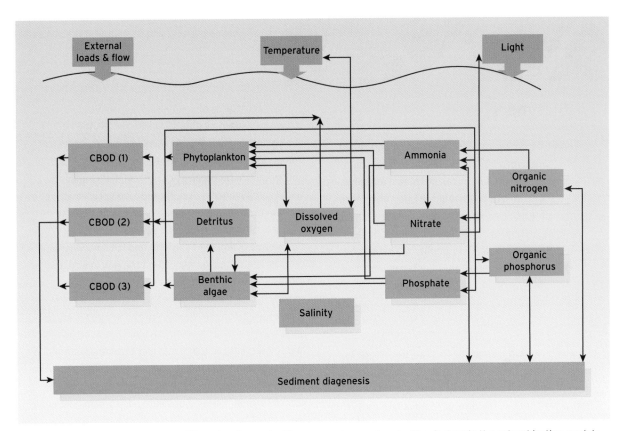

Figure 13.32 Interactions between different water controlling parameters and controlling factors in the eutrophication module within the WASP (Water quality Analysis Simulation Program) model. CBOD is the carbonaceous biochemical oxygen demand and is represented by three different compartments within this model. CBOD is measured in the same way as BOD (see Glossary) apart from the activity of bacteria that oxidize nitrogen, which is suppressed in the water sample in the determination of CBOD. (Source: United States Environmental Protection Agency (USEPA))

limits for polluters that are licensed to discharge pollution into watercourses. An example of this type of model is the WASP (Water quality Analysis Simulation Program) model which was developed by the United States Environmental Protection Agency (USEPA) to predict the effect of natural phenomena and human-made pollution on water quality. The processes of advection, dispersion, point and diffuse pollutant inputs and exchange between river reaches are simulated in the model. Based on these, WASP can predict concentrations of many water quality parameters, including different forms of nitrogen and phosphorus, dissolved oxygen, algae, mercury and organic chemicals. The interactions between different water quality parameters in the eutrophication model within WASP are shown in Figure 13.32. The WASP model was applied by Rose and Pedersen (2005) to assess the fate in river water and sediment of oxytetracycline (OTC) used in fish hatcheries to minimize the occurrence of fish disease and pathogens. The results suggested that

approximately 38% of the OTC used in fish hatcheries would be transferred to the river system. The fate of the OTC in the river environment was found to be primarily controlled by sedimentary processes. Concentrations of OTC in the river sediments in the immediate vicinity of the hatchery discharge, but not downstream, occasionally may be sufficiently high to affect the types of species present in the aquatic ecosystem.

Reflective question

➤ A model is being developed to predict the effect of urbanization on solute concentrations in the river draining a catchment. What do you think are the key processes and solutes to include in the model, and why?

13.8 Summary

This chapter demonstrates the importance and relevance of solute hydrology to physical geography and environmental science. Chemical, biological and physical factors affecting solute behaviour include solute form, pH, redox potential, temperature and pressure, particulates and interactions with microbiological activity. The behaviour of solutes in all components of the catchment hydrological system is important in determining the spatial and temporal variation of solute flux and solute concentration. Solute flux is the concentration of a solute multiplied by the discharge. Precipitation, evaporation/evapotranspiration, interception, soil water, groundwater, and river and lake processes are important in determining both solute concentrations and fluxes. In addition the interaction between solutes and hydrological pathways controls the downstream solute patterns. Different solute concentrations are contained and modified by different hydrological pathways. Slow-moving water that has a long soil or substrate residence time will tend to have very different solute contents and concentrations from water that has moved more quickly over or through the catchment. This faster-moving water may have solute concentrations that closely match that of the input precipitation. It is because of this that solutes are often used as flow tracers in catchments in order to identify the pathways through which the water has travelled.

Spatial patterns of solutes at the global and regional scales are caused by differences in climate, topography, geology, soil, vegetation and land use and management. Temporal patterns of solutes at the storm event, annual and long-term timescales are the result of changes in the relative dominance of runoff production processes within the region or catchment of study and by changes to inputs or stores of solutes within the system. Antecedent conditions (e.g. soil moisture content) and land management or environmental change may affect the way in which water moves across the catchment as well as the volume of water moving across the catchment. In addition inputs of fertilizer or changes in forestry practice may change the amount of certain solutes available for movement and removal. Modelling solutes within catchments and individual water bodies is required for management, planning and operational purposes. While many models help us with our predictions, a complete understanding of solute and contaminant behaviour cannot be attained without appreciating the interaction of geology, climate, soil, biotic and human activities with catchment hydrology. Such an understanding is necessary to advance our understanding of how the world works and to predict the effects of land management and climate change on solute and contaminant hydrology. Furthermore, adverse and unexpected consequences for human health, environmental resources and terrestrial and aquatic ecosystems can occur when our well-intentioned management interventions attempt to manipulate solute behaviour without taking account of these interactions.

Further reading
Books

Kendall, C. and McDonnell, J.J. (1998) (eds) *Isotope tracers in catchment hydrology*. Elsevier Science, Amsterdam, see: http://wwwrcamnl.wr.usgs.gov/isoig/isopubs/itchinfo.html.
The definitive textbook on this topic, beginning with chapters that explain the fundamentals of isotope geochemistry, followed by chapters examining the use of isotopes in all compartments of the catchment hydrological system.

Novotny, V. (2003) *Water quality: Diffuse pollution and watershed management*. John Wiley & Sons, New York.
A comprehensive and authoritative tome on diffuse pollution; a book to dip into, rather than to read from cover to cover.

Further reading
Papers

Cole, J.J., Prairie, Y.T., Caraco, N.F. *et al.* (2007) Plumbing the global carbon cycle: integrating inland waters into the terrestrial carbon budget. *Ecosystems*, 10, 171–184.
This paper demonstrates that freshwaters are not passive pipes conducting carbon from terrestrial environments to the ocean but make an important contribution to carbon dioxide emissions in the global carbon budget.

Johnes, P.J. (2007) Uncertainties in annual riverine phosphorus load estimation: impact of load estimation methodology, sampling frequency, baseflow index and catchment population density. *Journal of Hydrology*, 332, 241–258.
This paper examines the uncertainties in estimating river fluxes, including different calculation methods, using a relevant case study.

Meybeck, M. (2003) Global analysis of river systems: from earth system controls to Anthropocene syndromes. *Philosophical Transactions of the Royal Society B*, 358, 1935–1955.
A global overview of the effects of human activities on rivers is given in this paper.

Sumpter, J.P. and Johnson, A.C. (2005) Lessons from endocrine disruption and their application to other issues concerning trace organics in the aquatic environment. *Environmental Science & Technology*, 39, 4321–4332.
This paper provides an excellent critical overview of the knowledge (or lack of it) concerning endocrine disruptors in the water environment.

Whitehead, P.G., Wilby, R.L., Battarbee, R.W., Kernan, M. and Wade, A.J. (2009) A review of the potential impacts of climate change on surface water quality. *Hydrological Sciences Journal*, 54, 101–123.
A thoroughly referenced review which demonstrates that climate change is likely to cause deterioration in water quality.

CHAPTER 14

Dryland processes and environments

David S.G. Thomas

School of Geography, Oxford University Centre for the Environment, University of Oxford

Learning objectives

After reading this chapter you should be able to:

➤ understand what aridity is and what causes drylands and deserts

➤ describe where dryland areas are located today and the diversity of environments that drylands represent

➤ understand the main processes that have shaped drylands and operate within them

➤ understand why drylands are important regions for people and how their importance may change in the future

14.1 Introduction

Dryland environments are diverse (Figure 14.1), comprising much more than the sand-dune-covered landscapes that frequently feature in the media. The image of deserts as barren, ochrous sand seas, so often seen in magazine and television advertisements for cars and perfumes, reflects only a small part of the diverse environmental conditions present in the world's drylands. This diversity reflects variations in moisture availability, structural settings, parent rocks, ecological characteristics and local geomorphic conditions. Nor

is it the case that deserts and drylands are devoid of human interest and occupation (Figure 14.2). Over 5000 years ago, early Egyptian civilizations developed and flourished in a desert setting, as did those of Mesopotamia. Stone tools, found in many dryland contexts, are testimony of even earlier human (or hominid) engagement in these regions. This is exemplified by the incredible extensive stone tool scatters found in parts of the Kalahari Desert, Botswana (Figure 14.3), and the discovery in 1969 of the remains of 'Mungo man' in the Willandra Lakes region of dry interior Australia, and dated to *c*. 40 000 years ago.

The key to all these 'occupations' is an association with water: either in terms of now-dry former lake basins, testimony to past wetter climates or major inflows from wetter regions; or, as with the civilizations of Mesopotamia and Egypt, major rivers bringing water from wetter regions were the key to the development of societies in otherwise hostile regions. More generally, the ability to find strategies to cope with seasonal and longer moisture deficits contributed to the successful and sustainable development of migratory pastoral lifestyles throughout North Africa, the Middle East and central Asia and, in the waterless Kalahari of southern Africa, to the evolution of a hunter–gatherer lifestyle by the San, or bushmen. These persisted as the major forms of human use for many thousands of years, and in some cases are the dominant lifestyle today.

Figure 14.1 Three different dryland landscapes in the space of a few hundred metres, Namib Desert, looking due south. At the northern edge of the Namib sand sea, which contains mega-dunes that can reach heights of over 300 m (background), there is an abrupt transformation to a rocky and stoney plain, where processes such as salt weathering operate on the largely bare rock surfaces (foreground). This sharp transition occurs because of the east–west-orientated Kuiseb Valley. This valley, which has flowing water in it for no more than a few days a year, is a linear oasis fed by rains that fall in the highlands to the east of the desert. The small annual flow of the river is sufficient to inhibit the northward extension of the sand dunes, and therefore to preserve a distinction between the southerly sand desert and northerly stoney desert. There is enough seepage water in the valley sediments to support plants all year round.

Figure 14.2 Innovative living in drylands: a troglodyte house in the Matmata loess plateau, Tunisia. The even-grained but massive wind-blown dust deposits of loess have provided an ideal medium in which to excavate underground houses. The rooms that lead off this central courtyard maintain cool, even temperatures even at the hottest times of the year. There are many examples of human ingenuity being used over the millennia to cope with the harsh environmental conditions that can exist in dryland regions.

Explorers and travellers from western Europe began increasingly to visit African and Asian drylands regions from the eighteenth century onwards, although first arrivals were several centuries earlier. Dryland areas were frequently viewed with disdain and fear, and sometimes as barriers in the quest for more hospitable environments, as illustrated in the writings of Thomas Mitchell, following his travels in central Australia in the mid-1830s: 'After surmounting the barriers of parched deserts and hostile barbarians, I had at last the satisfaction of overlooking from a pyramid of granite a much better country' (Mitchell, 1837). In contrast to well-watered and well-vegetated temperate areas, the deficit of water often proved a major obstacle to movement, while the paucity of plant cover presented desert landscapes as spectacular and bizarre. This contributed to reports about deserts that emphasized the unusual aspects, which continue to fuel the images portrayed in the media today.

During the twentieth century, technologies evolved so that today, in the twenty-first century, drylands support over 2 billion people, almost a third of the global population. Drylands possess many cities with populations in excess of 2 million, including the extensive urban areas of

(a)

(b)

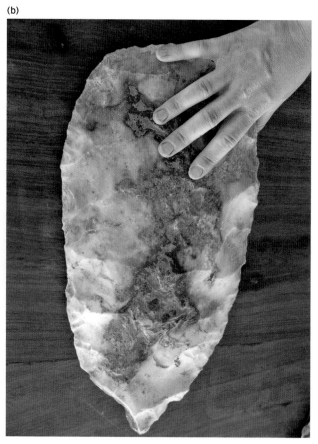

Figure 14.3 (a) Extensive stone tool scatters occur in parts of the Kalahari Desert, Botswana, including in the vicinity of the dry lake bed of former Palaeolake Makgadikgadi, a feature representing the past occurrence of a lake of up to 66 000 km² in this dryland context (Burrough et al., 2009). The tools are of uncertain age at present but are likely dominated by Middle Stone Age material, possibly from ~90-40 000 years ago. (b) Not only are there scatters of 'regular' artefacts in this basin, but several giant 'handaxes', amongst the largest on Earth, have been found.

such metropolises as Los Angeles and Beijing. Half of Africa's total population lives in drylands, while dryland countries and regions are amongst those with the highest annual population growth rates today. For example, the population growth rate is 4.77% in Afghanistan and 1.45% in Mongolia, compared with a global average of 1.15% per annum.

There is clearly an important social dimension to understanding how dryland environmental systems function. The widespread availability of satellite imagery from the 1970s onwards (Figure 14.4), together with a marked increase in systematic environmental research in drylands, have contributed both to a more comprehensive view of the diversity of deserts and drylands on the one hand, and, on the other hand, to a recognition that the processes that occur in these areas are not unique, differing from those in other environments in the frequency and magnitude of occurrence, rather than in type. This chapter explores dryland characteristics, landforms and the dominant processes that have shaped, and continue to shape, these extensive and important regions.

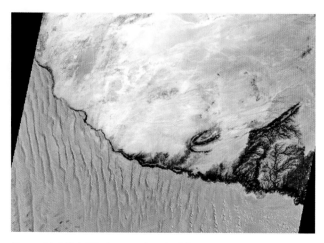

Figure 14.4 Satellite image of part of the Namib Desert. In the southern part of the image, the Namib sand sea is abruptly truncated at its northern margin by the Kuiseb Valley, north of which lies the stoney-gravelly Namib. Figure 14.1 shows conditions in this boundary zone from the ground. (Source: courtesy of NASA)

14.2 Aridity

The main characteristic of deserts and drylands is clearly a lack of available moisture. This does not mean that precipitation does not occur. Rather, it is erratic: it can be spread unevenly through the year, contributing to a long 'dry season', can be patchy in where it occurs in a given region, and can have a high propensity not to fall in any given year, contributing to uncertainty. Deserts and drylands are not necessarily hot (another common image), though some certainly are. Winters can be cool, even cold, and for example in parts of interior North America and Asia, winter snowfall can be an important precipitation source. Clear cloudless skies can lead to high diurnal temperature ranges of several tens of degrees Celsius. Of all the climatic characteristics of these regions, it is the annual overall net negative moisture balance, or **aridity**, that is the defining characteristic of deserts and drylands.

A number of methods exist for measuring the annual moisture balance at any location, all of which require meteorological data, preferably collected over a number of years so that mean values and trends can be determined. Data on monthly precipitation (P) and monthly potential evapotranspiration (PET) are required for the moisture balance to be calculated. Direct measurements of PET are not widely available, so that methods for its determination from other climate data have been used, even data from weather stations that collect only rudimentary information such as mean monthly temperature and the number of daylight hours.

Moisture balance values of less than 1.0 indicate an annual excess of PET over P, a moisture deficit. In practice, areas where values fall below 0.5 are regarded as drylands. Since the 1950s and the work of Perivail Meigs (1953), areas that are too cold for seasonal crop growth have been excluded from the classification of drylands. This convention was introduced by UNESCO in its arid lands programme, which was concerned with areas that support human populations, however sparsely. This means that tundra and extremely low-latitude areas, where actual precipitation amounts are very low, are not included in the classification. For example, the Dry Valleys area of Antarctica is extremely arid but temperatures are so low that this area does not meet the crop growth criterion mentioned above. Where P/PET values are below 0.5, a number of subdivisions have been introduced since the late 1970s by organizations such as the United Nations in its Environment Programme (UNEP), so that regions with different degrees of aridity can be distinguished.

14.2.1 Drylands

The driest, or **hyper-arid** areas, which can be regarded as the 'true deserts', have P/PET values of less than 0.05. These areas, which also have periods in excess of 12 months without rainfall being recorded, include large tracts of the central Sahara, the Arabian Peninsula and the coastal Atacama and Namib Deserts (Figure 14.5). Arid areas have P/PET values between 0.05 and 0.2 and include areas of central Australia and the northern and southern fringes of the Sahara. Annual rainfall may be up to 200 mm where it occurs predominantly in winter months and up to 300 mm where it is principally a summer occurrence. Much of the Sahel belt of Africa, large areas of the western interior of North America, north-east Brazil and areas of Mediterranean Europe are semi-arid. P/PET values fall between 0.2 and 0.5, and rainfall may be up to 500 mm a year in winter rainfall areas and up to 800 mm in summer rainfall areas. Whether hyper-arid, arid or semi-arid, dryland regions are all susceptible to large year-to-year variations in precipitation (Figure 14.6), and are susceptible to drought events. As a rule of thumb, the human occupation and agricultural use of drylands increases from hyper-arid to semi-arid, just as do natural vegetation amounts (see Chapter 18). In the early 1990s UNEP added dry–subhumid areas (P/PET = 0.5–0.65) to the areas that it regarded as drylands, because such regions experience similar problems, including susceptibilities to drought, as semi-arid areas (see Middleton and Thomas, 1997). These areas include the Canadian Prairies, parts of Kenya and southern Russia.

Figure 14.5 Barren rocky landscape in part of the coastal Atacama Desert, Chile. This is one of the driest places on Earth, where the most effective geomorphic agent is neither the work of water nor the wind, but rather it is the effect of the rapid uplift of the Andes region due to tectonic processes. This induces some slope instability, with Earth movement and gravity being the most effective geomorphic agents operating to shape the long-term development of the landscape.

Figure 14.6 In 2009, abnormal dry-season rains saw over half of the mean annual rainfall of the Kalahari, Botswana, fall in two days in June. Consequently water dominated the landscape of the 'dry season' in that year, filling parts of the Makgadikgadi Basin in central Botswana, leading to dry-season vegetation growth and irregular wildlife migrations.

Considered together, the different types of dryland cover over 37%, or 47% if dry–subhumid areas are included, of the Earth's land surface (Figure 14.7). As well as aridity, one of the principal characteristics of drylands is uncertainty about whether precipitation will occur. High inter-annual rainfall variation is therefore common. Since convection is a major cause of dryland rainfall, this variability can occur at a very local scale so that annual rainfall totals can vary dramatically over distances of a few kilometres. Perhaps the only certain thing about drylands is that all regions are

drought-susceptible and any dryland area has the potential to experience hyper-arid conditions in any year. Dryland environments represent a significant, but difficult to habitate, part of the globe. Droughts in Africa, and the resultant social consequences, are frequently reported in the media, for example through the 1970s and 1980s in the Sahel belt of Africa and in 2006 from Dafur Province, Sudan. When precipitation fails, consequences can be dramatic even in drylands in the developed world. For example, the Province of Saskatchewan, in the Canadian Prairies, experienced in 2002 its driest recorded year ever, with both winter snowfall and summer rains at very low levels. This had drastic effects on this agriculturally important region, both for crop production and for farming communities. Understanding the causes of aridity, and the processes operating in drylands, is therefore more than of academic interest.

14.2.2 Causes of aridity

The distribution of drylands shown in Figure 14.7 yields some clues to explain their distribution. At the global scale there are four main locations in which drylands occur: in subtropical regions (see Chapter 5); at considerable distance from the oceans with an interior continental location (see Chapter 5); in the lee of mountain zones (see Chapter 6); and on certain western coastlines of land masses. Some dryland areas meet more than one of these criteria. For example, the central Sahara is subtropical, has high continentality and is affected by mountain zones such as the Tibesti Massif and the Tamgat Mountains, while the Namib Desert is both coastal and subtropical. Thus causative factors may be

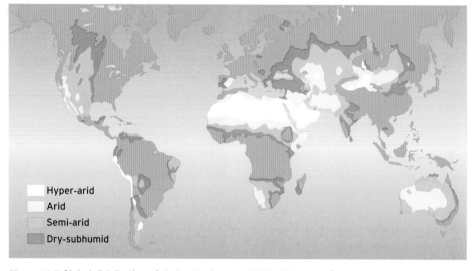

- Hyper-arid
- Arid
- Semi-arid
- Dry-subhumid

Figure 14.7 Global distribution of drylands. (Source: UNEP, ISRIC, CRU/UEA, after UN, 1992)

combined or cumulative, which may lead to moisture deficits that are greater than if only one factor applied.

Each of the four factors described above contributes to climatic conditions that result in low levels of precipitation. Subtropical areas are, in climatological terms, regions of stable, descending air, giving rise to the subtropical high-pressure belts. Such stable, descending air masses are not conducive to precipitation, although the seasonal movements of the **intertropical convergence zone** (ITCZ; see Chapters 4 and 5) can lead to rainfall affecting marginal areas. Subtropical drylands, which include the large dryland belt of North Africa and Arabia, comprise extensive hyper-arid/arid core areas, which are barely affected by ITCZ movements, and smaller marginal areas with semi-arid and dry–subhumid conditions. In these areas rainfall is unreliable and unpredictable, with a high incidence of droughts. This is illustrated by the problems experienced by the Sahel belt which lies on the southern side of the Sahara Desert.

Continentality contributes to aridity because areas that are a considerable distance from the oceans are not penetrated effectively by moisture-bearing weather systems, even when they fall outside the subtropics. Continentality is a major cause of dryland conditions in the interior of North America and central Asia. As well as having low precipitation levels, such areas also tend to experience very cold winter months owing to their great distances from warmer ocean conditions. Snow can be an important feature of winters in continental drylands while overall PET rates are lower than those of subtropical drylands.

Areas on the lee side of mountain barriers can have rain shadow dryland conditions generated by the orographic rainfall effects of the mountains (see Chapter 6). This usually enhances the level of aridity that would otherwise be caused by subtropical or continentality effects. For example, the Rocky Mountains, in the United States, create a rain shadow on their easterly side that enhances continentality effects, while the Great Divide in eastern Australia adds to interior aridity due to continentality/subtropical impacts.

The narrow, north–south-orientated, hyper-arid Namib and Atacama Deserts, respectively in Namibia and Chile, are a result of the impact of cold ocean currents that bring Antarctic waters to the ocean surface offshore of south-west Africa and South America. The lower atmosphere is cooled by the cold oceans, suppressing sea surface evaporation and therefore rainfall. Often fogs form in these conditions. These can drift onshore and make up the major precipitation source in these extremely dry deserts. Similar effects, but to a lesser extent, affect south-western Australia.

Table 14.1 A simple classification of dryland climates

Temperature regime	Percentage of global drylands	Mean monthly temperature (°C) Coldest months	Warmest month
Cold winter	24	<0	10–30
Cool winter	15	0–10	10–30
Mild winter	18	10–20	10–30
Hot	43	10–30	>30

The causes of aridity and dryland conditions are therefore relatively simple to understand, but for any dryland area the interaction of factors may be complex. Different combinations of P and PET levels can give rise to similar overall aridity levels, which is further complicated by the various degrees of seasonality that occur in different areas. Drylands caused by cold ocean currents and subtropical effects tend to have lower seasonal temperature contrasts than continental drylands, where winters can be extremely cold and summers very hot. Given that subtropical drylands are relatively close to the equator, seasonal contrasts are related more to the ITCZ movements (which affect rainfall) than to temperature. It can therefore be more appropriate in these cases to talk of wet and dry seasons, rather than of summer and winter. Where continentality adds to subtropical effects such as in the Kalahari Desert of interior southern Africa, both seasonal temperature and precipitation contrasts can be marked. Table 14.1 attempts to show the approximate percentages of the world's drylands with different temperature regimes. This both complements the classification based on aridity type and further illustrates the great range of background environmental conditions that are embraced by the term 'dryland' or 'desert'. Table 14.1 shows just how climatically diverse drylands can be.

Reflective questions

➤ How may drylands be defined and characterized?

➤ How is aridity calculated?

➤ Dryland areas embrace a range of climatic regimes. What is their uniting characteristic?

➤ How do different mechanisms that generate aridity operate?

➤ Can you provide some examples of dryland areas worldwide that are caused by these different mechanisms?

14.3 Dryland soil and vegetation systems

As well as having their own ecological significance, plant communities act at the interface between the land surface and the elements of climate and weather that contribute to geomorphic processes. In drylands, the variability and uncertainty of moisture availability, high seasonal, and also diurnal, temperature ranges, and the deficiencies of nutrients that some soil systems have are all major stresses upon plant growth. We have already considered the broad climatic conditions present in drylands, so that it is now necessary to review desert soils prior to moving on to a consideration of plant systems.

14.3.1 Dryland soils

Table 14.2 shows the principal characteristics, relevant to plant growth, of the main dryland soil types recognized in the United States Department of Agriculture (USDA) soil classification system (see Chapter 10). Entisols are little more than sedimentary material and therefore are dependent on the nature of the parent material for any nutrient content that they possess. Aridisols and alfisols are moisture deficient though the latter may be seasonally able to support

Table 14.2 Characteristics of main dryland soil orders

Soil order	Characteristics
Aridisol	All-year-round dryness and/or salinity
Alfisol	Moderate base saturation, some seasonal moisture
Entisol	Little-altered sedimentary material lacking horizonation
Mollisol	Base and organically rich
Vertisol	Deep-cracking clay soil

Figure 14.8 Green silcrete outcrop, Makgadikgadi Basin, Botswana.

plant growth. The heavy clay vertisols are either extremely dry or saturated, both conditions that are not conducive to most plants. Mollisols are more suited to plant growth owing to their higher organic and base content, but they are relatively scarce in occurrence.

An additional factor that potentially impacts upon plant growth is the susceptibility of soils to crusting in drylands. This crusting occurs through the effects of concentration of minerals in surface layers due to high evaporative rates, through raindrop impact during high-intensity storm events, or in the form of biological crusts. Salt crusting can be a common feature of dryland soils, where evaporation and limited flushing by rains can lead to the accumulation of, for example, gypsum and halite (see Chapters 7 and 10), which respectively can result in the development of crusts known as **gypcretes** and **salcretes** (Watson and Nash, 1997). Hard-to-see biological crusts, formed by algae or cyanobacteria, are also now recognized as an important surface feature in some drylands (Thomas and Dougill, 2007). Subsurface enrichment of dryland soils has also been widely noted, particularly but not exclusively by calcium carbonate (forming **calcretes**) and silica (forming **silcretes**, Figure 14.8). Both the processes of formation and precise chemical composition of these features, which collectively are known as **duricrusts**, can be complex (see Nash, 2011).

14.3.2 Dryland vegetation

Despite climatic and soil stresses in drylands, such areas are not always as devoid of plants as common perceptions suggest. Hence it is better to consider dryland plant communities in terms of their adaptations to moisture and temperature stresses rather than in terms of their absence.

While hyper-arid areas often appear to be largely plant-less, even they can offer niches that can favour the development of well-adapted plant communities, including lower plant orders such as algae and lichens. At the other end of the dryland environment spectrum, dry subhumid and semi-arid areas can, particularly during the wet season, have almost total plant coverage of the landscape (see Chapter 18). Figures 14.9, 14.10 and 14.11 show examples of different degrees of plant cover in dryland environments.

14.3.2.1 Characteristics of dryland plants

While general hot desert biome characterization is discussed in Chapter 18, particular attention is devoted here to plant survival strategies in such regions. At a very simple level all plants can be classified as **hydrophytes**, **mesophytes** or **xerophytes**. Hydrophytes are wetland plants that can tolerate permanently saturated soils, and therefore they are not found in drylands other than in very specific contexts such as the Okavango Delta of northern Botswana. Mesophytes exist in environments with 10–20% soil moisture. This is between the field capacity and the wilting point (see Chapter 10, especially Figure 10.3). These are the plants commonly found in temperate regions and also in the 'wetter' dry–subhumid and semi-arid dryland areas. Xerophytes are the true dryland plants, being able to tolerate extreme moisture deficiencies and to survive for very prolonged periods in situations with less than 20%, and as little as 5%, soil moisture.

Xerophytes possess a number of coping strategies to withstand seasonal and longer droughts. Lower plants such as lichens and algae are able to tolerate desiccation by entering a dormant state when moisture is absent, responding rapidly to an active state when moisture returns. Many grasses, whether annual or perennial species, have bulbs and rhizomes (see Chapter 18) and avoid drought by confining growth and reproduction stages to the wet season and lying dormant in the dry season and during droughts. Larger plants such as trees and shrubs, and succulents that include cacti, can remain active during dry seasons and droughts through having a range of strategies for evading, resisting or enduring moisture deficiencies. Table 14.3 summarizes the range of drought adaptations present among dryland plants.

The drought strategies of plants have a significant effect on the degree of plant cover that is afforded to the ground surface, and the temporal variability of that cover. Plant spacing allows both moisture and nutrients to be used opportunistically. Various classification systems have been produced in an attempt to capture the variability of dryland vegetation systems. This variability can be

Figure 14.9 *Welwichia mirabalis*, a rare plant adapted to the hyper-arid conditions of the Namib Desert. The spacing between plants reflects the scarcity of water and the extensive root networks that draw moisture from a wide area.

Figure 14.10 Clumped perennial grasses in the dune landscape of the arid south-west Kalahari Desert.

Figure 14.11 Mixed tree-bush savanna in dry-subhumid western Zimbabwe. The tree in the foreground is a specimen of *Acacia karoo*.

Table 14.3 Dryland plant adaptations to drought stress

	Strategy			
	Escape drought	Evade drought	Resist drought	Endure drought
Plant types	Annuals and ephemerals	Perennials	Perennials	Perennials
Plant examples	Grasses and some herbs	Trees	Cacti and other succulents	Shrubs
Characteristics	Dormancy/death during dry season, surviving as seeds	Tap deep water by extensive root networks	Store water in stems and roots	Reduce transpiration, for example by small leaf area/waxy leaf surfaces

extremely complex since moisture, temperature and nutrient factors can combine in a myriad of different ways. One simple but effective scheme that embraces the principal influences divides dryland vegetation systems into three categories: savanna, desert and extreme desert.

Savannas occur in semi-arid regions, with 10–30% cover of shrubs that may often be in dwarf form. Perennial grasses may provide extensive cover during wet periods but may die back to little more than root stock in times of stress. In some regions, succulent species are also an important component of the plant communities that are present. Desert systems possess perennial vegetation that rarely exceeds 10% cover and may comprise a mix of shrubs and grasses. A flush of herbaceous and grass annual growth follows rainfall events, but the total ground cover is unlikely to exceed 50%. Contracted or extreme desert systems possess vegetation in only the most favourable locations, such as ephemeral channel floors, where deep-rooted or salt-tolerant plants tap groundwater to depths that may be many tens of metres.

An important outcome of dryland plant adaptations to stress, which has been widely recognized only since the 1990s, is that Clementsian succession principles (see Chapters 19 and 20) do not necessarily apply. Biomass does not necessarily increase through time in these systems and dryland ecosystems are not necessarily stable, even under natural conditions. Ecosystems in drylands can 'crash' in response to disturbance which may include a drought event. Nevertheless the plants present tend to have the ability to recover once the disturbance has ceased. The manner in which drought escapers (Table 14.3) lie dormant during stress periods, or the way in which they have set seed prior to death, means that they have the ability to recover when moisture availability increases. A consequence of this is that

Figure 14.12 Heavily grassed dunes in the south-west Kalahari, photographed in 2006 after an abnormally rainy wet season, which promoted the effective growth of annual grasses. Contrast the density of vegetation cover with that shown in Figure 14.10, taken no more than 100 km away but during a time of 'normal' rains.

dryland landscapes can vary dramatically over time in terms of the degree of plant cover they support, with variations due to differences in antecedent rainfall amounts (Figure 14.12). Savanna systems commonly display this type of instability. Such instability is natural but has sometimes been mistaken as the result of negative human disturbances. See Chapter 18 for further information on dryland biome characteristics.

Reflective questions

➤ What are the characteristics of plants that can survive in arid conditions?

➤ What are the main soil types that are found in dryland areas, and what are their principal characteristics?

14.4 Geomorphological processes in drylands

14.4.1 Dryland landscapes

Dryland landscapes are as diverse as those found in any other global climatic zone. This reflects the range of structural (tectonic and geological) settings found in the regions that experience dryland climatic conditions. It also reflects the interactions between climate and landforms, which determine both the specific geomorphic processes that occur and their rates and effectiveness of operation. Some researchers have attempted to quantify the main landform types present in different areas, with the results providing useful illustrations of the diversity of dryland landscapes and the relative importance of different geomorphic processes (see Chapter 1 in Thomas, 2011). Three contrasting dryland regions are illustrated in Table 14.4.

The south-west United States, known as the basin and range region, is a high relative relief environment, which is reflected in the large extent of mountainous areas and intervening desert flats. The relatively sparsely vegetated mountain surfaces can generate high runoff rates when rain falls. Therefore alluvial fans are an important feature at the junction between mountains and flats. The Australian dryland interior has altogether much flatter terrain, and in contrast to the south-west United States is a tectonically 'quiet' region. Ephemeral river systems, which often have their headwaters in wetter areas next to the central deserts, are common. The abundant availability of sediment for transport by wind has generated extensive sand-dune systems. The Sahara and eastern Arabia are major dryland regions with significant landscape diversity. These include recently tectonically active areas that have contributed to the development of mountains (Figure 14.13), and extremely dry continental basins that have accumulated eroded sediments that have been reworked under extremely arid conditions into sand dunes.

The landforms in any particular dryland region may also incorporate a legacy from the operation of geomorphic processes under past climatic regimes during the Quaternary and even Tertiary periods (see Chapter 22). Thus in the south-west United States, dry lake beds, or playas (see Chapter 9), represent the sites of previously permanent lakes that existed during wetter climatic regimes in the late Quaternary. Even though the vegetation cover of the Australian dune systems is often relatively sparse, the dunes owe much of their construction to periods when conditions were even drier and windier than at present.

The examples in Table 14.4 show that dryland landscapes are composed of features that result from the effective operation of a range of geomorphic processes. These processes are not in themselves unique to dryland regions. However, the manner in which rock weathering, slope and channel processes and aeolian (wind-blown) processes operate is greatly influenced by dryland climatic conditions, both directly and through their effect upon plant cover. We will now consider these three sets of geomorphic processes, with particular emphasis on the elements important to drylands.

Table 14.4 Percentage estimates of different landscapes within three dryland areas

	SW USA	Australia	Sahara
Mountains	38	16	43
Alluvial fans	31	0	1
River plains and channel systems	5	13	3
Playas/dry lakes	1	1	1
Sand dunes	1	40	28
Undifferentiated flats	21	18	10
Other	3	12	14

(Source: adapted from Thomas, 1997a)

Figure 14.13 Rapid uplift since the mid-Tertiary period has formed the northern Hajar Mountains in the Mausandam Peninsula of Oman and Ras al Khaimah, United Arab Emirates. This has created a situation where, despite being extremely arid, runoff has generated deep wadi development (middle ground of photograph). A wadi is a river bed that remains dry except in wet weather.

14.4.2 Rock weathering in drylands

The presence of bare or relatively bare rock surfaces in drylands hills, mountains and plains (Figure 14.1), the high diurnal temperature ranges that drylands can experience, and the excess of evapotranspiration over precipitation, which can lead to salts becoming concentrated in surface locations, have all been considered as significant factors controlling dryland rock weathering regimes. Flaking rock surfaces are widely seen in drylands and have long been regarded as ample evidence of the favourable conditions that drylands present for rock weathering. However, it must not be forgotten that different rock types and structures vary in their weathering potential and susceptibility (see Chapter 7 for more detailed treatment of this topic). In recent years field measurements and experimental laboratory simulations have been enhancing our understanding of how these factors may influence weathering processes.

Insolation, or thermal weathering, has been widely cited as a major form of desert rock weathering, occurring effectively under the influence of high rock surface diurnal temperature ranges, which may exceed 50°C. This is thought to cause the expansion and contraction of rock particles and minerals, with different minerals experiencing different expansions and contractions, creating stresses that weaken rocks and loosen particles. The role of temperature change alone as the principal cause of this type of weathering,

which often takes the form of **granular disaggregation**, has been questioned. This is because the presence of moisture, even in extremely small quantities, may be vital in contributing to rock breakdown through its influence upon the movement and crystallization of salts and in extreme cases on the occurrence of interparticle freeze–thaw processes. Even in the extremely dry Namib Desert, fog (which contributes more than rainfall to total precipitation in parts of the Namib) and dew fall are common and can provide sufficient moisture to effect such processes.

Salt weathering is also a significant process in drylands. Salt accumulation in surface sediments and rocks is favoured by both high evaporation rates and limited leaching opportunities, supplemented in some cases by wind-borne salt (Figure 14.14). Salt weathering is a mechanical process that occurs principally through three mechanisms: crystallization when temperature increases lead to the growth of salt crystals, hydration when moisture inputs cause the salt volume to increase, and thermal expansion when salts increase in volume on heating. Different salts have different susceptibilities to these processes. Sodium sulphate is a good example of a salt that is both common and has a solubility that declines rapidly in line with falling temperatures such that crystallization occurs. It also experiences a substantial increase in volume upon wetting (Goudie, 1997). The overall impacts of salt weathering and the occurrence and importance of different salt weathering

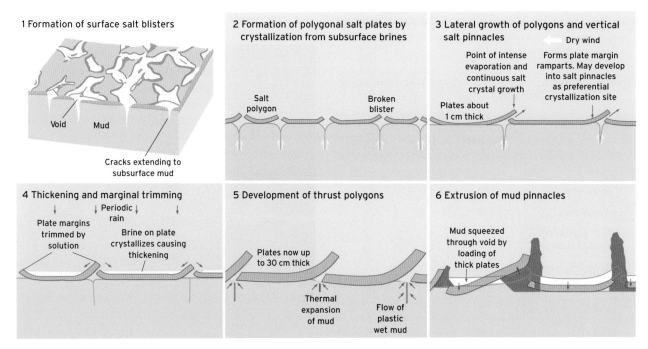

Figure 14.14 Salt crust development sequences on a playa surface. (Source: after Shaw and Thomas, 1997, based on a diagram originally in Krinsley, 1970)

processes have been the subject of much experimentation and debate. What is certain is that temperature and moisture changes are closely interlinked with salt weathering, and that insolation weathering and salt weathering effects are not necessarily discrete processes.

As well as contributing to the overall weathering of rock surfaces, these processes can be important in the development of particular weathering landforms. Cavernous and tafoni (honeycomb) weathering features have often been ascribed to salty weathering agencies. However, these landforms tend to be restricted to relatively even-grained rock types that permit the movement of salt-laden waters, or to the zone of capillary rise. Salt precipitation on flat, drying surfaces such as the floors of ephemeral dryland lakes and pools can contribute, through crystallization, to surface disruption and cracking that persists and develops until the next rainfall and flooding event (Figure 14.15).

14.4.3 Hillslope and channel processes

Although rainfall totals are, by definition, low in drylands, two factors mean that water can be a very effective medium for downslope sediment transport in drylands. Firstly, rainfall events are frequently intense. For example, rainfall in subtropical drylands is often associated with high rates of convection and thunderstorms. Secondly, vegetation cover is often only partial and at the end of the dry season when rains begin, biomass and ground cover are usually at their

Figure 14.15 The saline surface of the Chott el Jerid, a playa lake in southern Tunisia. The extensive flat floor of this salt lake may experience shallow inundation by water during the winter months when rain occurs. Evaporation during spring and summer soon dries the basin out, with the result that salt crusting occurs on the surface of the silt and clay floor. Palaeoenvironmental investigations show that there was an extensive water body during the late last glacial at this location when climate was both wetter and cooler in the region.

lowest levels. Runoff is therefore rapidly generated in many dryland environments (Figure 14.16), but is rarer, for example, in areas where sandy sediments dominate and infiltration rates are high.

In the areas where runoff is readily generated, it is usually in the form of infiltration-excess overland flow (see

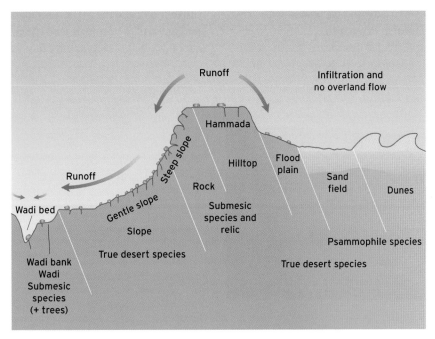

Figure 14.16 Elements of the hydrological cycle in drylands and the ability of different sub-environments to generate runoff. A hammada is a flat rocky area blown free of sand by the wind. (Source: after Shimda *et al.*, 1986)

411

Chapter 11). At one extreme this occurs when the ground surface is bare rock. However, the effect also applies to soft unconsolidated sediments where raindrop impact can further enhance the effect by promoting surface sealing and crusting. Sealing and crusting usually result from the breakdown of soil aggregates under raindrop impact, and the washing of the finer particles into the voids between larger ones (see Chapter 8). Overall infiltration rates tend to be higher where larger plants are present, not only because leaves and leaf litter on the ground act to intercept raindrops and thereby prevent sealing, but also because roots increase the presence of macropores in soils and sediments.

Although overland flow may be readily generated owing to high rainfall intensities, rainfall totals are usually low in individual storm events, but exceptions occur, as noted in Figure 14.6. Furthermore, evaporation rates are high and surface vegetation cover is patchy. Therefore during many storm events overland flow is discontinuous. Often only in the case of higher-magnitude or longer-duration rainfall events does overland flow actually connect to channel systems. Dryland soils are often poorly developed and stony. Slopes with almost non-existent soils are commonly mantled by coarse rock debris. Recent research has shown that the macro-roughness of slopes, which includes both the distribution of plants and that of stones, is a critical determinant of the continuity and nature of overland flow. Even when flow does not reach channel systems, rill development is a common feature of dryland hillslopes (see Chapter 8). In sediments and rock types that are susceptible to very efficient overland flow generation, gullies and even extensive badland systems can develop such as that shown in Figure 14.17.

The relative bareness of hillslopes means that in global terms sediment yields in drylands are the greatest of all

environments. This was illustrated by the ground-breaking work of the American geomorphologists Langbein and Schumm in 1958 (Figure 14.18). Rapid runoff means that flood hydrographs for dryland rivers peak early, with the highest-magnitude events capable of doing considerable structural damage.

The key features of dryland channel systems are summarized in Table 14.5. With the exception of perennial rivers in drylands that have their sources in neighbouring humid areas, channel flow is usually of short duration, irregular and frequently does not result in flow from the channel head to the end of the system. Two factors account for this. Firstly, the localized nature of dryland rainfall events means that a flow event may only receive a contribution from a small part of the catchment. Secondly, high infiltration losses through the bed may occur owing to unsaturated conditions prior to flooding. Owing to the presence of relatively bare slopes in dryland catchments, sediment transport, unlike in humid regions, is usually **transport limited** rather than sediment **supply limited**. There is, however, a major downside to this efficiency: dams and other impoundment structures can have relatively short life spans compared with their counterparts in wetter regions.

Owing to the irregular nature of dryland channel flow and high seepage losses, many such river systems do not terminate at the coastline. Inland-draining river systems are called **endoreic**. Some of these systems may end at inland salt pans, or playas, for example at Lake Eyre in Australia (Croke, 1997) and the Makgadikgadi pans in the Kalahari (see Figure 14.6, Burrough *et al.*, 2009). Although surface water may fill these basins after flow events or at the end of

Figure 14.17 Gullied and eroded loess deposits near Matmata, central Tunisia. The fine silty sediments are susceptible to flocculation (forming woolly, cloud-like aggregations) when wetted during rainfall events, which favours rapid dispersal and removal.

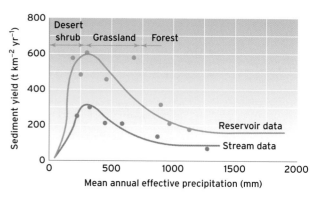

Figure 14.18 The Langbein and Schumm (1958) curve, relating mean sediment yield to mean effective rainfall (effective rainfall is that which produces runoff and is not lost to evapotranspiration or the soil moisture store). Drylands generate the highest sediment yields according to this model, which was based on the average values of measurements of sediment yields in different environments. (Source: after Langbein and Schumm, 1958)

Table 14.5 Some major characteristics of ephemeral dryland channel systems

Flow features	Geomorphic and ecological implications
Irregular and unpredictable	Wide channels, short-distance changes in channel form
High event-to-event variability	Channel systems may change rapidly and channel systems in a region may vary markedly
High water losses to evaporation and through the stream bed	Flow-limited transport rates
	Flow may not travel far down the channel system
	Subsurface moisture may be important for subsequent plant growth
Peaked hydrograph	Rapid flow onset during rainfall events. Flow may also end quickly
High velocities (>4 m s^{-1}) common	High sediment loads (bed load and/or suspended load)

the wet season, evaporative losses eventually result in drying out and the concentration of salts at the surface.

14.4.4 Aeolian processes and forms

The wind can be an effective agent of sediment transport, provided that sediment suitable for aeolian entrainment is available. The limited vegetation cover of drylands and the propensity for weathered sediment to accumulate in desert basins, because of the disjointed and ephemeral nature of flow in fluvial systems, explain why aeolian processes can be effective in these areas. The modern scientific investigation of aeolian processes and sand-dune development has its roots in the work of Brigadier Ralph Bagnold. He began observations in the North African deserts in the 1930s and 1940s. His seminal work on the physics of blown sand and desert dunes is still regarded as the baseline against which modern investigations are compared (Bagnold, 1941).

Only about 20% of drylands are covered by aeolian sand deposits and dunes. Many dunes occur in more vegetated semi-arid and dry–subhumid areas and are often relict from past drier climatic conditions at those sites. Silt, which has a mean particle diameter of 0.002–0.063 mm, is finer than sand, which has a mean particle diameter of 0.063–2.0 mm. Being much smaller and lighter than sand, aeolian silt (or dust) is often lifted several kilometres up into the atmosphere (Figure 14.19) and is therefore frequently transported beyond dryland areas and ultimately deposited in the oceans or other land areas. Aeolian dust presents a range of hazards within drylands, but is now also seen to be a critical part of a number of major global processes (see Box 14.1). In some regions thick deposits of wind-blown silt, known as **loess**, have accumulated, often on the margins of dryland areas and frequently banked up against hills. Somewhat paradoxically, however, silt can be less readily entrained by the wind than larger sand particles, principally because the small grains pack down better on the ground surface, thus offering more resistance to the forces of entrainment. Sand transport processes are described in Box 14.2.

Figure 14.19 Space Shuttle hand-held camera photograph, taken in 1992, of a major dust storm front in the Sahara, near the border between Algeria and Niger. (Source: courtesy of the NASA JSC Digital Image Collection)

HAZARDS

DESERT DUST: A BENEFICIAL HAZARD

Silt-sized particles, transported in the atmosphere by suspension as dust, are a significant component of dryland geomorphic systems. Locally in drylands dust is widely documented as a hazard, and dust storms are known to disrupt transport by road and air, through reducing visibility and clogging mechanical parts, as well as being a health hazard. The highest recorded annual frequencies of dust-storm days are recorded in parts of central Asia, where over 100 dust-days a year have been recorded in parts of Iran and Uzbekistan.

Dry lake beds are a particularly important source of dust to the atmosphere, because these are fluvial system end points where fine suspended sediment has accumulated. Lake beds may be exposed to the atmosphere because of climate change. For example, major lakes are known to have existed in parts of the Sahara, Kalahari and the south-west United States during the past 20 000 years. Lake beds may also be exposed to the atmosphere because of human activity. In recent decades, the drying-out of the Aral Sea due to over-exploitation of its river inflows for irrigation is a notable example of human actions that have had major negative human consequences. The shrinking of the sea to less than 30% of its former area has exposed 36 000 km^2 of former lake sediments, heavily polluted with chemical fertilizer residues, to the wind. Dust-storm frequencies are high, with high rates of lung disease and cancer incidences in Uzbekistan at least partially attributable to the 'Aral Sea dust effect' (O'Hara et al., 2001).

The ability to monitor the movement of dust in the atmosphere by TOMS (Total Ozone Mapping Spectrometer) satellite data has, since the 1990s, revolutionized the understanding of long-distance dust movement from drylands. It has been possible to identify major persistent transport pathways and the Earth's major dust sources in drylands. The Sahara produces approximately 66% of all the dust present in the Earth's atmosphere, with the Bodele Depression, in the dry Lake Chad Depression in the central Sahara, responsible for about half of this (Figure 14.20). The Bodele is particularly favoured for dust generation because of its topographic setting, whereby strong dust-generating winds are focused onto the dry, erodible, former lake bed.

Recently, the export of dust from the Sahara has been attributed a major role in the complex functioning of the Earth's environmental system. Using satellite data, Koren et al. (2006) have calculated that 40 million tons of dust is transported by the wind from the Sahara to the Amazon Basin each year, with this dust being a critical source of nutrients for the fertilization of the Amazon rainforests.

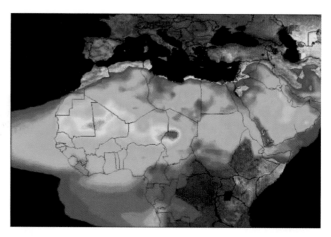

Figure 14.20 Total Ozone Mapping Spectrometer annual mean aerosol index data for 1980 to 1992. Red areas are of greatest dust concentrations, with the red area in the central Sahara being the Bodele Depression. (Source: Image courtesy of Sebastian Engelstaedter; Background image courtesy of NASA from Blue Marble: Land, Surface, Shallow Water, and Shaded Topography, http://visibleearth.nasa.gov/view_rec.php?id=2433)

BOX 14.1

FUNDAMENTAL PRINCIPLES

SAND TRANSPORT

Sand is transported primarily by saltation (a hopping motion) and creep (a rolling motion) (Figure 14.21). A process called **reptation** is also sometimes referred to (e.g. Anderson and Haff, 1988), which represents grains set into a low hopping motion due to the high-velocity impact of a descending saltating grain. Sand movement is affected by the particle size (or mass), the wind strength and any forces (including gravity and pore space moisture) that resist entrainment (see Chapters 8 and 9). Additional to these forces are ground surface conditions. However dry and sandy a surface is, aeolian entrainment will not take place even under high wind velocities if the surface is protected (e.g. by vegetation).

However, the relationship between surface cover and entrainment is complex, as noted by Wiggs *et al.* (1993) and others. This is because air moving over a surface develops a velocity profile whereby the surface imparts a frictional drag on air immediately adjacent to the surface, including a very thin (often less than 1 mm thick) zone of zero wind velocity, known confusingly as the aerodynamic roughness length or z_0. It is

necessary for these frictional effects to be overcome for particles to be mobilized, picked up by the wind, and moved vertically into faster-moving air away from the immediate surface. This process may be enhanced when air flow is not laminar but is turbulent (see Figure 9.3 in Chapter 9 for a diagrammatic explanation of laminar and turbulent flow).

Turbulence is now seen as a vital part of the entrainment process (Weaver and Wiggs, 2011) with small variations in surface topography, introduced by factors including variations in particle size and the occurrence of a partial vegetation cover, and variable surface heating by the Sun, contributing to the development of a turbulent velocity profile that enables particle entrainment to be effected.

For 2 mm diameter sand particles on a bare, dry surface, a wind velocity of about 5 m s^{-1} is sufficient to generate shear velocities (usually labelled as u^* and a term used to describe the gradient of the velocity profile) that enable entrainment to occur. As shear velocities increase, the sand transport rate rises exponentially. Once entrainment has commenced, the impact of descending saltating grains contributes further to particle movement on the surface. Sandy surfaces are usually rippled during aeolian transport events, which is a reflection of the impact of saltating grains on the ground surface. Ripples are usually transient features, with their size and spacing changing frequently and rapidly in response to gustiness and subtle directional changes in the wind.

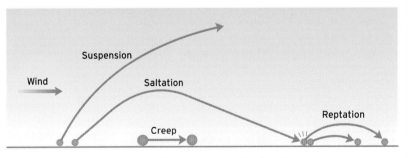

Figure 14.21 The processes of sediment movement by the wind.

BOX 14.2

While sand deposits, usually in the form of dunes, can occur as individual features, the vast majority of aeolian sand deposits are found in spatially extensive (greater than 2 km^2) accumulations known as sand seas (or ergs) (Figure 14.22). These occur at the end of regional-scale sand transport pathways. This reflects the resultant direction of sand-transporting winds, and topographic factors. Although the Sahara Desert may be the largest dryland area on Earth, its topography means that there are a number of discrete sand

seas within it. The relatively low relief of much of the Arabian Peninsula and the southern African and Australian interiors results in larger overall sand seas in these areas, with the Kalahari sand sea extending to almost 2.5 million km^2.

A sand dune forms when, at a specific location, the rate of arrival of sand in entrainment by the wind exceeds the rate of loss. This can occur when moving sand meets an obstacle that disrupts wind flow or reduces wind velocities, such that the capacity to transport sand is reduced. Obstacles

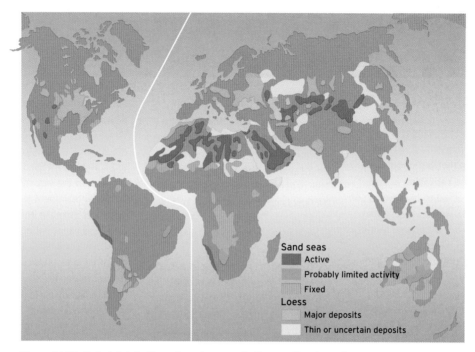

Figure 14.22 Global distribution of sand seas and other aeolian deposits. (Source: after Snead, 1972, and Thomas, 1997b)

that cause this to occur can range from hills, against which a dune may bank up, to a small plant or rock located within the transport pathway. Once sand accumulates, the accumulation itself becomes the obstacle, such that a feedback occurs in the aeolian system that enhances deposition (Figure 14.23). If sand transport direction changes frequently, the accumulation may not persist and may be destroyed and reworked, but if there is a single dominant direction, or if there are different directions but each has a sufficiently long duration, the accumulation will grow. As this occurs, the feature will begin to intrude into the lower atmosphere, which can lead to modification of air flow patterns and strengths. Of particular note is that wind velocities often increase (through compression of flow lines) as dune intrusion occurs. This allows sediment to be transported up the windward dune slope but eventually can limit the vertical accretion of the dune. Wind velocities can eventually create sufficient shear to move all of the sediment arriving at the dune crest through the system. As this occurs, steepening of the leeside of the dune can lead to air flow being separated from the dune surface, the development of a cell of reversed air flow, and the formation of a steep slip face (Figure 14.24), at the angle of repose for dry sand, as shown in Figure 14.23.

Dunes can form rapidly, in a matter of hours, or slowly over days, weeks or years. The rate depends on the sand transport capacity of the wind and on sediment availability.

Dune size is partially dependent on sediment availability and is also a function of dune type and the duration of accumulation (Livingstone and Warren, 1996). Sand dunes occur in a variety of forms, and have a variety of names

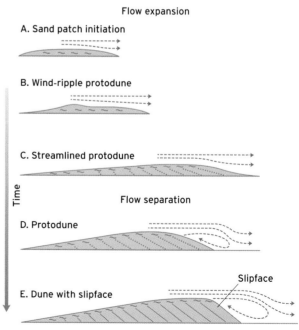

Figure 14.23 The stages of dune initiation and development. (Source: after Kocurek *et al.*, 1990)

Figure 14.24 The slip face of a transverse dune in the United Arab Emirates.

Figure 14.25 Barchan dunes, approximately 4 m high, formed in the interdune area between major linear dune ridges, in the Central Namib sand sea.

Figure 14.26 Giant linear dune ridges, with seif-like crests rising over 100 m above the interdune areas, in the Central Namib sand sea.

that reflect both scientific factors and local knowledge. For example, *barchan* is Turkish for 'active dune', and is widely applied to mobile crescent-shaped transverse dunes (Figure 14.25); *seif* is Arabic for a 'curved sword', and is sometimes applied to linear dunes with sinuous crests (Figure 14.26). In some environments sand transport occurs from a single dominant direction, in others, seasonal changes occur. This is a principal factor in determining the type of dune that develops. Chapter 9 details the main dune types (see for example Figure 9.11). Often dunes are found in extensive dune fields within sand seas (Figure 14.27). This is probably due to the area of dune development possessing more than a single obstacle that encourages sand deposition. It is also because a single dune modifies the wind environment, creating downwind flow perturbations that encourage further deposition.

Of the main dune types identified in Figure 9.11 and discussed in Chapter 9, star dunes usually attain the greatest

Figure 14.27 High dune ridges forming part of the Namib sand sea. The ridges, though orientated into quasi-linear patterns in this eastern part of the dune field, also have star-like peaks that rise over 300 m above the interdunes. The complexity of pattern reflects the seasonal changes in the wind regime in the area, and shows that, in reality, sand dunes often do not present themselves as neat 'textbook' examples. At the bottom of the image the dunes extend to the floor of Sossus Vlei, which occasionally experiences flowing water emanating from rains in the neighbouring Naukluft Mountains. (Source: image courtesy NASA/GSFC/MITI/ERSDAC/JAROS, and US/Japan ASTER Science Team, 2002)

size (Figure 14.28). This is because they develop in the depositional centres of sand seas, where net sediment accumulation and sand-transporting wind directional variability are greatest. Linear dunes in the Namib Desert may exceed 100 m in height but in other regions such as the Kalahari and central Australia, they are more usually 10–15 m high. Sediment availability is an important factor in determining the size of these features, which can in some situations extend uninterrupted for tens, and sometimes hundreds, of kilometres. Transverse dunes are more mobile than other dune types, since they form in more unidirectional wind environments and are therefore able to migrate forward more readily. Sand supply is again an important factor in respect of the maximum size attained by dunes in any area, and this in turn plays a major role in the migration rate. All other factors being equal, a larger dune will take longer to 'roll' forward. Net migration rates that have been recorded include over 60 m yr^{-1} for a 3 m high barchan dune in Mauritania (Sarnthein and Walger, 1974), 18 m yr^{-1} for a 17 m high dune in the same environment, and 20 m yr^{-1} for a 6 m high dune in the Algodones dune field, California (Norris, 1966). Movement may be highly seasonal and dominated by changes over a few days or weeks of the year.

Figure 14.28 Star dune, Sossus Vlei, Namib sand sea.

Not all sand sea surfaces are devoid of vegetation. A partial vegetation cover may even enhance sand transport since the turbulence of lower wind layers is enhanced over rough surfaces, and turbulent winds can be more effective in initiating sediment entrainment. Various attempts have been made to establish the vegetation cover limits on sand movement. The issue has relevance not only to understanding dune dynamics but also to understanding the potential risks of wind erosion from agricultural fields. Factors including the spacing of individual plants, plant height and structure all influence vegetation–wind–entrainment interactions, as well as ambient wind strengths. Grasses, and plant litter, are more effective than shrubs and bushes in limiting sand transport. About 90% of sand movement occurs in the lowest 50 cm of the atmosphere. Given the complex array of plant and surface variables that can affect sand transport, a single simple threshold percentage plant cover separating entrainment-susceptible surfaces from stable surfaces does not exist. However, sand transport (assuming winds exceed threshold velocities) tends to increase rapidly once plant cover falls below around 15%, but can still take place to a limited degree with covers of about 30%.

Sand and dust particles in entrainment can act as an effective abrasional agent, creating wind erosion features when they come into contact with immovable objects that are less resistant than the moving particles themselves. When abrasion is persistent, wind-sculpted landforms can result (Figure 14.29). These may range from smoothed surfaces of individual stones (termed **ventifacts**) to the smoothing and rounding of whole hills. Since sand particles largely move in saltation, the abrasive effect is limited to the maximum height of saltation and is usually concentrated in the lower 50–100 cm of rock surfaces. Buildings located in sand transport pathways are especially susceptible to wind blasting, such as that shown in Figure 14.30. Dust particles are therefore largely responsible for the wholesale smoothing of hills into features that are known as yardangs. As dust can be transported over long distances, extensive yardang fields, streamlined in the direction of sediment transport, can result, and have been observed in satellite imagery both in the Lut Desert of Iran and on the Tebesti Plateau in the Sahara Desert.

Figure 14.29 Wind-eroded soft sediments on the northern edge of Chott El Jerid, Tunisia. These features are commonly called yardangs, and are formed by the abrasive effects of saltating sand and suspended dust.

Figure 14.30 Wind erosion of a building. The height of the erosion notch represents the height of grain saltation, Kolmanskop, Namibia.

Reflective questions

➤ What influences the large-scale nature of dryland landscapes?

➤ How significant is insolation weathering in drylands?

➤ If drylands are moisture deficient, why do we need to consider the work of water in sediment transport?

➤ Why can the wind be an effective geomorphological agent in drylands?

14.5 Environmental change in drylands

Drylands are diverse, dynamic environments. Seasonal climatic variability and the occurrence of drought events are normal elements of these regions. Longer, more persistent changes have also affected these regions as a result of the major climatic shifts the Earth has experienced during the Quaternary period (see Chapter 22). Drylands have expanded and contracted in the past (Figure 14.31). For example, during the last glaciation many dryland regions were more extensive for periods lasting several hundreds to thousands of years. A colder climate meant a more arid climate. Presently inactive, and in many cases well vegetated, sand seas in the northern Kalahari, central Australia and in the Sahel were once dynamic features. Dust emissions from the Sahara were greater than at present, as evidenced by significant aeolian dust concentrations in ocean sediment cores from the mid-Atlantic Ocean and the Caribbean. Not all drylands were more extensive at this time, since the global circulation changes associated with major climatic shifts led, for example, to more effective precipitation falling over the presently arid south-western United States and in the currently hyper-arid Atacama Desert of Chile. During the Holocene (the last 10 000 years) evidence from the American mid-west, which is presently highly productive agriculturally, shows that arid episodes occurred on a number of occasions that were of sufficient duration to permit marked sand-dune development in some localities. Box 14.3 describes the types of evidence for expanding and contracting dryland conditions.

Today drylands are susceptible to change brought about by the combined impacts of human activity and climate change. Human activities, particularly in the face of growing dryland populations, attempt to increase natural resource and ecosystem use, and manage the inherent uncertainty and variability of drylands. Both these affect the operation of environmental processes. For example, expansions of agriculture in drylands often rely on the extraction of groundwater, which is then used to support livestock or in irrigation systems that support crop growth. This attempts to overcome the temporal variability in rainfall but, as well as positive effects, also results in a lowering of groundwater tables and a concentration of salts (through evaporative effects) in the soil. An outcome is often that such agricultural systems are unsustainable; water resources become depleted and the soil unsuitable for crop growth. This is a particular problem in a range of dryland areas such as south-east Spain, where irrigation schemes implemented in this semi-arid area in the 1970s and 1980s are now proving unproductive. These sorts of problem can lead to enhanced **desertification**. Desertification is described in Box 14.4.

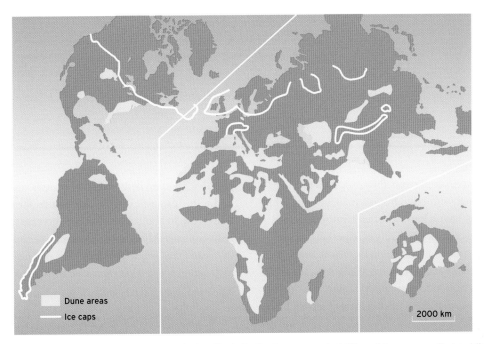

Figure 14.31 The maximum extent of major sand seas during the Late Quaternary period. When this map was first published in 1978 it was assumed that all these sand seas attained their maximum extent at the last Glacial Maximum. However, better dating control today indicates a more complex picture of the timing of the maximum activity of some of these sand seas. (Source: reprinted with permission from *Nature*, Sarnthein, M., 1978, Sand deserts during the last glacial maximum and climatic optimum, *Nature*, 272, Fig. 1b. Copyright 1978 Macmillan Magazines Ltd)

TECHNIQUES

INDICATORS OF LONG-TERM ENVIRONMENTAL CHANGE IN DRYLANDS

Past environmental changes, over hundreds and thousands of years, can be determined from a range of available data sources. These include the characteristics of preserved sediments and the occurrence of landforms that today are either inactive or out of equilibrium with present environmental processes. The timing of changes can be established either in a relative sense ('younger than' or 'earlier than') or numerically through the application of a radiometric dating technique. These include radiocarbon dating,

luminescence dating and uranium–thorium dating (see Chapter 20 for further information about these techniques). Different methods are applicable over different time periods. For example, radiocarbon dating can be applied to organic remains spanning the past 30–40 000 years, while luminescence dating can be applied to quartz sands over a time range in excess of 200 000 years and possibly up to 1 million years old. The following are examples of data sources that indicate (a) the greater extent of dryland conditions in the past and (b) the expansion of more humid conditions in the past into presently arid areas, though their interpretation can be complex and

controversial (Thomas and Burrough, 2011):

(a) Evidence of expanded dryland conditions:

- Stabilized, vegetated or degraded desert sand dunes.

- Dunes crossing river valleys, now breached by more recent fluvial action.

- Wind-blown dust accumulations in offshore ocean sediment cores.

- Wind-eroded hills in presently humid environments.

- Evaporite deposits in lake sediment sequences.

BOX 14.3 ➤

➤

(b) Evidence of contracted dryland conditions:

- Dry valley networks in present drylands.

- Valleys crossed by active sand dunes.

- Flow stone development in caves in drylands.

- Humid plant species' pollen preserved in dryland sediments.

BOX 14.3

Although most dryland rivers flow only ephemerally, they are increasingly subject to human management. This management may have a number of motivations including attempting to control flood surges, especially where rivers enter urban areas, and to regulate flow in order to make water, via dams and reservoirs, available in the dry season. The high suspended load of dryland rivers during flow periods means that dams often have a relatively brief effective life span before siltation causes reservoir capacity to diminish significantly. Furthermore, the changes in base level brought about by a dam can increase upstream flood risks considerably. This problem can be further enhanced in urban areas where the natural peakedness of flood hydrographs is further increased by the rapid rates of runoff delivered from paved surfaces (see Chapter 11).

ENVIRONMENTAL CHANGE

DESERTIFICATION

Dryland environments are increasingly subject to human pressures while at the same time being vulnerable to the impacts of naturally varying climatic conditions, especially drought. The expansion and intensification of agricultural systems in drylands, particularly from the mid-twentieth century, have led to a marked increase in environmental pressures and stresses, notably during periods of drought. The result of these pressures can be desertification. This is an often misunderstood and abused term, but is used to describe land degradation in drylands. Misunderstanding often arises because natural environmental (especially plant system) responses to drought, from which recovery usually occurs, have been confused with longer-term and more persistent negative changes.

The United Nations Convention to Combat Desertification (UNCCD) defines the issue as: 'land degradation in arid, semi-arid and dry-subhumid areas [the "susceptible drylands"] resulting from various factors, including climatic variations and human actions'. This definition recognized the natural environments in which desertification occurs, its human agency, and that the propensity for human actions to cause degradation is often enhanced during droughts when stresses are at their greatest. This is illustrated by the fact that desertification received wide global attention during the late 1970s when human pressures in the environment had been exacerbated in the Sahel region by a decade of drought.

Desertification is closely associated with unsustainable land management practices which affect both the soil and vegetation. Vegetation degradation includes the loss of natural plant cover through the lowering of water tables, and the replacement of palatable grasses used by livestock with unpalatable weeds or bushes, owing to excessive grazing levels. Distinguishing natural dryland ecosystem variability from longer-term changes is, however, difficult. Soil degradation in drylands takes on two main forms: erosion and internal changes. Wind and water erosion can both be enhanced by land-use practices, particularly if natural vegetation systems are disturbed, leaving slopes vulnerable to water erosion during storm events or sandy sediments vulnerable to wind action. Internal degradation embraces physical and chemical changes. The former includes soil crusting and compaction, which can again result from vegetation removal that increases the effect of raindrop impact. The latter includes the processes of nutrient depletion. This is a particular problem in developing world

BOX 14.4 ➤

➤

drylands where chemical fertilizers are expensive and often beyond the means of subsistence farmers. Salinization also affects irrigated lands whereby the irrigation waters evaporate, leaving salts behind which in turn render the soil intolerant to plants.

Establishing the extent of desertification has proved problematic, particularly at the global level. What is now known for certain is that the image, widely used in the 1970s and 1980s to portray the problem, of mobile sand dunes advancing over productive land is misleading, and only locally applicable (Thomas and Middleton, 1994). The Global Assessment of Soil Degradation (GLASOD) commissioned by the UN in the late 1980s and early 1990s was the first systematic attempt to establish soil degradation worldwide, and has been used to evaluate the extent of dryland degradation (Middleton and Thomas, 1997). GLASOD identified 1035 million ha out of a total susceptible dryland area of 5170 million ha (about 20%) as degraded. Of this, water erosion was the dominant degradation process in 45% of the area affected, wind erosion 42%, chemical degradation 10% and physical degradation 3%. Figure 14.32 shows the approximate global extent of human-induced soil degradation in drylands.

The GLASOD survey also assessed the severity of the problem, noting that 80% and 92% respectively of water and wind erosion was only light or moderate. Notwithstanding the limitations of the survey, this does suggest that dryland degradation may have been overestimated or confused with natural environmental variability in the past, and that rather than being an extensive problem, it most severely affects localized degradation hot spots.

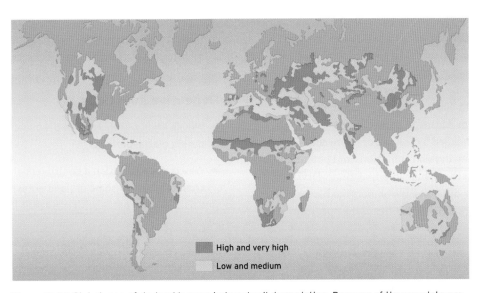

High and very high

Low and medium

Figure 14.32 Global map of dryland human-induced soil degradation. Because of the way data are cartographically represented, the extent of the degradation can appear greater than the extent of the affected areas (see Thomas and Middleton, 1994).
(Source: UNEP, ISRIC, CRU/UEA, after UNEP, 1997)

BOX 14.4

Human-induced global warming also impacts on drylands. Global climate model (GCM) predictions are by no means uniform, but 'double atmospheric CO_2 models' do provide an insight into possible major changes during the next 50–100 years. One approach to overcome the effects of using different GCMs is to compare the outputs of different models, which allows areas of agreement and disagreement to be highlighted. Table 14.6 shows possible changes for three dryland areas, derived from the integrated analysis of three GCMs. These changes show that not only will the direct effect of precipitation changes impact on dryland areas, but in many cases the effect of temperature change on evapotranspiration rates enhances the predicted moisture deficits. Climate change models can also be used to drive other models of how dryland landscapes may change in the twenty-first century. Two recent studies have done this for African drylands: one (Thomas *et al.*, 2005) using GCM outputs to see how the currently vegetated stable dune systems of the Kalahari may respond to climate change, the other (de Wit and Stankiewicz, 2006) focusing on how drainage systems, and thereby people's access to water, may alter. In the

Table 14.6 Global climate model predictions of dryland climate change under doubled atmospheric CO_2 levels. Predictions are integrated from three major GCMs

	Southern Europe drylands bordering the Mediterranean*	Southern African drylands	Australia*
Summer temperature change (°C)	+4-6	+2-4	+2-4
Winter temperature change (°C)	+2-6	+2-4, +4-6 in the Karoo region	+2-4
Summer precipitation change	Decrease	Decrease	Varied, mainly increased
Winter precipitation change	Varied, decrease	Slight decrease	Increase in east, decrease in west
Soil moisture change	Significant decrease	Significant decrease	Increase in east, decrease in west

*'Varies' refers to situations where models disagree.

(Source: adapted from Williams and Balling, 1995)

first case, the net drying trends predicted for interior southern Africa in most GCMs, coupled with increased wind transport energy, lead to a major reactivation of dune systems to an extent greater than experienced at any time in the past 14 000 years. In the second, major changes in drainage system density in parts of north and southern Africa are modelled by the end of the twenty-first century, enhancing dryland conditions and potentially impacting negatively on human populations. Drylands, which are already difficult environments to live in given the temporal and spatial variability of the operation of environmental processes, may well become more uncertain environments before the end of this century.

Reflective questions

➤ Why is aridity likely to have been more extensive during glacial periods?

➤ Why is desertification a controversial characteristic of some dryland areas?

14.6 Summary

The chapter demonstrates that far from being simply sandy wastelands, deserts and drylands are complex and diverse environments that are subject to a wide range of geomorphological processes and are increasingly subject to human impacts. The chapter has explored the definition of drylands and the measurement and causes of aridity in order better to understand and explain dryland diversity. Deserts and drylands are not necessarily hot. Aridity is a result of a range of factors which include subtropical anticyclonic conditions, mountain lee orography and continentality. Drylands vary in their water balance (precipitation/potential evapotranspiration). They are often classified as hyper-arid, arid and semi-arid.

Dryland soils are typically subject to severe moisture deficits and with high evaporation rates often develop surface crusts. Vegetation systems in drylands are naturally adapted to the conditions. Often this means that growth and seedling dispersal are done within very short periods whenever water is available and then dormancy might ensue for the majority of the time. A range of other plant adaptions can be identified in dryland environments which allow conservation of moisture.

Dryland landscapes are characterized by high rates of rock weathering via salt, temperature and moisture processes. These weathering processes result in a great deal of readily erodible material that can be transported by wind and water. However, because rainfall events tend to be of short duration, there is a surplus of sediment supply over sediment transport. Nevertheless, when dryland rivers flow they tend to have high sediment yields which create problems with reservoir infilling. Infiltration rates and overland flow are highly variable, depending on localized crusting or vegetation cover. Wind transport, deposition and erosion are important features of drylands and distinctive dune and abraded rock forms can be identified. Drylands are dynamic landscapes both over long timescales due to Quaternary climatic changes and over short timescales related to natural sediment and vegetational processes. However, humans are impacting dryland landscapes via a range of mechanisms. Changes to local vegetation and soil structure and abstraction of groundwater are localized impacts but anthropogenically enhanced global warming may also alter dryland landscapes and their spatial extent.

Further reading
Books

Agnew, C. and Anderson, E. (1992) *Water resources in the arid realm*. Routledge, London.
This book considers the issue of water availability and use in drylands, covering the main physical aspects and issues relating to human use.

Middleton, N.J. and Thomas, D.S.G. (1997) *World atlas of desertification*, 2nd edition. Edward Arnold, London.
Written for UNEP, this volume explains desertification and its measurement via the GLASOD Project, as well as providing detailed background coverage to the problem and case studies.

Thomas, D.S.G. (ed.) (2011) *Arid zone geomorphology: Process, form and change in drylands*, 3rd edition. John Wiley & Sons, Chichester.
This is a detailed textbook that covers all aspects of geomorphology in drylands, including the role of vegetation and the nature of, and evidence for, long-term environmental change.

Thomas, D.S.G. and Middleton, N.J. (1994) *Desertification: Exploding the Myth*. John Wiley & Sons, Chichester.
Explains the complexities and controversies associated with explaining and assessing desertification.

Further reading
Papers

Hulme, M. and Kelly, M. (1993) Exploring the links between desertification and climate change. *Environment*, 35, 5–45.
This is a sensible investigation of the links between natural and human-induced environmental changes in drylands.

CHAPTER 15

Coasts

Gerhard Masselink

School of Marine Science and Engineering, University of Plymouth

Learning objectives

After reading this chapter you should be able to:

➤ understand the spatial extent of the coastal zone and its importance for society

➤ appreciate the significance of considering coastal land-forms as 'morphodynamic systems', in which coastal processes and morphology mutually interact at a variety of temporal and spatial scales

➤ identify the main factors that control and drive coastal processes and landforms in a variety of settings

➤ describe the dominant coastal types and understand the key processes responsible for their development

➤ evaluate the suitability of a range of coastal management strategies and be aware that natural coastlines are more resilient to sea-level rise than those that are managed

15.1 Introduction

Coastal environments are arguably the most important and intensely used of all areas settled by humans. The most obvious use of coastal environments is providing living space and the coast is clearly a preferred site for urbanization. For example, 23% of the global population currently live within 100 km of the coast and less than 100 m above sea level, population density in coastal areas is three times larger than average and projected population growth rates in the coastal zone are the highest in the world (Small and Nicholls, 2003). In addition, 21 of the 33 mega cities (with more than 8 million people; projected top 5 for 2015: Tokyo, Mumbai, Lagos, Dhaka and Karachi) can be considered coastal cities (Martinez *et al.*, 2007). Perhaps most disconcertingly, Nicholls and Mimura (1998) predicted that 600 million people will occupy coastal floodplain land below the 1000 year flood level by 2100.

The coast is a very dynamic environment and this presents many challenges to coastal communities. The most serious threat is that posed by sea-level rise and it is appropriate to discuss this hazard at the start of the chapter. At present, sea level is rising at a rate of approximately 3 mm yr^{-1} and the predicted rise in sea level during this century is expected to be between 18 and 59 cm (Box 15.1). The two most obvious consequences of rising sea levels are coastal flooding and erosion, and the key factor that determines how severely coastal environments and communities are affected is the rate of sea-level rise. Therefore, many coastal

ENVIRONMENTAL CHANGE

SEA-LEVEL RISE

It has become clear over the past few decades that global sea level is rising at an accelerated rate. Figure 15.1 shows the change in global sea level over the last 150 years reconstructed by Church and White (2006) from an extensive analysis of tide-gauge records and satellite altimeter data. The sea-level curve shows a modest rate of rise of about 1 mm yr^{-1} around 1850, increasing to about 3 mm yr^{-1} more recently.

The Intergovernmental Panel on Climate Change (IPCC) has published its Fourth Assessment Report (AR4) which addresses at length the causes and implications of the current and future rise in sea level (IPCC, 2007a, 2007b). Four main contributing factors to the current sea-level rise have been identified: (1) thermal expansion of the ocean water due to an increase in the water temperature; (2) melting of mountain glaciers and ice caps; (3) melting of Greenland ice sheets; and (4) melting of Antarctic ice sheets. These contributions have been quantified and are listed in Table 15.1. For the observed sea-level rise from 1961 to 2003 (1.8 mm yr^{-1}) the cumulative effect of these four factors still left about 0.7 mm yr^{-1} unexplained. However, the use of more sophisticated and comprehensive monitoring has brought to light that melting of Greenland and Antarctic ice sheets has contributed considerably to the observed sea-level rise from 1993 to 2003 (3.1 mm yr^{-1}) leaving only 0.3 mm yr^{-1} unexplained. The reduction in the ice volume of these ice sheets is believed to be the result of increased flow speeds of outlet glaciers (especially for the Antarctic ice sheet) and because losses due to melting have exceeded accumulation due to snowfall. The uncertainty surrounding the contribution to sea-level rise of the melting of ice sheets is considerable, but it has now been established with some confidence that about 60% of the sea-level rise that occurred over the

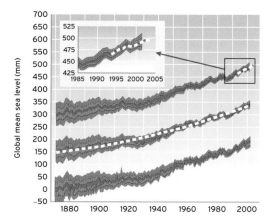

Figure 15.1 Reconstructed global mean sea level from 1870 to 2004. The three lines shown in the diagram are offset by 150 mm and represent, from bottom to top: (1) the monthly global average; (2) the yearly average with the quadratic fit; and (3) the yearly averages with satellite altimeter data superimposed. The inset shows a zoomed-in section of the most recent data with the satellite data represented by the thick white dashed line. The dark and light shading represent, respectively, the one and two standard deviation error estimates. (Source: after Church and White, 2006)

Table 15.1 Observed rate of sea-level rise and estimated contributions from different sources. The sea-level data prior to 1993 are from tide gauges and after 1993 from satellite altimetry

Source of sea-level rise	Rate of sea-level rise (mm yr^{-1})	
	1961-2003	1993-2003
Thermal expansion	0.42 ± 0.12	1.6 ± 0.5
Glaciers and ice caps	0.50 ± 0.18	0.77 ± 0.22
Greenland ice sheets	0.5 ± 1.2	2.1 ± 0.7
Antarctic ice sheets	1.4 ± 4.1	2.1 ± 3.5
Sum of individual climate contributions	1.1 ± 0.5	2.8 ± 0.7
Observed sea-level rise	1.8 ± 0.5	3.1 ± 0.7
Difference (observed minus sum of estimated climate contributions)	0.7 ± 0.7	0.3 ± 1

(Source: IPCC, 2007)

BOX 15.1 ➤

➤

period 1993 to 2009 can be attributed to land ice loss (Nicholls and Cazenave, 2010).

Using sophisticated computer models, the effect of different emission scenarios on climate change and sea-level rise by 2100 has been predicted (Table 15.2). The rise in global sea level by 2100 will be in the range from 18-38 to 26-59 cm depending on the emissions scenario. However, the predictions do not take into account any future rapid changes in ice flow, which are unpredictable and can potentially have very dramatic impacts. The actual sea-level rise may therefore be larger than predicted by the IPCC and recent estimates suggest that the rise in sea level by 2100 may be 1.8 m higher than at the start of this century (Nicholls and Cazenave, 2010).

Assessing the impacts of sea-level rise on our society, and formulating sustainable strategies to manage these, are obviously of great importance. An initial, and admittedly rather crude, approach is to quantify the extent of the land inundated and the number of people affected by rising sea level. Using a Digital Elevation Model (DEM) of the Earth's surface, Rowley *et al.* (2007) have determined that for a sea-level rise of 1 m, an area of 1.1 million km^2 will be inundated, affecting 108 million people. For a sea-level rise of 2 m, these numbers increase to 1.3 million km^2 and 175 million people.

Table 15.2 Projected globally averaged surface warming and sea-level rise by 2100 for different emission scenarios. The scenarios listed are described in the IPCC report (see also Section 24.2.3 in Chapter 24)

| Case | Temperature change by 2100 | | Sea-level rise by 2100 |
	Best estimate (°C)	Likely range (°C)	Model-based range (m)
B1 scenario	1.8	1.1-2.9	0.18-0.38
A1T scenario	2.4	1.4-3.8	0.20-0.45
B2 scenario	2.4	1.4-3.8	0.20-0.43
A1B scenario	2.8	1.7-4.4	0.21-0.48
A2 scenario	3.4	2.0-5.4	0.23-0.51
A2F1 scenario	4.0	2.4-6.4	0.26-0.59

(Source: IPCC, 2007)

BOX 15.1

researchers are concerned with quantifying how fast sea level is presently rising and whether this trend is likely to continue in the future. Sea-level rise is one of the main topics addressed by the Intergovernmental Panel on Climate Change (IPCC), which is a very large international and interdisciplinary group of leading scientists whose remit is to report on the environmental and societal causes and effects of climate change.

To manage the effects of sea-level rise, and also other anticipated consequences of climate change, such as increased storminess and changes to the prevailing wave direction, we need to have a good understanding of coastal processes. Of particular importance is the notion that different types of coastline respond differently to rising sea levels (Bird, 1993) and that coastal environments, especially those unaffected by humans, have a capacity to deal with the impacts of sea-level rise. This ability to respond to the consequences of sea-level rise is referred to as **resilience** and many natural features contribute to coastal resilience by providing ecological buffers (coral reefs, **salt marshes** and **mangrove** forests) and morphological protection (sand and gravel beaches, **barriers** and coastal dunes). A critical role in determining the resilience is played by the **sediment budget**. A coastline with a positive sediment budget may build up, rather than erode, under rising sea-level conditions. For example, the sediment deposition rate in salt marshes and tidal flats often exceeds the rate of sea-level rise; therefore, these environments may be able to 'keep up' with rising sea levels. Coral reefs are also examples of coastal environments that may be able to build up vertically faster than sea level rises provided they are in their natural state.

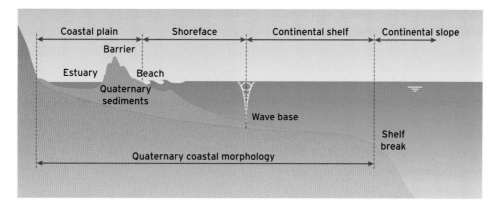

Figure 15.2 Spatial boundaries of the coastal zone. (Source: after Masselink and Hughes, 2003)

The overall objective of this chapter is to present an up-to-date overview of the main types of coastal environments and their governing processes. Before we start, however, it is important to indicate what we, as physical geographers, mean by the terms 'coasts' and 'coastal', because these terms are rather ambiguous and mean different things to different people. For most holidaymakers the coast is synonymous to the beach; for birdwatchers the coast generally refers to the intertidal zone; while for cartographers the coast is simply a line on the map separating the land from the sea. The spatial boundaries of 'our' coastal zone are defined in Figure 15.2 and its boundaries correspond to the limits to which coastal processes have extended during the Quaternary geological period. During this period, which lasted from approximately 2.4 million years ago until present, sea level fluctuated many times over 100 m vertically due to expansion and contraction of ice sheets (see Chapter 22). The landward limit of the coastal system therefore includes the coastal depositional landforms and the marine erosion surfaces formed when the sea level was high (slightly above present-day sea level) during warm interglacial periods. During cold glacial periods sea level was low and so coastal processes were close to the edge of

the continental shelf (see Chapter 3). The seaward limit of the coastal system is therefore defined by the edge of the **continental shelf**, which typically occurs in water depths of approximately 150 m.

15.2 Coastal morphodynamics

Coastal landforms and processes can be considered over a variety of temporal and spatial scales (time and space), ranging from the response of wave ripples to large wave groups on a timescale of minutes, to the infilling of estuaries following the drowning of river valleys due to sea-level rise over millennia. Regardless of the scale involved, a vital element in the coastal response is the presence of strong feedback between form and process. The morphodynamic approach, introduced by Wright and Thom (1977), formalizes this feedback by considering a coastal morphodynamic system comprising three linked elements (Figure 15.3):

1. *Processes* – this component includes all coastal processes that affect sediment movement. Hydrodynamic (waves, tides and currents) and aerodynamic (wind) processes are important. Weathering contributes

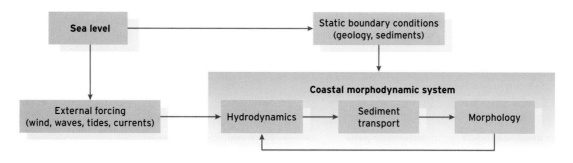

Figure 15.3 Coastal morphodynamic system with its energy input and boundary conditions.

significantly to sediment transport along rocky coasts, either directly through solution of minerals, or indirectly by weakening the rock surface to facilitate further sediment movement. Biological, biophysical and biochemical processes are important in coral reef, salt marsh and mangrove environments (Masselink *et al.*, 2011).

2. *Sediment transport* – a moving fluid imparts a stress on the bed, referred to as 'bed shear stress', and if the bed is mobile this may result in the entrainment ('picking up') and subsequent transport of sediment (see Chapter 9). The resulting pattern of erosion and deposition can be assessed using the sediment budget (see Box 15.2). If more sediment enters a coastal region than leaves it, then the sediment balance is positive and deposition will occur and the coastline may advance. A negative sediment balance will occur if more sediment leaves a coastal region than enters it and net erosion will ensue, with possible coastline retreat.

3. *Morphology* – the surface of a landform or assemblage of landforms such as coastal dunes, **deltas**, **estuaries**, beaches, coral reefs and **shore platforms** is referred to as the morphology. Changes in the morphology are brought about by erosion and deposition.

As the coastal system evolves over time, its evolution is recorded in the sediments (clay, silt, sand and gravel) in the form of the **stratigraphy**. It is important to realize that stratigraphic sequences are a record of the depositional history and that erosional events are represented only by gaps in the stratigraphic record.

Coastal systems exhibit a certain degree of autonomy in their behaviour, but they are ultimately driven and controlled by environmental factors, often referred to as 'boundary conditions'. The three most important boundary conditions are geology, sediments and external forcing (wind, waves, storms and tides), with sea level serving as a meta-control by determining where coastal processes operate. When contemporary coastal systems and processes are considered, human activity should also be taken into account. In fact, along many of our coastlines, human activities, such as beach nourishment, construction of coastal defences and land reclamation, are far more important in driving and controlling coastal dynamics than the natural boundary conditions, and cannot be ignored (Figure 15.4). It can even be considered that, through climate change, humans are altering the boundary conditions themselves!

A characteristic of coastal morphodynamic systems is the presence of strong links between form and process

Figure 15.4 The coastal frontage of Treport in north France is an excellent example of a coastline much affected by human activities.

(Cowell and Thom, 1994). The coupling mechanism between processes and morphology is provided by sediment transport and is relatively easy to comprehend. There is, however, also a link between morphology and processes to complete the morphodynamic feedback loop. For example, under calm wave conditions sand is transported on a beach in the onshore direction, resulting in beach accretion. As the beach builds up, its seaward slope progressively steepens and this has a profound effect on the wave breaking processes and sediment transport. At some stage during beach steepening, the hydrodynamic conditions may be sufficiently altered to stop further onshore sediment transport. Owing to the close coupling between process and form in morphodynamic systems it is often not clear whether the morphology is the result of the hydrodynamic processes, or vice versa. This makes it very difficult to predict coastal development over long timescales.

The feedback between morphology and processes is fundamental to coastal morphodynamics, and can be negative or positive. **Negative feedback** acts to oppose changes in morphology. For example, beach erosion during storms generally results in the development of an offshore bar. Wave breaking on the bar significantly reduces the amount of wave energy reaching the shoreline, thereby limiting further beach erosion. However, many coastal systems do not always behave that predictably and often the feedback between morphology and processes is positive, rather than negative. **Positive feedback** pushes a system away from equilibrium by modifying the morphology such that it is even less compatible with the processes to which it is exposed. A morphodynamic system driven by positive feedback seems to have a 'mind of its own' and exhibits

FUNDAMENTAL PRINCIPLES

SEDIMENT BUDGETS

Morphological change directly results from sediment transport processes. Sediment budgets help us to understand the different sediment inputs (sources) and outputs (sinks) involved (Figure 15.5). A sediment budget involves accounting for the sediment volumes (m^3) rather like you would account for money. Key components of the sediment budget are the sediment fluxes, which represent the direction and amount of sediment transport by certain processes and which are expressed as the quantity of sediment moved per unit of time ($kg\ s^{-1}$, $m^3\ yr^{-1}$). If the sediment

fluxes are known, sediment budgets can be used to predict how the morphology changes through time in a quantitative fashion.

Consider, for example, an estuary with a surface area of 1 km^2 (this is the same as 1 000 000 m^2) that receives an annual input of sediment from marine and fluvial sources of 100 000 m^3 per year. If it is further assumed that this sediment is evenly spread over the estuary floor, then the depth of the estuary will decrease by:

$$\frac{sediment\ input}{surface} = \frac{100\ 000\ m^3\ yr^{-1}}{1\ 000\ 000\ m^2}$$
$$= 0.1\ m\ yr^{-1}$$

If the average depth of the estuary is 10 m, then the estuary will be infilled in:

$$\frac{depth}{accretion\ rate} = \frac{10\ m}{0.1\ m\ yr^{-1}}$$
$$= 100\ years$$

Of course, this simple illustration assumes that the amount of sediment entering the estuary does not change while the estuary is infilling. It therefore ignores feedback between morphology and process which is one of the main principles of morphodynamic systems. Nevertheless, such simple calculations can still tell us a lot about environmental change in coastal zones.

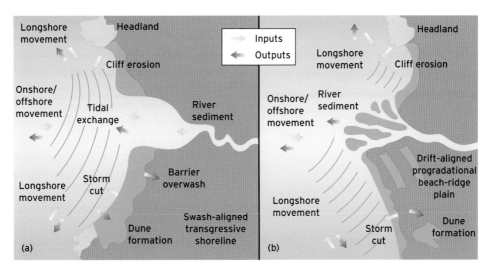

Figure 15.5 Qualitative examples of coastal sediment budgets. (a) Riverine sediment may be deposited into an estuarine embayment, which may also receive sediment from the seaward direction through tidal processes. (b) On a deltaic coast, riverine sediment contributes to the coastal sediment budget, and is moved up the drift-aligned coast, giving rise to a beach-ridge plain. (Source: adapted from Carter and Woodroffe, 1994)

BOX 15.2

self-forcing behaviour. An example of positive feedback is the infilling of deep estuaries by marine sediments due to uneven flows between an incoming tide and an outgoing tide. In a deep estuary, flood currents are stronger than ebb currents and this tidal 'asymmetry' results in a net influx of

sediment and infilling of the estuary. As the estuary is being infilled, the tidal asymmetry increases even more as friction effects are enhanced by the reduced water depths. In turn, the increase in tidal asymmetry speeds up the rate of estuarine infilling. This constitutes positive feedback between the estuarine morphology and the tidal processes, resulting in rapid infilling of the estuary. Eventually, tidal flats and salt marshes start developing in the estuary and this marks a reversal in feedback. As the intertidal areas become more extensive, the flood asymmetry of the tide progressively decreases so that the estuarine morphology approaches steady state as sediment imports and exports balance each other.

Adjustment of coastal morphology to changing conditions involves a redistribution of sediment which requires time. The time required for the adjustment to occur is known as the **relaxation time**. The relaxation time strongly depends on the volume of sediment involved in the adjustment of morphology and is related to the size of the landform. Large coastal landforms, such as coastal barriers, have longer relaxation times than small morphological features, such as **beach cusps**. Generally, the relaxation time exceeds the time between changes in environmental conditions. It is therefore unlikely that a 'steady-state equilibrium' (see Chapter 1) is ever reached, particularly for large coastal landforms.

Reflective questions

➤ How do negative and positive feedback affect our ability to make long-term predictions of coastal development?

➤ What does a morphodynamic approach to coasts involve?

15.3 Coastal processes: waves

Ocean waves are the principal agents for shaping the coast and driving **nearshore** sediment transport processes. Of course, wind and tides are also significant contributors, and are indeed dominant in coastal dune and estuarine environments, respectively, but the action of waves is dominant in most settings. Waves are generated by wind and the stronger the wind, the larger the waves. Not surprisingly, therefore, there is a strong climatic control on the global distribution of wave heights, with the largest waves found at the stormiest latitudes around the 'roaring forties' (40° north and south) (Figure 15.6).

It is important at the outset to make the distinction between regular and irregular (or random) waves. The motion of regular waves is periodic, that is the motion is repetitive over space (Figure 15.7a) or through fixed periods of time (Figure 15.7b). Regular waves can be described in terms of a single representative **wave height** H, **wavelength** L and **wave period** T. The wave height is the difference in elevation between the **wave crest** and the **wave trough**, the wavelength is the distance between successive crests (or troughs) and the wave period is the time it takes for the wave to travel a distance equal to its wavelength. Of these three parameters, the wave period is the easiest to determine in the field: simply count and time the passage of a large number of waves (at least 10) past a fixed point and divide the time by the number of waves. The **wave steepness** is also an important parameter and given by the ratio of wave height to wavelength H/L.

Natural waves are, however, highly irregular, and a range of wave heights and periods are present (Figure 15.8). To properly describe the wave conditions of irregular waves in

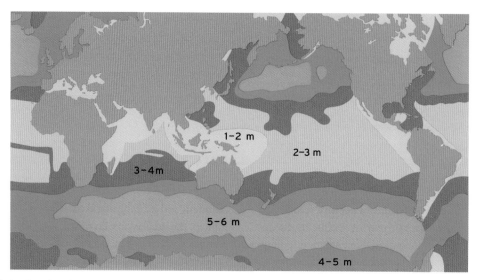

Figure 15.6 Global values for significant wave height which is exceeded 10% of the time. (Source: after Young and Holland, 1996)

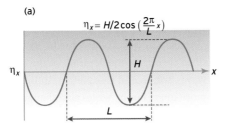

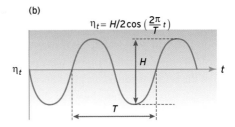

Figure 15.7 Schematic showing a regular wave train. (a) The spatial variation in water level η_x is measured at a single moment in time along the direction of wave travel. From such data the wavelength L can be derived. (b) The temporal variation in water level η_t is measured at a single location in space over a representative time period. Such data enable the determination of the wave period T. The wave height H can be derived from both types of wave data. (Source: Masselink and Hughes, 2003)

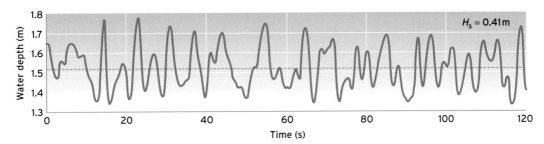

Figure 15.8 Example 2 minute time series of water depth showing 'real' waves measured just outside the surf zone in 1.5 m water depth on a sandy beach. These highly irregular and asymmetric waves are characterized by a significant wave height H_s of 0.41 m and a period of 4-5 s.

quantitative terms, statistical techniques are required. For example, a widely used measure of the wave height is the **significant wave height** H_s, which is defined as the average of the highest one-third of the waves. The significant wave height can be obtained by recording a time series of water depth and multiplying the associated **standard deviation** by four.

15.3.1 Linear wave theory

The behaviour of ocean waves can be estimated using **linear wave theory**. The equations associated with this theory are widely used (Komar, 1998). Based on the ratio of water depth h to deep-water wavelength L_o, we can identify three different wave regions, each characterized by different water particle motions under the waves (Figure 15.9):

1. *Deep water* ($h/L_o > 0.5$) – as waves travel across the sea surface, the water particles beneath undergo an almost closed circular path. The particles move forward under the crest of the wave and move seaward under the trough of the wave. The diameter of the orbits decreases with increasing depth until at some distance below the water surface, referred to as the **wave base**, the wave motion ceases. The wave base is thus defined as the depth below which wave motion cannot stir bed sediment. In deep water, the wavelength L is given by $gT^2/2\pi$ and the wave speed or celerity C is given by $gT/2\pi$, where g is the gravitational acceleration.

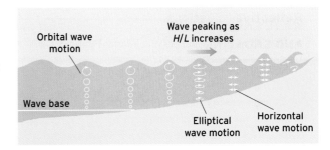

Figure 15.9 Motion of water particles under waves according to linear wave theory.

2. *Intermediate water* ($0.5 < h/L_o < 0.05$) – with decreasing water depth, the wave motion extends to the seabed and the surface waves are much affected by the presence of the seabed. As a result, the water particles now follow an elliptical path with the ellipses becoming flatter and smaller as the seabed is approached. At the seabed, the water particles merely undergo a horizontal to-and-fro motion. In intermediate water depths, L and C are not easily computed and fall between the values obtained by using the deep- and shallow-water equations.

3. *Shallow water* ($h/L_o < 0.05$) – very close to the **shore**, all water motion consists of horizontal movements to-and-fro which are uniform with depth. In shallow water, L and C are given by $T\sqrt{(gh)}$ and $\sqrt{(gh)}$, respectively.

The deep- and shallow-water wave equations are straightforward to apply. For example, think about a wave with a period of 10 s. In deep water, the wavelength and wave velocity are only dependent on the wave period and application of the relevant equations gives $L = 156$ m and $C = 15.6$ m s^{-1}. In shallow water, the water depth also needs to be considered. For a water depth of 1 m and a wave period of 10 s the relevant equations yield $L = 31$ m and $C = 3.1$ m s^{-1}. The example highlights a fundamental characteristic of waves: as waves propagate from deep to shallow water, they get closer together and slow down.

15.3.2 Wave processes in intermediate water

Wave **shoaling** is the process whereby the waves change in height as they travel into shallower water and start feeling the seabed. This process is particularly pronounced just before wave breaking at the seaward edge of the **surf zone**, where the waves are suddenly seen to 'pick up' and become steeper. The increase in wave height due to wave shoaling can be explained through consideration of the energy associated with wave motion and is explained in Box 15.3.

Perhaps more importantly, waves also change their shape during shoaling. In deep water, waves are characterized by a **sinusoidal** shape (as shown in Figure 15.7) and the water particle velocities associated with the wave motion are symmetrical, meaning that the onshore velocities are of equal strength and duration as the **offshore** velocities. However, as the waves enter shallower water, they become increasingly asymmetrical and develop peaked crests and flat troughs (Figure 15.10; see also Figure 15.8). The associated flow velocities also become asymmetric with the onshore phase of the wave being stronger, but of shorter duration than the offshore phase of the wave. The development of **wave asymmetry** is very important from a sediment transport

FUNDAMENTAL PRINCIPLES

WAVE ENERGY FLUX AND SHOALING

The rate at which wave energy is carried along by moving waves is known as the **wave energy flux** and is the product of the amount of energy associated with the waves and the speed at which this energy travels. The total wave energy E is given by:

$$E = \frac{1}{8}\rho g H^2 \qquad (15.1)$$

where ρ is the density of water, g is gravity and H is the wave height. The wave energy is expressed as the amount of energy per unit area (N m^{-2}) and is more appropriately referred to as the wave energy density. Wave energy depends on the square of the wave height. A doubling of the wave height therefore results in a four-fold increase in wave energy. The speed by which this energy is carried along is not the same as the speed of the individual waves, but is given by Cn, where C is the wave speed and n is given by:

$$n = \frac{1}{2}\left[1 + \frac{2kh}{\sinh(2kh)}\right] \qquad (15.2)$$

where sinh is sine hyperbolic. The parameter n increases from 0.5 to 1 from deep to shallow water. This means that deep-water waves travel at twice the speed of the energy ($n = 0.5$), whereas shallow-water waves travel at the same speed as the energy ($n = 1$). The wave energy flux P is then given by:

$$P = ECn \qquad (15.3)$$

As waves travel from deep to shallow water, the change in the wave height due to shoaling can be calculated by examining the wave energy flux P. Assuming that energy losses due to bed friction can be ignored, the wave energy flux remains constant during wave propagation. This can be expressed as:

$$P = (ECn)_1 = (ECn)_2 = \text{constant}$$

where the subscripts 1 and 2 indicate two different locations along the path of wave travel. Substituting the wave energy in this equation and rearranging the result produces equation (15.4):

$$H_2 = \left(\frac{C_1 n_1}{C_2 n_2}\right)^{1/2} H_1 \qquad (15.4)$$

If we would like to compute the wave height in shallow water, where $n = 1$, from the wave height in deep water, where $n = 0.5$, the equation can be simplified to:

$$H_2 = \left(0.5\frac{C_1}{C_2}\right)^{1/2} H_1 \qquad (15.5)$$

After inserting the appropriate deep- and shallow-water wave speed, this equation becomes:

$$H_2 = \left(0.5\frac{(gT/2\pi)}{\sqrt{gh}}\right)^{1/2} H_1 \qquad (15.6)$$

Assuming a deep-water wave height of 1 m and a period of 12 s, the wave height in 2 m water depth becomes 1.45 m, representing a significant increase.

BOX 15.3

Figure 15.11 Wave refraction across a bay.

Figure 15.10 Asymmetric waves in shallow water just prior to breaking characterized by peaked wave crests and broad wave troughs. Note also a rip current (see Section 15.3.4) in the centre of the image and the surfers for scale.

point of view because it promotes the onshore transport of sediment particles. In fact, without the development of wave asymmetry, beaches would not exist because there would be no mechanism to transport sand and gravel to the shore.

Wave **refraction** is another important process that takes place during shoaling. It occurs because the part of the wave that is in shallower water travels slower than the part of the wave in deeper water. This results in a rotation of the wave crest with respect to the bottom contours, or, in other words, a bending of the wave rays (Figure 15.11). The resulting change in wave direction causes the wave crests to be more aligned parallel to the bottom contours, reducing the wave angle. Irregular seabed topography can cause waves to be refracted in complex ways and produce significant variations in wave height and energy along the coast. Spreading of the wave rays, referred to as **wave divergence**, occurs when waves propagate over a localized

area of relatively deep water (e.g. depression in the sea floor) and this causes a reduction in the wave energy and wave height. Focusing of the wave rays is known as **wave convergence** and occurs when waves travel over a localized area of relatively shallow water such as a shoal on the sea floor and causes an increase in the wave energy and wave height.

15.3.3 Wave processes in shallow water

When the water depth becomes too shallow for a stable wave to exist, the wave breaks. **Wave breaking** occurs when a wave becomes too steep and the horizontal velocities of the water particles in the wave crest exceed the velocity of the wave and the wave disintegrates into bubbles and foam. Waves break in a depth slightly larger than their height and the region on the beach where waves break is known as the surf zone. Wave breaking is an important process, because when waves break, their energy is released and is used for generating nearshore currents and sediment transport. A continuum of breaker shapes occurs in nature; however, three main breaker types are commonly recognized (Figure 15.12):

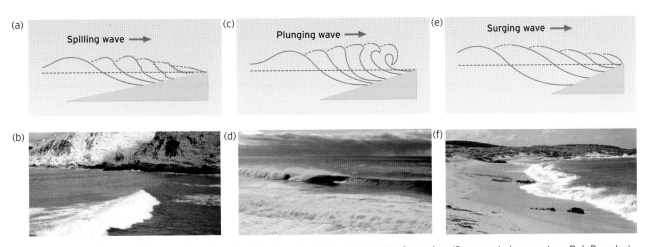

Figure 15.12 The three main types of breakers: (a, b) spilling, (c, d) plunging and (e, f) surging. (Source: photos courtesy Rob Brander)

1. *Spilling breakers* are associated with gentle beach gradients and steep incident waves. A gradual peaking of the wave occurs until the crest becomes unstable, resulting in a gentle forward spilling of the crest (Figure 15.12a and b).
2. *Plunging breakers* occur on steeper beaches than spilling breakers, with waves of intermediate steepness. The shoreward face of the wave becomes vertical, curling over, and plunging forward and downward as an intact mass of water. Plunging breakers are the most desirable type of breaker for surfers, because they offer the fastest rides and as they plunge over they produce 'tubes' (Figure 15.12c and d).
3. *Surging breakers* are found on steep beaches with low-steepness waves. The front face and crest of surging breakers remain relatively smooth and the wave slides directly up the beach without breaking (Figure 15.12e and f).

Breaker type can be predicted based on a consideration of the wave characteristics and the gradient of the beach using the Iribarren number, also known as the surf similarity parameter (Battjes, 1974):

$$\xi = \frac{\tan \beta}{\sqrt{H_b/L_o}} \tag{15.7}$$

where $\tan \beta$ is the gradient of the beach and the subscripts 'b' and 'o' indicate breaker and deep-water conditions, respectively. Small values for ξ are attained when the beach has a gentle gradient and the incident wave field is characterized by a large wave height and a short wavelength (or a short wave period). Large values of ξ are found when the beach is steep and the incident wave field is characterized by a small wave

height and a long wavelength (or a long wave period). Spilling breakers occur for $\xi < 0.4$; plunging breakers for $\xi = 0.4$ to 1; and surging breakers for $\xi > 1$.

For large values of the Iribarren number ($\xi > 1$), the incident wave energy is not dissipated by breaking, but is reflected at the beach, very much like light is reflected off a mirror. The proportion of reflected energy increases with ξ and is generally modest on beaches, unless they are very steep. When waves encounter the vertical face of a seawall, however, wave **reflection** approaches 100%. This may give rise to standing wave motion in front of the seawall and/or complicated criss-cross wave patterns (Figure 15.13).

When waves break, they produce **wave set-up**, which is a rise in the mean water level above the still water elevation of the sea (Figure 15.14). In popular terms, wave set-up is

Figure 15.13 Reflection of breaking waves at the vertical face of a seawall giving rise to a criss-cross wave pattern.

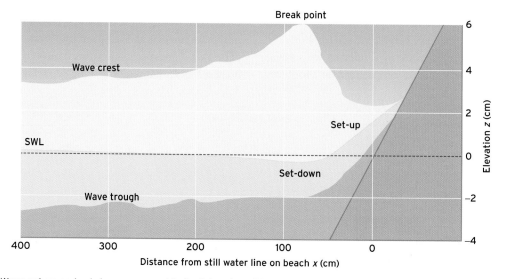

Figure 15.14 Wave set-up and set-down measured in the laboratory. Measurements were obtained using a wave height of 6.45 cm, a wave period of 1.14 s and a beach gradient of 0.082. SWL refers to still water level. (Source: after Bowen *et al.*, 1968)

conceived as a piling up of water against the shoreline due to the waves and is caused by the breaking waves driving water shoreward. As a general rule of thumb, the set-up at the shoreline is 20% of the offshore significant wave height. At the shoreline, surf zone waves propagate onto the 'dry' beach in the form of **swash**. Swash motion consists of an onshore phase with decelerating flow velocities (uprush) and an offshore phase characterized by accelerating flow velocities (**backwash**). For a number of reasons such as infiltration effects, **advection** of sediment and turbulence from the surf zone into the **swash zone**, the uprush is a more efficient transporter of sediment than the backwash, and this onshore swash asymmetry is responsible for maintaining the beach gradient.

15.3.4 Nearshore currents

In the surf zone, incident wave energy is lost, or dissipated, due to wave breaking. Much of this energy is used for generating nearshore currents and sediment transport, ultimately resulting in the formation of distinct coastal morphology. Nearshore currents derive their energy from wave breaking and the intensity of these currents increases with increasing incident wave energy level. The strongest currents are therefore encountered during storms. There are three types of wave-induced currents and these systems dominate the net water movement in the nearshore (Figure 15.15):

1. **Longshore currents** are shore-parallel flows within the surf zone. They are driven by waves entering the surf zone with their crests aligned at oblique angles to the shoreline. Longshore currents increase with the incident wave energy level and the angle of wave approach, and may reach velocities in excess of 1 m s^{-1}. They are also affected by alongshore winds and can be particularly strong when strong winds are blowing in the same direction as the longshore current.

2. The **bed return flow**, often somewhat misleadingly referred to as **undertow**, is an average flow near the bed flowing offshore. The current is part of a circulation of water characterized by onshore flow in the upper part of the water column and seaward flow near the bottom (Figure 15.15b). Measured bed return velocities are typically 0.1–0.3 m s^{-1}, but under extreme wave conditions may reach values of up to 0.5 m s^{-1}.

3. **Rip currents** are strong, narrow currents that flow seaward through the surf zone in channels and present a significant hazard to swimmers (Figures 15.10, 15.15c

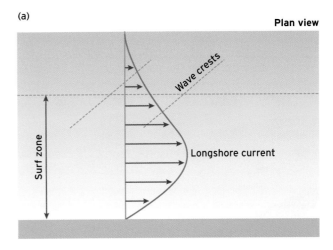

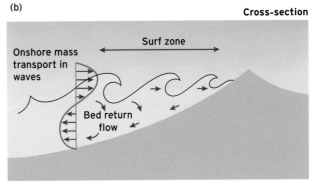

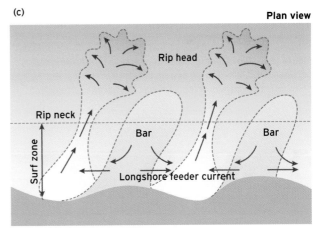

Figure 15.15 Quasi-steady currents in the surf zone: (a) shore-parallel longshore currents due to obliquely incident waves, (b) vertically segregated bed return flow or undertow and (c) horizontally segregated rip currents as part of the nearshore cell circulation system. (Source: after Masselink and Hughes, 2003)

and 15.16). They consist of onshore transport of water between rip currents, longshore feeder currents, and offshore transport of water in the rip itself. Maximum current velocities associated with circulation of water

(a)

(b)

Figure 15.16 Rip currents: (a) a single rip current flowing out between sand bars and (b) two rip currents present at either side of an intertidal rock outcrop. (Source: photo (a) courtesy Rob Brander)

in rip currents may reach up to 2 m s^{-1} under extreme storm conditions when so-called 'mega-rips' form. Typical rip current velocities are 0.5–1 m s^{-1} and flows are generally stronger during low tide than during high tide. Recent research into rip currents has revealed that under some conditions the longshore feeder and rip current develop into a larger rotating surf zone eddy as shown in Box 15.4.

Nearshore currents are capable of transporting large quantities of sediment. This is partly due to their often significant flow velocities, but also because sediment entrainment is considerably enhanced by the stirring motion of the breaking waves. The amount of sediment transported by longshore currents is known as the **littoral drift** and is generally of the order of 10 000 to 100 000 m^3 yr^{-1}. Such large transport rates have a major effect on coastal morphology and shoreline change.

Reflective questions

➤ Why is it important to distinguish between different types of breaking waves?

➤ What are the changes that occur to ocean waves when they travel from deep to shallow water?

➤ What is the best course of action when caught in a rip current?

TECHNIQUES

GPS DRIFTERS AND RIP CURRENT CIRCULATION

Global Positioning Systems (GPS) are now routinely used in coastal research for surveying coastal landforms and recording morphological change. Due to advances in GPS technology, which have led to a reduction in GPS size and an increase in their positional accuracy, they are now also widely used in society (e.g. SatNavs in cars). An innovative use of the latest generation of GPS devices is the measurement of rip currents.

Rip currents are the main hazard for surf zone users and are responsible for about 80 and 100 drownings per year on Australian and Florida beaches, respectively. However, it has proven difficult to characterize the flow pattern associated with rip currents due to their spatial and temporal variability. MacMahan et al. (2009) and Austin et al. (2010) used specially designed drifters equipped with GPS units (Figure 15.17a) to characterize rip circulation. By repeatedly releasing large numbers of these GPS drifters into the surf zone and combining the multiple drifter tracks, a representative characterization of

BOX 15.4 ➤

➤ the current patterns associated with rip currents can be obtained.

An example of the output produced by this research on a macrotidal beach in the south-west of England is shown in Figure 15.17. Contrary to the rip flow pattern illustrated by Figure 15.15c, which shows rip currents that flow straight out to beyond the surf zone, the GPS drifters suggest the presence of a large anticlockwise rotating eddy contained within the surf zone. By conducting similar experiments under a range of beach, wave and tide conditions valuable data can be collected to improve our understanding of rip currents. In turn, this information can be disseminated to the lifeguarding communities so they can provide a better service and reduce the number of rip-related drownings on beaches.

(a)

Figure 15.17 (a) Twelve GPS drifters awaiting deployment on Perranporth Beach in the south-west of England. (b) Rip circulation pattern obtained from a 2 h deployment of the 12 drifters, representing a total of 39 drifter tracks. The black arrows represent flow patterns obtained from at least five drifter tracks, whereas the red arrows represent isolated drifter tracks. The sinuous white band represents the region of wave breaking obtained using a video camera and the contours show bathymetric elevation. (Source: images courtesy of Martin Austin)

(b)

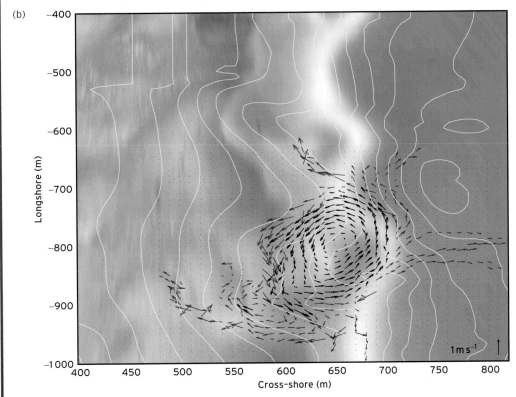

BOX 15.4

15.4 Coastal processes: storm surge, tides and tsunami

Although ocean waves are the principal agents for shaping the coast and driving nearshore sediment transport, the contributions of storm surge, tides and **tsunami** are also significant and require consideration. These three processes represent changes in the water level of the order of metres, but their timescales vary considerably. Storm surges generally last several days, tides occur daily or twice daily and tsunami represent water-level fluctuations of around 10 minutes.

15.4.1 Storm surge

During severe storms, the water level near the shore can be significantly elevated compared with tidal predictions, and the difference between the measured and the predicted water level is the storm surge (Figure 15.18). Storm surge levels can be in excess of 5 m under extreme conditions, such as during hurricanes or cyclones, leading to extensive coastal flooding and erosion. The maximum storm surge associated with the 2005 hurricane Katrina appears to have been in excess of 8 m. The amount of storm surge at the coast depends on three main factors:

1. *Low pressure* – sea level will rise approximately 1 cm for every 1 millibar fall in air pressure. Storms are always characterized by low pressure and hence the water level under a storm is always raised.

2. *Onshore wind* – if the wind is directed shoreward it can pond water against the coast, causing an increase in the water level.

3. *Coastal topography* – the effect of the storm surge on the coast depends greatly on coastal configuration. Relatively low-gradient, funnel-shaped coastal settings are particularly prone to extreme surges. Examples of these are the Bay of Bengal or the North Sea. Straight coastlines and promontories are generally less sensitive to storm surge.

15.4.2 Tides

The tidal rise and fall of the ocean surface are barely noticeable in the deep ocean, but on shallow continental shelves, along coastlines and within estuaries, tidal processes can be the dominant morphological agent. The two driving forces for ocean tides are the gravitational attraction of the Earth–Moon system and the Earth–Sun system, with the latter being almost half that of the former. The theory of tides is rather complicated and it is more useful to describe how tides are manifested along our coasts, than concern ourselves about how they are generated. In fact, the 'proper' theoretical explanation of tides involving bulges of water at either side of the Earth, known as the equilibrium tide theory, does not provide much insight into the actual characteristics of ocean tides other than their dominant periods (12.5 h for the lunar component and 12 h for the solar component).

According to the more practical explanation of tides, the dynamic tide theory, the global tidal water motion is broken up into a large number of tidal systems constrained by the coastal topography. In these systems, known as **amphidromes**, the tide travels around the centre of the amphidrome as a wave. Owing to the Coriolis force (see Chapter 4), the tidal wave travels clockwise in the southern hemisphere and anticlockwise in the northern hemisphere. The difference in water level between high and low water is the **tidal range**, and is practically zero at the centre of the amphidrome and maximum at the edge. Figure 15.19 shows the three amphidromes in the North Sea. At a particular location, the observed tidal range depends on which is the influential amphidrome and how far it is away from the centre. It should be noted that even the situation shown in Figure 15.19 is a simplification, because interaction between the tidal wave and the local topography can significantly modify the tidal water motion.

Along most coasts, tides cause a twice-daily rise and fall in the water level (some coasts have mainly once-daily tides, but these are relatively rare). The tidal range varies greatly

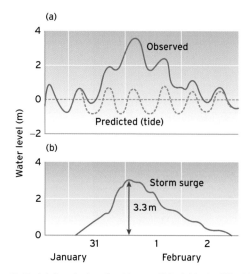

Figure 15.18 (a) Graph showing the predicted (dashed line) tidal water level and observed (solid line) water level at the Hook of Holland, January to February, 1953. (b) Graph showing that the residual tide, which is attributed to storm surge, reached 3.3 m. (Source: American Society of Civil Engineers)

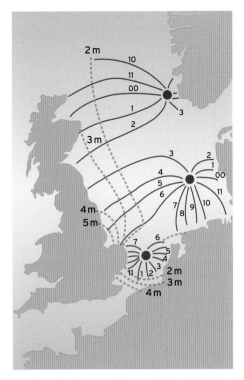

Figure 15.19 Map showing amphidromes in the North Sea. The dashed lines are lines of equal tidal range, whereas the solid lines show how the tidal wave propagates in an anticlockwise direction around the centre of the amphidrome. The numbers indicate where the tide is at the same stage in the tidal cycle (in hours). Along a line labelled '1', for example, the tide is one hour into the tidal cycle. (Source: after Pethick, 1984)

such as semi-enclosed basins like the North Sea and Irish Sea, funnel-shaped bays like the Bay of Fundy, Canada, and regions with wide continental shelves and island groups such as around north-west Australia.

The tidal range also varies over time, mainly due to the interaction between the tidal forces of the Earth–Moon system and the Earth–Sun system. When the Earth, Moon and Sun are all aligned, during either a full or new moon, the tidal forces of the Moon and the Sun are combined, resulting in extra-large tides, known as **spring tides**. When the Moon is at a right angle to the Earth with respect to the Sun, however, the tidal forces of the Moon and the Sun are competing, resulting in extra-small tides, known as **neap tides**. The Moon revolves around the Earth in 28 days; therefore, a spring-to-spring tidal cycle (or neap-to-neap tidal cycle) takes 14 days to complete (Figure 15.21).

Associated with the rising and falling tides are **tidal currents** whose strength increases with the tidal range. At any one place, ebb- and flood-tidal currents vary in strength and duration. This is especially the case for estuaries (see Section 15.7.3), but also along open coasts, in deep water and even on beaches. Therefore, over time there exists a tidal current (the tidal movements do not balance out over time) and this is known as the **residual tidal current**. These currents are important because they induce a net movement of nearshore sediment that may contribute significantly to coastal morphological development.

15.4.3 Tsunami

Often a tsunami is referred to as a 'tidal wave'. However, this term is inappropriate because tides and tsunami differ

around the world and it is common to distinguish between micro- (< 2 m), meso- (2–4 m) and macro-tidal (> 4 m) ranges (Figure 15.20). The largest tidal ranges are generally associated with rather complex coastal configurations,

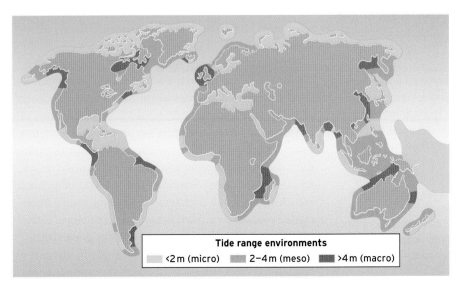

Figure 15.20 World distribution of mean spring tidal range. (Source: after Davies, 1980)

Tide range environments
<2 m (micro) 2–4 m (meso) >4 m (macro)

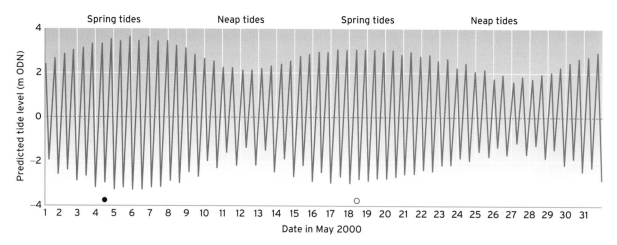

Figure 15.21 Predicted high and low tide levels for Immingham, England, over a one-month period (May 2000). The phases of the Moon are indicated by the solid (new moon) and open circle (full moon). Spring tides occur around the new and full moon, whereas neap tides occur halfway between these Moon phases. ODN refers to Ordnance Datum Newlyn, which is approximately mean sea level in the United Kingdom.

from each other in many respects, particularly in the way that they are generated. Tides are generated by astronomical forces involving the Earth, Moon and Sun, whereas a tsunami is generated as a result of the displacement of a large water mass by any of the following three mechanisms (Masselink *et al.*, 2011):

1. A displacement of the seabed by a submarine earthquake. The 2004 Sumatra earthquake produced a tsunami that killed more than 100 000 people.
2. A large landslide into the ocean, perhaps during a volcanic eruption. The 1883 Krakatau eruption produced a tsunami that killed more than 36 000 people.
3. An impulse generated by a meteorite impact striking the ocean. This generation mechanism has not occurred in recent historical time, but mega-tsunami events recorded in the geological record have been attributed to meteorites (Bryant, 2001).

In the open ocean tsunami typically have a wavelength of a few hundred kilometres and a height less than a metre. They travel at the shallow-water wave speed given by $\sqrt{(gh)}$, which for a typical water depth of 5 km implies a speed of around 700 km h^{-1}. When they cross the edge of a continental shelf, however, they rapidly begin to shoal, and become shorter in wavelength and larger in height. This process is identical to the shoaling process of wind waves (see Box 15.3). By the time tsunami reach the coastline, they can be several tens of metres high and in a period of only a couple of hours they can cause enormous property damage and loss of life. Chapter 3 provides more detail on tsunami processes.

Reflective questions

➤ How are neap and spring tides generated?

➤ What factors control the magnitude of a storm surge?

15.5 Coastal classification

Coastal landforms have often been classified to provide useful ways to help assess the different forcing factors and controls such as sea-level history, geology, climate, waves and tides that lead to the great variety of coastal landforms we encounter (Bird, 2000). Most early classification schemes were based on the realization that coastal landforms are largely the product of sea-level variations. Such classifications distinguish between submerged and emerged coasts. Typical submerged coasts are drowned river and glacial valleys, often referred to as rias and fjords, respectively. Coastal plains are characteristic of emerged coasts. Another type of classification distinguishes between primary and secondary coasts. Primary coasts have a configuration resulting mainly from non-marine processes and include drowned river valleys and deltaic coasts. Secondary coasts, on the other hand, are coasts that have a configuration resulting mainly from marine processes or marine organisms. Examples of such coasts are barrier coasts, coral reefs and mangrove coasts.

The main shortcoming of these early classifications is that the emphasis on geological inheritance and sea-level

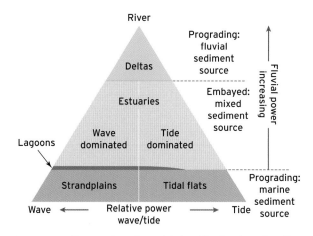

Figure 15.22 Process-based coastal classification based on the relative importance of wave, tide and river processes. (Source: after Boyd *et al.*, 1992)

history leaves only limited concern for the hydrodynamic processes. The morphology of depositional coastal environments (those consisting of mud, sand and gravel, rather than eroding rocky shores) responds to the relative dominance of river, wave and tidal factors (Boyd *et al.*, 1992). A diagram can be constructed that expresses the relative importance of river outflow, waves and tidal currents (Figure 15.22). In this diagram, deltas are positioned at the fluvial point of the triangle because a fluvial sediment source dominates, while prograding, non-deltaic coasts are located on the opposite wave–tide side of the triangle, because sediment is moved onshore by waves and tides. Estuaries occupy an intermediate position, because they have a mixed sediment source and are affected by river, wave and tidal factors. Different coastal typologies can be identified, reflecting the degree of fluvial, wave or tide dominance, and it is this classification that will be used for the remainder of this chapter. Specifically, the next three sections will discuss the dynamics of wave-, tide- and fluvial-dominated coasts, followed by rocky coastlines.

Reflective question

➤ What is the point of classifying coastlines in distinct types?

15.6 Wave-dominated coastal environments

The most easily recognized landform of wave-dominated coasts is the beach. However, beaches are only one component of wave-dominated coasts. The underwater slope that

lies seaward of the beach, known as the shoreface, is also dominated by wave processes and actually occupies a much larger area than the beach. Coastal dunes behind beaches are also common elements of wave-dominated coastal environments. Dunes, beaches and shorefaces are strongly linked by sediment transport pathways and collectively they make up 'coastal barriers', which are considered the basic depositional elements of wave-dominated coasts (Roy *et al.*, 1994). We will first discuss the dynamics of barriers and then move on to beaches and dunes.

15.6.1 Barriers

A large variety of barrier types exist, including **barrier islands** separated by tidal inlets and fronting wide shallow lagoons as along the coast of the Netherlands (Figure 15.23), continuous barrier systems that are backed by a coastal plain, **lagoon** or estuary, and mainland beaches on steep coasts with very little additional barrier morphology. Most barrier systems are made up of sand, but gravel barriers are also frequently found, especially at higher latitudes where glacial processes over the last few hundred thousand years have produced vast quantities of gravel-size material.

It is useful to distinguish between two fundamentally different types of barrier coasts: swash-aligned barriers and drift-aligned barriers. Swash-aligned barriers are oriented parallel to the crests of the prevailing incident waves (Figure 15.24a). They are closed systems in terms of longshore sediment transport and are characterized by a curved planform shape. The long-term average shoreline configuration of swash-aligned barriers is relatively constant since the net littoral drift is zero. However, short-term changes in incident wave conditions such as seasonal or inter-annual changes in the prevailing wave direction will induce minor adjustments in the planform shape.

Drift-aligned barriers are oriented obliquely to the crest of the prevailing waves (Figure 15.24b). The shoreline of drift-aligned coasts is primarily controlled by longshore sediment transport processes (Masselink and Hughes, 2003) and such coasts are sensitive to changes in littoral drift rates. A **spit** is a classic example of a drift-aligned barrier and is a narrow accumulation of sand or gravel, with one end attached to the mainland and the other projecting into the sea or across the mouth of an estuary or a bay. Spits grow in the littoral drift direction and can exist only through a continuous longshore supply of sediment. If this sediment supply ceases, the spit will eventually subsume itself and disappear.

Transgressive barriers are those that migrate landward under the influence of rising sea level and/or a negative

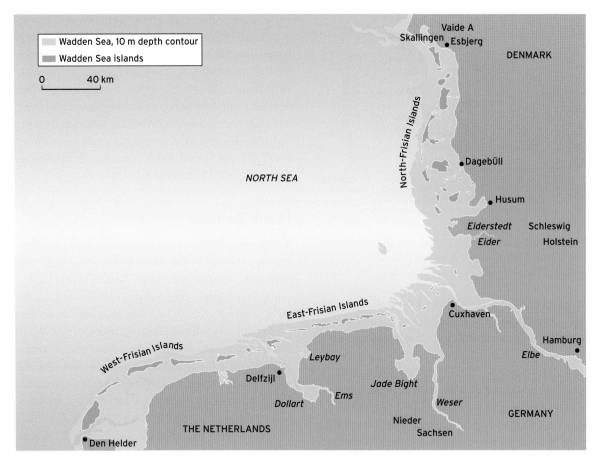

Figure 15.23 Barrier islands of the Wadden Sea in north-western Europe. (Source: after *The Physical Geography of Western Europe*, Oxford University Press (Hofstede, J. in Koster E.A. (ed.) 2005), and The Common Wadden Sea Secretariat (CWSS))

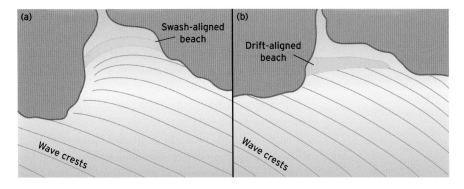

Figure 15.24 Two main types of coastal alignment: (a) swash alignment and (b) drift alignment. (Source: adapted from Davies, 1980)

sediment budget (Figure 15.25a). Transgressive barriers consist mainly of tidal delta and/or washover deposits, and are underlain by back-barrier estuarine or lagoonal deposits. Sediments deposited in seaward environments end up on top of sediment that originated in more landward environments (transgressive sequence). **Regressive barriers** or strandplains are those that develop under the influence

of a falling sea level and/or a positive sediment budget (Figure 15.25b). Here landward sediments are deposited on top of more seaward ones (regressive sequence). The barrier is generally overlain by windblown sand, below which there is beach and nearshore sand underlain by silt and clay that had been deposited on what was formerly the continental shelf.

(a)

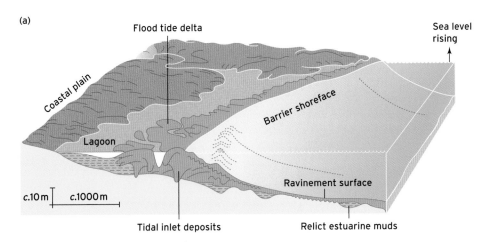

Figure 15.25 (a) Model of transgressive barrier during sea-level rise. Transgressive barriers are almost entirely composed of tidal delta and washover deposits. The barrier migrates into estuarine and lagoonal environments as sea level rises. (b) Model of regressive barrier during sea-level fall. The surface of the regressive barrier forms a wide strandplain, generally without estuaries. (Source: after Roy *et al.*, 1994)

(b)

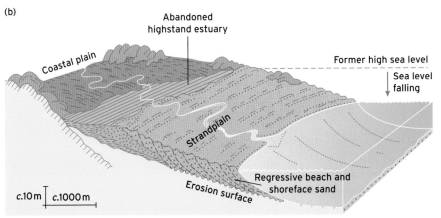

15.6.2 Beaches

A beach is a wave-deposited sand or gravel landform found along marine, lacustrine and estuarine shorelines. It represents the upper part of the shoreface and is generally characterized by an overall concave-upward shape. On most beaches, deviations to the concave profile may occur in the form of smaller-scale features and it is the presence of these features that gives beaches their distinctive morphology (Figures 15.26 and 15.27).

Berms result from the accumulation of sediment at the landward extreme of wave influence by swash processes (Figure 15.27b). They protect the back of the beach and coastal dunes from erosion under extreme wave conditions. The seaward part of the berm is often steep and is termed the beachface. Its equilibrium slope is thought to reflect the balance between differences in uprush and backwash. Onshore uprush forces are often greater than backwash forces. This results from energy losses due to bed friction and infiltration of water into the beach during the uprush, which is promoted by coarse and permeable sediments. The equilibrium beachface gradient is therefore positively correlated with the beachface sediment size and the steepest gradients occur on gravel beaches.

Beach cusps are rhythmic shoreline features formed by swash action and may develop on sand or gravel beaches. The spacing of the cusps is related to the horizontal extent of the swash motion and may range from about 10 cm on lake shores to 50 m on exposed ocean beaches. Beach cusps are considered self-organizing features, resulting from feedback between beachface morphology and swash processes as described in Box 15.5.

The surf zone on energetic beaches is often characterized by a nearshore bar morphology which can be rhythmic in planform. The development of rhythmic bar morphology has also been attributed to self-organization (Box 15.5). On beaches subjected to large tidal ranges, the intertidal zone is often flat and featureless. But when the upper part of the intertidal zone is relatively steep, because it is composed of coarse sediments for example, a distinct break in slope is generally found separating the steep upper part from the low-gradient, low-tide terrace.

Beaches respond to changing wave conditions and of greatest significance is the exchange of sediment between the upper beach and the surf zone, and the development of berm and bar profiles (Figure 15.28). Under calm conditions, sediment transport in the nearshore zone tends to be in the onshore direction, resulting in a steepening of the beach

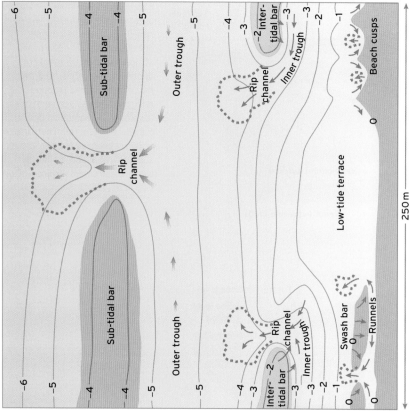

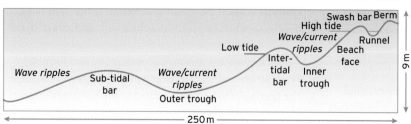

Figure 15.26 Schematic showing dominant morphological features on a beach. The top panel represents a contour plot of the beach morphology, whereas the bottom panel shows a typical cross-shore beach profile. High-tide level on this beach is at an elevation of 0 m and the tidal range is 2 m. (Source: after Masselink and Hughes, 2003)

Figure 15.27 Features of a beach: (a) large intertidal current ripples associated with rip feeder currents exposed at low tide; (b) actively developing berm near the high-tide level.

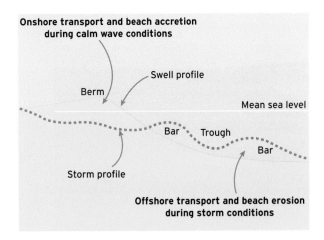

Onshore transport and beach accretion
during calm wave conditions

Swell profile

Berm

Mean sea level

Bar   Trough

Bar

Storm profile

Offshore transport and beach erosion
during storm conditions

Figure 15.28 Idealized beach profiles with and without bars. Storm conditions induce offshore transport, beach erosion and the formation of a nearshore bar. Calm wave conditions result in onshore sediment transport, beach accretion and the formation of a berm. (Source: Aagard and Masselink, 1999)

profile. If bars are present, these tend to migrate onshore and become part of the beach, resulting in the development of a steep beach with a pronounced berm. Such beaches are referred to as **reflective beaches**, because a significant part of the incoming wave energy is reflected back from the shoreline. In contrast, energetic wave conditions induce offshore sediment transport with prolonged high-wave conditions, resulting in the destruction of the berm and the formation of a flat beach with subdued bar morphology. The surf zone is likely to be wide with multiple lines of spilling breakers. The majority of the incoming wave energy is dissipated during the wave-breaking process and these beaches are known as **dissipative beaches**. Most beaches fall within these two extremes and are characterized by nearshore bar morphology over which a significant amount of wave energy is being dissipated due to wave breaking. The upper part of intermediate beaches is, however, rather steep and reflective. Therefore these beaches are referred to as **intermediate beaches**.

Depending on the wave conditions, beaches tend to move from one beach type to the other (Wright and Short, 1984). Along some coastlines, stormy conditions in the winter and calm conditions during summer give rise to a seasonal cycle of beach change comprising a winter profile with a nearshore bar and a summer profile with a berm, although this will vary in any given year. Depending on the wave conditions, beaches tend to move from one beach type to the other (Wright and Short, 1984). The occurrence of different types of beach morphology can be parameterized by the dimensionless fall velocity Ω given by:

$$\Omega = \frac{H}{w_s T} \qquad (15.8)$$

where H is the wave height, w_s is the sediment fall velocity (the speed at which a sediment particle falls through still water) and T is the wave period. Reflective beaches tend to develop when $\Omega < 1.5$, intermediate beaches are characterized by $\Omega = 1.5-5.5$, and dissipative beaches form when $\Omega > 5.5$.

NEW DIRECTIONS

SELF-ORGANIZATION

Morphodynamic systems often display a sequence of positive feedback driving the system towards a new state, followed by negative feedback, which stabilizes the system, resulting in equilibrium. This is referred to as self-organization. The result of this process is a rather orderly arrangement of sediments and landforms.

An example of self-organization is the formation of beach cusp morphology. Beach cusps are rhythmic shoreline features formed by swash action and are characterized by steep-gradient, seaward-pointing cusp horns and gentle-gradient, seaward-facing cusp embayments (Figure 15.29). The self-organization theory of beach cusp formation considers beach cusps to be the result of feedback between morphology and swash flow (Werner and Fink, 1993). Positive feedback causes small topographic depressions on the beachface to be amplified by attracting and accelerating water flow, thereby promoting erosion. At the same time, small positive relief features are enhanced by repelling and decelerating water flow, thereby promoting accretion. The sequence of positive feedback is followed by negative feedback processes, which inhibit erosion and accretion on well-developed cusps, and maintain equilibrium. The important feature is that the morphological regularity arises from the internal dynamics of the system.

The notion of self-organization has now become well established in a wide range of disciplines (Werner, 1999), including geomorphology (Murray *et al.*, 2009), and a range of coastal features are now interpreted as self-organizing features, including rhythmic features such as wave ripples, beach cusps, bar morphology and cuspate shoreline features (Coco and Murray, 2007).

BOX 15.5 ➤

Figure 15.29 Beach cusps on a gravel beach. The longshore spacing of the cusps is approximately 3 m.

BOX 15.5

15.6.3 Coastal dunes

Coastal dunes are common features in wave-dominated coastal environments and their dynamics are closely linked to that of the beach (Sherman and Bauer, 1993). Their formation requires an energetic wind climate and a suitably large supply of sand. Onshore winds capable of inducing sediment transport must occur for a significant amount of time. Dunes protect the coast from erosion by providing a buffer to extreme waves and winds. Extreme storm activity inevitably results in elevated water levels and beach erosion, and may lead to coastal flooding. However, well-developed dune systems dissipate the energy of storm waves through dune erosion. The sand eroded from the dune system will be transported offshore, but will eventually return to the beach under fair weather conditions. As the sediment is returned to the beach, wind processes may result in renewed dune development. Maintenance of coastal dune systems is thus an important component of coastal protection and management (see Box 15.6).

HAZARDS

EFFECT OF COASTAL DUNES ON IMPACT OF HURRICANES OF SANDY BARRIERS

A key aspect of the response of sandy barrier systems to extreme storm events is the destruction of foredunes. As long as the foredunes remain proud above the storm run-up level, sediment will be transported offshore from the seaward face of the dunes and upper beach area, to be deposited lower on the beach. But if the foredunes are relatively low, they will be overtopped and overwashed, resulting in onshore sediment transport from the upper beach area and dunes, and in the development of overwash deposits at the back of the barrier. The latter barrier response pattern characterizes transgressive barrier systems and is illustrated in Figure 15.30 which documents the response of a 2 km section of St. George Island, a barrier

BOX 15.6 ➤

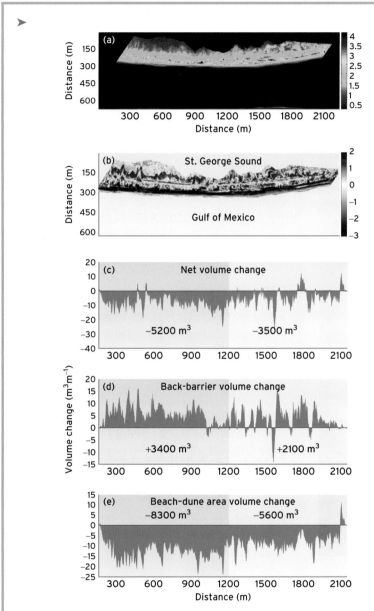

Figure 15.30 (a) Digital elevation model (DEM) of 2 km section of St. George Island State Park before the storm (May 2004). (b) Elevation change map (ECM) indicating the morphological change that occurred due to hurricane Dennis. The morphological change is also expressed in the net volume changes per unit metre width for the whole barrier (c) and for the landward (d) and seaward (e) parts of the barrier. The western part of the barrier (shaded) experienced increased sub-aerial erosion due to lower foredune elevations compared with the eastern part. The colour bars in (a) and (b) are in metres elevation and elevation change, respectively. (Source: images courtesy Anthony Priestas; modifed from Priestas and Fagherazzi, 2010)

island in Florida, to hurricane Dennis (Priestas and Fagherazzi, 2010).

Dennis made landfall as a category 3 hurricane on 10 July 2005, near Pensacola, Florida, 280 km west of St. George Island on the Gulf of Mexico. Sustained winds from Dennis at landfall were reported at 51–53 m s^{-1} and the storm surge at St. George Island was reported to be 2.5 m. Washover deposits as a result of the high storm surge translated large amounts of sediment to the back barrier, and where St.

George Island is narrow those deposits extended into St. George Sound.

Using elevation data obtained using LiDAR (Light Detection and Ranging), a widely used technique for rapidly assessing storm impact along the hurricane-prone south-east coast of the United States, an accurate picture of the morphological change and sediment volumes involved was obtained (Figure 15.30). The digital elevation model (DEM) from before the storm shows fragmented foredunes along the western part of the study area, but a well-developed foredune ridge with up to 4 m high dunes along the eastern part (Figure 15.30a). By comparing the DEMS from before and after hurricane Dennis, elevation change maps (ECMs) can be constructed, which provide an accurate picture of the storm response (Figure 15.30b). The ECM indicates that most of the dunes were flattened during the storm (up to 3 m erosion; blue areas in Figure 15.30b) with deposition of up to 2 m occurring in the back-barrier region (red areas in Figure 15.30b). Some deposition also occurred on the lower beach along the eastern part of the study area.

The erosion of the seaward part of the barrier (beach and dunes) and the deposition on the landward part (back-barrier region) can be quantified by considering the change in the sediment volumes (Figure 15.30c–e). What is particularly noteworthy is that the sediment losses from the beach–dune area and the gains in the back-barrier region for the western part of the study, where the dunes were least developed, are significantly greater than for the eastern area with higher and more continuous foredunes. These results convincingly demonstrate the benefits of maintaining well-developed foredunes: not only do foredunes limit the net sediment lost due to storms, but they also inhibit the development of overwash and barrier retreat.

BOX 15.6

Coastal dunes generally begin to develop around the drift line above the spring high-tide line. Here, tidal litter (seaweed, driftwood) represents an obstacle to the wind, promoting the formation of **shadow dunes** with tails stretching out downwind (Figure 15.31a). Shadow dunes cannot reach elevations higher than that of the obstacle, but ongoing accumulation of sediment can occur following the establishment of pioneer plant species. Pioneer plants are all characterized by a high tolerance to salt, elaborate root systems that can reach down to the freshwater table and rhizomes that grow parallel to the upper dune surface. The sand-trapping ability of the pioneer plants enables the shadow dune to grow upwards and outwards into incipient foredunes, which are 1–2 m high vegetated mounds of sand (Figure 15.31b). Given suitable conditions (onshore winds and adequate sand supply) and sufficient time, the incipient foredunes will coalesce, forming a **foredune ridge** (Figure 15.31c). Foredunes can grow quickly reaching a height of several metres over a period of 5–10 years.

(a)

(b)

(c)

Figure 15.31 Different stages in the formation of foredunes: (a) isolated shadow dunes associated with clumps of vegetation; (b) incipient foredune in front of older dune cliff; and (c) terrace-like foredune ridge backed by older dunes. (Source: Anthony Priestas)

Reflective questions

➤ Use the Google Maps search engine to compare the barrier islands of the Dutch Wadden Sea in the Netherlands with those of the New Jersey coast in the United States. Why are the Dutch barrier islands shorter?

➤ What is the difference in stratigraphy between transgressive and regressive barrier systems?

➤ Given the large wave energy level and the smoothing action of tides that beaches are exposed to, why are most beaches not flat and featureless?

➤ In what way are coastal dunes different from terrestrial desert dunes described in Chapters 9 and 14?

15.7 Tide-dominated coastal environments

Estuaries are the main tide-dominated coastal landform and represent zones of mixing between fluvial and marine processes (Figure 15.32). The development of present-day estuaries started when coastal river valleys were flooded as sea level rose following ice melt at the end of the last glacial period. Following stabilization of the sea level around 6000 years ago, infilling of the estuaries occurred as a result of the influx of sediments from both marine and terrestrial sources. Most estuaries can be divided into three zones: the inner zone, central zone and outer zone (Figure 15.33). These three zones

Figure 15.32 Panoramic view of the mouth of the Avon estuary, South Devon, UK, illustrating where the river meets the sea. (Source: photo courtesy Roland Gehrels)

are unique with respect to their energy regime, sediment type and morphology (Masselink *et at.*, 2011). River processes dominate at the head of the estuary and their influence decreases towards the mouth of the estuary. Marine processes are most important at the mouth and their role decreases towards the head. The energy regime in the inner zone is therefore river-dominated, in the outer zone it is marine-dominated (waves and tides) and in the central zone it is mixed (tide and river processes). There are many types of estuaries and a popular distinc-tion is that between wave- and tide-dominated estuaries (Dalrymple *et al.*, 1992).

15.7.1 Wave- and tide-dominated estuaries

Wave-dominated estuaries are found in coastal regions sub-jected to relatively high levels of wave energy (Figure 15.34). The outer zone consists of a barrier system and a tidal inlet, often with an ebb-tide delta and a flood-tide delta at the landward and seaward side, respectively. The outer zone

(a)

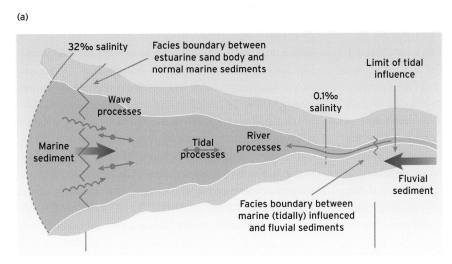

(b)

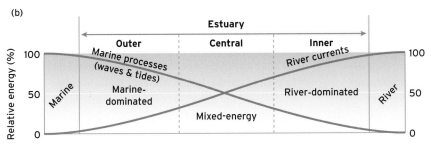

Figure 15.33 (a) Plan view of an estuary showing sediment and hydraulic boundaries. (b) Chart showing the changing mix of wave, tide and river processes along the estuary axis. (Source: after Dalrymple *et al.*, 1992, SEPM (Society for Sedimentary Geology))

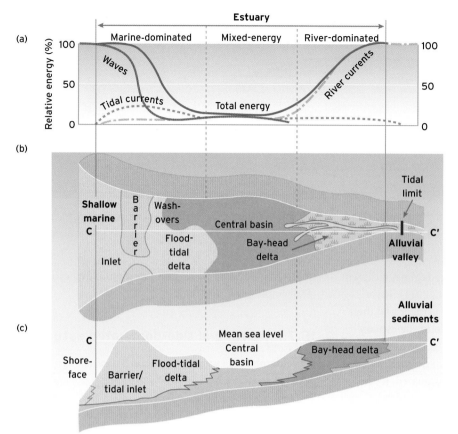

Figure 15.34 (a) Chart showing the change in energy regime along the axis of a wave-dominated estuary. (b) Plan view of the estuary showing positions of principal morphological features. (c) Section view along the estuary axis showing stratigraphy. (Source: after Dalrymple *et al.*, 1992, SEPM (Society for Sedimentary Geology))

is dominated by wave processes, but their effects decline rapidly with distance from the inlet due to wave breaking over both the barrier and tidal deltas. The tidal energy also decreases away from the coast, because the narrow tidal inlet restricts the tidal water motion between estuary and sea. A distinguishing characteristic of wave-dominated estuaries is the very low energy level in the central zone. If the estuary is relatively young, a deep central mud basin accumulates the finest sediments; if the estuary is mature, then the central zone is infilled and dominated by salt marshes or mangrove flats also composed of predominantly muddy sediments (Masselink *et al.*, 2011). If the river entering the estuary at the head carries significant amounts of sediment, a **bay-head delta** can be found in the inner zone. Wave-dominated estuaries infill through seaward progression of this bay-head delta and landward extension of the flood-tide delta. Eventually, fluvial and marine sands bury the central basin muds.

Tide-dominated estuaries are found in coastal regions experiencing relatively large tidal ranges and therefore strong tidal currents (Figure 15.35). The scouring action of the tidal currents keeps the entrance of the estuary relatively open and

gives tide-dominated estuaries their typical funnel shape. The strong tidal currents in the outer zone shape the sediments into linear sand bars separated by tidal channels. Waves are of secondary importance, but their influence can sometimes extend further into a tide-dominated estuary due to its funnel shape. Wave energy in the central basin is insignificant, whereas tidal energy is still relatively high. Therefore, the central zone of tide-dominated estuaries is more energetic than that in wave-dominated estuaries (Masselink *et al.*, 2011). A single meandering channel is commonly found in the central zone and this channel is tide-dominated for most of the time, but is significantly influenced by river processes during times of high discharge. The central zone is again a sink for fine sediment and includes extensive intertidal morphology, such as tidal flats and salt marshes. Because tidal currents remain strong, even up to the head of the estuary, fluvial sediments become progressively mixed with estuarine sediments in the inner zone and there is no discrete fluvial delta. The infilling of tide-dominated estuaries is often rapid and occurs through a steady seaward migration of the inner, central and outer zones along the drowned valley.

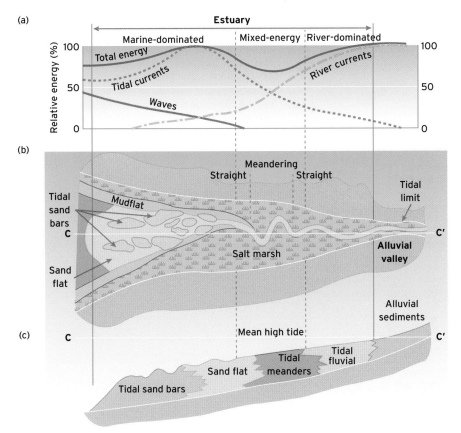

Figure 15.35 (a) Chart showing the change in energy regime along the axis of a tide-dominated estuary. (b) Plan view of the estuary showing positions of principal morphological features. (c) Section view along the estuary axis showing stratigraphy. (Source: after Dalrymple *et al.*, 1992, SEPM (Society for Sedimentary Geology))

15.7.2 Estuarine mixing

An important process occurring in estuaries is the mixing of salt- and freshwater masses that are delivered to the estuary by tide and river flows, respectively. The mixing is accomplished by the turbulence associated with river and tidal flows and is opposed by the density difference between the two water masses. On the basis of the mixing process, we can identify three types of estuaries (Dyer, 1998) as described below.

15.7.2.1 Stratified estuaries

Stratified estuaries tend to have fresh river water sitting on top of saline seawater, separated by a sharp interface, known as the **halocline** (Figure 15.36a). The saltwater near the bed is in the form of a salt wedge, which becomes thinner towards the landward end of the estuary. Stratified estuaries commonly occur along micro-tidal coasts with low to intermediate river discharge. The river discharge should be sufficient to develop a fresh surface water mass, but should not be so large as to expel the saltwater from the estuary. If the tidal or riverine flows become too dynamic, turbulence is generated and the stratification breaks down.

15.7.2.2 Partially mixed estuaries

In **partially mixed estuaries** there is a more gradual salinity gradient, but still with freshwater at the surface and saltwater near the bed (Figure 15.36b). However, there tends to be a more mixed layer in the central part of the water column. The salinity decreases towards the landward end of the estuary both at the surface and at depth. Partially mixed estuaries develop if the tidal energy is sufficient to cause mixing of the fresh- and saltwater, and are typical of meso- to macro-tidal coasts.

15.7.2.3 Well-mixed estuaries

In **well-mixed estuaries** the mixing is so effective that the salinity gradient in the vertical direction vanishes entirely. If such mixed estuaries are sufficiently wide, the Coriolis force (see Chapter 4 for an explanation of this process) starts to play a significant role in affecting the water circulation. Its effect on outflowing river water is to push the flow to the margin of the estuary (Figure 15.36c) and may result in a horizontal separation of river water and seawater. This segregation is best developed on the flooding tide when the fresh- and saltwater masses are opposed and the effect of the Coriolis force is to separate the two flows.

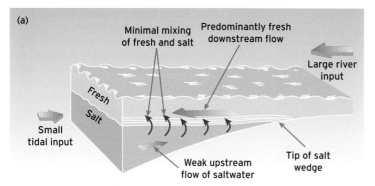

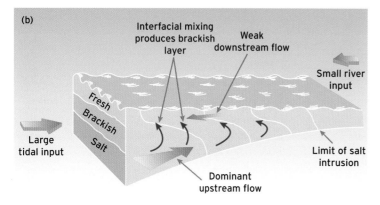

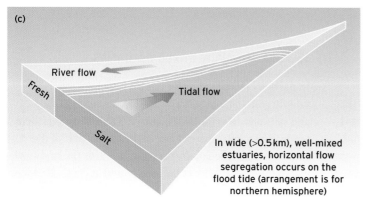

Figure 15.36 Diagram illustrating three main types of estuaries based on density stratification: (a) stratified; (b) partially mixed; and (c) well-mixed estuary. Water masses are indicated by shading on the front face of each block. Salinity contours (arbitrary scale) are indicated by thin lines on the side face of each block in (a) and (b) and on the top face of the block in (c). In (a) and (b) vertical mixing is indicated by thin arrows and non-tidal currents are indicated by thick arrows. (Source: after Pethick, 1984)

Wave-dominated estuaries are usually stratified, whereas tide-dominated estuaries are either partially mixed or well mixed.

15.7.3 Ebb- and flood-dominance

The total volume of water entering an estuary on the flooding tide is called the **tidal prism** and can be estimated by multiplying the tidal rise in water level in the estuary during high tide by the surface area of the estuary. Ignoring the contribution of the freshwater discharge and evaporation, this same volume of water leaves the estuary during the falling tide. The duration and strength of the flood and ebb flow, however, are significantly different in most estuarine

systems. Estuaries or channel sections that display a flooding tide that is larger in velocity magnitude and shorter in duration than the ebbing tide are said to be flood-dominant, whereas those that display an ebbing tide that is largest in magnitude and shortest in duration are said to be ebb-dominant.

Flood- or ebb-dominance often translates directly to net landward or seaward sediment transport, respectively (Friedrichs and Aubrey, 1988). Even a small difference in the velocity magnitude between the flood and ebb tide can lead to a large difference in the total amount of sediment transported by each, and therefore a net sediment transport. In general, flood-dominant estuaries tend to infill their entrance channels by continually pushing coastal sediment landward and as a result are often intermittently closed, whereas ebb-dominant estuaries tend to flush sediment seaward from their entrance channels and as a result are often stable (Masselink *et al.*, 2011). There are a number of mechanisms that can generate tidal flow asymmetry, and hence ebb- or flood-dominance. The two most important are tidal distortion (which is the change of the tidal wave shape due to shoaling, typical of long estuaries, and leads to flood-dominance) and a high proportion of intertidal areas such as salt marshes and tidal flats (which are more typical of small, wave-dominated estuaries and lead to ebb-dominance).

15.7.4 Salt marsh and mangroves

The lower intertidal zone in most estuaries is devoid of vegetation due to excessive bed shear stress preventing seedlings taking anchor in the sediment. The upper intertidal zone is less energetic, however, and in temperate environments is colonized by salt-tolerant grasses and reeds known collectively as salt marsh (Figure 15.37), and in

(sub)tropical environments by mangroves (see Figure 9.22 in Chapter 9). Both salt marshes and mangroves exhibit a distinct zonation across the upper intertidal zone with the most salt-tolerant species (the 'pioneers') lowest in the tidal frame, and the least salt-hardy species near the Highest Astronomical Tide (HAT) level.

Despite the obvious difference between these two types of ecosystems, their functioning from an estuarine evolutionary point of view is very similar: both significantly affect the tidal flows and through a variety of mechanisms enhance sediment deposition of clastic material, mainly silts and muds. In addition, in both systems there is a steady supply of organic detritus (roots, stems, leaves, branches) which also contributes to sedimentation. Vertical accretion rates are highly variable, both spatially and temporally, and are of the order of millimetres per year. Generally, sedimentation rates exceed current and even projected rates of sea-level rise, enabling these intertidal environments to keep up with rising sea level. Salt marshes and mangroves thus provide a good example of natural resilience to sea-level rise.

Figure 15.37 Well-developed salt marshes found along the inner bends of the tidal channel of the Avon estuary, South Devon, UK. (Source: photo courtesy Roland Gehrels)

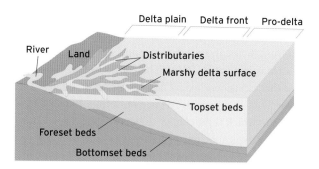

Figure 15.38 The three generic morphostratigraphic units found in all fine-grained deltas: delta plain (topset beds), delta front (foreset beds) and pro-delta (bottomset beds). (Source: after Haslett, 2000)

15.8 Fluvial-dominated coastal environments

Coastal deltas are accumulations of sediment deposited where rivers enter into the sea. River sediments may also accumulate at the head of coastal embayments if the coastline is drowned (bay-head delta), but these deposits are also controlled by estuarine processes and were briefly discussed in the previous section. In deltas, the amount of sediment delivered into the coastal margin by a river system outpaces the ability of the marine processes (waves and tides) to remove these sediments, causing the coastline to advance seaward. Since deltas are mainly associated with large river systems, they occupy only a relatively small proportion of the world's coastline. In Europe, for example, there are only six delta systems of significance (Danube Delta in Romania, Ebro Delta in Spain, Po Delta in Italy, Rhine/Meuse Delta of the Netherlands, Rhone Delta in France, Volga Delta in Russia). Nevertheless, deltas are very important from a societal point of view because they are characterized by relatively large population densities.

Although the detailed morphology of deltas varies from one example to the next, depending on the delta regime, there are three morphological units common to almost all deltas (Figure 15.38). The **delta plain** is the sedimentary platform that covers recent seaward advance, the **delta front** represents the seaward front of the delta that is located in relatively shallow water and is being reworked by wave and tidal processes, and the **pro-delta** is situated at the toe of the delta front in relatively deep water and is generally out of reach of wave processes (Masselink *et al.*, 2011). The delta builds out because the river continuously delivers sediment to the coastline. The coarsest sediments are deposited close to the river mouth and the finer sediments settle out further seaward. As the delta front progrades horizontally, the delta plain aggrades vertically. The resulting sediment distribution through the delta wedge is then a general fining in the seaward direction and a general coarsening upwards in the

vertical direction (Reading and Collinson, 1996). Based on the relative magnitude of river, wave and tide power, Galloway (1975) proposed a classification scheme of deltas with three end-members (Figure 15.39). This model has recently been extended by Hori and Saito (2008).

Fluvial-dominated deltas are associated with large catchments, river discharge into protected seas with minimal nearshore wave energy, and a small tidal prism. The freshwater river effluent is generally less dense than the saltwater in the receiving basin, and the river water flows out on top of the receiving water. If the outflowing river water is denser than the water in the receiving basin, for example due to extremely high suspended sediment concentrations, the river water will flow out along the seabed. Buoyancy limits the mixing between river and basin waters, allowing the river sediment to be transported further into the receiving basin before settling to the bed. As a result, the morphology of river-dominated deltas is characterized by pronounced seaward protrusions, a classic example of which is the so-called 'bird foot' of the Mississippi Delta (Figure 15.40).

Wave-dominated deltas are typically found in open-coast settings with a steep shoreface gradient, causing the deltaic coastline to be exposed to energetic waves. The action of waves at the river mouth induces strong mixing between the river flow and the receiving water. There are therefore limited buoyancy effects and the sediment transporting capacity decreases rapidly away from the mouth of the river, resulting in the formation of a distributary mouth bar across the river mouth. When waves approach with their crest parallel to the coastline they are refracted symmetrically around the distributary mouth bar. When they approach obliquely, however, they cause longshore transport and spit growth, with the river entrance constantly migrating downdrift. Regardless, the characteristic feature of wave-dominated deltas is a relatively straight, only weakly protruding coastline.

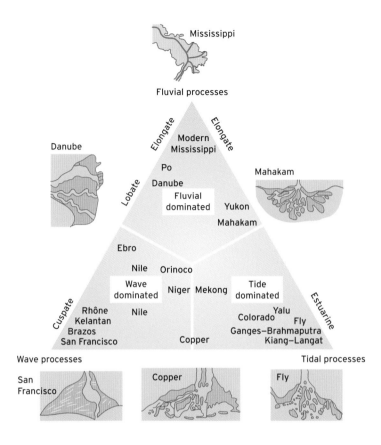

Figure 15.39 Classification of fine-grained deltas proposed by Galloway (1975). The end-member types are fluvial-, tide- and wave-dominated. (Source: from Briggs *et al.*, 1997)

Figure 15.40 The Mississippi Delta is shaped like a 'bird's foot' and is the classic example of a river-dominated delta. The scene shown in this false-colour image covers an area of 54 km by 57 km and represents the currently active delta front of the Mississippi known as the Balize. (Source: NASA)

Tide-dominated deltas develop when the tidal prism is larger than the fluvial discharge and are commonly found along meso- and macro-tidal coastlines. In such settings, strong tidal currents flow in and out of the river mouth, also causing strong mixing between the river effluent and the receiving water. Elongated bars in the mouth of the delta often result, separated by tidal channels. A fundamental difference between tide-dominated deltas and tide-dominated estuaries is that in the former case the bottom contours at the mouth of the river bend out (indicating advance of the landform out to sea), whereas in the latter case they bend in (indicating retreat of the landform inland).

Delta development relies heavily on an active sediment supply by the river. If this supply is less than the sediment removal and dispersal by wave and tidal processes, the deltaic shoreline will erode. A natural cause for this to happen is through **delta switching**, which occurs when the active region of coastal accumulation switches from one location on the delta to another. The interval between switching varies from hundreds to thousands of years, depending on the size of the delta, and many contemporary deltas have gone through several stages of delta switching during the past few thousand years. For example, the Mississippi Delta has had four major delta lobe switches over the past 5000 years,

457

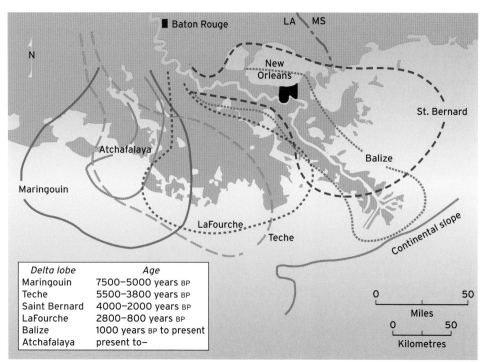

Figure 15.41 Outlines of the five delta lobes that make up the Mississippi Delta. The currently active delta lobe, the Balize, is shown in Figure 15.40. The youngest delta lobe is the Atchafalaya, which is not yet fully occupied by the river. (Source: adapted from Pilkey, 2003)

Delta lobe	Age
Maringouin	7500–5000 years BP
Teche	5500–3800 years BP
Saint Bernard	4000–2000 years BP
LaFourche	2800–800 years BP
Balize	1000 years BP to present
Atchafalaya	present to–

the last of which occurred approximately 800 years ago (Figure 15.41). Delta switching occurs due to an overextension of the presently active **distributaries**, which reduces their competency to convey sediment across the delta plain. At some stage, the existing channel (or channel network) is abandoned in favour of another more competent channel (or channel network) cut through a shorter, steeper section of the delta plain. When this happens, the former area of active delta formation becomes completely inactive with respect to fluvial processes and may undergo subsequent erosion by waves and tides.

Humans can modify the sediment supply in a river through actions on land such as building of dams, sediment abstraction and water abstraction for irrigation. This can lead to delta erosion since these interventions all reduce the sediment supply to the delta and may cause the deltaic shoreline to erode. Erosion problems in deltas are exacerbated by the fact that delta systems are generally subsiding due to the weight of the deltaic deposits on the Earth's crust and hence experience a **relative sea-level** rise. Not surprisingly, most deltaic shorelines are currently displaying large erosion rates, with the Nile Delta being the best-documented example (Stanley and Warne, 1998). Deltas are amongst the most vulnerable coastal environments to sea-level rise and extreme storm events as demonstrated by the devastation caused by hurricane Katrina in 2005.

Reflective questions

➤ What is the difference between a tide-dominated delta and a tide-dominated estuary?

➤ What are the distinguishing morphological features of fluvial-, wave- and tide-dominated deltas?

15.9 Erosive coasts

Rocky coasts are continually being cut back by the sea and are characterized by erosional features. The erosive nature of rocky coasts results in often stunning coastal scenery (Figure 15.42), but makes it difficult to deduce their evolutionary history, because the different evolutionary stages are not preserved in the stratigraphy. Along depositional coasts we can usually observe relatively quick morphological changes using measurements, maps and aerial photographs. However, in rocky coastal settings there tends to be a very slow rate of change. This slow rate of change averaged over long periods does not mean that changes cannot be dramatic and sudden. Moreover, rocky coastal features, especially when carved into resistant rocks, are often polygenetic (i.e. the product of more than one sea level) and rocky coast morphology can rarely be explained solely in terms of contemporary processes and sea level (Trenhaile, 2010).

Figure 15.42 Examples of rocky coastlines: (a) plunging limestone cliffs at the Bill of Portland, Dorset, UK; (b) boulder beach emplaced on shore platform and backed by cliffs along the high-energy west coast of Ireland; (c) cliffed coastline with embayed beaches, south Devon, UK; and (d) sloping platform near Minehead, Somerset, UK.

Sunamura (1992) categorized rocky coast morphology into three main types: sloping shore platform, sub-horizontal shore platform and plunging cliff as shown in Figure 15.43.

15.9.1 Rocky coast processes

Rocky coast erosion is accomplished by a wide range of processes, often working together (see Trenhaile, 1987, for a comprehensive overview of these processes). In terms of their function in controlling rocky coast morphology, these processes can be grouped into three main types: mass movements, rock-breakdown processes and marine rock-removal processes.

Mass movements are common along rocky coasts owing to the prevailing steep, and therefore unstable, slopes. A spec-

trum of mass movements can occur on rocky coasts, depending primarily on the properties of the rock (lithology and structure). They include rockfalls which are characteristic of hard rocks, landslides which typically occur in thick, fairly homogeneous deposits of clay, shale or marl, and flows which are mass movements that involve movement of material with a high liquid content. All types of mass movements are episodic and occur more commonly in winter than in summer due to increased rainfall and undercutting of the base of the rocky slopes by wave processes. The principal roles of mass movements are the downwearing of cliffs and the introduction of cliff material into the nearshore zone.

There are a host of physical, chemical and biological processes that weaken and loosen rock material, which then becomes available for removal by marine processes. Their relative importance depends principally on wave energy level,

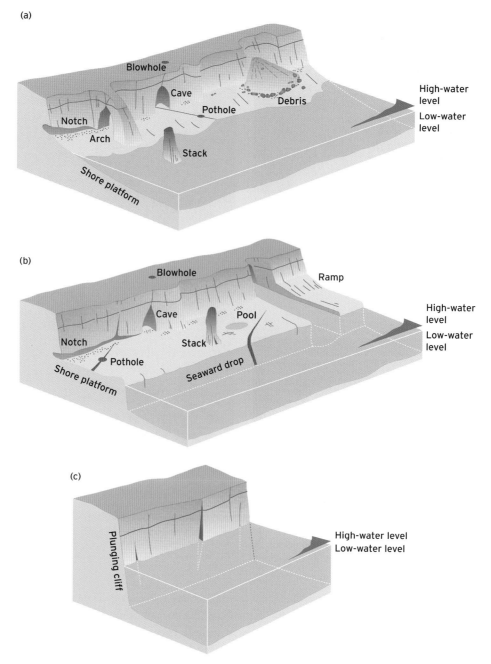

(a)

Blowhole

Cave

Pothole

Debris

Notch

Arch

Stack

Shore platform

High-water level

Low-water level

(b)

Blowhole

Ramp

Cave

Pool

Notch

Stack

Pothole

Shore platform

Seaward drop

High-water level

Low-water level

(c)

Plunging cliff

High-water level

Low-water level

Figure 15.43 Three major morphologies on rocky coasts with their characteristic erosional features: (a) sloping shore platform, (b) sub-horizontal shore platform and (c) plunging cliff. (Source: after Sunamura, 1992)

climate and rock type. Mechanical wave action (abrasion and hydraulic action) is the main erosional agent in most swell and storm-wave environments. In sheltered areas and on particularly susceptible rocks, weathering (see Chapter 7) is probably the major erosive mechanism along rocky coasts. Physical weathering breaks down rock through the formation and subsequent widening of cracks in the rock. These can occur due to frost action, alternating cycles of wetting and drying and the growth of salt crystals. Chemical weathering

of rocks is most significant in hot, wet climates. Finally, bio-erosion is the removal of rock by organisms and is most important in tropical regions due to the enormously varied marine biota and the abundance of calcareous substrates. Chapter 7 provides more detail on weathering processes and Chapter 8 deals with slope stability.

Mass movements and weathering weaken the rock and produce loose rock material that becomes available to marine processes for removal. Without the export of this material, rocky coasts stop evolving and simply develop into weathered terrestrial slopes. The larger the waves, the greater the efficiency of the cross-shore and longshore sediment transport processes to remove loose material and keep rocky coasts 'fresh'.

15.9.2 Coastal cliffs

Coastal cliffs can be defined as 'steep slopes that border ocean coasts' and occur along approximately 80% of the world's coastlines. A bewildering variety of cliff profiles are found in nature and this reflects the large number of factors involved in the development of coastal cliffs. In addition to rock type and sea-level history, the relative roles of marine or land processes are crucial in determining cliff morphology. If the ability of marine processes to remove the debris exceeds the supply of material by mass-wasting and weathering processes, sediment will not accumulate at the base of the cliff. In this case, the angle of the cliff profile depends mainly on the structure and lithology of the rock. If, on the other hand, the supply of debris exceeds the capacity of removal at the base of the cliff, the material accumulates into a talus slope (see Chapter 8). The resulting angle of the cliff profile is the **angle of repose** (see Chapter 8) of the debris. Between these two extremes lies an infinite range of slope forms, each of which depends on the relative rates of sediment supply and removal at the shoreline (Pethick, 1984).

The main factor that controls cliff erosion is the hardness of rock. A rock's resistance to erosion is mainly determined by its lithology, although factors such as wave energy and cliff height are also considered significant. As an example, typical cliff erosion rates for granite, shale and glacially deposited materials (glacial till) are <0.001 m yr^{-1}, 0.01–0.1 m yr^{-1} and 1–10 m yr^{-1}, respectively (Sunamura, 1992). It is telling to translate these recession rates to predictions of cliff retreat over the next 100 years: granitic cliffs can expect to retreat by less than 0.1 m, while glacial till cliffs are likely to be cut back by 100 to 1000 m. The number for the glacial till cliffs may appear excessive. However, the glacial till cliffs along the Holderness coast in eastern England have retreated by almost 3 km since Roman times, representing a cliff retreat rate of 1.5 m yr^{-1}. It is further noted that cliff retreat is a highly episodic process and that cliff recession rates represent long-term averages. In other words, cliffs do not erode at a uniform rate, but can go through phases of little erosion followed by a sudden phase of erosion.

15.9.3 Shore platforms

Shore platforms are erosional features that develop when erosion of a rocky coast and the subsequent removal of the debris by waves and currents leave behind an erosional surface, forming a horizontal or gently sloping rock surface in the intertidal zone (Figure 15.42d). They are often referred to as 'wave-cut platforms', but this term should not be used because it assumes that shore platforms result from wave action, which is not always true because weathering processes also significantly contribute to shore platform formation (Stephenson, 2000). The rate of platform lowering depends mainly on the hardness of the rock and averages between 0.1 and 2 mm yr^{-1} (Trenhaile, 1987). Shore platform formation is intrinsically linked with cliff erosion, and the rate of vertical lowering of the cliff base (top of the shore platform) is 2–5% of the horizontal cliff recession. The junction between the shore platform and the cliff is usually close to the high-tide level and sometimes a high-tide beach is present at this location. Shore platforms also form along cohesive coasts where they front clay cliffs. These shore platforms are generally covered by a thin layer of sand/gravel material, but become exposed when severe storms remove this veneer.

Reflective question

➤ What are the main factors controlling coastal cliff erosion and shore platform lowering?

15.10 Coastal zone management

Coastlines are the most important and intensely used of all areas settled by humans for a number of reasons, including historical settlement, trading or political linkages, climate, availability of fertile alluvial soils, proximity to fish stocks and, more recently, aesthetic and recreational reasons (Carter, 1988). From a human point of view, the coastal zone is a resource to be used and exploited, whereas from an environmental perspective, the coastal zone is an environment often adversely affected by human activities (French, 1997). The coastal zone is used for various activities, ranging from nature conservation to waste disposal, with most coasts supporting multiple activities. Interactions inevitably occur between two or more coastal uses, and management is required to plan and coordinate the different uses of the coastal zone to avoid conflicts (see also Chapter 27). In the past, coastal management was mainly concerned with single issues that could be dealt with by a single authority. This is no longer the case. The increased complexity of coastal management issues, and the varying spatial and temporal scales at which they operate, bring in many different organizations with an interest in the management of a coastline. These organizations typically include administrative authorities (councils, government agencies, environmental organizations), industry and other interest groups (residents, tourism). For effective management of the coast, an integrated approach should be adopted and the term '**integrated coastal zone management**' (ICZM) is used to indicate this approach.

Any ICZM initiative requires sustainability so that human activities should be non-destructive and the resources we exploit should be renewable. It is clear therefore that many coastal practices are not sustainable. However, Kay and Alder (1999) noted that sustainability is not a set of prescriptive actions but a 'way of thinking' about our use of the coastal zone and the resulting impacts. Generally the concept of sustainability to coastal management has resulted in a management approach with a longer-term view and more holistic perspective.

The output of ICZM consists of both legal policies and advisory initiatives. The latter are generally in the form of coastal management plans or shoreline management plans, which chart out a course for the future development of a stretch of coast and/or assist in resolving current management problems. Legally binding initiatives are a very powerful means to direct practices in the coastal zone. For example, the Dynamic Preservation Strategy adopted by the Dutch national government in 1991 (Koster and Hillen, 1995) included a legal provision that prescribed that the Dutch coastline be maintained at its 1990 position, irrespective of uncertain future developments. In other words, land losses due to coastal erosion are considered unlawful and have to be compensated for by beach nourishment. On a local level, councils can use by-laws to control activities in the coastal zone.

From a geographical point of view, the main issue associated with ICZM is to protect the coast from erosion and flooding. Both these aspects are particularly relevant at the moment, because 70% of our sandy coastlines are eroding (Bird, 1985) and sea level is rising at an increasing rate. There are four principal management options available to cope with coastal erosion and flooding due to sea-level rise:

1. *No active intervention* – this option is viable only if the coastline under question is undeveloped and nothing is at stake by giving up the land to coastal erosion.
2. *Managed realignment* – this option involves the relocation of coastal communities and industry, with a prohibition on further development. In this strategy, risks are minimized and costs of protection are avoided. However, social and economic costs associated with relocation and compensation are potentially high. The retreat option requires a strong governmental role with supportive legislation.
3. *Accommodation* – this allows continued occupancy and use of vulnerable coastal areas by adapting to, rather than protecting fully against, adverse impacts. It means learning to live with the sea-level rise and coastal flooding. Accommodation options include elevating buildings, enhancing storm and flood warning systems, and modifying drainage. The accommodation option can also involve changing activities, such as changing farming practices to suit the new environment, or simply accepting the risks of inundation and increasing insurance premiums. The accommodation option requires high levels of organization and community participation.
4. *Hold the line* – this option involves physically protecting the coast through **hard engineering** with structures such as seawalls and **groynes** or **soft engineering** through beach nourishment, for example. A summary of coastal protection measures and the problems associated with their implementation is given in Table 15.3. Protection has clear social, economic and political advantages, because assets and investments are safeguarded while economic activity can continue largely unhindered. Protection is the most expensive option to implement and maintain, and is only economically justifiable if the land to be protected is of great value.

The first three strategies are based on the premise that increased land losses and coastal flooding will be allowed to occur and that some coastal functions and values will be changed or lost. On the other hand, these strategies help to maintain the dynamic nature of the coast and allow it to adjust to rising sea levels naturally. It is beneficial to allow as many coastal regions as possible to retreat naturally, because erosion of these natural areas will liberate sediments, which may lessen the impact of sea-level rise on those areas that are not allowed to retreat naturally. The overall outcome is an increase in the resilience of the coastal system to sea-level rise. Hence, the first three options are most sustainable from a geomorphological point of view (although not necessarily from a socio-economic perspective). Certainly in developed countries there seems to be an increased push by national governments to pursue these more sustainable coastal protection strategies.

Notwithstanding the desire to maintain the dynamic nature of coasts, there will always remain a large role for coastal protection measures for the simple reason that many coastal areas are too valuable to be given up. When properly designed and constructed, hard engineering structures do

Table 15.3 A selection of coastal engineering works

Management issue	Engineered solution(s)	Types	Description	Problems	Illustration
Cliff erosion	Sea walls	Vertical wall	A wall constructed out of rock blocks, or bulkheads of wood or steel, or simply semi-vertical mounds of rubble in front of a cliff	Rock walls are highly reflective, bulkheads less so. Loose rubble, however, absorbs wave energy	
		Curved wall	A concrete constructed concave wall	Quite reflective, but the concave structure introduces a dissipative element	
		Stepped	A rectilinear stepped hard structure, as gently sloping as possible, often with a curved wave-return wall at the top	The scarps of the steps are reflective, but overall the structure is quite dissipative	
		Revetment	A sloping rectilinear armoured structure constructed with less reflective material, such as interlocking blocks (tetrapods), rock-filled gabions and asphalt	The slope and loose material ensure maximum dissipation of wave energy	
Coastal inundation	Sea walls	Earth banks	A free-standing bank of earth and loose material, often at the landward edge of coastal wetlands	May be susceptible to erosion, and overtopped during extreme high-water events	
	Tidal barriers		Barriers built across estuaries with sluice gates that may be closed when threatened by storm surge	Extremely costly, and relies on reliable storm surge warning system (e.g. Thames Barrier)	
Beach stabilization	Groynes		Shore-normal walls of mainly wood, built across beaches to trap drifting sediment	Starve downdrift beaches of sediment	
	Beach nourishment		Adding sediment to a beach to maintain beach levels and dimensions	Sediment is often rapidly removed through erosion and needs regular replenishing; often sourced by dredging coastal waters	
Offshore protection	Breakwaters		Structures situated offshore that intercept waves before they reach the shore. Constructed with concrete and/or rubble	Very costly and often suffer damage during storms	
Tidal inlet management	Jetties		Walls built to line the banks of tidal inlets or river outlets in order to stabilize the waterway for navigation	The jetties protrude into the sea and promote sediment deposition on the updrift side, but also sediment stavation and erosion on the downdrift side	

(Source: after Haslett, 2000)

(a)

(b)

(c)

(d)

Figure 15.44 Examples of hard coastal engineering structures: (a) vertical seawall with curved top; (b) short wooden groynes as usually deployed on gravel beaches. (c) Jetties are designed to prevent the silting up of tidal inlets and harbour entrance channels by blocking the littoral drift. In the photo, the littoral drift direction is from right to left. (d) Clay embankments or sea dikes installed as part of land reclamation efforts near the spring high-tide level seem inoffensive enough. However, the land they are protecting from flooding is at the same time deprived of the influx of sediment that would otherwise enable it to keep up with rising sea level. In this photo, the land to the right of the embankment, which is flooded during spring high tide, is at least 0.5 m higher than the land to the left. (Source for (c): Photo courtesy of Aart Kroon)

serve an important purpose: **storm surge barriers**, such as constructed across the Thames and in the south-west of the Netherlands, have prevented serious flooding on several occasions; **sea walls** and **breakwaters** protect coastal development from damage during extreme wave events; and groynes are successful to some extent in trapping sediments and maintaining a beach (Figure 15.44). The 'side effects' of hard engineering are considerable, however, and it is well established that, following the construction of hard coastal structures, erosion problems on the downdrift unprotected coastline are often exacerbated (or even created). Soft engineering practices, in the form of **beach nourishment** or beach recharge, largely circumvent the main problem associated with hard engineering. The artificial placement of a large amount of sediment, either on the underwater slope or on the beach itself, protects not only the recharged coast, but also the neighbouring coastline, because sediment transport processes will redistribute

the nourished sediment. This redistribution also represents a major downside of beach nourishment and treatment will have to be repeated at regular intervals. To reduce sediment losses following beach nourishment, groynes may be placed at the boundaries of the nourished area. On the whole, beach nourishment is more aligned with sustainable coastal management and is now very widely used (Bird, 1996).

Reflective questions

➤ What are the main factors involved in deciding on the best coastal management strategy for a particular stretch of coast?

➤ What are the problems associated with using hard engineering structures for coastal protection?

15.11 Summary

Coastal environments are arguably the most important and intensely used of all areas settled by humans. At the same time, they are currently at great risk from coastal erosion and flooding due to climate-induced sea-level rise. Over the last few decades, global sea level has been rising at a rate of 3 mm yr^{-1} and over the next 100 years this rate may increase to 6 mm yr^{-1}. Coastal morphologies are controlled and driven by a set of environmental boundary conditions, including sea-level change, geology, sediment supply and external forcing (wind, waves and tides), and changes in these boundary conditions will modify coastal processes and morphology. The current paradigm of coastal research is the so-called 'morphodynamic approach', which considers coastal systems and their dynamics to be mutually linked by negative and positive feedback. The presence of positive feedback between coastal form and process makes it difficult to predict coastal evolution, especially over longer timescales.

At the most basic level, coastal environments can be divided into clastic (comprising mud, sand and gravel) and rocky coasts. Clastic coastal environments are depositional and their morphology responds to the relative dominance of wave, tidal and fluvial factors. Barriers are the basic depositional elements of wave-dominated coasts and comprise the underwater shoreface and the above-water beach and dunes. They can be swash-aligned or drift-aligned, and respond strongly to changes in the sediment supply and sea level. Estuaries are the dominant tide-dominated coastal landform and represent zones of mixing between fluvial and marine (wave and tidal) processes. On a geological timescale, estuaries are rather short-lived features because they are characterized by relatively rapid infilling by fluvial and marine sediments. Deltas are accumulations of sediment deposited where rivers enter into the sea. Waves and tides significantly affect river outflow processes and play a significant role in determining the delta morphology. Rocky coasts are eroding coasts and the main factor controlling the erosion rate is the strength of the rocks relative to that of the eroding waves. The two dominant rocky coast landforms are (co-evolving) cliffs and shore platforms.

Effective management of the coast requires an integrated approach that considers all coastal users and stakeholders, at the same taking a long-term view to come up with sustainable solutions and policies. The term 'integrated coastal zone management' (ICZM) is used to indicate this approach. An important remit of ICZM is to address problems associated with coastal erosion and flooding, and there are two fundamentally different types of management approaches available. On the one hand, there are strategies that allow some loss of coastal land, functions and values, but help maintain the dynamic nature of the coast. On the other hand, there are strategies that protect the coastline using hard and soft engineering techniques. Allowing coastal regions to retreat naturally increases the resilience of the coast to sea-level rise and is preferable. However, in many instances the coastline under threat is simply too valuable to be sacrificed and coastal protection must be sought.

Further reading
Books

Davidson-Arnott, R.G.D. (2010) *Introduction to coastal processes and geomorphology.* Cambridge University Press, Cambridge.
Very recent coastal text with comprehensive treatment of all coastal environments and processes.

Dyer, K.R. (1998) *Estuaries: A physical introduction.* John Wiley & Sons, Chichester.
This text provides an in-depth, but accessible, introduction to tidal dynamics, stratification and mixing in estuaries.

French, P.W. (2001) *Coastal defences – Processes, problems and solutions.* Routledge, London.
A reasonably up-to-date and comprehensive overview of the different types of coastal defences.

Masselink, G., Hughes, M.G. and Knight, J. (2011) *Introduction to coastal processes and geomorphology,* 2nd edition. Hodder Education, London.
This text is co-written by the author of this chapter and provides an excellent continuation for further studies into coastal processes and morphology.

Pilkey, O.H. (2003) *A celebration of the world's barrier islands.* Columbia University Press, New York.
This book may be difficult to get hold of, but is a true gem for anyone interested in barrier islands. This is the only coastal coffee table book available, owing to the inclusion of many original batiks by the American artist Mary Edna Fraser, written at an advanced academic level.

Trenhaile, A.S. (1987) *The geomorphology of rock coasts.* Oxford University Press, Oxford.
A must for anyone interested in learning more about rocky coasts.

Viles, H. and Spencer, T. (1995) *Coastal problems.* Arnold, London.
This is an excellent text dealing specifically with environmental problems in the coastal environment.

Woodroffe, C.D. (2003) *Coasts, form, process and evolution.* Cambridge University Press, Cambridge.
This is an advanced-level text on coastal dynamics largely written from a geological perspective.

Further reading
Papers

Harris, P.T. and Heap, A.D. (2003) Environmental management of clastic coastal depositional environments: inferences from an Australian geomorphic database. *Ocean & Coastal Management,* 46, 457–478.
Very readable paper illustrating the application of classification models to coastal zone management.

Nicholls, R.J. and Cazenave, A. (2010) Sea-level rise and its impact on coastal zones. *Science,* 328, 1517.
Very readable paper with the latest on current sea-level rise and its impact on society.

CHAPTER 16

Glaciers and ice sheets

Tavi Murray

Department of Geography, Swansea University

Learning objectives

After reading this chapter you should be able to:

➤ describe different types of ice masses and their characteristics

➤ explain the concepts of glacier mass balance

➤ explain how glaciers move downslope, and understand glacier thermal regimes and their effect on glacier dynamics and glacier water systems

➤ describe how water moves through a glacier and understand why the water system is important in glacier dynamics

➤ explain how glaciers and ice sheets are reacting to climate change

➤ explain how glaciers erode their beds and modify sediments, and understand the differences in the sediment deposited via a range of mechanisms

➤ describe typical landscapes of glacial erosion and deposition

16.1 Introduction

Over the past 2.6 million years there has been repeated expansion and contraction of glaciers and continental-scale ice sheets in the middle latitudes. Today glaciers cover about 10% of the Earth's land surface. These ice masses store large volumes of water on land, thereby reducing the amount in the oceans. During phases of maximum ice advance, global sea levels fell by 130 m. The amount of water stored in glacial ice today is equivalent to approximately 75 m of global sea-level rise. Most of the water is stored in the great ice sheets of Antarctica and Greenland. For example, the West Antarctic and Greenland ice sheets individually contain the equivalent of about 5 and 7 m of sea-level rise. During the twentieth century global sea level rose by 12–22 cm, and the rate of sea-level rise accelerated: it is currently approximately 0.31 cm per year (IPCC, 2007b). Much of the sea-level rise is due to simple expansion of the water contained in the seas and oceans as it has warmed, but about 40% is attributed to the melting of glaciers and ice sheets.

Glaciers can be used as valuable resources. For example, glacial runoff is used to generate electricity in a number of countries such as Norway, Canada and Switzerland and in many regions it is an important water source for agriculture. In the Karakoram Himalayas, glacial meltwater makes a major contribution to the Indus and Yarkand Rivers and hence to the livelihood of 130 million people (Hewitt, 1998). An advantage of glacially fed water systems is that the water supply actually increases during hot summers, which are generally times of increased need. In addition skiing is a popular pastime, and in many areas this occurs on the surface of glaciers. The engineering of ski lifts and safe operation of these ski areas requires knowledge of glacier dynamics. Increasingly oil, gas and mineral extraction is taking place in glaciated environments. Furthermore icebergs are calved from glaciers, meaning that world shipping is affected by changes in ice dynamics at major outlet glaciers.

Changes in glacial extent have also affected the Earth's climate, for example by altering the Earth's albedo, and glaciers have made major modifications to much of the Earth's landscape by eroding, transporting and depositing sediment. These processes have helped shape the landscape over much of the middle latitudes in places where today there are no glaciers or ice sheets. Where they do occur, these landscape change processes operate at a variety of rates and over a range of scales. This chapter discusses glacial systems, glacial processes and the effects of glaciation on the landscape.

16.2 Glaciology

The aim of this section is to develop an understanding of glaciers and ice sheets, the controls on glacier extent and the effect of climate change on them, the evacuation of water through glacier hydrological systems and the processes and rates of ice flow.

16.2.1 Types of ice mass

There are a wide variety of ice masses including ice sheets, ice caps and valley glaciers. The largest ice masses are known as

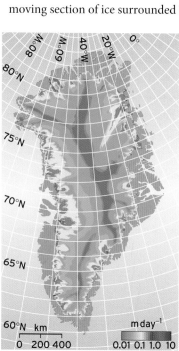

Figure 16.1 Schematic diagram of a marine-based ice sheet such as the West Antarctic ice sheet. Most ice from the interior is fed through the fast-flowing ice streams to the ice shelves where the mass is lost largely through iceberg calving and some basal melt beneath the ice shelves. (Source: after Alley, 1991)

ice sheets, which cover continental-size regions (Figures 16.1 and 16.2). Typically an ice sheet is 1–3 km thick. The Antarctic ice sheets cover approximately 11.97×10^6 km^2 and the Greenland ice sheet covers around 1.74×10^6 km^2. The bases of the Greenland and East Antarctic ice sheets are above sea level and these are known as terrestrial ice sheets, whereas a marine-based ice sheet has its base below sea level such as the West Antarctic ice sheet (Figure 16.1).

Ice sheets flow relatively slowly and most of the mass reaches the ice sheet margins through smaller, fast-flowing **ice streams** or outlet glaciers as shown in Figure 16.1. An ice stream is a fast-flowing 'river' of ice within more slowly moving ice sheet walls whereas an outlet glacier is a fast-moving section of ice surrounded by rock. An example of

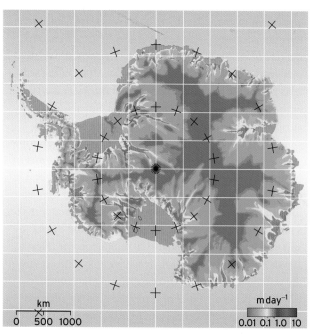

Figure 16.2 Antarctic and Greenland ice sheet balance velocities. The balance velocity of an ice sheet is the velocity required to evacuate the snow accumulated and maintain the current geometry. Calculating the balance velocity requires knowledge of the surface accumulation and surface slope as well as the bed geometry. (Source: courtesy of Jon Bamber/Adrian Luckman)

an ice stream is Whillans ice stream, West Antarctica, which flows at approximately 850 m yr^{-1}. In Antarctica, the ice streams feed **ice shelves** which are formations of floating ice as shown in Figure 16.1. Ice is lost from the ice shelves by melting into the ocean below or when **iceberg calving** occurs (Figure 16.1). This is when large chunks of ice are released from the ice shelf and float into the ocean.

Ice caps are smaller than ice sheets, but still cover considerable areas. For example, Vatnajökull, situated in south-east Iceland, covers some 8100 km^2. Many mountain regions are glaciated with valley glaciers, which are much smaller in area. Different regions and different types of ice mass have characteristic flow rates, thermal and hydrological systems.

16.2.2 Where do glaciers occur?

Glaciers are formed wherever the snow that falls in winter does not melt over the subsequent summer. This suggests that there are two fundamental requirements for the formation of glacier ice: precipitation in the form of snow (the percentage of precipitation in an area falling as snow is known as the **nivometric coefficient**) and low temperatures. Neither is sufficient on its own. For example, 23 000 years ago the majority of Canada and the northern United States was glaciated (see Chapter 22), but Alaska remained largely free of glacier ice because the precipitation in this area was very low (a cold arid region). Further factors that control the distribution of snow and ice include latitude, altitude, relief, aspect and continentality or the distance from the ocean. Glaciers differ greatly in locations where different factors are dominant. For example, the ice masses of Antarctica are characterized by very low surface temperatures and low precipitation. These are examples of polar glaciers, which occur at high latitudes. The glaciers of the European Alps occur in a relatively warm climate characterized by very high precipitation rates. These are examples of alpine glaciers, which exist because of their high altitude. Table 16.1 gives a summary of the current glacial coverage.

16.2.3 Glacier mass balance

Examination of glaciers in late summer generally shows that the **snowline** (where fresh snow still lies) has retreated up the glacier and that the lower region of the glacier is bare ice. An example can be seen in Figure 16.3. This lower region is known as the **ablation zone** and despite its becoming snow covered in winter it experiences net mass loss during the year (see also Figure 16.8 below). Ice may be

Table 16.1 Present-day glacial coverage. There are glaciers on every continent on Earth. Total world coverage is approximately 14 953 945 km^2

| | Area | |
	Km^2	%
South polar region	12 588 000	84.2
North polar region	2 137 274	14.2
North America	76 880	0.5
South America	26 500	0.2
Europe	7 410	< 0.1
Asia	116 854	0.8
Africa	12	< 0.1
Pacific/Australasia	1015	< 0.1

(Source: after Sugden and John, 1976, updated from Williams and Ferrigno, 1999)

Figure 16.3 Photograph of a typical alpine glacier at the end of summer. The snowline has retreated up the glacier leaving exposed ice close to the glacier margin. The lower part of the glacier, the ablation zone, is melting, and has lost mass over the year. Downslope of the ablation zone the previous extent of the glacier can clearly be seen in the sediment and erosion patterns. The upper part of the glacier, the accumulation zone, has retained snow and has gained mass over the year.

lost by melting at the glacier surface and by iceberg calving. Most ice loss in Antarctica occurs through iceberg calving. The region of the glacier that experiences net mass gain through the year is known as the **accumulation zone** (Figures 16.3 and 16.4). In this zone some or all of the snow that falls in winter survives the summer season to form **firn**, which is wetted and compacted snow more than a year old.

Figure 16.4 Ice cliff (~55 m high) showing annual layers of snow and ice in the accumulation zone of the Quelccaya ice cap in Peru. The layers are approximately 0.75 m thick. Each layer is demarked by a dirty layer of snow from the dry season. Typically the winter layers of ice are light in colour while summer layers are darker because the partial melting and lower summer accumulation rate produces higher concentrations of impurities. The age of each layer can be calculated by counting from the glacier surface. By identifying such layers in the accumulation zone the mass input to the glacier can be estimated, and such layers are used to date ice cores. (Source: photo courtesy of Lonnie G. Thompson, The Ohio State University)

Table 16.2 Typical densities of snow and ice. Snow is largely unaltered since deposition. Firn is wetted snow that has survived at least one summer. Firn becomes ice when there are no longer any interconnecting air passages

Material	Density (kg m^{-3})
New snow	50-70
Damp or settled snow	100-300
Depth hoar	100-300
Wind-packed snow	350-400
Firn	400-830
Very wet snow or firn	700-800
Glacier ice	830-910
Water	1000

(Source: from Cuffey and Paterson, 2010)

Slowly this firn becomes denser and forms ice when all the connecting air passages are closed off. Often annual layers of ice can be detected as a result of the differences in ice formation during the winter snow season and summer dry season as shown in Figure 16.4.

The division between the accumulation and ablation zones occurs at the **equilibrium line** (see Figure ·16.8 below). The mass of water that is added to or lost from a glacier over a year is known as the **mass balance**. A glacier is said to be in positive mass balance if it is gaining in mass over a year and in negative mass balance if it is losing mass. For a typical alpine glacier, mass balance is primarily a function of winter precipitation and temperature, and summer solar radiation and temperature.

When calculating mass balance it is better to use the mass of water gained or lost rather than the volume of change because of the extreme differences between the densities of snow, firn and ice (Table 16.2). Despite this, several techniques of estimating the mass balance actually measure the volume balance, because the mass change of a glacier is very difficult to measure (see Box 16.1). In many regions, such as Svalbard in the Norwegian High Arctic, small mountain glaciers are losing mass to the ocean. For two glaciers at Svalbard an acceleration in mass loss has been demonstrated (Kohler *et al.*, 2007). At Midtre Lovénbreen, for example (see Box 26.3), the thinning rate for 1936–1962 was 0.15 m yr^{-1}, while for 2003–2005 it was 0.69 m yr^{-1}. The values are given as depths because the volume loss is averaged over the glacier's area. In other words, the glacier is losing 0.69 m depth of ice per year averaged over its whole area.

The accumulation area of a glacier can be broken down into a number of zones although not all of these zones are present on all glaciers. The **dry-snow zone** is an area where there is no surface melt, even in summer. Most alpine glaciers do not have a dry-snow zone as some melting occurs, but much of Antarctica and the central region of the Greenland ice sheet are characterized by no melting. The region where the entire snowpack is saturated at the end of the summer

WEIGHING THE ICE SHEETS FROM SPACE

Measuring the mass balance of the major ice sheets is a difficult task, mainly because of their large size (Figure 16.5). The Gravity Recovery and Climate Experiment (GRACE), funded by NASA and the German Aerospace Center, is an exciting method for monitoring ice sheet mass balance from space. GRACE consists of two satellites orbiting the Earth (Figure 16.5a), which are separated by a distance of around 220 km. The distance between them varies slightly as the satellites pass over anomalies in the Earth's gravity field (Figure 16.5a). The gravity field allows calculation of changes in ice sheet mass (Figure 16.5b) once correction for glacial isostatic rebound and other factors is made.

(a)

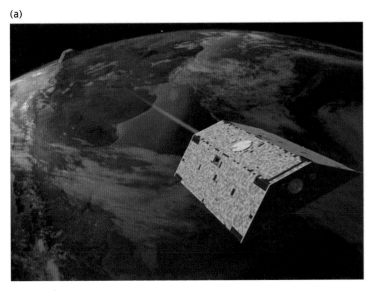
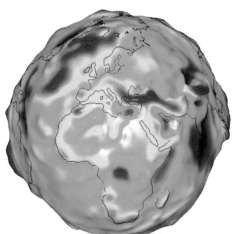

(b)

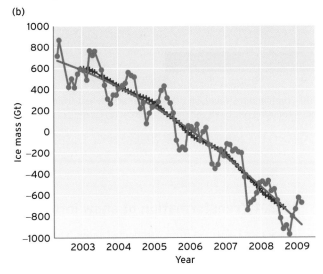

Figure 16.5 (a) Schematic diagram of the GRACE (Gravity Recovery and Climate Experiment) satellites. The two satellites orbit the Earth and variations in the gravity field cause small changes in their distance apart. This allows the production of a gravity field anomaly map shown on the globe. (b) Time series of ice mass changes for the Greenland ice sheet estimated from GRACE monthly mass solutions for the period from April 2002 to February 2009. (Source: (a) University of Texas Center for Space Research/NASA; (b) Velicogna, 2009)

BOX 16.1 ➤

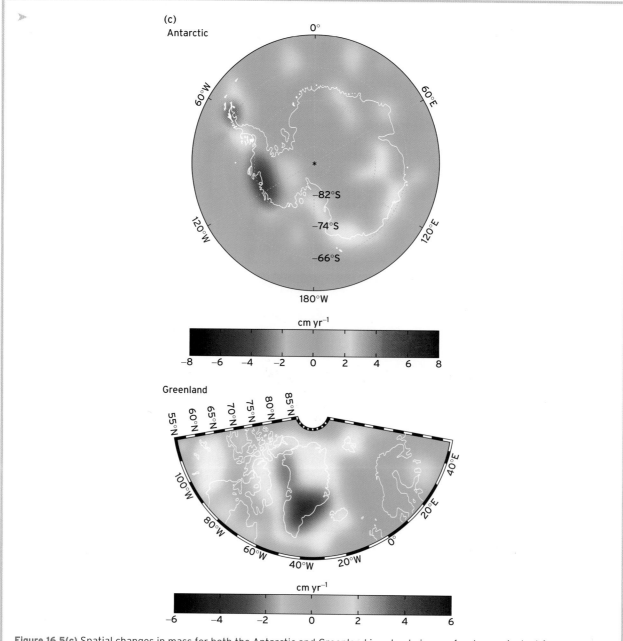

Figure 16.5(c) Spatial changes in mass for both the Antarctic and Greenland ice sheets in cm of water equivalent from GRACE during the period April 2002 to January 2009. Post-glacial rebound has been removed. (Source: (c) Cazenave and Chen, 2010)

BOX 16.1

is known as the **wet-snow zone**. Water that refreezes at the base of the snowpack is known as **superimposed ice**. Because this ice has formed by refreezing rather than compression, it has slightly different chemical and physical properties than glacier ice. Superimposed ice formation is an important component of glacier mass balance in High Arctic glaciers such as those in Svalbard and the Canadian Arctic.

16.2.4 Transformation of snow into ice

Glacier ice is formed from compacted snow. The processes that result in this transformation depend on whether there is water present in the snowpack. Changes are slow in the absence of water and result from packing changes and settling, changes in the ice crystal size and shape resulting from sublimation, and deformation of the crystals. The presence of water results in much faster changes, because melting, percolation and refreezing occur in the snowpack. The transformation of snow to ice may take only a few years where the snowpack becomes saturated (e.g. three to five years at the Upper Seward Glacier, Yukon Territory), but more than 100 years in Antarctica or Greenland. See Cuffey and Paterson (2010) for further details of the transformation processes.

16.2.5 Glacier thermal regime

There is an important distinction between cold and warm ice. The melting point of ice reduces with increasing pressure. At atmospheric pressure the melting point is 0°C, whereas beneath 1 km of ice the melting point is about −0.7°C. Therefore, the greater the thickness of an ice mass, the more likely it is to be at the pressure melting point. Warm ice is at the pressure melting point and contains water; cold ice is at temperatures below the pressure melting point and does not. The 'thermal regime' of a glacier exerts a fundamental control over the water system and over the range of processes that can operate at the bed of the glacier. Furthermore, the properties of warm and cold ice vary strongly. For example, the deformation rate of ice increases by about five times between −25 and −10°C. Many other properties of the ice are temperature or water-content dependent, such as the speed and attenuation of acoustic or electrical waves through the ice. This allows us to detect the thermal structure within glaciers using radar. An example of the thermal structure of a glacier is shown in Figure 16.6.

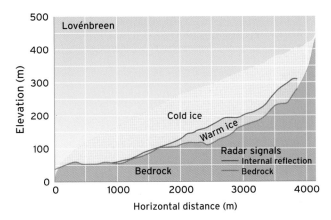

Figure 16.6 The internal structure of Midtre Lovénbreen, a glacier in Svalbard, based on radar. Warm ice near the glacier bed contains water and scatters the radar producing a reflection of the radar waves (indicated as an internal reflection on the figure). Note that in this glacier the downslope end is made of cold ice throughout. Because this glacier has both warm and cold ice, it is called a polythermal glacier. (Source: after Bjørnsson et al., 1996)

Generally three types of thermal regime are recognized. The first is a **temperate glacier**, which consists of warm ice throughout except for a layer, between 10 and 15 m in thickness, that is seasonally warmed and cooled by temperature variations at the surface (this layer is analogous to the active layer in permafrost, see Chapter 17). Typical examples of temperate glaciers are those in the European Alps, the south island of New Zealand and the Canadian Rockies. The second thermal regime is a **cold glacier**, which consists entirely of cold ice. A typical example might be Meserve Glacier in Antarctica. The final type is a **polythermal glacier**, which comprises both warm and cold ice; an example is given in Figure 16.6.

It is the temperature of the bed that is critical. If the bed of a glacier is cold then there will not be a basal water system and the occurrence of sliding and deformation of basal sediments will be greatly reduced. **Geothermal heat** (from inside the Earth) and heat generated by basal friction may warm the bed of a glacier, producing an active basal water system even in regions where the surface temperature is very cold such as beneath the Whillans ice stream, West Antarctica.

16.2.6 Glacier water systems

Water is produced at the surface of many glaciers in summer from the melt of both snow and ice. Small amounts of water are also produced by geothermal heating or as the glacier slides over its bed or deforms sediments. Additional water may be input at the surface directly due to rainfall. For many alpine glaciers the most important water source is from surface melt. Some of these sources are seasonal (such as surface melt), whereas other sources will persist throughout the year (such as basal melting).

Unless it refreezes, **supraglacial** (surface) meltwater will percolate downwards through any snow or firn and will flow at the surface of the glacier ice. The rate of water flow through snow or firn is fairly slow. Once the ice surface is reached, water will flow downslope along the ice surface and will emerge at the snowline as a series of supraglacial channels (Figure 16.7a). These channels (Figure 16.7a) transport water efficiently owing to their smooth ice walls. The water may continue to be routed to the glacier front across the glacier surface, which is common for High Arctic glaciers, or may enter the glacier to flow **englacially** (within the glacier) (Figure 16.8).

Although a small amount of water may percolate directly through the ice in small veins along the boundaries of individual ice crystals, the rate of this percolation is negligible. Most water flowing through a glacier does so through a system of englacial channels as shown in Figure 16.8. Water typically enters the glacier through **moulins** (Figure 16.7b) that tend to form where crevasses (fracture cracks in the ice)

(a)

(b)

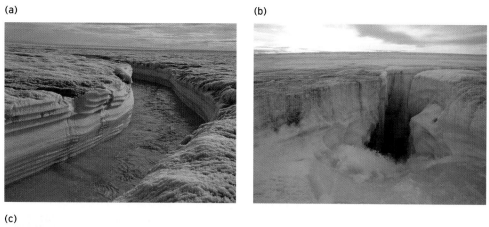

(c)

Figure 16.7 (a) Deeply incised supraglacial stream at the surface of the Greenland ice sheet. (b) A moulin on the Greenland ice sheet. Supraglacial water flows into the body of the glacier down moulins that lead to englacial channels. Eventually the water may reach the bed. (c) The water outlet at many alpine glaciers is a single or a small number of channels. This suggests that the lower region of this type of glacier is often drained by a channelized drainage system. (Source: photos (a and b) courtesy of Adam Booth; photo (c) courtesy of Mike Crabtree)

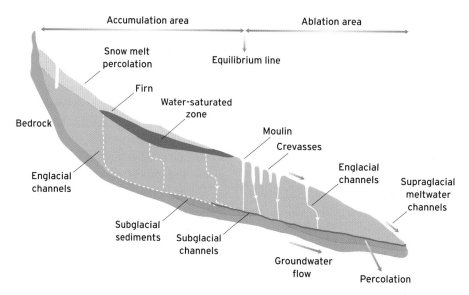

Figure 16.8 Schematic showing a typical temperate glacier hydrological system.

cut the drainage path of supraglacial water. Moulins usually descend rapidly to depth as semi-vertical shafts linked by short horizontal channels.

Englacial channels may intersect the glacier bed and flow along the floor of the glacier. Direct studies beneath glaciers are difficult and so many of our ideas about the configuration of **subglacial** (beneath the glacier) water systems are derived from largely theoretical considerations. Four main morphologies have been proposed, namely: (i) flow in a thin sheet; (ii) flow through a network of channels; (iii) flow through linked cavities; or (iv) flow through a braided system of 'canals' at the glacier bed (Figure 16.9).

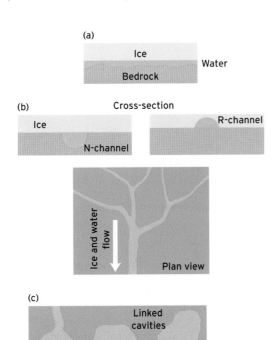

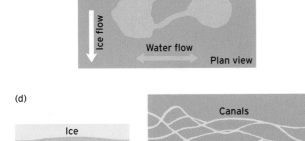

Figure 16.9 Conceptual forms of the basal water system. Water may flow at the bed of a glacier (a) through a sheet of water; (b) through a channel system eroded either down into the bed (Nye channels) or up into the ice (Röthlisberger channels); (c) through a tortuous series of linked cavities; or (d) through canals, which are thought to be braided in their plan form.

Study of subglacial drainage is an important branch of glaciology, because glacial meltwater is used to generate electricity and for agriculture, and because the water system has a profound effect on glacier dynamics.

16.2.6.1 Sheet flow

One evacuation route for basal water is thought to be through a thin film or sheet, just a micrometre (thousandth of a millimetre) to a millimetre in thickness (Figure 16.9a). The thickness of the sheet is thought to vary, being thinnest over bumps in the bed because of the higher pressure at their upstream side. Thicker films cannot develop because enhanced melting would result in the formation of subglacial channels. Flow through sheet flow is typified by very high water pressures, and research on bedrock at Blackfoot Glacier, USA, showed that sheet flow occurred over about 80% of the glacier bed (Walder and Hallet, 1979). The role of sheet flow is thought to be small for most alpine glaciers, perhaps transporting water into a channelized system. Sheet flow is thought to be dominant only beneath ice masses where the majority of basal water is derived from subglacial melting due to geothermal heating, sliding or bed deformation rather than surface meltwater.

16.2.6.2 Channelized flow

Water that descends from the glacier surface in englacial conduits will arrive at the bed at discrete locations and is unlikely to become dispersed into a thin sheet. This water will tend to flow at the bed in conduits or channels (Figure 16.9b). Evidence for channelized systems comes from the rapid rate of transfer of dye from moulins to the outlet stream at some glaciers and from the observation that many alpine glaciers have one or a few major outlet channels emerging at their margins (Figure 16.7c). Glacier outburst floods from subglacial sources (see Box 16.2) occur through efficient channelized water systems.

Two types of channels are thought to exist beneath glaciers, namely Nye (N) channels and Röthlisberger (R) channels (Figure 16.9b). N-channels are incised into the bedrock, and R-channels are melted upwards into the ice. N-channels are fixed in space and can be preserved on exposed bedrock surfaces. However, they may be destroyed by subglacial erosion due to ice sliding and may be slow to form because water erodes rock inefficiently. When they are seen they tend to form short channels approximately parallel to ice flow, rather than exposing entire water systems.

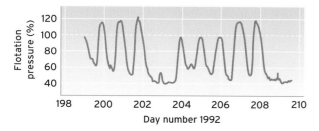

Figure 16.10 Water pressure variations measured by a basal water sensor during summer at Trapridge Glacier. Figure shows the percentage of the flotation pressure (which is the pressure at which the ice will float on its own water). Day 198 is 17 July. Note the diurnal pressure peaks corresponding to high-water input during the late afternoon.

R-channels are melted upwards into the glacier ice, which can be much more rapidly thermally eroded than the bedrock. R-channels form because the ice melts owing to frictional heat generated by water flow. These channels can close up because of **ice creep**, especially in winter. An important attribute of R-channels is that an increase in water flux decreases the water pressure in steady-state conditions. This inverse relationship between pressure and flux suggests that large R-channels will 'capture' small R-channels. Thus, water in R-channels will be concentrated in a small number of large channels and hence there is a tendency to form a tree-shaped channel network (Figure 16.9b).

Variations in surface melt drive variations in basal water pressure on both diurnal and longer timescales (Figure 16.10). Diurnally, water pressures are typically highest in the afternoon and lowest in the early morning as a result of melt cycles. In particular, abundant surface melt is routed to the glacier bed during spring, resulting in high water pressures until the water system evolves sufficiently to evacuate the water. Conversely, the beginning of winter is characterized by low basal water pressures. These characteristics result in an increase of sliding rate in spring and early summer at most alpine glaciers and lowest sliding rates during the winter.

16.2.6.3 Linked cavities

An alternative model of water flow across a hard bed is through linked cavities (Figure 16.9c), formed downstream of bedrock bumps as the glacier slides over its bed. These consist of a tortuous system of basal cavities, typically less than 1 m high and from less than 1 m to 10 m in length, linked by narrow connections less than 0.1 m in height. While the cavities have a large volume, water flow through the system is at high pressure and is very slow, so thermal erosion is inefficient. This means that the water system cannot increase its discharge rapidly in response to increased input.

16.2.6.4 Canals

The models of a glacier hydrological system described so far have assumed that there is hard and impermeable bedrock beneath the glacier. However, as most glaciers retreat they expose extensive glacial sediments, suggesting that many glaciers may be underlain by soft and potentially deformable beds (e.g. Figure 16.11). The wide coverage of glacial till in the middle latitudes suggests that the beds of many major ice sheets in the past were soft. Furthermore, soft beds have been identified beneath a number of contemporary glaciers, including Breiðamerkerjökull, south-east Iceland, Storglaciären, Sweden, and Trapridge Glacier, Yukon Territory.

Sediments are not just deformable; many also have greater permeability than rocks. This means that water can

Figure 16.11 Extensive subglacially eroded sediments at the margin of Trapridge Glacier, Yukon Territory, Canada. These sediments used to lie beneath the ice. The presence of such sediments is thought potentially to affect both the glacier dynamics and the morphology of the water system.

drain through the bed. However, for most ice masses the sediment is too fine to drain the volume of water produced, in which case some form of additional drainage system must develop. Two drainage systems are postulated to form over soft sediments. The first is essentially the same as the R-channel system that forms over a hard bed. The second system is analogous to N-channels over bedrock; however, in this case the channels are eroded downwards into the basal sediments (Figure 16.9d). These channels are wide and shallow and are known as 'canals'. Sediment deformation will tend to close canals, which must be balanced by erosion of the bed if channels are to remain open. Unlike in R-channels, water within a large canal is at a higher pressure than in a small canal. There is therefore no tendency for flow to concentrate in a few large channels. As a result a canal

drainage system is thought to consist of shallow, wide channels distributed more or less evenly across the bed and connected in a braided pattern. Canals are thought to be most likely to form where low-slope glaciers overlie soft sediment.

16.2.7 Glacier dynamics

In order to transfer ice formed from snow that accumulates in the upper reaches of all glaciers to the ablation zone it must flow downslope. To maintain constant glacier geometry, the mass transferred down the glacier should equal the mass lost by melting or calving. The **balance velocity** is the velocity required to maintain the ice mass in equilibrium (i.e. without change to its geometry; Figure 16.2). Those glaciers where the actual and balance velocities are similar are said to be in

HAZARDS

GLACIER LAKE OUTBURST FLOODS

Lakes are common at the margins of retreating temperate glaciers. Lakes can be dammed by glacier ice, by glacial moraines, or the water can be created underneath the glacier ice by a volcanic eruption or geothermal heating. These lakes tend to drain rapidly and cause a glacier lake outburst flood. A typical glacial lake outburst flood has a total discharge of water and debris up to 50 million m³ and a floodwave that may be up to 10 m high. These floods may cause widespread damage and destruction downstream, in some cases for hundreds of kilometres.

Lakes dammed by glacier ice can form when a glacier surges (see Section 16.2.7.5) or advances. One such flood occurred when Hubbard Glacier advanced and blocked Russell Fjord in Alaska in 2002. This formed Russell Lake, which rose

24 m over the summer before the moraine dam finally failed in August 2002, causing the second largest glacial lake outburst flood in historical times (Figure 16.12).

Many rapidly retreating mountain glaciers are forming lakes behind moraine dams, threatening villages downstream. An example of a glacier lake outburst flood occurred in Bhutan in the Himalayas where there are more than 2000 glacial lakes, many of which are increasing in volume as glaciers melt. In 1994, the Lugge Lake in northern Bhutan drained, releasing 18 million m³ of water. The floods killed 17 villagers and affected 91 households downstream.

Glacial outburst floods from a subglacial lake occur regularly from beneath Vatnajökull in south-east Iceland. Vatnajökull is a temperate ice cap of about 8100 km², and the volcanic fissure system of the Mid-Atlantic Ridge (see Chapters 2 and 3) lies beneath the western

portion of the ice cap where it causes substantial subglacial melt. One of the largest floods from Vatnajökull occurred in 1996, after an earthquake of magnitude 5 on the Richter scale. Subsidence bowls 100 m deep and cracks in the glacier surface showed that extensive melting was occurring at the base of the glacier along a fissure 5-6 km long and an eruption cloud emerged a few days after that extending to a height of 3 km. Meltwater from the eruption flowed into the 10 km diameter Grímsvotn subglacial lake at a rate of 5000 m³ s⁻¹.

The water stored in Grímsvotn produced an outburst flood, with a 4 m high wave crossing the floodplain. During the course of the flood two bridges were destroyed or badly damaged as were phone and power lines. Large chunks of ice were broken from the glacier margin and carried across the floodplain. The water peak flow was estimated to be 45 000 m³ s⁻¹ and the estimated damage was $15 million to roads, bridges and other infrastructure.

BOX 16.2 ➤

Figure 16.12 Outburst flood from Russell Lake, formed by the advance of the Hubbard Glacier to block the Russell Fjord. (Source: USGS)

BOX 16.2

balance with the current climate, and are unlikely to experience major changes in flow rate unless the climate changes.

There are three mechanisms by which glaciers flow. These are internal deformation or creep, sliding and bed deformation. The driving force for glacial flow is gravity, which is resisted by frictional forces at the sides and bed of the glacier. The gravitational driving force exerts stress on the ice, which can be divided into two components, the normal stress and the shear stress. Both increase with depth in the glacier (see Box 16.3).

16.2.7.1 Internal deformation of ice

Stress applied to ice causes it to deform, which is often known as creep (see Box 16.3). The rate of deformation depends on a wide variety of factors including temperature, crystal orientation and ice impurities. Creep is the dominant flow mechanism for cold glaciers, although the creep rate is much lower for cold ice than warm ice. Once the pressure melting point is reached at the glacier bed the processes of sliding and deformation of basal sediments can also occur.

16.2.7.2 Sliding at the glacier bed

If the bed of a glacier is cold, there is a strong bond between ice and bed that tends to prevent sliding. If, however, the bed is at the pressure melting point, this bond is not present and the presence of liquid water reduces friction. As a result the glacier can slide over its bed. Sliding is the process by which many of the geomorphic features we associate with glaciers are formed, and is the most efficient process for glacial erosion. The rate of sliding is controlled by the drag at the bed, which results from two factors. These are the bed roughness, which results in **form drag**, and the rock–rock friction that results from the interaction between sediment that is lodged in the basal ice (Figure 16.13) and the bed, which is known as **frictional drag**. The morphology of the basal water system and the basal water pressure also strongly influence the rate of glacier sliding by decoupling the glacier from its bed. High basal water pressures enhance the rate of glacier sliding.

Glacier beds are not smooth and basal sliding requires that ice is transferred around obstacles. Basal sliding occurs by two mechanisms under such rough conditions. These are **regelation**

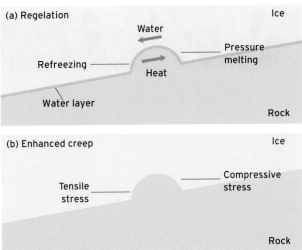

Figure 16.14 Sliding of glacier ice over a rough bed by (a) regelation and (b) enhanced creep. Regelation occurs where increased pressure causes melting on the upstream side of the obstacle. Water flows around the obstacle and refreezes on the downstream side. Heat released by refreezing is conducted upstream through the obstacle. Enhanced creep results because of enhanced compressive and tensile stresses around the obstacle.

Figure 16.13 Laminated basal ice exposed at the margin of Trapridge Glacier, Yukon Territory. The sequence shown is ~60 cm in height. Ice flow was from right to left.

and enhanced creep, which are shown in Figure 16.14. If ice flows around obstacles at the bed, most of the shear stress is supported on the upstream side of the obstacles. This results in excess pressure upstream of the obstacle and lowered pressure on the downstream side. Increased pressure lowers the ice melting point upstream of the obstacle. The melted water then flows around the obstacle to the low-pressure downstream side where it refreezes because the melting point is higher. This mechanism therefore allows the ice to slide past the obstacle. The downstream refreezing releases latent heat (see Chapter 4). The pressure drop across the obstacle results in a downstream temperature rise, so that heat is conducted upstream through the obstacle where it assists in melting. This process of melting, transfer of water and refreezing is known as regelation (Figure 16.14a). Regelation is limited by the rate of heat conduction upstream, which is most efficient for small obstacles. The regelation process is important in the formation of basal ice and the entrainment of debris at the ice base.

The second mechanism of basal sliding is known as enhanced creep. The presence of the obstacle causes an increased compressive stress on its upstream side. Glen's flow law (see Box 16.3, equation (16.3)) shows that the creep rate is proportional to the third power of the stress. Hence increased stress greatly increases the local creep rate (Figure 16.14b). The magnitude of stress enhancement is thought to be related to the size of the obstacle so that the mechanism works most effectively for large obstacles.

Because regelation is most efficient for small obstacles and enhanced creep for large obstacles there is a critical obstacle size which provides the majority of the resistance to glacial flow. This critical obstacle size is thought to lie between 0.05 and 0.5 m in diameter. The total sliding velocity is equal to the sum of the velocity due to regelation and that due to enhanced creep.

16.2.7.3 Deformation of basal sediments

Sliding theories usually assume that the glacier overlies undeformable and impermeable bedrock. However, where an ice mass overlies soft sediments, deformation of these sediments can contribute to surface flow. At Whillans ice stream, West Antarctica, high-resolution seismic surveys undertaken in the mid-1980s showed a 5–6 m thick sediment layer that was laterally continuous for at least 8 km. This layer was thought to be deforming throughout its thickness and this helps explain why the ice stream flows so fast despite its low-slope angle

FUNDAMENTAL PRINCIPLES

STRESS, STRAIN AND ICE DEFORMATION

Stress is defined as a force acting per unit area and has units of pascals (Pa). For a glacier, the normal stress, σ (which is the stress exerted at right angles, or 'normal', to the slope), can be calculated from:

$$\sigma = \rho g h \qquad (16.1)$$

where ρ is the density of ice (Table 16.2), g is the acceleration due to gravity (9.8 m s^{-2}) and h is the ice thickness. For ice 100 m thick, the normal stress is about 880 kPa.

The theoretical basal shear stress (the stress exerted at an angle parallel to the slope), τ, beneath an infinite parallel slab glacier is calculated from the bed slope (equal to the surface slope), α, and its thickness using:

$$\tau = \rho g h \sin \alpha \qquad (16.2)$$

For a typical glacier 100 m thick and with surface slope 4°, the shear stress will be about 57 kPa. Equation (16.2) allows us to predict certain geometric characteristics of a theoretical glacier. If we assume that the ice is perfectly plastic (does not deform until its **yield strength** but then has infinite **strain rate**) then we can set the basal shear stress in equation (16.2) to the yield strength of ice. This then

tells us that the thickness is controlled by the angle α, such that if α is large (the glacier is steep), then it will be thin. Conversely, if α is small and the glacier has a shallow surface and bed slope then the glacier will be thicker.

The stress applied to glacier ice causes it to deform, and the rate of deformation can be calculated from the flow law of ice (known as **Glen's law**). This flow law was determined in the laboratory and relates the strain rate, $\dot{\varepsilon}$, to the basal shear stress, τ:

$$\dot{\varepsilon} = A\tau^n \qquad (16.3)$$

The creep rate of ice is sensitively dependent on the temperature of the ice, and the value of A is about

1000 times greater at 0°C than it is at −50°C. The value of n determines how non-linear the response of ice is to an applied stress and is typically taken to be 3.

Glacier ice differs from the laboratory ice on which this result is based (Figure 16.15). Glacier ice is **polycrystalline** and the orientation of the crystals will vary. Glacier ice is inhomogeneous and can also contain water, air bubbles trapped as the snow becomes ice, and both soluble and insoluble impurities. These can all change the properties of the ice as can the **strain history** of the ice. In general, polycrystalline ice deforms less readily than a single crystal as reorientation must occur.

Figure 16.15 Thin section through glacier ice viewed through cross-polarizing filters. Each grain appears a different colour because of its different alignment. Also clearly seen are small air bubbles in the ice (seen as small round inclusions). These air samples provide a record of past atmospheric conditions. This thin section is from 333 m depth in the GISP2 core drilled at the summit of the Greenland ice sheet. (Source: photo courtesy of A. Gow)

BOX 16.3

(Alley *et al.*, 1986; Blankenship *et al.*, 1986). Furthermore, at Breiðamerkerjökull, south-east Iceland, deformation of the bed was shown to cause 88% of the total surface motion (Boulton and Hindmarsh, 1987).

16.2.7.4 Relative importance of glacier flow mechanisms

Glacier motion is made up from contributions of ice deformation, sliding and sediment deformation (Figure 16.16).

The relative importance of each at a particular site depends on the thermal regime, the availability and distribution of meltwater and the composition and morphology of the bed. At Trapridge Glacier, Yukon Territory, Canada, measurements have been performed using instruments installed through boreholes into the glacier bed designed to measure sliding and sediment deformation together with measurement of surface speed. The surface of Trapridge Glacier moves approximately 10 cm day^{-1}, of which sliding makes up about 4 cm day^{-1} and sediment deformation about

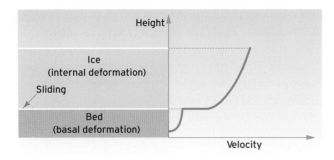

Figure 16.16 Mechanisms of glacial flow. The surface velocity of a glacier is made up from components due to internal deformation of the ice, sliding at the ice bed interface and deformation of subglacial sediments.

6 cm day^{-1}. The contribution at this site from internal ice deformation is less than 1 cm day^{-1} (Blake, 1992).

16.2.7.5 Glacier velocity, glacier surging and fast glacier flow

Glacier flow velocities vary considerably depending on the basal thermal regime, glacier mass balance, bed conditions and bed geometry (Clarke, 1987). Temperate valley glaciers typically flow at velocities of a few tens of metres per year, while cold-based continental glaciers may have maximum velocities of the order of 1–2 m yr^{-1}. The normal range of glacier velocities is considered to be 10–100 m yr^{-1} typically associated with basal shear stresses of the order of 40–120 kPa. For example, White Glacier, in the Canadian Arctic, flows at about 30 m yr^{-1}. Often glaciers can be quite inaccessible and so it is difficult to determine glacier velocity in many locations. However, using satellite technology it may be possible to measure glacier flow and glacier topography.

Box 16.4 provides information on one such technique. Even more inaccessible glaciers are found on other planets.

A smaller class of fast-flowing glaciers exists, with velocities of the order of 100 to over 1000 m yr^{-1}, which includes ice streams, tidewater terminating glaciers and surging glaciers. Jakobshavn Isbræ, West Greenland, thought to be Earth's fastest continuously flowing ice mass, flows at nearly 7000 m yr^{-1}, and several Greenland outlet glaciers have recently been measured to flow at rates of ~9000 m yr^{-1} at their calving fronts (Figure 16.17). Whillans ice stream, West Antarctica, flows at 827 m yr^{-1} and the Rutford ice stream, Antarctica, flows at 400 m yr^{-1}. Most fast-flowing glaciers are thought to do so by either fast sliding or fast sediment deformation. Several fast-flowing glaciers, such as Columbia Glacier, Alaska, and Whillans ice stream, West Antarctica, are discharging more ice than they accumulate. In contrast, Kamb ice stream, West Antarctica, which switched from a fast to slow flow mode approximately 250 years ago, flows at about 5 m yr^{-1}, and is currently gaining mass.

Surge-type glaciers switch between long periods of slow (or quiescent) flow and flow 10–1000 times faster (the surge). It is unclear whether the trigger that enables ice streams to switch between fast and slow flow is externally or internally controlled, but it appears that surging glaciers show internally driven switching, independent of external forcing. Most theories explain the change in velocity from slow to fast and vice versa as caused by a change from an efficient to an inefficient basal water system. Such a change can occur because the basal drainage system alters from a channelized to a linked cavity water system. If the glacier overlies a soft bed then the paths through sediments by which water is evacuated are thought to be destroyed by

Figure 16.17 Calving front of Helheim Glacier in south-east Greenland. Ice flow is from left to right, and the ice cliff marks the change between the glacier and the fjord, which is covered by icebergs. The glacier is about 5 km wide and the calving front is up to 100 m high. (Source: photo courtesy of Tim James)

shear deformation, which traps pressurized water at the bed causing high water pressures and fast flow.

Glacier surging involves the release of ice stored in the upper portion of a glacier (reservoir zone), allowing it to flow down the glacier rapidly into the receiving zone. For example, Variegated Glacier, Alaska, flowed up to 65 m day^{-1} during its 1982–1983 surge (Kamb *et al.*, 1985). Glaciers surge repeatedly, with a periodic and relatively constant surge cycle between 20 and 500 years. The fast flow or active phase is short, typically 1–15 years. The surge often

results in rapid advance of the front margin of the glacier as shown in Figure 16.18, sometimes by several kilometres in a few months or years. For example, the Hispar Glacier in the Karakoram Himalayas is reported to have advanced '2 miles in 8 days' and the Hassanbad Glacier in the same region advanced '9.7 km in 2.5 months' (Hewitt, 1998). Strand lines are often left marking the former pre-surge ice surface in the reservoir zone (Figure 16.18c) and the ice surface typically becomes intensively and chaotically crevassed (Figure 16.18d). Tributary glaciers can be sheared off, resulting

Figure 16.18 The effects of surging. (a) Aerial photograph showing the glacier Abrahamsenbreen, Svalbard, in 1969. The glacier surface is uncrevassed and tributary glaciers can be seen compressing the main trunk. (b) The same glacier in 1990 after a major surge. The glacier surface is highly crevassed and the moraine loops have been extended down the flow. The margin has advanced by ~3 km. (c) Glacier surges cause downdraw in the upper region of a glacier (the reservoir zone). At this glacier, Sortebræ in East Greenland, downdraw was measured to be up to 200 m, and (d) large crevasses opened in the lower part of the glacier. (Source: (a) air photo 569 1493 and (b) air photo 569 3134, © Norwegian Polar Institute; (c) and (d) Danish Lithospheric Centre)

TECHNIQUES

MEASURING GLACIER DYNAMICS FROM SPACE

Glaciers are often situated in remote and inaccessible regions, making remote sensing an attractive method for studying their dynamics. If repeated imagery is available then the movement of features such as crevasses on the glacier can be tracked automatically between the images. An area of the glacier is chosen in the first image and then the equivalent area is searched for in the second image (Figure 16.19). By repeating this process across the whole glacier surface and with knowledge of the time gap between the images, a map can be built up of the glacier's dynamics.

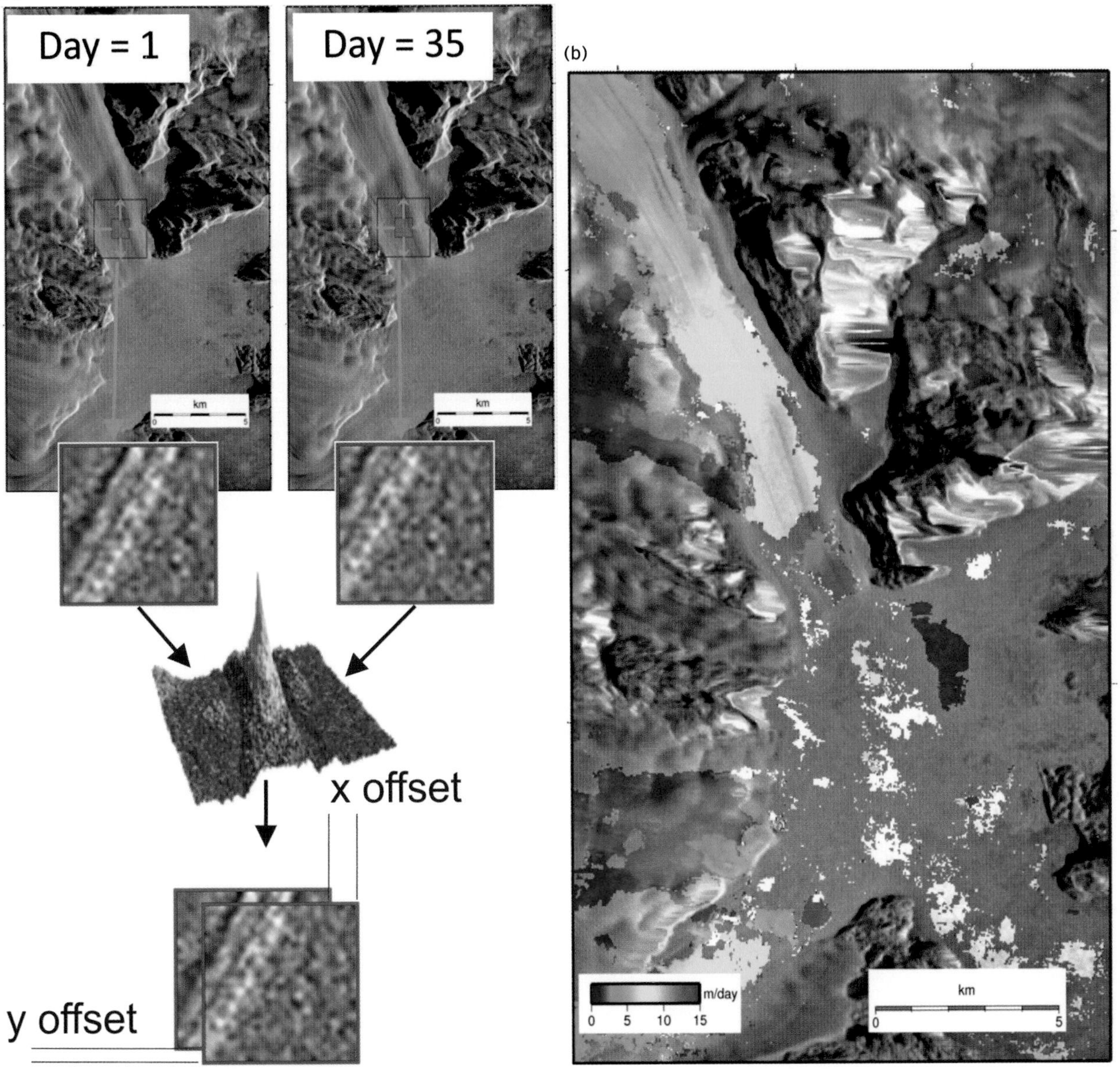

Figure 16.19 (a) Figure showing conceptually how tracking between two satellite images works. The rock (non-moving) areas in two images separated in time are co-registered (aligned spatially). Patches from one image are then matched in the second image allowing the displacement of features such as crevasses to be measured. (b) Speed map of the glacier Glydenlove in south-east Greenland from tracking on SAR imagery. (Source: courtesy Suzanne Bevan)

BOX 16.4

NEW DIRECTIONS

SUB-ICE-SHEET LAKES IN THE ANTARCTIC

Around 150 lakes have been mapped beneath the Antarctic ice sheet using radio-echo sounding (Siegert *et al.*, 2005) and seismic surveys (Woodward *et al.*, 2010a). As the lakes fill and drain under the ice sheet the surface elevation of the ice changes and this has been mapped using satellite techniques (e.g. Wingham *et al.*, 2006; Fricker *et al.*, 2007), and shows that the subglacial water

system beneath the ice sheet is very dynamic. The water in these lakes is in complete darkness, the ambient pressure is very high, and the waters may be isolated over long periods of time, so that the lakes are thought to be unique biological habitats. Lake Vostock was the first subglacial lake to be identified from radar and satellite imagery. The lake is the largest discovered, around 250×50 km^2 in size and up to 800 m deep, and is situated beneath 4 km of ice in East Antarctica. Bacteria were found in

refrozen ice at the bottom of the Vostok ice core (Karl *et al.*, 1999) and drilling is being undertaken to sample the lake waters. Subglacial Lake Ellsworth in West Antarctica (Figure 16.20) has been mapped using radar and seismic surveys: the lake has been shown to be around 15 km by 3 km in size and varies in depth from 52 to 156 m. Drilling is planned for Lake Ellsworth in 2012–2013 (see www.geos.ed.ac.uk/research/ellsworth).

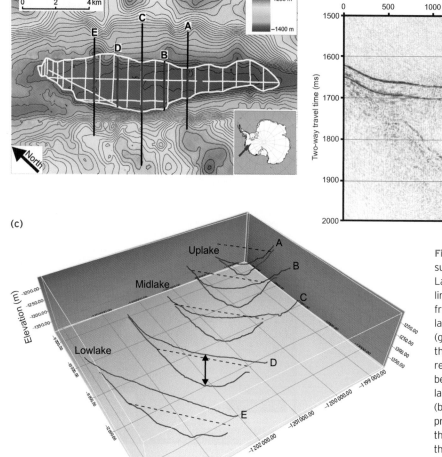

Figure 16.20 (a) Location map of and subglacial topography at subglacial Lake Ellsworth. (b) Example seismic line across the lake showing reflections from both the lake surface and the lake bed. The second line in each case (ghost) is energy that travelled up from the shot to the ice surface and then reflected down to the bed, so should be ignored. (c) A visualization of the lake surface (red lines) and lake bed (blue lines) identified from five seismic profiles. Black dashed lines represent the critical pressure boundary (ice thickness of ~3170 m) for each seismic line. View is uplake (into ice flow). (Source: from J. Woodward *et al.*, 2010)

BOX 16.5

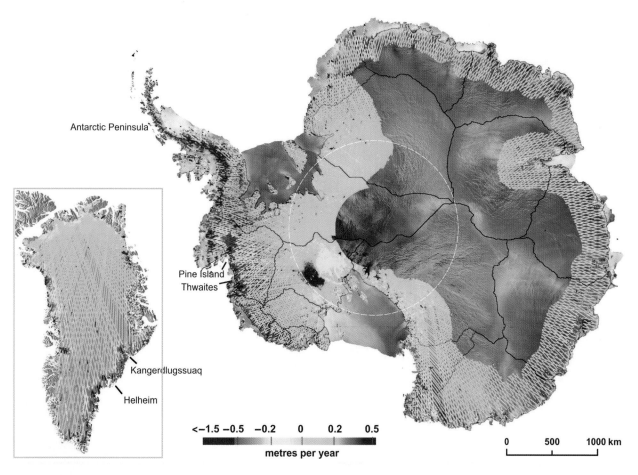

Figure 16.21 Elevation change on the Antarctic and Greenland ice sheets measured using ICESAT satellite altimetry. The central part of Antarctica cannot be imaged because there are no data; however, where imaged, the central portions of the ice sheet are thickening because of increased precipitation whereas the peripheries are thinning. The data show rapid thinning on the Greenland outlet glaciers (see also Boxes 16.1 and 16.6) as well as in West Antarctica (especially the Pine Island and Thwaites Glacier catchments) and the Antarctic Peninsula. Overall, it is likely that both Antarctica and Greenland are contributing to sea-level rise. (Source: Pritchard *et al.*, 2009)

in looped patterns of medial **moraines** (Figure 16.18b and c). The slow-flow phase is often characterized by frontal retreat and wasting ice. During slow-flow periods the ice builds up in the reservoir zone ready for the next surge, which often results in the formation of a bulge of ice.

Although only about 1% of Earth's glaciers are surge-type they tend to be more common in certain regions including Iceland, Svalbard, Karakoram, Pamirs, Tien Shan, Yukon–Alaska, Greenland and the Andes. These environments range from continental to maritime and the glacier thermal regimes from temperate to subpolar. Only cold-based glaciers have not been observed to surge.

16.2.7.6 Glaciers in a warming climate

Global climate is warming, although the spatial pattern is not uniform (see Chapter 24) and this warming is

causing widespread melting and retreat of glaciers. This loss of ice is resulting in sea-level rise (see Section 16.1). Mass loss is not happening everywhere, however. In some regions atmospheric warming has increased precipitation, resulting in mass gain. The major ice sheets are in general thickening in their central region, because of the increased snowfall, and thinning around their periphery (Figure 16.21). Some basins of the West Antarctic ice sheet which are grounded below sea level are thinning rapidly, as are many outlet glaciers in Greenland (see Box 16.6). Furthermore, a number of ice shelves on the Antarctic Peninsula have collapsed and the glaciers feeding into them have accelerated. Many outlet glaciers in southern Greenland have also speeded up. In most mountainous regions small glaciers are retreating and contributing to sea level despite the increase in precipitation in some areas.

CASE STUDIES

ACCELERATING GLACIERS IN GREENLAND

Greenland is the largest island on Earth and its ice sheet contains around 7.2 m of sea-level rise equivalent. Around half of the mass loss from the Greenland ice sheet is meltwater, which mainly runs off from the surface, but in some areas lakes form that may drain through the glacier to the bed (Figure 16.7a and b) and cause the ice sheet's flow to speed up. The remainder of the mass loss occurs through fast-flowing tidewater glaciers that calve icebergs into glacial fjords (Figure 16.17). In west Greenland, Jakobshavn Glacier was seen to accelerate in 1997 at the same time as ocean temperatures increased over the continental shelf (Holland *et al.*, 2008). In south-east Greenland, the two largest glaciers are Helheim and Kangerdlugssuaq (Figure 16.22). Both of these glaciers speeded up in 2004–2005 by 60 to 100%. This speed up was accompanied by rapid thinning of 70–100 m and a frontal retreat of several km. Subsequently in 2006, the glaciers slowed and partially readvanced, however, they remain at speeds faster than in 2000. Other glaciers in the south-east region also appear to have responded in the same manner and it is likely that changes in ocean temperature are responsible.

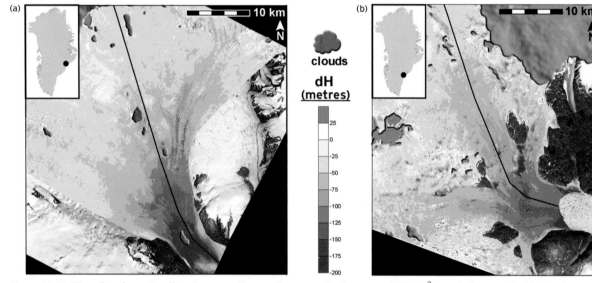

Figure 16.22 Map of surface elevation change on Kangerdlugssuaq Glacier over a 1750 km² area between July 2001 and July 2006. (b) Map of surface elevation change on Helheim Glacier (see also Figure 16.17) over a 1040 km² area between June 2002 and August 2005. (Source: from Stearns and Hamilton, 2007)

BOX 16.6

Reflective questions

➤ A mining company suggests disposing of waste down crevasses of a temperate glacier, arguing that the material would be held for many thousands of years before being released into the environment. Do you agree with this assessment?

➤ Can you explain the major differences between a warm- and cold-bedded glacier in terms of thermal structure, hydrology and dynamics?

➤ A ski company decides that in order to prolong the ski season it will move snow from the accumulation zone of a glacier to the ablation zone. Can you comment on this proposal with a view to the long-term future of the ski resort?

➤ In the Delta River Valley, Alaska, the Alaska oil pipeline passes within hundreds of metres of glaciers, including several surge-type glaciers. Why might this be a problem?

16.3 Glacial geological processes and glacial sediments

The aim of this section is to develop an understanding of the main geomorphic processes operating beneath glaciers, concentrating on the themes of erosion and deposition, and emphasizing the characteristics of the sediments deposited.

16.3.1 Processes of glacial erosion

Glacial erosion has resulted in some of the most spectacular scenery on Earth. Deep, previously glaciated valleys and fjords and lofty alpine ridges and horns are the typical scenery brought to mind. Enormous volumes of rock have been removed from these regions, transported by ice and water and deposited in lower-lying areas. This section outlines the processes by which these landscapes of erosion are formed. These processes have long been poorly understood, partly because of the inaccessibility of the subglacial environment and partly because of the complex interactions between the ice dynamics, thermal regime and hydrological system, which strongly affect the processes of erosion; however, recently remote sensing through the ice has revealed some of these sub-ice landscapes in detail for the first time.

16.3.1.1 Glacial crushing

Glacial crushing is the direct fracturing of bedrock because of the weight of ice above it. The thickest ice that has existed on Earth was around 5 km thick, which results in a normal stress of about 44 MPa (Box 16.3). However, most rocks are stronger than this. For example, granites have unconfined compressive strengths of 140–230 MPa, sandstones between 70 and 210 MPa, and even shales and tuffs have strengths greater than 10 MPa. It would appear that only the thickest glaciers overlying the weakest rocks will cause fracturing. Yet there is direct evidence that crushing does occur. This can result from the effects of stress concentration because of stones in the basal ice, the exploitation of pre-existing weaknesses and joints in the bedrock, and from repeated cycles of loading and unloading. Freeze–thaw weathering (see Chapter 7) in front of the glacier can fracture rock that the glacier subsequently advances over. Finally the removal of large amounts of rock by glacial erosion can cause fracturing due to pressure release. Bedrock crushing is enhanced by: (i) thick ice; (ii) the presence of particles entrained within the basal ice; (iii) cold patches at the bed; and (iv) large fluctuations in the basal water pressure. The products of bedrock crushing are typically large, angular rocks.

16.3.1.2 Plucking and quarrying

Once the bedrock is crushed it can be entrained (picked up) into the glacier ice. The process by which a glacier removes large chunks of rock from its bed is known as plucking or quarrying. Entrainment can result from freeze-on at the bed, by ice regelation around the rock or by incorporation into the ice along faults. Freeze-on can occur at the downstream side of obstacles at the bed where water refreezes as the glacier slides by regelation.

16.3.1.3 Glacial abrasion

Glacial abrasion occurs when glaciers slide relative to the material beneath them. Rock particles held within basal ice (Figure 16.13) are dragged over the glacier bed. This slowly scratches and wears the surface, rather like a piece of sandpaper wears wood. The rate of erosion due to abrasion is controlled by three factors. These are: the contact pressure between ice and its bed; the rate of sliding; and the concentration and nature of particles within the basal ice.

Abrasion rates increase as particle concentration within basal ice increases. Furthermore, fresh rough particles are more efficient tools of erosion than smoothed particles. Hence abrasion rates are higher in locations where basal melting brings a continual supply of fresh particles descending towards the bed. The relative hardness between entrained debris and the underlying bedrock will influence abrasion rates. For example, a glacier that flows from a hard to soft rock will have entrained hard rock pieces which will then efficiently abrade the softer rock downslope. Finally, erosion is more effective in locations where the fine sedimentary product is removed by meltwater flushing. Otherwise the fine sediment can clog or coat particles that are acting as abrasion tools. The sediment resulting from abrasion is very fine and when suspended in water is known as glacial flour. Glacial abrasion causes **striations** (see below) and smooths particles.

16.3.1.4 Mechanical and chemical erosion by basal meltwater

Meltwater at the bed of a glacier can cause erosion by mechanical or chemical processes. The rate of both is highly dependent on the nature of the glacial water system, the flux through it and the sediment or solute loading. Subglacial mechanical erosion occurs by the same processes as erosion in a surface channel flowing over bedrock. This operates by corrasion (mechanical wearing and grinding), cavitation (pressure changes due to bubble collapse in turbulent flow) and corrosion (chemical weathering). Glacial meltwaters are efficient at erosion, because they typically flow at high velocities and often have high viscosity because of their large suspended sediment load and cold temperature. Mechanical erosion forms smoothed bedrock, potholes and N-channels. Chemical erosion by meltwater at the bed results in the decomposition of minerals into their ionic constituents. The processes operating include: solution, hydrolysis, carbonation, hydration, oxidation and reduction, and cation exchange (see Chapters 7 and 13). All of these processes act at the surface of particles. The rate of chemical erosion at the bed is enhanced by the influx of fresh surface water, by the presence of freshly abraded surfaces and by the presence of dissolved CO_2, which has an enhanced solubility at low temperatures.

16.3.1.5 Erosion rates

The rate at which glaciers and ice sheets denude bedrock varies greatly owing to changes in basal temperature, glacier velocity and properties of the bedrock. In one experiment, marble and metal plates were installed beneath valley glaciers to measure the rate at which they were abraded (Boulton, 1979). The plates were abraded at rates between 0.9 and 36 mm yr^{-1}. Other estimates are derived from measurements of the sediment output from glacial systems in streams and in basal ice. These measurements suggest total erosion rates of between 0.073 and 165 mm yr^{-1}. The best global estimate is approximately 1 mm yr^{-1} which is equivalent to 1 km per million years. However, these estimates are for valley glaciers and do not include ice sheets. It should be emphasized that in some situations ice may have a protective role and bedrock may be eroded at much slower rates than would have otherwise been the case. Indeed, recent evidence from cosmogenic isotope data suggests that denudation by large ice sheets was minimal for large areas at least during the advance of the last glacial (Fabel *et al.*, 2002).

16.3.2 Entrainment and transport

Freeze–thaw and other processes result in material falling from rock slopes onto the surface of glaciers where it is transported downstream by the ice (Figure 16.23). If the material falls onto the glacier in the accumulation zone it will become buried, and either will be transported within the body of the glacier (englacially) or may descend to the bed. If material falls onto the surface in the ablation zone it will remain at the surface and be transported passively as supraglacial sediment. Material that is only buried partially will also be transported passively as englacial sediment and may emerge in the ablation zone as a result of ice melt. Sediment is also transported within the basal ice layer (Figure 16.13). Basal material may also become entrained within faults close to the margin or basal crevasses (Figure 16.24a). Medial and lateral moraines are made up of supraglacial material

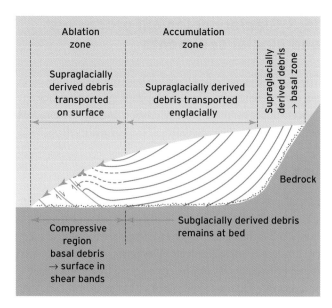

Figure 16.23 Transport paths of debris through glacial system. (Source: after Boulton, 1978)

and the particles are typically coarse and angular (Figure 16.24b). Particles that are eroded from the bed, or descend to the bed, become rounded and worn through abrasion (Figure 16.24a).

The glacial system can be thought of as a conveyor belt transporting ice and sediment downslope, modifying the sediment particles as they are transported (Figure 16.23). As particles travel down the glacier they are abraded for progressively longer time periods and break down to finer and finer particles. However, experiments in grinding mills show that the breakdown of sediment particles ceases once particles reach their **terminal mode**, the size of which depends on the mineralogy (Haldorsen, 1981). As the material is further abraded more of the particles reach this terminal mode, but the material does not break down to finer particles. As a result of abrasion the particle size distribution of basal sediments often becomes bimodal (two peaks in the distribution; in this case there are lots of pebble-sized material and lots of fine-clay-sized material, but with less of all the other size fractions) (Dreimanis and Vagners, 1971) (Figure 16.25).

16.3.3 Deposition

It is important to understand the processes and nature of glacial deposits, because they cover about 70–80% of mid-latitude regions, which represents some 8% of the Earth's total land surface. Deposition of particles from the 'glacier conveyor belt' is an active process and alteration of the sediment properties continues as the particles are deposited. The sediment consequently develops characteristics that result from the particles' source, transport route through the

(a)

(b)

Figure 16.24 (a) Basally derived material entrained in a fault in glacier ice. Basal material is actively transported and becomes rounded and finer. Ice flow was from right to left. (b) Supraglacial sediment is typically coarse and angular. Both supra- and englacial material are transported passively and so little modification occurs. Medial and lateral moraines are made up of supraglacial sediment.

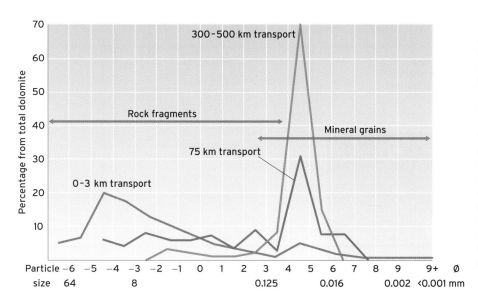

Figure 16.25 A typical particle size distribution for glacial till is bimodal or multimodal. The ϕ index is shown (see Chapter 9) on the x-axis along with values in mm. The two peaks in the distribution of the sediment sizes collected can be seen. There is more fine-grained (0.1 mm diameter) and pebble-sized (50 mm diameter) material than other particle sizes. This develops from the processes of glacial erosion that occur progressively with transport distance. With progressive distance from the ice mass centre, successively higher percentages of rock fragments are reduced to a smaller particle size. (Source: after Dreimanis and Vagners, 1971)

system and their mode of deposition. The resulting sediments are often referred to by the process of their deposition, as this has the strongest influence on their properties. This is known as a **genetic classification**. The generic term for sediment deposited directly from a glacier is a **till**.

16.3.3.1 Lodgement till

Lodgement of particles occurs when the frictional drag between a particle and the glacier bed exceeds the shear stress resulting from the moving ice (Box 16.3). The frictional drag on a particle depends on the contact pressure between it and the bed. This means that lodgement is most likely to occur beneath thick ice, where basal water pressures are low, on the upstream sides of obstacles, and in regions where basal melt rates are high. Boulder clusters will form because of preferential lodgement on the upstream side of obstacles at the bed. Lodgement tills form at the ice base from material that has been in transport in the basal zone. As a consequence the particles are typically rounded, are often striated and the sediment size distribution is often bimodal (see above).

The processes of lodgement may be size selective. A glacier slides over its bed by regelation and by enhanced creep. Since regelation is most efficient when obstacles are small, whereas enhanced creep is most efficient when obstacles are large, these processes will selectively keep particles of the critical size entrained within the glacier ice. The forces keeping small particles in motion within ice are relatively low because the glacier regelates around small particles easily, and the forces keeping large particles in motion are low because the ice creeps around the particles easily.

16.3.3.2 Deformation till

Many glaciers are thought to overlie deforming soft sediments. The processes operating beneath such glaciers are recorded in the tills that they deposit. Deformation rearranges particles and reorientates them. Examination of the tills can reveal structures that allow the strain history to be reconstructed (e.g. Figure 16.26). At low strain minor folding and faulting can occur. As the strain increases, the features formed can be quite spectacular. Compressive features include folds and some faults, whereas extension results

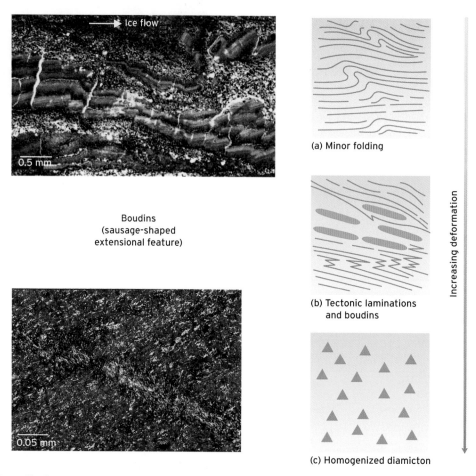

Figure 16.26 Schematic diagram showing the effect of deformation on tills. (a) At low strain minor folding occurs. Photograph: small-scale faulting in clay layers that shows that this sediment was deformed only to very low strain. (b) At intermediate strain dramatic features can be formed, depending on whether the strain is compressive or extensive. (c) At high strain the till can appear to be completely homogenized and massive. Photograph: shear zones in sediment that has no features apparent on the macroscale. Two directions of shearing are evident. Matrix-supported till from Criccieth, North Wales. (Source: after Hart and Boulton, 1991; photos courtesy of (a) Sarah J. Fuller and (c) Andy J. Evans)

in the formation of **boudins** (Figure 16.26b). As the strain increases still further, the features become progressively attenuated and tectonic laminations can be formed. Finally at high strains, likely to be typical where deformation is a significant component of ice flow, the till can appear homogenized. However, the sediment may still have many structures visible at the microscopic scale (Figure 16.26). Deformation tills can have high porosity (% of void space) because **dilation** (expansion) occurs as the sediment deforms. Particle **fabrics** (the alignment of coarse particles) are typically stronger for intermediate strains than for low or high strain. Often the fabric has both a flow parallel and a transverse component. The sediment typically comprises basal material, containing rounded and striated pebbles and rocks, and if the sediment has experienced high strain there is likely to be a wide range of lithologies because material will have travelled long distances. However, it can be difficult to distinguish a deformation till from a lodgement till, and there is controversy about the origin of several deposits.

16.3.3.3 Meltout till

Meltout till forms when ice surrounding sediment melts. In general, the term is used to describe deposits from subglacial meltout. Supraglacial meltout usually results in high water contents and therefore reworking of the deposit often occurs. Meltout till usually forms beneath stagnant ice masses. Meltout till can inherit properties from the basal ice from which it forms, often being weakly stratified and maintaining the fabric (Figure 16.27). This is most likely to occur where the basal ice has a high-sediment and low-ice content, and where water produced when the ice melts can drain away freely. The extent of meltout till therefore depends on the distribution and concentration of sediment in basal ice, which is in turn controlled by the thermal regime and strain regime. Basal ice layers are thickest beneath polythermal glaciers and where a compressive flow regime occurs. Surging glaciers often develop thick basal ice layers.

16.3.3.4 Flow till

The proglacial environment is very active. In summer, there is often a continuous supply of water from melting ice. Because many glacial sediments have low permeability, they drain poorly. As a result sediment flows are common. Such flows alter sediment properties, realigning particles, and cause particle sorting (Figure 16.28). The resulting deposit is known as flow till. It is estimated that up to 95% of the marginal area in front of a glacier may comprise flow till.

Figure 16.27 Meltout till at the margin of Trapridge Glacier, Yukon Territory.

16.3.3.5 Deposition from water

Sediments entrained into subglacial water flows may be transported considerable distances from the ice margin before deposition. As a result of the differences in time taken for particles of different sizes to settle through water (Table 16.3), there is potential for strong sorting of the deposits, both spatially and temporally. Spatial sorting occurs because the coarse particles settle close to the ice margin whereas the finer particles can remain in suspension and can travel long distances. Temporal sorting occurs because the outputs of both water and sediment vary on a variety of timescales (e.g. diurnally and seasonally). If we consider the seasonal variation, flows of sediment and water are highest in the early summer and lowest in the winter. The coarse particles are deposited first, in summer, and the finer particles progressively through the year. This forms repetitive sequences of deposited sediment that fine upwards; where such sequences represent annual inputs they are known as **varves**.

Figure 16.28 Mudflow (right to left) at the margin of Trapridge Glacier, Yukon Territory. Note that the flow has sorted the particles, with coarse particles migrating to the regions of least strain at the sides, surface and front of the flow. In the background a debris-rich basal ice sequence can be seen.

Table 16.3 Time taken for single grain to settle through 10 m of still water at 0°C. The very large differences in time between fine and coarse particles lead to strong sorting both spatially and temporally from a source

	Particle diameter (μm)	Time
Very fine sand	125	21 minutes
Coarse silt	31.25	5.7 hours
Fine silt	7.81	3.8 days
Coarse clay	0.98	7.7 months
Fine clay	0.06	175 years

Reflective questions

➤ What are the main processes of glacial erosion?

➤ Can you explain the differences between erosion beneath a warm- and a cold-bedded glacier in terms of the processes operating and their relative rates?

➤ What are the key differences that you expect to find in glacial sediments that allow the differentiation of: (i) supraglacial from basal material; and (ii) sediments deposited in water from those deposited on land?

16.4 The record of glacial change

The aim of this section is to develop an understanding of the effects of ice on the landscape at all scales from global to microscale, concentrating on the geomorphic features that result from the processes and environments discussed in Sections 16.2 and 16.3.

16.4.1 Ice sheet reconstruction

16.4.1.1 Geomorphology of regions of erosion

Regions dominated by the processes of glacial erosion form some of the most dramatic mountain scenery on Earth. Examples are the Southern Alps of New Zealand, the Rocky Mountains in North America and the Alps in Europe. In general erosion occurs most efficiently beneath a warm-based ice mass that is sliding over its bed. Furthermore, erosion requires that the glacier base is in contact with bedrock rather than sediments. This is more likely to be the case close to ice divides, at the centre of ice sheets or ice caps, and beneath local glaciers in alpine areas (Figure 16.29). The features formed by erosion are usually grouped by scale. At the macroscale, glaciers erode great **U-shaped valleys**, leaving ridges (or **arêtes**) and **horns** (Figure 16.30). Often side valleys and spurs are truncated, leaving **hanging valleys** (or hanging troughs) and **truncated spurs**. When the glaciers retreat, lakes often form in the overdeepened basins, which slowly fill with sediment over time. If the base of the glacial trough lies below sea level then after retreat it becomes inundated and forms a fjord. Less extensive glaciers often form **corries** (or cirques) which when they melt leave a depression that can be filled with water and are known as **tarns** (Figure 16.30). Figure 16.30 also shows some depositional features such as moraines that are discussed below.

Snowdonia, in North Wales, UK, is presently unglaciated but contains evidence of widespread erosion by glaciers. At around 23 000 years ago an ice sheet approximately 1000 m thick formed which was centred on North Wales. The accumulation area of this ice sheet was approximately at its maximum extent about 18 000 years ago. This ice sheet lingered until about 12 000 years ago. Then a second short cold spell refilled the area with ice between 11 000 and 10 000 years ago. Figure 16.31 shows the morphology of Cwm Idwal. This was one of the first sites where past glaciation was recognized in the 1840s. The corrie shows both erosional and depositional features. The overall corrie shape is governed by erosion, it is overdeepened and is currently occupied by a lake. Depositional moraines can be found within the valley (see below).

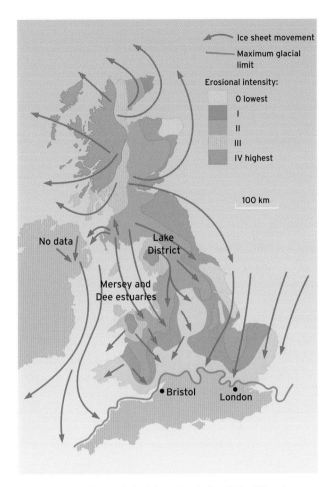

Figure 16.29 Pattern of glacial erosion in the United Kingdom. The erosion is greatest in the highest regions that formed the ice centres for major ice sheets and were glaciated by local glaciers during shorter or warmer stadials.

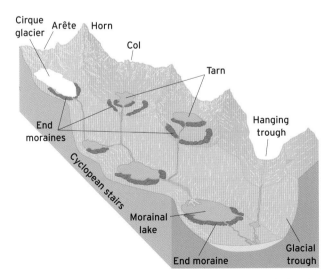

Figure 16.30 A landscape of glacial erosion. (Source: Easterbrook, D.J., *Surface processes and landforms*, 1st edition, © 1993. Reprinted with permission of Pearson Education, Inc., Upper Saddle River, NJ)

Mesoscale features are formed by the erosive power of both ice and water. Features formed by the sliding of debris-rich basal ice over the surface include **stoss-and-lee forms**, **roche moutonnées**, **whalebacks** and **crag and tail** features. Stoss-and-lee forms are streamlined features that have a gently sloped, glacially smoothed upstream side and a steeper, plucked, downstream lee side (Figure 16.32a). The stoss side is often striated. The features vary in size from centimetres to many metres in length. Small stoss-and-lee forms are often called roche moutonnées. Stoss-and-lee forms provide evidence of warm-based ice conditions and the formation of cavities downstream of obstacles at the bed. Whalebacks are a similar shape to stoss-and-lee forms but their steep side faces upstream and their tapered end downstream. Whalebacks probably form beneath warm-based ice where the sliding rate is slow enough that extensive cavitation does not occur behind obstacles at the bed. Crag and tail features form where resistant rock is left standing proud of the surface. Small cavities form behind the more resistant rock either protecting the tail of softer rock from erosion or allowing infilling to form a rock-cored **drumlin** (see Box 16.7 below). Steep-sided and smoothed channels are also found on bedrock, providing evidence of warm-based ice and a drainage system at least partly drained by N-channels.

Microscale features (Figure 16.32b) include those formed directly by erosion due to the ice, such as striations. These are formed by stones and rocks in the basal ice sliding over bedrock and leaving scratches on the surface. The presence of striations implies a warm-based ice mass, sliding over its bed, and provides a reliable indication of the direction of ice flow. Features related to striations include repetitive **chatter marks**, which show that the ice moved over its bed with a stick–slip motion. Larger **crescentric gouges**, scars and fractures also form, also typically concave down the glacier, but are less repetitive than chatter marks. All of these features form by non-uniform slip over bedrock. Benn and Evans (2010) provide further details on these features. Glacial flow direction can also be determined using mini crag and tail features that form on inhomogeneous bedrock. They consist of tails of uneroded bedrock preserved behind small, more resistant, grains on the surface of a rock. Smooth channels (glacial grooves) and depressions are also found on bedrock surfaces.

16.4.1.2 Geomorphology of areas of deposition

The geomorphology of areas of deposition is typically more subdued than regions where erosion dominates. Features may be formed by the direct action of ice such as moraines,

Figure 16.31 Cwm Idwal, Snowdonia, North Wales. This small valley was occupied by a valley glacier around 13 000 years ago. The glacier advanced to the terminal moraine (which forms at the front edge of a glacier). A complex suite of other moraines formed, which are discussed in the text. (Source: James Davies)

(a)

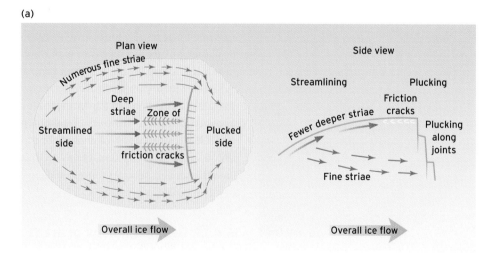

Figure 16.32(a) Medium- and small-scale ice erosional features, stoss-and-lee form; this example shows the typical features expected on a roche moutonnée, which is the name given to small versions of stoss-and-lee forms.

(b)

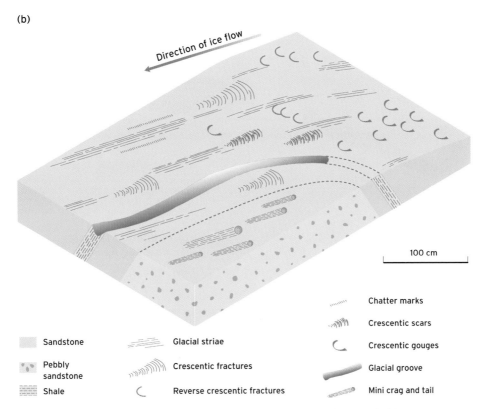

Direction of ice flow

100 cm

Chatter marks

Crescentic scars

Crescentic gouges

Glacial groove

Mini crag and tail

Sandstone

Pebbly sandstone

Shale

Glacial striae

Crescentic fractures

Reverse crescentic fractures

Figure 16.32(b) Microscale features of glacial erosion. (Source: (b) after Prest, 1983)

drumlins, **flutes**, **crevasse-fill ridges** and **kettle holes**. Except for kettle holes, these features all encode information about the dynamics of a retreating ice mass. Other features are formed by the action of meltwater, such as **eskers** and **kames**. These features may provide evidence for the nature of the basal water system of a former ice mass. Figure 16.33 shows a typical land system that forms at the margin of a retreating glacier.

Push moraines form when a glacier bulldozes into existing sediment at the glacier margin and raises it to form a ridge. Push moraines often contain evidence of glaciotectonic deformation within them and consist of glacially derived material. If the material is not glacially derived, the term **ice-pushed ridge** is used. Major push moraines often mark the maximum extent of glacial advance and allow us to map ice extent. See Figure 22.7 in Chapter 22 for an example of a map of the extent of the Laurentide ice sheet that covered large areas of North America, which is based on push moraine evidence. Smaller annual moraines can be formed in front of a retreating glacier marking seasonal re-advances of the glacier front.

Dump moraines are ridges formed approximately transverse to flow from material delivered to the margin of a glacier by ice flow. They mark the stationary position of an ice margin. The size of dump moraines depends on the rate of ice flow and the debris content of the ice. **Lateral**

moraines form parallel to the sides of glaciers from dumped material and frost-shattered material from the valley walls. **Hummocky moraines** form from the meltout of supraglacial or englacial material. These landforms tend to consist of irregular mounds of material.

Figure 16.34(a) shows arcuate ridges of sediment. The amplitude of these features, known as **de Geer moraine**, is about 2–15 m in height. These features are formed approximately transverse to flow where a retreating ice mass borders on a glacial lake. The features are thought to form by deposition at the margin of active ice and mark the position of the winter margin.

Drumlins are streamlined features aligned along the direction of ice flow (Box 16.7): drumlins and other elongate sedimentary bedforms have now been imaged underneath Rutford ice stream in Antarctica (Figure 16.34) where the bed comprises soft sediments and the ice is experiencing fast flow. Drumlins allow the direction of ice flow to be determined from their shape as they typically form with a blunt end pointing upstream. Eskers are often found together with drumlins and are thought to form in R-channels at the glacier bed, although they can also form in sub-aerial channels.

Glacial flutes form parallel to flow. Flutes have a much higher elongation ratio than the drumlins discussed above. Flutes are thought to form by the infilling of cavities on the

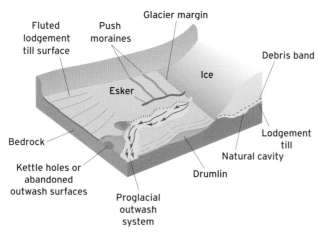

Figure 16.33 Subglacial-proglacial land system showing typical geomorphological features that form at the margin of a retreating ice mass.

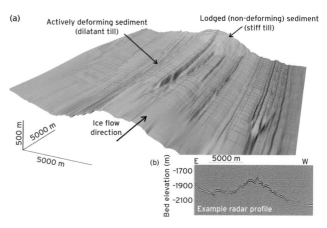

Figure 16.34 (a) Subglacial landscape beneath Rutford ice stream, West Antarctica, reconstructed from multiple radar lines sounded beneath ~2 km of ice. (b) Example single radar line. (Source: adapted from King *et al.*, 2009)

CASE STUDIES

FORMATION OF DRUMLINS

Drumlins are streamlined, elongated hills that form approximately parallel to ice flow. They are large-scale features and typical dimensions of a drumlin are 80–100 m high and 1–2 km long. Drumlins tend to have a blunt upstream end and a longer tapered downstream end. They are typically found in groups known as drumlin fields. Drumlins can be formed from either bedrock or sediment, and they may have a core of different composition at their upstream end. There are a number of theories as to the formation of drumlins. Not all theories can explain all types of drumlins. For example, theories that suggest that drumlins form by deposition of sediment cannot explain rock drumlins. Although both sediment drumlins and rock drumlins have similar morphologies it is possible that they form by different processes. Two contrasting theories of drumlin formation are the formation of drumlins by deformation of a soft bed and the formation of drumlins as a result of megafloods.

The idea that drumlins might form because of changes in the thickness and properties of a deforming bed was proposed by Boulton (1987) and Menzies (2000). They suggested that subglacial deformation can explain a continuum of forms from transverse features such as **rogen moraines**, through less elongate flow parallel forms such as drumlins, to very elongate forms such as flutes. The elongation ratio would increase with distance downstream from an ice mass centre.

The **Boulton–Menzies theory** suggests that a drumlin is formed by deposition in the lee of a slowly moving obstacle in the deforming layer and potentially by streamlining of the obstacle. This obstacle forms the core of the drumlin, and may be composed of bedrock, thermally frozen material or material that is better drained or less dilated than the surrounding material. The drumlin will slowly migrate down the glacier. As well as sedimentary evidence (Figure 16.35), proponents of this theory cite drumlins currently

forming at sites where deformation is known to occur, such as at the margin of Breiðamerkurjökull, south-east Iceland. Furthermore, time-lapse geophysical surveys have shown a drumlin forming from deforming sediments at Rutford ice stream (Smith *et al.*, 2007).

It has been suggested (J. Shaw, 1989, 1994) that all drumlins, including rock drumlins, were formed by a common agent, namely meltwater sheets during megafloods. Shaw used equivalence of form to argue that because drumlins have the same morphology as smaller-scale erosional scour features, they form by similar processes. He argued that horseshoe vortices form downstream of obstacles when megafloods occur (Figure 16.36a). These secondary flows cause erosion upwards into the ice and downwards into the bedrock. Rock drumlins are formed as the direct result of this pattern of erosion. Sediment drumlins then form either in a similar manner or at the end of flows when the flood waters become hyper-concentrated

BOX 16.7 ➤

➤ in sediment and the ice cap acts like a mould and settles back onto its bed (Figure 16.36b). Most of Shaw's work has considered drumlins that formed beneath the Laurentide ice sheet that covered Canada and parts of the United States. The megafloods that Shaw postulates involved a large amount of water. For example, for one drumlin field Shaw cited, 8×10^4 km^3 of water would be released over 16 to 162 days. This would have resulted in global sea-level rise of 0.23 m. Shaw suggests that the water required for this was probably stored in a depression beneath the ice sheet held back by frozen margins. There are currently about 150 lakes beneath the Antarctic ice sheet that store 4000–12 000 km^3 of water (see Box 16.5). Evidence to back up Shaw's theory includes oxygen isotope records (see Chapter 20), sedimentary cores from the ice sheet margin and flood myths in a number of human cultures.

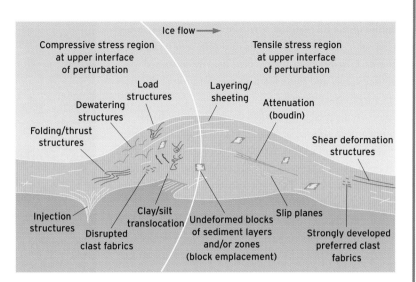

Figure 16.35 Sedimentary structures found in drumlins that would support their formation from an active deformable layer of sediment beneath a glacier via the Bouton-Menzies theory. Evidence thought to support this theory includes the presence of shear zones and other deformation structures in some drumlins. (Source: after Menzies, 1989)

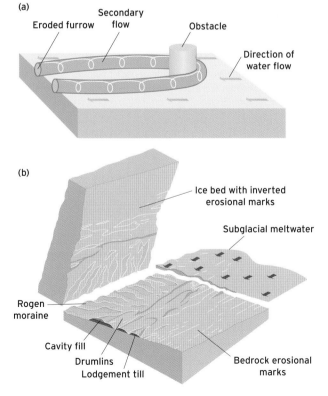

Figure 16.36 Shaw's theory of drumlin development as a result of megafloods. (a) During megafloods horseshoe vortices form downstream of obstacles resulting in erosion of the ice above and the rock or sediment below. (b) Drumlins form by this erosion or during the settling of the now drumlin-shaped ice sheet back onto soft sediments. Other bedforms such as rogen moraines may also form by the same process. (Source: (a) after Shaw and Sharpe, 1987; (b) after Shaw, 1994a)

BOX 16.7

downstream side of obstacles at the bed (Figure 16.37). The size of the flute is probably controlled by the ice velocity and the nature of the till material. Crevasse-fill ridges form when a glacier with deep crevasses sinks into soft, water-saturated sediments. Glaciers typically become heavily crevassed when they surge, and these features are common at the margins of surge-type glaciers.

Eskers are formed in ice-walled channels, either basal R-channels or in sub-aerial channels cut through ice blocks. An esker consists of a long, narrow ridge of sands and gravels which overlies till. Eskers are 20–30 m high, and range from a few to 500 km in length. Eskers formed in R-channels, which unlike sub-aerial channels may flow uphill as well as down, represent the relict form of a channelized water system. They are hence important in understanding the basal water system beneath past ice masses. Eskers are often quarried as a source of gravel and sand. A kame is a mound of sediment formed when a hole in a melting, stagnant ice mass becomes filled with sediment, whereas kettle holes are closed topographical depressions formed from melting stagnant ice, and are common in sediments downvalley from a glacier.

Reflective questions

➤ What are the main landforms of glacial erosion and glacial deposition?

➤ Which glacial features allow the direction of former ice flow to be determined?

➤ It is suggested that the Martian ice caps have previously been much larger than they are at present. What features would you look for in order to reconstruct past glacial extent on Mars?

Figure 16.37 Small-scale fluting formed at the margin of a modern glacier. The features formed by the infilling of cavities formed at the downstream side of each boulder. The largest boulder is about 0.5 m in diameter. Ice flow is away from viewer. (Source: John Shaw)

16.5 Summary

This chapter outlines the wide range of ice masses and their characteristics, and describes how they modify the landscape and how they are responding to climatic change. There are a range of feedbacks between the glacier water system, glacier dynamics and glacial thermal regimes. In summer, ice and snowmelt produce large volumes of water in many regions. This water must be removed on, within, or beneath the ice, often impacting on glacier dynamics by increasing rates of basal motion.

Glaciers move by three processes: ice deformation, basal sliding and the deformation of soft sediments beneath them. Of these, the last two are the most important in modifying the landscape: as the glacier sole scrapes over bedrock this bedrock is eroded by crushing, plucking and abrasion. The glacier then acts as a conveyor belt, moving this sediment downslope, modifying and finally depositing it. There are two transport routes through the glacial system: the high-level route via supra-glacial and englacial transport, which is passive and results in coarse, angular deposits; and the low-level route via basal transport, which is active.

Sediments that have been transported in the base of the glacier are typically bimodal and the coarser material becomes rounded and sometimes striated. Upland areas are typically source areas for ice where erosion domi-nates and this forms dramatic alpine landscapes. Lowland areas become covered with thick deposits of the resulting sediments. Understanding of the effects of glaciation on the landscape can be used when reconstructing past gla-cial extent and conditions.

Further reading
Books

Benn, D.I. and Evans, D.J.A. (2010) *Glaciers and glaciations*, 2nd edition. Hodder Arnold Education, Abingdon.
This recently updated book provides excellent and encyclopaedic coverage of all aspects of glaciers, glacial sediments and glaciated landscapes. It is suitable for courses throughout an undergraduate degree.

Bennett, M.R. and Glasser, N.F. (2009) *Glacial geology, ice sheets and landforms*, 2nd edition. John Wiley & Sons, Chichester.
This is an easy-to-access textbook covering most aspects of glacial systems.

Cuffey, K.M. and Paterson, W.S.B. (2010) *Physics of glaciers*, 4th edition. Butterworth–Heinemann, Oxford.
This is the glaciology 'bible' covering many topics comprehensively, albeit omitting landscapes and sediments. It is aimed at more advanced undergraduate courses or postgraduates.

Drewry, D. (1986) *Glacial geologic processes*. Edward Arnold, London.
This book is out of print but it is a unique approach to glacial geology. It contains many useful summary tables.

Hambrey, M.J. (2003) *Glacial environments*. Routledge, London.
Hambrey provides a concise introductory text with excellent photographs.

Hooke, R.L. (2005) *Principles of glacier mechanics*. Cambridge University Press, Cambridge.
A more advanced text for those interested in the details of glacier dynamics.

Hubbard, B. and Glasser, N.J. (2005) *Field techniques in glaciology and glacial geomorphology*. John Wiley & Sons, Chichester.
This book is worth delving into if you are interested in how measurements are made in the field in glacial environments and will certainly be helpful if you are planning such fieldwork for a project.

Post, A. and LaChapelle, E.R. (2000) *Glacier ice*. University of Washington Press, Washington, DC.
There are beautiful photographs of a wide range of glacial features and landscapes in this book. It contains mainly Alaskan examples.

Further reading
Papers

Clarke, G.K.C. (1987) Fast glacier flow: ice streams, surging and tidewater glaciers. *Journal of Geophysical Research*, 92(B9), 8835–8841.
This is a classic review paper covering concepts of balance velocity and fast glacier flow.

Fountain, A.G. and Walder, J.S. (1998) Water flow through temperate glaciers. *Reviews of Geophysics*, 36, 299–328.
This paper is the best review of glacier hydrology in temperate glaciers.

König, M., Winther, J.-G. and Isaksson, E. (2001) Measuring snow and glacier ice properties from satellite. *Reviews of Geophysics*, 39, 1–27.

This is an excellent review of remote sensing in glaciology and of relevance to Box 16.4.

Murray, T. (1997) Assessing the paradigm shift: deformable glacier beds. *Quaternary Science Reviews*, 16, 995–1016.
This is the author's own review of deforming glacier beds.

Quincey, D.J. and Luckman, A. (2009) Progress in satellite remote sensing of ice sheets. *Progress in Physical Geography*, 33, 547–567.
Contains useful summaries and examples of recent results from remote sensing.

Willis, I.C. (1995) Interannual variations in glacier motion – a review. *Progress in Physical Geography*, 19, 61–106.
This is a review of variability in glacial flow and the interactions between the glacial water system and glacier dynamics.

CHAPTER 17

Permafrost and periglaciation

Tavi Murray

Department of Geography, Swansea University

Learning objectives

After reading this chapter you should be able to:

➤ explain the differences between permafrost and periglacial environments and describe their main characteristics

➤ describe how ground temperatures in permafrost zones vary with time and define the active layer, and continuous and discontinuous permafrost

➤ describe and explain the main geomorphological features that form in permafrost and periglacial regions and their relict forms, as well as those geomorphological features that allow differentiation between these two environments

17.1 Introduction

Periglacial environments are defined as those that are cold but non-glacial, regardless of their spatial proximity to glaciers. These are environments in which freeze–thaw processes drive geomorphic change. Therefore they develop distinctive geomorphological features. **Permafrost** refers to soil or bedrock that is perennially frozen over long time-scales (greater than a minimum of two years). The presence of permafrost means that a periglacial environment exists

and periglacial processes are occurring. However, it is obviously possible to have a periglacial environment without permafrost. Permafrost usually contains ice within pore spaces (Figure 17.1). These ice features in the ground cause slow movement of the ground itself through expansion and contraction processes. Because the air temperature usually exceeds 0°C during summer in permafrost zones, a thin unfrozen layer (the **active layer**) can be present at the ground surface even when there are frozen layers below this. Currently permafrost affects approximately 26% of the Earth's surface (Washburn, 1979), while 35% is affected by freeze–thaw processes (Williams and Smith, 1989). The climate change that resulted in the advance and retreat of the major ice sheets during the past 2.6 million years (see Chapters 16 and 22) also resulted in past periglacial conditions far beyond the maximum extent of the ice sheets themselves. It is likely that an additional 20–25% of the Earth's surface has experienced periglacial conditions at some time in the past (French, 1996). These past periglacial conditions have left their mark on the landscape by producing periglacial landforms.

Freezing and thawing are important geomorphological processes as they drive fundamental changes in the ground's mechanical and hydrological properties. Both processes result in a volume change of the ground and there is typically a 9% increase in volume on freezing (Williams and

Figure 17.1 Ice layers from the top of the permafrost in Longyearbyen, Svalbard. (Source: Ole Humlum, University Centre, Svalbard)

Smith, 1989). This change in volume can result in frost shattering of rock. It is also partly the cause of **frost heave**, although other processes are also responsible. In fact, most of the ground movement caused by frost heave results because water migrates to the freezing zone as the ground freezes.

Frost heave is the vertical lifting of the soil surface and it disrupts structures and road surfaces in many parts of the world susceptible to permafrost conditions (Figure 17.2). Its understanding is therefore of great practical importance.

Most of the Earth's permafrost exists at temperatures just a few degrees below 0°C (IPCC, 2007a) and is thus highly susceptible to changes in climate or surface conditions. A slight warming or small changes to the ground surface, such as changes in vegetation cover, might cause dramatic change to permafrost zones. Particularly important for engineering in permafrost regions is the thawing that can occur owing to the increased insulation of the ground around a built structure, which can lead to subsidence (frozen ground often has 10–20% more water within it than unfrozen drained soil; see Figure 17.1). It is thus necessary for modern construction methods to take account of permafrost processes. This chapter is therefore of relevance to a large part of the Earth's land surface in terms of understanding landform development and human interactions with the landscape. It will begin by discussing permafrost processes before moving on to discuss permafrost and periglacial landforms and their formation.

Figure 17.2 Building subsidence in Dawson, Klondike, Yukon, Canada as a result of frost heave. (Source: Tony Waltham/Geophotos)

17.2 Permafrost processes

17.2.1 The distribution of permafrost

Today permafrost affects 82% of Alaska, 50% of Canada (Prowse and Ommaney, 1990; Trenhaile, 1990) and large parts of northern Siberia (Figure 17.3). This is known as **polar permafrost** and occurs because of low temperatures at high latitudes. **Alpine permafrost** occurs because of low temperatures at high altitudes. The major region of alpine permafrost occurs on the Tibetan Plateau, where some 2 million km^2 are affected (Figure 17.3). Alpine permafrost also occurs in the European Alps and the Rocky Mountains, North America. Permafrost is furthermore found beneath the sea, where it is known as **subsea permafrost**. The extent of subsea permafrost is not well known. Subsea permafrost usually occurs as a remnant of past colder temperatures and rising sea levels which drown frozen ground.

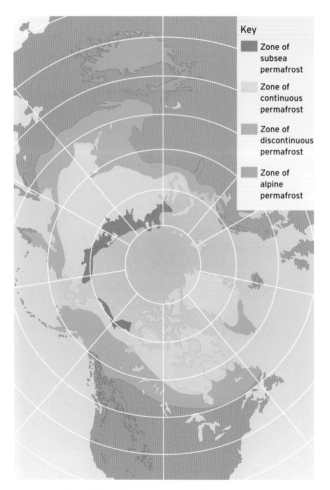

Figure 17.3 The distribution of permafrost in the northern hemisphere. Permafrost covers approximately 26% of the Earth's surface, the large majority of which is in the northern hemisphere. (Source: after Péwé, 1991)

A transect south across Canada from the Arctic Ocean towards the United States, as shown in Figure 17.4, demonstrates differences in the thickness and extent of permafrost. These differences correlate approximately with latitude. In the far north, the permafrost is laterally continuous and 1000 m in thickness. This is the **continuous permafrost zone**, where the frozen ground is broken only beneath lakes, rivers, glaciers and other thermal disturbances. Further south in the **discontinuous permafrost zone** the permafrost is not laterally continuous, and it varies from 1 to 10 m or more in thickness (see also Figure 17.3). The transition between the two zones occurs approximately at the mean annual air temperature of the –6 to –8°C isotherm (line of equal temperature) and often occurs at about the same location as the **treeline**. The southernmost extent of discontinuous permafrost is often taken to coincide with the –1°C isotherm, although isolated relict patches of permafrost occur further south, often in peatlands; this is sometimes known as sporadic permafrost. In both the continuous and discontinuous permafrost zones the ground is subject to annual and sometimes diurnal freeze–thaw close to the surface. This forms the active layer (Figure 17.5). Thawing of the active layer in summer causes considerable difficulties for travel in permafrost regions when thawed and water-logged terrain can become almost impassable. The active layer varies from a few tens of centimetres in the continuous permafrost zone to 15 m or more in the discontinuous permafrost zone. Unfrozen regions within the permafrost are known as **taliks**. Taliks are known as open if they are in contact with the active layer and closed if they are completely surrounded by permafrost.

Some measured permafrost depths are shown in Figure 17.5. At Resolute, Nunavut, Canada (74°N), the mean annual air temperature is –16.4°C. The permafrost thickness is greater than 400 m and the active layer is thin, only around 0.45 m thick. At Norman Wells, North West Territories, Canada (65°N), the permafrost is about 45 m thick with a moderate 1–2 m thick active layer. At Hay River, North West Territories, Canada (61°N), the mean annual air temperature is –4.4°C, the permafrost is discontinuous, approximately 12 m thick, and there is a 2–3 m thick active layer (Williams and Smith, 1989).

Fundamental changes in ground properties occur when it freezes. These changes include the ground mechanical properties (e.g. density, heat capacity, **thermal conductivity** and saturated hydraulic conductivity; see Chapter 11) and electrical properties such as **electrical conductivity**. These changes mean that geophysical survey techniques can be successfully used to map the extent of permafrost and massive ice within the ground. These techniques include seismic

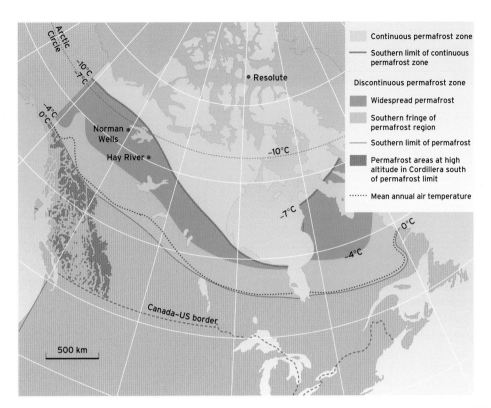

Figure 17.4 Permafrost zones in Canada. The continuous permafrost zone corresponds approximately to the −6 to −8°C isotherm of mean annual air temperature. (Source: adapted from Williams and Smith, 1989)

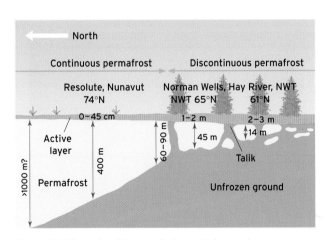

Figure 17.5 Permafrost in a north-to-south transect across Canada. The boundary between the continuous and discontinuous permafrost zones corresponds approximately with the treeline. Taliks are the unfrozen part of the ground within the permafrost. (Source: adapted from Williams and Smith, 1989)

reflection surveying, ground-penetrating radar (see Chapter 26) and electromagnetic induction. These techniques are useful when trying to establish the best locations and methods for engineering structures. The importance of permafrost for large engineering projects is discussed using the example of the trans-Alaska pipeline in Box 17.1.

17.2.2 Ground temperatures and permafrost thickness

Temperatures measured in permafrost boreholes are typically lowest close to the ground surface, excluding the active layer, and increase with depth, reaching the melting point at the base of the permafrost as shown in Figure 17.7 (Isaksen *et al.*, 2000). This trend may vary in the upper permafrost layer (see Box 17.2 for a discussion of how temperature varies in the ground in this upper layer). However, in general Figure 17.7 shows warming with depth. The main controls on permafrost thickness are the mean annual surface temperature, ground conductivity and the **geothermal heat flux**. Box 17.3 describes how to estimate permafrost thickness and the time taken to develop or degrade it. Surface albedo (see Chapter 4) can also play a role in permafrost processes. Dark vegetation cover may absorb more shortwave solar radiation and thus increase local ground temperature. Furthermore, vegetation cover or buildings can insulate the ground and cause thinning of permafrost.

The temperature at the Earth's surface varies between warmer summers and colder winters. However, deeper in the soil profile these variations are not so discernible. The **ground diffusivity** and the temperature fluctuation frequency control the depth to which surface temperature

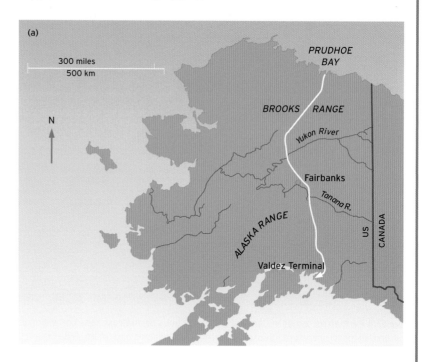

CASE STUDIES

THE TRANS-ALASKA PIPELINE

A large portion of the oil and gas reserves in North America lies along the northern coast of Alaska. The Alyeska oil pipeline (trans-Alaska pipeline) was built to transport oil 1300 km from Prudhoe Bay to Valdez at a cost of over $7 billion (Figure 17.6a). The pipeline consists of a 120 cm steel pipe that carries oil at a temperature of about 65°C; oil at lower temperatures is too viscous to transport through a pipeline. About 75% of the route overlies permafrost, which is about 600 m thick near Prudhoe Bay. The hot temperature of the pipeline would result in thawing of the surrounding ground if the pipe were buried or at the ground surface and this would result in subsidence and damage to the pipeline itself. The design chosen was to elevate the pipeline above the ground, thus minimizing thawing of the permafrost (Figure 17.6b). The pipe itself expands as it warms and contracts as it cools. To allow the pipe to do this as its temperature changes, the pipeline was built with bends. The pipe can move laterally by up to 4 m and vertically on its supports (Figure 17.6b). Furthermore, the vertical supports are equipped with thermal devices that help cool the permafrost during winter and prevent summer thawing. Provision was made for animals to cross the pipeline in certain places, which meant burying the pipeline for short distances with refrigeration units to prevent thawing of the ground. Despite its design the pipeline suffered two ruptures in 1977 and 1986 caused by unexpected settlement of the ground beneath (Williams, 1986).

Figure 17.6 The trans-Alaska pipeline: (a) the route of the trans-Alaska pipeline crosses both the continuous and discontinuous permafrost zones; (b) the design of the trans-Alaska pipeline allows it to move laterally and vertically with expansion and contraction. The supports are designed to prevent melting of the permafrost by the pipeline. (Source: Tony Waltham/Geophotos)

BOX 17.1

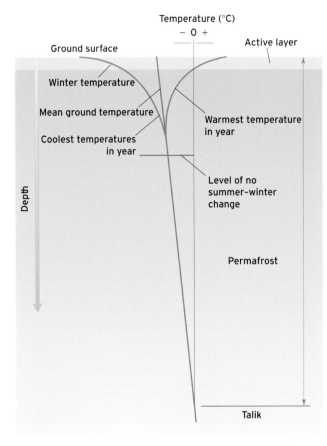

Figure 17.7 Theoretical temperature profile through material with constant thermal conductivity in a region in equilibrium with present conditions. (Source: after Isaksen *et al.*, 2000)

fluctuations are propagated. Box 17.2 describes the variation of ground temperature with depth in permafrost zones.

17.2.3 Reconstructing climate change from permafrost temperatures

Permafrost is ultimately a climatic phenomenon and there will be major changes in its distribution if climate warms as predicted over the twenty-first century. Because large quantities of carbon are stored in the permafrost, especially in peatlands and tundra regions, changes in distribution of frozen ground and the thickness of the active layer may result in the release of large amounts of greenhouse gas (e.g. methane) into the atmosphere (see below). Permafrost responds to climate change only slowly by aggrading or degrading over thousands of years. It therefore damps out short-period oscillations and records major climatic change of several degrees over long timescales (Figure 17.9). Under equilibrium conditions the temperature profile within the ground is linear with depth (assuming a constant ground thermal conductivity). The ground's temperature profile can therefore be used to extrapolate past surface conditions, particularly in tectonically stable areas. Permafrost is generally increasing in thickness in those areas with low surface temperatures and recent tectonic uplift (e.g. Melville Island, Canadian High Arctic) or glacial retreat and degrading in areas with warmer surface temperatures or where the thickness of snow accumulation is increasing (Zhang, 2005). In

FUNDAMENTAL PRINCIPLES

SEASONAL VARIATION OF GROUND TEMPERATURE

Figure 17.8 shows the air and ground temperatures measured at Barrow, Alaska (71ºN). Compare the variation in air temperature with the temperature at different depths in the ground. The air temperature is the forcing mechanism and the ground temperature is the response to this forcing. In general, the response is damped (the variation in temperature is less) at greater depths in the ground. For example, the air tempera-

ture shows a double peak in July/August, which can also be seen at 0.6 m depth in the ground but delayed by approximately one month. Below this depth this detail cannot be seen. It becomes more difficult to see the maxima and minima at greater depth in the ground and at 18.2 m there is almost no variation in temperature from summer to winter. The graphs also show the progressive delay in timing of the maxima and minima as they propagate into the ground. For example, at 9.1 m depth the maximum temperature occurs in March delayed

by some 8 months from the air temperature change that caused it.

The wavelength of a temperature variation, λ (such as diurnal or annual temperature variation), is given by:

$$\lambda = \sqrt{4\pi D/f} \qquad (17.1)$$

where D is the diffusivity of the ground (Table 17.1) and f is the frequency of the perturbation. The thermal diffusivity varies with soil type as shown in Table 17.1. At a depth of one wavelength, the amplitude of the temperature change is reduced

BOX 17.2 ➤

➤

by a factor of about 0.002. Diurnal fluctuations are thus attenuated rapidly and affect only the upper 1 m or so, whereas annual fluctuations propagate to depths of 15 m or more. The timing of the maximum and minimum temperature becomes progressively lagged with depth and the lag increases with decreasing frequency; the velocity of propagation of maxima and minima into the subsurface is $4D\pi f$.

It is possible to estimate the depth of the active layer at the site shown in Figure 17.8. The base of the active layer occurs where the ground temperature remains below freezing throughout the year. The graphs show that at 0.6 m depth in the ground the temperature is above freezing during July and August. At 2.4 m depth the temperature remains below freezing throughout the year; the active layer is therefore between

0.6 and 2.4 m in thickness. A final observation is that at 0.6 m depth the ground temperature remains constant at 0°C during October and November despite the air temperature becoming progressively colder during this period. This is because of the release of latent heat of freezing, which prevents the temperature of the ground dropping below 0°C until all the water at this depth in the soil is frozen.

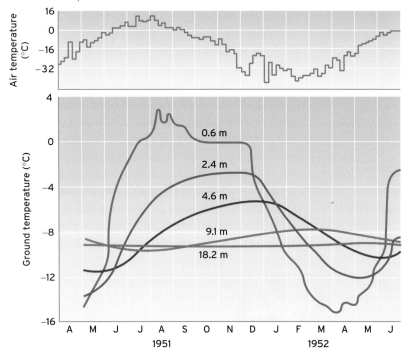

Figure 17.8 Typical air and ground temperatures at Barrow, Alaska. Surface temperature fluctuations take longer to have an impact for deeper layers and so the lag time between surface temperature increase and ground temperature increase is greater with depth. The magnitude of the change is also much greater near the surface and declines with depth. (Source: after Price, 1972)

Table 17.1 Typical thermal properties of soils and constituent components

Material	Water content (m³ m⁻³)	Density (kg m⁻³)	Mass heat capacity (J kg⁻¹ K⁻¹)	Thermal conductivity (W m⁻¹ K⁻¹)	Thermal diffusivity (×10⁻⁶ m² s⁻¹)
Sandy soil	Dry	1600	800	0.3	0.2
	0.2	1800	1180	1.8	0.9
Clay soil	0.2	1800	1250	1.2	0.5
Water (0°C)		1000	4180	0.6	0.1
Ice (0°C)		917	2100	2.2	1.2
Air		1.2	1010	0.03	20.6

(Source: after Williams and Smith, 1989)

BOX 17.2

ESTIMATING PERMAFROST THICKNESSES

As discussed above, the controls on permafrost thickness are the mean annual surface temperature (T), ground conductivity (k) and the geothermal heat flux (G) and these parameters can be used to estimate the equilibrium permafrost thickness, H:

$$H = \frac{Tk}{G} \qquad (17.2)$$

An estimate of the time t taken to form a thickness of permafrost H from unfrozen sediment at 0°C can be obtained from:

$$H = \sqrt{\frac{2kTt}{\rho L n}} \qquad (17.3)$$

where ρ is the density of the ground, L is the latent heat of fusion of ice and n is the porosity of the soil (Lock, 1990). Equation (17.3) is applicable to the early stages of growth of ice-rich permafrost from zero initial thickness. Large discrepancies in actual thickness compared with this simple model (equation (17.2)) can occur, mainly because of the presence of surface thermal disturbances such as water bodies and because of the long timescales for the response of permafrost to climate change.

BOX 17.3

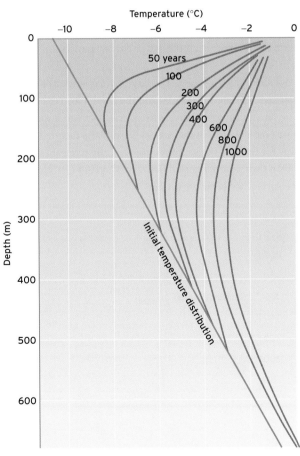

Figure 17.9 The effect of climate change on ground temperatures. The theoretical ground temperature curves show the effect of climate change (in this case a step climate warming) on the measured temperature in the ground. The figure ignores the active layer. (Source: reprinted with permission, after *Proceedings of the Second International Conference on Permafrost, Yakutsk, USSR* (1973) by the National Academy of Sciences, courtesy of the National Academies Press, Washington, DC)

Svalbard warming of the permafrost has been measured to 60 m depth at rates of ~0.04–0.07°C/year (Isaksen *et al.*, 2007), and in Alaska warming has been up to 3–4°C (Osterkamp, 2007). The presence of snow can make the ground surface cooler because of its high surface albedo, although the presence of seasonal snow more usually raises the ground temperature by insulating the ground from cold winter temperatures.

Permafrost is a globally significant carbon store, which is highly sensitive to climate change. In some regions, such as Siberia, permafrost contains large stores of carbon-rich material including animal bones, roots and peats, which are relics of tundra ecosystems from the last glacial period. As the permafrost melts, this carbon has the potential to be converted rapidly to methane and hence cause climate warming in a positive feedback (Zimov *et al.*, 2006).

17.2.4 Gas hydrates

A gas hydrate is a crystalline solid in which molecules of gas are combined with molecules of water. The hydrate of methane is stable at high pressures and low temperatures and so occurs commonly in ocean sediments with smaller amounts in permafrost regions. Global estimates of the methane stored in hydrate are around 10^{16} kg, which represents one of the largest sources of hydrocarbons on Earth: there is about twice as much carbon stored in gas hydrates than in all other fossil fuels put together. Increases in temperature or decreases in pressure may result in the hydrate becoming a mixture of gas and ice or water. The instability of hydrates may be a hazard during drilling operations, resulting in blow-outs due to gas build-up following warming via the drilling mechanism (Yakushev and Chuvilin, 2000). Hydrates of carbon dioxide are stable in the Martian ice caps, which exist at very low temperatures (140–155 K or –133 to –118°C). The flooding that appears to have scarred the landscape of Mars in the past may have resulted from catastrophic breakdown of these hydrates and associated release of greenhouse gases (Kastner *et al.*, 1998).

Hydrates may become commercially viable as a natural resource. Significant deposits are thought to exist in the Messaryakha gas field in western Siberia, the Mackenzie Delta and Arctic Islands and the Alaska North Shore. However, hydrates may contribute to future global warming (Kvenvolden, 1995) both via human extraction and combustion, and as a feedback response to global warming. With climate warming methane gas will be released from permafrost regions as they defrost. Methane is a strong greenhouse gas, and so this may result in positive feedback, leading to warmer global temperatures and release of further gas to the atmosphere. Release of methane at the end of the last glacial period may have played an important role in the rapid melting of the major ice sheets (Buffet, 2000).

17.2.5 Hydrology in permafrost regions

17.2.5.1 Groundwater flow

Groundwater movement within permafrost regions is often restricted by the presence of frozen ground that acts as a barrier to flow. At temperatures significantly below 0°C, the hydraulic conductivity of the ground is greatly reduced, exerting a retarding influence on groundwater flow. In such regions the freezing of the active layer introduces a seasonal aquitard (see Chapter 11) and the underlying permafrost represents a perennial aquitard to vertical flow, restricting aquifer recharge and water flow. In these regions permafrost can act as a barrier to contaminant transport. However, at temperatures close to zero there may be little difference between the conductivity of frozen and unfrozen regions of the ground (Anderson and Morgenstern, 1973), allowing water flow through frozen ground and aquifer recharge. In the discontinuous permafrost zone, water can flow through open taliks. Furthermore, in some regions seasonal freezing of the active layer does not reach the depth of the permafrost, allowing groundwater flow through a residual thaw layer above the permafrost confined below the upper frozen part of the active layer. Water can also flow beneath the permafrost in both discontinuous and continuous permafrost zones. The occurrence and flow of groundwater in permafrost regions are governed by the same physical processes as in more temperate regions (see Chapter 11).

17.2.5.2 River flow

River regimes in periglacial regions are typically very seasonal with large discharges resulting from the melt of winter snow cover during spring. This snowmelt usually occurs over a short period of time, typically 2–3 weeks, resulting in a short-lived flood event. In many areas between 25 and 75% of the total runoff is concentrated in a few days. Furthermore, because permafrost retards downward percolation, runoff is often rapid following a snowmelt or rainfall event. In glaciated catchments the large and variable flow continues into summer owing to ice melt and changes in the glacier hydrological system as shown in Figure 17.10.

Small rivers in periglacial regions flow only during the summer. However, despite cold winters major rivers continue to flow all year round under an ice cover. Thus, a distinction should be made between these large rivers and smaller rivers. The large rivers in the Arctic are often fed from a mixture of snowmelt plus input from non-periglacial regions and from deep springs within the discontinuous permafrost zone. Their discharge characteristics are therefore less peaked than smaller rivers which are fed solely from snow- and ice melt. The larger rivers, such as the Mackenzie River in Canada, and the Lena, Ob, Yenesei, Kolyna and Indigirka Rivers in Siberia, form an important transport network in the Arctic. Despite the fact that they flow for most of the year they are typically navigable for only a short annual period between May and October when the surface ice layer breaks up.

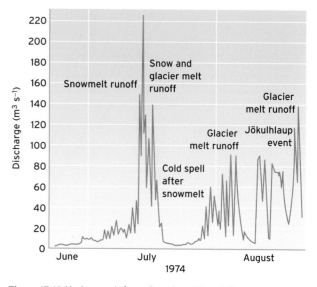

Figure 17.10 Hydrograph from Sverdrup River, Ellesmere Island, Nunuvut, Canada, showing a major peak from snowmelt and subsequent peaks from glacier melt and a glacier outburst flood known as a jökulhlaup. The basin area is 1630 km². (Source: after Woo, 1986)

The more peaked runoff characteristics of smaller rivers result in a high sediment transport rate compared with rivers with similar total discharges but which have discharge more evenly distributed during the year. For example, the River Mecham in Arctic Canada is fed largely by snowmelt. As a result 80–90% of its annual flow is concentrated into a 10 day period during which peak velocities may reach 4 m s^{-1} (Summerfield, 1991). Total sediment yield from the catchment is estimated to be 22.1 t km^{-2} yr^{-1} (French, 1996).

17.2.5.3 Seasonal features

Seasonal icings are mounds of ice that form in topographic lows during winter in locations where groundwater reaches the surface (Figure 17.11a). These are essentially zones where return flow (a component of saturation-excess overland flow, see Chapter 11) occurs and freezes. Icings also occur in river channels that freeze to their beds. Icings are common features downslope of warm-based glaciers (see Chapter 18) in Svalbard, Norway. Icings may form either below or above the ground surface. The ice forming these features is often stratified and may comprise **candle ice**, which consists of vertically orientated crystals of over 1 m in length (Figure 17.11b). In some regions, for example Yakutia, Siberia, large icings may form that do not fully melt in the summer and so survive from year to year.

(a)

(b)

Figure 17.11 (a) Icing formed downslope of a warm-based glacier in Yukon Territory, Canada. (b) The icing is made up of stratified ice and candle ice such as these elongated crystals. The black lens cap provides some idea of the scale of the crystals which can be over 1 m in length.

> ### Reflective questions
>
> ➤ What are the controls on the distribution of permafrost?
>
> ➤ How are surface temperature changes reflected by ground temperatures in permafrost zones?
>
> ➤ Aklavik, a town in Canada situated in a permafrost region, was planned to be replaced in 1954 because of flooding, subsidence and difficulties in construction due to increasing depth of the active layer. Inuvik was built to replace Aklavik (although in fact many residents stayed in Aklavik). What considerations would you use to choose the most suitable site for a new town like Inuvik?
>
> ➤ What are the characteristics of river regimes in periglacial environments?
>
> ➤ Why are permafrost regions particularly sensitive to climate change and why might they contribute to further global warming?

17.3 Geomorphology of permafrost and periglacial environments

In this section we discuss the geomorphological features that form in periglacial and permafrost environments (see Section 17.1 for the distinction). In particular we will emphasize those features that allow differentiation between the two. Cold climate environments develop distinctive geomorphology because of three basic processes:

1. The 9% expansion of water on freezing which causes frost shattering and in turn adds to scree development.
2. The contraction and cracking of rapidly freezing soils which forms **ice wedges** and polygonally shaped surface features.
3. The migration of water to the freezing front by suction which causes the formation of segregated ice.

Furthermore, permafrost and periglacial features form distinctive relict forms, collectively known as **thermokarst** because of their similarity to **karst** (cavernous limestone) features. Thermokarst features are extremely important in reconstructing past climate and the extent of former glaciers. They form in cold but non-glaciated terrain and are thus found at the margins of glaciers and can be used to demarcate former glacial boundaries. The presence of thermokarst features within the United Kingdom and Europe implies widespread permafrost conditions during the past (Figure 17.12). A range of active and relict permafrost and periglacial landforms are described in the following sections.

17.3.1 Ground ice features

Very large lenses of ice may slowly build up in soil that is frozen as a result of the migration of water to the freezing front in permafrost regions. These bodies of ice typically form only in the upper 5–6 m of ground and are known as **segregated ice**. Segregated ice may form thin bands. Where the bands are thick, sometimes up to several metres, they are known as **massive ice** (Figure 17.13). In both cases the concentration of ice can exceed 50% by volume. Clearly, the melting of such ground will produce a large volume of excess water. Such melting can cause seasonally impassable roads or the collapse of structures. Coarse deposits such as gravels and sands are highly permeable but have low potential to retain water, whereas finer deposits such as clay have low permeability but high water retention potential. This means that intermediate grain sizes, such as silt, have the greatest potential to form segregated or massive ice within the ground and are most susceptible to ice heave. When ice-rich ground thaws, **involutions** often form. Such features are shown in Figure 17.14. They are disruptions to the sedimentary structure of the ground and these features are often used as a diagnostic for past permafrost conditions. The melting of massive ground ice also affects the ground surface topography and produces thermokarst consisting of small irregularly shaped thaw lakes and depressions known as **alas**, which form when these thaw lakes drain. These features cover large areas in North America and Siberia. Similar features have been reported on the planet Mars (see Box 17.4).

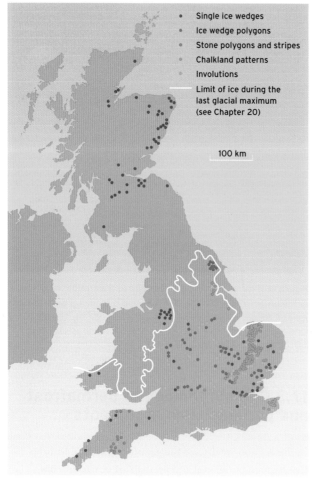

Figure 17.12 The distribution of periglacial patterned ground in the United Kingdom. Widespread periglacial conditions occurred beyond the margin of the Quaternary ice sheets. (Source: after Sparks and West, 1972)

Legend for map:
- Single ice wedges
- Ice wedge polygons
- Stone polygons and stripes
- Chalkland patterns
- Involutions
- Limit of ice during the last glacial maximum (see Chapter 20)

100 km

Figure 17.13 An example of banded massive ice, Peninsula Point, near Tuktoyaktuk, NWT, Canada. (Source: photo was taken by Julian Murton at Peninsula Point near Tuktoyaktuk, NWT, Canada)

Figure 17.14 Involutions (frost-disturbed structures) in Pegwell Bay, southern England. The involutions are the white/grey-coloured features protruding upwards into the brown soil and indicate the former presence of permafrost at this location. The structures are related to loading and density differences in water-saturated sediments, probably during the thawing of ice-rich permafrost. (Source: Dr Julian Murton/University of Sussex)

Ice wedges are V-shaped bodies of ground ice up to 1.5 m in width that can extend some 3–4 m into the permafrost (Figure 17.16a). Ice wedges develop because at low temperatures (less than approximately –15°C) frozen ground contracts as it is further cooled. If this occurs rapidly then cracking can occur as shown in Figure 17.17. The cracking of the feature is thought to occur in early winter, and the crack fills with water in spring and summer, which then freezes. The ice wedge, once developed, creates a weakness that tends to reopen annually as the ground contracts and hence the ice wedge grows. Ice wedges often exist in a network of **ice wedge polygons** (Figure 17.16b) which currently cover millions of square kilometres of the Earth's surface (Williams and Smith, 1989). Such polygons can also be identified on Mars (Box 17.4). Ice wedges actively grow in the continuous permafrost zone, forming only in perennially frozen ground, although some wedges may persist in the discontinuous permafrost zone. When ice wedges melt they often leave behind a landform known as an ice wedge

NEW DIRECTIONS

PERMAFROST AND PERIGLACIATION ON MARS

On the planet Mars, the mean annual temperature is −60°C and the planet has a dry periglacial-type climate. Permafrost currently extends over the planet's entire surface. Early in Martian history there is evidence for warmer climates and liquid water on the planet surface. At this time freeze-thaw processes were probably common. Certainly the surface of the planet provides evidence for polygonal-patterned ground similar, but at a larger scale, to that formed on Earth (Figure 17.15). Furthermore, the presence of apparent thermokarst features such as rampart craters, which resemble those resulting from the melting of ground ice on Earth, suggest high ice contents within the permafrost. On Earth, microbial life

has developed strategies to cope with extreme conditions and can survive within permafrost and basal ice beneath glaciers. It is thus possible that similar life forms have developed on Mars.

Figure 17.15 Taken from NASA's Pheonix spacecraft, northern Mars, showing the surface of Mars. The patterns should be compared to similar features on Earth in permafrost regions such as shown in Figure 19.22(a) and (b). (Source: NASA/ JPL-Caltech/University of Arizona)

BOX 17.4

(a)

(b)

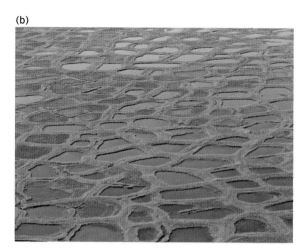

Figure 17.16 Ice wedge landforms: (a) ice wedge in Svalbard; (b) ice wedge polygons seen from the air. (Source: Alfred-Wenger-Institute)

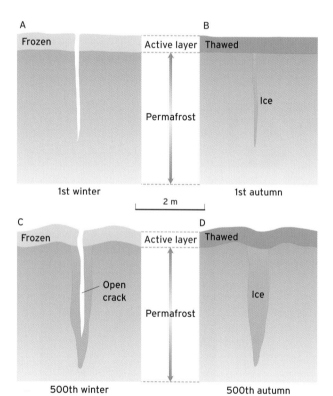

Figure 17.17 Formation of ice wedges. Soil will crack if it is cooled quickly (A). In summer this crack will fill with water (B), which subsequently freezes and expands the crack. Repeated thaw and freeze events (C) lead to the formation of an ice wedge (D). (Source: after Lachenbruch, 1962)

Figure 17.18 Ice wedge cast structures in the soil on the north end of the airstrip at Massacre Bay on Attu Island. (Source: USGS)

cast, as shown in Figure 17.18. The ice is replaced by sediment, occasionally forming polygonal or linear troughs. Such features are easily recognized in sediment sections that were marginal to the ice sheets in the United Kingdom and Europe and they are reliable thermokarst features for identifying past permafrost conditions.

Pingos are ice-cored mounds up to 55 m high and 500 m in length which form in permafrost zones. Examples are shown in Figure 17.19. The mounds can be either conical or elongated and they contain some segregated ice and a core of massive ice described as a lens. The top of the mound often becomes cracked as the ice core within the pingo grows. Two types of pingos are recognized: hydrostatic pingos (Figure 17.19a; formerly known as closed-system or Mackenzie Delta pingos) and hydraulic pingos (Figure 17.19b; formerly known as open-system or east Greenland pingos).

Hydrostatic pingos are caused by the doming of frozen ground as a result of the freezing of water expelled during talik elimination and the growth of permafrost beneath a former lake or other water body (Figure 17.20a). The features are usually isolated landforms found in continuous permafrost regions, predominantly in areas of low relief. Pingos formed over drained lakes are usually circular in shape, whereas pingos over old river channels may be linear

(a)
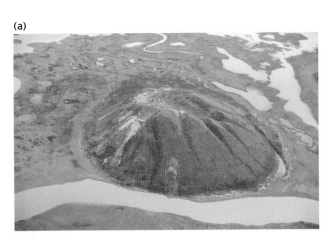

(b)

Figure 17.19 (a) Aerial view of a hydrostatic pingo in NWT, Canada. The Ibyuk pingo is 49 m high and 300 m long. (Source: Paolo Koch/ Science Photo Library Ltd.) (b) Hydraulic pingo in Svalbard. (Source: Ole Humlum, University Centre, Svalbard)

(a)
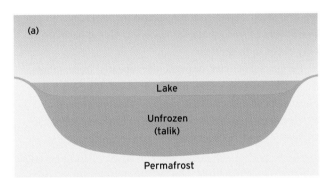

(b)
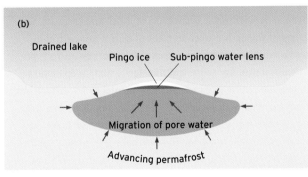

(c)
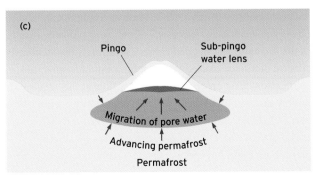

Figure 17.20 Formation of a hydrostatic pingo: (a) talik develops beneath a thaw lake; (b) lake drainage leads to refreezing of the talik; (c) progressive refreezing of the talik leads to pingo growth. (Source: after Mackay, 1983)

in form. After the drainage of a water body, a closed talik may form within the ground, which traps water within the freezing sediments. Migrating water moves to the freezing front, forming pore and segregated ice (Figure 17.20b). Very high pore water pressures develop that force the ground upwards creating the pingo (Figure 17.20c). The growth rate of pingos in the Mackenzie Delta has been measured to be initially quite fast at approximately 1.5 m yr^{-1} (Mackay, 1973, 1979) and to decrease with age so that some of the largest pingos are probably around 100 years old and growing at a rate of 2.3 cm yr^{-1} (Mackay, 1986). Similar results have been determined in Siberia. As pingos grow, their sides become steeper, eventually opening radial cracks at their summits (Figure 17.19), which expose the ice core and the pingo begins to subside. Hydrostatic pingos are common on the Mackenzie Delta, where they are found across the delta following switches in the course of the river. Hydrostatic pingos have also been reported on the floor of the Beaufort Sea.

Hydraulic pingos form at the foot of slopes and result from the inflow and freezing of groundwater seeping from upslope (Figure 17.21). These features are most likely to

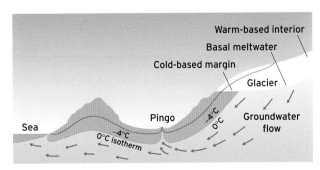

Figure 17.21 Formation of a hydraulic pingo. This figure shows a pingo forming from drainage from beneath a warm-based glacier in Svalbard. (Source: after Liestøl, 1977)

occur in thin or discontinuous permafrost regions. In Svalbard, Norway, the water supply for pingos often flows from warm-based glaciers upslope as in the example given in Figure 17.21. Hydraulic pingos are usually round or elliptical in form and typically smaller than hydrostatic pingos, rarely exceeding 35 m in height. They often consist of multiple mounds (unlike hydrostatic pingos) because as one begins to subside the continuing water supply will tend to form another. These features are common in east Greenland, central Yukon, Alaska and Svalbard.

The melting of the ice core of a pingo initially leaves a small lake, but the final relict form is a central depression with sediment ramparts, sometimes termed a **pingo scar**. Such features have been identified at various locations in Europe including the United Kingdom and the Netherlands.

Palsas are low mounds, 1–10 m high, that form in peat in permafrost zones (Gurney, 2001). They form where snow is thin or discontinuous and have a core of segregated ice generated through suction of unfrozen water migrating to the freezing front. The lack of snow allows the ground to freeze to a greater depth than the surroundings and this ice then survives the subsequent summer, insulated by the peat above. Repeated winters result in frost heave, and a mound forms. This topography causes a positive feedback as snow is more likely to blow off, resulting in even deeper freezing depths. Changing vegetation cover can also play a role in palsa formation. When palsas melt, mounds and small lake-filled hollows are formed. **Frost blisters** are small ice-cored mounds that develop over just a single winter as a result of groundwater that freezes and uplifts the ground surface.

The ground surface in periglacial regions is often characterized by metre-scale organization of topography, vegetation or particle size in regular geometric patterns. There are two types of patterned ground. These are sorted patterned ground, such as circles, polygons or stripes, and unsorted patterned ground defined by topography, such as unsorted circles and stripes, or by alternation of vegetated and non-vegetated ground.

Sorted stone circles are typically arranged so that the fine material occurs in the centre of an area of lowered relief and coarse material forms an uplifted outer rim as shown in Figure 17.22(a) and (b) (Hallet *et al.*, 1988). Typically, polygons or circles occur on flat surfaces, and sorted stripes that are elongated downslope form on low-angle slopes (Figure 17.22c). On slopes greater than about 30°, mass movement prevents the formation of patterned ground. The features probably form by a variety of processes. One hypothesis involves convection within the ground. In

summer, saturated soil close to the ground surface warms during the day whereas water at depth remains colder. Since water is densest at 4°C the colder water at depth is less dense than the water close to the surface, and this drives convection. Descending warm water can then melt the frozen surface below resulting in an undulating interface between frozen and unfrozen ground that is reflected in the surface topography. Sorting of the soil particles can occur if the soil particles convect with the soil water (Figure 17.23).

17.3.2 Slope processes

Many of the features that form in lowland areas also occur in alpine environments above the treeline. However, these alpine environments are often characterized by exposed, hard bedrock that has been eroded by the action of glacier ice, and steep slopes. Mountain tops are often covered by **blockfields** of frost-shattered material. Alpine permafrost may occur in these areas.

Periglacial activity, in particular repeated freeze–thaw, causes the formation and modification of slope deposits. Four types of processes cause **mass wasting** in periglacial environments: slopes evolve owing to fracture, debris (see Chapter 8) and **solifluction** flows, creep processes and **nivation**. Solifluction is especially likely to occur in regions underlain by permafrost. During the summer the active layer melts forming a mobile water-saturated layer. The process results in the formation of lobes and terraces. Slow creep of the active layer, at rates of only a few centimetres a year, occurs by processes of **frost creep** and a type of solifluction known as **gelifluction**. Frost creep occurs because freezing expands the soil normal to its surface but thawing results in settlement vertically resulting in a net downward movement. Gelifluction is this slow creep of water-saturated material. In seasonally frozen soil, movement typically occurs during spring as the ground thaws from the surface downwards. Because the underlying permafrost is largely impermeable, mass movement may occur on very low slopes (as low as 1°). Slow creep typically results in the formation of stepped ridges. Nivation is the localized erosion of a slope caused by a combination of frost action, gelifluction, frost creep and meltwater flow at the edges and underneath snow patches. Nivation commonly occurs in periglacial regions and is accentuated in permafrost-free areas. The combination of processes causes the development of nivation hollows as the snow patches sink into the hillside.

While most of these processes are not unique to periglacial environments and Chapter 7 provides details of their operation, mass wasting is probably most efficient

(a)

(b)

(c)

Figure 17.22 Organized topographic features: (a and b) stone polygons, Svalbard; (c) stone stripes, Svalbard. (Source: photos courtesy C. Fogwill, Exeter University)

in periglacial conditions (Table 17.2). The formation of screes from freeze–thaw fracturing of near-vertical rock faces occurs mainly in resistant rocks that are permeable only along fractures. Snow avalanches may move significant volumes of rock debris. Flows of unfrozen material are promoted in permafrost regions because the seasonal melting of the active layer forms an upper layer of high water content.

Periglacial slope processes result in a range of landforms particular to periglacial slopes. **Protalus (or pronivial) ramparts** are linear mounds of coarse sediment that form a small distance from the base of a slope (Figure 17.24). Snow persists at the foot of these slopes, particularly in the shade, which means that when a rockfall occurs boulders tend to slide across the snow and come to rest just beyond the

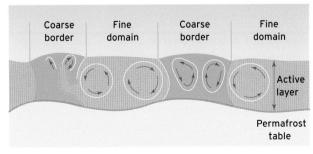

Figure 17.23 Inferred soil circulation patterns resulting in sorted circles. (Source: after Hallet *et al.*, 1988)

515

Figure 17.24 Fossil example of a pronivial (protalus) rampart from the Valldalen area of southern Norway. (Source: photo courtesy R. Shakesby)

Table 17.2 Rates of mass wasting by different processes at Karkevagge, Lapland, 1952–1960

Process	Rate (t km^{-1} yr^{-1})	Denudation (mm yr^{-1})
Transport of solutes by running water	26	0.010
Debris slides and flows	49.4	0.019
Slush avalanches, rock debris transport	14	0.005
Rockfalls	8.7	0.003
Solifluction	5.4	0.002
Talus creep	1.5	0.001

(Source: adapted from Rapp, 1960)

Figure 17.25 Ploughing boulder. A prow has built up in front of the boulder and a trough has formed upslope of it.

snow bank. **Ploughing boulders** can also be seen on slopes in periglacial regions as shown in Figure 17.25. The boulders move slowly downslope, leaving a trough upslope and forming a prow of sediment downslope. The movement of the boulders, typically a few millimetres per year, is thought to occur because of the different thermal conditions beneath the boulder compared with its surroundings.

Rock glaciers, as shown in Figure 17.26, typically consist of angular debris and have the form of a small glacier. They often show evidence of flow although there is no evidence of glacier ice at their surface. Within a rock glacier there is ice within the pore spaces (**interstitial ice**), and the slow flow of these features downslope is due to this. Some rock glaciers are probably formed from relict glaciers that have become debris covered, whereas other features are thought to originate as snow and rock avalanches or ice-cemented rock debris in permafrost regions. Relict rock glaciers have been identified in the United Kingdom, including one in North Wales (Harrison and Anderson, 2001).

Figure 17.26 Rock glacier in Svalbard forming from glacier ice covered by debris. Clean ice can still be seen in the upper part of this rock glacier.

17.3.3 Loess and aeolian activity

Many permafrost regions are characterized by extreme aridity and are defined as polar deserts. In these regions there is abundant evidence for the transport of material by wind and for erosion by abrasion of the wind-blown material. Dust clouds are characteristic of many permafrost regions in autumn (Sparks and West, 1972). Material is deposited when the wind velocity drops or when precipitation falls, forming a silt deposit known as loess. During the past 2.6 million years, deposits of fine sediment formed on the outwash plains of the ice sheets, providing a plentiful source of material for transport. Intense frost action broke down larger material and provided additional material. Stronger winds occurred during this period. As a result erosion by wind-borne sediment and wind entrainment has played a major role in the middle latitudes. Erosion formed wind-modified pebbles and blocks as well as **stone pavements**.

Deposition resulted in loess which covers tens of thousands of square kilometres and is more than 100 m thick in some places. Loess deposits occur in large areas of North America and Europe that were situated on the southern margin of the ice sheets (see Chapter 22).

Reflective questions

➤ How can pingos and ice wedge casts be used to make climatic reconstructions?

➤ What are the main landforms associated with ground ice and how do they form?

➤ How do slopes evolve in periglacial conditions?

➤ What are the unique geomorphological features found on periglacial slopes?

17.4 Summary

This chapter describes both permafrost and periglacial environments, their distribution and the processes that occur within them. Periglacial processes currently affect about 35% of the Earth's surface, and during the past their impact was even more widespread. This past activity has left a range of landforms that are characteristic of such environments and that can therefore be used to reconstruct past climates. Understanding the processes operating and features formed in these is important for engineering in permafrost and periglacial regions. These regions are extremely sensitive to climate change as well as being the possible source of greenhouse gases that may drive further climate change through gas hydrate and soil carbon release.

Permafrost can be found in high latitudes and at high altitudes. In the coldest locations it can be over 1000 m deep but in more marginal climates it can be discontinuous and thinner. The upper ground layer often melts during the summer and this is known as the active layer. Surface temperature fluctuations propagate down through the ground so that there is a delay in response. The lag time increases with depth and the magnitude of change decreases with depth. Deeper layers respond only very slowly to surface temperature change so that they can be used to identify former temperature regimes at that site.

Both permafrost and periglacial regions have a unique geomorphology with a range of landforms that develop owing to ice formation and melt at and below the ground surface. These melt and thaw processes cause expansion and contraction of the surface layers which results in surface collapse and cracking. As cracks fill with water and freeze they can expand and eventually large ice wedges can form. These are manifest at the surface by polygonal features. Freezing of water beneath thermal disturbances such as lakes can cause trapped water at high pressure which can result in features such as pingos. Slope features include rock glaciers and ploughing boulders. Periglacial slopes are dominated by mass wasting and gelifluction and frost creep processes.

Further reading
Books

Ballantyne, C.K. and Harris, C. (1995) *The periglaciation of Great Britain.* Cambridge University Press, Cambridge.
This is an excellent book giving detailed explanations of relict periglacial features, their distribution and the processes that formed them. Although it is related to Great Britain the discussion can equally be applied to other sites.

French, H.M. (2007) *The periglacial environment*, 3rd edition. John Wiley & Sons, Chichester.
A new edition of this clearly written textbook with an excellent range of diagrams and good use of examples.

Trenhaile, A.S. (2003) *Geomorphology: A Canadian perspective*, 2nd edition. Oxford University Press, Oxford.
This is a book describing the physical geography of Canada. Many of the descriptions are of features developed in permafrost and periglacial regions.

Williams, P.J. (1979) *Pipelines and permafrost: Physical geography and development in the circumpolar north.* Longman, Harlow.
This is an excellent short book describing the making of the Alaska pipeline and other related topics.

Williams, P.J. and Smith, M.W. (1991) *The frozen Earth: Fundamentals of geocryology.* Cambridge University Press, Cambridge.
This is the permafrost 'bible', although much is at quite a high level. If you want more information on processes in permafrost regions this is the book to look at.

Further reading
Papers

Buffet, B.A. (2000) Clathrate hydrates. *Annual Reviews of Earth and Planetary Science*, 28, 477–507.
This is a paper providing further (but technical) details on hydrates.

Zhang, T. (2005) Influence of the seasonal snow cover on the ground thermal regime: an overview. *Reviews of Geophysics*, 43, RG4002, doi:10.1029/2004RG000157.
An investigation of the science looking at how snow impacts ground temperature.

PART V

Biogeography and ecology

Figure PV.1 A poisonous mushroom, *Amanita muscaria*, often known as the fly agaric. Fungi such as these serve a vital role in helping to decompose dead matter to recycle nutrients so that they are ready to be taken up by new plants.

Part contents

Scope

The various features of the biosphere can be associated with all aspects of geography, from the climate system, oceanography, geology, hydrology, social issues and even global tectonics. These interactions result in distinctive regions of plants and animals (biomes) that differ depending upon the controlling variables, and often phase into one another along a gradient of change. Chapter 18 discusses the characteristics of these various biomes. It is not enough, however, simply to describe their features, but it is necessary to explain their nature and form. Chapter 19 is therefore concerned with studying the processes behind the spatial distribution of plants and animals and their change over time.

Any study in biogeography is inextricably linked to ecosystem processes which consider not only transfers of energy and matter but also individual species characteristics. Chapter 20 delves deeper into the characteristic features that underpin any understanding of modern ecology, such as the nature and role of ecosystems, habitats, communities, life strategies and the environmental niche at all scales down to an individual tree. By studying ecosystem processes, we become aware of the very dynamic nature of ecosystems and the checks and balances operating as drivers of ecological change. The planet's biomes are not static. Closer observation shows that important links between plants, animals and soils are related to processes involving the movement of energy and organic and inorganic materials through the system so that ecosystems are in a constant state of change.

Chapter 21 deals with living things in freshwater. Freshwater bodies hold around 6% of the Earth's species of plants and animals despite covering less than 1% of the Earth's surface. Factors that control the spatial variability of aquatic life are outlined within Chapter 21 as are the feedback effects between aquatic life, water quality and the geomorphology of freshwater aquatic systems.

The alteration and fragmentation of major biomes by non-sustainable practices of exploitation have been features of the human impact on the environment for thousands of years. However, the increasing density of human populations and improvements in technological capabilities put added pressure on the biosphere including freshwater bodies. Many of these pressures and their impacts are discussed within the chapters that form Part V of this book. The requirement of environmental managers to balance the preservation of ecosystem stability with the human needs for the ecosystem necessitates an understanding of the major ecological, biogeographical and environmental processes that characterize the living portion of our planet. The biogeographical system also plays an important role in the climate system and there are two-way interactions between climate and vegetation. However, discussion of these climate change processes will be dealt with mainly in Part VI of this book which deals with environmental change. More specifically the whole of Chapter 25 is devoted to the topic of vegetation and environmental change.

CHAPTER 18

The biosphere

Hilary S.C. Thomas
C2C Research Centre, University of Glamorgan

Learning objectives

After reading this chapter you should be able to:

➤ **define and explain the important terms 'biosphere', 'biome' and 'biogeographical realm'**

➤ **understand the main influences leading to patterns in plant and animal distributions**

➤ **describe the location and main characteristics of the major global biomes**

➤ **explain some of the factors leading to change in these regions**

18.1 Introduction

The **biosphere** is usually taken as referring to the surface of the Earth, together with those parts above and below it that maintain life. Its field of interest therefore includes parts of the atmosphere as well as the soils and waters of the Earth. Studies of the biosphere are linked with geology, ecology, soils, atmospheric processes and climates, and oceans. The biosphere is dependent on other components of our planet for its functioning as illustrated in Figure 18.1. As geographers, we are aware of the interactions between the atmosphere, biosphere and lithosphere but are fascinated by the patterns of distinctive regions of plants and animals that result on the Earth's surface. We are interested in not only the patterns but also how humans influence the biosphere through a range of deliberate and inadvertent practices. The following quote illustrates *one* viewpoint of the role of humans in the biosphere:

> *The biosphere comprises the natural world in which man has been placed, and which, thanks to his mental capacities, he is able to regard objectively, thus raising himself above it. . . . It is not the sole calling of man to use nature to his own ends. He also bears the responsibility for maintaining the Earth's ecological equilibrium, of tending and preserving it to the best of his ability. If he is to do this and to avoid exploiting the environment in a way which in the long run jeopardises his own existence, he has to recognise the laws of nature and act upon them.*

(Walter, 1985, p. 2)

At the beginning of the twenty-first century, as both the density of human population and its technological abilities increase (see Chapter 24), we are also aware of the extent to which, for good or ill, we are able to alter these patterns. Directing **ecological succession** and the management of energy flows through ecosystems allows commercial harvesting of resources to feed a rapidly growing population. The increase in disposable income and leisure time within

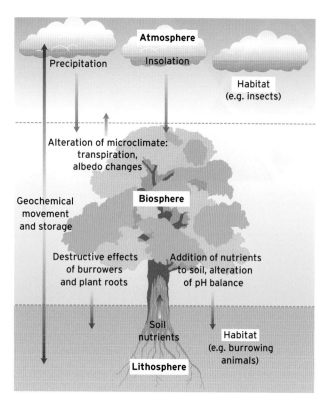

Figure 18.1 Interactions between the atmosphere, biosphere and lithosphere.

richer areas of the world, together with increasing appreciation of the aesthetic value of all forms of wildlife, has led to the use of the biosphere for a variety of forms of recreation. This varies from whale-watching, where the activity is dependent on the character of distinct locations, to projects such as those taking place in Cornwall, UK, as part of the Eden Project where tourists are given the opportunity to experience a variety of artificially created 'biomes' (see Box 18.2 below).

Biomes are major (global-scale) zones with characteristic life forms of plants and animals. The alteration and fragmentation of major biomes (such as the evergreen woodlands of the Mediterranean) by non-sustainable practices of exploitation has been a feature of human impact on the environment not just in recent decades, but for thousands of years. This has given rise, in some locations, to new biomes. However, biomes and the biosphere itself are in constant, if usually slow, change in response to fluctuations of climate. Modern geographers have the advantage of technology such as remote sensing and numerical modelling techniques to enable them to track changes in vegetation cover, growth and type (see Box 18.4 and Chapters 25 and 26). These academic efforts provide data and additionally provide a conceptual framework to aid those working towards conservation and restoration of environments.

This chapter aims to describe the fundamental properties and characteristics of the biosphere. The main factors leading to both spatial and temporal patterns within the biosphere are outlined, before focusing upon these patterns, at global and continental scales. Finally, ways in which elements of the biosphere may change over time are suggested and consideration is given to the role of the biosphere in human leisure activity. This chapter and the two following it are integrated by the concepts used in common by both biogeographers and ecologists. Cross-referencing is therefore made between them, to indicate where an aspect in a section is examined in more detail within one of the other two chapters. Occasionally Latin names are used for plants and animals during these three chapters. This is the standard practice and in many cases there are no common (or English) names for certain species. However, where common names do exist these are provided. You do not need to remember all the plant and animal names since it is more important to understand the processes and characteristics of the environments discussed. Nevertheless the species names are provided as examples.

18.2 Functions and processes within the biosphere

Although functions and processes within the biosphere are more properly dealt with as part of Chapter 19, it is worth noting here that these stem, for the most part, from the unique relationship between living things and their environment. Both are capable, to a greater or lesser degree, of affecting the other. These interactions can be observed to take place within a generally hierarchical structure, within which flows of energy, nutrients and matter operate at a variety of spatial and temporal scales.

18.2.1 Characteristics of the biosphere

The biosphere, like the atmosphere and oceans, is distinguished by movement and dynamism. For the biosphere this dynamism is of its life forms. The main characteristics of these fluxes are discussed in greater detail in Chapter 20. It is, however, worth noting here that this dynamism may be directional, as in the slow post-glacial return of species to high-latitude forests, or of a more cyclic nature as in regeneration after repeated fire damage within semi-arid zones.

Boundaries within the biosphere may result from a variety of factors, often relating to **environmental gradients** (e.g. climate, altitude, soil type), but increasingly in some areas to intensity of human usage. All geographers should be aware that a boundary depicted as 'a line on the map', especially at an atlas scale, needs considerable caution in interpretation. For example, we must think about how wide an area any line on the map of the tropical biomes (see Figure 18.5 below) covers on the ground, and how constant it is over time and space. Many terrestrial boundaries do not imply a total and sudden change to biomes. The resulting transitional areas are described as **ecotones** and can vary hugely in extent. Ecotones are expected to be early indicators of change in response to climate. Evidence exists to show that this has been the case in the past. For example, the northern Sahara was regarded as a major grain-providing area for the Roman Empire and the shifting patterns of desertification along the southern Sahelian boundary continue to provoke widespread concern.

The Earth's biosphere is not the same throughout, but has developed over time a pattern of distinctive regions at all scales from the global to the very local. These regions have characteristic energy flows, **biomass**, trophic levels and rates and types of nutrient cycling activity, and a number of factors influence these processes.

18.2.2 Major factors producing regions within the biosphere

18.2.2.1 Temperature regime

Both the actual temperature and its seasonal pattern may be of critical importance to plant or animal life. While four seasons are the norm for the middle latitudes, two seasons, or sometimes three, are more frequent on a global scale (see Chapter 5) and are of particular relevance to understanding biomes. The growing season for most plants creates an effective baseline of food for other creatures (see Chapter 20) and is taken to be the length of time when the average monthly temperature is above 10°C. This growing season may vary considerably within a small geographical range. The length of growing season has particular implications for **herbivores** (animals that just eat plants). Herbivores must adapt to the changing availability of food resources through the seasons. This is often done by becoming dormant (e.g. hibernation) for part of the year or by migration, sometimes over very considerable distances. High temperatures may exacerbate the effect of low rainfall. Vegetation change in response to temperature regimes may also affect the capacity for the plant–soil complex to retain moisture.

18.2.2.2 Moisture availability

Moisture availability is related to the local rainfall regime, in particular to the length of the dry season. It is also dependent upon the potential effectiveness of rainfall in comparison with losses of water via evaporation and accessibility of river or groundwater. A range of other factors will act locally to produce diversity within this framework. In particular, the roles of the soil type, geology, slope and altitude are often fundamental in providing zones of increased (e.g. in hillslope hollows) or decreased (e.g. on very coarse substrate) moisture for plants and animals. These major factors, together with others discussed below, can be classed as zonal or azonal in their effects.

18.2.2.3 Zonal factors

The regional macroclimate (equatorial, monsoonal, desert and so on – see Chapter 5) creates characteristically favourable or unfavourable conditions for plants and animals. The level of cold or drought may determine dormancy or deciduousness and can lead to distinctive life forms. It may also affect the relative speed of nutrient cycles. The major world biomes therefore correspond very closely to the major climatic zones (see also Chapter 25, especially Section 25.2). Zonal soil types, such as brown forest soils, have, for example, distinctive mechanical characteristics and types and rates of nutrient supply (see Chapter 10).

18.2.2.4 Azonal factors

Azonal factors have the ability to disrupt the otherwise climatically controlled pattern of biogeographical regions (biomes, ecotones, 'realms' – see below). Geomorphology will affect drainage, provide varying aspect and hence influence the receipt of solar radiation. Geology provides a varying substrate in terms of properties such as pH, nutrient availability and soil texture. The level and frequency of human influences, whether ancient or modern, 'managed' or unintentional, will have effects that are often highly disruptive to the zonal pattern. For example, the Bronze Age forest clearances led to high levels of soil erosion around the Mediterranean, similar to those seen much more recently in parts of the tropics (Evans, 1999).

18.2.3 The major biogeographical realms

The factors discussed above have together resulted in what are described as biogeographical zones or realms, which in turn are divided into biogeographical regions or biomes. The biomes, while often described as major terrestrial vegetation communities, defined by the similarity of the dominant plants (Archibold, 1995), also include characteristic animal communities. Some workers have divided the Earth into realms in relation to vegetation, while others

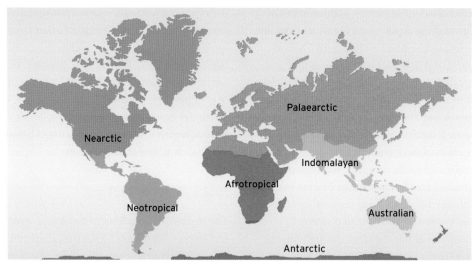

Biogeographical realms
NB Holarctic = Nearctic + Palaearctic

Figure 18.2 A classification of biogeographical realms. The Holarctic consists of the Nearctic and Palaearctic while the Palaeotropic consists of the remaining regions.

use a faunal basis (Bradbury, 1998). The classification discussed below draws together these two themes but also indicates where there are widely held differences of opinion. It should be noted, however, that although the geographical distribution of any single species within a realm may well coincide with the boundaries of that realm, it is usually the combination of species that will show a strong and characteristic association with it.

Figure 18.2 shows a typical classification of biogeographical realms. These fall into two main regions, which are then subdivided. The first region is the Holarctic region which although based on vegetation also comprises the two faunal realms of the Palaearctic (Eurasia) and the Nearctic (North America and Greenland). This region approximates to the ancient plate tectonic region of **Laurasia**, which was a large continent of the northern hemisphere that existed around 200 million years ago (see Chapter 2). The other major region may be regarded as being formed from **Gondwanaland**, the equally ancient southern continent. These regional similarities therefore indicate the importance of plate tectonics in determining the global distribution of species. The southern region is known as the Palaeotropic region and consists of further subdivisions, although there is often debate as to how these subdivisions should be made. The first subdivision is considered to be the Afrotropical realm (south of the Sahara). Many people think of the Indian (or south-east Asian) realm as a separate region as shown in Figure 18.2. The Asian (or Indomalayan) realm is similar to the African realm on floral grounds but there are distinct faunal differences.

There are also three further southern hemisphere realms that are considered to be distinct. The Neotropical realm refers to South America while the Australian realm sometimes excludes New Zealand on floral grounds (Takhtajan, 1969) but other people often include New Zealand, southern South America and southern South Africa within this realm (Walter, 1985). A further Antarctic realm is often distinguished which again, as in Figure 18.2, is considered to comprise southern South America, southern South Africa and New Zealand, all of which are related to the southern part of the old supercontinent of Gondwanaland. It should be noted that this classification has ignored the ocean realm. The divisions above result from the work of scientists usually interested in terrestrial distributions and as such the full richness of the world's biogeographical regions is sometimes underestimated (see Chapter 3).

Understanding past climatic and tectonic changes is important for understanding the patterns of plant and animal geography that can be identified today. These long-term climate and tectonic changes offer in many cases the clearest explanation for the variety of biodiversity to be found across the planet. The groupings in Figure 18.2 are of great use for purposes such as **taxonomy** (species classification). However, many geographers find that more detailed classification based on climatic controls is of greater usefulness. This is because it allows correlation with factors that have immediate effects on biomass, species dominance and, sometimes, human land use. Most workers agree that there is a very strong correlation between these biomes, the structural characteristics of their vegetation and the major world climate zones (Grime, 1997). Thus, these biomes form the main structure of the remainder of this chapter.

Since vegetation cover tends, in most cases, to be the determining factor in both the distribution of animals and the potential and actual human activity in the regions involved, the naming of the biomes described in the following sections of this chapter is based on a generally accepted classification based on the predominant vegetation type. As Cox and Moore (2005), who include several marine biomes within their classification, point out, there is no real agreement among geographers about the number of biomes in the world. For convenience, they are discussed in this chapter within three major climatic zones. Remote sensing techniques can be used to distinguish vegetation patterns and change as described in Box 18.1.

Reflective questions

➤ What are the main factors that produce regions of the biosphere and how do they interact?

➤ What is the difference between a biogeographical realm and a biome?

18.3 The tropical biomes

The tropical biomes include the complex and biomass-rich equatorial and tropical rainforests, the tropical woodland–grassland mix known as the savanna and the low-biomass region of the hot desert biome. Table 18.1 and

TECHNIQUES

REMOTE SENSING TO MONITOR LAND COVER CHANGE

Remote sensing is technology that has developed rapidly, from its initial military applications to a tool that enables environmental managers to obtain 'real-time' and 'multi-temporal' images of the globe, focused on specific regions or themes of investigation. Many satellites have instruments on board that enable them to detect different types of land cover. For example, Landsat's Thematic Mapper and Multispectral Scanner and NASA/ National Oceanographic and Atmospheric Administration are frequently used to provide consistent and reliable information. See Chapter 26 for a detailed exposition of remote sensing.

Vegetation change is one of the more easily distinguishable forms of land cover change. The very high reflective characteristics within the near-infrared part of the spectrum are distinctive to vegetation. Remote sensing from space can be used to

'see' these reflective characteristics. Further differentiation is possible using factors such as water content, vegetation density and a range of distinctive vegetation structures, even down to individual plant shapes. These factors combine to give a wide range of reflectance signatures, which can be translated into precise vegetation data. For example, deciduous leaves reflect more strongly than evergreen needles, and both ripening and wilting crops have associated reflectance changes. Each land cover can be assigned specific signatures, allowing mapping of very large areas and of regions not easily accessible. Results can be checked by ground-truth sampling rather than surveying the whole globe, reducing time, expense and hazard for field staff. Data are therefore more readily accessible for some developing countries than traditional field surveys.

An example of a remote sensing classification is shown in Figure 18.3, taken from the Global Land Cover Facility, provided by the University of Maryland. Although there is loss

of information at the scale shown in Figure 18.3, you should be able, in the light of information within the rest of this chapter, to identify many of the major biomes. Comparison of this figure with the other biome maps in this chapter (Figures 18.5, 18.14 and 18.22) demonstrates that the potential natural vegetation has often been supplanted by human land use.

Monitoring of biomes depends upon reliable, objective base line data and such data are often heavily reliant upon remotely sensed information. Considerable work is therefore ongoing into the identification of indicators of early change. This type of monitoring can also be used to check compliance with management plans and with regional agricultural and environmental legislation. It can be used to check the relative impacts of human- and climate-induced change (see Box 18.4 below). You may wish to visit NASA's Earth Observatory website which provides many images of geographical interest: http:// earthobservatory.nasa.gov.

BOX 18.1 ➤

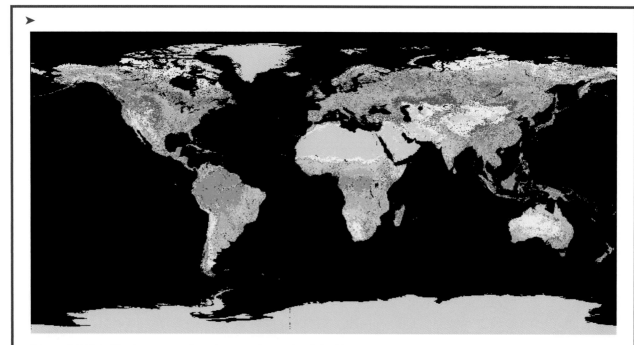

Figure 18.3 Global land cover map based on remote sensing data. (Source: Hansen *et al.*, 1998)

BOX 18.1

Table 18.1 Potential primary production of the Earth by climate zone

Climate zones	Area (km² × 10⁶)	Phytomass Total (t × 10⁹)	Phytomass Average (t h⁻¹)	Primary production Total (t ha⁻¹ × 10⁹)	Primary production Average (t ha⁻¹ yr⁻¹)
Polar	8.05	13.8	17.1	1.33	1.6
Boreal	23.2	439	189	15.2	6.5
Temperate					
Humid	7.39	254	342	9.34	12.6
Semi-arid	8.10	16.8	20.8	6.64	8.2
Arid	7.04	8.24	11.7	1.99	2.8
Subtropical					
Humid	6.24	228	366	15.9	25.5
Semi-arid	8.29	81.9	98.7	11.5	13.8
Arid	9.73	13.6	14.9	7.14	7.3
Tropical					
Humid	26.5	1166	440	77.3	29.2
Semi-arid	16.0	172	107	22.6	14.1
Arid	12.8	9.01	7.0	2.62	2.0
Geo-biosphere					
Land area	133.0	2400	180	172	12.8
Glaciers	13.9	0	0	0	0
Biohydrosphere					
Lakes and rivers	2.0	0.04	0.2	1.0	5.0
Oceans	361.0	0.17	0.005	60.0	1.7

(Source: Schultz, 1995)

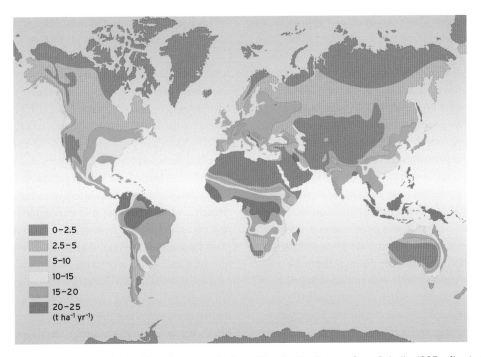

Figure 18.4 Map showing distribution of potential primary production of the Earth. (Source: from Schultz, 1995, after Leith, 1964)

Figure 18.4 show how the potential **primary productivity** (amount of energy fixed by plants during photosynthesis) is distributed across the Earth. Comparison of Figures 18.4 and 18.5 shows how the tropical biomes include regions of both large potential productivity and small potential productivity.

18.3.1 Equatorial and tropical forests

As a general rule tropical forest biomes are located where the climate is hot with mean annual temperatures around 25°C with little seasonal variation, and where there is around 2000 mm of rainfall per year. The rainfall needs to

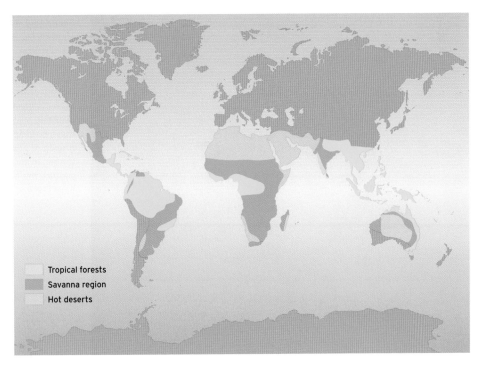

Figure 18.5 The location of the tropical biomes.

be consistent throughout the year, or at least any dry season should not extend longer than five months and should still produce rainfall exceeding 120 mm in every month. Soils reflect high rates of **biogeochemical** activity which occurs at three to four times the rate of other mature forests. This leads to deep but relatively infertile soils, such as **oxisols**. In these environments there may be a lack of recently weathered rock fragments, which would normally release new supplies of nutrients for plant uptake. Fertility therefore depends upon continual leaf fall and rapid cycling of nutrients through the vegetation, leaf litter and root system. Despite the high productivity, as a result of the constant nutrient demands made by the forest's rapid and non-seasonal growth, plant litter does not accumulate to great depths. With increasing distance from the equator, the more seasonal rainfall may produce soils with impeded drainage through drying and wetting cycles and this can affect the distribution of both vegetation and animals. These forests tend to be characterized by high levels of vegetative biomass (Figure 18.4), dependent upon the length of dry season. It is estimated that these forests provide 40% of the world's terrestrial **net primary productivity**. Many of the available nutrients are stored within the biomass and, in normal circumstances, little escapes from the system into rivers.

These tropical forest biomes are often referred to as rainforests (Figure 18.6). Tall trees are the dominant vegetation. Considerable variation of forest type exists within regions as well as between the major realms. Most tropical forest trees, however, are broad-leaved and evergreen species. Compared with other biomes, there is seldom a good survival reason for an organism to germinate or have its reproductive period at a specific season, since there is little climatic advantage to be gained. This has the effect of producing the characteristic luxuriant and continuous vegetation cover, with high levels of primary productivity and biomass, which benefit lower species throughout the system. Some plants shed all their leaves over a short period, in a similar fashion to those of the deciduous forests in Europe and North America, while others are able to maintain a continuous process of leaf fall and replenishment within the individual plant throughout the year. One effect of this is clearly seen in Figure 18.7 which emphasizes the nature of the closed canopy. However, with increasing distance from the equatorial areas, an increasing seasonality in primary productivity may be observed. Thus the non-seasonal equatorial rainforests can be distinguished from the tropical rainforests, which have one or two short dry seasons, the length of which increases at higher latitudes towards desert biomes.

The main issue for the vegetation is the competition for light. This leads to stratification of canopies, and proliferation of climbers (Figure 18.8) and **epiphytes** (plants which grow above the ground surface using other plants for support and that are not rooted in the soil). There may be

Figure 18.6 A rainforest showing tall, slender trees rapidly growing near a riverbank where older trees have fallen. In the background the density of the forest can be seen. As soon as light becomes available around fallen trees there is competition for rapid growth of new trees to reach the top of the canopy and capture the light. These trees have few low branches.

Figure 18.7 A rainforest floor in Sri Lanka. The struggle for light, often the main limiting factor in this otherwise highly favourable environment for vegetation in particular, can be seen here. Even a small break in the canopy is immediately utilized by climbers and by juvenile trees. (Source: photo courtesy of Lindsay Banin)

Figure 18.8 The strangler fig, Borneo. This climber grows up existing trees using them to support its growth and then eventually kills its host by blocking out sunlight. (Source: photo courtesy of Nicholas Berry)

discontinuous shrub and herb layers, giving perhaps five layers in all. Species with higher light requirements, such as **lianas** (a type of climbing vine) or epiphytes, display typically adaptive lifestyles. Lianas climb rapidly and will often assume a shrub form in clearings, but will not usually form leaves until sufficient light is available, at the canopy level. Epiphytes attach themselves to the trunks or branches of trees to obtain a similar advantage. The competition for light ensures that the canopy is continuous and dense, even down to ground level at river banks. This gives rise to the illusion of the 'impenetrable rainforest'. The tree layers are distinguished by a huge diversity of animals and birds, reflecting the availability of food sources. Species often restrict their range to a single stratum although they may travel extensively throughout the forest within that layer. Animal adaptations include prehensile tails (tails capable of grasping), grasping feet and clawed or suction-padded toes (Figure 18.9). Birds tend to be fruit rather than seed eaters.

The upper canopy contains the dominant, mature species, often 25–35 m above ground, with still taller 'emergent' individuals that may reach a height of 50 m. These individuals tend to have few branches below canopy level and fairly slender trunks in relation to their height, but may be supported by buttress roots (Figure 18.6). The second (~10–15 m) and third (~3–5 m) tree canopies are composed, to a large extent, of immature individuals of the dominant species. Second-canopy individuals are often triangular upward in shape, to take advantage of what light is available, and the second and third canopies are increasingly discontinuous, depending on the pattern of light availability.

The effectiveness of this competition ensures that in undisturbed forest there is relatively little understorey vegetation, since plants must be extremely shade-tolerant to survive. This space near the forest floor with limited vegetation, however, does provide room for large animals, such as wild pigs and jaguars, together with rodents, amphibians and reptiles. Nevertheless, most of the animal biomass is found in the soil fauna, including termites and beetles.

When natural clearings appear in the forest owing to fire, wind or water damage or resulting from the death of

Figure 18.9 A rainforest lizard with suction padded toes. (Source: photo courtesy of Nicholas Berry and Despina Psarra)

large individual trees, the saplings, climbers and other species contained within the lower canopies compete to obtain access to the increased light, fighting for this newly available niche. The regeneration process, if undisturbed, is usually regarded as having three distinct phases: gap, building and mature. Although complete regeneration can take around 250 years, the canopy is often closed once more after 5–20 years. Therefore the incidence of soil erosion and decreases in soil nutrient flows are minimized. Humans have mimicked this type of natural clearance for centuries within tropical forests via low-intensity shifting cultivation, otherwise known as slash and burn. When carried out at low intensity, shifting cultivation accommodates this natural regeneration process. However, increasing intensification of clearance, such as that which accompanies large-scale commercial agriculture or forestry operations, does not allow regeneration to take place and coarse grass such as *Imperata cylindrica* replaces the previously complex forest vegetation habitat.

In intact equatorial forests, there is seldom a problem in obtaining sufficient soil water, unlike the competition for light to enable photosynthesis to take place, and evergreen species usually dominate deciduous ones until the length of the dry season becomes appreciable further away from the equator. Some plants display modifications aiding the removal of surplus water, such as the drip-tips of *Ficus* species (e.g. rubber plants). Raunkiaer life forms are classes of plants based on how the new tissue (buds and shoots) develops and grows as shown in Figure 18.10. In terms of

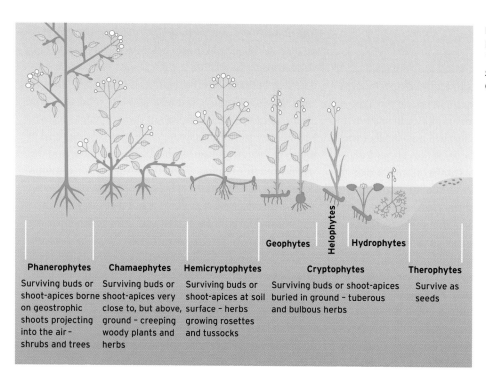

Figure 18.10 Raunkiaer's plant life form classification. (Source: *The life forms of plants and statistical plant geography* by Christen Raunkiaer, 1934)

Raunkiaer life forms, the equatorial forest biome is characterized by a high incidence of **phanerophytes** (tall plants, visible all year, with buds high in the air) such as the palm family, together with ephiphytes.

Although the rainforests contain half of the world's faunal and floral species, stands of vegetation tend to be associations, rather than dominated by a single species. The large number of species is probably due to the complex structure of these forests, together with the stable prevailing climatic conditions and the length of time since radical climatic disturbance. Large islands of forest may have remained relatively undisturbed by global climate changes for hundreds or thousands of years. This equatorial and tropical forest biome is probably the most ancient and stable biome, avoiding the major stresses of the Quaternary glaciations discussed by Dawson (1992) which resulted in a reduction of diversity in the Nearctic and Palaearctic realms.

While the characteristic structure of the equatorial and tropical forest realms tends to remain, the actual species to be found vary between the realms. For example, leguminous trees such as *Dalbergia* species characterize the South American Palaeotropic, while the African region generally has fewer species with leguminous tree species represented by *Brachystegia*. In the Indian region, **dipterocarp** (generally tall and large) trees are more likely to be dominant. However, the conditions and characteristics outlined above may be dramatically altered by local factors. Under specific azonal factors, such as local geology, it may be that drainage and nutrient levels provide a more challenging environment. In the Guiana Highlands of South America, the coarse sandstones have encouraged soil acidity and a form of tropical heathland exists. Under conditions of inundation by water, such as along parts of the Indian and Brazilian coasts, swamp formations such as mangrove forest will develop.

In areas with more than five months of dry season, the tropical forest biome changes in response to the water stresses on plants and animals. First deciduous and then strongly adapted species, such as thorn woodland species, begin to dominate in each realm, such as in the mulga and brigalow scrub of Australia and the acacia thorn scrublands typical of Africa. In all cases, the structure of the forest is of a low, 4–10 m canopy with thin or non-existent shrub and ground layers beneath.

The ecotone representing the forest–savanna boundary has been the object of much research, since forest clearance has been a major signature of human impact on Earth (Atkins *et al.*, 1998). The ecotone is highly variable in width, ranging from 30 m in parts of West Africa to areas of 'savanna-in-forest' and 'forest-in-savanna' mosaic covering hundreds of kilometres elsewhere (Mannion, 2002).

18.3.2 Savanna

The temperature of the savanna biome is very similar to that of the rainforests, but the dry season is sufficiently long to result in seasonal vegetation. Figure 18.5 indicates the typical location of savanna. The structure of this biome is characterized by a more open canopy than the tropical forests and the penetration of light allows the growth of grass and other ground flora (Figure 18.11). Most areas share the climatic characteristics of a dry season in which rainfall is below 250 mm a month for longer than five months. Where the annual rainfall drops below around 625 mm, thorn scrub begins to take over. A gradient within the biome, as shown in Figure 18.12, can often be recognized, as the increasing length of dry season has more pronounced effects away from the equator moving from savanna woodland, through tree savanna, then sometimes to a thorn/shrub savanna before savanna grassland dominates.

Most workers now assume that the tropical savanna areas are almost always the result of **edaphic** (influenced by the soil) or biotic factors dominating climatic factors. Savanna regions tend to be found in areas of **continental shields** such as the Mato Grosso, Central Africa or central Australia. Here, the effects of resistant, often infertile, substrates are compounded by long-continued human occupancy and usage with increasingly intensive woodland clearance, often by fire, with little chance of recovery. Climatic change during the Quaternary has also altered the tree cover to a very large extent. In general, the Australian, *Eucalyptus*-dominated, savanna woodlands are considered the most likely to be related to an earlier warmer climate, while those savannas in other realms may well be anthropogenic in origin and exist as fire climaxes (see Chapter 19). Termites also form an important element of the savanna because of their effects on litter decomposition and soil character.

Savanna areas appear as a mosaic. This is a result of diversity of origin and effectiveness of other factors. Large-scale diversity of structure within the biome is related to differences in the availability of water and soil nutrients. There are usually few trees. Where there are trees this is usually linked either to where the water table is close to the surface at depressions in the ground, or along river valleys, or to where drainage is enhanced within otherwise impermeable areas (e.g. the uplands of Trinidad). The density of tree spacing is dependent upon the level of competition for water. Woody species such as the giant baobab, with its

Figure 18.11 A savanna landscape in Kenya. The landscape is characterized by spiny shrubs and a tree cover that typically becomes sparser as average annual rainfall declines. The trees typically have a flat top and often lower browse line. (Source: Shutterstock.com: Oleg Znamenskiy)

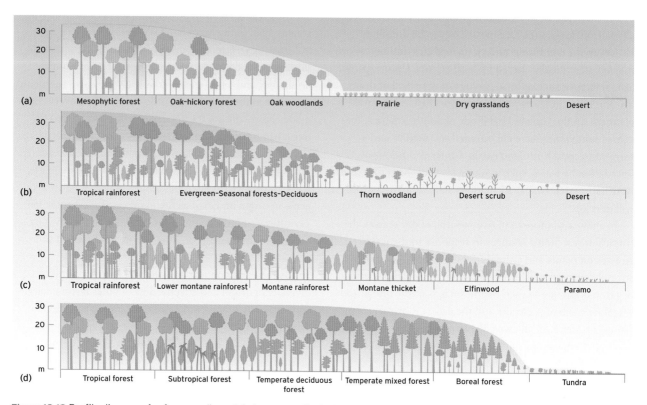

Figure 18.12 Profile diagrams for four ecoclines: (a) along a gradient of increasing aridity from moist forest in the Appalachians westwards to desert in the USA; (b) along a gradient of increasing aridity from rainforest to desert in South America; (c) along an elevation gradient up tropical mountains in South America from tropical rainforest to the alpine zone; and (d) along a temperature gradient from tropical seasonal forest northwards in forest climates to the Arctic tundra. (Source: after Whittaker, 1975)

thick bark, short season in leaf and water-retentive trunk, tend to be strongly adapted to withstand fire and drought. Many savanna trees are low with branches forming an umbrella-shaped crown. This is in distinct contrast to the form of typical rainforest species (Hoffmann *et al.*, 2003).

The fruiting period of savanna trees often takes advantage of 'fire seasons' (large fires occur every few years across parts of the savanna) by dropping fruit at the end of the fire into temporarily nutrient-rich soil. Other common adaptations include a deep root system, which allows suckers to respond to the wet season, thorns or spines to reduce water loss and deter grazers, and a general paucity of leaves and branches. Most trees are deciduous as an adaptation to the drought stresses of the dry season, such as *Brachystegia* and *Julbernardia* that dominate the African zone, although some, such as *Acacia faidherbia albida*, keep their leaves. Others, including *Acacia karoo*, produce very large numbers of seeds, since losses of seeds to fire, termites or drought can be great.

The savanna vegetation is often characterized by **xeromorphic** (adapted to dry conditions) grassland species which may produce a canopy several metres high. Elephant grass, for example, can reach 3–4 m in height. These grasses often have **rhizomes** (lateral stems through the ground that send up new shoots) or densely tufted habits to protect against fire and drought. Grassland stands are often locally dominated by very few species and levels of biomass are highly variable within this biome, related to the amount of trees. Collinson (1997) suggested a biomass range from 150 t ha^{-1} for savanna woodland to 2 t ha^{-1} for sparse grassland without trees and emphasized the direct relationship between productivity and rainfall.

Large animal predators tend to occupy wooded areas that provide shelter, cover and a variety of food. They prey on herbivores such as antelope found in huge numbers within East and Central Africa or deer species and marsupials in South America and Australia respectively. Hyenas exemplify the types of adaptation that lead to success within this environment (they also forage in drier regions), scavenging usually at night, since nocturnalism not only reduces competition for prey, such as from vultures, but also conserves their water supply. Other animal adaptations within this biome include the extensive migration of large grazing mammals and their related predators. Since prey migrate, the Serengeti spotted hyenas form temporary and mobile social groups, rather than permanent groups that would defend a single area (Mills, 1989). They will track seasonal availability of both water and vegetation, which at times conflicts with human administrative boundaries.

18.3.3 Hot deserts

The dominant factor to which the hot desert biome is attuned is the climate, often summarized as hot all year, dry all year. The criterion taken here is of insufficient moisture for complete ground cover, leading to characteristic adaptations of the life forms occupying the biome, or traversing it as part of migratory journeys. Both plants and animals tend therefore to display characteristics controlling evaporation and maximizing the conservation of any water available. Areas such as the Atacama Desert, in South America, include locations that may avoid sporadic rainfall for 30 years, while others may have more regular rainfall but which is insufficient to counteract the locally high levels of evaporation. Fog and dew, however, provide important supplementary water. In addition animals are also challenged by temperature stress. Heat may be extreme, with large diurnal ranges, due to lack of cloud cover (see Chapter 5).

Most desert soils are poorly developed but increased amounts of nutrients and improved structure are usually found around the roots of shrubs. This reflects the increased organic matter provided both by the plant and from those animals using the plant as shelter.

The structure of the vegetation of hot deserts is varied, but low in height and always with very open stands, as shown in the oasis in Figure 18.13. The vegetation density is dependent primarily upon on access to, and competition for, water. This is seen in the dramatic increase in biomass around oases and watercourses. Increased aridity results not only in increasing space between individuals but also in increasing clustering of their distribution. This is more complex than a purely climatic response as it is dependent upon local geology and geomorphology; soil conditions are very poor in hot deserts where the upper soil is of a coarse texture. Where runoff is concentrated, the habitat improves.

Responses to the hot desert's challenging environment often take the form, in plants, of controlling transpiration rates by, for example, only opening their stomata (see Chapter 4) at night. Cacti such as the Brazilian xique-xique store water within their stems and may display other moisture-conservation features. Leaves are often replaced by thorns (e.g. *Euphorbia ingens*, common in Central and eastern Africa) or, in succulent species, by stems with water-storing cells that also have the ability to photosynthesize (e.g. *Opuntia* species in Mexico and *Aloe* species in Africa). Other adaptations include woodiness, which prevents collapse of plant material during wilting, or a small total leaf area. For grass species, their characteristic short, tufted nature helps protect against drought and heat stress. Photographs

Figure 18.13 An oasis in the Awbari Sand Sea, in the Libyan Desert. When water is available, trees are able to establish in what are otherwise extremely arid areas. (Source: Shutterstock.com: Patrick Poendi)

of some dryland species are provided in Chapter 14 and some adaptations are provided in Table 14.3.

These adaptations may be combined with, or replaced by, a lengthy dormant season. This can involve a total die-back of above-ground tissue, perhaps where a large root system allows access either to deep groundwater or, more commonly where the substrate is coarse, to tap a wide lateral zone. The record for shrub root depth would appear to be held by a tamarisk (a small, narrow-leaved tree) in Suez, with roots that measured 45.7 m. Seeds may have germination-inhibiting coatings that require significant weakening by moisture before germination. This prevents germination in conditions that will otherwise be too harsh for the young seedling.

Where adaptations do not exist, plants may have very rapid life cycles, responding to the period between rainfall and resumption of drought, thereby avoiding the environmental challenge. They may be 'annual' plants, which means that the plant grows from a seed, flowers, produces its own seed and then dies. The same plant will then not grow again but its offspring will survive where seeds germinate during the next rainfall event. This accounts for the phenomenon of the desert 'blooming' for a few weeks after rainfall. Inevitably, these annuals are usually of small biomass and produce very large quantities of seed.

In the hot desert of Death Valley, USA, instead of grass there is a **xerophytic** broad-leaved ground layer, often called cactus scrub but including mesquite, cottonwood

and tumbleweed, with creosote bush and saltgrass nearer saltpans where evaporated water has left concentrated salts behind. The vegetation in equivalent Australian areas is mallee scrub (*Eucalyptus* and *Acacia* species) which can include spinifex grassland but may also develop into mulga shrubland. The caatinga of north-east Brazil is characterized by thorny acacias, together with xerophytic grasses.

Animal adaptations may be either physiological or behavioural. These often slow the loss of body moisture, through excretion by dry faecal pellets or concentrated urine, or through sweat, which is especially dangerous for smaller animals because their surface area in relation to body mass is high. Many animals, like sidewinder rattlesnakes, are without sweat glands. Small animals, such as the jack-rabbit, are nocturnal, hiding in burrows during the day, or undergo aestivation which is a state of dormancy during the driest season. Many insects, like Namib beetles, are also nocturnal and **cryptozoic** (shelter-seeking). Namib beetles obtain moisture from dew. Larger animals are relatively unusual within the desert fauna but may, like camels, have hairy coats to enable sweat to evaporate and produce a cooling effect. Camels also have variable body temperature, storing heat (not water!) within the body during the day and releasing it at night. Other species rely upon the moisture released by mist, fog or dew, or scavenge, like the hyena, at night.

It will therefore be apparent that hot desert biomass is mainly underground, with plant life forms mainly geophytes or therophytes (see Figure 18.10) to avoid drought. Biodiversity is low and net primary production strongly related to rainfall.

18.4 The temperate biomes

18.4.1 The Mediterranean/chaparral biome

The location of the temperate biomes can be seen in Figure 18.14. The 'Mediterranean' biome covers a wider range of areas than those surrounding the Mediterranean Sea. The characteristic climate is warm all year but with low rainfall, characterized by summer drought with high evaporation rates (see Chapter 5). This biome forms a transitional zone, similar to the savanna, but in this context between the desert and the true temperate biomes. Regardless of parent material, the upper soil horizon becomes very dry in summer and soil water is drawn upwards. Climate changes over the past few thousand years have tended to increase the aridity of these areas, producing an environment to which plants and animals would appear to have had to adapt rapidly.

The Mediterranean biome was once **sclerophyllous** (hard, tough-leaved plants) mixed woodland, with species such as the cork oak and the maritime pine. Largely as a result of continued human impact, working against woodland regeneration, the region is now dominated by maquis. Maquis is scrubland in which the canopy may reach 3 m, and typically contains gorse, broom, myrtle, arbutus or olive, together with aromatic herbs, such as rosemary and sage (Figure 18.15). Where water shortage is more intense, as on limestone substrate, the maquis tends to be replaced by garrigue vegetation. The actual species may be the same but garrigue vegetation tends to have a lower canopy than maquis with wider-spaced individual plants, which are mostly evergreen. There are, however, some deciduous species which lose leaves during dry periods.

In North America, the chaparral has similar, sclerophyllous life forms and canopy height. Typical species include *Eriodictyon tomentosum*, which has fine hairs on the leaves to decrease transpiration, and the creosote bush. Any trees tend to have thick bark, or other fire-resistant qualities. Animals include deer, elk and bears. The 'soft chaparral',

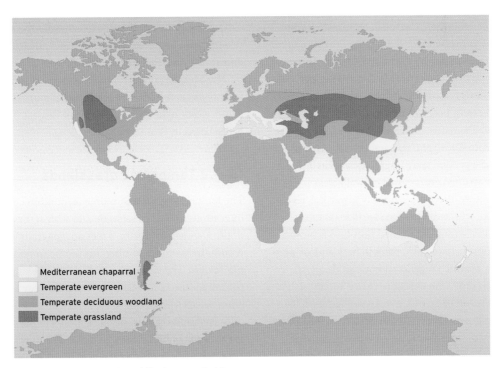

Figure 18.14 The location of the temperate biomes.

Figure 18.15 Maquis-type scrubland in Malta. (Source: Shutterstock.com: Baldovina)

found nearer the coast, with sage brush and *Salvia melliflora*, is literally softer and less sclerophyllous in nature.

The South African Mediterranean biome region has a distinctive flora and where the warm temperate forest remains, its structure is simpler than in the northern hemisphere. This is exemplified by the Southern Cape Province, where there is a mainly coniferous *Podocarpus* upper canopy and a lower-level flora dominated by tree ferns, comparable with the southern beech domination of similar regions of New Zealand, Tasmania and Chile. In Australia, mallee scrublands contain *Eucalyptus* species, with an *Acacia* shrub understorey and a ground layer of grasses such as *Trioda*, together with marsupial grazers such as wallabies.

Overall, therefore, there is a tendency towards a xerophytic scrubland, with strongly developed adaptations against the relatively frequent natural fires of this biome. These adaptations include thick, smooth bark and/or deep roots from both of which regeneration may take place. Some seeds open only after exposure to fire and there is a strong representation of cryptophytes (see Figure 18.10), surviving difficult periods as bulbs or rhizomes. As a result, a distinctive post-fire regeneration cycle exists, moving through domination by annuals, then herbaceous species before the scrubland returns. It should be noted, however,

that species representing the scrubland are usually present throughout the earlier stages and that the disturbance created by fire therefore increases species diversity within the earlier stages. Animals adapt to, or avoid the stresses of, drought and fire, often by speed (e.g. kangaroos, goats and emus) or by burrowing (e.g. mice).

The Mediterranean biome is one of several biomes that have been recreated as an educational tourist attraction in south-west England called the Eden Project. This attraction is discussed in Box 18.2.

18.4.2 Temperate grasslands

The temperate grassland biome is found in continental interior areas of the Holarctic and within the eastern region of the Neotropical realm. Trees are generally absent, except within the ecotones joining this biome to that of the deciduous forests. The vegetation is dominated by grasses, usually perennial (the same plant surviving for year after year) and often xerophyllous. It has, especially in its most climatically favoured regions, suffered intensive human impacts. Large herds of herbivores such as bison were characteristic, although are now dramatically reduced as a result of human activity. The extinction of the passenger pigeon, *Ectopistes*

THE EDEN PROJECT – BIOGEOGRAPHY AS LEISURE?

The Eden Project in Cornwall, south-west England (Figure 18.16), is a botanic garden focused on education about the world we live in and the ways humans use and abuse the world. Rather than presenting taxonomic collections, the site represents three of the world's biomes, with more planned for the future. The most famous icon of the project is the huge covered tropical biome, capable of showing rainforest trees at a mature size. This contains representations of the flora and land use of four regions: Guyana, Malaysia, Cameroon and Oceanic Islands. Within this largest 'biome', temperature and humidity are maintained at tropical levels throughout the year and plants are now reaching their full extent (Figure 18.17). Popularity for this venue is now huge and growing. A range of educational activities are focused upon this biome. There is also a covered warm temperate biome showing South Africa, California and the Mediterranean Basin. The outdoor displays show the temperate zone, covering northern Europe but also including displays that relate to other countries such as northern United States and Chile.

The plant collections are designed to show regional landscapes and land-use practices, so each biome shows both wild plants and crops appropriate to the region. These cannot be full ecological simulations of the different

Figure 18.16 The huge scale of the Eden Project can be appreciated in this picture. This is just one of the three triple-domed structures on the site. The walls of the old china-clay quarry can be identified in the background, together with the pre-industrial land surface.

BOX 18.2 ➤

Figure 18.17 Inside part of the tropical dome at the Eden Project.

Figure 18.18 Thought-provoking inscriptions are displayed throughout the Eden Project site.

zones and they will be managed displays rather than living landscapes. For example, although some invertebrates are contained within the displays it is impossible to represent the full range of species and, for many species, processes such as pollination or propagation are carried out by the staff.

Nevertheless, thought has gone into representing the zones as faithfully as possible. For example, the soils of the warm temperate region have been made to be drought prone and nutrient poor to encourage the plants to adopt typical growth forms. The plantings aim for a representation of the typical plant assemblages found in those regions and aim to mimic seasonal displays such as the flush of annuals found in South Africa following rain. An important aim is to infuse the displays with thought-provoking items such as using a log to reflect on island biogeography (Figure 18.18).

The 'outdoor biome' has more opportunity to represent authentic ecosystems and in an area called 'Wild Cornwall' examples of Cornish heaths, Atlantic coastal woodlands and farm and fieldbank communities are being created, using habitat restoration techniques. The educational displays are thus linked to research into restoration methodology that can be fed into real projects both within the region and globally.

BOX 18.2

migratorius, once observed in huge flocks in the temperate eastern grasslands of North America, was less directly but equally certainly due to human influences. Quammen (1996) suggested that its extinction followed hunting that resulted in numbers dropping below that compatible with its social ecology, even though numbers were still very large. The bird depended upon benefits of crowding to identify food sources during foraging flights and to avoid surprise attacks by predators. It is an example of a species where humans grossly underestimated the minimum size of the viable population.

The dry season of temperate grasslands tends to be lengthy, with annual precipitation usually less than 500 mm. Precipitation may take the form of snowfall in winter, while the intense heat of the continental interiors in summer leads to convectional rainfall and high evaporation. As a response to greater continentality, there is a gradual change from the moister, less continental conditions of neighbouring biomes, as shown in Figure 18.19. The geomorphological and glacial heritage of extensive, gently undulating areas with rich soils has led to regions such as the prairies and the steppes becoming important producers of cereal crops. Chernozem molisols together with vertisols and andisols are typical soils over these areas (see Chapter 10). Where it is damper there may be brown earth soils.

Adaptations, such as the height of the grassland, are similar to that of the forest biome and relate to maximizing effectiveness of precipitation and to minimizing damage due to natural fires. These adaptations include the shallow, turf-forming dense roots of many of the grasses typical of the moister areas (e.g. *Agropyron* species) or the tussocks typical in drier areas (e.g. *Poa* and *Festuca* species). These features provide fire resistance and water trapping. Lack of a protective cover from predators has caused animals to develop speed (e.g. antelope and deer), bulk (e.g. elk and bison) or a burrowing habit (e.g. mice and voles). The single vegetation layer results in a relatively limited diversity of birds but plant diversity is often as high as in forest formations. The tendency to summer fires means that systems at or below ground level that allow survival, such as bulbs, rhizomes or tubers (**perennating systems**), are important advantages (see Figure 18.10).

Where protection from fire is combined with increased soil moisture, then scrub and woodland have been seen to develop, notably in the South African veld, in the tussock grassland of New Zealand and in Australia. Sucker-producing trees such as aspen are among the first to appear, along watercourses. This has given rise to some controversy as to the origin of the wetter areas of this biome.

18.4.3 Temperate deciduous forest

Sometimes described as temperate deciduous woodlands, this biome is found only in the northern hemisphere, since in the southern hemisphere the vegetation of the equivalent climatic areas is predominantly evergreen. This is probably a result of the relatively late development of the deciduous habit, after the Laurasia–Gondwanaland split (see above). In all areas, however, there has been very considerable human interference, including transfer of species between regions,

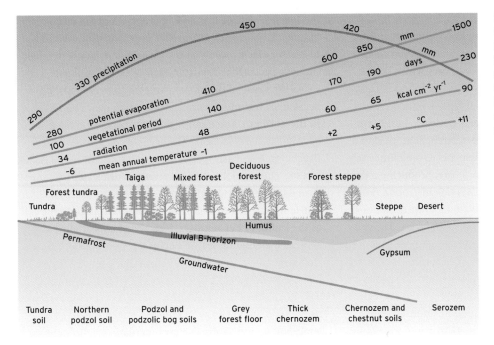

Figure 18.19 Schematic climate, vegetation and soil profile of the east European lowlands from north-west to south-east. See Chapter 10 for discussion of soil properties and Chapter 17 for discussion of permafrost processes. (Source: after Walter, 1990)

Figure 18.20 *Calluna vulgaris* (ling heather) in flower across a heathland.

Figure 18.21 A deciduous woodland photographed in early spring. By flowering early the ground layer plants such as snowdrops shown here enable the main part of their reproductive cycle to be completed before the tree canopy closes.

sometimes with disastrous effects. The effect of these introductions is discussed within Chapter 19. There is often a fairly sharp boundary between the deciduous forest and the more northerly boreal forest biome. The climate in the deciduous forest biome is moist but temperate all year, with few months having a mean temperature above 10°C.

Brown earth molisols (see Chapter 10) are typical of many areas, with a rich soil fauna providing mixing of nutrients. Where soils are under conifers, on naturally acidic rocks or have become heavily leached, podzols may be the characteristic soil. In response to this, heathland may develop, where common heather (more correctly called 'ling') and bell heather dominate, with a canopy of less than 60 cm (Figure 18.20).

Other components of the formation frequently include grasses and rushes such as *Nardus stricta*, *Molinia caerulea*, *Eriophorum vaginatum* and *Juncus effusus*, which form part of a typical heather cycle, producing a mosaic of age, species composition and structural types within the wider moorland habitat. On lowland heaths, such as Lüneberg Heath, northern Germany, junipers, birch, Scots pine and gorse may also form important elements. Over a lime-rich substrate, however, beech may dominate, producing a thick litter layer with a corresponding decrease in ground flora.

In the most favourable areas, the temperate deciduous biome displays a structure with four main layers. The canopy, where the trees with rounded deep crowns reach 8–30 m in height, is underlain by a shrub layer usually below 5 m in height and a field layer that includes many grass species as well as a ground layer of mosses and liverworts. The extent of the lower layers is dependent upon the nature of the species making up the canopy.

A characteristic of this biome is the very marked seasonality of the vegetation, with a corresponding effect upon wildlife. Since the growing season is restricted, areas may display a succession of dominants. This may begin with corms and other ground flora (e.g. snowdrops and dog's mercury; Figure 18.21), followed by species such as violets, primroses and bluebells, all of which have a flowering period before the canopy of new leaves closes and limits access to sunlight. To succeed in this environmental niche, they must initiate growth very early in the warmer period, often as soon as daylight hours increase, since their effective growing season is determined by the development of the species forming the upper canopy. As in the rainforest biome, climbers such as traveller's joy, ivy, wild rose and, to a lesser extent, ephiphytes such as mistletoe use the main canopy species for support in their quest for sunlight.

As the name of the biome suggests, trees are typically deciduous, including various oak species, beech and ash. In the northern American regions, maples will often also be important, replaced by hickory in more southern areas. Beech, birch and ash species are also important in the Asian forests. The autumnal leaf fall reduces both water loss and frost damage during the winter. Leaf loss also closes the nutrient cycle, returning absorbed nutrients back to the soil as the leaf litter decomposes. New leaves appear almost as soon as the growing season begins. Flowers appear before the leaves in tree species such as hazel and some willows but also in flowering plants such as coltsfoot and butterbur. This adaptation both maximizes the time available for fruit to ripen before the onset of the next winter and assists wind pollination. Animals may hibernate (e.g. hedgehogs) or burrow (e.g. rabbits) to avoid the challenges of the winter. Deer are found in all regions and there was once an abundance of other large mammals such as bears. Large mammals are now considerably reduced in both numbers and range in temperate deciduous forests as a result of human impact.

18.4.4 Southern hemisphere, evergreen temperate forest

This is the variant form of the temperate deciduous biome and usually displays a structure of two tree canopies and a shrub layer, having climbers and epiphytes but less frequently a ground flora. Southern beech are an important component, with conifers, of the podocarp family, found towards the boundaries with warmer areas. The New Zealand region has especially rich bird life, including carnivorous (meat-eating) parrots, compensating for a lack of native mammals.

Reflective questions

➤ What are the similarities and differences between the origins of Mediterranean and savanna biomes? How might their effective management differ?

➤ What are the main plant adaptations in temperate biomes?

18.5 The cold biomes

18.5.1 Taiga

The location of the cold biomes, comprising the taiga and the tundra, are shown in Figure 18.22. The lack of any extensive regions with these types of biogeographical character within the southern hemisphere should be noted. The taiga biome is often termed boreal forest. The taiga–tundra boundary often reflects both the post-glacial recovery, as a result of which biomes are continuing their slow poleward migration, and the effects of commercial forestry.

The climate of the taiga is cool all year, with relatively little rainfall, since it is located for the most part within continental interiors (see Chapter 5). There is, therefore, a summer maximum of precipitation, often via convection. The biome is usually taken as extending polewards from where there are less than five months with air temperatures above a mean of 10°C, until only a single month fulfils this criterion. Since most plants will cease to grow when temperatures fall below this value, there is a very limited growing season which has a major influence upon the faunal (animal) as well as the floral (plant) elements of the biome (Figure 18.23).

Soils tend to be variants of the podzol family (see Chapter 10) and vast areas have been glacially eroded. Characteristics include slow nutrient cycles and the litter layer may contain three to five times the annual accumulation and strong vertical layering. This is caused, in part, by the acidic

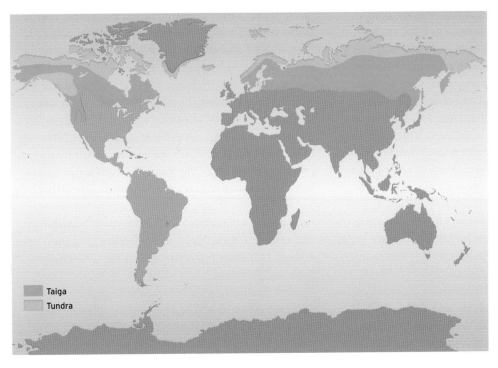

Figure 18.22 The location of the cold biomes.

Taiga
Tundra

Figure 18.23 The taiga in Brooks Range Mountains, Alaska, with spruce willows and rugged landscape. The trees have wider branches nearer the foot of the tree tapering to narrower branches at the top. This minimizes potentially damaging snow accumulation. (Source: Getty Images: Christopher Gruver)

nature of the leaf litter but also by the relative lack of soil fauna leading to very slow decomposition, a response to a cold and short growing season and poor drainage. In those areas where glacial effects include impeded drainage, peat bogs may develop rather than forest.

A very wide ecotone exists between the temperate and boreal forests but it tends to be represented by mosaics rather than stands of mixed deciduous trees and conifers. These are often related to local conditions. For example, spruce prefers loamy soils and in the most advantageous areas oaks may survive, while pines are found on sandy outwash plains. In the warmer and wetter taiga areas, forests are characterized by a structure of unbroken stands of tall (24–30 m) conically shaped trees, a discontinuous shrub and herb layer in response to the lack of sunlight penetrating the canopy and a thick resinous leaf litter and well-developed moss ground layer. The true central taiga, however, has lower and more discontinuous vegetation, with greater dominance of ground layer lichens. The trees tend to become more sparse in areas where the growing season diminishes. This ecotone is sometimes referred to as 'sub-arctic parkland' with elements similar to that characterizing the rainforest–savanna boundary.

Most Eurasian forests are dominated in the west by Scots pine and Norway spruce giving way to larch, but those in North America are characterized by the lodgepole pine and Alpine fir in the west and white spruce, black spruce

and balsam fir further east. Not all trees are coniferous, and larch, birch and alder are frequently of great regional importance. *Larix dahurica* dominates the world's most northerly forest in Siberia. Its deciduous nature enables it to withstand the combination of intense cold and strong winds. Tufted grasses and heathland plants such as crowberry and bilberry are typical of the ground layer. In the colder areas, as tree cover diminishes, the ground is often carpeted with lichens. A 100-year-old tree may be only 1.5 m high, with a 7 cm basal diameter. The deer family is well represented in North America, Europe and Asia, and is usually migrant within the forest, although both caribou (North America) and reindeer (Europe) also utilize the tundra (see below). Weasels (*Mustela* species) successfully utilize the more open areas both in the taiga and in neighbouring biomes. The fur of northern species turns white in winter, providing camouflage. The weasels' success is mainly due, however, to the advantages provided by their small but long and thin bodies. These include the ability to follow prey such as mice and rabbits into their burrows. Weasels may also move into these burrows after feeding upon the original owners. This offers protection against the harsh environment and predators.

In the brief spring period available in taiga biomes, evergreen species are able to begin immediate photosynthesis as sunlight and temperature increase. This maximizes the growing season. Where the effective soil layer is shallow as a result of the local geology or permafrost (see Chapter 17), plants may respond with fan-shaped root systems, allowing take-up of water as soon as the spring thaw occurs, achieving a similar result. The typical form of trees in this biome is of a tapering single trunk with the lowest branches being the longest, minimizing potentially damaging snow accumulation (Figure 18.23). Under the closed canopy the lower limbs die and in mature trees live branches often only exist above 6–10 m on the trunk.

Many taiga species display an increased concentration of sugar in their sap during the winter. This offers protection against both cold and lack of water, since plant roots absorb water from the soil far less efficiently than during warmer conditions. A sugar increase, combined with a decrease in water content of the plant cells, means that water absorbed from the soil can enter but not leave the vegetation. Similar adaptations are displayed where transpiration rates are lowered by stomatal closure (see Chapter 4), in the most severe conditions, or by thickened leaves or bark.

18.5.2 Tundra

Tundra regions provide, in many ways, the most challenging environments for plants and animals. Note that the boundary shown in Figure 18.22 within each continental land mass is further north in the west than the east, indicating the warming effect of maritime winds (see Chapter 5). The equatorward limit is taken to be a temperature of 10°C for the warmest month. The 'summer' period, with temperatures rising above freezing, may only last for two months of the year and winter temperatures may plunge to −50°C. Strong, cold, dry winds are generally prevalent, with precipitation most likely during the summer. Precipitation is seldom substantial but snow accumulates from year to year in many sheltered sites. Lowland tundra areas are also characterized by continuous permafrost (see Chapter 17). This results in low levels of soil fauna and limits viable rooting depths.

Tundra soils (cryosols) are characterized by a litter layer of partly decomposed, often highly acidic plant material, commonly up to 10 cm thick but reaching over 1 m in boggy locations. This rests upon a gleyed horizon (see Chapter 10), which in turn lies upon a permanently frozen layer. Geomorphology has a stronger than usual immediate influence on this biome. Glacial erosion and contemporary fluvial activity have left shallow soils at best, but with underlying sands and gravels in valley floors. This provides more opportunities for trees within such valleys, and where there may be a deeper layer that is not permanently frozen. The vegetation generally forms a mosaic closely related to local geomorphological and microclimatic features, in competition for the least stressful sites (Figure 18.24). This has the effect that, although on a microhabitat scale the number of species is often very small, in a regional context there may be very considerable diversity (Matthews, 1992).

In the most favoured localities there may be a characteristic three-fold structure comprising a low shrub layer above tussocky grasses and cushion-form herbs, underlain by a final layer of mosses and lichens. As in other biomes, this structure becomes progressively simpler and lower as conditions become more severe. Vegetation growth throughout this biome is slow. A 400-year-old juniper trunk may

Figure 18.24 Tundra is the treeless Arctic region that lies between the polar regions of perpetual snow and ice and the northern limit of tree growth. Only the most favourable areas are utilized by plants. These usually occur where protection from the wind is combined with the availability of surface water. Tundra vegetation consists mostly of mosses, lichens and small dwarf shrubs. (Source: Alamy Images: Phil Degginger)

measure only 25 mm in diameter. Therefore the colonization of bare ground may take many decades. Similarly, the **metamorphosis** of some tundra insects such as *Gynaephora groendlandica* can occupy several years.

The low productivity of the vegetation means that large areas are required to support the migrating herds of large mammals, such as reindeer. Migratory birds, such as terns, utilize the mosquito and other summer insect populations and are in turn preyed upon by hawks, falcons and owls. Plant-eating lemmings are very important creatures in the tundra zone. They are preyed upon by a variety of species, such as owls, foxes and weasels. Changes in lemming populations therefore result in changes in predator populations.

The tundra is limited floristically, probably because throughout the Quaternary this biome was destroyed for long periods during glacial periods as the ice covered the land masses of these regions (see Chapter 20). In the most severely challenging tundra areas, the plant associations are usually composed of sparse mosses, lichens and tufted grasses, all of which may become dormant when necessary for survival. Most plants of the tundra are perennial with perennating tissue at or below the surface. As climatic conditions become increasingly extreme, the importance of various means of vegetative reproduction also increases. Underground runners, rhizomes and bulbs dominate, as seeding becomes less reliable during the short growing season. Some species are self-fertile and do not require a partner for pollination, and others are wind rather than insect pollinated. The dominant plants overall are grasses and sedges although this is highly variable. In the most favoured areas there are patches of dwarf trees, such as the dwarf birch, *Betula nana*, but these form a shrub layer structurally. Rigid sedge and alpine meadow grass together with crowberry, mosses and lichens characterize this variant of tundra. As conditions deteriorate polewards, *Sphagnum* moss is replaced by hardier mosses, such as *Distichium capillaceum*. Flowering plants include members of the buttercup, poppy and saxifrage families, often with bright, insect-attracting flowers.

Heath dominates on the coarser-grained substrates, and is especially important over much of tundra Greenland. Heathland species include members of the *Vaccinium* and *Erica* families such as the arctic blueberry. In marshy areas such as the Mackenzie Delta, grasses, sedges and rushes are more characteristic, with *Sphagnum* moss and willow cotton grass where waterlogging occurs. Dwarf trees such as alder and willow may also exist as a shrub layer. Dwarf forms of vegetation provide several vital benefits arising from the

microclimates created. These include a relative calm, since wind speeds are reduced by friction near the ground, which helps reduce evapotranspiration. The insulating effect of a tussock or cushion habit both provides warmer temperatures during the growing season and conserves moisture. Willows may reach only 0.3 m and will occupy the most sheltered areas. A semi-horizontal growth form benefits from the lower wind speeds and also protects shoots to some extent from frost. Many herbs are present here only in cushion or rosette form. Growth can begin from within the cushion or as part of the compressed rosette as soon as the snow melts, giving a slightly longer growing season. The next season's flowering buds are often pre-formed.

Dormancy allows plants to survive cold and drought. Those trees and shrubs able to survive in the most sheltered tundra areas tend to be deciduous where shoots are likely to be killed by frost. This creates a disadvantage, however, since new leaves must form within the short growing season. As in the taiga biome, the high sugar content within sap offers frost protection but also increases nutritional value.

Animal behavioural adaptations include hibernation and migration. Species such as caribou and reindeer, Arctic foxes, wolves and lemmings are all known to migrate between the taiga and tundra, although musk oxen and some caribou species will remain within the tundra biome. Burrowing animals include voles and snowshoe hares. Animals and birds may have insulating fur or feathers, such as the ptarmigan, with feathers on the soles of its feet and also white camouflage in winter.

Despite the actual constituents of the plants and animals differing within the tundra biome, their habit and structure in this highly challenging environment are actually very similar throughout the biome, across the realms. This is often referred to as an example of 'convergent evolution' whereby the species in different areas have all evolved in the same way to cope with the conditions. Box 18.3 provides information on surveying and protecting Arctic biodiversity.

Reflective question

➤ What are the main differences between tundra and taiga biomes and how do plant adaptations reflect these?

ENVIRONMENTAL CHANGE

ARCTIC LIFE

The UNEP suggests that the Arctic region contains the biosphere's 'largest continuously intact ecosystems' (Johnsen *et al.*, 2010). The Arctic environment, of intense cold, severe winds and low biomass, has discouraged human settlement, except for the well-adapted ethnic groups thinly dispersed across the region and a few intensely focused primary industry bases. Global pressures impacting on the region include climate change (demonstrated by heat stressors on animal species as varied as polar bears and caribou), pollution (threatening ecosystem integrity and stability), and increasing development infrastructure (transport, extractive industry and urban settlement). Because the bio-region crosses many national boundaries, collaboration is essential to address these global pressures effectively, especially with respect to climate change, and biodiversity conservation.

The Arctic states comprise Canada, the United States (Alaska), the Russian Federation, Finland, Sweden, Norway, Denmark (Greenland and the Faroe Islands) and Iceland, as shown on Figure 18.25. Governance of the Arctic biomes must therefore endeavour to reflect the priorities of a wide range of disparate stakeholders, requiring patient, protracted negotiation to seek consensus. However, biome protection has been increasing rapidly, with 11% of the region now having protected status in some form, ahead of many of the Earth's eco-regions (see Box 18.5).

The Arctic Circle is an inappropriate southern boundary for the Arctic bio-region and defining an agreed limit has generated much discussion based on land cover, temperature regimes, and so on. Published research before the impacts of climate change were studied prioritized the geomorphological aspects of the Arctic environment, rather than dynamic biosphere issues. The Arctic's complex transnational structure also demonstrates the negotiation difficulties that arise in managing biodiversity conservation effectively, amid competing national, commercial and environmental interests. Figure 18.26 illustrates the range of institutions with an active interest in the Arctic environment. Although similar constraints in terms of administration affect most areas attempting to conserve biodiversity, such as the Biosphere Reserves (see Chapter 20), these are added to in the Arctic context by those specific to each of the eight nations involved and the effects upon relatively straightforward

Figure 18.25 The Arctic region.

BOX 18.3 ➤

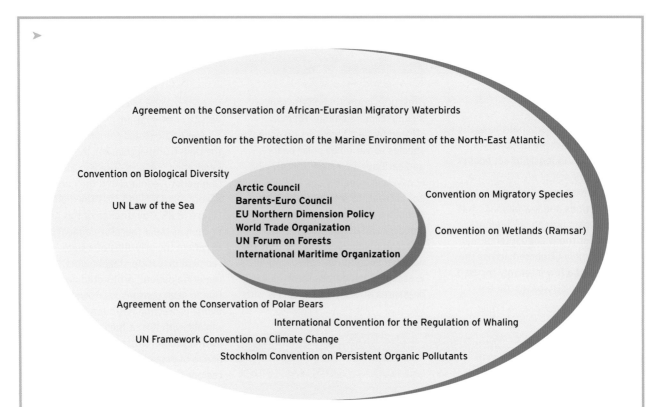

Figure 18.26 Important international organizations and Multinational Environment Agreements relating to the Arctic.

practical projects can appear overwhelming and work against rapid responses to change.

Arctic ecosystems respond to the long daylight hours of the Arctic summer with a brief surge of high productivity, inducing much inward migration. Species travel immense distances, especially birds and cetaceans. Hump-backed whales arrive from winter feeding grounds in the Caribbean, while bar-tailed godwits migrate between Australasia and the Siberian tundra. Seasonal movements of caribou reaching 1300 km have been observed within North America. Adverse habitat changes along these migration routes – such as a loss of feeding grounds or an increase in predation along flyways – can threaten the success of the migrating species. Where degradation in the Arctic region harms summer visitors, impacts will arise in warmer biomes along the migration routes, a reminder that one way in which ecosystems exhibit both interdependence and vulnerability is through species migration.

To identify and address knowledge gaps in relation to this environment, the Arctic Biodiversity Assessment was set up in 2006 and the Arctic Council aspires towards having a minimum of 12% of each ecozone with some form of protected status, in line with Millennium Development Goal 7 (see Box 18.5). One of the recent products of this research has been the Arctic Biodiversity Trends published in 2010 (Johnsen *et al.*, 2010). Key findings include the reduction in habitats such as sea ice and thermokarst ponds (see Chapter 17) as well as the effects of the poleward movement of biomes, leading to incursion of trees into tundra zones and the degradation of those peatlands which are based on permafrost conditions. As elsewhere, changing patterns of local temperatures have led to a less strong correlation of food availability with migration or reproduction periods. Local economic use of biome resources is also challenged with the prospect of unsustainability, mitigated for some by the influx of potential replacement species poleward. The trends analyzed reinforce the fact that those species with specialized habitat requirements or with limited distributions, such as the polar bear, are those most vulnerable to the effects of Arctic warming. As sea routes become more available, an increase in environmental disruption and pollution, such as that already emanating from the nickel-copper smelters in the Russian Federation, is regarded as inevitable.

The Circumpolar Biodiversity Monitoring Programme is intended

BOX 18.3 ➤

to monitor and respond to these changes, providing data sets including those from traditional knowledge to inform management policies for both conservation and sustainable development of the region. The type of information being collated for future use is wide-ranging in its nature and relevance. Evidence is growing that some species such as the mountain aven display local adaptations which may increase resilience to climate change. Tree-ring studies of *Salix lanata* show increased annual growth across both the European and Siberian tundra and taiga over the last few decades, in part at least due to lengthened growing seasons, which will impact upon carbon cycling in the affected ecosystems. The traditional methods of predicting seasonal changes are losing status as weather events become more erratic, impacting upon those who still attempt to gain a living in traditional ways within the Arctic. Elsewhere, studies on Ellesmere Island have provided ground-based support for the widespread satellite evidence of tundra 'greening', with increased abundance of evergreen shrubs on permanent plots – although in this study, species diversity appeared unchanged. Work in the Yakutia region of Siberia (Diekmann *et al.*, 2007), close to the classic palaeobotany sites on the coast, is proving useful in the interpretation of palaeobotanic data. Some localities appear to have present-day plant distributions which can be regarded

as relict steppes, co-existing with 'arctic alpine vegetation mainly with *Kobresia* meadows and alpine pioneer plants'. This study faced the common problem, slowly being overcome by organizations such as the International Association for Vegetation Studies, of finding new formats to combine data from the different traditions of vegetation description and analyses that are the norm in each group of Arctic states. New methods of study are helpful in such remote areas. Studies based on Middleton Island show effective use of GPS tracking of the kittiwakes (*Rissa tridactyla*) nesting on the island. This study investigated the flight paths and duration, allowing analysis of foraging trips, suggesting that short trips under 10 km were made

for chick-feeding purposes as opposed to those for adult-feeding purposes which could cover over 40 km.

Research aimed at conserving the special characteristics of the Arctic biomes (Figure 18.27) is therefore proceeding on a range of fronts, supported by both national and transnational institutions and organizations. Although the challenges, both long- and short-term, are undisputed, this region of the biosphere, perhaps because of its governance by technologically advanced nations, is turning these into opportunities for biodiversity and other research projects with results and methodologies applicable elsewhere.

(This box was produced with assistance from G. Ovens)

Figure 18.27 Moving towards the Arctic summer in a National Wildlife Refuge in the taiga/tundra transition zone – heavy snow cover persists. Less exposed areas show deciduous shrubs with occasional evergreens. (Source: photo courtesy of US Fish and Wildlife Service)

BOX 18.3

18.6 Mountain biomes

As with climate (see Chapter 6), there are regional modifications to the underlying structure of biomes. For example, topography can play a local role in altering regional patterns. High mountain areas are usually sufficiently unique to be considered as separate biomes to their surrounding regions. They may also form significant barriers to the

dispersal of life forms. The effect of aspect on local-scale biodiversity tends to be important within any mountainous area and it can also be influenced by local land-use patterns. Since mountain biomes tend to be unique to the area within which those mountains are found, these mountain biomes will be discussed by use of a case study. The Andes mountain biome is therefore described in Box 18.4.

CASE STUDIES

THE MOUNTAIN BIOMES – THE ANDES

As you ascend the Andes foothills, precipitation, cloud and fog increase and this is especially important in the arid regions of Peru and Ecuador. In higher zones above significant cloud cover, diurnal temperature ranges can exceed the normal annual ranges of the region especially within tropical areas of the Andes. This results in increased temperature stress for life forms. There are therefore altitudinal differences in mountain biome characteristics. Those for the Andes are described below from the lower slopes to the higher slopes and the zones are indicated in Figure 18.28. The lowest slopes called *Tierra calienta* generally have vegetation characteristics similar to the surrounding lowland biome and so are not discussed here.

Lower slopes: *Tierra templada*

The lower slopes consist of tropical pre-montane forest, mostly within the eastern inland foothills. Two distinctive formations include:

- Lomas: fog/cloud-dependent vegetation within the otherwise hot desert biome of the coastal lowlands. For example, *Tillandsia paleacea* takes in moisture via hairlike structures on leaf surfaces.

- Matorral: evergreen shrubs with an open canopy usually less than 5 m high in the equivalent of the Mediterranean biome, with *Lithaea* and *Acacia* species.

Middle slopes: the upper *Tierra templada* and the lower zones of *Tierra fria*

Temperate rain- or cloud-forests exist in most areas of the middle Andes slopes. This is mixed forest in northern areas which is mostly evergreen, although deciduous beech species become more important with increasing altitude and further south where conditions become cooler. Where locally drier conditions exist, savanna-type structures are found.

Upper slopes: *Tierra fria* and *Tierra helada*

Above the treeline, species are often xerophytic (e.g. *Stipa*) and there may

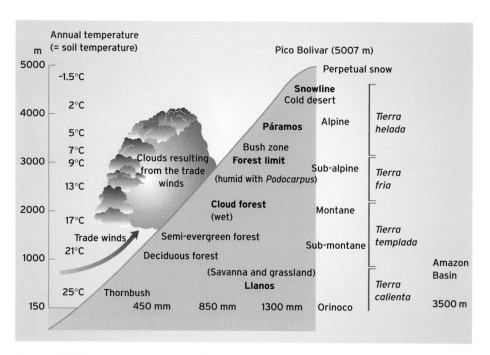

Figure 18.28 Schematic representation of the altitudinal belts in Venezuela, with mean annual precipitation and temperature.
(Source: after Breckle, 2002)

BOX 18.4 ➤

➤

be giant forms of plants such as giant groundsel and giant lobelia. These giant forms may be due to increased ultraviolet light or to lower oxygen levels. Grasses are found mixed with small evergreen shrubs where conditions are sufficiently moist but tend to be sparse and of cushion habit. Llama and alpaca grazing may be important controls on these areas (Figure 18.29).

A 'tropical alpine' zone exists before the true tundra zone, composed of low perennials but with slightly different species from the equivalent tundra genera and with more flowering plants, as opposed to mosses and lichens. Spring flowering is an important adaptation, to allow a rapid beginning to the growing season, as soon as the snow melts and/ or temperatures rise. Adaptations against the cold nights are found, such as within *Espeletia* species, where dead leaves shelter the stem/ trunks, or the woolly-leaved *Senecio* species. Tundra-like conditions and life forms occur in the highest zone between the treeline and the snowline. Plant adaptations are similar to those of other tundra biomes, especially a low habit to escape the strong cold and dry winds. Animals at this height, such as llamas, are adapted to low atmospheric oxygen by modified cardiovascular systems and may migrate to lower areas in winter. However, seasonal differences are not as extreme as those encountered on a diurnal basis.

In the highest regions, expanses of rock face or areas with gravels and **scree** are common. Their rapid drainage can lead effectively to drought conditions and very low biomass is the norm. However, sheltered areas may support lichens, together with some insects and birds.

Figure 18.29 The Andean altiplano, despite its bleak nature, produces sufficient vegetation to support a significant large mammal population. Like the tussocky habit of the grassland, these llamas are adapted to withstand the extreme climate conditions of this altitude. (Source: photo courtesy of Wouter Buytaert)

BOX 18.4

Reflective question

➤ Why might spatial maps of the biomes be complicated by topography?

18.7 Changing biomes

The biomes described above are depicted as having specific characteristics, either functional or structural, but it is clear that they are all dynamic in nature. With intensification of resource usage and climate change, the ability to regenerate and hence the degree of permanence of any biome cannot be guaranteed. 'Maritime versus continental climates, soil effects and fire effects can shift the balance between woodland, shrubland and grassland types' (Whittaker, 1975). Such natural causes of change have probably always existed, but the effects of human intervention are now equally important. Multinational bodies have been attempting to establish the consequences of biome and ecosystem changes not only for the ecosystem itself but also for human well-being.

18.7.1 The Millennium Ecosystem Assessment

The Millennium Ecosystem Assessment was conducted under the auspices of the United Nations, running from 2001 to 2005, with contributors from 95 countries. Its aim was 'to assess the consequences of ecosystem change for human well-being and to establish the scientific basis for actions needed to enhance the conservation and sustainable use of ecosystems and their contributions to human well-being' (www.maweb.org).

In the context of this chapter, the UN wished to assess the state of the biomes and the consequences of any major changes to them, for whatever reason, for humanity. Those controlling the exercise included non-governmental organizations and indigenous peoples as well as businesses, governments and international institutions. The work makes clear how the health and stability of the biosphere are intimately connected with human social and economic well-being. Changes to the biosphere and to biodiversity were seen as having importance to humankind's economy and well-being and needed to be monitored and, if possible, directed.

Many of the findings are highly significant. Biodiversity is being reduced. This is in part due to intensification of human land management. Often species regarded as without value, such as many insect species, are not protected and where species are considered to be a threat to the local economy, whether elephants or viral infections, they are removed. Increasing globalization has also increased the potential for invasion of previously isolated ecosystems by alien species (see Chapters 19 and 20), threatening further groups of species. The greatest immediate threat to biodiversity within the next half-century would appear to come not from climate change but from land-use change, although ultimately climate change will become the greater driver of change.

The Millennium Ecosystem Assessment found that although the rate of conversion of 'natural land' to arable or other intensive human use was slowing in biomes such as the temperate woodlands and Mediterranean zones, this was largely due to the fact that, within these regions, land suitable for development was running out. Van Vuuren *et al.* (2006) believed that tropical biomes such as savanna grasslands and shrublands, and tropical woodlands and forests, are the most likely to show marked changes to their vegetation. These will, inevitably, cascade changes throughout the rest of the ecosystems involved.

The Millennium Ecosystem Assessment came up with four potential future scenarios, relating to different types of human development drivers (Figure 18.30):

1. The *techno-garden* form of development, where environmentally sound technology, within a globally connected world, provides highly engineered and monitored ecosystems. Management would constantly be monitoring these ecosystems, enabling them to be proactive should any problem arise, with the technology and global connectivity available to put strategies into place quickly.

2. The *global orchestration* scenario would focus on high global economic growth, connections and trade.

3. The *adapting mosaic* is where local societies are stronger than global ones and people develop ecosystem management strategies appropriate to their own region, with local monitoring being a strong driver of policy-making. Managers would be very aware of the wider global initiatives but may pre-empt these by local action.

4. The *order from strength* situation is where, if a problem is seen to be developing in their region, then institutional and governmental action will certainly be taken, especially if their regional markets are likely to be affected.

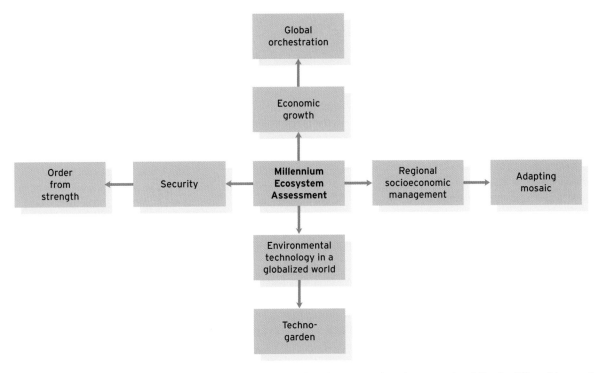

Figure 18.30 The four potential future scenarios produced by the Millennium Ecosystem Assessment, relating to different types of human development drivers.

It should be emphasized that these four scenarios were not specific states that the UN expected to come into being but were intended to guide discussion by providing a framework of possibilities against which research could be assessed. They are being used (as in the work of van Veeren *et al.*, 2006) to predict, for example, habitat change and species loss. Each scenario has, of course, been the result of scientific model building and discussions and relies upon global development being focused, at least in part, on one of the four main drivers for change. You can obtain more details on each of the scenarios by visiting the Millennium Ecosystem Assessment website at www.maweb.org.

18.7.2 Biome resilience

Deliberate and accidental introductions of species, which act as weed species without their natural controls of predators, alter local balances. Vast numbers of species were introduced by European colonists into, for example, New Zealand and the United States (see Chapters 19 and 20). The majority of biomes are identified by the presence of species that share characteristics allowing them to colonize particular geographical and climatic regions. The associations between these 'life zones' and the distribution of species throughout them present a way of monitoring and

perhaps ameliorating environmental and climatic change by ecosystem management on a massive scale. Approaches to resource utilization are changing, in the light of changed priorities and values. This is reflected in the popularity of venues such as the Eden Project discussed in Box 18.2.

How resilient is the biosphere in its response to human and natural agents of change? A backward glance in time suggests that the biosphere has the capacity to accommodate vast fluctuations in climatic and environmental conditions and changes in the numbers and density of the species which it supports. However, much of the reassurance gained from this hindsight lies in evidence from the fossil record with, inevitably, gaps in our knowledge. For the biosphere to maintain the capacity to support the current great diversity of life, some form of management is essential. The following two chapters suggest some of the means by which biogeographers and ecologists are attempting to address such issues.

Reflective question

➤ What might cause changes to biome characteristics and what might cause changes to biome location?

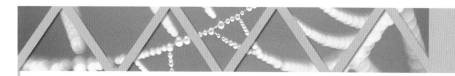

MILLENNIUM DEVELOPMENT GOAL 7 AND ITS IMPACTS

The Millennium Declaration, signed by 189 heads of state in the year 2000, aimed to 'free our fellow men, women and children from the abject and dehumanizing conditions of extreme poverty, to which more than a billion of them are currently subjected'. Eight goals were produced to achieve this aim by, it was hoped, 2015. The goals are ambitious, especially in relation to the timescale set, but are frequently used as a framework for international aid and as a spur to both local and national government efforts. Millennium Development Goal 7 is intended to 'ensure environmental sustainability', through the attainment of four targets:

7(a) Integrate the principles of sustainable development into country policies and programmes; reverse loss of environmental resources.

7(b) Reduce biodiversity loss, achieving, by 2010, a significant reduction in the rate of loss.

7(c) Reduce by half the proportion of people without sustainable access to safe drinking water and basic sanitation.

7(d) Achieve significant improvement in lives of at least 100 million slum dwellers, by 2020.

The second target in particular has great significance for the future of the biosphere and humankind's impact upon it, the original goal for 2010 having been missed (although the rate of deforestation in particular shows signs of decreasing) with targets now extended until 2015. Low levels of environmental sustainability, especially of biodiversity, will inevitably hamper efforts to meet other Millennium Development Goals (MDGs), such as ending hunger.

The Millennium Development Goal's Report (2010) observed that, despite increased investment in conservation planning and action, the major drivers of biodiversity loss were not yet being sufficiently addressed. Although nearly 12% of the planet's land area and nearly 1% of its sea area are currently under protection, other areas critical to the Earth's biodiversity are not yet adequately safeguarded. In addition, while certain areas may be officially 'protected', this does not mean that they are adequately managed or that the coverage provided is sufficient to effectively conserve critical habitats and species.

GEF Projects to address MDG7

The United Nations Development Programme is heavily involved in biodiversity projects, mainly to 'maintain and enhance the beneficial services provided by natural ecosystems in order to secure livelihoods, food, water and health security, reduce vulnerability to climate change, sequester carbon, and avoid greenhouse gas emissions'. The UNDP Global Environment Facility (GEF) funds very many (>6000) small grants in developing countries for biodiversity projects using local NGOs and community-based organizations. The scope of these projects will usually include building capacity, increasing investment, extending the protection of under-represented ecosystems, adapting to, and mitigating the effects of, climate change. Since the pressure for economic survival is greatest in rural areas with few job prospects, projects are geared towards balancing the needs of the local human society with those of the global and local environment. In Uganda, almost 200 small projects have been funded to date and a typical one, begun in 2005, is outlined below.

Mabira Green Ventures - a free-standing GEF Project

The Mabira Forest Reserve (Figure 18.31) lies along the main road between the capital, Kampala, and Jinja - an industrial centre which is expanding as a tourism resort, due to its location on the shores of both Lake Victoria and the Nile. Considerable pressure exists along the boundaries of the forest reserve, in part due to its potential for sugar plantations and charcoal production as well as for urban development, with a major roadside market already existing at Najembe serving the buses running between Kampala and Jinja. Several villages existed within the reserve area and illegal use of the forest for a variety of uses, from the collection of medicinal plants to illegal logging, continues at intervals together with attempted land claims from developers, sometimes with a political motive.

The rainforest reserve, regarded as a remnant of the ancient Guinea-Congo forest, has been protected in some form since 1932. Before that, the area suffered from intensive timber, banana and coffee production, to the extent that considerable areas were reduced to secondary forest. The topography means that whilst the matrix of the reserve comprises

BOX 18.5 ➤

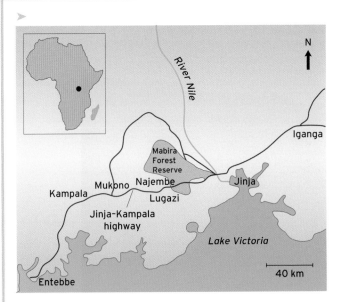

Figure 18.31 The Mabira Forest Reserve region. (Source: derived by C. Bennett)

tropical deciduous forest, some of the valleys are characterized by papyrus (*Cyperus papyrus*) swamps (Figure 18.32) and some drier, grassland areas also exist. The combination of habitats, especially the old forest stands within the reserve, make it one of the richest areas for bird diversity within Uganda.

Management, whilst including extensive replanting of the native species, aims to maintain the diversity of habitat that would naturally occur within a predominantly forest zone. The Green Ventures project is focused upon biodiversity and the conservation of this large block (almost 30 000 hectares) of complete semi-deciduous forest ecosystems and operates through the 'Mabira Forest Integrated Community Organisation'. There is little history of local community involvement in forestry projects and the intention is to develop a community-based ecotourism and conservation project, linked to Mabira Forest Lodge, already catering for visitors. As is usual in this type of project, much of the funding needs to be spent on building the capacity of local people for relevant forest, visitor and project management. Integrated conservation and business development will form part of the project outcomes, ensuring both local collaboration with forest issues and also sustainable development as increasingly the forest reserve is seen as providing a source of income rather than inconvenience.

Figure 18.32 Papyrus swamp, characteristic of riverside areas in the Mabira district.

BOX 18.5 ➤

Figure 18.33 Slope on the Mount Elgon foothills in Bududa, showing steep gradients and old land-slide scars and gullies, as well as continuing clearance for farm plots, mostly for small-scale banana or coffee production.

GEF small grants as part of a wider scheme

From 2011, a significant tranche of GEF Ugandan funding will be allocated to projects within the Greater Mbale TACC (Territorial Approach to Climate Change) Region, which comprises Bududa and Manafwa districts as well as that of Mbale. TACC is an approach to building regional resilience against climate change, working at the local rather than national level, on the assumption that the most efficient means of resilience are likely to be those related to local rather than national conditions. The Greater Mbale region has been selected as one of 10 TACC pilot projects, each of which is linked with a region from the developed world, in this case Wales. Major funds will come from the UNDP, including GEF projects tailored to the ITCP (see below). Welsh expertise will support the Ugandan region with exchange visits and joint projects. An holistic Integrated Territorial Climate Plan will be one of the outcomes, developed in response to the production of a series of vulnerability maps relating to the specific environmental and economic challenges likely to be felt in this region as a result of climate change. A very wide range of stakeholders are involved, not least because of the inclusion of a sector of the Mount Elgon National Park within the boundary (see Box 19.1 in Chapter 19) and one of the most important Arabica coffee producing areas of Uganda. The scoping mission has already identified tree planting and an increase in canopy cover as major responses to the regional environmental stressors in the east of the region, including an increase in landslides – a response to local geology combined with the need to farm ever-more marginal areas as population increases (Figure 18.33).

Further information on topics relevant to this box can be found at: www.endpoverty2015.org/en/goals www.un.org/millenniumgoals www.undp.org/biodiversity

BOX 18.5

18.8 Summary

Within this chapter the main features of the biosphere, especially of the major biomes, have been put forward. Any one of these biomes could form a chapter, or even a book, of its own. The current objective, however, is to draw out common themes to allow geographical comparisons to be made. The themes have been the typical climates, soils and wildlife of all types. Since vegetation tends to underpin most other biogeographical activity, the characteristic vegetation for each biome has been an important part of the descriptions provided for each one. Specific mention has also been made of the plant and

animal adaptations to the challenges of each biome. The factors underlying the geographical patterns may then, in many cases, be reduced to the effects of temperature and available moisture. A useful summary diagram is given in Figure 18.34. However, local human factors can alter biome characteristics. Furthermore, there are often no distinct boundaries between biomes. Instead a biome may slowly grade into another biome; this graded area is often called an ecotone.

The tropical biomes range from high-productivity, high-biomass rainforests to low-productivity, low-biomass deserts. Within each of these, species are adapted to compete for local resources. In tropical forests with fast

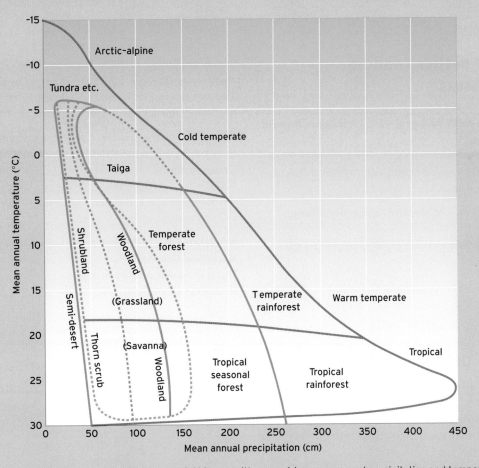

Figure 18.34 The distribution of the major terrestrial biomes with respect to mean annual precipitation and temperature. Within regions delimited by the dashed lines a number of factors, including proximity to oceans, seasonality of drought and human land use, may affect the biome type that develops. Boundaries are, of course, approximate. (Source: after Whittaker, 1975)

nutrient cycling, rapid bacterial decomposition and plentiful water, species compete for light resources and at ground level there may be hardly any light penetration. However, in savanna grasslands and hot deserts, species are adapted to compete for water resources. Often this means only being active for short periods when water is available and developing methods for conserving water, including nocturnal activity and daytime dormancy.

The temperate biomes vary from deciduous and evergreen forests to seasonal grasslands with perennial plants and large grazing herbivores. Rainfall, temperature and sunlight are seasonal and Mediterranean and temperate grassland biomes are subject to natural fires to which plants and animals are adapted (e.g. by burrowing or rhizomes). The cold biomes of the tundra and taiga are mainly found in the northern hemisphere and have slow

nutrient cycles, a lack of soil fauna and acidic leaf litter. Tree and shrub growth can often be stunted and where water resources are frozen for part of the year, plants may develop wide fanning root systems to take up as much water as possible as soon as it thaws. Evergreen species maximize productivity potential by allowing photosynthesis and growth to take place as soon as spring sunlight becomes available. Hibernation and burrowing allow species to cope with long cold winters. The tundra is dominated by sparse vegetation cover and the low productivity of vegetation means that large areas are required to support migratory mammals such as reindeer. In highly stressful environments only mosses and lichens may be able to survive and the tundra has a very low diversity of plants.

Further reading
Books

Archibold, O.W. (1995) *Ecology of world vegetation.* Chapman and Hall, London.
This has a North American focus and many useful photographs and diagrams. It also provides an unusually large list of references that can be followed up.

Bradbury, I.K. (1998) *The biosphere.* John Wiley & Sons, Chichester.
This is a standard undergraduate text which is good for the general principles.

Cox, B. and Moore, P. (2005) *Biogeography: An ecological and evolutionary approach.* John Wiley & Sons, Chichester.
This is an excellent and classic undergraduate text which is of relevance to this chapter and for Chapters 9 and 10.

Walter, H. (2002) *Vegetation of the Earth and ecological systems of the geo-biosphere,* translated by Owen Muise. Springer-Verlag, Berlin.
This classic book clearly shows the author's love of his subject and the results of extensive personal research. Therefore there is a slightly patchy coverage.

Whittaker, R.H. (1975) *Communities and ecosystems.* Macmillan, New York.
This is an American classic text, written by the proponent of environmental gradients.

Further reading
Papers

Hill, G.B. and Henry, G.H.R. (2011) Responses of high Arctic wet sedge tundra to climate warming since 1980. *Global Change Biology,* 17, 276–287.
Examples of climate change impacts in the Arctic relevant to Box 18.3.

Hudson, J.M.G. and Henry, G.H.R. (2009) Increased plant biomass in a High Arctic heath community from 1981 to 2008. *Ecology,* 90, 2657–2663.

CHAPTER 19

Biogeographical concepts

Hilary S.C. Thomas
C2C Research Centre, University of Glamorgan

Learning objectives

After reading this chapter you should be able to:

➤ understand key concepts within biogeography

➤ understand processes leading to changes in biogeographical distributions over time and space

➤ describe the basic principles of landscape ecology

➤ understand the theory of island biogeography and how it can be used to inform management

➤ show an awareness of current issues in biogeography

19.1 Introduction

Biogeography is an important area of science and informs global environmental policy as well as local land management practice. Biogeography is the study of the distribution and patterns of life on Earth. Biogeography is therefore usually focused upon the study of the current distribution of living things. However, it also recognizes that to understand, predict and manage changes in biogeographical patterns it is necessary to examine the processes that create these distributions. The links with ecology are strong. Figure 19.1 shows the range of topics shared by both disciplines. Those with the greatest interest for biogeographers are located

to the right of the diagram. Most of these biogeographical topics are therefore addressed within this chapter. The various types of spatial biogeographical distributions will be explained, together with factors that may lead to change over time such as climate change or the introduction of 'alien' species. Factors more closely allied to ecology are located to the left of Figure 19.1. Many topics such as island biogeography and landscape ecology are part of both research areas but will form important elements of this chapter. Concepts such as biomes, **succession** and conservation are very important to both and are shown in bold at the centre of Figure 19.1. These topics are dealt with in Chapters 18, 19, 20 and 25. Of course, with biomes being so fundamental to geography there is a separate chapter on this topic (Chapter 18).

Of the many factors and processes that underpin our present understanding of biogeography, certain groups are especially important. These include those relating to the concept of succession, the strategies used by plants, animals, birds and insects as they utilize the biosphere, and the effects of habitat disturbance. Common themes for all these factors include the role of time and how geological history can help explain plant and animal distributions, the relationships between species and their environment and how ecological processes operate to produce biogeographical distributions. These fundamental themes will be explored in this chapter.

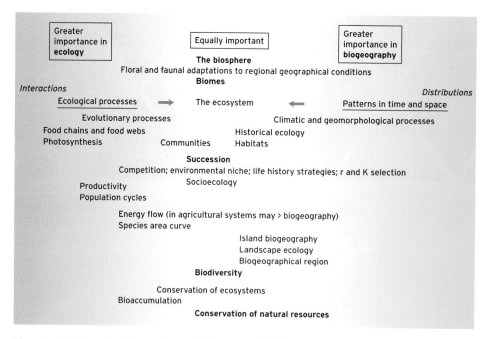

Figure 19.1 A comparison of ecology and biogeography in terms of the emphasis placed on selected topics important to both disciplines. Note that an emphasis on interactions tends to characterize ecology, while a greater emphasis on distributions is usually found within biogeography.

19.2 Succession

Classic studies within biogeography have focused on vegetation succession, where the sequential changes to be seen over considerable lengths of time have been used to develop a number of concepts that underlie modern biogeographical thinking. The distribution of living things within the biosphere is rarely constant, at whatever scale it is studied. Birds may migrate vast distances, insects crawl from one blade of grass to another and the seeds of plants may drift with the wind, roll down slopes or be moved through the

agency of animals. To begin to understand the processes involved it is worthwhile taking a snapshot view that is revisited over time. When this is done it usually becomes clear that directional change is taking place. Such a change tends to lead towards increased complexity of the **community** structure and increased biodiversity. This is seen, for example, in studies of bare ground such as on a new volcanic island or on areas abandoned from other land uses (Figure 19.2). In nearly all cases the disturbed or recently exposed ground is colonized by animals and vegetation. This process is described as **primary succession**.

Figure 19.2 Newly colonized land on a former mining waste heap.

Change is fairly slow at first and then becomes very rapid, before becoming much slower again. This final slow phase generally continues for as long as the area's development is undisturbed.

Disturbance usually causes the whole process to begin once more, shifts the succession back one or more stages, or may cause it to move forward in a different direction. Succession is therefore the result of a range of environmental and ecological processes and has immense importance within biogeography as well as in ecology. The following section discusses the early influential schools of thought on succession while the topic is given further treatment in Chapter 20.

19.2.1 The development of succession theory

While Cowles (1899) did pioneering work on dune succession on the shores of Lake Michigan, which demonstrated how plants, soil, water, climate and topography interact, there were two other major contributors to the concept of succession as it is used today: Clements and Gleason. Clements (1916, 1928) based his work around the concept that plant distributions formed recognizable communities that became increasingly complex in their interactions within themselves. He observed that each community operated almost as a single developing organism. As a community develops, the populations of plants and other species change and replace each other and may also alter the environment. This may occur, for example, through a reduction in wind speed and daily temperature ranges as vegetation increases in height. These changes lead to climatically controlled **climax communities**. During the development towards the climax community, the community passes through recognizable **seral stages** such as from an initial cover of lichen to communities dominated by mosses, then by grasses, then by shrubs and finally by forest. At this final stage a dynamic equilibrium (see Chapter 1) would keep the community basically the same despite any temporary local or minor disturbances. A division was seen between primary succession from bare ground and **secondary succession** that would take place after disturbance where a seed-bank, remnant roots and other materials were likely to remain and possibly influence the character of the resulting community. In either case, Clements assumed that six distinctive processes were in operation, controlling the development of the community:

1. *Nudation* – creation of the initial bare surface.
2. *Migration* – seed or vegetative spread.
3. *Ecesis* – establishment, to the extent that a complete life cycle can occur.
4. *Competition* – for space, water and so on.
5. *Reaction* – modification of the habitat by the vegetation.
6. *Stabilization* – usually the end product or climax community related to the regional climate.

A widely accepted amendment to Clements's idea that there would be a single final community in any one area was made by the British ecologist Tansley. He believed that where there were strong local environmental influences, these would control the type of final community. There might be, for example, 'topographic' climax communities, controlled by the local geomorphology, or 'edaphic' climax communities where the major influences were the local soil conditions.

Much modern work, however, from glacier forelands to roadside verges, indicates that communities, rather than becoming locally characteristic over time, frequently become increasingly different. This suggests that neither a **monoclimax** (single final community) nor a set of **polyclimaxes** (several co-existing final communities) may be the end result of succession. Studies continue on this theme because the ability to manipulate succession is of great value, in applications relating to economic usefulness to humans or where maximized conservation value is required.

In what appears, at first, to be the exact opposite of Clements's ideas, Gleason (1926) assumed that communities existed by chance, as a result of their individual response to the environment in which they found themselves. He suggested that the distribution of species would be controlled by factors such as migration opportunities, environmental selection and frequency of disturbance. He also maintained that there was no true directional succession, since change was random and disturbance was so frequent that a stable final assemblage of species was not possible. His ideas, with others, are shown for easy comparison in Table 19.1.

Whittaker (1953), however, considered that undisturbed communities could be generally observed to change slowly from one type of community to another along habitat gradients such as light or water availability. As a result, he regarded climax communities as a pattern of communities that reflected the pattern of environmental gradients. Within this there would be a central, most extensive community type that would represent the prevailing or climatic climax. Whittaker's succession model is regarded as a climax pattern model, which he considered to be self-maintaining and 'potentially immortal if not disturbed'. He also introduced the term '**ecocline**', for a combination of environmental variants that change together through space. For example, both temperature and exposure change simultaneously as a result of altitude. Matthews (1992)

Table 19.1 Some early ideas on succession important to biogeographical prediction of patterns

Author's initial publication	Idea	End result
Clements (1916)	Like a single developing organism	A climatically controlled climax community
Tansley (1920)	Some divergence of outcome	Mosaic of polyclimax communities, reflecting local environmental factors
Gleason (1917)	Individualistic plant responses	No stable, characteristic final assemblage
Whittaker (1953)	Environmental gradients	Gradual changes in response to gradients of environmental change

pointed out, however, that despite Whittaker's proposal being more complex and including elements of Gleason's individual responses, it still provided no mechanism. There is therefore a need to move towards more process-based approaches when thinking about succession. As the study of succession moved forward, the emphasis changed towards the search for mechanisms. This helps to increase the practical application of a concept that has become, while still contested, one of the most important elements in the explanation of changes in biogeographical pattern over time. Such successional processes are discussed in Chapter 20.

Reflective questions

➤ How does the concept of succession alter our perception of 'restoration' in degraded or disturbed areas?

➤ Can you summarize succession theories?

19.3 Spatial patterns and processes

A variety of processes lie behind the patterns seen in the distribution of plants and animals. Geographical factors are often very closely allied with those relating to ecology. For example, where the local climate leads to the development of environmental gradients, the result will often determine the range of available **niches** for plants and animals. Biogeographical processes will vary in importance according to the scale or resolution of the region selected by an investigator. It will become apparent that varying scales of study are important for the development of biogeographical understanding.

19.3.1 Global-scale patterns

19.3.1.1 Climate

On a global scale the patterns of species distribution can be seen to reflect those of the major climatic zones (see Chapter 18). The length of the growing season, for example, provides a constraint upon the distribution of species. It would, however, be simplistic to assume that one factor, even at this scale, dominates biogeographical distributions. Whittaker (1975) noted that biomes should be defined by their structure, but in practice they have to be defined by combining this with their environment. The availability of water, for example, is often related to climate, but may also be a response to local geology and topography and hence, perhaps, soil conditions.

19.3.1.2 Geological factors

Tectonic movement, including the break-up of continents, has led to opportunities for species to spread but also to the creation of barriers to spreading, such as mountain ranges or open water. Lengthy geographical isolation, for whatever reason, may produce distinctive communities at a variety of scales. Perhaps the most famous biogeographical concept arising from this is that of 'Wallace's line'. Alfred Russel Wallace was impressed by the sudden difference in bird families (and other fauna and flora) he encountered when he sailed from the island of Bali and landed on Lombok, 30 km to the east, in the 1850s (Figure 19.3). The birds on Bali were clearly related to those of the larger islands of Java and Sumatra and mainland Malaysia. On Lombok, however, the birds were instead related to those of New Guinea and Australia. He marked the channel between Bali and Lombok as the divide

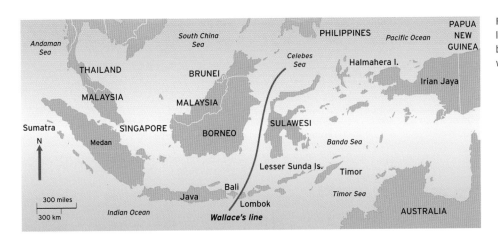

Figure 19.3 Wallace's line. This line marks a distinct change in biogeography and is coincident with an ocean trench.

between two biogeographical regions, the Oriental and Australian. In his honour this dividing line as shown in Figure 19.3 is still referred to as Wallace's line. This line on the map corresponds roughly to the location of an ocean trench separating global plates. Thus the islands to the north of the line were formed from the Asian continent and the islands to the south were originally part of the Australian continent and they had been separate continents for over 200 million years. Of course it is only since the 1960s that we have known about plate tectonics (see Chapter 2) so Wallace was way ahead of his time.

19.3.1.3 Ecological factors

Ecological factors affecting global-scale biogeographical patterns and distributions include habitat conditions such as the presence or absence of predators and prey, competition for nutrients and living space, and the ability to adapt and migrate (Huggett, 2004). Ecological factors have been the focus of much research and there are a range of terms to describe the geographical patterns of living creatures (Table 19.2). The distribution of some species is regarded as cosmopolitan since they are found worldwide when suitable habitats exist, but this is not the usual situation (Spellerberg and Sawyer, 1999). Geological processes can result in barriers to movement and lead to certain species being **endemic** (confined to a particular area), such as the Hawaiian hawk, *Buteo solitarius*. If species have never been known to occur in other regions their distribution is regarded as being of **primary endemism** such as the Australasian marsupials. However, certain endemic species are only endemic because extinctions of those species have occurred in the other places where they used to survive (such as the mammals of the West Indies). These species are therefore described as having a **secondary endemic** distribution. If a species is quite widely distributed but with

Table 19.2 Commonly used terms to describe patterns in the distribution of species and other taxonomic groups

Type of distribution	Definition
Cosmopolitan	A species or taxonomic group that is distributed widely throughout the world
Primary endemic	A species or taxonomic group in a particular region native only to that region
Secondary endemic	A species or taxonomic group in a particular region whose distribution has contracted so that it is now native only to the region in which it is found
Rare	A species or taxonomic group that is restricted geographically or is widespread but never found in abundance
Disjunct	A species that occupies areas that are widely separated and scattered (species with a discontinuous distribution)
Indigenous or native	A species that originates in a particular place or which has arrived there entirely independently of human activity

(Source: from Spellerberg *et al.*, 1999)

Table 19.3 Biogeographical features of three oceanic island systems

	Hawaiian Islands (Pacific)	Galapagos Islands (Pacific)	Mauritius (Indian Ocean)
Vascular plants	900 of which 850 are endemic	540 of which 170 are endemic	878 of which 329 are endemic
Threatened native plants	343	82	269
Introduced plant species	4000	195	-
Animals known to have gone extinct	86	5	41
Extinct endemic plants	108	2	24
Species of land snails	c.1000, all endemic	90 of which 66 are endemic	109 of which 77 are endemic
Extinct snail species	29	1	25
Unusual wildlife	High endemism, superspecies-rich taxa, e.g. fruitflies	High endemism, unusual species, e.g. marine iguanas	High endemism, once home to the dodo

(Source: adapted from Jeffries, 1997)

large gaps between regions its biogeographical distribution is described as **disjunct**. The mountain avens, *Dryas octopetala*, for example, is found in the Burren Hills, Ireland, although it is usually considered to be a species of high mountain regions.

The isolation of groups of organisms (such that they do not interbreed and the gene flow from the larger population is suppressed) can lead to the evolution of new species, or **speciation**. This particular mechanism is termed allopatric (or geographic) speciation. A good example of such speciation is the finches from the Galapagos Islands, specimens of which were collected by Charles Darwin and studied back in London by John Gould along with Darwin. The finches probably contain 13 different species, within the subfamily Geospizinae. It is normal for genetic drift to take place within populations such as the gradual change in height or in hair colour over generations, especially if these changes are adaptive to the local environment (e.g. providing better camouflage against predators). In small populations, such as within the Galapagos Islands, the founder population, which may consist of a very few individuals that arrive in a new location (or are left in an almost destroyed old one), may become increasingly diverse over time. The Galapagos finches consist of species with different-shaped beaks (to take advantage of different food sources) and some that are vegetarian and others that are carnivorous. Oceanic islands and other isolated regions are very likely to support such endemic

species, as shown by Table 19.3. The table also shows that these islands are also highly susceptible to disturbance and extinction.

Living creatures vary tremendously in their mobility. This is considered to be part of their **vital attributes** and may determine the most likely combinations of species to be found in new or disturbed localities. Most species spread very gradually, as opportunity arises. This movement may take centuries to cover a few tens of kilometres but is still highly effective. It allows species (such as oak trees), but not necessarily individuals, to disperse. For example, species may disperse away from an area that is becoming increasingly cold or nutrient poor. Individuals in cooler margins eventually die but their offspring preferentially succeed in the warmer margins of the local distribution. This may especially be the case where ecological niches are vacated by other species with greater warmth requirements.

A far more rapid change is possible where, for example, plants such as dandelion (Figure 19.4) or cotton grass have mobile seeds with feathers that help them drift in the wind. Many species have taken advantage of the increased mobility offered by human forms of transport and are found in clusters around, for example, rural railway stations. Seeds and insects also inadvertently take the train, bus or car from one locality to another and may develop new areas of distribution if the conditions in the new locality are suitable in terms of habitat and a vacant or incompletely filled ecological

(a)

(b)

Figure 19.4 Many species are highly mobile. (a) Dandelion seed head. The seeds can travel long distances even in a gentle breeze. (b) Raspberry seeds are carried by birds and other animals who eat the fruit and then drop the seeds elsewhere either while eating or after passing through the animal's digestive system.

niche. Often, however, the dispersed individuals are unable to succeed as they cannot overcome competitive exclusion from local resources.

19.3.2 Small-scale patterns

When regions are investigated there are often smaller-scale nested patterns. These may relate to the availability of light, water or nutrients, to the dampness or dryness of the site, or to the intensity of disturbance by humans. Humans may create an environmental gradient that we could call a 'disturbance gradient'. This may range from virtually untouched wilderness to a suburban back garden. Four points along such a gradient have been identified by West-hoff (1983) as shown in Table 19.4. It should be noted, however, that there are very few areas of the world that today would truly belong to the 'natural' end of the gradient, as identified in his classification.

19.3.3 Landscape ecology

In 1939, Troll described **landscape ecology** as 'the study of the entire complex cause–effect network between the living communities and their environmental conditions which prevails in a specific section of the landscape … and becomes apparent in a specific landscape pattern'. The living landscape can show clear structure, functions and change and may be studied at a huge variety of scales. At probably the finest scale, a suburban garden can provide evidence of the effects of microclimatic and pedology on the pattern of living things. On a coarser scale (e.g. a national park), a landscape system will display within it a hierarchy of smaller-scale patterns. Too coarse a scale of approach, however, may often lack the detail required for many biogeographical purposes. Equally, too fine a resolution such as concentrating upon a single tree may obscure patterns.

Table 19.4 Degrees of naturalness

Natural	A landscape or an ecosystem not influenced by humans
Subnatural	A landscape or an ecosystem influenced to some degree, but still holding to the same formation or vegetation type from which it derived, such as grassland
Semi-natural	A landscape or an ecosystem in which the flora and fauna are largely spontaneous but in which the vegetation structure is altered in such a way that it now constitutes a different type, such as from woodland to heather moorland
Cultural	A landscape or an ecosystem in which the flora and fauna are essentially influenced by humans, such as on arable land or within plantation forests

(Source: after Westhoff, 1983)

Landscape ecology is a major growth area in biogeography, and in land planning, conservation and ecology. The results of landscape ecological investigations are used to inform planning and management decisions in many countries (see Box 19.1 for example). Troll developed his ideas when working with air photographs in Africa. Present-day geographers are still likely to use these, but they will also use a **geographical information system** to help map and visualize different effects such as geology, geomorphology, climate, soil, biogeography, economic activity, settlement, culture and social structure. These collectively give a landscape its identity and help determine its biogeographical character. Landscape ecology is of particular value to the biogeographer when defining and helping to explain the distributions seen in groups of species. It helps when identifying species habitats and functions within a landscape and attempting to measure, through analysis of landscape stability, how likely it is that a pattern may change over time. The International Association for Landscape Ecology puts forward several core themes for its activity:

- the spatial pattern or structure of landscapes, ranging from wilderness to cities;
- the relationship between pattern and process in landscapes;
- the relationship of human activity to landscape pattern, process and change;
- the effect of scale and disturbance on the landscape.

Any landscape, regardless of the scale of an investigation, can be analyzed in terms of its landscape ecological structure. The typical **landscape patches**, **landscape matrices** and **landscape corridors** together define the character of that landscape and hence help define the 'sense of place' vital in preserving the heritage of an area.

19.3.3.1 Landscape patches

Landscape patches are distinctive elements within the wider landscape, such as ponds, woods or towns. These patches may have value put upon them by those living or working in the area. The spread of a town may be described as 'urban sprawl' and be locally unpopular, while the cutting down of part of a woodland might be considered disastrous. The objective analysis of landscape patches usually deals with the patch characteristics of shape, frequency, origin and stability.

The shape of a patch may indicate its vulnerability to change as a result of outside influences. The greater the proportion of 'edge' to 'interior' (Figure 19.5a), the more likely the patch is to be influenced by external factors. For

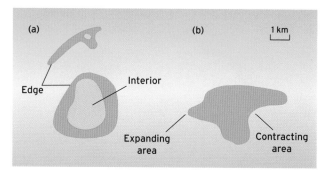

Figure 19.5 Landscape patches: (a) the proportion of edge to interior is an important factor in determining whether a patch will be influenced by external factors; (b) convex edges indicate contracting patches while concave edges indicate an expanding part of a landscape patch.

example, a small wooded area will be more exposed to light entering the understorey from surrounding land than a more substantial wooded area, where the central areas may be sheltered from sunlight and wind. This will allow species from outside the smaller patch to colonize more easily. Figure 19.5(a) shows a patch which is elongated and therefore has a greater proportion of edges to interior. The figure also shows a much wider, more circular patch which therefore has a smaller proportion of edges to interior and will be less susceptible to external disturbance. Patches with significant concave boundary sections tend to demonstrate that in those sectors the patch is contracting, as opposed to areas of expansion, indicated by convex sections as shown in Figure 19.5(b).

The origin of a landscape patch may also determine other characteristics (Figure 19.6). Classic geographical sites, such as a wet point (e.g. a spring) or dry point (e.g. a hummock), also produce related landscape patches. In each case, the change in local environmental conditions tends to result in distinctive biogeographical and land-use patterns. These may then become altered or reinforced by other processes such as the bare ground that results from trampling and intensive grazing around waterholes in semi-arid areas. Some introduced patches originate from management, such as the field patterns of arable land, **polders** (land reclaimed from the sea through embankments) or intentionally burned patches of heather moorland or rainforest. Other patches may be inadvertent such as those created by cattle trampling at the approaches to field gateways. Many patches result from natural disturbance such as within a forest where a dominant tree within the canopy dies and falls, creating a new area for regeneration. Of course, most human settlements form patches of varying scale. Within the built-up area of a major city, biogeographically important patches

Figure 19.6 A hillslope in northern England where a variety of landscape patches can be identified. The upper slopes are cool and wet and only suitable for short moorland species. Lower down the slope woodland patches can grow. Other bright green pasture patches have been created by humans who have fertilized the land and concentrated sheep grazing using walls and fences.

of parks, lakes and gardens exist. Perhaps the most important patches on a global scale in the twenty-first century will be remnants of biomes that have otherwise vanished and thus act as a haven for species that would otherwise be extinct.

19.3.3.2 Landscape matrices

The matrix of the landscape is usually regarded as that element of the landscape that occupies a greater area than any patch type within it. In general, it plays a dominant role in the dynamics and character of the local biogeography, and also in other elements of the local geography, so helping to create a distinctive sense of place. The matrix contains within it the other landscape elements (patches, corridors) and a measure of its stability can be obtained by a study of the extent to which patches appear and develop within it, creating 'porosity'. Slight porosity

is usual but if this reaches a very high level, the integrity of the matrix may be threatened. For example, clearance of a forest for agriculture initially produces a forested landscape containing patches of farmland. These patches of farmland may increase in size or number over time to the point at which the landscape is better described as a farmland containing patches of forest. There may be an interim scenario, where two or more matrices contribute to make up a particular landscape. Such a situation may occur across the transition between biomes, seen classically in the forest–savanna boundary of tropical regions, as discussed in Chapter 18.

19.3.3.3 Landscape corridors

Landscape corridors are narrow strips of land that differ from the matrix that exists on either side of the strip. They may be isolated strips but are usually attached in

Figure 19.7 Hedgerows acting as landscape corridors. The hedges support diverse plants communities and shelter animals.

some manner to a patch, often of similar character. For example, hedges may unintentionally link woodland, rivers may run from lakes, and roads join built-up areas. The key characteristics of corridors for the biogeographer relate to their connectivity both with similar corridors and with other landscape features and to the fact that there are often sharp microclimatic and soil gradients from one side of a corridor to another. An example is shown in Figure 19.7 of connected hedgerows acting as landscape corridors. In this artificial situation natural, parallel corridors such as embankments, verges, carriageways, central reservations and drainage zones, each with their own characteristics, exist within the major landscape feature. Increasingly, corridors such as hedgerows or streams are considered to be of immense potential conservation value, by giving species an escape route from threatened patches, and perhaps providing migration opportunities. Corridors are also geographically important in that they are frequently the element of the landscape that both inhabitants and visitors may regard

as typifying the area and giving it its character, such as drystone walls, or small, irregular hedges dividing up the landscape. A new corridor, such as a line of electricity pylons, may attract intense criticism and, despite its small size, be regarded as 'ruining the area'. It may also form a significant barrier or filter to movement for a whole range of species.

Reflective questions

➤ What are the major factors influencing global-scale biogeographical patterns?

➤ Can you provide examples of each of the Westhoff landscape types from within your country?

➤ What are the distinctive landscape patches and corridors in your local landscape?

CASE STUDIES

WORLD BIOSPHERE RESERVES

Biosphere Reserves form one of the most important UNESCO responses to global environmental concerns, working at the regional landscape scale. Biosphere Reserves are areas of terrestrial and coastal ecosystems promoting solutions to reconcile the conservation of biodiversity with its sustainable use. They are internationally recognized, nominated by national governments and remain under sovereign jurisdiction of the states where they are located. Biosphere Reserves serve in some ways as 'living laboratories' for testing out and demonstrating integrated management of land, water and biodiversity. Collectively, Biosphere Reserves form a world network: the World Network of Biosphere Reserves (WNBR). Within this network, exchanges of information, experience and personnel are facilitated. There are over 500 Biosphere Reserves in over 100 countries.

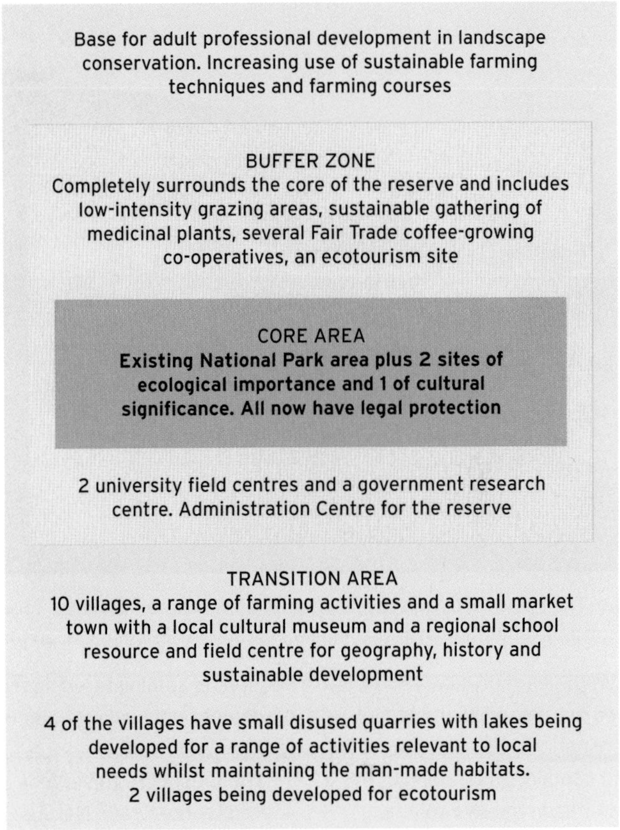

Each Biosphere Reserve has distinctive characteristics and these landscapes and ecosystems are monitored and conserved with respect to both regional and global issues. Economic development that is culturally and ecologically sustainable in the context of that reserve is encouraged in a spirit of community co-operation and management, with the intention of developing a high level of focused knowledge and research on those aspects which made the reserve worthy of its initial designation. This information may then be shared with other Biosphere Reserves. Management increasingly addresses education and appropriate, sustainable livelihood issues alongside those of conservation, especially in the buffer

Base for adult professional development in landscape conservation. Increasing use of sustainable farming techniques and farming courses

BUFFER ZONE
Completely surrounds the core of the reserve and includes low-intensity grazing areas, sustainable gathering of medicinal plants, several Fair Trade coffee-growing co-operatives, an ecotourism site

CORE AREA
Existing National Park area plus 2 sites of ecological importance and 1 of cultural significance. All now have legal protection

2 university field centres and a government research centre. Administration Centre for the reserve

TRANSITION AREA
10 villages, a range of farming activities and a small market town with a local cultural museum and a regional school resource and field centre for geography, history and sustainable development

4 of the villages have small disused quarries with lakes being developed for a range of activities relevant to local needs whilst maintaining the man-made habitats. 2 villages being developed for ecotourism

Figure 19.8 A typical UNESCO Biosphere Reserve model.

regions where human impact is tolerated and even encouraged to preserve as far as possible the core regions of highest conservation value as shown in Figure 19.8.

Where a regional landscape has had Biosphere Reserve designation for more than 10 years, the designation is subject to review, giving an opportunity to respond to new research and guidelines, perhaps by expansion. The UK, for example, currently has seven reserves: Beinn Eighe, Dyfi, Loch Druidibeg, Moorhouse–Upper Teesdale, North Devon, North Norfolk and the Cairnsmore of Fleet and Silver Flowe–Merrick Kells Biosphere Reserve.

Most of these were originally designated in the 1970s. Recent moves have been towards the establishment of regional networks and transboundary reserves, assisting management of shared landscapes and ecosystems. From 2010 to 2017, the BREESE programme (Biosphere Reserves for Environmental and Economic Security) will use the 'living laboratory' role of the reserves as a means of addressing poverty issues in the light of climate change, focusing on South-East Asia.

Two contrasting case studies of Biosphere Reserves are presented below: the Lower Moravia Reserve is well established and was extended in

BOX 19.1 ➤

Figure 19.9 The Pavlov Hills and the small town of Mikulov which is the focus of vine-growing and tourism. The hills have a thin cover of soil and vegetation over the limestone. (Source: Ronald A. S. Johnston)

2003 from the earlier Palava Reserve designated in 1986, while the Mount Elgon transnational Biosphere Reserve still has to complete its transition to a single unit from the two adjoining Mount Elgon Biosphere Reserves recently set up in Kenya and Uganda.

The Lower Moravia Biosphere Reserve, Czech Republic

The Lower Moravia UNESCO Biosphere Reserve area, in the south of the Czech Republic near Brno, is one of the driest parts of the Czech Republic and is dominated by Jurassic limestone hills (Figure 19.9). The main range, the Pavlov Hills, gave their name to the original Biosphere Reserve, set up in 1986. The reserve contains a range of nature reserves and other protected areas, both ecological and cultural. There is evidence of settlement in the area from the Stone Age with pasture and coppicing as land uses. The historic small town of Mikulov, from where the Biosphere

Reserve is administered, is the focus of vine-growing villages and is also of tourism importance.

The Biosphere Reserve is administered by the Czech Nature Conservation Authority, which works with other institutions, especially the Masaryk and Mendel Universities in the nearest city, Brno. Like most Biosphere Reserves, it is a working landscape with emphasis on management to maintain the current balance between humans and nature, to protect the distinctive landscape and to facilitate environmental research and education.

The area is at a biogeographical crossroads, containing species from alpine as well as northern and southern continental zones, and contains many endangered species, such as the nationally scarce Old World swallowtail (*Papilio machaon*). Although the potential natural vegetation is mixed oak forest, the type of oak forest alters with height

and includes species-rich shrub and herb layers. Dry grassland areas are also important elements of the landscape. Some of these are naturally occurring 'karstic forest steppes' but others reflect the intensive grazing of the area in the past. Although grazing has a long history in the region, in communist times the state ran game reserves in the area, with mouflon and goats stocked at higher rates than are regarded as acceptable today. Other problems stem from mining, with some hills almost destroyed, and from soil erosion. Stock management in parts of the reserve has changed, especially from 1996, but intensive grazing persists in some areas.

The Mount Elgon transboundary Biosphere Reserve, Kenya and Uganda

In 2003, Kenya designated its Mount Elgon region as a biosphere

BOX 19.1 ➤

➤ reserve, with Uganda responding by designating the Ugandan portion of the massif in 2005, each country already having its own 'Mount Elgon National Park'. This huge extinct volcanic area, with the world's largest **caldera**, straddles the international border between south-east Uganda and south-west Kenya (Figure 19.10). The massif is of immense importance to both countries and beyond as a generator of weather patterns and a catchment area feeding the Nile, Lake Victoria and other regionally significant river systems (Figure 19.11). For both nations, Mount Elgon has great cultural significance, the distribution of indigenous tribes long predating

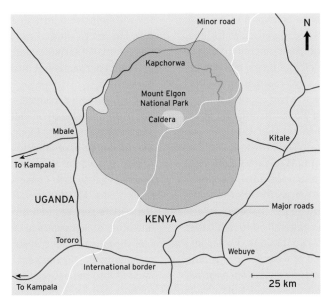

Figure 19.10 Map of the Mount Elgon region. (Source: original map devised by C. Bennett)

Figure 19.11 The steep slopes of the Mount Elgon region, together with one of the waterfalls created by the volcanic scarps. Tropical montane vegetation begins to displace the deciduous rainforest that would be the naturally occurring vegetation of the surrounding area. Sites such as this are slowly becoming the focus of ecotourism and Biosphere Reserve designation should help develop this in a sustainable manner.

BOX 19.1 ➤

> the colonial boundaries. The region is renowned in particular for its very high plant diversity. Since the end of colonial times, management priorities and regimes have differed between the two countries, for example elephants existing on the Kenyan rather than Ugandan slopes at present. Until recently, National Park wardens in each country were unable to pursue either poachers or game across the international boundary and exchange of information was generally difficult. Other problems continue to arise from the existence of a dense rural population of mainly subsistence farmers but also small growers utilizing the rich soils and modified mountain climate for coffee, banana and vegetable production. Population is increasing rapidly as mortality rates fall with improved infrastructure. This has led to widespread use of unsustainable farming methods and periodic encroachment into national park areas.
>
> Considerable ecological research has been undertaken on individual sites within the Mount Elgon region, although little in relation to the region as a whole. The Mount Elgon Regional Ecosystem Conservation Programme (MERCP) is working with the wildlife services of both countries and other stakeholders to develop a realistic capacity for transboundary management. Management initiatives must fit into the national strategies for both countries and equality of decision-making has to be demonstrated, but activities such as those under MERCP are reducing levels of poaching and illegal harvesting.
>
> Transboundary reserves 'are rare because of lack of trust and transparency between countries, unclear benefits and political considerations that hamper co-operation between nations' (Sankey, 2003). Nevertheless, progress towards a formal Ugandan/Kenyan transboundary reserve continues, with dialogue between institutions in both countries. It has been agreed by UNESCO that the way forward will be in two stages, with the initial establishment of designated Mount Elgon Biosphere Reserves on each side of the border now achieved. This will be followed by the development of joint project proposals and other cross-border activities. The final stage will be a single functional Biosphere Reserve.
>
> BOX 19.1

19.4 Temporal patterns and distributions

19.4.1 Geological time

The role of geological processes in biogeography via plate movements and barrier development has already been discussed above and in Chapter 18. However, the geological record also suggests that there have been periods of the past when large-scale mass extinctions occurred. Over the past 600 million years there are believed to have been five mass extinction phases. A good example is the mass extinctions associated with the loss of the dinosaurs 65 million years ago. This extinction resulted in 40% of the terrestrial invertebrate species being wiped out. A range of causes for mass extinctions have been proposed and these include asteroid impact, rising sea levels, prolonged glaciations and other environmental changes (e.g. global atmospheric gas changes following periods of extended volcanism).

19.4.2 Post-glacial change

Over a lesser but still extended period of time, climatic fluctuations over the past 2.6 million years have been important determinants of biogeographical change. Many regions of the world have experienced major climatic change to which plants and animals have been forced to adjust in order to survive. Many of these changes were relatively slow such that the distribution patterns of most plants were able to respond to these changes. Areas such as southern Europe became refuges for more northern species as conditions became increasingly colder (see Chapter 22). Here they survived during the coldest periods and some species were in a position to recolonize as conditions improved. The British Isles, however, are relatively species-poor for their latitude. The Irish Sea and the English Channel became deeper and wider following the last Glacial Maximum. As the ice retreated from northern Europe and sea levels rose, those species with dispersal mechanisms unable to operate over extensive water bodies were unsuccessful colonizers. Box 19.2 describes biogeographical change in northwest Europe over the past 10 000 years. It should be clear from this box that such regions have not had natural stable biogeographies and that the landscape is in a constant state of change.

19.4.3 Migratory patterns

The migratory movement of species creates biogeographical change over time. This type of change, however, forms a recurring pattern over long periods and may be an important element of a region's biogeography. Species may be present in a region at one time of year and absent at another. Therefore, if a site is being studied at the same time every year (which is often done to record change and avoid the effects of changing seasons) migratory species may be missed from the survey. Figure 19.13 shows that the 'summer' and 'winter' distributions of a migratory species may be not only very distant from each other, rarely just following temperature changes, but also quite different in scale. For Kirtland's warbler the range is restricted to a small area between Lakes Michigan and Huron during the summer, whereas during the winter the range is much larger covering the entire Bahamas region. Important considerations for migratory species include suitable areas for breeding and the availability of food. Some species travel great distances

ENVIRONMENTAL CHANGE

POST-GLACIAL BIOGEOGRAPHICAL CHANGE IN NORTH-WEST EUROPE

Around 11 000 to 10 000 years ago north-west Europe was dominated by a tundra biome (see Chapter 23). This consisted of grasses and sedges, often rosette and tussock species with dwarf trees and shrubs such as willow and birch together with alpine species of herbs. Many of these species were survivors from isolated pockets that were maintained during the glacial periods in warmer areas of southern Europe such as around the Bay of Biscay.

By 9000 to 7500 years ago, groves of birch, aspen and juniper and, further south, more substantial blocks of woodland began to develop. As warming progressed, lowlands became increasingly wooded, up to about 1000 m altitude. Species varied with oak, elm, hazel and lime common in the south, with some beech and yew where the land was chalky. Pine, birch and hazel existed further north. The more cold-tolerant plants were gradually pushed into the mountain areas and to the moorlands.

Between about 8000 and 5000 years ago the climate was warm and wet. The beginnings of peat formation were seen in many regions, together with the inundation of some formerly wooded areas by sea (e.g. south-east England fenlands). The North Sea together with the English Channel rapidly widened as sea levels rose in response to further melting of the world's ice caps. This reduced the chances of species migration into mainland Britain and Ireland. As a result, while Britain at present has around 1500 native species the climatically equivalent French zones contain nearer 6000. The Atlantic period also saw a decline in elm all over north-west Europe. Suggestions for this include a response to climate, the effect of humans, a rise in the occurrence of the weed species *Plantago lanceolata*, or possibly due to disease.

The period around 5000 to 3000 years ago was even warmer and drier and evidence for the beginnings of agriculture can be found in many regions during this period. There is evidence of some decline in woodland, together with an increase in heathland. The period from 3000 to 2000 years ago was ge nerally a

wetter and cooler period, with moorland vegetation developing in many areas. Cereals were introduced to agriculture and there is increasing evidence of use of trees by humans for shelter, cooking and fodder.

Thus we need to think quite carefully about what we mean when we talk about natural landscapes and 'native' species (Figure 19.12). Often in areas disturbed by humans there is a perception that we should restore the landscape to its natural state. However, this natural state is actually one of continuous change. Most species present in north-west Europe today were not present 12 000 years ago. The forests of northern Europe have not been around for thousands of years. If an area has been affected by humans for 2000 years, what should we restore it to? Should we try to restore species that were present 2000 years ago on the site or should we introduce species that we *think* should be there if humans had not affected the land? Historical biogeography can be a very useful tool in allowing us to place contemporary biogeographical distributions into context and in allowing us to understand contemporary patterns.

BOX 19.2 ➤

Figure 19.12 Moorland vegetation in the English Pennines. The vegetation is periodically burnt and grazed to control its growth and yet these landscapes are considered to be ones of 'outstanding *natural* beauty'.

BOX 19.2

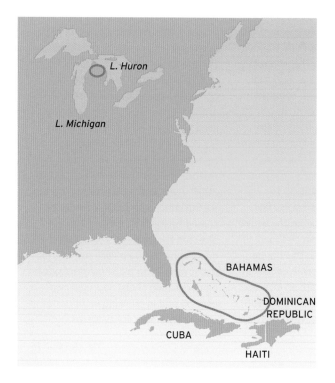

Figure 19.13 The summer and winter range of Kirtland's warbler, *Dendroica kirtlandii*. The bird is restricted to a small area between Lakes Huron and Michigan in the summer while the winter range is much greater with the birds dispersed througout the Bahamas. (Source: Van Tyne, 1951)

to enable them to utilize areas for part of the year which for permanent residents would require dormancy, or other adaptations.

19.4.4 Alien introductions

Alien species are any species not part of the 'native' bio-geography. Their interest for the biogeographer is the manner in which they are able to arrive and in many cases establish and spread. This colonization can sometimes occur at astounding rates, far outcompeting local species. The aliens may be introduced accidentally or on purpose. Certain communities and regions seem more susceptible to invasion than others. Many continental islands, such as Tasmania, Ireland and mainland Britain, were once connected to a continent by land that, although it once formed a bridge, is now submerged as a result of post-glacial sea-level rises. These islands therefore tend to be relatively species-poor (see above), which makes them more likely to be susceptible to invasions. Box 19.3 describes an example of mammalian invasion of islands and the effect on other species.

Firstly, environmental niches may not be completely filled in these locations so that these areas can actually support a much greater diversity of species than at present. Secondly, in the absence of certain hosts, parasites, predators and diseases may also be absent from such species-poor communities or they may have adapted to utilize different species. Thirdly, certain forms of habitat management, especially those related to farming or to conservation, may also create an environment that is capable of favouring alien species. In such environments it is easy to envisage the easy integration of newly arriving species which are able to occupy some of these biogeographical gaps. Human colonizers over the years have deliberately brought alien species into areas that were suitable for agriculture or even for nostalgia. In many cases, such as the British rabbits intro-

duced into the New Zealand grasslands, they may now be considered as pests (see Chapter 10). The edible dormouse, *Glis glis*, originally from Hungary, has reached pest status in parts of southern Britain, even though the rate of its spread is slow compared with that of other introductions. Shown in Figure 19.14 is the spread of the American grey squirrel, within mainland Britain, at the expense of the native red squirrel. Although the spread of the grey squirrel has been well documented since the deliberate releases in the late nineteenth and early twentieth centuries, the reasons for its success are still disputed. However, the grey squirrel was adapted to fierce competition in its native North America, so it is not a surprise that it has spread so quickly in the much less hostile British deciduous woodlands (Yalden, 1999). Some species have been targeted for eradication after their populations exploded or became problematic following introduction. These include copyu in Britain and limu seaweed introduced around Hawaii which is threatening native coral by blocking out sunlight. Box 19.3 describes the problem of Japanese knotweed in the United Kingdom. Further examples of aquatic species invasions are provided in Section 21.2.4 of Chapter 21.

Figure 19.14 The effectiveness of spread among introduced species: the grey squirrel. Since the 1920s the grey squirrel has taken over the habitats of the red squirrel which has consequently been outcompeted and declined. (Source: after Yalden, 1999)

TRESPASSERS

Trespassers in the Lord Howe Islands, Australia

The black rat (*Rattus rattus*) is regarded as one of the 'world's worst' invading species by the World Conservation Union. Its main adverse effect arises from its opportunistic predation, as opposed to the effects on habitats of other species. The effect of the arrival of rats on isolated communities such as islands has been observed worldwide, although a common problem in quantifying this has been the lack of data on conditions before rats arrived. A typical example is found on the geographically remote Lord Howe Island group, off the east coast of Australia, which has had the status of World Heritage Site since 1982. The black rat reached many of Australia's offshore islands in the nineteenth century and is known to have arrived on the main Lord Howe Island in 1918, after which it spread rapidly. Anecdotal and qualitative records of its effects are now being supplemented by more formal surveys, related to the work of

'Biosecurity Australia' (see Box 20.6 in Chapter 20). The arrival and spread of the black rat is likely to have been the tipping factor leading to population decline in some species on the Lord Howe Islands. For example, the Lord Howe Island wood-feeding cockroach, *Panesthia lata*, is now extinct through predation by the black rat on the main island and only survives in small numbers on the smaller islands, such as Roach Island, where the rats have not yet arrived. The Lord Howe Island gecko, *Christinus guentheri*, has also suffered from the impact of the rats, again through predation, in this case of eggs and young geckos. The geckos are sensitive to such predation because each female lays very few eggs.

Trespassers in the United Kingdom

Japanese knotweed, *Fallopia japonica* var. japonica (Figure 19.15), is probably the most invasive UK plant species at the beginning of the twenty-first century. Originally used in the United Kingdom as a garden ornamental plant, it was introduced in the nineteenth century and spreads

rapidly though its underground rhizomes. It also hybridizes with similar species such as *F. sachalinensis* and *F. japonica* var. compacta. The spread of this alien is frequently assisted by movement of topsoil by deliberate human action and from material accidentally dropped by construction traffic as well as through natural processes. Even small sections of rhizome will form a new plant and the dense clumps formed after two or three years can be 9 m high, substantially altering local habitats and out-competing local species. As a result its distribution can be strongly linear along roads and footpaths, riverbanks, beaches and railway lines, as well as on construction sites.

A range of control methods are used against Japanese knotweed. However, it takes several years to destroy a rhizome system using herbicide and removal of the plant requires not just the root system to be taken out but also the surrounding soil to ensure that no fragments of rhizome remain. Cutting above-ground material, with the intention of exhausting the plant, is another long-term technique but this creates the problem of disposal of the cut material.

It is now a criminal offence to grow the species in the wild or to knowingly spread it in the United Kingdom. Japanese knotweed is classed as 'controlled waste' under the UK Environmental Protection Act 1990 and so may only be taken to licensed landfill sites. The Environment Agency's website (www. environment-agency.gov.uk) provides advice for those needing to deal with the species and many counties have a 'Knotweed Forum'. A multi-agency

Figure 19.15 Japanese knotweed. (Source: Alamy Images: Mambo)

BOX 19.3 ➤

➤

team is working to identify natural predators of the plant that may be suitable for release (after quarantine testing) in the United Kingdom as an aid to elimination or control of Japanese knotweed.

In March 2010, as a result of these tests, Defra (UK government agency) and the Welsh Assembly gave approval for the limited release of *Aphalara itadori*, a psyllid louse, into closely monitored knotweed-infested sites and it is likely that the fungus *Mycosphaerella polygoni-cuspidati* (Figure 19.16) may also be released as a biological control. The release of *A. itadori* forms a landmark within the UK, being the first deliberate import of a biocontrol species. Twentieth-century deliberate releases in Australia of imported species, such as the attempts to control the prickly pear cactus, led to a chain reaction of other uncontrolled alien species destroying local ecosystems (see Chapter 10). The British release has therefore raised considerable opposition. A review undertaken by Kabat *et al.* (2006) stated that there was poor data accessibility to existing control and eradication methodologies to give an objective picture against which the results for other controls could be assessed. The potential cost-effectiveness of a biological solution is likely to have been a major incentive for its adoption. It has been estimated that to treat a single 30 m^2 plot with current methods can cost building developers over £50 000 and to treat the UK nationally would cost in the region of £1.5 billion (Cabi, 2009).

The spread of alien species results in the main from the absence of natural predators and stressors in the invaded locality. The results of the studies on Japanese knotweed have shown that, within its original

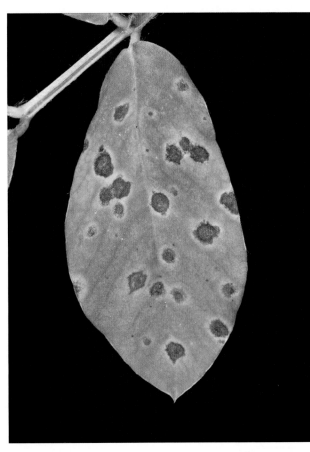

Figure 19.16 *Mycosphaerella* fungus may slow or halt the spread of Japanese knotweed. (Source: Getty Images: Nigel Cattlin)

ecosystems, an important stressor is *M. polygoni-cuspidati*, a leaf-spot fungus, which is a co-evolved natural enemy, having a severe impact on the fitness of infected plants (Kurose *et al.*, 2010). It is probable (Murrell *et al.*, 2011) that part of the knotweed's invasive capacity derives from **allelopathy** which can be reduced by regular mechanical destruction of the rhizomes. *A. itadori*, especially in its juvenile stage, sucks the sap from Japanese knotweed, causing significant damage. It could therefore be possible to replicate these stressors, through mechanical means plus biocontrols in the UK environment.

It seems logical that future control of alien species within ecosystems is likely to come from a combination of methods, such as outlined here. The costs of preliminary research are sufficiently high to make these solutions at present likely to be applied only where there are severe economic pressures. As the true costs of ecosystem health become clearer with the increase in research on ecosystem services (see Chapter 20), this may change in the future, but the costs of rash importation of species as biocontrols are likely to remain immense.

BOX 19.3

Although it is often the lack of predators in the new locality that may explain the success of a new species, it should be noted that even should this biogeographically fortunate state exist, it may not be prolonged. The Guernsey fleabane (a type of daisy) has spread since the 1980s throughout London and appears to be a rival for the niches occupied by the butterfly bush species (English Nature, 2002). However, it is now consumed by the insect *Nysius senecionis*, which is another alien species first recorded in Britain in 1992 some nine years after the arrival of the Guernsey fleabane.

Reflective questions

➤ What role have tectonic and glacial processes played in the biographical patterns and processes that we see today?

➤ Why might you have to design a biogeographical research project carefully in terms of its timing during the year?

➤ What factors determine the success or failure of an invading species?

19.5 Biogeographical modelling

In this section two very different examples are provided. The first, island biogeography, contains perhaps the most famous use of the term 'biogeography' and has been the basis of much pure research and applied management techniques. It has also fuelled fierce controversy over the years. Like the ideas of Clements, Gleason and Whittaker discussed above, the concepts arising from the island biogeography model of species change are almost unconsciously used as part of the language of biogeography and ecology. In comparison, the second example of biogeographical climate modelling is far more recent. The ability to predict where and when a species is likely to achieve pest status can be a valuable warning device for a wide range of users including farmers and environmental and economic planners.

19.5.1 Island biogeography

The study of isolated areas such as islands has provided knowledge and understanding of huge importance. Islands often provide, through their clearly defined boundaries and geographical isolation, as near a situation to a scientific laboratory that the biogeographer is likely to encounter. On many islands the biogeography is simplified owing to a lack of external factors, thus allowing us to examine individual processes more clearly.

MacArthur and Wilson (1967) considered a balance between the rates of immigration of species to an island and the rates of extinction. Their theory of island biogeography suggested a relationship between the species richness of an island and its size and isolation. The opportunities for species to arrive at a newly created island and be able to sustain themselves depend upon ease of access and the nature of the habitat encountered. Islands close to a mainland would usually be more accessible, while more remote ones would be disadvantaged (Figure 19.17c), but might develop a range of species more slowly, perhaps if they were part of a chain of islands, where species might have the opportunity to cross between islands. If the most favourable habitat was already occupied, later immigrant species might have to adapt to survive. Larger islands or those with a greater variety of habitats might be able to support a greater range of species (Figure 19.17b). Most data sets support the idea that larger islands tend to contain a greater number of species than those that are smaller.

The rate of extinction of species inhabiting the island would initially be low, since competition for resources would be low. As the number of species increased, however, and pressure on resources increased, the rate of extinction would rise. Figure 19.17(a) expresses the curves for extinction and immigration and suggests that there will be an equilibrium number of species when the rate of immigration is matched by that of species extinction. The role of distance from immigration source and size of island is also shown on the figure as it alters the slope of the lines on the graphs (Figure 19.17b and c).

Several important extensions of island biogeography theory are used in biogeography and elsewhere to explain patterns of distribution. Firstly, if an island is created by the loss of a land bridge to the mainland, such as Tasmania, following a sea-level rise, the new 'continental' island might initially be species-rich. Not all species might be able to be supported by the restricted resources of the new island, so in this case, an increase in extinction rates and a drop in species richness might reasonably be expected over time. This is opposite to the circumstances governing the developing species richness of a new 'oceanic' island, such as Hawaii.

The second extension of the theory is to encompass within it 'virtual' islands, or landscape patches, such as

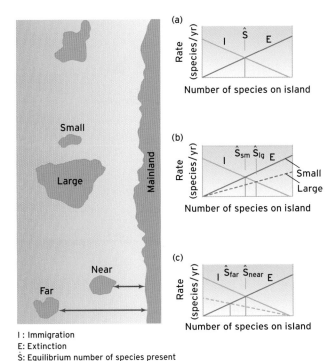

I : Immigration
E: Extinction
Ŝ: Equilibrium number of species present

Figure 19.17 The MacArthur and Wilson (1967) theory of island biogeography. (a) The island has immigration of new species from the mainland and extinctions of species. The number of species on the island should be in balance since if there are too many then extinctions will be greater than the rate of immigration of new species and if there are too few for the environmental niches available then new immigrants will find it easy to survive and extinction rate will be low. (b) Smaller islands have fewer environmental niches and so can support fewer species than larger Islands. (c) Islands closer to the mainland species' source will have a greater number of species as immigration rates will be greater than for distant islands.

woodland clearings, ponds or an isolated marshy area. This element of the island biogeography model has had extensive application in determining the most suitable areas for conservation. This has resulted in modelling to determine whether it is better to have a single large patch that can preserve a greater number of species or several small patches. Several smaller areas when added together may have the same total area as one large patch. However, these two situations (lots of small patches and one large patch) do not produce the same diversity even though they have the same total area (May, 1975). The 'single large' school of thought is supported by Wilson (1994) and by Diamond (1975), Diamond and May (1976) and Diamond and Veitch (1981). Diamond supports the use of a few, large reserves, such as Yellowstone National Park, especially if the conservation of large mammals or those requiring migration routes is involved. Studies monitoring species loss have shown that single large reserves such as Yellowstone and similar

very large 'islands' have lost fewer species than the smaller American national parks. It is also contended that neither initial species diversity nor latitudinal range (both of which are used as gauges of habitat diversity) are as important as the size of the reserve in predicting species diversity. A major political disadvantage, however, is that agencies have used this school of thought to downgrade the protection of certain reserves on the grounds that they must be too small to be viable.

Others such as Simberloff (1983) and van der Maarel (1997) regard habitat diversity as the most important factor influencing the number of species existing in a given reserve. In this light, it can be argued that 'small' is not only acceptable at times but can also provide insurance against site loss by providing replication. In other words, having species in lots of small sites means that if species were lost from one site there would still be others preserving them. However, if we only protected one large site and a species was lost from that, then the species would be lost completely. In many countries there is simply insufficient land or too great a demand for land to allow the creation of large nature reserves.

19.5.2 Species distribution modelling

There is often an economic as well as a scientific reason for learning about the preferred 'geography' of a species. This may be in order to increase the efficiency of its production, such as breeding varieties of sheep able to cope well in specific conditions. It may also be to control the spread of a pest species. Using climate change models to predict how species distributions might change is becoming very popular today. The range of many insects will expand or change, and new combinations of pests and diseases may emerge as natural systems respond to altered temperature and precipitation profiles. It will be important to predict such problems in advance in order to prepare for and mitigate against them.

Any species will have, for each environmental variable, not only a preferred ecological niche where conditions are ideal, but also a less favourable wider area within which it is still able to survive and reproduce. Outside this area conditions are so stressful that the species is unlikely to be found. Such areas can be measured indirectly by observation of population abundance. These areas can then be related to the local climatic conditions. It should then be possible to predict where species might be distributed for a given climate regime. This can first be tested by applying the predictions to other areas where the population has been observed but not measured.

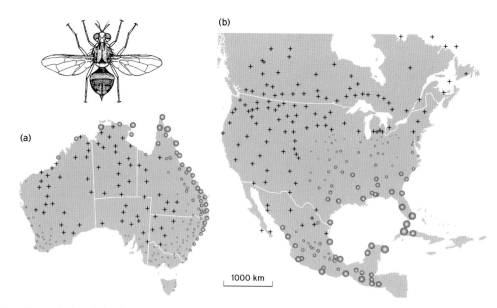

Figure 19.18 Climatic modelling of the Queensland fruit fly for (a) Australia and (b) North America. The area of each circle is proportional to the Ecoclimatic Index of the fly; crosses indicate that the fly could not permanently colonize that area. (Source: courtesy of CSIRO Australia, © CSIRO Australia 1991)

For example, in many parts of Australia, fruit growing has become important as the basis for agriculture, both for home use and for export. A major pest is the Queensland fruit fly, *Bactrocera tryoni*. Sutherst and Maywald (1985) developed a model allowing the potential distribution of this fruit fly to be predicted. They developed an 'Ecoclimatic Index' to allow precise evaluation of areas suited to the Queensland fruit fly to be carried out. The evaluation is based upon factors such as the stress created for the species as a result of localities where temperatures or moisture levels are either too high or too low. The resulting model can predict distribution of the fly, based upon the Ecoclimatic Index, not just for Australia but for other areas where there is the potential for inadvertent introduction of the species by accident (e.g. poor hygiene). The Ecoclimatic Index is shown for Australia in Figure 19.18(a) and shows those areas most favourable to colonization. The potential of the fly as a pest for North America has also been predicted using the model and predictions are shown in Figure 19.18(b). This has been done by matching the climate of North America with that of the fly's native range. The figure shows that accidental transport of the fly to North America could lead to large areas of potential colonization. Most of Mexico and the east coast of the United States are vulnerable. Canada and the western United States, however, are unlikely to support a permanent colonization. Thus it is possible to calculate which regions of a continent are at risk and therefore which regions should have their imports more carefully checked.

Reflective questions

➤ What is the theory of island biogeography?

➤ How is island biogeography theory relevant to places that are not oceanic islands?

➤ What benefits are there in modelling species habitats in a world of changing climate and mass transportation of foodstuffs?

19.6 Biogeography and environmental management

Many aspects of biogeography have great relevance and impact today. It has enabled the development of techniques for managing a range of environmental factors within an increasingly technological environment, where the need for land makes multiple land-use demands common (Figure 19.19). The management of wild or semi-natural animals and plants may have a variety of objectives, methods and levels of intensity. Four important management aspects of biogeography are those of:

1. agriculture – sometimes taken as being a form of 'directed' succession;

Figure 19.19 A picnic area in the upper Taff Valley, Wales, managed as an amenity area while controlling access to a reservoir and helping to minimize sedimentation flows into the water.

2. conservation – either of a particular species or group of species, or as a means of increasing biodiversity;
3. recreation/amenity environments;
4. environmental tools – plants or animals may be a means to an end rather than being present for their intrinsic value. This can include the planting of grasses to increase the stability of new slopes or the keeping of fierce guard dogs.

The management aims of agricultural systems are focused upon optimizing conditions for one particular species of plant or animal, often through the control and direction of the normal successional processes. This may involve the reduction of competition by removing unwanted 'weed' species. Other methods include increasing nutrient supplies in combinations particularly appropriate to the selected species (e.g. via fertilizers). The reduction or removal of predators, pests and parasites is also a feature of this type of biogeographical management. These unwanted species, however, may be regarded by conservationists as endangered species, or by agriculturists elsewhere as valued potential crops. A weed is often merely a plant in the wrong place at the wrong time. The same is true of many faunal 'pests'. A further complication arises when, for example, a generally unwelcome species, such as bracken, provides a habitat for species with a high conservation value, such as the high brown fritillary butterfly. Some management practices may seem severe, such as the burning of heather moorland, and often they are carried out as a tradition rather than being based on best possible practice (Holden *et al.*, 2007). Often the biogeographical research has not yet been done to establish best practice or the most efficient techniques. It is also important to understand the resistance or **resilience** (ability to 'bounce back') of biogeographical systems to make management practices efficient (Box 19.4).

Often a wide range of biogeographical techniques have to be adopted which are directed towards economic goals such as tourism and sporting activities. These may be combined with those of conservation and the preservation of cultural values. Frequently there are multiple land-use requirements made upon management (see Chapter 27) and priorities change as seasonal demands on the site alter. As a result, development of multifaceted plans for parks and reserves is increasingly common, often including biodiversity action plans. Amenity management can include

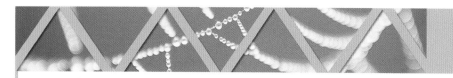

RESILIENCE AND PANARCHY

The biosphere is dynamic and often has the ability to 'bounce back' or to recover from stresses applied to it. This capacity to return to its previous state is described as resilience. In Figure 19.20, the area of grassland on the headland has remained for many years, despite the stressors of the coastal environment and from trampling by visitors along the coastal path. Unlike straightforward physical resilience, such as within a rubber band, ecological resilience is highly complex and may involve a range of short- and longer-term adaptations. It is problematic to measure or to pre-dict the ecological limits of any one locality. With increasingly intensive human land use, coupled with climate and other environmental changes, it is important to be able to measure ecological resilience and to find the tipping point. In other words it is important to determine how much stress or disruption can be accommodated before a change to a new state is inevitable. This new state may result in biogeographical boundary changes and the processes involved in the adaptive capacity of ecosystems are therefore important.

The term 'resilience', in this context, was first used by C.S. Holling in 1973 and the measurement of resilience is becoming an important tool in ecosystem management. The concept should be treated with care, however, as it can be problematic if taken too literally. It is sometimes equated, if on a simplistic level, with 'sustainability' or with successful environmental management. Ecological succession (see Chapter 20) can be regarded as a series of stable states, separated by dynamic, transitional conditions. Some workers, who regard ecological resilience as both a functional and philosophical tool, consider it to be the determinant of movement between stable states and that a system's adaptive capacity can act as a buffer against movement

Figure 19.20 Tintagel, Cornwall, England. An example of an area where habitats are displaying their adaptive capacity to both environmental and human-related stresses.

BOX 19.4 ➤

from one state to another (grassland and desert, for example) but that adaptive capacity may also be the medium of change form one state to another.

The natural environment, however, is 'noisy' and the complexity of relationships within and between, for example, trophic levels is becoming increasingly recognized, their measurement and analysis appearing to go beyond the traditional hierarchical structures used in ecology and biogeography. This complexity has given rise to a concept which is old in philosophical terms (1860), but relatively new in terms of its ecological meaning. This is the concept of 'panarchy' which is upheld by many of the workers supporting resilience theory. It claims that, to fully understand environmental change, it is necessary to consider long- and short-term interactions. These interactions might occur both across and within trophic levels adapting to change and creating resilience. Panarchy equates these interactions with the evolutionary development of ecosystems but in a more random way than is traditionally accepted. Instead of ecological hierarchies, there are panarchies that incorporate adaptive cycles, signifying their dynamism. With this concept there is an emphasis on the connectivity between panarchy levels.

BOX 19.4

extreme forms of management. The creation of appropriate turf conditions for international standard soccer pitches may involve massive inputs of energy and biogeographical and ecological expertise at a microscale. Even civic amenities such as parks and playing fields require the control of vegetation succession.

Biogeographical management may involve land restoration or land reclamation. Land reclamation suggests that the land can be used again whereas land restoration is about returning the site to its former state. Large areas of many countries may have initially unpromising environments that make either of these proposals seem very difficult. Figure 19.21 lists major characteristics of industrial wastes that limit vegetation development. However, an understanding of biogeography suggests that some of these harsh areas may in fact act as sanctuaries where species unable to tolerate the competition found in more usual environments may be able to flourish. Thus the harsh conditions shown in Figure 19.21 may not be considered as limiting but actually as enabling. For example, it has been possible to locate areas such as those with high levels of lead in the soil by their distinctive vegetation. Thus,

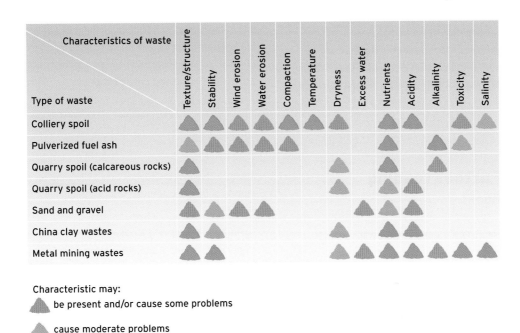

Type of waste	Texture/structure	Stability	Wind erosion	Water erosion	Compaction	Temperature	Dryness	Excess water	Nutrients	Acidity	Alkalinity	Toxicity	Salinity
Colliery spoil	▲	▲	▲	▲	▲	▲	▲		▲	▲		▲	▲
Pulverized fuel ash	▲	▲	▲	▲	▲				▲		▲	▲	
Quarry spoil (calcareous rocks)	▲						▲		▲		▲		
Quarry spoil (acid rocks)	▲						▲		▲	▲			
Sand and gravel	▲	▲	▲	▲			▲	▲	▲				
China clay wastes	▲	▲					▲		▲				
Metal mining wastes	▲	▲					▲	▲	▲	▲	▲	▲	▲

Characteristic may:

▲ be present and/or cause some problems

▲ cause moderate problems

▲ cause major problems

Figure 19.21 Major characteristics of industrial wastes which limit soil and vegetation development. (Source: after Wheater, 1999)

where expenditure for restoration is prohibitive or the land has been affected too severely, other uses for that land may still provide environmental opportunities. Flooded gravel pits may provide valuable wetland nature reserves, for example.

Reflective question

➤ What can an understanding of biogeography offer environmental managers?

19.7 Summary

This chapter has shown the importance of a range of processes in determining biogeographical patterns in time and space. Large-scale factors such as climate and plate tectonics may play a role in the distribution of species and species groups. Ecological factors such as succession, evolution, extinction, species mobility and the immigration of alien species are also vital in determining the biogeographical patterns we see today. On a smaller scale local topography or pedology may control the spatial distribution of species. Landscape ecology provides a useful concept for exploring regional and small-scale biogeographical distributions. This theory suggests that landscapes are made up of a matrix which contains patches and corridors. The patches provide oases surrounded by the more usual distributions and the corridors provide pathways for species dispersal and movement between patches.

Overriding most of this chapter is the element of change in the distribution of living things. Current biogeography can produce only a snapshot of a constantly changing situation. Habitats develop and change in response to gradual changes, such as in a regional climate, or more dramatic disturbance through fire or the agency of humans. In turn the community occupying the habitat may respond by losing or gaining species. The theory of island biogeography suggests that islands (or landscape patches) are more likely to have a greater number of species if they are larger because the habitats within them are more likely to be diverse. This is because any given area will be subject to immigration of new species over time and extinctions of species as the increased number of species results in increased competition. There is thus an equilibrium point for the number of species any given island area can hold. This sort of theory, while it has its critics, also has important practical applications in the management of the living elements of the physical environment, whether for economic gain or for conservation. Biogeography has a wide range of applications from land reclamation on industrial sites to predicting the influence of climate change on the spatial distribution of crop growth potential and crop pests.

Further reading
Books

Bradbury, I.K. (1998) *The biosphere.* John Wiley & Sons, Chichester.
This is a standard undergraduate text which is good for the general principles and discusses evolutionary aspects of biographical pattern.

Cox, B. and Moore, P. (2005) *Biogeography: An ecological and evolutionary approach.* John Wiley & Sons, Chichester.

Another classic text which is especially good on the evolutionary aspects of biogeographical patterns, going back far into geological time.

Forman, R.T.T. (1995) *Land mosaics – The ecology of landscapes and regions.* Cambridge University Press, Cambridge.
Forman is not a supporter of island biogeography but this is still one of the best texts on landscape ecology and is written in a very accessible style.

Huggett, R.J. (2004) *Fundamentals of biogeography.* Routledge, London.

Huggett has written several textbooks in this area, all worth exploring. This one is full of illustrative examples.

Kent, M. and Coker, P. (1992) *Vegetation description and analysis – A practical approach.* **Belhaven Press, Chichester.**
This is a good hands-on guide to many of the techniques needed in fieldwork design and the analysis of data collected. It provides a range of useful case studies.

Matthews, J.A. (1992) *The ecology of recently deglaciated terrain: A geoecological approach to glacier forelands and primary succession.* **Cambridge University Press, Cambridge.**
This excellent text is aimed at a higher level and will be suitable for those interested in either succession or harsh upland or semi-arctic environments.

Quammen, D. (1996) *The song of the dodo: Island biogeography in an age of extinctions.* **Hutchinson, London.**
An excellent example of narrative non-fiction is provided by Quammen. It works better for the fast reader since it is a chunky volume, but you might want to dip into it for fascinating accounts of interviews with a range of well-known scientists.

Spellerberg, I.F. and Sawyer, J.W.D. (1999) *An introduction to applied biogeography.* **Cambridge University Press, Cambridge.**
This book provides a southern hemisphere view and covers far more than the title suggests. It includes an overview of the discipline and an extensive section on the application of landscape ecology techniques.

Whittaker, R.J. and Fernández-Palacios, J.M. (2007) *Island biogeography: Ecology, evolution, and conservation.* **Oxford University Press, Oxford.**
A detailed overview of the theory of island biogeography and the role of islands as ecological hot spots, the threats to species and measures to protect species.

Further reading
Papers

Global Ecology and Biogeography, Volume 9 (2000)
This particular volume of the journal includes several papers discussing the need for 'a new paradigm of island biogeography'.

CHAPTER 20

Ecological processes

Hilary S.C. Thomas
C2C Research Centre, University of Glamorgan

Learning objectives

After reading this chapter you should be able to:

➤ understand important ecological characteristics

➤ describe ecosystem processes of energy, nutrient and material flows

➤ describe spatial and temporal ecological processes

➤ understand how humans influence spatial and temporal ecological processes and how they can create unique ecological communities

20.1 Introduction

The study of ecology is focused upon the interactions of living things with their environment. The study of these interrelationships has led to the development of ideas that are seen as increasingly important in the twenty-first century. In 1989, the British Ecological Society, as part of its 75th anniversary celebrations, published the results of a poll of its members. The aims had been to determine the key ideas in ecology and those areas of the subject that were considered to have contributed most to our understanding of the natural world. The ideas that most ecologists agreed upon

included: the ecosystem, **succession**, energy flow, conservation of resources, competition, niche, materials recycling, the community, life-history strategies and ecosystem fragility, along with a range of other themes that will be discussed in this chapter. Cherrett (1989) statistically analyzed this survey to see if there were clusters of responses that might indicate groups of ecologists with similar priorities. The analysis resulted in Table 20.1 which indicates how certain important ecological concepts formed four distinct groups. There has been little change to these priorities since the original survey and so the concepts addressed in this chapter will be similar in character to the four groups shown in Table 20.1.

The characteristic features underpinning any understanding of modern ecology, such as the nature and role of ecosystems, habitats, communities and the environmental niche, will be discussed in this chapter. The chapter will consider spatial patterns created via ecological processes at a range of scales including the scale of an individual tree. It will then move on to consider temporal patterns and distributions, such as the effects of climate change. In particular, the variety of cycles and flows of nutrients, energy and matter that underlie these patterns and distributions are considered, since the study of these interrelationships is the defining feature of ecology. The checks and balances operating within these patterns lead to a consideration of

Table 20.1 Important areas of ecological study, based upon Cherrett's (1989) analyses

The ecosystem	Studies in 'holistic' theories
Energy flows	
Geochemical cycles	
Succession	
Ecosystem fragility	Practical/applied studies
Conservation of resources	
Habitat restoration	
Sustainability	
Life-history strategies	Studies of biotic influences
Predator-prey relationships	
Ecological adaptations	
Plant-herbivore interactions	
Environmental heterogeneity	Studies related to external influences
Stochastic processes	
Natural disturbance	

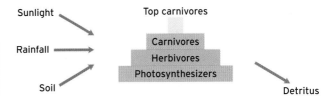

Figure 20.1 A simple ecosystem. The photosynthesizers use the energy of the Sun, and nutrients and water within the soil, to help produce organic matter. This plant matter is then eaten by herbivores (plant eaters) which in turn are eaten by carnivores (meat eaters). Some of these carnivores may be preyed upon by other carnivores. During this process waste is produced. Some of this waste may be recycled back into the soil to be taken up by plants at a later stage, whereas other wastes may be lost from the ecosystem by wind or water movement.

the drivers, both natural and anthropogenic, of ecological change. Some of the resulting issues, relating to the conservation and rehabilitation of ecologically sensitive areas and to the effects of climate change, are discussed towards the end of the chapter. They are indicators of areas where ecology has great relevance and impact today.

20.2 The functions and characteristics of ecology

The following section briefly introduces major themes in ecology that will be elaborated on throughout the rest of this chapter.

20.2.1 The ecosystem

Huggett (2004) suggested that 'ecosystems are communities together with the physical environment that sustains them'. An ecosystem may be a pond, a catchment basin, a biome or the Earth's biosphere. Selecting the appropriate boundary is dependent on the problem and on the timescale (Waring, 1989). However, more important than scale issues is the idea that ecosystems are about the characteristics of and interrelationships between the properties of the environment and its living inhabitants, permanent or temporary, and of the flows of a variety of materials between these

elements. This characteristic of interconnection means that if one part of an ecosystem changes then this will affect other parts of the ecosystem too.

Most ecosystems are taken to be 'open systems' where there is an element of input to and/or output from the system, rather than complete self-containment. At its simplest level as shown in Figure 20.1 the major input is usually taken to be energy, in the form of solar radiation. Outputs include energy loss at a variety of levels. In the simplified structure of Figure 20.1 sunlight, rainfall and soil are used to indicate typical ecosystem inputs, and various types of **detritus** (waste) typify ecosystem outputs. Within it, an ecosystem can be divided into several nutrient levels. In reality, as indicated by Figures 20.6 and 20.7 below, both inputs and outputs may be far more complex, making the realistic modelling of many ecosystems and ecological processes problematic (Murray, 1968).

20.2.2 The habitat

The physical environment provides the distinctive habitat or habitats of the ecosystem. This is the living area for a species or group of species and may be very small, such as a single branch of a tree for a lichen, or very large, such as the areas needed by predators of grazing animals in tropical grassland biomes (see Chapter 18). Sometimes the terms habitat and ecosystem are used incorrectly, as if they were the same, but the term habitat does not include the large-scale interrelationships that are an essential part of the description of an ecosystem. An ecosystem may contain a range of habitats within it. For example, a sand-dune ecosystem with dunes and slacks provides a wide range of habitats, at a variety of scales.

20.2.3 Populations

Populations of species vary over time often in response to changing environmental conditions (Beeby and Brennan, 1997). Chain reactions are the norm such as a decrease in predators leading to an increase in prey species and vice versa. Sometimes more indirect changes may occur owing to the removal of one important species upon which lots of other species and processes were dependent. Limits in population size arise from the limited resources available within the ecological niche and from the balance between births and deaths within the population as a whole. Some of the commonly used models of population growth and decline are shown in Figure 20.2. If there were no limits then population growth would occur as indicated in Figure 20.2(a). Population curves are produced for situations where limiting factors include finite amounts of nutrients or space (Figure 20.2b), or where altitude limits population growth owing to more severe climate conditions (Figure 20.2c). Population curves can also be produced for a range of other scenarios including those where disease in one species allows increased survival of another species, or of younger generations of the same species. Disturbance of a species may impact on the population, and models such as that shown in Figure 20.2(d) help indicate the typical recovery time (or relaxation time – see Chapter 1) and what may happen if disturbances occurred more frequently than the full time needed for total population recovery. Therefore these population models are of enormous benefit for management planning.

20.2.4 Ecological communities

A community is made up of groups of individuals that may be of any living organisms, although most familiar are often those defined by plants or animals, which occupy a specific area. From the geographer's viewpoint, the community is therefore a mappable entity. Communities operate at a variety of scales with boundaries that may be sharp or very gradual (e.g. Figure 20.3) but their character is determined by the 'core' of the community, which may be some definite species, such as the Tasmanian blue gum which often co-exists with other species. Within the community, there may well be competition and change over time, so that an early community may evolve into a very different later one. This may be through effecting changes within the habitat or by responding to changes imposed upon it from outside.

Communities are often referred to by the Latin classification term for their dominant species. A *Callunetum* is therefore a community dominated by *Calluna vulgaris* or

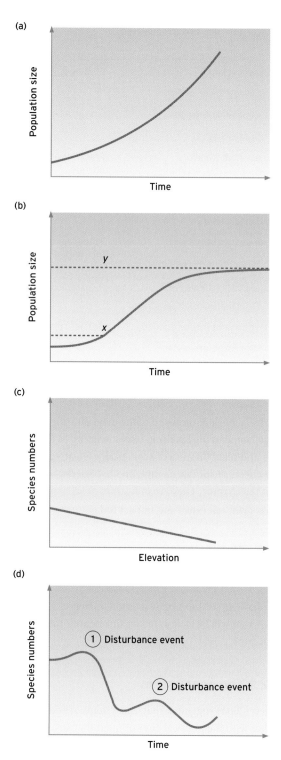

Figure 20.2 Some typical simple population curves: (a) rapid growth in the absence of limiting factors; (b) growth of population after a certain size (*x*) comes under increasing environmental friction until the carrying capacity (*y*) of that location is reached. Population will fluctuate around the carrying capacity. (c) A typical environmental limiting factor. The number of species declines with increasing elevation. (d) Effects of repeated disturbance. Note the importance of sufficient recovery time between events.

Figure 20.3 Effect of exposure on treeline near Grey Mare's Tail waterfalls, Moffat, southern Scotland. Small changes in slope, shelter and altitude can create changes to the plant association. In this instance, trees are able to colonize those lower areas that are also the most sheltered with heather and bracken in fairly well-sheltered zones. Note how clear some of these boundaries appear and the differences between the two sides of the valley.

the common heather. Where other species are also present in certain subtypes of that community, they are described as **differentials**. These terms provide a useful way of describing the likely features and processes operating within an area. The ecological community is also characterized by its interactions. For example, the heather in a *Callunetum* provides shelter, food and access to water for a range of creatures including birds, insects and mammals.

20.2.5 Ecological functions

Within ecology, the study of the global functions of the biosphere has particular relevance for humans (e.g. see Box 20.1). A classification of those functions purely in terms of their utility for humans (van der Maarel, 2000) is suggested below:

- A carrier – where the local ecology provides space and surface for human activities.
- Production – as a supplier of matter and energy from natural and agricultural resources (Figure 20.4).

Figure 20.4 One of the functions ecosystems serve from a human perspective is to supply food.

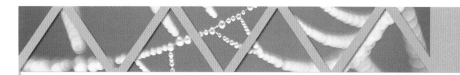

ECOSYSTEM SERVICES

In part an attempt to place a monetary or other value on the natural environment for decision-makers, the Millennium Ecosystem Assessment (see Chapter 18) was one of the first attempts to identify and put values on the benefits provided by the environment. Two years later, the British Government (Defra, 2007) adopted this approach to support policy decision-making. Internationally the topic is of growing importance. Major international organizations such as Conservation International now use the ecosystem services concept to help prioritize which conservation projects to take forward. The analysis of both elements of and complete 'ecosystem services' is now of major importance. This is often described as an 'ecosystem approach' where land, water and living resources are managed in an integrated fashion, but it should be noted that it is almost always focused upon the impact and benefits for humans. The description 'ecological economics' is also apt. Essentially, ecosystem services are the benefits that people obtain from ecosystems. Since the rural poor are often the most directly dependent upon natural capital, either daily or in times of hardship, the sustainable management of ecological services is an important element of poverty reduction policies and the Millennium Development Goals (TEEB, 2010).

The main elements of ecosystems in terms of the services they provide to humanity are usually placed into four groups, as shown in Figure 20.5. The support services, not always readily apparent to the non-specialist, are those which underpin a resilient ecosystem, such as the role of invertebrates in maintaining soil fertility or of photosynthesis in relation to primary production and the food webs within the ecosystem. Without these services, synthetic or other nutrients might need to be bought in to maintain agricultural production. Regulatory benefits include those where, for example, the presence of trees (Figure 20.6) allows the impact of heavy rainfall on the soil to be moderated by the barrier provided by leaves and branches, while allowing a longer period for soil water replenishment. Tree roots also stabilize a roadside verge or mangrove roots protect a section of coastline. Value may be seen here in terms of the amount of money needed to provide equivalent protection to that supplied by the trees. The provisioning and cultural services are usually the most obvious to those living within or near a specific ecosystem. The outputs of food, medicinal and other economic products such as timber are visible as they are transported or sold. Equally, the importance of many landscapes in terms of their cultural value gives rise to evocative terms such as 'The Lake District', and has often been used in the past to justify protected area status, despite the difficulty of assigning a monetary value to what is being given such status. Perhaps this is most closely approached in the difference in house prices with and without a desirable view.

A current problem is that of calibration. Many of the services and the importance of their contribution are not yet measured in a manner suited for objective calculation, nor are there widely accepted examples of best practice or national standards.

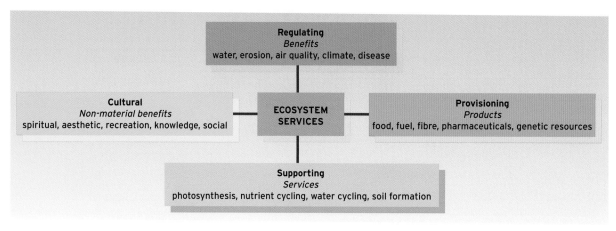

Figure 20.5 Categories of ecosystem services.

BOX 20.1 ➤

It is, however, clear that the cost of providing equivalent services on a commercial basis would be huge. In South Africa (eThekwini Municipality Biodiversity Report, 2007) an open space of about 63 000 ha in the Durban region was given a value of R3.1 billion in terms of the environmental goods and services it supplied, compared with the municipal budget of R6.5 billion.

In Australia local authorities in Canberra expect to accrue benefits of over US$ 50 million over a 4 year period through planting 400 000 trees (Brack, 2002). The services provided by this planting include carbon storage, micro-climate remediation and reductions in air pollution and in energy costs related to air-conditioning.

Glaves and Egan (2010) rightly point out that it might also be appropriate to consider ecosystem disservices, such as increased fire risk for residents near restored heathland.

Figure 20.6 Trees as ecosystem service providers on the Mbale Kumi road, reducing soil loss, providing shelter, taking up carbon from the atmosphere and regulating climate, maintaining fertility and reducing downstream floods.

BOX 20.1

- Information – for orientation, aesthetic appreciation, philosophical identification, research, education, recreation and, through monitoring, to detect changes (Figure 20.7).
- Regulation – based on ecosystem stabilization, for example via chemical cycles and damping of climatic change.

The environmental function of production is fundamental to human existence. Humans survive by eating foods and certain crops can also be used for other purposes. These include building materials, fuel and clothing. Ecosystems also provide a regulatory function that is fundamental for ecology as a modern science and one of great modern relevance. They regulate atmospheric carbon dioxide concentrations, for example by promoting photosynthesis, absorbing carbon dioxide and releasing oxygen back into the atmosphere. However, the human alteration of ecosystems is of great concern as this may result in change to these regulatory systems. Regional alterations could have global implications. The inevitable interdependence and interaction between individuals, species and their environments provides ecosystems and communities with both strength and potential fragility. This is drawn in part from their habitats but also from their living components.

The functions and characteristics described above form important components of ecological research and will be elaborated on within the following sections.

(a)

(b)

Figure 20.7 Gravel pits now serving as a leisure, education and flood sacrifice area in the Rhone Valley just outside of the city of Lyons. The site provides countryside park leisure facilities for picnics, swimming, fishing, boating, painters and photographers and there are birdwatching hides. Some zones are managed for threatened and newly re-establishing species. Educational interpretation boards are found in the area as shown in (b).

Reflective questions

➤ What is the difference between an ecosystem and a habitat?

➤ What are the main ecological functions from a human perspective?

➤ What are the benefits and disadvantages of putting an economic value on ecosystem services?

20.3 Ecological processes

20.3.1 Energy and nutrient flows

Most plants and some other fairly simple life forms such as certain bacteria and algae take energy from sunlight. Through the process of photosynthesis they convert this to sugars, which then become available for animals, birds or insects to use for nutrition. The life forms involved in this initial process within the ecosystem are described as producers or **autotrophs** and they form the lowest level of the food chain and the first trophic level (or nourishment level). As shown in Figure 20.8, primary producers can include a range of vegetation types. The purpose of eating is to generate energy in order to live and to collect nutrients in order to grow. The consumers or **heterotrophs** feeding on the autotrophs in the example shown in Figure 20.8 are grasshoppers, rabbits and mice. These creatures, usually herbivores, feed directly on the producers and are described as first-order consumers and form the next trophic level. The energy gained by consuming the autotrophs is used by these herbivore consumers for a range of functions such as digestion, movement, reproduction and respiration or is lost as heat. As a result, there is relatively little energy passed on, perhaps only 10%, from one trophic level to the next. Therefore a consumer requires a large amount of biomass from lower trophic levels. For example, rabbits must spend a considerable amount of their time grazing. Herbivores are consumed by the **carnivores** (such as the snake) and **omnivores** of the next trophic level. These, in turn, are likely to form the food and therefore energy base of higher-level carnivores or omnivores such as the hawk shown in Figure 20.8. When they die, their remains will probably form the diet of the decomposer **saprovores**, such as maggots, becoming part of a detritus-based, rather than vegetation-based, food chain. This final stage releases the last of the energy as heat, and is also important in the recycling of inorganic nutrients that have been passed through the food chain. These include materials such as nitrogen and phosphorus, which entered the system as nutrition for

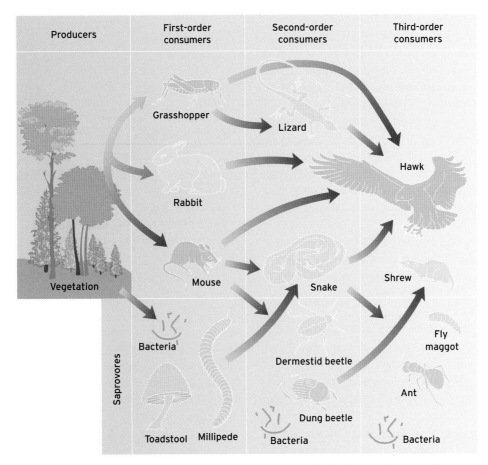

Figure 20.8 A simplified example of nutrient flows within a temperate ecosystem, indicating trophic levels. (Source: after Murray, 1968)

those original producers, perhaps taken up from the soil water (see Chapter 10).

Food chains are not always simple and, as Figure 20.9 shows, many animals are able to operate at a variety of levels within what is better described as a food web. This is less confusing if translated into human dietary terms.

The wealthy human who is primarily and perhaps preferentially a carnivore will still, in most cases, add vegetables to their diet (e.g. steak and chips/French fries). The less wealthy may subsist for the most part on vegetables such as beans and rice but will augment this with meat or fish if possible.

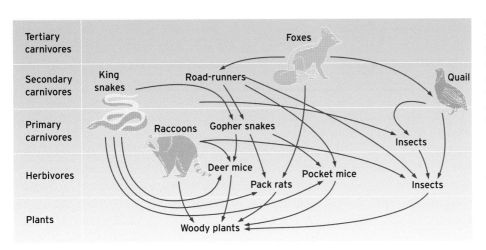

Figure 20.9 The complexity of a typical food web. This example is from the chaparral scrub of California. The various trophic levels indicate the animals of different groups and it is noticeable that not all animals are confined to one trophic level. Raccoons, for example, are both herbivores and carnivores. However, they are ultimately dependent on the plants.

Since not all of the biomass in any one trophic level is eaten, terrestrial food webs are often shown, as in Figure 20.1, as biomass 'pyramids'. The lower trophic levels tend to have the greatest bulk, not all of which will be consumed. This is because there are always some animals that die of old age rather than being caught and eaten by predators. However, in aquatic systems the trophic pyramid is not always accurate. Some forms of algae, for example, are both rapid reproducers and highly nutritious, so they may form the diet of a larger biomass than exists within their own trophic level. This is like a hospital patient attached to a 'drip' of such highly nutritious material that very little is required to keep the patient alive.

In addition to energy transfers, material is transferred through an ecosystem. The material within ecosystems follows complex paths that often include temporary storage for varying amounts of time. The difference in time for chemicals held within the Atlantic Ocean compared with those held within a single potato is enormous. **Biogeochemical** cycles involve links with the local variations in geology and climate. There are many types of such cycles but for simplicity the main components of the phosphate form of the phosphorus cycle are shown in Figure 20.10 in order to illustrate the biogeochemical cycling process.

Many rocks contain phosphates. When weathering takes place, the water-soluble phosphate goes into solution, often as part of the soil water or in water bodies such as lakes and streams. Here it becomes available for producers within local food webs, becoming incorporated into cell membranes. Subsequent consumers may also incorporate the phosphate into their skeletons (Figure 20.10). However, a subcycle may occur if a **mycorrhizal** fungi–plant root relationship is able to take place. Here, the plant roots extract the phosphate from the fungus, so making the plant in this instance a consumer rather than producer. In exchange, the fungus is a supplier of sugars. Back within the main cycle, the phosphate moves on from its storage within the body when released either as a component of defecation or upon death. At this stage saprovores will aid decomposition and the phosphate will once more become part of the soil or water. The phosphorus cycle can have almost endless repetition until it is leached out of a soil into a watercourse where it could be used for an aquatic ecosystem. On a larger timescale phosphorus can be incorporated as part of deep ocean sediments. Over time the sediment could be **lithified** and form sedimentary rock. As geological processes continue the rock may eventually reach the surface again. At this stage it may become weathered and the phosphate cycle will continue. Other

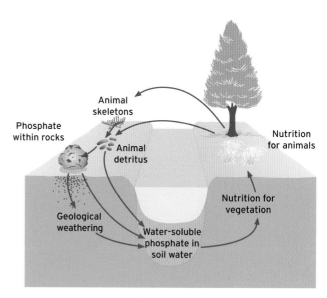

Figure 20.10 An example of a biogeochemical cycle.

chemicals may undergo very similar or very different pathways through geology, soil, water, plant and animal transfer processes. Thus ecosystems consist not only of complex food webs but also of complex energy and material cycling processes.

20.3.2 Bioaccumulation

Through the biogeochemical cycling process there may be locations where chemicals build up or become concentrated. Some chemicals have far greater potential than others to accumulate in zones where they become bioavailable (e.g. within soil water). Producers and consumers take up nutrients from these zones and store them for future use. However, less desirable materials may also be taken up and stored if they too are locally bioavailable. As a result, toxins may accumulate in specific parts of an ecosystem, usually in the higher levels of food chains or webs (e.g. Figure 20.11). This can have severe effects on an ecosystem once a given threshold of toxin storage is reached or to higher-level consumers (e.g. humans) who may suffer health problems as a result.

Figure 20.11 illustrates for an aquatic ecosystem how **bioaccumulation** can occur. A contaminant such as mercury can build up in the sediment and then be taken up by mussels. Each mussel may have only a small amount of mercury. However, as small fish eat a lot of these mussels the mercury becomes more concentrated inside the fish. Then, as other predators eat lots of these fish the mercury can accumulate to high levels within them. This

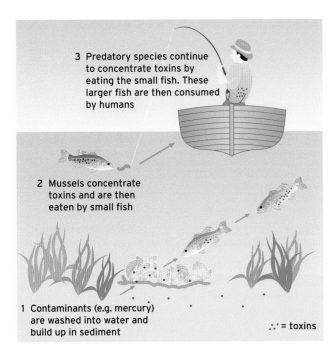

3 Predatory species continue to concentrate toxins by eating the small fish. These larger fish are then consumed by humans

2 Mussels concentrate toxins and are then eaten by small fish

1 Contaminants (e.g. mercury) are washed into water and build up in sediment

∴ = toxins

Figure 20.11 An example of bioaccumulation in an aquatic ecosystem. The toxin is found in low concentrations in mussels but as these are consumed in large quantities by a fish then the toxin concentrations are much greater in that fish. The toxin is concentrated even further higher up the food chain.

can have toxic consequences for humans who may eat the larger predatory fish.

To be able to 'bioaccumulate', materials must not merely be stored in the consumer, but must be stored in those parts of the individual that will be consumed. Therefore, those materials stored in 'meat' are far more likely to become bioaccumulated than those stored in bone. Water-soluble toxins are less likely to produce this type of problem, since they tend to be lost as part of the more general water loss of the consumer and, unless the toxin is very long-lived, may not be taken up again. In the immediate period following the Chernobyl nuclear explosion, soil and water pollution occurred in northern Europe as the local winds spread the radioactive materials over large areas. This resulted in long-term bioaccumulation in vegetation. These toxins were then consumed by grazing animals in northern Europe during the following year, spreading the negative effects to the wider ecosystem. **Bioconcentration** levels, which are the extent to which a material is found in tissue compared with background levels, are generally site specific. Therefore the effects of the Chernobyl fallout were not found in Australia.

Reflective questions

➤ Why do herbivores such as rabbits or sheep need to spend a significant amount of their time grazing?

➤ Where is energy gained and lost in an ecosystem?

➤ What is bioaccumulation?

➤ Why may small, seemingly harmless, concentrations of a chemical found at the lower end of the food chain eventually prove harmful to the ecosystem?

20.4 Spatial patterns and distributions in ecology

20.4.1 The ecological niche

As with so much in ecology, understanding the spatial patterns of living things comes from studying not just the individual requirements of a species but their interrelationship with both their physical environment and the other species occupying it. In any locality, the status and character of populations are connected, in particular, with the level of competition for the ecological space available: the environmental niches. Competition is related to the type and amount of resources available at any one location and the ecological niche is the combination of those resources required by the individual organism. The niche is not the same as the habitat, since the ecological niche relates more to the distribution of the species within the wider community and is explained further below.

The resource requirements for most living things usually centre upon provision of appropriate climate, shelter, food and water. The ecological niche, in the forms described below, is the basis of most ecological patterns. In the simplest of situations, where there are no competitors for any of the resources required by an individual or species, the **fundamental niche** can be occupied, which contains ideal conditions. This is rather like your being able to use your educational institution's library resources as the only client for the duration of your course. Most species, however, have to utilize a **realized niche** that is the result of competitive interaction between several species attracted to the resources. This is like taking the only textbook left on the library shelf as other books are not available because the keenest students took the best books out of the library at the beginning of term. You therefore have to hope that at least one of these other books becomes available before the examination. This competition is usually strongest between similar species, since their ecological niches are likely to overlap. The species able to survive on the lowest amount of the limiting resource will be advantaged.

There are many instances of niches that are used by different groups at different points in time. For example, daytime and night-time predation or the use of scarce water resources may allow resources to become partitioned. Different parts of a resource, such as a tree, are often used by different species. Observations in a British wood showed a characteristic tree-use pattern by blue, great and marsh tit birds (Gibb, 1954) which was later confirmed by ornithologists elsewhere. Blue tits were far more likely to spend feeding time among the leaves during the summer and on the twigs and buds in the winter, than elsewhere on the tree. Great tits obtained their prey from the ground beneath the trees, in winter especially, although they also used the leaves in the summer. Marsh tits used a far greater proportion of the trees throughout the year.

20.4.2 Competition

Both interspecific and intraspecific **competition** are important factors determining ecological distributions. Interspecific competition arises in situations such as at an African waterhole in an otherwise arid area. To be able to drink the water, herbivores and other vulnerable species may need to wait until carnivores have drunk their fill and moved away. The disadvantage of the long wait is compensated by the knowledge that eventually the herbivores too will be able to drink. The species in this case share the same spatial distribution, but not at the same time. Intraspecific competition can be seen where a number of squirrels need to collect acorns from a small area within a city park. Intraspecific competition may lead to the exclusion of weaker individuals and explain the patterns of territories which control both feeding and reproduction opportunities for the squirrels.

20.4.3 Life strategies

A further factor in the patterns and distributions found in ecology relates to the life strategies of species and individuals (Drury and Nisbet, 1973). These are usually related to elements of the life cycle of a species, particularly its means of dispersal. Understanding a species' life strategy may help to predict or to explain its ecological distribution (Table 20.2). There are two main types of strategies, known as **'r' and 'K' selection**. Both are likely to occur within the wider landscape but the 'r' strategists are more likely to be found in new or disturbed sites, since they have good colonizing ability. They include many weed species and are often described as **ruderal**. The second type of strategist is the 'K' strategist.

Table 20.2 Characteristics of 'r' and 'K' life strategists

'r'-selected	'K'-selected
Rapid development	Slow development
Short-lived	Long-lived
Small body size	Larger body size
Have many offspring – tend to overproduce	Have few offspring
Single reproduction	Repeated reproduction
Do not care much for individual youngsters	Care for their offspring
Population not regulated by density – boom and bust	Population stabilizes near a capacity level
Opportunistic – will invade new territory	Maintain numbers in stable ecosystems
Lax competition	Keen competition

These species are more likely to do well in a less disturbed environment. Table 20.2 illustrates the typical characteristics of 'r' and 'K' strategists.

Both stress and disturbance may alter a community. Stress on species, perhaps as a result of recurrent drought, affects productivity. Disturbance, such as a forest fire, destroys biomass. The 'r' strategists are likely to succeed in conditions of high disturbance but low stress. 'K' strategists are more likely to outcompete other species in conditions of low disturbance and low stress. This is an area of ecological research which is very important given potential climate change and human impacts on stress and disturbance.

20.4.4 Biodiversity: patterns of species richness

Figure 20.12 illustrates the links to biodiversity in terms of its definition, why we should be bothered about it, the potential threats and solutions. In recent decades there have been increasing interest in and concern about 'biodiversity'. This is a measure of species richness. It is the variation among living organisms and ecosystems of which they are part. There are a number of diversity indices and some are described in Box 20.2. Some areas may have many more species than others. For example, the tropical forests are believed to provide habitats for 40% of the species on Earth. There have therefore been increasing attempts both to identify biodiversity 'hot spots' and to conserve biodiverse areas. Concerns exist because of increased extinction in areas of

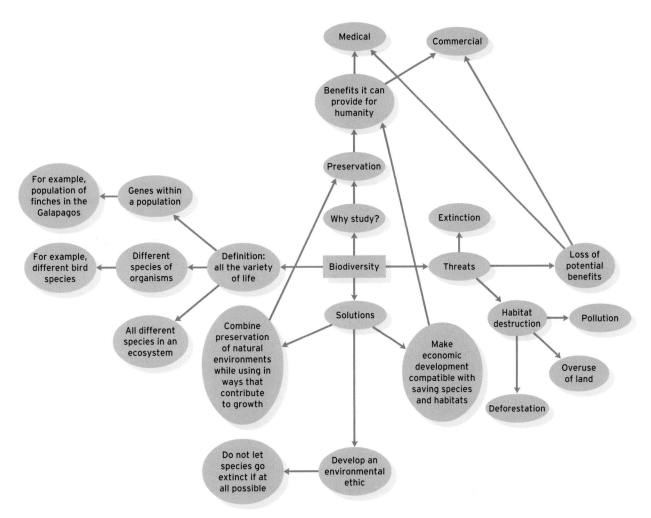

Figure 20.12 A biodiversity concept map. (Source: 'A biodiversity concept map', Michael J. Robinson, from the website of the 1999 Summer Biology Institute on Diversity, produced by the Leadership Programme for Teachers at the Woodrow Wilson National Fellowship Foundation (Princeton, NJ, USA: Woodrow Wilson National Fellowship Foundation 1999). http://woodrow.org/teachers/bi/1999/projects/group9/Robinson/conceptmap.html. Used by permission of the Woodrow Wilson National Fellowship Foundation, 2004)

altered land use such as tropical forest regions. In these areas, species may become extinct before their existence (and potential human function, e.g. for medicine) is known. Therefore it is important to recognize factors that may lead to enhanced species richness. Global-scale areas of species richness are usually the result of factors including lack of disturbance and lack of isolation.

Tropical areas have the greatest species diversity (Wilson, 1992). This is, to a great extent, a result of the relatively minor impact (lack of disturbance) of the Quaternary glaciations on tropical areas. Where a region's ecology has been uninterrupted, in terms of major climatic or other disturbance for lengthy periods, species are less likely to have become locally extinct. These areas may also benefit if species from disturbed areas arrive as immigrants. Barriers such as high mountains or oceans will lessen immigration

opportunities. While many isolated areas are species-poor, they may contain species now unique to the area that were originally more widespread. Regional and local patterns of species richness may result from short-term disturbances or habitat diversity. Short-term disturbances such as natural fires are likely to encourage high levels of species richness, as in Mediterranean regions (see Chapter 18). A mosaic of patches of differing age-since-disturbance often develops, each with a distinctive community. This gives the area as a whole a high species richness. Some environments are able to provide a range of habitats within a local area. Woodland tends to have a range of light and shelter conditions, providing a diversity of habitats within it as well as those created by the woodland edges, where species from both the woodland and the surrounding area may partition the resources available.

DIVERSITY INDICES

It is often important to measure the species richness of a site or habitat. Measurement might be very general: concerned with numerous vegetative, insect, animal or bird species; or quite specific: confined, for example, to rodents, or mosses, or a particular family of birds. Quantifying the abundance and diversity of species in any taxonomic group can be done using a diversity index.

Diversity indices are a useful way of objectively measuring species abundance or diversity in almost all terrestrial and aquatic habitats. They can be used to describe spatial changes (differences in species between different habitats or between different areas of the same site), or to establish temporal change (the impact of pollution or management by repeated sampling of the same site periodically over time). One of the most common, and most flexible, diversity indices is the Shannon–Wiener Diversity Index (also known as the Shannon index or the Shannon–Weaver index). This is often used to compare two or more situations as the index is of most use as a comparative measure. It is important to compare 'like with like'. In other words, comparing the diversity of wetland birds between two areas or the diversity of grasses in a field at specific time intervals is fine, but comparing the diversity of moths at one site with the diversity of fish at another is meaningless. In addition, care must be taken to sample a similar area between sites or monitor the site for a similar length of time in order that the index is not biased.

The Shannon–Wiener index has a complicated-looking formula but is quite simple to work out by hand or on a computer if you take things one stage at a time:

$$H = -\sum^{S}[p_i \times \ln(p_i)] \qquad (20.1)$$

where H is the Shannon–Wiener index, S is the total number of species at a site, and p_i is the proportion of a particular species. $\sum$ indicates the sum of, while ln is the natural logarithm, which can be calculated on most scientific calculators or in most computer spreadsheet packages. For ease, this can be expanded to:

$$H = -\sum[N_i/N_{tot} \times \ln(N_i/N_{tot})] \qquad (20.2)$$

where N_i is the abundance of an individual species and N_{tot} is the total abundance. A worked example is provided below.

Table 20.3 provides data on wetland bird species from two marshes. In this case, five species were found at Marsh 1 and only three species were found at Marsh 2, but the number of individuals found at each site was the same (70). This is for ease in this example and need not be the case.

The first step is to calculate p_i for each species at each site, then calculate $\ln(p_i)$, and $p_i \times \ln(p_i)$. Thus for Marsh 1 the values are given in Table 20.4 with values for Marsh 2 shown in Table 20.5. In this example the calculations have been done to two decimal places. Once this has been completed you should then sum all the individual results (one for each species) from the formula $p_i \times (\ln p_i)$. For Marsh 1 this gives the result −1.54. Finally, take the absolute of this value to give the Shannon–Wiener score: $H = (−1.54) = 1.54$. For Marsh 2, $H = 0.61$.

The most important thing to remember when interpreting values for the Shannon–Wiener index is that they are relative values. This means that if one area has an index value of 3 this is not double (twice as species-rich) as another area with an index of 1.5. In this instance all that can be said is the first site has a higher species richness or is more species diverse than the second. Shannon–Wiener values typically range from 1.5 to 3.5 but (as here) can range from 0 to over 4.5 depending on the number of species involved. In the above example, Marsh 1 has a higher (but not double) diversity of wetland

Table 20.3 Data on wetland bird species from two marshes

Marsh 1	Marsh 2
Species 1 = 20	Species 1 = 4
Species 2 = 5	Species 2 = 9
Species 3 = 14	Species 3 = 57
Species 4 = 15	
Species 5 = 16	

BOX 20.2 ➤

➤
bird species than does Marsh 2. This might be as a result of location, greater habitat diversity, better management or many other site-specific factors. In terms of setting conservation priorities, Marsh 1 has a greater species richness and might be considered as being of higher conservation importance than Marsh 2. Conversely, it might be considered that more conservation strategies are needed at Marsh 2 to increase its species diversity.

Source: Material courtesy of Anne Goodenough, University of Gloucestershire.

Table 20.4 Calculations to help work out the Shannon-Wiener index for Marsh 1

Formula	Species 1	Species 2	Species 3	Species 4	Species 5
$N_i \div N_{tot}$ (gives p_i)	$20 \div 70 = 0.29$	$5 \div 70 = 0.07$	$14 \div 70 = 0.20$	$15 \div 70 = 0.21$	$16 \div 70 = 0.23$
$\ln(p_i)$	-1.24	-2.66	-1.61	-1.56	-1.47
$p_i \times \ln(p_i)$	-0.36	-0.19	-0.32	-0.33	-0.34

Table 20.5 Calculations to help work out the Shannon-Wiener index for Marsh 2

Formula	Species 1	Species 2	Species 3
$N_i \div N_{tot}$ (gives p_i)	$4 \div 70 = 0.06$	$9 \div 70 = 0.13$	$57 \div 70 = 0.81$
$\ln(p_i)$	-2.18	-2.04	-0.21
$p_i \times \ln(p_i)$	-0.17	-0.27	0.17

BOX 20.2

Reflective questions

➤ What is the difference between a fundamental niche and a realized niche?

➤ What is the difference between intraspecific and interspecific competition?

➤ Think about endangered species that you have heard of: are they mostly 'r' strategists or 'K' strategists? What type of strategists are humans?

20.5 Temporal change in ecological patterns and distributions

20.5.1 Succession

Many ecologists have noted that from the community to the ecosystem scale, the composition, complexity and biomass of an ecological unit change over time (e.g. Connell and Slatyer, 1977; Tilman, 1985; Wilson, 1992). Most of these observers have concluded that such change is seldom completely random in nature and is more often directional, moving towards or away from a particular state or character. This form of change is known as succession. Important schools of thought have developed concerning succession and these are discussed in Chapter 19. However, more process-based assessments have also been developed and these are discussed in Box 20.3.

20.5.2 Human influence

A wide range of ecological stressors may initiate changes to ecological systems or to habitats, which may alter the population and community flows and possibly lead to the extinction of certain species. Such changes have always been part of the natural processes operating within the environment. However, humans have been responsible for a variety of deliberate alterations. Species-selective agricultural production or clearance of an area followed by urban construction is usually intentional. However, the after-effects may well be unintentional and uncontrolled.

20.5.2.1 Agriculture and its effects upon ecosystems

Agriculture provides the most obvious example of intentional human manipulation of natural ecological patterns. Its effects cover huge areas of the Earth's surface. It is

ENVIRONMENTAL CHANGE

ECOLOGICAL OLYMPICS – THE SUCCESSION RACES

For many years, ecologists have accepted succession as a module to work from. The changes over time could be measured but research continued in order to determine why these changes occurred. Table 20.6 provides a list of ideas that attempt to describe the processes involved in succession. Egler (1954), in searching for a mechanism for succession, studied individual species. He looked at initial invasion of a site through to species disappearance from the local community. From this analysis he developed the concept of **relay floristics**. This suggested that as one group of plant species establishes it is replaced by another and then they are replaced until a stable state is achieved. Many ecologists have studied the importance of evolutionary 'strategies' such as life-cycle or physiological strategies as a succession mechanism. Noble and Slatyer (1980) developed the concept of vital attributes. These could be characteristics such as the speed of reaching reproductive state and would determine the place of that species in the vegetation replacement series.

Three different models of succession mechanisms were compared by Connell and Slatyer (1977) and each achieved wide ecological acceptance under different situations. These are facilitation (the relay floristics model), tolerance and inhibition and are shown in Figure 20.13. The facilitation model suggests that a species may colonize and then change the local environment in terms of pH, moisture availability, structure and shading, for example. This may mean that the conditions are facilitated for another species. Therefore when another species comes along and finds the conditions more to its liking, it can out-compete the first species.

The tolerance model proposes that most species are present from the start. The fastest growers are the early dominants. These have little or no effect on slow growers or later colonizers, although late successional species will suppress the early species. There is thus a competitive hierarchy. Several attempts have been made to distinguish between species on the grounds of their competitive abilities under different circumstances. As such, lists similar to league tables have been produced for a range of species. Tilman (1985) linked this competitive hierarchy to the resources of soil and light and how these might change during succession. Where there is a single limiting nutrient for survival or reproduction, the species that can be successful at the lowest availability level of that resource will win.

Finally the inhibition model proposes that colonists capture space and inhibit colonization of later species. It therefore requires disturbance of pioneers before later species can establish.

Table 20.6 Ideas on the processes involved in succession

Authors	Ideas
Egler (1954)	Relay floristics
Noble and Slatyer (1980)	Vital attribute
Drury and Nisbet (1973), Grime (1997)	Life strategies
Connell and Slatyer (1977)	Facilitation, tolerance and inhibition
Tilman (1985)	Resource ratio/allocation

BOX 20.3 ➤

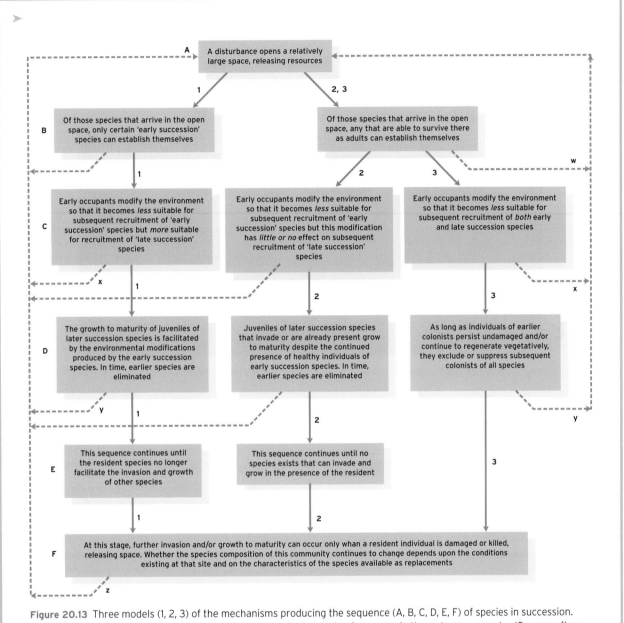

Figure 20.13 Three models (1, 2, 3) of the mechanisms producing the sequence (A, B, C, D, E, F) of species in succession. The dashed lines represent interruptions to the process, in decreasing frequency in the order w, x, y and z. (Source: after Connell, J.H. and Slatyer, R.O., 1977, Mechanisms of succession in natural communities and their role in community stability and organization, *American Naturalist*, 111, p. 1121, Fig. 1, © University of Chicago Press. Reprinted by permission of the University of Chicago Press)

BOX 20.3

estimated that as much as 70% of the developed areas of the world might already be in a state of an agroecosystem, which is where the agriculture is the dominant part of the system. In the developing world agroecosystems are a major driver of landscape change. Many of the more successful agroecosystems mimic elements of the original ecosystem such as the replacement of prairie grasslands by large-scale cereal cropping. However, our increasing technical abilities can produce far more original landscapes. In Figure 20.14 commercial forestry has replaced the moorland which can still be seen in the foreground.

Agricultural systems can be regarded as examples of ecosystems with managed inputs and outputs of energy and nutrients with controlled species diversity. The unintended

Figure 20.14 Softwood plantation above upland heather moorland and blanket bog in northern Scotland.

results may, however, act as a trigger to a variety of unwelcome patterns and distributions. Intentional nutritional increases to crops may be leached (see Chapter 7) into areas downslope of the original application, into water systems or to the sea. This may cause eutrophication, giving rise to secondary problems such as a phytoplankton bloom (NEGTAP, 2001). The main effects of agricultural expansion may be seen as an increase in cropland, but other known effects include an increase in savanna woodland as a result of the overgrazing of grasses.

Slash and burn or shifting agriculture allows the 'useless' but difficult-to-clear species to remain and also allows those species regarded locally as useful to remain. These species therefore remain on site while crops are planted around them. The land will remain in agricultural use until either weed growth or dropping yields become unacceptable and then a move to a new plot is made. In the Apo Kayan area of Borneo, for example, temporary long houses and grain stores are located in the centres of areas of shifting cultivation. Intermixed with areas in current use are unused plots of varying ages, which slowly revert to secondary forest after about 20 years. The old plot is therefore traditionally allowed to regenerate. However, too often population or other pressures mean that secondary regrowth, with the accompanying return of soil nutrient levels, will not be complete before the plot is cleared once more. Over time

this process creates a situation where the seed bank is not large enough to allow regeneration of those intermediate species lost during each clearing phase. An altered pattern of species and competition therefore arises over time.

More intensive agriculture, both before and after sowing, may remove unwanted species either through selective herbicide, reinforced by the use of selective fertilizing, or, where organic means are preferred, through manual or mechanized weeding. The end result, of decreased biodiversity, remains the same.

20.5.2.2 Urban ecology

The growth of urban areas is often regarded as ecologically destructive; 50% of the world's population lives in urban areas. It is accompanied by domestic animals, pests, parasites and other species that are able to compete successfully in these relatively new ecological environments. The resulting urban ecosystems, while varying in response to wider-scale geographical patterns such as latitude, are characterized by: **encapsulated countryside** (e.g. Figure 20.15) (either ancient habitats or previously managed land – both likely to suffer ecosystem degradation unless management is provided); highly managed places such as gardens; and abandoned land including post-industrial habitats that may now be sustaining unique communities that are a mixture of native and alien species.

Figure 20.15 Ancient woodland on the sides of many South Wales valleys is now often encapsulated by later development. A combination of public appreciation and local conservation interests acts to preserve many of these for the future. Those shown here now form a 'campus woodland reserve' for Glamorgan University.

The distinctive character of urban microclimates includes, in most instances, increased temperature. Local air flows may be modified by the physical structure of the city surface and the differences in temperature regime with the surrounding rural areas (see Chapter 6). Air pollution is often a problem, with high levels of particulates at certain times of day. Absolute humidity and the incidence of fog, including photochemical smog, are also likely to be increased (Oke, 1987). Water quality in urban areas may be affected, depending upon constructional and meteorological conditions (see Chapter 15). A general decrease in vegetation cover, coupled with an increase in impermeable surfaces, creates increased likelihood of flooding coupled with lower soil moisture and groundwater levels. Increased runoff may stress local aquatic ecosystems in terms of water flow, sedimentation and pollutants, which may include road salts during winter periods (see Chapters 13 and 21). Land quality is generally regarded as being diminished after urbanization with problems relating to exhausted soils, altered slopes and the concentration of wastes such as at landfill sites.

Probably the greatest interest in urban ecology is the 'urban wildlife'. Certain species have found the new combination of resources advantageous and have made significant dietary and/or habitat changes to enable them to fill vacant urban niches. The cliff-ledge habits of the rock dove have transferred easily to the window-sills utilized by feral pigeons. The wide nutritional range provided by urban waste dumps attracts a range of birds and other species. House mice and house spiders, together with the brown rat, have long been the prey of domesticated cats and dogs. More exotic pets intentionally or accidentally released are often able to survive, if at the edge of their ecological niche, such as terrapins which are now regarded with affection as long-term residents in an urban lake in Cardiff, Wales. Inevitably, the existence of these smaller creatures has allowed higher trophic-level predators to move into towns and cities, such as birds of prey. Foxes moved into many British urban areas during the early part of the twentieth century when improved transport systems led to the building of suburban housing in once rural areas. Rural foxes quickly urbanized, taking advantage of the food and shelter provided in gardens from compost heaps, bird tables and garden buildings. Following the destruction of hedgerows and woods, food and shelter resources may now actually be better for some species within towns and cities. Garden escapes of a range of plants have also increased local biodiversity and

Figure 20.16 The Hottentot fig, originally from southern Africa. This photo shows a garden escape now dominating native species along the cliffs at Lizard Point, Cornwall, England.

competition (Figure 20.16). Urbanization of species may create problems for humans, such as the urban raccoons of Washington, DC, and New York that are believed to have been responsible for a rabies outbreak (Chang *et al.*, 2002).

20.5.2.3 Recombinant communities

Recombinant communities are working communities of plants and animals that are a mixture of native and alien species. The arrival and integration of introduced species into the specific recombinant community may be relatively recent or may stem from very ancient introductions, such as the sycamore into the British countryside. Such communities are not restricted to urban or to industrial areas, although this is often their most common occurrence. It has been suggested that, in European cities, some 20–35% of the flora may be composed of recently established species and discussion is under way about how we categorize such systems.

20.5.3 Ecosystem fragility

The fragile nature of many ecosystems subject to exacerbated change is the object of much current research and debate (e.g. Solé and Montoya, 2001). The complexity of many ecosystems makes them robust under a range of conditions, such as the random removal of species. However, it also makes them vulnerable when highly connected **keystone species** are removed. The nodes in Figure 20.17 represent individual species within a specific food web. The responses to species removal include both secondary extinction of other species and the fragmentation of originally complex food webs that were always assumed to provide effective buffering, into discontinuous smaller webs. Those systems where the network is somewhere between a regular and a totally random distribution of connections are able to withstand perturbations relatively well and to respond rapidly to change. However, the remaining ecosystems are much more fragile. Decay of the original web into disconnected species may result in ecosystem collapse. Species that are highly connected with the rest of the web are those

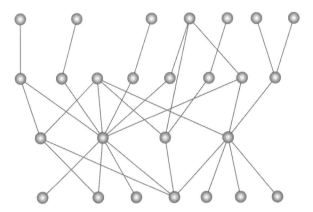

Figure 20.17 Schematic representation of ecological relationships between species. The links may include not just predator-prey links but many other links including seed dispersal and shelter. Few species have many links and many have just one or two links. If one of the species is removed then because of the many links with the other species connected to it they will also be affected. Those species that are highly connected to other species are known as keystone species. (Source: after Solé and Montoya, 2001)

whose elimination is likely to be most destructive. These are known as keystone species. Keystone species may be damaged if bioaccumulation results in toxic conditions for those species. These species may also be those posing a perceived threat to humans. If such a species is removed then the resulting chain of effects might include a population explosion of previous prey, or the extinction of species dependent upon the removed species for nutrition or seed dispersal. Organisms other than the top trophic-level predators may be keystone species. In many food webs, species may appear at several trophic levels as shown in Figure 20.9.

The keystone relationship, often vital to successful conservation planning, may be derived from a variety of characteristics seen in the examples below:

- Elephants are generally regarded as grassland keystone species in tropical Africa. Their browsing deters succession from changing these habitats into shrubland or forest.
- A subspecies, the forest elephant, is key in maintaining some West African forests through its role in differential seed dispersal.
- In temperate grasslands, prairie dogs are a major keystone species as a result of their high level of linkages. They appear as grazers, predators, prey or, through their burrows, as providers of microhabitats for other species.
- The starfish *Pisaster ochraceus* underpins the stability of many rocky shoreline ecosystems on the North American coast, through its preferred diet of Californian mussels, which are otherwise the competitive dominants of the community.

20.6 Ecological processes and environmental management

20.6.1 Conservation and sustainability

Conservation is almost universally regarded as a good thing but definitions as to what conservation means are varied. Conservation always implies intervention and value-judgements have to be made as to what should be conserved and what should be ignored. This forms an uneasy alliance with the objectivity implicit in ecological science. Protected conservation areas are intended to maintain functioning ecosystems, minimizing known vulnerabilities and maximizing biodiversity. Often management is required to increase local diversity or protect what is considered to be a rare and important species (Box 20.4). For example, the Kenfig National Nature Reserve, South Wales, consists of a dune system which provides a habitat for rare plants such as the fen orchid. The main problem at the site is that a lack of fresh sandy sediment supply has led to an increased rate of succession. During soil development, increasing amounts of organic matter and nutrients and decreasing pH have created opportunities for rapid-growing species to replace species characteristic of the dune habitat. Thus a programme of removal of litter was introduced to remove organic matter from the system. Artificial bare areas (blow-outs) were also created in the dunes to promote plant communities of earlier stages as shown in Figure 20.18 and sheep and rabbit grazing is actively encouraged. The results were that levels of diversity increased dramatically.

However, there are questions about whether we should strive to maintain ecological communities that are now no longer self-sufficient (see Chapter 25). The Kenfig system will need continuous management intervention if biodiversity is to be maintained at a high level. There may also be more interesting and distinctive recombinant communities which are sometimes the only refuge for species with

Figure 20.18 Kenfig National Nature Reserve, South Wales. There are a variety of habitats within the dune system with shrubby vegetation in some of the drier slacks. This photo shows ponding in an artificially created dune slack which has been created to try to establish earlier succession species. (Source: Kenfig National Nature Reserve)

unusual niche requirements. The idea of sustainability is related not only to the ecosystem we are trying to manage but also to how we behave in the world as a whole and how sustainable our use of biosphere resources actually is. Calculations of sustainability (and unsustainability) are discussed in Box 20.5.

Shaw *et al.* (1998) showed for urban landscapes in the Tucson metropolitan area that among the 33 land cover categories, golf courses and neighbourhood parks had the highest vegetative cover (93% and 77% respectively). Schools (45%) and moderate density housing (2–8 houses per hectare) had a higher vegetative cover (44%) than natural open space in this semi-arid region which had only 22% cover. The study concluded that the single most important strategy for integrating conservation into planning for Tucson's growth would be to protect an interconnected matrix of habitats. Networks of reserves with corridors maintained between them are often regarded as a potential conservation solution in areas of fragmented landscapes.

UNESCO Biosphere Reserves (such as the one described in Box 19.1) are areas of terrestrial and coastal ecosystems promoting solutions to reconcile the conservation of biodiversity with sustainable use. They are internationally

recognized sites for testing out and demonstrating integrated management of land, water and biodiversity. The emphasis is on *conservation* rather than *preservation*. There is also an importance placed upon the use of the land by the local human population and potentially a much wider population. The emphasis on the human dimension appeals to most geographers and is a distinctive feature of the reserves. Each reserve contains a core area, surrounded first by a buffer zone and then by a zone of transition. The transition zone is the one in which the human impact is intended to be the greatest but also where there are likely to be areas within which sustainable economic and social development of the region will take place.

The idea that social and economic considerations need to be taken into account when thinking about ecological management suggests that multidisciplinary concepts are required. **Social ecology**, for example, is having increased influence on ecological thought. This is where conflicts between social values and ecological sustainability may be addressed. For Brunckhorst (1995) all environmental problems were seen as social problems. He put forward as a potential solution a framework known as cultural **bioregional theory**. This is based on landscape ecology (see Chapter 19) and integrates the arts and culture. Landscape

CASE STUDIES

CONSERVATION OF THE HAWAIIAN GOOSE

Ecology and historic decline

The Hawaiian goose (*Branta sandvicensis*), or nene (pronounced nay-nay), is a small brown goose endemic to the Islands of Hawaii and has the smallest biogeographical range of any living goose (Figure 20.19). The state emblem of Hawaii, the nene was a common species on many of the larger islands with an estimated historic population of 25 000 birds. However, after colonization of the islands by Europeans in 1778, the population began to decline dramatically. This decline was fuelled by direct threats including hunting and egg-collecting, and also by the indirect threat posed by habitat loss as the landscape became increasingly farmed. The single most serious threat, however, was the introduction of non-native predators to the islands. Most devastating was the predation of adults, chicks and eggs by the small Indian mongoose (*Herpestes auropunctatus*) which was

introduced to the islands in 1883. The combination of these human-induced problems of hunting, habitat change, and introduction of alien predators resulted in massive population decline: a demonstration of the interactions between a species and its biotic and abiotic environment. In 1907, the species was listed as a protected species but this had little effect, and by the late 1940s, only about 20 to 30 wild birds remained in Hawaii.

The nene's decline demonstrates the vulnerability of island populations to environmental change. In the case of the nene, the situation was made all the more serious by its endemic status: extinction in Hawaii meant extinction from the biosphere. Moreover, the nene is a keystone species, dispersing seeds and playing an important role in vegetation development, particularly in early successional communities on lava slopes. This illustrates how the decline of one species can have consequences for many more.

Conservation: captive breeding and reintroduction

Because of the critical nature of the species' population, several birds were captured during 1949 and 1950 and a captive breeding programme was started with two main bases, one on Hawaii and one in England at the Wildfowl and Wetland Trust at Slimbridge, Gloucestershire. Fortunately, the species adapted well to captive conditions and this conservation strategy probably saved the species from extinction.

A reintroduction programme of captive-bred individuals was started on three Hawaiian Islands (Hawaii, Kauai and Maui) in 1960 and over 2300 birds have now been released. The wild population has increased significantly as a result of this reintroduction programme and there are now about 1000 birds in eight populations. However, at least five of these populations are not self-sustaining, still requiring continued release of captive-bred birds.

Although it is probably only the captive breeding and subsequent reintroduction programme that have saved the species from extinction, there are still problems with these conservation approaches. Recent DNA analysis of the captive breeding nene populations suggests captive populations show low genetic variation. This is not surprising given that the population went through such a bottleneck (i.e. the number of individuals fell so low) meaning that all birds alive today come from a very few breeding individuals. Such chronic inbreeding may have led to inbreeding depression in the reintroduced population, although this is

Figure 20.19 The Hawaiian goose, ringed for identification purposes. (Source: photo courtesy of Anne Goodenough)

BOX 20.4 ➤

➤ very difficult to test scientifically. There are also concerns that reintroduced birds are showing altered patterns of behaviour and a lack of adaptability to change as a result of being captive-bred. These problems demonstrate the value of *in situ* conservation approaches whenever possible, such as suitable habitat management and control of alien species, rather than captive breeding and reintroduction. Of course, this can only be done well before the situation becomes critical.

The current situation

Because of its small and non self-sufficient population, the nene is still listed on the IUCN Red List as 'vulnerable' and is federally listed as 'endangered'. The current population dynamics of, and conservation strategies for, the nene in Hawaii are summarized in Figure 20.20. Many of the current threats are the same as those which caused the historic decline, but the difference is that most of these are being addressed through habitat management and

other conservation strategies. In addition, environmental education programmes seek to inform people about the nene and reduce illegal poaching, and extensive research programmes to establish and inform future conservation are continuing in the hope that the wild populations of nene can one day become self-sustaining.

Material courtesy of Anne Goodenough, University of Gloucestershire.

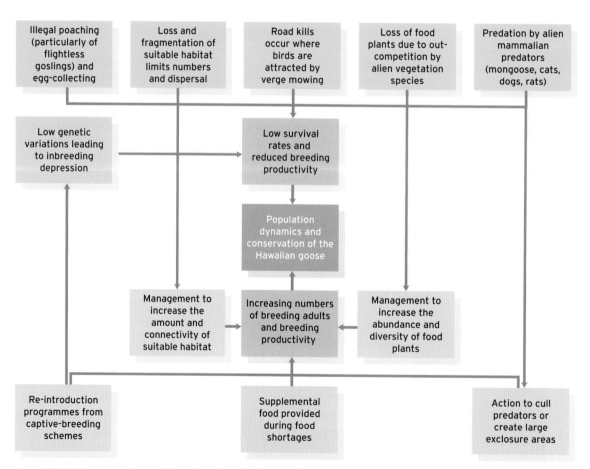

Figure 20.20 The current population dynamics of wild Hawaiian geese in Hawaii. Purple boxes indicate factors causing population decline; green boxes show management and conservation strategies. (Source: courtesy of Anne Goodenough)

BOX 20.4

and conservation values usually develop from the cultural history of the locality. In the belief that such a bioregional framework for planning should include both economic and ecological issues, it is similar to the philosophy behind the UNESCO Biosphere Reserves (see Box 19.1 in Chapter 19). Bioregional theory suggests that there are unified ecosystems defined by geophysical indicators (local watersheds, flora and fauna, geology and weather patterns) and cultural indicators (history of local human habitation, evolution as a community, presence and arrangement of material constructions and socio-economic structures). A bioregional approach attempts to understand and value the interconnected and sustainable workings of these individual components.

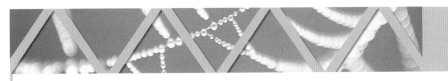

NEW DIRECTIONS

THE ECOLOGICAL FOOTPRINT

The ecological footprint is a recently developed sustainability index. It can provide estimates of the amount of ecological resources in hectares per person that are used by individuals, companies, specific activities and so on. The footprint is a measure of the amount of biologically productive area needed to both produce resources used and to absorb waste created. The measure reflects the intensity of land use and population densities as well as levels of resource use. When the index is used some startling results can be seen.

Table 20.7 shows how, of the countries listed, only Australia and Canada are below capacity. All of the other countries use more resources than they can sustain. It has been estimated that at current rates the annual consumption of resources by the world's population requires 14 months to be renewed and that many countries are running at an ecological deficit. The footprint can therefore be a potential measure of sustainability. Goals can be set for reducing local or national footprints by decreasing consumption or increasing productivity. Technology can be utilized to improve current productivity such as the use of roof gardens and solar panels, creating multiple use of space in urban areas. Farming methods such as slope terracing or multiple harvesting from individual fields can also be adopted. Consumption can be reduced through the use of recycled materials and using public rather than private transport and energy-efficient materials and products. You can calculate your own ecological footprint on the web; see the list of web resources on the Companion Website.

Table 20.7 Examples of ecological footprints calculated for a range of developed countries at the end of the twentieth century

	Population (millions)	Footprint*	Current capacity*	±*
World	6210.1	21.9	15.7	−6.2
Australia	19.7	79.1	110.2	31.2
Canada	31.2	83.0	85.9	2.9
Japan	127.2	53.2	8.8	−44.4
Netherlands	16.1	69.1	8.0	−61.1
South Africa	44.2	40.6	20.8	−19.8
USA	288.3	109.0	20.4	−88.6
UK	60.2	62.6	10.5	−52.1

*Global hectares per capita.

(Source: from the Redefining Progress Ecological Footprint Accounts for 2001)

BOX 20.5

20.6.2 Climate change

Root *et al.* (2003) considered data sets extracted from 143 earlier ecological or biogeographical studies, where temperature information was available as well as species occurrence and location. The intention was to identify spatial trends over time. Some **phenological** data (timing of events such as leaf fall and buds appearing) were also incorporated into the study. Overall, their results identified a consistent temperature-related shift. Where mobile, many species in the regions studied have moved either polewards or to higher altitudes over the past century. There has been a discernible impact of recent global warming on animals and plants. Walker and Steffan (1997), in their overview of the implications of global change for natural and managed terrestrial ecosystems, consider changes in the species composition and structure of ecosystems and how these might affect the system's functions. They predict

that the biosphere will be generally weedier and structurally simpler, with fewer areas in an ecologically complex old-growth state. Biomes will not shift as intact entities and the terrestrial biosphere may become a carbon source rather than a sink. Future research should consider the interactive effects of changes in carbon dioxide, nitrogen and temperature rather than discrete elements of change. The effects of changes in land use will probably continue to reduce species richness and an increasing number of 'natural' ecosystems will move towards production systems for human use.

20.6.3 Biosecurity

Isolated regions such as Australia and New Zealand have always been vulnerable to the arrival and establishment of alien species. For example, Figure 20.21 was produced

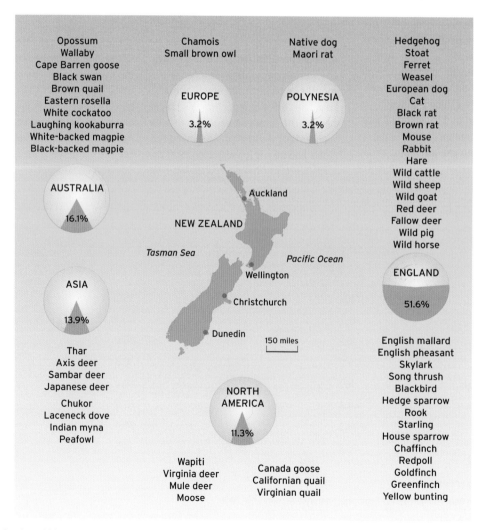

Figure 20.21 Introduced birds and mammals that have established populations in New Zealand, with their countries/continents of origin. The percentage from each origin location is shown. (Source: after Wodzicki, 1950)

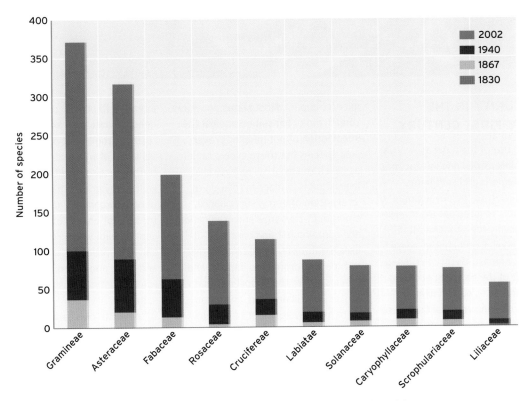

Figure 20.22 The 10 largest families of naturalized flora in New Zealand on four dates.
(Source: from Williams and Cameron, 2006)

in 1950 and shows the alien birds and mammals that had successfully established in New Zealand at the time. Therefore concern over the introduction of species is not a new phenomenon. Figure 20.22 shows the 10 largest groups of flora which have established in the wild in New Zealand since 1830, showing the enormous impact that invasions have had on the environment. Biosecurity is one response to this, as shown in Box 20.6. New Zealand, for example, has had a Minister of Biosecurity since 1997 and New Zealand's systems for protecting its borders are acknowledged to be among the best in the world. It is estimated that the costs of dealing with invasive pest species in New Zealand are currently US$4 billion per year. The most obvious measures to people arriving from abroad are those taken by the Quarantine Service. Ships, aircraft and mail are searched and sanitized as a matter of routine. Stopping alien species on arrival or dealing with their impact once recognized within the New Zealand

environment are reactive measures. However, more proactive measures include requiring ships travelling between Australia and New Zealand to empty their water-ballast tanks mid-voyage and to ensure the water is treated before release. This reduces the risk of importing known invasive species such as the Northern Pacific starfish.

Reflective question

➤ Think of your local environment: what conservation projects are taking place and to what extent does social ecology play a role in those projects?

NEW DIRECTIONS

BIOSECURITY IN THE TWENTY-FIRST CENTURY

As human use of the natural environment has intensified and interactions between regions and continents become easier, so the development of factors liable to interrupt the stability (and especially the economic stability) of ecological units has increased. Biosecurity measures often originate as a result of the threat of economic loss or damage to human health. Recent global and regional biosecurity threats include avian influenza, swine fever and 'foot and mouth' disease. Biosecurity attempts to secure the boundaries of an ecological unit against generally known threats. Just as ecosystems operate at a very wide range of scales, so biosecurity can operate in the context of a pond, field or farm breeding unit, a nature reserve, a nation state or globally (e.g. quarantine measures in relation to space research). While biosecurity often relates to an 'invading species', a distinguishing feature of biosecurity is that it is generally taken to be a set of human control processes. It is associated with the perceived value of species or habitats and their use by humans. The level of potential damage to the ecosystem may be minor, such as where fencing stops sheep straying into woodland and destroying saplings, or may have potentially disastrous impacts, such as those relating to avian influenza.

How does biosecurity work?

Risk assessment is fundamental to biosecurity. This involves calculating the probability and likely impact of an event and balancing these against the costs of implementing measures to counteract that event or to mini-mize its impact. Most states now have some form of legislation against the importation of potential or proven pest species but these evolve rapidly as new hazards arise. In Australia legislation is in place in relation to 'prevention of entry, surveillance, emergency response and eradication, and containment and control'. This uses a database divided into permitted species, prohibited imports and species where the import risk analysis process is still ongoing. The establishment and development of the powers of 'Biosecurity Australia' is a typical example of national ecological hazard minimization, where regulation is constantly revised and developed in the light of new information. This includes a 'hotline' for the public to report invasive species, such as fire ants (*Solenopsis invicta*), and surveillance programmes for known highly invasive species not yet observed in Australia. In a similar manner, 'Biosecurity New Zealand' is the lead agency in New Zealand, working in a 'whole of system' leadership role including economic, environmental, social and cultural issues.

Avian influenza

In the United Kingdom, as the possibility of a 'bird flu' epidemic has increased, with the potential for mutation into a human disease, a series of government guidelines have been produced. The Great Britain Poultry Register was set up to provide centralized information in relation to commercially bred poultry. The growth of various forms of 'animal passports' also continues to increase, incorporating monitoring schemes and, as required, restrictions on animal (and animal product) movement. These schemes are also intended to bolster consumer confidence in agricultural products. As a result of the threat of avian flu, surveillance of wild birds has been added to the existing monitoring of poultry welfare by government agencies. A general recommendation is that poultry be kept under cover to minimize contact with wild birds. However, this can raise problems with small-scale and free-range production units, where existing housing may not be suitable.

When an outbreak occurs or is suspected, a range of buffer zones around potentially infected sites are set up. Once the disease is confirmed, culling of infected and other individuals is usually required. In 2007, turkey chicks being reared in certain poultry sheds, forming part of a major production and processing unit in southern England, were exterminated, while the Canadian Food Inspection Agency, in response to an outbreak of the H7N3 avian flu strain in 2004, slaughtered 17 million birds within a 70 km^2 buffer zone. Beyond the immediate zone of infection, it is usual to have a zone where poultry movement is prohibited, in turn surrounded by one where only restricted movement is permitted. In the United Kingdom, should these measures prove insufficient to curb the spread of disease, the Government has also acquired millions of doses of avian flu vaccine. Such expensive interventions are often not politically popular but, especially where the economic and human impact is clear and where the threat is restricted to certain species, it may be the only means to allow trade to continue. Doses of vaccine would also be available, from

BOX 20.6 ➤

a separate stockpile, for the preventive vaccination of zoo birds, in the interests of conservation.

Community awareness and involvement

Costs and benefits of biosecurity are not always quantifiable and measures may be seen as heavy-handed by the public or organizations affected. In such situations, public education and awareness are vital elements of hazard control. A successful bio-security campaign with high levels of community awareness and involvement is the 'didymo' campaign in New Zealand. *Didymosphenia geminata* is a non-native freshwater alga that, by attaching itself to the beds of streams, rivers and lakes, can form a thick brown mat on rocks, plants and other materials, destroying habitats. *D. geminata* was first identified during a routine survey of algal growth and has since been found in many rivers in New Zealand's South Island. Since the algae are most likely to be spread by humans moving items between waterways, people using rivers and lakes must comply with control measures or face penalties of up to five year's imprisonment or a heavy fine. These measures focus upon a regime of 'check, clean, dry', especially when moving between waterways. Posters, websites and other public information attempt to reinforce the message.

BOX 20.6

20.7 Summary

This chapter has presented a range of ecological concepts and processes. While an ecosystem may appear to be static, closer observation shows that important links between plants, animals and soil are related to processes involving movement of energy, organic and inorganic materials through the system, so that ecosystems are in a constant state of change. There are a range of trophic levels which result in inefficient energy transfers between a hierarchy of species. In most situations, rather than a simple linear or circular route for processes there are more likely to be a range of options at various stages. Hence food chains are often better described as food webs with the same species appearing in a range of stages. Biogeochemical cycles may incorporate reservoirs for matter or nutrients. These reservoirs may form a vital buffer against change resulting from pulses of increased flow or the removal of a source of material.

Often, if a substance is bioavailable it can be taken up and stored for future use. Sometimes toxic substances are also bioavailable and these can accumulate and become concentrated within certain parts of an ecosystem. This may result in a range of problems for the system as a whole if the bioaccumulation results in severe toxicity to a species feeding on it. This may be very problematic if the species being adversely affected is a keystone species. These keystone species are very important to the whole ecosystem because they are so well connected to many other parts of the system. Identifying such keystone species is an important goal for environmental managers. Losing keystone species may result in ecosystem collapse and loss of biodiversity. Biodiversity is positively related to the number of habitats within an ecosystem. Some ecosystems may be more fragile than others, depending on the nature of the linkages within the ecosystem. The role of environmental niches, competition and a range of life strategies all play a part in biodiversity and the succession of an ecosystem.

Human disturbance of ecosystems through agriculture and urbanization provides important areas of research. While urbanization may seem a negative factor for ecosystems it can result in new recombinant communities. Environmental management requires consideration not only of ecosystem processes such as population dynamics and succession but also of human needs for the ecosystem. In other words, ecosystems can be considered to serve a range of functions. Ecological thought must therefore also consider socio-economic, value-laden ideas within the area of management.

Further reading
Books

Beeby, A.H. and Brennan, A.M. (1997) *First ecology*. Chapman and Hall, London.
This is a relatively easy read with some useful boxed studies, self-assessment exercises (with answers) and a nice glossary section.

Bradbury, I.K. (1998) *The biosphere*. John Wiley & Sons, Chichester.
This book provides a good overview of many of the ecological concepts discussed in this chapter and provides detail regarding taxonomy and the functions of ecosystems.

Ganderton, P. and Coker, P. (2005) *Environmental biogeography*. Prentice Hall, Harlow.
Strong, accessible introduction to the area.

Quammen, D. (1996) *The song of the dodo: Island biogeography in an age of extinctions*. Hutchinson, London.
An excellent example of narrative non-fiction is provided by Quammen. It works better for the fast reader since it is a chunky volume, but you might want to dip into it for fascinating accounts of interviews with a range of well-known scientists.

Ladle, R.J. and Whittaker, R.J. (eds) (2011) *Conservation biogeography*. Wiley-Blackwell, Chichester.
This text looks closely at conservation planning

van der Maarel, E. (1997) *Biodiversity: From Babel to biosphere management*. Opulus Press, Grangarde, Sweden.
This is, essentially, the inaugural lecture at Groningen of one of Europe's most distinguished vegetation scientists and is written in an entertaining, accessible and at times intentionally provocative style.

Further reading
Papers

Egoh, B., Rouget. M., Revers, B. *et al.* (2007) Integrating ecosystem services into conservation assessments: a review. *Ecological Economics*, 63, 714–721.
The authors interview conservationists and review how much the ecosystem services concept has become embedded into decision-making.

Salles, J.M. (2011) Valuing biodiversity and ecosystem services: why put economic values on Nature? *Comptes Rendus Biologies*, 334, 469–482.
A review and discussion about ecosystem services and monetary values.

CHAPTER 21

Freshwater ecosystems

Lee E. Brown

School of Geography, University of Leeds

Learning objectives

After reading this chapter you should be able to:

➤ understand some of the key scientific concepts underpinning the study of life in rivers and lakes

➤ recognize some of the major groups of freshwater plants and animals

➤ appreciate the major differences between flowing and still water ecosystems

➤ describe some of the ways in which freshwater ecosystems have been altered by human environmental change

21.1 Introduction

Although rivers and lakes constitute only an estimated 0.01% of the world's water resources and cover approximately 0.8% of the Earth's surface, these habitats have a disproportionately high diversity of plants and animals with *at least* 6% (or >100 000) of known species estimated to be found in freshwaters (Dudgeon *et al.*, 2006). Over 10 000 fish species are known from freshwaters with some 90 000 species of invertebrates described, the richest groups being the insects, crustaceans, molluscs and mites. Other well-studied groups of freshwater organisms include amphibians, mammals, birds, macrophytes (plants) and algae (e.g. diatoms, phytoplankton) (Figure 21.1). A diverse assemblage of bacteria, fungi, protozoa and rotifers is also found. However, knowledge of freshwater diversity remains incomplete and new species are identified every year.

Organisms inhabiting freshwaters can be grouped according to their role within aquatic food webs. For example, producers or **autotrophs** are the plants and algae that synthesize biomass from inorganic compounds and light. Producers can be attached to surfaces such as rocks or other plants (e.g. filamentous algae or macrophytes), be rooted in loose sediments (macrophytes), or be free living in the water column (e.g. phytoplankton). Those species that exist within the aquatic ecosystem and directly provide energy to the aquatic food web are termed **autochthonous** producers. **Heterotrophs** are organisms that obtain organic matter by consuming autotrophs, other heterotrophs or detritus. Members of this group can be considered either as herbivores (consumers of attached algae, plants and phytoplankton), detritivores (consumers of dead organic matter) or predators (consumers of living heterotrophs), although some organisms are omnivorous, feeding on a variety of resources. The diets of many aquatic heterotrophs are also subsidized with organic materials from adjacent terrestrial ecosystems (e.g. leaf litter from

Figure 21.1 Some examples of the diverse life forms associated with aquatic ecosystems. (a) insect larvae (blackfly); (b) juvenile fish (dolly varden); (c) insect larva (cranefly) (d) amphibian (Pyrenean brook newt); (e) adult aquatic insect (mayfly); (f) filamentous algae; (g) frog; (h) single-celled algae (diatoms); (i) macrophytes (water lilies); (j) fish (tench). (Source: (d) Getty Images: Mark Ledger)

trees, terrestrial insects). Resources originating externally to the river are termed **allochthonous**.

Further grouping of freshwater organisms can be made based on where they spend the majority of their existence. **Benthic** organisms (or the **benthos**) are those living on, in, or near the bed sediments of rivers or lakes. **Nektonic** organisms (or the **nekton**) are known collectively as organisms that can actively move around within the water column, in contrast to **planktonic** organisms that are suspended and passively float or drift around in the water column. A fourth group exists predominantly on, or just beneath, the surface of water bodies and these organisms are collectively termed the **neuston**.

The focus of this chapter is river and lake freshwater ecosystems. Wetlands (areas where the water table reaches the surface and persists long enough to support aquatic plants) can also be considered as freshwater ecosystems (see Dobson and Frid, 2009) but these systems are not considered in this chapter. The study of any 'ecosystem' unavoidably requires some knowledge of *both* the living organisms and their effective environment (see Chapter 20). Where necessary, this chapter makes reference to the aquatic environment to place understanding of the biota into relevant context, but the reader is directed to Chapters 7 to 13 for more detailed information on hillslope processes, sediments, catchment hydrology, fluvial geomorphology and solutes.

21.2 Running waters: rivers and streams

The terms *stream* and *river* are often used interchangeably when referring to flowing waters, because in reality there is no obvious distinction and the latter term is typically used when describing a 'larger' running watercourse. Some authors use the term **lotic** to describe running water systems of any size. The term 'river' is used for consistency in this chapter. Rivers are extremely diverse in their geomorphological form and physicochemical characteristics, which in turn influences the remarkable diversity of organisms that we find in flowing water ecosystems. Describing the characteristics of an individual river ecosystem can be difficult owing to this diversity of characteristics. However, it is helpful to learn about river ecosystems based upon: (i) their hierarchical organization, which considers the nested scales at which components of river ecosystems can be observed; (ii) the interactive pathways, or spatial dimensions, of interest, including upstream–downstream changes, land–water interfaces and surface–subterranean water mixing zones; and (iii) the temporal scale over which observations are made. Human influences have had a major effect on river ecosystem structure and functioning. In many environments such changes are obvious, but even the most remote and apparently pristine river environments will have undergone some degree of alteration due to the effects of atmospheric pollution (Moss, 2010).

21.2.1 River ecosystem geomorphological units

A widely used spatial framework used to aid understanding of nested river ecosystem units is that of hierarchical organization (Frissell *et al.*, 1986). Spatial units include the whole catchment, river segments, river reaches, mesohabitats and patches (or microhabitats) (Figure 21.2). The environment varies considerably at all of these scales, with larger units exerting significant influence over those at smaller spatial scales. This variability is reflected in the plants and animals that inhabit these different spatial units. The most influential environmental variables at each spatial scale can be considered as 'filters', for which species must possess appropriate biological traits enabling them to disperse to, and exist in, that unit (see Box 21.1 for further details). For example, at the catchment scale geology and vegetation are generally important natural factors influencing which organisms are present. With respect to vegetation, dense forest cover is likely to restrict the presence of in-stream producers and their consumers but can be beneficial to detritivores. In individual **river segments**, water quality variables such as stream temperature, or the flood or low-flow disturbance regime, may be important determinants of species distribution.

Closer examination of river segments reveals distinctly different **reaches** at the scale of tens of metres. For example, a river may flow through a relatively narrow, deep slower section (e.g. Figure 21.2c) before entering a wider reach where the wetted channel is separated from the bordering riparian vegetation by deposits of sediment (e.g. Figure 21.2d). Consequently, there may be reach-scale differences in shading affecting the level of primary production, for example. Each reach is typically composed of smaller units known as **mesohabitats**, including deeper, flatter, slower-flowing pools with lower dissolved oxygen and finer sediment accumulations compared with shallow, steeper, faster-flowing riffle sections (Table 21.1). Within mesohabitats there are clear differences in habitat at the patch (**microhabitat**) scale, which can play important roles in the temporal dynamics of stream communities (Townsend *et al.*, 1989), often by providing refugia from flow disturbances. Differences in flow velocity or bed sediments lead to small-scale variability in accumulations of detritus or algal growth, which in turn can influence the abundance of heterotrophs. Alternatively, small-scale differences in flow velocity may influence the abundance of aquatic invertebrates depending on, for example, their ability to adhere to the sediment or their morphology (e.g. whether streamlined or not).

It is obvious from these examples that river habitat can be extremely varied. Therefore, there are major differences in both the abundance of individual species as well as the composition of the stream biotic community depending on the level of examination across these nested spatial scales. The variability that can be observed across even the smallest spatial scales is important for scientists to consider

Figure 21.2 The hierarchical organization of river ecosystems, showing features at nested spatial scales: (a) the catchment (1000 m+) is composed of (b) various river segments (100 m). Each segment is composed of (c and d) different reaches (10 m). Closer inspection at the reach scale reveals (e) different mesohabitats (1 m). Mesohabitats are themselves composed of patch-scale (0.1 m) components, for example (f) collections of leaf litter and algal tufts on an individual boulder, (g) sand–silt film over cobbles, (h) a thin algal covering on a partially exposed rock and (i) moss attached to a small rock. (Source: photo (e) courtesy of Sandy Miner)

Table 21.1 Mesohabitat types and defining features as typically found in rivers with pool-riffle sequences

Mesohabitat	Features
Riffle	Sloping bed, shallow depth, poorly defined **thalweg** (i.e. line of greatest water depth)
Run	> Depth but < slope compared with riffles. Thalweg often well defined
Pool	Deep, surface slope almost zero, often located at the outside of meander bends
Glide	Located immediately downstream of pools, where the stream bed slopes to become progressively shallower

(a)

(b)

Figure 21.3 Collection of river benthic macroinvertebrates. (a) A Surber net. The operator places the quadrat onto the bed of the stream and uses a hand to disturb a small patch of stream bed in a fixed area (in this case 0.1 m²). Water is directed into the net by the orange side panels. Typically between five and ten samples are collected from each stream reach using this method. (b) A kick net. The operator places the net onto the bed of the stream and then disturbs the sediments immediately upstream of the net with one foot. The operator typically moves to a different part of the stream reach or mesohabitat every 30 seconds, sampling for a total of three minutes per site. Nets for both sampling methods typically have mesh size of between 250 μm and 1 mm. (Source: (a) photo courtesy of Jonathan Carrivick)

when studying river ecosystems or attempting to manage degraded systems. Multiple patch-scale samples will be necessary to characterize the local species pool of just one mesohabitat. River ecologists often have to accept some level of sampling error as a consequence of the exceptional level of habitat variability, although careful sampling can ensure comparable data sets for different river sites. For example, if the aim was to identify changes along an entire river segment, a typical approach for sampling aquatic invertebrates would be to focus on riffle mesohabitats at several points along the upstream–downstream gradient. Sampling may then be undertaken using a device such as a Surber sampler (Figure 21.3a), or a composite sample may be collected from several patches over a fixed time period by taking a kick sample (Figure 21.3b).

21.2.2 Spatial variability of river ecosystems

The concept of hierarchical organization in river ecosystems recognizes that stream habitats, and the biota they support, are strongly influenced by the surrounding catchment (Hynes, 1975). When considering the importance of these spatial perspectives on river ecosystem pattern and processes, it is also important to acknowledge three interactive spatial pathways, or dimensions (Ward, 1989): (i) the upstream–downstream continuum, or longitudinal dimension; (ii) exchanges of matter between the river channel and the riparian zone/floodplain, or the lateral dimension; and (iii) interactions between the channel and groundwater, or the vertical dimension.

21.2.2.1 Longitudinal dimension

The longitudinal (upstream–downstream) dimension is the best studied of the three spatial dimensions in rivers. At the largest scale, river biota change markedly with distance downstream from the headwaters linked to changes

FUNDAMENTAL PRINCIPLES

BIOLOGICAL TRAITS AND ENVIRONMENTAL FILTERS

Biological traits are variants of a phenotypic character (e.g. size, shape, life-history attributes, dispersal mechanism, diet) that may be inherited or environmentally determined (or both). Biological traits have been used to classify organisms since the early twentieth century when the Saprobian system was developed to assess the extent of organic pollution of rivers based on species' oxygen requirements (see Statzner and Bêche, 2010). Cummins (1973) later developed a classification of river invertebrates based on their dominant mode of feeding, so-called functional feeding groups (FFGs). The concept of FFGs was subsequently adopted by Vannote *et al.* (1980), forming the theoretical foundations of the hugely influential river continuum concept. More recently, freshwater ecologists have developed methods based on characterizing multiple biological or ecological traits of species, which can potentially be used to better understand how assemblages vary across environmental gradients compared with more traditional taxonomic approaches.

While predicting the abundance of species as a function of environmental features has been a long-standing goal in river ecology, the complexities associated with spatial habitat variability and frequent disturbance episodes (particularly in rivers) mean this has proven difficult to achieve (Poff, 1997). However, freshwater habitats with similar environmental features should theoretically have species assemblages with comparable biological trait attributes, even for

two water bodies in different regions of the world, because traits respond to environmental selection regimes regardless of biogeographic boundaries (e.g. Townsend and Hildrew, 1994). Freshwater species can be grouped on the basis of numerous biological traits broadly categorized as follows: (i) life history, (ii) mobility, (iii) morphology and (iv) ecology (Table 21.2). Individual traits can be related to habitat gradients; for example, species inhabiting frequently disturbed habitats will often possess traits that confer resilience (i.e. an ability to recover quickly from the disturbance) such as early age at reproduction, short reproductive cycles (multivoltine), high adult mobility and high fecundity. Alternatively,

taxa may possess resistance (ability to withstand the disturbance) traits such as clinging habit, streamlined or flattened body shapes or life-cycle forms such as egg stages or **diapause** and life stages outside the river.

A knowledge of species traits allows researchers to consider the selective environmental forces operating at different spatial scales (region to patch), each of which acts as a 'filter' reducing the total species pool of a region to the different assemblages found in habitat patches (Figure 21.4). For example, consider a regional species pool containing 100 different species. Natural factors such as barriers to dispersal (e.g. waterfalls, prevailing wind direction) may restrict 20 species from establishing in a given catchment

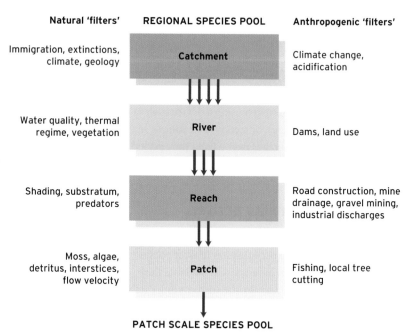

Figure 21.4 Examples of some natural and anthropogenic 'filters' at hierarchical spatial scales, which act upon species traits. Filters operating at the catchment scale restrict the occurrence or abundance of traits at smaller spatial scales, and so on. As river ecosystems are dynamic, the ways in which the environmental filters restrict the species pool can change in space and time. (Source: modified from Malmqvist, 2002)

BOX 21.1 ➤

Table 21.2 Some indicative biological traits of invertebrates and fish

Trait category	Indicative biological traits
Life history	Voltinism (e.g. number of generations per year)
	Maximum age (e.g. fish)
	Development (e.g. seasonal, non-seasonal for insects)
	Synchronization of adult insect emergence (e.g. none, poor, good)
	Adult life span (e.g. days, weeks, months)
	Fish reproductive strategy (migratory, broadcaster, simple nester, complex nester–guarder and bearer)
	Resistance forms (e.g. eggs, diapauses)
	Fecundity (number of eggs laid)
	Egg type (single or multiple batches)
Mobility	Adult insect dispersal distance
	Adult insect flying strength
	Travel distance (drift, crawling, swimming, flight)
	Swimming ability (strong, weak, none)
	Locomotion (cruisers, accelerators, manoeuverers, benthic high-velocity huggers, benthic low-velocity creepers)
Morphology	Shape (streamlined or not for invertebrates; torpedo, arrow, disc, arched, teardrop, elongate for fish)
	Morphometric characteristics of fish (e.g. body length/depth, height of caudal fin, head width)
	Mode of respiration (gills, tegument, plastron, spiracle)
	Size at maturity (body length or biomass)
Ecology	Habitat (e.g. erosional, depositional, sediment types, mesohabitat (fish))
	Thermal preference (cold, cool, warm)
	Invertebrate habit (e.g. burrower, clinger, swimmer)
	Fish stream size preference (expressed in m)
	Feeding behaviour (e.g. FFGs for invertebrates; herbivores, detritivores, planktivores, invertivores and carnivores for fish)
	Trophic classification (e.g. mouth position, teeth, gut size for fish)
	Water body type (river, lake, wetland, estuary)
	Elevation distribution
	Salinity tolerance
	Oxygen requirements
	pH tolerance

because they lack the necessary traits for dispersal to that catchment. Within that catchment, if we sampled an individual river in its entirety then we might find only 40 of the 80 species, perhaps due to a cool thermal regime selecting against species preferring warm water. Within an individual reach, only 20 of the 40 species may be able to colonize because shading by riparian vegetation means there is too little algal production to support grazing invertebrates. Finally, an individual habitat patch may then have only 10 of the remaining 20 species because the high flow velocity selects against species without streamlined bodies. In addition to natural factors, researchers also have to consider the effect of habitat filters imposed by the actions of human interference which are widespread in river systems, and which themselves operate across different spatial scales (Figure 21.4).

The concept of environmental filtering is based on the idea that the species pool at each spatial scale is filtered by the environment. It is important to realize that the actual species pool at each scale is also influenced by dispersal constraints, such that some species may be absent even though the habitat is suitable for their existence. Biotic interactions should also be taken into account, because a given species may be absent due to a lack of food or the existence of a predator, even if the abiotic environment is suitable.

BOX 21.1

619

in physical properties such as increased river size (width/depth), lower gradient, downstream fining of river bed sediments and a typically warmer, less variable thermal regime. One of the most influential research ideas in the history of river ecology is the river continuum concept (RCC; Vannote *et al.*, 1980), which generalized the changes in basal resource supply, the functioning of stream macroinvertebrates, with reference to their functional feeding groups (FFGs; Table 21.3) and fish communities that are typical in temperate forested river systems from headwater streams downstream to larger rivers (Figure 21.5).

In headwater or upland reaches, the river continuum concept considers rivers to be net heterotrophic (i.e. respiration > production). Heavy shading by trees limits light and thus instream primary production. Coarse particulate organic matter (CPOM; predominantly leaf litter but also small pieces of wood >1 mm) from deciduous trees and plants acts as a main source of energy for aquatic food webs. Leaf litter is colonized by microbes and fungi making it more palatable for aquatic invertebrates. The physical action of the river's current and movement of sediment can also break leaf litter into smaller pieces. This organic matter is consumed by invertebrates classified into a functional feeding group called shredders, which includes organisms such as freshwater shrimp (Gammaridae), water louse (Asellidae), some stoneflies (e.g. Leuctridae) and some caddis fly larvae (e.g. Limnephilidae). The processes of shredding and instream

physical breakdown break the coarse material into fine particulate organic matter (FPOM; < 1mm).

FPOM easily washes downstream, thus serving as an upstream–downstream connection in the aquatic food web, hence the use of the term 'continuum'. Fine particulates are considered to be a particularly important food source in the mid reaches of rivers. FPOM can be collected from the water column by organisms such as blackfly larvae (Simuliidae) which possess filter-feeding adaptations (collector-filterers) or gathered from deposits on the riverbed by collector-gatherers. However, the increasing width of the river channel means that shading of the river by riparian vegetation is largely restricted to the margins. Therefore, more light can reach the stream allowing for an increase in primary production and a shift to net production. Abundant algal growth supports a greater abundance of grazers or scrapers (e.g. snails, some mayflies) which directly consume producers growing on surfaces. In the deeper, wide, slow-flowing lowland reaches of rivers, light may be unable to penetrate to the river bed due to the high level of suspended sediment and organic material transported from the upper reaches of the river system. Therefore primary production falls and these reaches will be net heterotrophic. FPOM washed from upstream is a dominant food source, and the community will be almost exclusively composed of collectors. Throughout the whole river, shredders, collectors and grazers are themselves consumed

Table 21.3 Functional feeding groups (FFGs), their feeding mechanisms, food sources (FPOM/CPOM = fine/coarse particulate organic matter, respectively) and typical size range of particles ingested

FFG	Feeding mechanism	Main food source	Particle size range (mm)
Collector-filterer	Suspension feeders – filtering particles from the water column	FPOM, detritus, algae, bacteria, fungi	0.01-1.0
Collector-gatherer	Deposit feeders – collecting deposited particles of FPOM or sediment	FPOM, detritus, algae, bacteria, fungi	0.05-1.0
Grazer	Graze surfaces of rocks, plants and wood	Producers such as attached algae, and associated detritus, bacteria and fungi	0.01-1.0
Predator	Capture prey and either engulf whole or pierce and ingest body fluids	Living animals	>0.5
Shredder	Chewing	CPOM or live plant tissue	>1.0

(Source: modified from Merritt *et al.*, 2008)

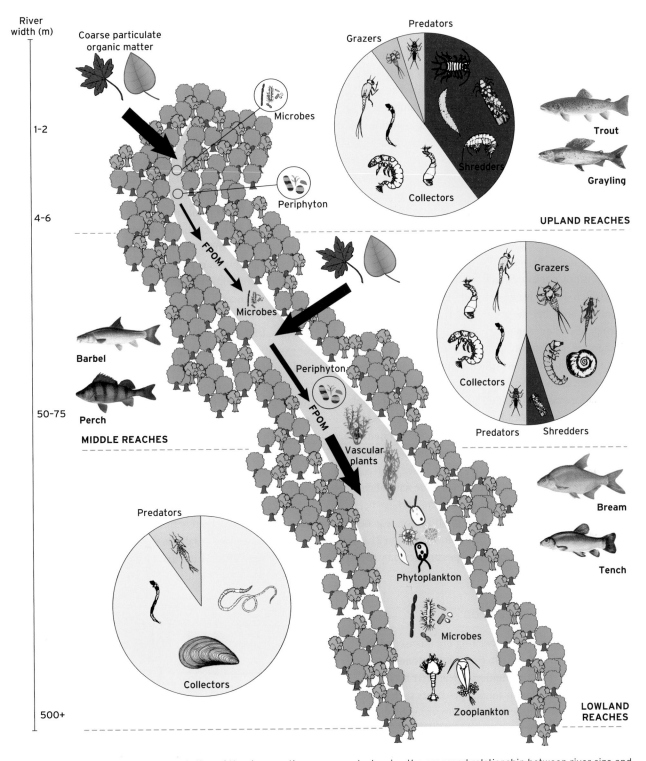

Figure 21.5 Diagrammatical representation of the river continuum concept, showing the proposed relationship between river size and the downstream changes in structural and functional attributes of the river ecosystem. (Source: after Vannote *et al.*, 1980)

by **predators** such as invertebrates (e.g. some stoneflies; Perlidae), fish and amphibians.

There is also a major microbial component to aquatic food webs throughout a river's length, fuelled largely by

dissolved organic carbon (DOC) sourced from the decomposition of organic matter in soils and the stream (Meyer, 1994). Carbon and other nutrients in river water can be absorbed and assimilated by plants and microbes or adsorbed to sediments. When these nutrients are eventually released back to the water, some may be recycled immediately into the biota but most move some distance downstream with the flow. The combined action of nutrient cycling and downstream movement can be conceptualized as a spiral. This idea forms the basis of another important piece of work that is built around the longitudinal dimension of rivers, the nutrient spiralling concept (Newbold *et al.*, 1982).

It should be realized that some organisms cannot be easily placed into functional feeding groups as diet can vary with life stage and body size, while some species are omnivorous, consuming a variety of different food resources. It is also important to appreciate that the RCC describes only an idealized sequence of ecosystem change along undisturbed, temperate forested rivers. The downstream sequence of changes is not 'continual' and can be interrupted by tributaries which deliver water, sediment, food sources and organisms, and alter the geomorphology. Riparian forest clearings also allow reaches of high primary production in otherwise net heterotrophic sections of river. The RCC cannot be applied to rivers which naturally lack abundant forest cover, such as those in Arctic or alpine regions (see Box 21.2, for example), and it is not considered applicable to streams in New Zealand due to an absence of shredders, because many streams are unstable due to the frequent overriding influence of rainfall and snowmelt-induced flood disturbances, and because many streams lie above the treeline (Winterbourn *et al.*, 1981). Additionally, the natural characteristics of many rivers and their catchments have been altered by a long history of human modification (e.g. deforestation, dam building, channelization; see Section 21.2.4) meaning that the downstream changes predicted by the RCC no longer apply.

21.2.2.2 Lateral dimension

As the RCC suggests, river-dwelling organisms and the adjacent terrestrial ecosystem (or **riparian zone**) can be linked through the consumption of leaf litter by shredders. However, the diet of aquatic predators such as fish and some invertebrates can also be supported by inputs of terrestrial invertebrates (e.g. ants, beetles, caterpillars) which fall into the water and are washed downstream. In this sense it can be said that the terrestrial ecosystem feeds, or subsidizes, the aquatic ecosystem (Nakano *et al.*, 1999). Inputs of terrestrial invertebrates to rivers vary seasonally, with peaks typically occurring from spring to early autumn in temperate zones. Some studies have shown that the diet of river-dwelling fish can be made up of >50% terrestrial invertebrates (Baxter *et al.*, 2005).

The flow of resources and energy between the terrestrial and aquatic systems is not, however, a one-way process. Recent research indicates that there are subsidies from the river back to the terrestrial ecosystem (Baxter *et al.*, 2005). Aquatic insects develop in the stream as larvae (nymphs) but towards the end of their life cycle they emerge as adult flies with a terrestrial aerial stage. The number of flies emerging from rivers varies widely but in some studies >150 000 per m^2 have been measured. The emergence of adult aquatic insects is typically concentrated into a small part of the year in temperate regions (peaking in early summer) whereas in tropical areas it may occur year round. These insects provide food for many different terrestrial organisms including spiders, bats, birds, amphibians and beetles. High abundances of these predatory organisms can be found along some river channels at times of peak aquatic insect emergence.

Terrestrial animals can also benefit from fish populations in rivers. In the Pacific north-west region of North America, for example, the annual salmon runs are often characterized by thousands of fish moving upstream in each river to their spawning grounds. During and after this migration, the salmon die leading to accumulations of carcasses in the stream, entangled in bankside vegetation and washed up on river banks (Figure 21.6). These salmon provide abundant food resources for insects, scavenging birds and mammals including bears. Bears also actively predate live fish, as the many TV documentaries attest. Bear populations can be up to 80 times denser in areas where salmon are abundant compared with fishless rivers (Gende *et al.*, 2002). Nutrients from salmon carcasses may also fertilize terrestrial vegetation such that growth may be increased significantly in areas where bears deposit carcass remains and excrete waste.

Figure 21.6 Pink salmon carcass in a coastal river in Alaska, with evidence of consumption by a bear (part-eaten head).

21.2.2.3 Vertical dimension

Rivers were traditionally viewed as components of the landscape where water was flowing over the surface. However, river scientists now understand that the spatial extent of rivers does not cease at the upper surface of the river bed. At small scales (up to metres), river water flows within the openings of river bed sediments (called **interstices**). These spaces are important habitats for invertebrate and fish eggs and larvae, offering refuge from swift-flowing surface currents and some predatory organisms. The slower flow of water through the interstices, compared with the water column, is important for nutrient cycling and organic matter respiration by microbial organisms because it results in greater sediment–water contact times. At larger spatial scales (up to kilometres) river ecosystems can extend downwards and outwards where subsurface and river bank sediments are permeable and surface water interacts with subsurface groundwater (Ward *et al.*, 2002). The boundary between the river water and groundwater, known as the **hyporheic zone**, can have steep physical, chemical and biological gradients and is considered to be particularly important for nutrient cycling, a permanent habitat for some species and as a refuge from disturbance for organisms that typically inhabit the upper layers of the stream bed.

21.2.3 Temporal variability of river ecosystems

In addition to the three spatial dimensions discussed above, a fourth dimension, time, exerts significant influence on river ecosystems (Ward, 1989). It is useful to consider temporal river ecosystem dynamics either as those that are relatively regular and predictable allowing the biota to adapt or persist, or those that are discrete disturbances, defined as 'an event in time that is characterized by a frequency, intensity and magnitude that is outside a predictable range and that disrupts ecosystem, community or population structure' (Resh *et al.*, 1988, p. 433).

Predictable changes in river ecosystems occur over diurnal timescales, where there can be, for example, significant changes in water temperature due to solar radiation receipt, causing fish to migrate temporarily to cooler reaches. In mountain rivers fed by snow and ice there are usually diurnal discharge variations due to melt (see Chapter 11), which can influence the behaviour of macroinvertebrates, causing them to move down into bed sediments to avoid higher flow velocities, or to migrate downstream as drifting individuals. At longer timescales there are seasonal changes in the composition of river ecosystems due to life-cycle progression and migrations to avoid/take advantage of wet or dry seasons. At inter-annual timescales, researchers have found evidence that the abundance and diversity of fish and invertebrates respond to climatic cycles associated with phenomena such as the El Niño Southern Oscillation and the North Atlantic Oscillation (Hurrell *et al.*, 2003; see also Chapters 4 and 10). Long-term changes have also been documented in some river ecosystems as a consequence of climate warming and the retreat of glaciers (Box 21.2).

Less predictable or **stochastic** disturbances in river ecosystems are typically associated with floods (the most common form of disturbance in rivers) or droughts, although major changes to river ecosystems can also result from disturbances associated with freezing, landslides, high

ENVIRONMENTAL CHANGE

RIVER ECOSYSTEM RESPONSE TO GLACIER RETREAT

Glaciers are distributed worldwide, not only at high latitudes but at lower latitudes where they are located in mountainous areas (e.g. the European Alps and the Pyrenees). Glacier melt contributes significantly to river flow and water resources across the globe, and rivers with glacial meltwater inputs provide habitat for fisheries and a number of rare and endemic aquatic invertebrate species. Glacier-fed rivers are considered to be among the most vulnerable to climate change because of strong interconnections between atmospheric forcing, snow-packs, glacier mass balance (see Chapter 16), stream flow, water quality and fluvial geomorphology and river biota (Milner *et al.*, 2009).

Glacier retreat is frequently attributed to a warmer climate and will lead to major shifts in water sourcing of Arctic and alpine rivers, with glacier

BOX 21.2 ➤

and snowmelt reductions and changes in river hydrological and geomorphological dynamics. These changes are likely to have significant, widespread consequences for plants and animals of alpine stream ecosystems which are strongly influenced by river channel stability and water temperature, two variables determined by the amount of meltwater contribution and valley geomorphology (Milner *et al.*, 2001). The non-biting midge larvae (Chironomidae) genus *Diamesa* typically dominates European glacier-fed rivers where maximum water temperature is <2°C and river channel stability is low (Figure 21.7). Further downstream from glacier margins, river channels become

more stable and water temperature increases allowing mayfly, stonefly and caddisfly larvae to become increasingly dominant along with blackfly (Simuliidae) and other groups of chironomid larvae.

In Glacier Bay, south-east Alaska, glacial retreat has been occurring since around 1750 (the end of the Little Ice Age), opening up vast areas of deglaciated terrain, and creating hundreds of kilometres of new rivers that subsequently undergo colonization and primary succession by biotic communities. Stream ecosystem response to the loss of glacial ice masses has been studied in detail since 1978 at Wolf Point Creek.

Progressive increases in stream temperature, riparian vegetation cover and instream habitat complexity over time have been accompanied by a diverse group of macroinvertebrate colonizers (Milner *et al.*, 2008). In 1978 when catchment glacial cover was ~70%, just five taxa (all Chironomidae of the subfamilies Diamesinae and Orthocladiinae) were found. The first mayflies and stoneflies were found at ~50–60% glacier cover in 1986, the first non-insect taxa (Oligochaete worms) at ~30% in 1992, and Dytiscidae beetles and Corixidae (pond skaters) in 2000 and 2003, respectively, after ice masses had almost vanished.

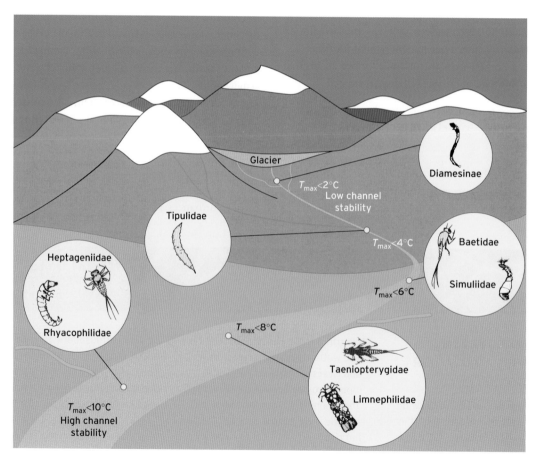

Figure 21.7 Diagram outlining the likely first appearance of macroinvertebrate taxa along a longitudinal continuum from the ice margin for European glacier-fed rivers. (Source: based on data from Milner *et al.*, 2001)

BOX 21.2 ➤

Further loss of snow and ice masses in the future will alter the spatial and temporal dynamics of river basin runoff through changes in the relative proportion of snowmelt, glacier melt, hillslope and groundwater contributions. If the climate continues to warm, glacial runoff may increase initially but eventually it will decrease in the long term as the ice disappears. Seasonal discharge patterns will therefore shift, with earlier and higher spring peaks and lower summer flows, and diurnal meltwater cycles may be reduced with day-to-day variability increasing with a greater disposition to flooding. These hydrological changes will alter the habitat within which freshwater organisms exist, necessitating adaptation or potentially causing species losses (Brown et al., 2007).

BOX 21.2

winds depositing large numbers of trees into rivers, or even anthropogenic pollution events. High-flow events may be devastating for the biota of small, steep upland streams owing to large sections of river being affected by sediment erosion and deposition. In lowland rivers, floods can be beneficial because of the potential for floodwater to expand across the floodplain and transfer energy and nutrients between the terrestrial and aquatic systems, whilst providing slower-flowing refuge areas for mobile biota (Junk et al., 1989). During droughts, the contraction of wetted area can lead to river biota becoming concentrated in pools and short reaches, increasing the risk of predation.

21.2.4 Human alterations to river ecosystems

Humans have severely changed river ecosystems through hydrological (e.g. water abstraction, dams), geomorphological (e.g. channelization, culverts, dams, gravel extraction), water quality (e.g. acidification, organic pollution, nutrients, thermal pollution) and biological alterations (e.g. introduced species; see Vörösmarty et al., 2010). These anthropogenic alterations span the range of river ecosystem spatial scales (see Box 21.1 and Mason, 2002). At the catchment scale and more widely, climate change is expected to drive changes in the thermal regime of freshwaters (Woodward et al., 2010). Warming increases metabolic rates of ectothermic organisms (i.e. those reliant on the environment to maintain their own body temperature), and above certain limits can induce stress on physiological systems, leading to the loss of species. Thermal regime changes can also occur from afforestation or deforestation, as well as the release of heated effluent from factories and power stations into streams and rivers. Warmer waters hold less dissolved oxygen and this will consequently impact respiration. Climate change effects on river ecosystems may also be indirect, with changes in precipitation leading to more frequent and severe droughts in some areas but more severe and unpredictable flooding in others. Many organisms have adapted their life cycles and morphological traits to the natural flow regime over long timescales. Changes to the magnitude, timing, frequency, predictability or duration of extreme flow events are therefore likely to induce major changes in river ecosystems (Lytle and Poff, 2004).

At the scale of individual river systems, major alterations can occur following land-use changes, in particular the conversion of forest to grassland or arable farmland and the urbanization of catchments. Agricultural development leads to increased diffuse nutrient fluxes to rivers (see Chapter 13), potentially increasing primary production which in turn may lead to changes in assemblages of invertebrates and fish (Allan, 2004). Nutrient enrichment can alter rates of leaf litter breakdown by microbes and fungi, leading to changes in the metabolic balance of freshwaters. Sediment eroded from bare agricultural fields can smother river bed sediments, reducing algal production and, by filling in the voids between larger sediments, affecting the spawning habitats of fish. Globally, regulation of rivers by flow control structures (typically dams) has caused major negative impacts on aquatic biodiversity by changing river flow regimes, altering habitat and impeding the migration of organisms. In the United States alone, there are an estimated 2.5 million structures controlling river flow, and only 2% of rivers are considered to have no impacts from flow control structures (Lytle and Poff, 2004).

There is a long history of humans discharging waste products to rivers from industrial processes and sewage treatment works via point source pipes or culverts (Figure 21.8). Sewage works were historically associated with inputs of organically rich effluent which led to severe oxygen decreases in affected river reaches downstream of the outlets (Mason, 2002). The severe lack of oxygen caused by organic enrichment typically leads to losses of most invertebrates and fish with the exception of a few tolerant groups

Figure 21.8 Point source discharge of sewage effluent to the River Calder, UK.

such as worms and some midge larvae. However, in the last two decades there have been major improvements in river water quality (particularly in the UK) linked to investment in enhanced sewage treatment technologies. Some problems still remain, though, owing to the inadequacies of sewage treatment processes in removing the active compounds contained in human pharmaceuticals, recreational drugs and personal care products such as toothpaste, shower gels and shampoo. Research in the last couple of decades has, for example, found that male roach (*Rutilus rutilus*) have developed female characteristics in many UK rivers due to steroidal estrogens and chemicals that mimic estrogens reaching the river in sewage effluent (Tyler and Jobling, 2008). The ecological effects of the many hundreds of other synthetic chemicals released into rivers remain to be studied.

While many rivers have been physically or chemically altered by humans, there are also biological changes caused by the intentional or accidental introduction of 'exotic' species. The rainbow trout (*Oncorhynchus mykiss*), for example, originates from North America, but it has been spread intentionally across the world for the purposes of sport angling and as a food source. When individuals escape they can compete for food and habitat with other species of fish, often to the detriment of the native species (Allan and Castillo, 2007). Similarly, introduction of the brown trout (*Salmo trutta*) to New Zealand has lead to dramatic declines in populations of native galaxiids (McIntosh *et al.*, 2010). Another example of biological change in rivers concerns the American signal crayfish (*Pacifastacus leniusculus*) which was brought to the UK because it is much larger than the native white-clawed crayfish (*Austropotamobius pallipes*), and was therefore more appealing as a species to farm for food supply. While the signal crayfish was initially introduced to selected fish farms, some individuals escaped into river systems and over time these have bred and expanded their range. The signal crayfish are better competitors for food and habitat, and they carry a fungus which causes mortality among the white-clawed crayfish, meaning they are at risk of being lost from the few rivers where they still exist (Dunn *et al.*, 2009). Chapter 19 contains more on alien species invasions and Chapter 20 describes biosecurity measures (e.g. Box 20.6).

21.3 Still waters: lakes and ponds

Lakes and ponds are distinct bodies of 'still' water that form in geomorphological settings with outflows small relative to the size of the water body. These water bodies are not completely still, with currents driven by wind, thermal currents and the slow movement of water from the inflow to outflow. However, they have relatively longer **residence times** than rivers, and some authors use the term **lentic** rather than referring interchangeably to still waters, ponds or lakes. The term 'lake' is used for consistency in this chapter because most of the examples refer to larger bodies of water. Lakes and ponds are typically fed by rivers, although some may be more isolated in the landscape and fed by rain, groundwater seepage/springs or inundation during times of flooding (Dobson and Frid, 2009). Similar to rivers, the study of lakes cannot be undertaken without considering the effects that humans have had on many of these systems. Describing the characteristics of an individual lake ecosystem can be difficult owing to the diversity of morphological and physico-chemical characteristics. However, it is helpful to learn about lake ecosystems based upon: (i) their geomorphological origin, and (ii) the tendency or otherwise to stratify, either thermally or chemically. Within individual lakes it is then possible to identify (iii) spatially discrete habitat zones. Each of these classification systems is detailed below.

21.3.1 Classification of lake ecosystems

21.3.1.1 Geomorphology and processes of formation

Seventy-six types of lakes were classified by Hutchinson (1957) and the list has been extended to over 100 lake types

in the intervening period. However, 10 distinct groups can be recognized (Table 21.4), with the most common being formed by glacial activity (Figure 21.9). The geomorphological characteristics of lakes and their surrounding landscape are particularly diverse. These characteristics control the nature of a lake's drainage, nutrient inputs and residence time, which in turn have strong influences on physical, chemical and biological characteristics (Wetzel, 2001).

21.3.1.2 Stratification of lake ecosystems

Some lakes, such as those that are shallow and strongly influenced by the actions of the wind, are well mixed. In contrast, many lakes become subject to **stratification** at certain times of the year, when water that is not readily mixed forms layers with distinct physical and chemical properties (in particular, differences in water temperature). The absorption of solar radiation from the Sun decreases exponentially with water depth (Moss, 1998) and, as a result, there should theoretically be an associated temperature decrease. However, the effect of wind at the lake surface is to mix the water and, if this mixing is insufficient to reach the bed of the lake (i.e. because it is too deep relative to the mixing depth), then a warmer upper layer of water (termed the **epilimnion**) can form. Where distinct layers occur and temperatures drop rapidly the 'boundary' is referred to as the **thermocline**. Lakes can also stratify in winter when ice cover develops at the surface, resulting in a warming with depth and a reverse thermocline close to the surface.

The classification system proposed by Hutchinson (1957) is still used today to describe differences in lake stratification. Lakes that never stratify are termed **amictic**; these lakes are characteristically found at high latitudes or altitudes where surface ice cover persists year-round and prevents wind from mixing the water. **Meromictic** lakes mix incompletely and form where lakes are either very deep, preventing complete turnover, or where inflows are denser than lake water owing to high solute concentrations, causing them to sink to the lake bottom. A third group of lakes are **holomictic**, and these mix completely. Holomictic lakes can be subdivided into: (i) **monomictic** lakes which have a single season of mixing, or overturn (e.g. warm, temperate lakes, or some tropical lakes that have large temporal changes in depth); (i) **dimictic** lakes which stratify in the summer and winter (e.g. cold temperate lakes, Figure 21.10); (iii) **polymictic** lakes, such as warm tropical lakes which may stratify as often as each day, albeit slightly; and (iv) **oligomictic lakes**, which may be continually stratified except for when irregular storms induce mixing. During

Table 21.4 Groups of lakes, with a brief description of processes by which they were formed

Group	Typical processes of formation	Example
Tectonic basins	Depressions formed by movement of the Earth's crust that infill with water. The major types are caused by faulting of the crust	Lake Baikal, eastern Siberia. Covers 31 500 km² with a maximum depth of 1620 m Also lakes in the rift valley of East Africa
Volcanic	Depressions and cavities in cooling magma may collect water and form lakes. Crater lakes are often found occupying inactive volcanoes. Small depressions where magma has been ejected and overlying material has collapsed. Caldera lakes form where the rook of partially emptied magmatic chamber collapse. Lava flows may create basins or block existing rivers	Lake Toba, North Sumatra; 100 km long × 30 km wide, maximum depth of 505 m
Landslides	Sudden movements of unconsolidated material form new depressions or block existing rivers. Many lakes formed in this way are transitory, lasting for only a few weeks or months because the dams are usually susceptible to erosion and failure	Hunza River, north-west Pakistan, formed in January 2010 following a massive landslide
Glacial	These are the most numerous type of lake. Gradual erosional and depositional activities of glacial ice create depressions. With the retreat of the last **Pleistocene** glaciers (see Chapter 22) many small lakes were formed. Lakes can also form in, on or under ice masses, or behind terminal moraines in river valleys where glaciers still exist	Windermere, Lake District, England. Main fish species include trout, char, pike and perch. Much of the pioneering research into lake ecology was undertaken here by the Freshwater Biological Association
Solution lakes	Slow chemical weathering of soluble rock creates depressions. Commonly found on limestone terrain. Typically very circular and conical	Many lakes in Florida, USA, lie on limestone and have formed by dissolution
River activity	Plunge pool lakes – excavated at the foot of waterfalls along previous river channels	Grand Coulee plunge pool system, Washington State, USA
	Lateral lakes form where mainstem rivers deposit large amounts of sediment overbank, blocking tributaries	Lake Chicot, Arkansas, USA, is the largest oxbow lake in North America and was originally part of the Mississippi River system
	Several types of floodplain lakes form where rivers inundate the riparian zone	
	Oxbow lakes form where river meanders become isolated from the main channel (see Chapter 12)	
Wind-formed	Form in arid regions where wind-blown sediments form dunes that block rivers	Common in dry regions such as inner Australia, South Africa and arid regions of the USA
	Deflation basins form where sediments are blown away and depressions form	
Shoreline activity	Longshore currents can deposit marine sediments across the mouth of bays, separating the bay from the sea (see Chapter 15)	Slapton Ley, England
Biological origin	Lakes formed by the damming action of plant growth or detritus build up. Can also include beaver dams	
Human	Reservoirs formed behind dams	

(Source: after Wetzel, 2001)

Figure 21.9 Four contrasting glacially formed lakes: (a) Latnajaura, near Abisko, Sweden, a lake formed by glacial scour, and which is frozen for up to 10 months per year; (b) a small cirque lake in south-east Alaska, in the depression formed at the head of a former glacier; (c) Lake Wakatipu, near Queenstown, New Zealand, a moraine dammed lake; and (d) kettle lakes, formed through the melting of ice blocks deposited in unconsolidated sediment by retreating glaciers.

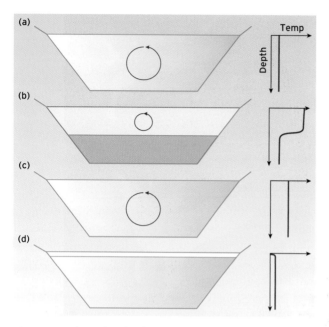

Figure 21.10 Annual cycle of stratification for a cold temperate dimicitic lake. Circular arrows denote zones of mixing. Panels to the right show the temperature changes that occur with increasing distance from the surface. (a) In spring, there is complete mixing, whereas (b) in summer a thermocline develops. (c) The thermocline breaks down in autumn and the lake becomes thoroughly mixed, until (d) in winter stratification can occur for a second time due to surface ice cover resulting in a reverse thermocline at the surface. (Source: modified from Dobson and Frid, 2009)

stratification, the upper layer is typically referred to as the **epilimnion** and the lower layer the **hypolimnion**. The zone characterized by the thermocline (zone of rapid temperature change) is the **metalimnion**. It is important to bear in mind that these general patterns may be altered by, for example, local variations in climate, lake geomorphology and water movement.

21.3.1.3 Nutrient status

Lakes are often described on the basis of their nutrient status with two common descriptions being oligotrophic and eutrophic, although there are various other trophic states (e.g. dystrophic, mesotrophic, hypertrophic). Various methods exist for determining the trophic status of lakes, and although there are no universally accepted definitions, some quantitative differences have been proposed (Table 21.5). These include taking measurements of total phosphorus concentration, phytoplankton biomass or chlorophyll-*a* concentration in a water sample, primary production rate measured as the uptake of carbon, and water transparency measured using a device called a Secchi disc (Figure 21.11).

21.3.2 Spatial variability of lake ecosystems

21.3.2.1 Edges and bottom

Lake habitats can be considered within a simple framework (Figure 21.12). The **littoral zone** is the part of a lake nearest the shore. It is the zone of both shallow and deep lakes where light can penetrate to the bed allowing the growth of diverse macrophytes (plants) and benthic algae assemblages. Primary producers can dominate the energy base of aquatic food webs in these areas. In small and shallow lakes, the littoral zone may constitute a major habitat due to its areal dominance. The edges and bottom of lakes are considered to be relatively productive and diverse perhaps because of the light penetration, combined with abundant sediment nutrient stores and the spatial heterogeneity of these habitats (Moss, 2010).

Table 21.5 Approximate values for four parameters used to categorize the trophic status of lakes

	Total phosphorus (µg L^{-1})	Maximum chlorophyll-*a* (µg L^{-1})	Primary production (mg C m^{-2} day^{-1})	Maximum Secchi depth (m)
Ultra-oligotrophic	<4	<2.5	<30	>6
Oligotrophic	4-10	2.5-8	30-100	3-6
Mesotrophic	10-35	8-25	100-300	1.5-3
Eutrophic	35-100	25-75	300-3000	0.7-1.5
Hypertrophic	>100	>75	>3000	<0.7

(Source: modified from Dobson and Frid, 2009)

The shallow edges and bottom in the littoral zone have an assortment of sediment sizes supporting a diverse assemblage of organisms. Waves act to regularly disturb the lake bed and create a mosaic of sediments (ranging from silts and mud in calmer waters, to bare rock and gravel where wave action is intense) which in turn determine the composition of biotic communities. Wave disturbance is a major factor influencing the distribution of benthic organisms in the littoral zone, with wave-impacted rocky shorelines only covered with attached algae, bacteria, limpets and occasional caddisfly or mayfly larvae. More sheltered areas of the shoreline may have gravel deposits but this moves readily and therefore has only a small permanent community. Where wave action is less considerable, finer sand deposits can build up and be colonized by aquatic plants and diverse communities of bacteria, protozoa or epipelic (living on, or in, fine sediment) algae. There may also be an array of emergent, submerged and floating plants, fish, invertebrates, and both herbivorous and piscivorous birds and mammals (Moss, 1998). The calm waters around these plants can be particularly beneficial for the larval development of nuisance species such as mosquitoes (Box 21.3).

The zone where light penetrates the water column is known as the **euphotic zone**. Below this is the **profundal zone** and here the lack of light prevents photosynthesis. Thus, any organisms inhabiting this area permanently are dependent on detritus (dead organic matter) as a food source, either supplied from the overlying water column or washed from the littoral zone. The profundal zone typically supports a lower diversity

Figure 21.11 Measurement of water transparency using a Secchi disc. The disc is 20 cm in diameter with quarters painted black and white. The operator lowers the disc into the water until it is no longer visible. The Secchi depth is then recorded from the attached tape measure. (Source: photo courtesy of Mel Bickerton)

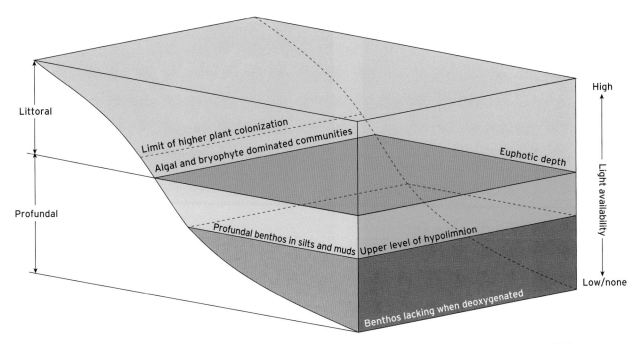

Figure 21.12 A general scheme of littoral and profundal habitats in a freshwater lake. (Source: modified from Moss, 1998)

HAZARDS

MOSQUITOES AND INFECTIOUS DISEASES

Mosquitoes are familiar as the small flying insects which can deliver painful bites. The name mosquito does not refer to one individual species; in fact, there are more than 3500 known species of mosquito from around the world. Many species are found only in tropical and subtropical regions but some have adapted to life in temperate and subarctic regions. While these insects spend the majority of their life cycle as adult flies (between 4 and 8 weeks), three of their four life-cycle stages (egg, larvae and pupa) are completely reliant on freshwater (with the exception of a few species which can develop in salt marshes). Adult females lay their eggs in still waters such as ponds, lakes and wetlands and also water butts, buckets, or water-filled hollows of plants. These eggs develop into larvae and subsequently pupae over a

period of up to two weeks depending on the environmental temperature and food availability.

In the larval stage (Figure 21.13a), the mosquito has a well-developed head with mouth brushes that are used for feeding on phytoplankton and suspended bacteria. The larva's body is segmented and lacking legs but has a distinctive eighth abdominal segment called a spiracle, through which it breathes. To breathe, it must hang beneath the water surface with the spiracle piercing the surface film. When disturbed, the larvae swim to depth by undulating the body or using hairs around the mouth for propulsion. Following the larval stage, each individual forms a pupa which also breathes from the water surface but it does not feed and is far less active than the larvae. After a few days the adult fly emerges from the pupal cocoon.

During the adult stage of the life cycle (Figure 21.13b) both male and

female mosquitoes feed on plant nectar. However, the female flies of most mosquito species also require a blood meal which provides the protein and iron required for the development of egg masses. When these blood meals are obtained from humans, the mosquito can deliver a painful bite as its proboscis pierces the skin and injects anti-coagulant saliva to prevent localized blood clotting while it feeds. The bite itself typically causes a red itchy spot in most people that subsides quickly. However, mosquitoes pose major health hazards in many tropical and subtropical areas because they are vectors for several life-threatening diseases.

The most commonly known mosquito-borne disease is malaria which is spread to humans by mosquitoes of the genus *Anopheles*. Malaria is caused when humans become infected by parasitic protists of the genus *Plasmodium* which are carried by the mosquito. These

(a)

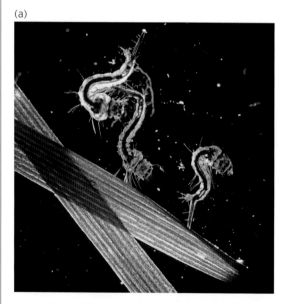

(b)

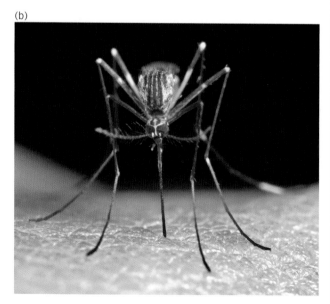

Figure 21.13 (a) Larval mosquito hanging from the water surface layer, and (b) adult mosquito with the distinctive proboscis used to pierce the flesh of its prey. (Source: (a) Getty Images: Harold Taylor, (b) Paul Bertner)

BOX 21.3 ➤

parasites have a complex life cycle, partly in the mosquito and partly in vertebrate animals. When a mosquito bites an infected person, malarial parasites can pass into the mosquito with the blood. These parasites then develop in the mosquito over a period of about one week until they are subsequently injected into another victim with the mosquito's anti-coagulant saliva. The *Plasmodium* parasites then develop in the human liver over a period of at least two weeks before they start

to multiply and malarial symptoms arise. Symptoms of malaria can include severe fever and headache as well as hallucinations, coma and even death if left untreated. While a variety of anti-malarial drugs are available these are not accessible to many of the world's poorest people, particularly in developing countries. Consequently there are thought to be more than a million deaths per year across the world due to malaria. Additionally, some species of mosquito carry a parasitic worm that

causes elephantiasis (severe swelling of different body parts). The mosquito species *Aedes aegypti* spreads the viral diseases yellow fever, dengue fever and Chikungunya, whilst other mosquitoes spread Ross River fever, West Nile virus and Japanese encephalitis. An estimated 700 million people worldwide are affected by mosquito-transmitted diseases each year, and the scale of the problem is expected to increase, with northward shifts of the disease projected due to climate change.

BOX 21.3

---

of organisms than the littoral zone, and is often characterized by simple communities of bacteria (including Cyanobacteria or blue–green algae), protozoa, invertebrates and fish (Moss, 1998). Often the zone below the hypolimnion will become deoxygenated, restricting the biotic community to anaerobic bacteria and protozoa.

21.3.2.2 Open water

The region of open water above the profundal and out from the littoral is called the **pelagic zone**. The open-water habitat of lakes contains a complex assortment of organisms, although in reality many of these are invisible to the naked eye except for where they form immense colonies, or blooms. The planktonic organisms that are primary producers are collectively known as **phytoplankton**, and range in size from 1 μm (picoplankton), <5 μm (ultraplankton), 5–60 μm (nanoplankton) and >60 μm (net plankton). This group includes a hugely diverse array of bacteria, green, yellow–green and golden-yellow algae, diatoms, cryptophytes, euglenoids and dinoflagellates (Moss, 1998). The growth of phytoplankton is strongly influenced by light and dissolved nutrients such as nitrogen and phosphorus. Phytoplankton can be sampled by collecting water samples (usually 1 L) from the water surface, from depth using a pump and hose, or by deploying sealed bottles which are submerged to the required depth before being opened remotely. Alternative methods include the use of plankton nets (Figure 21.14) which can be towed behind moving boats to filter larger volumes of water, or which can be lowered from the side of a vessel and then slowly pulled upwards.

Figure 21.14 Sampling of lake plankton from a small boat. The fine mesh net is lowered into the water column to the desired depth then pulled up to the surface to sample a column of water. (Source: photo courtesy of Mel Bickerton)

Rotifers, protozoa and crustaceans such as the cladoceran *Daphnia* form the major groups of freshwater **zooplankton**, an assortment of organisms which comprises heterotrophs feeding on phytoplankton, suspended organic particles and sometimes other smaller zooplankton. Other less common constituents of the zooplankton include freshwater jellyfish, some flatworms and mites (Moss, 1998). Some groups of zooplankton such as the *Cladocera* are much more active than the phytoplankton, being able to move actively through the water column, form shoals and undertake migrations through the water column either in response to diurnal cycles or to avoid predators. Larger, faster-moving zooplankton groups include shrimp-like mysids (up to 2 cm in length) which are predators of other zooplankton. Sampling zooplankton can be tricky due to their spatial aggregation and ability to move actively. Plankton nets are the favoured method because bottle samples of lake water are unlikely to capture many individuals.

A familiar third group of open-water-dwelling organisms, with major socio-economic importance, is the fishes. Over 10 000 species, or ~40% of the world's known fishes, are from freshwaters (Dudgeon *et al.*, 2006). Most freshwater fish species are found in the tropics, perhaps as result of shorter generation times in warmer waters permitting faster evolution, coupled with the long history of permanent water bodies in these regions (Moss, 1998). The fishes have particularly diverse morphological, behavioural and ecological traits, meaning that they inhabit a range of lake habitats as well as the open water, and they have considerable dietary differences with some being herbivorous or detritivorous, whilst others are predatory. Many fish are omnivorous

(feeding on both plants and animals) and their diet changes as they develop from larvae to fry to adults. Fish have different spawning needs and thermal tolerances but being highly mobile they are able to move between different areas of lakes.

21.3.3 Human influences on lake ecosystems

Lakes suffer from many of the same problems as rivers (see Section 21.2.4 and Mason, 2002), particularly those related to hydrological, water quality and biological alterations. The hydrological behaviour of many lakes throughout the world has been modified by human abstractions for water supply and irrigation. Some lakes have been modified morphologically by the construction of dams to raise the water level (e.g. Malham Tarn, UK) which alters the configuration of the shoreline and littoral zone. Diffuse pollutants from the wider catchment, and point source pollutants entering either directly through wastewater pipes or indirectly via inflowing streams and rivers, are particularly major problems. In many lakes around the world, nutrients (nitrogen and phosphorus in particular) contained in diffuse runoff from surrounding agricultural land as well as point source discharges have led to major changes in the trophic status of lakes, leading to the widespread problem of eutrophication (see Box 21.4 and Moss, 2010). Some lakes are used as direct

CASE STUDIES

ANTHROPOGENIC EUTROPHICATION

While eutrophic lakes (high nutrients, high productivity) can occur naturally, the term eutrophication is more synonymous with the problem of anthropogenic eutrophication. Anthropogenic eutrophication is caused by an increase in nutrient concentrations (nitrogen and phosphorus) in lakes, usually from treated sewage effluent, raw sewage (from sewer overflows, farm runoff or fish farm effluent), and by leaching of fertilizers from arable farmland in the surrounding catchment (Moss, 1998).

The accumulation of phosphorus and nitrogen in previously oligotrophic (low-nutrient, low-productivity) lakes can cause a dramatic shift along the nutrient and productivity spectrum towards a eutrophic state. The problem was particularly prevalent in the 1960s to 1970s when the water quality and fisheries of many lakes across Europe and North America deteriorated as a consequence of massive algal blooms associated with eutrophication. Eutrophication is now increasing in countries such as China which are undergoing major economic development.

The problem of eutrophication is recognizable by increased growth of aquatic plants and phytoplankton. Occasionally these blooms contain species of blue-green algae that produce substances toxic to mammals, in particular affecting cattle using the lake margins as a source of drinking water or dogs swimming in the water. Algal blooms also clog filters where water is extracted from lakes for human consumption, increasing the cost of treatment and sometimes altering the taste of the water (Moss, 1998). In extreme cases the high densities of algae can lead to deoxygenation of lakes

BOX 21.4 ➤

Figure 21.15 Lake 226 on 4 September 1973. The yellow line is the top of a vinyl curtain reinforced with nylon, which was sealed into the lake bed sediments and fastened to the lakeside bedrock. The lower part of the lake (north-east basin) received additions of phosphorus and quickly developed a bloom of blue-green algae as can be seen from the light green colour of the water. (Source: Fisheries & Oceans, Canada: E. Debruyn)

because night-time algal respiration, coupled with the decomposition of detritus originating from the increased phytoplankton bloom, consumes dissolved oxygen from the water column more quickly than it can be replaced by diffusion from the atmosphere. Consequently, there can be major losses of fishes and other aquatic organisms during these deoxygenation episodes. If the hypolimnion becomes deoxygenated, sulphides can be released from bed sediments causing further water quality problems for water supply companies and aquatic ecosystems.

Solving the problem of eutrophication can involve treating the symptoms, by either physically removing aquatic plants and phytoplankton or altering the mixing regime of lakes. However, these interventions can be costly in the long term and do nothing to address the root cause of the problem. Laboratory studies carried out in the 1960s and 1970s suggested that phosphorus, nitrogen and carbon were all responsible for eutrophication. However, experimental studies in lakes showed that the problem could be controlled successfully by restricting the input of phosphorus alone. In a series of experiments, scientists from the Canadian Fisheries and Marine Service manipulated the nutrient content of lakes in north-western Ontario. Figure 21.15 shows an image of Lake 226, which has two basins of similar size separated by a narrow constriction (Schindler, 1974). The constriction provided an opportunity to physically separate the basins using a barrier. Both basins were treated with nitrogen and carbon throughout the year but only the north-east basin received additions of phosphorus. The north-east basin quickly developed a major bloom of the blue-green alga *Anabaena spiroides* whilst the phytoplankton community of the adjacent basin did not differ from the pre-treatment period. The findings of this and other experiments led to a realization that the problems of eutrophication could be mitigated by making efforts to: reduce phosphorus loads in sewage treatment effluent discharges (either by phosphate stripping or by reducing phosphorus in household and industrial detergents), control the application of phosphorus to farmland, reduce livestock densities in vulnerable lake catchments, or manipulate the aquatic food web (Scheffer *et al.*, 1993).

BOX 21.4

disposal routes for sewage effluent which can cause problems related to nutrient enrichment, organic pollution and the introduction of pharmaceuticals. For example, Windermere, England's largest natural lake, receives effluent from two sewage treatment works and while these remove a large component of the potential pollutant and nutrient load, some pollutants remain in the effluent. Additionally, releases of untreated waste are common from overflow pipes during heavy rain.

One of the major ways in which lakes provide goods and services to humans is through fisheries. In many regions of the world, lake fisheries provide fish that comprise the major dietary component. However, a typical problem is overfishing which in turn often leads to the restocking of fish to replace those which have been over-exploited. In some places commercial fish farms are common on lakes, with fish being reared in large cages (see Figure 3.23 in Chapter 3). Inevitably, some fish escape and these can be particularly problematic if they are non-native species because they often out-compete native species. Other invasive species that have caused major problems in lakes are zebra mussels (*Dreissena polymorpha*). These bivalve molluscs originate from the Caspian Sea which borders northern Iran and southern Russia, but they were spread to western Europe in the nineteenth century, North America in the late 1980s, and to other parts of the world, most likely in the ballast waters of cargo ships. Since being introduced

to North America they have spread quickly across the Great Lakes and into Canada. Dispersal is rapid because high numbers of larval offspring are produced by each adult zebra mussel (e.g. Johnson and Carlton, 1996), and these are distributed by lake currents. Adult mussels are also transported between lakes attached to boats, and their unrestricted spread has caused major problems with the fouling of vessels, docks and blocking of water supply pipes. Their mode of feeding by filtering detritus and phytoplankton from the water column has also led to increased clarity of lakes, meaning increased sunlight penetration and algal growth with knock-on effects throughout the lake food web.

Reflective questions

➤ Can you draw the typical zones of a lake ecosystem?

➤ Consider a lake that you have seen recently. By what process is it likely to have been formed and what classification might it be accorded based on the stratification process described by Hutchinson (1957)?

➤ How have the activities of humans altered lake ecosystems?

21.4 Summary

This chapter has provided an introduction to some of the diverse habitats and biotic groups that can be found in rivers and lakes, the differences between these two types of freshwater system and some of the alterations imposed by human activities. River systems can be examined at different scales and the chapter has shown the importance of longitudinal, lateral and vertical spatial dimensions as well as some of the important changes that occur in river ecosystems over timescales ranging from days to years. The river continuum concept has been a very influential theory over the past three decades and describes

the aquatic zones within river systems from headwater streams downstream to larger rivers. Connectivity is provided by flowing water carrying, for example, fine particulate organic matter or nutrients as part of the freshwater nutrient spiral. Human modifications to river and lake ecosystems have included the introduction of alien, invasive species, changes to chemical and thermal characteristics through pollution, modification to flow regimes through river and lake control structures (e.g. dams), and climate change impacts (e.g. changes to glacial meltwater release, floods or drought). Lake ecosystem classification systems including geomorphological, stratification and nutrient status have been outlined. An individual lake ecosystem can be described by a simple framework of zones dependent on physical and chemical characteristics which interact with biota. However, it is important to remember that rivers and lakes are particularly diverse ecosystems, and the ideas discussed within this chapter for the most part provide general frameworks, rather than a blueprint for

how individual river or lake systems can be expected to work. The reader is encouraged to consult the recommended further reading, key research papers, references and weblinks on the Companion Website to build up a more detailed understanding and knowledge of freshwater ecosystems.

Further reading
Books

Allan, D.J. and Castillo, M.M. (2007) *Stream ecology: Structure and function of running waters*. Springer, Dordrecht, Netherlands.
This is a very thorough text providing a detailed review of research studies on the subject of stream ecology. It is suitable for more advanced undergraduate study but also has introductory information on stream hydrology, water chemistry and human modifications of river systems.

Closs, G., Downes, B. and Boulton, A. (2004) *Freshwater ecology: A scientific introduction*. Blackwell Publishing, Oxford.
An easily accessible introductory text to freshwater ecosystems aimed at those who have not previously studied freshwater ecology. The book includes an overview of scientific methodology, reviews of several key concepts and overviews of relevant case studies, with each chapter based around a key question in aquatic ecology. Content includes material on lakes and rivers.

Hauer, F.R. and Lamberti, G. (2006) *Methods in stream ecology*. Academic Press, London.
Confused by a crazy experiment you've read about in a journal or book? Unclear exactly how a piece of research has been

carried out? Check this book out. Each chapter has an overview of the scientific rationale and applications of most common methods used by stream ecologists, then provides more technical information on how the methods can be used and the resultant data analyzed.

Mason, C.F. (2002) *Biology of freshwater pollution*. Pearson Education, Harlow.
Detailed text covering many of the ways in which humans have altered freshwater ecosystems through pollution, habitat modification and introducing invasive species.

Moss, B. (2010) *Ecology of freshwaters: A view for the twenty-first century*. Wiley-Blackwell, Chichester.
A comprehensive book covering lakes and wetlands but including some chapters on streams and rivers. The book provides a detailed overview and is aimed at students who wish to gain an integrated view of freshwaters.

Wetzel, R.G. (2001) *Limnology: Lake and river ecosystems*. Academic Press, London.
This book offers an extremely detailed review of just about all aspects of lake ecosystems, with some comparative assessment of rivers for good measure. The material is more suited to advanced study rather than introductory overviews.

Further reading
Papers

Carpenter, S.R., Kitchell, J.F. and Hodgson, J.R. (1985) Cascading trophic interactions and lake productivity. *BioScience*, 35, 634–639.
Links between parts of lake ecosystems are explored in this paper.

Hutchinson, G.E. (1961) The paradox of the plankton. *American Naturalist*, 95, 137–145.
A classic paper which discusses why so many species of phytoplankton are found in lakes.

Junk, W.J., Bayley, P.B. and Sparks, R.E. (1989) The flood pulse concept in river–floodplain systems. *Canadian Special Publication of Fisheries and Aquatic Science*, 106, 110–127.
A classic paper outlining the importance of flood pulses for the existence and productivity of lowland rivers.

Nakano, S., Miyasaka, H. and Kuhara, N. (1999) Terrestrial-aquatic linkages: riparian arthropod inputs alter trophic cascades in a stream food web. *Ecology*, 80, 2435–2441.
A paper that reminds the reader about the importance of linkages in and out of the aquatic system with surrounding landscape ecological processes.

Resh, V.H., Brown, A.V., Covich, A.P. *et al.* (1988) The role of disturbance in stream ecology. *Journal of the North American Benthological Society*, 7, 433–455.
Natural and human disturbance impacts on stream ecosystems are illustrated.

Scheffer, M., Hosper, S.H., Meijer, M.L., Moss, B. and Jeppesen, E. (1993) Alternative equilibria in shallow lakes. *Trends in Ecology and Evolution*, 8, 275–279.
Steady states and thresholds of change in lake systems.

Schindler, D.W. (1974) Eutrophication and recovery in experimental lakes: implications for lake management. *Science*, 184, 897–899.
A classic paper on lake eutrophication of relevance to Box 21.4.

Vannote, R.L., Minshall, G.W., Cummins, K.W., Sedell, J.R. and Cushing, C.E. (1980) The river continuum concept. *Canadian Journal of Fisheries and Aquatic Sciences*, 37, 130–137.
This is the paper that proposed the highly influential concept in river ecosystem studies.

Ward, J.V., Tockner, K., Arscott, D.B. and Claret, C. (2002) Riverine landscape diversity. *Freshwater Biology*, 47, 517–540.
An excellent review of landscape features in river corridors, landscape evolution, ecological succession, connectivity and biodiversity.

PART VI

Environmental change

Figure PVI.1 A core taken from a peat bog can tell us a lot about environmental change over the past few thousand years. It contains within it a record of the plants and microscopic animals that lived on the site in the past from which we can infer changes in environmental conditions over time.

Scope

This final part deals with environmental change, its monitoring, modelling and management. The first two chapters tackle our understanding of global environmental change over the past 2.6 million years. This period is known as the Quaternary. The main part of this period is known as the Pleistocene epoch for which a range of evidence suggests that the Earth's climate has warmed and cooled several times resulting in massive expansion and contraction of the world's ice sheets and an associated rise and fall of global sea level. These natural environmental changes are driven by a range of processes, some of which are related to external factors such as Earth's changing distance from the Sun and others driven by internal factors associated with surface-ocean-atmosphere-biosphere interactions. Such natural changes have been observed over timescales of hundreds of thousands of years down to decadal. The evidence from history suggests that the Earth's climate system can sometimes change very rapidly. Studying the past can provide us with greater foresight and lessons for the future. Important and complex feedback mechanisms have been observed involving interactions between oceans, ice sheets, the biosphere and the atmosphere. The discovery of these interrelated adjustments helps explain more about present-day landforms, plant and animal distributions, and the climate system and how it may behave in the future. Chapter 22 therefore provides this context. The last 11 700 years of the Quaternary are known as the Holocene epoch and this is dealt with in Chapter 23. Evidence of environmental change during the Holocene is widespread because it is the most recent period in the Earth's history and therefore deposits containing evidence of change are easy to get to. Evidence from human archive material is also available. The Holocene is also the time in history when humans suddenly had a major impact on the landscape initially through farming, but also through population growth and later through the development of infrastructure.

Chapter 24 examines contemporary climate change and demonstrates how this is occurring at an unprecedented rate with higher atmospheric CO_2 concentrations than at any other time in the past 2.6 million years. It describes how human populations have changed the landscape and altered the concentration of natural and human-made chemicals in the atmosphere, land and waters of the Earth. Much of the change has been a result of human requirements for food and for fuel. Around 7 billion humans now consume or dominate 40-50% of the land's biological production. They have progressively attained the capacity to alter the Earth's physical systems (such as the carbon and nitrogen cycles) in drastic and potentially permanent ways. The chapter discusses predictions of global climate models and how we might mitigate against climate change. The chapter also

discusses the carbon cycle and the role of humans in altering the amounts of carbon stored in different parts of the cycle.

Chapter 25 deals with how climate change is driving changes in vegetation distributions and processes, but also establishes further feedbacks between vegetation and the environment. The chapter presents a series of case studies from different environments to provide evidence that vegetation is responding to human-induced climate change. Biodiversity loss and its importance are discussed in this chapter, which also shows how human modification of the world's biomes is having large impacts on climate and how integrated climate-vegetation modelling techniques are allowing us to predict what might happen in the future.

It is necessary to measure contemporary environmental change so that we can be aware of the global and local nature of such change. Many global and local-scale changes are difficult to measure from the ground and many areas remain inaccessible. However, types of remote sensing technology (such as satellite imagery) help provide quick, cost-effective and new methods of monitoring global change at all scales. Remote sensing now plays a pinnacle role in monitoring and detecting environmental change, such as alterations in vegetation cover and type, ice caps, global and oceanic temperatures, cloud cover and the ozone layer. The significance of this vital commodity will no doubt continue to grow, and for this reason it is important to understand the basic techniques involved. Chapter 26 gives a valuable overview of the methods employed.

A basic premise of environmental science is the desire to apply the knowledge we attain to help manage the environment. Elements of environmental change are described in every chapter in this book. In addition, environmental management issues are also discussed in every chapter in this book. Therefore it is necessary for the final chapter of this book to describe some of the general tools and pitfalls associated with managing environmental change that can be applied to all of the areas that physical geography touches. Arguably, the management of environmental change, species threatening as it is, is one of the most important aspects of science and management today. With improved physical geography it is often possible to foresee and perhaps to forestall the consequences of our actions. As demonstrated in the accompanying chapters of Part VI, change is not a single or simple concept: complex relationships exist between the factors that promote change and the nature of the change in terms of its direction, rate and reversibility. Accordingly, management of environmental change is not infallible and requires both scientific understanding and an ability to judge and make difficult decisions. The situation is further compounded by the economic and political imperatives that often take precedence. Nevertheless, Chapter 27 seeks to show how physical geography can be tailored to a form that is useful in supporting environmental management.

CHAPTER 22

The Pleistocene

Joseph Holden and Ian Lawson

School of Geography, University of Leeds

Learning objectives

After reading this chapter you should be able to:

➤ describe the major climate changes of the Pleistocene

➤ understand the orbital forcing hypothesis and a range of feedback mechanisms

➤ describe the main types of evidence for Pleistocene environmental change

➤ list the main dating techniques and explain issues surrounding correlation of different types of evidence from different locations

➤ evaluate the role of modelling in palaeoclimatic reconstruction

22.1 Introduction

The Pleistocene is the period of geological time which covers the period from 2.6 million years ago until 11 700 years ago. It forms the first **epoch** of the Quaternary (Figure 22.1). The onset of the Quaternary (and hence the Pleistocene) is defined by a change in the Earth's climate from a period of general stability (a very slow cooling over 50 million years with some minor fluctuations) to a period of great instability.

The Pleistocene epoch is characterized by frequent fluctuations of warming and cooling of the Earth's atmosphere and the growth and retreat of major ice sheets. Defining the exact time when the Quaternary period started is problematic because the change from the warmer earlier Tertiary period to the colder Quaternary was gradual. However, Gibbard and Head (2010) more precisely suggest the Quaternary began 2.588 million years ago. Figure 22.1 provides the subdivision of Earth history within which we can place the Quaternary and Pleistocene with approximate dates before present (BP). It should be noted that although 2.6 million years might seem a long time, in the context of Earth's history the Quaternary period is actually very short. Since we live in the Quaternary, we are actually living in the ice ages. Before the Quaternary the world was much warmer, although as Figure 22.1 shows, there were probably four other major cold periods in earlier Earth's history. The most recent of these was around 280 million years ago.

The second epoch of the Quaternary is known as the Holocene. The Holocene epoch is the past 11 700 years during which the Earth's climate has been relatively warm and the major ice sheets that covered large parts of the land surface retreated. The Holocene is called an **interglacial interval** as it follows a colder period with larger ice sheets (a **glacial interval**). The Holocene is dealt with in Chapter 23. Here we concentrate on the Pleistocene.

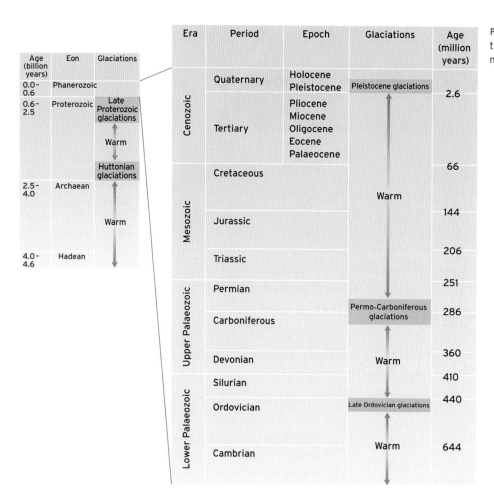

Figure 22.1 Earth's geological timescale. The Quaternary is the most recent period.

One of the major challenges that Pleistocene scientists face is in putting together the evidence for past environmental change. Many natural systems are affected by climate and so evidence of these past systems might give us information on past climate. There are many types of evidence for Pleistocene environmental change including the constituents of gas bubbles trapped in ice cores, ocean floor sediments and their organic and inorganic constituents, glacial deposits and erosional features, periglacial and glacio-eustatic features, wind-blown deposits (see Chapters 9, 14, 16 and 17), mineral deposits, pollen found in peat and lake deposits that have built up over time, and other plant and animal remains. There are two main challenges with these types of data: the first is in establishing how these types of evidence actually relate to a particular state of the environment; the second is in determining which period of time they relate to and how they fit into the global picture. This is particularly problematic because there are often no appropriate absolute dating methods that can be employed.

It is useful to understand the environmental changes that have taken place during the Pleistocene period so that we can explain more about present-day landforms and

plant and animal distributions. It also helps us to explain how the climate system is behaving today and how it is likely to behave in the future. Understanding Pleistocene environmental change allows us to place present-day climate changes into a much longer-term context so that we can judge their importance and potential implications. For example, we know that carbon dioxide (CO_2) concentrations have fluctuated between high and low in the past. We have also learnt that current post-industrial CO_2 concentrations are already much greater than any known during the past 2.6 million years (see Chapter 24). By studying the past we can see how the Earth–ocean–atmosphere system behaved under naturally high CO_2 conditions and learn how the system might change if humans continue to increase atmospheric CO_2 concentrations. As this chapter will show, there are important and complex feedback mechanisms that must be taken into account. While some climatic changes can take thousands of years to take place, others can be sudden and occur over just decades. It is therefore crucial that we understand Pleistocene environmental change as there are major implications for our present-day world.

22.2 Long-term cycles, astronomical forcing and feedback mechanisms

The gradual cooling during the Cenozoic (Figure 22.1) may have been related to the movement of the Earth's plates which resulted in changing ocean and atmospheric circulation patterns through continental drift and mountain building. Plate tectonics (see Chapter 2) created the suitable conditions for the beginning of the ice ages by positioning Antarctica in a thermally isolated position over the South Pole. The northern hemisphere continents were also huddled around the Arctic Ocean. This then allowed other factors to become important including an important external driving factor that helps partly explain the glacial cycles that have occurred during the Pleistocene. This is known as orbital forcing or the **Milankovitch theory**.

22.2.1 Orbital forcing theory

From 1912, Milutin Milankovitch, a mathematician, followed up the nineteenth-century work of astronomers Joseph Adhemar and James Croll by computing the radiation received at the top of the Earth's atmosphere over time as the Earth's orbit varied. The basis of the Milankovitch theory is that the amount of energy reaching different parts of the Earth from the Sun varies as the shape of Earth's orbit around the Sun, the angle of tilt of the Earth's axis, and the direction that the axis of rotation points change over time. These three parameters, illustrated in Figure 22.2, vary in a regular and predictable way to determine the amount and distribution of solar radiation at the Earth's surface.

The first variable orbital parameter is called the **eccentricity of orbit**, which is where gravitational forces cause the shape of the Earth's orbit around the Sun to change from almost circular to more elliptical and back again over a period of 100 000 years. As the eccentricity increases, the difference in the Earth's distance from the Sun at the orbit's closest and furthest points also increases. This affects the severity of the seasons. If winter (in the northern hemisphere) occurs when the distance between the Earth and the Sun is at its maximum, the winter season will be more severe. The greater the eccentricity of the orbit (the more elliptical it is), the stronger this effect can be. The second orbital factor is the **tilt of the Earth**. The tilt of the Earth's axis varies from approximately 21° to 24° and back over a period of 41 000 years. The greater the tilt, the more intense the seasons in both hemispheres become; summers get hotter and winters colder. The third variable is axial

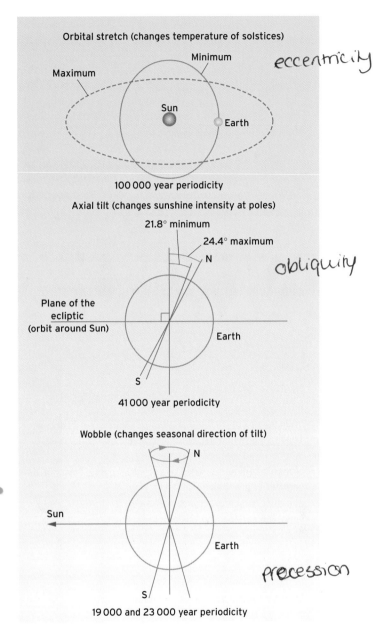

Figure 22.2 Milankovitch orbital forcing mechanisms.

'wobble', which causes a phenomenon called **precession of the equinoxes**. The gravitational pull exerted by the Sun and the Moon cause the Earth to wobble on its axis like a spinning top, where the end of the axis describes a circle in space and determines where in the orbit the seasons occur, and most importantly the season when the Earth is closest to the Sun. This happens over two cycles of 19 000 and 23 000 years. This factor governs the interplay between the first two factors since precession determines whether summer in a given hemisphere occurs at a near or distant point in the orbit around the Sun. For example, at the present time the Earth reaches its furthest point from the Sun during the southern hemisphere winter. Therefore southern hemisphere winters are slightly colder than northern hemisphere winters. Southern hemisphere summers are also slightly warmer than in the northern hemisphere.

22.2.2 Evidence that orbital forcing causes climate change

In the 1950s the first continuous record of the Pleistocene came from marine sediment cores, when Emiliani (1955) analyzed the isotopic composition of fossil calcium carbonate skeletons of small, single-celled marine organisms called

foraminifera. This record consisted of a long core of ocean sediment that had been deposited over time. The sediment on the ocean floor builds up slowly as the remains of marine creatures sink to the bottom when they die. The nature of the oxygen isotope ratios from the skeletons of foraminifera taken from different layers of the core can tell us how much of the world's water was locked up in glaciers. The expanding ice sheets result in a fall of global volumetric sea level (this type of sea-level change is often called eustasy). The chemistry of the ocean water changes depending on how much water is left in the oceans compared with that stored as ice. The mechanism is explained in Box 22.1.

FUNDAMENTAL PRINCIPLES

EVIDENCE FROM OCEAN CORES

On the deep ocean floor, sediments have been accumulating relatively undisturbed for millions of years. These sediments are a mixture of land-derived material and marine biogenic sediments. The terrestrial sediments are mainly sand, silt, clay and dust, which reach the ocean via wind, **ice rafting** and fluvial inputs. The biogenic sediment is composed of calcareous and siliceous skeletal remains of microorganisms that lived in the ocean waters. Figure 22.3 shows a typical section of sediment in a core being analyzed while Figure 22.4 illustrates the remains of planktonic foraminifera which are surface-dwelling organisms and are indicators of the prevalent sea surface temperature at the time. Their shells are coiled left in temperatures below 7°C and right in temperatures greater than 7°C. The different species that make up foraminiferal assemblages also indicate the temperature, salinity and nutrient availability in the oceans.

Figure 22.3 Evidence from sediments deposited over time being investigated in this part of a core. (Source: photo courtesy of Alice Milner)

BOX 22.1 ➤

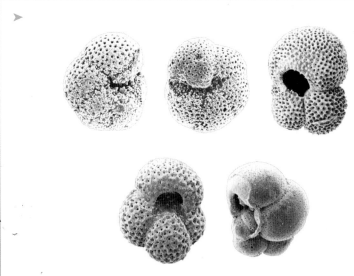

Figure 22.4 Fossil shells of foraminifera. Species assemblages can tell us about sea temperatures and productivity and the chemistry of their shells can tell us about the chemistry of the ocean water when the microorganisms were alive. (Source: Thomas M. Gibson/USGS)

heavier isotope is also precipitated out more easily before it can get there). Therefore, during glacial periods the proportion of ^{16}O to the heavier ^{18}O isotope in the oceans decreases. During interglacials when the water is returned to the oceans the proportion of ^{16}O increases. Therefore, the ratio of ^{18}O to ^{16}O in the ocean sediment provides a long-term proxy for global ice volume during the Pleistocene. This record has been divided up into stages, each of which corresponds to either a glacial or interglacial period (Figure 22.5; Figure 22.6e). Odd numbers represent warm periods and even numbers represent cold periods. More minor climatic events (stadials and interstadials) are sometimes given a letter coding or decimal place (e.g. 5e = 5.5) after a number (Figure 22.5).

Shackleton and Opdyke (1973) formalized the use of oxygen isotope stratigraphy to indicate changes in global sea level. Oxygen isotope analysis can be performed on foraminifera, and is typically done on bottom-dwelling (**benthic**) species. Their skeletons contain some of the chemical constituents of the seawater when they were formed. During glacial periods more of the world's water is locked up in ice sheets and less is in the ocean. The lighter isotope of oxygen which is contained in water (H_2O) is evaporated more easily and more readily reaches polar regions to be stored as ice (the

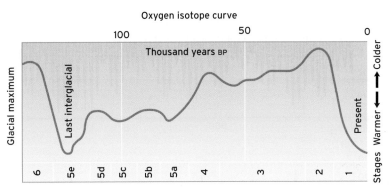

Figure 22.5 The oxygen isotope curve for the last glacial and interglacial period. The curve acts as a proxy for sea-level change and ice build-up. Note that the ice seems to build up slowly and then very quickly melt away.

BOX 3.1

The results from ocean core work showed that there were many fluctuations in the Earth's climate during the Pleistocene that caused the growth and retreat of ice sheets. These results seemed partly to match the predictions provided by the Milankovitch model as shown in Figure 22.6 (Broecker and Denton, 1990). Figure 22.6(a), (b) and (c) shows the pattern of eccentricity, tilt and precession during the past 1.6 million years. Figure 22.6(d) shows the cumulative effect of the three forcing mechanisms on radiation received at 65°N in June. Figure 22.6(e) shows the volume of the Earth's ice sheets determined from foraminiferal oxygen isotope ratios in ocean sediments for the same period. Significantly, there is an approximate match between Figure 22.6(d) and (e). For example, there are eight large glacial build-ups over the past 800 000 years on an approximately 100 000 year cycle, each coinciding with minimum eccentricity. Smaller decreases or surges in ice volume have come at intervals of approximately 23 000 years and 41 000 years in keeping with the precession and tilt frequencies.

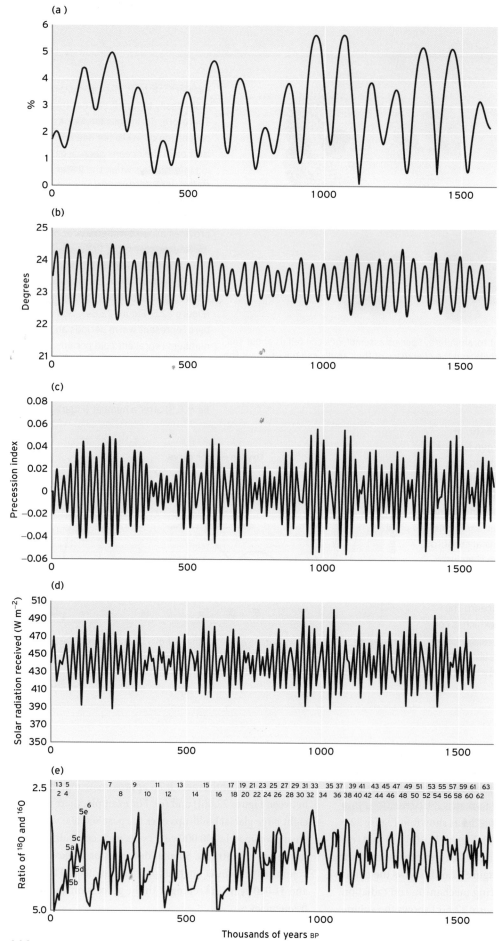

Figure 22.6 Orbital forcing and global ice volume for the past 1.6 million years with calculated effect of orbital forcing parameters on solar radiation received at 65°N: (a) eccentricity; (b) tilt; (c) precesion; (d) cumulative effect of all orbital forcing parameters at 65°N during June; (e) the actual oxygen isotope record of global ice volume. (Source: (e) after Bridgland, 1994. Compiled from data published in Ruddiman *et al.*, 1989)

Other evidence for the length of the glacial cycles comes from the record of sea-level change provided, for example, by coral reefs in Barbados or Papua New Guinea (Chappell and Shackleton, 1986). Since water is stored on the continents as ice during glacials, and is released during interglacials when ice sheets melt, sea levels rise and fall by over 100 metres during climatic cycles. During the last glacial period (the peak of which was reached only 19 000 years ago) it was possible to walk from present-day England to mainland Europe, as the North Sea and English Channel were dry land. Coral growth is linked to sea level and different growth periods for the coral can be dated up to about 160 000 years ago. This dating can be done by determining the ratio of uranium to thorium in the coral. Newly formed coral is rich in uranium whereas old coral has less uranium and more thorium (see below). Again the evidence seems to match that from the ocean sediments and the predictions of the Milankovitch cycles as to the timing of glacial cycles (e.g. Chappell and Shackleton, 1986). High sea-level events coincide with the interglacial intervals, and also with the predicted timing of these from Milankovitch calculations.

Today it is recognized that there have been dozens of alternating cold and warm phases during the Pleistocene. The cold phases are called glacial intervals and the warm phases are known as interglacial intervals. During glacial periods ice sheets and glaciers extended to lower latitudes from the poles. Less intense and shorter variations in global temperature that occur during these main cold and warm periods are called **stadial** (cold) intervals and **interstadial** (warm) intervals.

22.2.3 Problems with orbital forcing theory

There are a range of problems with orbital forcing theory and the use of marine oxygen isotope records. Firstly, it has recently emerged that benthic foraminifera oxygen isotope records (Box 22.1) do not strictly represent a pure sea-level record and hence a record of ice advance and retreat (Skinner and Shackleton, 2005, 2006; Shackleton, 2006). The oxygen isotope record is also affected by changes in deep ocean temperature and this signal can vary between oceans as the local hydrology of the oceans reacts differently to changes in atmospheric temperature and ice growth or retreat. For example, Skinner and Shackleton (2005) showed that the isotope record for the end of the last glacial appears to be about 2200 years earlier than the likely change in ice volume, which is further 1700 years earlier than the deep Pacific isotope record. This offset in timing creates a problem because it is now no longer possible to simply assume that all sediments showing a change from cold to warm from different oceans are the same age.

Secondly, the expected changes in temperature caused by changes in insolation via orbital forcing are too small to explain the large temperature changes required for the vast ice expansions and retreats recorded. In fact there appears to be a 4–6°C shortfall. So something other than orbital changes must have also been acting to cause changes in Pleistocene environments.

Thirdly, the 100 000 year cycle, according to orbital calculations, should have a much weaker effect on incoming solar radiation than the shorter cycles. However, as Figure 22.6(e) shows, the 100 000 year cycle is dominant at least from about 800 000 years ago. Finally, the Milankovitch cycles should show a smooth rise and fall as in Figure 22.6(e) whereas the ice curve in Figure 22.6(e) is sawtooth in pattern. Over tens of thousands of years ice sheets built up several kilometres thick and scoured and scarred the landscape as far south as central Europe and Midwestern United States (e.g. Figure 22.7). But each cycle ended abruptly. Within a few thousand years the ice sheets melted back to present-day patterns.

So while orbital forcing mechanisms seem to be a good 'pacemaker' of Pleistocene environmental change (Imbrie and Imbrie, 1979) it is necessary to look at other processes too. The next section details how internal feedback mechanisms may help explain some of the patterns that orbital forcing theory alone cannot account for.

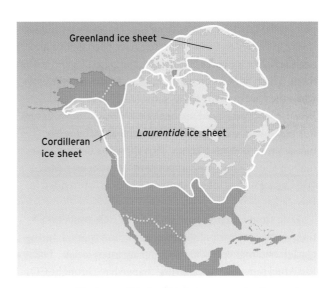

Figure 22.7 The Laurentide ice sheet grew very slowly over tens of thousands of years and then melted within a few thousand years.

22.2.4 Internal feedback mechanisms

22.2.4.1 Albedo and sea-level change

Once ice sheets start to form, the albedo of the Earth's surface increases because ice is highly reflective of the Sun's energy. This increased albedo results in a further drop in temperatures, allowing ice sheets to expand further. A further positive feedback mechanism comes from the drop in sea level induced by ice sheet growth. This would make it easier for ice to flow out from the land further onto the continental shelves. Therefore the ice sheets can expand even further, and albedo would be greater, allowing the Earth to cool further. This might help explain some of the extra global cooling we have seen during the Pleistocene that could not be explained by the changes in insolation predicted by the Milankovitch theory alone. However, general circulation models (GCMs) suggest that although albedo and sea-level change may be important they still would not account for the full magnitude of cooling seen during the Pleistocene glacial periods (Broecker and Denton, 1990).

There are two potentially more important internal feedback processes that have been the focus of research since the 1990s: these relate to the combined effect of (i) changes in the ocean circulation and atmospheric CO_2 concentrations and (ii) the nature of ice sheet dynamics.

22.2.4.2 The missing CO_2 link: oceans and ice sheet dynamics

The centres of some very large ice sheets do not melt even during interglacials (that is why we still have ice over the poles today). Long ice cores from the centre of very old ice sheets in Greenland and Antarctica have been drilled and analyzed. The ice cores contain bubbles of gas locked within them. These bubbles are representative of the air contents when the ice originally fell as snow (Alley *et al.*, 1993). Because ice layers build up during each year's snow season, it is possible to calculate how old the air within the ice bubbles is by counting layers. This allows us to gain information about global air composition from thousands of years ago. Further information on ice cores is provided in Box 22.2.

Data from ice cores have shown that CO_2 concentrations were very low (~180 parts per million (ppm)) during glacials and higher (~280 ppm) during interglacials. Today fossil fuel burning and land-use change have brought CO_2 concentrations above 380 ppm (see Chapter 24). CO_2 is an important greenhouse gas and therefore would be expected partly to control the temperature of the Earth. However, this poses an interesting problem: why should CO_2 levels rise and fall during the Pleistocene?

The oceans hold around 60 times as much CO_2 as the atmosphere. The gas readily diffuses between the ocean surface and the atmosphere. Therefore, its concentration in surface waters regulates the atmospheric concentration. Changes in CO_2 concentration in the ocean surface waters may affect atmospheric concentrations, which may then cause the Earth's climate to warm or cool. Interestingly, however, the CO_2 concentration changes appear to lag behind the temperature changes.

While there are many ocean currents, often driven by surface trade winds (see Chapter 3), there is one very important deep ocean current. This is known as the thermohaline circulation system. The current is driven by temperature gradients (thermo) and salt concentration gradients (haline). This ocean circulation is described in detail in Chapter 3, which should be referred to in order to comprehend fully the remainder of this section. This strong deep current acts as a pump that can transfer CO_2 and nutrients from the surface of the oceans to the deeper waters and return them to the surface again. Today there are sensitive zones where such downwelling and upwelling occur (see Chapter 3).

It is, however, living things in the oceans that partly control the concentration of CO_2 in the surface waters (Broecker and Denton, 1990). Tiny green plants (plankton) in the upper sunlit layers of the ocean take up CO_2 from the water as part of photosynthesis and this helps to form the plant tissue itself. When the plants die, their debris falls to the bottom of the ocean where bacteria oxidize it back to CO_2. Today the thermohaline circulation system allows CO_2 from the bottom of the oceans to be stirred up and taken back to the surface again. However, when the circulation slows, deep carbon stores are not circulated back to the surface as quickly. Therefore, when the plants in the ocean waters die and fall to the bottom of the ocean, taking their absorbed CO_2 with them, this CO_2 is not returned to the surface in as great a quantity as it is being sent down into the deep. Therefore when the thermohaline circulation system slows down the surface waters of the oceans have less CO_2 (since less is pumped back to the surface by currents) and therefore less is returned from the water into the atmosphere. Hence, atmospheric concentrations of CO_2 fall as the ocean plants continue photosynthesis. This results in global cooling. Evidence for changes in ocean circulation has come from faunal and chemical studies of deep-sea sediments. These have indicated that the production of deep water in the Atlantic, the driving force of present-day circulation systems (see Chapter 3), was reduced greatly during past glacials.

TECHNIQUES

THE SEARCH FOR THE OLDEST ICE ON EARTH

Ice builds up incrementally on the surface of ice sheets. Typically the winter layers of ice are light in colour while summer layers are darker because the partial melting and lower summer accumulation rate produce higher concentrations of impurities. This layering and X-ray analysis allow each seasonal and annual layer of ice to be identified. Figure 22.8 shows an ice core being collected. It contains annual increments that will allow a relative chronology of the ice to be developed. Each of the layers can then be analyzed for their chemical properties. Major ice drilling programmes have been carried out in Greenland, the Canadian Arctic and the Antarctic as well at sites in Peru and Tibet. The longest cores are over 3 km deep.

Ice cores contain an abundance of highly detailed climate information. There are three main approaches to climatic reconstruction based on ice cores. These are analysis of: (i) oxygen isotopes; (ii) gases from air bubbles trapped in the ice; and (iii) dissolved and particulate matter in the ice.

- *Isotopic analysis:* this is partly based on the same approach as discussed in Box 22.1 for ocean sediments. The isotopic composition of the ice is partly controlled by the isotopic composition of seawater. In addition, atmospheric temperature controls the isotopic ratio in the precipitation and thus the relative amount of the heavier ^{18}O isotope reveals the temperature at the site when the snow fell.

Figure 22.8 Scientists collecting an ice core. (Source: Getty Images: Fred Hirschmann)

- *Gas analysis:* the mixing time for atmospheric gases over Earth is around one to two years, meaning that changes are rapidly diffused throughout the lower atmosphere. Bubbles trapped in the ice contain records of past atmospheric composition. It is possible, by analyzing the bubbles, to determine how the past concentrations of CO_2, methane and other gases have varied over time. The ice cores have revealed dramatic increases in CO_2, methane and nitrous oxide during interglacial episodes and decreases during glacial episodes. There are also many shorter-term fluctuations in gas concentrations over millennial and even centennial timescales (e.g. Luthi *et al.*, 2008). However, a fundamental problem is that air in ice bubbles is always younger than the age of the surrounding ice. As snow becomes buried and transforms to firn and then ice, the air between the snow crystals remains in contact with the

atmosphere until the pores of air become fully sealed upon ice formation. The sealed bubbles therefore contain air that is representative of the conditions long after snow deposition. This needs to be taken into account when producing high-resolution records.

- *Chemical content:* the presence of dust and trace chemicals in the ice can be determined. Evidence for volcanic episodes or periods of increased aridity can be traced. The types of dust can also be analyzed to determine their sources and hence prevailing atmospheric circulation patterns.

The oldest ice core that scientists have sampled extends to 800 000 years ago, but models suggest that there should be ice that is 1.5 million years old in parts of the East Antarctic ice sheet. This could yield excellent data on climate change over that period. However, finding ice that is so old requires better mapping and

BOX 22.2 ➤

Changes in the strength of the ocean circulation system changed the energy transfers between the equator and poles. In the North Atlantic, for example, a reduction in the strength of the thermohaline system today would cause western Europe to cool by several degrees changing local climates quite dramatically (Paillard, 2001). Figure 22.9 shows predicted changes in mean annual temperature 30 years after a collapse of the thermohaline circulation.

An explanation is needed for why the ocean circulation strength should change in the first place. One of the answers might lie in the formation of large ice sheets. Experiments with GCMs suggested that the topographical effects of ice sheets could explain a lot of the extra cooling not accounted for by the Milankovitch mechanism alone. A small amount of cooling caused by orbital

forcing can cause the growth of some ice sheets. These thick ice sheets (several kilometres) could change local air currents that were deflected around the ice domes which might contribute further to cooling and ice sheet growth. Changing air currents may result in changes to ocean currents. For example, MacAyeal (1993) argued that the topography of the huge ice sheets altered the North Atlantic trade winds. This, in combination with a cooler climate, reduced the evaporation in sensitive areas of deep-water formation in the North Atlantic, thereby reducing the strength of the thermohaline circulation system. This occurred because evaporation controls the saltiness of the water left behind; increased evaporation leads to increased salinity. Saltier water is denser and will sink, allowing the circulation 'pump' to remain strong. A reduction in evaporation would result in a slowing of

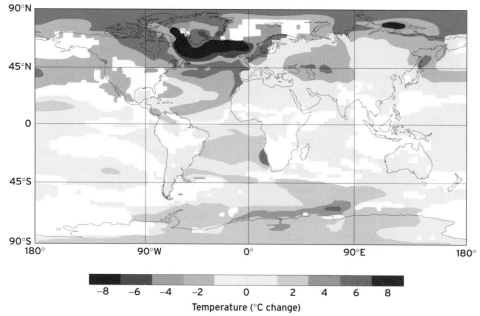

Figure 22.9 Predicted change in annual temperature 30 years after a collapse of the thermohaline circulation. (Source: from Cramer *et al.*, 2001)

the thermohaline circulation as there would no longer be intense downwelling of dense salty water. Return of CO_2 to the atmosphere would slow, since the upwelling amounts are reduced. Hence the northern hemisphere ice sheets could cause changes to climate worldwide, allowing southern hemisphere ice sheets to advance and global cooling to be greater than if simply caused by orbital forcing alone.

One problem with this theory is that extra ice formation at the poles and particularly in the Antarctic Ocean would cause more downwelling. This is because as ice forms it excludes the salt, allowing the remaining ocean water to become denser and to sink as a strong downwelling pump. This would increase the strength of the circulation system and lead to a negative feedback as a strong thermohaline circulation brings heat from the equator to high latitudes. Furthermore, a colder ocean would mean lower biological productivity and thus more CO_2 in the atmosphere. A colder climate will also reduce the amount of precipitation available to supply glaciers. Again these are negative feedbacks which suggest that there may be some self-regulation to the Earth's climate system. However, it seems that the North Atlantic is the crucial and most sensitive part of the entire system.

The ocean–atmosphere circulation system is extremely important but very complex and as yet we do not fully understand all the processes involved. In addition to long-term Quaternary changes, the thermohaline circulation system may play a role in short-term climate changes including those that may occur in the very near future. These processes are discussed further below.

22.3 Short-term cycles

22.3.1 Glacial instability

Ocean cores provide long continuous records of climate change but because sedimentation rates are so slow they are of lower temporal resolution than ice cores. Ice cores provide high-resolution data through annual layering of snow accumulation (Box 22.2). However, the ice records do not extend back very far into the Pleistocene. The most important evidence to emerge from the ice records is that Pleistocene climate changes were often very rapid with significant warm interstadial episodes and cool stadial episodes occurring just a few centuries apart. Glacials appear to have been very unstable with several short warming and cooling episodes. Interglacials, however, were more stable with some millennial-scale variability but this was more subdued than during glacials.

Some 20 interstadial events have been identified in the Greenland ice cores during the period 80–20 000 years BP during which temperatures fluctuated by 5–8°C. These events are known as **Dansgaard-Oeschger (D-O) events** and last for no more than 500–2000 years. D-O events are characterized by abrupt changes in temperature (gradual cooling followed by abrupt warming), dust content, ice accumulation rate, methane concentrations and CO_2 concentrations (Broecker, 1994). The abruptness is of the order of a few decades. Bond *et al.* (1993) grouped these D-O events together into larger cycles which contain a long cooling trend followed by an abrupt warming (Figure 22.10). Ocean sediments from the North Atlantic revealed that at the coldest part of these longer **Bond cycles** vast discharges of icebergs floated

Reflective questions

➤ Can you explain Milankovitch orbital forcing theory?

➤ How do the predictions of the Milankovitch cycles differ from the evidence for climate change during the Pleistocene?

➤ What are the main positive and negative feedbacks if solar radiation decreases?

➤ What are the main positive and negative feedbacks if solar radiation increases?

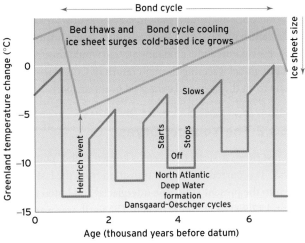

Figure 22.10 Schematic representation of the relationship between Dansgaard-Oeschger events, Bond cycles and Heinrich events.

southwards across the North Atlantic from the North American and European ice sheets. These releases of ice are known as **Heinrich events**. Heinrich events were immediately followed by abrupt warming. Evidence for Heinrich events comes from the fact that some distinctive layers of sediments taken from North Atlantic sediment cores contain coarse angular sand grains that have been brought by icebergs from Canada and Europe. They are named after Helmut Heinrich who described and numbered them and explained how the layers came to be there (Heinrich, 1988).

Several scientists (e.g. MacAyeal, 1993; Hunt and Malin, 1998) have proposed that Heinrich events were caused by inherent instabilities of large ice sheets. The great Laurentide ice sheet grew over North America so that it became very thick during glacials. This meant that under the sheer mass of the ice the sediments beneath could weaken and meltwater would be produced in greater quantities at the ice base owing to the extra pressure. This ice loading and failure of the substrate may then have caused parts of the ice sheet to flow much more quickly. This would result in the release of large quantities of icebergs from the Canadian coast. The icebergs would carry with them sediment from the Canadian land mass that had been plucked and scoured by the ice and then as the iceberg melted the sediment would be deposited on the ocean floor. Once the excess ice has been released, the ice sheet may stabilize again and start to grow once more; the ice rafting would cease. This mechanism is known as the **binge-purge model** of ice sheet development.

Ice rafting events in the North Atlantic caused climatic changes much further afield. MacAyeal (1993) proposed that the decrease in ice sheet volume changed the wind action around the ice sheet. This change returned the wind to its pre-glacial formation with enhanced sea surface evaporation. This would result in enhanced downwelling and a return of the strong thermohaline circulation system. This reinvigorated thermohaline system would then result in rapid global warming through return of CO_2 to surface waters from the deep and enhanced heat transport.

22.3.2 The Younger Dryas

The last short, major, cold event is known as the 'Younger Dryas'. It occurred during the transition from the last glacial into the present Holocene interglacial and lasted from around 12 900 to 11 700 years ago. The Earth was warming from about 19 000 years ago. This warming trend was interrupted by several cold reversals, the most pronounced of which was the Younger Dryas. A suggested explanation is that the meltwater of retreating continental ice masses was

released into the sensitive parts of the North Atlantic (from the melting North American ice sheet) where it substantially reduced the density of the ocean surface water (Broecker and Denton, 1990). As the ice sheets melted, a switching of drainage of the Laurentide ice sheet from the Caribbean towards the North Atlantic, via the Gulf of St Lawrence, led to an input of fresh meltwater to surface ocean layers. Being fresh (less dense than seawater) this meltwater input slowed the downwelling of water in the North Atlantic and thereby slowed down the deep-water formation. The result of this was to slow down the thermohaline ocean circulation system which had carried warm tropical waters to the north. Without this source of heat Europe and North America began to cool again (Figure 22.9) and the ice sheets started to re-advance. Many of the glacial landscapes that we can see in northern Europe today are those created by that sudden cooling and re-advance of glaciers. This series of events demonstrates that climatic feedback effects can be strongly non-linear; global warming led to a sudden cooling.

> ### Reflective questions
>
> ➤ Why might glacial periods be more unstable than interglacial periods?
>
> ➤ Why do results from Younger Dryas research have implications for predictions of the future impacts of current global warming caused by humans?

22.4 Further evidence for environmental change

Evidence from deep-sea and ice core sediments has already been discussed in this chapter but a range of other types of evidence can be used to reconstruct Pleistocene and Holocene environments. Table 22.1 summarizes the main evidence types used. The value of using multiple lines of evidence lies in the richness of the information that they bring. For example, if data from one site are mapped with data from other locations then a regional synthesis can be produced, providing greater insight into palaeoclimate and former circulation patterns. Local environmental gradients and information on precipitation and aridity, for example, can be reconstructed. This information helps us understand more about the processes involved in environmental change and shows us which factors need to be manipulated in models predicting future global environmental change. Multiproxy studies provide a diversity of information, spatial coverage,

Table 22.1 Main sources of data used to reconstruct Quaternary environments

Proxy data source	Variable measured	Proxy data source	Variable measured
Geology and geomorphology – continental:		Biology and biogeography – continental:	
Relict soils	Soil types	Tree rings	Ring-width anomaly, density
Closed-basin lakes	Lake level		Isotopic composition
Lake sediments	Varve thickness	Fossil pollen and spores	Type, relative abundance and/or absolute concentration
Aeolian deposits – loess, desert dust, dunes, sand plains	Composition, depth, layering	Plant macrofossils	Age, distribution
Lacustrine deposits and erosional features	Age	Plant microfossils	Age, distribution
Evaporites, tufas	Stable isotope composition	Vertebrate fossils	Age, distribution
Speleothems	Composition, layering, dated layers	Invertebrate fossils: mollusca, ostracods	Age, distribution
Geology and geomorphology – marine:		Diatoms	Type, assemblage abundance
Ocean sediments	Ash and sand accumulation rates	Insects	Refuges
	Fossil plankton composition	Modern population distributions	Relict populations of plants and animals
	Isotopic composition of planktonic and benthic fossils	Biology and biogeography – marine:	
Continental dust	Mineralogical composition and surface texture	Diatoms	Age, distribution, geochemistry
Biogenic dust: pollen, diatoms, phytoliths	Geochemistry	Foraminifera	Faunal and floral abundance
Marine shorelines	Coastal features	Coral reefs	Morphological variations
Fluviatile inputs	Reef growth	Archaeology:	
Glaciology:		Written records	
Mountain glaciers, ice sheets	Terminal positions	Plant remains	Age, distribution
Glacial deposits and features of glacial erosion	Composition, orientation, shape	Animal remains, including hominids	Age, distribution
Periglacial features	Composition	Rock art	Age, distribution, content
Glacio-eustatic features	Shorelines	Hearths, dwellings, workshops	Age, distribution, geochemistry
Layered ice cores	Oxygen isotope concentration	Artefacts: bone, stone, wood, shell, leather	Composition, age, distribution
	Physical properties (e.g. ice fabric)		
	Trace element and microparticle concentrations		

(Source: from Williams *et al.*, 1998)

temporal coverage, a diversity of environments recorded and a completeness of records. They enable us to obtain information about a range of different environments and within each environmental setting they enable us to obtain a record of different aspects of that environment. A selection of evidence types and their uses is discussed in the following sections and in Chapters 9, 14, 16, 17, 23, 24 and 25.

22.4.1 Landforms

Over 20 major variations in climate are recognized in the undisturbed records of the deep oceans, but the terrestrial (land-based) record of climatic change preserves far fewer, because processes of erosion make the record much more fragmentary. For example, the last glacial period allowed ice sheets to expand across much of northern Europe, eroding earlier sedimentary records of Pleistocene landscapes, plant and animal life. Nevertheless there are still many geomorphological features providing ample evidence for climatic change. These include landforms of former glacial erosion and deposition, former periglacial landforms, river terraces, cave deposits and wind-blown sediments. Data can be obtained from the sediments themselves by relating observations of present depositional environments to features present in the stratigraphic record. Furthermore, since many deposits contain fossils, inferences can be made about the type of environment and climate experienced at that point in the sedimentary sequence.

22.4.1.1 Glacial and periglacial landforms

Sedimentary evidence for cold climates includes glacial tills, fluvioglacial sands and gravels, and glaciolacustrine (from water bodies associated with glaciers) silts and clays. Such proxy evidence can be augmented by other landform evidence, including, for example, moraines, eskers, kames, drumlin fields and meltwater channels and periglacial landforms (see Chapters 16 and 17). The unvegetated landscape of periglacial areas is often subject to intense wind erosion and the abundant supply of fine-grained sediment often leads to the deposition of significant thicknesses of fine-grained silt (termed **loess**) and sand deposits (termed **coversands**). Erosional features such as U-shaped valleys and rock **striations** (Figure 22.11) can also be used to infer former landscape processes.

Mapping the spatial distribution of these sediments and landforms (by both remote sensing techniques (see Chapter 26) and ground mapping) allows the reconstruction of glacial limits (e.g. Clarke *et al.*, 2000). Figure 22.12 shows an example of geomorphological mapping in northern Europe,

Figure 22.11 Parallel grooves scoured into a rock indicating the former direction of ice flow. These grooves are known as striations.

allowing ice sheets to be reconstructed through analysis of landforms. The direction of ice movements can also be determined through analysis of morphological features such as striations and moraines. This type of mapping work, however, requires understanding about the genesis of different landforms and the complexity of such landscapes and is often subjective. Therefore different people may produce different reconstruction maps depending on the way they view landscape features.

Technological advances now allow maps to be produced that can increase insights into past landscapes through improved visualization techniques. For example, a detailed glacial landform map of Britain can be downloaded from: www.shef.ac.uk/geography/staff/clark_chris/britice. The map is available with information in different layers and formats and has been produced in digital format so that all of the information sources to the 20 000 features mapped can be traced back to the original investigations or publications. The main limitation of the map is a lack of consistency and reliability of some of the features since the data sources span 150 years of scientific effort. Some moraines are shown for an area around York in northern England in

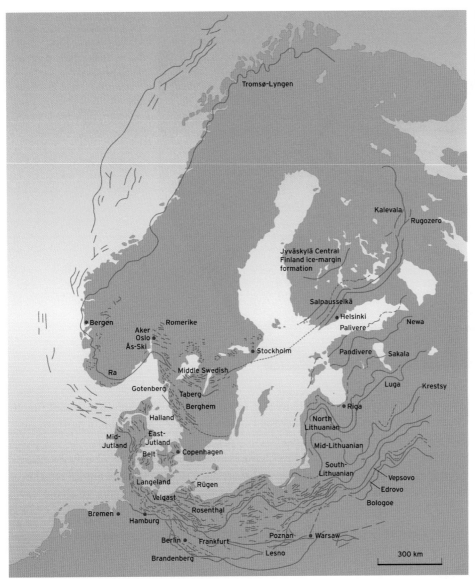

Figure 22.12 Ice sheet reconstruction on the basis of morphological mapping over Scandinavia during the last glacial maximum. (Source: after Lowe and Walker, 1997)

Figure 22.13 and have been mapped showing the underlying topography. It is also possible to use modern computer technology to 'fly through the landscape' and view mapped glacial landforms from different angles and this may help gain insights into landform processes.

22.4.1.2 River terraces

River valleys throughout the world contain evidence of past climates. River terraces often exist along valley sides (Figure 22.14) or on the floodplain and sometimes exist in vertical steps such as on the River Meuse (Figure 22.15). Terrace sequences reflect both incision and lateral migration of a river channel. Often the highest terrace is the oldest, and lower terraces are younger. Sometimes the sequence is more complex than this where, for example,

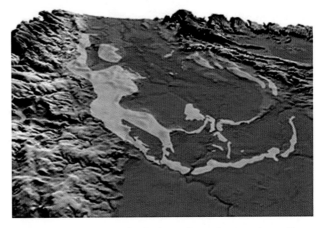

Figure 22.13 Some moraine features (brown) mapped over the landscape topography near York in northern England. (Source: Chris Clark)

Figure 22.14 River terrace series formed by river incision on the Pahoemeroi River, Idaho, USA. (Source: Ken Hamblin)

older terraces have become buried by younger alluvial deposits. Terraces can develop through a range of processes including: changes in sea level where a fall in sea level causes a river to incise further; changes in precipitation regimes and vegetation cover resulting in changes to water and sediment supply; tectonic processes; and human activity (e.g. forest clearance can result in increased runoff and erosion and increased sedimentation in downstream areas).

River terraces are of great interest because they contain sediments that are often rich in plant and animal remains. Often the nature of the environment under which the

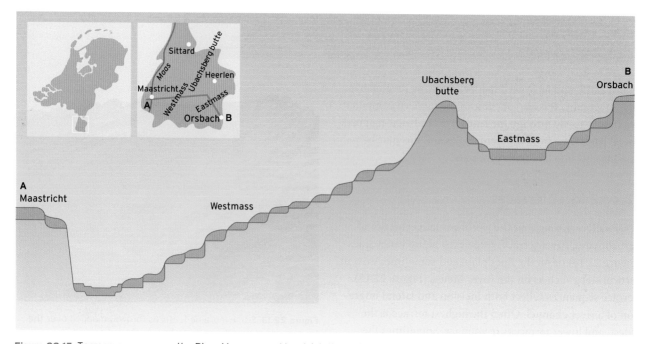

Figure 22.15 Terrace sequence on the River Meuse, near Maastrict, the Netherlands. (Source: after Ruegg, 1994)

terraces evolved can be determined from the deposits. The terraces of the River Thames in England are some of the most studied in the world and fossil assemblages have allowed glacial, interstadial and interglacial sequences to be determined and correlated with terrace sequences throughout lowland England (Gibbard, 1994; Bridgland, 1994). This allows a comprehensive picture of local environmental change to be built up involving changes in vegetation and animal communities through time.

22.4.1.3 Cave deposits

Cave systems and **rock shelters** (shallow niches in the hillside) can also yield useful information about palaeoclimates. Caves form excellent traps for sediment as they are protected from surface weathering and erosion. Cave detritus can often contain skeletal parts of animals and many caves and rock shelters were occupied by humans resulting in rich archaeological deposits. For example, the rock shelters of northern Greece have yielded evidence of human tools and debris dating back to 50 000 years ago. Interspersed with these human artefacts are sediments of faunal, floral, fluvial and mineralogical processes providing

rich detail on local palaeoenvironments (Woodward and Goldberg, 2001).

These cave systems can also contain **speleothems**, which are mineral deposits formed in limestone regions by water dripping from the ceiling or walls of a cave (e.g. Figure 22.16). Often these are formed as stalactites and stalagmites. They are primarily composed of calcium carbonate which is precipitated from groundwater that has percolated through the adjacent carbonate rock. Certain trace elements such as uranium can be used to determine the ages of layers of speleothems. Cessation of speleothem growth can be detected where adjacent layers yield very different ages. Such growth restriction is likely to be climatically driven. For example, cold conditions could both prevent groundwater percolation and reduce biotic activity (resulting in less carbonate in solution). Periods of maximum speleothem growth in Britain and Tasmania correspond well with warm periods (interglacial and interstadial episodes) in the ocean oxygen isotope record (Atkinson *et al.*, 1986; Goede and Harmon, 1983). In addition, oxygen isotopes can be extracted from some speleothems that are representative of local surface temperatures, providing further evidence for local climatic conditions.

Figure 22.16 Speleothems in Harrison's Cave, Barbados. (Source: Tony Waltham/Geophotos)

Figure 22.17 Layers of loess (yellowy) formed during cold dry periods alternating with layers of soil (brown) formed during warmer, wet periods. (Source: Tony Waltham/Geophotos)

22.4.1.4 Wind-blown sediments

Deposits in many arid parts of the world are dominated by wind-blown sediment (see Chapters 12 and 16). Loess and coversands can accumulate to great thicknesses and they cover around 10% of the Earth's land surface. The deposits of the Loess Plateau in China, for example, cover around 276 000 km^2. The deposits often contain evidence of climate change such as layers of soil formed during warm periods, found between layers of wind-blown sediments deposited during cold phases (Figure 22.17). Indeed work by Guo *et al.* (2002) showed that the alternating layers in the Chinese loess seemed to correspond well to Milankovitch cycle timelines. The sediments also provide useful evidence of prevailing wind direction, atmospheric circulation patterns and gradients in the past (e.g. Figure 22.18).

22.4.2 Plants

Plant remains have provided a major source of information on Pleistocene environmental change. Much research uses the principle that the environment where

particular species or species assemblages survive today is similar to that in the past where such plant remains have been found. Macrofossils of plant remains such as leaves, tree stumps and seeds can be identified in sedimentary deposits. Microfossils of minute biological remains can also be identified using microscopes. These microfossil remains include fungal spores, algae, seeds and pollen. Pollen analysis, or **palynology,** is the most widely adopted method of reconstructing Pleistocene environmental change. Box 23.2 in Chapter 23 provides further information on palynology.

Using pollen analysis combined with other techniques it has been possible to map how individual species have migrated over time and Figure 22.19 presents information for deciduous oak trees in Europe. During the last Glacial Maximum at 18 000 years BP, the oaks survived only in isolated pockets in the south and were not present elsewhere in Europe. During the post-glacial period, as the ice melted, the oaks spread northwards. However, the sudden onset of the Younger Dryas lasting about 1000 years caused them to retreat southwards again before further warming allowed full expansion. It

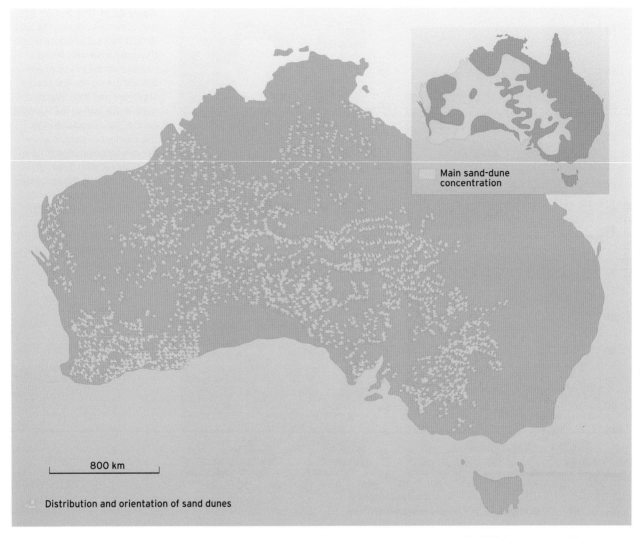

Figure 22.18 Fossil dune fields of Australia illustrating palaeowind directions. (Source: Wasson *et al.*, 1988, Large-scale patterns of dune type, spacing and orientation in the Australian continental dunefield. *Australian Geographer*, **19**, 89–104, Taylor & Francis Ltd and www.informaworld.com)

seems that the isolated pockets in southern Europe where the oak trees persisted during glacial periods allowed the species to survive and provided a reservoir of oak trees which expanded again to cover northern Europe once the climate warmed. These sites, known as **refugia**, are crucial to the preservation of particular species through cold episodes. Indeed, Tzedakis (1993) showed for a 430 000 year record at Ioannina, in north-west Greece, that this site has acted as a long-term refugium for the oak trees throughout the Pleistocene. At such sites it appears that moisture availability was the critical climatic factor that allowed trees to survive with temperature having a supporting role. Tzedakis *et al.* (2002) showed that some of these sites are refugia for several species. Thus, refugia may be areas of special value for maintaining long-term biodiversity.

22.4.3 Insects

A range of insects have been found fossilized in Pleistocene deposits and as with vegetation these allow reconstruction of local temperature and moisture regimes based on comparison with contemporary habitats. Of all the insects it is the remains of beetles that have proved the most useful in Pleistocene research. This is because their exoskeletons are well preserved and allow good identification of species. Beetles (**Coleoptera**) account for around 25% of all the species on Earth and they occupy almost every terrestrial and freshwater zone. Many species survive only within narrow temperature ranges or very specific habitats. They evolve very slowly so that over the timescale of the Pleistocene they have been stable; their environmental tolerances have remained stable. Therefore, it is possible to reconstruct with some precision the past

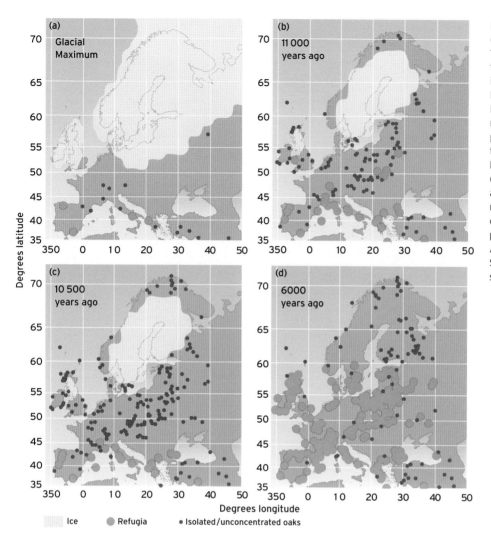

Figure 22.19 Distribution of deciduous oaks during the past 18 000 years. The figure shows that oaks were only found in isolated pockets in southern Europe during the last Glacial Maximum and then migrated northwards during the warming period. The changes during the Younger Dryas can also be seen by comparing (b) with (c). (Source: after Taberlet, P. and Cheddadi, R. 2002, Quaternary refugia and persistence of biodiversity. *Science*, **297**, pp. 2009–10. Copyright 2002 AAAS. Illustration: Katharine Sutcliff. Reprinted with permission from AAAS)

environmental settings for the Coleopteran fossils found within sedimentary sequences. They offer tremendous potential for palaeoclimatic research as they provide information on short-term environmental fluctuations because they respond immediately to climatic change (by migrating to remain in their favoured environmental range). They can be used to reconstruct seasonal and mean annual temperature maps for different periods. Indeed, they can also be used to tell us about likely vegetation assemblages that they were associated with.

22.4.4 Other animal remains

A range of other fossil remains are used in reconstructing Pleistocene environmental change, such as terrestrial and marine molluscs, diatoms, non-biting midges (chironomids), bivalved crustaceans (ostracods), foraminifera, animal bones and teeth, coral polyps, fungal remains and testate amoebae. The last, for example, are found in many peats and are controlled by hydrological variations that affect peat formation. They can be used to reconstruct past moisture regimes. Care should be taken when interpreting animal and plant remains and inferring climate because some species may or may not be present owing to ecological processes rather than climatological ones. For example, species competition or migrational isolation might mean that some species are not present when at other times with a similar climate they were present. Among Pleistocene deposits there is evidence for large animals in many places where they no longer occur and many of which are now extinct. Examples of these are described in Box 22.3.

Reflective questions

➤ Why is it beneficial to use several types of evidence when reconstructing Pleistocene environmental change at any given location?

➤ What are the advantages and disadvantages of using pollen grains to infer environmental change?

ENVIRONMENTAL CHANGE

PLEISTOCENE MEGAFAUNA EXTINCTIONS

During the Pleistocene many large mammals, birds and reptiles became extinct. In fact 85% of large mammal species disappeared. Most of these extinctions occurred between 50 000 and 10 000 years ago. In North America large animals that disappeared include the American lion, ground sloth, dire wolf, several species of mammoth and the short-faced bear. In Africa and Asia several sabre-toothed cat species disappeared along with giant tapir, giant apes, giant hartebeest, pigmy hippos and several types of elephant. In Australia extinctions included giant kangaroos, snakes and several types of wombat. South America lost the glyptodon (Figure 22.20) and the three-toed litoptern. In Europe woolly mammoths, woolly rhinoceros, cave lions and Neander-thals were among the casualties. The extinction of megafauna (large animals) during the late Pleistocene was the second largest of the last 55 million years.

It is not entirely clear why the rate of extinction of megafauna was

Figure 22.20 The glyptodon lived in South America and was the size of a car. It became extinct around 10 000 years ago. It is thought that humans once used their huge shells as shelters in bad weather.

so great during the late Pleistocene. However, the spread of humans for the first time to new areas may have played a role in hunting some species to extinction, introducing new species, introducing disease and in changing habitats (e.g. through fire) by fragmenting megafaunal ranges. Human impact *combined* (i.e. not on its own) with the late Pleistocene climate changes appears to be the most plausible explanation for megafaunal extinction (Koch and Barnoski, 2006).

BOX 22.3

22.5 Dating methods

Reliable dating is highly desired by Pleistocene scientists. There are three types of technique: age estimation, age equivalence and relative dating methods (Lowe and Walker, 1997). Some examples of these techniques are listed in Figure 22.21. Often the techniques used are valid for only limited parts of the Pleistocene. There are no perfect techniques and each has a range of problems which lead to difficulties when interpreting data.

22.5.1 Age estimation techniques

There are two types of age estimation techniques. The first comprise the **incremental methods** which involve measurements of regular accumulations of sediment or biological matter through time. These include the use of tree rings for dating (**dendrochronology**), annual layers of ice on a glacier and analysis of varves (layered seasonal accumulations of sediment often found in glaciofluvial settings). The layers can be counted back through time. The second types of estimation techniques comprise the **radiometric methods**

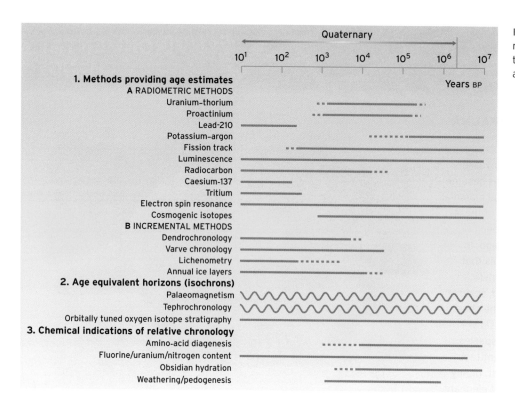

Figure 22.21 Some Quaternary dating techniques and their time range. (Source: after Lowe and Walker, 1997)

that are based on the natural radioactive properties of particular isotopes. Some isotopes decay over time to produce a more stable atomic form. Because radioactive decay is time dependent and the rate of decay over time is constant, the age of a sample can be determined. Some elements decay in a few seconds whereas others transform slowly over millions of years. Typical methods include potassium–argon dating, uranium series dating, carbon dating and *in-situ*-produced cosmogenic isotope analysis.

Carbon dating is explained in Box 23.1 in Chapter 23. Potassium–argon and uranium series dating involve measuring the products of radioactive decay. Potassium–argon dating is used to date the age of volcanic rocks. Uranium dating is used to date rocks and carbonate deposits such as speleothems, corals, bone and molluscs. Care must be taken to ensure that samples are not contaminated by materials from different ages (e.g. by detritus). Luminescence dating allows the amount of time that sediment has been buried to be calculated. Materials exposed to radiation accumulate a **thermoluminescent** property over time. Sediments everywhere contain low concentrations of uranium, thorium and potassium which produce, over geological time periods, a constant flux of ionizing radiation. The ionizing radiation is absorbed and stored by surrounding sediments and, with stimulation (heating), this stored dose can be released, producing an emission of light (the thermoluminescence). The amount of luminescence emitted is proportional to the accumulated dose. If the annual dose (how much ionizing radiation the sediment is exposed to per year) can be estimated then an age for the sediments can be determined. Sunlight removes the luminescence signal and so resets the 'time clock'. However, the major problem lies in calculating the annual dose, which means that errors are often greater than ±10% of the age of the material.

Cosmogenic isotope analysis produced *in situ* is now becoming the standard technique for dating landforms and some types of terrestrial deposits (Cockburn and Summerfield, 2004). The Earth is under constant bombardment by cosmic radiation. Some of this radiation comes from the Sun, some from other locations within our galaxy, while the highest-energy particles originate from outside our galaxy. These high-energy particles, which consist mainly of neutrons, interact with elements in a shallow surface layer when they reach the Earth's surface. This interaction produces extremely small quantities of cosmogenic nuclides. Measurements of the amounts of these cosmogenic nuclides accumulated over time can provide valuable information on the age and rate of change of the land surface. They are particularly useful because, unlike ^{14}C, the cosmogenic isotopes have very long half-lives, ranging from thousands to millions of years. For more information on these techniques see the further reading at the end of this chapter.

22.5.2 Age equivalent labels

Distinctive horizons can often be found in Pleistocene deposits which are found in more than one place. If the horizon can be dated at just one of the locations where it is found (e.g. by **radiometric** or incremental methods) then this allows dates to be extended to other places where this distinctive horizon occurs. Distinctive horizons are produced by layers of volcanic ash spread across a large proportion of the planet after a major volcanic eruption. They are also produced by reversals and strength changes of the Earth's magnetic field (see Chapter 2).

One commonly used age equivalence technique uses the marine oxygen isotope curve as the basis of a globally applicable stratigraphic scheme. The oxygen isotope stages (shown in Figure 22.6e), established in deep-ocean sediment records, can be correlated with sediment sequences (both marine and terrestrial) throughout the world. The ages of each stage are assigned by matching the stage with the predicted age for that stage as determined by Milankovitch orbital forcing mechanisms. This 'tuning' of course only provides a relative age and assumes the deep-sea record of climate change should be perfectly tuned to the orbital cycles. This is unlikely to be the case because of a large number of lags and feedback mechanisms operating in the climate system (Elkibbi and Rial, 2001).

22.5.3 Relative chronology

Sediments and rocks are affected by chemical reactions that take time to occur. The amount of weathering or organic decomposition may provide a basis for relative dating. Bones and shells contain proteins even hundreds of thousands of years after death. The protein undergoes transformations over time, which changes, for example, the ratios of certain amino acids. Thus a relative chronology can be developed by examination of the types and amount of transformation (**diagenesis**) that has taken place. Commonly this involves analysis of amino acids. The rate of the chemical reactions concerned is directly related to temperature.

Consequently, amino acid diagenesis will proceed more slowly at cooler sites than at warmer sites. Samples from some mid-latitude regions, for example, provide a resolution of 20 000–30 000 years with a useful range of only approximately 2 million years since diagenesis proceeds rapidly; Arctic samples provide a resolution of 100 000 years with a useful range of 5 to 6 million years, because diagenesis is slower here (Bradley, 1999). Fossil bones can also be assessed for their content of elements absorbed into them from the surrounding sediments over time. For example, bones often absorb fluorine and uranium from groundwater over time.

Reflective question

➤ Why do we need to develop and use several different dating techniques?

22.6 Pleistocene stratigraphy and correlation

In order to assess environmental change through time it is necessary to analyze sequences of sediments preserved in a range of contexts on land, in ice and beneath the ocean floor. Two aspects of this work are ordering the record at any one location into a time sequence (stratigraphy) and determining how the evidence at one site relates to the evidence at another (correlation). Since only a short period of time is recorded in the terrestrial deposits at any one site it is rare to find even one complete cycle of glacial and interglacial sediments on the land surface of northern Europe or North America. Therefore sequences from different sites must be pieced together to form a complete picture. Only in deep oceans, deep lake deposits and thick loess are long stratigraphic sequences preserved. At most places, where the record is fragmentary, careful analysis is required before sequences can be ordered at one place and then related to those at another.

Often stratigraphic methods are descriptive as it is the visible features that allow formal subdivision of the sequence (Figure 22.22). Classification of subdivisions can be done in a number of ways. For example, sequences can be classified on the basis of fossil evidence found within them, with each **biostratigraphic** unit having a distinctive fossil assemblage. Traditionally, pollen assemblages have been used in the subdivision

(a)

(b)

Figure 22.22 Stratigraphic deposits: (a) a short terrestrial deposit which dates back approximately 18 000 years; (b) deposits made visible by cliff erosion on the Norfolk coast, England, dating back 25 000 years.

of interglacial stages of North America and northern Europe. Other methods involve the relative dating of landforms present or inferring changes in climate from the sedimentary structure. Where a particular stratigraphic unit is very clear and well recorded, and where its lower boundary can be well defined, this site may be designated a **stratotype** or typesite. This site becomes the reference point and the place where a particular stratigraphic subdivision is officially defined. Then other sites where the record of that equivalent unit is only partly present or poorly preserved can be compared with the stratotype that is the standard reference site. Often the location of these stratotype sites provides the names for apparent events in the Pleistocene record. However, because of the spatial and temporal variation of Pleistocene environments often the stratotypes are only locally important.

Nevertheless, it is important to try to correlate evidence at one site with that of others so that the extent and spatial variability of environmental change can be determined and so that we can piece together a long continuous record. However, the repeated cooling and warming during the Pleistocene has meant that similar depositional features may be preserved that are actually of very different ages. This makes correlation difficult. Figure 22.23 presents how stratigraphic schemes established using evidence from different regions have been related to each other on a large scale. There is often a great deal of debate about whether a particular sequence really represents similar changes recorded elsewhere. In some places there is no evidence in the stratigraphic record for events that have been recorded

elsewhere. It is now common practice to try to relate terrestrial stratigraphic units to the marine isotope stratigraphy (Box 22.4), but many of the marine stages have yet to be identified in the terrestrial record. This may be because the event did not have an impact on local processes at a given site or region, but it is more likely to be because sedimentary evidence of the event has not been preserved.

Even at a small scale problems can arise in correlating sedimentary sequences. This may happen because during one glacial period the ice may extend over a particular site. However, even if the next glacial period was just as cold and ice volumes just as great, the same site may not be subject to the same glacial action. It may, for example, be part of a fluvioglacial outwash plain during the next glacial period. There are also problems caused by the erosion or reworking of old sediments which are then redeposited on top of a younger layer of sediment. Thus, fossils and other sediments may be incorrectly associated with a younger period of time than they actually belong to because of their erroneous stratigraphic position.

Deep-ocean records appear to be relatively undisturbed, unlike many terrestrial sequences, and therefore represent a good long-term global timeframe of events. In fact the ocean record is now used as the standard of reference for most other stratigraphic sequences. Producing a good correlation between terrestrial and marine sequences is one of the key areas of Pleistocene research (Box 22.4). However, some terrestrial sequences preserve temporal detail that is as good as the deep-ocean sedimentary records and reveal similar patterns of change. These include undisturbed sedimentary basins or lakes

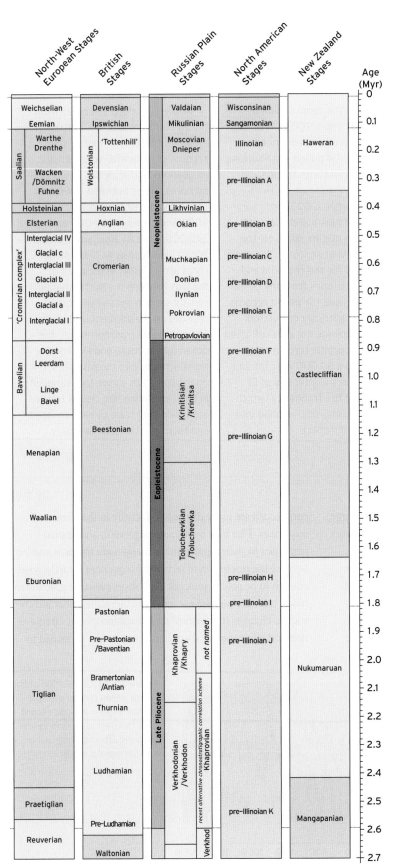

Figure 22.23 Quaternary stratigraphy correlations from different regions. (Source: Phil Gibbard)

NEW DIRECTIONS

DIRECT MARINE AND TERRESTRIAL STRATIGRAPHIC CORRELATION

In order to get around the problems of making sure that a land sequence really was deposited at the same time as an apparently equivalent ocean record some workers have used deep-ocean sediment cores obtained near land margins (e.g. Heusser *et al.*, 2006). This is because the sediment deposited at these sites often contains pollen and dust from local terrestrial sources. Marine pollen records of the vegetation that grew on land can be correlated directly with proxy evidence of the marine environment and the oxygen isotope stratigraphy

preserved in the same sediment (e.g. Heusser and Oppo, 2003). This also makes sedimentary sequences on land with similar pollen distributions easy to correlate with the marine sequences.

Kershaw *et al.* (2003), for example, analyzed pollen and charcoal records from marine cores off the northern coast of Australia. The radiocarbon dates for the charcoal and the oxygen isotopes from the ocean sediment allowed the pollen records to be more rigorously correlated. Vegetation change not only reflected Milankovitch orbital forcing patterns but reflected a 30 000 year fluctuation change in the intensity of El Niño induced fire frequency (which produced the charcoal washed into the ocean sediments).

Analysis of deep-sea cores from the western Portuguese margin provides continuous, high-resolution records of millennial-scale climatic oscillations and work has been done on cores dating between 9000 and 65 000 years BP (Roucoux *et al.*, 2001) and between 180 000 and 345 000 years BP (Roucoux *et al.*, 2006). Pollen analysis of the same cores allows direct assessment of the lags between the North Atlantic climate system and the vegetation changes on the adjacent land mass. The pollen was transported into the ocean by the Douro and Modego Rivers which flow into the Atlantic. Work on the cores has shown that variability in tree population size closely tracked both millennial-scale climate variability and Milankovitch-scale variability.

BOX 22.4

(e.g. Funza, Colombian Andes; Lake Biwa, Japan; Lake George, Australia; Carpathian Basin, Hungary) where sediments have been consistently accumulating for millions of years. For example, Hooghiemstra and Sarmiento (1991) showed that tree pollen at Funza in the Colombian Andes correlated well with the marine isotope record (Figure 22.24).

Shorter-term changes such as D–O events and even shorter decadal to century-scale events recorded in the ice

core record are more difficult to identify in the terrestrial sequences. This may be because vegetation and animal responses to such rapid climate changes are too slow and/or because the resolution of terrestrial sequences is rarely good enough to be able to pick out such short events (e.g. sedimentation rates are not fast enough). Furthermore, other effects such as changes in moisture regimes, atmospheric circulation and environmental gradients may not be identified in

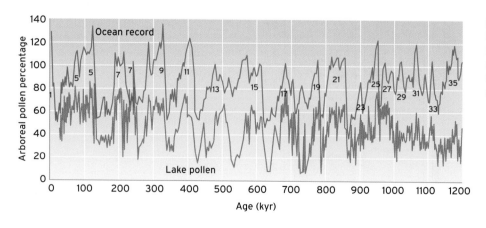

Figure 22.24 Correlation of the Funza pollen record with the marine oxygen isotope record over the past 1.2 million years. (Source: after Hooghiemstra and Sarmiento, 1991)

the record until the vegetation has had time to respond and migrate or *in situ* decline can take place. However, the vegetation surrounding tree population refugia (e.g. in southern Europe) seems to respond rapidly and therefore registers the D–O cycles in its pollen records (Tzedakis *et al.*, 2002).

Reflective questions

➤ What are the main problems and benefits of correlating evidence for Pleistocene environmental change?

➤ Why is using deep-sea cores close to land margins an exciting prospect for future Pleistocene research?

22.7 Palaeoclimate modelling

The use of data from a number of sources and use of high-speed numerical simulation models has benefited Pleistocene research enormously. At a simple level, wave modelling (or spectral analysis) has allowed the fluctuations of the ocean oxygen isotope record to be analyzed in more detail to determine more exact matches and deviations from the Milankovitch predictions (Elkibbi and Rial, 2001). However, it is concern over the consequences of human activity on atmospheric greenhouse gases that has driven the development of models of the climate system. GCMs have been run with pre-industrial levels of CO_2 (as indicated by ice core evidence) and then with double CO_2 levels, or with sequential changes in CO_2, to examine projected climatic changes. However, such models, which are based on the present state of the Earth system, should be questioned for reliability in light of what we have learned about the Pleistocene. For example, it is questionable how well such a model can work for a future climate state in which boundary conditions (sea level, solar radiation, extent of ice, etc.) are very different from today (just as different as some of the climatic states that occurred during the Pleistocene). In order to get around this problem many of these models have been used to predict past climates using boundary conditions that are known to have occurred in the past. If models can reproduce past climatic conditions that we know about from their proxy environmental record we may be more confident in their ability to predict future climates.

This modelling approach, which is concerned with improving future predictions, has also benefited Pleistocene research. The models provide insight into potential forcing mechanisms and the interactions between different parts of the climate system (oceans, ice sheets, atmosphere, biosphere, land surface). They can also indicate potentially unreliable data or areas where further research is needed. For example, many models are unable to produce glaciations with relatively warm tropical sea surface temperatures and this goes against some data from ocean sediments.

Several types of models are used for palaeoclimatology, including energy balance models, radiative convective models and GCMs. Energy balance models simply consider energy exchange between zones of the Earth and can incorporate latitudinal, longitudinal and altitudinal transfers. These simple models allow the effects of feedbacks such as albedo, CO_2 concentrations and solar input to be investigated relatively quickly. Radiative convective models examine atmospheric radiation processes and have been used to examine the effects of aerosols and clouds on temperature. Often these models have been used to look at changes in one parameter at a time (e.g. methane concentration), manipulating them to see which factors are most important and how individual factors might act as feedbacks in the system (Bradley, 1999).

For GCMs the Earth's surface is divided into a grid of boxes which also extend vertically into the atmosphere. The verticals are sliced up into boxes for each ground-based cell (Figure 22.25). Equations involving conservation of energy, mass and momentum are then solved at each grid point and for each vertical level for every time interval required. This requires an enormous amount of computer time to run, and the higher the spatial and temporal resolution, the more calculations that have to be performed and the longer it takes. Nevertheless, research is demonstrating that even more complex models that combine a series of smaller models are required to simulate the climate system. Ocean circulation models are often coupled to atmospheric circulation models. Ice dynamic models are also added in to demonstrate the impacts on ocean circulation and atmospheric circulation of ice sheets growing and melting. Coupling these models to biosphere (Chapter 25) and land surface models will add complexity but may aid understanding of the whole system behaviour. GCMs can also be used to trace the pathways of materials within the climate system. For example, desert dust pathways have been modelled and predictions tested against observed changes in the dust content of ice cores in order to determine the direction of prevailing circulation patterns (Mahowald *et al.*, 1999). Sources of moisture supplying precipitation to ice sheets have also been modelled in order to help understand how source regions differed in the past, which helps with interpretation of ice core geochemistry.

Given the strong evidence in the climate record for sudden and dramatic changes in the climate system indicating

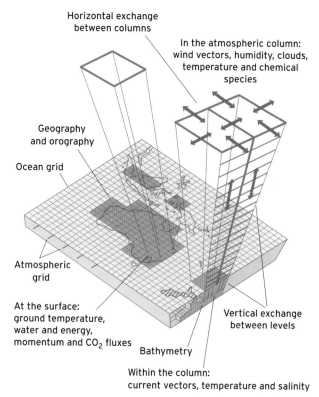

Horizontal exchange
between columns

In the atmospheric column:
wind vectors, humidity, clouds,
temperature and chemical
species

Geography
and orography

Ocean grid

Atmospheric
grid

At the surface:
ground temperature,
water and energy,
momentum and CO_2 fluxes

Bathymetry

Vertical exchange
between levels

Within the column:
current vectors, temperature and salinity

Figure 22.25 Putting grid cells over the Earth. The atmosphere and oceans are split into columns and each calculation is performed for each box. (Source: after McGuffie and Henderson-Sellers, 1997)

thresholds in the system, it is important that such thresholds are incorporated into palaeoclimate models. However, it is often difficult to work out what these thresholds should be. For example, we do not really know how great a change in evaporation or meltwater input is required in the North Atlantic to shut down the thermohaline circulation system, or whether the amount required changes through time as other processes operate (e.g. changes in ocean circulation due to plate tectonics affecting the shape of the ocean basins). Some models are able to simulate key features of the climate record but uncertainties about the processes, the role of feedbacks and boundary conditions still need to be resolved. Even the most complex models today are still too simple to represent climate processes and many of the feedbacks in detail. Nevertheless, even simple models can produce interesting results that suggest new avenues for research. Such an example is given at the start of this book in Chapter 1 (Box 1.2).

Reflective questions

➤ What are the possibilities and limitations of palaeoclimate models?

➤ Why are palaeoclimate models useful for making predictions about future climate change?

22.8 Summary

The Pleistocene is a dynamic period of Earth's history which has seen oscillation upon oscillation in climatic conditions. Huge ice sheets have expanded and retreated over tens of thousands of years. Short-term, but dramatic, changes in global temperature superimposed on top of this have occurred on a timescale of just a few decades. Evidence for Pleistocene environmental change comes from a range of sources on land, in ice and on the ocean floor. The terrestrial evidence is detailed but fragmentary, the ocean evidence is continuous but low resolution (and is restricted in that it tells us only about the ocean and ice volume) and the ice core evidence is high resolution but relatively short in comparison with the ocean sedimentary record.

There are several problems in correlating different types of evidence from different locations. One problem is that it is often difficult to date a piece of evidence for past environmental change precisely. Other problems include the fragmentary nature of some records and the fact that in some places a climatic change may not have a local impact whereas in other places it does.

The evidence for climatic change is always indirect evidence (a proxy, e.g. vegetation or animal assemblages). Therefore, even if proxy data are well dated they will be of little use for climatic reconstruction unless the climatic dependency of that proxy can be clearly established. There is room for considerable improvement in this field because it is often not appropriate to use modern analogues. For example, species can adapt to new environments (Chapters 18–20) and environmental gradients can vary over time. Nevertheless a rich amount of detail can be obtained on ecological responses to climate, the physical extent of former ice sheets, and former

> atmospheric and ocean circulation patterns, using proxy environmental and climatic data.

This chapter has shown that there are a range of factors exerting strong controls on Pleistocene climate. These factors include external orbital forcing processes and a range of internal positive and negative feedback effects. The complex interactions between oceans, ice sheets, the biosphere and the atmosphere mean that there are no simple cause–effect relationships. The Earth's environment appears to have certain thresholds which, when crossed, can result in sudden changes to the climate system. Pleistocene research is of use to those involved in predicting future climate change. It has suggested that slow changes of climate in one direction can result in sudden changes in a different direction. Modelling of climate change allows us to make predictions of future climates while modelling past climates serves to test the models' sensitivity. It also allows us to determine where we need more detailed information in order to improve the quality of these predictions. It seems evident that a great deal more research is required before our predictions include the full complexity of real environmental processes and the feedbacks between them.

Further reading
Books

Bradley, R.S. (1999) *Paleoclimatology: Reconstructing climates of the Quaternary*, 2nd edition. Academic Press, London.
This is a very detailed (and sometimes technical) textbook on techniques and the advantages and disadvantages of different types of evidence.

Ehlers, J. (1996) *Quaternary and glacial geology*. John Wiley & Sons, Chichester.
Translated by Phil Gibbard, the latter part of this book contains useful examples of stratigraphic analysis and provides regional synopses of local environmental change inferred from the evidence.

Imbrie, J. and Imbrie, K.P. (1979) *Ice ages: Solving the mystery*. Macmillan, London.
This book is rather dated now but is very relaxed bed-time reading about the development of Quaternary research.

Lowe, J.J. and Walker, M.J.C. (1997) *Reconstructing Quaternary environments*. Pearson Education, Harlow.
This is a very comprehensive overview of types of evidence and methods of their analysis. This book contains lots of useful examples and explanations and is an excellent textbook.

Walker, M. (2005) *Quaternary dating methods*. John Wiley & Sons, Chichester.
This book clearly explains the basics of dating, the techniques available and the strengths and weaknesses of each technique. While the title might sound daunting, the book is very accessible and the techniques are described in a very understandable way.

Walker, M. and Bell, W. (2005) *Late Quaternary environmental change: Physical and human perspectives*. Pearson Education, Harlow.
Another clearly written book, but this time concentrating on the more recent period of the Quaternary over the past 25 000 years during which time the ice sheets have retreated.

Further reading
Papers

Baker, V.R. (2003) Icy Martian mysteries. *Nature*, 426, 779–780.
The orbital forcing theory operating on Mars.

Bridgland, D., Maddy, D. and Bates, M. (2004) River terrace sequences: templates for Quaternary geochronology and marine-terrestrial correlation. *Journal of Quaternary Science*, 19, 203–218.
The issues of stratigraphic correlation are outlined here.

Koch, P.L. and Barnosky, A.D. (2006) Late Quaternary extinctions: state of the debate. *Annual Review of Ecology, Evolution, and Systematics*, 37, 215–250.
An excellent review of the Pleistocene megafaunal extinctions.

Lockwood, J.G. (2001) Abrupt and sudden climatic transitions and fluctuations: a review. *International Journal of Climatology*, 21, 1153–1179.
John Lockwood presents a straightforward review summarizing the main debates and issues surrounding rapid climate oscillations.

Loehle, C. (2007) Predicting Pleistocene climate from vegetation in North America. *Climate of the Past*, 3, 109–118.
This paper shows why care is needed when using vegetation to reconstruct climate.

Paillard, D. (2001) Glacial cycles: toward a new paradigm. *Reviews of Geophysics*, 39, 325–346.
Paillard's paper is a discussion about the problems with orbital forcing and the need to think about feedback mechanisms and thresholds.

CHAPTER 23

The Holocene

Ian Lawson
School of Geography, University of Leeds

Learning objectives

After reading this chapter you should be able to:

➤ describe how the Holocene began

➤ outline the main drivers of climate change during the Holocene

➤ explain the difference between local and global sea-level changes during the Holocene

➤ describe the key changes in global ecosystems during the Holocene

➤ show an awareness of the way human societies developed and spread during the Holocene and their long-term impacts upon the landscape

23.1 Introduction

The Holocene is a geological **epoch** which began 11 700 years ago and continues to the present day (Figure 23.1). The name derives from the Greek for 'wholly recent'. In the older literature the Holocene is sometimes called 'the Recent' (with a capital R), 'the Postglacial', or by a local name (e.g. 'Flandrian' by British researchers). The Holocene epoch follows the Pleistocene epoch which extends from 2.6 million to 11 700 years ago (Chapter 22), Together, the Holocene and Pleistocene epochs make up the Quaternary **period**.

Why does the Holocene deserve a whole epoch name – and chapter of this book – to itself? After all, it represents just a tiny fraction of the vast 4.6 billion-year history of planet Earth. In many ways, the Holocene is a quite unexceptional period of Earth history, a period of moderate warmth and relatively stable climate. Nonetheless, it stands out from the rest of the geological record for three reasons.

Firstly, the Holocene is the most recent period of Earth history. Studying the Holocene allows us to reconstruct the immediate history of the world around us. Studying more distant periods of time can be fascinating (we would all like to discover the remains of a dinosaur), but the Holocene has a special relevance in offering explanations of why the physical environment is the way it is. Landscape features as diverse as soils, peat bogs, salt marshes, raised beaches, river floodplains, lake basins and coral atolls are often (though not always) geologically recent phenomena whose history began since the last glaciation. Furthermore, to properly explain these phenomena, we have to understand their history (and mostly that means their history during the Holocene). For example, in order to explain why Irish ecosystems are less diverse than British or mainland European ones, it is necessary to know that at the start of the Holocene most temperate animals and plants were still sheltering from the harsh climate of the last glacial period

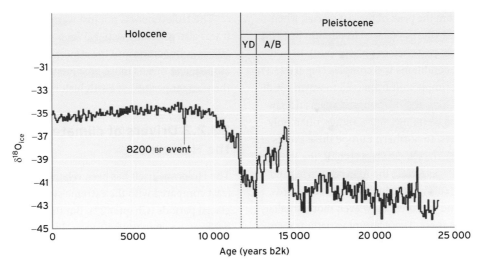

Figure 23.1 A familiar portrayal of late glacial and Holocene climatic change: this is the upper part of an oxygen isotope record from a Greenland ice core (NGRIP). It can be interpreted as a temperature curve (warm towards the top). The Holocene/Pleistocene boundary occurs at 11 700 years b2k (before AD 2000). Two key events towards the end of the Pleistocene are YD - the Younger Dryas and A/B - the Allerød/Bølling interstadial. Many smaller events have also been described (Walker *et al.*, 1998). The 8200 years BP event appears as a relatively small cool spike in the Holocene record. (Source: you can derive your own plots from publically available data at www.ncdc.noaa.gov/paleo/icecore/greenland/ngrip/ngrip-data.html)

in southern Europe. These animals and plants made their way across the continent as the Holocene progressed, and some did not make it in time before the land bridges connecting Britain and Ireland to the European continent were drowned as the ice sheets melted and sea level rose. This is just one example of the way in which studying the past offers explanations about the present.

The second reason why the Holocene is special is that it can be studied in a different way from most of the rest of Earth history. Holocene deposits are much more widespread than those of any previous period, because there has not been time for erosion to remove them; a large proportion of the Earth's surface is mantled by Holocene sediments and soils. As they tend to be at or near the surface, they are easy to access, and like most Quaternary deposits they are usually soft enough to be sampled without recourse to the traditional geologist's hammer: a spade or trowel will do instead. And because Holocene deposits have not usually been buried very deeply or for very long, they often escape the squeezing, deformation and chemical alteration that can make more ancient deposits difficult to interpret. The Holocene is thus very often the easiest part of the geological record to study, and its deposits are capable of yielding an unusually rich and well-resolved record of the past.

The third key characteristic of the Holocene is the importance of people within it. Hominins, the group of species including our own species *Homo sapiens*, have been evolving separately from chimpanzees and other primates since 5–8 million years ago. *Homo sapiens* as a species is at

least 120 000 years old. For most of their history, hominins fitted into ecosystems rather like other large animals. They were relatively rare and appear to have lived in quite small groups; they ate plants and, later, scavenged and hunted animals. Gradually their use of technology became more sophisticated: fire and tool-making gave them the ability to manipulate the environment in ways that other animals could not. But the beginning of the Holocene coincided with, arguably, the greatest step change in the relationship between humans and the environment in the whole of our history: the beginnings of agriculture. As we will see later in this chapter, the ramifications of this change included a huge increase in human populations, the emergence of complex civilizations, and an ever-greater ability to modify the physical geography of the Earth. We will begin, however, by looking at the backdrop for this human story: the climatic history of the Holocene and natural processes of ecosystem and landscape change.

23.2 Holocene climatic change

23.2.1 How the Holocene began

The Holocene is currently formally defined as a geological epoch whose onset coincides with the rapid transition from the Younger Dryas stadial to the interglacial conditions of the Holocene, as preserved in the NGRIP (North GReenland Ice-core Project) ice core in Greenland (Walker *et al.*, 2009).

The transition from the peak of the last glacial, when ice sheets were largest (the last Glacial Maximum or LGM, ending around 19 000 years before present (BP)), to fully interglacial climatic conditions was complex (Figure 23.1). Rather than being a simple shift from cold to warm conditions, the transition took several thousand years and saw a series of short-lived warmings and coolings, often only lasting a few centuries. In northern Europe these events are, in order (oldest first): the Oldest Dryas (cold), the Bølling (warm), the Older Dryas (cold), the Allerød (warm) and the Younger Dryas (cold); some highly resolved records, such as the Greenland ice cores, show even more, smaller, events nested within these. These abrupt variations seem to have been caused indirectly by the melting of the great ice sheets in Fennoscandia and North America as the Earth slowly warmed: the influx of fresh meltwater into the North Atlantic disrupted the delicate balance of ocean water salinity that mediates the **thermohaline circulation**, causing the currents that bring warm water from the tropics into the North Atlantic to switch on or off (Broecker *et al.*, 2010; see Chapter 22). The end of the Younger Dryas marks a convenient place to define the lower boundary (i.e. the start) of the Holocene, although more minor perturbations of the ocean currents by the melting ice sheets continued well into the Holocene, the last known event of this kind occurring about 8200 years ago.

Palaeoclimate researchers have tended to focus on Europe and North America, but if research had begun on the other side of the planet we might have chosen a different starting point for the Holocene. This is because the effect of ocean circulation disruptions differed around the world. The regions around the North Atlantic were most strongly affected by the temperature changes in the North Atlantic itself, which is why the Greenland record shows such a clear transition between the Younger Dryas and the Holocene. In much of the northern Pacific, where the ocean circulation was more stable, the changes were more muted. Most of the southern hemisphere, on the other hand, experienced exactly the opposite sequence of abrupt climatic changes during the glacial–interglacial transition: whenever it was warm in the northern hemisphere it was cold in the southern hemisphere, and vice versa. This so-called 'bipolar see-saw' (Broecker, 1998) arose because if the currents in the North Atlantic fail to draw warm water up from the tropics into the high latitudes of the North Atlantic, then all the heat in that warm water has to go somewhere else, and so much of the southern hemisphere warms instead. Ice core records of abrupt climate events from Antarctica during the run-up to the Holocene are almost mirror images of those from Greenland, although the overall trend (from cold to warm) is the same.

The Holocene was not just warmer than the last glacial, it was also generally wetter. Water evaporates more readily from warm oceans into the atmosphere; more water in the atmosphere means more precipitation, so, as a general rule, the warmer periods of Earth history are also the wetter ones.

23.2.2 Drivers of climate change during the Holocene

The Holocene itself has been relatively climatically stable, at least compared with the extreme swings in climate during glacial periods (Chapter 22). But that is not to say that there has been no change whatsoever. Indeed, the subtleties of Holocene climate change promise to teach us a lot about the modern climate system and the possible shape of future climate change, although they are often very difficult to detect using the palaeoenvironmental techniques at our disposal today. Our knowledge is increasing but this remains a key research frontier. At present we know of several different mechanisms by which Holocene climate has changed, each of which will be discussed in turn below.

23.2.2.1 Meltwater-driven ocean circulation changes

The continental ice sheets that had built up in northern Europe and North America during the last glacial were so large that, even once the global climate had warmed to temperatures similar to those of today, it took several thousand years for the last traces of the ice to disappear. Meltwater probably continued to upset North Atlantic ocean circulation on several occasions. The largest and best known of these meltwater release events occurred approximately 8200 years BP (Alley *et al.*, 1997). At that time the North American (or 'Laurentide') ice sheet had reduced in size considerably and fragmented, but still presented a barrier to northward-flowing runoff from the continent. Consequently a very large lake, named Lake Agassiz by geologists, ponded up behind the ice sheet in an area centred approximately on the Great Lakes (as it had done on several previous occasions during the deglaciation). As the ice holding back this water gradually thinned, eventually the water burst through, catastrophically. The enormous flood channels and boulder-sized sediments left by the ensuing outburst flood have been identified in eastern Canada (Figure 23.2). The sudden draining of Lake Agassiz into the North Atlantic and the subsequent reorganization of the river systems may have caused the thermohaline circulation (see Chapters 3 and 22) to slow markedly for the next two centuries: a strong cooling event is clearly recorded in the

Figure 23.2 Giant current ripples formed by huge outburst floods during the end of the last glacial in the north-west USA. A similar flood event is widely thought to have caused the 8200 years BP cold event. (Source: Tom Foster)

Greenland ice, and is visible in many of the more sensitive records from North America and Europe.

23.2.2.2 Orbital forcing

Milankovitch cycles (Chapter 22), although slow, have continued to have an important effect on Holocene climate. In Chapter 22 it was explained that variations in summer insolation in the northern hemisphere are thought to be the main driver of the balance between ice accumulation and ablation: the greater the summer insolation, the greater the extent to which the winter snow is melted completely away. Northern hemisphere summer insolation peaked during the early Holocene, although the ice sheets were slow to respond and only reached their present size around 7000 years BP. The early Holocene was, for the northern hemisphere, a period of warm summers and cold winters (the opposite was true for the southern hemisphere); across much of Europe the summer temperature might have been a degree or two higher than today on average. In the northern hemisphere subtropics this had the important effect of increasing the strength of the monsoon systems (Chapter 5), whereby intense heating of

the land surface in summer brings heavy rainfall. Thus for many parts of the subtropics the early Holocene was considerably wetter than today. Perhaps the most striking effect of this is that much of the area now occupied by the Sahara Desert was wet enough to support savanna ecosystems; large game animals such as giraffe and antelope, and the humans who hunted them, roamed across much of the eastern Sahara until the mid-Holocene (see Section 23.4.2, below).

Since the early Holocene, northern hemisphere summer insolation has slowly declined, northern summers have become cooler, the monsoons have weakened and the Sahara has dried up. Until the theory of anthropogenic global warming became firmly established in the 1980s, many palaeoclimatologists thought that the insolation trend might be leading us, slowly but inevitably, into the next glacial period.

23.2.2.3 Shorter-term climatic cycles: solar forcing and internal oscillations

Milankovitch theory explains changes in climate over thousands or tens of thousands of years as the result of changes in the shape of the Earth's orbit around the Sun,

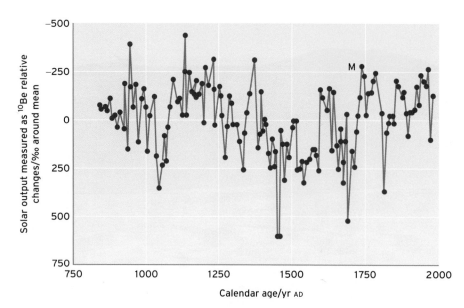

Figure 23.3 A ^{10}Be record from the South Pole for the last ~1200 years. Greater solar irradiance is towards the top of the y-axis. M marks the Maunder minimum. (Source: from Wolff, 2005; after Bard *et al.*, 1997, 2000)

which affect the distribution of incoming solar radiation (or insolation) across the Earth's surface in space and time. There is a growing appreciation, however, that the amount of heat emitted by the Sun varies over time. The reason for this is not well understood but probably reflects persistent, chaotic variations in the way in which hot plasmas circulate within the Sun itself. These variations in the Sun's outputs may cause climatic changes (Bard and Frank, 2006), which would then be described as due to '(direct) solar forcing', as opposed to 'orbital forcing' for Milankovitch-driven changes.

Measurements of solar forcing variations on a day-to-day basis are now being carried out by satellites, and for some centuries astronomers have kept records of sunspot activity, which is correlated to solar output. Measuring changes deeper in the past is also possible using a small number of recently developed techniques. Most rely on the apparent correlation between solar activity and the strength of the cosmic ray flux: the stream of photons (particles of light) generated by the Sun intercepts some of the cosmic radiation reaching the Earth from the rest of the universe, and the stronger the stream of photons (in other words, the greater the Sun's output), the smaller the flux of cosmic rays reaching the Earth. Cosmic rays (protons) collide with atoms in the Earth's atmosphere and can cause nuclear reactions, sometimes producing unusual and often unstable, radioactive isotopes. The most well known of these is carbon-14 (^{14}C or radiocarbon). Variations in the past production of ^{14}C can be measured and used to infer changes in solar output, although there are complicating factors (see Box 23.1). Some other isotopes, notably beryllium-10 (^{10}Be) which is preserved in ice cores, are less problematic, and the

^{10}Be record is regarded as a good proxy for past solar output variations (Figure 23.3).

A combination of direct and proxy measurements has established that there are at least three significant, more or less periodic cycles in solar output: the relatively well-known 11-year 'Schwabe' sunspot cycle, a 78-year 'Gleissberg' cycle and the 200-year 'Suess' cycle. Several Holocene climate records suggest that there might be two longer cycles with periods of about 1500 years and 2200 years, but researchers have proposed a large number of other possible cyclicities in Holocene climate, unfortunately often on the basis of palaeoclimatic data from a single site. These apparent cyclicities may not be real, and even if they are, they may not be due to changes in solar output. This uncertainty about the importance of solar forcing is essentially due to three factors. Firstly, the actual magnitude of changes in solar output is very small – of the order of ± 0.1% for the 11-year sunspot cycle – so they are difficult to measure. Secondly, these small changes in solar output translate into only very small climatic changes, even with the help of positive feedbacks (Figure 23.4). These climatic changes are difficult to detect using existing palaeoclimatic proxy techniques and their interpretation is usually ambiguous; it is a real challenge to find records sensitive enough to pick up such small climatic variations. Thirdly, in order to detect cycles on decadal and century scales we need very accurate knowledge of the age of the deposits being analyzed. A typical radiocarbon date might itself have an analytical uncertainty, after calibration, of centuries. Generating the kind of detailed, precise age models required in solar forcing studies is sometimes possible, but it is very time consuming and expensive to do so.

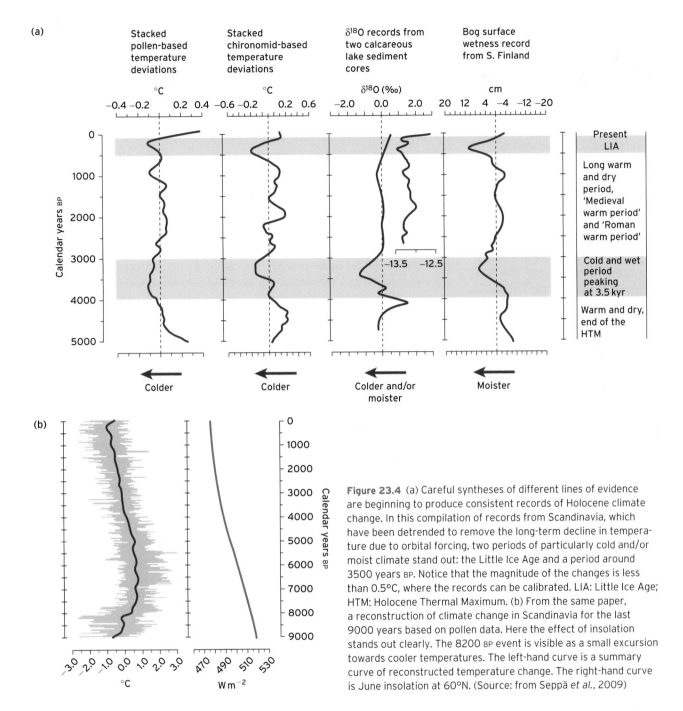

Figure 23.4 (a) Careful syntheses of different lines of evidence are beginning to produce consistent records of Holocene climate change. In this compilation of records from Scandinavia, which have been detrended to remove the long-term decline in temperature due to orbital forcing, two periods of particularly cold and/or moist climate stand out: the Little Ice Age and a period around 3500 years BP. Notice that the magnitude of the changes is less than 0.5°C, where the records can be calibrated. LIA: Little Ice Age; HTM: Holocene Thermal Maximum. (b) From the same paper, a reconstruction of climate change in Scandinavia for the last 9000 years based on pollen data. Here the effect of insolation stands out clearly. The 8200 BP event is visible as a small excursion towards cooler temperatures. The left-hand curve is a summary curve of reconstructed temperature change. The right-hand curve is June insolation at 60°N. (Source: from Seppä et al., 2009)

The main alternative explanations of cyclical climatic change during the Holocene are what are usually called 'internal oscillations'. That means they are internal to the climate system, an inherent part of it. Perhaps the best-known internal climatic oscillation is the **El Niño Southern Oscillation** (ENSO, Chapters 4 and 6). ENSO, which has a periodicity of around seven years, is essentially caused by a time delay in the response of one part of the Pacific Ocean to temperature and circulation changes in another part of it. This delay drives a semi-periodic cycle in sea surface temperatures, which in turn cause further

changes in air temperature and moisture load, eventually affecting the climate of a large part of the planet.

Systems theorists have shown that such oscillations are quite common in complex systems where feedbacks from one part of the system to another are delayed, and climate scientists thus suspect that there might be many more internal oscillations in the climate system operating on a variety of timescales. Two others operating on timescales of decades and known from instrumental data are the North Atlantic Oscillation (NAO) and Arctic Oscillation (AO)

TECHNIQUES

RADIOCARBON DATING

In reconstructing the story of Holocene environmental change we always need to know the age of the deposits we are studying. Many dating techniques have been developed, some of which were introduced in Chapter 22. In Holocene research, undoubtedly the most important is radiocarbon (^{14}C) dating. This technique makes use of the fact that the radioactive form of carbon, ^{14}C, is continuously formed in the upper atmosphere: cosmic rays collide with ^{14}N atoms and create ^{14}C atoms, which are then oxidized and incorporated in carbon dioxide (CO_2) molecules. Carbon dioxide is mixed throughout the atmosphere and absorbed by plants, which incorporate it into new cellular growth. This means that all living plants, and the animals that eat them, incorporate some ^{14}C, as well as the far more common ^{12}C and ^{13}C isotopes (Figure 23.5).

^{14}C is radioactive, which means that, over time, it decays back to nitrogen. The rate at which it does so is constant. We say that ^{14}C has a half-life of 5570 years: if your body contains 1×10^{-14}% radiocarbon today, then in 5570 years your skeleton, if safely preserved out of harm's way, will only contain half as much radiocarbon (0.5×10^{-14}%). By the time it is 11 040 years old (two half-lives), it will only contain a quarter as much radiocarbon as it does today (Figure 23.6).

By the same logic, if we take a sample of material containing carbon from a Holocene deposit and measure its radiocarbon content, we can infer its age. When radiocarbon dating was invented in the 1950s its impact was tremendous: for the first time geologists and archaeologists were able to put a firm figure on the age of a sample. Willard Libby, who did most to develop the technique, was awarded a Nobel Prize for his work.

Radiocarbon dating rests on many assumptions. The most important of these is that we know the proportion of radiocarbon in the atmosphere. In the early days of radiocarbon dating it was assumed that this had not changed over time, and the resulting dates seemed acceptable. However, eventually it became apparent that the proportion of radiocarbon in the atmosphere had in fact changed quite substantially, for two main reasons. The first is to do with the production of new radiocarbon in the upper atmosphere, which is normally approximately in balance with the rate at which radiocarbon decays, so that the overall proportion of radiocarbon in the atmosphere stays roughly constant. In fact, the rate of production does change over time. It depends on the strength of the cosmic ray flux, which is linked to the intensity of the Sun's output. This is known to change on a variety of

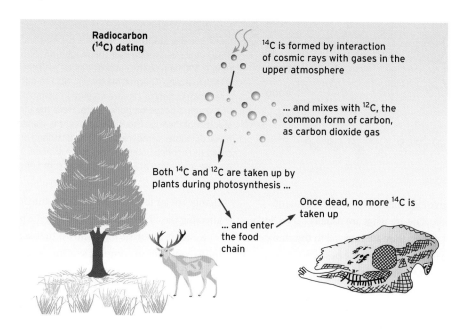

Radiocarbon (^{14}C) dating

^{14}C is formed by interaction of cosmic rays with gases in the upper atmosphere

... and mixes with ^{12}C, the common form of carbon, as carbon dioxide gas

Both ^{14}C and ^{12}C are taken up by plants during photosynthesis ...

Once dead, no more ^{14}C is taken up

... and enter the food chain

Figure 23.5 The principles of radiocarbon dating.

BOX 23.1 ➤

➤

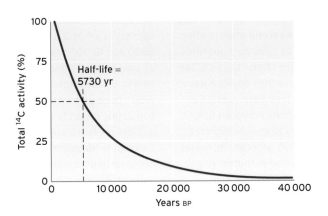

Figure 23.6 The radiocarbon decay curve.

timescales, the best documented of which is the 11-year Schwabe sunspot cycle, but there is evidence that the strength of the Sun's output varies on longer timescales too. So the amount of new radiocarbon produced varies semi-periodically on a range of timescales. The second reason that the proportion of radiocarbon in the atmosphere changes over time is that a substantial amount of CO_2 is stored in the oceans, potentially for thousands of years, and the rate of exchange between the oceans and the atmosphere varies. Sometimes large amounts of CO_2 are released from the oceans back into the atmosphere quite suddenly. This happened, for example, at the very beginning of the Holocene, when ocean overturning in the North Atlantic restarted after the Younger Dryas. Large amounts of carbon had been locked up in stagnant deep water, losing radiocarbon all the time. When the North Atlantic deep-water circulation started again, much of this ^{14}C-depleted CO_2 was released back into the atmosphere, causing the proportion of radiocarbon in the atmosphere to drop markedly. The net result of these changes in the proportion of radiocarbon in the Earth's atmosphere is that simply calculating radiocarbon ages using the half-life

does not yield an accurate estimate of the true age of the sample. The discrepancy tends to increase with age, so that a sample from the very beginning of the Holocene, 11 700 years ago, will give an apparent radiocarbon age of about 10 000 years. This would be called an uncalibrated date.

One reason we know that uncalibrated radiocarbon dates are inaccurate is that scientists have, for many decades, been assembling sequences of modern and sub-fossil wood. The width of rings laid down annually by trees changes from year to year depending on the weather during the growing season. Year-to-year changes in weather leave distinctive patterns of variation in ring widths,

which means that the beginning of one sequence of tree rings in a sample of sub-fossil wood can be matched with the end of a sequence in a second, older sample. In this way several continuous sequences of tree rings have been compiled from various parts of the world, some of which extend all the way from the present day into the last glacial. And because we can count each annual ring, we know the precise actual (or calendar) age of each tree ring. The science of estimating ages from tree rings is called dendrochronology (*dendron* means tree and *khronos* means time in Greek).

Tree rings also provide the solution to correcting or calibrating radiocarbon dates. Samples of wood from individual tree rings, from thousands of samples, have been painstakingly radiocarbon dated. Plotting all of these dates on a graph with the radiocarbon age on one axis and the actual age of the sample (determined by counting the tree rings) yields what is called a calibration curve. The most commonly used calibration curve at the time of writing, INTCAL09, has been assembled by an international team of scientists and refined over many years (Figure 23.7). To calibrate a radiocarbon date, we look up the calibration

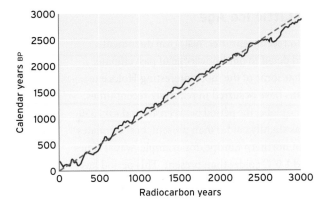

Figure 23.7 A radiocarbon calibration curve.

BOX 23.1 ➤

curve for the apparent radiocarbon age of our sample and read off the corresponding calendar age of the tree ring that yields the same apparent radiocarbon age. A complicating factor is that, often, tree rings of several different ages give the same apparent radiocarbon age, so in practice statistical techniques are used to cope with this uncertainty and calibrated ages are often quoted as a range of ages within which the true age of the sample is likely to fall, with (usually) 95% confidence.

In practical terms, this means that you need to be careful when reading the literature as calibrated and uncalibrated ages are often used by different authors (and they do not always specify which they have used!). Generally, publications published before 2000 are likely to use uncalibrated dates. The offset generally increases with age. As a rule of thumb, you should memorize the following approximately equivalent dates: true calendar age (years BP): 0, 6000, 11 650 (start of Holocene) are equiva-

lent to the uncalibrated radiocarbon age (years BP) of 0 (i.e. AD 1950), 5500 and 10 000 respectively. Bear in mind that 'BP' (before present) means, by convention, 'before AD 1950' (because that is when radiocarbon dating began to be used). Many authors prefer to use AD and BC dates. To simplify the arithmetic involved in converting from BP to AD/BC, some authors are now starting to quote dates as 'b2k', which is shorthand for 'before the year AD 2000'.

BOX 23.1

(see Chapters 6 and 11). Much longer oscillations have been identified in numerical models of oceanic circulation and ice sheet behaviour, and have been suggested as the cause of **Dansgaard-Oeschger** climate cycles during the last glacial (Chapter 22). Some argue that internal oscillations such as these could account for most, if not all, of the quasi-periodic climatic variability in the Holocene, without the need to invoke solar forcing.

Although it is a considerable challenge to collect data that are good enough to test alternative explanations, the study of solar output variations and internal climatic oscillations, and their effects on climate, is a growing field. It is also an important one, because the relatively fast pace of many of these cycles means that they may help to predict future climate change on timescales that are relevant to people.

23.2.3 The Little Ice Age

As we move into recent centuries, historical documents increasingly add detail to our knowledge of past climate. It so happens that some of the most interesting Holocene climate variations have occurred in relatively recent times. Over much of the globe, the period between around AD 1350 and AD 1900 was slightly colder than present; best estimates are that much of northern Europe, for example, was at times between 0.5 and 1.0°C colder than present. This period has become known as the 'Little Ice Age' (Grove, 2004; some authors use the term to refer only to the period between AD 1570 and AD 1900, when temperatures were at their lowest).

Some of the geomorphological effects of the Little Ice Age are clearly visible. For example, many glaciers in

northern Europe have relict terminal and lateral moraines (Chapter 16) which demonstrate that they were larger in the recent past (Figure 23.8). But perhaps the particular fascination of the Little Ice Age is that it demonstrably had an effect on people. During the most severe decades, upland farms throughout Europe were abandoned and farm production declined. The Arctic sea ice made incursions much further south than it does today, and polar bears carried by the ice became a regular menace to the people of Iceland. Rivers and lakes in Britain and Holland froze much more regularly and thickly than they do today; people held fairs on the ice on the River Thames in London, the farmers of

Figure 23.8 Little Ice Age moraines at Gígjökull, Iceland. The glacier is surrounded by debris (foreground) left by its Little Ice Age advance. A muddy meltwater lake has ponded up behind the moraine. This picture was taken in 2008; the eruption of Eyjafjallajökull just south of Gígjökull in 2010 reduced the size of the glacier considerably.

East Anglia became famous for their prowess in ice-skating on their frozen ditches, and the Scots developed the sport of curling, played on the ice of frozen lakes. The cultural memory of the Little Ice Age lingers on: British Christmas cards often portray snowbound scenes that would have been much more commonly a part of the Christmas experience in Victorian London than they are today.

Some palaeoclimatologists interpret the Little Ice Age as marking the beginning of the transition from the Holocene interglacial into the next glacial, resulting ultimately from the gradual Milankovitch-driven downward trend of northern hemisphere summer insolation. Ruddiman *et al.* (2005) used climate modelling experiments to suggest that, if it had continued, the Little Ice Age would have allowed ice sheets to accumulate in northern Canada. Positive feedbacks such as the ice–albedo feedback (Chapter 22) would have amplified the effect of the gradual summer insolation decline. According to Ruddiman *et al.* (2005), only the release of CO_2 by fossil fuel burning reversed this long-term cooling.

Reflective questions

➤ To what extent do you think patterns of Holocene climate change might suggest patterns and mechanisms of *future* climate change?

➤ Why was the climate generally more stable during the Holocene than during the previous glacial? (Compare with Chapter 22.)

23.3 Holocene geomorphological change

23.3.1 Retreating ice sheets

The general warming at the start of the Holocene had the most obvious direct geomorphological effect of causing most of the Earth's ice sheets to diminish or disappear, a process that took thousands of years. The largest were the Laurentide (North American) and Fennoscandian (northern European) ice sheets, but smaller ice caps centred on mountain ranges such as the Alps and the Rockies also shrank and fragmented. In most cases these regions now host only a limited number of valley glaciers. Only two significant ice sheets survived into the Holocene, in Greenland and Antarctica. The Greenland ice sheet is much smaller than it was during the last glacial, but the Antarctic ice

sheet is almost the same size: its growth during glacials is limited by the fact that it terminates in the ocean, and during interglacials its weather systems remain separated from the rest of the planet by strong oceanic and atmospheric currents that encircle and isolate the Antarctic continent.

The loss of most of the Earth's ice sheets opened up large areas of land in North America and Eurasia, which together with the change towards a warmer and wetter climate meant a fundamental change in the kind of geomorphological processes acting over large parts of the planet. Periglacial processes, permafrost, and glacial erosion and deposition all became less widespread.

In some places the loss of ice sheets and valley glaciers caused a delayed relaxation of the landscape to fit the new conditions. For example, where glaciers carve out a U-shaped valley, the valley sides can become very steep, even where the bedrock is quite soft or mechanically weak, because the ice supports the valley walls and prevents them from collapsing. When the ice disappears, the support is removed, and the slopes can fail through a series of landslips until the valley sides are at a more stable, shallower angle. Many previously glaciated valleys show evidence of this kind of mass movement, which often dates to the first few centuries or millennia of the early Holocene. Particularly good examples are to be found on the western side of the Trotternish Peninsula on the Isle of Skye (Figure 23.9). This is an example of **paraglacial geomorphology,** which produces a landscape feature that owes its existence to (the previous presence of) ice, even though the ice is only indirectly involved (Ballantyne, 2002).

Another slow response of landscapes to the removal of the ice sheets is **isostatic rebound**. The Earth's crust rests on the mantle, which is partly molten and behaves as a very viscous fluid. The additional weight of the 3 km thick Laurentide and Fennoscandian ice sheets pushed the crust into the mantle. Once the ice sheets were removed, the land began to rise again. The response is extremely slow and indeed is still continuing; close to the former centre of the Fennoscandian ice sheets, in the northern Gulf of Bothnia (Sweden/Finland), the land is still rising at about 9 mm yr^{-1}.

As the ice sheets decayed and periglacial disturbance diminished, soils began to form through the chemical and physical weathering of newly exposed and stabilized regolith (Chapter 7). In wetter areas, peat built up. The extensive peatlands that now occupy large parts of Canada, Scandinavia and western Siberia are globally important because, on the one hand, they represent a long-term store of carbon and, on the other hand, the slow decay of organic matter under waterlogged conditions in the peat leads to emissions

Figure 23.9 A paraglacial landscape on the Trotternish Peninsula, Isle of Skye, Scotland. During the last glacial the cliff on the left of the photograph was supported by the edge of an ice sheet. When the ice melted, the support was removed, and the oversteepened cliff edge collapsed in a series of rotational landslips.

of methane (CH_4), a powerful greenhouse gas. Ice core records show that atmospheric methane concentrations increased markedly at the start of the Holocene, probably reflecting the expansion of northern peatlands. The greenhouse effect of this additional methane acted as a positive feedback to the initial warming, accelerating the pace of deglaciation.

23.3.2 Rising seas

As the ice sheets melted, they returned a huge quantity of water – some 44 million km^3 – to the oceans. This meant that global average sea level (sometimes called **eustatic sea level**, to differentiate it from local sea-level changes that can be caused by changes in the elevation of the land – see below) dropped to around 120 m below present at the height of the last glacial. The eustatic sea level began to rise again around 19 000 years BP, and only reached approximately its present level around 7000 years BP. In some parts of the world, particularly close to the previous ice sheets, sea-level history has been more complex because, at the same time that global sea level has been rising, so too has the land been rising due to isostatic rebound. In parts of northern Europe, for example, there are a series of old coastlines which have been successively uplifted during the Holocene to form what are called **raised beaches**, outpacing the general rise in sea level.

In most of the world, rather than the uplifting of land from the sea, the opposite has occurred: a great deal of land was drowned by the eustatic sea-level rise during the first 5000 years of the Holocene. This has some substantial implications. The majority of global coastlines are only a few thousand years old: most of our coastal geomorphological features, such as salt-marshes, beaches and spits (Chapter 15), are relatively young. In some regions the outlines of continents and islands have changed substantially. One of the best-studied areas in this respect is the southern North Sea, between Britain and the Netherlands. Today this is a relatively shallow sea, often only 20 m deep, with occasional shallows and sandbanks that are dangerous to shipping. At the start of the Holocene the whole area was dry land, a region that palaeogeographers call 'Doggerland' after the Dogger Bank, the largest of the remaining sandbanks (Figure 23.10). Trawlers operating in the North Sea regularly dredge up the traces of this lost landscape, such as the teeth of woolly mammoth and stone tools left by hunter–gatherers. As the sea invaded this largely flat landscape during the early Holocene, the geography must have changed quickly; even within a human lifetime the advance of the sea would have been obvious.

There are cases where the effect of the rising sea level on geography may have been even more dramatic. Some palaeogeographers think that the Bosporus Strait, which separates the Mediterranean from what is now the Black Sea, was

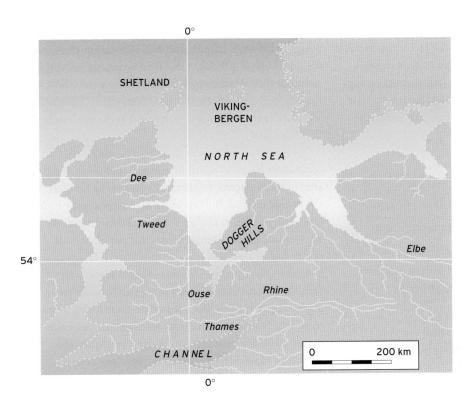

Figure 23.10 A reconstruction of the coastline of Britain during the early Holocene, showing the area now under the North Sea known as Doggerland. (Source: from Coles, 2000)

reached by the rising sea around 9400 years BP (Ryan *et al.*, 1997). The Black Sea was then a relatively small freshwater lake (Figure 23.11). The rising sea eventually overtopped the sill at the Bosporus Strait. Geomorphological evidence gained by seismic survey of the bed of the Black Sea suggests that this may have initiated a catastrophic flood, accelerating as the flood waters eroded the base of the sill; one of the more extreme estimates is that, at its peak, the Black

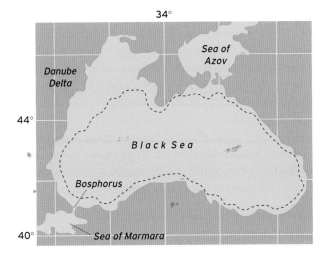

Figure 23.11 The current coastline of the Black Sea, and its much smaller extent prior to the early Holocene 'megaflood' proposed by Ryan *et al.* (1997). (Source: after Giosan *et al.*, 2009)

Sea rose at a rate of several centimetres *per day*. The abrupt flooding could have meant the loss of 100 000 km² of land in as little as a year, an event so cataclysmic for local populations that some people suggest that it inspired myths and legends of catastrophic floods in many Near Eastern cultures, persisting to this day in the biblical story of Noah.

The effects of the rising sea level were not just limited to the coasts. River systems are adjusted to the base level of the sea (Chapter 12). As sea level rose, so the gradients of river systems declined. This meant that there was less energy to transport sediment from inland to the oceans. Consequently, during the Holocene the general pattern has been for river systems at lower latitudes to deposit sediment in their lower reaches through a process of **aggradation**. However, river geomorphology is controlled by a number of other factors, including changes to the runoff regime (largely driven by climate) and the extent of sediment supply (which, in temperate regions at least, is often a function of the degree of vegetation cover, which protects soils from erosion). The interaction of these and other factors means that rivers have had strikingly different and often complex Holocene histories in different parts of the world (Chapter 12; see also, for example, Gibbard and Lewin, 2002).

In this section we have seen that the geomorphological history of the Holocene has been dominated by the response to the major shift in climate during the transition from the last Glacial Maximum to the warmth of the interglacial, and the subsequent equilibration of landscapes. There has also, of course, been climatic change *within* the Holocene, and at times this has caused substantial changes in geomorphology: for example, the re-expansion of many glaciers during the Little Ice Age. But, as we shall see later, the most important cause of further geomorphological change within the Holocene was human action.

Reflective questions

➤ Why is there a difference between eustatic and local sea-level history?

➤ Why do geomorphological systems often respond to climate change with a lag of hundreds or thousands of years?

➤ In what ways might geomorphological systems continue to react to the last glacial-Holocene transition in the future?

23.4 Holocene ecosystem change

23.4.1 Responses of ecosystems to the end of the last glacial

Just as in the case of geomorphology, the long-term pattern of change in ecosystems during the Holocene has largely been one of a slow and continuing adjustment to the warmer and wetter conditions of the Holocene after the last glacial. The biogeography of the Earth during the last glacial was very different from today. The lower temperatures meant that ecosystems such as temperate forests tended to be found much closer to the equator, and/or at lower altitudes in mountain ranges. Tundra and arid steppe grasslands were much more widespread. In some parts of the tropics, a lack of moisture meant that rainforests contracted. There were also some ecosystems that we would not recognize today. Pollen evidence (see Box 23.2) shows that much of northern Europe was dominated by species such as sea buckthorn (*Hippophaë rhamnoides*), which today is associated with the dry, salty conditions of sandy shores. Its abundance inland during the last glacial, along with other halophilic (salt-tolerant) species, is attributed to aridity causing soil salinization (Chapters 9 and 14) widely across the continent. There were also many species that no

longer exist on Earth, especially large animals such as the woolly mammoth. Large herbivores like mammoths were so abundant that it is thought that their browsing, and fertilization by their dung, helped to maintain grass-dominated, nutrient-rich ecosystems that have been called 'mammoth steppe'; direct analogues are not found in the modern world (Guthrie, 2001).

As temperatures increased during the late glacial, so ecosystems began to adjust to warmer conditions; at the start of the Holocene, following the end of the Younger Dryas cold interval, the process of adjustment really took hold. In many cases, finding a new equilibrium with the warmer interglacial climate meant that populations of plants and animals had to move hundreds of kilometres away from the equator. This was less problematic for animals than for plants, which can generally only 'move' by dispersing their seeds. For a few of the large forest-forming trees, which often produce large seeds that cannot easily be dispersed by the wind but must instead be carried by animals, and whose generations can be many decades apart, this was a very slow process. In a few cases (e.g. spruce – *Picea abies*: see below) the post-glacial spread may still be continuing. With each new immigrating species, ecosystems had to adjust through competition towards equilibrium, a process that can take many generations; again, thousands of years in the case of some long-lived tree species.

Climatic changes during the Holocene have also driven major changes in ecosystems. We have already encountered one example, the greening of the Sahara during the early Holocene and its subsequent desiccation. Another is the hypothesis that the spread through Europe of some relatively cool-adapted trees, such as silver fir (*Abies alba*), may have been aided by the 8200 BP cold spell, which gave it a 200 year window in which to out-compete less cold-tolerant species and gain a foothold in the forests. Another important factor has been the development of soils, which is itself a slow process that is closely coupled to vegetation development. Changes in animal populations have also been important.

All of these different factors have combined in unique ways in each region of the world. To give some flavour of the diversity of ecosystem changes that have occurred during the Holocene, we will look at three case studies:

1. Ecosystems in tropical Africa and the Sahara.
2. European ecosystems.
3. The particular ecological patterns found on isolated islands worldwide.

(handwritten, left margin, vertical): No decomposition: bacteria can't respire

TECHNIQUES

POLLEN ANALYSIS AND PALAEOECOLOGY

Pollen analysis (sometimes also called palynology) is one of the most powerful techniques available for reconstructing past environments. Many plants produce tiny pollen grains, each typically 10–100 micrometres in diameter, as part of the reproductive cycle (Figure 23.12). Pollen, which contains genetic material, is released from the male organs of a flower and transported by wind or by insects, hopefully to collide with and fertilize the female organ of a flower of the same species. The vast majority of pollen grains never reach their goal, but drop to the ground – or sometimes fall on a lake, a peat bog, or some other environment where they can be buried and preserved in oxygen-poor conditions. Different species of plant produce pollen grains of differing shapes and sizes, so examining the pollen preserved in a sample of peat of a certain age reveals which plants were growing in the environment at the time, and in roughly what proportions.

Pollen analysis was developed in the first half of the twentieth century and has remained the bedrock of Holocene palaeoecology, for several reasons. Firstly, pollen analysis tells us something that is very useful. Vegetation dominates ecosystems in terms of structure and biomass, and it is important to animals and humans, so it is often the first thing we want to know about. Secondly, we can use knowledge of the vegetation to infer other properties of the environment, such as climate. It stands to reason, for example, that a pollen assemblage made up of Arctic species implies an Arctic climate; the climatic preferences of individual species can often be used to pin down the palaeoclimate more precisely. Thirdly, pollen analysis can be widely applied. Pollen grains are chemically very tough and will survive in most waterlogged environments; with the exception of hot deserts, pollen analysis has been used in most parts of the world with great success.

Nonetheless, pollen analysis, like all indirect measures of past environmental conditions, has its problems. For example, some species produce practically identical pollen. Most members of the grass family (Poaceae) produce almost identical pollen grains – spherical, with a single round hole called a pore – which means that, in a pollen diagram, we cannot distinguish between such ecologically different plants as *Phragmites australis* (a reed, common in nutrient-rich lakes, often up to 2 m in height), *Molinia caerulea* (a medium-sized tussock grass typical of nutrient-poor, acidic peatlands and moors), and *Festuca rubra* (a small grass common on sub-Arctic heaths). Fortunately, some of the most significant grasses for humans, cereals such as *Triticum* (wheat), *Hordeum* (barley) and *Zea* (maize), can, with care, be recognized at least to genus level. Even distantly related plants sometimes, by chance, produce identical pollen: a classic example is *Corylus avellana*, the hazel, a common shrub or small tree found in woodlands throughout much of Europe. Its pollen is almost undistinguishable from that of *Myrica gale* (bog myrtle), a small shrub of a completely different family that is found on peat bogs. Only when the preservation of the pollen grains is absolutely perfect can the two be distinguished (Blackmore *et al.*, 2003). Another serious constraint of pollen analysis is that it is rarely possible to tell exactly where the pollen in a deposit has come from. Despite a lot of work on the theory and prediction of pollen dispersal, transport and deposition over the last 50 years, pollen analysts generally have trouble in saying exactly what area of landscape their pollen data represent. Pollen can literally be blown

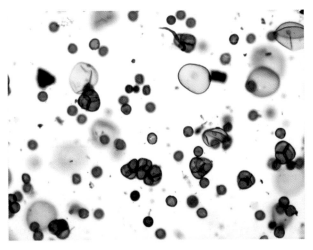

Figure 23.12 Pollen grains under the microscope. A range of pollen shapes and sizes can be seen, each indicating the presence of a different species. (Source: photo courtesy of John Corr)

BOX 23.2 ➤

around the world: in a famous early experiment in pollen trapping it was recovered from the air in the middle of the North Atlantic using a vacuum cleaner on the deck of an ocean liner (Erdtman, 1937)! Pollen from American and European tree species regularly turns up in, for instance, lake sediment samples from Iceland. So we cannot infer the past presence of a particular species close to our site from a find of one or two pollen grains. Pollen analysts thus have to use a lot of caution in interpreting their data.

The development of pollen analysis led the way for many other palaeoecological techniques focusing on particular groups of microfossils, which can generally tell us more specific things about more restricted parts of the environment. Some examples of these include:

- Diatoms: single-celled algae (plants) which live mainly in lakes and the ocean; they can be used to infer lake depth, pH, salinity, and nutrient status.
- Ostracods: small crustaceans which live in lakes; again the species assemblages reflect variations in lake depth, salinity, and nutrient status.
- Foraminifera: tiny oceanic animals; different species and growth forms occur in different temperature and salinity conditions.
- Chironomids: a type of midge; the head-capsules of chironomid larvae can be abundant in lake sediments, and different species prefer different water temperatures.
- Testate amoebae: single-celled animals which are particularly common in peat and which are believed to reflect the wetness of the peat.

BOX 23.2

23.4.2 Tropical Africa and the Sahara

In general, the temperature difference between glacials and interglacials is greatest at the poles and least at the equator. This often means that ecosystem change during glacial–interglacial transitions is least marked in tropical areas. Certainly, temperatures during the last glacial must have remained warm enough that rainforest taxa survived in at least parts of lowland sub-Saharan Africa. Pollen evidence shows that, overall, the area of rainforest declined, but rainforest taxa persisted in quite sizeable areas (Lezine and Cazet, 2005). At the start of the Holocene the forests expanded, at the expense of the surrounding savanna. This history is still reflected in patterns of biodiversity in the modern rainforests: biodiversity tends to be greatest in the areas where the rainforest survived during the last glacial (so-called 'refugia'). This has led to debate over whether the high biodiversity is a result of enhanced **speciation** in these small refugial areas, or merely the result of enhanced survival of species that were never successful in recolonizing adjacent areas when the climate improved.

Further north, in what is now the central and eastern Sahara, the effect on the landscape was very dramatic. The early Holocene was much wetter than today because of the enhanced monsoon system. Lakes filled and much of the desert was covered with a lush savanna (the western Sahara probably always remained a desert). Populations of humans made their living by fishing for perch and catfish in the lakes, while keeping a wary eye out for hippos and crocodiles; they hunted antelope across the savanna, grazed domesticated cattle and produced decorative pottery,

Figure 23.13 Rock art from an early Holocene settlement site in the Sahara. This site in Ennidi Massif lies in the arid desert; during the early Holocene, human hunter-gatherer pastoralists lived here in a savanna landscape, hunting antelope and fishing from now dried-up lakes. (Source: Getty Images: DEA / C Sappa)

in areas that are now absolutely barren, stony desert (Figure 23.13; see also Figure 14.3 in Chapter 14). This enhanced wetness extended into tropical West Africa in what is known as the 'African Humid Period'. Eventually, however, the monsoon began to weaken; the Sahara became a desert, the savanna and rainforest belts contracted, and West Africa adopted its current, markedly drier, climate.

In addition to this long-term drying trend, there is evidence of substantial variations in West African climate within the last few thousand years. A lake-level record from Lake Bosumtwi in Ghana shows that lake levels have fluctuated several times over the last few millennia, indicating phases of drought that lasted often for many decades. This is thought to indicate that the position of the **intertropical convergence zone** (ITCZ), a boundary between air masses which controls precipitation in West Africa, is quite variable and delicately balanced.

23.4.3 European ecosystems

In Europe during the last glacial, tundra and grassy steppe dominated over large areas, possibly with some conifers and cold-tolerant trees north of Alps; to the south, most of the warmth-demanding species that dominate European forests today were restricted to small refugial areas. At the start of the Holocene the warmth-demanding species began the slow spread northwards, one generation at a time, spreading their seed by wind, water or with the help of animals and birds. The pace at which forest-forming species spread varied considerably (Huntley and Birks 1983). Birch (*Betula* sp.) has very light seeds with thin wings which help it to travel on the wind. At the start of the Holocene, birch spread extremely quickly and had reached the northern parts of Britain within a few centuries. Most trees, such as pine (*Pinus*), oak (*Quercus*) and elm (*Ulmus*), took a little longer, up to a few thousand years, to expand from southern Europe to the north of Britain. A few species, such as spruce and perhaps beech (*Fagus sylvatica*), are probably still spreading. Others have probably reached their high-water mark and are retreating: hazel (*Corylus*), for example, occurred a little further north of its current limits in Scandinavia under the slightly warmer conditions of the early Holocene.

What this means is that, if an observer were able to stand in one location in northern Europe and watch the surrounding ecosystem develop around them over the course of the Holocene, they would see a succession of radically different vegetation communities. Typically, in southern Britain, this would consist first of thin, open birch woodland, then pine would invade; then both would be replaced by a darker, denser woodland of oak and elm; this in turn would be invaded and enriched by later arrivals such as alder (*Alnus*) and hornbeam (*Carpinus*). All the while the soil would be becoming deeper and more fertile (though where rainfall is high, fertility peaks then declines as soluble nutrients are leached away). Eventually the whole forest might be replaced by beech, a relatively recent immigrant to southern Britain but an extremely aggressive one, capable of out-competing all other trees in certain circumstances to form the monospecific woodlands that are now so characteristic of parts of southern Britain.

More ancient interglacials were populated by a fantastic range of animals, many of which we now associate with the tropics. Famously, river deposits excavated from Trafalgar

Figure 23.14 A hippopotamus skeleton recovered from last interglacial deposits at Barrington, near Cambridge, and now on display at the Sedgwick Museum in Cambridge.

Square in London dating to the last interglacial, also known as the Ipswichian or Eemian (*c*.127 000–117 000 years BP), contained fossil material from animals including lion, elephant and hippopotamus (Franks, 1960; Figure 23.14). Many of these animal species did not become widespread in Europe during the Holocene, probably due to hunting by people (though lions were still being hunted in classical Greece, and a small breed of hippos survived into the Holocene on Cyprus). Nonetheless, early Holocene environments were faunally richer than they are today, with species such as lynx, wolf, bear, beaver and the extinct aurochs (a creature similar to cattle) common even in Britain. In fact some of these species survived until relatively recent times: the last wolf in Scotland was shot about AD 1750. There have been attempts to reintroduce some of these species to restore the faunal diversity of various parts of Europe. The absence of beavers from Britain is particularly significant because these animals have the habit of making dams which block rivers (they are often referred to as 'ecosystem engineers'). This activity creates new wetlands and encourages flooding. Beavers have recently been reintroduced on a trial basis to small parts of Scotland and if they prosper as some people hope, we can expect the landscape to change considerably.

23.4.4 Island ecosystems

The rise in sea level during the early Holocene was particularly important for ecosystems on islands. Many islands were drowned and new ones were created by the rising seas. Animals and plants living on islands sometimes benefit from the protection that insularity brings: many of Europe's largest seabird colonies, for example, are established on islands from which predators such as rats are absent. But at the same time, island ecosystems can be vulnerable because resources

19ka
)

are limited and small populations of plants and animals are prone to local extinction if they are disturbed. The early Holocene rise in sea level made many islands steadily smaller, which put their populations under pressure.

A strange feature of islands is that very often, over long periods of time, the animals that live on them change size. Large animals tend to become smaller. This is because, if they are lucky enough to be stranded on an island without predators, there is little evolutionary pressure for them to maintain a large body size as a defensive measure; and there is selective pressure for them to be small, so that they are less vulnerable to food shortages and can grow to reproductive maturity more quickly. Many of the new islands formed by the Holocene sea-level rise saw spectacular examples of this 'island dwarfing'. One well-documented case is a group of islands off the north coast of Siberia. Here woolly mammoths became stranded as the sea level rose and, on each island where they persisted, underwent dramatic dwarfing. On Wrangel Island, where the mammoths persisted until just 3500 years BP, the last individuals were only about 2 m high at the shoulder (Guthrie, 2004). (On the Channel Islands of California, during the last glacial, isolated populations of mammoths became even more dwarfed, to only 1.5 m at the shoulder.) Eventually, however, the last of the dwarf mammoths died out, probably due to human predation.

> **Reflective questions**
>
> ➤ Which factors might allow some species to react quickly to climate change, and not others? Some things to consider: reproduction rate; mobility; degree of specialization to any one habitat; whether they are plants or animals; and so on.
>
> ➤ Should we be making more of an effort to reintroduce locally extinct animals into areas with a long history of impoverishment, such as Europe, as well as trying to preserve biodiversity in less disturbed areas?

23.5 The rise of civilizations

23.5.1 Humans at the end of the last glacial

While the boundary between the last glacial and the Holocene was significant for geomorphology and for ecosystems in general, it was perhaps even more significant for humans. At or close to this transition, the first steps were taken towards constructing a wholly new relationship between people and the environment: the development of agriculture. By the end of the last glacial, our species, *Homo sapiens*, appears to have been the only species of hominin left on the planet. On present evidence, for much of the last glacial we had shared the Earth with at least two and probably three other species of hominin: the neanderthal *Homo neanderthalensis*, which went extinct about 30 000 years BP; the 'hobbit', *Homo floresiensis*, first discovered in 2004 on the island of Flores and thought to have survived until at least 18 000 years BP (Brown *et al.*, 2004); and a probable third species, at the time of writing still unnamed, known only from DNA recovered from a finger bone dated to 30 000–48 000 years BP found in a cave in the Altai Mountains of Siberia in 2008 (Krause *et al.*, 2010). For one reason or another – habitat loss, competition for resources, or perhaps direct hunting by modern humans – these three species, and perhaps others we do not yet know about, disappeared. *Homo sapiens* was left as the only survivor of 5 million years of hominin evolution and diversification.

At the same time, modern humans were spreading around the globe, entering Eurasia from Africa around 120 000 years BP, reaching Australasia around 50 000 years BP and taking advantage of the waning Laurentide ice sheet to enter the Americas for the first time during the late glacial (Figure 23.15). The spread of humans was to continue during the Holocene, with the remote islands of the Pacific being colonized progressively (the last, the Chatham Islands off New Zealand, were only colonized around AD 1000). Despite their wide spread, the impact of humans on ecosystems was, as yet, relatively minor by modern standards: people lived by hunting, fishing and gathering plant foods, and most societies were nomadic, so their impacts were spread thinly over the landscape. Estimates of prehistoric population size are notoriously difficult to make, but few would argue that at the end of the last glacial there were more than 10 million people on Earth – about the same as the population of London, Paris or Chicago today. The most important environmental impact of people at that time was their probable role in the extinction of many species of large animal (see Box 22.3 in Chapter 22).

23.5.2 The beginnings of agriculture

Agriculture can be considered as a way of life that involves two key components: cultivating plants and herding animals. The animals and plants in question have in most cases undergone a process called domestication, which means that humans have taken control of the reproduction of the individual plants and animals concerned. The process of

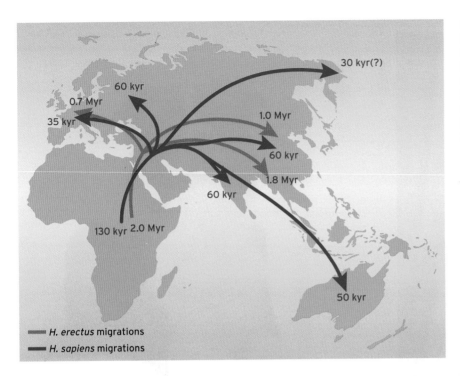

Figure 23.15 The migration of hominins: the most widely accepted theory is that most hominin lineages arose in Africa and spread across Eurasia during two migration periods. The first involved *Homo erectus* over a million years ago. The second was the expansion of our own species, *Homo sapiens*, out of Africa around 130 000 years ago.

domestication probably differed for each individual species that came to be cultivated or herded, so it is perhaps easiest to understand by examining a few case studies.

The first animal to be domesticated was the dog, possibly as early as 135 000 years ago according to recent genetic evidence, and there are good reasons why it was the first. The ancestors of dogs, various species of wolf, share key behavioural properties with humans. Wolves live in social groups; they have a strong social structure, with a hierarchy and an inherent willingness to submit to the pack leader. Wolf pups taken from the wild today can be tamed and, sometimes, trained to submit to humans, though they are usually still dangerously aggressive. Wolves are also pack hunters, used to sharing the task of chasing down prey and sharing the proceeds. For humans, tamed wolves would have made very useful hunting tools: wolves are faster and better armed with teeth and claws than humans, so using tamed wolves to help with hunting would make up for some of the biological inadequacies of humans as carnivores.

Taming an animal from the wild is one thing; domestication requires the further step of intervening in the reproduction of individuals. In the case of tamed wolves, some might have preferable characteristics: they might be a little more docile and more easily controlled than others, for example. These wolves might be kept, and allowed to breed; the more aggressive might be killed at an early age. If only docile individuals are allowed to breed and if docility is a genetically controlled trait, then they will tend to

have docile offspring. Over many generations this process of conscious or unconscious selection of certain traits has produced not just several types of hunting dogs – spaniels, Afghan hounds, terriers – but also breeds prized for intelligence (sheepdogs), speed (racing dogs), companionship (many small breeds) and even aesthetics: the fact that most dog breeds have much shorter muzzles than wolves has been suggested to reflect unconscious selection of individuals with flatter, more 'human-like' faces (Figure 23.16).

The development of domestication is very significant in human history because it marks a qualitative change in the relationship between people and the environment. Domestication is a special kind of symbiosis, a kind of co-evolution between humans and other species. The new life forms created by domestication were effectively a new kind of 'tool' that humans could use to increase the efficiency and effectiveness with which they could control their environment. We have seen that the domestication of dogs helped people to transcend their biological limitations as hunters. In the Holocene, the domestication of many other animals – goats, sheep, horses, cattle – gave humans the ability to convert biomass they could not eat (grass) into food they could (meat and milk); new technologies for keeping warm (wool); and the ability to travel much faster than human anatomy alone will allow and a ready source of superhuman motive power (horses and other beasts of burden). Similarly, the development of crop plants gave us a way to turn more of the available sunlight, through photosynthesis, into

(a)

(b)

(c)

(d)

Figure 23.16 Selective breeding has transformed wild animals into valued domestic animals. The transformative power of selective breeding is never more obvious than in the case of the transformation of the wolf into the many breeds of dogs that now fulfill many functions, from guarding sheep, helping with hunting, racing and companionship: (a) wolf, (b) sheepdog, (c) bichon frise, (d) greyhound. (Source: (a) Creatas, (b) Shutterstock.com: Erik Lam, (c) Shutterstock.com: Sue C, (d) Shutterstock.com: Eric Isselee)

the things we find useful: food of course, but also textiles, fuel, wood and medicines.

The dog is rather exceptional in having been domesticated so early. Almost all domestication occurred during the Holocene (or shortly beforehand, during the Younger Dryas; Blockley and Pinhasi, 2011). Most of the earliest sites of domestication are in the Near East, in the so-called 'fertile crescent'. Here, around 12 000 years ago, hunter–gatherers lived by collecting the large seeds of certain wild grasses, the ancestors of wheat and barley (Figure 23.17). At some point they must have realized that planting the seed in the soil led to a more predictable crop the following year. They favoured the plants with the largest seeds, and with non-shattering seed heads (most grasses drop their seed to the ground when ripe, but a few genetic mutants in any population will retain the seeds on the head where they can

more easily be collected). Consequently, over time, the average seed size of the wheat and barley they grew increased; and as the seeds remained on the seed heads, these new strains could not reproduce without human help. The first domesticated crops had been developed by the first cultivators.

In these first farming communities in the Near East the new varieties of cereals were joined by other crops – peas, lentils, beans – and by domesticated animals, notably goats in the first instance. Over time, more crops and animals were added to the Near Eastern agricultural package. Elsewhere in the world, and often entirely independently, people began domesticating plants and animals as the Holocene progressed. Domestication seems to have been concentrated in a few key areas (known as 'Vavilov centres' after the first person to identify them). From South and Central America, for example, we now have potatoes, maize, squash, chilli peppers, cotton, tobacco, turkeys and llamas, while from South-East Asia we have rice and chickens.

23.5.3 Social and environmental consequences of agriculture

The beginning of agriculture was important not just in its own right, as a notable innovation in human behaviour, but for its repercussions, both for *Homo sapiens* as a species and for the relationship between people and the environment. The two are inextricably linked, because the invention of agriculture seems to have triggered a series of further innovations in human behaviour which, in turn, affected the environment.

One important consequence of agriculture was that the populations of farming communities grew rapidly. The gap between offspring in modern hunter–gatherer societies is often around four years. In farming communities, where the food supply can be more assured, where more food is available from a given area of land, and where the settled way of life means that infants are not such a burden as they are to nomadic hunter–gatherers, the gap between births is typically around two years. Although there are exceptions to this pattern, groups dependent on farming generally grew rapidly. That is one reason why there are as many as 7 billion people on Earth today.

Farming also appears to have enabled technological innovation. Nomadic hunter–gatherers tend to own few possessions, and their societies tend to be non-hierarchical and unspecialized: everyone contributes more or less equally to food collection (although there can be strong gender biases). For farming societies, which are rooted to the land they cultivate, there is nothing to prevent them

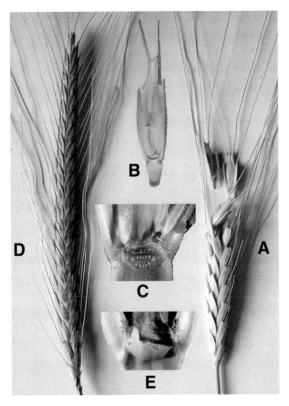

Figure 23.17 Differences in the shape of wild and domestic einkorn wheat. Wild einkorn (a) disperses its seeds (b) naturally, leaving a smooth scar (c) on the base of each spikelet. By contrast, the domesticated variety of einkorn (d) has a 'non-shattering' head: the seeds are firmly attached. This makes it much easier for farmers to harvest the ripe grain. When the seeds are separated from the ear during threshing, the spikelets break off leaving a characteristically rough edge (e) which makes it easy for archaeologists to tell whether individual grains come from wild or domesticated varieties. (Source: from Tanno and Willcox, 2006)

from accumulating goods, even if those goods are large and heavy. Furthermore, because many farming communities can produce more food than they need, they are able to support some individuals who do little or no food production. These people can specialize in using and developing other technologies, such as pottery, woodwork and, in the second half of the Holocene, metalwork. Because individuals are able to accumulate wealth in the form of land and material goods, some become more wealthy than others. This encourages the development of stratified societies, where power is concentrated in the hands of a few people. They are then often able to mobilize others to contribute to large-scale social projects: constructing irrigation systems, building city walls, or raising monuments and temples. It is impossible to envisage the complex society we live in today being supported by a hunter–gatherer way of life. By permitting technological innovation and enabling social stratification, the transition to agriculture paved the way for the long-term increase in humankind's ability to shape and transform the planet. We will explore some of these transformations in the next section.

Reflective questions

➤ Summarize the differences between natural and artificial selection.

➤ Why do you think that so many species of plants and animals have *not* been successfully domesticated?

23.6 Human interaction with physical geography

23.6.1 Out of Eden?

One possible interpretation of environmental history, and perhaps a common one among western peoples, is that before the Industrial Revolution (*c.* AD 1800) people lived in relative harmony with nature. This view can be traced back to the Romantic movement of the nineteenth century, which was critical of the changes that were being wrought on the landscape in the name of progress, the 'dark satanic mills', and that harked back to an idealized, gentler past, where people farmed and hunted in a timeless, changeless world. There is no doubt that the invention of the steam engine and subsequent methods of energy production and transforma-

tion, like the invention of agriculture, marked a step change in the behaviour of people. The impacts of post-industrial human activity on the natural environment are beyond the scope of this chapter. But academic investigation of the pre-industrial past is counteracting the Romantic myth of a 'golden age' of human–environment relations. Hundreds of case studies have shown that pre-industrial people had significant and far-reaching effects on the environment: they deforested large tracts of land, reduced biodiversity in many areas and introduced new species to others, eroded the soil, altered drainage patterns, and moved earth and reshaped the land to a surprising extent, given that it was almost all done with human and animal muscle-power alone.

That pre-industrial humans changed the world is beyond doubt. A more subtle debate concerns the extent to which this change was for the worse. Environmental historians and archaeologists have perhaps tended to emphasize the negative aspects of human impact: the detrimental loss of biodiversity, the loss of soils, the unintended and unfortunate consequences of deforestation. This echoes the growing concern of environmentalists over the last 50 years or more that modern society is facing an environmental crisis. There is a tendency to look for signs of similar crises in the past, which perhaps says more about the preoccupations of modern society than about the realities of the past. Indeed, one justification that environmental historians sometimes use for their work is that a better understanding of how people caused and reacted to past crises, which often take decades or centuries to play out, can help us to understand our present situation. For example, there are many case studies of societies that 'collapsed' in the past due at least in part to ecological pressure on their environment. The most famous example is that of the Easter Islanders, whose society appears to have collapsed due in part at least to deforestation (see Box 23.3). If an ecological crisis through over use of a key ecological resource can cause an island society to collapse, perhaps this proves that a similar collapse is a real threat to global civilization today.

While the investigation of environmental crises is certainly an important aspect of this field, we can make a strong argument that the story of past human impacts is largely a positive one. For every example of resource depletion there is another example of successful husbandry of resources over the course of millennia. For every degraded and ravished landscape, there is another that has been transformed by generations of farmers into a rich, productive, benign home for people. Studying the past can give us a deeper appreciation of the extent to which our civilization depends on this accumulated long-term effort, and can help us to maintain that legacy far into the future.

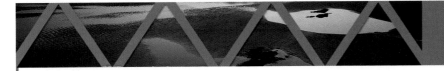

CASE STUDIES

EASTER ISLAND

Easter Island, or Rapa Nui, is the most isolated inhabited island on Earth, lying some 2250 km from Pitcairn Island, its nearest neighbour in the South Pacific (Figure 23.18). Almost from its first discovery by Europeans on Easter Day, AD 1722, the island has been the cause of much discussion and mystery. The first European visitors encountered an island almost bare of trees, mostly covered by grassland, and populated by a few thousand Polynesian farmers growing potatoes and plantains and keeping chickens. The most striking feature of the island was the presence of hundreds of huge statues, called *moai*, carved out of the volcanic rock, up to five metres in height, often with a large rounded stone perched on top like a hat, and arranged in rows on well-crafted masonry platforms. Many of them had fallen over by the time the Europeans arrived. Captain Cook,

who visited the island in 1774, marvelled at the statues in his account of his voyage: 'We could hardly conceive how these islanders, wholly unacquainted with any mechanical power, could raise such stupendous figures, and afterwards place the large cylindric stones before mentioned upon their heads' (Cook, 1777). Cook assumed that the statues had been made by the ancestors of the contemporary Easter Islanders, as he noted that the production of new statues had stopped and no attempt had been made to raise the fallen ones. This led to speculation that the Easter Islanders had undergone a social collapse, leading to the loss of the art of statue-making.

At just 22 km by 11 km, the resources on Easter Island are inherently limited. This has led many researchers to suspect that the decline of the native society was due to a lack of prudence in con-

serving resources. In the 1980s the palaeoecologist John Flenley and colleagues offered new data to support this idea (see Flenley *et al.*, 1991). They studied the pollen in sediment sequences from the three volcanic craters on the island. All three pollen records suggested that, until about AD 1500, the island was covered in woodland dominated by a species of large palm tree (Figure 23.19). This conclusion was shortly afterwards confirmed by archaeological finds of fossil fruits similar to those of the palm *Jubaea chilensis*. Flenley and his colleagues, and later workers, speculated that deforestation would have caused soil erosion, undermining agriculture. Deforestation might also have caused local aridity by reducing recycling of water by evapotranspiration and increasing exposure to desiccating winds. The loss of trees in itself would have constituted a major loss of resources: without timber,

Figure 23.18 Location of Easter Island in the South Pacific.

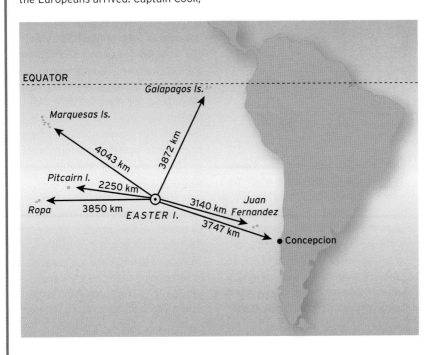

BOX 23.3 ➤

➤

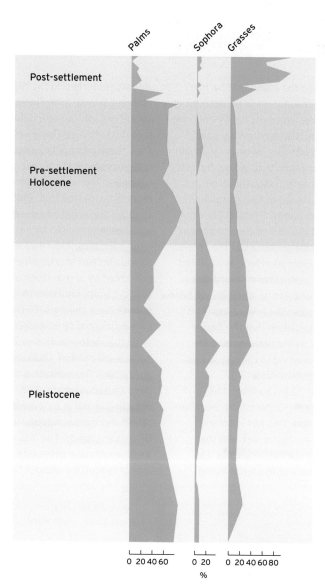

Post-settlement

Pre-settlement
Holocene

Pleistocene

0 20 40 60   0 20   0 20 40 60 80
%

Figure 23.19 Summary pollen data from Rano Raraku, Easter Island. (Source: based on data presented in Flenley *et al.*, 1991)

canoes for fishing could not be built and houses could not be repaired, and the loss of palm fruits would have removed a valuable food source.

The causes of this deforestation have been much debated. Some researchers see Easter Island as a paradigmatic example of ecological disaster brought about by careless resource use. One scenario is that the Easter Islanders became driven by inter-tribal competition which was expressed in the building of *moai* (Figure 23.20), which symbolized prestige or ownership of land (perhaps in the same way that expensive cars and grand buildings symbolize wealth and power in modern western society). They may have needed large timbers to construct sledges and rollers to help move the statues from the quarries to their final platforms, and then to lever them upright. Perhaps the compulsion to build *moai* outweighed the need to conserve palm trees; perhaps, as Diamond (2005) envisaged, someone consciously took the decision to chop down the last tree, knowing that there would be no more.

This vision of ecological suicide is compelling to many environmentalists who fear that modern society might be facing a similar situation on a grander scale. Perhaps what happened on Easter Island is a harbinger of the self-destruction that might occur for the whole population of our planet, if we keep polluting the atmosphere, changing the climate, replacing forests with fields, and driving animals and plants to extinction.

Other authors are more sceptical. They point out that even in the case of Easter Island, which is probably the best-documented example of 'ecological overshoot' (Tainter, 2006) in the palaeoecological canon, the case for such thoughtless self-destruction is far from proven. The society of Easter Island might indeed have become simpler over time, but not to the point of total collapse: there were still several thousand Easter Islanders even at the time of European contact, making a reasonable living from agriculture, with food to spare for trading. Moreover, the extinction of the palm trees might not simply have been due to intentional deforestation by people. Every one of the fossil palm fruits that have been recovered from archaeological contexts shows signs of having been gnawed by a small species of rat, probably *Rattus exulans* which is common on Pacific Islands and was probably introduced by the first Polynesian settlers on Easter Island (who probably arrived around AD 1200; Hunt and

BOX 23.3 ➤

Figure 23.20 Examples of *moai*, the giant statues raised in prehistory by the Easter Islanders. (Source: photo courtesy of Ty Parker)

Lipo, 2006). As the rat population increased, more and more of the palm seeds would have been eaten before they could germinate. So even if the decline of Easter Island's society can be blamed on ecological degradation, that degradation might have been accidental (by the unintended introduction of the rats) and, once under way, unstoppable: still bad news for Easter Islanders, and perhaps still with important parallels in the modern world, but much easier to accept than that people should behave so illogically as to seal their own fate by a deliberate act of environmental suicide. These arguments are still being hotly debated: see, for example, Hunt and Lipo (2009), Mieth and Bork (2010) and Rull *et al.* (2010).

BOX 23.3

In the following sections we will look at three key facets of human impact on past environments: deforestation, soil erosion and drainage.

23.6.2 Deforestation

Having evolved from a lineage of apes that was specialized in consuming fruits, which are relatively rich in sugars, human beings have unusually limited digestive systems. Unlike most herbivores, we cannot digest cellulose, which makes up the bulk of plant biomass on the planet; and, unlike most carnivores, we find uncooked meat relatively difficult to digest (meat-eating seems to have been a relatively late innovation in hominin evolution). Only in the tropics are fruits readily available all year round, and even then they make up a small proportion of the available biomass. The natural vegetation of temperate zones is even less edible from the point of view of humans. One great liberating consequence of farming is that it enables us to turn much more of the available energy from the Sun into a form we can eat. We replace inedible forests with fields of grain, or, on poorer soils, with grassy pastures stocked with sheep and cattle which convert the cellulose in grass into edible (cooked) meat and milk. Thus agriculture necessarily involves the conversion of natural, relatively unproductive vegetation – and over most of the planet, that means forest – into a cultural landscape where a much greater proportion of the Sun's energy ultimately becomes stored in a form that is useful to humans.

Destruction of woodland by colonizing farmers is thus a common feature of the past, and pollen diagrams from many regions of the Earth mark the coming of farmers by a decline in tree pollen abundance. The actual process of deforestation probably varied from place to place. Trees can be cleared deliberately for agriculture by axe, though this is hard work. In drier environments, woodland can be burned, as it is in the subtropics today. Less rapid, but much easier, is to allow your animals to clear the woodland for you. Goats and sheep will eat saplings and pigs will root them up; over the course of decades to centuries the woodlands around a pastoral settlement will tend to thin out unless grazing is carefully kept in check. Collection of relatively small amounts of wood for fuel and timber for construction can, over long periods, have a similar effect.

The consequences of converting forest to what Rackham (1986, p. 72) called 'an imitation of the dry open steppes of the Near East' – the natural environment of the ancestors of most of our crops and animals – go beyond the obvious loss of plant and animal diversity. After all, we are talking about a major change in the structure of the ecosystem. Forests are three-dimensional structures: they harbour animals and understorey plants that are simply unable to survive in the disturbed, unshaded, two-dimensional landscape of fields and pastures. With clearance, the total biomass declines, and as the trees rot or are burned their stored carbon is released back into the atmosphere. Carbon emissions from modern deforestation are a significant contributor to global warming, but Ruddiman (2003) pointed out that the same process has been occurring for at least 8000 years. He argued that atmospheric CO_2 concentrations have been elevated for most of the Holocene as a result (though other climate scientists argue that pre-industrial deforestation was slow enough for the excess CO_2 to be absorbed by the oceans, rather than accumulating in the atmosphere). A canopy of trees also intercepts a lot of rainfall; loss of trees therefore has consequences for soil erosion and hydrology (see below).

All this talk of destruction might seem rather negative, so it is worth reiterating that the loss of forests is not a 'bad thing' in every respect. After all, without cleared space for farmland, most of us would not be here. Furthermore, western people today have a tendency to romanticize forests. Just two centuries ago in Europe and North America, attitudes were very different: forests were dark, ominous wildernesses, unproductive and harbouring bandits and dangerous animals; they made a landscape claustrophobic and difficult to traverse. To an eighteenth-century pioneer in North America, our current fondness for forest would seem difficult to understand.

People have also had a more positive relationship with woodland. In most environments, efforts have been made to retain some woodlands because they provide valuable resources: fuel-wood and timber, some seasonal nuts and fruits, deer and other animals for (recreational) hunting, and forage for animals such as pigs. These remaining woodlands were often much more carefully managed than they are today, with regular cycles of cutting (coppicing and pollarding) and legislation to prevent over-exploitation that can seem surprisingly modern in its concern for conservation and sustainability. Many European woodlands still show the signs of past management in the form of, for example, banks and ditches designed to control animals and delineate parcels of land, overgrown coppice stools that have produced a sustainable harvest of wood for centuries, and deliberate plantings of favoured species such as sweet chestnut.

23.6.3 Soil erosion and impoverishment

For farming communities, the health of the soil is absolutely paramount. Most crops grow best in relatively thick soil with good water-retention properties and the capacity to store and provide nutrients. Unfortunately, in many places one long-term consequence of farming has been to erode or otherwise impoverish the soil resource (Chapter 10). Often this comes as a consequence of deforestation. A forest canopy limits the amount of rainfall that reaches the soil, which limits the erosive effects of rainfall impact and slope-wash. So little rainfall reaching the soil means that soluble nutrients such as some forms of phosphorus are not lost to leaching. Tree roots bind the soil physically together. Leaf litter adds a constant supply of organic material to the soil, which improves water retention and nutrient status. When the forest is removed by deliberate clearance, fire or grazing, all of these benefits are lost.

The results of deforestation on soils can be quite spectacular. Physical soil erosion is particularly important on

Figure 23.21 Degraded hillslopes in Spain. The loss of forest cover has led to soil erosion on a large scale across much of the limestone terrain of the Mediterranean. (Source: photo courtesy of Katy Roucoux)

sloping terrain. For example, across wide areas of Britain, rivers have deposited large amounts – often a metre or more – of reworked soil since the Bronze and Iron Age; these deposits are often referred to as the 'Romano-British Silt', although their origins can be older than the Roman period. In more mountainous regions, such as the Mediterranean Basin (Figure 23.21), the removal of forests mainly through over-grazing has caused the almost total loss of the thin soils that had built up on limestones over many thousands of years.

Archaeological sites of Greek or Roman age are frequently buried under many metres of river-transported soils. In Iceland, where the soil is made up of very fine volcanic ash, the removal of birch woodland by Viking colonists allowed the soil to be simply blown away by the wind into the North Atlantic (Figure 23.22). It is estimated that 40% of Iceland's soil cover has been stripped since the settlement began around AD 870.

Of course, not all soils have been lost. In fact eroded soils are often transported into basins or re-deposited on river floodplains, where they can make exceptionally fertile farmland: one region's loss can be another's gain. In flatter areas, or where the soil contained enough clay to make it sticky and resist erosion, the original soil cover may still be more or less intact. Nonetheless, even where it remains *in situ*, the quality of the soil can change. In the uplands of the UK, the removal of trees and the consequent increase in the amount of rainfall reaching the soil has meant that soluble nutrients

have tended to be washed away more quickly. This accelerated leaching tends to make the soil not just less fertile, but also more acidic, which encourages colonization by bog plants such as *Sphagnum* moss. *Sphagnum* absorbs water and makes the soil much wetter, and therefore more likely to be anaerobic, so plant litter decays more slowly. Under the right circumstances, this can lead to the beginning of peat growth. Blanket peats across much of the British and

Figure 23.22 A degraded landscape in Iceland. The fine volcanic soils of Iceland are particularly prone to wind erosion. Once the woodland cover is damaged the wind begins to eat into the soil from above, and then sideways. Up to 40% of Iceland's soil has been stripped from the glacial substrate since the settlement of the island around AD 870. (Source: photo courtesy of Mike Church)

Irish uplands probably owe their origin, at least in part, to prehistoric deforestation.

23.6.4 Irrigation and drainage

Plants need water to survive, and in arid areas farmers need to find ways to direct freshwater onto their land, and to keep it there. On the other hand, waterlogged basins can be too wet: because most of the ancestors of our crops are adapted to a semi-arid climate, they will not flourish in wet conditions. For these reasons, a range of technologies have been developed over millennia to allow farmers to control the amount of water their plants receive.

Irrigation and drainage have a long history in many parts of the world. In Britain, the focus has been on drainage, particularly in the low-lying areas of eastern England. The Romans began extensive drainage of the fenland of East Anglia, a process that was accelerated in the seventeenth and eighteenth centuries when first wind power and then steam were used to pump water out of the low-lying peaty fens. Rivers were straightened so that they would drain more quickly to the sea. Earth banks were used to constrain the floodplains of rivers and to keep the sea at bay. Every field had its own ditches and drains to encourage the water to leave as quickly as possible. The drained soil, made up mostly of rich fen peat or (towards the coast) silty sediments, made superb farmland, and the region remains a cornerstone of British agriculture. One unexpected consequence of the drainage, however, was that the soil began to disappear. Peat can be more than 90% water by weight, and the loss of water causes it to shrink. The dry peats become vulnerable to wind erosion. Worse still, as the peat dries out, so oxygen levels rise, encouraging rapid aerobic decay, so the peats are gradually being oxidized to carbon dioxide gas. In some places, the vertical thickness of peat lost is impressive. In 1851 an iron post was driven into the peat until its top was flush with the surface at Holme Fen in Cambridgeshire (Figure 23.23). Today, the top of the post stands more than 4 m above the peat surface. Embanked rivers often flow high above the level of the surrounding land. Artificial coastal defences and constant pumping are all that keep the Fens from being reclaimed by the sea.

The East Anglian Fens are an extreme case of a general tendency across the temperate world to drain boggy land and convert it to farmland. Studying an early map will often show that, just a century ago, the landscape was much more watery than it is today. Thousands of ponds, fens, marshes and wet woodlands have been reclaimed by agriculture. Preserving and restoring wetlands, for the benefit of the plant

Figure 23.23 The Holme Fen Post. In 1851 this post was driven into the peat at Holme Fen so that its top was level with the surface. Now, due to drainage and peat shrinkage, it stands 4 m tall.

and animal diversity that they support, is now a conservation priority.

In the drier parts of the world, by contrast, the problem has always been to *increase* the water supply. Like drainage, irrigation has a very long history. Some of the most famous early civilizations developed on the basis of their command of the water supply, such as the so-called 'hydraulic civilizations' of Mesopotamia (Iraq) (Figure 23.24). From around 10 000 years ago, groups of farmers came together on the floodplains of the Tigris and Euphrates Rivers and developed sophisticated systems of flood-farming, diverting the courses of rivers and embanking and arranging their fields so that water could be retained and made to flow from one field to another. The combination of a hot climate, fertile floodplain soils, and a carefully controlled water supply made these farming techniques very productive, and empires were built on the surpluses they produced. In the long run, however, problems could emerge. Disputes over water rights, and the political difficulty of building and maintaining such large shared resources,

Figure 23.24 The Ziggurat of Ur. This ancient, mud-brick temple has been largely reconstructed to give a sense of the achievement of its builders. Mesopotamian cities including Ur flourished between 6000 and 1500 years ago, but failure to maintain complex irrigation networks, coupled with problems such as soil salinization, means that today they stand in the desert.

could lead to conflict. In such arid areas, excessive evaporation in irrigated fields can draw salts upwards through the soil, eventually making it toxic to plants and ruining the land. On the whole, though, the transformation of Mesopotamia from semi-desert to the bread-basket of the Near East was an outstanding example of landscape engineering that allowed complex civilizations to flourish for many millennia.

Reflective questions

➤ To what extent have past human–environment relationships been sustainable?

➤ How do human–environment interactions today differ from those of the pre-industrial past?

23.7 Summary

Although it is the most recent epoch of Earth history, and although we have learned an astonishing amount about it in recent decades, there is much still to be found out about the Holocene. In particular, although it would still be fair to say that the Holocene has been climatically rather stable by comparison with most of Quaternary time, the true complexity of Holocene climate variability is only just beginning to emerge, spurred on by developments in palaeoclimatological techniques. Scientists are still debating the very existence of some climatic events that lie close to the detection limits of current methods. For example, there are hints in various palaeoclimatic records that the later Holocene was punctuated by short-lived cold or dry events reminiscent of, but less extreme and less widespread than, the 8200 BP event; one such drought event, 4200 years ago, is thought to have led to the collapse of the Akkadian Empire in the Near East and of the Old Kingdom in Egypt. The exact character and magnitude of this and many other events and cycles remain to be properly established. There are also other possible causes of climate change, such as volcanic eruptions, which remain to be explored in detail.

There is also a lot left to be discovered about the effects of Holocene climate change on geomorphology and ecology. For instance, efforts have long been under way to determine the history of glaciers and ice caps within the Holocene; the results are beginning to help us to put the present changes in the planet's ice into context, and to predict what to expect in the warming world of the twenty-first century. Similarly, even in areas where Holocene palaeoecological research has a long history, such as northern Europe, new techniques and new data sets are constantly refining our understanding of the sensitivity of ecosystems to environmental change. In less-explored parts of the world, notably the tropics, palaeoecological research is still in its infancy.

In the last two sections this chapter examined the rise of civilizations, and some of the ways in which pre-industrial societies interacted with and affected their environment. Living as we are in an age where technological progress is rapid, the human population is expanding dramatically, and 'natural' ecosystems seem to be under increasing pressure, it is important to place our modern situation in the longer-term context. Many of the environmental problems that society is struggling with today, such as deforestation, soil erosion and the loss of biodiversity, have a much longer history than we commonly recognize. We can also appreciate that much of what we value about the environment is, in fact, the result of human activity. Management of natural resources has not always been detrimental: the countryside and landscapes that we enjoy today are more often than not the result of coppicing, hedging, ploughing, ditching, drainage, planting and herding over centuries or millennia: a huge investment of human labour and ingenuity. Understanding how these landscapes came about is key to making the most of this legacy for the future.

Further reading
Books

Anderson, D.E., Goudie, A.S. and Parker, A.G. (2007) *Global environments through the Quaternary: Exploring environmental change.* Oxford University Press, Oxford.
Perhaps the best of the current textbooks about Quaternary science in general.

Diamond, J.M. (1997) *Guns, germs and steel: A short history of everybody for the last 13,000 Years.* Jonathan Cape, London.
A classic and very readable account of the history of the human race from an environmental perspective.

Diamond, J.M. (2005) *Collapse: How societies choose to fail or survive.* Allen Lane, London.

An excellent account of the interaction between society and environment during the Holocene.

Fagan, B. (2000) *The Little Ice Age.* Basic, New York.
A popular account of the Little Ice Age.

Grove, J.M. (2004) *Little Ice Ages: Ancient and modern,* 2nd edition. Routledge, London.
This is the classic account of the Little Ice Age, drawing on a wide range of sources to paint a fascinating picture of this climatic event and its impacts on society.

Roberts, N. (1998) *The Holocene: An environmental history,* 2nd edition. Blackwell, Oxford.
This is a good, well-illustrated textbook on the Holocene.

Further reading
Papers

Ballantyne, C.K. (2002) Paraglacial geomorphology. *Quaternary Science Reviews*, 21, 1935–2017.
A good review of this subject.

Bard, E. and Frank, M. (2006) Climate change and solar variability: what's new under the sun? *Earth and Planetary Science Letters*, 248, 1–14.
A sceptical look at direct solar forcing.

Flenley, J.R., King, A.S.M., Jackson, J., Chew, C., Teller, J.T. and Prentice, M.E. (1991) The Late Quaternary vegetational and climatic history of Easter Island. *Journal of Quaternary Science*, 6, 85–115.

This is a classic paper on the palaeoecological story of Easter Island. It very carefully presents and considers the available evidence and constructs an elegant argument. The three more recent papers on Easter Island below show how these basic data can be debated and re-interpreted:

Hunt, T.L. and Lipo, C.P. (2009) Revisiting Rapa Nui (Easter Island) 'ecocide'. *Pacific Science*, 63, 601–616.

Mieth, A and Bork, H.-R. (2010) Humans, climate or introduced rats – which is to blame for the woodland destruction on prehistoric Rapa Nui (Easter Island)? *Journal of Archaeological Science*, 37, 417–426.

Rull, V., Canellas-Bolta, N., Saez, A., Giralt, S., Pla, S. and Margalef, O. (2010) Paleoecology of Easter Island: evidence and uncertainties. *Earth-Science Reviews*, 99, 50–60.

CHAPTER 24

Climate change: an unprecedented environmental challenge

John Grace
School of GeoSciences, University of Edinburgh

Learning objectives

After reading this chapter you should be able to:

➤ discuss the evidence for anthropogenic global warming

➤ understand how the use of fossil fuels has impacted upon the climate

➤ describe how the carbon cycle has been perturbed

➤ appreciate how humankind has created environmental problems and perceive how they may be solved

24.1 Introduction

Environmental change on a global scale first became a matter of public concern in the 1960s. Before then, the perceived environmental problem was urban pollution, which affected human health and the quality of life of so many people. Although urban pollution became acute during the Industrial Revolution, it was not new. The smelting of toxic metals such as copper and lead was a health hazard in ancient Rome, as revealed by analysis of hair samples from the preserved corpses of Roman soldiers found in bogs, and from traces of metal in Greenland ice cores. Coal was used in London in the thirteenth century. Coal contains not only carbon but also 1–4% sulphur and traces of heavy metals, and therefore its combustion releases a multitude of pollutants as well as carbon dioxide (CO_2). With the onset of the Industrial Revolution in western Europe, around 1780, the use of coal increased dramatically and cities like London became heavily polluted with smog, a mixture of fog and smoke. Domestic coal burning was a major contributor to smog, and the industrial regions around Birmingham in England became known as the Black Country and even non-industrial Edinburgh was known as Auld Reekie, referring to the smell of coal burning.

Diseases such as bronchitis and tuberculosis were widespread following the Industrial Revolution, and nearly a quarter of deaths in Victorian Britain (1837–1901) were from lung diseases. In one week of December in 1952, 4000 Londoners were killed by a particularly severe episode of smog. The ensuing public outcry resulted in the Clean Air Act of 1957, which restricted coal burning and resulted in the use of cleaner energy sources such as oil, gas and electricity. Other coal-burning cities of the world such as Pittsburgh in the United States have a similar history. Problems were greatly exacerbated by the growth in use of the motor car, especially in regions receiving high solar radiation, such as Los Angeles, Athens and Mexico City, where the ultraviolet radiation reacts with uncombusted hydrocarbons and oxides of nitrogen from exhausts of cars to yield photochemical smog, irritating the eyes, nose and throat. An important milestone in the awakening of environmental concern was prompted by the widespread use of the persistent pesticides that were introduced after the Second World War and the publication of Rachel Carson's book *Silent Spring* in 1962 (see also Chapter 27). *Silent Spring* warned against the dangers of pesticides, especially to songbird populations, indicating how persistent chemicals might spread in food chains as well as in the atmosphere, and ultimately damage non-target species. At the same time, other scientists were demonstrating that the pesticide DDT could be found in rainwater in the Antarctic, and that pesticides were responsible for eggshell-thinning in wild birds, threatening especially those species at the end of food chains such as raptors. Thus, the idea of environmental change *on a global scale* soon became a permanent part of western culture, and part of the international research agenda.

The global scale of human influence on the planet is today felt even more strongly, but not because of fears of widespread pollution of the land and sea by pesticides. The global environmental challenge that we face now is climate change, and that is the main focus of this chapter.

24.2 Climate change

24.2.1 Long-term change

The climate has always fluctuated, but usually over very long timescales. There are many sources of information that help in the reconstruction of past climates. These include historical records, evidence from the annual growth rings of trees, deposits of pollen in lakes and bogs, isotopes and fossils. The picture that emerges is quite complex, showing cyclic fluctuations on several scales (Figure 24.1). The long-term cyclic trends in the Earth's temperature, associated with periods of glaciation known as 'the Ice Ages', were attributed by Scottish geologist James Croll and Serbian astronomer Milutin Milankovitch to the irregularities in the orbit and tilt of the Earth, which influence the energy received from the Sun (see Chapter 22). These are known as the Milankovitch cycles. In contrast to these gradual changes, there have also been catastrophic events causing mass extinctions on a global scale. For example, in the Late Permian (245 million years ago) about half the families of marine animals were lost. At the boundary of the Cretaceous and Tertiary (known as the KT boundary, some 65 million years ago) 15% of marine families were lost, perhaps 75%

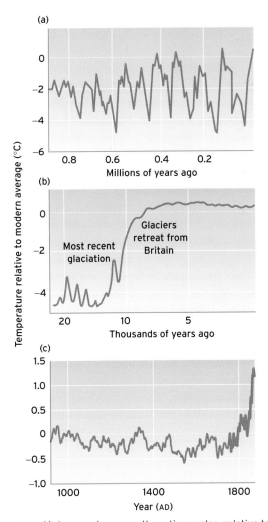

Figure 24.1 Air temperature over three timescales, relative to modern records: (a) the past million years; (b) the past 20 000 years; (c) the past millennium. (Source: after Brohan *et al.* 2006; Hegerl *et al.* 2007)

of plant species, and (most famously) dinosaurs became extinct. Such events are now usually attributed to the impact of comets, asteroids or large meteorites, which would have thrown up debris into the atmosphere and greatly reduced the penetration of solar radiation, causing widespread cooling, a reduction in photosynthesis and collapse of food chains. The KT boundary is considered to have been caused by an asteroid 10 km in diameter, which impacted at Chicxulub, north-east Yucatan, Mexico. Geologists recognize five such mass extinction events in the fossil record, all of them global in extent, taxonomically broad, and most evident in marine invertebrates (for which the fossil record is relatively complete). It is against this background that we examine the changes in the climate system which are currently occurring, and their link to anthropogenic activity.

24.2.2 Recent climate change and its causes

Over the past century the Earth has warmed by about 0.7°C, particularly over the last decade (Figure 24.2). Apart from this modern instrumental record from meteorological stations, there are a number of independent sources of information to demonstrate the phenomenon of climate warming: glaciers have been receding, snow cover has declined, polar ice has been melting, sea levels have been rising and spring has been earlier.

Many authors describe the present-day temperatures as 'unprecedented'. We know the temperatures and the concentrations of CO_2 and CH_4 that have occurred over the last 650 000 years from analysis of deep cores taken from polar ice, and we can compare these with those being experienced now. Although the 650 000 year record does contain large fluctuations, associated with the ice ages and the warm interglacial periods, we see that today's temperatures and concentrations of CO_2 and CH_4 are much higher. Moreover, the rate of increase in temperature is faster now than previously.

A causal association between greenhouse gases and temperature is inevitable, ever since the demonstration in 1859 by the Irish scientist John Tyndall that CO_2 absorbs infrared

radiation. In the Earth's atmosphere, CO_2 and a range of other gases, including water vapour, absorb some of the infrared radiation that would otherwise stream directly out to space, thus causing a heating effect known as the greenhouse effect (the name arises because glass also absorbs infrared radiation and so the glass panes in a greenhouse have exactly the same effect on a local scale). The amounts of three of these gases, CO_2, CH_4 and N_2O (nitrous oxide), have risen sharply in recent times, and the extent of warming to be expected from these rises can be calculated (see the right-hand axis of Figure 24.3). The rise in heat supply to the Earth's surface caused by these gases, known as radiation forcing (see Chapter 4), amounts to 2–3 watts per square metre ($W\ m^{-2}$). This is much smaller than the average incoming solar radiation (averaging about 230 $W\ m^{-2}$ at the Earth's surface) but enough to increase the global temperatures.

We know for sure that humans have emitted vast quantities of CO_2 by burning fossil fuels and biomass, thus interfering with the global carbon cycle. The rise in CH_4 concentrations can be attributed to increases in various types of human activity. Only about 45% of all CH_4 emissions are produced naturally: from wetlands, termites, the ocean and from the decomposition of gas hydrates. The remainder is anthropogenic: from energy production, rice fields, landfills, ruminant livestock, waste treatment and biomass burning. The rate of increase in CH_4 has in fact been levelling off in the last few years. As for N_2O, the causes of its increase are somewhat less clear. It is produced naturally by microbial activity in the nitrogen cycle (see Chapter 25), and at a much faster rate when land is 'improved' by the use of nitrogen fertilizer. It is estimated that about one-third of the global emissions of N_2O are anthropogenic.

Other processes influence global temperatures (see Figure 4.23 in Chapter 4). Some are less well understood, and are the subject of current research. One such case is the influence and general behaviour of aerosols. These are particles in the atmosphere, including fumes and smoke from industrial processes and transport, and naturally produced particles such as pollens and spores. To some extent, they shield the planet from solar radiation, absorbing, scattering and reflecting solar radiation. The aerosol 'haze' which we see in the clear sky (especially in the northern hemisphere) effectively reflects part of the incident solar energy back into space, contributing to a cooling effect and therefore offsetting the warming effect of greenhouse gases (see Chapter 4). Periodic changes in the aerosol content of the atmosphere, for example by major volcanoes and by periods of heavy industrialization or biomass burning, have the capacity to change the temperature of the planet. Marine **phytoplankton**

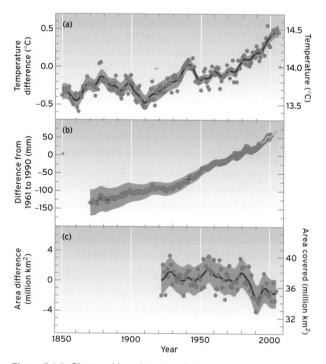

Figure 24.2 Observed trends on (a) global average temperature, (b) global average sea level and (c) northern hemisphere snow cover for the past century. (Source: IPCC 2007a)

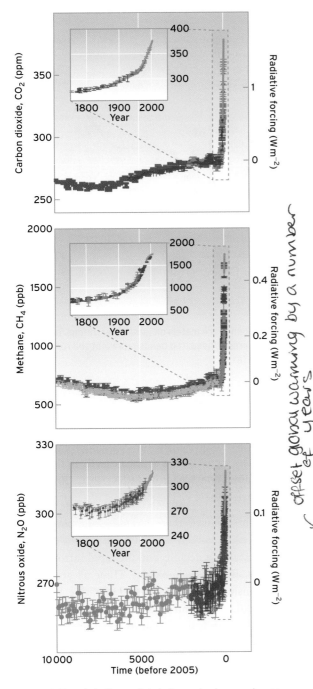

Figure 24.3 Trends in three global atmospheric greenhouse gases over the past 10 000 years. Different colours denote different studies. The inset box shows the period since 1750 in more detail. The left-hand axis shows the concentration of each gas and the right-hand side shows the radiative forcing that the concentration implies. (Source: IPCC, 2007a)

and the vegetation itself contribute to the aerosol content of the atmosphere, by emitting certain volatile organic compounds which form aerosols (Meir *et al.*, 2006).

The uncertainty inherent in estimating the radiative forcing of aerosols comes from the recognition that not all of them behave in the same way. Aerosols from biomass burning (black carbon) and the 'brown clouds' that come from urban sources may have the opposite effect. Ramanathan *et al.* (2007) flew small unmanned aircraft in brown clouds over the Indian Ocean and showed that these low-elevation clouds had a warming effect.

Variations in the Sun's energy output is sometimes proposed as a possible cause of global warming. Although we talk of the energy incident on the Earth as measured outside the atmosphere as the 'solar constant' (and assign it the value of 1366 W m^{-2}), it is not quite constant. The most conspicuous variations are associated with sunspots, which appear as dark marks on the solar surface and arise because of variations in the magnetic properties of the Sun. They occur in an 11 year cycle, but there is a possibility of less conspicuous longer-term trends (see Chapter 23). Changes in the solar output caused by sunspots are only about 0.1% of the solar constant. According to estimates in IPCC (2007a) the changes in solar irradiance since 1750 have caused a radiative forcing of only +0.12 W m^{-2}, and Lockwood and Fröhlich (2007) have shown that the changes in solar radiation over the past 20 years have been in the opposite direction to that required to explain the observed rise in global mean temperatures.

Volcanic eruptions are sometimes large enough to have a short-term impact on the climate (see Box 4.7 in Chapter 4). The June 1991 eruption of Mount Pinatubo injected large amounts of aerosols into the stratosphere. Over the following months, the aerosols formed a reflective layer of sulphuric acid haze and global temperatures dropped by about 0.5°C. Likewise, the April 1815 eruption of Mount Tambora in (modern-day) Indonesia is believed to have been the cause of the exceptionally cold conditions everywhere in the world in the following year: 1816 is known as 'the year without a summer'. Significant volcanic eruptions in recent times were Mount Agung in Bali in 1963 and El Chichonal in Mexico in 1981.

The scientific consensus is, overwhelmingly, that the production of greenhouse gases by humans is the primary cause of recent global warming, as outlined in various reports from the IPCC (www.ipcc.ch). One of the most compelling lines of evidence is that global climate models (GCMs), in which production of these gases is simulated, show the same pattern of global warming as that observed, and in all parts of the world (Figure 24.4). When run without adding anthropogenic production of greenhouse gases, GCMs show no appreciable global warming.

Understanding the causes of climatic variation is still an area of intense research, drawing upon expertise from many scientific disciplines (see Chapters 4, 22 and 23). One

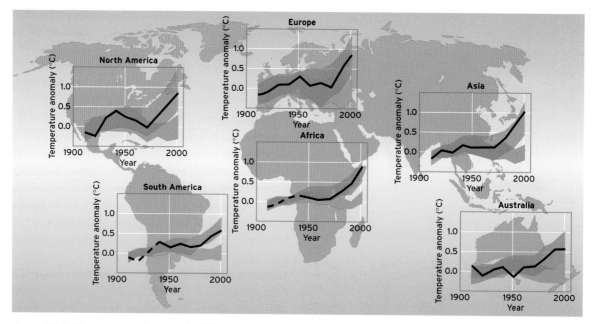

Figure 24.4 Observed warming rates in different regions of the world. Black lines are meteorological observations, blue bands represent the range of model results when no anthropogenic effects are included, red bands represent the range of model results when anthropogenic effects are included. (Source: IPCC, 2007a)

important issue is the behaviour of a myriad of negative feedbacks which tend to dampen any instability, and the extent of the influence of positive feedbacks which might cause run-away warming. For example, as the climate warms and the polar ice melts, the overall albedo of the planet will decline (land and sea absorb more energy than ice and snow). A decline in albedo will make the planet's surface absorb more solar radiation, and thus further warming will occur. In this way, warming may give rise to further warming, an example of a positive feedback loop. Some examples of positive and negative feedbacks are given in Box 24.1.

24.2.3 Predictions from global climate models (GCMs)

Global climate models have developed from global circulation models (both abbreviated GCM), which in turn sprang from the application of numerical methods to weather forecasting. Predictions of the climate for the next century are made by running GCMs with specified prior assumptions about the pattern of greenhouse gas emissions. In reality, these patterns will depend on social, political and economic development in the world, and they are patterns which we cannot foretell. So researchers define them as 'scenarios' or 'storylines', each storyline

having a particular pattern of greenhouse gas emissions, and use GCMs to investigate the consequence of each scenario. The scenarios are defined exactly in the Special Report on Emission Scenarios, SRES (IPCC, 2000) and are summarized here.

A1: In this scenario there is rapid economic growth, an increasing human population until mid-century and thereafter a decline, and the rapid introduction of more efficient technologies. Three A1 groups are distinguished: A1FI is fossil fuel intensive, A1T uses non-fossil energy and A1B uses a mixture of the two. A1B corresponds to what most traditionalists imagine will happen. A1FI leads to the CO_2 concentration rising from its present 380 ppm to around 960 ppm while A1B results in 710 ppm by 2100.

A2: Here, the world develops in a more heterogeneous way with emphasis on self-reliance. Fertility patterns have regional characteristics and converge slowly; economic growth and technological uptake are more fragmented. In this scenario, the CO_2 concentration rises from its present 380 ppm to 860 ppm by 2100.

B1: Like A1, scenario B1 is a convergent world with a population that peaks in the mid-century but with a strong evolution of a service and information technology, with reductions in material intensity and clean

ENVIRONMENTAL CHANGE

CLIMATE FEEDBACK

Anthropogenic activity is believed to be enhancing climate change and encouraging the planet to warm. However, there may be a range of feedbacks that result in different responses to human activity. Positive feedbacks on the climate system will accelerate global warming, whilst negative feedbacks will suppress warming. There are a whole range of interlinked processes that suggest we need to look at environmental change taking a whole-system viewpoint. This box lists some of the hypothesized climate feedbacks which global modellers are investigating.

Positive feedbacks

- Warming will cause release of CO_2 from increased biomass decomposition, primarily in the forest regions of the world but also in the tundra, thus accelerating warming.

- As the sea surface warms, more water vapour enters the atmosphere. H_2O is a strong absorber of infrared radiation and so this increased water vapour is expected to increase the rate of warming.

- Warming will melt snow and ice, decreasing albedo and thus increasing warming, melting even more snow and ice.

- Tropical deforestation will cause warming and drying, itself causing a decline in the rainforests of the world.

- Increased cover of woody vegetation in the high latitudes, caused by warming, will decrease the reflectance of the land surface, and thus accelerate warming.

- Warming will increase the decomposition rate of gas hydrates (see Chapter 17) leading to a release of the potent greenhouse gas methane; this will increase warming.

Negative feedbacks

- Deforestation will lead to an increase of soil erosion, atmospheric aerosols will increase and solar radiation at the surface will decline, causing cooling.

- Replacement of coniferous forest by warmth-loving broadleaved forests and by agriculture will decrease planetary reflectance, causing cooling.

- Increased transpiration in a warm world will lead to more clouds, cooling the planet.

- Increased precipitation and ice melt will result in increased runoff into sensitive parts of the oceans altering the balance between freshwater and saline water thereby resulting in a slowing of ocean circulation and allowing northern high latitudes to cool (see Chapters 22 and 23).

BOX 24.1

technologies. In B1 global solutions are found to economic, social and environmental sustainability. In this scenario, the CO_2 concentration rises from its present 380 ppm to 540 ppm by 2100.

B2: In B2 local solutions to economic, social and environmental sustainability are found; the population growth rate is slower than A2, with intermediate levels of economic development, and there is less rapid technological change than in A1 and B1. In this scenario, the CO_2 concentration rises from its present 380 ppm to 615 by 2100.

When the GCMs are run, we see a warming by 2100 ranging from 1.8°C in the B1 scenario to nearly 4°C in the A1B scenario (Figure 24.5). There are associated changes in rainfall. In the A1B scenario the rainfall patterns are substantially different from those today with more rain

falling in the polar regions while the mid-latitudes will become drier (Figure 24.6). The Mediterranean region of Europe and Central America will both become especially dry according to this prediction.

These changes are profound, especially so as the models suggest a more variable climate with an increasing frequency of extreme events. News reports of storms, droughts and hurricanes are increasingly shown in the media but these alone should not be taken as evidence of a link between global warming and extreme events. Analyses of reliable long-term records and model predictions are the proper evidence that must be considered. Emanuel (2005) investigated data on hurricanes and found the total power dissipated (longer storms and more intense storms) has increased markedly since the mid-1970s. Moreover, model predictions do indeed show

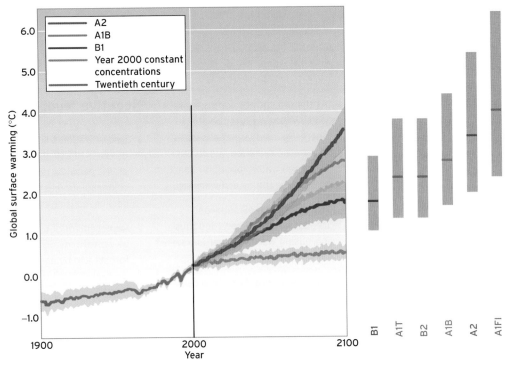

Figure 24.5 Predictions of global warming for different scenarios (see text for an explanation). The coloured lines show the assumed socio-economic scenarios and the bands around the lines show the range of model behaviours. Note: the orange line shows the effect of keeping the concentration constant from the year 2000. (Source: IPCC, 2007a)

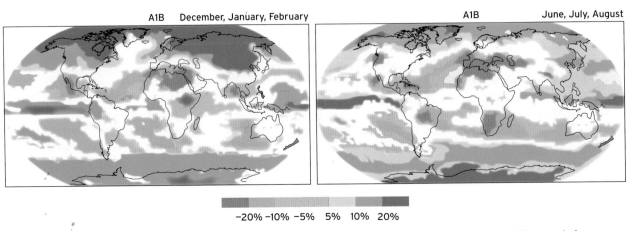

Figure 24.6 Relative changes in precipitation for the period 2090-2099, relative to 1980-1999, assuming the A1B scenario for December to February (left) and June to August (right). White areas are where less than 66% of the models agree on the sign of change. (Source: IPCC, 2007)

an increased variability with an increased frequency of extreme events (IPCC, 2007b). Heatwaves, for example, are expected to increase (Figure 24.7). Results such as these prompt economic analysis and receive the attention of the public and of politicians. Insurance companies can no longer base their premiums on the analysis of past data when the climate system is so clearly changing its behaviour. Such 'extremes' in temperature, rain and wind will all cause appreciable damage, and the cost of repairing the damage will ultimately consume much of the wealth of the world, as emphasized by the report of Sir Nicholas Stern made to the UK government (Stern, 2006). Box 24.2 provides some examples of potential impacts of global warming.

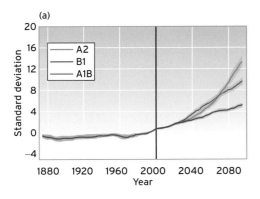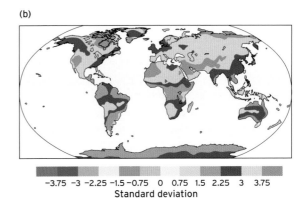

Figure 24.7 Increase in heatwaves over the rest of the century, expressed as the standard deviation of temperature: (a) trends from runs of the GCM for three scenarios; (b) global pattern for A1B. (Source: IPCC, 2007a)

NEW DIRECTIONS

IMPACTS OF GLOBAL WARMING

Most current models suggest a 2–4°C increase in global temperature over the next century. This rate is 10 times faster than the warming experienced over the past 10 000 years, and substantial impacts on human societies are anticipated. The following is a summary of some of the main effects that have been widely discussed in various publications.

- Extinction rates, already very high, will be increased even more. Species would have to be capable of very fast migration to keep up with the isotherms which would move northwards at about 10 km per year. A recent study suggests that 13–37% of species will be 'committed to extinction' by 2050 (Thomas *et al.*, 2004).

- Although cold countries such as Russia may enjoy an increase in agricultural productivity as a result of warming, at least in the short term, many of the world's

main food-producing regions may become too hot and dry for crops to grow. This would include major 'breadbasket' regions such as central and southern Europe and North America.

- Low-lying ground, including many major cities of the world and some entire small island states, may be inundated as a result of thermal expansion of the ocean and the melting of ice. Rates of rise in sea level are likely to be in the range 0.20–0.86 m from 1990 to 2100 (IPCC, 2001).

- Some geographical regions will suffer more than others. Temperature rises are currently especially high in the high latitudes (leading to melting of ice). Models show an increase in the extent of El Niño, with high rates of warming and drying in some of the tropical regions, causing replacement of the rainforest of Brazil by savanna.

- Diseases are likely to spread from the tropics to the

temperate and northern regions as the climate warms. Outbreaks of pests may become more extreme, as the natural biological control processes may not always be present. Of particular concern is the northward spread of insect pests which damage crops or transmit disease. One such example is Lyme disease, a life-threatening disease carried by ticks and found to be more prevalent in the United Kingdom during warm years.

- The cost of repairing damage caused by extreme events will escalate and occupy a major proportion of the world's economic production.

Although there is considerable uncertainty in model predictions, partly because the models do not incorporate many of the likely feedbacks that derive from the vegetation itself, there is now agreement that the countries of the world must co-operate to reduce the emissions of greenhouse gases.

BOX 24.2

24.2.4 Critical evaluation of the state-of-the-art in GCMs

GCMs have significant weaknesses which are frequently highlighted by sceptics. However, according to the IPCC, most climatologists agree that better models would not materially influence the conclusions of the model runs. Some of these perceived weaknesses are mentioned for consideration here, as follows.

24.2.4.1 Resolution

Spatial resolution may be too coarse. For example, in the HADCM3 model, the GCM used at the Hadley Centre in the United Kingdom, the grid cells for the global runs made for the IPCC are 2.5 × 3.75 degrees in latitude × longitude, and the time steps are half-hour. There is a practical limitation on spatial and temporal resolution imposed by the speed of the supercomputer as it takes a long time to run a GCM for a 100 year or more simulation. As computing power increases over the next few years there will be improvements in resolution. Some features of the climate system of course have a characteristic size which is small in relation to the grid cells (hurricanes and even clouds have to somehow be represented).

24.2.4.2 Biological and chemical coupling

Attempts to represent the impact of the biology and chemistry in GCMs are in their infancy. To some extent this bottleneck relates to the lack of process understanding. For example, how should we model the effect of warming on the respiratory production of CO_2 by the soil microbes? In general, our understanding of the carbon cycle is incomplete, and arguably we are not yet ready to represent it in global climate models, yet its behavior clearly has the potential to generate 'surprises' in the form of new and substantial sinks and sources of carbon. This is touched upon in Section 24.3 below. Similar remarks could be made in the realm of atmospheric chemistry and aerosol science.

24.2.4.3 The behaviour of ice

It is very difficult to represent properly the melting of ice, and the existing GCMs fail to deal specifically with the consequences of the possible melting of the Greenland ice sheet. This would cause a massive influx of meltwater into the North Atlantic, changing the ocean circulation patterns and therefore profoundly altering the distribution of heat over the Earth's surface. Such abrupt events may have occurred before, as the Heinrich events, which are evident during the last glacial period (Rahmstorf, 2000; see Chapter 22).

24.2.4.4 The human dimension

No one has attempted to incorporate models of the social and economic life into this type of climate model. The most significant change in land use is currently tropical deforestation. When forest is replaced by pasture, as for example in Amazonia, the land surface becomes more reflective, the pattern of evapotranspiration becomes more seasonal, the cloud cover is reduced, and the surface becomes aerodynamically smoother (Figure 24.8). These changes, on the scale of the Amazon Basin (over 4 million km^2), have the potential to change the climate not only in Brazil but elsewhere.

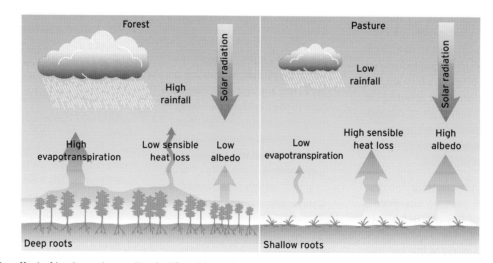

Figure 24.8 The effect of land-use change (tropical forest to pasture) on the energy and water balance of the landscape. Forests absorb more solar radiation and have a higher evapotranspiration rate than pastures.

Reflective questions

➤ What evidence is these that global warming is caused by human action?

➤ What might be the impacts of climate change associated with enhanced greenhouse gas concentrations?

➤ What are the benefits and limitations of GCMs?

24.3 The carbon cycle: interaction with the climate system

The carbon cycle is the circulation of carbon atoms between the ocean, land and atmosphere by physical, chemical and biological processes. Living organisms are of great importance in the carbon cycle, because life on planet Earth is made of carbon compounds such as proteins, carbohydrates and lipids. Thus, the cycle is one of many **biogeochemical cycles**, whereby organisms reuse vital elements such as carbon, nitrogen, sulphur and phosphorus.

The principal processes involved in transfer from the atmosphere are the dissolution of CO_2 in the oceans and the uptake of CO_2 by the balance between photosynthesis and respiration of living organisms. The processes involved in the return to the atmosphere are the release of CO_2 from the ocean in regions where the ocean upwells (see Chapter 3), and the breakdown of organic matter by respiration or fire. We can thus envisage the carbon cycle as a set of fluxes between major pools as shown in Figure 24.9. Greenhouse gases have 'sources' and 'sinks'. In the case of CH_4 the main sink is the photochemical process whereby it is oxidized in the troposphere by the radical OH, and its lifetime in the atmosphere is well known (9.6 years). In the case of CO_2 the sinks are many and some of them are long-lived compounds such as lignin and the carbonates, which makes it much more difficult to estimate the lifetime.

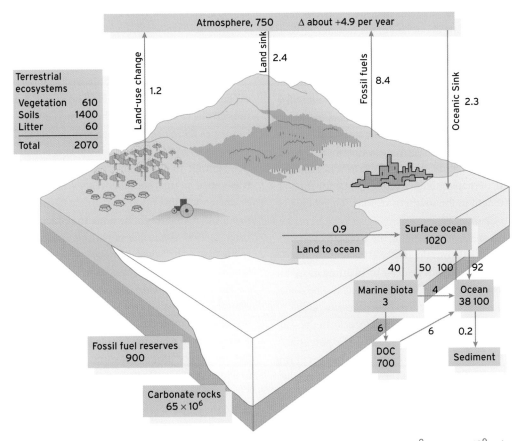

Figure 24.9 The global carbon cycle, updated for 2010 (units are gigatonnes, i.e. billions of tonnes, 10^9 tonnes or 10^9 petagrams). The term for fossil fuels includes emissions from cement manufacture. Data for emissions have the following uncertainty: fossil fuels 8.4 ± 0.5, land-use change 1.2 ± 0.7. Land and ocean sinks may vary from year to year, especially the land sink. The land sink is the net of very large gains by photosynthesis (around 120 gigatonnes of carbon per year) and losses of a similar magnitude by respiration of plants and other organisms. Data for stocks on the land and in the ocean are very uncertain. 'DOC' is an abbreviation for dissolved organic carbon.

An understanding of the carbon cycle is fundamental to our understanding of life itself, as all biomass is carbon-based. The principal biochemical constituents of cells (carbohydrates, proteins, lipids) have a high carbon content, and the overall carbon content of dry biomass is in the range 45–55%. The carbon cycle has become especially topical in recent years, since the realization that warming is large enough to force the cycle out of equilibrium, whereby the concentration of the gas in the atmosphere is rising. As we have seen above, CO_2 is by far the most important of the several gases that absorb infrared radiation emitted from the planetary surface, and its continuing rise is capable of causing additional global warming.

The quantity of carbon emitted in 2010 by burning fossil fuels was 8.4 Gt C yr^{-1} and clearing forests in the tropics released about 1.2 Gt C yr^{-1}, making a total anthropogenic burden on the atmosphere of 9.6 Gt C yr^{-1}. However, the concentration of CO_2 in the atmosphere is increasing by only about 4 Gt C yr^{-1}. The CO_2 not appearing in the atmosphere is being taken up by terrestrial photosynthesis or dissolving in the ocean. It seems that the terrestrial vegetation may be absorbing about half of the missing carbon, and the rest is dissolving in the ocean (Freidlingstein *et al.*, 2010). This is reflected in the fluxes shown in Figure 24.9. However, the ocean and terrestrial sinks might not continue. According to some models, the terrestrial sink will diminish and then become a source as a result of the impact of high temperatures and droughts in the tropics (Cox *et al.*, 2000). On the other hand, as warming occurs there is a reasonable expectation that the sinks in the northern regions will strengthen, perhaps enough to compensate for the loss of sink strength in the southern regions. This theme is returned to in Chapter 25.

There are prospects of enhancing the strength of the sinks in order to slow down the rate of climate warming. On the terrestrial side, this might be done by planting more forests, or by protecting existing forests. It could also be done by modifying agricultural practices (less ploughing, for example) in order to conserve the carbon stocks in the soil or to protect peatlands and other wetlands. The potential of enhancing the terrestrial sink by land-use changes of this kind is considerable (IPCC, 2007b). The ocean sink might also be managed by fertilizing the ocean. The scientific basis for this proposition is the observation that phytoplankton in the deep ocean are short of the micronutrient iron. When iron is added as ferric ions the productivity of phytoplankton is increased. This provides more food for zooplankton, and more food for the fish that eat the zooplankton, so there should be an enhanced stream of dead biota and carbonate shells that sink to deeper layers of the ocean. This, in turn,

should enable more CO_2 to dissolve in the surface waters. No one really knows whether this will work on a large scale. Environmentalists have generally opposed all suggestions of increasing the strength of ocean sinks, and sometimes sinks in general, arguing that the mechanisms and processes are imperfectly understood, and the sinks cannot be depended upon in the long term. They argue that there is no alternative but to reduce emissions by reducing consumption and finding alternative 'clean' sources of energy.

Reflective questions

➤ Can you draw a diagram of the carbon cycle and explain it?

➤ Why is the total increase of CO_2 in the atmosphere currently less than the CO_2 emitted by human actions?

24.4 Mitigation

The question of 'what can be done?' to avoid dangerous climate warming has to be addressed at a global scale. If one country alone were to apply stringent measures at great expense to reduce fossil fuel emissions while other countries go ahead and increase theirs, then no one would gain. Countries therefore must engage in debate and decide on the actions required before it is too late. Following the 1992 'Rio Summit' to discuss environmental change, many countries joined an international treaty – the United Nations Framework Convention on Climate Change (UNFCCC) – to consider what might be done to reduce global warming. Later, in 1997, a number of nations approved an addition to the treaty: the Kyoto Protocol, which imposes powerful (and legally binding) measures to reduce emissions.

The Kyoto Protocol is an international agreement to limit the emission of greenhouse gases. Six gases are mentioned in the protocol, of which CO_2 is the most important contributor to warming (Table 24.1). They differ greatly in their residence time in the atmosphere, and in the extent to which they are effective in absorbing infrared radiation. These two factors together are incorporated into an index called the global warming potential (GWP), which measures the relative effectiveness of the gas, on a per molecule basis, in causing global warming over a century.

Only Parties to the Convention that have also become Parties to the Protocol (by ratifying, accepting and approving) are bound by the Protocol's commitments; 175 countries have ratified the Protocol to date. Of these, 36 countries and the countries of the European Union are required to reduce greenhouse gas emissions below levels specified for each of them in the treaty. The individual targets for these Annex I countries are listed in the Kyoto Protocol (see the Protocol at unfccc.int/2860.php). These add up to a total cut in greenhouse gas emissions of 5.2% by 2012 from 1990 levels, but for these countries only. In the meantime, some major emitting countries which are not Annex I members (China, India) have increased their emissions sharply. In China's case, much of the industrial activity has been in the manufacture of goods for the west, but the emissions are counted according to where they are incurred not to where the resulting goods are consumed.

At an individual level, some people opt to take personal responsibility for their carbon emissions. In a developed western society it is possibly to make substantial savings in this way, as Reay (2006) has pointed out. But only a few people have so far taken direct control over their 'carbon footprints' by changing their lifestyles. Most people in the developed world, and an increasing number in the developing countries, have a substantial component of emissions from travel; they often ignore options to reduce these by selecting low-carbon-emitting modes of transport (train not plane, bike not car). Currently, the cheapest and most convenient mode of transport is usually not the one with the lowest carbon emissions. In a recent survey of the travelling habits of the people of Edinburgh a linear relationship was found between income and transport emissions (Figure 24.10), suggesting an inevitable association between wealth and travel emissions. The challenge for governments is now to enable modes of travel which are efficient and affordable yet which do not incur such high emissions of greenhouse gases.

At national level, some countries have set themselves carbon emission reduction targets which go well beyond those prescribed by the Kyoto Protocol. Various countries have passed legislation to limit their own emissions. For example, the UK government's Climate Change Act states: 'It is the duty of the Secretary of State to ensure that the net UK carbon account for the year 2050 is at least 80% lower than the 1990 baseline.' Other European countries have similar legislation, passed or in preparation. These goals will only be achieved by quite radical and potentially unpopular changes which will have to include: carbon taxes, new technologies (especially, the burial of CO_2 in geological strata) and a move back to nuclear power. The use of renewable energy sources such as wind power and biomass energy can make a significant contribution. Countries such as China that are developing economically at a very fast rate, and have a large reserve of easily accessible coal, may take several decades to control their emissions effectively.

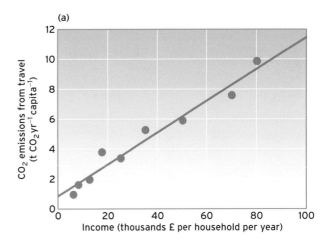

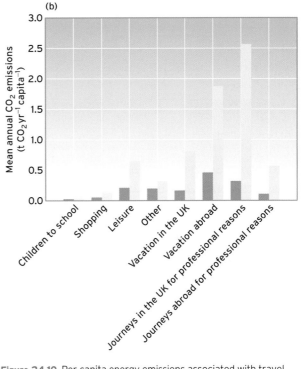

Figure 24.10 Per capita energy emissions associated with travel for people of Edinburgh, Scotland, in 2005: (a) relationship with income; (b) type of travel (blue bars are deprived social group, grey bars are affluent social group). (Source: from Korbetis et al., 2006)

Reflective questions

➤ Can you list some positive and negative climate feedbacks and some of the impacts of climate change that could be associated with increased greenhouse gas concentrations?

➤ Using Internet tools work out your own personal carbon footprint and reflect how you might make it smaller. Would this impact on your quality of life?

Table 24.1 Greenhouse gases in the Kyoto Protocol, lifetime in the atmosphere and global warming potential (GWP) on a scale where $CO_2 = 1$ (see text for definition)

Gas	Sources	Lifetime (years)	GWP	Comment
Carbon dioxide (CO_2)	Fossil fuel burning, cement manufacture	Sometimes assumed to be 100 years but probably longer	1	On the increase still, but Annex 1 countries* likely to reduce emissions
Methane (CH_4)	Wetlands, rice fields, burning, oil wells, ruminants, termites	12	21	
Nitrous oxide (N_2O)	Land disturbance, use of nitrogen fertilizers	120	310	
Hydrofluorocarbons	Industry, refrigerants	1-300	140-11 700	Substituting for CFCs. They contain only hydrogen, fluorine and carbon, and do not damage the ozone layer (CF_3CFH_2, CF_3CF_2H, CHF_3, CF_3CH_3 and CF_2HCH_3)
Perfluorocarbons (CF_4, C_2F_6)	Industry, electronics, firefighting, solvents	2600-50 000	6500	Also substituting for CFCs

* Annex 1 countries are 37 developed countries and economies in transition who had committed themselves to reducing emissions to 1990 levels by 2000. The term 'Annex 1 countries' is still used despite the original date having lapsed. Annex 1 countries pledged in the Kyoto Protocol to reduce emissions to 5.2% below 1990 levels in the 2008-2012 period.

(Source: Woodward et al., 2004)

24.5 Destruction of the ozone layer by chlorofluorocarbons (CFCs)

Electromagnetic radiation from the Sun reaches the Earth's atmosphere at a rate of about 1366 W m^{-2}. Much of it is scattered back into space, and only a fraction reaches the surface; it drives photosynthesis and evaporation, and warms the planetary surface. The radiation contains ultraviolet (waveband 100–400 nm) radiation, visible radiation, which happens also to be the photosynthetic waveband (waveband 400–700 nm), and near-infrared radiation (waveband 700 nm to a few nanometres). Ultraviolet radiation is absorbed by the DNA of all organisms, causing damage to the genetic code and consequently interfering with protein synthesis and the control of cell division. In humans the most common effects include reduction in the immune system (all races), skin cancer in Caucasian-type humans and damage to the eyes.

For the last billion years, the Earth has been shielded from damaging ultraviolet radiation by ozone (O_3) in the stratosphere. This protection has enabled life to develop on the land. Now, the ozone layer is diminishing as a result of a chain of chemical reactions that begins with totally

human-made chemicals, the chlorofluorocarbons (CFCs). These CFCs are synthetic non-reactive gases and liquids first made in 1930 and used as refrigerants (later as propellants in spray cans). Being inert under normal conditions, they persist in the atmosphere, and slowly make their way to the stratosphere. Laboratory studies in 1974 established that CFCs could catalytically break down ozone in the presence of ultraviolet radiation to form highly reactive radicals such as ClO and OClO. It is these radicals that catalyze the breakdown of O_3 to O_2. A ground-based survey of stratospheric ozone was started in Antarctica in 1956, followed by satellite surveys in the early 1970s. In 1985 a British team based in Antarctica reported a 10% drop in the ozone level during the spring, which they attributed to CFCs and oxides of nitrogen. A similar decline was also seen in data from NASA's Nimbus 7 satellite carrying TOMS (Total Ozone Mapping Spectrometer; see Chapter 26), and it is now evident that a steady decline is occurring over Antarctica whilst a decline has been detected over the Arctic (Figure 24.11). In the 1980s Australians sunbathed much less than before, and sales of sunhats and skin creams to protect against ultraviolet radiation increased. Plants and animals are less able to take protective measures.

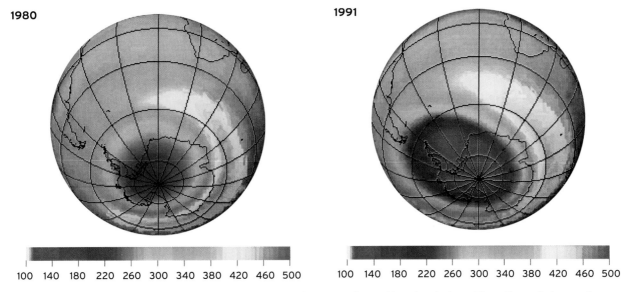

1980   **1991**

100 140 180 220 260 300 340 380 420 460 500   100 140 180 220 260 300 340 380 420 460 500

Figure 24.11 Images of the thickness of the stratospheric ozone layer over the southern hemisphere. The units are Dobson units named after G.M.B. Dobson, an early investigator. Normal thickness is 300 and the scale is linear. (Source: Centre for Atmospheric Science, University of Cambridge, UK)

The Montreal Protocol of 1987, in which nations agreed to phase out the use of CFCs, has undergone several modifications. Trade sanctions on CFCs have been imposed and a total phasing out is due in 2030. In March 1989 environmental ministers of the European Community announced a total phase-out of CFCs in Europe by the year 2000. More recently, related chemicals which do not significantly destroy ozone have been introduced: these are the hydrofluorocarbons (HFCs) and the perfluorocarbons (PFCs). Unfortunately, these gases are powerful greenhouse gases (Table 24.1), although their concentration in the atmosphere is very low and so they do not presently contribute much to global warming.

Reflective questions

➤ Why is the ozone layer crucial to life on Earth?

➤ Has the Montreal Protocol of 1987 been more successful than the Kyoto Protocol of 1997?

24.6 The future

Fossil fuel continues to be the main source of energy. Moreover, the developing world, which consists of about five-sixths of humankind, will continue to increase its population and its fossil fuel burning for many years after the rich countries have stabilized and decreased their dependency on fossil fuels. Some poor countries have neither fossil fuels nor any other supply of energy, and so cannot develop. Even fuelwood is in short supply.

Nuclear power was developed enthusiastically by many countries in the 1950s, and 29 countries were running 437 nuclear power plants by 1998. Early optimism about development of an energy economy from nuclear fission faded, following nuclear accidents and leakages such as Chernobyl in the USSR in 1986 and Fukishima in Japan in 2011. Many environmentalists believe that the risks that are inherent in nuclear fission are quite unacceptable. Power from nuclear fission is very expensive, once the costs of handling radioactive waste and decommissioning old power stations are taken into account. Despite all this, many governments are in favour of continuing and even expanding their nuclear power programmes, and for many it is the only practical way to reduce carbon emissions.

There are, however, some reasons for optimism. In the period since 1960 considerable progress has been made towards developing alternative sources of energy to replace carbon-based fuels (coal, oil and gas). Governments in the richer countries are setting ambitious targets to decarbonize their energy economies (which means they will use less fossil fuels to produce energy) and are pushing for investment in renewable forms of energy such as wind and biofuels. Solar cells are likely to become increasingly important. This technology was first developed during the exploration of space in the 1960s. Silicon solar cells convert solar energy to electrical energy with an efficiency of 20%, and the energy may be stored and transported in fuel cells. The construction cost of a solar cell is rather low, as the main elemental ingredient, silicon, is abundant. Unlike wind turbines and wave power generators there are no moving parts and consequently maintenance costs are extremely low. Even in winter in northern countries, solar cells can provide useful quantities of solar energy. In the future, roof tiles incorporating solar cells may be used in all new housing construction, and solar energy 'farms' may cover large areas of deserts.

The major problems facing the world are related to each other: climate change, energy supply, poverty, and the tensions arising from a disparity of living standards between different people. It is difficult to foresee what kinds of environmental change are just around the corner, and therefore hard to plan for the future. A hundred years ago, the problems of today were invisible and quite unpredicted. But there is some evidence from recent history (Box 24.3) that we are at last beginning to grasp the nature of the human–environment interaction. For 200 years humans have been inadvertently damaging the life-support system of the planet. In the last 30–40 years we have realized what is happening, and governments are beginning to take remedial action.

Reflective question

➤ Despite the challenges ahead, why should we be optimistic about dealing with climate change?

ENVIRONMENTAL CHANGE

IMPORTANT RECORDED EVENTS

The most important dates in the relationship between humans and the global environment are listed below. These events and discoveries changed our perceptions of the world we inhabit, and contributed to global environmental change. From 1500 to 2007 the human population increased from 0.5 billion to 6.6 billion, and the world gross domestic product (GDP) rose from $240 billion to $30 000 billion.

1400–1450	Chinese fleets arrive in Middle East and East Africa; Arab fleets cross Indian Ocean. Intercontinental trade begins.
1450	Johannes Gutenberg, Germany, establishes printing press, enabling humans to communicate efficiently and to learn from each other more effectively than ever before.
1490–1500	Decade of European maritime exploration of Asia, Africa and America, paving way for settlement, slavery, trading, further spread of economically important plant and animal species.
1610	Galileo Galilei, Italy, uses the telescope to observe behaviour of the Moon and planets. Other scientific instruments for Earth observation were developed: microscope (1618), thermometer (1641) and barometer (1644).
1780–1820	Industrial Revolution. Dramatic increase in the use of coal. Western Europe sees rapid technological, social and economic transformation, driven largely by the steam engine fuelled by coal. Widespread urban pollution, exploitation of workforce, occupational diseases. Humans begin to alter the composition of the global atmosphere.
1796	First blast furnace opens in Gleiwitz, Poland. Manufacture of iron and steel followed in urban centres: Belgium, Germany, Great Britain. Respiratory diseases increase.
1798	Thomas Malthus, English clergyman, writes *Essay on the principle of population,* pointing out the natural tendency of human populations to grow exponentially, outrunning the food supply.
1821	William Hart obtains natural gas (methane) from a 9 m well in New York, and provides street lighting.
1827	Jean Baptiste Joseph Fourier (France) first proposed that 'light finds less resistance in penetrating the air, than in repassing into the air when converted into non-luminous heat'. Possibly the first articulation of the greenhouse effect.
1838	Birth of John Muir at Dunbar, Scotland. First person to call for conservation of wilderness areas, arguably the father of the modern conservation movement.

BOX 24.3 ➤

1839	Antoine-Cesar Becquerel, France, discovers the photoelectric effect, demonstrating that sunlight can be converted into electricity. But practical solar cells were not developed until 1954.
1842	John Lawes, England, invented superphosphate fertilizer.
1851	James Young, Scotland, discovers how to extract hydrocarbons from oil shale, and develops process of refining oil. He establishes a paraffin industry in Scotland (paraffin is called kerosene in the USA) and is nicknamed 'Paraffin Young'.
1859	Edwin Drake strikes oil at 20 m in Pennsylvania, USA. Oil was soon discovered in North and South America, Mexico, Russia, Iran, Iraq, Romania, Japan, Burma and elsewhere. Oil soon plays its part in the industrialization of the world.
1859	Charles Darwin publishes *The origin of species*, proposing the theory of evolution by natural selection.
1859	Irish scientist John Tyndall discovers that H_2O and CO_2 absorb selective wavebands of infrared radiation, and suggested a role for these gases in the regulation of the Earth's temperature.
1864	James Croll, Scotsman, proposed theory of long-term climate change to account for the Ice Ages (see also the reference to Milutin Milankovitch, 1895)
1866	German engineers Langen and Otto patent the internal combustion engine. The manager in Otto's factory, Daimler, makes the first petrol engine in 1884.
1868	In Japan, the Meiji Restoration. Japan opens to the west and large-scale industrialization spreads to Asia.
1885	German chemist Robert Bunsen discovers how to make a very hot flame from gas, by mixing it with air before combustion. Gas burners were thereafter much more efficient.
1895	Serbian astrophysicist Milutin Milankovitch describes theory of long-term climate change. Essentially it is the same theory that Croll had proposed in 1864, but Croll's work was ignored by his peers.
1896	Arrhenius, Swedish chemist, advances theory that CO_2 emissions will lead to global warming, and postulates the ocean as a CO_2 sink.
1901	Italian Guglielmo Marconi invents radio, achieves first transatlantic radio message.
1903	Henry Ford, USA (1863-1947), establishes the Ford Motor Company, makes Model-T Ford cars in 1908. Others would follow Ford's idea of mass production, and car ownership would increase rapidly.
1903	Orville and Wilbur Wright, USA, demonstrate a flying machine based on the internal combustion engine. Rapid intercontinental travel by air would follow in 50 years.
1907	Henry Ford completes first tractor, a machine that was to revolutionize agriculture.
1909	Fritz Haber, Germany, shows how to synthesize ammonia from N_2 in the atmosphere, and Karl Bosch uses this process for mass production of nitrogen fertilizer. This was to greatly increase the capacity of humans to grow crops.
1915	German geophysicist Alfred Wegener publishes his controversial hypothesis of continental drift, in a book entitled *The origin of continents and oceans*.
1917	Chainsaw manufactured for the first time. Enables rapid deforestation.
1926	John Logie Baird, Scotland, invents television, but TV broadcasts did not start until 1936 (Britain) and 1941 (USA), and TV sets were not widespread until the 1950s. Television has allowed ordinary people to see how others live, and thus to understand better their own place in the world.
1927	Alexander Fleming, Britain, discovers the antibiotic effect of the fungus *Penicillium*. Much later (1940) Florey and Chain, working in the USA, discover how to make the antibiotic penicillin in bulk. This launches golden age of medicine. Infant mortality declines and people live longer.
1928	Mohandas Karamchand Gandhi, India leader, questions the sustainability of the industrial age: 'God forbid that India should ever take to industrialism after the manner of the West. If it took to similar economic exploitation, it would strip the world bare like locusts.'

BOX 24.3 ➤

➤

1930 Thomas Midgely, USA, invents the gas freon. It was the first of the chlorofluorocarbons (CFCs) which much later (1970s) became widely used and caused thinning of the stratospheric ozone layer.

1938 Guy Stewart Callendar (UK) predicted global warming at a rate of 0.03°C per decade.

1943 Primitive electronic computers constructed: Harvard Mk I (USA) and Colossus (Britain).

1944 Pilotless planes and rockets developed in Germany for use in warfare. Later, the technology formed the basis for space exploration and Earth observation.

1945 Atomic bomb dropped on Hiroshima. *Guardian* newspaper (UK) comments 'man is well on the way to mastery of the means of destroying himself utterly'.

1948 Agricultural efficiency increases dramatically in the developed world, as a result of mechanization, fertilizers, pesticides, plant breeding and managerial skill. Crop yields increase. Later (1960s and 1970s) the new agricultural technology is taken to the developing countries, and becomes known as the Green Revolution.

1948 First operational stored-program computer, known as Manchester Mk I (Williams, Kilburn and Wilkes, Britain).

1951 Age of nuclear power starts with first commercial nuclear reactor at Idaho, USA. Later there are significant accidents: fire at Windscale, UK, in 1957; meltdown at Three Mile Island, Michigan, USA, in 1979; meltdown and large release of fission products at Chernobyl, USSR, in 1986.

1952 British jet airliner, De Havilland Comet, begins regular intercontinental travel.

1954 Chapin, Fuller and Pearson develop silicon solar cell capable of converting solar energy into electrical energy with a conversion efficiency of 15%.

1957 First spacecraft, Sputnik, USSR; to be followed by first man in space, Yuri Gagarin, USSR, in 1961 and first man on the Moon, Neil Armstrong, USA, 1969.

1958 Charles Keeling, of the Scripps Institute in the USA, begins the first reliable measurements of atmospheric CO_2 at Mauna Loa in Hawaii.

1960 Soviet engineers begin large-scale irrigation using rivers flowing to the Aral Sea, the world's fourth largest lake. Within 40 years the lake would almost disappear, possibly the greatest hydrological change yet engineered by humankind.

1962 *Silent spring* by Rachel Carson, USA, warns of dangers of pesticides to wildlife. This best-seller inspired a whole generation of environmentalists.

1968 Satellite remote sensing starts. Pictures of Earth from deep space, Apollo 8 mission, USA; followed in 1972 by Earth Resources Satellite ERTS-1 carrying multispectral sensors later called Landsat.

1969 In the USA, the Advanced Research Projects Agency (ARPA) begins the ARPANET. Soon, global communication by e-mail and Internet would become possible.

1970 Establishment of Environmental Protection Agency (EPA), USA.

1971 Formation of Greenpeace. A group of activists sail their small boat into a US bomb-test zone near Alaska to draw attention to the environmental dangers of nuclear war.

1971 Swedish scientists demonstrate long range transport of sulphur as the cause of acidification of Swedish lakes, and predict that acid rain will damage fresh water ecosystems and forests.

1972 In the UK, publication in *The Ecologist* of 'A blueprint for survival', warning of the extreme gravity of the global situation and criticizing governments for failing to take corrective action.

1972 Publication of *The limits to growth* by the Club of Rome, dealing with computer simulation of global environmental change. Fails to identify the threat of global warming; points to resource depletion and pollution as the major threats.

1972 First international conference on the environment, Stockholm, leading to the establishment of the United Nations Environment Programme (UNEP). Acid rain is widely publicized, especially in relation to forest decline, but since then the developed world has been moving to low-sulphur fuels.

BOX 24.3 ➤

➤	
1972	The anchovy fishery of Peru collapses because of overfishing and bad weather. Other fish stocks decline sharply, and management of marine resources becomes an important issue.
1973	Organization of Petroleum Exporting Countries (OPEC) restricts the supply of oil, forcing its price to rise five-fold and threatening the global economy.
1979	James Lovelock proposes the Gaia hypothesis (see Chapter 23).
1985	Farman, Gardiner and Shanklin, a British team working in the Antarctic, report thinning of stratospheric ozone, attributable to CFCs.
1986	Nuclear accident at Chernobyl, USSR, creates radioactive fallout everywhere in the northern hemisphere, reminding people that environmental problems cross political boundaries. The expansion of nuclear power in the west falters.
1987	First appearance of the word biodiversity in the scientific literature (by E.O. Wilson, USA).
1987	Ice core from Antarctica, taken by French and Russian scientists, reveals close correlation between CO_2 and temperature over the last 100 000 years.
1987	Montreal Protocol signed, an agreement to phase out CFCs.
1987	United Nations World Commission on Environment and Development produce the Brundtland Report, dealing with definitions of sustainability.
1988	Intergovernmental Panel on Climate Change (IPPC) is established.
1990	IPPC's first Scientific Assessment Report, linking greenhouse gas emissions to warming.
1992	Implementation of the International Geosphere Biosphere Programme (IGBP) to predict the effects of changes in climate, atmospheric composition and land use on terrestrial ecosystems; and to determine how these effects lead to feedbacks to the atmosphere.
1992	Earth Summit, Rio de Janeiro. Leaders of the world's nations meet in Rio and set out an ambitious agenda to address the environmental, economic and social challenges facing the international community. Heads of state sign the UN Framework Convention on Climate Change. The UNFCCC was one of three conventions adopted. The others – the Convention on Biological Diversity and the Convention to Combat Desertification – involve matters strongly affected by climate change.
1997	Kyoto Protocol is drafted, the first international agreement to limit greenhouse gas emissions.
1997–1998	Particularly severe El Niño causes drought and widespread forest fires in Indonesia, Malaysia, Brazil and Mexico. In SE Asia the fires affect 10 000 km^2 of forest.
1998	The warmest year of the century, and probably of the millennium.
2000	International Coral Reef Initiative reports that 27% of the world's corals reefs are lost, mainly a consequence of climate warming.
2001	US President George Bush announces that the USA will not ratify the Kyoto Protocol.
2002	As warming of Antarctica proceeds, some 3200 km^2 of the Larsen B ice shelf collapses.
2002	Schools in Seoul, S Korea, are closed when a dust storm originating from China sweeps over the country.
2003	Gates of China's Three Gorges Dam are shut, and the world's largest hydropower reservoir is created, destroying archaeological sites and forcing the relocation of nearly 2 million people.
2003	European heatwave causes premature death of 35 000 people.
2004	Indian Ocean earthquake causes large tsunami, killing a quarter of a million people.
2005	Kyoto Protocol comes into force, 16 February.
2005	Hurricanes sweep the US Gulf Coast, causing widespread damage and loss of life. New Orleans evacuated.
2005	Worst Amazon drought in the last 100 years.
2006	The film *An Inconvenient Truth* (Director Davis Guggenheim, Presenter Al Gore) presents the science of global warming in a manner accessible to non-scientists. It is a box-office success and wins awards.

BOX 24.3 ➤

➤

2006	Nicholas Stern, an economist, in a report for the UK Government, suggests that global warming will cause the greatest and widest-ranging market failure ever seen, and proposes environmental taxes as the best remedy.
2006	China becomes the top CO_2 emitter, overtaking the USA.
2007	IPCC Fourth Assessment Report predicts that temperatures will rise by 1.8–4.0°C by the end of the century.
2008	Forests of western North America devastated by epidemic of mountain pine beetle.
2009	Worse wildfires in Australian history, 181 killed in the state of Victoria.
2009	Global fossil fuel emissions fell by 1.3% to 8.4 Pg C per year, as a result of the economic recession; emissions in China and India rose, nevertheless, by 8% and 6% respectively.
2009	United Nations Climate Change Conference 'Copenhagen Climate Summit' fails to reach legally binding agreement on curbing greenhouse gas emissions or reducing deforestation. In the resulting Copenhagen Accord many governments pledge to reduce emissions by 2020, including the USA.
2010	Deepwater Horizon oil leak discharges the equivalent of 4.9 million barrels of oil into the Gulf of Mexico, causing a 6500 km^2 oil slick and harming fisheries. It became known as the BP oil disaster.
2010	Severe Amazon drought, thought to be the second (after 2005) most severe drought on record.
2010	Arctic sea ice now covers its smallest area since records began.
2011	Two papers in the journal *Nature* present compelling evidence that an increase in the frequency of heavy rain and floods in the northern hemisphere is linked to the increase in greenhouse gases in the atmosphere.

BOX 24.3

24.7 Summary

World population growth has been associated with increased utilization of the land for agriculture. Increased domestication of plants and animals and use of wood for fuel have resulted in vast amounts of deforestation, particularly over the past 200 years when the human population has increased from 1 billion to 6.9 billion. Humans now appropriate 40–50% of the land's biological production. Large-scale change in the land surface from forest to farm influences regional and global climates in many ways that are still not properly understood. Conversion of forest to pasture involves release of CO_2 to the atmosphere, as well as changes to albedo, air movements, and the water, carbon and nitrogen cycles.

In addition, the burning of fossil fuels and the creation of human-made chemicals such as CFCs and techniques of creating nitrogen fertilizer all have an impact on the environment. The enhanced greenhouse effect appears to be causing accelerated climate change which will have major impacts on humans and global ecosystems. However, there are a number of positive and negative feedback effects that global circulation models are only just being able to predict and model. Atmospheric pollution from industry, combustion engines and agricultural practice is impacting on human health and biodiversity. Cities often experience harmful smogs, and the ozone layer which protects the Earth from vast amounts of harmful ultraviolet radiation is suffering severe damage due to CFC use.

However, humans are an inventive species and technological improvements are continuously being made that may help us cope with and mitigate environmental change. The development of safe nuclear fusion and increased use of solar cells may allow us to harness the world's resources in a more sustainable manner. In addition, the international recognition that global environmental change is taking place has been achieved and there is a willingness around the world to try to combat environmental problems.

Further reading
Books

Carson, R. (1962) *Silent spring*. Houghton Mifflin, Boston, MA.
This is a classic and seminal book, essential reading for those who are interested in the environmental movement.

Evans, R.L. (2007) *Fueling our future: An introduction to sustainable energy*. Cambridge University Press, Cambridge.
Outline of alternative energy supplies, including nuclear energy.

Houghton, J.T. (2004) *Global warming: The complete briefing,* 3rd edition. Cambridge University Press, Cambridge.
An authoritative and lucid account. If you want to read only one book about global warming then it should be this.

Leggett, J.K. (2000) *The carbon war: Global warming and the end of the oil era*. Penguin, London.
This is excellent bedtime reading, a first-hand story about the international negotiations surrounding reductions in carbon emissions.

Lovelock, J.L. (1979) *Gaia: A new look at life on Earth*. Oxford University Press, Oxford.
This is a very readable book, and a seminal work, proposing that the biosphere behaves as a homeostatic system. In his recent book, *The revenge of Gaia*, published in 2006, the author argues that it is now too late to avoid substantial global heating which will make large regions of the Earth inhospitable.

McNeill, J. (2000) *Something new under the Sun*. Allen Lane & Penguin, London.
This is a brilliant and readable account of environmental history.

Moore, P.D., Challoner, W. and Stott, P. (1996) *Global environmental change*. Blackwell, Oxford.
This provides a good overview of the subject and is well illustrated.

Wilson, E.O. (2002) *The future of life*. Abacus, London.
This is a thoughtful and highly informed set of speculations on what will happen next.

Further reading
Papers

Brohan, P., Kennedy, J.J., Harris, S.F.B. *et al.* (2006) Uncertainty estimates in regional and global observed temperature changes: a new dataset from 1850. *Journal of Geophysical Research,* 111, doi: 10.1029/2005JD006548.

Observation of temperature change.

Friedlingstein, P., Houghton, R.A., Marland, G. *et al.* (2010) Update on CO_2 emissions. *Nature Geosciences,* 3, 811–812.
Recent estimates on sources and sinks of CO_2.

Hegerl, G.C., Crowley, M. Allen, W. *et al.* (2007) Detection of human influence on a new 1500-year temperature reconstruction. *Journal of Climate,* 20, 650–666.

Pulls out human impacts on recent temperature record.

CHAPTER 25

Vegetation and environmental change

John Grace
School of GeoSciences, University of Edinburgh

Learning objectives

After reading this chapter you should be able to:

➤ understand the way vegetation responds to climatological variables, and appreciate some of the underlying mechanisms of this response

➤ outline the evidence that shows how temperature and water supply have a leading role in determining the global patterns in vegetation

➤ appreciate how researchers are using models to predict future vegetation patterns

➤ discuss how human activities interact with climatic impacts, in both 'good' and 'bad' ways

➤ understand the relationship between climate models, vegetation models and observations, and realize that much remains to be understood about the impact of climate on vegetation and vice versa

25.1 Introduction

Vegetation responds to a large number of factors in the physical environment, such as temperature, the available solar energy, and the supply of nutrients and water via the soil. Humans have been discovering the nature and extent of this response for a very long time, at least since biblical times when people first began to grow crops. The impact of drought, for example, is woven into the history of tribes and whole civilizations. In modern times, the study of the relationship between plant life and the climate falls within several scientific disciplines. **Plant physiology** is concerned with understanding the functioning of plants, and this includes the response of plants to their environment and the acquisition of resources by plants. From plant physiology we learn how species have different environmental requirements as a result of differing structural and biochemical make-up. **Agronomy**, **horticulture** and **silviculture** are applications of plant physiology in the service of humankind to provide field crops, garden plants and wood products. Ecology, on the other hand, looks more broadly at vegetation, and has a focus on plant distribution; here we learn that plant distribution is not only constrained by climate and soil, but also influenced by factors such as **competition**, herbivory, fire and disturbance. **Palaeoecology** is about ancient distributions (see Chapters 22 and 23), often inferred from deposits of pollen in a few places where material is well preserved, such as peat bogs and lake sediments. Every year, the surface of the peat or the bed of a lake is added to by a 'rain' of pollen grains. Cores can be extracted from the material, and slices can be acid-digested to remove organic debris

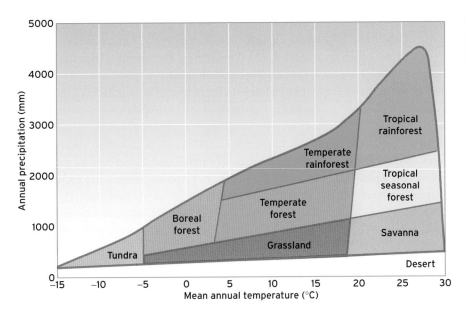

Figure 25.1 Approximate distribution of biomes in relation to precipitation and mean annual temperature. (Source: from Woodward *et al.*, 2004)

leaving pollen grains. The temporal patterns (with some assumptions) can be correlated with changes in temperature. Likewise, patterns of growth over long periods can also be measured in some organisms, such as corals and trees. **Dendrochronology** is the study of climate patterns from the annual growth rings of trees, and the related **dendroecology** is the study of tree rings to investigate ecological processes. In old trees these growth records extend over hundreds of years, and by joining together the records of past generations of trees, for example using trees found preserved in bogs or the timbers of ancient buildings, it is possible to obtain a record over thousands of years. Finally, **phenology** is the study of the annual variations in the timing of key events in the life cycle of plants, such as the date when specific tree species open their leaves, the date when flowers appear, or the date of autumn colouring when leaves lose their chlorophyll and photosynthesis ceases. There are long-term records of these phenological events, and they can be related to trends in the climate.

In this chapter we will draw upon work from all these disciplines, noting at the same time that to synthesize knowledge over a range of disciplines often requires some kind of mathematical model that incorporates knowledge and understanding, and that can be run from historical climate data or from data generated by climate models. In fact, we use predictions of what the climate may be in the future to estimate how the vegetation may change in the future.

We may also note in passing that there is an inevitable relation between climate and the native vegetation, as discussed in the classic work by Wladimir Köppen, a German climatologist, in the a 1931 paper, and Holdridge, an ecologist, in 1947. We emphasize 'native' vegetation, because much of the world's vegetation is affected by humans and transformed or removed so that the land can be used for agriculture and forestry. It turns out that for the native vegetation, the most important variables determining distribution on a global basis are average annual and monthly temperatures, and precipitation. A modern expression of these relationships can be seen in a paper by Woodward *et al.* (2004), reproduced here as Figure 25.1. Here, the land cover is represented as only nine '**biomes**', each biome occupying a particular region of climate space. From such relationships we see how a warmer world might shift particular locations from one biome to another.

25.2 Fundamentals of how plants respond to climatic variations

25.2.1 Light

About 50% of the dry weight of plants is carbon, which has been accumulated through the process of **photosynthesis** (from the Greek *phos* = light and *synthesis* = combination) summarized as follows:

$$6CO_{2(gas)} + 12H_2O_{(liquid)} + \text{photons} \rightarrow C_6H_{12}O_{6(aqueous)} + 6O_{2(gas)} + 6H_2O_{(liquid)} \quad (25.1)$$

carbon dioxide + water + light energy → glucose + oxygen + water

Green leaves achieve this by capturing photons (energy as light from the Sun) with a set of pigments, of which the most important are the chlorophylls (green pigments); then they use the captured energy to drive a series of chemical reactions that result in CO_2 being absorbed from the atmosphere and converted into simple sugars, such as glucose. Subsequently, glucose is made into storage compounds, the most common being starch, and structural compounds such as cellulose (of which cell walls of leaves and roots are made) and lignin (a component of the cell walls of woody stems). From this discussion, it is clear that green plants need light, and that the more light they have, the more growth can be expected. It is worth pointing out that the process of photosynthesis has been going on for nearly 4 billion years, and that the by-product of photosynthesis, oxygen, enables all aerobic life including our own.

From a biochemical perspective, different types of photosynthesis are recognized. The majority of plant species are found to fix CO_2 into 3-carbon compounds, triose phosphates. However, some other species have a different enzyme system and they make a 4-carbon compound instead, oxaloacetic acid. The former condition is known as C_3 photosynthesis and the latter as C_4 photosynthesis. C_4 plants are relatively recent. We see them in the fossil record only 20–25 million years ago and they spread remarkably only 8 million years ago. Most of them are grasses, some are sedges, a few are herbs and shrubs, and only one of them

is a tree. When they evolved from C_3 ancestors, fire was on the increase because of a drying climate, and browsing increased with the evolution and spread of large mammals in the Oligocene and Miocene.

Only a few per cent of species are currently C_4 but they can utilize high levels of solar radiation more effectively and may account for as much as 30% of global photosynthesis, mostly in the tropics and mainly in savannas. There are many differences between C_3 and C_4 but here we need note only a few of them. C_4 plants utilize water more efficiently and so they tolerate periods without rain. C_4 photosynthesis works best in warm climates, and so we can reasonably expect C_4 plants to be favoured in the future, warmer climate (Figures 25.2 and 25.3). They are among the faster growing crops of the world, including maize, sorghum, millet and sugar-cane. They dominate some of the most productive natural **ecosystems** in the world, from the floodplains in Brazil where we find a C_4 aquatic grass *Echinochloa polystachya* to the lakes in Africa where monospecific **stands** of *Cyperus papyrus* occur, the sedge used by ancient civilizations to make 'papyrus'. They are widespread as the C_4 grasses in tropical savannas, where they co-exist with C_3 trees. A third type of photosynthesis, first discovered in the plant family Crassulaceae, is associated with leaves that are fleshy and accumulate malic acid during the night. This condition is termed crassulacean acid metabolism (CAM). In contrast to all other plants, CAM plants

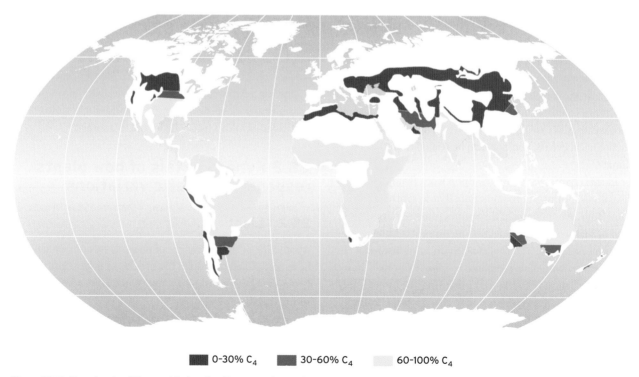

■ 0-30% C_4　　■ 30-60% C_4　　□ 60-100% C_4

Figure 25.2 Grasslands of the world, showing the percentage of the grasses that are C_4. (Source: from Ehleringer *et al.*, 2005)

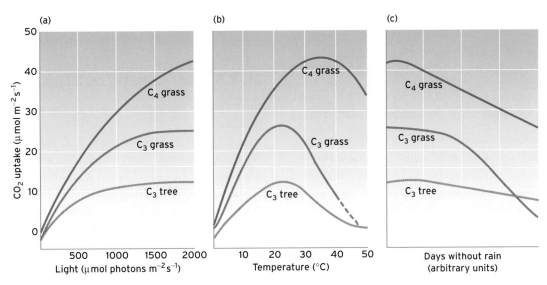

Figure 25.3 Environmental responses of photosynthetic CO_2 uptake by leaves of three hypothetical species: a C_4 grass, a C_3 grass and a C_3 tree. (a) Under different amounts of light, the C_4 grass has higher rates of CO_2 uptake and is able to continue to exploit the incident solar energy even in bright sunlight; C_3 grasses and herbs often have a higher rate of photosynthesis than C_3 trees. (b) The performance of the three species at high temperature; we see how the C_4 grass has a higher optimum temperature for photosynthesis. (c) How the hypothetical species might respond to a period with no rain: C_4 species have a higher water use efficiency than C_3 species (they lose only 250-300 g of water for every gram of CO_2 fixed, while C_3 lose 400-500 g water per g CO_2); grasses have near-surface roots and they die down in drought but trees often have deep roots and can exploit stored water more effectively (grasses will regrow leaves when the rain comes).

keep their stomata closed in the day and hence avoid water loss; but they open them at night using a different enzyme system to capture CO_2 and form malic acid. Then, during the day the malic acid is broken down to yield CO_2 which is fixed into glucose in the same way as we find in C_3 plants. As well as members of the Crassulaceae, CAM is found in cacti, the pineapple family (Bromeliaceae) and orchids.

25.2.2 Water

Inside the plant cell, water is the medium in which all the important biochemistry takes place. For land plants water is additionally important as they are mechanically supported by the water pressure inside them. Take away water from a plant and it becomes flaccid (it wilts). So, rain (strictly, the supply of water) is an important variable in controlling the distribution of plants, and determining their growth rates. Only a few plants can survive desiccation. These are the so-called resurrection plants that can disassemble the photosynthetic machinery when drought comes and then reassemble it later when water is abundant. Almost all land plants benefit from an extra supply of water, but species vary hugely in how well they can tolerate dry periods. Some tolerate drought by shedding leaves or by having special characteristics such as a thick and waterproof coating (cuticle) on their leaf surfaces, or by having small thick

leaves and special organs to store water. Plants with adaptations to dry conditions are called **xerophytes**, and are found especially in deserts. Changes in water supply are expected as a component of climate change: as we saw in Chapter 24 we expect some regions to become wetter and others to be drier; we may therefore expect vegetation to change in some places and patterns of food production to alter. For example, it has been predicted that the Amazon rainforest will be replaced by a more xerophytic vegetation within 100 years as El Niño events become more extreme and possibly more frequent (Cramer *et al.*, 2001).

25.2.3 Temperature

The biochemical reactions involved in photosynthesis and growth all require warmth. Most plants photosynthesize and grow best between 10 and 30°C although there are important variations. Some organisms can even thrive in the extreme conditions of hot springs and others live in the coldest places on Earth. But these extremophiles are mostly bacteria, not vascular plants. Much of the planet's surface is too cold for many plant species, and the long winters of the boreal zone limit photosynthesis and prevent cell division. As climate warms we expect northern and mountain regions to become greener, to photosynthesize more rapidly and to grow faster. Indeed, there is evidence from historical photography and satellite imagery that this is already happening (Nagy *et al.*, 2003). We expect that in the north especially, there will be widespread changes in the distribution of plants, as the length of the 'growing season' increases.

25.2.4 Carbon dioxide concentration

The atmospheric concentration of CO_2 has risen from 270 parts per million (ppm) before industrialization to over 390 ppm today. As photosynthesis involves the diffusion of CO_2 molecules from the atmosphere through the stomatal pores to the active sites inside the leaf where photosynthetic fixation occurs, it is to be expected that any increase in the external concentration of CO_2 will increase photosynthesis. This is because, according to Fick's law, the rate of diffusion increases in proportion to the concentration gradient. The increase in photosynthesis might also increase the growth rate and final yield of the plants. In fact, commercial growers have often used high CO_2 concentrations in greenhouses to speed up the production of crops like tomato and lettuce. The increase in photosynthesis may not be as clear as expected, however, because plants raised at high CO_2 develop fewer stomata and their stomata become somewhat closed at high CO_2. Moreover, the products of photosynthesis may not be readily translated into growth and yield. For example, there may be a shortage of a vital nutrient such as nitrogen, or the temperature may be too low for cell division to occur.

The extent to which these expectations about the effect of high CO_2 are true can be tested using a type of experimentation called FACE (Free Air enriChment Experiment). In FACE, a set of supply tubes surrounds a small circular field in which crops, small trees or other plants are being raised. Then, CO_2 is fed through the tubes to enrich the atmosphere. Sensors measure the concentration and adjust the supply rate. Usually the experiment is aimed at doubling the CO_2 concentration, to try to predict what may happen in a future high CO_2 world. The results of such experiments show that the increase in photosynthesis, productivity and yield is rather smaller than people have imagined. A doubling of external CO_2 will almost double the diffusion gradient, so in a simple physical system we might expect this to translate into a near doubling of CO_2 uptake. However, in a review of recent data, Leakey *et al.* (2009) reported a mean stimulation in biological production of 17% and in yield of only 16%. This type of experimentation cannot easily be done on major biomes such as the tropical and boreal forest, which collectively are acting as a carbon sink. So far there is just one experiment on mature trees, and a few on trees enclosed in large chambers. Consequently we conclude

that the extent of the stimulation of photosynthesis by CO_2 is one of the unknowns in global models of vegetation.

25.2.5 Other climatic variables

Other climatic variables are important too, especially on a local basis. Photosynthesis responds strongly to changes in humidity, as leaves tend to close their stomata in dry air. Wind is an important factor also. It plays a direct role in some regions of the world, where it may limit the extent to which forest or woodland can develop. In all environments wind is important because it determines the relation between the climate and the microclimate at the surface of leaves. In this regard, the influence of wind on heat transfer between leaves and the atmosphere determines surface temperatures, and plants that are sheltered are generally a few degrees warmer by day than those that are exposed in the same location. Wind may also convey adverse materials, including salt spray and pollutants.

Reflective questions

➤ Should we take into account *extremes* of climatological variables or do we work merely with mean values?

➤ Is the temperature of a leaf the same as the temperature of the air?

➤ How will global food production change with global warming?

25.3 Observational studies: how we know for sure that vegetation responds to a changing climate

In this section we review examples of field observations that show how natural vegetation frequently changes as a result of climate change.

25.3.1 The forest/savanna boundary in southern Amazonia

The Amazon rainforest occupies about 6 million km^2 and holds around one-third of the global **biodiversity**. However, the world's most famous rainforest may not have been as extensive just a few thousand years ago. Records obtained from lake sediments in Bolivia in the southern part of Amazonia suggest that the boundary with the savanna may have undergone substantial changes over the past few thousands of years (Mayle *et al.*, 2000). The results are presented as a pollen diagram (Figure 25.4), in which we see the fluctuations of some major groups of plants found as pollen from a 3 m core taken from the sediment at the bottom of a deep lake. In such studies, the age of the samples is determined by the ^{14}C dating (see Box 23.1 in Chapter 23) of the strata from which the samples have been taken. At present, the lake is surrounded by rainforest and the pollen 'rain' into the lake is dominated by pollen of the families Moracaceae and Urticaceae which contain predominantly trees. There is also a signal from *Cecropia*, a tree genus. Grass pollen from members of the Poaceae is relatively uncommon, and so are *Mauritia* and *Mauritiella*, both palms of wet places.

However, just 4000 years ago the situation was very different. The site was evidently much drier. We can tell the water level in the lake must have been much lower because we see pollen of *Isoetes* (quillwort), a plant that lives in the shallows at lake margins (Figure 25.4). Most significantly, there is a strong component of grass pollen and relatively little tree pollen, showing this area to have been savanna at the time. There is also charcoal, an indication of dry conditions and human presence. The conclusion, that rainforest is quite recent, and preceded by grassland vegetation, can be supported by other evidence. Mayle *et al.* (2000) point to a regional increase in rainfall at the time tree pollen began to be common: it is known that the water level in Lake Titicaca (located in the Andes on the border of Peru and Bolivia) rose after 3200 BP, reaching modern levels by 2100 BP. This change in climate was not anthropogenic, although

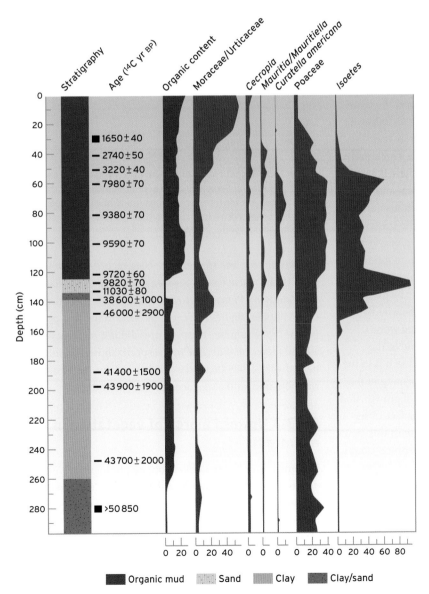

Figure 25.4 Vegetational history of southern Amazonia seen in the pollen record. (Source: Mayle *et al.*, 2000, from F.E. Mayle, R. Burbridge & T.J. Killeen, Millennial-scale dynamics of Southern Amazonian rain forests, *Science* 290: 2291-2294, 2000. Reprinted with permission from AAAS.)

the savannas were probably occupied and fire was used as a tool, but rather the consequence of a southward shift in the **inter tropical convergence zone** (see Chapter 4), itself triggered by Milankovitch forcing. A few similar studies have been conducted at other tropical and subtropical regions, for example the Sahara Desert which was formerly much wetter and vegetated, with a considerable human population (see Chapter 23).

The work is relevant to ideas of how tropical vegetation will change in the future. Some climate models predict substantial drying in Amazonia and elsewhere, and it is thought likely that part of the region will revert to savanna. Today, the changes might well be accelerated by humans who are inclined to use fire for clearing. When the forest is dry there can be large-scale destruction of forests by fire as we have seen in many parts of the world in recent years. This removal of forest by fire is one of the mechanisms by which savanna replaces forest.

25.3.2 The northern tree line

Trees are excluded from cold places. In most older texts it is claimed that trees are absent from sites wherever the temperature of the warmest month is less than 10°C, following a line of thinking by Köppen. This is not entirely true: for some Andean sites the tree line coincides with a maximum summer temperature of only 6°C. However, there is a consensus that summer temperatures are important and that they must be sufficient for adequate rates of photosynthesis and cell division. The consequence of this is the existence of a phytogeographical boundary where trees give way to dwarf shrubs, found at northern latitudes and on mountains all over the world. Around the boundary, known as the tree line, the trees grow slowly and are often stunted with contorted stems. The German word to describe the woodland composed of these trees is *krummholz* (crooked wood).

Climate warming has so far been more rapid in the extreme northern regions than elsewhere, and so it is in the north that we might expect to see a sign of warming in the vegetation, for example increases in the rates of growth, and northerly advances in the distribution of trees. In fact, observations from several regions of the world suppport this view. Gamache and Payette (2004) sudied sub-Arctic spruce–tundra in northern Québec, Canada. Here, the black spruce (*Picea mariana*) is normally rather shrubby. The authors measured growth rates of the plants along a 300 km latitudinal transect. Height growth decreased with increasing latitude, and they found that height growth had increased in the northern plants since the 1970s, so that the northern trees were growing almost as well as the southern trees (Figure 25.5). Similar

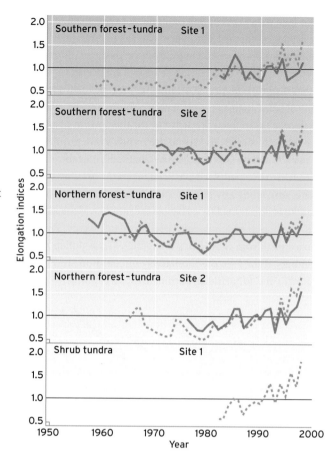

Figure 25.5 Recent spurts of growth of black spruce in northern forest and tundra. Growth is measured as elongation of the shoots. Data are for trees or saplings found at the tree line (solid lines) and above the tree line (dashed lines). (Source: from Gamache and Payette, 2004)

changes in tree line species have been recorded in other parts of the world too. In Sweden, Kullman (2005) has been studying *Pinus sylvestris* since 1973. He finds the size of the population has increased by 50%. In the first period (1973–1987) summer temperatures did not increase at all and there was a decline in the species; however, in the period 1988–2005 there was a strong increase in growth and reproduction, and also a decrease in winter damage.

25.3.3 Upward march of vegetation in mountains

In some places it is possible to obtain photographic evidence of shifts in vegetation, either by repeatedly revisiting the same place, comparing aerial photographs (often, they go back to the 1940s), or simply as photographs taken by ecologists and naturalists on the ground. One such case is the Montseny Mountains of Catalonia in north-east Spain (Peñuelas and Boada, 2003), where ground-based images were available from 1945. In these mountains the beech forest has moved up the slope by 70 m at the highest altitudes. At medium altitudes the existing beech forest is becoming degraded, partially defoliated and is not regenerating itself as it formerly did. Nearby, heathland is being replaced by the more drought and heat-tolerant holm oak (Figure 25.6). Changes like this have been reported from the European Alps as well. There, the existence of many small alpine species is threatened with local extinction as their particular habitat declines in area as a result of global warming (Pauli *et al.*, 2007). This general phenomenon, first recognized two decades ago, has prompted the establishment of long-term sample plots, which are revisited every few years for enumeration of the plants. The project has established a global network of sites in mountain regions, known as GLORIA (Global Observation Research Initiative in Alpine Environments, www.gloria.ac.at).

▓ *Quercus ilex* young forest	▓ *Erica scoparia* heathland
▓ *Calluna vulgaris* heathland	▓ *Pteridium aquilinum* fernland
▓ Grassland and *Cytisus scoparius*	

MS Meteorological station

Figure 25.6 Rapid vegetational change attributed to climate warming. Panels on the left-hand side are from photographs in 1969, those on the right-hand side are from 2001. (Source: from Peñuelas and Boada, 2003)

25.3.4 Changes in the timing of flowering

The recording of the annual cycle of growth and development of plants has interested amateur naturalists and professional meteorologists for over 100 years (Jeffree, 1960; Sparks *et al.*, 2000). In some countries, 'phenological gardens' have been established to monitor such things as the date of bud-break every spring, the date of first flowering and the date on which trees shed their leaves. These data sets are a rich archive of information, linking biology and meteorology. In some cases these records are very long ones, broken only by interruptions by war or failure of funding. In other cases, they are the result of one person's painstaking efforts over a lifetime. One of the best examples is provided by the naturalist Richard Fitter, who recorded the data of first flowering for a set of 557 wild plant species over a period of 47 years in south-central England (Fitter *et al.*, 1995; Fitter and Fitter, 2002). These authors found a very strong dependency of first flowering date on temperature in the previous months. Only 24 out of the 243 species selected showed no dependency. More recently, Menzel *et al.* (2006) have completed an analysis of more than 100 000 time series of phenological stages across Europe. They found that nearly all the phenological events, and also farmers' activities, were related to temperature (Figure 25.7). In Figure 25.7, the recordings represent the date of onset of the event. A negative correlation coefficient therefore means that warming is associated with an earlier date of onset. Nearly all phenological events occur earlier as a result of warming in the spring. Overall, the authors calculate that as a result of warming, spring in Europe is arriving 2.5 days earlier per decade. A few events have positive correlations with temperature, meaning that warming is associated with a later occurrence of the event. For example, warm summers are associated with later leaf colouring in the autumn.

We may safely conclude that, at least for temperate perennial plants, the life cycle is to a large extent set by the temperature, within certain limits. The cycle of growth and development may be advanced or retarded, and the growth period shortened or prolonged, according to the temperature pattern of a particular year.

From such data, it is possible to model what might happen with a few degrees of warming. To go further and to predict the ecological consequences is somewhat more difficult. If insect-pollinated plants flower early, for example, they will not be fertilized unless the relevant insect pollinators are available at the same time. Also, if flowering is too early there might be some increased risk of damage by frost. Clearly, this work has enormous economic relevance

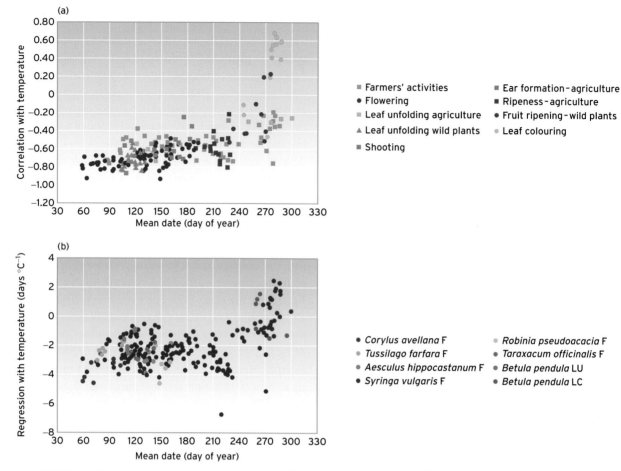

Figure 25.7 The relationship between phenological events and temperature measured on different days of the year: (a) the correlation coefficient (negative coefficient means high temperatures on that day of the year produce earlier events); (b) the sensitivity of the bud opening to temperature on that day of the year for a range of six trees and two perennial herbs. For example, a regression of −4 on day 120 means that one degree of warming stimulates early bud-opening by 4 days. F = flowering, LU = leaf unfolding, LC = leaf colouring. (Source: from Menzel *et al.*, 2006)

and it assists growers to plan for the future and also to evaluate risks.

Today, satellite sensors can be brought to bear on the problem of relating climate change to the cycles of plant life, especially in relation to the greening up of the land in spring. However, satellites look at huge swathes of landscape and it is usually hard to disaggregate the signal of natural vegetation from the frequent changes in agricultural practice. So, there remains a role for amateur phenologists, and interest in phenological gardens is currently as high as it has ever been.

Reflective question

➤ What evidence is there that vegetation is responding to climate change?

25.4 Models for prediction

In the preceding sections we saw something of how plants respond to climatic variables. There are substantial differences between species: how long they live, their growth rates and physiology, and this in general is why different species occupy different regions of the land surface. In principle, it would be possible to determine a set of parameters for each species, and then to use a model to determine where the species could possibly occur within a global climatic envelope like the example shown in Figure 25.1. The task would be large, as even closely related species respond differently to environmental variables, and in any case there are important other influences, to which we have already alluded, such as competition between species and the tendency for dominant species to modify the conditions for other species. It would not be practical to screen the characteristics of all extant species of vascular plants, as there are about 400 000 of them. The size of this task is one of the bottlenecks in the development of predictive models. The research community is still discussing the best way to link knowledge of species to knowledge of biomes; however, there is a consensus that some coarse classification of species is required so as to capture the essence of what makes one vegetation type different from another. Ideally, this description should have a 'meaning' across several disciplines, so that remotely sensed data on reflectance of the land cover can be related to what ecologists can discern as a more or less homogeneous entity. To this end, and after a number of international conferences on the subject, the idea of plant functional types (PFTs) emerged.

A PFT is a group of plants with similar traits and which are similar in their association with environmental variables. Each PFT is defined by a variety of optical, morphological and physiological parameters. At present there is no real consensus beyond this, but in Table 25.1 we show one set of PFT definitions from Woodward *et al.* (2004). From this starting point we can make mathematical models of how vegetation responds to a changing climate. Most models keep track of the flow of carbon as well as the PFTs and

biomes. The important processes which need to be modelled in that case are as follows.

1. *The rate of photosynthesis, and its dependency on environmental factors and the supply of nutrients.* The rate of photosynthesis on an ecosystem scale is termed gross primary productivity (GPP); it is the CO_2 uptake in the daytime period, adjusted by an estimate of how much carbon is simultaneously lost by plant respiration.

2. *Allocation patterns.* The allocation of the carbon acquired in photosynthesis to different plant parts (leaf, stem and root) is required and so is the respiration rate required to build complex molecules such as proteins and lignin from the simpler ones like glucose. If we know how this process of allocation depends on environment, and we also know the rate of photosynthesis, then we can compute the growth rate. On an ecosystem scale this is known as net primary productivity (NPP), related to GPP as follows:

$$GPP = NPP + R_a \tag{25.2}$$

where R_a is the respiration that the plant expends, known as autotrophic respiration. There are some established theoretical relationships that help us to find R_a. For example, the

Table 25.1 A plant functional type classification adopted by Woodward et al. (2004)

Classification	Comment
Evergreen needleleaf forests	Lands dominated by trees with a canopy cover of more than 60% and height exceeding 2 m. Almost all trees remain green all year. Canopy is never without green foliage
Evergreen broadleaf forests	Lands dominated by trees with a canopy cover of more than 60% and height exceeding 2 m. Almost all trees remain green all year. Canopy is never without green foliage
Deciduous needleleaf forests	Lands dominated by trees with a canopy cover of more than 60% and height exceeding 2 m. Consists of seasonal needleleaf tree communities with an annual cycle of leaf-on and leaf-off periods
Deciduous broadleaf forests	Lands dominated by trees with a canopy cover of more than 60% and height exceeding 2 m. Consists of seasonal broadleaf tree communities with an annual cycle of leaf-on and leaf-off periods
Mixed forests	Lands dominated by trees with a canopy cover of more than 60% and height exceeding 2 m. Consists of tree communities with interspersed mixtures or mosaics of the other four forest cover types. None of the forest types exceeds 60% of the landscape
Closed shrublands	Lands with woody vegetation less than 2 m tall and with shrub canopy cover more than 60%. The shrub foliage can be either evergreen or deciduous
Open shrublands	Lands with woody vegetation less than 2 m tall and with shrub canopy cover between 10 and 60%. The shrub foliage can be either evergreen or deciduous
Woody savannas	Lands with herbaceous and other understorey systems, and with forest canopy cover of between 30 and 60%. The forest cover height exceeds 2 m
Savannas	Lands with herbaceous and other understorey systems, and with forest canopy cover between 10 and 30%. The forest cover height exceeds 2 m
Grassland	Lands with herbaceous types of cover. Tree and shrub cover is less than 10%

Source: Woodward *et al.*, 2004

synthesis of cellulose requires the energy obtained from the breakdown of glucose as well as a supply of glucose molecules, so the formation of 1 g cellulose is associated with the release of a specific quantity of respiratory CO_2. These are 'classical' issues in plant physiology, discussed by Amthor (2000).

3. *Birth, death, phenology.* Plant parts are formed and they are shed, often on an annual cycle. Plants as a whole are 'born' every time a seed germinates and they eventually die. Some are annuals (programmed to die after a few weeks or months), others are biennials (the life cycle takes two years, sometimes a little longer) and perennials (most grasses and all trees are good examples). Generally, in global models we have to provide some 'rules' for each PFT. As in all models, we have to ignore the exceptions and abide by simple rules. For conifer trees, for example, leaves are retained for several years whereas for deciduous trees they are shed on an annual cycle. The resulting 'litter' from shedding and death is incorporated into the soil, and decomposed by microbes. The microbial respiration gives rise to further efflux of CO_2 from the ecosystem, R_h, known as heterotrophic respiration. The resulting net flux of CO_2 between the ecosystem and the environment is the net ecosystem production (NEP), related to the previous terms as:

$$GPP = NEP + R_a + R_h \qquad (25.3)$$

In a world with a constant environment, all the GPP would be respired by plants or microbes and so NEP would be zero. In the real world there are additional losses of carbon caused by natural and human influences. To take these into account we would need to insert a new carbon loss, the disturbance flux R_d.

4. *Population-level processes.* Models need to represent the process of plant **succession**, whereby species colonize the land and develop towards an equilibrium or near-equilibrium state. To do this, the attributes of each PFT are needed, and some further rules are required to determine in what circumstances one species succeeds another.

Models are run using a sequence of climatological data, in time steps that vary from months to minutes, according to the scheme of Figure 25.8. How good are such models? They can be tested in several ways: (i) by examining whether the model calculations can reproduce today's vegetation by running the models from historical climate data; (ii) by investigating whether they produce the carbon fluxes that are measured in field studies; and (iii) by comparing them with other models, developed in other laboratories more of less independently.

An important comparison of six models was reported by Cramer *et al.* (2001). They computed the 'expected' vegetation on the basis of a world divided into regions of 3.75° longitude and 2.5° latitude, given the climatic scenario. In this scenario of climate change, temperature increases by 4.5°C over the next 100 years, and the CO_2 concentration rises from 380 to 800 ppm by the end of the century. Although the six models do not all agree in detail,

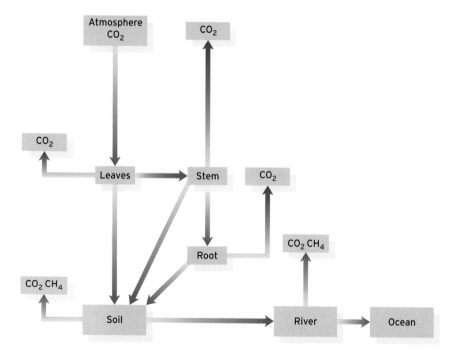

Figure 25.8 Structure of a very basic model of the carbon flows associated with photosynthesis, respiration, growth and storage of carbon. 'Losses' of CO_2 from leaves, stems and roots (autotrophic respiration) are associated with maintenance and growth of the plant tissues; losses from the soil are either from microbial respiration (heterotrophic respiration) or in the drainage water. Such a model might be configured for an individual plant or for a set of PFTs in an ecosystem. More advanced models include flows of water and nutrients, and their interaction with carbon flows.

and although they do not faithfully reproduce the current vegetation, there is a general consensus on how vegetation change will occur (Figure 25.9). The principal changes are:

- Transitions from forest to savanna. Such transitions are predicted to occur in the Amazon, the central part of the American continent and in South-East Asia.
- Evergreen forest will replace 'grassland' in parts of North America and north Europe.

- Parts of the Mediterranean grasslands will become savannas.

As for the carbon balance, the models produce an interesting trend, and there is some agreement between models. The NPP is stimulated by warmer conditions (felt especially in the cold northern regions) and also by the elevated CO_2 (Figure 25.10). However, the heterotrophic respiration is increased by warming to an even larger extent, and the

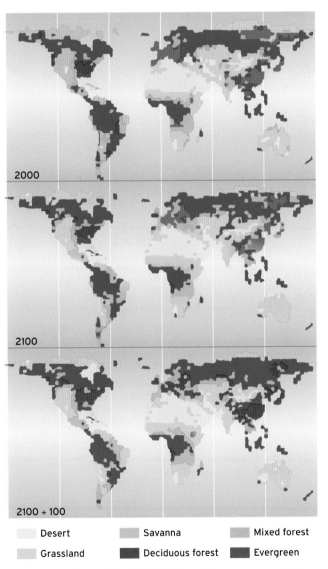

Figure 25.9 The predicted distribution of biomes as a result of climate change and elevated CO_2 according to Cramer *et al.* (2001). The models stopped simulating climate and CO_2 change in 2100. Therefore changes from the middle to bottom panel are related to vegetation dynamics as the system equilibrates. See text for further interpretation. (Source: from Cramer *et al.*, 2001)

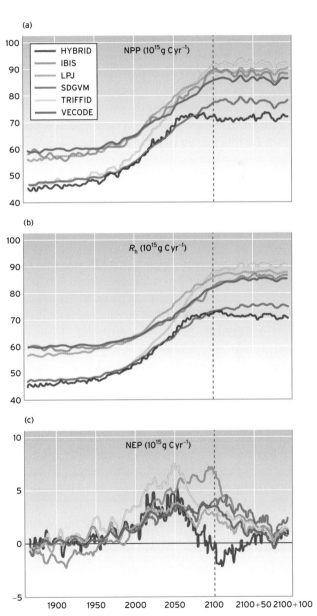

Figure 25.10 Predicting the carbon balance of terrestrial vegetation for the next century by means of six models: (a) the net primary productivity (NPP); (b) the heterotrophic respiration (R_h microbes and animals); (c) the overall carbon flux made by subtracting R_h from NPP. Each line is a different model, as shown in (a). (Source: from Cramer *et al.*, 2001)

consequence is a large rise in global respiration over several decades. It is interesting to put these numbers into perspective as follows. The heterotrophic respiration from all the microbes (heterotrophic respiration of all terrestrial ecosystems) is currently around 50–60 gigatonnes of carbon per year (Gt C yr^{-1}), which completely dwarfs the fossil fuel emissions of about 6.5 Gt C yr^{-1}.

Large though R_h is, it is more than offset by photosynthetic production, so the NEP is positive. In other words the terrestrial ecosystems are collectively a 'sink' for carbon. In fact, the current terrestrial sink strength of 1–2 Gt C yr^{-1} is predicted to rise for a few decades before taking a downward turn as the effect of temperature on respiration is increasingly felt. There is a substantial predicted downturn in NEP so that by 2100 the sink has diminished and, in one of the models, has turned into a source.

The most significant weakness of these models is that they are not coupled to a model of the climate system, so the CO_2 effect of a vegetational source or sink on the temperature is not apparent. In the 'real world' the atmospheric CO_2 and therefore the global warming rate would be influenced by the vegetation. Such feedbacks may well be important. In recent years, simple vegetation models have, however, been coupled to global circulation models (GCMs) in an attempt to capture the essential feedbacks. Here, we refer to a synthesis study (Friedlingstein *et al.*, 2006) in which 11 models were compared. The results are quite variable and the models do not all show the same general trend (Figure 25.11). They clearly differ in their sensitivity to climate change. The most sensitive result of all is shown by the Hadley Centre Model where the vegetation becomes a progressively weakening carbon sink in the next few decades and then moves into carbon deficit rather in the same way as the uncoupled models discussed in the previous paragraph. For the other models in the Friedlingstein study, there is a weakening of the sink except in one case where the sink continues to intensify. Further work is clearly needed before we can make further estimates of the future based on models.

One of the more difficult aspects of predicting the future from models is that humans interact with climate change and this interaction is hard to predict. In truly managed ecosystems the economically relevant outputs are the ones that are recognized, monitored and increased largely as a result of applying science. In modern UK agriculture, for example, the agriculture has become more intensive since the Second World War, and has enjoyed substantial government subsidies, yet a recent study showed widespread recent carbon losses from the soil in England and Wales, thus contributing inadvertently to global warming.

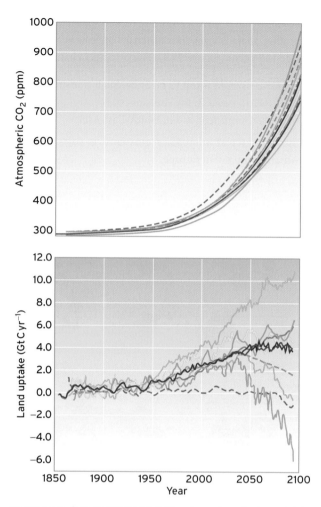

Figure 25.11 Carbon uptake from the atmosphere by terrestrial ecosystems (units: Gt = gigatonnes = 10^9 tonnes = 1 billion tonnes). The calculations are made after coupling several GCMs to simple models of the terrestrial biosphere. Each line is the result from a different model. The current terrestrial carbon 'sink' is somewhere between 0 and 4 Gt of carbon per year. Some models predict that the land sink will become a source. (Source: from Friedlingstein *et al.*, 2006)

Reflective questions

➤ Models are only as good as the understanding built into them. What are the gaps in our understanding of vegetation–climate models?

➤ How well have the vegetation models been tested?

➤ What is a plant functional type?

➤ What important processes need to be modelled in PFT, biome and carbon flow vegetation models?

25.5 The complex interaction between human activities and climate change

25.5.1 Does atmospheric pollution sometimes benefit plants?

As a result of human activities, notably agriculture and the driving of motor vehicles, much more nitrogen in a chemically active form (ammonium and nitrate especially) is deposited to the land surface than hitherto (Box 25.1). If there is too much N-deposition, ecosystems may show signs of 'nitrogen saturation', a condition whereby the land surface may 'leak' nitrogen to the drainage water and give off nitrous oxide, another greenhouse gas. If, on the other hand, the deposition rate is below a certain threshold level, there may be a fertilizer effect for many ecosystems, especially those that are otherwise N-deficient. This would be seen as a stimulation of photosynthesis and possibly an increase in growth rates

and a strengthening of the carbon sink. Indeed, many model calculations of the impact of environmental change on global vegetation contain a term to allow for this. The N-effect and climate change interact in ways that are not understood very well, as highlighted by Magnani *et al.* (2007). Data on the C-fluxes over forests in Europe and North America were collected, and attempts were made to relate the GPP and R_h to temperature. In fact, both showed a remarkable linear relation with temperature; but the NEP (i.e. GPP $- R_{eco}$, where R_{eco} is the total ecosystem respiration) is rather a weak function of temperature (Figure 25.12). However, NEP is strongly related to the deposition of anthropogenic nitrogen from the atmosphere. Hence, we may suggest that pollution of the atmosphere with nitrogen (principally from vehicles, agricultural systems) is at a level that stimulates production. This is of course a somewhat controversial claim, as most people link N-deposition to nitrogen saturation of ecosystems or, worse, with the production of acid rain that has deleterious impacts on forests.

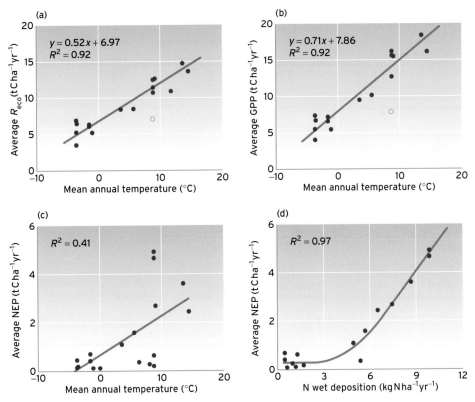

Figure 25.12 Measured carbon fluxes over 18 forests in Europe and North America: (a) ecosystem respiration, R_{eco}; (b) gross primary productivity, GPP; (c) and (d) the overall carbon flux as directly measured as NEP, showing a carbon sink of between 0 and 5 t C ha^{-1} yr^{-1}. Note that the overall carbon flux NEP is a fairly weak function of temperature but a strong function of the deposition of nitrogen from the atmosphere. (Source: Magnani *et al.*, 2007, reprinted by permission of Macmillan Publishers Ltd: *NATURE*, F. Magnani, M. Mencuccini, M. Borghetti, P. Berbigier, F. Berninger, S. Delzon, A. Grelle, P. Hari, P.G. Jarvis, P. Kolari, A.S. Kowalski, H. Lankreijer, B.E. Law, A. Lindroth, D. Loustau, G. Manca, J.B. Moncrieff, M. Rayment, V. Tedeschi, R. Valentini, J. Grace, The human footprint in the carbon cycle of temperate and boreal forests, vol. 447: 848-850. Copyright 2007.)

ENVIRONMENTAL CHANGE

THE NITROGEN CYCLE AND ANTHROPOGENIC PERTURBATIONS

Nitrogen is an important constituent of proteins and nucleic acids, and so is essential for life. Nitrogen exists primarily as an unreactive gas in the atmosphere as dinitrogen, N_2. It constitutes 79% of the air we breathe. Very important reactive forms of nitrogen also exist as gases and ions. The gases are ammonia (NH_3) and the oxides NO, NO_2 and N_2O, all present in trace concentrations. The main ions, found in soils and water, are ammonium NH_4^+, nitrate NO_3^- and nitrite NO_2^- ions (see Chapter 10). Two natural processes convert nitrogen to reactive forms that can be taken up by the roots of plants: lightning and biological nitrogen fixation (BNF). Very

small amounts of N_2 are reacted with O_2 during lightning, to form the gas nitric oxide, NO, which eventually reaches the ground as nitrate. BNF is quantitatively more important than lightning as an agent of nitrogen fixation: bacteria living in the soil fix N_2 to make the reactive forms ammonium NH_4^+ and nitrate NO_3^- which can be used by plants. Some of these nitrogen-fixing bacteria are free-living, but others form **symbiotic** relationships with plants, especially those of the pea family Leguminosae. Many members of this family are used in agricultural systems as a 'free' source of nitrogen fertilizer (examples are clover, lucerne, groundnuts, soybeans, alfalfa and lupins). Such plants have root nodules containing populations of the nitrogen-fixing bacteria. As a consequence of the free-living nitrogen-fixers and the symbiotic

nitrogen-fixers, 130–330 Gt N yr^{-1} are made available in the soil solution as ammonium NH_4^+ and nitrate NO_3^- and can thus be taken up by plants to make protein and other biochemical constituents. Herbivorous animals obtain their protein by consuming plant material in prodigious quantities. Dead plants and animals decompose in the soil, and some of the nitrogen is acted upon by denitrifying bacteria. Nitrogen is thereby returned to the atmosphere as N_2 or N_2O. The process is thus cyclic (Figure 25.13).

Humans perturb the nitrogen cycle in fundamental ways. The industrial fixation of N_2 was developed in 1909 by Haber and Bosch, and provides a supply of nitrogen fertilizer, estimated to be 78 Gt N yr^{-1}. This is applied to the soil, but much of it is released to the atmosphere as nitrous oxide N_2O, a greenhouse gas. Some of it enters

Figure 25.13 The nitrogen cycle.

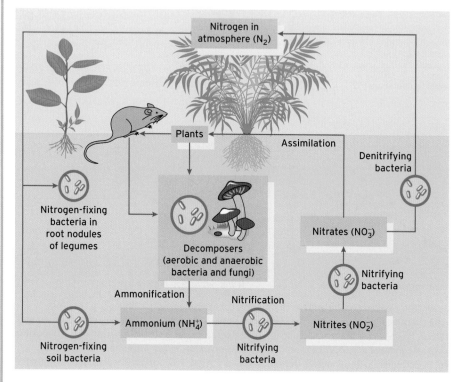

BOX 25.1 ➤

➤

drainage water, and causes excessive growth of algae in streams and rivers (see Chapter 13). There are other anthropogenic sources of nitrogen. The internal combustion engine and some other fuel-burning devices are responsible for emissions of oxides of nitrogen to the atmosphere, through the combination of atmos-pheric oxygen and nitrogen inside the combustion chamber. Moreover, cultivation and disturbance of the land results in emissions of nitrous oxide. Finally, animal rearing is associated with the emission of NH_3, produced by the decomposition of urine and faeces.

The consequence of increased formation of NH_3 and oxides of nitrogen is an enhanced rate of ammonium and nitrate deposition to land and waters. Nitrogen depo-sition rates are now much higher in populous regions of the world than they were in pre-industrial times (Figure 25.14). This deposition occurs as dry deposition of the gases themselves, and also as nitric acid,

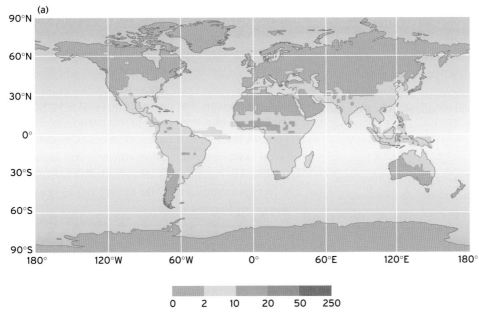

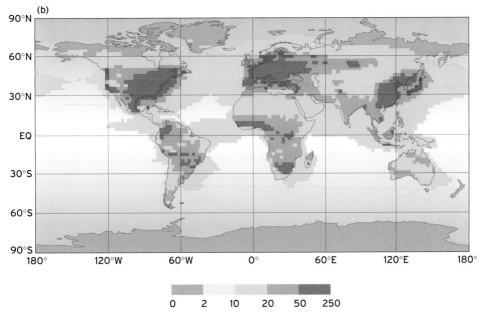

Figure 25.14 Deposition of reactive nitrogen to the Earth's surface (in mmol N m^{-2} yr^{-1}): (a) pre-industrial levels; (b) current atmospheric depo-sition. Note the enhanced deposition in densely populated areas. (Source: after Galloway *et al.*, 1995)

BOX 25.1 ➤

735

> HNO$_3$, a contributor to acid rain. The growth of plants is generally limited by the availability of nitrogen in the soil, and so the enhanced deposition may be increasing plant growth, and contributing to a widespread increase in the rate at which trees grow. On the other hand, in some areas, the imbalance in nutrients in acid rain may cause damage to forests (see Chapter 10).

BOX 25.1

25.5.2 How does fire interact with climate change?

Many predictions regarding climate change suggest that some areas of the tropics will become dryer as a result of an increased frequency and harshness of El Niño events. As we saw above, this increase in drought is one of the causes of the expected conversion from forest to savanna. However, as humans encroach upon the rainforest, some researchers believe the effect will be amplified by the creation of forest edges in a fragmented forest, and the use of fire (Laurance and Williamson, 2001). Dense forests have a microclimate characterized, for example, by high humidities and daytime temperatures that are usually lower than those measured at the top of the canopy. At forest edges the situation is different, with free horizontal ventilation and mixing of canopy air with air from outside. When drought occurs, relatively dry air penetrates the canopy, and reductions in humidity and increases in plant mortality have been measured at up to 100 m from the canopy's edge. Humans light fires, and these fires are likely to ignite more easily and spread more rapidly in the dry conditions of the forest edge. Hence, the forest is damaged and possibly destroyed at a faster rate to what would occur in the absence of humans. The processes involved are quite complex and collectively amount to a positive feedback (Figure 25.15). In the Laurance–Williamson model, deforestation causes less evaporation, which in turn leads to less rainfall and hence droughts are exacerbated. Logging can also be important as it thins the canopy and increases vulnerability to combustion.

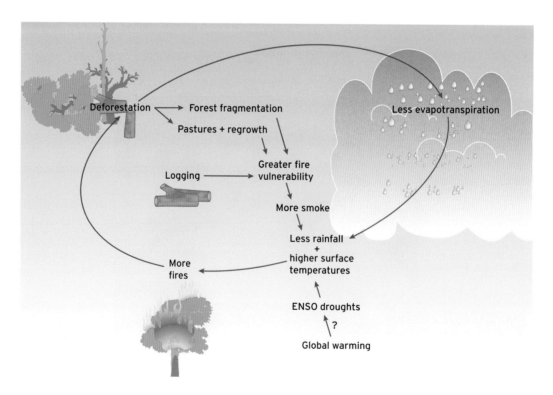

Figure 25.15 The Laurance-Williamson model of how droughts and human impacts combine to degrade and destroy rainforests. (Source: Laurance and Williamson, 2001)

25.6 Loss of biodiversity

Biodiversity is defined as the number and variety of species in ecological systems, at local, regional and global scales (see Chapters 19 and Chapter 20). There is concern that both human activities and climate change are causing a decline in the number of species. Given global warming, the importance of climate change is likely to become progressively more important, as pointed out in the recent report by the IPCC (IPCC, 2007b).

There are at least two reasons why we may expect biodiversity to alter under climate change. In the first place, the pattern of land use will change as the demand for agricultural land and biofuel increases. Thus, species-rich habitats such as rainforest and savanna in the tropics will decline and at some critical point the species they contain will be lost. The second, and more subtle, reason for loss of biodiversity is that species differ greatly in their sensitivity to warming and to water supply. Thus, particular species in an ecosystem may migrate, leaving a functional void in their original location. In the new ecosystems that they come to occupy they are 'invasive species' and may be strong competitors, ousting some of the more delicate species. Thus, ecosystems which are more or less in a species equilibrium may be destabilized with unknown consequences.

Although it is impossible to know the total number of species in the world (only 1.5 million are known but many are yet undiscovered), extinctions themselves are generally well documented. Since 1600, a minimum of 490 plant and 580 animal species have become extinct. Some groups, such as mammals and birds, have suffered more than others. In geological time there have of course been catastrophic mass extinctions. The natural or background extinction rate can be estimated from the fossil record. For example, in mammals the background rate is about one in 400 years, but this is much lower than the observed rate. The number of species that are threatened far exceeds our capacity to protect those species, and so conservationists have identified 'hot spots' where conservation effort and resources should be greatest (Myers *et al.*, 2000). It is particularly important to protect areas with a high degree of endemism (an endemic species is one restricted to a particular region). When this exercise was carried out there were some surprises, as shown in Box 25.2. For example, the natural vegetation of the tropical Andes heads the list. Its vegetation has been reduced to 25% of its original extent, yet it contains 6.7% of all plant species in the world and 5.7% of all vertebrate animals.

ENVIRONMENTAL CHANGE

HOT SPOTS AND CLIMATE CHANGE

Myers *et al.* (2000) reported that 44% of all species of vascular plants are confined to 25 hot spots constituting only 1.4% of the land surface. The authors urged these to be singled out for the attention of conservationists, to attempt to protect them. The Myers *et al.* (2000) map of hot spots is reproduced here, along with projections of temperatures from the IPCC (2007b) 4th Assessment Report (Figure 25.16). Figure 25.16(b) shows the warming predicted from the A2 scenario (see Chapter 24). Note: all hot spots will be warmer.

Over the next century there will be further extinctions, and climate change will be an important driver, along with land-use change. Everyone agrees that species should be

BOX 25.2 ➤

➤
protected, and that natural environments have an inherent value. Indeed, the idea that humans are the guardians of nature is deeply embedded in Christian and other religions. There is also much folklore that evokes the conservation ethic. The author's grandmother used to say:

if you wish to live and thrive
let a spider run alive.

This appears to be a reference to a 'keystone species', as spiders are voracious predators, required to control populations of small flying insects which often carry disease. When keystones are removed the ecosystem is in trouble.

At the moment there is a plethora of international agreements to protect species and habitats and many individuals subscribe to organiza-

tions such as Friends of the Earth or Greenpeace. The challenge for nature conservation is to protect natural ecosystems but traditional conservation and protection against human encroachment is clearly not sufficient: it does not protect species and ecosystems from the impact of human-made climate change.

(a)

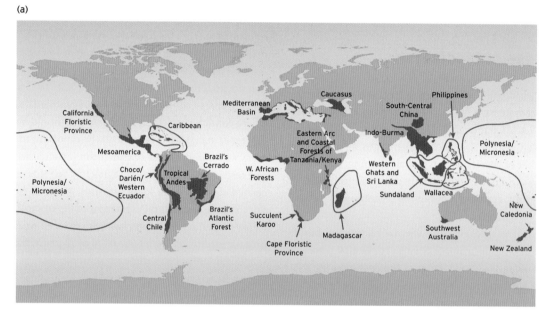

(b)

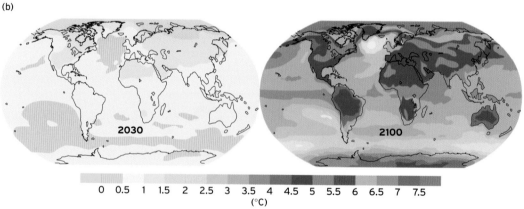

Figure 25.16 (a) Twenty-five biodiversity hot spots identified where 44% of all species of vascular plants and 35% of all species in four invertebrate groups are confined to the hot spots comprising only 1.4% of the land surface of the Earth. (b) Projected temperature increases for 2030 and 2100 from model scenario A2. (Source: (a) from Myers *et al.*, 2000, reprinted by permission of Macmillan Publishers Ltd: *NATURE*, Norman Myers, Russell A. Mittermeier, Christina G. Mittermeier, Gustaro A.B. da Fonseca & Jennifer Kent, Biodiversity hotspots for conservation priorities, vol. 403: 853–858: Copyright 2000; (b) from IPCC, 2007a)

BOX 25.2

25.7 Summary

Plant growth is influenced by climatological variables, especially light, temperature and moisture. For natural vegetation it is predominantly temperature and moisture that determine the type of land cover on a global scale. Both temperature and moisture are changing and they are expected to change rapidly in the next 100 years. Observations in the field, conducted over the last few decades, support the general view that the vegetation is changing in response to global climate change. Over the next century, it is likely that substantial changes in vegetation will result, and that these are likely to be so important as to impact upon the carbon cycle and the climate system. Specifically, regions of the world such as the humid tropics, which are now believed to be a carbon sink, may become a source; conversely, cold northern regions will become warmer and therefore more favourable for the growth of plants and especially for trees. They may become a sink for carbon. Predictions are, however, based on state-of-the-art model calculations, and there are still many uncertainties. One of the largest unknowns in the system is the behaviour of humans. Their behaviour determines the rate of global warming and the nature of the land cover, and also modifies the response of vegetation on a global scale through the use of agents such as cultivation, fertilizers and fire.

Further reading
Books

Beerling, D.J. and Woodward, F.I. (2001) *Vegetation and the terrestrial carbon cycle. Modelling the first 400 million years.* Cambridge University Press, Cambridge.
Original and insightful if you are interested in long (geological) timeframes.

Lovejoy, T.E. and Hannah, L. (2005) *Climate change and biodiversity.* Yale University Press, New Haven, CT.
A good overview book.

Malhi, Y. and Phillips, O. (2005) *Tropical forests and global atmospheric change.* Oxford University Press, Oxford.
A research-level enquiry into one of the most important issues raised in this chapter.

Schulze, E.D., Beck, E. and Müller-Hohenstein, K. (2001) *Plant ecology. Springer-Verlag, Berlin.*
An excellent compendium of plant physiology as it relates to ecology; a good place to learn about photosynthesis.

Further reading
Papers

Leakey, A.D.B., Ainsworth, E.A., Bernacchi, C.J., Rogers, A., Long, S.P. and Ort, D.R. (2009) Elevated CO_2 effects on plant carbon, nitrogen and water relations: six important lessons from FACE. *Journal of Experimental Botany*, 60, 2859–2876.
Useful review of state-of-the-art experimentation.

Menzel, A., Sparks, T.H., Estrella, N. *et al.* (2006) European phonological response to climate change matches the warming pattern. *Global Change Biology*, 12, 1969–1976.
A research paper showing how temperature appears to control phenological changes.

CHAPTER 26

Remote sensing of environmental change

Timothy D. James
Department of Geography, Swansea University

Learning objectives

After reading this chapter you should be able to:

➤ appreciate the role of remote sensing as an indispensable tool in physical geography for monitoring and understanding the processes of environmental change

➤ understand the nature of electromagnetic radiation and how it interacts with the physical environment

➤ evaluate different methods of remote sensing for collecting environmental change data

➤ describe methods of digital image processing that enhance the quality and interpretation of remotely sensed data

26.1 Introduction

In physical geography, a great many subjects are studied that cover a wide range of scales. Geographers study processes that occur over the macroscale (e.g. global climates, Chapter 5), the mesoscale (e.g. glaciers and ice sheets, Chapter 16) and the microscale (e.g. soils, Chapter 10). At each of these levels, scientists need to collect suitable data as quickly and efficiently as possible using cost-effective methods and without damaging or interfering with the environments they seek to understand. Whatever the scale, many of these data collection requirements are met through the methods of remote sensing. Over the last few decades, data collected using remote sensing have significantly increased our ability to measure the physical, chemical, biological and cultural characteristics of the Earth's surface.

The American Society of Photogrammetry and Remote Sensing defines remote sensing as any technique whereby information about objects and the environment is obtained from a distance. Remote sensing data can be acquired from one of three platforms: terrestrial, airborne and space-borne. Terrestrial sensors often include simple handheld instruments such as a regular 35 mm camera handheld or mounted on a gantry, or sensors that are placed close to the ground for detecting features beneath the Earth's surface. Airborne sensors are mounted in helicopters or small aircraft for low-altitude missions, and more specialized aircraft for higher-altitude missions. Spaceborne sensors are mounted on orbiting vehicles such as satellites or the Space Shuttle.

Most often, remotely sensed information comes in the form of images that can be interpreted and analyzed. There are several characteristics of remote sensing that make it ideal for use in physical geography and detecting environmental change. Firstly, it minimizes the need for field visits. This is important when studying environments that are either dangerous (e.g. natural hazards, political strife), isolated and difficult to reach (e.g. remote islands or glaciers), or fragile (e.g. periglacial and dune ecosystems). Secondly, within some limitations remote sensing instruments can be positioned as far from, or as near to, a surface as required. Thus, they can collect information over a very large area as in satellite imagery or over a very small area such as a camera mounted on a microscope. Finally, remote sensing offers the ability to repeat data collection over relatively short periods thus adding a fourth dimension to geographical studies and allowing environmental change to be detected. As a result, the methods of remote sensing have become essential for research in physical geography.

The subject of remote sensing is exceptionally extensive as it covers a wide range of instruments and applications. To address the subject in one chapter can scarcely do it justice. Where necessary, you are therefore encouraged to consult the further reading and references for more detailed coverage of the topics herein. With this in mind, the aim of this chapter is two-fold. The first is to provide you with a general introduction to the various components of remote sensing. The second is to act as a starting point from which you can explore more comprehensive references on any of the subjects discussed in the following sections. In this chapter you will be introduced to important image characteristics, the foundations of remote sensing, the various data sources and digital image processing.

26.2 Image characteristics

26.2.1 Types of image

Although remote sensing data are not exclusively in the form of imagery, this is by far the most common format and the most useful to physical geography. Therefore, this section discusses the various characteristics of imagery produced by remote sensing techniques. There are basically two types of images in remote sensing: analogue and digital. **Analogue images** include photographic negatives, **diapositives** and prints, each of which can be described as continuous tone images. Even in this digital age, analogue images still represent a common image format. However, in the face of improving technology and decreasing costs, digital equivalents are slowly replacing their analogue predecessors.

Wolf and Dewitt (2000) defined a **digital image** as a computer-compatible, pictorial rendition divided into a fine, two-dimensional grid of **pixels**. The term pixel comes from a contraction of 'picture elements' and each represents

Table 26.1 Bit scale and grey levels

Bit scale (depth)	Range of DNs	Number of grey levels
1-bit (2^1)	0-1	2
6-bit (2^6)	0-63	64
7-bit (2^7)	0-127	128
8-bit (2^8)	0-255	256
24-bit (2^{24})	0-16 777 215	16 777 216
32-bit (2^{32})	0-4 294 967 296	4 294 967 297

a finite area in the image. A **digital number (DN)** is assigned to each of these pixels to summarize the average reflection that was recorded for that unit area of surface. In the displayed images, the DNs determine the colour of each pixel according to the **bit scale** of the recorded image and the **colour palette** used for its display. A colour palette is simply a record of predefined colours each linked to one or more DNs. Colour palettes can range from only two colours (i.e. black and white) to billions of colours. The bit scale (or depth) of the image determines the number (or range) of DNs that are used in an image. A simple example is a 1-bit image, called a binary image, composed only of the numbers 1 and 0. Thus, the number of colours used in an image is determined by the bit scale as shown in Table 26.1. An example of the DNs in an image is given in Figure 26.1. Here, a simple 8-bit image using 256 grey levels is shown with small sections of increasing magnification to show how the DNs combine to form an image.

26.2.2 Image orientation, scale and resolution

For both analogue and digital images, there are a variety of image characteristics that are essential for their use in geographical research. An appreciation of these characteristics is central to understanding remote sensing and therefore they are addressed here. The first three characteristics discussed are variables that are determined prior to data collection. They are **image orientation**, **scale** and **resolution**.

A general assumption is that remote sensing instruments are always pointed downwards (**nadir**-looking) from some position above the surface of interest. However, this is not necessarily the case. Although vertical imagery

98	98	98	92	88	72	55	48	44	41	40	38	37	45	59	48
99	98	97	92	83	62	49	44	42	39	39	39	39	42	55	46
97	99	94	87	69	52	46	47	52	54	49	45	43	42	46	42
97	97	92	79	61	51	83	138	460	157	113	68	48	47	46	38
98	94	89	71	55	62	151	212	197	198	198	138	60	53	50	39
94	92	81	61	53	70	172	196	113	103	175	191	92	58	56	40
90	86	73	52	47	59	154	206	128	81	149	201	110	59	54	40
89	81	62	45	42	51	94	187	198	174	196	194	94	62	55	40
87	71	49	44	44	46	51	88	141	176	171	118	64	63	50	38
83	61	45	42	44	45	46	46	54	69	68	55	54	68	46	36
73	50	42	41	42	44	46	47	46	48	44	43	52	72	48	38
33	27	26	26	25	32	41	47	49	48	45	42	51	61	44	38

Figure 26.1 Construction of a digital image. As the magnification increases, what appears to be an image of continuous tone breaks down into units called pixels, each of which is represented by a digital number. In this case the image is 8-bit greyscale which has 256 shades of grey.

is more common, the orientation of a remote sensing instrument can theoretically be in any direction: vertically (up or down), horizontally or obliquely (at an angle other than vertical or horizontal). The orientation of the imagery depends on the characteristics of the available sensor and the data requirements of each application. Although more difficult to analyze quantitatively, oblique imagery can be more cost effective for qualitative applications as a greater area can be covered than in one vertical image. This is illustrated in Figure 26.2, which shows a vertical and oblique image of the flooded Ouse River in York, England, in November 2000. Mapping water extent and flood risk management (Box 26.4 below) is an important application of remote sensing.

The scale of an image describes the relationship between a linear distance on the image and the corresponding linear horizontal distance on the ground. Thus, scale is usually expressed as a ratio such as 1 : 50 000 which describes an image–ground relationship such that 1 cm on the image equals 50 000 cm (or 500 m) on the ground. It is important to note that where images contain relief or are captured obliquely, image scale will change across an image. In any case, images or maps can be described as being small (least

detail), medium or large scale (most detail) and although the limits are not standardized, a rough guide is 1 : 50 000 and smaller for small scale and 1 : 12 000 and larger for large scale. The importance of the scale of an image lies in that it largely dictates the usefulness of an image for a particular application. For example, a weather satellite image covering all of western Europe would be of little use for studying hillslope processes, whereas an aerial photograph of the English Channel would not be useful for studying global ocean circulation.

The resolution of an image also plays an important role in determining its suitability for a given application. In general it refers to the ability of a system to separate a scene into constituent 'parts'. In remote sensing, resolution is divided into three components: spatial, temporal and spectral. Spatial and temporal resolution are discussed here, whereas spectral resolution is more appropriately addressed later in this chapter.

In most cases, the resolution of an image describes its spatial resolution. This refers to the degree to which a system can isolate a unit of area on a surface as being separate from its surroundings. Resolution is expressed as the size of the smallest individual component of an image in surface

(a)

(b)

Figure 26.2 Vertical (a) and oblique (b) aerial photographs of the November 2000 flood of the River Ouse in the city of York, England. These images were used to map the extent of the flood waters for Britain's Environment Agency. Notice how much more of the flood-plain is visible in the oblique image compared with the vertical image. (Source: photos © the UK Environment Agency)

measurement units. Although there are many factors that influence resolution during image capture, ultimately the resolution is determined by the size of individual film grains in photographs or by the size of individual pixels in digital images. These in turn are largely dependent on scale. Typically, the smaller the scale of the imagery, the lower the resolution. For example, the size of the smallest unit in a satellite image taken from 800 km above the Earth's surface will be much larger (smaller scale, lower resolution) than the smallest unit in an aerial photograph taken from 1000 m (larger scale, higher resolution).

With temporal resolution, the ability of the system to separate a scene into constituent parts refers not to parts over space but over time. Thus, a system that can capture many images over a unit period of time will have a higher temporal resolution than one that captures only a few. Clearly temporal resolution is highly dependent on the sensor used to collect the data. For terrestrial systems, the resolution can be very high. For example, with a 35 mm camera on a tripod or gantry, photographs can be taken separated only by a few tenths of a second. Such instruments are useful for recording events of environmental change that occur very quickly, such as avalanches and volcanic activity. However, with satellites that have to orbit the Earth several times before revisiting the same scene, the temporal separation of images will be of the order of days. Coarse temporal resolution can be very useful for recording environmental changes that occur over longer periods of time, such as deforestation, sea ice dynamics and coastal processes.

26.2.3 Characteristics of image content

There are a number of image characteristics that form the basis of image interpretation. They do not describe characteristics of the imagery as a whole but rather the characteristics of image content and thus are used to extract information from the images through human image interpretation and computer-aided analysis. They are dependent on the illumination of the scene and the characteristics of the surface being imaged. They include shape, size, pattern, association, tone and texture.

The first four elements, shape, size, pattern and association, are fairly self-explanatory. They describe the spatial characteristics of features in an image and their relationship to each other. The tone or colour describes the relative brightness of the surface as detected and interpreted by the sensing instrument. Given the context of an image, many features can be identified based largely on their tone. For example, green tones in an image are associated with parks and trees, brown tones with crops and soil, and grey tones with buildings and roads. However, tone is of little use in isolation.

Texture can be described as the tonal variation in an image as a function of scale. Thus, where tone describes the spectral information in an image, texture describes the spatial variation of the spectral information in an image. In essence, texture is the effect created by an agglomeration of features in an image that are too small to be detected individually. The size of these texture features determines the coarseness or smoothness of the texture and it is one of its most important defining characteristics of features in an image. It might be argued that tone is more important, but why then is it so easy for humans to identify objects in a greyscale image where the role of tone is greatly diminished? This shows how central the role of texture is in human vision and in our ability to identify objects around us and, therefore, in image interpretation.

26.3 Foundations of remote sensing

All objects on the Earth's surface are capable of reflecting, absorbing and emitting energy called **electromagnetic radiation**. The foundations of remote sensing lie in the nature of this energy; how it interacts with our atmosphere and the surface of the Earth; and in our ability to detect and record it using remote sensing instruments. These are discussed in the following sections.

26.3.1 Electromagnetic radiation

The objective of remote sensing is to detect, measure and analyze electromagnetic radiation. When a sensing instrument is pointed at a surface from a remote location, it is the electromagnetic radiation intercepting the sensor from the surface of interest that is being measured and recorded. Sometimes called **electromagnetic energy** or just simply radiation, electromagnetic radiation refers to a form of energy in transit where both the electric and magnetic fields vary simultaneously. Visible light is just one form of electromagnetic radiation. In a vacuum, this energy travels in a straight line at the speed of light (c = approximately $300\,000$ km s^{-1}) and can only be detected when it interacts with some form of matter.

Electromagnetic radiation can be described either as a wave or as a stream of particles travelling through space. For the applications of physical geography, the characteristics of electromagnetic radiation are best described using the wave model (Avery and Berlin, 1992). However, the particle model is important for understanding and describing how electromagnetic radiation interacts with objects. Therefore, both models should be explored and these are addressed in Box 26.1.

26.3.2 Electromagnetic spectrum

Equations (26.1) to (26.3) in Box 26.1 describe functions where wavelength and frequency are continuous rather than discrete variables. This implies that electromagnetic radiation exists over a continuum. This continuum is called the **electromagnetic spectrum**. As shown in Box 26.1, the way in which radiation interacts with the atmosphere and with objects on the Earth's surface is dependent on wavelength and thus on its position in the spectrum. Therefore, the electromagnetic spectrum has been divided up into discrete categories of wavelengths sharing similar properties called **spectral bands**. The spectral bands and their location on the electromagnetic spectrum, shown in Figure 26.5, provide a convenient system of reference for describing electromagnetic radiation of similar properties.

From Figure 26.5, it can be seen that radiation ranges from lethal gamma rays (short wavelength, high frequency, high energy) to harmless television and radio waves (long wavelength, low frequency, low energy). The most familiar range of the spectrum for humans is the visible band that

FUNDAMENTAL PRINCIPLES

WAVE AND PARTICLE MODELS OF ELECTRO-MAGNETIC RADIATION

When electromagnetic radiation is described in terms of the wave model, reference is made to its wavelength (λ) and **wave frequency** (f). Wavelength describes the linear distance between successive peaks or troughs of the energy wave and frequency refers to the number of peaks (or troughs), called cycles, that pass a fixed point in space in a given period of time (Figure 26.3). The relationship between the wavelength, frequency and velocity of the radiation is described as:

$$c = f\lambda \quad \text{or} \quad f = \frac{c}{\lambda} \quad (26.1)$$

where c is velocity, equal to the speed of light, $\sim3.0 \times 10^8$ m s^{-1}, λ is wavelength, in metres (m), and f is frequency, in cycles per second or hertz (Hz). This inverse relationship between wavelength and frequency is illustrated in Figure 26.4.

The particle model takes a different approach by describing radiation in terms of a discrete unit called a **photon**, which is a quantum of radiation. The energy of a quantum of radiation varies directly with its frequency (f) and inversely with its wavelength. This is given by:

$$E = hf \quad (26.2)$$

where E is the energy of one quantum, in joules (J), and h is Planck's constant, 6.626×10^{-34} J s^{-1}.

Using equation (26.1), we can link together the wave and particle models with:

$$E = \frac{hc}{\lambda} \quad (26.3)$$

This shows that the energy content of radiation is dependent on its wavelength and therefore has important consequences on how the radiation will interact with and affect any object it encounters. This will become increasingly obvious as this chapter progresses. Further details about radiation are also provided in Chapter 4.

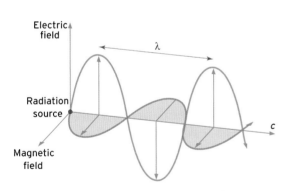

Figure 26.3 Illustrating the wavelength (λ) of the magnetic and electric fields of electromagnetic radiation. c denotes the speed of light equal to $\sim3.0 \times 10^8$ m s^{-1}.

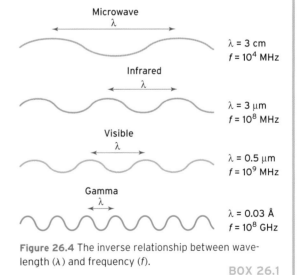

Figure 26.4 The inverse relationship between wavelength (λ) and frequency (f).

BOX 26.1

quite literally includes all the colours of the rainbow. This can be found between the ultraviolet wavelengths at 0.4 μm and the near-infrared wavelengths at 0.7 μm. The visible band is the range of radiation to which human vision has adapted. However, important information can also be obtained about objects by measuring and analyzing radiation from other parts of the spectrum. This is discussed in later sections of this chapter.

26.3.3 Atmospheric and terrestrial interactions

Because the characteristics of electromagnetic radiation vary from one end of the spectrum to the other, it follows that the way it interacts with matter will also vary. The implications of this are that not all bands of the spectrum are suitable for remote sensing applications. One property

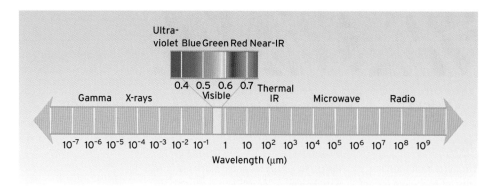

Figure 26.5 The electromagnetic spectrum showing a magnified section of the visible bands to which human vision is adapted. (Source: after Lillesand and Kiefer, 2000)

of energy that is constant throughout the spectrum is that the energy that intersects an object, whether it be an atmospheric molecule, a leaf or a road surface, must be in balance with the energy that leaves the object. This conservation of energy is called the first law of thermodynamics, which states that the energy that enters a system must equal the energy that leaves the system. This can be represented by:

$$E_i = E_o \qquad (26.4)$$

where E_i is incident energy and E_o is output energy.

When energy intercepts matter in the atmosphere such as gases, water molecules or particulate matter, it will be scattered, absorbed or transmitted (Figure 26.6; see also Chapter 4). Therefore, we can write equation (26.4) as:

$$E_i = E_s + E_a + E_t \qquad (26.5)$$

where E_s is scattered energy, E_a is absorbed energy and E_t is transmitted energy.

Scattering is divided into three types: Rayleigh, Mie and non-selective. **Rayleigh scattering** is caused by atmospheric

molecules and particles whose diameters are much smaller than the wavelength of the incident radiation. As there is an inverse relationship between wavelength and the degree of scatter, shorter wavelengths are most affected by Rayleigh scattering. This accounts for the blue colour of the sky. **Mie scattering** is caused by atmospheric molecules that are about equal in diameter to the wavelength of the incident radiation. This includes water molecules and dust particles and most strongly affects longer wavelengths. Conversely, **non-selective scattering** is not dependent on wavelength and scatters all wavelengths between the visible and mid-infrared. Non-selective scattering is caused by atmospheric particles such as water droplets and ice crystals that are much larger in diameter than the incoming wavelength.

Matter in the atmosphere that scatters and absorbs energy prevents the energy from the Earth's surface from reaching the scanner and is therefore a hindrance to remote sensing. As this **interference** is often wavelength-dependent, some wavelengths of radiation will pass through the Earth's atmosphere virtually unobstructed while others will be almost completely scattered or absorbed. This yields bands of the spectrum called **transmission bands** (or atmospheric windows) and **absorption bands**. The various transmission and absorption bands of the atmosphere and the various contributing atmospheric molecules are shown in Figure 26.7.

Equations (26.4) and (26.5) must also hold true for terrestrial interactions with incoming electromagnetic radiation. The way in which natural and artificial objects on the Earth's surface distribute incident energy varies greatly. When incident radiation strikes a surface, it is reflected, absorbed or transmitted. In terrestrial interactions the broader term 'reflection' is used rather than scattering. In this context, reflection is subdivided into two types: specular and diffuse (scattering) reflection. **Specular reflection**

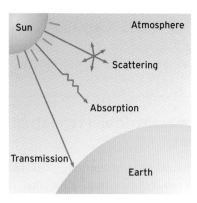

Figure 26.6 Schematic diagram illustrating the division of incoming solar radiation into the three components: scattered, absorbed and transmitted energy.

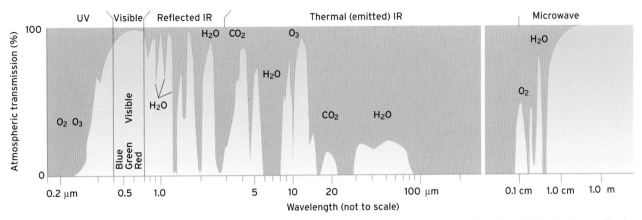

Figure 26.7 Atmospheric transmission and absorption bands of electromagnetic radiation. Orange shaded area indicates transmitted energy while green indicates absorbed energy. (Source: after Scherz and Stevens, 1970)

occurs when radiation hits a surface that is smooth relative to the radiation's wavelength. In this case, the radiation is redirected in a predictable direction such as when the Sun glints off a body of water. In contrast, **diffuse reflection** occurs when radiation encounters a surface that is rough relative to its wavelength. Here, the radiation is reflected randomly in many directions like the soft light on a cloudy day. Thus equation (26.5) can be expressed as:

$$E_i = E_r + E_a + E_t \qquad (26.6)$$

or:

$$E_i = (E_{st} + E_{dr}) + E_a + E_t \qquad (26.7)$$

where E_r is reflected energy, E_{st} is specular reflection and E_{dr} is diffuse reflection. Albedo, discussed in Chapter 4, is the term used to describe the ability of an object to reflect incoming radiation. Some typical albedo values for some common surfaces are given in Table 4.1 in Chapter 4.

Equation (26.7) suggests that specular and diffuse reflection, called directly reflected radiation, is the only source of radiation that reaches the sensor. However, there is also indirectly reflected radiation to consider. Indirectly reflected radiation occurs after incoming radiation is absorbed (E_a) by an object, converted to internal heat energy and then subsequently emitted at longer wavelengths. Since the emitted radiation has a longer wavelength than its solar source, this energy is called long-wave radiation, whereas the incoming solar radiation and directly reflected energy is called short-wave radiation. Therefore, equation (26.7) holds true either only for short-wave radiation or if the term E_r includes both types of reflected radiation. In any event, both directly and indirectly reflected radiation are important in remote sensing for characterizing the reflectance characteristics of a surface.

The reflectance characteristics of an object or surface across the electromagnetic spectrum are called its **spectral signature**. A surface's spectral signature describes to what extent electromagnetic radiation is transmitted, absorbed and reflected at different wavelengths and, therefore, to some extent is an indication of its chemical composition and physical state. Thus, in many ways a spectral signature is analogous to a fingerprint. Every object reflects natural and artificial radiation in different ways. Just as fingerprints can be used to identify people, we can use spectral signatures to identify surface features. Figure 26.8(a) gives the average spectral signatures of various types of common surface covers. Notice the high reflection of vegetation and the absorption of water in the near-infrared. Figure 26.8(b) shows how different types of vegetation reflect differently. Notice how much overlap there is between the curves except in the near-infrared. This is a good example of the value of performing remote sensing in bands outside the visible spectrum.

Having introduced the concept of spectral signatures, it is now possible to revisit the issue of spectral resolution. Consider that at every wavelength in the spectrum a surface will interact in a predictable way. The spectral resolution of a sensor refers to its ability to define sections of the spectrum by wavelength and to provide a measurement of the radiation at each section. For example, a camera using black and white film takes one measurement for the entire visible portion of the spectrum (0.4–0.7 μm). From this no information about how the surface reflects in the rest of the spectrum will be available. Such a system has a low spectral resolution. However, a camera using colour film essentially takes a measurement for three separate bands (one for each of blue, green and red) thereby providing more information about the reflectance characteristics of the surface. Therefore, this

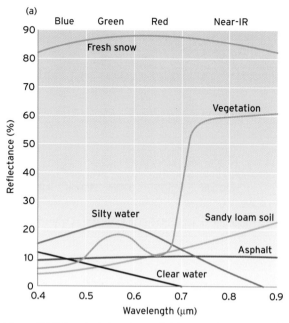

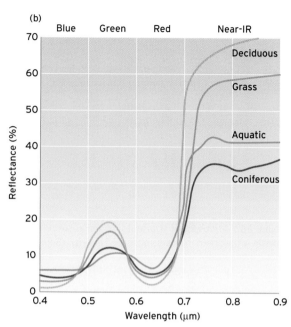

Figure 26.8 Spectral signatures of: (a) several common surface covers; and (b) vegetation types for the blue to near-infrared (0.4 and 0.9 μm) portion of the spectrum. (Source: after Avery and Berlin, 1992)

system would have higher spectral resolution. If you were to split up the entire spectrum into small slivers and take a measurement for each sliver you would have a very high spectral resolution indeed.

In the following sections, two classifications of sensing instruments will be discussed: passive and active. Passive sensors measure the naturally occurring radiation that is reflected or emitted from a surface. A common 35 mm camera, when used without a flash, is a good example of a passive system. Alternatively, active sensors provide their own source of illumination. They emit a radiation pulse and measure any radiation that is reflected back to the source. A 35 mm camera used with a flash is a good example of an active sensing system. The main divisions of remote sensing instruments as they are addressed in the following sections are: (i) camera sensors; (ii) electro-optical sensors; and (iii) microwave and ranging sensors.

26.4 Camera sensors

One of the most versatile remote sensing instruments is the analogue photographic camera and its digital equivalent. The difference between analogue and digital cameras lies in how each system captures and stores images. Whereas traditional analogue cameras use photographic film to capture images, and diapositives, glass plates and prints to store them, digital cameras use electronic photosensitive devices to capture images and electronic media for storage (e.g. computer hard disk). These technological advances have necessitated the expansion of the traditional definition of 'camera' to include digital imaging devices. However, although the performance and cost of digital cameras are improving, analogue cameras are still heavily used, especially for aerial applications.

Cameras used in remote sensing range from standard single lens reflecting (SLR) cameras to highly specialized, large-format aerial cameras. Whether analogue or digital, these are divided into non-metric, semi-metric and metric cameras. In metric cameras, the camera geometry is closely monitored (camera calibration), enabling precise measurements to be made from the imagery. With semi-metric and non-metric cameras, access to this information is limited. These cameras are less expensive to operate than metric cameras, but measurements will consequently be less precise.

With digital cameras, the electronic photosensitive devices usually record imagery within a predefined portion of the spectrum and allow some flexibility over collection parameters such as image size, resolution and so on. Conversely, analogue cameras depend on different types of film. Thus, as modern image processing methods require digital imagery as input, analogue images are digitized using scanners which range in quality from standard desktop scanners to highly specialized metric scanners, which aim to preserve the geometric and radiometric qualities of the analogue photographs.

Common film formats for analogue cameras include standard 35 mm, medium-format and large-format film. The last is the most common format used in airborne remote sensing and is designed for both high-resolution and high image quality. An example of a standard, large-format aerial photograph is given in Figure 26.9, which shows the fiducial marks in the four corners (used to define a coordinate system in image space) and flight data recorded around the border of the image. This image was taken using black and white **panchromatic** film, which is sensitive to all the colours of the visible spectrum and uses shades of grey between black and white to record them. In addition to panchromatic film, black and white film that is also sensitive to infrared light is in common use. Figure 26.10 gives a comparison of a panchromatic and black and white infrared photograph. Take care to note the differences between these two images and how the vegetation and water reflect differently.

The obvious alternative to black and white film is colour film to which human vision is better adapted. However,

Figure 26.9 Typical aerial photograph showing the fiducial marks and flight data including the contractor contact details, date and time of flight, aircraft elevation, image number, film and frame number and level indicator. This image was taken over Upper Wharfedale in the Yorkshire Dales National Park in northern England on 17 May 1992. North is roughly to the right and the image measures 3.2 km across. (Source: photo © the UK Environment Agency)

(a)

(b)

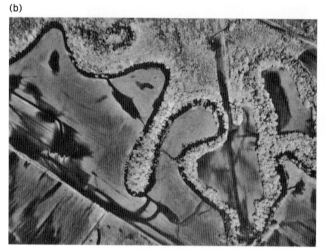

Figure 26.10 Examples of (a) a panchromatic and (b) a black and white infrared photograph. Notice the difference in reflection of the water and the vegetation between the two images. These 1 : 9000 images show flooding of Bear Creek in north-west Alabama, USA. (Source: from Lillesand and Kiefer, 2000)

(a)

(b)

Figure 26.11 Examples of (a) a normal colour and (b) a colour infrared aerial photograph. These images show the campus and stadium of the University of Wisconsin. Notice that the vegetation appears red in the colour infrared image and how the artificial turf in the stadium is green in both. (Source: from Lillesand and Kiefer, 2000)

26.4.1 Photogrammetry

A common application of camera-based aerial imagery and increasingly of satellite imagery (discussed in the next section) is in a process called photogrammetry. The American Society for Photogrammetry and Remote Sensing defines photogrammetry as the recording, measurement and interpretation of both photographic images and recorded radiant electromagnetic radiation and other phenomena (Wolf and Dewitt, 2000). However, the main application of photogrammetry is the generation of **digital elevation models (DEMs)**. A DEM is a digital file that stores the three-dimensional coordinates of an array of points that correspond to a real surface. They are often oriented in a regular grid so they can be easily manipulated in standard image processing packages where they can be useful for modelling Earth surface processes that are dependent on surface gradient topography such as erosion (Chapter 8) and for monitoring and measuring environmental change such as glacier mass balance (Chapter 16).

the processing of colour film tends to be more expensive. With colour film, the printed colours need not necessarily correspond to the real colour of the scene as this can be controlled by film type and processing method. As a result, infrared-sensitive colour film can be displayed in a false-colour image where colours are assigned so that green light is recorded as blue, red light as green and near-infrared as red. This is much easier to interpret over black and white infrared images. Like Figure 26.10, Figure 26.11 provides a comparison between a normal colour and colour infrared image. Again, take note of the differences between the images. As predicted by Figure 26.8, the bright vegetation in both infrared images and the dark water in Figure 26.10(b) indicate how strongly vegetation reflects and how much water absorbs in the near-infrared.

To derive three-dimensional (3D) topographic co-ordinates from remote sensing data requires stereo imagery. **Stereo images** are images that have been captured in overlapping pairs, strips or blocks (Kasser and Egels, 2002). Overlapping images allow a surface to be observed from two different positions. This produces a phenomenon called **parallax**, which refers to the apparent change in position of a stationary object when viewed from two different positions. To demonstrate, hold your thumb up in front of you at arm's length and look at it with one eye shut. Now switch to the other eye and observe the change of position of your finger relative to the background. The magnitude of this change in position is dependent on several factors including the distance between your eyes (i.e. between the points of observation), the distance from your finger to your eyes and the distance from your finger to the background. Because of the parallax, the position of any point that has been imaged in at least two aerial photographs can be determined if information such as the flying height and attitude of the aircraft and the distance between the photographs is known (Wolf and Dewitt, 2000). Therefore, the goal of photogrammetry is to reconstruct the exact geometry of the film, camera (sensor), **platform** (usually an aircraft) and ground at the time each photo was captured. The geometrical relationship between these components is shown in Figure 26.12. Traditionally, this was accomplished using large, highly specialized instruments called stereoplotters.

However, nowadays digital photogrammetry is the norm where the re-creation of these geometrical relationships is carried out entirely in a digital environment. In either case, the variety of high-precision measurements made in the camera, on the photographs and on the ground are used to produce a model of the system geometry (Figure 26.12) at the time each photo was taken using a process called a **least-squares adjustment**. This is a mathematical method for fitting a model to data so as to minimize errors between the observed values and the model predicted values. With a good fit of data to the model, 3D positional measurements can be made from the photographs.

As a result, photogrammetry represents an important source of topographic data for a variety of applications. For example, Lane *et al.* (2000) applied digital photogrammetry on 1 : 3000 scale aerial photography to an area of complex topography in the coniferous-forest-covered Glen Affric catchment in Scotland. The results showed that the precision of defining the surface topography was largely governed by photogrammetric data quality (camera calibration, base : distance ratio, ground control), combined with either scanning density or digital image resolution. The effect of the vegetation itself could be digitally removed using the stereo images, allowing the ground surface beneath to be mapped. Another example of the use of photogrammetry in the monitoring of environmental change is the work of Andreassen *et al.* (2002) who used historical aerial photography to track volume changes of a series of Norwegian glaciers.

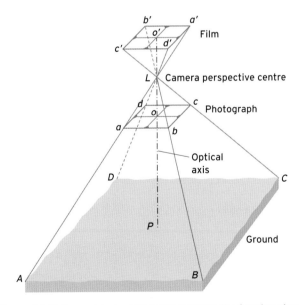

Figure 26.12 Geometrical relationship between the four imaging components of photogrammetry: film, camera perspective centre, photograph and ground. (Source: after Wolf, P.R. and Dewitt, B.A., *Elements of Photogrammetry with Applications in GIS*, 3rd edition, 2000, McGraw-Hill, Fig. 6.1, reproduced with permission of the McGraw-Hill Companies)

Reflective questions

➤ What are the advantages and disadvantages between the use of conventional film-based cameras and digital cameras for remote sensing?

➤ In addition to elevation, what other information about a surface can be derived from a DEM?

26.5 Electro-optical scanners

Scanners use electro-optical detectors to measure incoming radiation. An important difference between camera-type instruments and scanners is the higher spectral resolution that can be achieved. Scanning sensors are capable of measuring reflectance over the entire electromagnetic spectrum but most commonly operate between about 0.3 and 14 μm (blue, green, red, near-, mid- and thermal infrared) and can measure radiation in numerous very narrow bands of the spectrum, thus providing high spectral resolution. Conversely, cameras tend to operate in one broad band between about 0.3 and 0.9 μm (ultraviolet, blue, green, red and near-infrared) although they tend to provide a higher spatial and temporal resolution than scanners.

Scanning instruments form a family of sensors that produce two-dimensional (2D) digital images by collecting data continuously under a **swath** beneath a moving platform. This differs from cameras, which use a lens and a shutter to capture an entire image simultaneously. However, the data can similarly be described in terms of resolution, bit scale and so on. What is very different between these images is how they are acquired. There are two main types of scanning instruments: **across-track** and **along-track sensors**. Although both types of instruments measure the incoming radiation from a swath below a moving platform, in across-track sensors the scanner's line of sight (also called the instantaneous field of view, IFOV) is directed in a sweeping motion at right angles to the direction of travel by a rotating or oscillating mirror. The forward motion of the platform causes the field of view to move forward, thus covering the entire 2D swath beneath the platform. The operation of these sensors is depicted in Figure 26.13(a). Because of this motion, across-track sensors are often referred to as whiskbroom scanners.

Along-track sensors produce 2D images using a linear array of charge-coupled devices (CCDs) oriented perpen-

dicularly to the direction of travel covering one side of the swath to the other. As the platform moves forward, the field of view of the sensor array moves forward, thus producing a continuous 2D image of the swath. The operation of along-track sensors, often referred to as pushbroom scanners, is illustrated in Figure 26.13(b).

Scanners are most often mounted on airborne or spaceborne platforms and there are a variety of different types and configurations, which are discussed in the following sections.

26.5.1 Multispectral, thermal and hyperspectral instruments

Using either whiskbroom or pushbroom mechanisms, **multispectral scanners** produce digital images with more bands than is possible with camera-type sensing instruments. With technological advances, these instruments have given way to **hyperspectral scanners**, also called **imaging spectrometers**, which are similar in principle to multispectral scanners except that they can image with hundreds of bands, each of which is very narrow (approximately 2 nm). This produces images over virtually a continuous spectrum between the visible and thermal infrared regions of the spectrum. This provides very high spectral resolution and therefore offers a very detailed spectral signature for surface materials in the image. In hyperspectral imagery, surface objects can be more accurately identified and characterized remotely. These types of data have innumerable

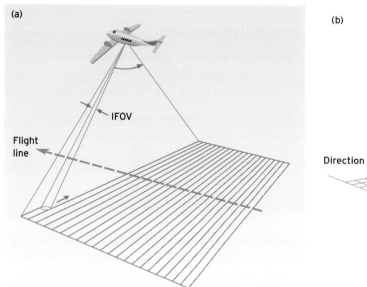

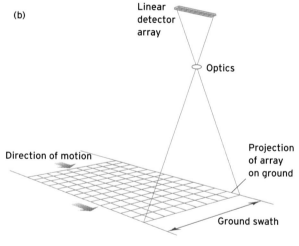

Figure 26.13 The operation of (a) an across-track (whiskbroom) sensor and (b) an along-track (pushbroom) sensor. (Source: after Lillesand and Kiefer, 2000)

applications such as detecting changes in land use (Section 26.2), water quality monitoring, and assessing vegetation health (Box 26.3 below). To show that they are made up of hundreds of spectral bands, hyperspectral images are often displayed graphically as an image stack or image cube as shown in Figure 26.14.

Thermal scanners measure radiation in the same way as multispectral and hyperspectral scanners but target the thermal infrared band of the spectrum. All objects with a temperature above $0\,\mathrm{K}$ ($-273.15°\mathrm{C}$) emit thermal or radiant energy as a function of their internal temperatures (see Chapter 4). This energy can be measured, interpreted and analyzed in the same way as in multispectral and hyperspectral scanners. Some applications of thermal remote sensing are qualitative, thus requiring only interpretation of relative differences between surface objects. Other applications may require absolute temperature difference to be measured.

To do so requires that the **emissivity**, which is measured by the sensor, is converted to absolute temperatures. Thermal remote sensing is useful for a variety of applications in physical geography including studying ocean circulation patterns, geological structure, soil mapping and for assessing volcanic activity. Figure 26.15 shows a good comparison of the Chiliques Volcano in northern Chile imaged in the visible and thermal wavelengths.

(a)

(b)

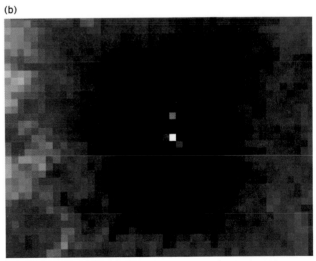

Figure 26.15 The Chiliques Volcano in northern Chile has been dormant for the past 10 000 years and, as shown in this pair of images, is now coming back to life. These images were captured by the Advanced Spaceborne Thermal Emission and Reflection Radiometer (ASTER) and show the volcano in the visible and near-infrared (left) and in the thermal infrared (right). Note the hot spots around the volcano's crater caused by magma appearing just below the surface. (Source: images courtesy of NASA/GSFC/MITS/ERSDAC/JAROS, and US/Japan ASTER Science Team)

Figure 26.14 This image is an example of a hyperspectral image stack. The instrument used to capture the data was the Airborne Visible/Infrared Imaging Spectrometer (AVIRIS) which collects data in 224 bands of about 10 nm width between 0.40 and 2.45 mm. This particular image was captured over Konza Prairie, Kansas, in late August. (Source: from Lillesand and Kiefer, 2000)

Table 26.2 Landsat mission characteristics

Satellite	Launched	Decommissioned	RBV bands	MSS bands	TM bands	Repeat period (days)	Altitude (km)
1	23 July 1972	6 January 1978	1-3	4-7	None	18	917
2	22 January 1975	25 February 1982	1-3	4-7	None	18	917
3	5 March 1978	31 March 1983	A-D[a]	4-8[b]	None	18	917
4	16 July 1982	15 June 2001[c]	-	4-7	1-7[d]	16	705
5	1 March 1984	TM operational, MSS powered off August 1995	-	4-7	1-7	16	705
6	5 October 1993	Failed on launch	-	-	ETM 1-7	16	705
7	15 April 1999	Still operational	-	-	ETM+ 1-8	16	705

[a]Single-band panchromatic images from two cameras.
[b]Band 8 failed after launch.
[c]Science operations ended 14 December 1993.
[d]TM data transmission failed in August 1993.

26.5.2 Spaceborne instruments

Remote sensing of the Earth from space has undergone many developments since the days of the Mercury, Gemini and Apollo missions in the early 1960s. Today, there are many satellite programmes in operation that provide endless volumes of image and other data about the Earth's surface. A discussion of the wealth of remote sensing data and their satellite platforms that are now available would be a textbook in itself. Therefore, this section provides a detailed overview of two important and well-established Earth observation programmes, the Landsat and SPOT missions. This is followed by a short introduction to a number of other satellite sensors that are important in monitoring environmental change. While the resolution of satellite imagery is always increasing – several satellites offer imagery in the 1 m pixel range – the scale of interest for environmental monitoring tends to favour the lower-resolution satellites and thus the discussion focuses on these. Owing to the fast pace of change in this field, it is important to refer to the further reading (especially Lillesand *et al.*, 2008) sections at the end of this chapter and the Companion Website for more comprehensive coverage of available satellite platforms and the most up-to-date information.

26.5.2.1 Landsat

The Landsat programme, which has been generating data of the Earth's surface since 1972, is very prominent in Earth observation research since the early days of remote sensing. The programme has become even more important with the release of a large part of the Landsat data archive to the public. The characteristics of each Landsat mission are summarized in Table 26.2. Overall, five sensor instruments have been used in these missions. They are the Return Beam Vidicom (RBV), the Multispectral Scanner (MSS), the Thematic Mapper (TM), the Enhanced Thematic Mapper (ETM) and the Enhanced Thematic Mapper Plus (ETM+). Table 26.3 summarizes some of the specifications of these five sensors. Enhancements included improvements to spatial resolution (80 m to 30 m), temporal resolution (18 versus 16 days to complete coverage) and spectral resolution (four to eight bands).

The MSS, TM and ETM+ sensors aboard Landsat 5 and 7 are all based on the whiskbroom scanner design and, unlike their predecessors, are still very much in use and represent an important source of Earth observation data. Landsat 5 was launched in a circular, Sun-synchronous, near-polar orbit at an altitude of 705 km. The use of **Sun-synchronous orbits** means that satellites pass over any given latitude at the same local time each day. This ensures that the same solar illumination conditions prevail for each pass. The repeat period, the time it takes for a satellite to revisit a point on the ground, is completed every 16 days but the orbits were offset to reduce this period to 8 days when Landsat 4 was still operational. As Landsat 7 was designed in order to maintain data continuity with Landsat 4 and 5, the same specifications (orbit, bands, resolution and swath

Table 26.3 Sensors of the Landsat programme

Sensor	Mission	Sensitivity (μm)	Resolution (m)	Swath width (km)	Bit scale
RBV	1, 2	0.475-0.575	80	185	n/a
		0.580-0.680	80		
		0.690-0.830	80		
	3	0.505-0.750	30	183	
MSS	1-5	0.5-0.6	79/82*	185	6-bit
		0.6-0.7	79/82*		
		0.7-0.8	79/82*		
		0.8-1.1	79/82*		
	3	10.4-12.6	240		
TM	4, 5	0.45-0.52	30	185	8-bit
		0.52-0.60	30		
		0.63-0.69	30		
		0.76-0.90	30		
		1.55-1.75	30		
		10.4-12.5	120		
		2.08-2.35	30		
ETM	6	TM bands	30 (120 in thermal)	185	8-bit
		0.52-0.90	15 (panchromatic)		
ETM+	7	TM bands	30 (120 in thermal)	185	8-bit
		0.52-0.90	15 (panchromatic)		

*79 m pixels for Landsat 1 to 3 and 82 m for Landsat 4 and 5.

(Source: from Lillesand *et al.*, 2000)

width) were used with the exception of an added high-resolution 15 m panchromatic band. A sample of the five non-thermal bands that are the most frequently used TM bands (Bands 1 to 5) is given in Figure 26.16. The information that can be extracted from images has been used in innumerable applications of physical geography and in detecting environmental change, including studies of vegetation change (Chapters 18, 19, 20 and 25), soil and coastal erosion (Chapters 10 and 15), sediment and pollution movements in watercourses (Chapters 9, 11, 12, 13 and 21) and glacial studies (Chapter 16). The Landsat Data Continuity Mission (LDCM), set for launch in December 2012, is currently underway to ensure this remarkable resource continues into the future.

26.5.2.2 SPOT

The SPOT programme, an acronym of the Système Pour l'Observation de la Terre, was undertaken by the French Government in 1978 with early collaboration with Belgium and Sweden and has now become a successful international endeavour. The SPOT satellites, of which there are five, use the pushbroom sensor design. Their mission characteristics are given in Table 26.4. As the goal of the SPOT programme was long-term data continuity, the characteristics of various satellites are similar. All the SPOT satellites are in identical circular, Sun-synchronous, near-polar orbits at an altitude of 832 km. Although the SPOT satellites take much longer to cover the globe than Landsat (i.e. 26 days), the optical

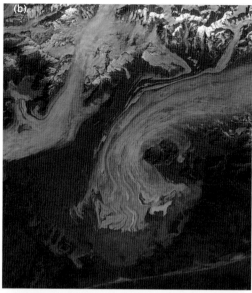

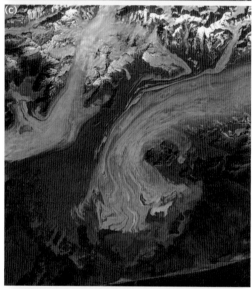

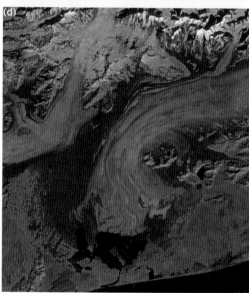

Figure 26.16 This figure provides a sample of the five non-thermal Landsat TM bands, each in 8-bit greyscale: (a) Band 1 (0.45–0.52 mm blue); (b) Band 2 (0.52–0.60 mm green); (c) Band 3 (0.63–0.69 mm red); (d) Band 4 (0.76–0.90 mm near-infrared); and (e) Band 5 (1.55–1.75 mm mid-infrared). This Landsat 5 image was captured over the Bering Glacier, Alaska, on 25 September 1986 at 11:03 local time. Notice the large difference in reflectance between the vegetation along the bottom in the image and the glacier in Bands 4 and 5. North is up and the images are roughly 53 km across. (Source: NASA Landsat Program, 11/08/1999, Landsat TM scene, p064r18_5t860925. Geocover Ortho, MDA Federal, Rockville, 09/25/1986. Data courtesy of Global Land Cover Facility, www.landcover.org, images courtesy of the Global Land Cover Facility, and Adrian Luckman)

Table 26.4 SPOT mission characteristics

SPOT satellite	Launched	Decommissioned	Repeat period (days)	Altitude (km)	High-resolution stereoscopic
1	22 February 1986	31 December 1990	26	822	No
2	22 January 1990	29 July 2009	26	822	No
3	26 September 1993	14 November 1997	26	822	No
4	24 March 1998	Still operational	26	822	No
5	4 May 2002	Still operational	26	822	Yes

devices in the sensors are movable to permit off-vertical imaging. This decreases the 26 day repeat period to less than 5 days. It also enables stereoscopic viewing, enabling the imagery to be used easily in photogrammetric applications. Sensor systems also include vegetation monitoring instruments that provide 1 km resolution multispectral imagery and daily global coverage. On the newer SPOT 5, imagery is provided at 5 m resolution. Plans are afoot for SPOT 6 and 7 satellites to extend the mission well into the future. For more detailed information about the SPOT instruments including band sensitivity visit the SPOT technical information link at www.spotimage.com.

26.5.2.3 NASA's Earth Observing System Program

The Earth Observing System (EOS) is a program of coordinated polar-orbiting satellites aimed at providing long-term global observations of the land surface, biosphere, solid Earth, atmosphere and oceans. The goal of EOS is to improve our understanding of the Earth–Sun system and its response to natural and human-induced changes. The first two satellites of the EOS program are of particular relevance to the monitoring of environmental change. The Terra satellite, launched on 18 December 1999, was designed to monitor the state of the Earth's environment, how it is changing and to determine the consequences of these changes on life. It was followed by the Aqua satellite, which was launched on 4 May 2002. As the name might suggest, the Aqua satellite's primary focus is the Earth's water cycle. However, both satellites carry a suite of instruments designed to complement each other and provide important data for a number of interrelated scientific questions. Here we focus on only two sensors of the Terra/Aqua pair which are commonly

used in environmental monitoring: the Moderate Resolution Imaging Spectro-radiometer (MODIS) and the Advanced Spaceborne Thermal Emission and Reflection Radiometer (ASTER). For more information about the suite of EOS satellites and sensors refer to the NASA EOS website (http://eospso.gsfc.nasa.gov/).

The MOSIS instrument, which is mounted on both the Terra and Aqua satellites, is a 36-band imaging sensor with a resolution between 250 and 1000 m depending on wavelength. Like all EOS instruments, it is designed for long-term, continuous monitoring of the Earth's surface for detecting even subtle changes in land, ocean and atmospheric processes simultaneously. Its 'moderate' resolution imagery can cover the globe in multispectral imagery every two days! Each of the 36 bands was chosen for specific applications which, along with their spatial resolution, are given in Table 26.5. Because of its resolution, MODIS is most suitable for macroscale applications such as the measurement of sea surface temperatures, monitoring drought and mapping deforestation in the Amazon. An interesting characteristic of the MODIS product is that its true- and false-colour imagery have been made publically available online via the Rapid Response System (http://rapidfire.sci.gsfc.nasa.gov) in almost real time. The NASA MODIS page provides details about the MODIS sensor as well as some stunning examples of MODIS imagery and other data products (http://modis.gsfc.nasa.gov).

Unlike the MODIS sensor, the ASTER instrument is only mounted on the Terra satellite and, as a much higher-resolution sensor, has been described as a zoom lens for MODIS. It also differs from MODIS in that the ASTER instrument is made up of three separate systems that operate in different spectral bands: Visible and Near Infrared (VNIR), Short Wave Infrared (SWIR) and Thermal Infrared (TIR). Their channels and bandwidths are given in

Table 26.5 MODIS spectral bands and their primary uses

Primary use	Band	Bandwidth (μm)	Resolution (m)
Land/cloud/aerosols boundaries	1	0.620–0.670	250
	2	0.841–0.876	250
Land/cloud/aerosols properties	3	0.459–0.479	500
	4	0.545–0.565	500
	5	0.1230–0.1250	500
	6	0.1628–0.1652	500
	7	0.2105–0.2155	500
Ocean colour/phytoplankton/biogeochemistry	8	0.405–0.420	1000
	9	0.438–0.448	1000
	10	0.483–0.493	1000
	11	0.526–0.536	1000
	12	0.546–0.556	1000
	13	0.662–0.672	1000
	14	0.673–0.683	1000
	15	0.743–0.753	1000
	16	0.862–0.877	1000
Atmospheric water vapour	17	0.890–0.920	1000
	18	0.931–0.941	1000
	19	0.915–0.965	1000
Surface/cloud temperature	20	3.660–3.840	1000
	21	3.929–3.989	1000
	22	3.929–3.989	1000
	23	4.020–4.080	1000
Atmospheric temperature	24	4.433–4.498	1000
	25	4.482–4.549	1000
Cirrus clouds water vapour	26	1.360–1.390	1000
	27	6.535–6.895	1000
	28	7.175–7.475	1000
Cloud properties	29	8.400–8.700	1000
Ozone	30	9.580–9.880	1000
Surface/cloud temperature	31	10.780–11.280	1000
	32	11.770–12.270	1000
Cloud top altitude	33	13.185–13.485	1000
	34	13.485–13.785	1000
	35	13.785–14.085	1000
	36	14.085–14.385	1000

(Source: adapted from http://modis.gsfc.nasa.gov/about/specifications.php, NASA)

Table 26.6 ASTER sub-instruments and their spectral bands

Sub-instrument	Band	Bandwidth (μm)	Resolution (m)
VNIR	1	0.52–0.60	15
VNIR	2	0.63–0.69	15
VNIR	3	0.76–0.86	15
VNIR	3	0.76–0.86	15
		(backward-looking)	
SWIR	4	1.600–1700	30
SWIR	5	2.145–2.185	30
SWIR	6	2.185–2.225	30
SWIR	7	2.235–2.285	30
SWIR	8	2.295–2.365	30
SWIR	9	2.360–2.430	30
TIR	10	8.125–8.475	90
TIR	11	8.475–8.825	90
TIR	12	8.925–9.275	90
TIR	13	10.25–10.95	90
TIR	14	10.95–11.62	90

(Source: from http://asterweb.jpl.nasa.gov/characteristics.asp, NASA)

Table 26.6. Uniquely, the VNIR measures its three spectral bands using two telescopes, one looking down (nadir-looking) and the other looking behind the sensor at an angle of 27.7° off-vertical. As discussed in Section 26.4.1, this has the benefit of providing stereo imagery for the generation of three-dimensional topographic data. The globe is covered by the VNIR system every four days whereas the repeat period for coverage in all 14 ASTER bands is 16 days. The relatively high resolution and stereo abilities of the ASTER imagery make it ideal for monitoring changes in glacier elevation (Howat *et al.*, 2008). The high resolution also makes these data valuable for studying surface processes and phenomena that have high spatial variability like mineral mapping and tracking changes in land cover and land use. The NASA ASTER web pages provide more detail about the ASTER sensor and give many examples of interesting applications (http://asterweb.jpl.nasa.gov).

26.5.2.4 High-resolution sensors

At the turn of the twenty-first century, the age of high-resolution satellites began with the launch of the IKONOS satellite in late 1999. This commercial system was the first to offer 1 m resolution imagery (1 m panchromatic, 4 m multispectral) to the public. Since then, a number of commercial undertakings have started to provide ultra-high-resolution imagery of the Earth's surface. For example, DigitalGlobe, a US-based company, has three sub-metre satellites in orbit, of which the first one, QuickBird, is one of the highest-resolution satellites that are publically available (60 cm panchromatic, 2.4 m multispectral). Its other satellites which are of limited availability to the public include WorldView-1 (50 cm panchromatic) and WorldView-2 (46 cm panchromatic, 1.84 m multispectral). Several other systems exist which can be explored in the further reading section below. While high-resolution imagery enables us to monitor the Earth's surface in unprecedented detail, its use for meso- and macroscale applications is impractical. Imagine looking at the 1.7 million km^2 Greenland ice sheet in 1 m resolution imagery!

Regardless of the sensor, the application of multispectral data depends on many factors including time of day, specific bandwidths, imaging conditions (e.g. weather), resolution and experience of the interpreter. However, some general guidelines exist for interpreting the different bands of the spectrum which can be applied to any sensing instrument. These are highlighted in Table 26.7.

Table 26.7 Principal applications of bands commonly used in remote sensing

Spectral band	Principal applications
Blue	Good water penetration so suitable for coastal mapping and bathymetry. Also useful for soil/vegetation/forest-type discrimination, cultural feature identification. Sensitive to atmospheric haze
Green	Has some ability to penetrate water but sensitive to turbidity. High reflectance from vegetation so useful for vegetation discrimination and vigour assessment. Also cultural feature identification. Sensitive to atmospheric haze
Red	High chlorophyll absorption in vegetation and high reflection in soils so good for differentiating between soil and vegetation. Also good for delineating snow cover and cultural features
Near-IR	High reflection from vegetation and absorption in water. Best band for discriminating between vegetation types and vigour. Good for delineating water bodies and soil moisture
Mid-IR	In shorter wavelengths, good for vegetation and soil moisture content and for discriminating between snow and clouds. In longer wavelengths, good for discriminating between mineral and rock types and also moisture content
Thermal	Vegetation stress analysis, soil moisture and thermal mapping

(Sources: after Avery and Berlin, 1992; Lillesand and Kiefer, 2000)

Reflective questions

➤ What types of electro-optical scanners are there?

➤ What is the advantage of high spectral resolution instruments over lower spectral resolution instruments such as the Landsat instruments?

➤ How do sensor mechanics affect the output image?

➤ From Figure 26.5, can you determine the colour bands of the VNIR? Will these bands produce a colour or false-colour image?

➤ Compare the ASTER sensor bands to those of MODIS (Table 26.5). Can you see how the two instruments might complement each other?

26.6 Microwave and ranging sensors

In this section we investigate the sensors of the microwave wavelengths as well as the ranging sensors including radar, sonar and laser altimetry (lidar). Although the ranging sensors use vastly different sources of illumination, they have many operational similarities and therefore it is appropriate to discuss them here.

26.6.1 Microwave sensors

The microwave portion of the electromagnetic spectrum can be found between the wavelengths of 0.001 and 1 m. Radiation in this portion of the spectrum is more commonly characterized in terms of its frequency (3.0×10^5 to 3×10^9 Hz). Microwave remote sensing has two important advantages. Firstly, it is well suited for penetrating through the atmosphere in conditions that typically interfere with other wavelengths, such as cloud, dust, rain, snow and smoke. Secondly, the interaction of microwaves with surfaces tends to be quite different from those of other common wavelengths such as the visible and infrared bands. This provides researchers with unique insights into the surfaces under investigation.

There are two main types of remote sensing instruments that operate in the microwave portion of the electromagnetic spectrum. The first is a passive instrument and is called a microwave radiometer. The second, a more widely used instrument, is an active microwave sensor called radar. Remember that, unlike an active sensor, a passive sensor does not provide its own source of radiation. Thus, microwave radiometers detect only the very low levels of naturally occurring microwave radiation that are emitted from all objects on the Earth's surface. Because all objects emit these microwave signals, the interpretation of passive microwave data can be very difficult since the signal for any given point is the sum of four component sources of radiation: (i) emitted radiation from the object of interest; (ii) emitted radiation from the atmosphere; (iii) reflected radiation from another source such as the Sun; and (iv) transmitted energy from the subsurface. Thus, the applications of passive systems are somewhat limited.

Radar, however, is very widely used and has many applications. It is an acronym of **ra**dio **d**etection **a**nd **r**anging, although microwaves are now used in place of radio waves. As an active sensor, it generates a microwave pulse that is aimed at the surface of interest and any returned energy, called the echo or **backscatter**, is measured by an antenna. Two common non-imaging radar systems are Doppler radar and the plan position indicator (PPI). Doppler radar is used to measure the velocity of remote objects such as cars. PPI is used to plot the planimetric position of large objects such as aircraft around an airport or ships at sea. It involves the use of a rotating antenna that continuously updates measured positions of objects on a circular screen.

Imaging radar systems were developed after the Second World War and were based on PPI technology primarily for peering over enemy lines. Radar systems were mounted in aircraft to look sideways deep into enemy territory while flying safely in friendly skies. This technology was called side-looking radar (SLR) or side-looking airborne radar (SLAR) for airborne systems. Like the Doppler and PPI radar, the instrument sends out a short, high-energy pulse as shown in Figure 26.17. The energy from this pulse reaches the ground and reflects off the surfaces it encounters (buildings, trees, soil and so on). The first object the pulse encounters will produce a return signal that will be the first to reach the antenna. The travel time of this return is recorded. All subsequent objects that the pulse encounters will produce returns that will be detected by the antenna.

Since the pulse travels at the speed of light, *c*, and the travel time is known (*t*), the distance (*d*) of the return producing objects from the aircraft is given by:

$$d = \frac{ct}{2} \tag{26.8}$$

As the aircraft moves forward a series of pulse returns can be combined to form an image. The tone of a point in a radar-generated image measures the intensity of the backscatter from that point.

The resolution of an SLR image is determined by several factors. One is the length of the pulse and the other is the **beamwidth** (Figure 26.17). Two ways to improve the resolution of the system are to use shorter pulses and narrower beamwidths. Shorter pulses can be achieved by shortening the time the instrument emits the outgoing signal. Instruments can control beamwidths either by increasing the physical length of the antenna or by simulating a longer antenna length. The former are called brute force or real aperture systems. Those that simulate antenna length are called synthetic aperture radar (SAR). Thus, SAR systems enable the use of very narrow beamwidths without the physical requirements of a long antenna. These instruments represent one of the most important sources of remote sensing data in the microwave portion of the spectrum.

Overall, the reflection characteristics of the radar pulse can be attributed to a surface's geometric properties such as surface roughness and orientation, and electrical properties, which are largely influenced by the presence of water. Although these interactions are complex, they have enabled the successful application of radar to a wide variety of

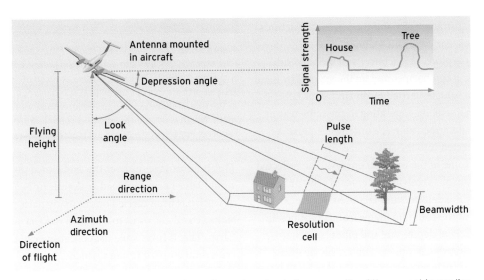

Figure 26.17 Operation of a side-looking radar system. Inset shows a profile of the current image line.

applications such as geological mapping, discriminating among vegetation types and sea ice studies.

Radar sensors are ideally suited for use on satellite platforms. As they are unaffected by clouds and do not require sunlight, they can provide imagery of the Earth's surface with far greater regularity than optical instruments. As such, radar instruments have been in use on satellite platforms since the late 1970s and there are now a number in orbit. The European Space Agency's ERS-2 satellite has been in operation since its launch in April 1995, despite having a projected service life of only 3 years. An example of a radar image is given in Figure 26.18 used for studying the Monaco Glacier in Svalbard, Norway. ENVISAT, a European follow-on mission to the ERS programme, was launched in early 2002. Data collected using instruments aboard ENVISAT on atmospheric aerosols are discussed in Box 26.2. In late 1995 the Canadian Space Agency launched its first remote sensing satellite RADARSAT-1, which is still in operation after an expected life span of 5 years. The RADARSAT-2 follow-on mission was launched in December 2007. The Japanese also have a radar satellite in orbit called ALOS, which has been in operation since January 2006. The newest trend in radar remote sensing is the development of high-resolution sensors, such as the German TerraSAR-X (launched in June 2007) which offers imagery up to 1 m resolution. The applications of such a remarkable data set are only beginning to unfold.

Another use of radar technology, which has many environmental uses, is ground-penetrating radar (GPR). It is a technology that has been used since the 1920s when it was first applied to a glacier in Austria (Stern, 1929, 1930). Unfortunately, the method was virtually forgotten until the late 1950s. Since then, GPR has become an invaluable tool for a whole host of applications including locating buried objects, delineating stratigraphy and the internal structures of glaciers, bedrock, concrete and sediment.

GPR is used to explore, characterize and monitor subsurface structures. It can be operated on the ground by hand or in some type of vehicle, carried by an aircraft or helicopter and even from a satellite. Operating in the same way as the radar systems described above, a microwave pulse is generated and directed into the subsurface. Structures in the subsurface cause some of the waves of energy to be reflected back to the instrument's receiving antenna as echo or backscatter where they are detected and recorded. Figure 26.19 gives an example of the resulting image from a peat bog in northern England. Peat depths and soil pipes (see Chapter 11) across the survey catchment could easily be measured, producing a continuous profile (Holden et al., 2002).

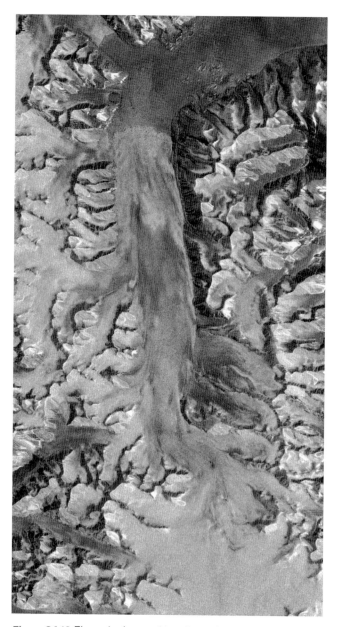

Figure 26.18 The radar image shown here shows Monacobreen (the Monaco Glacier) and surrounding mountains in north-west Svalbard, Norway. This image was captured in June 1995 by the ESA's Earth Resources Satellite (ERS-2). Notice the distortion in the mountain peaks due to the look angle of the radar instrument. North is up and the image is roughly 23 km across. (Source: image courtesy of Tavi Murray and Adrian Luckman)

This is advantageous over the traditional point data that would have been produced with rods or by coring. This type of work has been carried out in many parts of Finland and Norway and allows us to establish the amount of peat on Earth. This is invaluable data because peatlands are a very important store of carbon, which if released may enhance global warming (Holden, 2005b).

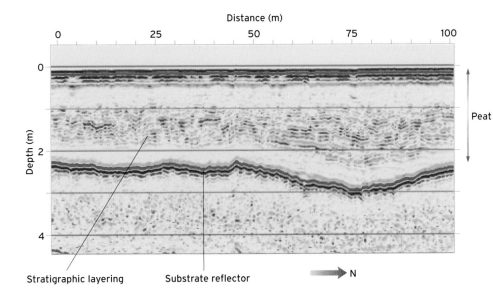

Distance (m)

Peat

Stratigraphic layering   Substrate reflector   N

Figure 26.19 A GPR radargram from a peatland survey in the North Pennines, England. The image clearly shows the peat depth and some of the layers within the peat, and how it changes across the transect. The 'substrate reflector' is the bedrock below the peat.

CASE STUDIES

IMPROVING OUR UNDERSTANDING OF THE EARTH SYSTEM AND CLIMATE CHANGE: MEASURING AEROSOL CONCENTRATIONS FROM SPACE

Aerosols, either natural or anthropogenic, are fine particles suspended in the atmosphere (see Chapter 4, Section 4.9.3.2). They play an important role in the Earth's energy balance and thus in how our climate changes over time.

Generally, aerosols are thought to exert a cooling effect on the Earth's climate. However, the precise role of aerosols in the Earth's climate system is not well understood and therefore is a significant source of uncertainty for scientists making future predictions of climate change.

Unfortunately atmospheric aerosols are difficult to measure. Unlike greenhouse gases, which are distributed fairly evenly around the globe and can remain in the atmosphere for up to 100 years, aerosols are very heterogeneous in their spatial and temporal distribution. The Advance Along-Track Scanning Radiometer (AATSR) aboard the European Space Agency's (ESA) ENVISAT satellite was originally designed for measuring surface temperatures of the world's oceans but has also been utilized to measure aerosols (Bevan *et al.*, 2011). When a satellite looks down at the Earth it 'sees' scattered radiation from the atmosphere as well as the Earth's surface. Since this sensor is able look in both the nadir (down) and forward directions and since we have knowledge of how surface reflectance changes with view angle at different wavelengths, it is possible

to separate scattering from these two sources and infer from the atmospheric component the nature and concentration of aerosols. With a spatial resolution of only 1 km x 1 km the AATSR sensor enables the measurement of aerosol concentrations across the globe.

Figure 26.20 shows two maps of global June/July/August and September/October/November concentrations of aerosols derived using the AATSR sensor. The sensor can also differentiate between fine and course aerosols, which is indicative of the aerosol's source. Typically, fine aerosols tend to come from anthropogenic sources like smoke and industrial pollution whereas coarse particles tend to be of natural origin like dust and ocean salt. In Figure 26.20(a), high concentrations of fine aerosols are seen over China and India due to industrial pollution

BOX 26.2►

➤

and over Alaska from the forest fire season of the northern hemisphere summer. In Figure 26.20(b), as the southern hemisphere summer approaches, aerosols from the burning of the rainforests in South America become prominent. The most striking feature in both seasons is the high concentration of coarse aerosols associated with the Sahara and Arabian Deserts.

(a)

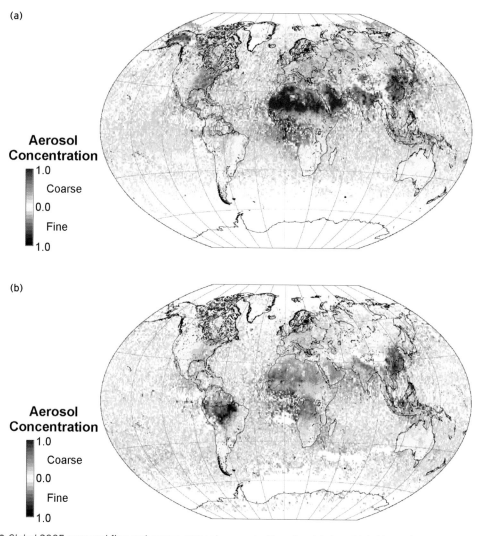

(b)

Figure 26.20 Global 2005 seasonal fine and course aerosol concentrations for: (a) June/July/August; and (b) September/October/November. Red and blue indicate coarse and fine aerosol concentration, respectively. Concentration increases with colour intensity and is on a relative scale. (Source: after Bevan et al., 2011)

BOX 26.2

26.6.2 Sonar

Although sonar employs a very different type of energy to 'illuminate' a surface, it is operationally very similar to radar. Sonar is an acronym of **so**und **n**avigation and **r**anging and is a form of active remote sensing that uses sound waves or acoustic energy propagated through water to detect objects and surfaces. There are two groups of sonar systems. The first are non-imaging systems used to measure water depths and include echo-sounding profilers or bathymetric sonar (for measuring underwater topography). The second produce images or sonographs and are called sidescan imaging sonar.

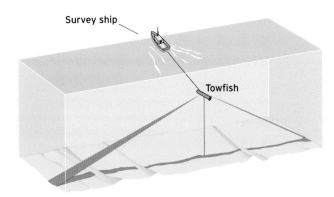

Figure 26.21 Operation of a sonar instrument. (Source: after Avery and Berlin, 1992)

Figure 26.21 illustrates how a sonar system is deployed. A torpedo-like vessel called a towfish is towed at a known depth behind a survey ship joined by a cable. Within this cable is the link between the recording instruments on board the ship and the sending and receiving instruments (called transducers) in the towfish. The transducers send out fan-shaped acoustical pulses perpendicular to the direction of travel. Upon receiving and recording the echo from surface features, a sonograph is produced. The topographic features on the sea floor, for example, must be thought of as reflectors in the same way features are with microwave energy (Avery and Berlin, 1992). Features that face the transducers will be better reflectors than those that face away from the transducers. Furthermore, smooth surfaces (e.g. sand and mud) will tend to act as specular reflectors of acoustic energy and thus will return no signal. Alternatively, rough surfaces such as gravel and boulders will reflect much of the energy back to the sensor. The main application of sonar in physical geography is the mapping of the ocean sediments, river and lake beds, and the ocean floor topography.

26.6.3 Lidar

Lidar technology has been used since the 1970s and is also called laser scanning or laser altimetry. It is an acronym of **l**ight **d**etection **a**nd **r**anging. Instead of using microwaves or acoustical energy to 'illuminate' a subject, lidar employs a highly focused beam of light called a laser. The primary function of lidar is to produce DEMs. A laser scanner measures the time it takes for the laser pulse to travel from the scanner to the surface and back to the scanner. If the direction, velocity and travel time of the laser pulse are known, the distance and position of the surface point relative to the platform (e.g. aircraft or helicopter) can be determined. However, in order to turn this relative position

into absolute coordinates, the exact position and attitude of the platform must be precisely known at a frequency that matches that of point measurement (approximately 15 000 pulses s^{-1}). Until very recently this locating technology was not available but today laser scanning systems rival and often exceed the performance and efficiency of digital photogrammetric systems in producing DEMs.

A modern system requires three components as illustrated in Figure 26.22. The first is the laser scanner to determine the distance and direction of each measured point from the platform. The second is a **global positioning system (GPS)** used to determine the position of the aircraft using a constellation of satellites that orbit the Earth. This requires a static reference GPS receiver (called a **base station**) on a known point on the ground, and a mobile GPS receiver (called a **rover**) in the aircraft. This system is called **differential GPS**. The third component is an inertial navigation system (INS), which measures any unwanted rotation of the aircraft. With these three systems operating in synchronization with each other, a very accurate 3D position of a point on the ground can be achieved. Modern instruments now also record the strength or intensity of each laser return signal. The strength of the return is heavily dependent on surface characteristics in the same way the strength of reflected light is heavily dependent on the surface from which it was reflected. Therefore, this intensity can be used in the form of an image, like a photograph, to assist in the interpretation for the lidar data. This is demonstrated in Box 26.3.

One of the great benefits of lidar systems is their ability to record multiple returns for the same point. As the laser beam travels towards the surface, it experiences some divergence such that it produces a **footprint** on the ground whose diameter (usually around 25 cm) is dependent on the height of the aircraft. The first object within this footprint that the beam encounters will produce a return. Likewise, all objects within this footprint that the beam encounters will produce a return. Ideally, the first return will represent the top of a vegetation canopy and the last return will represent the ground. This enables the modelling of canopy heights while at the same time offering an accurate representation of the surface. However, sufficiently dense vegetation and solid objects such as buildings do present a problem for accurate representation of the ground surface. Thus, sometimes DEMs that include non-surface objects are called digital surface models (DSMs) and those representing only the ground surface are called digital terrain models (DTMs). Lidar segmentation, the process by which lidar points (often called hits) that represent non-surface objects are separated from hits which represent the surface, is one of the principal challenges of laser altimetry. For example, lidar data used

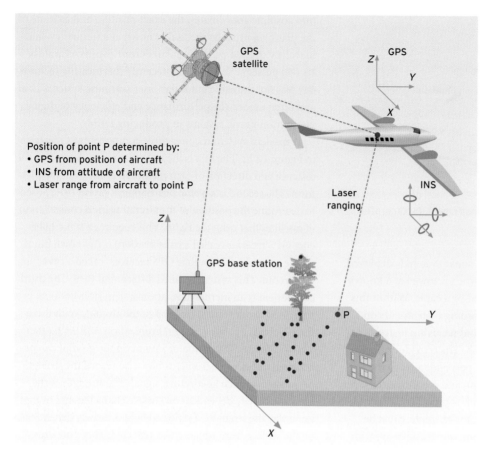

Figure 26.22 Operation of a lidar system. The combination of an accurate differential GPS position of the platform/instrument, an INS measurement of the attitude of the platform/instrument (rotation on X, Y and Z axes) and the laser range gives an accurate position in X, Y and Z of point P.

Position of point P determined by:
• GPS from position of aircraft
• INS from attitude of aircraft
• Laser range from aircraft to point P

CASE STUDIES

AIRBORNE LIDAR FOR MEASURING CHANGES IN GLACIER GEOMETRY, EXTENT AND MASS BALANCE IN SVALBARD, NORWAY

Mountain glaciers constitute only about 3% of the glacierized area on Earth. However, under current climate change predictions, they are expected to contribute significantly to sea-level rise because of their heightened sensitivity to climate change (IPCC, 2007a). Unfortunately, estimates of the contribution of these glaciers to future sea-level rise are uncertain owing largely to the global shortage of long-term glacier mass balance observations (see Chapter 16). Glacier mass balance is an important measure of the response of a glacier to changes in climate variables, especially summer temperatures and winter precipitation. Of more than 160 000 glaciers worldwide only about 40 have mass balance records of more than 20 years. In the high latitudes mass balance records are even worse because glaciers in these areas tend to be very remote. Unfortunately, it is at these high latitudes where climate change is expected to have the greatest impact and where long-term mass balance measurements are needed most.

There are several challenges to measuring glacier mass balance:

• Glaciers are usually found in remote locations.

• They are often associated with extreme topography.

• They can be very large, which makes it difficult to ensure that a small number of measurements are representative of the whole glacier.

BOX 26.3 ➤

➤ Airborne remote sensing methods such as lidar (Section 26.6.3) can provide an ideal solution to these challenges and, as a result, have become important tools for measuring glacier mass balance.

The figures in this box demonstrate one recent application of lidar for mass balance monitoring in the heavily glaciated Arctic archipelago of Svalbard, which lies about 700 km off the north coast of Norway and about 1100 km south of the North Pole. Data presented here were collected over the summers of 2003

and 2005 by the Airborne Remote Survey Facility (ARSF) of the UK Natural Environment Research Council (NERC). Figure 26.23 shows a photograph taken of Midtre Lovénbreen from the NERC aircraft with a handheld digital camera. The corresponding lidar data in Figure 26.24 is presented with a similar perspective to the photograph in Figure 26.2 for ease of interpretation. Compare the DEM with the photograph to appreciate the incredible surface detail that the lidar DEM has captured. The DEM of difference between the two study

years in Figure 26.25 shows the greatest thinning towards the glacier terminus with more than 5 m of elevation loss over the 2 year period. This decreases to 0 m in the accumulation zone. The blue colour of the surrounding bedrock indicates no change between the data sets as expected. The average mass balance over the whole glacier area during this 2 year period was found to be −121 180 tonnes per year, which equals −0.6 m of water equivalent per year. More details of this study can be found in Kohler *et al.* (2007).

Figure 26.23 Handheld digital photograph of the Svalbard glacier Midtre Lovénbreen looking south from an aircraft. The glacier is about 4 km long and 1 km wide.

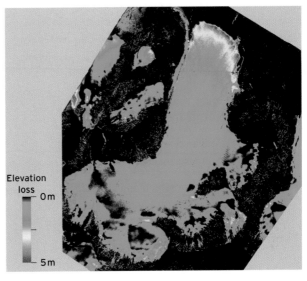

Elevation loss 0 m — 5 m

Figure 26.25 DEM of difference between the 2005 lidar data shown in Figure 26.24 and a similar lidar data set that was collected (also by NERC) during summer 2003. North is up and the DEM of difference is approximately 4 km across.

(a)

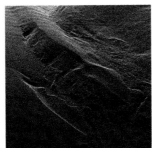

(b)

Figure 26.24 The 2005 lidar DEM of Midtre Lovénbreen: (a) with intensity overlay; and (b) in shaded relief. The latter clearly highlights the surface detail that can be captured by this high-resolution instrument. Both DEMs were interpolated from the raw lidar point cloud (inset) to a 5 m grid. Thanks to 3D imaging software, QT Modeler, courtesy of Applied Imagery, for use of the software to produce this image.

BOX 26.3

in mass balance modelling (Box 26.3) and coastal management (Chapter 15) must only model points that represent the ground surface whereas land-use and vegetation mapping (Chapter 25) may require that all surface features are modelled. Compare this to a more complex scenario such as floodplain mapping for flood management which requires the modelling of the ground surface but also of any surface feature that would act as a barrier to slow flood waters (i.e. walls and hedges but not buildings or trees). With lidar becoming increasingly important for monitoring environmental change, the lidar segmentation issue has become an important area of research (Sithole and Vosselman, 2006).

A very exciting application of laser altimetry is its operation from satellite platforms. This is exactly the intention of ICESat (Ice, Cloud and land Elevation Satellite), part of NASA's Earth Observation Program, which was launched in January 2003. Despite instrument problems throughout, which forced limited usage time, and total instrument failure in October 2009, the instrument proved invaluable especially for the monitoring of the ice sheets of Antarctica and Greenland. Plans are afoot for a possible successor to the ICESat mission.

Reflective questions

➤ What are the advantages of microwave remote sensing over optical sensing?

➤ When collecting lidar data, if any of the three instruments (scanner, GPS and INS) are in error or out of synchronization, are the resulting lidar measurements reliable?

➤ Can you explain the patterns in Figure 26.18?

26.7 Digital image processing

Regardless of the source of remotely sensed data, photographs and images are rarely in a state suitable for immediate analysis. Most often, owing to radiometric or geometric errors, images require some form of manipulation or interpretation before they are ready for analysis. The manipulation and interpretation of digital images are called digital image processing. There are many procedures in digital image processing that are frequently used for detecting environmental change. Important procedures include image rectification, image enhancement and image classification.

26.7.1 Digital images

Digital images are made up of a 2D array of DNs that usually represent the brightness of the corresponding area on the surface. A pixel can also be defined in terms of its position in a coordinate system. In the simplest case, pixel space coordinates are used where the top left corner is defined as the origin (0, 0). To link an image to an absolute coordinate system (called **georeferencing**), a pixel can be assigned an x and y coordinate that corresponds to some ground reference system (e.g. latitude/longitude). Images that are georeferenced can be joined together to create an **image mosaic**.

As digital images are made up of a series of numbers, their characteristics can be expressed in terms of a **histogram** and summary statistics. Figure 26.26 provides a histogram of the greyscale range of DN values for an image. Images can be subjected to a variety of processes based on the manipulation of their DNs and histograms.

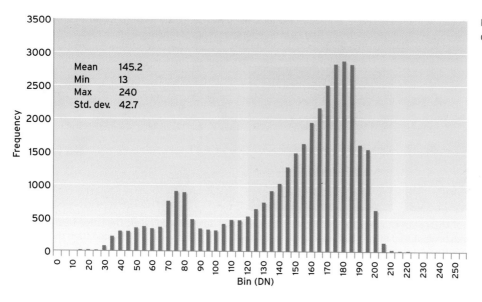

Figure 26.26 Histogram and descriptive statistics of an image.

Mean	145.2
Min	13
Max	240
Std. dev.	42.7

26.7.2 Image rectification

The purpose of image rectification is to attempt to compensate for any image distortions or degradations that may have resulted from the image acquisition process. The potential sources of such artefacts are many, as are the procedures for their correction. Image rectification processes are divided into two types: geometric and radiometric.

When an image requires geometric correction, inconsistencies in the relationship between the sensor and the surface have occurred during data collection. These inconsistencies depend greatly on the sensor used but in general they can be caused by changes in the sensor's altitude, attitude or velocity, changes in the surface such as the rotation and curvature of the Earth, and effects of the atmosphere. Some of these errors are predictable (e.g. Earth's rotation and curvature) and can be modelled mathematically to remove their effects. These are called **systematic errors**. Other errors that are unpredictable cannot be modelled mathematically. These are called **random errors**. To account for the effects of random errors, **ground control points** (GCPs) are used to tie the imagery to known points on the ground. The location of many GCPs in the distorted images is measured. Then, the coordinates in both ground and image systems are used to determine the transformation equations to correct the position of each pixel. Orthorectification is a specific type of rectification for aerial photography where the effects of elevation are removed from the imagery using not only GCPs but also information about the internal geometry of the camera to compensate for these errors.

Unlike geometric correction, radiometric correction does not change the position of the pixels, only the value of their DNs. However, like geometric correction, sources of radiometric errors are also largely dependent on the sensor used. There is a long list of factors that can interfere with the radiance measured by an instrument. Some examples include atmospheric conditions, illuminance (time of year/day), instrument sensitivity and viewing geometry. To correct an image radiometrically, the DNs are adjusted based on some model to correct the pixel brightness to represent the true radiance characteristics of that scene. The three most common adjustments are: noise removal, Sun-angle and haze correction.

Noise removal is necessary when electronic noise is recorded in an image that is unrelated to the radiance of the scene. It appears either as random or periodic errors found throughout the image. In many cases, these errors are caused by the malfunction of the sensing instrument. The errors can be removed by interpolating new DN values from unaffected pixels around the errors. The effect of a noise removal routine can be seen in Figure 26.27.

Remote sensing imagery is collected at various times of the year and different parts of the day. The position of the Sun at any given point in time will have an influence on how much sunlight reaches the surface and therefore on the radiance that is recorded by the instrument. This only becomes a problem when working with groups of images with different **illuminations**. Thus, compensation for solar position, called Sun-angle correction, is often required. This is easily achieved by applying a correction factor to each DN based on the illumination angle. Another problem is that atmospheric scattering (called haze) can be recorded by the sensing instrument. Haze has the effect of making an image appear foggy and

(a)

(b)

Figure 26.27 Image showing the effects of noise removal. (a) An image containing noise which can be seen as a random white speckle distributed throughout the image. In (b) a noise removal algorithm has been applied. Although the noise has been removed, notice how the image appears less sharp than previously.

it tends to mask the radiance characteristics of the surface. Haze correction can be applied when a feature in the scene is known to have a reflection at or near zero, such as water or shadows. Haze makes dark features appear more grey rather than black. Thus, an estimate of the effects of haze can be made and subtracted from the rest of the image.

26.7.3 Image enhancement

Image enhancement refers to any form of image manipulation that attempts to redisplay the information in the image in a way that better represents image characteristics or features of interest. Image enhancement applications are infinite. However, the most common applications include contrast stretching, spatial filtering and band ratioing.

Contrast stretching refers to a group of processes that are used to redistribute the range of DNs of an image to make better use of the image's bit scale. The image in Figure 26.28(a) and its histogram in Figure 26.28(b) show a rural scene dominated by pastures all of fairly uniform tone and very little texture. In this image, only the grey levels between 55 and 230 are being used to display the image, leaving many unused grey levels on either end of the histogram. To improve the interpretability of the information available in this image, it is sometimes beneficial to make use of all the grey levels available on the bit scale (in this case 0–255) and thus we use contrast stretching. A linear stretch is simplest and uniformly expands the range of DNs that are used in the image to include the entire range of values. A histogram equalization stretch also makes use of the entire range of DN values. However, unlike a linear stretch, a histogram equalization takes into account the frequency of DN occurrences. This means that in the output image, more DNs are reserved for brightness values in the image that occur frequently. For Figure 26.28, for example, this histogram equalization stretch will increase the contrast of the pasture pixels more than the urban pixels since the former occur more frequently in the image. The histogram equalization stretch is well suited for improving the interpretability of this image and the results of its application are given in Figure 26.29.

Another type of commonly used image enhancement is spatial filtering. Spatial filtering is a broad term used to refer to a variety of applications for the enhancement of the spatial variation in image tone (texture). In spatial filtering, texture is often described in terms of frequency.

(a)

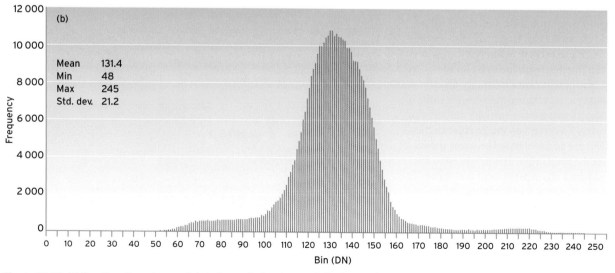

Figure 26.28 (a) Small section of an aerial photograph showing a relatively poor image contrast in the pastures. (b) The histogram and descriptive statistics of the image.

A smooth texture (small variations in tone) has a low frequency, whereas a coarse texture (large variations in tone) has a high frequency. This type of operation is called filtering because, like a filter, it selectively passes or preserves certain spatial frequencies, thereby enhancing some spatial frequencies and suppressing others. Low-pass filters produce a smoother image by enhancing low-frequency features and suppressing high-frequency features, whereas high-pass filters do the reverse and produce a sharper image. To illustrate this Figure 26.30 shows the results of a low- and high-pass filter when applied to the image in Figure 26.28.

Spatial filtering is a local operation, which means that the pixels of the filtered image only reflect the conditions in the immediate neighbourhood of each pixel. This is accomplished using a process called convolution. Convolution describes the process where a small moving window, usually 3×3, 5×5 or 7×7 pixels in size, passes over each pixel in an image. For each pixel some operation is computed using only the pixels within the window to arrive at a new DN value for the central pixel. For the low-pass filter described previously, the convolution filter (also called a kernel) determines the mean of the values in the moving window and assigns this new value to the central pixel of the window before moving to the next pixel to repeat the calculation. This process is illustrated in Figure 26.31. Thus, it can be seen that by changing the values in the kernel, a great number of spatial filtering operations can be carried out.

Another set of spatial filters are edge enhancement filters. These are also conducted using convolution and are used to exaggerate abrupt changes of DNs in an image that are usually associated with an edge or boundary of some feature of interest. Because edges are an example of a high-frequency feature, these filters tend to produce sharper images similar to the high-pass filter. Edge enhancement filters can be used simply to highlight all edge pixels, to remove non-edge pixels or to identify edges that lie only in a certain direction.

Finally, band ratioing is a process where the DN values of one band are divided by those in another to reveal the subtle variations between bands that would otherwise be masked by differences in illumination across the scene.

(a)

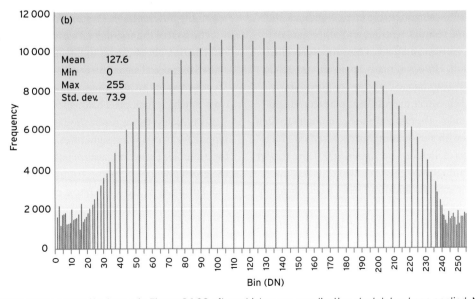

(b)

Mean 127.6
Min 0
Max 255
Std. dev. 73.9

Figure 26.29 (a) This figure shows the image in Figure 26.28 after a histogram equalization stretch has been applied. Notice how the improved contrast highlights the small variance in reflection in the pastures. (b) The histogram and descriptive statistics of this image are given. The image now uses all the 256 available grey levels and the mean and standard deviation have increased accordingly.

(a)

(b)

Figure 26.30 The image in Figure 26.28 is shown after applying (a) a low-pass filter which suppresses high-frequency variation in the image and (b) a high-pass filter which highlights the high-frequency variation in the image.

One very common example of band ratioing is the calculation of a normalized difference vegetation index (NDVI) image which provides an important tool for monitoring vegetation. It is calculated using the red and near-infrared bands of a sensor which represent wavelengths of high absorption and high reflection, respectively, and is the subject of Box 26.4.

26.7.4 Image classification

When humans look at an image, the brain automatically orders the scene into categories (e.g. trees, buildings and roads) based on the colours, textures and other spatial relationships of pixels in an image. This process is called pattern recognition. However, many applications such as mapping and land-use change detection require a more numerical approach where the results can be stored and manipulated digitally. Therefore we use computers to identify relationships with surrounding pixels based on image tones and patterns (Lillesand and Kiefer, 2000) and to sort patterns into meaningful categories called classes. This process is called image classification. Since a classification usually involves the use of several bands of data, it is often referred to as multispectral classification and it is really the sum of a three-component process: training, defining signatures and the decision rule.

Before the computer can classify an image, it has to be trained to recognize data patterns. This training can be carried out either automatically (unsupervised training) or under the guidance of the user (supervised training).

In unsupervised training the computer identifies intrinsic statistical clusters in the data that often do not correspond to typical land-use classes. In supervised training the user defines the clusters of data by choosing groups of pixels in the image that are representative of the desired classes. These pixel groups are called training samples while their physical locations in the image are the training sites. Supervised training tends to yield more accurate results than unsupervised training since there are many tests that can be applied to the training samples to determine how representative they are of the classes they characterize and how spectrally differentiable they are from the other classes.

The goal of the training process is to provide spectral information about classes in the image from which a representative spectral signature can be developed for each class. If each class in the image has a unique signature, then the class to which each pixel belongs can be easily determined. Unfortunately, this is rarely the case since many land cover types have overlapping signatures. Take the visible spectrum, for example. Trees and pasture both have a high reflection in the green portion of the spectrum and a low reflection in the red. Fortunately, the input into a classification is not limited to the visible portion of the spectrum. In fact, multispectral classification is not even limited to bands of surface reflectance but can include elevation, and statistical and map data layers. Additional layers added to a classification are called ancillary layers and the number of layers used in a classification is referred to as its dimensionality. The results of a classification can be considerably improved by increasing its dimensionality (Tso and Mather, 2001) but there is a trade-off

(a)

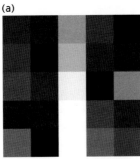

Original image showing
a vertical edge

(b)

-1	2	-1
-1	2	-1
-1	2	-1

Vertical edge
detection kernel

3	2	10	4	1
4	1	12	3	2
5	3	16	1	9
2	2	18	5	3
7	1	17	7	5

(c)
(-1*3)+(2*2)+(-1*10)+(-1*4)+(2*1)+(-1*12)+(-1*5)+(2*3)+(-1*16) = -38

Since the value is negative the new pixel is assigned a value of 0.

(d)

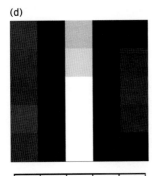

14	0	54	0	0
12	0	62	0	8
10	0	77	0	10
16	0	83	0	8
12	0	84	0	0

Image after convolution showing
a highlighted vertical edge

Figure 26.31 This figure provides an example of the application of convolution. For illustration purposes, a vertical edge detection kernel is applied to a simple 5 × 5 image. The original image is shown in (a) and the edge detection kernel is given in (b). In essence, the kernel is applied by centring the kernel window over each pixel in the image and multiplying the overlapping pixels. The value of the new pixel is achieved by summing the results of the nine cells. In this case, a value of 0 is given to all negative results. (c) An example of the calculation performed when the window is centred over the highlighted pixel in (a). (d) The results of the vertical edge detection kernel applied to the whole image. Notice how the vertical edge in the original image has been enhanced in the filtered image.

ENVIRONMENTAL CHANGE

LANDSAT TM FOR MEASURING AND MONITORING THE EARTH'S GREEN VEGETATION

The use of remote sensing data has proved very successful for assessing the type, extent and condition of vegetation over the Earth's surface. Such data have allowed researchers to estimate the area of different crop types, to assess the impact on vegetation from natural or human-made stresses (e.g. pests, fire, disease and pollution) and to delimit boundaries between various types of vegetation cover. It has also been an important tool in assessing and publicizing the extent of desertification (Chapter 14) and deforestation (Chapter 25).

Satellite imagery can be enhanced to improve its interpretability through a variety of image processing techniques. One such technique that is applicable to vegetation monitoring is a band ratioing application called the normalized difference vegetation index (NDVI). Like most other vegetative indices, it is calculated as a ratio of measured reflectivity in the red and near-infrared portions of the electromagnetic spectrum. These two spectral bands are chosen because they are most affected by the absorption of chlorophyll in green leafy vegetation and by the density of green vegetation on the surface. Also, in the red and near-infrared bands, the contrast between vegetation and soil is at a maximum.

The NDVI is a simple ratio of the difference between these two bands to their sum and it has been applied to the study of seasonal vegetation variations, leaf area index measurements and tropical deforestation. It can be calculated by:

$$NDVI = \frac{NIR - RED}{NIR + RED} \qquad (26.9)$$

where NIR is the reflectivity in the near-infrared and RED is the reflectivity in the red portion of the spectrum.

For Landsat data the red band is Band 3 and the near-infrared band is Band 4. Figure 26.32 shows the red and near-infrared bands of a Landsat scene of the Wirral Peninsula in north-west England, which is separated from the city of Liverpool by the River Mersey. Applying the NDVI to these bands gives the image in Figure 26.33. Notice how the vegetated parts of the Landsat scene stand out in the NDVI image compared with either the red or infrared bands independently.

(a)

Figure 26.32 A magnified view of (a) Band 3 (0.63–0.69 mm red) and (b) Band 4 (0.76–0.90 mm near-infrared). The image was captured over Liverpool, England, on 19 December 2000 at 12:15 in the afternoon. The city of Liverpool can be clearly seen in the top-centre portion of the image on the north side of the Mersey. (Source: images (a) and (b) courtesy of MIMAS)

BOX 26.4 ➤

(b)

Figure 26.32 *Continued.*

Figure 26.33 Landsat NDVI image which is generated using Landsat Bands 3 and 4 shown in Figure 26.32. Notice how clearly the city of Liverpool stands out against the parks and surrounding farmland as well as the brightness of the marshlands on the north shore of the Dee in the south-west corner of the image. (Source: image courtesy of MIMAS)

BOX 26.4

because the inclusion of redundant layers can strain storage and processing resources (Ohanian and Dubes, 1992).

The third stage of the classification process is the decision rule, which is the mathematical algorithm that uses the signatures to assign each pixel to a class. In an unsupervised classification, the three stages (training, signatures and the execution of the decision rule) are usually carried out automatically in one continuous process. Thus, it is the most automated form of image classification and requires minimal input from the user. The most common decision

rule used in unsupervised classification is called the Iterative Self-Organizing Data Analysis Technique (ISODATA) and is based on the spectral distance of each pixel from the cluster means as determined automatically by the computer from the statistics of the image (Tou and Gonzalez, 1974). The number of classes used is defined by the user. Initially, arbitrary means are assigned to each class, and a classification is performed. The means are adjusted to the resulting classes and the process is repeated until the means closely fit the classes in the data. As a result, the final classes rarely coincide with

(a)

(b)

(c)

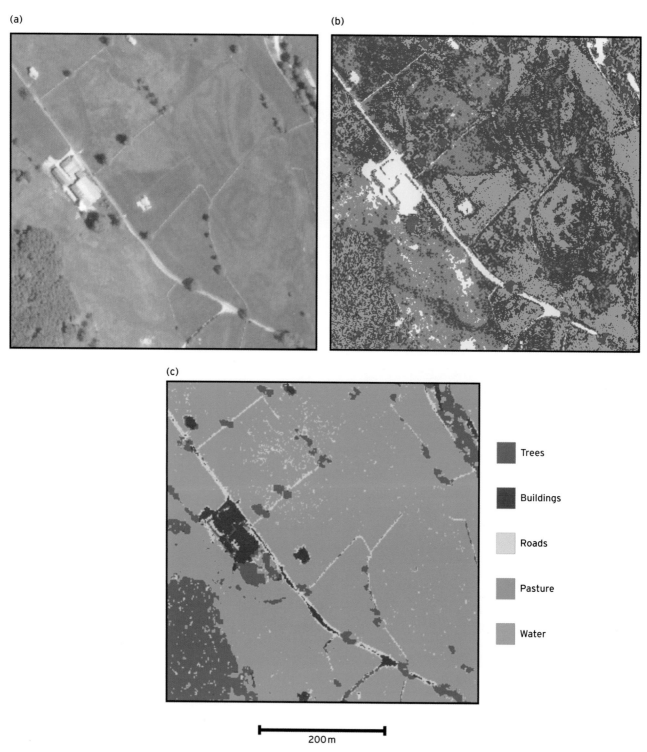

Trees

Buildings

Roads

Pasture

Water

200 m

Figure 26.34 Image classification: (a) a subsection of an aerial photograph that was taken over Upper Wharfedale, northern England, in May 1992; (b) the same image classified by an unsupervised classification using five arbitrary classes; (c) a supervised classification applied using five predefined classes: trees, buildings, roads, pasture and water. Notice the confusion in the supervised classification image between walls, roads and buildings owing to their similar reflective characteristics. Also, notice how the unsupervised classification identified the heterogeneity of the pastures and trees while the supervised classification did not. Each image is about 440 m across.

land-use classes on the ground. Nonetheless, such information is useful when little is known about the surface in question or as a guide for subsequent supervised classification.

Unlike unsupervised classification, supervised classification requires much more input from the user and the decision rule stage is no exception. There are a variety of methods available but these are the three most common methods:

- Minimum distance – assigns the pixel to the class whose spectral mean is closest to that of the pixel.
- Mahalanobis distance – the same as minimum distance except that it takes into account the variability of each class. A highly variable class will have pixels assigned to it with greater spectral distances than one with low variation.
- Maximum likelihood – assigns pixels to classes based on the probability of their membership (Tso and Mather, 2001).

To illustrate the types of classification results that can be expected, Figure 26.34 gives an image that has been classified using an unsupervised classification (Figure 26.34b) and a maximum likelihood supervised classification (Figure 26.34c), both of which used five classes. Notice the difference between the results for the two classifications. For Figure 26.34(b) the intrinsic statistical divisions in the image can be seen whereas Figure 26.34(c) shows land cover divisions.

Reflective questions

➤ Which image processing routine would you apply if you wanted to highlight the location of roads, hedges or fences?

➤ Which image processing routine would you apply to remove the effects of atmospheric scattering?

26.8 Summary

This chapter has introduced remote sensing in the context of its application to physical geography and environmental change. There is not one major component of physical geography covered by the chapters in this textbook that has not benefited from remote sensing. Digital images are made up of pixels, each of which has a digital number. This number is a code for the colour of that pixel. Important image characteristics include orientation, scale and resolution as well as content characteristics such as tone, texture, shape, size and pattern. These characteristics determine the quality and type of information that can be gained from the image. Photogrammetry can be used to produce digital elevation models and digital terrain models from photographs and if repeated images are taken over time then environmental change such as landscape erosion can be measured remotely.

The properties of electromagnetic radiation are fundamental to remote sensing because it is not only visible light that can be detected and recorded by sensors but the full range of the spectrum. Microwave sensors and ranging sensors such as radar, sonar and lidar provide additional information on environmental change.

Reflection, transmission, absorption and scattering processes are important in determining the type and amount of energy received by sensors. Every object reflects radiation in different ways and thus has a characteristic spectral signature. These signatures can be detected by passive and active sensors. Active sensors use their own source of radiation and measure the quantity of emitted radiation reflected back. Scanners can be used to detect radiation across narrow wavelength bands or over the full spectrum. A range of spaceborne, airborne and ground scanners are used to create a wealth of spatial and temporal information on environmental change.

Often it is not sufficient to use raw remote sensing images and it is necessary to process these images in some way. This processing may involve rectification which allows us to compensate for any image distortions or degradations that may have resulted during image acquisition. Image enhancement attempts to redisplay image information in a way that better represents image characteristics or features of interest. Methods include contrast stretching, spatial filtering and band ratioing. Computer classification of image features is also an important tool and aids automation of remote sensing data analysis.

Further reading
Books

Kasser, M. and Egels, Y. (2002) *Digital photogrammetry.* **Taylor & Francis, London.**
This is a good photogrammetry reference that specifically addresses issues of digital photogrammetry.

Lillesand, T.M., Kiefer, R.W. and Chipman, J. (2008) *Remote sensing and image interpretation*, **6th edition. John Wiley & Sons, Toronto.**
This is an invaluable reference providing an in-depth look at the principles of remote sensing and includes a comprehensive review of the Earth observation satellites currently in use. It also explores the links between remote sensing and geographical information systems. Appendix B on p. 732 provides an extensive list of remote sensing data sources with web addresses.

Russ, J.C. (2011) *The image processing handbook*, **6th edition. CRC Press, Boca Raton, FL.**
This is a well-respected user manual of digital image processing. It presents a high-quality discussion of the techniques of image processing and when/why they are applicable. It is up to date and draws on examples from a variety of real-world applications such as medicine, microscopy and remote sensing.

Wolf, P.R. and Dewitt, B.A. (2000) *Elements of photogrammetry with applications in GIS.* **McGraw-Hill, New York.**
This new edition of the principal photogrammetry reference provides a thorough and modern perspective on photogrammetry in the twenty-first century. It includes sections on principles of photography, coordinate systems, GIS, topographic mapping, digital image processing and project planning.

Further reading
Papers

Howat, I.M., Smith, B.E., Joughin, I. and Scambos T.A. (2008) Rates of southeast Greenland ice volume loss from combined ICESat and ASTER observations. *Geophysical Research Letters*, **35, L17505, doi:10.1029/2008GL034496.**
A paper showing glacier elevation changes.

Kohler, J., James, T.D., Murray, T. *et al.* **(2007) Acceleration in thinning rate on western Svalbard glaciers.** *Geophysical Research Letters*, **34, L18502, doi: 10.1029/2007GL030681.**
An important paper relevant to Box 26.3.

CHAPTER 27

Managing environmental change

Adrian T. McDonald

School of Geography, University of Leeds

Learning objectives

After reading this chapter you should be able to:

➤ understand environmental change and define and categorize such change

➤ appreciate the dependence of environmental management on good physical geography and describe how physical geography can be applied

➤ understand basic environmental management tools and management processes

➤ evaluate problems and processes involved in implementation of environmental interventions

27.1 Introduction

Four issues stand out from the rest of the field as the major global challenges of the twenty-first century: food security, energy security, water security and climate change. These are highly interrelated. Indeed Professor Sir John Beddington, the UK Government's Chief Scientist, raised the unwelcome possibility of a 'perfect storm' in which several of these issues coalesce to create a serious economic, welfare and unsustainable situation across many parts of the world (Beddington,

2009). The hypothesis is that as population grows globally, and as China and India continue their economic expansion, they start to buy into the global food market driving up prices, but if this happens at a time of depressed yields due to climate change and at a time when many food crops have been replaced by energy crops, there may simply be insufficient food to satisfy world markets and even more people will go hungry. Does this seem like scaremongering? Consider the words of Zheng Chunmiao, a leading water resource research manager at Peking University: 'The government must adopt a new policy to reduce water consumption. The main thing is to reduce demand. We have relied too much on engineering projects ... We must reduce food production ... It would be more economical to import' (quoted in the *Guardian*, 28 June 2011). See Box 27.1 to discover more for yourself.

Ten years ago the management of the environment was viewed as important by only a modest number of resource-dependent companies. The B&Q do-it-yourself stores stood out from the pack as a supplier of certified timber (from well-managed forests to internationally agreed standards), for example. Today almost every global company has environmental policies, action plans and responsible managers. For companies like Unilever, BASF, Crown Technologies and others, managing carbon and carbon footprint, carbon trading and offsetting, and corporate social responsibility activities are standard. The pre-existing approach of CATNAP (Cheapest Available Technology Narrowly Avoiding Prosecution) is simply unacceptable in the board room and in its place is the presence of environmental performance as part of a balanced scorecard (performing sufficiently across a range of important indicators) and indeed as a marketing requirement. Tesco (a major grocery store in Europe) has three key aims of:

- helping its customers by making green choices easier and more affordable;
- setting an example by measuring and making big cuts in Tesco's greenhouse gas emissions around the world;
- working with others to develop new low-carbon technology throughout the supply chain.

It has made a number of specific commitments (Leahy, 2007), for example:

- to use a 50 : 50 bio-diesel mix in its fleet;
- to add 5% bio-ethanol to its petrol mix in 300 stations;
- to halve the price of energy-efficient bulbs;
- to label air-freighted food and to reduce this to 1% of products.

Now compare this situation with that of 2004 when the first edition of this book was written and when it expressed concern that environmental management was all talk and no action. Perhaps this was driven by pessimism derived from the outcomes of the major Johannesburg conference on climate change in 2003 which seemed to have very few actions resulting from it. However, the growing strength of a market-led approach to environmental performance seems to have created optimism so that progress may be made. However, activists such as George Monbiot continue to argue that the rate and scale of a changed lifestyle are too slow to make a difference and so we still need to imagine a world without life. This might be a world in which the biosphere had been silenced by excessive and careless use of bioaccumulating pesticides (see Chapters 3, 10, 13 and 18). This was the future offered by the courageous ecologist and author Rachel Carson when she published, in 1962, her landmark book *Silent spring* in an effort to communicate the dangers of DDT (see also Chapter 25). This was a pesticide sprayed on crops and used around the world from the 1930s. The problem with this pesticide is that it contained toxins that stayed in the food chain and accumulated in soils, plants and animals. These toxins became more and more dangerous as they built up over time. Therefore DDT was banned over most of the world in the late 1970s. There is no case for complacency despite the obvious improvements in the developed world at least. Ibuprofen

finds its way into our rivers already populated by more female fish thanks to release of hormones derived from the birth control pill. Metaldahyde, a chemical derived from slug pellets, is also found in our rivers and water supplies and is challenging to treat and remove. But at least these issues are being addressed. Of more concern perhaps is the transfer of pollutants to the Third World and BRIC countries as production is moved to more economic and less regulated places.

Silent spring was probably the warning that finally awoke the world to the significance of the environment and the degradation that was in progress. People began to realize that humans were degrading the landscape and damaging the environment. Surprisingly, the subsequent increase in environmental awareness, investigation and popular reporting has been slow to impact on our thinking. When environmental degradation first began, with such actions as forest clearance for timber and agricultural activities, it was at a small scale and the overall effects of the actions were not collated or understood. That cannot be the case today. We have a global picture of change (see Chapter 26) and some appreciation of the ramifications of such changes (see Chapter 24). We also have the tools and management skills with which to start to cope or respond and it is the purpose of this chapter to describe these management tools.

A fundamental question must be answered before we continue in this chapter. Why, in a book about physical geography, is there a chapter on environmental management? We study physical geography for two basic reasons. The first relates to intellectual curiosity, which drives us to understand better our world in all its diversity and the processes through which the landscapes in which we live have become as they are. The second relates to the effect we have on our environment by living in it and drawing natural resources from it. To manage the way in which we impact on the environment, to foresee and perhaps to forestall the consequences of our actions require that we understand the processes which govern our environment. After all, Rachel Carson was first and foremost a practising scientist who understood the processes operating in the environment. Society needs good physical geographers so that process-based science can provide the information necessary for environmental management. Table 27.1 provides, for each of the other chapters in this book, just two examples of environmental management issues that were discussed within those chapters. It is clear that all areas of physical geography have relevant application to environmental management.

Arguably, the management of environmental change, species threatening as it is, is one of the most important aspects

Table 27.1 Two examples of environmental management issues from each of the other chapters in this textbook

Chapter	Subject	Two example management issues
1	Approaching physical geography	Channel avulsion, rock weathering on building materials
2	Earth geology and tectonics	Earthquakes, volcanic eruptions
3	Oceans	Overfishing, pollution
4	Atmospheric processes	Drought, aerosols
5	Global climate and weather	Hurricanes, monsoons
6	Regional and local climates	Haze, shelter belts
7	Weathering	Rock fractures around infrastructure, building stone decay
8	Hillslopes and landform evolution	Mass movements, erosion/landscape degradation
9	Sediments and sedimentation	Dams, mining
10	Soils and the environment	Soil erosion, soil pollution
11	Catchment hydrology	Flooding, salinization
12	Fluvial geomorphology and river management	Bank erosion, biodiversity
13	Solutes	Water pollution/contamination, lake acidification
14	Dryland process and environments	Desertification, water resources
15	Coasts	Coastal erosion, sea-level rise
16	Glaciers and ice sheets	Water resources and energy, skiing/tourism
17	Permafrost and periglaciation	Infrastructure stability, gas hydrates
18	The biosphere	Fire, deforestation
19	Biogeographical concepts	Alien introductions, recreation needs
20	Ecological processes	Conservation, bioaccumulation
21	Freshwater ecosystems	Fisheries, water quality
22	The Pleistocene	Greenhouse gases, species extinctions
23	The Holocene	Agriculture, deforestation
24	Climate change	Mitigation, pollution
25	Vegetation and environmental change	Biodiversity, food supply
26	Remote sensing of environmental change	Land-use change, flooding

of science and management today. Managing and promoting change in the environment are full of uncertainties. Yet action needs to be taken, often at an early stage before the picture is clear. This is known as the **precautionary principle**. This is action taken as a precaution before we understand the outcome, but being aware that some aspects of the possible outcomes cannot be tolerated. However, taking such action also involves risk, both a physical risk of the wrong outcome developing and an economic risk of taking expensive action that is not required. The scientific research discussed in the earlier chapters of this book provides information to help reduce uncertainty about how the environment will respond to a management action. It is, however, the management tools themselves that will be discussed in this chapter. Environmental management is the practical application of physical geography. It has become a discipline in its own right. This chapter seeks to show how physical geography informs environmental management and to outline some of the key techniques used to tailor understanding of physical geography to a form that is useful for management.

27.2 Understanding environmental change

The previous five chapters of this book outline issues surrounding environmental change over different timescales. Chapters 22 and 23 discuss environmental changes over the past 2.6 million years, whereas Chapter 24 provides information on more recent human-induced environmental change and related problems. Chapter 25 describes interactions between vegetation and climate change while Chapter 26 discusses contemporary techniques for monitoring present global and local environmental change. The following section, however, provides more general ideas about the characteristics of change of relevance to environmental management.

27.2.1 Characteristics of change

In order to manage environmental change we need to have some understanding of change itself, both the 'upstream' drivers and the 'downstream' consequences. This more holistic approach is gathered in the DPSIR process. This is a framework that assumes a cause and effect relationship between the various interrelated parts (spatial, temporal and component) of social, economic and environmental systems:

Driving forces of environmental change (e.g. greater agricultural production)

Pressures on the environment (e.g. release of chemicals to the atmosphere)

State of the environment (e.g. urban air quality)

Impacts on population, economy, ecosystems (e.g. health impacts, respiratory problems)

Response of the society (e.g. clean air acts, vehicle emission control)

DPSIR is extremely flexible and although created with developed country economies and systems in mind, it has already proven effective in rural environmental management in many developing countries. Figure 27.1 shows an example result from Ghana when applied to an area suffering from **desertification** and declining crop yield. The adoption of the DPSIR framework in northern Ghana allowed the interconnectedness of the various factors to be understood, communicated and applied to policy formulation.

Perhaps the most important aspect of change concerns the relationship between factors that promote change and the nature of the change. The nature of change is summarized in Table 27.2. If the change is irreversible or quick and unstable, for example, it is a much more problematic type of change than one that is reversible, slow and stable. Change can be considered to be driven by a factor or group of factors. For example, increased acid rain may be driven by the sulphur released in the burning of coal for energy (see Chapters 10 and 24). However, the nature of the relationship between coal burning and acid rain and the resulting management implications will be constrained by the mix of good and problematic attributes of the relationship. This is generalized in Figure 27.2 which shows the range of relationships that exist. These can be reversible or

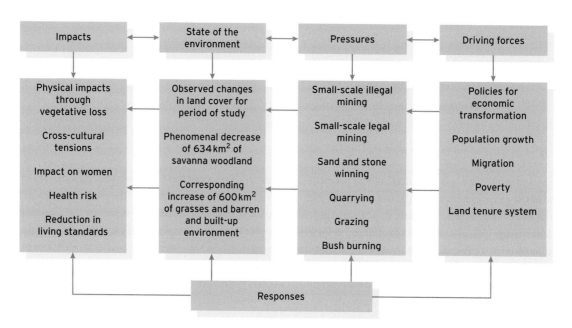

Figure 27.1 DPSIR framework analysis results from northern Ghana. (Source: Isaac Agyemang)

Table 27.2 The nature of change. Change is not simple. If a circumstance or situation involves all the attributes listed as 'good' then it is certainly simpler than a situation involving some or all of the 'problematic' attributes. It is a good management exercise to consider where, in the spectrum of change, a particular case lies

Good attributes	Problematic attributes
Stable	Unstable
Reversible	Irreversible
Slow transformation	Rapid transformation
Limited	Unlimited
Within tolerances	Beyond tolerances
Linear	Non-linear
Small scale	Global

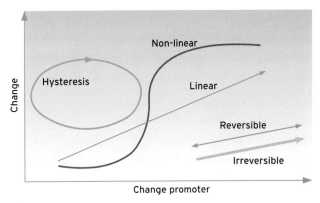

Figure 27.2 Some characteristics of change represented graphically.

irreversible, simple (linear) or complex (non-linear) or even contain **hysteresis**. If we assumed that a relationship was linear and reversible then we might put in place a management strategy to reverse this change. However, it may be the case that the change was not linear or reversible and so this management strategy might make the problem worse.

CASE STUDIES

UNDERSTANDING THE PERFECT STORM

This box is designed to get you thinking creatively. Design a diagram to illustrate the connections between the four main elements of the Perfect Storm (Figure 27.3). These elements are food, water security, energy security and climate change. For example: (i) food security might be enhanced by fertilizers but their use may compromise energy security (as fertilizers take a lot of energy to produce) and increased energy use may (depending on the energy source) impact on climate change or (ii) food security might be improved by irrigation but that might impact on water security elsewhere. One possible skeleton of such a diagram is given below. Should there be a fourth or fifth 'Security' box or is everything else subsumed in these headings? Is climate change the only controlling driver or should the central box have several other items? Is the economy, for example, so fundamental that it should appear with climate change? Should there be daughter cells around each security issue and should these be generic like demand reduction and supply enhancement, or specific like nuclear, biofuels and wind power? Alternatively, might the newer 'measurement' concepts such as carbon footprints, virtual water, food miles and so on have a place? Finally, might the whole diagram be made optimistic by concentrating on the options

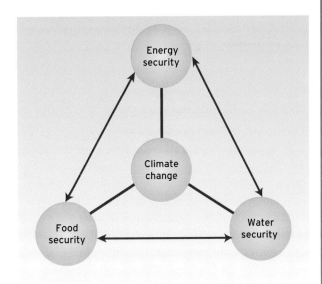

Figure 27.3 Four components of the 'Perfect Storm'.

or strategies for remediating these issues such as carbon trading, water efficiency and recycling, slow cities and so on?

BOX 27.1

Figure 27.4 Drainage ditch in a peatland which has eroded to become a wide channel. The exposure of the bare peat to the Sun has caused it to dry out and crack. (Source: photo courtesy of Alona Armstrong)

A good example of this comes from many peatlands which have been drained. The digging of ditches into a peatland often causes the soil to dry out and to crack (the change) as shown in Figure 27.4. The ditches themselves often lead to faster water movement when it rains, causing more flooding, and also to more land erosion because of the channelling of flow along the unvegetated drains (associated change). The management solution is to block up the drains by placing dams into them. This fills up the drains with water when it rains, rather than letting water run along their floors. However, in many cases this sends more water through the newly created cracks on the floors and sides of the drains, which results in more erosion of the soil below the surface (Holden *et al.*, 2006). The cracking is an irreversible change because when peat dries out it changes its properties so that it can no longer hold as much water (Holden and Burt, 2002b). Therefore the management solution may result in even faster change in the landscape by encouraging more turbulent flow through soil cracks causing erosion. So the simple definition of change, a transition from one state to another, can hide the risks, uncertainty and complexity of change and it is therefore

important to be aware of the attributes of change as described in Table 27.2 and Figure 27.2.

27.2.2 Rate of change

The rate of onset of change is a vital consideration because, if slow, (i) it may offer species the opportunity to adapt or shift their range, (ii) it may provide the environmental manager with the time required to devise and implement a remediation strategy, and (iii) it may offer the politicians sufficient time to perceive the potential damage and provide the resources needed by the manager. Of course in this last case, slow onset may also be a disadvantage in that it might not develop to be a serious problem within the limited time horizon that some politicians possess.

Non-linearity of, and rapid alteration in, rate of onset of change is a further, and in many ways frightening, issue. If we consider Figure 27.5 we can see that it consists of a surface which slowly curves from one stable location (look at either the upper or lower part of the surface). Forecasts based on a statistical assessment of the trend may indicate a slow onset change (relatively flat surface) allowing plenty of time for a considered response or perhaps an equilibrium situation in which a large change in the driver causes little response in the factor of concern. Unfortunately, the next section of the surface, the 'cusp', shows a very rapid response to a new stable state. In other words, a very slow change suddenly

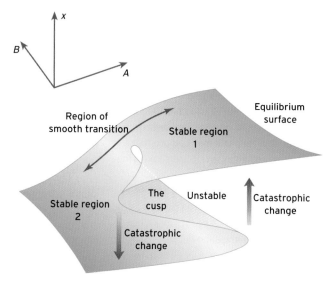

Figure 27.5 Catastrophe theory. The surface shown here is in three dimensions. *A* and *B* are the changes in environmental conditions and *x* is the change in the response. In the real environment there are many more than two variables but for simplicity just two have been drawn. Catastrophe occurs at the two arrows, the region of instability at the cusp. Rapid change occurs which is not reversed by a reversal in system conditions.

becomes a very rapid change as if the system has reached a threshold and it suddenly jumps out of one stable state. This figure summarizes **catastrophe theory** (Thom, 1968). A sudden catastrophic change occurs and then the system reaches a new stable (or equilibrium; see Chapter 1) position. So the theory describes cases when we have a sudden jump from one stable mode of operation to another, but where any changes at first may be very slow or hardly noticeable.

A good example of such a problem is provided by the thermohaline circulation system discussed in Chapters 3, 22 and 23. This is the deep-ocean circulation system that is today a very strong system for transporting heat away from the equator towards the poles. It is a strong system because of the saltiness of the water in certain sensitive locations. This salty water is dense and so sinks, thereby forcing water at the bottom of the oceans up in a large circulation system. However, with global warming, more of the world's glaciers are melting and so rivers flowing into oceans such as the North Atlantic are producing more freshwater (Dickson *et al.*, 2002). As river water is not very salty, these inputs may prevent the deep sinking that previously took place. This may shut down the ocean circulation system (Paillard, 2001). This change may be very sudden, and there is evidence from the past that the climate system has suddenly jumped from one stable warm mode to a stable very cold mode because of changes in ocean circulation (Broecker and Denton, 1990). The immediate result would be a

dramatic cooling of Europe by several degrees and a consequent growth of glaciers. These glaciers would increase albedo and thereby cool the planet even further and we would enter a glacial period. Thus, while we are in a stable warm period we may have a very slow change (slow warming) that eventually results in a rapid and catastrophic shutting down of the ocean circulation system which then sends the system into a stable cold phase. This of course would be irreversible on human timescales. Thus the concept of catastrophe theory, of a non-linear switch between stable states without the prospect of a reverse, is real and possible.

27.2.3 Environmental tolerance

Many definitions of the environment are somewhat limited in that they focus primarily on an ecological definition of the environment relating to organisms. However, environmental change relates to Earth surface–ocean–atmosphere–biosphere environments at any point on the global to local scale. The environmental manager needs to encompass the ecological perspective but must also include the physical characteristics and processes of the Earth's surface, the oceans and the atmosphere. At a global scale this perspective merges with the concept of **Gaia** (Box 27.2) at least to the extent of recognizing highly complex interactions of all elements of the Earth surface and that the response to a stimulus will be complex and unpredictable, almost like a giant organism (Lovelock, 1979).

ENVIRONMENTAL CHANGE

GAIA HYPOTHESIS

James Lovelock (1979) presented his Gaia theory which suggested that the Earth is a superorganism. It was argued that the maintenance of conditions suitable for life and particularly the maintenance of the Earth's atmospheric chemistry was derived from the self-regulation and feedback mechanisms that exist on the planet. The biological systems and the landscape-ocean-atmosphere systems are highly coupled and operate as a single living entity. A justification for this hypothesis is that the composition of the Earth's atmosphere would be radically different if there were not life on the surface. Without flora and fauna, the atmosphere would be mostly carbon dioxide, with very little nitrogen or oxygen. However, with the addition of life, in combination with the Earth's other subsystems, the Earth's various aspects constitute a feedback system that seeks an optimal environment to sustain life. In its most basic form, the Earth acts to regulate flows of energy and cycling of materials. The unlimited input of energy from the Sun is captured by the Earth as heat or photosynthetic processes, and returned to space as long-wave radiation. At the same time, the material possessions of the Earth are limited. Thus, while energy flows through the Earth (Sun to Earth to space), matter is recycled within the Earth.

The idea of the Earth acting as a single system as put forth in the Gaia hypothesis has stimulated a new awareness of the connectedness of all things on our planet and the impact that humans have on global processes. It means we cannot think of separate components or parts of the Earth; there are too many interconnections for this to be the case.

BOX 27.2 ➤

Therefore what humans do on one part of the planet can affect other parts of the planet. The most difficult part of this idea is how to determine whether these effects are positive or negative. If the Earth is indeed self-regulating, then it will adjust to the impacts of humans. However, these adjustments may act to exclude humans, much as the introduction of oxygen into the atmosphere by **photosynthetic bacteria** millions of years ago acted to exclude **anaerobic bacteria**. This is the crux of the Gaia hypothesis.

Lovelock argued that the Earth is not 'alive' in the normal sense but that seeing the Earth as a living organism is just a convenient way of organizing facts about the Earth. He argued that a tree processes sunlight and water to grow and change, but this happens so imperceptibly that in practical terms it often appears unchanged, and wondered if the Earth might have the same characteristics. Many people will reject Gaia as one step too far. However, if we accept that Earth is an integration between all elements of a system at all scales and if we seek to understand how all components of the system relate to each other and avoid excluding any factors, then we start to approach Gaia.

BOX 27.2

An environmental tolerance is a valuable concept within the ecological community. Each species has a range of environmental conditions within which it can operate. This might, for example, be a temperature range such as $-2°C$ to $+30°C$. In reality all species will have a series of 'ranges' within which they can survive, controlled for example by soil moisture content, oxygen saturation, temperature and pH. When conditions fall outside any *one* of these values for a critical period of time (seconds or hours depending on the species and determinant) the organism cannot survive. That just *one* of a series of environmental factors is crucial at a particular time is known as the **law of limiting factors**. The same law or concept appears in other areas of management under titles such as 'bottleneck theory' and 'threshold theory'. The generalized concept of environmental tolerance is measured and made specific for an individual determinant through measures such as the $LD_{50/24}$ measure, which is the dose of a determinant which would be lethal to 50% of the organisms under consideration within 24 h (Lawrence *et al.*, 1998).

An ecosystem contains a set of species which have individual tolerances but there is also a tolerance of the ecosystem as a whole. The tolerance of the ecosystem will be greater than that of the individual species that form it. Thus the scale of environmental change will have to be greater to destroy ecosystems than to destroy an individual species. Of course, ecosystems may be more degraded if the individual species that is destroyed is a keystone species (see Chapters 20, 21 and 25). Figure 27.6 shows that the tolerance range of the ecosystem (the range beyond which none of the ecosystem species survive) is considerably greater than that of individual component species (McDonald, 2000). Under 'normal' conditions nine of the species survive, although species number 8 is at the extreme of its range. About half the species will survive changes to unusual conditions but note that the species that survive an upward unusual change are different from those that survive a downward unusual change. Only two species are shown to survive swings to either extreme. This is one reason why the promotion of diversity is important. Diverse ecosystems tend to be more resilient (Bradbury, 1998). Box 27.3 provides a case study of managing beetle infestations around the Upper Fraser River, British Columbia, which have come about due to climate change and change in local forest management.

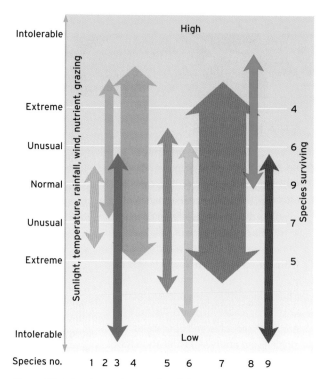

Figure 27.6 Species and ecosystem tolerance. (Source: after McDonald, 2000)

MANAGING THE CONSEQUENCES OF CLIMATE CHANGE – THE PINE BEETLE IN PRINCE GEORGE, BRITISH COLUMBIA

Prince George lies in the geographical centre of British Columbia. It is a small city on the banks of the Upper Fraser River and is the administrative centre of northern British Columbia. Its origins are as a logging town and it is surrounded by coniferous forest for hundreds of kilometres in every direction. What is strange about Prince George is that it appears from the trees as if it is autumn all year. The trees are brown and light filters through to the forest floor. However, these are coniferous trees that should stay green all year round. This is wholesale destruction by the pine beetle in epidemic proportions. Seven million trees are infected (Figure 27.7) with areas where the tree kill is 100%.

So there are three (generic) questions that any manager will pose in such a situation:

1. What is the cause?
2. What are the consequences?
3. What can be done?

Causes

The pine beetle is a natural grazer and occupier of conifers (Figure 27.8). Normally it is in balance with the growth of the vegetation. However, beetle numbers are increasing substantially and damaging the tree cover. There appear to be two basic causes of the spread of this beetle. The first relates to climate change. The occurrence of long-duration, severe cold periods in winter has declined. Temperatures reach only −20°C for a few days at a time whereas

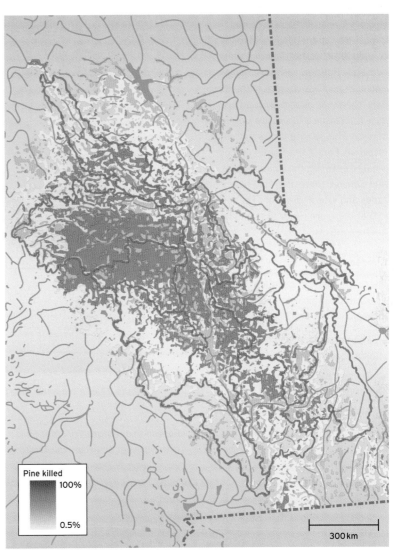

Figure 27.7 A map showing the total proportion of pine trees killed by pine beetle between 1999 and 2005 in British Columbia. (Source: Courtesy of BC Ministry of Environment)

in the past such temperatures were exceeded for weeks at a time. Beetles are therefore surviving the winter in much larger numbers than before. An added factor may be the drier years that have occurred since 1995 which may have weakened the mature pines, making them more susceptible to pine

beetle attack. As is so often the case it is the larvae and the fungal infections carried by the adult that destroy the trees – not the beetle itself.

The second cause is more directly human induced. Fire is an integral part of the ecological processes that operate in the forests of British Columbia.

BOX 27.3 ➤

787

The fires occur naturally from lightning strikes and are usually relatively low-temperature burns that reduce the numbers of beetles in an area. However, as forests become more important commercially, as fires start to arise from deliberate or careless human actions and as more homes and communities develop in the forests, there has been a growing pressure to provide a forest fire-fighting service. This service is now extremely effective. Forest fire-fighting is coordinated between the United States and Canada, there are rapid response times and effective action. The result is that there are fewer fires and fewer beetle controls.

Consequences

There are many consequences, some direct and tangible, some indirect and intangible:

- The immediate result is that there is a growing area of dead forest.

- Timber not extracted in a 5 year period will have limited value.

- A rapid increase in timber volume reaching market has an influence on price.

- Replanting with pine is a high-risk option.

- Loss of leaf cover alters snow capture, trunk climate and snowmelt.

- Loss of cover means direct rainfall to the forest floor, increasing erosion.

- Loss of habitat and loss of biodiversity.

- Raised fire risk from dead trees (and accumulated unburnt brash).

- Loss of forest as a cultural resource to first nations.

Intervention

Many companies offer advice on detecting the pine beetle and treating the

Figure 27.8 A mountain pine beetle. (Source: Forestry Images: Ron Long)

infection at a suburban or small-plot scale. However, the economic treatment of very large swathes of landscape by chemicals is unrealistic. The options are:

- species change occurring naturally;

- the exploitation of the trees at a younger age (as it is a disease of mature trees);

- the relaxation of the burning control regimes to promote large areas of younger trees;

- harvesting of the damaged trees (luckily wood prices are high at the moment) followed by, legally required, replanting.

Alternatively we may opt to accept this as a natural occurrence and so accept:

- changed flow regimes in the rivers;

- more flooding downstream;

- more erosion;

- loss of habitat;

- loss of employment;

- increased risk of wildfires.

An excellent video of the pine beetle problem can be found on the British Columbia government website www. for.gov.bc.ca/hfp/mountain_pine_ beetle/video.htm and the substantial costs of replanting and community fire protection are given in www2. news.gov.bc.ca/news_releases_2005- 2009/2005OTP0108-000832.htm.

BOX 27.3

27.2.4 The 'duty' and need to manage change

Some ecologists take comfort in ecosystem resilience to change. However, the environmental manager can take much less comfort because economic and political imperatives always appear to have the highest priorities. In other words, money and power often come before environmental management issues. The human species is nearly always placed at the top in any environmental management decision. However, there are both ancient and modern values that do not place the human species higher than others in the biosphere. These ideas form part of 'environmental ethics' and suggest that we might have a duty to care for other species and the Earth itself.

Environmental ethics is the idea that different people might have different viewpoints (e.g. religious, cultural) on environmental management. One viewpoint is that we should treat other species and environmental processes as equal to ourselves. It is a philosophy that places humans on an equal level with other species. Therefore, advocates of environmental ethics would suggest that we should think carefully about managing the environment for the environment's sake and not just for our own selfish uses of the environment.

Environmental ethics is not a new concept but it has been dismissed in industrialized western societies over the past few centuries. These societies have tended to have a technocentric perspective on the environment. This is a view that suggests that with technological developments we can 'conquer' the environment. In western society today an appreciation of the importance of environmental ethics has awakened. Many people around the world believe that we have a duty to look after and care for the environment. It is therefore vital that the environmental manager appreciates the very different values and perspectives that can be held by different communities (Hargrove, 1989).

When managing the environment we must carefully think about the actions of others and not just about our own actions. A good example of this is what has been termed the **tragedy of the commons** (Hardin, 1968). This involves a field that anyone can use (common land). It is to be expected that each farmer will try to keep as many cattle as possible on the field. However, the logic of the commons will bring tragedy. This is because, as a rational being, each farmer seeks to maximize their gain. The farmer will ask 'what is the benefit or disadvantage of adding one more animal to my herd?' However, this action has one negative and one positive component. The positive component is a function of the increment of one animal. Since the farmer receives all the proceeds from the sale of the additional animal, the positive utility is nearly +1. However, the negative component is a function of the additional overgrazing created by one more animal. Since, however, the effects of overgrazing are shared by all the farmers using the field, the negative utility for any particular decision-making farmer is only a small fraction of -1.

Adding together the component parts, the rational farmer will conclude that the only sensible course to pursue is to add another animal to the farmer's herd on the field. This will be followed by the addition of another and maybe many more. However, this is also the conclusion reached by each and every rational farmer sharing a field. Therein is the tragedy. Each farmer exists within a system that encourages them to increase their herd without limit but in a world with limited resources. The field will be massively overgrazed and all of the vegetation will be removed. Therefore none of the animals will survive and the farmers will be ruined. So a society that believes in freedom of the commons may be a society that brings its own downfall.

This analogy, of course, is almost the same as allowing all people the freedom to pump pollutants into the atmosphere or oceans when and where they want and in whatever quantity they want as if the atmosphere and oceans were unlimited resources. It is also almost the same as individuals thinking 'what will it matter if I throw just one more plastic carton in the bin; surely one more will not do any harm?' The analogy fits environmental problems throughout the world. We have finite resources and we must therefore manage those resources. Allowing people freedom to do what they want may bring only degradation unless they can agree to collective management. Management is therefore required in order to identify the range of feasible alternative outcomes in response to changing environmental conditions, the identification of the costs, risks and uncertainties associated with pursuing each alternative, and the organization and mediation of the resources needed to implement a chosen alternative (Lawrence *et al.*, 1998).

27.2.5 Types of change to be managed

There are broadly three types of environmental change that can be 'managed':

1. Responding to natural environmental change.
2. Controlling anthropogenic environmental change.
3. Implementing local change.

We do not have the ability to alter or reverse natural environmental change and certainly not in a sustainable manner (see below) and not at any scale beyond the local. For example,

Table 27.3 Some east coast flooding responses in England

Flood response	Advantages	Disadvantages
Sea walls and similar hard engineering responses	Rapid implementation	Ugly
	Defined action	Promotes flood potential
	Retain existing use	Costly
Beach maintenance (adding sand)	Rapid implementation	Regular action needed
	Retains scenic value	Questionable sustainability
Land use	Long-term viability	Change in use
	Recognizes risk	Loss of existing business
Insurance	No capital cost	Availability?
	Not a government expenditure	Cost
	Individual responsibility	
Managed retreat (allowing the sea to advance inland)	A sustainable defence	Change in use
	Provides biodiversity	Outcome not easy to predict
	Reduced expenditure?	
No active response	No capital cost	Lack of coordination
	Individual responsibility	Lack of expertise

the isostatic rebound of the land masses in the north of Britain following the thousands of years under the weight of the ice and the resultant settling of the southern half of Britain as part of that response has increased coastal flood risk in southern Britain and will continue to do so. We cannot raise the south-east of England and so all our responses are forms of '**coping strategies**' (see Chapter 15). Table 27.3 lists some of the coping strategies to flooding on the east coast of England and lists the advantages and disadvantages of each option. The lists of options, advantages and disadvantages are by no means complete but the table provides examples to show the sorts of responses we may be able to perform.

When considering the management of environmental change we must consider how long we want to manage the system for. An action taken today to cope only with the current situation may not be adequate in the long term as, for example, sea levels continue to rise. This necessity for long-term management is often subsumed within the term '**sustainable development**' incorporating a requirement to consider not just fair outcomes for the present population but outcomes that are satisfactory for future generations: 'Sustainable development is development that meets the

needs of the present without compromising the ability of future generations to meet their own needs' (World Commission on Environment and Development, 1987). Sustainable development also incorporates the broader vision of the social and economic environment.

Anthropogenic change, or change resulting from human activity, is usually more rapid than similar processes operating through natural phenomena. Examples of anthropogenic change are widespread (acid rain, overfishing, deforestation, river diversion, and so on) and further cases involve a blend of anthropogenic and natural change such as desertification. In most cases the change is one directional rather than cyclical, with broadly irreversible consequences. For the manager there are two main sets of options: (i) using the coping strategies available to manage natural change; and (ii) limiting the drivers that are creating the environmental change. Of course to limit the drivers assumes that the process is reversible or at least reversible at the stage which the situation has reached. It is also assumed, more importantly, that the political, legal and economic will is present to commit to curtailing the activity that is giving rise to the environmental change. This 'will' is only forthcoming if the environmental scientist can prove clearly and simply that the relationship between the initiating activity and the consequent change exists and that the change is very serious and must be addressed. Thus the disciplines of the physical geographer and the environmental manager must never become competitive. Each must be aware of the contribution of the other.

At a local scale we now have the power to make changes to the environment. For example, for a hundred years or more (thousands in the case of the Rivers Amu Darya, Syr Darya, Tigris, Euphrates and Nile at least) we have 'trained' rivers to follow the course most convenient to our developing use of land. However, such power has often not been accompanied by an equal measure of knowledge and so, as discussed in Chapter 12, river channelization techniques have often made flood problems worse by sending the water downstream in larger volumes over shorter periods of time (e.g. by shortening the length of a river by removing meander bends). They have also resulted in a series of other problems related to geomorphological processes. For example, straightening a section of river causes it to erode its bed and banks in the straightened section because the average slope of the river along that section has increased (shorter river channel distance over the same fall in altitude) and so stream power is greater. This undermines the river channel engineering structures themselves and also adds more sediment to the downstream part of the river. This can cause the downstream river channel to infill with sediment and so its capacity to carry water is reduced and it will overflow more easily, causing increased flooding. It is therefore important to improve and incorporate process understanding into environmental management plans.

Reflective questions

➤ Why are diverse ecosystems more resilient than individual species?

➤ What is catastrophe theory?

➤ Can you think of an example of human action that is analogous to the tragedy of the commons?

27.3 Tools for management

The needs of the environmental manager are simple. The manager needs to be able to identify the nature, scale and timing of the impacts of any actions. This can be the action of others or humankind in general. The manager must be able to demonstrate that remedial action is required either

to pre-empt or reverse other actions or to create change for which there is demonstrably no adverse reaction. Hence the first need of managers is to be able to forecast. Statistical and numerical modelling can help make predictions about environmental change and the impacts of management strategies. Such modelling tools are discussed in Chapter 1 and elsewhere in this book.

27.3.1 Hazard assessment

Studies of tectonics (Chapter 2), slope processes (Chapter 8), solutes (Chapter 13) and many other areas of physical geography covered in this book provide us with information that allows us to gain some idea of the risk of an environmental hazard. This risk is measured in terms of the likely recurrence interval (e.g. 1 in 1000 years event) or when the event might occur and how much damage the hazard might cause. Thus, for environmental management it is often necessary to be aware of environmental hazards and be able to predict the risks that those hazards pose. There are a range of tools for assessing and predicting hazards but most of these are site or hazard specific such as models for diagnosing urban areas against the risk of earthquakes, or slope stability equations for assessing when slope failure might occur.

These techniques are often spatial in their approach and it may be of great benefit to map areas of risk associated with a given environmental problem. For example, Figure 27.9 shows a map of soil erosion risk in Europe. Once these hazard assessments have been made it is then possible to show the results to local or international governors in order to gain the money (and legislation if necessary) to spend on management strategies for those areas considered to be most risky. For example, those areas with a high risk should be protected by reducing grazing levels or planting trees, or using some other management technique (Kirkby, 2001). These hazard assessments can then justify environmental management policies such as spending money on trying to prevent soil erosion in certain sensitive areas, or by moving people out of an area that is very risky (e.g. in terms of slope failure or contaminants).

27.3.2 Impact assessment

Making an environmental management decision is rarely simple. Environmental systems are complex. Small changes in one component may have unforeseen consequences elsewhere. To reflect the complexity of possible impacts, managers have developed a series of assessment techniques that seek to report fully the impacts of an action. Initially

Estimated annual erosion (t ha^{-1} yr^{-1})

- 0
- 0–0.5
- 0.5–1
- 1.0–2
- 2.0–5
- 5.0–10
- 10.0–30
- 30.0–50
- 50.0–100

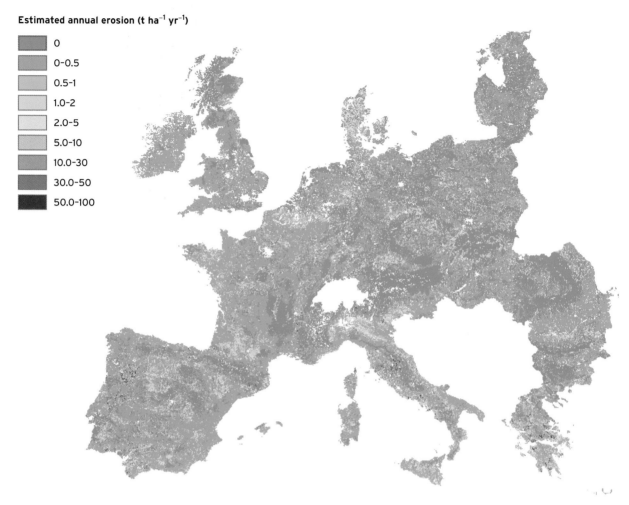

Figure 27.9 Soil erosion map for Europe showing estimated rates of erosion. (Source: courtesy of Brian Irvine and the EU PESERA Project)

these were called **environmental impact assessments (EIAs)**. They are vital tools in the arsenal of the manager and are legally required in many countries depending on the characteristics of a proposed action (Petts, 1999). However, the focus on the physical environment to the exclusion of other elements in early forms of EIAs and related legislation has led to other forms of impact assessment such as **environmental technology assessment** (a new technology is assessed for its potential environmental, economic and social impacts), **social impact assessment** and **health impact assessment (HIA)**. Since the manager must seek to identify, quantify and aggregate the impacts regardless of the 'arena' in which they occur, it is clear that a scoping study is required to identify the type of impact assessment required.

All forms of impact assessment, environmental, social, strategic and health, have a generic similarity. They all have screening and scoping elements to determine whether the assessment is required in the first place and to define the boundaries of the investigation. They should all be prospective (occur before the proposed management activity) in nature and are therefore a form of forecasting tool. They are relatively simple and are driven by checklists and matrices to try to ensure that all elements are considered and that the significance of the impact in terms of scale, intensity, cost, irreversibility, longevity and so on are addressed. The key common stages are shown in Figure 27.10. Screening assesses whether an EIA is required and, if so, at which scale (short/outline or full). Scoping involves defining the boundaries of the investigation in spatial, regulatory and ramification terms. The third stage requires the definition of the type of impact assessment (strategic, social, environmental, health, etc.) while the fourth step is the implementation of the EIA. Implementation is usually promoted through some form of checklist which prompts the analyst to consider each possible impact element. For example, a prompt could concern the nature of any pollutant: liquid, gas, particulates and so on. The fifth step is the reporting

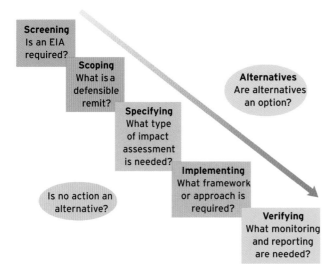

Figure 27.10 The main stages in an environmental impact assessment.

and monitoring stage. In total all these form a series of steps that provide a consistent framework for making environmental management decisions. It allows results between sites to be compared and for decisions to be made in a clear way so that other people can see how a given decision has been arrived at.

HIAs are a newer form of EIA and feature more prominently on the public agenda than normal EIAs. They have gained wide acceptance by governments and are a combination of procedures, methods and tools by which a policy, programme or project may be judged as to its potential effects on the health of a population, and the distribution of those effects within the population (World Health Organization, 1999). These methods have global endorsement. Policies to address environmental change at anything beyond the local scale are almost, by default, major policies and so are likely to require an HIA. However, 'health' in this context is not limited to the absence of disease but includes the presence of well-being: mental, physical and social. The HIA has applied impact analysis to a broadly interpreted health agenda in a process that incorporates the principles of sustainable development. Because the HIA has only recently been developed it incorporates a more explicit reflection of sustainable development principles. Many economic, social and environmental (as well as inherent biological) factors influence the well-being of individuals and communities. For example, personal circumstances, lifestyles and the social and physical environment influence health (e.g. education, income, employment, behaviour, culture, social support networks, community participation, air quality, housing, crime, civic design and transport). In addition HIAs tend to have a moral and ethical dimension with an

aim of trying to make sure that everyone can enjoy the same health quality of life.

27.3.3 Life costing

Putting impact assessment and forecasting together allows the development of **life-cycle analysis (LCA)** and raises an interesting question about the perceptions of environmental managers. Do managers have the vision to see all the ramifications of an action or do they tend to see only the immediate consequences and limited future effects irrespective of the perception they may hold? In the light of many studies which identify the shortcomings of a manager's vision, LCA has become an important tool in allowing a full 'cradle to grave' assessment of management actions or intervention. LCA is the examination of everything that happens in the manufacture, use and disposal of a product, from the time the raw materials are taken from the Earth to the time the product is thrown away and added to the ecosystem (Figure 27.11). The basic idea of LCA is to identify and evaluate all the environmental impacts of a given product (Ciambrone, 1997).

This is similar to producing an ecological footprint, which is where a calculation is performed as to what area of biosphere is required to sustain an individual, a company/organization or a country (see Box 20.5 in Chapter 20). The challenge of LCA, however, lies in the word 'evaluate'. We are aware that many natural systems have a capacity to absorb impacts. What is needed is a measure of this capacity. Thus we need detailed physical geography science to be carried out to provide us with such information. Sometimes we can use, as in the case of the ecological footprint, a measure of land area to indicate the capacity of a system. We can also use other indices such as the amount of carbon used. Most of the time, however, we are using multiple resources (fossil fuels, nutrients, etc.) and can create substances that may last for thousands of years without being recycled into the ecosystem (e.g. nuclear waste). Environmental managers, like managers in all other sectors, are seldom free to do what they like. The actions that they take need resources, and these need to be justified. Furthermore, the boundaries within which managers can exercise choice must be agreed. Since all actions take place from within a limited budget and since money spent on the environment will inevitably divert resources away from other deserving causes, it is often helpful to express impact and effects in monetary values to promote communication with decision-makers. This has expanded to a major discipline in its own right, known as **environmental economics**. When we merge LCA with the financial expression of consequences the resulting approach is called **whole-life costing**.

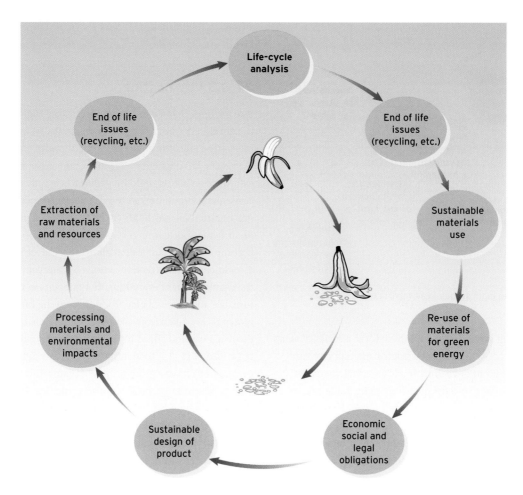

Figure 27.11 Life-cycle analysis.

Reflective questions

➤ What is an environmental impact assessment?

➤ What is life-cycle analysis?

27.4 Implementation

Without effective implementation, environmental science and management approaches discussed above are futile. There are two key elements to implementation: (i) building a consensus and (ii) managing the project. The following section deals with these two elements.

27.4.1 Stakeholder involvement

Building a consensus for action is vital but is usually time consuming. It requires that everyone with an interest is consulted or feels consulted. The identification of precisely 'everyone with an interest' is not simple. Each discipline has its own perspective and, certainly in the past, engineers would tend to consult engineers and politicians would negotiate with other politicians. Today we seek to identify '**stakeholders**' through a process called stakeholder analysis. Stakeholder analysis is simply a sequence of brainstorming events at which a group of project managers seek to identify everyone with a reasonable interest in a project. It is important to appreciate that this goes far beyond simply the list of statutory bodies that a politician might first identify and beyond the owners, businesses, consumers and liability holders that a lawyer might identify. Box 27.4 provides an example of stakeholder analysis.

Making an adequate definition of stakeholders is the foundation of consensus building. Agreement cannot arise from people who are not part of the agreement process. This may seem self-evident but the Chinese Government, in November 2003, paid compensation to contractors who were unable to proceed with the development of the Hong Kong harbour. This was after the courts decided that protestors who disagreed with the proposed development

TECHNIQUES

STAKEHOLDER ANALYSIS

The following example is hypothetical and illustrates the main themes that must be thought about in terms of stakeholder analysis. The Columbia River has its headwaters in the Rocky Mountains of southern Canada in the province of British Columbia. It then flows into the United States (Washington State). There are several dams on the US part of the river. However, it is proposed that further developments are to take place on the Canadian section of the Columbia River where a large tributary is to be dammed for hydroelectric power (Figure 27.12). It is therefore necessary to draw up a comprehensive list of stakeholders for this situation. We would start by listing the different groups that might have an interest. It is usually better to attempt this systematically.

Often it is good to start at the largest scale and work down and thus we might identify governments, regional administrations, businesses, pressure groups, native first peoples and local communities. Note that each of these can be subdivided. For example, governments subdivide into US Government and Canadian Government and the International Joint Commission. Pressure groups might subdivide into global interest groups such as Greenpeace, Friends of the Earth, the Sierra Club and local pressure groups. (You might like to complete this process through additional web research.) The list then has to be tested to determine whether, for example, pressure groups that had an interest in some of the earlier British Columbia and Washington State (two more stakeholders) water developments would retain an

Figure 27.12 Map of the Columbia River, North America.

interest in further Columbia River developments.

When you have completed the list, look at Table 27.4 which lists some of the stakeholders involved with the nearby Vancouver Island power generation project at Nanaimo (a smaller project; see Figure 27.12). You may now feel that you need to add more stakeholders to your list. It is very important to be inclusive of stakeholders and to make sure they are adequately represented as many people become very sensitive about their involvement in decision-making. For example, the following extract is from a letter sent in May 1998 to the British Columbia Premier about membership of a local watershed council from members of the Federation of British Columbian Naturalists (FBCN):

> Your letter was addressed to Denis and June Wood, with no mention of the groups on whose behalf we speak. My constituency is naturalists who are organised under the umbrella of the Federation of B.C. Naturalists. The FBCN is comprised of over 50 naturalist clubs province-wide, representing over 5300 members. The Federation has provided a unified voice for naturalist clubs since 1969 and our input is valued and respected throughout British Columbia as reasoned and informed.

The FBCN resigned from the watershed council because it felt it was not being respected by the council as a full stakeholder, and this quote illustrates how people can become upset if they feel their organizations are not given sufficient recognition as a stakeholder in an environmental management decision. Without people being fully involved with project decisions it is then very difficult to reach a consensus and therefore to gain agreement about an environmental management decision.

BOX 27.4 ➤

Table 27.4 Some of the stakeholders contacted regarding the Vancouver Island generation project at Nanaimo in British Columbia, Canada

Government	Businesses and community groups
Municipal/regional government	Vancouver Island Real Estate Board
Islands Trust	Fire departments
Fire Department	Harmac Mill
City councillors	General hospitals
Recreation directors	E&N Railway Company
Tourism directors	Chambers of Commerce
Environmental service director	Community Futures Development Corporation of Central Island
Trails coordinator	Nanaimo Rotary
Liquid waste manager	Vancouver Island building trades
Planner manager	Nanaimo Port Authority
Public works officer	Tourism Association of Vancouver Island
Roads and traffic office	Nanaimo fisheries
Airport manager	
Mayor	
Economic development officer	
Provincial government	
Environmental assessment office	
Aboriginal relations advisor	
Ministry of Health Services	
Ministry of Health Planning	
Ministry of Sustainable Resource Management	
Ministry of Transportation	
Ministry of Water, Land and Air Protection	
Federal government	
Mortgage and Housing Corporation	
Department of Fisheries and Oceans	
Environment Canada	
Royal Canadian Mounted Police	

BOX 27.4

Table 27.5 Levels of governance and styles of participation

Style of management	Role of stakeholder	Form of interaction
Facilitative	Initiator	Interactive
Co-operative	Co-operating partner	
Delegating	Co-decision-maker	
Participating	Adviser	
Consultative	Consultant	Non-interactive
Open authoritative	Information receiver	
Closed authoritative	None	

(Source: from van Ast and Boot, 2003)

had not had an opportunity to register their disagreement and had not been able to engage in a proper debate about the proposals with those in favour of the scheme.

However, even if you do manage to identify all of the stakeholders it is very unlikely that all the stakeholders will agree with a single inflexible proposal. Indeed such an approach is almost inevitably going to cause opposition. If, as an environmental manager, this 'single solution' approach is adopted it will need to be resolved through the legal system or through some national or international arbitration process. Of course any such approach is costly and may have unfortunate consequences:

- For example, a project such as a runway development, abandoned for environmental considerations, may reduce employment potential. The same project permitted to go ahead will result in claims of loss of property value.

- Future opposition to similar projects is likely to be stronger and more organized. This will occur because opponents will have learned key lessons and will be strengthened by their past success or hardened by their previous failures.

- The legal process is costly.

- Arbitration and legal processes are lengthy. Such processes can result in problems where local development opportunities are lost because people do not know which way a region, area or project will develop.

- International disagreement can result in stalemate. Within a country there is likely to be a mechanism for dispute resolution. In the international arena, however, there are few such mechanisms and, even if apparent agreement is reached, one side can simply decide to ignore the agreement, which is effectively how the United States views the Kyoto Carbon Treaty.

Therefore the process of offering a single option destroys rather than builds on the prospects of future consensus. In place of such an approach there should be an intelligent negotiation of alternatives. In reality, however, these two approaches (offering a single solution and offering a choice of options) are simply two points on a spectrum of types of participation in decision-making identified by van Ast and Boot (2003). The full spectrum is given in Table 27.5 which shows how the style of management and role of stakeholder may interact. The two approaches discussed above correspond to 'open authoritative' and 'participating/delegating' categories within the table.

Alternative dispute resolution is a process that avoids any winners and losers. It does this by presenting a range of choices and uses the stakeholder groups to participate in the analysis of which options are viable and which are not. The process of encouraging all stakeholders to take an active part in the decision process is called **participatory analysis** while the identification of key criteria through which to evaluate options is called **options analysis**.

27.4.2 Project management

The second part of implementation involves project management with an extended period of monitoring. Project management is the delivery of an agreed objective to an agreed timescale. It follows then that there are two requirements that are fundamental to effective project management: (i) a clearly specified, agreed, feasible objective and (ii) the resources to reach that objective. An example is provided in Box 27.5. Project management requires the subdivision of the project into a series of discrete tasks. Each task will have a clear specification, resources allocated, a time duration to complete the task and a person or agency responsible for the

CASE STUDIES

LOCAL KNOWLEDGE FOR PROJECT MANAGEMENT AT THE NATIONAL SCALE: THE CASE OF FLOOD DEFENCE

Floods cause damage and threaten lives and livelihoods and so need to be managed. Flooding is a wide-spread national problem which needs national policy. This case study deals with the management of flooding in England.

- 8% of England is at risk of river flooding.

- 1.7 million dwellings are at risk.

- 130 000 commercial premises are at risk.

- 4.5 million people are at risk.

- 12% agricultural land (river and sea floods) is at risk.

In summer 2007 there was major flooding across many parts of England. For example, Figure 27.13 shows flood locations in the Yorkshire region in northern England. This lead to a review of flood project management. Flood management requires considerable *local* knowledge and one of the lessons of the Pitt Report on the management of flooding in England and Wales (Pitt, 2008) is that a more planned local focus is required in which the integration with adjacent flood management agencies

is effective. To achieve this a new set of strategic flood defence partnerships has been established in which all the relevant local authorities are represented. In addition the local water companies are represented. In parallel to these assemblies of local knowledge, the Environment Agency Regional Flood Defence Committees and Environment Agency staff provide the planning and professional flood management needed. These committees have an enlarged remit under the Pitt proposals and are now called Flood and Coastal Management Committees. Two important changes have occurred since 2007: (i) the explicit addition of the 'coast' and (ii) the replacement of the word

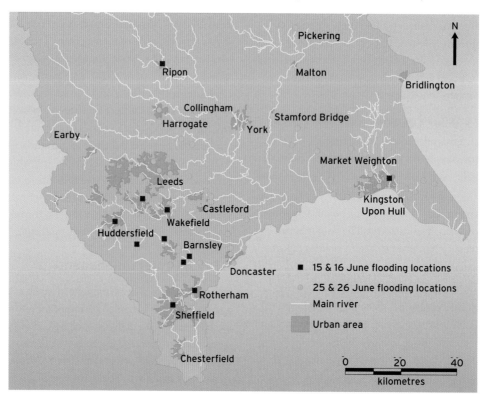

Figure 27.13 Flooding in the Yorkshire region of northern England during summer 2007. (Source: Environment Agency, 2007)

BOX 27.5 ➤

➤ 'defence' with the word 'management'.

The coast, water companies and local authorities have been added to flood defence committees because there are four different types of flood: (i) pluvial (rainfall flood without river flooding), (ii) fluvial (river over-bank), (iii) coastal and (iv) sewage and so all the agencies with primary responsibility have to be gathered into the same project management fora in order to ensure that decision-making takes account of the various factors involved and so that the responsible bodies fully communicate with each other to manage flood defence projects. Hence local knowledge, co-operation and investment takes place as part of the national flood defence project.

There are two funding sources for flood management investment in England. The first is from central government and is called Grant in Aid. The second is locally derived and is called Local Levy. Use of Local Levy is more flexible and restriction free while Grant in Aid has to follow specific allocation metrics. The allocation and use of these funds is carefully tracked. Sometimes, even after extensive deliberations and agreements, the actual implementation can be delayed. Since this is, in effect, not protecting a community in a timely manner, the flood protection managers 'allow an element of over-programming', which simply means they embark on more projects than they can afford because experience has shown that a proportion will always be delayed. This tracking is shown in Figure 27.14 for Yorkshire Regional Flood Defence as a whole. Of course individual projects are also tracked.

A new set of six 'outcome measures' have been developed in order to determine which projects should receive investment and to facilitate the management of projects:

- the whole-life cost-benefit ratio;
- numbers of households at less flood risk;
- numbers of households at less coastal erosion risk;
- extent of habitat improvement;
- proportion receiving flood warning direct service;
- proportion of planning decisions reflecting Environment Agency flood advice.

These can be found in more detail and with calculator tools at **www. environment-agency.gov.uk/ research/planning/122070.aspx**. The tools and outcome measures are designed so that allocations are fair across the country.

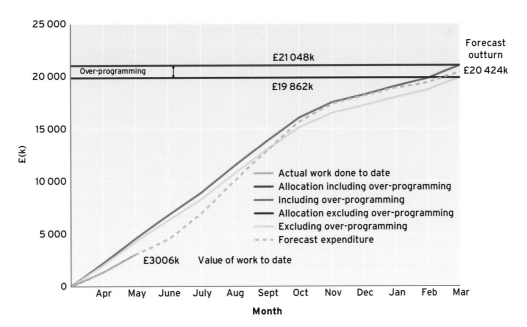

Figure 27.14 Flood defence project cost tracking for Yorkshire in northern England for 2011. The symbol k means £1000, so a value of £3000k is equivalent to £3 million. (Source: Environment Agency, 2011)

BOX 27.5

delivery. There must also be an identification of the relationships between tasks because in some cases one task may not be completed without another task being completed. For example, a structure such as a weir could not be added to a river without permission from the planning agency and that permission might not be forthcoming without a formal EIA. Not all tasks must be performed in sequence and many might be performed at the same time. However, in any large project a network of tasks is necessary and the track through the network which has the highest cumulative duration is the critical path. This is the path that will take the longest to complete as it relies on one thing being completed before the next can be taken forward. A delay in any of the tasks on this track will delay the whole project and conversely a saving on this track will speed up the whole project. Larger projects also need definitions of milestones and deliverables. In effect these are simply key stages at which major steps are realized and at which identifiable items can be 'delivered' to the end user or next developer.

For the management of environmental change, however, the project management as outlined above is too simple or at least too clearly defined. Any project that involves managing environmental change will have a large element of uncertainty and so there will be a need to monitor outcomes to ensure that the anticipated outcomes are being realized.

However, there are usually too many outcomes to report effectively to a monitoring group, particularly as many members of that group will not be familiar with the science or the terminology.

For example, we may want to restore urban air quality that has been degraded by pollution. Urban air quality will be measured in several ways such as particulate concentrations, low-level ozone, sulphur dioxide concentrations, visibility and so on, and each of these will be related to one or more sources of pollution. For each pollution source there may be several options for solution or reduction, and the implementation of these options would constitute a project. Some sources, causes and solution options are given in Table 27.6. The environmental manager must reflect on the solution options, such as declaring a smoke-free zone, or incentivize use of renewable energy such as wind power or solar panels (Figure 27.15) and consider the best way forward. Nevertheless, the question still remains as to how a non-technical project management or oversight group can measure progress. At the simplest level an 'improving' trend is the clearest signal of progress. However, this would ignore the rate of improvement unless an annual target had been set or a satisfactory concentration had been defined. In addition, a large number of 'trends' with no recognition of differences in the significance or

Table 27.6 Options to limit urban air quality degradation. SO_x and NO_x are forms of sulphur oxide and nitrous oxide respectively and PM_{10} is the concentration of particles that are less than 10 μm in diameter

Measure of degradation	Cause of degradation	Solution option
SO_x	Domestic emissions	Smoke-free zone
		Punitive taxation on domestic fuel
		Home insulation support
		Combined heat and power facilities
NO_x	Transport emissions	Congestion charging
		Odd- and even-day access
		Parking restrictions
		Subsidized public transport
		Pedestrianization
		Zero-emission vehicles
PM_{10}	Industrial emissions	More aggressive regulation
		Changed process
		Resiting
		Incentivize use of renewable energy

(a)

(b)

Figure 27.15 Renewable energy generation: (a) wind power and (b) a building clad in solar panels.

meaning of the trends will make the derivation of information from data difficult.

Therefore, oversight groups need indicators. These may be selected key indicators or may be aggregate indicators. The indicators may be 'normalized' so that all measures are within similar numeric values or they may be transformed to word format (e.g. improving, satisfactory, deteriorating, unsatisfactory, unchanged) or symbols (e.g. tick/cross, traffic light symbol) to improve ease of communication to a diverse group.

Reflective questions

➤ What are the main issues associated with gaining agreement on environmental management?

➤ What are the main general problems associated with project management?

27.5 Summary

Environmental management is about the management of change. The objective may be to stop a detrimental change or to encourage a responsible change. However, change is not simple and predictable, yet it is vital to be able to forecast change. Environmental managers therefore rely on process understanding provided by physical geographers to offer scenarios of the possible changes that might take place and the uncertainty that is associated with each scenario. Change may be linear or non-linear, reversible or irreversible, and hysteretic. Slow change may cause a system eventually to cross a threshold and then dramatically alter (e.g. river channels suddenly switching their course) and such changes may be illustrated by catastrophe theory.

Environmental impact assessments are important environmental management tools and allow us to determine the relative merits of a variety of management options. It is often possible to estimate the financial advantages and disadvantages of management options but most of the time such schemes do not incorporate full life-cycle analysis or costing of the wider effects of a given management activity or technology.

The money to pay for environmental management will be provided only if politicians are willing to provide such finance. The political stance has to reflect the consensus for support for a given action and such consensus can only be found through all stakeholders finding common ground. Implementation of the actions that are agreed requires monitoring. The monitoring results eventually need to be reported to the non-technical (e.g. the non-scientist) groups that are paying for the management or that are stakeholders within the project. Therefore the results must often be expressed as a series of non-technical indicators.

Further reading
Books

Asante-Duah, D.K. (1998) *Risk assessment in environmental management.* John Wiley & Sons, Chichester.
This book describes the nature of contaminant problems, and demonstrates how you can calculate risk in different ways. There is occasionally some complex theory and a little maths.

Barrow, C.J. (1997) *Environmental and social impact assessment: An introduction.* Arnold, London.
This book describes the development, role, processes and methods of EIAs. There are good boxes with case studies.

Barrow, C.J. (2006) *Environmental management for sustainable development.* Routledge, London.
This is a clear text which also covers law, standards and economics in addition to the topics covered in this chapter.

Erickson, S.L. and King, B.J. (1999) *Fundamentals of environmental management.* John Wiley & Sons, Chichester.
This book relates a lot of its discussion to regulations in North America. There is a very useful appendix which contains checklists for environmental management audit exercises and a good glossary.

Nath, B., Hens, L., Compton, P. and Devuyt, D. (eds) (1998) *Environmental management in practice.* Routledge, London.
This book has three volumes and is very detailed. Volume 2 is good for information on the management of physical components such as soils, water quality, transport, sustainable agriculture, tourism, fisheries, and so on, whereas volume 3 deals with ecosystems of the world.

Owen, L. and Unwin, T. (eds) (1997) *Environmental management: Readings and case studies.* Blackwell, Cambridge, MA.
This book is aimed at undergraduate level and is a compilation of important papers and chapters by different authors. It is an excellent resource for case studies which could be used in essays and exams.

Further reading
Papers

McDonald, A.T., Lane, S.N., Haycock, N.E. and Chalk, E.A. (2004) Rivers of dreams: on the gulf between theoretical and practical aspects of an upland river restoration. *Transactions of the Institute of British Geographers*, 29, 257–281.
A good case study of stakeholder engagement and project management issues.

A

Ablation The action of removing and carrying away a superficial material; it primarily relates to the *sublimation*, melting and evaporation which remove snow and ice from the surface of a glacier or snowfield. It can also be applied to the wearing away of rock by water and the removal of salt or sand from a surface by the action of wind.

Ablation zone The region of a glacier that experiences net loss of mass throughout the year (i.e. ablation exceeds accumulation).

Absolute zero The temperature at which molecules have no internal energy and are at a complete rest. This is equivalent to zero kelvin or −275.15 degrees Celsius.

Absorption bands Sections of the electromagnetic spectrum that interfere with the passage of radiation to the Earth's surface.

Abyssal plain The flat deep-sea floor extending seawards from the base of the continental slope and continental rise, reaching depths of 4–6 km between mid-ocean ridges and trenches.

Accumulation zone The region of a glacier that experiences net gain of mass throughout the year (i.e. accumulation of snow, *firn* and ice).

Across-track sensors Scanning instruments that collect data by directing the line of sight in a sweeping motion at right angles to the direction of travel by a rotating or oscillating mirror. These are also called whiskbroom scanners.

Active layer The thin top layer of the ground surface that is seasonally frozen and unfrozen above *permafrost*.

Adiabatic The expansion of a body of air without loss or gain of heat.

Adsorbed When a specific gas, liquid or substance in solution adheres to the exposed surface of a material with which it is in contact, usually a solid, e.g. the adsorption of anions to the surface of clay minerals by electrostatic attractions.

Adsorption Attachment of a substance in solution to a solid.

Advection The horizontal circulation of an ocean body or air mass.

Advective solution The process in which solutes are removed and transported through the soil in a flow of water (e.g. runoff and percolating water); it is very effective at removing solutes from near the surface; similar to leaching.

Aeolian Pertaining to the processes, Earth materials and landforms that involve the role of wind.

Aeration The ventilation of soil.

Aerosols Minute particles suspended in the atmosphere that interact with the Earth's radiation budget and climate.

Aggradation The raising of a surface caused by the accumulation of material deposited by various geomorphological agents (e.g. wind, water or wave).

Aggregate A grouping of soil particles adhered together and separated from surrounding aggregates by voids in the soil. Also known as *ped*.

Agronomy Subject of utilizing plant processes for crop growth.

Air mass An extensive body of air possessing relatively uniform conditions of temperature and moisture that is in contact with the ground.

Alas Small irregularly shaped lakes and depressions caused by the melting of massive ground ice.

Albedo The proportion of radiation reflected from a surface. Surfaces such as snow have a high albedo.

Alkaline A substance which has properties of an alkali, a pH greater than 7.

Alkalinity The capacity of water to neutralize acid, determined by the quantity of base *cations* (Na^+, K^+, Mg^{2+}, Ca^{2+}) in the substance. Measured by titration of the sample with a strong mineral acid.

Allelopathy A biological occurrence whereby one plant species can limit the growth of another through the release of inhibiting metabolic chemicals into the local environment.

Allochthonous A material or resource which is not indigenous to the location in which is it found and has therefore travelled to reach the said location.

Allogenic Pertaining to a change in system dynamics caused by the influence of an external environmental factor, i.e. in relation to river channel adjustment, an allogenic change involves a move away from equilibrium conditions in response to an alteration in the sediment and water regime of the river.

Alluvial fan A fan-shaped landform composed of alluvium, deposited where a tributary stream loses momentum on entering a more gently sloping valley.

Along-track sensors Scanning devices that collect data using a linear array of instruments oriented perpendicularly to the direction of travel covering one side of the *swath* to the other. These are also called pushbroom scanners.

Alpine permafrost Permafrost that occurs locally owing to low temperatures at high altitudes, e.g. the Rocky Mountains.

Alternative dispute resolution The process of settling an environmental dispute through mediation and arbitration (avoiding litigation); it avoids strict winners and losers by presenting a range of choices and encouraging *participatory analysis*.

Amictic Amictic comes from amixis, which means a lack of mixing. Amictic lakes are those which do not circulate, usually because they are covered in ice for the majority of the year and therefore wind is not able to mix the water; they are predominantly found in high-latitude and high-altitude locations.

Amphidrome Points in the oceans where there is zero tidal range due to cancelling out of tides. Tides radiate out from amphidromes.

Anaerobic Functioning in the absence of oxygen.

Anaerobic bacteria Microorganisms that survive in environments containing no free or dissolved oxygen; they obtain oxygen through the decomposition of chemical compounds, such as nitrates.

Analogue image An image composed of continuous tone.

Anastomosing Pertaining to the tendency for certain rivers to divide and reunite, producing a complex pattern of channels with large, stable islands between the channels.

Angle of repose The maximum slope gradient at which unconsolidated material will remain stable without collapse.

Anions Negatively charged *ions*, i.e. an atom which has gained one or more negatively charged electrons, e.g. the chloride ion (Cl⁻).

Anoxic Depleted of oxygen; in water usually a result of bacterial oxygen consumption and other respiration in areas of restricted circulation.

Antarctic Bottom Water (ABW) A body of water formed along the edge of the Antarctic continent. Very dense water created by the very cold, saline conditions is forced to sink and flow north underneath the *North Atlantic Deep Water (NADW)*; together they power the *thermohaline circulation* of the world's oceans.

Anticyclogenesis A condition in which a zone of descending air results in high pressure at ground level and air to circulate slowly outwards from the descending zone. This results in anticyclonic conditions.

Anti-dune A type of small-scale cross-bedding feature formed from a sand deposit on a river bed. It develops from a 'normal' dune when the flow velocity increases in a highly loaded river; erosion from the downstream slope throws material into saltation and suspension more easily than it can be replenished from upstream, causing upstream migration of the bedform feature.

Aquifer A layer of rock with sufficient porosity to absorb and store water and permeable enough to allow water to pass freely through as groundwater.

Aquitard An *aquifer* which has been confined between impermeable rock layers and only open for recharge and discharge at certain locations.

Archimedes' principle Any object wholly or partially immersed in a fluid will experience a buoyant force (or upthrust) equal to the weight of the fluid displaced.

Arête A steep knife-edge ridge that divides the steep walls of two adjacent cirques in a mountainous region.

Argillic A term to describe a soil horizon characterized by clay accumulation.

Aridity A state of lacking in moisture, when evapotranspiration exceeds precipitation. It can be defined by the annual overall net negative moisture balance of a particular environment.

Armoured layer The coarser stoned surface layer of a mixed gravel-bed river, protecting the finer material beneath.

Artificial neural network (ANN) A type of parallel computing in which memory is distributed across a number of smaller processing units that process information in a parallel manner.

Aspiration The act of drawing air.

Asthenosphere The ductile layer of the Earth's mantle located 100–400 km below the surface, on which the rigid lithospheric plates glide.

Atmometer An instrument for taking direct measurements of evaporation; by connecting a water supply to a porous surface the amount of evaporation over a given time is measured by the change in water stored.

Atolls Coral reefs that surround a central lagoon; most are found in the Indian and Pacific Oceans.

Autochthonous A material or resource which is indigenous to the location in which it is found.

Autogenic Pertaining to a change in system dynamics caused by the influence of an internal, self-produced factor, i.e. in relation to river channel adjustments, an autogenic change involves a fluctuation about an equilibrium condition.

Autotrophs Those life forms that acquire their energy from the Sun via the process of photosynthesis. They form the first *trophic level* by creating a source of energy for other animals, birds and insects.

Available water The water available in soil for plant growth after excess water has drained owing to gravitational forces, i.e. the water retained between the states of *field capacity* and *wilting point*.

Avulsion The process whereby a channel shifts, abandoning its old course for a new course, and leaving an intervening area of floodplain intact.

B

Back-scar The upslope section of the wall from which a landslide has occurred, creating a scar.

Backscatter The return signal of a radiation pulse from an active remote sensor. This is also called the echo.

Backwash The seaward return pulse of water from a breaking wave along the shoreline, moving under the force of gravity.

Badlands A deeply eroded barren landscape characterized by very irregular topography with ridges, peaks and mesas resulting from wind and water erosion of sedimentary rock. Badlands originally referred to the heavily eroded arid region of south-west South Dakota and north-west Nebraska in the United States but is now a more generic term.

Balance velocity The flow velocity required by a glacier to maintain the ice in equilibrium; the mass transferred down the glacier should equal that lost by melting in the ablation zone.

Bank-full Condition when the river channel is full of water.

Bank-full discharge The level of discharge at which any more water would cause the river to spill out of the channel onto adjacent low-lying land.

Barchan dunes Isolated crescentic sand dunes with a shallow windward and a steep lee side whose horns point in the direction of dune movement (usually forming under conditions of limited sand supply).

Barrier beaches Elongated offshore banks of coarse granular debris (sand, gravel) lying parallel to the coastline that are not submerged by the tide (see *barriers*).

Barrier islands Elongated offshore islands of coarse granular material, lying parallel to the coastline, similar to barrier beaches but larger in scale and forming behind barrier beaches (see *barriers*).

Barriers Elements of a beach planform located just offshore that involve the accumulation of landward-migrating sand shoals running parallel to the coastline that achieve surface elevation as they roll inland (see *barrier beaches* and *barrier islands*).

Bars Ridges of coarse sediment deposited on a stream bed where the stream velocity drops, especially mid-stream and on the inside of meanders.

Base Pertaining to substances with a pH above 7 (notably calcium, magnesium, potassium and sodium) or substances that release hydroxide ions (OH^-).

Base saturation The percentage of base cations that make up the total exchangeable cations in soil.

Base station A stationary *global positioning system* receiver positioned over a known point that continuously collects data from satellites used to correct the recorded positions of the roving global positioning system receivers.

Baseflow The stable portion of a river's discharge, contributed by *groundwater* transfers.

Batholith A large irregular shaped mass of *igneous rock*, often granite; it is formed through the intrusion of magma into the strata at depth, where it melts the strata. The igneous rock is then exposed when the less resistant overlying rock is eroded away over time.

Bathymetry The study and mapping of ocean floor topography.

Bay-head delta A delta at the head of an estuary or a bay into which a river discharges. They typically occur if the river carries large amounts of sediment, or where the coastline is being submerged.

Beach cusps Crescent-shaped accumulations of sand or shingle surrounding a depression on a beach; they are always found in combination and formed when outgoing *rip currents* and incoming waves combine to create circular water movements.

Beach nourishment A *soft engineering* technique implemented to reduce the impact of beach erosion. The method involves adding sediment to the beach to maintain the beach profile.

Beamwidth The width of a radar pulse in the direction of travel. A narrow beamwidth means higher resolution.

Bed load Sediment grains transported in water by rolling along the bed surface or through *saltation*.

Bed return flow The average return flow of water offshore near to the bed, after it has been brought onshore in the upper water column. Large amounts of beach material can be removed from the beach in the bed return flow.

Bedform A morphological feature developed by fluid flow across the surface of soft sediment, involving the entrainment or deposition of sediment.

Benthic Pertaining to organisms dwelling on the river or sea floor.

Benthos Benthic organisms (see *Benthic*).

Bergeron process The formation of precipitation described by the Bergeron–Findeison theory. Ice crystals fall from the upper part of a cloud, leading to aggregation of crystals and accretion of supercooled water. Ice crystals grow preferentially by *sublimation* at the expense of surrounding water droplets because the relative humidity above an ice surface is greater than a liquid surface and hence the saturation vapour pressure over water is greater than ice, causing a pressure gradient towards the ice.

Berms Prominent ridges at the back of a beach with a steep seaward face and flat top, marking the limit of the swash zone.

Best Management Practices (BMPs) Methods of minimizing diffuse pollution. BMPs consist of two types: structural, e.g. wetlands, and procedural, e.g. handling methods for polluting chemicals.

Bifurcation point Bifurcation theory is a branch of chaos theory, and the bifurcation point refers to an occurrence within this theory. The bifurcation point is the point of change in a non-linear system where there is a branching off into different paths, which can lead to differing outcomes. Along each branch of the system new influences will determine the eventual outcome which may result in feedback loops or descent into chaos. The bifurcation point marks a sudden change as opposed to a slow one which would allow for gradual evolution.

Bimodal distribution A statistical term signifying that the frequency curve of a distribution of data has two maxima (two modal classes).

Binge-purge model A model of ice sheet development related to inherent instabilities of large ice sheets; ice loading due to the growth of an ice sheet increases basal pressure causing substrate failure and greater meltwater production, thereby increasing ice flow velocity and ice rafting. The release of this excess ice would then stabilize the ice sheet again. Proposed as a possible cause of *Heinrich events*.

Bioaccumulation The accumulation of toxins in specific parts of the ecosystem (usually at the higher levels of food chains) due to the greater ability of some chemicals to accumulate in zones where they become bioavailable, and are taken up and stored by producers and consumers. The materials must be stored in those parts of the individual that will be consumed. It may result in severe adverse effects on an ecosystem once a threshold level of toxin storage is reached.

Biochemical oxygen demand (BOD) A measure of the amount of biochemically degradable organic matter in water that is widely used in water pollution assessments.

Bioconcentration The level of concentration of accumulated toxins found in plant or animal tissues (via the process of *bioaccumulation*) compared with background natural levels.

Biodiversity The number and variety of taxonomic groups (usually species) of plants and animals at a site or within a region.

Biogenous Pertaining to material derived from organisms.

Biogeochemical Pertaining to the chemical relationships between the geology of an area and its plant and animal life.

Biogeochemical cycles These cycles are the pathways though which the elements necessary for life travel in the biotic and abiotic environment. Along these pathways there are changes in the fluxes of the element depending on the residence time in varying reservoirs and the transformations of the elements which occur.

Bioherm An ancient mass of rock formed by sedentary organisms, such as corals.

Biological sediments Sediments derived from organic materials, either remains of dead organisms (e.g. shells, plants) or framework organisms (e.g. coral reefs).

Biomass The total dry weight of living organic matter, usually measured per unit area over a particular time interval. Tends to include dead parts of organisms when referring to soils.

Biome A coarse unit of ecosystem classification based on what the land cover and ecosystems look like.

Bioregional theory A concept first put forward by Brunckhorst (1995) as a framework for regional planning based on a firm understanding of the sustainable and interconnected elements of both the geophysical and the cultural indicators that unify an *ecosystem*, therefore valuing both the natural *landscape ecology* and the conservation values developed from local cultural beliefs.

Biosphere All the organisms on the planet, viewed as a system of interacting components making a thin film on the planet's surface, and including parts of the atmosphere, hydrosphere and *lithosphere*.

Biostratigraphic Pertaining to the division of sedimentary deposits based upon their fossil evidence, each biostratigraphic unit having a distinctive fossil assemblage, e.g. the use of fossilized pollen assemblages for studying European *interglacial intervals*.

Bit scale Refers to the number of colours used to quantify brightness values in a digital image.

Black body An ideal body or surface that absorbs and emits all radiant energy dependent on its absolute temperature.

Black smokers Hydrothermal springs lying along the rift valley of *mid-ocean ridges*. Seawater that seeps into fissures in the basaltic lava becomes superheated and chemically interacts with the basaltic rock to create a black precipitate of metal sulphides.

Blockfield A continuous spread of angular rock fragments across a high mountain or plateau in a periglacial environment; formed *in situ* by frost shattering (occasionally transported and deposited by saturated material in *gelifluction*).

Bond cycles A grouping of *Dansgaard-Oeschger (D-O) events* together into a larger cycle with a long cooling trend followed by an abrupt warming.

Bottomset beds Horizontally layered sediment beds deposited in front of a delta as it prograde seawards. They become covered and end up at the bottom of a stack of deltaic sediments (below the *foreset beds* and *topset beds*).

Boudins Bands of connected debris-rich ice lenses in a glacier, once connected but broken under pressure (forming a sausage shape; *boudin* is French for sausage).

Boulder-controlled slope A scree slope at the base of a cliff in which the scree material is removed as quickly as new material is added to it by rockfall creating a thin covering maintained at the angle of repose. The landform (cliff and boulder-controlled slope) retreats at an almost constant ratio.

Boulton-Menzies theory A theory of *drumlin* formation suggesting drumlins are formed by deposition in the lee of a slowly moving obstacle in the deforming layer of a glacier, therefore streamlining the deposition.

Boundary conditions The physical conditions at the boundaries of a system. They are particularly used in modelling work and for example would refer to the impermeable nature of the floor of an aquifer in the model.

Boundary roughness The roughness of the river channel bed and the submerged bank.

Braided channel A river channel consisting of separate, but interlinked, migrating channels flowing either side of active unvegetated bars that change position owing to bed load transport.

Breakwater A coastal management feature in which a submerged artificial barrier offshore acts to break incoming waves or create new diffraction patterns, protecting the shoreline from wave action in the process.

Bulk density The weight per unit volume of a solid particulate as it is normally packed (including solids and pore spaces). Usually expressed as lb/ft^3 or $g\ cm^{-3}$.

C

Calcareous ooze Fine-grained deep-ocean *biogenous* sediment containing at least 30% skeletal remains of marine organisms based on calcium carbonate ($CaCO_3$).

Calcrete A *duricrust* composed mainly of calcium carbonate.

Caldera A large, steep-sided, land surface depression containing volcanic vents. Formed by large-scale subsidence as the parent magma chamber cools and contracts following a major volcanic eruption.

Calibre The size of sediment particles.

Candle ice Ice consisting of vertically orientated crystals often over 1 m in length.

Capillarity Also known as capillary action. It is the process whereby a liquid can act against the force of gravity and flow upwards in a narrow tube such as those found in a porous material. It is dependent upon the surface tension of the liquid and the adhesion of the molecules.

Capillary water Water that remains in small pores in the soil against the forces of gravity; the major source of water available for plant uptake.

Carbon sequestration The uptake of carbon by a system. Carbon dioxide can be absorbed by plants from the atmosphere and then this is converted into solid plant material. The carbon is then part of the terrestrial system and has been 'taken up' from the atmosphere.

Carnivores Organisms (usually an animal) that consume meat and therefore occupy a high *trophic level* in the *ecosystem*.

Catastrophe theory A theory in which non-linear interactions within a system cause a threshold to be crossed which then leads to a sudden and dramatic change to a new stable model of operation. Before the threshold is crossed changes may be slow and barely noticeable.

Catchment An open system defined as the area of land drained by a particular stream or river; it represents a fundamental unit in hydrology and is usually topographically well defined. A catchment may be composed of a series of subcatchments.

Catena The sequence of soils occupying a slope transect from the topographical divide to the bottom of the adjacent valley that have developed from similar parent material but vary in profile characteristics owing to the differing topographical and drainage conditions under which they formed.

Cation exchange The process of interchange between a *cation* in soil solution and another on the surface of a soil colloid.

Cation exchange capacity (CEC) The overall net negative charge of clay minerals per unit mass of soil, usually expressed as milliequivalents (meq) per kg of oven-dried soil.

Cations Positively charged *ions*, e.g. the sodium ion (Na^+).

Causal inference The process in which a cause is linked to observations under the assumption that every event must have a cause. It is a key element in the scientific method.

Cavitation A process of fluvial erosion, characteristic of waterfalls and rapids. Constriction of channel flow raises the flow velocity, thereby reducing water pressure and leading to the formation of air bubbles. As the stream widens again and the velocity decreases the air bubbles collapse and the shock waves place considerable stress on the channel walls.

Channel planform The form of channels when viewed from above.

Channel sinuosity A measure of the degree of curvature in channels with meandering *channel planforms*.

Channelization The artificial modification of natural river channels for the purposes of flood alleviation, land drainage or relocation. It may involve channel widening, deepening, straightening, stabilizing (using concrete or piling) or embanking.

Chatter marks Microscale erosional features that appear as crescentic scars.

Chelates A stable compound formed between organic molecules and metallic cations in which more than one bond links the two components (also see *complex*). Chelates are especially important for the behaviour of aluminium and iron in the soil.

Chelation The process whereby *chelates* are formed.

Chemical mixing model A chemical mass balance model in which it is assumed that the concentration of an *ion* in solution consists of the mixture of concentrations and flows from different sources.

Chemical sediments Sediments produced by chemical processes, formed predominantly as a result of precipitation of minerals directly from a water body.

Chinook A warm, dry local wind that blows east down the lee slopes of the Rocky Mountains. The wind is subject to warming by adiabatic compression on descent and is warmer in absolute terms at any given altitude than on its windward ascent.

Chroma A measurable variable of soil colour describing the purity or strength of the colour (a chroma of 0 is natural grey).

Clastic sediments Sediments composed of grains of rock which have been weathered and eroded from a pre-existing bedrock material; they are dominated by those grains most resistant to weathering.

Clay A soil mineral particle within the *fine earth* fraction, having an upper limit of 2 μm (two-millionths of a metre) in diameter; very important in determining soil properties.

Clay skin A thin film of clay which has lined an area of soil.

Climax communities A community of plants and animals in steady-state equilibrium with prevailing conditions in the physical environment, seen as the self-perpetuating terminal stages of *ecological succession*.

Coefficient of friction The ratio of the frictional force between two surfaces sliding across one another to the normal force acting perpendicular to the surfaces; it depends primarily on the nature of both surfaces in contact.

Coefficient of thermal expansion When heated, most substances will expand and on cooling will contract. It is the response of the substance to this alteration in temperature which is the coefficient of thermal expansion.

Cohesion The force by which a homogeneous substance is held together owing to attraction between like molecules.

Cold glacier A glacier in which ice remains at very low temperatures, tens of degrees below freezing, with no appreciable surface melting. The absence of meltwater causes the glacier to remain largely frozen to the underlying substrate.

Coleoptera A large and important order of insects, distinguished by anterior wings converted into hard sheaths covering the other pair when not in use, i.e. beetles.

Collector-filterers These organisms exist in the water column and feed by filtering out fine particulate organic matter in the water column.

Collector-gatherers These organisms can be found in the slower flowing areas of a river, they are more likely to feed from the bottom of the water column as they wait until the fine particulate organic matter has fallen out of the water column due to the reduced energy in the water flow and then gather it up to feed on.

Colloid A substance in which very small particles (0.1-10 μm (millionths of a metre) in diameter) are held in a state midway between a solution and a suspension.

Colour palette Colour reference table for displaying the *digital number* of an image.

Community (in ecological terms) The total living biotic component of an ecosystem (plants, animals and microbes).

Competition (in ecological terms) Negative interaction between organisms caused by the need for a common resource such as light, water or nutrients.

Complex A compound formed between organic molecules and metallic cations by a single bond (also see *chelates*).

Compressive stress The action of a force pushing inwards to the centre of an object along its cross-sectional axis.

Concentration The mass of substance of interest per unit volume, for solutes normally expressed as mg L^{-1}.

Conditionally unstable Instability in the atmosphere that is conditional upon an air parcel becoming saturated, which leads to a shift from cooling via the dry adiabatic lapse rate to the saturated adiabatic lapse rate. This causes the air to become warmer than the surrounding air and ascend more rapidly, leading to strong upward convection.

Conduction Transfer of energy between two bodies in contact.

Congruent dissolution This is the process whereby a whole solid dissolves into its constituent elements and there is no secondary solid phase. An example of such a process is the dissolution of calcite into calcium, water and carbon dioxide.

Continental shelf The zone bordering a continent extending from the line of permanent immersion to the depth at which there is a marked increase in the downward slope which descends to the deep-ocean floor.

Continental shield The ancient, stable, low-relief interior of continents; composed primarily of Precambrian crystalline rocks, some as old as 2 to 3 billion years.

Continuous permafrost zone A region in which *permafrost* occurs everywhere in the ground surface except beneath large bodies of water or ice.

Contour scaling This often occurs on urban stone and is associated with salt crystallisation in subsurface layers of the stone. The result is a layer of stone up to several centimetres thick blistering and then completely falling away.

Control A single test performed within a larger set of experiments, whereby no variables are altered from the norm; it acts to monitor the quality of the experimentation and ensure that no unaccounted variables are influencing the results.

Convection Transfer of energy through a fluid (liquid or gas) by molecular motions.

Convergent plate boundary The boundary between two lithospheric plates in which one plate descends below the other, resulting in the consumption of *lithosphere* via the process of *subduction*.

Coping strategies Strategies used to reduce stress by mitigating the influences of an event/situation as opposed to attempting to change the situation.

Coriolis effect As a result of the Earth's rotation any moving object or fluid is deflected towards the right in the northern hemisphere and to the left in the southern hemisphere.

Corrasion The mechanical breakdown of rock due to wearing and grinding caused by material carried in transport across the rock surface.

Corries Basins excavated into a mountainside by the erosive power of a cirque glacier possessing defining features: steep retaining rock walls, a gently inclined rock basin, abundant signs of glacial scour and a terminal moraine. Also known as cirques.

Corrosion A weathering process involving the breakdown of solid rock by means of chemical reactions.

Cosmogenous Pertaining to material that originated extraterrestrially, e.g. meteor fragments and cosmic dust.

Coversands An extensive sand sheet (generally thin and lacking bedforms), formed by wind action in an unvegetated periglacial environment adjacent to an ice sheet. Similar to *loess* but coarser grained.

Crag and tail A streamlined ridge consisting of a resistant rock mass (the 'crag') with an elongated body of less resistant glacial till (the 'tail') on the lee side; a result of preferential erosion around a ridge below a glacier.

Crescentric gouges Crescentric fractures formed by irregular rolling of boulders carried at the base of a glacier; similar to *chatter marks* but larger in scale and less repetitive.

Crevasse-fill ridge A short, linear ridge of glaciofluvial material deposited as meltwater debris in a crevasse of a previous glacier (similar to an *esker* but shorter and less sinuous).

Critical rationalism A form of *positivism* involving the use of *deductive* reasoning. A theory is first adopted, leading to the formation of a hypothesis; this hypothesis is then tested in an attempt to falsify it.

Crusting The process in which a crust (a hard coating) is formed on the ground surface, caused either by concentration of minerals in surface layers due to high evaporative rates, or through high-intensity raindrop impact. It is most common in dryland environments.

Cryptozoic Pertaining to organisms which seek shelter as the preferred ecological niche.

Cultural landscape A landscape fashioned by human intervention, normally over a long period. Typically, a cultural landscape will be dominated by vegetation whose structure and species composition is different from what would be expected under pristine conditions. An obvious example would be farmland; a less obvious example might be a moorland created by felling a forest and maintained by grazing by sheep.

Curie point The temperature above which a metal is no longer attracted to a magnet.

Current ripples Small unidirectional ridges (up to 5 cm in height and 30 cm in wavelength) formed on a beach or on a sandy river bed by the motion of water.

Cyclogenesis A condition in which high-level air divergence is greater than low-level convergence such that air is able to rise. This results in low pressure and the development of cyclonic conditions.

D

Dansgaard–Oeschger (D–O) events *Interstadial* episodes during the Quaternary lasting no more than 500-2000 years involving an abrupt change in temperature of the order of 5-8°C as quickly as a few decades (gradual cooling followed by rapid warming). Large differences are also recorded in atmospheric dust content, ice accumulation rate, methane concentration and CO_2 concentrations. They are interpreted to be the response of internal feedback mechanisms.

Darcy's law A physical law to determine the flow of water through the matrix of a porous medium (e.g. groundwater through an aquifer). The relationship holds only for laminar (non-turbulent) flow of fluids in homogeneous porous media and does not hold well for well-jointed limestone with numerous fissures. According to Darcy's law the discharge of water through permeable material is equal to the cross-sectional area times the slope of the water surface times the coefficient of permeability.

Davisian cycles of erosion A theoretical sequence of 'weardown' processes and forms that occur between the initial uplift of a land mass and its erosion to a *peneplain* (first codified by William Davis in 1899). Its utility within real situations has often been challenged, particularly in relation to the effect of frequent global climatic changes that render any simple sequence of forms improbable.

de Geer moraine An arc-shaped ridge of glaciofluvial material (2-15 m in height) formed transverse to flow at the margin of a retreating glacier where the ice mass borders a glacial lake.

Debris avalanche A type of sudden and very rapid mass movement common on steep mountain slopes mobilized by gravity

and commonly originating from a rockslide. It is composed of an unsorted mass of rock and soil that disintegrates during movement into a range of fragment sizes. Movement is characterized by flowage in a dry or wet state; momentum is maintained by a bouncing layer of grain-to-grain collisions at the ground surface.

Debris flow A highly destructive mass wasting process involving a slurry-like flow composed of rock grains, sediment and water with a wide range of sediment size grades and little internal stratification on deposition.

Deductive Pertaining to the process of drawing a conclusion via observational and measurement testing of a principle/hypothesis that is already assumed, i.e. inference by reasoning from general laws to particular instances (opposed to *inductive*).

Degrees of freedom Pertaining to the capability of variation within a system. The number of degrees of freedom in a particular system refers to the number of independent variables that can be freely changed to bring the system to a new equilibrium without altering the phases of the system.

Delta A sedimentary landform where the mouth of a river reaches another water body such as an ocean or lake.

Delta front The limit of the accumulation zone of a delta; it forms from the settling of finer-grained sediment carried furthest into the basin by the decelerating currents.

Delta plain The flat surface of the delta over which the river channel migrates and channel sands accumulate.

Delta switching The process whereby the position of a river delta changes from one site to another. This can occur when the delta becomes so infilled with sediment that the river changes its course dramatically to allow the water to drain more freely into the ocean.

Dendrochronology The study of annual growth rings in certain tree species for dating of the recent past.

Dendroecology The study of tree rings to understand ecological processes.

Denitrification The process whereby nitrate is reduced to produce nitric oxide to then produce nitrous oxide to then produce nitrogen. It occurs in anaerobic conditions where the microbes use nitrate for respiration in place of oxygen during the decomposition process. The result of denitrification is the loss of nitrogen into the atmosphere.

Denudation chronology The process of attempting to determine the history of a landscape by establishing which stage of the *Davisian cycles of erosion* it represents.

Desert varnish A tough dark layer covering *scree slopes* found in arid areas, produced by weathering of the interior of the scree slope boulders.

Desertification The spread of desert-like conditions and land degradation in arid and semi-arid environments as a result of mainly human influence or climatic change.

Detritivore An organism that feeds off litter breaking down dead plants and animals or their waste.

Detritus Waste from living organisms, including dead organisms and cast-off fragments.

Deuterium A stable isotope of hydrogen containing two neutrons in the nucleus.

Diagenesis Minor, non-destructive changes in the mechanical or chemical properties of rock shortly after deposition, particularly cementation and compaction (associated with the final stages of lithification (see *lithified*)).

Diapause A phenomenon which is often associated with insects; it is a period in the life cycle of an insect where its development is suspended. It can be instigated by environmental conditions.

Diapositives A positive photograph developed on plastic or glass with high dimensional stability, rather than on paper, to minimize distortions with shrinkage or expansion of the photographic media.

Diatoms Microscopic single-celled marine or freshwater plants with silica skeletons that contribute to the formation of sedimentary deposits when they die.

Differential GPS A global positioning system that uses one or more roving receivers along with one stationary global positioning system receiver positioned over a known point that continuously collects data from the satellites. This information can be used to correct errors in the global positioning system signals received by the rovers to produce high-quality positional measurements.

Differentials Those species present as a subtype of an ecosystem community that is dominated by another species.

Diffuse pollution The release of contaminants over a large area, e.g. leaching of nitrates from cultivated fields, wash off of oil from highway surfaces.

Diffuse reflection The redirection of radiation off a rough surface such that the radiation is redirected in many random directions.

Digital elevation model (DEM) A digital representation of a three-dimensional surface, where pixel *digital numbers* represent elevation rather than brightness.

Digital image An image composed of an array of discrete pixels with a numerical assignment to define its tone.

Digital number (DN) The numerical value assigned to a pixel in a digital image, the range of which is defined by the image depth or *bit scale*.

Dilation In general terms, the expansion of material.

Dimictic Dimictic lakes are those which have two mixing periods per year. During summer and winter the lake is likely to be stratified due to thermal differences at the top and bottom of the lake. The mixing periods are likely to be in the spring and autumn when the temperature of the lake is the same throughout.

Dinoflagellate Unicellular organisms which exhibit a great diversity of form; the most dramatic effect on surrounding life is in marine ecosystems during 'bloom' periods.

Dipterocarp A family of large tropical trees that dominate South-East Asian rainforest ecosystems.

Discontinuous permafrost zone A region in which frozen ground occurs but is not laterally continuous.

Disjunct When two related groups of organisms are separated geographically by a large distance.

Dispersive grain stress A force acting to lift particles in a debris avalanche, caused by grain-to-grain collisions which bounce the particles along the base of the flow.

Displacement flow The method in which soil water at the bottom of a slope is rapidly pushed out of the soil by new infiltrating water entering at the top of a slope, contributing directly to storm hydrographs.

Dissipative beaches This type of beach is also known as a high-energy beach. Due to the energetic waves sediment is transported offshore. The resultant beach is flat with a wide surf zone. On these beaches most of the wave energy is lost in the surf zone.

Distributaries Separate river channels that are created when a river splits and does not rejoin the main channel.

Divergent plate boundaries The boundary between two lithospheric plates which are moving apart, resulting in the formation of new *lithosphere*.

Domestication The process whereby the evolution of a plant or animal species becomes controlled by humans. The deliberate breeding of dogs from their tamed wolf ancestors is an example.

Drainage basin That part of the landscape which is drained by a unitary river system.

Drainage density The measure of total stream channel length per unit area of drainage basin (stream length divided by drainage area).

Drainage divide The perimeter boundary of a *drainage basin*.

Drumlin A depositional bedform characterized by elongated accumulations of till streamlined in the direction of ice flow (they can reach 1 km in length, 500 m in width and 50 m in height). Debate exists as to the process of their formation; most accept they are deposited when the competence of a glacier overloaded with sediment reduces.

Dry adiabatic lapse rate The rate at which rising air is cooled as it expands when no condensation is occurring: 9.8°C km^{-1}.

Dry-snow zone A zone within the accumulation zone of a glacier in which there is no surface melt, even in summer.

Dump moraines Ridges formed at the margin of a glacier from material delivered by the ice flow; they lie transverse to the flow direction.

Dunes Migrating ridges of sediment, either as terrestrial deposits of sand formed by aeolian processes, or as small stream bed deposits of sand and clay common in streams of high velocity. The steeper front of the dune is termed the 'stoss' side and the gentler the 'lee' side.

Duricrust A hard, crystalline crust found on arid land surfaces. Evaporation and limited flushing by rains can lead to the accumulation of minerals on the surface or subsurface as capillary action transports minerals from underlying soils and rocks towards the surface.

E

Eccentricity of orbit The shape of the Earth's orbit around the Sun changes from more circular to more elliptical and back again over a 100 000 year period owing to gravitational forces. An increase in eccentricity causes the seasons in one hemisphere to become more intense while the seasons in the other are moderated.

Ecocline A biogeographical term introduced by Whittaker (1953) to describe the combination of environmental factors changing together through space along a gradient, e.g. the simultaneous change in temperature, exposure and soil type resulting from a change in altitude.

Ecological succession The mixture of processes that produce a gradual directional change in ecosystem structure and community at a given site over time, involving progressive habitat modification. Clements (1916) first described ecological succession as a sequence of plant communities characterized by increasing complexity of life form.

Ecosystem An organized open system consisting of biotic (plants and animals) and abiotic (environmental) components interconnected through flows of energy and materials.

Ecotone A zone of transition between two plant communities that is generally characterized by plant competition and can have special significance for more mobile animals due to edge effects.

Edaphic Pertaining to the characteristics of soil, i.e. those environmental conditions influencing a terrestrial ecosystem that are determined by the physical, chemical and biological properties of the soil.

Eh–pH stability field A plot showing the Eh–pH conditions in which the aqueous species of an element occur in a system at equilibrium (see also *redox potential*).

El Niño Southern Oscillation (ENSO) A reduction in the trade wind strength over the equatorial Pacific Ocean causes the westward-driven equatorial ocean current to falter. This leads to the cessation of the typical upwelling of cold deep water off the South American Pacific. The result is the appearance of unusually warm weather and the disturbance of pressure and precipitation systems throughout the southern hemisphere. Literally, 'The Christ Child' for its periodic occurrence every few years commencing during the Christmas season.

Elastic creep Deformation caused by strain forces; the method of movement of solid mantle rocks.

Elastic limit This is the limit to which a material can be stretched without irreversible alterations being made to the dimensions and form of the material.

Electrical conductivity The degree to which a substance conducts an electric current.

Electromagnetic energy/radiation A type of energy in transit (or radiation) in which electric and magnetic fields vary simultaneously.

Electromagnetic spectrum The continuum comprising the entire range of wavelengths of electromagnetic energy.

Electrons Within an atom electrons are the negatively charged particles; as a whole these electrons create a negative charge which is balanced by the positively charged protons. Electrons can be found on the shells surrounding the nucleus of an atom. Electrons are a key feature in the bonds which exist between atoms.

Eluviation The removal of solid or dissolved material from one soil horizon.

Emissivity The rate of emission of energy from a surface per degree of temperature difference between the surface and surrounding substances.

Encapsulated countryside A type of urban ecosystem that is likely to suffer ecosystem degradation in the absence of human management, either ancient habitats or previously managed land.

Endemic Referring to a plant or animal species that is indigenous to one particular location or region, i.e. a consequence of geological isolation and allopatric (populations become more isolated from one another) *speciation*.

Endoreic Pertaining to an inland drainage system which does not terminate at the coastline; predominantly found in dryland environments ending in salt pans or *playas*.

Englacial Inside a glacier between the surface and the bed.

Entrainment The process in which small sediment particles are mobilized from a bed surface and transported in fluid suspension.

Entropy The degree of disorder or uncertainty in a system.

Environmental economics A branch of economics that involves the analysis and expression of the impacts of an activity on the environment in monetary values in order to promote communication with decision-makers and providing a more straightforward view of the impacts.

Environmental gradient The change in an environmental variable that acts as a control on plant and animal communities along a transect from one location to another, e.g. altitude, moisture, temperature, soil acidity. Environmental gradients vary in steepness, direction and the severity of their influence upon the community; they can lead to evolutionary branching and hence *speciation*.

Environmental impact assessment (EIA) An evaluation designed to identify and predict the impact of proposed action or project on the environment in order to ensure all possible impacts are considered before implementation. They are legally required in many countries.

Environmental lapse rate The actual rate at which temperature falls with increasing altitude in the local atmosphere.

Environmental technology assessment An evaluation designed to identify and predict the potential environmental, economic and social impacts of a new technology.

Ephemeral Short-lived.

Epilimnion The surface layer of water in a water body which is warmer and less dense than the water layer below which remains trapped as it is cooler and more dense than the water above.

Epiphyte A plant growing above the ground surface that is not rooted in the soil but uses other plants for support; commonly associated with tropical rainforests.

Epoch In geological nomenclature, an epoch is a division of geological time. Two or more epochs make up a *period*.

Equifinality This is the argument with respect to open systems that the end point, the formation of a landscape for example, can be achieved by a number of processes and it is not possible to attribute the outcome to just one input. Different processes may achieve the same end point.

Equilibrium line The boundary between the *accumulation zone* and the *ablation zone* of a glacier where the net mass balance is zero.

Ergodic method Studying the development of a process or object over time (i.e. the sequence of landforms) by evaluating areas that represent different stages of advance in the process, therefore substituting space for time. For example, the sequence of successional stages of a salt marsh over time can be observed by studying a horizontal transect through the marsh from the youngest to the oldest section.

Ergs Another term for *sand seas*.

Esker A narrow winding ridge of glaciofluvial sand and gravel deposited by a meltwater stream flowing at the bed of a glacier.

Essential elements Elements found in the soil without which, or in the wrong proportions, green plants cannot grow normally. There are 16 essential elements consisting of *micronutrients* and *macronutrients*.

Estuary The mouth of a river where it broadens into the sea and within which the tide ebbs and flows, leading to an intermixing of freshwater and seawater. Estuaries are usually sites of deposition, especially if the river charges more sediment than can be removed by the tidal current or wave action.

Eustasy Global change in ocean water level due to change in the volume of water in the oceans.

Eustatic sea level The mean global sea level.

Eutrophic zone The upper layer of water where there is enough light penetration for photosynthesis.

Eutrophication Enrichment of freshwater and marine water bodies with plant nutrients to the extent that plants in the water bloom at the expense of other aquatic organisms.

Evaporite A mineral or sedimentary rock composed of soluble salts resulting from the evaporation of a body of water.

Evapotranspiration The transfer of liquid water from the Earth's surface to water vapour in the atmosphere by means of evaporation and plant *transpiration*.

Exchangeable cations Cations attracted to the surface of clay minerals and adsorbed by electrostatic attractions which can be displaced by cations in the soil solution through *cation exchange*.

F

Fabric The orientation of particles within a rock or sediment.

Falling limb The section of a storm hydrograph depicting the decrease in river discharge after rainfall has ceased following a storm event.

Fatigue failure This is the result of low-magnitude stresses being applied to a material frequently over a prolonged period. There is often not any evidence of the stress occurring but it is the accumulation of the impacts of the stress which can cause the failure.

Fecundity The faculty of reproduction. An organism has a high fecundity if it reproduces quickly and in large numbers and can therefore recover its population quickly after a problem.

Feeder–seeder mechanism A type of *orographic* enhancement of precipitation. Adiabatic cooling of air forced to rise over mountains causes saturation of water vapour and cloud formation. The water vapour of this 'feeder' cloud is swept into the precipitation of a frontal 'seeder' cloud aloft, increasing the overall precipitation on the mountain.

Fiducial marks Marks exposed on imagery that act as a frame of reference for the x, y coordinate system of the image.

Field capacity A term to describe the state of the soil when all *gravitational water* has drained away, leaving only the *capillary water*.

Field drains Subsurface drainage system installed under agricultural land to reduce the soil moisture content.

Fine earth The fine fraction of soil mineral particles consisting of sand, silt and clay, less than 2 mm in diameter.

Firn Compacted granular snow with interconnecting air spaces in at least its second accumulation season, in the process of being transformed into glacier ice (with a density usually greater than 0.4 but less than 0.8 kg m^{-3}).

Fjord A long, deep basin, previously excavated by a glacier, and has since become inundated by the sea as a result of sea-level rise during deglaciation.

Flashy regime A term used to describe a stream with a fast hydrological response to precipitation events involving a rapid rise in channel water levels.

Flow duration curve A graphically plotted curve used in the analysis of river flow frequency. The frequency distribution of the mean flow of a river at a particular site is calculated and the percentage of time any particular discharge rate is reached and exceeded is then plotted. The slope of the curve indicates the magnitude of the flow.

Flow traction A type of wash process that involves the transport of sediment along the ground surface due to the stress applied by rainwater flowing downslope and the friction maintaining the rolling and bumping of particles within the moving water; a type of *rillwash*.

Flows A type of rapid mass movement in which different parts of the mass move over each other with differential levels of movement; see *debris flow*.

Fluidization The process in which water detaches an entire fine-grained sediment deposit and converts it into a mixture of water and sediment that takes on most of the properties of a fluid.

Flute An elongated ridge formed by the infilling of cavities on the downstream side of obstacles in the bed (more elongated than drumlins); the depositional equivalent of crag and tail features.

Föhn wind The European equivalent of the *Chinook* wind.

Footprint The area on the ground 'seen' by a sensor at an instantaneous moment in time.

Foraminifera An order of Rhizopoda: small, single-celled marine organisms with a shell of calcium carbonate.

Foredune ridge A ridge of sand on the seaward side of a dune system where several smaller dunes which are forming have coalesced.

Foreset bed A seaward-sloping sediment bed deposited at the advancing edge of a growing delta; the sediment accumulates underwater and constitutes the bulk of the delta (overlain by the *topset bed*).

Form drag A drag force that slows and curves a flow path over surface topography, e.g. the slowing of glacier flow due to obstacles on the glacier bed, or the slowing of air flow over a mountain range.

Frequency distribution A method of plotting numerical data to show the number of times different values of a variable occur within a sample.

Friction cracks A variety of rock fractures caused by the action of glacier ice passing over bedrock (including *crescentic gouges*).

Frictional drag The retarding drag force associated with the interaction between a particle moving across the surface of another particle.

Frost blisters Small ice-cored mounds that develop over a single winter as a result of *frost heave*.

Frost creep The slow downslope creep of the *active layer* on a slope due to freezing of the soil causing it to expand normal to the surface; subsequent thawing then permits the vertical settlement of the soil, resulting in a net downward movement.

Frost heave The vertical lifting of the soil surface into doming frost hillocks as a result of pressures caused by the freezing of groundwater under periglacial conditions.

Fundamental niche A *niche* containing the ideal conditions for the species requirements, only realized in simple situations where no other competitors for any of the resources exist.

G

Gabbro A type of intrusive igneous rock that crystallizes slowly at depth; its minerals (plagioclases and pyroxenes) are large and visible to the naked eye.

Gabions Wire rock-filled cages used to prevent river bank erosion and stabilize slumping hillsides.

Gaia A conceptual theory developed by James Lovelock (1979) that proposes that the Earth maintains conditions suitable for life by self-regulation and feedback mechanisms whereby all elements of the Earth are interlinked at all scales; the Earth acts almost as a conscious biological organism.

Gas hydrate An ice-like crystalline solid formed from a mixture of water and a gas, often methane.

Gelifluction A type of *solifluction* only occurring in areas of permafrost. When seasonally frozen soil in the *active layer* thaws during spring, water cannot percolate down owing to lower layers of permafrost, thereby creating a lubricating effect for the slow downslope creep of the water-saturated material.

Genetic classification A term used in association with the classification of sediment types by the mode of their deposition.

Geographical information system A system of hardware, software and procedures designed to support the capture, management, manipulation, analysis, modelling and display of spatially referenced geographical data for solving complex planning and management problems.

Geomaterials This is the cumulative name given to rocks and regoliths, materials which are derived from a geologic source.

Georeferencing Correcting a remote sensing image to remove geometric distortions caused by the motion of the sensor or by the motion of the Earth beneath the sensor so that objects in the image correspond to their positions on the ground.

Geostrophic wind High-level wind blowing parallel to isobars, in which the pressure gradient and Coriolis force are in balance.

Geothermal heat Energy derived from the heat in the interior of the Earth. This energy is found everywhere beneath the Earth's surface, although the highest temperatures are concentrated in regions of active or geologically young volcanoes.

Geothermal heat flux The amount of heat given off by the Earth's crust; generated by radioactive decay, chemical processes and friction at plate boundaries.

Glacial intervals An extended cold phase during the Quaternary in which ice sheets and glaciers extended widely from the poles to lower latitudes.

Glen's law A physical law referring to the rate of deformation of glacier ice due to shear stress; fundamental in the understanding of glacier flow.

Gleying The formation of a gley soil; a blue-grey soil or soil layer caused by the reduction of iron and manganese compounds in stagnant saturated conditions.

Global positioning system (GPS) A system consisting of a constellation of orbiting satellites and one or more global positioning system receivers on the ground (or in an air- or watercraft) that are used for precise positioning.

Gondwanaland A large ancient continent of the southern hemisphere made up of present-day South America, Africa, Australia and Antarctica.

Graben A structural rock mass that is downdropped by parallel faults on both sides, often forming a structural valley.

Grain roundness A textural property of sediments associated with the degree of angularity/roundness that provides environmental information, e.g. very rounded grains are formed via considerable abrasion and are likely to have undergone extended transport.

Granular disaggregation A process associated with salt weathering; as a result grains of the rock can become loose and lead to a rough uneven surface on the rock.

Gravitational water The water that drains from *macropores* in the soil following a period of soil saturation due to gravitational forces and is replaced by air.

Grazers (or scrapers) Organisms which feed on algae attached to stones in the water column by scraping or grazing from the surface of the stone. The abundance of these organisms in the river is dependent on the extent of light penetration which in turn determines algae production.

Greenhouse effect – enhanced Human disturbance within the last few centuries has caused the concentration of the major greenhouse gases to increase. Thus, the presence of higher concentrations of such gases may enhance the greenhouse effect, making the planet warmer. The Earth's average global temperature has been estimated to have risen by 0.3–0.6°C in the past 100 years, as a result of human influence.

Greenhouse effect – natural The atmosphere traps heat energy at the Earth's surface and within the atmosphere by absorbing and re-emitting long-wave radiation; 90% of long-wave radiation emitted back to space is intercepted and absorbed by greenhouse gases such as carbon dioxide and water vapour. Without the greenhouse effect the Earth's average global temperature would be −18°C.

Ground control points Points on the ground of known coordinates that can be identified in the imagery and can be used to remove distortions in a process called geocorrection.

Ground diffusivity Pertaining to the ability of the soil to propagate fluctuations in surface temperature to greater depths in the ground.

Ground-penetrating radar A geophysical technique involving the propagation of high-frequency electromagnetic waves into the ground which are then reflected back to the surface from boundaries at which there are electrical property contrasts. It allows high-resolution mapping of subsurface features such as soil pipes, bedrock and other geophysical anomalies.

Groundwater The portion of subsurface water stored in both soils and aquifer rocks below the water table (in the *saturated zone*).

Groynes Artificial structures positioned across a beach at right angles to the shore to trap sediment being transported by longshore currents, therefore inhibiting loss of sediment by longshore drift.

Gypcrete A *duricrust* composed of calcium sulphate.

Gypsum An *evaporite* mineral composed of calcium sulphate with water.

Gyre A large circular movement of water. It usually refers to the large oceanic gyres in the subtropical high-pressure zones where geostrophic currents rotate clockwise in the northern hemisphere and counterclockwise in the southern hemisphere.

H

Halocline A layer of water in which the salinity (saltiness) changes rapidly with depth.

Hanging valley A tributary valley which, at convergence with the trunk valley, has a higher ground level, resulting in a sharp drop in elevation (the result of glacial activity in the main valley).

Hard engineering Pertaining to coastal management practices that involve the construction of large-scale structures to protect the coastline (e.g. sea walls, breakwaters and *groynes*); most hard engineering practices change the local sediment dynamics.

Health impact assessment (HIA) An evaluation designed to identify and predict the health impacts on society of a proposed activity.

Heinrich events Referring to events occurring during the coldest points of *Bond cycles* in which vast amounts of icebergs were discharged into the North Atlantic, immediately followed by abrupt warming.

Helicoidal flow A process by which water flows in an outward direction when approaching a meander bend, causing water levels on the outside of meander bends to become superelevated. Water then flows inwards along the channel bed as a return flow. This results in individual water molecules corkscrewing around the meander bend.

Herbivore An organism that feeds off plants (primary producers) and therefore occupies the second *trophic level*.

Heterotrophs Organisms that cannot photosynthesize and therefore feed directly on *autotrophs* and other heterotrophs for their energy supply. They are described as first-order consumers and form the second *trophic level*.

Histogram A column diagram where values are divided into equal parts and the frequency of occurrences within each subdivision is summed and plotted.

Hjulström curve An empirical curve defining the threshold flow velocities (i) to initiate motion of sediment grains of different sizes on a stream bed, (ii) necessary to keep the sediment grains in transport and (iii) the depositional velocity.

Holocene The second epoch of the Quaternary, in which we live now, preceded by the *Pleistocene*. It began approximately 10 000 years ago. The Holocene is an interglacial period.

Holomictic This is a group of lakes which mix completely, which means that during the year there will be a time when the temperature of the lake is the same at the top as it is at the bottom and therefore mixing is possible. Holomictic lakes can be divided into four subgroups: monomictic, dimictic, polymictic and oligomictic.

Horizontal interception Formation of water droplets by condensation of atmospheric moisture on vegetation surfaces, contributing significantly to catchment precipitation inputs under vegetation canopies in conditions of high atmospheric humidity.

Horn A high, spire-shaped mountain summit with steep sides formed by the convergence of intersecting walls of several cirques.

Horticulture Subject of using plant processes for the purposes of garden development.

Hot spots A centre of volcanic activity and igneous rock production located away from plate margins, thought to be positioned over a rising mantle plume and related to convection processes which originate at the core–mantle boundary deep in the Earth.

Hue A measurable variable of soil colour that describes the dominant colour of the pure spectrum (usually redness or yellowness).

Hummocky moraine A type of *moraine* characterized by considerably undulating terrain; caused by deposition from meltout of *supraglacial* and *englacial* material in *kettle holes* and crevasses.

Humus A type of soil organic matter which is very resistant to decomposition.

Hydrates A compound which contains water but which can dissociate into water and another compound. This process is reversible and therefore the compound and the water can combine again.

Hydraulic geometry Referring to the river flow characteristics of a channel (such as discharge, depth, width and velocity) and their relationship to one another.

Hydraulic radius The ratio of the cross-sectional area of water flowing through a channel to the length of the *wetted perimeter*; it represents a measure of the efficiency of the channel at conveying water and the proportion of water subject to bed surface friction.

Hydraulic sorting The process by which particles of river bed material are sorted into sections of near uniform particle size due to the change in river competence throughout its journey from source to mouth.

Hydrogenous Pertaining to sediments derived from *ions* in seawater through geochemical processes, e.g. metal ions of iron and manganese are released from hydrothermal vents and oxidize or combine with silica to form metal-rich sediments.

Hydrograph A graph showing river discharge plotted against time for a point on the river channel network, displaying a characteristic shape during rainfall events.

Hydrophyte A plant that has adapted to grow in wet or water-logged conditions.

Hygroscopic water Soil water held as a tight film around individual soil particles and unavailable to plants because of the very strong attraction between the water and soil particle.

Hyper-arid Pertaining to extremely dry areas ('true deserts') that may go as long as 12 months without rainfall (e.g. central Sahara).

Hyper-concentrated flow Similar to a debris flow but with a higher water content, acting more like a liquid and with less viscosity; these flows behave like a sediment-rich stream maintained by forces of turbulence.

Hyperspectral scanners Scanning remote sensing instruments that record digital images using multiple narrow bands. Similar to *multispectral scanners* but with higher spectral resolution. These are also called imaging spectrometers.

Hypolimnion A cooler, lower layer of water in a water body which does not readily mix with the upper warmer layer as it is more dense and hence remains below the warmer, less dense layer.

Hyporheic zone This is the zone under a river channel where there is a mixing of channel water and groundwater. It exists where the river channel flows over a permeable substrate.

Hysteresis A process whose progress is determined by the direction in which the reaction is occurring. It is normally described by a bivariate plot in which the value of one variable is dependent on whether the other variable is increasing or decreasing.

I

Ice creep A slow, continuous movement of ice involving non-recoverable deformation of the ice owing to intergranular motion caused by internal pressure and the force of gravity.

Ice rafting The process by which glacially eroded debris is transported by floating ice (ice floes or icebergs); it may be transported great distances and deposited either on the sea floor when the ice melts or on beaches.

Ice segregation The formation of layers of ice in rocks or soils where there has been a movement of water into gaps which has then frozen.

Ice shelves Thick floating sheets of ice extending over the sea from a landward ice sheet, fed by the ice sheet and snow accumulation.

Ice streams Fast-flowing 'rivers' of ice within more slowly moving ice sheet walls.

Ice wedge polygons *Ice wedges* that have joined together owing to the annual reopening and expansion of the ice wedge.

Ice wedges V-shaped bodies of ground ice that extend into *permafrost* (up to 1.5 m in width and 3–4 m in depth). Under very low temperatures frozen ground contracts as it is further cooled, causing it to crack; water enters during spring and summer and then freezes into an ice wedge.

Iceberg calving The process in which a large mass of floating ice breaks away from an ice shelf; a major method in which mass is lost from ice sheets.

Ice-pushed ridge A ridge of material accumulated by the ploughing action of a glacier but composed of material that is not glacially derived (i.e. similar to a *push moraine* but not consisting of glacially derived debris).

Igneous rock Rock that has originated from a molten state such as lava from a volcano.

Illumination The degree to which a scene or object is lit, in this case by the Sun.

Illuviation The deposition of solid or dissolved material into a soil horizon.

Image mosaic A composite of remote sensing images to produce an image of greater coverage.

Image orientation The direction in which a sensor is pointed to capture an image. Images can be orientated vertically, horizontally or obliquely.

Imaging spectrometers Another term for *hyperspectral scanners*.

Imbrication The wedging of particles among others. Often small particles become trapped by larger ones so that even though the flow is great enough to entrain them, they cannot move until the larger particles are entrained.

Incremental methods Techniques for estimating the age of deposits based on the measurements of regular accumulations of sediment or biological matter through time, e.g. *dendrochronology*, analysis of *varves* and ice cores.

Indentation hardness The resistance of a material to deformation as a result of the application of compressive stress from a sharp object. The indentation hardness of a material can be determined by the amount of compressive force required to make an indentation to a certain depth or by the size of an indentation left as a result of a fixed size of compressive force.

Inductive Pertaining to the process of inferring a general law or principle from the observation of particular instances; by classifying and ordering unordered knowledge, regularities may be identified and general laws discovered (opposed to *deductive*).

Indurated Pertaining to soils and sedimentary rocks which have become hardened and compacted by post-depositional chemical and physical alterations.

Industrial Revolution A major shift of technological and cultural practices in the late eighteenth century and early nineteenth century in some western countries. It began in Britain and spread throughout the world and consisted of an engagement with energy generation through fossil fuel burning, construction, invention and mass transport systems.

Infiltration capacity The maximum rate at which water can enter soil under specified conditions.

Infiltration-excess overland flow A form of *overland flow* occurring when rainfall intensity exceeds the infiltration capacity and excess water is stored and transported on the surface (also known as Hortonian overland flow).

Infiltration rates The volume of water passing into the soil per unit area per unit time (i.e. the rate at which water added to the surface enters the soil).

Integrated coastal zone management A management approach where all parties concerned in coastal protection and development are involved; it considers the socio-economic and environmental issues which are present to achieve a sustainable outcome. Planning should be based on shared knowledge and long-term goals need to be identified.

Interception The process by which precipitation is prevented from reaching the ground by the vegetation layer.

Interception storage The storage of water on leaves and tree trunks when precipitation has been intercepted by vegetation en route to the ground surface.

Interference The fading, disturbance or degradation of a signal (in this case surface reflectance) caused by signals from unwanted sources (i.e. the atmosphere).

Interglacial intervals A long, distinct warm phase between glacial stages during the Quaternary; the Earth's glaciers become severely diminished owing to climatic amelioration (restricted to very limited locations with sufficient conditions).

Intermediate beaches This type of beach falls between the extremes of the high-energy (*dissipative*) and low-energy (*reflective*) beaches. They often feature near shore bars which dissipate some of the wave energy which reaches the beach. The upper part of an intermediate beach can be steep, however, and show characteristics of a *reflective beach*. Intermediate beaches can be dynamic and change their morphology.

Interstadial A short period of climatic amelioration and ice retreat within a glacial stage, less pronounced than an *interglacial interval*.

Interstices This is a small or narrow space between particles or objects.

Interstitial ice Ice crystals (individual or fused together) occupying the pore spaces of a soil or rock.

Intertropical convergence zone (ITCZ) The zone where the north-east trade winds from the northern hemisphere and the south-east trade winds from the southern hemisphere come together over the equatorial region. This zone is characterized by cloud bands which illustrate rising air yet it is not a continuous band around the Earth. The ITCZ migrates northwards and southwards across the equator with the seasons, so that it resides in the hemisphere which is experiencing summer.

Involutions Features caused by the deformation of unconsolidated surface materials (i.e. disruption to the sedimentary structure and soil profile) due to thawing of ice-rich ground; often used as a diagnostic for past permafrost conditions.

Ionic diffusion The upward movement of ions through the soil without the aid of water, due to the difference in concentration of ions from the base to the surface of the soil. The close proximity to parent material at the bottom of the soil profile results in a greater quantity of ions; the random movements of the ions will then form a general upward movement to an area of fewer ions.

Ions Positively or negatively charged atoms.

Island biogeography The study of the distribution and evolution of organisms on islands or even 'virtual islands' (resulting from some barrier other than the sea). More narrowly, island biogeography is the examination of MacArthur and Wilson's (1967) equilibrium theory of *speciation* in geographically isolated areas, whereby a relationship is identified between the species richness of an island and its size and isolation, among other characteristics.

Isobars Lines on a map joining points of equal atmospheric pressure.

Isohyets Contour lines connecting points of equal rainfall.

Isomorphous substitution During the formation of a clay mineral, the process in which one atom in the crystal lattice is replaced by another of similar size without disrupting the crystal structure. The replacing ion is generally of a lower positive charge, causing the clay mineral to become electrically negative.

Isostasy The principle by which the Earth's crust 'floats' upon the denser *mantle*, following Archimedes' law of hydrostatics. The thicker, more buoyant crust (continental regions) stands topographically higher than the thinner, denser crust (under the oceans) to create an equilibrium situation.

Isostatic rebound The process whereby, after a heavy weight (such as an ice cap) is removed, the Earth's lithosphere slowly relaxes and the surface rises to a new equilibrium level.

Isovels Contour lines connecting points of equal velocity.

J

Jet streams High-speed long, narrow winds in the upper atmosphere. These currents meander and reach speeds of 400 km h^{-1}.

Jetties *Hard engineering* coastal management structures built along the banks of a tidal inlet at a river mouth in order to stabilize unpredictable shifting channels for navigation purposes.

K

Kames Steep-sided isolated conical hills of bedded glaciofluvial materials deposited by meltwater along the sides or margins of a glacier.

Karst Referring to the ground surface depressions and extensive underground drainage network created by limestone solution.

Katabatic drainage Radiative cooling at night causes the air close to the ground to cool; this cooler air is slightly denser and slowly moves downslope to lower ground and into depressions. It is greatest in cloud-free and dry conditions with light winds (limited mechanical mixing of the air).

Kettle hole A closed depression found in glacial till deposits, formed by the melting of a large mass of ice that became incorporated and preserved in glacial till.

Keystone species Species that are highly connected to the entire food web; their loss may result in *ecosystem* collapse and huge loss of biodiversity.

Kinematic viscosity The ratio between the density and viscosity of a fluid.

L

Lagoon A coastal bay totally or partially enclosed and cut off from the open sea by a *barrier beach*, *spit*, shingle ridge or an offshore reef.

Lahars Flows of loose soil, rock, ash and water following a volcanic eruption.

Laminar flow One of the two ways in which water can flow; it involves all water molecules flowing in the same direction parallel to one another resulting in no mixing of water.

Landscape corridors Narrow strips of land that differ from the *landscape matrix* existing on either side; the key characteristic relates to their function in connecting different environments and the often sharp microclimatic and soil gradients from one side of a corridor to another.

Landscape ecology A concept used in exploring regional and small-scale biogeographical distributions. It refers to the analysis of the cause-effect relationships between the living community and the immediate environmental conditions, which have created the specific landscape pattern observed. The theory suggests the landscape consists of a matrix of patches and corridors providing oases and pathways for species dispersal and movement.

Landscape matrix The element of the landscape that contains within it other landscape components (patches and corridors) into a complex system that controls the local biogeography. The stability of the matrix is dependent upon the extent and development of *landscapes patches* and *landscape corridors*.

Landscape patches Distinctive elements within the wider landscape, such as ponds, woods or towns. Their analysis involves the influence of patch characteristics (shape, frequency, origin and stability) upon the local ecosystem. The community of a landscape patch may vary substantially to the surrounding landscape and be very vulnerable to its influences, thus having particular relevance to conservation ecology.

Landslides A mass movement process whereby a large coherent mass of material moves down a slope under the influence of gravity, remaining undeformed.

Lapse rate Rate at which temperature decreases with altitudinal increase.

Latent heat The amount of heat required to change the state of a substance, e.g. from a liquid to a gas, or vice versa.

Lateral moraine A ridge of glacial debris lying parallel to the sides of a glacier or lying along the sides of a valley formerly occupied by a glacier, consisting of dumped material and frost-shattered material from the valley walls.

Laterization The process in which high temperatures and heavy rainfall cause intense weathering and leaching of the soil, producing horizons depleted in base cations and enriched in silica and oxides of aluminium and iron.

Laurasia A large ancient continent of the northern hemisphere made up of present-day North America, Europe and Asia.

Law of limiting factors Pertaining to a species, the necessity for all the environmental factors that control its survival to be maintained within a range that the organism can tolerate; if just one of these controlling variables falls outside of the tolerance range the organism will not survive.

Laws of thermodynamics Laws pertaining to the conservation of energy. The first law of thermodynamics states that energy cannot be created or destroyed, only transformed from one form into another; thus energy is conserved. The second law of thermodynamics states that isolated systems become more disorganized over time.

Leaching Downward transport of soluble soil material in solution through the soil profile by percolating surplus water, depositing some in lower layers but removing the most soluble entirely.

Least-squares adjustment A mathematical method for fitting a model to data so as to minimize error between the observed values and the estimated values.

Lentic Term used to refer to things which are related to or inhabit still water bodies such as ponds and lakes.

Liana A woody vine supported on the trunk or branch of trees, usually tropical.

Life-cycle analysis (LCA) The evaluation of all the environmental impacts of a product from the time the raw materials are taken from the Earth to the time the product is thrown away and added to the ecosystem (including its manufacture, use and disposal).

Linear wave theory Main theory of ocean surface waves used in ocean and coastal engineering from which important equations are derived.

Lithified Pertaining to the transformation of unconsolidated sediments into a cohesive sedimentary rock mass through cementation, compaction and crystallization (lithification).

Lithogenous Pertaining to material derived from the physical and chemical breakdown of rocks and minerals.

Lithosphere The rigid outermost layer of the Earth, consisting of the crust and upper section of the mantle above the asthenosphere; characterized by brittle behaviour.

Litter A type of soil organic matter consisting of decomposing residues of plant and animal debris.

Littoral drift The transport of beach material along the coast, sometimes referred to as *longshore drift*. Waves surging along the beach at an oblique angle transport sediment up and along the beach in the *swash* followed by transport more perpendicular to the coast in the *backwash* (creating a zig-zag movement of sediment along the beach).

Littoral zone The part of the lake which is closest to the shoreline. It occurs in both shallow and deep lakes and is where light can penetrate to the bottom thereby allowing for a diverse array of plants and algae to grow.

Lobate Characterized by having a tongue-like shape, e.g. the ice lobe of an alpine glacier.

Loess A fine-grained (less than 50 μm (fifty-millionths of a metre)), commonly non-stratified and unconsolidated sediment. It is composed of quartz, feldspar, carbonate and clay minerals that have been transported by wind from arid land surfaces and deposited elsewhere, sometimes thousands of kilometres away.

Logical positivism A form of *positivism* in which *inductive* reasoning is used to form theory and acquire knowledge from experimentation.

Long profile A graphical curve displaying the longitudinal altitude profile of a river from source to mouth (height of the river plotted against distance from stream source). It illustrates the change in river gradient downstream.

Longshore currents A net movement of water parallel to a coastline. This occurs because waves surging along beaches at oblique angles are followed by more perpendicular transport out to sea resulting in a net water movement along the coastline.

Longshore drift Another term for *littoral drift*.

Long-wave radiation Radiation that has been emitted by a surface at a longer wavelength than its solar source. It is also called terrestrial radiation.

Lotic Term used to refer to things which are related to or inhabit fast-moving water bodies such as rivers.

Lumped model A catchment model in which catchment characteristics are assumed to be uniform across space.

Luvisols A group of soils produced by clay *eluviation* (also known as acid brown earths).

Lysimeter An instrument for taking direct measurements of *evapotranspiration*; by isolating a block of soil (with its vegetation cover), the weight of the block can be used to represent the quantity of water and its change over time can be calculated.

M

Macrogélivation The process whereby rocks are broken up into clast-size debris through the utilization of existing fractures and fissures in the rock.

Macroinvertebrates These are organisms which live in the water column and are greater than half a millimetre in size; they live in varying locations in the water column including on rocks and in aquatic plants.

Macronutrients The group of *essential elements* found in high concentration in plants (carbon, oxygen, hydrogen, nitrogen, phosphorus, sulphur, calcium, magnesium, potassium and chlorides).

Macropore Infrequent large opening or void in the soil (greater than 0.1 mm in diameter) that can promote rapid, preferential transport of water and chemicals, formed by structural cracks and fissures or by biological activity, e.g. earthworms, burrowing creatures and plant roots.

Macropore flow The movement of water through the soil within larger pores (*macropores*).

Magnetometer An instrument for measuring the strength of the Earth's magnetic field.

Main stream length The distance of the main river channel in a catchment from source to mouth (equating to the length of the *long profile*). Given in kilometres.

Mangroves A term applying to the variety of trees and shrubs which grow on saline mudflats in tropical coastal areas to form a dense swamp forest. Their roots trap silt which accumulates to form a swamp.

Mantle The zone within the Earth's interior lying between the partially molten core and the thin surface crust, containing 70% of the earth's total mass and composed principally of magnesium-iron silicates.

Mass balance The difference between the total accumulation and *ablation* of a glacier with time, i.e. a positive mass balance exists when accumulation exceeds ablation for a given period.

Mass movement The downslope movement of sediment, soil and rock material as a single unit (the individual fragments are in close contact); a number of mass movement processes can be identified including *debris flows*, *debris avalanches*, *slumping* and *landslides*.

Mass wasting The spontaneous downhill movement of surface materials (soil, *regolith* and bedrock) under the influence of gravity, without the active aid of fluid agents.

Massive ice Very thick bands of *segregated ice*, up to several metres thick.

Matrix flow The movement of water through the soil within very fine pores.

Meandering rivers Sinuous river channels that migrate downstream owing to river bank erosion on the outside of meander bends and deposition of bed material on the inner bank. Excessive meandering leads to oxbow lake formation.

Meromictic Derived from meromixis, this is where there is an incomplete mixing of a lake. Deep lakes are meromictic as their depth prevents a complete overturning circulation. Lakes which also have inflows of different density water can be meromictic as stratification can occur owing to the different densities.

Mesohabitats The smaller units which combine to define a reach of a river. Each mesohabitat within a river reach is determined by similar characteristics such as deeper, slower flowing pools and shallower, faster riffle sections.

Mesophyte A plant that requires a moderate climate in terms of temperature and precipitation in order to survive.

Metalimnion A thin layer of water in a lake or ocean where there is a zone of rapid temperature change with depth.

Metamorphic rock Rock which has altered its form through structural and mineralogical change due to heat and pressure from the surrounding conditions.

Metamorphosis (metamorphism) (biological) A change in the form, function or habits of a living organism by a natural process of growth or development, e.g. the change of a caterpillar into a butterfly.

Microgélivation The occurrence of ice crystallization within pores and fissures of a rock at the scale of grains and crystals can result in the formation of fine rock fragments.

Microhabitats The smallest area in an ecosystem which is home to an individual array of organisms and vegetation, microhabitats can include clumps of grass or fallen trees.

Micronutrients The group of *essential elements* found in small concentration in plants (iron, manganese, zinc, copper, boron and molybdenum).

Micropores Very small pores in the soil that can hold soil water (less than 0.1 mm in diameter).

Mid-ocean ridges The zones in which oceanic lithosphere is created by the spreading of *divergent plate boundaries*. The relative buoyancy of the newly formed oceanic crust causes the topography to be raised, creating a high-relief ridge.

Mie scattering The wavelength-dependent redirection or scattering of electromagnetic radiation at wavelengths of about the same magnitude as the size of the particles.

Milankovitch theory A hypothesis formalized by Milutin Milankovitch describing the external driving force behind the glacial cycles of the Quaternary. The amount of solar radiation reaching different parts of the Earth from the Sun varies as the eccentricity of the Earth's orbit, the obliquity of the axis of rotation and the precession of the equinoxes change over time in a regular and predictable way.

Mineralization The process of forming a mineral by combination of a metal with another element.

Mohorovičić discontinuity (Moho) The contact surface between the crust and the mantle; the zone in which seismic waves are significantly modified.

Mole The quantity of a substance that contains the same number of chemical units as there are atoms in exactly 12.000 grams of carbon-12.

Molten rock Rock in a state of a liquid; the rock has melted and flows as any liquid.

Monoclimax A theory of vegetation requiring that all sequences of *ecological succession* within a given climatic region converge on a single uniform stable climax community depending solely on regional climate.

Monomictic With respect to lakes, these lakes only have one season of overturning and mixing in a year.

Monsoon A system of winds that switch direction from ocean-continent to continent-ocean between summer and winter in response to the northerly and southerly movements of the intertropical convergence zone (ITCZ). The characteristics of a monsoon climate are most apparent in India and South-East Asia; the *jet stream* reverses from westerly to easterly, causing the north-east and south-west monsoon seasons that are responsible for the majority of inter-annual climatic change in the region.

Montmorillonite A soft mineral that forms as very small plate-shaped crystals. Two silicon tetrahedral sheets enclose an aluminium octahedral sheet in the structure. Considerable expansion can occur when water moves between the silica sheets.

Moraine An accumulation of glacial till that has been transported and deposited by a glacier or ice sheet; classifications of moraines are usually based on the mode of their formation: see *de Geer moraine*, *dump moraines*, *hummocky moraine*, *lateral moraine*, *push moraine* and *rogen moraine*.

Moulin A rounded vertical or steeply inclined hole within a glacier down which meltwater travels.

Multispectral scanners Scanning remote sensing instruments that record digital images using several, moderately narrow bands, typically between the ultraviolet and infrared portions of the spectrum. Similar to *hyperspectral scanners* but with lower spectral resolution.

Munsell colour chart A standard system for the description of soil colour based on three measurable variables (*hue*, *value* and *chroma*).

Mycorrhizal Pertaining to the nature of mycorrhiza, a fungus growing in or on a plant root involving a symbiotic relationship between the two.

N

Nadir The point below a point of observation. The nadir will be the point in the centre of an aerial photograph that is perfectly vertical.

Neap tide A tide that occurs at the first and third quarters of the Moon when the gravitational force of the Moon is opposed to that of the Sun, thereby producing a relatively small tidal range, and causing lower than average high tides and higher than average low tides. The velocity of tidal currents is slowed at this time.

Nearshore A process-based term for the area comprising the swash, surf and breaker zone; the area in which waves are forced to break owing to the shallowing of water closer to the shoreline.

Negative feedback An event or process resulting from another event that counteracts its effects.

Nekton A nektonic organism (see *nektonic*).

Nektonic The collective name for organisms which are active in the water column and move around in it rather than being restricted to the top or the bottom.

Net primary productivity (NPP) The amount of energy (carbohydrate) fixed by plants during photosynthesis subtracting that used in respiration; it represents the growth of the plant/ecosystem, measured in unit area per unit time.

Net radiation The difference between the total incoming radiation and the total outgoing radiation.

Neuston The name given to a group of organisms which exist mainly on the surface of the water or just below the surface.

Niche The position or role of an animal or plant species within its community in relation to its specific requirement of habitat resources and microclimatic conditions (i.e. climate, shelter, food, water). No two species with identical resource requirements can occupy the same niche (the principle of competitive exclusion applies).

Nitrification The process whereby ammonium is oxidized to produce nitrite which is then oxidized to form nitrate. In the natural environment the process is carried out by nitrifying bacteria which are able to gain energy through the oxidization of the compounds. Nitrification is an aerobic process which means it requires oxygen to occur.

Nitrifying bacteria Bacteria that oxidize ammonium to nitrite and thence to nitrate.

Nitrogen fixation Nitrogen in the form N_2 is not accessible to plants to use for their growth. Nitrogen fixation is the process whereby bacteria which exist in the soil or in nodules on the roots of plants (symbiotic relationship) are able to reduce nitrogen to ammonium thereby making it accessible to plants.

Nivation Localized erosion of a slope caused by the combination of frost action, *gelifluction*, *frost creep* and meltwater flow at the edges and beneath a snowpack; accentuated in permafrost-free zones during periodic freezing and thawing of constantly moistened ground.

Nivometric coefficient The percentage of precipitation falling as snow within a given area.

Non-selective scattering The wavelength-independent redirection or scattering of electromagnetic radiation caused by atmospheric particles that are much larger than the wavelengths of the light they scatter.

North Atlantic Deep Water (NADW) A body of water formed in the North Atlantic Ocean. Relatively saline water from the Gulf Stream cools when it moves rapidly north into the Norwegian Sea; it becomes denser and sinks, flowing back south to form a major component of the *thermohaline circulation* of the oceans.

Nuée ardente A cloud of superheated gas-charged ash that develops into a pyroclastic gravity flow following a very explosive volcanic eruption.

O

Occluded front The process in which a cold front of a depression overtakes a warm front. The occluded front is classified as warm or cold depending on whether the air ahead of the warm front is colder or warmer than the air following the cold front.

Occult deposition The occurrence where the contact of mist or fog with buildings or vegetation can result in the deposition of pollutants on these surfaces.

Offshore A morphological term for the area below the wave base, just beyond the shoreline and foreshore.

Oligomictic With reference to lakes, these lakes may be permanently stratified unless there is some perturbation to induce mixing, such as a storm event.

Omnivore Organism that feeds on both plants and animals.

Ophiolites A layer of oceanic crust created at mid-oceanic ridges and uplifted at convergent plate boundaries, now lying exposed above the water at continental margins.

Options analysis The identification of key criteria through which to evaluate management options when dealing with an environmental issue.

Organic Pertaining to any compound containing carbon, except simple compounds such as oxides and carbonates (which are considered inorganic).

Orogeny The process of mountain building, it is particularly related to the growth of mountains which occurs as a result of the deformation of the Earth's crust through folding actions and compressional forces.

Orographic Pertaining to mountains; for example, orographic precipitation is caused by the forced ascent of air over high ground/mountain barrier.

Oscillatory flow Currents that oscillate backwards and forwards such as wave currents.

Overland flow The motion of a surface layer of water as sheet flow (unchannelled).

Oxidation A chemical weathering process involving the combination of free oxygen with minerals to form oxides with a positive electrical charge.

Oxidation state The electronic state of an atom in a particular compound; equal to the difference between the number of electrons it has compared with a free atom, e.g. in calcium chloride ($CaCl_2$), calcium has the oxidation state of +2 (Ca^{2+}) and chlorine has the state −1 (Cl^-).

Oxisols A soil order found in the tropics consisting of old, extremely weathered soils which have been highly leached and consequently become infertile with a low base status.

P

Palaeo- Spelt 'Paleo-' in American English, derived from the Greek for 'old', and often used as a prefix to mean 'past'. Thus 'palaeoecology' is the ecology of the past; 'palaeoclimatology' is the study of past climates; etc.

Palaeoecology The study of ancient plant and animal distributions and processes.

Palaeomagnetism The study of the magnetism of igneous rock; the strongly magnetic particles of magnetite in igneous rock become permanently orientated in the direction of the Earth's magnetic field at the time of the lava cooling.

Palsas Low permafrost-cored mounds, 1–10 m high, formed in peat of both continuous and discontinuous permafrost zones; caused by differential frost heaving linked to the thermal conductivity of peat.

Palynology The branch of science concerned with the study of living or fossil pollen and spores; often used in the reconstruction of palaeoenvironments via analysis of pollen types preserved in peat, organic soils and lake muds.

Panchromatic Sensitive to all colours of the visible spectrum.

Pangaea The Earth's most recent supercontinent formed during the Permian by the coalescence of most continental plates (Gondwanaland and Laurasia, among other smaller continents) and rifted apart in the Jurassic.

Paraglacial geomorphology The study of landscape features which owe their existence to the presence of ice, albeit indirectly and often in the past.

Parallax The apparent change in position of a stationary object when viewed from two different positions.

Parent material The material upon which soil is developed and constitutes the main input of soil material through the process of weathering. It may be the weathered surface of exposed unconsolidated *in situ* rock surfaces, or unconsolidated superficial material transported and deposited by gravity, water, ice and wind.

Partial contributing area concept The idea that *infiltration-excess overland flow* will often occur only in spatially localized parts of the hillslope as opposed to the entire catchment (as originally postulated by Horton).

Partially mixed estuaries Estuaries that are highly influenced by tidal currents, causing greater mixing (*advection* and *diffusion*) of fresh- and saltwater and a more gradual salinity gradient in the water column.

Participatory analysis The process of encouraging all stakeholders to take an active part in the decision-making process when deciding on an environmental management strategy where conflicting interests are involved.

Particle movement The physical transportation of material down a hillslope where grains move one, or a few, at a time and do not significantly interact with one another, as opposed to a *mass movement*.

Peat A type of predominantly dark organic soil derived from partially decomposed compacted plant materials that accumulate under waterlogged conditions.

Pedogenesis The process of soil formation.

Peds Clumps or structural units of soil separated by small natural voids.

Pelagic sedimentation Sediments formed in an open-ocean environment by the slow background sedimentation of fine-grained material (usually marine organisms and red clays) falling through the water column to the seabed.

Pelagic zone The zone at the top of the water column where light abundance allows for photosynthesis to take place; it is also related to open-water areas and is therefore only found away from the banks of a water body (*littoral zone*).

Peneplain A low-relief plain that is the theoretical end product of erosion in the absence of tectonic activity (following *Davisian cycles of erosion*).

Perennating system A group of plants that persist for several years, usually with new growth from a perennating part of the plant, e.g. bulbs, *rhizomes* or tubers.

Period In geological terms, a period is a formal division of geological time. The Quaternary is an example of a geological period. Two or more periods make up an era.

Permafrost A condition existing below the ground surface, in which the soil or bedrock material remains perennially frozen, below 0°C for a minimum of two years. Currently permafrost affects approximately 26% of the Earth's surface.

Phanerophytes Woody perennials (trees and shrubs) with visible buds on upright perennial stems high above the ground, e.g. the palm family.

Phenological Pertaining to the timing of recurring natural phenomena, such as the timing of events such as leaf fall and buds appearing on plants.

Phenology The study of the timing of natural phenomena in relation to climate. For example, the appearance of the first flower of spring.

Photic zone The layer of the surface ocean which receives enough sunlight to enable photosynthesis to occur. Also known as the euphotic zone.

Photochemical oxidation A process which occurs in the presence of light and results in the chemical change of a substance through the loss of electrons.

Photon A quantum of electromagnetic radiation. Units of light or other electromagnetic radiation, the energy of which is proportional to the frequency of the radiation.

Photoreduction Chemical reduction of a substance caused by ultraviolet radiation, e.g. in sunlight.

Photosynthesis The process of converting light energy to chemical energy and storing it as sugar. This process occurs in plants and some bacteria and algae. Plants need only light energy, and to make sugar.

Photosynthetic bacteria Bacteria that are able to carry out *photosynthesis* (light is absorbed by bacteriochlorophyll), e.g. blue-green algae.

Phytoplankton Photosynthesizing plants, often microscopic, that live in saline and freshwaters and are the foundation for the aquatic food chain.

Pillow lava The name for lava that erupts from vents underwater and cools rapidly forming rounded structures surrounding the vent.

Pingo An ice-cored mound (up to 55 m high and 500 m long) found in permafrost areas; derived from an Inuit word meaning 'hill'.

Pingo scar A relict periglacial feature formed by the melting of the ice core of a *pingo*, leaving a central surface depression with sediment ramparts.

Pipeflow The movement of water through the soil within *soil pipes*.

Pixel A contraction of 'picture element' that refers to the smallest unit of a *digital image*.

Planktonic A group of organisms which reside in the water column in suspension and move through drifting and floating.

Plant physiology The study of the functioning of plants, and this includes the response of plants to their environment and the acquisition of resources by plants.

Platform The stationary (i.e. gantry) or moving (i.e. aircraft) position on which remote sensors are mounted.

Playa A depression in the centre of an inland desert basin; the site of occasional temporary lakes; high levels of evaporation often create alluvial flats of saline mud.

Pleistocene The first epoch of the Quaternary, preceded by the Pliocene and succeeded by the *Holocene*. Lasting from approximately 1.8 million to 10 000 years before the present (when the Earth was most extensively glaciated).

Ploughing boulders Boulders found on periglacial slopes that slowly move downslope owing to different thermal conditions beneath the boulder compared with the surroundings. They leave a trough upslope and form a sediment prow downslope.

Point source pollution Release of contaminants from a clearly identified point, e.g. a pipe from a factory.

Polar front The surface of contact between a cold polar air mass and a warm tropical air mass.

Polar permafrost Extensive *permafrost* that occurs owing to low temperatures in high-latitude areas, e.g. Alaska.

Polder Land reclaimed from the sea via the development of embankments.

Polyclimaxes A theory of vegetation allowing the co-existence of several final *climax communities* for a given type of area, all of which rank equally rather than being subordinate to a single climatic climax community (as required by the *mono-climax* theory). Instead of total convergence into a single community type, succession therefore produces partial convergence to a mosaic of different stable communities in different habitats.

Polycrystalline Referring to a crystalline structure in which there is a random variation in the orientation of different parts.

Polymictic With reference to lakes, polymictic lakes tend to be shallow and are predominantly continually circulating. They have limited stability although there may be a point in each day where thermal stratification is able to occur.

Polythermal glacier A glacier composed of both warm and cold ice.

Pore spaces The voids between solid soil particles.

Pore water pressure The pressure exerted by water in the pores of soil and aquifer rocks which may force particles apart during saturated conditions.

Porosity The pore space of a substrate (i.e. the factor controlling soil and rock permeability).

Positive feedback An event or process resulting from another event or process which exacerbates or magnifies the original effect.

Positivism A traditional philosophy of science, originally attempting to distinguish science from religion by ensuring the application of a unitary scientific method of observation, involving direct and repeatable experimentation on which to base theory. The underlying premise is that a firm empirical basis will lead to the identification of scientific laws which become progressively unified into a system of knowledge and 'absolute truth' about the natural world.

Potential evapotranspiration The *evapotranspiration* that would occur from a vegetated surface with an unlimited water supply.

Potholes Circular depressions found on bedrock surfaces. In reference to rivers they are scoured out by the effect of a pebble rotating in an eddy.

Precautionary principle An approach to decision-making which states that where there are threats of serious or irreversible damage, lack of scientific certainty should not be used as an excuse to preclude preventative action. Action should be taken at an early stage before victims or negative impacts occur; 'better safe than sorry'.

Precession of the equinoxes The gravitational pull exerted by the Sun and the Moon cause the Earth to wobble on its axis like a spinning top, determining where in the orbit the seasons occur, and the season when the Earth is closest to the Sun.

Precipitation The deposition of water in a solid or liquid form on the Earth's surface from atmospheric sources (including dew, drizzle, hail, rain, sleet and snow).

Precipitation deficits The lack of precipitation in a water balance when considering the losses in the form of evaporation or losses through gravity-driven movement of water to riverflow or groundwater.

Predators Carnivorous organisms which exist by preying and feeding on other organisms.

Primary endemism When a species occurrence is unique to one specific area alone and unknown to any other region, e.g. Australian marsupials.

Primary minerals Minerals that have not changed from their original state since they were formed in magma (e.g. quartz, feldspars and micas).

Primary productivity The amount of biological material (biomass) produced by photosynthesis per unit area and unit time by plants.

Primary succession *Ecological succession* beginning on a newly constructed substrate previously devoid of vegetation (e.g. a new volcanic island); the recently exposed land is colonized by animals and plants.

Pro-delta The shelf area offshore of a river mouth which marks the intersection between the delta sediments and the adjacent basin.

Profundal zone A deep zone of water, usually in an ocean or lake where there is reduced light penetration. It is often found below the *thermocline* and due to the lack of light there is reduced biological diversity in this zone.

Pronivial ramparts Another name for *protalus ramparts*.

Protalus ramparts Linear ridges of coarse sediment found a small distance away from a slope base, formed from the accumulation of frost-shattered debris that, once fallen from a backwall, slides down a snowpack to its lower margin.

Push moraine A ridge of material accumulated at the glacier margin by the bulldozing action of a glacier front and consisting of glacially derived material.

R

'r' and 'K' selection A theory of two life strategies to cope with competition and stress. Natural selection may favour either individuals with high reproductive rates and rapid development ('r' selection) or individuals with low reproductive rates and better competitive ability ('K' selection).

Radiation Emitted electromagnetic energy.

Radiometric Of or pertaining to the measurement or representation of radiation.

Radiometric methods Techniques for estimating the age of deposits based on the time-dependent radioactive decay of particular radioactive isotopes found in sediments.

Raindrop impact The force exerted by a falling raindrop on a soil surface. The impact of the raindrop causes a shock wave which detaches grains of soil or small aggregates up to 10 mm in diameter and projects them into the air in all directions; the rate of detachment is roughly proportional to the square of rainfall intensity.

Rainflow In shallow overland flow, the transport of water resulting from a combination of detachment by raindrop impact and transportation by rainwater flowing downhill.

Rainsplash A type of soil erosion caused by *raindrop impact* in which sediment is transported through the air.

Rainwash The erosion of soil by overland flow processes; normally occurs in concert with *rainsplash*.

Raised beach A step-like feature along a coastline which marks the former position of the high tides and which once used to be a beach. Often, raised beaches are created by isostatic rebound, which lifts the beach out of reach of the waves.

Random errors Non-systematic errors that are unpredictable and cannot be removed from data and can only be estimated.

Rating equation An equation used to infer river discharge values from measured water levels at particular points along a river. The known discharges of the river at various different water levels are plotted and the equation for the line of best fit is calculated; the discharge at any water height can then be inferred (although there are inevitable errors in this process).

Rayleigh scattering The wavelength-dependent redirection or scattering of *electromagnetic radiation* caused by atmospheric particles that are much smaller than the wavelengths of the light they scatter.

Reaches (river) Sections of a river which show uniform characteristics such as flow depth, slope and area. This could be found between two morphological features such as debris dams or boulders. They are usually observed at a scale of up to 10 m.

Reagent A substance or compound which is added to another substance to initiate a chemical reaction or to determine whether a reaction will occur.

Realized niche A term to describe the *niche* more commonly utilized by most species whereby competitive interaction between several species attracted to the same resource has inhibited attainment of the *fundamental niche*.

Rebound hardness The hardness of a substance as measured by a hammer or other object which bounces off it. The height of the bounce from a given drop height is compared from substance to substance.

Recharge Replenishment of groundwater stores.

Redox potential The reducing or oxidizing intensity of a system, measured with an inert platinum half-cell and a reference half-cell calibrated against the hydrogen electrode. A measurement conducted in this manner is known as the Eh.

Reduction A chemical weathering process in which oxygen is dissociated from minerals creating a negative electrical charge; it usually occurs in anaerobic conditions.

Reductionist The assumption that the system under study is 'closed', i.e. all other variables within the system are held constant, allowing the direct relationship between two variables to be ascertained, and thereby eliminating reference to the potential influence of extraneous variables (e.g. *positivism* is reductionist).

Reflection The process by which a wave approaching a vertical or near-vertical object (e.g. sea cliff or sea wall) is rebounded from the object. If the angle of wave approach is parallel to the object, the wave will be reflected in the opposite direction to the line of approach. If the wave strikes at an angle of incidence other than parallel the wave is reflected in the tangent to the angle of approach.

Reflective beaches Also termed low-wave energy beaches, they often form in protected pockets on the lee side of rocks and can be identified by steep narrow beaches of coarse sand and a narrow surf zone.

Reflectivity The ability of a body to return energy.

Refraction The process by which a wave front bends and changes direction owing to a reduction in velocity as the wave enters shallow water.

Refugia Isolated habitats with distinctive ecological, geological, geomorphological or microclimatic characteristics that allow formerly widespread species to survive following a period of climatic change.

Regelation A two-fold process involving the melting of ice under pressure (the melting point of ice under pressure is lower than 0°C) and its subsequent transport and refreezing where the pressure is reduced; a major factor in the mechanism of downslope movement of a glacier.

Regime The seasonal variation in river flow which tends to be repeated each year is the river regime.

Regolith(s) The basal layer of soil overlying the bedrock composed of loose, unconsolidated weathered rock and gravel debris; it is the raw material from which soils are developed.

Regressive barriers Large mounds of sediment that have developed under the influence of a falling sea level and/or excess sediment supply. Landward sediments are deposited on top of more seaward ones.

Relative sea level Level of the sea relative to the land determined by *eustasy* and *isostasy*.

Relaxation time The amount of time that an environment/landscape takes to recover from a major event (e.g. a flood or landslide).

Relay floristics A concept relating to species invasion and disappearance from a local community; it suggests that as one group of species establishes itself it is replaced by another which is then replaced until a stable state is achieved.

Renaissance A period of change in culture in Europe when classical art and learning was re-examined and embraced. It began in the late fifteenth century in Italy and then spread to other European countries.

Reptation A method of sand transport in which grains are set into a low motion due to the high-velocity impact of a descending saltating grain.

Residence times The period of time a substance, e.g. nutrients, remains in a single location.

Residual tidal current Net movement of water due to tides occurring over long time periods. Tidal movements do not necessarily balance out over time thereby creating an overall water movement in a particular direction.

Resilience The ability of a system to recover from an event, change or shock.

Resolution Describes the ability of a system to separate a scene into constituent parts whether these parts be spatial, temporal or spectral.

Return flow Subsurface flow in the soil, either throughflow or macropore flow, that encounters a zone of soil saturation or lower hydraulic conductivity and is forced up through the soil profile to flow over the ground surface.

Rhizomes The underground lateral stems of certain plants that send up the new shoots (the rootstock).

Rhizosphere A zone approximately 1 mm wide surrounding the roots of a plant. The chemistry and biology of the soil in this zone are influenced by the plant root as a result of plant uptake and exudates.

Riffle A *bar* deposit found on the bed of river channels, usually spaced between 5 and 10 times the channel width apart. The height above the average bed surface causes fast-flowing, shallow and broken water under low- and medium-flow conditions.

Rill A small channel, formed by the merging of sheet wash into channelized flow, that acts as a conduit for water and sediment and is liable to collapse and change location between each runoff event.

Rillwash A hillslope erosion process that occurs when rainflow is deeper than 6 mm (generally in small channels carved out of the hillslope), rendering raindrop detachment ineffective; sediment detachment occurs when the downslope component of gravity and fluid flow traction overcome the frictional resistance of the soil.

Rip currents A strong seaward-directed current associated with water returning to the sea after being brought onshore by wave-breaking activity; an accumulation of water develops which pushes down the beach via a line of least resistance.

Riparian zone A region on either side of a stream or river which is characterized by vegetation which differs from that outside of the riparian zone due to soil conditions found in the region. Riparian zones can provide a number of ecosystem services including acting as filters for overland runoff, providing habitats and also producing *allochthonous* resources for river ecosystems.

Rising limb The increase in river discharge in response to a rainfall event, as depicted in a storm hydrograph.

River reach See *Reaches (river)*.

River segments These are sections of a river which are the length of the river between two bends. They are usually studied at the scale of approximately 100 m.

Roche moutonnées Small *stoss-and-lee forms*.

Rock flour Fine-grained rock particles pulverized by glacial erosion.

Rock glacier A tongue-like body of angular debris resembling a small glacier but with no ice evident at the surface and only *interstitial ice* in the pore spaces between the debris. Their movement downvalley is very slow and many appear stagnant.

Rock shelters Shallow, sheltered niches in a hillside, smaller and less pronounced than a cave. Ancient human occupation often results in rich archaeological findings, in addition to other deposits indicative of past environmental conditions.

Rockfall A mass-wasting process whereby consolidated material falls and breaks up into a jumble of material at the base of a cliff or steep slope.

Rogen moraine A moraine characterized by a series of ribs of sediments lying transverse to the direction of ice advance, approximately 10-30 m in height.

Rossby waves Upper-air waves that undulate horizontally in the flow path of the jet streams and the westerlies.

Roughness length An indicator of the roughness of the ground surface and its impact upon surface winds, i.e. an urban surface has a much greater roughness length (up to 10 m for tall buildings) than agricultural crops (approximately 5-20 cm).

Rover A non-stationary global positioning system receiver that is used to collect three-dimensional position data over an area.

Ruderal Pertaining to species with a good colonizing ability, capable of growing on new or disturbed sites, e.g. weeds (also described as 'r' strategists).

S

Sabkha A salt-encrusted tidal flat environment; evaporation of groundwater draws in seawater which upon evaporation precipitates *gypsum* (e.g. the coasts of the Persian Gulf).

Safety factor The ratio of the sum of forces resisting movement to the sum of forces promoting movement of material down a slope; a value below 1 means movement will begin.

Salcrete A *duricrust* predominantly composed of sodium chloride (rock salt), a halite.

Salinization A process involving the accumulation of soluble salts of sodium, magnesium and calcium in the soil to the extent that the soil fertility is severely reduced.

Salt marshes Coastal marshes that develop on low-lying sheltered sections of coastlines (primarily in a lagoon, behind a spit or in an estuary). Specialized salt-tolerant vegetation (halophytes) traps silt particles and consolidates the environment through processes of vegetation succession.

Saltation A mechanism of sediment transport involving sediment grains being bounced along a bed surface.

Sand Sediment particles between 0.06 and 2 mm in diameter.

Sand seas Large areas of sand accumulations characterized by sand sheets and dunes; sediment grains are well rounded and typically quartz (e.g. the Sahara and Namibian Deserts). Also known as *ergs*.

Saprolite A soft, clay-rich, disintegrating rock found in its original place, formed by chemical weathering of igneous or metamorphic rock in humid, tropical or subtropical climates.

Saprovores An organism that survives on dead organic matter.

Saturated A term to describe the state of the soil when all soil pores are filled with water.

Saturated adiabatic lapse rate The rate at which temperature decreases in a rising parcel of saturated air.

Saturated hydraulic conductivity The rate of water movement through a porous medium when it is saturated (calculated using *Darcy's law*).

Saturated zone The zone under the surface, which lies beneath the *water table*, in which all pores of the aquifer rock or soil are filled with groundwater (i.e. saturated).

Saturation-excess overland flow A form of overland flow that occurs when all available soil pore spaces become full (i.e. the soil is saturated). Excess water is forced to flow over the surface.

Scale Describes the linear relationship between a linear distance on an image and the corresponding distance on the ground which determines how much detail is captured in the image.

Sclerophyllous Refers to plants with small, tough evergreen leaves which maintain a rigid structure at low water potentials thereby avoiding wilting. They are usually found in low-rainfall areas since the tough leaves help to reduce water loss.

Scrapers Another name for grazers.

Scratch hardness The resistance of a material, such as stone, to scratching by another known material. These known materials are assembled into a standard scale which is known as Moh's scale of minerals.

Scree Loose, angular, rocky material that has been loosened from a slope through weathering and deposited further down the slope.

Scree slope The area at the base of a hillside where loose angular sediment (*scree*) accumulates.

Sea walls Massive concrete, steel or timber structures built along the coastline, with a vertical or sometimes curved face. A *hard engineering* coastal management technique employed to protect local infrastructure from flooding or erosion.

Seamounts Individual volcanoes on the ocean floor whose origin is distinct from the plate boundary volcanic system of mid-ocean ridges or subduction zones, i.e. usually formed as a plate moves over a *hot spot*.

Sea-salt events Enrichment of precipitation with sea salts incorporated from sea spray in windy conditions.

Seasonal icings Mounds of ice formed in winter in topographic lows where groundwater reaches the surface, i.e. in areas where *return flow* occurs and freezes.

Secondary endemic A species becomes endemic through the extinction of those species occurring in other places where they once survived (e.g. mammals of the West Indies).

Secondary minerals Minerals formed by the breakdown and chemical weathering of less resistant primary minerals (e.g. clays and oxides of iron and aluminium).

Secondary succession *Ecological succession* beginning on a previously vegetated site that has been recently disturbed by natural agents (e.g. fire, flood and hurricanes) or by human activities (e.g. deforestation). Remnant seed banks and root systems may influence the character of the resulting community.

Sediment budget An account of the inputs, outputs and stores of sediment for a given system.

Sediment yield The amount of sediment, both in suspension and transported as bed load, that is lost from a catchment. Usually measured as tonnes per year or tonnes per year per unit catchment area.

Sedimentary rock Rock which has formed by the gradual accumulation of sediment through time which has then solidified.

Sedimentation The process in which sediment is deposited leading to its accumulation (e.g. at deltas).

Segregated ice Very large lenses of ice that have slowly built up in frozen soil as a result of the migration of water to the freezing front (typically only in the upper 5-6 m of ground).

Seif dunes Linear dunes formed where two dominant wind directions are present at approximately right angles to each other.

Sensible heat Heat that can be measured by a thermometer and felt by humans.

Seral stage A stage within the process of *ecological succession* which is characterized by a particular biotic community. A series of seral stages (and their associated biotic communities) successively follow one another in the path to the climax community; each community creates conditions more favourable for a succeeding community.

Sesquioxides An oxide containing three atoms of oxygen to two atoms (or radicals) of some other substance.

Sessile A term to describe benthic organisms attached to a substrate and hence immobile (fixed to the ocean bottom).

Shadow dunes Small wind-blown dunes that develop in coastal or dryland areas around obstacles such as driftwood, a rock or a dead animal.

Shear A condition or force causing two contacting layers to slide past each other in opposite directions parallel to their plane of contact.

Shear stress A stress that acts upon a particle in the same plane as the surface the particle is resting upon (i.e. opposed to normal stress acting in the direction of gravity), resulting in either movement or strain of the particle. In the context of river systems, shear stress is the velocity of flowing water; when a critical flow velocity (and hence a critical shear stress) is reached, frictional forces may be overcome and a particle lifted from the bed.

Shield areas Tectonically stable areas of exposed rock which date back to the Precambrian era, these areas are relatively flat and show limited evidence of tectonic activity such as mountain building which is more evident at the margins of shield areas.

Shield volcanoes Large, dome-like rarely explosive volcanoes with gentle slopes of 6-12° formed by alternate layers of runny basalt, e.g. the Hawaiian shield volcanoes.

Shifting cultivation A form of plant cultivation in which seeds are planted in the fertile soil prepared by cutting and burning the natural growth. Relatively short periods of cultivation are followed by longer periods of fallow to allow soil rejuvenation, returning to the site years later.

Shoaling A gradual shallowing of the seabed.

Shore The land bordering the sea between the water's edge at low tide and the upper limit of effective wave action.

Shoreline The water's edge where the shore and the water meet; it varies over time.

Shore platform An erosional surface of horizontal or gently sloping rock in the intertidal zone that has developed following erosion of a rocky coast.

Short-wave radiation Incoming radiation whose wavelength is unchanged from its solar source.

Shredders Organisms which feed on coarse particulate organic matter, breaking it down into fine particulate organic matter. These organisms are considered most important in temperate river systems and are predominantly found in forest streams where there is much plant and leaf litter.

Significant wave height The mean height of the top tenth of all wave heights recorded at a given location (used as an approximate measure of wave energy for that location).

Silcrete A *duricrust* predominantly composed of silicates.

Silt Sediment particles between 0.004 and 0.06 mm in diameter.

Silviculture Subject of utilizing plant processes to grow trees for harvesting.

Sinusoidal The mathematical shape of a curve of sines, i.e. a wave consists of a simple sinusoidal form.

Slaking A process that involves raindrops striking a soil surface and water being forced into a soil aggregate therefore compressing the air inside and causing the aggregate to explode into its constituent grains.

Slantwise convection Convection (vertical rise in an air parcel) is inhibited when the prevailing lapse rate is less than the appropriate adiabatic lapse rate. However, a poleward horizontal movement of an air mass may bring the air parcel into an environment denser than itself, thereby allowing the air parcel to rise through slantwise convection.

Slide *Mass movements* which involve a large mass of earth or rock essentially moving as a block as opposed to *flows*.

Slumping A mass movement process whereby saturated slope material moves downslope under the force of gravity and deforms upon movement.

Smelting The process of extracting a metal from its ores by heating.

Snowline The altitude marking the lower limit of permanent snow in upland or high-latitude areas, i.e. the line where the winter snowfall exceeds the amount removed by summer melting and evaporation.

Social ecology The study of the dynamics and diversity of social behaviour and social systems of animals; social ecological variables include measures of group composition, inter-male competition and habitat preference.

Social impact assessment An evaluation of the impact of a proposed activity on all the social aspects of the environment including: people's *coping strategies* (economic, social and cultural); use of the natural environment; the way communities are organized through social and cultural institutions; and the identity and cultural character of a community. It involves characterizing the existing state of these aspects of the social environment in addition to predicting how they might change.

Soft engineering Pertaining to more 'sensitive' management practices that involve methods more closely associated with geomorphological processes and local sediment dynamics; large *'hard engineering'* types of structures are avoided.

Soil A complex medium consisting of inorganic materials, organic matter (living and dead), and water and air variously organized and subject to dynamic processes and interactions. It forms the natural terrestrial surface layer that is the supporting medium for the growth of plants.

Soil biomass The living component of soil organic matter, it is the term given to a mass of organisms in a specified mass of soil. It is often used as an indicator of soil quality and includes organisms such as bacteria and earthworms.

Soil colloid Very small mineral particles (less than 0.002 mm in diameter) that stay suspended in water, the most important being clay minerals capable of remaining suspended in water indefinitely.

Soil colour A visible characteristic of the soil that allows the determination of soil properties such as organic matter content, iron content, soil drainage and soil aeration.

Soil creep The very slow, imperceptible, movement of material downslope under the force of gravity.

Soil horizons Distinctive horizontal layers within a *soil profile*, created primarily by the translocation of materials with water moving through the soil.

Soil organic matter Predominantly consists of carbon, but is also made up of other elements including nitrogen, phosphorus, oxygen and sulphur. It can be split into three groups - litter, humus and biomass - and is an important component of a healthy and productive soil. It has a number of functions including retaining moisture and organic pollutants and providing food for soil biomass.

Soil pipes Horizontal tube-like subsurface cavities within the soil; special forms of macropores greater than 1 mm in diameter. They are continuous in length such that they can transmit water, sediment and solutes through the soil and bypass the soil matrix.

Soil profile A vertical section through the soil from the ground surface down to the parent material; the profile characteristics determine the soil type.

Soil solution The water held in the soil pores that contains dissolved organic and inorganic substances and hence is not pure.

Soil structure The shape, size and distinctiveness of soil *aggregates*, divided into four principal types (blocky, spherical, platy and prismatic).

Soil texture The relative proportions of sand, silt and clay-sized fractions of a soil.

Solifluction Form of slow mass movement in environments that experience freeze-thaw action or highly variable warming and cooling of the surface. This results in a slow movement of soil material downslope.

Solum The portion of the soil where soil-forming processes are active and plant and animal life are mostly confined; the A, E and B horizons.

Solute load The total mass of material transported in solution by a flow.

Solvent A substance which dissolves another substance to produce a solution.

Sorting A measure of the spread, or standard deviation, of grain sizes within a sediment. In general the further a sediment deposit has been transported from its source, the greater the sorting of grains.

Speciation The evolution of new species involving the relatively gradual change in the characteristics of successive generations of an organism, ultimately giving rise to species different from the common ancestor. Most biologists accept Darwin's basic hypothesis of speciation from a common ancestor as a result of natural selection of those attributes best suited to survival in a given habitat with limited resources. Speciation can take a number of forms, whether sympatric (populations overlapping) or allopatric (populations become isolated from one another).

Specific conductance The ability of water to conduct an electric current, dependent on the concentration of ions in solution.

Specific heat The energy required to change the temperature of 1 gram of a substance by 1 degree Celsius. Water has a higher specific heat than air, requiring more energy to be absorbed for any given temperature change.

Spectral band A division of the electromagnetic spectrum that groups energy according to similarities.

Spectral signature Describes the reflectance characteristics of a surface across the electromagnetic spectrum.

Specular reflection The redirection of radiation off a smooth surface such that the radiation is otherwise unchanged.

Speleothems Structures formed in a cave by the deposition of minerals from water, e.g. a stalactite, stalagmite. They are primarily composed of calcium carbonate precipitated from groundwater percolating through carbonate rock, e.g. limestone.

Spits Narrow and elongated accumulations of sand and shingle projecting into the sea, usually with a curved seaward end caused by wave action. They grow out from the coastline when the shore orientation changes but longshore currents do not deviate and continue to transport and deposit along a projected coastline, e.g. at the mouth of an estuary.

Spring tides A tide that occurs at or near the new moon and full moon when the gravitational pull of the Sun reinforces that of the Moon producing a large tidal range, causing higher than average high tides and lower than average low tides.

Stable isotope Isotopes of an element possess the same number of protons in their nuclei but have different numbers of neutrons. A stable isotope does not break down by radioactive decay. For example, ^{12}C is a stable isotope and the most widespread form of carbon in the environment, but the radioactive isotope ^{14}C and the stable isotope ^{13}C also occur.

Stadial A short period of climatic deterioration within an interglacial period; glaciers advanced and periglacial conditions extended but in a less pronounced way than during a *glacial interval*.

Stakeholders A person or group who can affect or is affected by an action and therefore has a vested interest in the outcomes. Responsible decision-making requires consideration of the effects on all stakeholders.

Stand An area of more or less homogeneous vegetation.

Standard deviation A measure of how spread out the data are around the mean.

Stellate dunes Star-shaped dunes formed under conditions of variable wind direction with no one prevailing wind direction. These dunes do not migrate.

Stemflow The flow of water down the trunk of a tree or stems of other vegetation allowing water to reach the hillslope.

Stereo images Aerial photographs that have been obtained such that each photo overlaps another by a prescribed amount. Overlapping coverage provides two points of observation to provide *parallax* required for *digital elevation model* generation.

Stochastic A model that contains some random element in the operation or input data so that more than one, and usually a very large number of, outcomes are possible.

Stone pavements Accumulations of flat-lying boulders in a mosaic pattern at the ground surface in periglacial environments. Some argue they are formed by aeolian removal of fine surface particles, but it is more commonly argued they are displaced upwards as small particles fall into ground cracks created during freeze–thaw cycles while the larger boulders cannot.

Stormflow The peak flow that occurs during or immediately following a rainfall event occurring as a result of overland flow and rapid subsurface flow (e.g. pipeflow contributions may also be high).

Storm surge barriers A *hard engineering* technique with a main function to protect low-lying and coastal areas from flooding during storm events which can be exacerbated by the occurrence of rising sea levels.

Stoss-and-lee forms Streamlined elongated rock exposure formed by the sliding of debris-rich basal ice over the bedrock surface under a glacier; characterized by a gently sloping glacially smoothed upstream side and a steeper plucked downstream side (centimetres to metres in length).

Strain history The amount of deformation of a substance that has occurred owing to previous stress impact; it can affect present and future stress–strain relationships.

Strain rate The amount of deformation occurring over time for a given material (i.e. the rate of deformation). For glaciers, the strain rate for a given shear stress is determined by *Glen's law*.

Stratification (stratify) Division of water in deep lakes, reservoirs and stable water bodies into layers of differing density.

Stratified estuaries Estuaries with limited saltwater and freshwater mixing (via *advection* and *diffusion*) causing a lower layer of denser and saltier water with an upper layer of less dense freshwater; a salt wedge develops.

Stratigraphy The layering of sediments.

Stratosphere A layer of the atmosphere lying above the *troposphere* about 50 km above the Earth's surface.

Stratotype A particular stratigraphic unit with clear and well-recorded characteristics and boundaries. This site can become the point of reference for comparison with a more poorly preserved record. Also known as typesite.

Stream competence The maximum particle size a stream can transport.

Stream order Numbering of the drainage network according to the number of tributaries and stream network linkages.

Stream power The rate of energy supply in a river that is available for work to be done at the stream bed, measured in $W\ m^{-2}$.

Stress The force per unit area acting on a plane within a body due to application of an external load; six values are required to characterize the stress at a point completely (three normal components and three shear components).

Striations Microscale erosional features on rock surfaces, resembling a scratch.

Sub-aerial An object or a process which exists or occurs near to or on the surface of the Earth.

Subduction The process in which one lithospheric plate descends beneath another into the asthenosphere when the two plates converge.

Subglacial Pertaining to the environment at the base of a glacier.

Sublimate A change in the physical state of a substance directly from solid to gaseous form.

Sublimation The chemical process in which a solid changes directly into a gas.

Subsea permafrost *Permafrost* found beneath the sea; sometimes due to low temperatures at the bed, more usually a remnant of past colder temperatures and rising sea levels (drowning frozen ground).

Subsurface flow Pertaining to *throughflow* that occurs through *micropores*, *macropores* and *soil pipes*.

Succession Changes over time in the structure or composition of an ecological community. These changes often follow a predictable pattern.

Sun-synchronous orbit The orbit of a satellite travelling around the Earth which is timed such that it passes over any given latitude at the same time at each pass so as to ensure that illumination conditions remain constant between subsequent images.

Super-adiabatic A term used for localized steep *lapse rates* that are greater than even the *dry adiabatic lapse rate* causing rapid local convection, e.g. strong radiational heating of the ground surface.

Superimposed ice The ice formed when water from melting snow comes into contact with the cold surface ice of a glacier at the base of the snowpack and refreezes.

Supply limited A transport process that is limited by the lack of sediment supply, not the capacity to transport sediment since more force is available than is being utilized. For example, rockfalls are limited by the amount of material that is loose enough to fall.

Supraglacial Pertaining to the environment at the surface of a glacier.

Surf zone A process-based term for the area within the *nearshore* zone where breaking waves approach the shore usually over a wide, low gradient.

Surface boundary layer A layer extending upwards from the Earth to a height that ranges anywhere between 100 and 3000 m. Here, almost all interactions between the atmosphere and humans take place.

Surface tension The resistance of the surface of a material to external forces. It is determined by the cohesive energy between the molecules which form the surface of the object. Molecules in the middle of an object are subject to equal forces on all sides whereas forces acting on the surface molecules are not in balance, however.

Suspended load The sediment transported in water when lifted from the bed surface and kept in suspension by turbulent fluid flow.

Sustainable development Development (any form of development from an action, project, strategy or legislation) that meets the needs of people today without compromising the ability of future generations to meet their own needs.

Swash The thin sheet of water that travels up the beach following the breaking of a wave.

Swash zone Process-based term for the area within the *nearshore* zone where broken waves travel up the beach as *swash* and return as *backwash*.

Swath The area on the ground covered by the motion of a remote sensing instrument.

Symbiotic Living together in a mutually beneficial relationship.

Systematic errors Predictable errors that can be modelled and removed from the data.

T

Tafoni These are cavities in a rock face which are the result of weathering. They are elliptical in shape and are most commonly found on vertical and sloping rock faces. They usually exist in honeycomb-like groups and can be found on a variety of rock types.

Talik A Siberian word for an unfrozen pocket within permafrost; for example, beneath a lake or warm-bedded glacier.

Talus An accumulation of angular rock debris from rockfalls found at the base of a slope.

Tarn A depression located at the site of a melted *corrie* glacier; a lake usually forms in the centre.

Taxonomy The study, description and systematic classification of living organisms (plant and animal) into groups based on similarities of structure or origin. Synonymous with systematics.

Telemetry The process of obtaining measurements in one place and relaying them for recording or display at a different site.

Temperate glacier A glacier formed in temperate climates where the temperature of the entire glacier is at the pressure melting point except for the surface 10–20 m (which fluctuates with the season); considerable quantities of meltwater are generated causing high rates of glacier movement and erosion.

Temperature inversion A reversal of the normal environmental temperature *lapse rate*; air temperature increases with altitude.

Tensile strengths The amount of force which is required to pull an object to the point of fracture.

Tensile stress The action of a force pulling an object along its cross-sectional axes in an outwards motion from its centre.

Terminal mode The final form of a particle of glacial sediment in which the particle will not break down into a finer form even with prolonged transport in the glacial system.

Terminal velocity The velocity at which the frictional drag forces acting on a falling object are equal to the driving forces of gravity, resulting in a constant fall rate (neither accelerating nor slowing down).

Thalweg The line of maximum water velocity down the path of a river.

Thermal conductivity The degree to which a substance transmits heat.

Thermal scanner A remote sensing instrument similar to a *multispectral scanner* but that can only sense radiation in the thermal infrared portion of the spectrum.

Thermocline The depth at which the temperature gradient of the water column rapidly changes in the vertical dimension, marking the contact zone between water masses of markedly different temperatures. Also known as the metaliminion.

Thermohaline circulation Large-scale circulation of the world's oceans, involving the vertical movement of large bodies of water, driven by water density differences. Cold, salty water sinks in 'downwelling zones', particularly at high latitudes in the North Atlantic, and flows slowly southward along the bottom of the Atlantic and into the Pacific, where it rises again mainly in an 'upwelling' zone of western South and Central America. It then flows back as a surface current. The thermohaline circulation is estimated to take 2000 years to complete one revolution. It is very important in transporting heat through the Earth system.

Thermokarst A term referring to the ground surface depressions which are created by the thawing of ground ice (and subsequent water erosion) in periglacial areas, e.g. *pingos*.

Thermoluminescent Pertaining to luminescence (an emission of light) resulting from exposure to high temperature; used as a means of dating ancient material.

Thiessen polygon The spatial influence of a particular data point calculated using arithmetic spatial averaging techniques on a network of data points.

Thixotropic Pertaining to the property of becoming fluid when agitated but recovering its original condition upon standing; viscosity decreases as the rate of shear (relative movement) increases.

Throughfall Water reaching the ground surface after dripping or bouncing off overlying vegetation.

Throughflow The downslope movement of water draining through the soil.

Through-wash A wash process involving the movement of regolith particles through the pores between grains in the regolith; the particles must be at least 10 times smaller than the grains they are passing between, and the process is therefore only significant in washing silt and clay out of clean sands.

Tidal currents Currents produced by the rise and fall of the tide; either the movement in and out of an estuary or bay, or the movement of water between two points affected by different tidal regimes (especially common in straits).

Tidal prism The volume of water that moves in or out of an area such as an estuary during a tidal cycle.

Tidal range The difference in water level between high and low water during a tide.

Till The generic term for sediment deposited directly by glacier ice.

Tillage erosion An anthropogenic soil erosion process (similar to creep) which is the result of ploughing either up- and downslope or along the contour. The turning over of soil produces a direct downhill movement. Whatever the ploughing direction, the process is faster than natural soil creep.

Tilt of the Earth The Earth's axis lies at an angle that varies from approximately 21° to 24° and back again in a 41 000 year cycle. The greater the tilt, the more intense the seasons in both hemispheres become.

Topset bed A horizontal bed of coarse sediment deposited by braided streams crossing a *delta plain*; it represents the sub-aerial part of the delta.

Topsoil The upper section of the soil that is most important for plant growth (usually the A horizon or plough horizon).

Total stream length The combined length (km) of all components of the channel network.

Trade winds The prevailing winds in the tropics blowing from high pressure at the tropics to low pressure at the equator. The winds do not blow directly north-south because the rotation of the Earth deflects them to the left in the southern hemisphere and to the right in the northern hemisphere.

Tragedy of the commons A term coined by Garret Harding in 1968 that refers to the excessive exploitation of a communal resource to a point of degradation due to the selfish nature of rational people who will use more than their fair share of the resource; no one person will take responsibility for something owned by all. It is often used to demonstrate the mistake in allowing a growing population to increase steadily its exploitation of the ecosystem which supports it.

Transform faults Major strike-slip faults occurring where two plates slide past each other in the horizontal plane. They are capable of causing major destructive earthquakes, e.g. the San Andreas Fault.

Transgressive barriers Accumulations of sediment just offshore running parallel to the coastline which have formed under the influence of rising sea level and/or a negative sediment budget. They tend to consist mainly of tidal delta and/or washover deposits, and are underlain by estuarine or lagoonal deposits. In this instance sediments deposited in seaward environments end up on top of sediment that originated in more landward environments.

Translocation The transport of dissolved ions and small particles through the soil within the *soil solution*, to surface water and groundwater.

Transmission bands Sections of the electromagnetic spectrum that allow radiation to pass unobstructed. These are also called atmospheric windows.

Transpiration The loss of water to the atmosphere through the process of evaporation from leaf pores and plants.

Transport limited A transport processes that can only move material a limited distance from the source despite the plentiful supply of material, e.g. rainsplash.

Transporting capacity The maximum amount of material which the transport process can carry.

Transverse dunes Linear dunes with a shallow windward side and steep lee slope (similar in structure to dunes and ripples formed below water).

Treatments A single test within a larger experiment where a single variable has been altered from the control situation by a known quantity and applied to the principal substrate in order to ascertain its effect.

Treeline The altitudinal upper limit of tree growth; affected by latitude and local factors such as slope, soil, aspect and exposure.

Trophic level A functional or process category describing the position of an organism or group of organisms in a food chain. Primary producers are at the first trophic level, those that feed on primary producers are at the second trophic level.

Tropopause The boundary between the *troposphere* and the *stratosphere*.

Troposphere The lowermost layer of the atmosphere extending to approximately 11 km above the Earth's surface.

Truncated spur A valley side spur that has been abruptly cut off and steepened at its lower end by the erosive action of a glacier.

Tsunami A large sea wave generated by submarine seismic activity (earthquakes, slides, volcanic activity) or meteor impact in the ocean. These waves can be extremely destructive, especially in the Pacific Rim, contributing to the coastline development of these areas.

Turbidite current A density current involving mixtures of sediment and water which, owing to their increased density relative to seawater, flow down and along the bottom of the oceans transporting sand and clay-sized sediment from shelf slopes to deeper oceanic environments.

Turbulent flow One of the two ways in which water can flow. It involves water molecules moving in many directions, with an overall net flow in one direction; as a result the flow is well mixed.

U

Undertow See *bed return flow*.

Unidirectional flow Currents flowing in one dominant direction (e.g. rivers and wind).

Uniformitarianism A practical principle of modern science concerning the method in which scientists explain phenomena; it advocates the use of the simplest explanation which is consistent with both evidence and known scientific laws. It is primarily related to James Hutton's demonstration in 1788 that the simplest explanation of the development of the Earth's landscape is through the observed processes of erosion and uplift acting gradually over time, rather than catastrophic landform development through divine intervention.

Urban boundary layer The section of the atmosphere directly overlying the *urban canopy layer*, subject to *urban heat island* effects through the entrainment of air from the urban canopy layer and anthropogenic heat from roofs and chimneys.

Urban canopy layer The section of the atmosphere immediately overlying an urban area, subject to heavy localized *urban heat island* effects, including greater daytime heat storage, anthropogenic heat release from buildings and decreased evaporation.

Urban heat island Urban landscapes adjust the local microclimatic processes, resulting in an 'island' of warmer air surrounded by cooler rural air; two distinct regions of atmospheric modification are observed (the *urban canopy layer* and the *urban boundary layer*). The effect is most apparent at night, owing to slower cooling of the urban landscape, and during light winds.

U-shaped valley A wide valley with steep sides formed by glacial erosion of a V-shaped valley, involving the formation of truncated spurs during the valley straightening.

V

Valent A substance which has valence, often used to refer to the potential of an atom to combine with another atom. This is dependent on the number of atoms which can be shared, lost or gained through combining.

Value A measurable variable of soil colour describing the degree of darkness or lightness of the colour (a value of 0 represents black).

Variable source area concept The idea that the area of a catchment that produces *saturation-excess overland flow* will vary through time, i.e. during a rainfall event a greater proportion of the catchment will begin to contribute saturation-excess overland flow as time progresses, and the catchment becomes more saturated.

Varve A thin laminar bed of glacial sediment deposited by a proglacial stream in a repetitive annual sequence; coarse particles are deposited in summer and the finer particles progressively throughout the year.

Ventifacts The smoothed surfaces of individual stones eroded by sand and dust particles entrained in the wind.

Venturi effect Wind forced to funnel between two buildings causes the local wind speed to increase.

Vital attributes The critical physical characteristics of plants that determine their ability to survive disturbance, including their methods of persistence, conditions for establishment and timing of life stages.

W

Wadati-Benioff zone The band of rock (20 km thick) which dips from the trench region under an overlying plate in a subduction zone. It is the location of earthquake foci that are associated with descending lithospheric plates.

Wandering gravel-bed river A gravel-bed river channel characterized by an irregularly sinuous thalweg that is frequently split around vegetated islands and low-order braiding within complex bar deposits where the river is laterally unstable.

Water balance Pertaining to the cyclical movements of volumes of water within a drainage basin per unit time. It relates to the various inputs, storage and outputs of water within the drainage basin system and controls the nature of river discharge.

Water table The upper boundary of the zone of groundwater saturation (the *saturated zone*). Its level varies with the amount of precipitation, evapotranspiration and percolation.

Watershed Another name for *catchment*.

Wave asymmetry The nature of waves that is not symmetrical on either side of the wave crest. As waves enter shallower water they develop peaked crests and flat troughs. In addition to the asymmetrical shape, the water flow velocities also become asymmetric with the onshore side of the wave being stronger but of shorter duration than the offshore side of the wave.

Wave base The point of a wave below which there is no orbital movement of water.

Wave breaking The destruction of a wave when it becomes too steep and disintegrates.

Wave convergence The focusing of wave rays so that they come together increasing in energy and height.

Wave crest The peak of the curve of a wave.

Wave divergence The separation of wave rays so that waves move apart. Typically waves will become shorter and less energetic.

Wave energy flux The rate of transfer of energy by waves.

Wave frequency The number of *wave crests* which pass a fixed point over a set timescale.

Wave height The vertical distance between the *wave trough* and the *wave crest*.

Wave period A measure of wave speed; the time taken for two successive *wave crests* to pass a fixed point.

Wave set-up Wave breaking results in water piling up against the shore. This results in a slope of water with higher water pushed nearer the shore and this 'set-up' is sufficient to oppose the shoreward wave stresses.

Wave steepness Wave height divided by wavelength.

Wave trough The base of the curve of a wavelength.

Wavelength The distance between a *wave crest* to the next wave crest, or between trough to trough.

Weathering The breakdown of rocks and minerals by the physical and chemical processes of erosion.

Well-mixed estuary An estuary in which mixing is so effective that the salinity gradient in the vertical direction vanishes entirely. If the estuary is wide enough then the Coriolis force pushes the flow of the outflowing river to the margin of the estuary and may result in a horizontal separation of riverwater and seawater.

Wet-snow zone The region of a glacier in which the entire snowpack becomes saturated at the end of the summer.

Wetted perimeter The contact area between the channel bed and water when viewed in cross-section. Bank-full wetted perimeter is calculated as the estimated contact zone when the channel is completely full with water.

Whaleback A streamlined elongated rock exposure formed by the basal sliding action of a glacier; similar in shape to *stoss-and-lee forms* but the steep side faces upstream and the tapered end downstream.

Whole-life costing Expressing the results of *life-cycle analysis* in financial terms, i.e. placing a monetary value on all the environmental impacts of a product from its manufacture to disposal.

Wilting point The condition of a soil when plants cannot withdraw the necessary water for growth as the only remaining soil water is that held tightly to soil particles by hygroscopic forces and is unavailable for plant use.

Wind shear A change in wind speed or direction with altitude in the atmosphere.

X

Xeromorphic The possession, by a plant, of features adapted to conditions of limited moisture availability.

Xerophyte A plant that has adapted to grow in very arid conditions with restricted water availability by minimizing water loss and maximizing water efficiency.

Xerophytic Pertaining to having the character of a *xerophyte*, i.e. an organism adapted to growth in conditions of limited water availability.

Y

Yardangs Large-scale dryland features; the erosive power of dust carried in the wind leads to the smoothing of entire hills streamlined in the direction of sediment transport.

Yield strength The stress at which a material exhibits a deviation from the proportionality of stress to strain, to produce a specified amount of plastic deformation, i.e. below the yield strength the material acts as an elastic and above as a viscous material.

Young's modulus Also known as the modulus of elasticity, it is a measure of the rate of change of stress in a material as a result of the strain applied, often in the form of compressional forces. On a stress–strain diagram it is the straight-line part of the graph.

Z

Zooplankton Microscopic organisms which consume other plankton. Zooplankton exist in a variety of forms including larval, immature stages of larger animals, single-celled organisms and tiny crustaceans.

A

Aagard, T. and Masselink, G. (1999) The surf zone. In: Short, A.D. (ed.), *Handbook of beach and shore morphodynamics*. Wiley Interscience, New York, 72-118.

Abernethy, B and Rutherfurd, I.D. (2000) The effect of riparian tree roots on the mass-stability of riverbanks. *Earth Surface Processes and Landforms*, **25**, 921-937.

Abrol, I.P., Yadav, J.S.P. and Messoud, F.I. (1988) Salt affected soils and their management. *FAO Soils Bulletin*, **39**.

Achard, F., Eva, H.D., Stibig, H.J., Mayaux, P., Gallego, J., Richards, T. and Malingeau, J.P. (2002) Determination of the deforestation rate of the world's humid tropical forests. *Science*, **297**, 999-1002.

Agnew, C. and Anderson, E. (1992) *Water resources in the arid realm*. Routledge, London.

Ahnert, F. (1976) Brief description of a comprehensive three-dimensional process-response model of landform development, *Zeitschrift für Geomorphologie, Supplement*, **25**, 29-49.

Ahrens, C.D. (2000) *Meteorology today*, 6th edition. West Publishing, Mineapolis, MN.

Ahrens, C.D. (2003) *Meteorology today – An introduction to weather, climate, and the environment*. Brooks/Cole, Pacific Grove, CA.

Allan, D.J. and Castillo, M.M. (2007) *Stream ecology: Structure and function of running waters*. Springer-Verlag, Dordrecht.

Allan, J.D. (2004) Landscapes and riverscapes: the influence of land use on stream ecosystems. *Annual Review of Ecology and Systematics*, **35**, 257-284.

Allen, J.R.L. (1985) *Principles of physical sedimentology*. George Allen and Unwin, London.

Allen, M.R., Frame, D.J., Huntingford, C. *et al.* (2009) Warming caused by cumulative carbon emissions towards the trillionth tonne. *Nature*, **458**, 1163-1166.

Allen, P.A. (1997) *Earth surface processes*. Blackwell Science, Oxford.

Alley, R.B. (1991) West Antarctic collapse – how likely? *Episodes*, **13**, 231-238.

Alley, R.B., Blankenship, D.D., Bentley, C.R. and Rooney, S.T. (1986) Deformation of till beneath ice stream B, West Antarctica. *Nature*, **322**, 57-59.

Alley, R.B., Meese, D.A., Shuman, A.J. *et al.* (1993) Abrupt increase in Greenland snow accumulation at the end of the Younger Dryas Event. *Nature*, **362**, 527-529.

Allsopp, D., Seal, K. and Gaylarde, C. (2004) *Introduction to biodeterioration,* 2nd edition. Cambridge University Press, Cambridge.

American Meteorological Society (1997) Meteorological drought – policy statement. *Bulletin of the American Meteorological Society*, **78**, 847-849.

Amthor, J.S. (2000) The McCree-de Wit-Penning de Vries-Thornley respiration paradigms: 30 years later. *Annals of Botany*, **86**, 1-20.

Anderson, D.E., Goudie, A.S. and Parker, A.G. (2007) *Global environments through the Quaternary: Exploring environmental change.* Oxford University Press, Oxford.

Anderson, D.M. and Morgenstern, N.R. (1973) Physics, chemistry and mechanics of frozen ground: a review. *North American Contribution, 2nd Permafrost International Conference*, US National Academy of Sciences, Washington, DC, 257-288.

Anderson, K. and Bows, A. (2008) Reframing the climate change challenge in light of post-2000 emission trends. *Philosophical Transactions of the Royal Society A*, **366**, 3863-3882.

Anderson, M.G. (ed.) (1988) *Modelling geomorphological systems*. John Wiley & Sons, Chichester.

Anderson, M.G. and Burt, T.P. (eds) (1985) *Hydrological forecasting*. John Wiley & Sons, Chichester.

Anderson, R.S. and Haff, P.K. (1988) Simulation of eolian saltation. *Science*, **241**, 820-823.

Andoh, R.Y.G. (1994) Urban runoff – nature, characteristics and control. *Journal of the Institution of Water and Environmental Management*, **8**, 371-378.

Andreassen, L.M., Elvehøy, H. and Kjøllmoen, B. (2002) Using aerial photography to study glacier changes in Norway. *Annals of Glaciology*, **34**, 343-348.

ANRA (2001) *Fast facts 21 – Dryland salinity in Australia*. Australian National Resources Atlas, Canberra.

Archer, D. (2005) Fate of fossil fuel CO_2 in geologic time. *Journal of Geophysical Research*, **110**, C09S05, doi:10.1029/2004JC0022625.

Archfield, S.A. and Vogel, R.M. (2010) Map correlation method: selection of a reference streamgage to estimate daily streamflow at ungaged catchments. *Water Resources Research*, **46**, W105134, doi:10.1029/2009WR008481.

Archibold, O.W. (1995) *Ecology of world vegetation*. Chapman and Hall, London.

Arnell, N. (2002) *Hydrology and global environmental change*. Pearson Education, Harlow.

Asante-Duah, D.K. (1998) *Risk assessment in environmental management*. John Wiley & Sons, Chichester.

Ashurst, J. and Dimes, F.G. (1998) *Conservation of building and decorative stone*. Butterworth-Heinemann, Oxford.

Atkins, P., Roberts, B. and Simmons, I. (1998) *People, land and time: An historical introduction to the relations between landscape, culture and environment*. Arnold, London.

Atkinson, T.C., Briffa, K.R., Coope, G.R. *et al.* (1986) Climatic calibration of coleopteran data. In: Berglund, B.E. (ed.), *Handbook of Holocene palaeoecology and palaeohydrology*. John Wiley & Sons, Chichester, 851–858.

Austin, M.J., Scott, T.M., Brown, J.W. *et al.* (2010) Temporal observations of rip current circulation on a macro-tidal beach. *Continental Shelf Research*, **30**, 1149–1165.

Avery, B.W. (1990) *Soils of the British Isles*. CAB International, Wallingford.

Avery, T.E. and Berlin, G.L. (1992) *Fundamentals of remote sensing and airphoto interpretation*. Macmillan, Toronto.

B

Bäckström, M., Karlsson, S., Bäckman, L., Folkeson, L. and Lind, B. (2004) Mobilisation of heavy metals by deicing salts in a roadside environment. *Water Research*, **38**, 720–732.

Bagnold, R.A. (1941) *The physics of blown sand and desert dunes*. Methuen, London.

Baillie, M.G.L. and Munroe, M.A.R. (1988) Irish tree rings, Santorini and volcanic dust veils. *Nature*, **332**, 344–346.

Baker, V.R. (2003) Icy Martian mysteries. *Nature*, **426**, 779–780.

Ballantyne, C.K. (2002) Paraglacial geomorphology. *Quaternary Science Reviews*, **21**, 1935–2017.

Ballantyne, C.K. and Harris, C. (1994) *The periglaciation of Great Britain*. Cambridge University Press, Cambridge.

Barber, D.C., Dyke, A., Hillaire-Marcel, C. *et al.* (1999) Forcing of the cold event of 8,200 years ago by catastrophic drainage of Laurentide lakes. *Nature*, **400**, 344–348.

Bard, E. and Frank, M. (2006) Climate change and solar variability: what's new under the sun? *Earth and Planetary Science Letters*, **248**, 1–14.

Bard, E., Raisbeck, G., Yiou, F. and Jouzel, J. (2000) Solar irradiance during the last 1200 years on cosmogenic nuclides. *Tellus*, **52B**, 985–992.

Bard, E., Raisbeck, G.M., Yiou, F. and Jouzel, J. (2000) Solar modulation of cosmogenic nuclide production over the last millennium: comparison between ^{14}C and ^{10}Be records. *Earth and Planetary Science Letters*, **150**, 453–462.

Bardgett, R.D. (2005) *The biology of soil: A community and ecosystem approach*. Oxford University Press, New York.

Barnes, R.T., Raymond, P.A. and Casciotti, K.L. (2008) Dual isotope analyses indicate efficient processing of atmospheric nitrate by forested watersheds in the north eastern U.S. *Biogeochemistry*, **90**, 15–27.

Barrow, C.J. (1997) *Environmental and social impact assessment: An introduction*. Arnold, London.

Barrow, C.J. (2003) *Environmental management: Principles and practice*. Routledge, London.

Barrow, C.J. (2006) *Environmental management for sustainable development*. Routledge, London.

Barry, R.G. (1992) *Mountain weather and climate*. Routledge, London.

Barry, R.G. and Chorley, R.J. (2009) *Atmosphere, weather and climate*, 9th edition. Routledge, London.

Bass, I.G. (1982) Ophiolites. *Scientific American*, **247**, 108–117.

Batjes, N.H. (1996) Total carbon and nitrogen in the soils of the world. *European Journal of Soil Science*, **47**, 151–163.

Battjes, J.A. (1974) Surf similarity. *Proceedings 14th International Conference on Coastal Engineering*, ASCE, Reston, VA, 466–480.

Baxter, C.V., Fausch, K.D. and Saunders, W.C. (2005) Tangled webs: reciprocal flows of invertebrate prey link streams and riparian zones. *Freshwater Biology*, **50**, 201–220.

Beatley, T., Brower, D.J. and Schwab, A.K. (2002) *An introduction to coastal zone management*. Island Press, Washington, DC.

Beddington, J. (2009) Food, energy, water and the climate: a perfect storm of global events?, available at: www.bis.gov.uk/go-science/news/speeches.

Beeby, A.H. and Brennan, A.M. (1997) *First ecology*. Chapman and Hall, London.

Beerling, D.J. and Woodward, F.I. (2001) *Vegetation and the terrestrial carbon cycle. Modelling the first 400 million years*. Cambridge University Press, Cambridge.

Bellamy, P.H., Loveland, P.J., Bradley, R.I., Murray Lark, R. and Kirk, G.J.D. (2005) Carbon losses from all soils across England and Wales 1978-2003. *Nature*, **437**, 245–248.

Beniston, M. (2006) Mountain weather and climate: a general overview and a focus on climatic change in the Alps. *Hydrobiologia*, **562**, 3–16.

Benn, D.I. and Evans, D.J.A. (1996) The interpretation and classification of subglacially-deformed materials. *Quaternary Science Reviews*, **15**, 23–52.

Benn, D.I. and Evans, D.J.A. (2010) *Glaciers and glaciations*, 2nd edition. Hodder Arnold Education, Abingdon.

Bennett, M.R. and Glasser, N.F. (2009) *Glacial geology, ice sheets and landforms*, 2nd edition. John Wiley & Sons, Chichester.

Berner, E.K. and Berner, R.A. (1996) *Global environment: Water, air and geochemical cycles*. Prentice Hall, Englewood Cliffs, NJ.

Betson, R.P. (1964) What is watershed runoff? *Journal of Geophysical Research*, **69**, 1541–1552.

Bevan, S.L., North, P.R.J., Los, S.O. and Grey, W.M.F. (2011, in press) A global dataset of atmospheric aerosol optical depth and surface reflectance from AATSR. *Remote sensing of environment*, in press.

Beven, K. (2007) Towards integrated environmental models of everywhere: uncertainty, data and modelling as a learning process. *Hydrology and Earth System Sciences*, **22**, 460-467.

Beven, K.J. (1989) Changing ideas in hydrology – the case of physically-based models. *Journal of Hydrology*, **105**, 157-172.

Beven, K.J. and Germann, P. (1982) Macropores and water flow in soils. *Water Resources Research*, **18**, 1311-1325.

Bigg, G.R. (2001) Back to basics: the oceans and their interaction with the atmosphere. *Weather*, **56**, 296-304.

Billett, M.F., Palmer, S.M., Hope, D. *et al.* (2004) Linking land-atmosphere-stream carbon fluxes in a lowland peatland system. *Global Biogeochemical Cycles*, **18**, GB1024, doi:10.1029/2003GB002058.

Bird, E. (2000) *Coastal geomorphology: An introduction.* John Wiley & Sons, Chichester.

Bird, E.C.F. (1985) *Coastline changes: A global review.* Wiley-Interscience, Chichester.

Bird, E.C.F. (1993) *Submerging coasts: The effects of a rising sea level on coastal environments.* John Wiley & Sons, Chichester.

Bird, E.C.F. (1996) *Beach management.* John Wiley & Sons, Chichester.

Birkeland, P.W. (1999) *Soils and geomorphology*, 3rd edition. Oxford University Press, Oxford.

Bjørnsson, H., Gjessing, Y., Hamran, S.-E. *et al.* (1996) The thermal regime of sub-polar glaciers mapped by multi-frequency radio-echo sounding. *Journal of Glaciology*, **42**, 23-32.

Blackmore, S., Steinmann, J.A.J., Hoen, P.P. and Punt, W. (2003) The northwest European pollen flora 65: Betulaceae and Corylaceae. *Review of Palaeobotany and Palynology*, **123,** 71-98.

Blake, E.W. (1992) The deforming bed beneath a surge-type glacier: measurement of mechanical and electrical properties. PhD thesis, University of British Columbia, Vancouver.

Bland, W. and Rolls, D. (1998) *Weathering: An introduction to the scientific principles.* Arnold, London.

Blankenship, D.D., Bentley, C.R., Rooney, S.T. and Alley, R.B. (1986) Seismic measurements reveal a saturated porous layer beneath an active Antarctic ice stream. *Nature*, **322**, 54-57.

Blockley, S.P.E. and Pinhasi, R. (2011) A revised chronology for the adoption of agriculture in the Southern Levant and the role of Lateglacial climatic change. *Quaternary Science Reviews*, **30**, 98-108.

Bodman, G.B. and Colman, E.A. (1943) Moisture and energy conditions during downward entry of water into soils. *Proceedings of the Soil Science Society of America*, **8**, 116-122.

Bond, G., Broecker, W., Johnsen, S. *et al.* (1993) Correlations between climate records from North Atlantic sediments and Greenland ice. *Nature*, **365**, 143-147.

Boulton, G.S. (1978) Boulder shapes and grain size distributions of debris as indicators of transport paths through a glacier and till genesis. *Sedimentology*, **25**, 773-799.

Boulton, G.S. (1979) Processes of glacier erosion on different substrata. *Journal of Glaciology*, **23**, 15-38.

Boulton, G.S. (1987) A theory of drumlin formation by subglacial sediment deformation. In: Menzies, J. and Rose, J. (eds), *Drumlin symposium*, Balkema, Rotterdam, 25-80.

Boulton, G.S. and Hindmarsh, R.C.A. (1987) Sediment deformation beneath glaciers: rheology and geological consequences. *Journal of Geophysical Research*, **92** (B9), 9059-9082.

Bowen, A.J., Inman, D.L. and Simmons, V.P. (1968) Wave 'set-down' and setup. *Journal of Geophysical Research*, **73**, 2569-2577.

Bowerman, N.H.A., Frame, D.J., Huntingford, C., Lowe, J.A. and Allen, M.R. (2010) Cumulative carbon emissions, emissions floors and short-term rates of warming: implications for policy. *Philosophical Transactions of the Royal Society A*, **369**, 45-66.

Boyd, P.W., Watson, A.J., Law, C.S. *et al.* (2000) A mesoscale phytoplankton bloom in the polar Southern Ocean stimulated by iron fertilization. *Nature*, **407**, 695-702.

Boyd, R., Dalrymple, R. and Zaitlin, B.A. (1992) Classification of clastic coastal depositional environments. *Sedimentary Geology*, **80**, 139-150.

Bradbury, I.K. (1998) *The biosphere.* John Wiley & Sons, Chichester.

Bradley, R.S. (1999) *Palaeoclimatology: Reconstructing climates of the quaternary.* Harcourt Academic Press, London.

Brady, N.C. and Weil, R.R. (2002) *The nature and properties of soil*, 13th edition. Prentice Hall, Upper Saddle River, NJ.

Brady, N.C. and Weil, R.R. (2007) *The nature and properties of soil*, 14th edition. Pearson International Edition, Harlow.

Brasher, R.E. and Zheng, X. (1995) Tropical cyclones in the southwest Pacific. *Journal of Climate*, **8**, 1249-1260.

Breckle, S.-W. (2002) *Walter's vegetation of the Earth: The ecological systems of the geo-biosphere*, 4th edition. Springer-Verlag. Berlin.

Bridge, J.S. (2003) *Rivers and floodplains.* Blackwell, Oxford.

Bridgland, D.R. (1994) *Quaternary of the Thames.* Chapman and Hall, London.

Bridgland, D., Maddy, D. and Bates, M. (2004) River terrace sequences: templates for Quaternary geochronology and marine-terrestrial correlation. *Journal of Quaternary Science*, **19**, 203-218.

Briggs, D., Smithson, P., Addison, K. and Atkinson, K. (1997) *Fundamentals of the physical environment*, 2nd edition. Routledge, London.

Broecker, W.S. (1994) Massive iceberg discharges as triggers for global climate change. *Nature*, **372**, 421-424.

Broecker, W.S. (1998) Paleocean circulation during the last glaciation. A bipolar seesaw? *Paleoceanography*, **13**, 119-121.

Broecker, W.S. and Denton, G.H. (1990) What drives glacial cycles? *Scientific American*, **262**, 43-50.

Broecker, W.S., Denton, G.H., Edwards, R.L., Cheng, H., Alley, R.B. and Putnam, A.E. (2010) Putting the Younger Dryas cold event into context. *Quaternary Science Reviews*, **29,** 1078-1081.

Broham, P., Kennedy, J.J., Harris, S.F.B. (2006) Uncertainty esti-mates in regional and global observed temperature changes: a new dataset from 1850. *Journal of Geophysical Research*, **111**, doi: 10.1029/2005JD006548.

Brookes, A. (1985) River channelisation, traditional engineering methods, physical consequences and alternative practices. *Progress in Physical Geography*, **9**, 44-73.

Brown, J.D. (2004) Knowledge, uncertainty and physical geog-raphy: towards the development of methodologies for questioning belief. *Transactions of the Institute of Physical Geographers*, **29**, 367-381.

Brown, L.E., Hannah, D.M. and Milner, A.M. (2007) Vulnerability of alpine stream biodiversity to shrinking glaciers and snow-packs. *Global Change Biology*, **13**, 958-966.

Brown, P., Sutikna, T., Morwood, M.J. et al. (2004) A new small-bodied hominin from the Late Pleistocene of Flores, Indonesia. *Nature*, **431,** 1055-1061.

Browning, K.A. and Hill, F.F. (1981) Orographic rain. *Weather*, **36**, 326-329.

Brunckhorst, D.J. (1995) Sustaining nature and scociety – a biore-gional approach. *Inhabit*, **3**, 5-9.

Brutsaert, W. (1982) *Evaporation into the atmosphere: Theory, history and application*. D. Reidel, Dordrecht.

Bryan, K. (1922) Erosion and sedimentation in the Papago County, Arizona. *USGS Bulletin*, **730**.

Bryant, E. (2001) *Tsunami: The underrated hazard*. Cambridge University Press, Cambridge.

Budyko, M.I. (1974) *Climate and life*. Academic Press, New York.

Buffet, B.A. (2000) Clathrate hydrates. *Annual Reviews of Earth and Planetary Science*, **28**, 477-507.

Bunte, K. and Abt, S.R. (2001) *Sampling surface and subsurface particle-size distributions in gravel- and cobble-bed streams for analyses in sediment transport, hydraulics and streambed monitoring*. USGS Rocky Mountain Research Station General Technical Report RMRS-GTR-74.

Burke, E.J., Brown, S.J. and Christidis, N. (2006) Modeling the recent evolution of global drought and projections for the twenty-first century with the Hadley Centre Climate Model. *Journal of Hydrometeorology*, **7**, 1113-1125.

Burrough, S.L., Thomas, D.S.G. and Bailey, R. (2009) Mega-Lake in the Kalahari. A 250 kyr record of Palaeolake Makgadikgadi. *Quaternary Science Reviews*, **28**, 1392-1411.

Burt, J., Barber, K. and Rigby, D.L. (2006) *Elementary statistics for geographers,* 3rd edition. Guilford, New York.

Burt, T.P. (1986) Runoff processes and solutional denudation rates on humid temperate hillslopes. In: Trudgill, S.T. (ed.), *Solute processes*. John Wiley & Sons, Chichester, 193-249.

Burt, T.P. (1988) A practical exercise to demonstrate the variable source area model. *Journal of Geography in Higher Education*, **12**, 177-186.

Burt, T.P. (1996) The hydrology of headwater catchments. In: Petts, G.E. and Calow, P. (eds), *River flows and channel forms*. Blackwell, Oxford, 6-31.

Burt, T.P. and Holden, J. (2010) Changing temperature and rain-fall gradients in the uplands. *Climate Research*, **45**, 57-70.

Burt, T.P., Heathwaite, A.L. and Trudgill, S.T. (1993) *Nitrate: Process, patterns and management*. John Wiley & Sons, Chichester.

Byrne, S. and Ingersoll, A.P. (2003) A sublimation model for Martian South Polar Ice features. *Science*, **299**, 1051-1053.

C

Campos, E.F., Fabry, F. and Hocking, W. (2007) Precipitation measurements using VHF wind profiler radars: measuring rainfall and vertical air velocities using only observations with a VHF radar. *Radio Science*, **42**, Art. No. RS3003.

Carpenter, S.R., Kitchell, J.F. and Hodgson, J.R. (1985) Cascading trophic interactions and lake productivity. *BioScience*, **35**, 634-639.

Carson, M.A. and Kirkby, M.J. (1972) *Hillslope form and process*. Cambridge University Press, Cambridge.

Carson, M.A. and Kirkby, M.J. (2009) *Hillslope form and process*. Cambridge University Press, Cambridge.

Carson, R. (1962) *Silent spring*. Houghton Mifflin, Boston, MA.

Carter, R.W.G. (1988) *Coastal environments*. Academic Press, London.

Carter, R.W.G. and Woodroffe, C.D. (1994) Coastal evolution: an introduction. In: Carter, R.W.G. and Woodroffe, C.D. (eds), *Coastal evolution*. Cambridge University Press, Cambridge, 1-31.

Cazenave, A. and Chen, J. (2010) Time-variable gravity from space and present-day mass redistribution in the Earth system. *Earth and Planetary Science Letters*, **298**, 263-274.

Chandler, T.J. (1965) *The climate of London*. Hutchinson, London.

Chang, H.-G.H., Eidson, M., Noonan-Toly, C. et al. (2002) Public health impact of reemergence of rabies, New York. *Emerging Infectious Diseases*, **8**, 909-913.

Chapman, D.V. (ed.) (1996) *Water quality assessments: A guide to the use of biota, sediments and water in environmental moni-toring*, 2nd edition. Chapman and Hall, London.

Chappell, J. (1980) Coral morphology, diversity and reef growth. *Nature*, **286**, 249-252.

Chappell, J.M.A. and Shackleton, N.J. (1986) Oxygen isotopes and sea-level. *Nature*, **324**, 137-138.

Chengrui, M. and Dregne, H.E. (2001) Silt and the future develop-ment of China's Yellow River. *The Geographical Journal*, **167**, 7-22.

Cherrett, J.M. (1989) *Ecological concepts*. Blackwell Scientific, Oxford.

Childs, E.C. (1969) *Introduction to the physical basis of soil water phenomenon*. John Wiley & Sons, London.

Chow, V.T. (1959) *Open channel hydraulics*. McGraw-Hill, New York.

Chow, W.T.L. and Roth, M. (2006) Temporal dynamics of the urban heat island of Singapore. *International Journal of Climatology*, **26**, 2243-2260.

Christopher, S.F., Mitchell, M.J., McHale, M.R., Boyer, E.W., Burns, D.A. and Kendall, C. (2008) Factors controlling nitrogen release from two forested catchments with contrasting hydrochemical responses. *Hydrological Processes*, **22**, 46-62.

Church, J.A. and White, N.J. (2006) A 20th century acceleration in global sea-level rise. *Geophysical Research Letters*, **33**, L01602, doi:10.1029/2005GL024826.

Ciambrone, D.F. (1997) *Environmental life cycle analysis*. Lewis, Boca Raton, FL.

CLAG (Critical Loads Advisory Group) (1994) *Critical levels of acidity in the United Kingdom*. Summary report prepared at the request of the UK Department of the Environment, Transport and Regions.

Clark, J.M., Bottrell, S.H., Evans, C.D. *et al.* (2010) The importance of the relationship between scale and process in understanding long-term DOC dynamics. *Science of the Total Environment*, **408**, 2768-2775.

Clarke, C.D., Knight, J.K. and Gray, J.T. (2000) Geomorphological reconstruction of the Labrador sector of the Laurentide Ice Sheet. *Quaternary Science Reviews*, **19**, 1343-1366.

Clarke, G.K.C. (1987) Fast glacier flow: ice streams, surging and tidewater glaciers. *Journal of Geophysical Research*, **92**(B9), 8835-8841.

Clarke, S. (2010) Ecosystem services: from theory to application. *In practice: Bulletin of Institute of Ecology and Environmental Management*, **68**, 4-7.

Clements, F.E. (1916) *Plant succession*. Carnegie Institute of Washington, Washington, DC.

Clements, F.E. (1928) *Plant succession and indicators*. H.W. Wilson, New York.

Clifford, N.J. (1993) Formation of riffle-pool sequences: field evidence for an autogenetic process. *Sedimentary Geology*, **85**, 39-51.

Clifford, N.J. (2002) Hydrology: the changing paradigm. *Progress in Physical Geography*, **26**, 290-301.

Clifford, N.J. and Valentine, G. (eds) (2003) *Key methods in geography*. Sage, London.

Clifford, S.M., Crisp, D., Fisher, D.A. *et al.* (2000) The state and future of Mars polar science and exploration. *Icarus*, **144**, 210-242.

Clifford, N.J., Holloway, S., Rice, S.P. and Valentine, G. (eds) (2008) *Key concepts in geography*. Sage, London.

Clifford, N.J., French, S. and Valentine, G. (eds) (2010) *Key methods in geography*. Sage Publications, London.

Closs, G., Downes, B. and Boulton, A. (2004) *Freshwater ecology: A scientific introduction*. Blackwell, Oxford.

Coco, G. and Murray, A.B. (2007) Patterns in the sand: from forcing templates to self-organisation. *Geomorphology*, **91**, 271-290.

Cockburn, H.A.P. and Summerfield, M.A. (2004) Geomorphological applications of cosmogenic isotope analysis. *Progress in Physical Geography*, **28**, 1-42.

Cole, J.J., Prairie, Y.T., Caraco, N.F. *et al.* (2007) Plumbing the global carbon cycle: integrating inland waters into the terrestrial carbon budget. *Ecosystems*, **10**, 171-184.

Cole, L., Bargett, R.D., Ineson, P. and Adamson, J. (2002) Relationship between enchytraeid worms (*Oligochaeta*), temperature, and the release of dissolved organic carbon from blanket peat in northern England. *Soil Biology and Biochemistry*, **34**, 599-607.

Coles, B.J. (2000) Doggerland: the cultural dynamics of a shifting coastline. In: Pye, K. and Allen, S.R.L. (eds), *Coastal and estuarine environments: Sedimentology, geomorphology and geoarchaeology*. Geological Society Special Publication No. 175. The Geological Society, London, 393-401.

Collier, C.G. (1999) Precipitation measuring weather radar. In: Herschy, R.W. (ed.), *Hydrometry: Principles and practice*. John Wiley & Sons, Chichester, 181-198.

Collinson, A.S. (1997) *Introduction to world vegetation*. Allen and Unwin, London.

COM (2002) *Towards a thematic strategy for soil protection*. Commission of the European Communities, Brussels.

Committee on Climate Change (2008) *Building a low-carbon economy - the UK's contribution to tackling climate change: The first report of the committee on climate change, December 2008*. TSO, London.

Connell, J.H. and Slatyer, R.D. (1977) Mechanisms of succession in natural communities and their role in community stability and organisation. *American Naturalist*, **111**, 1119-1144.

Cook, J. (1777) *A voyage towards the South Pole and round the world performed in His Majesty's ships the* Resolution *and* Adventure, *in the years 1772, 1773, 1774, and 1775*. W. Strahan and T. Cadell, London.

Cooke, R.U. and Smalley, I.J. (1968) Salt weathering in deserts. *Nature*, **220**, 1226-1227.

Cosby, B.J., Ferrier, R.C., Jenkins, A. and Wright, R.F. (2001) Modelling the effects of acid deposition: refinements, adjustments and inclusion of nitrogen dynamics in the MAGIC model. *Hydrology and Earth System Sciences*, **5**, 499-517.

Cotton, C.A. (1922) *Geomorphology of New Zealand*. Dominion Museum, Wellington.

Cowell, P.J. and Thom, B.G. (1994) Morphodynamics of coastal evolution. In: Carter, R.W.G. and Woodroffe, C.D. (eds), *Coastal evolution*. Cambridge University Press, Cambridge, 33-86.

Cowles, H.C (1899) The ecological relations of the vegetation on the sand dunes of Lake Michigan. *Botanical Gazette*, **27**, 95-117, 167-202, 281-308, 361-391.

Cox, B. and Moore, P. (2005) *Biogeography: An ecological and evolutionary approach*. John Wiley & Sons, Chichester.

Cox, P.M., Betts, R.A., Jones, C.D. *et al.* (2000) Acceleration of global warming due to carbon-cycle feedbacks in a coupled climate model. *Nature*, **408**, 184-187.

Cramer, W., Bondeau, A., Woodward, F.I. *et al.* (2001) Global response of terrestrial ecosystem structure and function to CO_2 and climate change: results from six dynamic global vegetation models. *Global Change Biology*, **7**, 357-373.

Cresser, M.S., Kilhma, K. and Edwards, A.C. (1993) *Soil chemistry and its applications*. Cambridge University Press, Cambridge.

Critchfield, H.J. (1983) *General climatology*, 4th edition. Prentice Hall, Englewood Cliffs, NJ.

Croke, J. (1997) Australia. In: Thomas, D.S.G. (ed.), *Arid zone geomorphology: Process, form and change in drylands*. John Wiley & Sons, Chichester, 563-573.

Cuffey, K.M. and Paterson, W.S.B. (2010) *Physics of glaciers,* 4th edition. Butterworth-Heinemann, London.

Culling, W.E.H. (1963) Soil creep and the development of hillside slopes. *Journal of Geology*, **71**, 127-161.

Cummins, K.W. (1973) Trophic relations of aquatic insects. *Annual Review of Entomology*, **18**, 183-206.

Curtis, L.F., Courtney, F.M. and Trudgill, S.T. (1976) *Soils in the British Isles*. Longman, London.

D

Dai, A., Trenberth, K.E. and Qian, T. (2004) A global dataset of Palmer Drought Severity Index for 1870-2002: relationship with soil moisture and effects of surface warming. *Journal of Hydrometeorology*, **5**, 1117-1130.

Dalrymple, R.W., Zaitlin, B.A. and Boyd, R. (1992) Estuarine facies models: conceptual basis and stratigraphic implications. *Journal of Sedimentary Petrology*, **62**, 1130-1146.

D'Arcy, B.J., Ellis, J.B., Ferrier, R.C., Jenkins, A. and Dils, R. (eds) (2000) *Diffuse pollution impacts: The environmental and economic impacts of diffuse pollution in the UK*. Terence Dalton, London.

Darwin, C. (1859) *The origin of species by means of natural selection.* John Murray, London.

Davidson-Arnott, R.G.D. (2010) *Introduction to coastal processes and geomorphology*. Cambridge University Press, Cambridge.

Davies, J.L. (1980) *Geographical variation in coastal development*, 2nd edition. Longman, London.

Davies, T.A. and Gorsline, D.S. (1976) Oceanic sediments and sedimentary properties. In: Riley, J.P. and Chester, R. (eds), *Chemical oceanography,* 2nd edition. Academic Press, Orlando, FL.

Davis, C.H., Li, Y.H., McConnell, J.R. *et al.* (2005) Snowfall-driven growth in East Antarctic ice sheet mitigates recent sea-level rise. *Science*, **308**, 1898-1901.

Davis, W.M. (1889) The rivers and valleys of Pennsylvania. *National Geographic Magazine*, **1**, 183-253.

Davis, W.M. (1954) *Geographical essays*, ed. D.W. Johnson. Dover, New York.

Dawson, A.G. (1992) *Ice Age Earth*. Routledge, London.

Dawson, J.C. and Smith., P. (2007) Carbon losses from soil and its consequences for land-use management. *Science of the Total Environment*, **382**, 165-190.

Dawson, J.J.C., Soulsby, C., Tetzlaff, D., Hrachowitz, M., Dunn, S.M. and Malcolm, I.A. (2008) Influence of hydrology and seasonality on DOC exports from three contrasting upland catchments. *Biogeochemistry*, **90**, 93-113.

Day, M.J. (1981) Rock hardness and landform development in the Gunong Mulu National Park, Sarawak, E. Malaysia. *Earth Surface Processes and Landforms*, **6**, 165-172.

DeConti, R.M. and Pollard, D. (2003) Rapid Cenozoic glaciation of Antarctica induced by declining atmospheric CO_2. *Nature*, **421**, 245-249.

Defra (2007) *Securing a healthy natural environment: An action plan for embedding an ecosystems approach*. Defra, London.

Dekker, L.W. and Ritsema, C.J. (1996) Preferential flow paths in a water repellent clay soil with grass cover. *Water Resources Research*, **32**, 1239-1249.

Delworth, T.L. and Dixon, K.W. (2000) Implications of the recent trend in the Arctic/North Atlantic Oscillation for the North Atlantic thermohaline circulation. *Journal of Climate*, **13**, 3721-3727.

Deosthali, V. (2000) Impact of rapid urban growth on heat and moisture islands in Pune City, India. *Atmospheric Environment*, **34**, 2745-2754.

de Wit, M. and Stankiewicz, J. (2006) Changes in surface water supply across Africa with predicted climate change. *Science*, **311**, 1917-1921.

Diamond, J. (2005) *Collapse: How societies choose to fail or survive.* Allen Lane, London.

Diamond, J.M. (1975) The island dilemma: lessons of modern biogeographic studies for the design of nature reserves. *Biological Conservation*, **7**, 129-146.

Diamond, J.M. (1997) *Guns, germs and steel: A short history of everybody for the last 13,000 years*. Jonathan Cape, London.

Diamond, J.M. and May, R.E. (1976) Island biogeography and the design of nature reserves. In: May, R.E. (ed.), *Theoretical ecology.* Sinauer Associates, Sunderland, MA.

Diamond, J.M. and Veitch, C.R. (1981) Extinctions and introductions in the New Zealand avifauna - cause and effect? *Science*, **211**, 499-501.

Dickson, B., Yashayaev, I., Meincke, J. *et al.* (2002) Rapid freshening of the deep North Atlantic Ocean over the past four decades. *Nature*, **416**, 832-837.

Dickson, R.R., Gmitrowicz, E.M. and Watson, A.J. (1990) Deepwater renewal in the northern North Atlantic. *Nature*, **344**, 848-850.

Diekmann, B., Wetterich, S. and Kienast, F. (2007) Russian-German cooperation Potsdam-Yakutsk: the Expedition Central Yakutia 2005. In: Scirrmeister, L. (ed.), *Expeditions in Siberia in 2005*. Reports on Polar and Marine Research. Alfred Wegener Institute for Polar and Marine Research, Bremerhaven, **550**, 243-289.

Dietrich, W.E. and Dunne, T. (1978) Sediment budget for a small catchment in mountainous terrain. *Zeitschrift für Geomorphologie, Supplement*, **29**, 191-206.

Dietrich, W.E. and Dunne, T. (1993) The channel head. In: Beven, K.J. and Kirkby, M.J. (eds), *Channel network hydrology*. John Wiley & Sons, Chichester, 175-219.

Dobson, M. and Frid, C. (2009) *Ecology of aquatic systems,* 2nd edition. Oxford University Press, Oxford.

Dorn, R.I. (1998) *Rock coatings*. Elsevier, Amsterdam.

Doran, J.W. and Parkin, T.B. (1994) Defining and assessing soil quality. In: Doran, J.W. *et al.* (eds), *Defining soil quality for a sustainable environment*. Special Publication 35. Soil Science Society of America, Madison, WI, 3-21.

Dougill, A.J., Reed, M.S., Fraser, E.D.G. *et al.* (2006) Learning from doing participatory rural research: lessons from the Peak District National Park. *Journal of Agricultural Economics*, **57**, 259-275.

Douglas, I. (1967) Man, vegetation and the sediment yield of rivers. *Nature*, **215**, 925-928.

Douglas, I. (1979) *Humid landforms*. MIT Press, Cambridge, MA.

Dreimanis, A. and Vagners, U.J. (1971) Bimodal distribution of rock and mineral fragments in basal tills. In: Goldthwait, R.P. (ed.), *Till, a symposium*. Ohio State University Press, Columbus, OH, 237-250.

Drewry, D. (1986) *Glacial geologic processes*. Edward Arnold, London.

Drury, W.H. and Nisbet, I.C.T. (1973) Succession. *Journal of the Arnold Arboretum*, **53**, 331-368.

Dudgeon, D., Arthington, A.H., Gessner, M.O. *et al.* (2006) Freshwater biodiversity: importance, threats, status and conservation challenges. *Biological Reviews*, **81**, 163-182.

Dunbar, R.B. (2000) Clues from corals. *Nature*, **407**, 956-957.

Dunn, J.C., McClymont, H.E., Christmas, M. and Dunn, A.M. (2009) Competition and parasitism in the native white clawed crayfish *Austropotamobius pallipes* and the invasive signal crayfish *Pacifastacus leniusculus* in the UK. *Biological Invasions*, **11**, 315-324.

Dunne, T. (1978) Field studies of hillslope flow processes. In: Kirkby, M.J. (ed.), *Hillslope hydrology*. John Wiley & Sons, Chichester, 227-293.

Dunne, T. and Aubry, B.F. (1986) Evaluation of Horton's theory of sheetwash and rill erosion on the basis of field experiments. In: Abrahams, A.D. (ed.), *Hillslope processes*. Allen & Unwin, London, 31-53.

Dyer, K.R. (1998) *Estuaries: A physical introduction*. John Wiley & Sons, Chichester.

Dyurgerov, M.B. and Meier, M.F. (2000) Twentieth century climate change: evidence from small glaciers. *Proceedings of the National Academy of Sciences, USA*, **97**, 1406-1411.

E

Easterbrook, D.J. (1993) *Surface processes and landforms*. Macmillan, London.

Edwards, C.A. and Bohlen, P.J. (1996) *Earthworm ecology of earthworms*, 3rd edition. Chapman and Hall, London.

Egler, F.E. (1954) Vegetation science concepts I. Initial floristic composition, a factor in old-field vegetation development. *Vegetatio*, **4**, 412-417.

Egoh, B., Rouget, M., Revers, B. *et al.* (2007) Integrating ecosystem services into conservation assessments: a review. *Ecological Economics*, **63**, 714-721.

Ehleringer, J.R., Cerling, T.E. and Dearing, M.D. (eds) (2005) *A history of atmospheric CO_2 and its effect on plants, animals, and ecosystems*. Springer-Verlag, New York.

Ehlers, J. (1996) *Quaternary and glacial geology*. John Wiley & Sons, Chichester.

Eicher, D.L. (1976) *Geologic time*. Prentice Hall, Englewood Cliffs, NJ.

Elkibbi, M. and Rial, J.A. (2001) An outsider's review of the astronomical theory of climate: is the eccentricity-driven insolation the main driver of the ice ages? *Earth-Science Reviews*, **56**, 161-177.

Emanuel, K. (2005) Increasing destructiveness of tropical cyclones over the past 30 years. *Nature*, **436**, 686-688.

Emiliani, C. (1955) Pleistocene temperatures. *Journal of Geology*, **63**, 538-575.

Emmett, B.A., Reynolds, B., Chamberlain, P.M. *et al.* (2010) *Countryside Survey: Soils Report from 2007*. Technical Report No. 9/07, NERC/Centre for Ecology & Hydrology (CEH Project Number: C03259).

Emmett, W.W. and Wolman, M.G. (2001) Effective discharge and gravel-bed rivers. *Earth Surface Processes and Landforms*, **26**, 1369-1380.

England, P. and Jackson, J. (2011) Uncharted seismic risk. *Nature Geoscience*, **4**, 348-349.

English Nature (2002) *Urbanities urbio: Urban diversity and human nature*. English Nature, Peterborough, Issue **1**, 14.

Ensign, S.H. and Doyle, M.W. (2006) Nutrient spiraling in streams and river networks. *Journal of Geophysical Research*, **111**, G04009, doi:10.1029/2005JG000114.

Environment Agency (2007) *Briefing to the regional flood defence committee on the summer 2007 floods in south and west Yorkshire*. Environment Agency NE Region, Newcastle.

Environment Agency (2011) *Report to regional flood defence committee on aggregate flood project expenditure*. Environment Agency NE Region, Newcastle.

Erdtman, G. (1937) Pollen grains recovered from the atmosphere over the Atlantic. *Meddelanden från Göteborgs Botaniska Trädgård*, **12**, 185-196.

Erickson, S.L. and King, B.J. (1999) *Fundamentals of environmental management*. John Wiley & Sons, Chichester.

eThekwini Municipality and Local Action for Biodiversity (2007) *EThekwini municipality biodiversity report 2007*. eThekwinin Municipality, Durban and Local Action for Biodiversity, Vlaeberg, South Africa.

European Commission (2008) Review of existing information on the interrelations betweens oil and climate change, available at: http://ec.europa.eu/environment/soil/pdf/climsoil_report_dec_2008.pdf.

European Commission (2010) Soil biodiversity: functions, threats and tools for policy makers, available at: http://ec.europa.eu/environment/soil/biodiversity.htm.

European Union (2000) Directive 2000/60/EC of the European Parliament and of the Council establishing a framework for the Community action in the field of water policy, available at: http://ec.europa.eu/environment/water/water-framework/index_en.html.

Evans, C. and Davies, T.D. (1998) Causes of concentration/discharge hysteresis and its potential as a tool for analysis of episode hydrochemistry. *Water Resources Research*, **34**, 129-137.

Evans, C.D., Reynolds, B., Hinton, C. *et al.* (2008) Effects of decreasing acid deposition and climate change on acid extremes in an upland stream. *Hydrology and Earth System Sciences*, **12**, 337-351.

Evans, D.J. and Kirkby, M.J. (eds) (2004) *Critical concepts in geomorphology, vol. 2 – Hillslope geomorphology.* Taylor & Francis, London.

Evans, J.G. (1999) *Land and archaeology: Histories of human environment in the British Isles.* Tempus Publishing, Stroud.

Evans, M.G., Burt, T.P., Holden, J. and Adamson, J. (1999) Runoff generation and water table fluctuations in blanket peat: evidence from UK data spanning the dry summer of 1995. *Journal of Hydrology*, **221**, 141-160.

Evans, R.L. (2007) *Fueling our future: An introduction to sustainable energy.* Cambridge University Press, Cambridge.

F

Fabel, D., Stroeven, A.P., Harbor, J. *et al.* (2002) Landscape preservation under Fennoscandian ice sheets determined from *in situ* produced [10]Be and [26]Al. *Earth and Planetary Science Letters*, **201**, 397-406.

Fagan, B. (2000) *The Little Ice Age.* Basic, New York.

Fagundo-Castillo, J.R., Carrillo-Rivera, J.J., Antigüedad-Auzmendi, I. *et al.* (2008) Chemical and geological control of spring water in Eastern Guaniguanico mountain range, Pinar del Rìo, Cuba. *Environmental Geology*, **55**, 247-267.

FAO/UNESCO (1974) *Soil map of the world, scale 1 : 5 000 000: Volume 1, legend.* UNESCO, Paris.

Ferguson, R.I. (1981) Channel form and channel changes. In: Lewin, J. (ed.), *British rivers.* George Allen and Unwin, London, 90-125.

Ferguson, R.I., Hoey, T., Wathen, S., and Werritty, A (1996) Field evidence for rapid downstream fining of river gravels through selective transport. *Geology*, **24**, 635-43.

Fisher, O. (1866) On the disintegration of a chalk cliff. *Geological Magazine*, **3**, 354-356.

Fitter, A.H. and Fitter, R.S.R. (2002) Rapid changes in flowering time in British plants. *Science*, **296**, 1689-1691.

Fitter, A.H., Fitter, R.S.R., Harris, I.T.B. and Williamson, M.H. (1995) Relationships between first flowering date and temperature in the flora of a locality in central England. *Functional Ecology*, **9**, 55-60.

FitzPatrick, E.A. (1983) *Soils: Their formation, classification and distribution.* Longman, Harlow.

Flenley, J.R., King, A.S.M., Jackson, J., Chew, C., Teller, J.T. and Prentice, M.E. (1991) The Late Quaternary vegetational and climatic history of Easter Island. *Journal of Quaternary Science*, **6**, 85-115.

Focazio, M.J., Kolpin, D.W., Barnes, K.K. *et al.* (2008) A national reconnaissance for pharmaceuticals and other organic waste-water contaminants in the United States – II: Untreated drinking water sources. *Science of the Total Environment*, **402**, 201-226.

Fookes, P.G., Gourley, C.S. and Ohikere, C. (1988) Rock weathering in engineering time. *Quarterly Journal of Engineering Geology*, **21**, 33-37.

Ford, D.C and Williams, P. (2007) *Karst hydrogeology and geomorphology.* John Wiley & Sons, Chichester.

Forman, R.T.T. (1995) *Land mosiacs – The ecology of landscapes and regions.* Cambridge University Press, Cambridge.

Fountain, A.G. and Walder, J.S. (1998) Water flow through temperate glaciers. *Reviews of Geophysics*, **36**, 299-328.

Fourier, J. (1824) Remarques generales sur les temperatures du globe terrestre et des espaces planetaires. *Annales de chimie et de physique*, **27**, 136-167.

Franks, J.W. (1960) Interglacial deposits at Trafalgar Square, London. *New Phytologist*, **59**, 145-152.

Freeman, C., Ostle, N. and Kang, H. (2001) An enzymic 'latch' on a global carbon store. *Nature*, **409**, 149-150.

French, H.M. (1996) *The periglacial environment*, 2nd edition. Longman, Harlow.

French, H.M. (2007) *The periglacial environment*, 3rd edition. John Wiley & Sons, Chichester.

French, P.W. (1997) *Coastal and estuarine management.* Routledge, London.

French, P.W. (2001) *Coastal defences – Processes, problems and solutions.* Routledge, London.

Fricker, H.A., Scambos, T.A., Bindschadler, R.A. and Padman, L. (2007) An active subglacial water system in west Antarctica mapped from space. *Science*, doi: 10.1126/science.1136897.

Friedlingstein, P., Houghton, R.A., Marland, G. *et al.* (2010) Update on CO_2 emissions. *Nature Geoscience*, **3**, 811-812.

Friedlingston, P., Cox, P., Betts, R. *et al.* (2006) Climate-carbon cycle feedback analysis: results from the C4MIP model intercomparison. *Journal of Climate*, **19**, 3337-3353.

Friedrichs, C.T. and Aubrey, D.G. (1988) Non-linear tidal distortion in shallow well-mixed estuaries: a synthesis. *Estuarine, Coastal and Shelf Science*, **27**, 521-545.

Frissell, C.A., Liss, W.L., Warren, C.E. and Hurley, M.D. (1986) A hierarchical framework for stream habitat classification: viewing streams in a watershed context. *Environmental Management*, **10**, 199-214.

Fu, Q., Johanson, C.M., Wallace, J. M. and Reicher, T. (2006) Enhanced mid-latitude tropospheric warming in satellite measurements, *Science*, **312**, 1179.

G

Gaillardet, J., Negrel, P., Dupre, B. and Allegre, C.J. (1997) Chemical and physical denudation in the Amazon River Basin. *Chemical Geology*, **142**, 141-173.

Galloway, J.N., Schlesinger, W.H., Levy, H. et al. (1995) Nitrogen fixation: anthropogenic enhancement and environmental response. Global Biogeochemical Cycles, 9, 235-252.

Galloway, W.E. (1975) Process framework for describing the morphologic and stratigraphic evolution of deltaic depositional systems. In: Broussard, M.L. (ed.), Deltas, models for exploration. Houston Geological Society, Houston, TX, 87-98.

Gamache, I. and Payette, S. (2004) Height growth response of tree line spruce to recent climate warming across the forest-tundra of eastern Canada. Journal of Ecology, 92, 835-845.

Gammons, C.H., Nimick, D.A., Parker, S.R., Cleasby, T.E. and McCleskey, R.B. (2005) Diel behavior of iron and other heavy metals in a mountain stream with acidic to neural pH: Fisher Creek, Montana, USA. Geochimica et Cosmochimica Acta, 69, 2505-2516.

Ganderton, P. and Coker, P. (2005) Environmental biogeography. Prentice Hall, Harlow.

Gaunt, J. et al. (2008) Soils within the Catchment Sensitive Farming Programme: project to deliver improvements in soil management. Defra project SP08014. Defra, London.

Gedzelman, S.D. and Austin, S. (2003) Mesoscale aspects of the urban heat island around New York City. Theoretical and Applied Climatology, 75, 29-42.

Geiger, R. (1965) The climate near the ground, 2nd edition. Harvard University Press, Cambridge, MA.

Geiger, R. and Pohl, W. (1953) Revision of Koppen-Geiger climate maps of the Earth. Justes Perthes, Darmstadt.

Geiger, R., Aron, R.H. and Todhunter, P. (2003) The climate near the ground. Rowman and Littlefield, Oxford.

Gellert, P.K. (1998) A brief history and analysis of Indonesia's forest fire crisis. Indonesia, 65, 63-85.

Gende, S.M., Edwards, R.T., Willson, M.F. and Wipfli, M.S. (2002) Pacific salmon in aquatic and terrestrial ecosystems. Bioscience, 52, 917-928.

Gerrard, J. (2000) Fundamentals of soils. Routledge, London.

Gibb, J. (1954) Feeding ecology of tits, with notes on treecreeper and goldcrest. Ibis, 96, 513-543.

Gibbard, P.L. (1994) Pleistocene history of the lower Thames. Cambridge University Press, Cambridge.

Gibbard, P.L. and Lewin, J. (2002) Climate and related controls on interglacial fluvial sedimentation in lowland Britain. Sedimentary Geology, 151, 187-210.

Gilbert, G.K. (1909) The convexity of hilltops. Journal of Geology, 17, 344-351.

Gildor, H. and Tziperman, E. (2001) A sea-ice climate switch mechanism for the 100-kyr glacial cycles. Journal of Geophysical Research, 106, 9117-9133.

Gilvear, D.J. (2003) Patterns of channel adjustment due to impoundment of the Upper River Spey, Scotland (1942-2000). River Research and Applications, 19, 1-17.

Gilvear, D.J. and Black, A.R. (1999) Flood induced embankment failures on the River Tay: implications of climatically induced hydrological change in Scotland. Hydrological Sciences Journal, 44, 345-362.

Gilvear, D.J. and Bradley, S. (1997) Geomorphic adjustment of a newly constructed ecologically sound river diversion on an upland gravel bed river, Evan Water, Scotland. Regulated Rivers, 13, 1-13.

Gilvear, D.J., Winterbottom, S.J. and Sichingabula, H. (2000) Mechanisms of channel planform change on the actively meandering Luangwa River, Zambia. Earth Surface Processes and Landforms, 25, 421-436.

Giosan, L., Filip, F. and Constatinescu, S. (2009) Was the Black Sea catastrophically flooded in the early Holocene? Quaternary Science Reviews, 28, 1-6.

GLASOD (1990) World map of the status of human-induced soil degradation. Winand Staring Centre, Wageningen.

Glaves, P. and Egan, D. (2010) Valuing and using ecosystem services in practice: findings from the east of England pilot studies. In practice: Bulletin of Institute of Ecology and Environmental Management, 68, 12-15.

Gleason, H.A. (1917) The structure and development of the plant association. Bulletin of the Torrey Botanical Club, 43, 463-481.

Gleason, H.A. (1926) The individualistic concept of plant association. Bulletin of the Torrey Botanic Club, 53, 7-26.

Goede, A. and Harmon, R.S. (1983) Radiometric dating of Tasmanian speleothems - evidence of cave sediments and climate change. Journal of the Geological Society of Australia, 30, 89-100.

Goldich, S.S. (1938) A study of rock weathering. Journal of Geology, 46, 17-58.

Goldstein, R.M., Engelhardt, H., Kamb, B. and Frolich, R.M. (1993) Satellite radar interferometry for monitoring ice stream motion: application to an Antarctic ice stream. Science, 262, 1525-1530.

Goller, R., Wilcke, W., Fleischbein, K., Valarezo, C. and Zech, W. (2006) Dissolved nitrogen, phosphorus, and sulfur forms in the ecosystem fluxes of a montane forest in Ecuador. Biogeochemistry, 77, 57-89.

Gong, D. and Wang, S. (1999) Definition of Antarctic oscillation index. Geophysical Research Letters, 26, 459-462.

Goodbred, S.L. and Kuehl, S.A. (1999) Holocene and modern sediment budgets for the Ganges-Brahmaputra river system: evidence for highstand dispersal to flood-plain, shelf, and deep-sea depocenters. Geology, 26, 559-562.

Goudie, A.S. (1973) Duricrusts of tropical and subtropical landscapes. Clarendon Press, Oxford.

Goudie, A.S. (1997) Weathering processes. In: Thomas, D.S.G. (ed.), Arid zone geomorphology: Process, form and change in drylands. John Wiley & Sons, Chichester, 25-39.

Goudie, A.S. and Viles, H.A. (1997) Salt weathering hazards. John Wiley & Sons, Chichester.

Gowing, G. (2003) Salinisation - one of our biggest environmental problems, available at: www.amonline.net.au/factsheets/salinisation.htm.

Grace, J. and Unsworth, M.H. (1988) Climate and microclimate of the uplands. In: Usher, M.B. and Thompson, D.B.A.

(eds), *Ecological change in the uplands*, Special Publication Number 7, British Ecological Society. Blackwell, Oxford, 137-150.

Graumann, A., Houston, T., Lawrimore, J. *et al.* (2005) Hurricane Katrina - a climtaological perspective. NOAA National Data Center Technical Report 2005-01, Ashville, NC.

Gregory, K.J. (1985) *The nature of physical geography.* Arnold, London.

Gregory, K.J. (1992) Vegetation and river channel process interactions. In: Boon, P.J., Calow, P. and Petts, G.E. (eds), *River conservation and management*. John Wiley & Sons, Chichester, 255-269.

Gregory, K.J. (2000) *The changing nature of physical geography.* Arnold, London.

Gregory, K.J. (2010) *The Earth's land surface.* Sage, London.

Gregory, K.J., Davis, R.J. and Downs, P.W. (1992) Identification of river channel changes due to urbanisation. *Applied Geography*, **12**, 299-318.

Griggs, D.T. (1936) The factor of fatigue in rock exfoliation. *Journal of Geology*, **74**, 733-796.

Grime, P. (1997) Climate change and vegetation. In: Crawley, M.J. (ed.), *Plant ecology*. Blackwell Scientific, Oxford.

Grimes, J.P. (1979) *Plant strategies and vegetation processes.* John Wiley & Sons, Chichester.

Grotzinger, J. and Jordan, T.H. (2010) *Understanding Earth.* W.H. Freeman, New York.

Grove, J.M. (2004) *Little ice ages: Ancient and modern*, 2nd edition. Routledge, London.

Guinotte, J.M., Buddemeier, R.W. and Kleypas, J.A. (2003) Future coral reef habitat marginality: climate change in the Pacific basin. *Coral Reefs*, **22**, 551-558.

Guo, Z.T., Ruddiman, W.F., Hao, Q.Z. *et al.* (2002) Onset of Asian desertification by 22 Myr ago inferred from loess deposits in China. *Nature*, **416**, 159-163.

Gupta, A. (2007) *Large rivers: Geomorphology and management.* John Wiley & Sons, Chichester.

Gurnell, A.M., Morrissey, I.P., Boitsidis, A.J. *et al.* (2006) Initial adjustments within a new river channel: interactions between fluvial processes, colonising vegetation and bank profile development. *Environmental Management*, **38**, 580-596.

Gurney, K.R., Law, R.M., Denning, A.S. *et al.* (2002) Towards robust regional estimates of CO_2 sources and sinks using atmospheric transport models. *Nature*, **415**, 626-630.

Gurney, S.D. (2001) Aspects of the genesis and geomorphology and terminology of palsas: perennial cryogenic mounds. *Progress in Physical Geography*, **25**, 249-260.

Guthrie, R.D. (2001) Origin and causes of the mammoth steppe: a story of cloud cover, woolly mammal tooth pits, buckles, and inside-out Beringia. *Quaternary Science Reviews*, **20**, 549-574.

Guthrie, R.D. (2004) Radiocarbon evidence of mid-Holocene mammoths stranded on an Alaskan Bering Sea island. *Nature*, **429**, 746-749.

H

Hack, J.T. (1960) Interpretation of erosional topography in humid temperate regions. *American Journal of Science*, **258**, 80-97.

Haines-Young, R.H. and Petch, J.R. (1986) *Physical geography: Its nature and methods.* Harper and Row, London.

Haldorsen, S. (1981) Grain-size distribution of subglacial till and its relation to glacier crushing and abrasion. *Boreas*, **10**, 91-105.

Hall, K. and Thorn, C. (2011) The historical legacy of spatial scales in freeze-thaw weathering: misrepresentation and resulting is misdirection. *Geomorphology*, **130**, 83-90.

Hallam, A. (1973) *A revolution in earth sciences.* Clarendon Press, Oxford.

Hallet, B., Anderson, S.P., Stubbs, C.W. and Gregory, E.C. (1988) Surface soil displacements in sorted circles, Western Spitzbergen. In: *Proceedings 5th International Conference on Permafrost*, Vol. 1. Tapir Publishers, Trondheim, 770-775.

Hambrey, M.J. (2003) *Glacial environments.* Routledge, London.

Hammer, C.U., Clausen, H.B., Freidrich, W.L. and Tauber, H. (1987) The Minoan eruption of Santorini dated to 1645 BC? *Nature*, **328**, 517-519.

Hansen, J., Nazarenko, L., Ruedy, R. *et al.* (2005) Earth's energy imbalance: confirmation and implications. *Science*, **308**, 1431-1435.

Hansen, J., Sato, M., Kharecha, P. *et al.* (2008) Target atmospheric CO_2: where should humanity aim? *The Open Atmospheric Science Journal*, **2**, 217-231.

Hansen, M., DeFries, R., Townshend, J.R.G. and Sohlberg, R. (1998) *UMD Global land cover classification, 1 kilometer, 1.0, 1981-1994*. Department of Geography, University of Maryland, College Park, MD.

Harding, G. (1968) The tragedy of the commons. *Science*, **162**, 1243-1248.

Harding, R.J. (1978) The variation of the altitudinal gradient of temperature within the British Isles. *Geografiska Annalar* A, **60**, 43-49.

Harding, R.J. (1979) Radiation in the British uplands. *Journal of Applied Ecology*, **16**, 161-170.

Harding, R.J., Hall, R.L., Neal, C., Roberts, J.M., Rosier, P.T.W. and Kinniburgh, D.G. (1992) *Hydrological impacts of broadleaf woodlands: Implications for water use and water quality*. National Rivers Authority R & D Note 56. National Rivers Authority, Bristol.

Hargrove, E.C. (1989) *Foundations of environmental ethics.* Prentice Hall, Englewood Cliffs, NJ.

Harré, R. (2002) *Great scientific experiments: Twenty experiments that changed our view of the world*. Dover, New York.

Harris, P.T. and Heap, A.D. (2003) Environmental management of clastic coastal depositional environments: inferences from an Australian geomorphic database. *Ocean & Coastal Management*, **46**, 457-478.

Harrison, S. and Anderson, E. (2001) A Late Devensian rock glacier in the Nantlle Valley, North Wales. *Glacial Geology and Geomorphology*, available at: http://boris.qub.ac.uk/ggg/papers/full/2001/rp012001/rp01/html.

Hart, J.K. and Boulton, G.S. (1991) The interrelation of glaciotectonic and glaciodepositional processes within the glacial environment. *Quaternary Science Reviews*, **10**, 335-350.

Harvey, A.M. (1997) The role of alluvial fans in arid zone fluvial systems. In: Thomas, D.S.G. (ed.), *Arid zone geomorphology: Processes, forms and change in drylands*. John Wiley & Sons, Chichester, 231-259.

Haslett, S.K. (2000) *Coastal systems*. Routledge, London.

Hauer, F.R. and Lamberti, G. (2006) *Methods in stream ecology*. Academic Press, London.

Hegerl, G.C., Crowley, M., Allen, W. *et al.* (2007) Detection of human influence on a new 1500-year temperature reconstruction. *Journal of Climate*, **20**, 650-666.

Heinrich, H. (1988) Origin and consequences of cyclic ice rafting in the Northeast Atlantic Ocean during the past 130,000 years. *Quaternary Research*, **29**, 142-152.

Helliwell, R.C. and Simpson, G.L. (2010) The present is the key to the past, but what does the future hold for the recovery of surface waters from acidification? *Water Research*, **44**, 3166-3180.

Henderson-Sellers, A. and Robinson, P.J. (1986) *Contemporary climatology*. Longman, London.

Herschy, R.W. (ed.) (1999) *Hydrometry, principles and practice*. John Wiley & Sons, Chichester.

Hermann, L., Anongrak, N., Zarei, M., Schuler, U. and Spohrer. K. (2007) Factors and processes of Gibbsite formation in Northern Thailand. *Catena*, **71**, 279-291.

Hess, H.H. (1962) History of ocean basins. In: Engel, A.E.J., James, H.L. and Leonard, B.F. (eds), *Petrologic studies: A volume in honor of A.F. Buddington*. Geological Society of America, Boulder, CO.

Heusser, L. and Oppo, D. (2003) Millennial- and orbital-scale variability in southeastern United States and in the subtropical Atlantic during Marine Isotope Stage 5: evidence from pollen and isotopes in ODP Site 1059. *Earth and Planetary Science Letters*, **214**, 483-490.

Heusser, L., Heusser, C., Mix, A. and McManus, J. (2006) Chilean and Southeast Pacific palaeoclimate variations during the last glacial cycle: directly correlated pollen and delta [18]O records from ODP Site 1234. *Quaternary Science Reviews*, **25**, 3404-3415.

Hewitt, K. (1998) Recent glacier surges in the Karakoram himalaya, south central Asia, available at: www.agu.org/eosrelec/97106e.html.

Hewlett, J.D. (1961) Watershed management. In: *Report for 1961 southeastern forest experiment station*. US Forest Service, Ashville, NC, 62-66.

Hey, R. (1996) Environmentally sensitive river engineering. In: Petts, G.E. and Calow, P. (eds), *River restoration*. Blackwell Scientific, Oxford, 80-105.

Hill, G.B. and Henry, G.H.R (2011) Responses of high Arctic wet sedge tundra to climate warming since 1980. *Global Change Biology*, **17**, 276-287.

Hoegh-Guldberg, O., Mumby, P.J., Hooten, A.J. *et al.* (2007) Coral reefs under rapid climate change and ocean acidification. *Science*, **318**, 1737-1742.

Hoffmann, A.A., Hallas, R.J., Dean, J.A. and Schiffer, M. (2003) Low potential for climatic stress adaptation in a rainforest *Drosophila* species. *Science*, **301**, 100-102.

Hofstede, J. (2005) Danish-German-Dutch Wadden Environments. In: E.A. Koster (ed.), *The physical geography of Western Europe*. Oxford University Press, Oxford, 185-205.

Holden, J. (2005a) Darcy's law. In: Lehr, J.H. and Keeley, J. (eds), *Water encyclopedia*, Vol. 5. John Wiley & Sons, New York, 63-64.

Holden, J. (2005b) Peatland hydrology and carbon cycling: why small-scale process matters. *Philosophical Transactions of the Royal Society A*, **363**, 2891-2913.

Holden, J. (2009) Flow through macropores of different size classes in blanket peat. *Journal of Hydrology*, **364**, 342-348.

Holden, J. (2011) *Physical geography: The basics*. Routledge, London.

Holden, J. and Burt, T.P. (2002a) Piping and pipeflow in a deep peat catchment. *Catena*, **48**, 163-199.

Holden, J. and Burt, T.P. (2002b) Laboratory experiments on drought and runoff in blanket peat. *European Journal of Soil Science*, **53**, 675-689.

Holden, J. and Gell, K.F. (2009) Morphological characterization of solute flow in a brown earth grassland soil with cranefly larvae burrows (leatherjackets). *Geoderma*, **152**, 181-186.

Holden, J. and Rose, R. (2011) Temperature and surface lapse rate change: a study of the UK's longest upland instrumental record. *International Journal of Climatology*, **31**, 907-919.

Holden, J. and Wright, A. (2004) A UK tornado climatology and the development of simple prediction tools. *Quarterly Journal of the Royal Meteorological Society*, **130**, 1009-1022.

Holden, J., Burt, T.P. and Cox, N.J. (2001) Macroporosity and infiltration in blanket peat: the implications of tension disc infiltrometer measurements. *Hydrological Processes*, **15**, 289-303.

Holden, J., Burt, T.P. and Vilas, M. (2002) Application of ground penetrating radar to the identification of subsurface piping in blanket peat. *Earth Surface Processes and Landforms*, **27**, 235-249.

Holden, J., Evans, M.G., Burt, T.P. and Horton, M. (2006) Impact of land drainage on peatland hydrology. *Journal of Environmental Quality*, **35**, 1764-1778.

Holden, J., Shotbolt, L., Bonn, A. *et al.* (2007) Environmental change in moorland landscapes. *Earth-Science Reviews*, **82**, 75-100.

Holling, C.S. (1973) Resilience and stability of ecological systems. *Annual Review of Ecology and Systematics*, **4**, 1-23.

Holt-Jensen, A. (1980) *Geography: History and concepts*. Sage, London.

Hooghiemstra, H. and Sarmiento, G. (1991) New long continental pollen record from a tropical intermontane basin: Late Pliocene and Pleistocene history from a 540 m core. *Episodes*, **14**, 107-115.

Hooke, J.M. (1980) Magnitude and distribution of rates of river bank erosion. *Earth Surface Processes and Landforms*, **5**, 143-147.

Hooke, R.L. (2005) *Principles of glacier mechanics*. Cambridge University Press, Cambridge.

Hori, K. and Saito, Y. (2008) Classification, architecture and evolution of large river deltas. In: Gupta, A. (ed.), *Large rivers: Geomorphology and management*. John Wiley & Sons, Chichester, 214-231.

Horton, R.E. (1933) The role of infiltration in the hydrological cycle. *Transactions of the American Geophysical Union*, **14**, 446-460.

Horton, R.E. (1945) Erosional development of streams and their drainage basins; hydrophysical approach to quantitative morphology. *Bulletin of the Geological Society of America*, **56**, 275-370.

Houghton, J. (1994) *Global warming: The complete briefing*. Oxford: Lion.

Houghton, J.T. (1997) *Global warming: The complete briefing* (paperback edition). Cambridge University Press, Cambridge.

Houghton, J.T. (2004) *Global warming: The complete briefing*, 3rd edition. Cambridge University Press, Cambridge.

Houghton, J. (2009) *Global warming: The complete briefing*, 4th edition. Cambridge University Press, Cambridge.

Howard, A.D. (1994) A detachment limited model of drainage basin evolution. *Water Resources Research*, **30**, 2261-2285.

Howat, I.M., Smith, B.E., Joughin, I. and Scambos T.A. (2008) Rates of southeast Greenland ice volume loss from combined ICESat and ASTER observations. *Geophysical Research Letters*, **35**, L17505, doi:10.1029/2008GL034496.

Hubbard, B. and Glasser, N.J. (2005) *Field techniques in glaciology and glacial geomorphology*. John Wiley & Sons, Chichester.

Hudson, J.M.G. and Henry, G.H.R. (2009) Increased plant biomass in a High Arctic heath community from 1981 to 2008. *Ecology*, **90**, 2657-2663.

Hudson-Edwards, K.A., Macklin, M.G., Miller, J.R. and Lechlar, P.J. (2001) Sources, distribution and storage of heavy metals in the Rio Pilcomayo, Bolivia. *Journal of Geochemical Exploration*, **72**, 229-250.

Hudson-Edwards, K.A., Heather, E., Jamieson, H.E., Charnock, J.M. and Macklin, M.G. (2005) Arsenic speciation in waters and sediment of ephemeral floodplain pools, Rios Agrio-Guadiamar, Aznalcóllar, Spain. *Chemical Geology*, **219**, 175-192.

Huggett, R.J. (1995) *Geoecology: An evolutionary approach*. Routledge, London.

Huggett, R.J. (2004) *Fundamentals of biogeography*. Routledge, London.

Hulme, M. and Kelly, M. (1993) Exploring the links between desertification and climate change. *Environment*, **35**, 5-45.

Hunt, A.G. and Malin, P.E. (1998) Possible triggering of Heinrich events by ice-load induced earthquakes. *Nature*, **393**, 155-158.

Hunt, T.L. and Lipo, C.P. (2007) Chronology, deforestation, and 'collapse': evidence vs. faith in Rapa Nui prehistory. *Rapa Nui Journal*, **21**, 85-97.

Hunt, T.L. and Lipo, C.P. (2009) Revisiting Rapa Nui (Easter Island) 'ecocide'. *Pacific Science*, **63**, 601-616.

Huntley, B. and Prentice, C.I. (1993) Holocene vegetation and climates of Europe. In: Wright, H.E. Jr. *et al.* (eds), *Global climates since the last glacial maximum*. University of Minnesota Press, Minneapolis, MA, 136-168.

Hurrell, J.W., Kushnir, Y., Ottersen, G. and Visbeck, M. (2003) The North Atlantic Oscillation: climatic significance and environmental impact. *Geophysical Monograph*. **134** American Geophysical Union, Washington, DC.

Hutchinson, G.E. (1957) *A treatise on limnology, Vol. 1. Geography, physics and chemistry*. John Wiley & Sons, New York.

Hutchinson, G.E. (1961) The paradox of the plankton. *American Naturalist*, **95**, 137-145.

Hutton, J. (1795) *Theory of the Earth*. Edinburgh, 2 volumes.

Hynes, H.B.N. (1975) The stream and its valley. *Verhandlungen Internationale Vereinigung für theoretische und angewandte Limnologie*, **19**, 1-15.

I

ICONA (1991) *Plan national de lutte contre l'érosion. Ministère de l'Agriculture, de la Peche et de l'Alimenation*. Institut National pour la conservation de la Nature, Madrid.

Ilaiwi, M., Abdelgawad, G. and Jabour, E. (1992) Syria: human induced soil degradation. In: *UNEP world atlas of desertification*. Edward Arnold, London, 43-45.

Imbrie, J. and Imbrie, K.P. (1979) *Ice ages: Solving the mystery*. Macmillan, London.

Imeson, A.C. and Kwaad, F.J.P.M. (1990) The response of tilled soils to wetting by rainfall and the dynamic character of soil erodibility. In: Boardman, J., Foster, I.D.L. and Dearing, J.A. (eds), *Soil erosion on agricultural land*. John Wiley & Sons, Chichester, 3-14.

Inkpen, R. (2004) *Science, philosophy and physical geography*. Routledge, London.

International Energy Agency (2010) *Key world energy statistics 2010*. IEA, Paris.

IPCC (1995) *IPCC second assessment - Climate change 1995*, A Report of the Intergovernmental Panel on Climate Change. IPCC, Geneva.

IPCC (2000) *Special report on emission scenarios*. Nakicenovic, N. and Swart, R. (eds). Cambridge University Press, Cambridge.

IPCC (2001) *Climate change 2001: The scientific basis*. Contribution of Working Group 1 to the Third Assessment Report of the Intergovernmental Panel on Climate Change. Houghton, J.T. *et al.* (eds). Cambridge University Press, Cambridge.

IPCC (2007a) *Climate change 2007: The physical science basis*. Contribution of Working Group 1 to the Fourth Assessment Report of the Intergovernmental Panel on Climate Change. Solomon, S. *et al.* (eds). Cambridge University Press, Cambridge.

IPCC (2007b) *Climate change 2007: Climate change impacts, adaptation and vulnerability*. Contribution of Working Group 2 to the Fourth Assessment Report of the Intergovernmental Panel on Climate Change. Adger, N. *et al.* (eds). Cambridge University Press, Cambridge.

Isacks, B., Oliver, J. and Sykes, L.R. (1968) Seismology and the new global tectonics. *Journal of Geophysical Research*, **73**, 5855-5899.

Isaksen, K., Mühll, D.V., Gubler, H., Kohl, T. and Sollid, J.L. (2000) Ground surface temperature reconstruction based on data from a deep borehole in permafrost at Janssonhaugen, Svalbard. *Annals of Glaciology*, **31**, 287-294.

Isaksen, K., Sollid, J.L., Holmlund, P. and Harris, C. (2007) Recent warming of mountain permafrost in Svalbard and Scandinavia. *Journal of Geophysical Research*, **112**, F02S04, doi:10.1029/2006JF000522.

J

Jarvis, M.G., Allen, R.H., Fordham, S.J., Hazelden, J., Moffat, A.J. and Sturdy, R.G. (1984) *Soils and their use in south east England*. Soil Survey of England and Wales. London.

Jauregi, E. (1990-91) Influence of a large urban park on temperature and convective precipitation in a tropical city. *Energy and Buildings*, **15-16**, 457-463.

Jeffree, E.P. (1960) Some long-term means from *The phenological reports* (1891-1948) of the Royal Meteorological Society. *Quarterly Journal of the Royal Meteorological Society*, **86**, 95-103.

Jeffries, M.J. (1997) *Biodiversity and conservation*. Routledge, London.

Jenkinson, D.S. and Powlson, D.S. (1976) The effects of biocidal treatments on metabolism in soil. 1. Fumigation with chloroform. *Soil Biology and Biochemistry*, **8**, 167-177.

Jenny, H. (1941) *Factors of soil formation: A system of quantitative pedology*. McGraw-Hill, New York.

Johannessen, O.M., Khvorostovsky, K., Miles, M.W. *et al.* (2005) Recent ice-sheet growth in the interior of Greenland. *Science*, **310**, 1013-1016.

Johnes, P.J. (2007) Uncertainties in annual riverine phosphorus load estimation: Impact of load estimation methodology, sampling frequency, baseflow index and catchment population density. *Journal of Hydrology*, **332**, 241-258.

Johnsen, K.I., Alfthan, B., Hislop, L. and Skaalvik, J.F. (eds) (2010) *Protecting arctic biodiversity*. United Nations Environment Programme, GRID-Arendal, available at: www.grida.no.

Johnson, L.E. and Carlton, J.T. (1996) Post establishment spread in large-scale invasions: dispersal mechanisms of the Zebra Mussel *Dreisenna polymorpha. Ecology*, **77**, 1686-1690.

Jones, A., Duck, R., Reed, R. and Weyers, J. (2000) *Practical skills in environmental science*. Prentice Hall, Harlow.

Jones, J.A.A. (1981) *The nature of soil piping: A review of research*, British Geomorphological Research Group Monograph Series 3. Geo Books, Norwich.

Jones, J.A.A. (1997a) *Global hydrology: Processes, resources and environmental management*. Pearson Education, Harlow.

Jones, J.A.A. (1997b) Pipeflow contribution areas and runoff response. *Hydrological Processes*, **11**, 35-41.

Jones, J.A.A. (2010) *Water sustainability: A global perspective*. Hodder Education, London.

Jouzel, J., Barkov, N.I., Barnola, J.M. *et al.* (1993) Extending the Vostok ice-core record of palaeoclimate to the penultimate glacial period. *Nature*, **364**, 407-412.

Judd, K.E., Likens, G.E. and Groffman, P.M. (2007) High nitrate retention during winter in soils of the Hubbard Brook Experimental Forest. *Ecosystems*, **10**, 217-225.

Junk, W.J., Bayley, P.B. and Sparks, R.E. (1989) The flood pulse concept in river-floodplain systems. *Canadian Special Publication of Fisheries and Aquatic Science*, **106**, 110-127.

K

Kabat, T.J., Stewart, G.B. and Pullin, A.S. (2006) Are Japanese knotweed (*Fallopia japonica*) control and eradication interventions effective? *Systematic Review No. 21. Collaboration for Environmental Evidence*. Centre for Evidence-Based Conservation, Birmingham.

Kamb, B., Raymond, C., Harrison, W. *et al.* (1985) Glacier surge mechanism: 1982-1983 surge of Variegated Glacier, Alaska. *Science*, **227**, 469-479.

Kaplan, M. (2002) Let the river run. *New Scientist*, **175**, 32-35.

Karavayeva, N.A., Nefedova, T.G. and Targulian, V.O. (1991) Historical land use changes and soil degradation on the Russian Plain. In: Brouwer, F.M., Thomas, A.J. and Chadwick, M.J. (eds), *Land use changes in Europe: Processes of change, environmental transformations and future patterns*. Kluwer Academic, Dordrecht, 351-377.

Karl, D.M., Bird, D.F. Bjorkman, K., Houlihan, T., Shackelford, R. and Tupas, L. (1999) Microorganisms in the accreted ice of lake Vostok, Antarctica. *Science*, **286**, 2144-2147.

Kasser, M. and Egels, Y. (2002) *Digital photogrammetry*. Taylor & Francis, London.

Kastner, M., Kvenvolden, K.A. and Lorenson, T.D. (1998) Chemistry, isotopic composition and origin of a methane-hydrogen sulfide hydrate at the Cascadia subduction zone. *Earth Planetary Science Letters*, **156**, 173-183.

Kay, R. and Alder, J. (1999) *Coastal planning and management*. Routledge, London.

Keller, E.A. and Macdonald, A. (1995) River channel change: the role of large woody debris. In: Gurnell, A. and Petts, G.E. (eds), *Changing river channels*. John Wiley & Sons, Chichester, 216-236.

Keller, E.A. and Swanson, F.J. (1979) Effects of large organic material on channel form and fluvial processes. *Earth Surface Processes*, **4**, 361-380.

Kellerhals, R. and Church, M. (1989) The morphology of large rivers: characterization and management. *Canadian Special Publication of Fisheries and Aquatic Sciences*, **106**, 31-48.

Kendall, C. and McDonnell, J.J. (1998) (eds) *Isotope tracers in catchment hydrology*. Elsevier Science, Amsterdam.

Kennedy, J. and Parker, D. (2010) Global and regional climate in 2009. *Weather*, **65**, 244-250.

Kent, M. and Coker, P. (1992) *Vegetation description and analysis - A practical approach*. Belhaven Press, Chichester.

Kernan, M., Battarbee, R.W., Curtis, C.J., Monteith, D.T. and Shilland, E.M. (eds) (2010) *Recovery of lakes and streams in the UK from the effects of acid rain: Acid Waters Monitoring Network 20 year interpretative report.* Environmental Change Research Centre, University College London.

Kershaw, A.P., van der Kaars, S. and Moss, P.T. (2003) Late Quaternary Milankovitch-scale climatic change and variability and its impact on monsoonal Australasia. *Marine Geology,* **201**, 81-95.

King, C. (1991) *The depositional environments,* Sedimentology Book 2. Longman, Harlow.

King, E.C., Hindmarsh, R.C.A. and Stokes, C.R. (2009) Formation of mega-scale glacial lineations observed beneath a West Antarctic ice stream. *Nature Geoscience,* **2**, 585-588.

King, J.A., Bradley, R.I., Harrison, R. and Carter, A.D. (2004) Carbon sequestration and saving potential associated with changes to the management of agricultural soils in England. *Soil Use and Management,* **20**, 394-402.

Kirkby, M.J. (1971) Hillslope process-response models based on the continuity equation. *Transactions of the Institute of British Geographers,* Special Publication No. 3.

Kirkby, M.J. (2001) Modelling the interactions between soil surface properties and water erosion. *Catena,* **46**, 89-102.

Kirkby, M.J. (2011) Hydromorphology, erosion and sediment. In: McDonnell, J.J. (ed.), *Benchmark volumes in hydrology.* IAHS Publications, Wallingford.

Kirkby, M.J., Naden, P.S., Burt, T.P. and Butcher, D.P. (1993) *Computer simulation in physical geography,* 2nd edition. John Wiley & Sons, Chichester.

Kirschbaum, M.U.F. (2000) Will changes in organic soil organic carbon act as a positive or negative feedback on global warming? *Biogeochemistry,* **48**, 21-51.

Koch, P.L. and Barnosky, A.D. (2006) Late Quaternary extinctions: state of the debate. *Annual Review of Ecology, Evolution, and Systematics,* **37**, 215-250.

Kocurek, G., Townsley, M., Yeh, E. *et al.* (1990) Dune and dune-field development on Padre Island, Texas, with implications for interdune deposition and water table controlled accumulation. *Journal of Sedimentary Petrology,* **62**, 622-635.

Kohler, J., James, T.D., Murray, T. *et al.* (2007) Acceleration in thinning rate on western Svalbard glaciers. *Geophysical Research Letters,* **34**, L18502, doi:10.1029/ 2007GL030681.

Komar, P.D. (1998) *Beach processes and sedimentation,* 2nd edition. Prentice Hall, Englewood Cliffs, NJ.

Kondolf, G.M. (1997) Hungry water: effects of dams and gravel-mining on river channels. *Environmental Management,* **21**, 533-551.

Kondolf, G.M., Piegey, H. and Landon, N. (2002) Channel response to increased and decreased bedload supply from land use change: contrasts between two catchments. *Geomorphology,* **45**, 35-51.

Kondolf, M. and Piegey, H. (2003) *Tools in fluvial geomorphology.* John Wiley & Sons, Chichester.

König, M., Winther, J.-G. and Isaksson, E. (2001) Measuring snow and glacier ice properties from satellite. *Reviews of Geophysics,* **39**, 1-27.

Korbetis, M., Reay, D.S. and Grace, J. (2006) New directions: rich in CO_2. *Atmospheric Environment,* **40**, 3219-3220.

Koren, I., Kaufman, Y.J., Washington, R. *et al.* (2006) The Bodélé depression: a single spot in the Sahara that provides most of the mineral dust in the Amazon forest. *Environmental Research Letters,* **1**, 014005.

Kortelainen, P., Mattsson, T., Ahtiainen, M., Saukkonen, S. and Sallantaus, T. (2006) Controls on the export of C, N, P and Fe from undisturbed boreal catchments, Finland. *Aquatic Sciences,* **68**, 453-468.

Koster, M.J. and Hillen, R. (1995) Combat erosion by law: coastal defence policy for The Netherlands. *Journal of Coastal Research,* **11**, 1221-1228.

Kraus, U. and Wiegand, J. (2006) Long-term effects of the Aznalcóllar mine spill - heavy metal content and mobility in soils and sediments of the Guadiamar river valley (SW Spain). *Science of the Total Environment,* **367**, 855-871.

Krause, J., Fu, Q.M., Good, J.M. *et al.* (2010) The complete mitochondrial DNA genome of an unknown hominin from southern Siberia. *Nature,* **464**, 894-897.

Krinsley, D.P. (1970) *A geomorphological and palaeoclimatological study of the playas of Iran.* US Geological Survey, Washington, DC.

Kullman, L. (2002) Rapid recent range-margin rise of tree and shrub species in the Swedish Scandes. *Journal of Ecology,* **90**, 68-77.

Kullman, L. (2005) On the presence of Lateglacial trees in the Scandes. *Journal of Biogeography,* **32**, 1499-1500.

Kumar, K.K., Rajagopalan, B., Hoerling, M. *et al.* (2006) Unraveling the mystery of Indian monsoon failure during El Niño. *Science,* **314**, 115-119.

Kurose, D., Evans, H.C., Djeddour, D.H., Cannon, P.F., Furuya, N. and Tsuchiya, K. (2010) Systematics of *Mycosphaerella* species associated with the invasive weed *Fallopia japonica,* including the potential biological control agent *M. polygoni-cuspidati*. *Mycoscience,* **50**, 179-189.

Kvenvolden, K.A. (1995) A review of the geochemistry of methane in natural gas hydrate. *Organic Geochemistry,* **23**, 997-1008.

L

Lachenbruch, A. (1962) *Mechanics of thermal contraction cracks and ice-wedge polygons in permafrost.* Special Report 70. Geological Society of America, Boulder, CO.

Lack, D. (1971) *Ecological isolation in birds.* Blackwell Scientific, Oxford.

Ladle, R.J. and Whittaker, R.J. (eds) (2011) *Conservation biogeography.* Wiley-Blackwell, Chichester.

Lal, R. (2004) Soil carbon sequestration impacts on global climate change and food security. *Science,* **304**, 1623-1627.

LaMarche, V.C. and Hirschboeck, K.K. (1984) Frost rings in trees as records of major volcanic eruptions. *Nature,* **307**, 121-126.

Lane, S.N. (2001) Constructive comments on D. Massey 'Space-time, "science" and the relationship between physical geography and human geography'. *Transactions of the Institute of British Geographers*, **26**, 243-256.

Lane, S.N. (2003) Numerical modelling in physical geography: understanding, explanation and prediction. In: Clifford, N.J. and Valentine, G. (eds), *Key methods in geography*. Sage, London, 263-290.

Lane, S.N., James, T.D. and Crowell, M.D. (2000) Application of digital photogrammetry to complex topography for geomorphological research. *Photogrammetric Record*, **16**, 793-821.

Lane, S.N., Brookes, C.J., Kirby, M.J. and Holden, J. (2004) A network-index-based version of TOPMODEL for use with high-resolution topographic data. *Hydrological Processes*, **18**, 191-201.

Langbein, W.B. (1964) Geometry of river channels. *Journal of the Hydraulics Division American Society of Civil Engineers*, **90**, 301-312.

Langbein, W.B. and Schumm, S.A. (1958) Yield of sediment in relation to mean annual precipitation. *Transactions of the American Geophysical Union*, **39**, 1076-1084.

Langmuir, D. (1997) *Aqueous environmental geochemistry*. Prentice Hall, Englewood Cliffs, NJ.

Laudon, H., Poléo, A.B.S., Vøllestad, L.A. and Bishop, K. (2005) Survival of brown trout during spring flood in DOC-rich streams in northern Sweden: the effect of present acid deposition and modelled pre-industrial water quality. *Environmental Pollution*, **135**, 121-130.

Laudon, H., Sjöblom, V., Buffam, I., Seibert, J. and Mörth, C.M. (2007) The role of catchment scale and landscape characteristics for runoff generation of boreal streams. *Journal of Hydrology*, **344**, 198-209.

Laurance, W.F. and Williamson, B. (2001) Positive feedbacks among forest fragmentation, drought and climate change in the Amazon. *Conservation Biology*, **15**, 1529-1535.

Laurance, W.F., Cochrane, M.A., Bergen, S. *et al.* (2001) Environment - the future of the Brazilian Amazon. *Science*, **291**, 438-439.

Lauscher, F. (1976) Weltweite Typen der Höhenabhängigkeit des Niederschlags, *Wetter und Leben*, **28**, 80-90.

Lawrence, E., Jackson, A. and Jackson, J. (1998) *Dictionary of environmental science*. Longman, Harlow.

Leahy, T. (2007) Tesco, carbon and the consumer, available at: www.tesco.com/climatechange/speech.asp.

Leakey, A.D.B., Ainsworth, E.A., Bernacchi, C.J., Rogers, A., Long, S.P. and Ort, D.R. (2009) Elevated CO_2 effects on plant carbon, nitrogen and water relations: six important lessons from FACE. *Journal of Experimental Botany*, **60**, 2859-2876.

Leatherman, S.P. (1983) Barrier dynamics and landward migration with Holocene sea-level rise. *Nature*, **301**, 415-418.

Leeder, M.R. (1999) *Sedimentology and sedimentary basins*. Blackwell Scientific, Oxford.

Leeks, G.J., Lewin, J. and Newson, M.D. (1988) Channel change, fluvial geomorphology and river engineering: the case of the Afon Trannon, Mid Wales. *Earth Surface Processes and Landforms*, **13**, 207-223.

Leggett, J.K. (2000) *The carbon war: Global warming and the end of the oil era*. Penguin, London.

Lehr, J.H. (ed.) (2005) *Water encyclopedia*. John Wiley & Sons, New York, 5 volumes.

Leith, H. (1964) *Versuch einer kartographischen Darstelung der Produktivitat der Pflanzendecke auf der Erde*. Geographische Taschebuch, Wiesbaden.

Leonard, J., Perrier, E. and de Marsily, G. (2001) A model for simulating the influence of a spatial distribution of large circular macropores on surface runoff. *Water Resources Research*, **37**, 3217-3225.

Leopold, L.B. and Maddock, T. Jr (1953) The hydraulic geometry of stream channels and some physiographic implications. *US Geological Survey Professional Paper* **252**.

Leopold, L.B. and Wolman, M.G. (1957) River channel patterns - braided, meandering and straight. *US Geological Survey Professional Paper* **282B**.

Lezine, A.-M. and Cazet, J.-P. (2005) High-resolution pollen record from core KW31, Gulf of Guinea, documents the history of the lowland forests of West Equatorial Africa since 40,000 yr ago. *Quaternary Research*, **64**, 432-443.

Li, X., Zhou, J. and Dodson, J. (2003) The vegetation characteristics of the 'Yuan' area at Yaoxian on the Loess Plateau in China over the last 12 000 years. *Review of Paleobotany and Palynology*, **124**, 1-7.

Liestøl, O. (1977) Pingos, springs and permafrost in Spitsbergen. *Norsk Polarinstitutt Årbok*, **1975**, 7-29.

Likens, G.E. and Bormann, F.H. (1995) *Biogeochemistry of a forested ecosystem*, 2nd edition. Springer-Verlag, New York.

Lillesand, T.M. and Kiefer, R.W. (2000) *Remote sensing and image interpretation*. John Wiley & Sons, Toronto.

Lillesand, T.M., Kiefer, R.W. and Chipman, J. (2008) *Remote sensing and image interpretation*, 6th edition. John Wiley & Sons, Toronto.

Linacre, E. and Geerts, B. (1997) *Climates and weather explained*. Routledge, London.

Lisci, M., Monte, M. and Pacini, E. (2003) Lichens and higher plants on stone: a review. *International Biodeterioration and Biodegradation*, **51**, 1-17.

Livingstone, I. and Warren, A. (1996) *Aeolian geomorphology*. Longman, London.

Lock, G.S.H. (1990) *The growth and decay of ice*. Cambridge University Press, Cambridge.

Lockwood, J.G. (1962) Occurrence of Föhn winds in the British Isles. *Meteorological Magazine*, **91**, 57-65.

Lockwood, J.G. (1965) The Indian Monsoon - a review. *Weather*, **20**, 2-8.

Lockwood, J.G. (1979) *Causes of climate*. Edward Arnold, London.

Lockwood, J.G. (1993) Impact of global warming on evapotranspiration. *Weather*, **48**, 291-299.

Lockwood, J.G. (1999) Is potential evapotranspiration and its relationship with actual evapotranspiration sensitive to elevated atmospheric CO_2 levels? *Climatic Change*, **41**, 193-212.

Lockwood, J.G. (2001) Abrupt and sudden climatic transitions and fluctuations: a review. *International Journal of Climatology*, **21**, 1153-1179.

Lockwood, J.G. (2009) The climate of the Earth. In: Hewitt, C.N. and Jackson, A.V. (eds), *Atmospheric science for environmental scientists*. Wiley-Blackwell, Chichester, 1-25.

Lockwood, M. and Fröhlich, C. (2007) Recent oppositely directed trends in solar climate forcings and the global mean surface air temperature. *Proceedings of the Royal Society A.*, doi: 10.1098/rspa.2007.1880.

Loehle, C. (2007) Predicting Pleistocene climate from vegetation in North America. *Climate of the Past*, **3**, 109-118.

Lovejoy, T.E. and Hannah, L. (2005) *Climate change and biodiversity.* Yale University Press, New Haven, CT.

Lovelock, J.L. (1979) *Gaia: A new look at life on Earth*. Oxford University Press, Oxford.

Lowe, J.J. and Walker, M.J.C. (1997) *Reconstructing quaternary environments*. Pearson Education, Harlow.

Luthi, D., Le Floch, M., Bereiter, B. *et al.* (2008) High-resolution carbon dioxide concentration record 650,000 years before present. *Nature*, **453**, 379-382.

Lytle, D.A, and Poff, N. (2006) Adaptation to natural flow regimes. *TRENDS in Ecology and Evolution*, **19**, 94-100.

M

MacArthur, R.H. and Wilson, E.O. (1967) *The theory of island biogeography*. Princetown University Press, Princeton, NJ.

MacAyeal, D.R. (1993) Binge-purge oscillations of the Laurentide Ice Sheet as a cause of the North Atlantic's Heinrich events. *Palaeoceanography*, **8**, 775-784.

MacKay, D.J.C. (2009) *Sustainable energy - Without the hot air.* UIT, Cambridge.

Mackay, J.R. (1973) The growth of pingos, western Arctic coast, Canada. *Canadian Journal of Earth Sciences*, **10**, 979-1004.

Mackay, J.R. (1979) Pingos of the Tuktoyaktuk Peninsula area, Northwest Territories. *Canadian Journal of Earth Sciences*, **15**, 461-462.

Mackay, J.R. (1983) Pingo growth and subpingo water lenses. In: *Permafrost: Fourth International Conference, Proceedings, Fairbanks, Alaska*. National Academy Press, Washington, DC, 762-766.

Mackay, J.R. (1986) Growth of Ibyuk Pingo, Western Arctic Coast, Canada and some implications for environmental reconstructions. *Quaternary Research*, **26**, 68-80.

Macklin, M.G., Brewer, P.A., Balteanu, D. *et al.* (2003) The long term fate and environmental significance of contaminant metals released by the January and March 2000 mining tailings dam failures in Maramures County, upper Tisa Basin, Romania. *Applied Geochemistry*, **18**, 241-257.

MacMahan, J., Brown, J. and Thornton, E., (2009) Low-cost handheld global positioning system for measuring surf zone currents. *Journal of Coastal Research*, **25**, 744-754.

Magnani, F., Mencuccini, M., Borghetti, M. *et al.* (2007) The human footprint in the carbon cycle of temperate and boreal forests. *Nature*, **447**, 848-850.

Mahowald, N., Kohfeld, K.E., Hansson, M. *et al.* (1999) Dust sources and deposition in the last glacial maximum and current climate: a comparison of model results with paleodata from ice cores and marine sediments. *Journal of Geophysical Research - Atmospheres*, **104**, 15895-15916.

Makepeace, D.K., Smith, D.W. and Stanley, S.J. (1995) Urban stormwater quality: summary of contaminant data. *Critical reviews in Environmental Science and Technology*, **25**, 93-139.

Malhi, Y. and Phillips, O. (2005) *Tropical forests and global atmospheric change*. Oxford University Press, Oxford.

Malkus, J.S. (1958) Tropical weather disturbances: why do so few become hurricanes? *Weather*, **13**, 75-89.

Malmqvist, B. (2002) Aquatic invertebrates in riverine landscapes. *Freshwater Biology*, **47**, 679-694.

Manning, R. (1891) On the flow of water in open channels and pipes. *Transactions of the Institute of Civil Engineers of Ireland*, **20**, 161-207.

Mannion, A.M. (2002) *Dynamic world land-cover and land-use change*. Arnold, London.

Marshall, G.J. (2003) Trends in the Southern Annular Mode from observations and reanalyses. *Journal of Climate*, **16**, 4134-4143.

Marshall, J., Kushnir, Y., Battistf, D. *et al.* (2001) North Atlantic climate variability: phenomena, impacts and mechanisms. *International Journal of Climatology*, **21**, 1863-1898.

Martinez, M.L., Intralawan, A., Vazquez, G., Perez-Maqueo, O., Sutton, P. and Landgrave, R. (2007) The coasts of our world: ecological, economic and social importance. *Ecological Economics*, **63**, 254-272.

Mason, C.F. (2002) *Biology of freshwater pollution*. Pearson Education, Harlow.

Mason, R.G. and Raff, A.D. (1961) Magnetic surveys off the west coast of North America, 32°N to 42°N latitude. *Geological Society of America Bulletin*, **72**, 1259-1266.

Masselink, G. and Hughes, M.G. (2003) *Introduction to coastal processes and geomorphology*. Hodder Arnold, London.

Masselink, G., Hughes, M.G. and Knight, J. (2011) *Introduction to coastal processes and geomorphology*, 2nd edition. Hodder Education, London.

Matthews, J.A. (1992) *The ecology of recently deglaciated terrain: A geoecological approach to glacier forelands and primary succession*. Cambridge University Press, Cambridge.

Matthews, H.D. and Caldeira, K. (2008) Stabilizing climate requires near-zero emissions. *Geophysical Research Letters*, **35**, L04705, doi: 10.1029/2007GL032388.

May, R.E. (1975) Island biogeography and the design of nature reserves. *Nature*, **254**, 277-278.

May, R.M., Lawton, J.H. and Stork, N.E. (1995) Assessing extinction rates. In: Lawton, J.H. and May, R.M. (eds), *Extinction rates*. Oxford University Press, Oxford, 1-24.

Mayle, F.E., Burbridge, R. and Killeen, T.J. (2000) Millennial-scale dynamics of Southern Amazonian rainforests. *Science,* **290**, 2291-2294.

McCabe, S., Smith, B.J. and Warke, P.A. (2007) Preliminary observations on the impact of complex stress histories on sandstone response to salt weathering: laboratory simulations of process combinations. *Environmental Geology*, **52**, 251-258.

McClatchey, J. (1996) Spatial and altitudinal gradients of precipitation in Scotland. In: Merot, P. and Jigorel, A. (eds), *Hydrologie dans les pays celtiques*. INRA, Paris, 45-51.

McClatchey, J., Runcres, A.M.E. and Collier, P. (1987) Satellite images of the distribution of extremely low temperatures in the Scottish Highlands. *Meteorological Magazine*, **116**, 376-386.

McDonald, A.T. (2000) Ecosystems and their management. In: Kent, A. (ed.), *Reflective practice in geography teaching*. Paul Chapman, London, 11-25.

McDonald, A.T. (2002) Environment at the crossroads. In: *Regional surveys of the world: The USA and Canada*. Europa Press, London, 22-28.

McDonald, A.T., Lane, S.N., Haycock, N.E. and Chalk, E.A. (2004) Rivers of dreams: on the gulf between theoretical and practical aspects of an upland river restoration. *Transactions of the Institute of British Geographers*, **29**, 257-281.

McGuffie, K. and Henderson-Sellers, A. (1997) *A climate modeling primer*. John Wiley & Sons, Chichester.

McIlveen, J.F.R. (2010) *Fundamentals of weather and climate*. Oxford University Press, Oxford.

McIntosh, A.R., McHugh, P., Dunn, N.R. *et al.* (2010) The impact of trout on galaxiid fishes in New Zealand. *New Zealand Journal of Ecology*, **34**, 195-206.

McKenzie, D.P. and Parker, D.L. (1967) The North Pacific: an example of tectonic on a sphere. *Nature*, **216**, 1276-1280.

McKnight, D.M. and Bencala, K.E. (1990) The chemistry of iron, aluminium, and dissolved organic material in three acidic, metal-enriched, mountain streams, as controlled by watershed and in-stream processes. *Water Resources Research*, **26**, 3087-3100.

McManus, J. and Duck, R.W. (eds) (1993) *Geomorphology and sedimentology of lakes and reservoirs*. John Wiley & Sons, Chichester.

McNeill, J. (2000) *Something new under the sun*. Allen Lane & Penguin, London.

McRae, S.G. (1988) *Practical pedology: Studying soils in the field*. Ellis Horwood, Chichester.

Meadows, D.H., Meadows, D.I., Randers, J. and Behrens, W.W. III (1972) *The limits to growth*. Earth Island, London.

Meigs, P. (1953) World distribution of arid and semi-arid homoclimates. *Arid zone hydrology*, UNESCO Arid Zone Research Series 1. UNESCO, Paris, 203-209.

Meir, P., Cox, P. and Grace, J. (2006) The influence of terrestrial ecosystems on climate. *Trends in Ecology and Evolution,* **21**, 254-260.

Melton, M.A. (1965) Debris-covered slopes of the southern Arizona desert. *Journal of Geology*, **73**, 715-729.

Menzel, A., Sparks, T.H., Estrella, N. *et al.* (2006) European phenological response to climate change matches the warming pattern. *Global Change Biology,* **12**, 1969-1976.

Menzies, J. (1989) Drumlins - products of controlled or uncontrolled glaciodynamic response? *Quaternary Science Reviews*, **8**, 151-158.

Menzies, J. (2000) Micromorphological analyses of microfabrics and microstructures indicative of deformation processes in glacial sediments. In: Maltman, A.J., Hubbard, B. and Hambrey, M.J. (eds), *Deformation of glacial materials*. The Geological Society, London, 245-257.

Merritt, R.W., Cummins, K.W. and Berg, M.B. (2008) *An introduction to the aquatic insects of North America*, 4th edition. Kendall/Hunt, Dubuque, IA.

Meybeck, M. (2003) Global analysis of river systems: from earth system controls to Anthropocene syndromes. *Philosophical Transactions of the Royal Society B*, **358**, 1935-1955.

Meyer, J.L. (1994) The microbial loop in flowing waters. *Microbial Ecology*, **28**, 195-199.

Micheli, E.R. and Kirchner, J.W. (2002) Effects of wet meadow riparian vegetation on streambank erosion: remote sensing measurements of streambank migration and erodibility. *Earth Surface Processes and Landforms*, **27**, 627-639.

Middleton, G.V. and Wilcock, P.R. (1994) *Mechanics in the Earth and environmental sciences*. Cambridge University Press, Cambridge.

Middleton, N.J. and Thomas, D.S.G. (1997) *World atlas of desertification*, 2nd edition. Edward Arnold, London.

Mieth, A. and Bork, H.-R. (2010) Humans, climate or introduced rats - which is to blame for the woodland destruction on prehistoric Rapa Nui (Easter Island)? *Journal of Archaeological Science*, **37**, 417-426.

Migo, P. (2006) *Granite landscapes of the world*. Oxford University Press. Oxford.

Migo, P. and Lidmar-Bergström, K. (2001) Weathering mantles and their significance for geomorphological evolution of central and northern Europe since the Mesozoic. *Earth-Science Reviews*, **56**, 285-324.

Millennium Ecosystem Assessment (2005) *Ecosystems and human well-being: Biodiversity synthesis*. World Resources Institute, Washington, DC.

Mills, M.G.L. (1989) The comparative behavioural ecology of hyenas: the importance of diet and food dispersion. In: Gittleman, J.L. (ed.), *Carnivore behaviour, ecology and evolution*. Chapman and Hall, London.

Milne, G. (1935) Composite units for the mapping of complex soil associations. *Transactions of the Third International Congress of Soil Science*, **1**, 345-347.

Milne, R. and Brown, T.A. (1997) Carbon in the vegetation and soil of Great Britain. *Journal of Environmental Management*, **49**, 413-433.

Milner, A.M., Brittain, J.E., Castella, E. and Petts, G.E. (2001) Trends of macroinvertebrate community structure in glacier-fed rivers in relation to environmental conditions: a synthesis. *Freshwater Biology*, **46**, 1833-1848.

Milner, A.M., Robertson, A., Monaghan, K., Veal, A.J. and Flory, E.A. (2008) Colonization and development of a stream community over 28 years; Wolf Point Creek in Glacier Bay, Alaska. *Frontiers in Ecology and the Environment*, **6**, 413-419.

Milner, A.M., Brown, L.E. and Hannah, D.M. (2009) Hydroecological effects of shrinking glaciers. *Hydrological Processes*, **23**, 62-77.

MISR (1984) *Organisation and methods of the 1:250000 Soil Survey of Scotland*. The Macaulay Institute for Soil Research, Aberdeen.

Mitchell, T.L. (1837) Account of the recent exploring expedition to the interior of Australia. *Journal of the Royal Geographical Society*, **7**, 271-285.

Moffitt, B.J. and Ratcliffe, R.A.S. (1972) Northern hemisphere monthly mean 500 millibar and 1000-500 millibar thickness charts and some derived statistics (1951-66). *Geophysical Memoirs*, 117. Meteorological Office, London.

Molochuskin, E.N. (1973) The effect of thermal abrasion on the temperature of the permafrost in the coastal zone of the Laptev Sea. In: *Proceedings of the Second International Conference on Permafrost, Yakutsk, USSR*, USSR Contribution. National Science Academy, Washington, DC, 90-93.

Monteith, J.L. and Unsworth, M.H. (1990) *Principles of environmental physics*, 2nd edition. Edward Arnold, London.

Montello, D.R. and Sutton, P.C. (2006) *An introduction to scientific research methods in geography*. Sage, London.

Montgomery, D.R. and Buffington, J.M. (1997) Channel-reach morphology in mountain drainage basins. *Geological Society of America Bulletin*, **109**, 596-611.

Moore, C.J., Moore, S.L., Leecaster, M.K., and Weisberg, S.B. (2001) A comparison of plastic and plankton in the North Pacific central gyre. *Marine Pollution Bulletin*, **42**, 1297-1300.

Moore, J.K. and Braucher, O. (2008) Sedimentary and mineral dust sources of dissolved iron to the world ocean. *Biogeosciences*, **5**, 631-656.

Moore, P.D., Challoner, W. and Stott, P. (1996) *Global environmental change*. Blackwell, Oxford.

Morgan, J.J. and Stumm, W. (1965) The role of multivalent metal oxides in limnological transformations, as exemplified by iron and manganese. In: *Advances in water pollution research, Proceedings of the Second International Conference on Water Pollution Research*. Pergamon Press, Oxford, **1**, 103-118.

Morgan, R.P.C. (ed.) (1986) *Soil erosion and conservation*. Longman, Harlow.

Moss, B. (1998) *Ecology of freshwaters: Man and medium, past to future*. Blackwell, Oxford.

Moss, B. (2010) *Ecology of freshwaters: A view for the twenty-first century*. Wiley-Blackwell, Chichester.

Muhweezi, A.B., Sikoyo, G.M. and Chemonges, M. (2007) Introducing a transboundary ecosystem management approach in the Mount Elgon region. *Mountain Research and Development*, **27**, 215-219.

Murray, A.B., Lazarus, E., Ashton, A. *et al.* (2009) Geomorphology, complexity, and the merging science of the Earth's surface. *Geomorphology*, **103**, 496-505.

Murray, J. (1968) *The world of life: The biosphere*. University of Colorado/Jarrold and Sons, Norwich.

Murray, T. (1997) Assessing the paradigm shift: deformable glacier beds. *Quaternary Science Reviews*, **16**, 995-1016.

Murrell, C., Gerber, E., Krebs, C., Parepa, M., Schaffner, U. and Bossdorf, O. (2011) Invasive knotweed affects native plants through allelopathy. *American Journal of Botany*, **98**, 38-43.

Myers, N., Mittermeier, R.A., Mittermeier, C.G. *et al.* (2000) Biodiversity hotspots for conservation priorities. *Nature*, **403**, 853-858.

N

Nägeli, W. (1946) Weitere Untersuchungen uber die Windverhaltnisse im Bereich von Windschutzanglen. *Mitteilungen Schweizerischen Anstalt für das forstliche Versuchswesen*, **24**, 659-737.

Nagler, P.L., Glenn, E.P. and Huete, A.R. (2001) Assessment of spectral vegetation indices for riparian vegetation in the Colorado River delta, Mexico. *Journal of Arid Environments*, **49**, 91-110.

Nagy, L., Grabherr, G., Körner, C. and Thompson, D.B.A. (2003) *Alpine biodiversity in Europe*. Springer-Verlag, Berlin.

Nakano, S., Miyasaka, H. and Kuhara, N. (1999) Terrestrial-aquatic linkages: riparian arthropod inputs alter trophic cascades in a stream food web. *Ecology*, **80**, 2435-2441.

Nash, D.J. (2011) Chapter 8: Desert crusts and rock coatings. In: Thomas, D.S.G. (ed.), *Arid zone geomorphology: Process, form and change in drylands*, 3rd edition. John Wiley & Sons, Chichester.

Nath, B., Hens, L., Compton, P. and Devuyt, D. (eds) (1998) *Environmental management in practice*. Routledge, London.

NEGTAP (2001) *Transboundary air pollution: Acidification, eutrophication and ground-level ozone in the UK*. Report by the National Expert Group on Transboundary Air Pollution (NEGTAP) on behalf of the UK Department for Environment, Food and Rural Affairs.

Netzband, A., Reincke, H. and Bergemann, M. (2002) The River Elbe: a case study for the ecological and economical chain of sediments. *Journal of Soils and Sediments*, **2**, 112-116.

Neuman, A.C. and MacIntyre, I. (1985) Reef response to a sea level rise: keep-up, catch-up or give-up. *Proceedings, 5th International Coral Reef Congress*, 105-110.

Newbold, J.D., O'Neill, R.V., Elwood, J.W. and Van Winkle, W. (1982) Nutrient spiralling in streams: implications for nutrient limitation and invertebrate activity. *American Naturalist*, **120**, 628-652.

Nicholls, N. (1991) The El Niño-Southern Oscillation and Australian vegetation. *Vegetatio*, **91**, 23-36.

Nicholls, R.J. and Mimura, N. (1998) Regional issues raised by sea-level rise and their policy implications. *Climate Research*, **11**, 5-18.

Nicholls, R.J. and Cazenave, A. (2010) Sea-level rise and its impact on coastal zones. *Science*, **328**, 1517.

Nichols, G. (2009) *Sedimentology and stratigraphy*, 2nd edition. Blackwell Science, Oxford.

Nicholson, S.E., Ba, M.B. and Kim, J.Y. (1996) Rainfall in the Sahel during 1994. *Journal of Climate*, **9**, 1673-1676.

Nilsson, J. and Grennfelt, P. (1988) *Critical loads for sulphur and nitrogen* (Report 1988 : 15). Nordic Council of Ministers, Copenhagen.

Noble, I.R. and Slatyer, R.O. (1980) The use of vital attributes to predict successional changes in plant communities subject to recurrent disturbances. *Vegetatio*, **43**, 5-21.

Norris, R.M. (1966) Barchan dunes of Imperial Valley, California. *Journal of Geology*, **74**, 292-306.

Novotny, V. (2003) *Water quality: Diffuse pollution and watershed management*. John Wiley & Sons, New York.

Novotny, V. and Harvey, O. (1994) *Water quality: Prevention, identification and management of diffuse pollution*. Van Nostrand Reinhold, London.

NSRI (2004) Spatial analysis of change in organic carbon and pH using re-sampled national Soil Inventory data across the whole of England and Wales. Defra project SP0545.

O

O'Connell, E., Ewen, J., O'Donnell, G. and Quinn, P. (2007) Is there a link between agricultural land-use management and flooding? *Hydrology and Earth Systems Sciences*, **11**, 96-107.

Ohanian, P.P. and Dubes, R.C. (1992) Performance evaluation for four class of texture features. *Pattern Recognition*, **25**, 819-833.

O'Hara, S.L., Wiggs, G.F.S., Wegerdt, J. *et al.* (2001) Dust exposure and respiratory health amongst children in the environmental disaster zone of Karakalpakstan, Central Asia: preliminary findings of the ASARD project. In: Brebbia, A. and Fayzieva, D. (eds), *Environmental health risk*. WIT Press, Southampton, 71-82.

O'Hare, G., Sweeney, J. and Wilby, R. (2005) *Weather, climate and climate change*. Prentice Hall, Harlow.

Oke, T.R. (1976) The distinction between canopy and boundary-layer urban heat islands. *Atmosphere*, **14**, 268-277.

Oke, T.R. (1987) *Boundary layer climates*. Routledge, London.

Oke, T.R. and East, C. (1971) The urban boundary layer in Montreal. *Boundary Layer Meteorology*, **1**, 411-437.

Oldeman, L.R., Wakkeling, R.T.A. and Sombroek, W.G. (1991) *World map of the status of human induced soil degradation*, Vol. 2: *Global assessment of soil degradation*. UNEP/International Soil Reference and Information Centre, Wageningen.

Olvmo, M., Lidmar-Bergström, K., Ericson, K. and Bonow, J.M. (2005) Saprolite remnants as indicators of pre-glacial landform genesis in Southeast Sweden. *Geografiska Annaler A*, **87**, 447-460.

Osterkamp, T.E. (2007) Characteristics of the recent warming of permafrost in Alaska. *Journal of Geophysical Research*, **112**, F02S02, doi:10.1029/2006JF000578.

Owen, L. and Unwin, T. (eds) (1997) *Environmental management: Readings and case studies*. Blackwell, Cambridge, MA.

Owens, P.N. (2005) Conceptual models and budgets for sediment management at the river basin scale. *Journal of Soils and Sediments*, **5**, 201-212.

Owens, P.N., Walling, D.E. and He, Q. (1996) The behaviour of bomb-derived caesium-137 fallout in catchment soils. *Journal of Environmental Radioactivity*, **32**, 169-191.

Owens, S. and Owens, P.L. (1991) *Environment, resources and conservation*. Cambridge University Press, Cambridge.

Oyarzún, C.E., Godoy, R., de Schrijver, A., Staelens, J. and Lust, N. (2004) Water chemistry and nutrient budgets in an undisturbed evergreen rainforest of southern Chile. *Biogeochemistry*, **71**, 107-123.

P

Paillard, D. (2001) Glacial cycles: toward a new paradigm. *Reviews of Geophysics*, **39**, 325-346.

Parker, D.E. and Alexander, L.V. (2002) Global and regional climate in 2001. *Weather*, **57**, 328-340.

Paterson, W.S.B. (1994) *The physics of glaciers*, 3rd edition. Pergamon Press, Oxford.

Pauli, H., Gottfried, M., Reiter, K., Klettner, C. and Grabherr, G. (2007) Signals of range expansions and contractions of vascular plants in the high Alps: observations (1994-2004) at the GLORIA master site Schrankogel, Tyrol, Austria. *Global Change Biology*, **13**, 147-156.

Pedersen, R.B., Rapp, H.T., Thorseth, H.I. *et al.* (2010) Discovery of a black smoker vent field and vent fauna at the Arctic Mid-Ocean Ridge. *Nature Communications*, **1**, 126.

Peel, D.A. (1994) The Greenland Ice-Core Project (GRIP): reducing uncertainty in climate change? *NERC News*, 26-30 April.

Peixoto, J.P. and Oort, A.H. (1992) *Physics of climate*. American Institute of Physics, New York.

Penck, W. (1924) *Morphological analysis of landforms*, trans. from the Czech, 1953. Macmillan, London.

Peñuelas, J. and Boada, M. (2003) A global change-induced biome shift in the Montseny mountains (NE Spain). *Global Change Biology*, **9**, 131-140.

Perry, C.T. and Taylor, K.G. (2004) Impacts of bauxite sediment inputs on a carbonate-dominated embayment, Discovery Bay, Jamaica. *Journal of Coastal Research*, **20**, 1070-1079.

Perry, C.T. and Taylor, K.G. (eds) (2007) *Environmental sedimentology*. Blackwell, Oxford.

Perry, C.T., Smithers, S.G., Palmer, S.E., Larcombe, P. and Johnson, K.G. (2008) A 1200 year paleoecological record of coral community development from the terrigenous inner-shelf of the Great Barrier Reef. *Geology*, **36**, 691-694.

Perry, M.C., Prior, M.J. and Parker, D.E. (2007) An assessment of the suitability of a plastic thermometer screen for climatic data collection. *International Journal of Climatology*, **27**, 267-276.

Pethick, J. (1984) *An introduction to coastal geomorphology*. Arnold, London.

Pettijohn, F.J., Potter, P.E. and Siever, R. (1987) *Sand and sandstone*. Springer-Verlag, New York.

Petts, G.E. (1979) Complex response of river channel morphology subsequent to reservoir construction. *Progress in Physical Geography*, **3**, 329-369.

Petts, G.E. (1984) Sedimentation within a regulated river. *Earth Surface Processes and Landforms*, **9**, 125-134.

Petts, G.E. and Calow, P. (eds) (1996a) *River flows and channel forms*, Blackwell Scientific, Oxford.

Petts, G.E. and Calow, P. (eds) (1996b) *River restoration*. Blackwell Scientific, Oxford.

Petts, J. (ed.) (1999) *Handbook of environmental impact assessment*. Blackwell Scientific, Oxford.

Péwé, T.L. (1991) Permafrost. In: Kiersch, G.A. (ed.), *The heritage of engineering geology: The first hundred years*. Centennial Special Volume 3. Geological Society of America, Boulder, CO, 277-298.

Philander, G. (1998) Learning from El Niño. *Weather*, **53**, 270-274.

Pilkey, O.H. (2003) *A celebration of the world's barrier islands*. Columbia University Press, New York.

Pimentel, D., Harvey, C., Resosudarmo, P. *et al.* (1995) Environmental and economic costs of soil erosion and conservation benefits. *Science*, **267**, 1117-1123.

Pinder, G.F. and Jones, J.F. (1969) Determination of the groundwater component of peak discharge from the chemistry of total runoff. *Water Resources Research*, **5**, 438-445.

Pinet, P.R. (1996) *Invitation to oceanography*. West Publishing, St Paul, MN.

Pinet, P.R. (2009) *Invitation to oceanography*, 5th edition. Jones & Bartlett, Sudbury, MA.

Pirrie, D., Power, M.R., Wheeler, P.D. *et al.* (2002) Geochemical signature of historical mining: Fowey Estuary, Cornwall, UK. *Journal of Geochemical Exploration*, **76**, 31-43.

Pitt, M. (2008) *The Pitt Review: Learning lessons from the 2007 floods*. UK Cabinet Office, London.

Playfair, J. (1802) *Illustrations of the Huttonian Theory of the Earth*. Reprinted 1964, Dover, New York.

Poff, N.L. (1997) Landscape filters and species traits: towards mechanistic understanding and prediction in stream ecology. *Journal of the North American Benthological Society*, **16**, 391-409.

Post, A. and LaChapelle, E.R. (2000) *Glacier ice*. University of Washington Press, Seattle, WA.

Power, E.T. and Smith, B.J. (1994) A comparative study of deep weathering and weathering products: case studies from Ireland, Corsica and Southeast Brazil. In: Robinson, D.A. and Williams, R.B.G. (eds), *Rock weathering and landform evolution*, John Wiley & Sons, Chichester, 21-40.

Press, F. and Siever, R. (1986) *Earth*. W.H. Freeman, New York.

Press, F., Siever, R., Grotzinger, J. and Jordan, T.H. (2004) *Understanding earth*. W.H. Freeman, New York.

Prest, V.K. (1983) *Canada's heritage of glacial features*. Miscellaneous Report 28, Geological Survey of Canada.

Price, J.S. (1992) Blanket bog in Newfoundland: Part 1. The occurrence and accumulation of fog-water deposits. Hydrological processes. *Journal of Hydrology*, **135**, 87-101.

Price, L.W. (1972) *The Preglacial environment, permafrost and man*. Commission on College Geography Resource Paper No. 14. Association of American Geographers, Washington, DC.

Price, M. (2002) *Introducing groundwater*. Nelson Thornes, Cheltenham.

Priestas, A.M. and Faherazzi, S. (2010) Morphological barrier island changes and recovery of dunes after Hurricane Dennis, St. George Island, Florida. *Geomorphology*, **114**, 614-626.

Pritchard, H.D., Arthern, R.J., Vaughan, D.G. and Edwards, L.A. (2009) Extensive thinning on the margins of the Greenland and Antarctic ice sheets. *Nature*, doi:10.1038/nature08471.

Prowse, T.D. and Ommaney, C.S.L. (1990) *Northern hydrology, Canadian perspectives*. NHRI Science Report 1.

Pye, K. (1987) *Aeolian dust and dust deposits*. Academic Press, London.

Q

Quammen, D. (1996) *The song of the dodo: Island biogeography in an age of extinctions*. Hutchinson, London.

Quincey, D.J. and Luckman, A. (2009) Progress in satellite remote sensing of ice sheets. *Progress in Physical Geography*, **33**, 547.

R

Rackham, O. (1986) *The history of the countryside*. Dent, London.

Rahmstorf, S. (2006) Thermohaline ocean circulation. In: Elias, S.A. (ed.), *Encyclopedia of quaternary science*. Elsevier, Amsterdam.

Ramanathan, V., Ramana, M.V., Roberts, G. *et al.* (2007) Warming trends in Asia amplified by brown cloud solar absorption. *Nature*, **448**, 575-578.

Ranasinghe, R., Symonds, G. and Holman, R.A. (2004) Morphodynamics of intermediate beaches: a video imaging and numerical modelling study. *Coastal Engineering*, **51**, 629-655.

Rapp, A. (1960) Recent development of mountain slopes in Karkevagge and surroundings, northern Sweden. *Geografiska Annaler*, **42**, 71-200.

Raschke, K., Vonder Haar, T.H., Bandeen, W.R. and Pasternak, M. (1973) The annual radiation balance of the earth-atmosphere system during 1969-70 from Nimbus 3 measurements. *Journal of Atmospheric Sciences*, **30**, 341-364.

Raunkiaer, C. (1934) *The life forms of plants and statistical plant geography*. Oxford University Press, Oxford.

Readers Digest Universal Dictionary (1987) Readers Digest, London.

Reading, H. (1998) *Sedimentary environments: Processes, facies and stratigraphy*, 3rd edition. Blackwell Scientific, Oxford.

Reading, H.G. and Collinson, J.D. (1996) Clastic coasts. In: Reading, H.G. (ed.), *Sedimentary environments: Processes, facies and stratigraphy*. Blackwell Scientific, Oxford, 154-231.

Reay, D.S. (2006) *Climate change begins at home: Life on the two-way street of global warming*. Macmillan Science, Basingstoke.

Redfield, A.C., Ketchum, B.H. and Richards, F.A. (1963) The influence of organisms on the composition of seawater. In: Hill, M.N. (ed.), *The sea*, Vol. 2. Interscience, New York.

Reed, M.S., Bonn, A., Slee, W. *et al.* (2010) Future of the uplands. *Land Use Policy*, **26**(1), 204-216.

Reid, J.M., McLeod, D.A. and Cresser, M.S. (1981) Factors affecting the chemistry of precipitation and river water in an upland catchment. *Journal of Hydrology*, **50**, 129-145.

Reiser, D.W., Ramey, M.P., Beck, S., Lambert, T.R. and Geary, R.E. (1991) Flushing flow recommendations for maintenance of salmonid spawning gravels in a steep regulated stream. *Regulated Rivers*, **3**, 267-276.

Renfrew, I.A. and Anderson, P.S. (2002) The surface climatology of an ordinary katabatic wind regime in Coats Land, Antarctica. *Tellus, Series A: Dynamic Meteorology and Oceanography*, **54**, 463-484.

Resh, V.H., Brown, A.V., Covich, A.P. *et al.* (1988) The role of disturbance in stream ecology. *Journal of the North American Benthological Society*, **7**, 433-455.

Rhoads, B.L. and Thorn, C.E. (eds) (1996) *The scientific nature of geomorphology: Proceedings of the 27th Binghampton Symposium in Geomorphology, 27-29 September*. John Wiley & Sons, Chichester.

Richards, K.S. (1996) Samples and cases: generalisation and explanation in geomorphology. In: Rhoads, B.L. and Thorn, C.E. (eds), *The scientific nature of geomorphology: Proceedings of the 27th Binghampton Symposium in Geomorphology, 27-29 September*. John Wiley & Sons, Chichester, 171-190.

Richards, K.S., Brookes, S., Clifford, N.J., Harris, T.R.J. and Lane, S.N. (1997) Theory, measurement and testing in 'real' geomorphology and physical geography. In: Stoddart, D.R. (ed.), *Process and form in geomorphology*. Routledge, London, 265-292.

Riehl, H. (1965) *Introduction to the atmosphere*. McGraw-Hill, New York.

Roberts, N. (1998) *The Holocene: An environmental history*, 2nd edition. Blackwell, Oxford.

Robertson, D.J., Taylor, K.G. and Hoon, S.R. (2003) Geochemical and mineral magnetic characterisation of urban sediment particulates, Manchester, UK. *Applied Geochemistry*, **18**, 269-282.

Robinson, P.J. and Henderson-Sellers, A. (1999) *Contemporary climatology*. Pearson Education, Harlow.

Root, T.L., Price, J.T., Hall, K.R. *et al.* (2003) Fingerprints of global warming on wild animals and plants. *Nature*, **421**, 56-60.

Rose, P.E. and Pedersen, J.A. (2005) Fate of oxytetracycline in streams receiving aquaculture discharges: model simulations. *Environmental Toxicology and Chemistry*, **24**, 40-50.

Rosenberg, N.J., Blad, B.L. and Verma, S.B. (1983) *Microclimate: The biological environment*. John Wiley & Sons, New York.

Rosenqvist, I.T. (1990) From rain to lake: water pathways and chemical changes. *Journal of Hydrology*, **116**, 3-10.

RoTAP (in press) *Acidification, eutrophication, ground level ozone and heavy metals in the UK*. Review of Transboundary Air Pollution (RoTAP), Defra, London.

Rotty, R.M. and Mitchell, J.M.J. (1974) *Man's energy and world climate*. Institute for Energy Analysis Report, Oak Ridge Associated Universities, Oak Ridge, TN.

Roucoux, K.H., Shackleton, N.J., Schonfeld, J. and Tzedakis, P.C. (2001) Combined marine proxy and pollen analyses reveal rapid Iberian vegetation response to North Atlantic millennial-scale climate oscillations. *Quaternary Research*, **56**, 128-132.

Roucoux, K.H., Tzedakis, P.C., de Abreu, L. and Shackleton, N.J. (2006) Climate and vegetation changes 180 000 to 345 000 years ago recorded in a deep-sea core off Portugal. *Earth and Planetary Science Letters*, **249**, 307-325.

Rowell, D.L. (1994) *Soil science: Methods and applications*. Longman, London.

Rowley, R.J., Kostelnick, J.C., Braaten, D. *et al.* (2007) Risk of rising sea level to population and land area. *Eos*, **88**, 105-107.

Roy, P.S., Cowell, P.J., Ferland, M.A. and Thom, B.G. (1994) Wave-dominated coasts. In: Carter, R.W.G. and Woodroffe, C.D. (eds), *Coastal evolution: Late Quaternary shoreline morphodynamics*. Cambridge University Press, Cambridge, 121-186.

Royal Commission on Environmental Pollution (1996) *Nineteenth report - sustainable use of soil*. HMSO, London.

Ruddiman, W.F. (2003) The anthropogenic greenhouse era began thousands of years ago. *Climatic Change*, **61**, 261-293.

Ruddiman, W.F., Raymo, M.E., Martinson, D.G., Clement, B and Backman, J. (1989) Pleistocene evolution: northern hemisphere ice sheets and North Atlantic ocean. *Paleoceanography*, **4**, 353-412.

Ruegg, G.H.J. (1994) Alluvial architecture of the Quaternary Rhin-Meuses River. *Geologie en Mijnbouw*, **72**, 321-330.

Rull, V., Canellas-Bolta, N., Saez, A., Giralt, S., Pla, S. and Margalef, O. (2010) Paleoecology of Easter Island: evidence and uncertainties. *Earth-Science Reviews*, **99**, 50-60.

Rusjan, S. and Mikoš, M. (2010) Seasonal variability of diurnal instream nitrate concentration oscillations under hydrologically stable conditions. *Biogeochemistry*, **97**, 123-140.

Russ, J.C. (2011) *The image processing handbook*, 6th edition. CRC Press, Boca Raton, FL.

Ryan, W.B.F., Pitman, W.C., Major, C.O. *et al.* (1997) An abrupt drowning of the Black Sea shelf. *Marine Geology*, **138**, 119-126.

S

Salles, J.M. (2011) Valuing biodiversity and ecosystem services: why put economic values on Nature? *Comptes Rendus Biologies*, **334**, 469-482.

Sánchez Goñi, M.F., Eynaud, F., Turon, J.L. and Shackleton, N.J. (1999) High resolution palynological record off the Iberian margin: direct land-sea correlation for the Last Interglacial complex. *Earth and Planetary Science Letters*, **171**, 123-137.

Sankey, T. (2003) Mount Elgon transboundary biosphere initiative: key stakeholder consultative workshop, Eldoret, Kenya, available at: www.natcomreport.com/uganda4/livre/mount.html.

Sarnthein, M. (1978) Sand deserts during glacial maximum and climatic optimum. *Nature*, **272**, 43-46.

Sarnthein, M. and Walger, E. (1974) Der aolische Sandstrom aus der W-Sahara zur Atlantikkuste. *Geologische Rundschau*, **63**, 1065-1087.

Scheffer, M., Hosper, S.H., Meijer, M.L., Moss, B. and Jeppesen, E. (1993) Alternative equilibria in shallow lakes. *Trends in Ecology and Evolution*, **8**, 275-279.

Scheidegger, A.E. (1973) On the prediction of the reach and velocity of catastrophic landslides. *Rock Mechanics*, **5**, 231-236.

Scherz, J.P. and Stevens, A.R. (1970) *An introduction to remote sensing for environmental monitoring*, Report No. 1, University of Wisconsin Madison.

Schindler, D.W. (1974) Eutrophication and recovery in experimental lakes: implications for lake management. *Science*, **184**, 897-899.

Schlosser, P., Bayer, R., Bonisch, G. *et al.* (1999) Pathways and mean residence times of dissolved pollutants in the ocean derived from transient tracers and stable isotopes. *Science of the Total Environment*, **238**, 15-30.

Schmugge, T.J. and Jackson, T.J. (1996) Soil moisture variability. In: Stewart, J.B., Engman, E.T., Feddes, R.A. and Kerr, Y. (eds), *Scaling up in hydrology using remote sensing*. John Wiley & Sons, Chichester, 183-192.

Schultz, J. (1995) *Die Okozonen der Erde*, 2. Aufl. UTB 1514. Ulmer, Stuttgart.

Schulze, E.D., Beck, E. and Müller-Hohenstein, K. (2001) *Plant ecology*. Springer-Verlag, Berlin.

Schumm, S.A. (1956) Evolution of drainage systems and slopes on badlands at Perth Amboy, New Jersey. *Geological Society of America Bulletin*, **67**, 597-646.

Schumm, S.A. (1964) Seasonal variations of erosion rates and processes on hillslopes in western Colorado. *Zeitschrift für Geomorphologie Supplement*, **5**, 215-238.

Schumm, S.A. (1977) Applied fluvial geomorphology. In: Hails, J.R. (ed.), *Applied geomorphology*. Elsevier, Amsterdam, 119-156.

Schumm, S.A. (1985) *The fluvial system*. John Wiley & Sons, New York.

Schumm, S.A. (1991) *To interpret the Earth: Ten ways to be wrong*. Cambridge University Press, Cambridge.

Schumm, S.A. and Lichty, R.W. (1965) Time, space and causality in geomorphology. *American Journal of Science*, **263**, 110-119.

Scottish Environment Protection Agency (2007) Significant water management issues in the Scotland River Basin District, available at: www.sepa.org.uk/water/water_publications/swmi.aspx.

Sear, D.A., Lee, M.W.E., Oakley, R.J., Carling, P.A. and Collins, M.B. (2000) Coarse sediment tracing technology for littoral and fluvial environments: a review. In Foster, I.D.L. (ed.), *Tracers in geomorphology*, John Wiley & Sons, Chichester, 21-55.

Seibert, J. and Beven, K.J. (2009) Gauging the ungauged basin: how many discharge measurements are needed? *Hydrology and Earth System Sciences*, **13**, 883-892.

Seigert, M.J., Carter, S., Tabacco, I., Popov, S. and Blankenship, D.D. (2005) A revised inventory of Antarctic subglacial lakes. *Antarctic Science*, **17**, 453-460.

Segar, D.A. (2007) *Introduction to ocean sciences*, 2nd edition. W.W. Norton, New York.

Selby, M.J. (1993) *Hillslope materials and processes*, 2nd edition. Oxford University Press, Oxford.

Sellers, W.D. (1965) *Physical climatology*. University of Chicago Press, Chicago.

Seppa, H., Bjune, A.E., Telford, R.J., Birks, H.J.B. and Veski, S. (2005) Last nine-thousand years of temperature variability in Northern Europe. *Climates of the Past*, **5**, 523-535.

Shackleton, N.J. (2006) Formal Quaternary stratigraphy - what do we expect and need? *Quaternary Science Reviews*, **25**, 3458-3461.

Shackleton, N.J. and Opdyke, N.D. (1973) Oxygen isotope analyses and palaeomagnetic stratigraphy of equatorial Pacific core V28-238: oxygen isotope temperatures and ice volumes on a 10^5 and 10^6 year scale. *Quaternary Research*, **3**, 39-55.

Shaw, E.M., Beven, K.J., Chappell, N.A. and Lamb, R. (2010) *Hydrology in practice*. Taylor & Francis, London.

Shaw, J. (1989) Drumlins, subglacial meltwater floods and ocean responses. *Geology*, **17**, 853-856.

Shaw, J. (1994a) A qualitative view of sub-ice-sheet landscape evolution. *Progress in Physical Geography*, **18**, 159-184.

Shaw, J. (1994b) Hairpin erosional marks, horseshoe vortices and subglacial erosion. *Sedimentary Geology*, **91**, 269-283.

Shaw, J. and Sharpe, D.R. (1987) Drumlin formation by subglacial meltwater erosion. *Canadian Journal of Earth Sciences*, **24**, 2316-2322.

Shaw, P.A. (1997) Africa and Europe. In: Thomas, D.S.G. (ed.), *Arid zone geomorphology: Process, form and change in drylands*. John Wiley & Sons, Chichester, 467-485.

Shaw, P.A. and Thomas, D.S.G. (1997) Pans, playas and salt lakes. In: Thomas, D.S.G. (ed.), *Arid zone geomorphology: Process, form and change in drylands*. John Wiley & Sons, Chichester, 294-317.

Shaw, W.W., Harris, L.K. and Livingston, M. (1998) Vegetative characteristics of urban land covers in metropolitan Tucson. *Urban Ecosystems*, **2**, 65-73.

Sherman, D.J. and Bauer, B.O. (1993) Dynamics of beach-dune systems. *Progress in Physical Geography*, **17**, 413-447.

Sherman, G.D. (1952) The genesis and morphology of the alumina-rich laterite clays. In: American Institute of Mining and Metallurgical Engineers, New York, *Problems of clay and laterite genesis: Symposium at Annual Meeting, St. Louis, Missouri, February 19-22, 1951*.

Shimda, A., Evenari, M. and Noy-Meir, I. (1986) Hot desert eco-systems: an integrated view. In: Evenari, M., Noy-Meir, I. and Goodall, D.W. (eds), *Ecosystems of the world*, Vol. 12B: *Hot deserts and arid shrublands*. Elsevier, Amsterdam, 378-388.

Shinn, E.A., Smith, G.W., Prospero, J.M. *et al.* (2000) African dust and the demise of Caribbean coral reefs. *Geophysical Research Letters*, **19**, 3029-3032.

Short, A.D. (1999) Global variation in beach systems. In: Short, A.D. (ed.), *Beach morphodynamics*. John Wiley & Sons, Chichester, 72-118.

Sibbesen, E. and Runge-Metzger, A. (1995) Phosphorus balance in European agriculture – status and policy options. In: Tiessen, H. (ed.), *Phosphorus in the global environment*. John Wiley & Sons, Chichester, 43-60.

Sigmundsson, F., Hreindóttir, S., Hooper, A. *et al.* (2010) Intrusion triggering of the 2010 Eyjafjallajökull explosive eruption. *Nature*, **468**, 426-430.

Simberloff, D.S. (1983) When is an island community in equilibrium? *Science*, **220**, 1275-1277.

Simon, A. and Castro, J. (2003) Measurement and analysis of alluvial channel form. In: Kondolf, M. and Piegey, H. (eds), *Tools in fluvial geomorphology*. John Wiley & Sons, Chichester, 291-322.

Simpson, C.J. and Smith, D.G. (2001) The braided Milk River, northern Montana, fails the Leopold Wolman discharge gradient test. *Geomorphology*, **41**, 337-353.

Sithole, G. and Vosselman, G. (2006) Bridge detection in airborne laser scanner data. *ISPRS Journal of Photogrammetry and Remote Sensing*, **61**, 33-46.

Skempton, A.W. and DeLory, F.A. (1957) Stability of natural slopes in London Clay. *Proceedings of the Fourth International Conference on Soil Mechanics and Foundation Engineering*, Vol. 2, 378-381.

Skinner, B.J., Porter, S.C. and Park, J. (2003) *The dynamic Earth: An introduction to physical geology*. John Wiley & Sons, New York.

Skinner, L.C. and Shackleton, N.J. (2005) An Atlantic lead over Pacific deep-water change across Termination I: implications for the application of the marine isotope stratigraphy. *Quaternary Science Reviews*, **24**, 571-580.

Skinner, L.C. and Shackleton, N.J. (2006) Deconstructing Terminations I and II: revisiting the glacioeustatic paradigm based on deep-water temperature estimates. *Quaternary Science Reviews*, **25**, 3312-3321.

Slingo, J. (1998) The 1997/98 El Niño. *Weather*, **53**, 274-281.

Small, C. and Nicholls, R.J. (2003) A global analysis of human settlement in coastal zones. *Journal of Coastal Research*, **19**, 584-599.

Smart, R.P., Holden, J., Dinsmore, K. *et al.* (2012) The dynamics of natural pipe hydrological behaviour in blanket peat. *Hydrological Processes*, in press.

Smith, A.M. and Murray, T. (2009) Bedform topography and basal conditions beneath a fast-flowing West Antarctic ice stream. *Quaternary Science Reviews*, **28**, 584-596.

Smith, A.M., Murray, T., Nicholls, K.W. *et al.* (2007) Rapid erosion and drumlin formation observed beneath a fast-flowing Antarctic ice stream. *Geology*, **35**, 127-130.

Smith, B.J. (2003) Background controls on urban stone decay: lessons from natural rock weathering. In: Brimblecombe, P. (ed.), *The effects of air pollution on the built environment*. Air Pollution Reviews, Vol. 2. Imperial College Press, London, 31-61.

Smith, B.J. (2009) Weathering processes and forms. In: Abrahams, A. and Parsons, A. (eds), *Geomorphology of desert environments*, 2nd edition. Springer-Verlag, Berlin, 69-100.

Smith, B.J. and McGreevy, J.P. (1988) Contour scaling of a sandstone by salt weathering under simulated hot desert conditions. *Earth Surface Processes and Landforms*, **13**, 697-706.

Smith, B.J. and Sanchez, B.A. (1992) Landuse change and erosion hazard in the Niteroi district of southeast Brazil. *Geography Review*, **6**, 37-41.

Smith, B.J. and Viles, H.A. (2006) Rapid, catastrophic decay of building limestones: thoughts on causes, effects and consequences. In: Fort González, R., Alvarez de Buergo, M. and Gómez-Heras, M. (eds), *Heritage weathering and conservation*. Taylor & Francis, London, 191-197.

Smith, B.J., Gomez-Heras, M. and McCabe, S. (2008) Understanding the decay of stone-built cultural heritage. *Progress in Physical Geography*, **32**, 439-461.

Smith, B.J., Gomez-Heras, M. and Viles, H.A. (2010a) Underlying issues on the selection, use and conservation of building limestone. *Geological Society of London Special Publication*, **331**, 1-11.

Smith, C.L., Richards, K.J. and Fasham, M.J.R. (1996) The impact of mesoscale eddies on plankton dynamics in the upper ocean. *Deep-Sea Research I*, **43**, 1807-1832.

Smith, F.D.M., May, R.M., Pellew, R. *et al.* (1993) Estimating extinction rates. *Nature*, **364**, 494-496.

Smith, L.C. (2010) *The world in 2050: Four forces shaping civilization's northern future*. Dutton, New York.

Smith, P. (2004) Soil as carbon sinks: the global context. *Soil Use and Management*, **20**, 212-218.

Smith, P., Mile, R., Powlson, D.S. *et al.* (2000) Revised estimates of the carbon mitigation potential of UK agricultural land. *Soil Use and Management*, **16**, 293-295.

Smith, P., Chapman, S.J., Scott, W.A. *et al.* (2010b) Climate change cannot be entirely responsible for soil carbon loss observed in England and Wales, 1978-2003. *Global Change Biology*, **13**, 2605-2609.

Smithson, P., Briggs, D., Addison, K. and Atkinson, K. (2002) *Fundamentals of the physical environment*, 3rd edition. Routledge, London.

Snead, R.E. (1972) *Atlas of world physical features*. John Wiley & Sons, New York.

Soar, P.J. and Thorne, C.R. (2001) *Channel restoration design for meandering rivers*. United States Army Corps of Engineers Coastal and Hydraulics Laboratory publication ERDC/CHL CR-01-1. USACE, Washington, DC.

Solé, R.V. and Montoya, J.M. (2001) Complexity and fragility in ecological networks. *Proceedings of the Royal Society B*, **268**, 1-7.

Soulsby, C. (1997) Hydrochemical processes. In: Wilby, R.L. (ed.), *Contemporary hydrology*. John Wiley & Sons, Chichester, 59-106.

Sparks, B.W. and West, R.G. (1972) *The Ice Age in Britain*. Methuen, London.

Sparks, T.H., Jeffree, E.P. and Jeffree, C.E. (2000) An examination of the relationship between flowering times and temperature at the national scale using long-term phenological records from the UK. *International Journal of Biometeorology*, **44**, 82-87.

Spellerberg, I.F. and Sawyer, J.W.D. (1999) *An introduction to applied biogeography*. Cambridge University Press, Cambridge.

Spellerberg, I.F., Sawyer, J.W.D and Whitten, T. (1999) *An introduction to applied biogeography*. Cambridge University Press, Cambridge.

Stamp, L.D. (1949) *The world: A general geography*. Longman, London.

Stanley, D.J. and Sheng, H. (1986) Volcanic shards from Santorini (Upper Minoan ash) in the Nile delta, Egypt. *Nature*, **320**, 733-735.

Stanley, D.J. and Warne, A.G. (1998) Nile delta in its destruction phase. *Journal of Coastal Research*, **14**, 794-825.

Stanners, D. and Bourdeau, P. (eds) (1991) *The DOBRIS Report: Europe's environment*. Earthscan, London.

Stanners, D. and Bourdeau, P. (1995) *Europe's environment: The Dobris assessment*. European Environment Agency, Copenhagen.

Statzner, B. and Bêche, L.A. (2010) Can biological invertebrate traits resolve effects of multiple stressors on running water ecosystems? *Freshwater Biology*, **55** (Supplement 1), 80-119.

Stearns, L. A. and Hamilton, G.S. (2007) Rapid volume loss from two East Greenland outlet glaciers quantified using repeat stereo satellite imagery, *Geophysical Research Letters*, **34**, L05503, doi:10.1029/2006GL028982.

Stephenson, W.J. (2000) Shore platforms: remain a neglected coastal feature. *Progress in Physical Geography*, **24**, 311-327.

Stern, N. (2006) *The economics of climate change: The Stern Review*. Cambridge University Press, Cambridge.

Stern, W. (1929) Versuch einer elektrodynamischen Dickenmessung von Gletschereis. *Gerlands Beiträge zur Geophysik*, **23**, 292-333.

Stern, W. (1930) Uber Grundlagen, Methodik und bisherige Ergebnisse elektrodynamischer Dickenmessung von Gletschereis. *Zeitschrift für Gletscherkunde*, **15**, 24-42.

Stewart, M. (2003) Using woods, 1600-1850 (i) the community resource. In: Smout, T.C. (ed.), *People and woods in Scotland - A history*. Edinburgh University Press, Edinburgh, 82-104.

Stidson, R.T., Dickey, C.A., Cape, J.N., Heal, K.V. and Heal, M.R. (2004a) Fluxes and reservoirs of trichloroacetic acid at a forest and moorland catchment. *Environmental Science & Technology*, **38**, 1639-1647.

Stidson, R.T., Heal, K.V., Dickey, C.A., Cape, J.N. and Heal, M.R. (2004b) Fluxes of trichloroacetic acid through a conifer forest canopy. *Environmental Pollution*, **132**, 72-84.

Stockdale, E.A., Watson, C.A., Black, H.I.J. and Philipps, L. (2006) *Do farm management practices alter below ground biodiversity and ecosystem function? Implications for sustainable land management.* JNCC report no. 364.

Stoddart, D.R. (1966) Darwin's impact on geography. *Transactions of the Institute of British Geographers*, **56**, 683-698.

Stow, D.A.V. (2005) *Sedimentary rocks in the field: A colour guide*. Manson Publishing, London.

Strahler, A.N. (1957) Quantitative analysis of watershed geomorphology. *Transactions of the American Geophysical Union*, **38**, 913-920.

Strakhov, N.M. (1967) *Principles of lithogenesis*, Vol. 1. T. M. Thomas, London.

Strangeways, I.C. (2006) *Precipitation: theory, measurement and distribution*. Cambridge University Press, Cambridge.

Sugden, D.E. and John, B.S. (1976) *Glaciers and landscape*. Edward Arnold, London.

Summerfield, M.A. (1991) *Global geomorphology*. Longman Scientific, Harlow.

Sumpter, J.P. and Johnson, A.C. (2005) Lessons from endocrine disruption and their application to other issues concerning trace organics in the aquatic environment. *Environmental Science & Technology*, **39**, 4321-4332.

Sunamura, T. (1992) *Geomorphology of rocky coasts*. John Wiley & Sons, Chichester.

Sutherst, R.W. and Maywald, G.F. (1985) A computerised system for matching climates in ecology. *Agriculture, Ecosystems and Environment*, **13**, 281-299.

Sverdrup, H.U., Johnson, M.W. and Fleming, R.H. (1942) *The oceans*. Prentice Hall, Englewood Cliffs, NJ.

Sverdrup, K.A. and Armbrust, E.V. (2008) *An introduction to the world's oceans*, 9th edition. McGraw-Hill, London.

T

Taberlet, P. and Cheddadi, R. (2002) Quarternary refugia and persistance of biodiversity. *Science*, **297**, 2009-2010.

Tainter, J.A. (2006) Archaeology of overshoot and collapse. *Annual Review of Anthropology*, **35,** 59-74.

Takhtajan, A. (1969) *Flowering plants, origins and dispersal*. Oliver and Boyd, London.

Tanno, K.-I. and Willcox, G. (2006) How fast was wild wheat domesticated? *Science*, **311**, 1886.

Tansley, A.G. (1920) The classification of vegetation and the concept of development. *Journal of Ecology*, **22**, 554-571.

Tansley, A.G. (1935) The use and abuse of vegetational concepts and terms. *Ecology*, **43**, 614-624.

Tarbuck, E. and Lutgens, F. (1991) *Sedimentary petrology: An introduction to the origin of sedimentary rocks*. Blackwell Scientific, Oxford.

Taylor, K.G. and Owens, P.N. (2009) Sediments in urban river basins: a review of sediment-contaminant dynamics in an environmental system conditioned by human activities. *Journal of Soils and Sediments*, **9**, 281-303.

Taylor, K.G., Boyd, N.A. and Boult, S. (2003) Sediments, porewaters and diagenesis in an urban water body, Salford, UK: impacts of remediation. *Hydrological Processes*, **17**, 2049-2061.

Taylor, K.G., Perry, C.T., Greenaway, A.M. and Machent, P.G. (2007) Bacterial iron oxide reduction in a terrigenous-sediment impacted tropical shallow marine carbonate system, north Jamaica. *Marine Chemistry*, **107**, 449-463.

TEEB (2010) *The economics of ecosystems and biodiversity: Maintaining the economics of nature: A synthesis of the approach, conclusions and recommendations of TEEB*. Progress Press, Malta.

Terzaghi, K. (1962) Stability of steep slopes on hard unweathered rock. *Géotechnique*, **12**, 251-270.

Thom, R.F. (1968) Une théorie dynamique de la morphogénèse. In: C.E.H. Waddington (ed.), *Towards a theoretical biology*, Vol. I. Edinburgh University Press, Edinburgh, 152-166.

Thomas, A.D. and Dougill, A.J. (2007) Spatial and temporal distribution of cyanobacterial soil crusts in the Kalahari: implications for soil surface properties. *Geomorphology*, **85**, 17-29.

Thomas, C.D., Cameron, A., Green, R.E. *et al.* (2004) Extinction risk from climate change. *Nature*, **427**, 145-148.

Thomas, D.S.G. (1997a) Arid environments: their nature and context. In: Thomas, D.S.G. (ed.), *Arid zone geomorphology: Process, form and change in drylands*. John Wiley & Sons, Chichester, 3-12.

Thomas, D.S.G. (1997b) Sand seas and aeolian bedforms. In: Thomas, D.S.G. (ed.), *Arid zone geomorphology: Process, form and change in drylands*. John Wiley & Sons, Chichester, 373-412.

Thomas, D.S.G. (ed.) (2011) *Arid zone geomorphology: Process, form and change in drylands*, 3rd edition. John Wiley & Sons, Chichester.

Thomas, D.S.G. and Burrough, S.L. (2012) Interpreting geo-proxies of late Quaternary climate change in African drylands: implications for understanding environmental and early human behaviour. *Quaternary international*, in press.

Thomas, D.S.G. and Middleton, N.J. (1994) *Desertification: Exploding the myth*. John Wiley & Sons, Chichester.

Thomas, D.S.G. and Shaw, P.A. (1991) *The Kalahari environment*. Cambridge University Press, Cambridge.

Thomas, D.S.G., Knight, M. and Wiggs, G.F.S. (2005) Remobilization of southern African desert dune systems by twenty-first century global warming. *Nature*, **435**, 1218-1221.

Thomas, M.F. (1994) *Geomorphology in the Tropics: A study of weathering and denudation in low latitudes*. John Wiley & Sons. Chichester.

Thompson, L.G., Mosley-Thompson, E., Davis, M.E. *et al.* (1995) Late Glacial Stage and Holocene tropical ice core records from Huascarán, Peru. *Science*, **269**, 46-50.

Thorne, C.R., Hey, R.D. and Newson, M.D. (1997) *Applied fluvial geomorphology for river engineering and management*. John Wiley & Sons, Chichester.

Thornes, J.B. (1994) Catchment and channel hydrology. In: Abrahams, A.D. and Parsons, A.J. (eds), *Geomorphology of desert environments*. Chapman and Hall, London, 257-287.

Thurman, H.V. (2004) *Introductory oceanography*. Prentice Hall, Upper Saddle River, NJ.

Thurow, C. (1983) *Improving street climate through urban design*. Planning Advisory Service, No. 376. American Planning Association, Chicago.

Tilman, D. (1985) The resource-ratio hypothesis of plant succession. *The American Naturalist*, **125**, 439-464.

Tipping, E., Smith, E.J., Bryant, C.L. and Adamson, J.K. (2007) The organic carbon dynamics of a moorland catchment in N.W. England. *Biogeochemistry*, **84**, 171-189.

Titus, T.N., Kieffer, H.H. and Christensen, P.R. (2002) Exposed water ice discovered near the South Pole of Mars. *Science*, **299**, 1048-1051.

Tomlinson, R.W. and Milne, R.M. (2006) Soil carbon stocks and land cover in Northern Ireland from 1939 to 2000. *Applied Geography*, **26**, 18-39.

Tou, J.T. and Gonzalez, R.C. (1974) *Pattern recognition principles*. Addison-Wesley, Reading, MA.

Townsend, C.R. (1989) The patch dynamics concept of stream community ecology. *Journal of the North American Benthological Society*, **8**, 36-50.

Townsend, C.R. and Hildrew, A.G. (1994) Species traits in relation to a habitat template for river systems. *Freshwater Biology*, **31**, 265-275.

Toy, T.J., Foster, G.R. and Renard, K.G. (2002) *Soil erosion: Processes, prediction, measurement and control*. John Wiley & Sons, Chichester.

Trenhaile, A.S. (1987) *The geomorphology of rock coasts*. Oxford University Press, Oxford.

Trenhaile, A.S. (2003) *Geomorphology: A Canadian perspective*, 2nd edition. Oxford University Press, Oxford.

Trenhaile, A.S. (2010) The effect of Holocene changes in relative sea level on the morphology of rocky coasts. *Geomorphology*, **114**, 30-41.

Troll, C. (1939) Luftbildplan und ökologische Bodenforschung. *Zeitschrift der Gesellschaft für Erdkunde zu Berlin*, **7/8**, 241-298.

Trudgill, S.T. (ed.) (1986) *Solute processes*. John Wiley & Sons, Chichester.

Tso, B. and Mather, P.M. (2001) *Classification methods for remotely sensed data*. Taylor & Francis, London.

Tucker, C.J., Dregne, H.E. and Newcomb, W. (1991) Expansion and contraction of the Sahara Desert from 1980 to 1990. *Science*, **253**, 299-301.

Tucker, G.E., Lancaster, S.T., Gasparini, N.M. and Bras, R.L. (2001) The channel-hillslope integrated landscape development model (CHILD). In: Harmon, R.S. and Doe, W.W.I. (eds),

Landscape erosion and evolution models. Kluwer Academic, New York, 349-388.

Tucker, M.E. (1981) *Sedimentary petrology: An introduction.* Blackwell Scientific, Oxford.

Tucker, M.E. and Wright, V.P. (1990) *Carbonate sedimentology.* Blackwell Scientific, Oxford.

Turbé, A., De Toni, A., Benito, P. *et al.* (2010) *Soil biodiversity: functions, threats and tools for policy makers.* Bio Intelligence Service, IRD and NIOO, Report for European Commission (DG Environment).

Turkington, A.V. (1998) Cavernous weathering in sandstone: lessons to be learned from natural exposures. *Quarterly Journal of Engineering Geology and Hydrogeology,* **31**, 375-383.

Turner, N. (1757) *An essay on draining and improving peat bogs; in which their nature and properties are fully considered.* Baldwin and Pew, London.

Tyler, C.R. and Jobling, S. (2008) Roach, sex, and gender-bending chemicals: the feminization of wild fish in English rivers. *BioScience,* **58**, 1051-1059.

Tzedakis, P.C. (1993) Long-term tree populations in northwest Greece through multiple Quaternary climatic cycles. *Nature,* **364**, 437-440.

Tzedakis, P.C., Lawson, I.T., Frogley, M.R. *et al.* (2002) Buffered tree population changes in a Quaternary refugia: evolutionary implications. *Nature,* **297**, 2044-2047.

U

Uchida, T., Asano, Y., Onda, Y. and Miyata, S. (2005) Are headwaters just the sum of hillslopes? *Hydrological Processes,* **19**, 3251-3261.

UK Department of Energy (1979) *Heat loads in British Cities,* Energy Paper No. 34. HMSO, London.

Ulén, B., Bechmann, M., Fölster, J., Jarvie, H.P. and Tunney, H. (2007) Agriculture as a phosphorus source for eutrophication in the north-west European countries, Norway, Sweden, United Kingdom and Ireland: a review. *Soil Use and Management,* **23**, 5-15.

UNEP (1997) *World Atlas of Desertification,* 2nd edition, Edward Arnold, London.

UNEP (2000) *Guidelines for erosion and desertification control management.* United Nations Environment Programme.

UNFCCC (2010) *Report of the Conference of the Parties on its fifteenth session, held in Copenhagen from 7 to 19 December 2009. Addendum. Part two: Action taken by the Conference of the Parties at its fifteenth session.* United Nations Office at Geneva, Geneva.

United Nations (1992) *United nations framework convention on climate change.* UN, New York.

Urban, F.E., Cole, J.E. and Overpeck. T. (2000) Influence of mean climate change on climate variability from a 155-year tropical Pacific coral record. *Nature,* **407**, 989-993.

V

van Ast, J.A. and Boot, S.P. (2003) Participation in European water policy. *Physics and Chemistry of the Earth,* **28**, 555-562.

Van Camp, L., Buarrabal, B., Gentile, A.R. *et al.* (eds) (2004) *Reports of the technical working groups established under the thematic strategy for soil protection.* EUR 21319 EN/1. European Commission.

Van der Berg, J.H. (1995) Prediction of alluvial channel pattern of perennial rivers. *Geomorphology,* **12**, 259-279.

van der Maarel, E. (1997) *Biodiversity: From Babel to biosphere management.* Opulus Press, Grangarde, Sweden.

van der Maarel, E. (ed.) (2000) *Succession and zonation on mountains, particularly on volcanoes.* Opulus Press, Grangarde, Sweden.

Van Tyne, J. (1951) The distribution of the Kirtland Warbler (Dendroica Kirtlandii). *Proceedings of the International Ornithological Congress,* **10**, 537-544.

Van Vuuren, D.P., Sala, O.E. and Pereira, H.M. (2006) The future of vascular plant diversity under four global scenarios. *Ecology and Society,* **11**, article 25.

Vannote, R.L., Minshall, G.W., Cummins, K.W., Sedell, J.R. and Cushing, C.E. (1980) The river continuum concept. *Canadian Journal of Fisheries and Aquatic Sciences,* **37**, 130-137.

Vaughan, D.G., Marshall, G.J., Connolley, W.M. *et al.* (2003) Recent rapid regional climate warming on the Antarctic Peninsula. *Climatic Change,* **60**, 243-274.

Velicogna, I. (2009) Increasing rates of ice mass loss from the Greenland and Antarctic ice sheets revealed by GRACE. *Geophysical Research Letters,* **36**, L19503, doi:10.1029/2009GL040222.

Velicogna, I. and Wahr, J. (2006) Acceleration of Greenland ice mass loss in spring 2004. *Nature,* **443**, 329-331.

Viles, H. and Spencer, T. (1995) *Coastal problems.* Arnold, London.

Viles, H.A., Goudie, A.S., Grab, S. and Lalley, J. (2010) The use of the Schmidt Hammer and Equotip for rock hardness assessment in geomorphology and heritage science: a comparative analysis. *Earth Surface Processes and Landforms,* **36**, 320-333.

Vincent, P. (1990) *The biogeography of the British Isles.* Routledge, London.

Vine, F.J. and Matthews, D.H. (1963) Magnetic anomalies over oceanic ridges. *Nature,* **199**, 947-949.

Vitousek, P.M., Ehrlich, P.R., Ehrlich, A.H. and Matson, P.A. (1986) Human appropriation of the products of photosynthesis. *Bioscience,* **36**, 368-373.

Vörösmarty, C.J., McIntyre, P.B. Gessner, M.O. *et al.* (2010) Global threats to human water security and river biodiversity. *Nature,* **467**, 555-561.

W

Walder, J.S. and Hallet, B. (1979) Geometry of former subglacial water channels and cavities. *Journal of Glaciology,* **23**, 335-346.

Waldron, S., Flowers, H., Arlaud, C., Bryant, C. and McFarlane, S. (2009) The significance of organic carbon and nutrient export from peatland-dominated landscapes subject to disturbance, a stoichiometric perspective. *Biogeosciences*, **6**, 363-374.

Walker, B. and Steffan, W. (1997) An overview of the implications of global change for natural and managed terrestrial ecosystems. *Conservation Ecology*, **1**, 2.

Walker, J.P., Houser, P.R. and Willgoose, G.R. (2004) Active microwave remote sensing for soil moisture measurement: a field evaluation using ERS-2. *Hydrological Processes*, **11**, 1975-1997.

Walker, M. (2005) *Quaternary dating methods*. John Wiley & Sons, Chichester.

Walker, M. and Bell, W. (2005) *Late quaternary environmental change: Physical and human perspectives*. Pearson Education, Harlow.

Walker, M.J.C., Johnsen, S., Rasmussen, S.O. *et al.* (2009) Formal definition and dating of the GSSP (Global Stratotype Section and Point) for the base of the Holocene using the Greenland NGRIP ice core, and selected auxiliary records. *Journal of Quaternary Science*, **24**, 3-17.

Walling, D.E. and Webb, B.W. (1986) Solutes in river systems. In: Trudgill, S.T. (ed.), *Solute processes*. John Wiley & Sons, Chichester, 251-327.

Walter, H. (1985) *Vegetation of the Earth and ecological systems of the geo-biosphere*, 3rd edition. Springer-Verlag, New York.

Walter, H. (1990) *Vegetationszonen und Klima*, 6th edition. Ulmer, Stuttgart.

Walter, H. (2002) *Vegetation of the Earth and ecological systems of the geo-biosphere*, translated by Owen Muise. Springer-Verlag, Berlin.

Ward, J.V. (1989) The four-dimensional nature of lotic ecosystems. *Journal of the North American Benthological Society*, **8**, 2-9.

Ward, J.V., Tockner, K., Arscott, D.B. and Claret, C. (2002) Riverine landscape diversity. *Freshwater Biology*, **47**, 517-540.

Ward, R.C. and Robinson, M. (2000) *Principles of hydrology*. McGraw-Hill, London.

Wardle, D.A. (2002) *Communities and ecosystems: Linking the aboveground and belowground components*. Monographs in Population Biology 34. Princeton University Press, Princeton, NJ.

Waring, R.H. (1989) Ecosystems: fluxes of matter and energy. In: Cherrett, J.M. (ed.), *Ecological concepts*. Blackwell Scientific, Oxford, 17-41.

Warke, P., Smith, B. and Savage, J. (2010) *Stone by stone*. Appletree Press, Belfast.

Washburn, A.L. (1979) *Geocryology: A study of periglacial processes and environments*. Arnold, London.

Wasson, R.J., Fichett, K., Mackay, B. and Hyde, R. (1988) Large-scale patterns of dune types, spacing and orientation in the Australian continental dunefield. *Australian Geographer*, **19**, 80-104.

Watson, A. and Nash, D.J. (1997) Desert crusts and varnishes. In: Thomas, D.S.G. (ed.), *Arid zone geomorphology: Process, form and change in drylands*. John Wiley & Sons, Chichester, 69-107.

Waugh, D. (1995) *Geography: An integrated approach*. Thomas Nelson, London.

Weaver, C.M. and Wiggs, G.F.S. (2011) Field measurements of mean and turbulent airflow over a barchan sand dune. *Geomorphology*, in press.

Webster, P.J. (1981) Monsoons. *Scientific American*, **245**, 109-118.

Wegener, A.L. (1915) *Die Entstehung der Kontinente und Ozeane*. Sammlung Vieweg, No. 23. F. Vieweg und Sohn, Braunschweig.

Wegener, A.L. (1966) *The origin of continents and oceans*, trans. from the 4th revised German edition of 1929 by J. Biram, with an introduction by B.C. King. Methuen, London.

Wemelsfelder, P.J. (1953) The disaster in the Netherlands caused by the storm flood of February 1, 1953. *Proceedings of the 4th Coastal Engineering Conference*, ASCE, 256-271.

Werner, B.T. (1999) Complexity in natural landform patterns. *Science*, **284**, 102-104.

Westhoff, V. (1983) Man's attitude towards vegetation. In: Holzner, W., Werger, M.J.A. and Ikusima, I. (eds), *Man's impact on vegetation*. Dr. W. Junk, The Hague, 7-24.

Wetzel, R.G. (2001) *Limnology: Lake and river ecosystems*. Academic Press, London.

Wheater, C.P. (1999) *Urban habitats*. Routledge, London.

White, I.D., Mottershead, D.N. and Harrison, S.J. (1992) *Environmental systems: An introductory text*, 2nd edition. Taylor & Francis, London.

White, R.E. (1997) *Principles and practice of soil science: The soil as a natural resource*, 3rd edition. Blackwell Scientific, Oxford.

Whitehead, P.G., Wilby, R.L., Battarbee, R.W., Kernan, M. and Wade, A.J. (2009) A review of the potential impacts of climate change on surface water quality. *Hydrological Sciences Journal*, **54**, 101-123.

Whittaker, R.H. (1953) A consideration of climax theory: the climax as a population and pattern. *Ecological Monographs*, **23**, 41-78.

Whittaker, R.H. (1975) *Communities and ecosystems*. Macmillan, New York.

Whittaker, R.J. and Fernández-Palacios, J.M. (2007) *Island biogeography: Ecology, evolution, and conservation*. Oxford University Press, Oxford.

Wiggs, G.F.W. (1993) Desert dune dynamics and the evaluation of shear velocity: an integratede approach. In: Pye, K. (ed.), *The dynamics and environmental context of aeolian sedimentary systems*. Special Publication 72. Geological Society of London, 37-46.

Wild, A. (1993) *Soils and the environment: An introduction*. Cambridge University Press, Cambridge.

Willgoose, G.R., Bras, R.L. and Rodriguez-Iturbe, I. (1991) Results from a new model of river basin evolution. *Earth Surface Processes and Landforms*, **16**, 237-254.

Williams, G.P. and Wolman, M.G. (1983) Downstream effects of dams on alluvial rivers. *US Geological Survey Professional Paper* 1286.

Williams, M.A.J. and Balling, R.C. (1995) *Interactions of desertification and climate*. Edward Arnold, London.

Williams, M.A.J., Dunkerley, D., Deckker, P. de *et al.* (1998) *Quarternary environments*. Edward Arnold, London.

Williams, P.A. and Cameron, E.K. (2006) Creating gardens: the diversity and progression of European plant introductions. In: Allen, R.B. and Lee, W.G. (eds), *Biological invasions in New Zealand*. Ecological Studies 186. Springer-Verlag, Berlin.

Williams, P.J. (1986) *Pipelines and permafrost: Physical geography and development in the circumpolar north*. Longman, Harlow.

Williams, P.J. and Smith, M.W. (1989) *The frozen Earth: Fundamentals of geocryology*. Cambridge University Press, Cambridge.

Williams, P.J. and Smith, M.W. (1991) *The frozen Earth: Fundamentals of geocryology* (paperback). Cambridge University Press, Cambridge.

Williams, R.S. and Ferrigno, J.G. (1999) Satellite image atlas of glaciers of the world, Chapter A: Introduction. *US Geological Survey Professional Paper* 1386-A.

Willis, I.C. (1995) Interannual variations in glacier motion – a review. *Progress in Physical Geography*, **19**, 61-106.

Wilson, E.O. (1992) *The diversity of life*. Harvard University Press, Cambridge, MA.

Wilson, E.O. (1994) *Naturalist*. The Penguin Press, London.

Wilson, E.O. (2002) *The future of life*. Abacus, London.

Wilson I. (1985) *The exodus enigma*. Weidenfeld and Nicolson, London.

Wingham, D.J., Shepherd, A., Muir, A. and Marshall, G.J. (2006) Mass balance of the Antarctic ice sheet. *Philosophical Transactions of the Royal Society A*, **364**, 1627-1635.

Winterbourn, M.J., Rounick, J.S. and Cowie, B. (1981) Are New Zealand stream ecosystems really different? *New Zealand Journal of Marine and Freshwater Research*, **15**, 321-328.

Withers, P.J.A., Edwards, A.C. and Foy, R.H. (2001) Phosphorus cycling in UK agriculture and implications for phosphorus loss from soil. *Soil Use and Management*, **17**, 139-149.

Wodzicki, K.A. (1950) *Introduced mammals of New Zealand*. Department of Scientific and Industrial Research, Wellington.

Wolf, P.R. and Dewitt, B.A. (2000) *Elements of photogrammetry with applications in GIS*. McGraw-Hill, New York.

Wolff, E.W. (2005) Understanding the past-climate history from Antarctica. *Antarctica Science*, **17**, 487-495.

Wolman, M.G. (1967) A cycle of sedimentation and erosion in urban river channels. *Geografiska Annaler*, **49**, 385-395.

Wolman, M.G. and Miller, J.P. (1960) Magnitude and frequency of forces in geomorphic processes. *Journal of Geology*, **68**, 54-74.

Woo, M.-K. (1986) Permafrost hydrology in north America. *Atmosphere-Ocean*, **24**, 201-234.

Wood, S.J., Jones, D.A. and Moore, R.J. (2000) Accuracy of rainfall measurement for scales of hydrological interest. *Hydrology and Earth System Sciences*, **4**, 531-543.

Woodroffe, C.D. (2003) *Coasts, form, process and evolution*. Cambridge University Press, Cambridge.

Woodward, F.I., Lomas, M.R. and Kelly, C.K. (2004) Global climate and the distribution of plant biomes. *Philosophical Transactions of the Royal Society B*, **359**, 1465-1476.

Woodward, G., Perkins, D. and Brown, L.E. (2010b) Climate change and freshwater ecosystems: impacts across multiple levels of organisation. *Philosophical Transactions of the Royal Society B*, **365**, 2093-2106.

Woodward, J., Smith, A.M., Ross, N. *et al.* (2010a) Location for direct access to subglacial Lake Ellsworth: an assessment of geophysical data and modelling. *Geophysical Research Letters*, **37**, L11501, doi:10.1029/2010GL042884.

Woodward, J.C. and Goldberg, P. (2001) The sedimentary records in Mediterranean rockshelters and caves: archives of environmental change. *Geoarchaeology: An International Journal*, **16**, 327-354.

Wooldridge, S.W. and Linton, D.L. (1939) *Structure, surface and drainage in South East England*. Philip, London.

Woolgar, S. (1988) *Science: The very idea*. Methuen, London.

World Commission on Environment and Development (1987) *Our common future*. Oxford University Press, Oxford.

World Health Organization (1993) *Guidelines for drinking water quality*, Vol. 1: *Recommendations*, 2nd edition. WHO, Geneva.

World Health Organization (1999) *Gothenburg consensus paper*. WHO, Geneva.

Worrall, F. and Burt, T.P. (2004) Time series analysis of long-term river dissolved organic carbon records. *Hydrological Processes*, **18**, 893-911.

Wright, L.D. and Short, A.D. (1984) Morphodynamic variability of surf zones and beaches: a synthesis. *Marine Geology*, **56**, 93-118.

Wright, L.D. and Thom, B.G. (1977) Coastal depositional landforms: a morphodynamic approach. *Progress in Physical Geography*, **1**, 412-459.

Wright, L.D., Chappell, J., Thom, B. *et al.* (1979) Morpho-dynamics of reflective and dissipative beach and inshore systems, South Australia. *Marine Geology*, **32**, 105-140.

Wyrtki, K. (1982) The Southern Oscillation, ocean-atmosphere interaction, and El Niño. *Marine Technology Society Journal*, **16**, 3-10.

Y

Yakushev, V.S. and Chuvilin, E.M. (2000) Natural gas and gas hydrate accumulations within permafrost in Russia. *Cold Regions Science and Technology*, **31**, 189-197.

Yalden, D. (1999) *The history of the British mammals*. Academic Press, London.

Yaron, G., Hall, N., Gaunt, J. and Brett, C. (2008) SP08014 Soils within the Catchment Sensitive Farming Programme: testing non-deterministic approaches to provision of advice on soil organic matter management, available at: http://randd.defra.gov.uk/Document.aspx?Document=SP08014_8258_TRP.pdf.

Yatsu, E. (1966) *Rock control in geomorphology*. Sozosha, Tokyo.

Yatsu, E. (1988) *The nature of weathering: an introduction.* Sozosha, Tokyo.

Yatsu, E. (1992) To make geomorphology more scientific. *Transactions of the Japanese Geomorphological Union*, **13**, 87-124.

Young, I.R. and Holland, G.J. (1996) *Atlas of oceans, wind and wave climate*. Elsevier, Oxford.

Z

Zhang, T. (2005) Influence of the seasonal snow cover on the ground thermal regime: an overview. *Reviews of Geophysics*, **43**, RG4002, doi:10.1029/2004RG000157.

Zhu, T.X. (1997) Deep-seated, complex tunnel systems - a hydrological study in a semi-arid catchment, Loess Plateau, China. *Geomorphology*, **20**, 255-267.

Zimmerman, R.C., Goodlet, J.C. and Comer, G.H. (1967) The influence of vegetation on channel form of small streams. Symposium on River Morphology, *International Association of Hydrological Sciences Publication* 75, General Assembly at Bern, 255-275.

Zimov, S.A., Davydov, S.P., Zimova, G.M. *et al.* (2006) Permafrost carbon: stock and decomposability of a globally significant carbon pool. *Geophysical Research Letters*, **33**, L20502, doi:10.1029/2006GL027484.

INDEX